WATER TREATMENT PLANT OPERATION

Fifth Edition
Volume I

A Field Study Training Program

prepared by

Office of Water Programs
College of Engineering and Computer Science
California State University, Sacramento

in cooperation with the

National Environmental Training Association
(Now National Environmental, Safety & Health Training Association (NESHTA))

★★★★★★★★★★★★★★★★★★★★★★★★★★★★★★★★★

Kenneth D. Kerri, Project Director

★★★★★★★★★★★★★★★★★★★★★★★★★★★★★★★★★

for the

California Department of Health Services
Sanitary Engineering Branch
Standard Agreement #80-64652

and

U.S. Environmental Protection Agency
Office of Drinking Water
Grant No. T-901361-01-0

2004

NOTICE

This manual is revised and updated before each printing based on comments from persons using this manual.

FIRST EDITION

    First Printing, 1983   7,000
    Second Printing, 1986   8,000

SECOND EDITION

    First Printing, 1989   8,000
    Second Printing, 1990   10,000

THIRD EDITION

    First Printing, 1992   12,000
    Second Printing, 1994   15,000
    Third Printing, 1996   20,000

FOURTH EDITION

    First Printing, 1999   20,000
    Second Printing, 2002   15,000

FIFTH EDITION

    First Printing, 2004   20,000

OPERATOR TRAINING MATERIALS

OPERATOR TRAINING MANUALS AND VIDEOS IN THIS SERIES are available from the Office of Water Programs, California State University, Sacramento, 6000 J Street, Sacramento, CA 95819-6025, phone: (916) 278-6142, e-mail: wateroffice@csus.edu, FAX: (916) 278-5959, website: www.owp.csus.edu.

1. *WATER TREATMENT PLANT OPERATION*, 2 Volumes,

2. *SMALL WATER SYSTEM OPERATION AND MAINTENANCE,**

3. *WATER DISTRIBUTION SYSTEM OPERATION AND MAINTENANCE,*

4. *UTILITY MANAGEMENT,*

5. *OPERATION OF WASTEWATER TREATMENT PLANTS*, 2 Volumes,

6. *ADVANCED WASTE TREATMENT,*

7. *INDUSTRIAL WASTE TREATMENT*, 2 Volumes,

8. *TREATMENT OF METAL WASTESTREAMS,*

9. *PRETREATMENT FACILITY INSPECTION,*

10. *OPERATION AND MAINTENANCE OF WASTEWATER COLLECTION SYSTEMS*, 2 Volumes, (Spanish edition available),

11. *COLLECTION SYSTEMS: METHODS FOR EVALUATING AND IMPROVING PERFORMANCE*, and

12. *SMALL WASTEWATER SYSTEM OPERATION AND MAINTENANCE*, 2 Volumes.

*   Other training materials and training aids developed by the Office of Water Programs to assist operators in improving small water system operation and maintenance and overall performance of their systems include the *SMALL WATER SYSTEMS VIDEO INFORMATION SERIES*. This series of ten 15- to 59-minute videos was prepared for the operators, managers, owners, and elected officials of very small water systems. The videos provide information on the responsibilities of operators and managers. They also demonstrate the procedures to safely and effectively operate and maintain surface water treatment systems, groundwater treatment systems, and distribution and storage systems. Other topics covered include monitoring, managerial, financial, and emergency response procedures for small systems. These videos are used with a *LEARNING BOOKLET* that provides additional essential information and references. The videos complement and reinforce the information presented in *SMALL WATER SYSTEM OPERATION AND MAINTENANCE*.

The Office of Water Programs at California State University, Sacramento, has been designated by the U.S. Environmental Protection Agency as a *SMALL PUBLIC WATER SYSTEMS TECHNOLOGY ASSISTANCE CENTER*. This recognition will provide funding for the development of training videos for the operators and managers of small public water systems. Additional training materials will be produced to assist the operators and managers of small systems.

PREFACE

The purposes of this water supply system field study training program are to:

1. Develop new qualified water treatment plant operators,

2. Expand the abilities of existing operators, permitting better service to both their employers and the public, and

3. Prepare operators for civil service and *CERTIFICATION EXAMINATIONS.*[1]

To provide you with the knowledge and skills needed to operate and maintain water treatment plants as efficiently and effectively as possible, experienced water treatment plant operators prepared the material in each chapter of this manual.

Water treatment plants vary from city to city and from region to region. The material contained in this program is presented to provide you with an understanding of the basic operation and maintenance aspects of your water treatment plant and with information to help you analyze and solve operation and maintenance problems. This information will help you operate and maintain your plant in a safe and efficient manner.

Water treatment plant operation and maintenance is a rapidly advancing field. To keep pace with scientific and technological advances, the material in this manual must be periodically revised and updated. *THIS MEANS THAT YOU, THE OPERATOR, MUST RECOGNIZE THE NEED TO BE AWARE OF NEW ADVANCES AND THE NEED FOR CONTINUOUS TRAINING BEYOND THIS PROGRAM.*

The Project Director is indebted to the many operators and other persons who contributed to this manual. Every effort was made to acknowledge material from the many excellent references in the water treatment field. Reviewers Leonard Ainsworth, Jack Rossum, and Joe Monscvitz deserve special recognition for their extremely thorough review and helpful suggestions. John Trax, Chet Pauls, and Ken Hay, Office of Drinking Water, U.S. Environmental Protection Agency, and John Gaston, Bill MacPherson, Bert Ellsworth, Clarence Young, Ted Bakker, and Beverlie Vandre, Sanitary Engineering Branch, California Department of Health Services, all performed outstanding jobs as resource persons, consultants and advisors. Larry Hannah served as Education Consultant. Illustrations were drawn by Martin Garrity. Charlene Arora helped type the field test and final manuscript for printing. Special thanks are well deserved by the Program Administrator, Gay Kornweibel, who typed, administered the field test, managed the office, administered the budget, and did everything else that had to be done to complete this project successfully.

KENNETH D. KERRI
PROJECT DIRECTOR

---

[1] *Certification Examination.   An examination administered by a state agency that operators take to indicate a level of professional competence. In most plants the Chief Operator of the plant must be "certified" (successfully pass a certification examination). In the United States, certification of operators of water treatment plants and wastewater treatment plants is mandatory.*

OBJECTIVES OF THIS MANUAL

Proper installation, inspection, operation, maintenance, repair, and management of water treatment plants have a significant impact on the operation and maintenance costs and effectiveness of the plants. The objective of this manual is to provide water treatment plant operators with the knowledge and skills required to operate and maintain water treatment plants effectively, thus eliminating or reducing the following problems.

1. Health hazards created by the production or output of unsafe water from the plant;

2. System failures that result from the lack of proper installation, inspection, preventive maintenance, surveillance, and repair programs designed to protect the public's investment in the plant;

3. Taste and odor complaints from consumers;

4. Turbid or colored waters that are unacceptable to consumers;

5. Corrosion damages to pipes, equipment, tanks, and structures at the water treatment plant and in the distribution system;

6. Complaints from the public or local officials due to the unreliability or failure of the water treatment plant to perform as designed; and

7. Fire damage caused by insufficient water at a time of need.

SCOPE OF THIS MANUAL

This manual on water treatment plant operation is divided into two volumes. Volume I stresses the knowledge and skills needed by an operator working in a conventional water treatment plant used for treating surface waters. Volume II emphasizes material needed by operators trying to control iron and manganese, softening hard waters, and trihalomethanes. Also contained in Volume II is information needed by all operators responsible for the administration and management of a water treatment plant, such as maintenance, instrumentation, safety, and laboratory procedures.

Volume I contains information on:

1. What water treatment plant operators do;

2. How to manage reservoirs and intake structures;

3. How to operate and maintain coagulation, flocculation, sedimentation, and filtration water treatment processes;

4. Disinfection of water;

5. Procedures for controlling corrosion;

6. Techniques for identifying the causes of taste and odor problems and suggestions for correcting such problems;

7. Procedures for operating, maintaining, and administering a water treatment plant; and

8. Basic laboratory procedures.

Volume II contains information on:

1. How to control iron and manganese;

2. Procedures for fluoridating water;

3. Techniques for softening water;

4. How to control trihalomethanes;

5. Techniques for treating dissolved solids in water;

6. Handling and disposal of process wastes;

7. Procedures for maintaining processes, equipment, and facilities;

8. How to maintain and troubleshoot instrumentation;

9. Techniques for recognizing hazards and developing safe procedures and safety programs;

10. Advanced laboratory procedures for analyzing samples of water;

11. Water quality regulations; and

12. Administrative considerations for supervisors and managers.

Material in this manual furnishes you with information concerning situations encountered by most water treatment plant operators in most areas. These materials provide you with an understanding of the basic operational and maintenance concepts for water treatment plants and with an ability to analyze and solve problems when they occur. Operation and maintenance programs for water treatment plants will vary with the age of the plant, the extent and effectiveness of previous programs, and local conditions. You will have to adapt the information and procedures in this manual to your particular situation.

Technology is advancing very rapidly in the field of operation and maintenance of water treatment plants. To keep pace with scientific advances, the material in this program must be periodically revised and updated. This means that you, the water treatment plant operator, must be aware of new advances and recognize the need for continuous personal training reaching beyond this program. *TRAINING OPPORTUNITIES EXIST IN YOUR DAILY WORK EXPERIENCE, FROM YOUR ASSOCIATES, AND FROM ATTENDING MEETINGS, WORKSHOPS, CONFERENCES, AND CLASSES.*

USES OF THIS MANUAL

This manual was developed to serve the needs of operators in several different situations. The format used was developed to serve as a home-study or self-paced instruction course for operators in remote areas or persons unable to attend formal classes either due to shift work, personal reasons, or the unavailability of suitable classes. This home-study training program uses the concepts of self-paced instruction where you are your own instructor and work at your own speed. In order to certify that a person has successfully completed this program, objective tests and special answer sheets for each chapter are provided when a person enrolls in this course.

Also, this manual can serve effectively as a textbook in the classroom. Many colleges and universities have used this manual as a text in formal classes (often taught by operators). In areas where colleges are not available or are unable to offer classes in the operation of water treatment plants, operators and utility agencies can join together to offer their own courses using the manual.

Cities or utility agencies can use the manual in several types of on-the-job training programs. In one type of program, a manual is purchased for each operator. A senior operator or a group of operators are designated as instructors. These operators help answer questions when the persons in the training program have questions or need assistance. The instructors grade the objective tests, record scores, and notify California State University, Sacramento, of the scores when a person successfully completes this program. This approach eliminates any waiting while papers are being graded and returned by CSUS.

This manual was prepared to help operators operate and maintain their water treatment plants. Please feel free to use the manual in the manner which best fits your training needs and the needs of your operators. We will be happy to work with you to assist you in developing your training program. Please feel free to contact:

Project Director
Office of Water Programs
California State University, Sacramento
6000 J Street
Sacramento, California 95819-6025

Phone: (916) 278-6142
FAX: (916) 278-5959

ENROLLMENT FOR CREDIT AND CERTIFICATE

Students wishing to earn credits and a certificate for completing this course may enroll by contacting the Office of Water Programs, California State University, Sacramento, 6000 J Street, Sacramento, CA 95819-6025, (916) 278-6142. If you have already enrolled, the enrollment packet you were sent contains detailed instructions for completing and returning the objective tests. Please read these important instructions carefully before marking your answer sheets.

Following successful completion of each volume in this program, a Certificate of Completion will be sent to you. If you wish, the Certificate can be sent to your supervisor, the mayor of your town, or any other official you think appropriate. Some operators have been presented their Certificate at a City Council meeting, got their picture in the newspaper, and received a pay raise.

INSTRUCTIONS TO PARTICIPANTS IN HOME-STUDY COURSE

Procedures for reading the lessons and answering the questions are contained in this section.

To progress steadily through this program, you should establish a regular study schedule. For example, many operators in the past have set aside two hours during two evenings a week for study.

The study material is contained in two volumes divided into 23 chapters. Some chapters are longer and more difficult than others. For this reason, many of the chapters are divided into two or more lessons. The time required to complete a lesson will depend on your background and experience. Some people might require an hour to complete a lesson and some might require three hours; but that is perfectly all right. *THE IMPORTANT THING IS THAT YOU UNDERSTAND THE MATERIAL IN THE LESSON!*

Each lesson is arranged for you to read a short section, write the answers to the questions at the end of the section, check your answers against suggested answers; and then *YOU* decide if you understand the material sufficiently to continue or whether you should read the section again. You will find that this procedure is slower than reading a normal textbook, but you will remember much more when you have finished the lesson.

Some discussion and review questions are provided following each lesson in some of the chapters. These questions review the important points you have covered in the lesson. Write the answers to these questions in your notebook.

After you have completed the last chapter, you will find a final examination. This exam is provided for you to review how well you remember the material. You may wish to review the entire manual before you take the final exam. Some of the questions are essay-type questions, which are used by some states for higher-level certification examinations. After you have completed the final examination, grade your own paper and determine the areas in which you might need additional review before your next certification or civil service examination.

You are your own teacher in this program. You could merely look up the suggested answers from the answer sheet or copy them from someone else, but you would not understand the material. Consequently, you would not be able to apply the material to the operation of your facilities nor recall it during an examination for certification or a civil service position.

YOU WILL GET OUT OF THIS PROGRAM WHAT YOU PUT INTO IT.

SUMMARY OF PROCEDURE

OPERATOR (YOU)

1. Read what you are expected to learn in each chapter (the chapter objectives).

2. Read sections in the lesson.

3. Write your answers to questions at the end of each section in your notebook. You should write the answers to the questions just as you would if these were questions on a test.

4. Check your answers with the suggested answers.

5. Decide whether to reread the section or to continue with the next section.

6. Write your answers to the discussion and review questions at the end of each lesson in your notebook.

ORDER OF WORKING LESSONS

To complete this program you will have to work all of the lessons. You may proceed in numerical order, or you may wish to work some lessons sooner.

SAFETY IS A VERY IMPORTANT TOPIC. Everyone working in a water treatment plant must always be safety conscious. Operators daily encounter situations and equipment that can cause a serious disabling injury or illness if the operator is not aware of the potential danger and does not exercise adequate precautions. For these reasons you may decide to work on the chapter on "Safety" early in your studies. In each chapter, *SAFE PROCEDURES ARE ALWAYS STRESSED.* See Chapter 20, "Safety," Volume II, for details.

SAFE PROCEDURES ARE ALWAYS STRESSED

TECHNICAL CONSULTANTS

John Brady   Jim Sequeira
Gerald Davidson   R. Rhodes Trussell
Larry Hannah   Mike Young

NATIONAL ENVIRONMENTAL TRAINING ASSOCIATION REVIEWERS

George Kinias, Project Coordinator

E.E. "Skeet" Arasmith   Andrew Holtan   Rich Metcalf
Terry Engelhardt   Deborah Horton   William Redman
Dempsey Hall   Kirk Laflin   Kenneth Walimaa
Jerry Higgins   Anthony Zigment

PROJECT REVIEWERS

Leonard Ainsworth   Jerry Hayes   Joe Monscvitz   Gerald Samuel
Ted Bakker   Ed Henley   Angela Moore   Carl Schwing
Jo Boyd   Charles Jeffs   Harold Mowry   David Sorenson
Dean Chausee   Chet Latif   Theron Palmer   Russell Sutphen
Walter Cockrell   Frank Lewis   Eugene Parham   Robert Wentzel
Fred Fahlen   Perry Libby   Catherine Perman   James Wright
David Fitch   D. Mackay   David Rexing   Mike Yee
Richard Haberman   William Maguire   Jack Rossum   Clarence Young
Lee Harry   Nancy McTigue   William Ruff

WATER TREATMENT PLANT OPERATION, VOLUME I
COURSE OUTLINE

WATER TREATMENT PLANT OPERATION, VOLUME II
COURSE OUTLINE

Other similar operator training programs that may be of interest to you are our courses and training manuals on the operation and maintenance of water distribution systems and small water systems.

WATER DISTRIBUTION SYSTEM OPERATION AND MAINTENANCE
COURSE OUTLINE

SMALL WATER SYSTEM OPERATION AND MAINTENANCE
(Wells, Small Treatment Plants, and Rates)
COURSE OUTLINE

CHAPTER 1

THE WATER TREATMENT PLANT OPERATOR

by

Ken Kerri

TABLE OF CONTENTS

Chapter 1.   THE WATER TREATMENT PLANT OPERATOR

OBJECTIVES

Chapter 1.   THE WATER TREATMENT PLANT OPERATOR

At the beginning of each chapter in this manual you will find a list of *OBJECTIVES*. The purpose of this list is to stress those topics in the chapter that are most important. Contained in the list will be items you need to know and skills you must develop to operate, maintain, repair, and manage a water treatment plant as efficiently and as safely as possible.

Following completion of Chapter 1, you should be able to:

1. Explain the type of work done by water treatment plant operators,

2. Describe where to look for jobs in this profession, and

3. Describe how you can learn to do the jobs performed by water treatment plant operators.

CHAPTER 1.   THE WATER TREATMENT PLANT OPERATOR

Chapter 1 is prepared especially for new operators or people interested in becoming water treatment plant operators. If you are an experienced water treatment plant operator, you may find some new viewpoints in this chapter.

1.0   NEED FOR WATER TREATMENT PLANT OPERATORS

People need safe water to drink. Many sources of water are not directly suitable for drinking purposes without treatment because of pollution and contamination by humans and nature. Before modern society and the intensive use of available water resources, sun, wind, filtration through soil, and time purified water. Today water treatment plants are built to provide us with safe drinking water. Thus, nature is given an assist by a team consisting of designers, builders, and treatment plant operators. Designers and builders occupy the scene for only a short time, but operators go on forever. Water treatment plant personnel operate, maintain, repair, and manage water treatment plants. These operators have the responsibility of producing safe and pleasant drinking water from their plants. Cities and towns need qualified, capable, and dedicated operators to do these jobs.

The need for *RESPONSIBLE* water treatment plant operators cannot be overstressed. You, as a water treatment plant operator, have the responsibility for the health and well-being of the community you serve. Yes, you are responsible for the drinking water of your community and any time you fail to do your job, you could be responsible for an outbreak of a waterborne disease which could even result in death. As an operator, you do not want the knowledge that you were negligent in

your duty and, as a result, were responsible for the death of a fellow human being.

QUESTIONS

Below are some questions for you to answer. You should have a notebook in which you can write the answers to the questions. By writing down the answers to the questions, you are helping yourself learn and retain the information. After you have answered all the questions, compare your answers with those given in the Suggested Answers section on page 11. Reread any sections you do not understand and then proceed to the next section. You are your own teacher in this training program, and *YOU* should decide when you understand the material and are ready to continue with new material.

1.0A   Why do many sources of water need treatment?

1.0B   Why is there a need for water treatment plant operators?

1.0C   Why must water treatment plant operators be responsible persons?

1.1   WHAT IS A WATER TREATMENT PLANT?

1.10   Conventional Surface Water Treatment Plant

The purpose of a water treatment plant is to produce safe and pleasant drinking water. This water must be free of disease-causing organisms and toxic substances. Also, the water should not have a disagreeable taste, odor, or appearance.

A water treatment plant takes raw water from a source, such as a stream or lake, and passes the water through a series of treatment processes. The raw water flows through tanks or basins where chemicals are added and mixed with it. Then the water slowly flows through larger tanks which allow the heavier suspended solids to settle out. Any remaining solids are removed by filtration and the water is disinfected. The size of a water treatment plant as well as the number and specific types of processes it uses will depend on several factors: (1) the impurities in the raw water, (2) water quality (purity) standards, (3) the demand for water by the population being served, (4) fire protection, and (5) cost considerations.

To describe a water treatment plant, we will follow a drop of water as it passes through a typical or conventional surface water treatment plant. Most surface waters receive this type of treatment. Figure 1.1 shows a flow diagram of water treatment plant processes and the purpose or function of each process. Figure 1.2 illustrates the flow pattern through a water treatment plant. In this figure both the plan (top view) and the profile (side view) are provided to help you visualize the appearance of a water treatment plant.

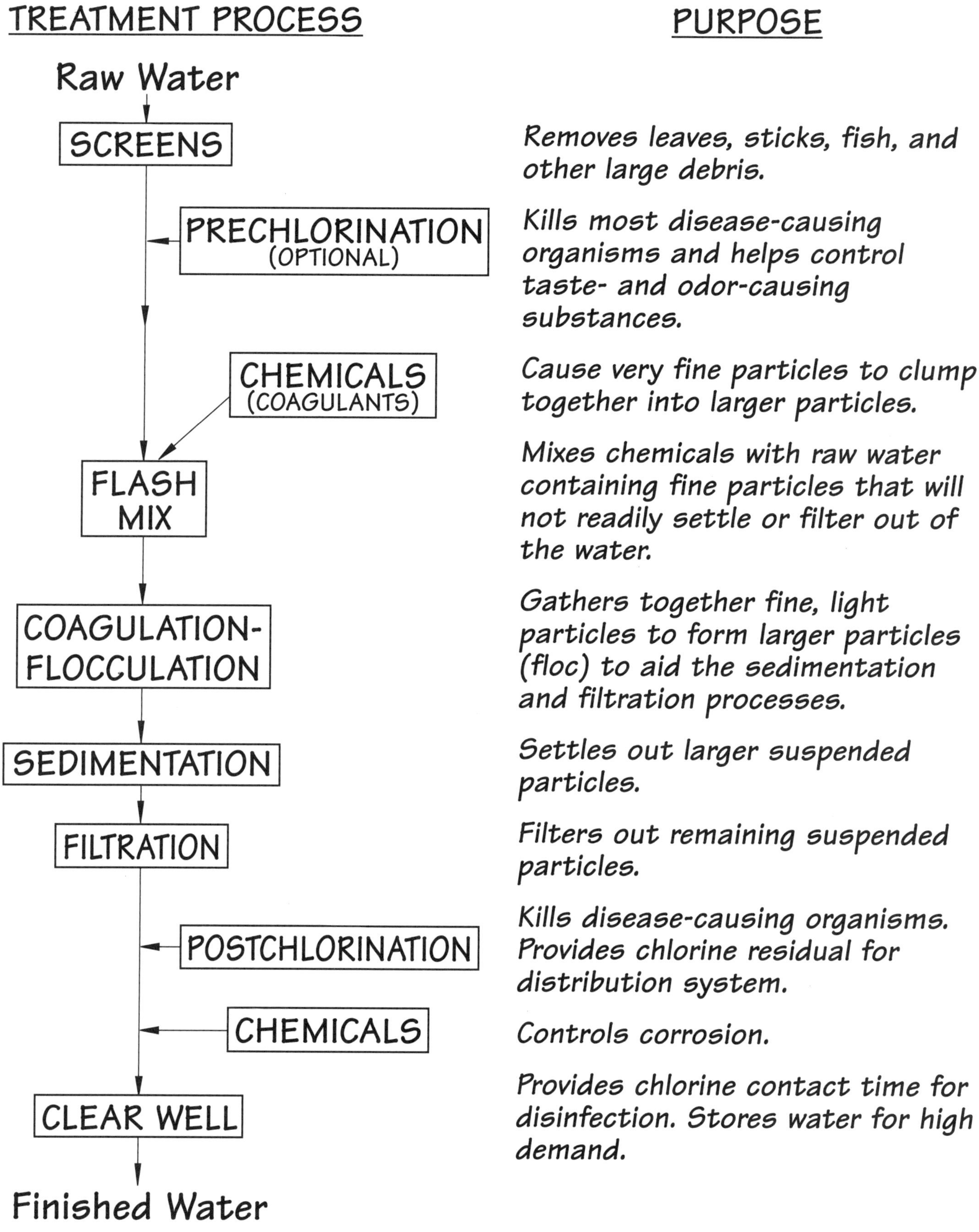

TREATMENT PROCESS

Raw Water

SCREENS

PRECHLORINATION
(OPTIONAL)

CHEMICALS
(COAGULANTS)

FLASH
MIX

COAGULATION-
FLOCCULATION

SEDIMENTATION

FILTRATION

POSTCHLORINATION

CHEMICALS

CLEAR WELL

Finished Water

PURPOSE

Removes leaves, sticks, fish, and other large debris.

Kills most disease-causing organisms and helps control taste- and odor-causing substances.

Cause very fine particles to clump together into larger particles.

Mixes chemicals with raw water containing fine particles that will not readily settle or filter out of the water.

Gathers together fine, light particles to form larger particles (floc) to aid the sedimentation and filtration processes.

Settles out larger suspended particles.

Filters out remaining suspended particles.

Kills disease-causing organisms. Provides chlorine residual for distribution system.

Controls corrosion.

Provides chlorine contact time for disinfection. Stores water for high demand.

Fig. 1.1   Flow diagram of conventional surface water treatment plant processes

Raw water usually enters a water treatment plant through some type of intake structure. The main purpose of the intake structure is to draw in water while preventing leaves and other debris from clogging or damaging pumps, pipes, and other pieces of equipment in the treatment plant. Various types of screens are often found in intake structures or in the suction line to raw water pumps.

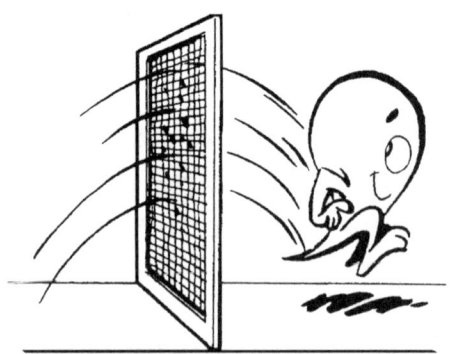

Chlorination at the beginning of a water treatment plant (prechlorination) can help control tastes and odors and also prevent the growth of algae and slimes in other treatment processes. Chlorine is added to water to kill pathogenic (disease-causing) organisms. Also, the use of prechlorination often reduces chlorine requirements for postchlorination. Some waters should not be prechlorinated because they contain substances which will react with chlorine and form cancer-causing compounds (trihalomethanes).

Coagulant chemicals such as alum are added to help remove light, fine particles and other materials suspended in the water. Coagulants cause these very fine particles to clump together into larger particles. A flash mixer is used to thoroughly mix the coagulating chemicals with the water being treated. Flocculation is the name of the treatment process in which paddles gently mix the water. The clumps of particles formed by coagulation come together and form larger and larger floc particles. These larger floc particles are easier to remove by sedimentation and filtration.

Sedimentation is an operation in which the water being treated flows very slowly through a large tank or basin. During this time the heavier floc particles gradually settle out of the water being treated. The flocs and settled solids that reach the bottom of the basin form a sludge that must be removed and either discharged to a sewer or disposed of in a landfill after being dried in drying beds. Treated water leaves the sedimentation basin by flowing over weirs (a flow control device) at the outlet end of the basin.

After sedimentation, the water passes through some type of filter to remove the remaining suspended impurities and flocs. The filter may be made of sand, anthracite coal, or some other type of granular material or a combination of these materials.

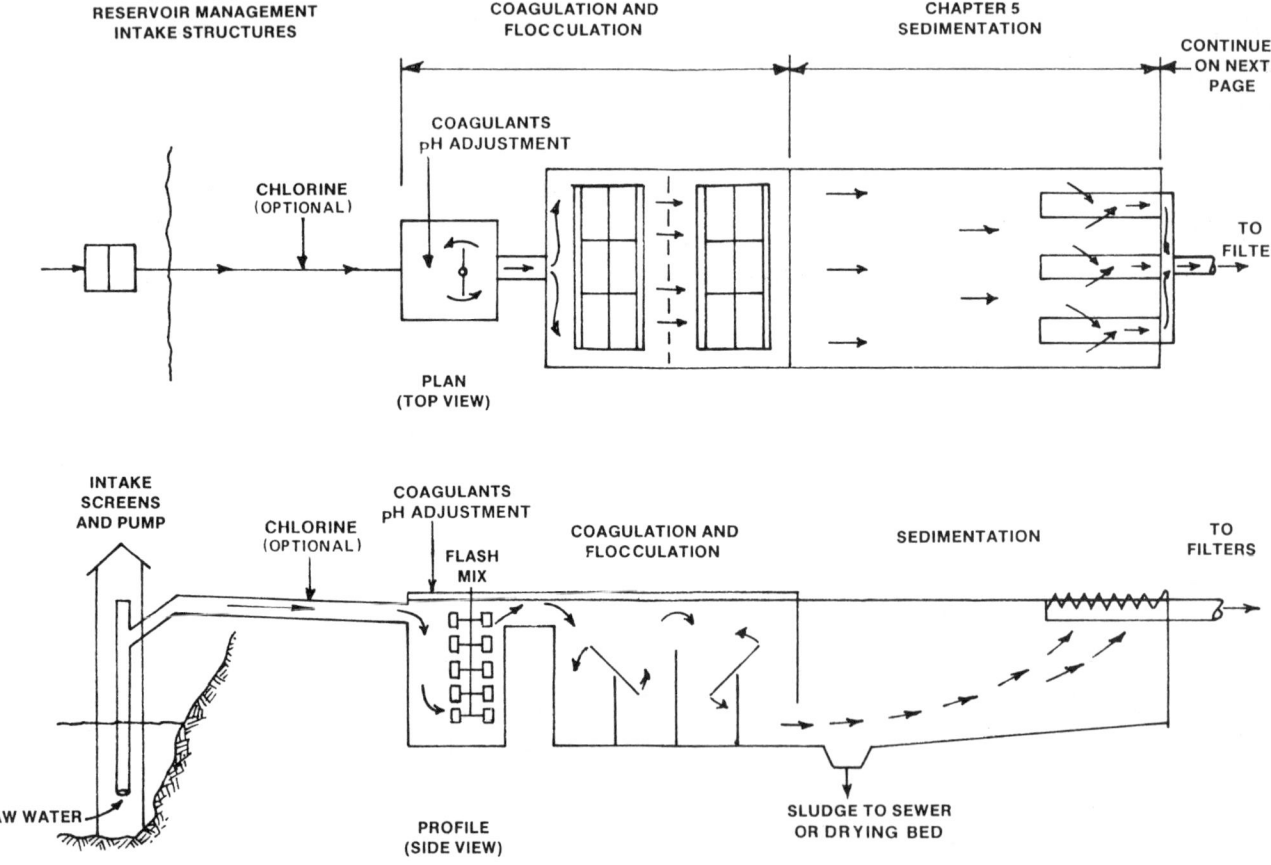

Fig. 1.2   Flow pattern through a conventional surface water treatment plant

After filtration the water is usually disinfected by some type of chlorination process. The purpose of disinfection is to kill the remaining disease-causing organisms in the water.

If the treated water is corrosive (capable of deteriorating metal pipe), chemicals should be added to reduce the corrosivity of the water or to prevent scale (rust) formation.

Treated water is stored in a large tank or basin (clear well) until it is pumped into the distribution system for use or to service storage during low demand periods for later use during periods of high demand. Storage also provides chlorine contact time for disinfection.

1.11 Softening

Some water treatment plants include processes for softening water. Waters are softened to remove excess hardness caused by calcium and magnesium. Extra soap is needed to clean or wash with hard water. Also hard waters will cause scale to develop in water heaters, pipes, and fittings.

1.12 Iron and Manganese Control

Iron and manganese are undesirable because they will cause undesirable color in water and also stain clothes and plumbing fixtures. Iron and manganese also can promote the growth of iron bacteria which can cause tastes and odors. Flow diagrams and treatment processes for removing hardness and also iron and manganese will be discussed in Volume II of this manual.

All of the chapters in Volume I of this manual deal with the processes shown in Figures 1.1 and 1.2. As you go through this manual, feel free to return to these figures so you will understand the location and purposes of these processes and how they relate to each other. *REMEMBER* that if the quality of the raw water changes, or any process fails to do its intended job, all of the downstream processes will be affected.

QUESTIONS

Write your answers in a notebook and then compare your answers with those on page 11.

1.1A What is the purpose of a water treatment plant?

1.1B Why do intake structures at water treatment plants have screens?

1.1C How is the sludge disposed of after it is removed from a sedimentation basin?

1.1D Why is excessive hardness removed from drinking water?

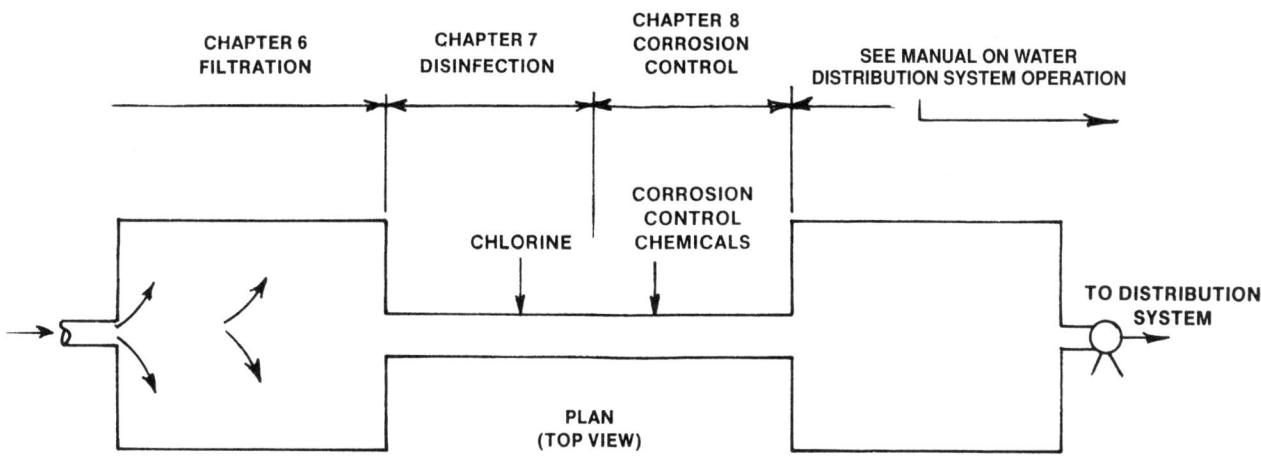

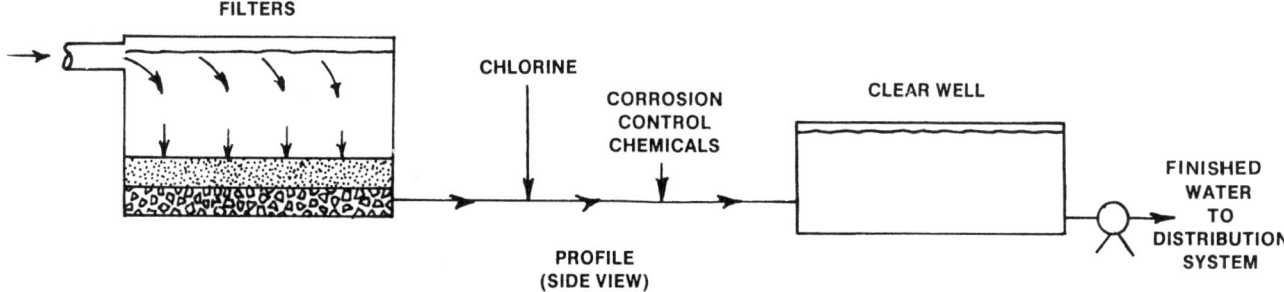

Fig. 1.2 Flow pattern through a conventional surface water treatment plant (continued)

1.2   WHAT DOES A WATER TREATMENT PLANT OPERATOR DO?

1.20   Operation and Maintenance

Simply described, water treatment plant operators keep a treatment plant operating to produce a safe, pleasant, and adequate supply of water. They monitor the raw water entering the plant and keep an eye on the water as it flows through all of the various treatment processes. Flows into the plant are adjusted according to conditions of the raw water and system demands for water. Equipment and facilities are maintained and repaired as necessary to keep the water flowing and the plant working today and into the future. Typical duties performed by water treatment plant operators are summarized in Table 1.1.

To start at the beginning, let's assume that the need for a new or improved water treatment plant has been recognized by the community. The community has voted to issue the necessary bonds to finance the project, and the consulting engineers have been requested to submit plans and specifications. In the best interests of the community and the consulting engineer, you should be present (or at least available) during both the design and construction periods in order to be completely familiar with the entire plant layout, including the piping, equipment, and machinery and their intended operation. This will provide you with the opportunity to relate your plant drawings to actual facilities. At this time you should gather together all the data and literature for the equipment in order to prepare a regular maintenance schedule. You and the engineer should discuss how the water treatment plant should best be run and the means of operation the designer had in mind when the plant was designed.

If the plant is an old one that is being remodeled, you may be in a position to offer excellent advice to the consulting engineer. Your experience provides valuable technical knowledge concerning the characteristics of the raw water and the limitations of the present facilities. Together with the consultant, you can be a member of an expert team able to advise your water utility.

1.21   Supervision and Administration

In addition to operation and maintenance duties for your water treatment plant, you may also be responsible for supervision of personnel. Chief operators frequently have the responsibility of training new operators and should encourage all operators to strive for higher levels of certification.

As a plant administrator, you may be in charge of record-keeping. In this case, you will be responsible for operating and maintaining the facilities as efficiently as possible, keeping in mind that the primary objective is to produce safe and pleasant drinking water from your plant. Without adequate, reliable records of the important phases of operation and maintenance, the effectiveness of your operation will not be properly documented (recorded). Also, accurate records are required by regulatory agencies for compliance with the Primary Drinking Water Regulations of the Safe Drinking Water Act.

Records are an excellent operating tool. Reference to past records can be quite helpful in adjusting treatment processes for various changes in raw water.

You may also be the budget administrator. Here you will be in the best position to give advice on budget requirements, management problems, and utility planning. You should be aware of the necessity for additional expenditures, including funds for plant maintenance and enlargement, equipment replacement, laboratory requirements, and personnel needs. You should recognize and define such needs in sufficient time to inform the proper officials to enable them to accomplish early planning and budgeting.

1.22   Public Relations

As an operator, you are in the field of public relations and must be able to explain the purpose and operation of your water treatment plant to visitors, civic organizations, school classes, representatives of the news media, and even to city council members or directors of your district. A well-guided tour for officials of regulatory agencies or other operators may provide these people with sufficient understanding of your

TABLE 1.1   TYPICAL DUTIES OF A WATER TREATMENT PLANT OPERATOR

1. Start up, shut down, and make periodic operating checks of plant equipment, such as pumping systems, chemical feeders, auxiliary equipment (compressors), and measuring and control systems.

2. Perform routine preventive maintenance, such as lubrication, operating adjustments, cleaning, and painting equipment.

3. Load and unload chemicals, such as chlorine cylinders, bulk liquids, powdered chemicals, and bagged chemicals, either by hand or using chemical-handling equipment such as forklifts and hoists.

4. Perform minor corrective maintenance on plant mechanical equipment, for example, chemical feed pumps and small units.

5. Maintain plant records, including operating logs, daily diaries, chemical inventories, and data logs.

6. Monitor the status of plant operating guidelines, such as flows, pressures, chemical feeds, levels, and water quality indicators, by reference to measuring systems.

7. Collect representative water samples and perform laboratory tests on samples for turbidity, color, odor, coliforms, chlorine residual, and other tests as required.

8. Order chemicals, repair parts, and use tools.

9. Estimate and justify budget needs for equipment and supplies.

10. Conduct safety inspections, follow safety rules for plant operations, and also develop and conduct tailgate safety meetings.

11. Discuss water quality with the public, conduct tours of your plant (especially for school children), and participate in your employer's public relations program.

12. Communicate effectively with other operators and supervisors on the technical level expected for your position.

13. Make arithmetic calculations to determine chemical feed rates, flow quantities, detention and contact times, and hydraulic loadings as required for plant operations.

plant to allow them to suggest helpful solutions to operational problems. One of the best results from a well-guided tour is gaining support from your city council and the public to obtain the funds necessary to run a good operation.

The overall appearance of your water treatment plant indicates to visitors the type of operation you maintain. If the plant looks dirty and run-down, you will be unable to convince your visitors that you are doing a good job. *YOUR RECORDS SHOWING THAT YOU ARE PRODUCING A SAFE DRINKING WATER WILL MEAN NOTHING TO VISITORS UNLESS YOUR PLANT APPEARS CLEAN AND WELL MAINTAINED.*

Another aspect of your job may be handling complaints. When someone contacts you complaining that their drinking water looks muddy, tastes bad, or smells bad, you have a serious problem. Whenever someone complains, record all of the necessary information (name, date, location, and phone number) and have the complaint thoroughly investigated. Be sure to notify the person making the complaint of the results of your investigation and what corrective action was or will be taken.

1.23 Safety

Safety is a very important operator responsibility. Unfortunately, too many operators take safety for granted. *YOU* have the responsibility to be sure that your water treatment plant is a safe place to work and visit. Everyone must follow safe procedures and understand why safe procedures must be followed at all times. All operators must be aware of the safety hazards in and around treatment plants. Most accidents result from carelessness or negligence. You should plan or be a part of an active safety program. Chief operators frequently have the responsibility of training new operators and safe procedures must be stressed.

Clearly, today's water treatment plant operator must be capable of doing many jobs—*AND DOING THEM ALL SAFELY!*

QUESTIONS

Write your answers in a notebook and then compare your answers with those on page 11.

1.2A Why should a water treatment plant operator be present when a new plant is being constructed?

1.2B What is the reason for keeping adequate, reliable records?

1.2C Why are well-guided tours for officials of regulatory agencies or other operators important?

1.2D Why is safety important?

1.3 JOB OPPORTUNITIES

1.30 Staffing Needs

The water treatment field is changing rapidly. New treatment plants are being constructed, and old plants are being modified and enlarged to meet the water demands of our growing population and industries. Towns, municipalities, special districts, and industries all employ water treatment plant operators. Operators, maintenance personnel, supervisors, managers, instrumentation experts, and laboratory technicians are sorely needed now and will be into the future.

1.31 Who Hires Water Treatment Plant Operators?

Operators' paychecks usually come from a city, water agency or district, or a private utility company. The operator also may be employed by one of the many large industries which operate their own water treatment facilities. As an operator, you are always responsible to your employer for operating and maintaining an economical and efficient water treatment plant. An even greater obligation rests with the operator because of the great number of people who drink the water from the water treatment plant. In the final analysis, the operator is really working for the people who depend on the operator to produce safe and pleasant drinking water from the treatment plant.

1.32 Where Do Water Treatment Plant Operators Work?

Jobs are available for water treatment plant operators wherever people live and need someone to treat water for their homes, offices, or industrial processes. The different types and locations of water treatment plants offer a wide range of working conditions. From the mountains to the seas, wherever people gather together into communities, water treatment plants will be found. From a single process operator or a computer control center operator at a complex municipal treatment plant to a one-person manager of a small town water treatment plant, you can select your own special place in water treatment plant operation.

1.33 What Pay Can a Water Treatment Plant Operator Expect?

In dollars? Prestige? Job satisfaction? Community service? In opportunities for advancement? By whatever scale you use, returns are mainly what you make them. If you choose a large municipality, the pay is good and advancement prospects are tops. Choose a small town and the pay may not be as good, but job satisfaction, freedom from time-clock hours, community service, and prestige may well add up to a more desirable outstanding personal achievement. If you have the ability and take advantage of the opportunities, you can make this field your career and advance to an enviable position. Many of these positions are or will be represented by an employee organization that will try to obtain higher pay and other benefits for you. Total reward depends on you and how *YOU APPLY YOURSELF.*

1.34   What Does It Take To Be a Treatment Plant Operator?

DESIRE. First you must make the serious decision to enter this fine profession. You can do this with a high school or a college education. While some jobs will always exist for manual labor, the real and expanding need is for *QUALIFIED OPERATORS.* You must be willing to study and take an active role in upgrading your capabilities. New techniques, advanced equipment, and increasing use of complex instrumentation and computers require a new breed of water treatment plant operator: one who is willing to learn today, and gain tomorrow, for surely your water treatment plant will move toward newer and more effective operation and maintenance procedures. Indeed, the truly service-minded operator assists in adding to and improving the performance of the water treatment plant on a continuing basis.

You can be a water treatment plant operator tomorrow by beginning your learning today; or you can be a better operator, ready for advancement, by accelerating your learning today.

This training course, then, is your start toward a better tomorrow, both for you and for the public who will receive better water from your efforts.

QUESTIONS

Write your answers in a notebook and then compare your answers with those on page 11.

1.3A   Who hires water treatment plant operators?

1.3B   What does it take to be a good water treatment plant operator?

1.4   PREPARING YOURSELF FOR THE FUTURE

1.40   Your Qualifications

What do you know about your job or the job you'd like to obtain? Perhaps a little, and perhaps a lot. You must evaluate the knowledge, skills, and experience you already have and what you will need to achieve future jobs and advancement.

The knowledge and skills required for your job depend to a large degree on the size and type of water treatment plant where you work. You may work in a large, complex water treatment plant serving several hundred thousand persons and employing 15 to 25 operators.

On the other hand, you may operate and maintain a small water treatment plant serving only a thousand people or even fewer. You may be the only operator and have other duties or, at best, you may have one or two helpers. If this is the case,

you must be a "jack-of-all-trades" because of the diversity of your tasks. For additional information about the types of jobs a water treatment plant operator performs, working conditions, wages, and the job outlook, refer to the Bureau of Labor Statistics (BLS) website: www.bls.gov.

1.41   Your Personal Training Program

Beginning on this page you are starting a training course which has been carefully prepared to help you to improve your knowledge and skills to operate and maintain water treatment plants.

You will be able to proceed at your own pace; you will have the opportunity to learn a little or a lot about each topic. This training manual has been prepared this way to meet the various needs of water treatment plant operators, depending on the size and type of plant for which you are responsible. To study for certification and civil service exams, you may have to cover most of the material in both Volumes I and II. You will never know everything about water treatment plants and the equipment, processes, and procedures available for operation and maintenance. However, you will be able to answer some very important questions about how, why, and when certain things happen in these plants. You can also learn how to manage your water treatment plant to produce a reliable output of safe and pleasant drinking water for your customers while minimizing costs in the long run.

This training course is not the only one available to help you improve your abilities. Some state water utility associations, vocational schools, community colleges, and universities offer training courses on both a short- and long-term basis. Many state, local, and private agencies have conducted training programs and informative seminars. Most state health departments can be very helpful in providing training programs or directing you to good programs.

Some libraries can provide you with useful journals and books on water treatment. Listed below are several very good references in the field of water treatment. Prices listed were those available when this manual was published; they will probably increase in the future.

1. *MANUAL OF INSTRUCTION FOR WATER TREATMENT PLANT OPERATORS (NEW YORK MANUAL).* Obtain from Health Education Services, PO Box 7126, Albany, NY 12224. Price, $36.00, includes cost of shipping and handling.

2. *MANUAL OF WATER UTILITIES OPERATIONS (TEXAS MANUAL).* Obtain from Texas Water Utilities Association, 1106 Clayton Lane, Suite 101 East, Austin, TX 78723-1093. Price to members, $22.85; nonmembers, $34.85; price includes cost of shipping and handling.

3. *WATER TREATMENT.* Obtain from American Water Works Association (AWWA), Bookstore, 6666 West Quincy Avenue, Denver, CO 80235. Order No. 1956. ISBN 1-58321-230-2. Price to members, $83.50; nonmembers, $97.50; price includes cost of shipping and handling.

4. *WATER QUALITY.* Obtain from American Water Works Association (AWWA), Bookstore, 6666 West Quincy Avenue, Denver, CO 80235. Order No. 1958. ISBN 1-58321-232-9. Price to members, $73.50; nonmembers, $106.50; price includes cost of shipping and handling.

5. *BASIC SCIENCE CONCEPTS AND APPLICATIONS.* Obtain from American Water Works Association (AWWA), Bookstore, 6666 West Quincy Avenue, Denver, CO 80235. Order No. 1959. ISBN 1-58321-233-7. Price to members, $83.50; nonmembers, $125.00; price includes cost of shipping and handling.

Throughout this manual we will be recommending American Water Works Association (AWWA) publications. Members of AWWA can buy some publications at reduced prices. You can join AWWA by writing to the headquarters office in Denver or by contacting a member of AWWA. Headquarters can help you contact your own state or regional AWWA Section. This professional organization can offer you many helpful training opportunities and educational materials when you join and actively participate with your associates in the field.

1.42   Certification

Certification examinations are usually administered by state regulatory agencies or professional associations. Operators take these exams in order to obtain certificates which indicate a level of professional competence. You should continually strive to achieve higher levels of certification. Successful completion of this operator training program will help you achieve your certification goals.

1.5   ACKNOWLEDGMENTS

Many of the topics and ideas discussed in this chapter were based on similar work written by Larry Trumbull and Walt Driggs.

SUGGESTED ANSWERS

Chapter 1.   THE WATER TREATMENT PLANT OPERATOR

You are not expected to have the exact answer suggested for questions requiring written answers, but you should have the correct idea. The numbering of the questions refers to the section in the chapter where you can find the information to answer the questions. Answers to questions number 1.0A and 1.0B can be found in Section 1.0, "Need for Water Treatment Plant Operators."

Answers to questions on page 4.

1.0A   Many sources of water need treatment because they have been polluted and contaminated by man and nature and are not directly suitable for drinking.

1.0B   Water treatment plant operators are needed to operate, maintain, repair, and manage the water treatment plants that provide us with safe drinking water.

1.0C   Water treatment plant operators must be responsible persons because they are responsible for the health and well-being of the community they serve.

Answers to questions on page 7.

1.1A   The purpose of a water treatment plant is to produce safe and pleasant drinking water.

1.1B   Intake structures at water treatment plants have screens to prevent leaves and other debris from clogging or damaging pumps, pipes, and other pieces of equipment in the treatment plant.

1.1C   Sludge removed from a sedimentation basin is disposed of by discharge to a sewer or is dried and then disposed of in a landfill or on land.

1.1D   Excessive hardness is removed from drinking water because hardness requires extra soap to clean or wash and may cause scale to develop in water heaters, pipes, and fittings.

Answers to questions on page 9.

1.2A   A water treatment plant operator should be present when a new plant is being constructed in order to be completely familiar with the entire plant layout, including the piping, equipment, and machinery and their intended operation.

1.2B   Adequate, reliable records are important to document (record) the effectiveness of your operation and are required by regulatory agencies.

1.2C   Well-guided tours for officials of regulatory agencies or other operators may provide these people with sufficient understanding of your plant to allow them to suggest helpful solutions to operational problems. Also well-guided tours will help gain the funds necessary to run a good operation.

1.2D   Safety is a very important operator responsibility. Most accidents result from carelessness or negligence. Safe procedures must be stressed at all times.

Answers to questions on page 10.

1.3A   Water treatment plant operators may be hired by cities, water agencies or districts, private utility companies, or industries.

1.3B   *DESIRE.* If you want to be a qualified water treatment plant operator, you can do it.

CHAPTER 2

WATER SOURCES AND TREATMENT

by

Bert Ellsworth

TABLE OF CONTENTS

Chapter 2.   WATER SOURCES AND TREATMENT

OBJECTIVES

Chapter 2. WATER SOURCES AND TREATMENT

Following completion of Chapter 2, you should be able to:

1. Describe the importance of water,

2. Identify various sources of water,

3. Outline the procedures of a sanitary survey,

4. Evaluate the suitability of a water source for drinking purposes and as a general water supply, and

5. Identify water quality problems and treatment processes to solve the problems.

PROJECT PRONUNCIATION KEY

by Warren L. Prentice

The Project Pronunciation Key is designed to aid you in the pronunciation of new words. While this key is based primarily on familiar sounds, it does not attempt to follow any particular pronunciation guide. This key is designed solely to aid operators in this program.

You may find it helpful to refer to other available sources for pronunciation help. Each current standard dictionary contains a guide to its own pronunciation key. Each key will be different from each other and from this key. Examples of the difference between the key used in this program and the *WEBSTER'S NEW WORLD COLLEGE DICTIONARY* "Key"* are shown below.

In using this key, you should accent (say louder) the syllable which appears in capital letters. The following chart is presented to give examples of how to pronounce words using the Project Key.

	SYLLABLE				
WORD	1st	2nd	3rd	4th	5th
acid	AS	id			
coliform	COAL	i	form		
biological	BUY	o	LODGE	ik	cull

The first word, *ACID*, has its first syllable accented. The second word, *COLIFORM*, has its first syllable accented. The third word, *BIOLOGICAL*, has its first and third syllables accented.

We hope you will find the key useful in unlocking the pronunciation of any new word.

Term	Project Key	Webster Key
acid	AS-id	aś id
coliform	COAL-i-form	kō′ lə fôrm
biological	BUY-o-LODGE-ik-cull	bī ə läj′ i kəl

---

* *The WEBSTER'S NEW WORLD COLLEGE DICTIONARY, Fourth Edition, 1999, was chosen rather than an unabridged dictionary because of its availability to the operator. Other editions may be slightly different.*

WORDS

Chapter 2. WATER SOURCES AND TREATMENT

ACID RAIN ACID RAIN

Precipitation which has been rendered (made) acidic by airborne pollutants.

APPROPRIATIVE APPROPRIATIVE

Water rights to or ownership of a water supply which is acquired for the beneficial use of water by following a specific legal procedure.

AQUIFER (ACK-wi-fer) AQUIFER

A natural underground layer of porous, water-bearing materials (sand, gravel) usually capable of yielding a large amount or supply of water.

ARTESIAN (are-TEE-zhun) ARTESIAN

Pertaining to groundwater, a well, or underground basin where the water is under a pressure greater than atmospheric and will rise above the level of its upper confining surface if given an opportunity to do so.

CAPILLARY FRINGE CAPILLARY FRINGE

The porous material just above the water table which may hold water by capillarity (a property of surface tension that draws water upward) in the smaller void spaces.

CISTERN (SIS-turn) CISTERN

A small tank (usually covered) or a storage facility used to store water for a home or farm. Often used to store rainwater.

CONTAMINATION CONTAMINATION

The introduction into water of microorganisms, chemicals, toxic substances, wastes, or wastewater in a concentration that makes the water unfit for its next intended use.

CROSS CONNECTION CROSS CONNECTION

A connection between a drinking (potable) water system and an unapproved water supply. For example, if you have a pump moving nonpotable water and hook into the drinking water system to supply water for the pump seal, a cross connection or mixing between the two water systems can occur. This mixing may lead to contamination of the drinking water.

DETENTION TIME DETENTION TIME

(1) The theoretical (calculated) time required for a small amount of water to pass through a tank at a given rate of flow.

(2) The actual time in hours, minutes or seconds that a small amount of water is in a settling basin, flocculating basin or rapid-mix chamber. In storage reservoirs, detention time is the length of time entering water will be held before being drafted for use (several weeks to years, several months being typical).

$$\text{Detention Time, hr} = \frac{(\text{Basin Volume, gal})(24 \text{ hr/day})}{\text{Flow, gal/day}}$$

DIRECT RUNOFF DIRECT RUNOFF

Water that flows over the ground surface or through the ground directly into streams, rivers, or lakes.

DRAWDOWN DRAWDOWN

(1) The drop in the water table or level of water in the ground when water is being pumped from a well.

(2) The amount of water used from a tank or reservoir.

(3) The drop in the water level of a tank or reservoir.

EPIDEMIOLOGY (EP-uh-DE-me-ALL-o-gee) EPIDEMIOLOGY

A branch of medicine which studies epidemics (diseases which affect significant numbers of people during the same time period in the same locality). The objective of epidemiology is to determine the factors that cause epidemic diseases and how to prevent them.

EVAPORATION EVAPORATION

The process by which water or other liquid becomes a gas (water vapor or ammonia vapor).

EVAPOTRANSPIRATION (ee-VAP-o-TRANS-purr-A-shun) EVAPOTRANSPIRATION

(1) The process by which water vapor passes into the atmosphere from living plants. Also called TRANSPIRATION.

(2) The total water removed from an area by transpiration (plants) and by evaporation from soil, snow and water surfaces.

GEOLOGICAL LOG GEOLOGICAL LOG

A detailed description of all underground features discovered during the drilling of a well (depth, thickness and type of formations).

HYDROLOGIC (HI-dro-LOJ-ick) CYCLE HYDROLOGIC CYCLE

The process of evaporation of water into the air and its return to earth by precipitation (rain or snow). This process also includes transpiration from plants, groundwater movement, and runoff into rivers, streams and the ocean. Also called the WATER CYCLE.

IMPERMEABLE (im-PURR-me-uh-BULL) IMPERMEABLE

Not easily penetrated. The property of a material or soil that does not allow, or allows only with great difficulty, the movement or passage of water.

INFILTRATION (IN-fill-TRAY-shun) INFILTRATION

The seepage of groundwater into a sewer system, including service connections. Seepage frequently occurs through defective or cracked pipes, pipe joints, connections or manhole walls.

MICROORGANISMS (MY-crow-OR-gan-IS-zums) MICROORGANISMS

Living organisms that can be seen individually only with the aid of a microscope.

NONPOTABLE (non-POE-tuh-bull) NONPOTABLE

Water that may contain objectionable pollution, contamination, minerals, or infective agents and is considered unsafe and/or unpalatable for drinking.

PALATABLE (PAL-uh-tuh-bull) PALATABLE

Water at a desirable temperature that is free from objectionable tastes, odors, colors, and turbidity. Pleasing to the senses.

PATHOGENIC (PATH-o-JEN-ick) ORGANISMS PATHOGENIC ORGANISMS

Organisms, including bacteria, viruses or cysts, capable of causing diseases (giardiasis, cryptosporidiosis, typhoid, cholera, dysentery) in a host (such as a person). There are many types of organisms which do *NOT* cause disease. These organisms are called non-pathogenic.

PERCOLATION (PURR-co-LAY-shun) PERCOLATION

The slow passage of water through a filter medium; or, the gradual penetration of soil and rocks by water.

POLLUTION POLLUTION

The impairment (reduction) of water quality by agricultural, domestic, or industrial wastes (including thermal and radioactive wastes) to a degree that has an adverse effect on any beneficial use of water.

POTABLE (POE-tuh-bull) WATER POTABLE WATER

Water that does not contain objectionable pollution, contamination, minerals, or infective agents and is considered satisfactory for drinking.

PRECIPITATION (pre-SIP-uh-TAY-shun) PRECIPITATION

(1) The process by which atmospheric moisture falls onto a land or water surface as rain, snow, hail, or other forms of moisture.

(2) The chemical transformation of a substance in solution into an insoluble form (precipitate).

PRESCRIPTIVE (pre-SKRIP-tive)   PRESCRIPTIVE

Water rights which are acquired by diverting water and putting it to use in accordance with specified procedures. These procedures include filing a request (with a state agency) to use unused water in a stream, river or lake.

RAW WATER   RAW WATER

(1)   Water in its natural state, prior to any treatment.

(2)   Usually the water entering the first treatment process of a water treatment plant.

RIPARIAN (ri-PAIR-ee-an)   RIPARIAN

Water rights which are acquired together with title to the land bordering a source of surface water. The right to put to beneficial use surface water adjacent to your land.

SAFE DRINKING WATER ACT   SAFE DRINKING WATER ACT

Commonly referred to as SDWA. An Act passed by the U.S. Congress in 1974. The Act establishes a cooperative program among local, state and federal agencies to ensure safe drinking water for consumers. The Act has been amended several times, including the 1980, 1986, and 1996 Amendments.

SAFE WATER   SAFE WATER

Water that does not contain harmful bacteria, or toxic materials or chemicals. Water may have taste and odor problems, color and certain mineral problems and still be considered safe for drinking.

SAFE YIELD   SAFE YIELD

The annual quantity of water that can be taken from a source of supply over a period of years without depleting the source permanently (beyond its ability to be replenished naturally in "wet years").

SANITARY SURVEY   SANITARY SURVEY

A detailed evaluation and/or inspection of a source of water supply and all conveyances, storage, treatment and distribution facilities to ensure protection of the water supply from all pollution sources.

SEWAGE   SEWAGE

The used household water and water-carried solids that flow in sewers to a wastewater treatment plant. The preferred term is WASTEWATER.

SHORT-CIRCUITING   SHORT-CIRCUITING

A condition that occurs in tanks or basins when some of the flowing water entering a tank or basin flows along a nearly direct pathway from the inlet to the outlet. This is usually undesirable since it may result in shorter contact, reaction, or settling times in comparison with the theoretical (calculated) or presumed detention times.

STRATIFICATION (STRAT-uh-fuh-KAY-shun)   STRATIFICATION

The formation of separate layers (of temperature, plant, or animal life) in a lake or reservoir. Each layer has similar characteristics such as all water in the layer has the same temperature.

TOPOGRAPHY (toe-PAH-gruh-fee)   TOPOGRAPHY

The arrangement of hills and valleys in a geographic area.

TRANSPIRATION (TRAN-spur-RAY-shun)   TRANSPIRATION

The process by which water vapor is released to the atmosphere by living plants. This process is similar to people sweating. Also see EVAPOTRANSPIRATION.

TRIHALOMETHANES (THMs) (tri-HAL-o-METH-hanes)   TRIHALOMETHANES (THMs)

Derivatives of methane, CH_4, in which three halogen atoms (chlorine or bromine) are substituted for three of the hydrogen atoms. Often formed during chlorination by reactions with natural organic materials in the water. The resulting compounds (THMs) are suspected of causing cancer.

TURBIDITY (ter-BID-it-tee)   TURBIDITY

The cloudy appearance of water caused by the presence of suspended and colloidal matter. In the waterworks field, a turbidity measurement is used to indicate the clarity of water. Technically, turbidity is an optical property of the water based on the amount of light reflected by suspended particles. Turbidity cannot be directly equated to suspended solids because white particles reflect more light than dark-colored particles and many small particles will reflect more light than an equivalent large particle.

TURBIDITY UNITS (TU)   TURBIDITY UNITS (TU)

Turbidity units are a measure of the cloudiness of water. If measured by a nephelometric (deflected light) instrumental procedure, turbidity units are expressed in nephelometric turbidity units (NTU) or simply TU. Those turbidity units obtained by visual methods are expressed in Jackson Turbidity Units (JTU) which are a measure of the cloudiness of water; they are used to indicate the clarity of water. There is no real connection between NTUs and JTUs. The Jackson turbidimeter is a visual method and the nephelometer is an instrumental method based on deflected light.

WASTEWATER   WASTEWATER

A community's used water and water-carried solids (including used water from industrial processes) that flow to a treatment plant. Storm water, surface water, and groundwater infiltration also may be included in the wastewater that enters a wastewater treatment plant. The term "sewage" usually refers to household wastes, but this word is being replaced by the term "wastewater."

WATER CYCLE   WATER CYCLE

The process of evaporation of water into the air and its return to earth by precipitation (rain or snow). This process also includes transpiration from plants, groundwater movement, and runoff into rivers, streams and the ocean. Also called the HYDROLOGIC CYCLE.

WATER TABLE   WATER TABLE

The upper surface of the zone of saturation of groundwater in an unconfined aquifer.

YIELD   YIELD

The quantity of water (expressed as a rate of flow—GPM, GPH, GPD, or total quantity per year) that can be collected for a given use from surface or groundwater sources. The yield may vary with the use proposed, with the plan of development, and also with economic considerations. Also see SAFE YIELD.

ZONE OF AERATION   ZONE OF AERATION

The comparatively dry soil or rock located between the ground surface and the top of the water table.

ZONE OF SATURATION   ZONE OF SATURATION

The soil or rock located below the top of the groundwater table. By definition, the zone of saturation is saturated with water. Also see WATER TABLE.

CHAPTER 2.   WATER SOURCES AND TREATMENT

2.0   IMPORTANCE OF WATER

For decades Americans have used water as though their supply would never fail. In recent years, drought conditions have forcibly brought the need to conserve and properly budget our water resources to the minds of water supply managers. Even in the driest years, though, rain across the country enormously exceeds water use. The trouble is that the nation's water resources are unevenly distributed. The Pacific Northwest has a big surplus. The agricultural states of the Southwest fight for the last salty drop of river water from the Lower Colorado River. The federal government has spent billions of dollars building and operating facilities to divert water for use in arid and water-short areas. Contamination is a problem, too. Mineral residues from irrigation have damaged once fertile soil. *ACID RAIN*[1] is killing the fish in mountain lakes. America's drinking water has been tainted with substances as exotic as trichloroethylene (TCE) and as commonplace as highway salt. Vast underground basins of water, deposited over many years, have been seriously depleted in a matter of decades.

All water comes as rain or precipitation from the sky, but 92 percent of the water either evaporates immediately or runs off eventually into the oceans. One-quarter of the water that irrigates, powers, and bathes America is taken from an ancient network of underground aquifers. In 1950, the United States took some 12 trillion gallons (45 billion cubic meters) of water out of the ground; by 1980 the figure had more than doubled.

Water shortages directly influence energy consumption. As groundwater levels fall, more energy is required to pump water from deeper levels in the basin. In several areas, vast water projects use large amounts of electricity to pump water many miles along the project. As energy becomes more expensive, the users of the water will see the increased cost reflected in their water rates. This link between water demands and pumping costs is a driving force behind water conservation programs.

Water is regarded as commonplace because it is the most plentiful liquid on earth and because of our familiarity with it. All of the tissues of our bodies are bathed in it. Whatever may be the thing which we call life on earth, it requires a water environment. Our foods must be suspended or dissolved in water solutions to be carried to the different parts of the body. Also, most waste products are eliminated from the body as water-soluble substances.

Both plant life and animal life depend upon water for survival. A plant receives the greater part of its food from the soil in water solutions and manufactures the rest of its food in the presence of water.

Water is present in almost all natural objects and in almost every part of the earth that people can reach. There is water vapor in the air, and liquid water in rocks and soil. In addition to the water that wets them, clay and certain kinds of rocks contain water in chemical combination with other substances.

Water may be commonplace, but useful water is not always readily available. Even before the discovery of America, one of the common causes of war between Native American tribes was water rights. Among the first considerations of any new land development is water. Useful water is only rarely free, and it is not very abundant in many parts of the United States. There are not many places left where a person can feel safe in drinking water from a spring, stream, or pond. Even the groundwater produced by wells must be tested regularly. In some areas, human activities have made it difficult to locate a safe water supply of any sort. In certain coastal areas, for example, overpumping from the ground has depleted the groundwater basins. As a result, the intrusion of seawater is ruining the basin for most useful purposes. Other sources of groundwater contamination include seepage from septic tank leaching systems and agricultural drainage systems, the improper disposal of hazardous wastes in sanitary landfills and dumps, and the entry of surface runoff into poorly constructed wells. Some human activities clearly pose a serious threat to life on this planet.

---

[1] *Acid Rain.   Precipitation which has been rendered (made) acidic by airborne pollutants.*

QUESTIONS

Write your answers in a notebook and then compare your answers with those on page 33.

2.0A   Why has it become necessary to conserve and properly budget our water resources?

2.0B   How are water shortages and energy consumption linked together?

2.0C   Name three ways groundwater may become contaminated.

2.1   SOURCES OF WATER

2.10   The Water (Hydrologic) Cycle (Figure 2.1)

All water comes in the form of precipitation. Water evaporates from the ocean by the energy of the sun at an overall rate of about six feet (1.8 m) of water annually. The water which is evaporated is salt-free water since the heavier mineral salts are left behind. This water vapor rises, is carried along by winds, and eventually condenses into clouds. When these clouds become chilled, the small particles of water collect into larger droplets which may precipitate over land or water. As the water falls in the form of rain, snow, sleet, or hail, it clings to and carries with it all the dust and dirt in the air. Needless to say, the first water that falls during a storm picks up the greatest concentration of contamination. After a short period of fall, the precipitation is relatively free of pollutants. A large part of the evaporated water is carried over land masses by the winds and the droplets that fall there make up our supply of fresh water. These droplets may  soak into the ground, fall as snow on the mountain tops, or collect in lakes, but in one form or another, all of the droplets eventually return to the ocean. This, in brief, is the framework of the water cycle.

2.11   Rights to the Use of Water

The rights of an individual to use water for domestic, irrigation, or other purposes varies in different states. Some water rights stem from ownership of the land bordering or overlying the source, while others are acquired by a performance of certain acts required by law.

There are three basic types of water rights:

1. *RIPARIAN*—rights which are acquired with title to the land bordering a source of surface water.

2. *APPROPRIATIVE*—rights which are acquired for the beneficial use of water by following a specific legal procedure.

3. *PRESCRIPTIVE*—rights which are acquired by diverting water, to which other parties may or may not have prior claims, and putting it to use for a period of time specified by statute.

When there is any question regarding the right to the use of water, a property owner should consult with the appropriate authority and clearly establish rights to its use.

2.12   Ocean

At some time in its history, virtually all water resided in the oceans. Water in the oceans is too salty for drinking and irrigation uses. Desalination technology is advancing and can make the water in the oceans suitable for other beneficial uses. By evaporation, moisture is transferred from the ocean surface to the atmosphere, where winds carry the moisture-laden air over land masses. Under certain conditions, this water vapor condenses to form clouds, which release their moisture as precipitation in the form of rain, hail, sleet, or snow. When rain falls toward the earth, part of it may re-evaporate and return immediately to the atmosphere. Precipitation in excess of the amount that wets a surface or evaporates immediately is available as a potential source of water supply.

2.13   Surface Water

2.130   Direct Runoff

Surface water accumulates mainly as a result of direct runoff from precipitation (rain or snow). Precipitation that does not enter the ground through infiltration or is not returned to the atmosphere by evaporation flows over the ground surface and is classified as direct runoff. Direct runoff is water that drains from saturated or *IMPERMEABLE*[2] surfaces, into stream channels, and then into natural or artificial storage sites (or into the ocean in coastal areas).

The amount of available surface water depends largely upon rainfall. When rainfall is limited, the supply of surface water will vary considerably between wet and dry years. In areas of scant rainfall, people build individual cisterns for the storage of rain which drains from the catchment areas of roofs. This type of water supply is used extensively in areas such as the Bermuda Islands, where groundwater is virtually nonexistent and there are no streams.

Surface water supplies may be further divided into river, lake, and reservoir supplies. In general, they are characterized by soft water, turbidity, suspended solids, some color, and microbial contamination. Groundwaters, on the other hand, are characterized by higher concentrations of dissolved solids, dissolved gases, lower levels of color, relatively high hardness, and freedom from microbial contamination.

---

[2] Impermeable (im-PURR-me-uh-BULL).   Not easily penetrated. The property of a material or soil that does not allow, or allows only with great difficulty, the movement or passage of water.

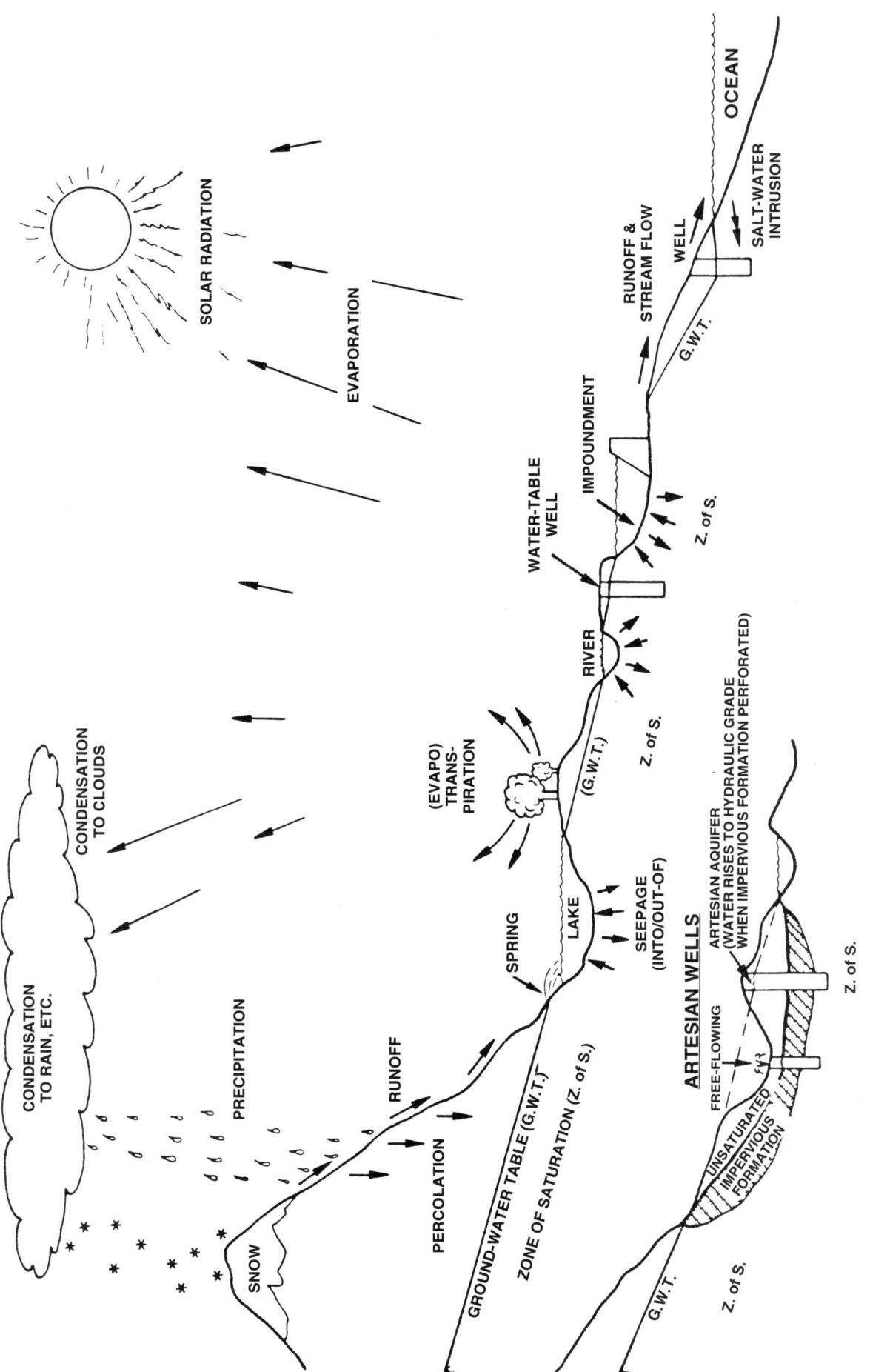

Fig. 2.1   Water (hydrologic) cycle as related to water supply
(Source: *BASIC WATER TREATMENT OPERATOR TRAINING COURSE I,*
by Leonard Ainsworth, by permission of California-Nevada Section, AWWA)

2.131 Rivers and Streams

Many of the largest cities in the world depend entirely upon large rivers for their water supplies. In using a river or stream supply, one should always be concerned with upstream conditions. Some cities draw drinking water from a stream or river into which the treated wastewater (sewage) from upstream cities has been discharged. This can present very serious problems in water treatment. Because of upstream pollution (wastewater, agricultural drainage, or industrial waste), the proper treatment of river and stream supplies is extremely important. Rivers and streams are also susceptible to scouring of the bottom, changing channels, and silting. Before the intake for a water supply is located in a river or stream, a careful study must be made of the stream bottom, its degree of scour, and the settling out of silt. Provisions must be made in the design of the intake to make sure that it can withstand the forces which will act upon it during times of flood, heavy silting, ice conditions, and adverse runoff conditions. Because of variations in the quality of water supplied by a river or stream, purification effectiveness must be continually checked. This is especially true if there are industries upstream from the intake which may dump undesirable wastes into the supply. Sudden pollutant loads might not be discovered unless constant monitoring of the raw water is maintained by the treatment plant operator.

2.132 Lakes and Reservoirs

The selection and use of water from any surface storage source requires considerable study and thought. When ponds, lakes, or open reservoirs are used as sources of water supply, the danger of contamination and of the consequent spread of diseases such as typhoid, hepatitis, dysentery, giardiasis, and cryptosporidiosis exists. Clear water is not always safe water and the old saying that running water "purifies itself" to drinking water quality within a stated distance is false.

The potential for contamination of surface water makes it necessary to regard such sources of supply as unsafe for domestic use unless properly treated, including filtration and disinfection. To ensure the delivery of a constant, safe drinking water to consumers also requires diligent attention to the operation and maintenance of the distribution system.

Lakes and reservoirs are subject to seasonal changes in water quality such as those brought about by *STRATIFICATION*[3] and the possible increase of organic and mineral contamination that occurs when a lake "turns over." In any body of water, the surface water will be warmed by the sun in spring and summer causing higher temperatures on the surface. Then in the fall, the cooler air temperatures cool the surface water until it reaches the same temperature as the subsurface waters. At this point, the water temperature is fairly uniform (the same) throughout the entire depth of the lake or reservoir. A breeze will start the surface water circulating and cause the lake to "turn over," thus bringing poor quality deeper water to the surface.

Lakes and reservoirs are susceptible to algal blooms, especially after fall or spring turnovers. The rapid growth of algae (blooms) will occur when the temperature is right and the water contains enough nutrients to support rapid algal growth. In any given body of water, blooms of various types of algae can occur several times during a season depending on what algae are present and whether the conditions are right for algal growth.

Water supplies drawn from large lakes and reservoirs through multiport intake facilities (openings at several depths) are generally of good quality since the water can be drawn from a depth where algal growths are not prevalent. A large lake or reservoir also dilutes any contamination that may have been discharged into it or one of its tributaries.

Large bodies of water are generally attractive recreation areas. If the water is also used for domestic supplies, however, it must be protected from contamination. This will require proper construction and location of recreation facilities such as boat launching ramps, boat harbors, picnic and camping areas, fishing, and open beach areas away from the intake area. The location and construction of wastewater collection, treatment, and disposal facilities must also be carefully studied to protect domestic water supplies from contamination.

QUESTIONS

Write your answers in a notebook and then compare your answers with those on page 33.

2.1A   What is the water (hydrologic) cycle?

2.1B   List the three basic types of water rights.

2.1C   What are the general water quality characteristics of surface water supplies?

2.1D   What are the general water quality characteristics of groundwater supplies?

2.1E   What items should be considered before selecting a location and constructing a water supply intake located in a river or stream?

2.1F   What water treatment processes are considered essential to reliably treat physically and bacteriologically contaminated surface waters for domestic use?

2.1G   How can provisions be made to allow recreation on water supply lakes and reservoirs without endangering water quality?

2.14 Groundwater

2.140 Sources *(refer to Figure 2.1, page 23)*

Part of the precipitation that falls infiltrates the soil. This water replenishes the soil moisture, or is used by growing

---

[3] *Stratification (STRAT-uh-fuh-KAY-shun). The formation of separate layers (of temperature, plant, or animal life) in a lake or reservoir. Each layer has similar characteristics such as all water in the layer has the same temperature.*

plants and returned to the atmosphere by *TRANSPIRATION*.[4] Water that drains downward (percolates) below the root zone finally reaches a level at which all of the openings or voids in the earth's materials are filled with water. This zone is known as the zone of saturation. Water in the zone of saturation is referred to as groundwater. The upper surface of the zone of saturation, if not confined by impermeable material, is called the water table. When an overlying, impermeable formation confines the water in the zone of saturation under pressure, the groundwater is said to be under artesian pressure. The name "artesian" comes from the ancient province of Artesium in France where, in the days of the Romans, water flowed to the surface of the ground from a well. However, not all water from wells that penetrate artesian formations flows to ground level. For a well to be artesian, the water in the well must rise above the top of the aquifer. (An aquifer, or water-bearing formation, is an underground layer of rock or soil which permits the passage of water.)

The porous material just above the water table may contain water by capillarity in the smaller void spaces. This zone is referred to as the capillary fringe. Since the water held in the capillary fringe will not drain freely by gravity, this zone is not considered a true source of supply.

Because of the irregularities in underground deposits or layers and in surface *TOPOGRAPHY*,[5] the water table occasionally intersects (meets) the surface of the ground at a spring or in the bed of a stream, lake, or the ocean. As a result, groundwater moves to these locations as seepage out of the aquifer (groundwater reservoir). Thus, groundwater is continually moving within aquifers even though the movement may be very slow (see Figure 2.1). The water table (artesian pressure surface) thus may slope from areas of recharge to lower areas of discharge. The total head difference represented by these slopes causes the flow of groundwater within the aquifer. Seasonal variations in the supply of water to the underground reservoir cause considerable changes in the elevation and slope of the water table and the artesian pressure level.

2.141 Wells

A well that penetrates the water table can be used to extract water from the groundwater basin (see Figure 2.1). The removal of water by pumping will naturally cause a lowering of the water table near the well. If pumping continues at a rate that exceeds the rate of replacement by the water-bearing formations, the "sustained yield" of the well or group of wells has been exceeded. The "safe yield" will be exceeded if wells ex-

tract water from an aquifer over a period of time at a rate that will deplete the aquifer and bring about other undesired results (such as seawater intrusion and land subsidence). This situation is a poor practice, but occurs quite frequently in many areas of the U.S.

2.142 Springs

Groundwater that flows naturally from the ground is called a spring. Depending upon whether the discharge is from a water table or an artesian aquifer, springs may flow by gravity or by artesian pressure. The flow from a spring may vary considerably; when the water table or artesian pressure fluctuates, so does the flow from the spring.

2.15 Reclaimed Water

The use of treated wastewater as a source of water for non-food crop irrigation is an established practice in many regions of the world. The type of crop which can be safely irrigated depends somewhat on the quality of wastewater and method of irrigation. At the present time, more than 20,000 acres (8,000 hectares) of agricultural lands in California are irrigated, all or in part, with reclaimed water. Two of the largest operations are at Bakersfield and at Fresno, California. The City of Bakersfield has used wastewater effluent for irrigation since 1912. At the present time, approximately 2,400 acres (1,000 hectares) of alfalfa, cotton, barley, sugar beets, and pasture are irrigated. Fresno irrigates 3,500 acres (1,400 hectares) of the same types of crops. These two operations use almost 30,000 acre-feet (37 million cubic meters) of reclaimed water per year. Almost 90 percent of the 2,000,000 acre-feet (2.5 billion cubic meters) of reclaimed wastewater in California is used for crop irrigation.

Other uses for reclaimed wastewater include:

1. Greenbelt (parks) areas,

2. Golf course irrigation,

3. Landscape irrigation,

4. Industrial reuse,

5. Groundwater recharge,

6. Landscape impoundments, and

7. Wetlands/marsh enhancement.

Reclaimed water can be used safely for any of these purposes, with the possible exception of groundwater recharge. Health experts have serious questions regarding organic compounds that are present in wastewater and about our ability to reduce them to safe levels. These doubts increase further when we realize that a large number of new and potentially toxic chemicals are developed each year. Laboratories are unable to adequately detect all of these chemicals without expensive monitoring programs. To protect groundwater resources, regulations require that "reclaimed water used for groundwater recharge of domestic water supply aquifers by surface spreading shall be at all times of a quality that fully protects public health." Proposed groundwater recharge projects must be investigated on an individual basis where the use of reclaimed water involves a potential risk to public health.

---

[4] *Transpiration (TRAN-spur-RAY-shun). The process by which water vapor is released to the atmosphere by living plants. This process is similar to people sweating.*

[5] *Topography (toe-PAH-gruh-fee). The arrangement of hills and valleys in a geographic area.*

Treatment of reclaimed water should be appropriate for the intended use. The greater the potential exposure to the public, the more extensive the treatment needs to be. Regulations often specify not only the degree of treatment for the usage of water, but also the reliability features that must be incorporated into the treatment processes to ensure a continuous high degree of finished water quality. Studies indicate that the average reclamation plant does not achieve the quality of treatment expected on a continuous basis. This is of concern to the water supplier and also the regulatory agencies that are charged with the responsibility of ensuring that the health of the public is protected.

Write your answers in a notebook and then compare your answers with those on page 33.

2.1H   What causes the flow of groundwater within an aquifer?

2.1I   How can the "safe yield" of an aquifer be exceeded?

2.1J   List some of the possible uses of reclaimed wastewater.

2.1K   How much treatment should reclaimed water receive before use?

2.2   SELECTION OF A WATER SOURCE

2.20   Sanitary Survey[6]

The importance of detailed sanitary surveys of water supply sources cannot be overemphasized. With a new supply, the sanitary survey should be made during the collection of initial engineering data covering the development of a given source and its capacity to meet existing and future needs. The sanitary survey should include the location of all potential and existing health hazards and the determination of their present and future importance. Persons trained in public health engineering and the *EPIDEMIOLOGY*[7] of waterborne diseases should conduct the sanitary survey. In the case of an existing supply, sanitary surveys should be made frequently enough to control health hazards and to maintain high water quality.

The information furnished by a sanitary survey is essential to evaluating the bacteriological and chemical water quality data. The following outline lists the essential factors which should be investigated or considered in a sanitary survey. These items are essential to (1) identify potential hazards, (2) determine factors which affect water quality, and (3) select treatment requirements. Not all of the items are important to any one supply and, in some cases, items not in the list could be found to be significant during the field investigation.

GROUNDWATER SUPPLIES:

a. Character of local geology; slope (topography) of ground surface.

b. Nature of soil and underlying porous material, whether clay, sand, gravel, rock (especially porous limestone); coarseness of sand or gravel; thickness of water-bearing stratum; depth of water table; location and *GEOLOGICAL LOG*[8] of nearby wells.

c. Slope of water table, preferably as determined from observation wells or as indicated by slope of the ground surface.

d. Extent of the drainage area likely to contribute water to the supply.

e. Nature, distance, and direction of local sources of pollution.

f. Possibility of surface-drainage water entering the supply and of wells becoming flooded.

g. Methods used for protecting the supply against contamination from wastewater collection and treatment facilities and industrial waste disposal sites.

h. Well construction: materials, diameter, depth of casing and concrete collar; depth to well screens or perforations; length of well screens or perforations.

i. Protection of well head at the top and on the sides.

j. Pumping station construction (floors, drains); capacity of pumps; storage or direct to distribution system.

k. Drawdown when pumps are in operation; recovery rate when pumps are off.

l. Presence of an unsafe supply nearby, and the possibility of *CROSS CONNECTIONS*[9] causing a danger to the public health.

m. Disinfection: equipment, supervision, test kits, or other types of laboratory control.

SURFACE WATER SUPPLIES:

a. Nature of surface geology; character of soils and rocks.

b. Character of vegetation; forests; cultivated and irrigated land.

c. Population and wastewater collection, treatment, and disposal on the watershed.

---

[6] *Sanitary Survey.   A detailed evaluation and/or inspection of a source of water supply and all conveyances, storage, treatment and distribution facilities to ensure protection of the water supply from all pollution sources.*

[7] *Epidemiology (EP-uh-DE-me-ALL-o-gee).   A branch of medicine which studies epidemics (diseases which affect significant numbers of people during the same time period in the same locality). The objective of epidemiology is to determine the factors that cause epidemic diseases and how to prevent them.*

[8] *Geological Log.   A detailed description of all underground features discovered during the drilling of a well (depth, thickness and type of formations).*

[9] *Cross Connection.   A connection between a drinking (potable) water system and an unapproved water supply. For example, if you have a pump moving nonpotable water and hook into the drinking water system to supply water for the pump seal, a cross connection or mixing between the two water systems can occur. This mixing may lead to contamination of the drinking water.*

d. Methods of wastewater disposal, whether by diversion from watershed or by reclamation treatment.

e. Closeness of sources of fecal pollution (especially birds) to intake of water supply.

f. Proximity to watershed and character of sources of contamination including industrial wastes, oil field brines, acid waters from mines, sanitary landfills, and agricultural drain waters.

g. Adequacy of supply as to quantity (safe yield).

h. For lake or reservoir supplies: wind direction and velocity data; drift of pollution; and algae growth potential.

i. Character and quality of raw water: typical coliform counts (MPN or membrane filter), algae, turbidity, color, and objectionable mineral constituents.

j. Normal period of *DETENTION TIME*.[10]

k. Probable minimum time required for water to flow from sources of pollution to reservoir and through the reservoir to the intake tower.

l. The possible currents of water within the reservoir (induced by wind or reservoir discharge) which could cause *SHORT-CIRCUITING*[11] to occur.

m. Protective measures in connection with the use of the watershed to control fishing, boating, landing of airplanes, swimming, wading, ice cutting, and permitting animals on shoreline areas.

n. Efficiency and constancy of policing activities on the watershed and around the lake.

o. Treatment of water: kind and adequacy of equipment; duplication of parts for reliable treatment; effectiveness of treatment; numbers and competency of supervising and operating personnel; contact period after disinfection; free chlorine residuals and monitoring of the water supply both during treatment and following treatment.

p. Pumping facilities: pump station design, pump capacity, and standby unit(s).

q. Presence of an unsafe supply nearby, and the possibility of cross connections causing a danger to the public health.

2.21 Precipitation

Precipitation in the form of rain, snow, hail, or sleet contains very few impurities. (However, there are exceptions such as acid rain and dust from dust bowl areas.) Trace amounts of mineral matter, gases, and other substances may be picked up by precipitation as it forms and falls through the earth's atmosphere. Precipitation, however, has virtually no microbial content.

Once precipitation reaches the earth's surface, many opportunities are presented for the introduction of mineral and organic substances, microorganisms, and other forms of contamination. When water runs over or through the ground surface it may pick up particles of soil. This is noticeable in the water as cloudiness or turbidity. This water also picks up particles of organic matter and microorganisms. As surface water seeps downward into the soil and through the underlying material to the water table, most of the suspended particles are filtered out. This natural filtration may be partially effective in removing microorganisms and other particulate materials; however, the chemical characteristics of the water usually change considerably when it comes in contact with underground mineral deposits.

The widespread use of synthetically produced chemical compounds, especially pesticides, has raised concern for their potential to contaminate water. Many of these materials are known to be toxic (poisonous), some cause cancer, and others have certain undesirable characteristics even when present in a relatively small concentration.

Agents which alter the quality of water as it moves over or below the surface of the earth may be classified under four major headings:

A. *PHYSICAL*—Physical characteristics relate to the sensory qualities of water for domestic use; for example, the water's observed color, turbidity, temperature, taste, and odor.

B. *CHEMICAL*—Chemical differences between waters include mineral content and the presence or absence of constituents such as fluoride, sulfide, and acids. The comparative performance of hard and soft waters in laundering is one visible effect.

C. *BIOLOGICAL*—The presence of organisms (viruses, bacteria, algae, mosquito larvae), alive or dead, and their metabolic products determine the biological character of water. These may also be significant in modifying the physical and chemical characteristics of water.

D. *RADIOLOGICAL*—Radiological factors must be considered because there is a possibility that the water may have come in contact with radioactive substances.

---

[10] *Detention Time.   (1) The theoretical (calculated) time required for a small amount of water to pass through a tank at a given rate of flow. (2) The actual time in hours, minutes or seconds that a small amount of water is in a settling basin, flocculating basin or rapid-mix chamber. In storage reservoirs, detention time is the length of time entering water will be held before being drafted for use (several weeks to years, several months being typical).*

$$\text{Detention Time, hr} = \frac{(\text{Basin Volume, gal})(24 \text{ hr/day})}{\text{Flow, gal/day}}$$

[11] *Short-Circuiting.   A condition that occurs in tanks or basins when some of the flowing water entering a tank or basin flows along a nearly direct pathway from the inlet to the outlet. This is usually undesirable since it may result in shorter contact, reaction, or settling times in comparison with the theoretical (calculated) or presumed detention times.*

Consequently, in the development of water supply systems, it is necessary to examine carefully all the factors which might adversely affect the water supply.

QUESTIONS

Write your answers in a notebook and then compare your answers with those on page 33.

2.2A   What is the purpose of a sanitary survey?

2.2B   How frequently should a sanitary survey be conducted for an existing water supply?

2.2C   When conducting a sanitary survey, what protective measures should be investigated regarding use of the watershed?

2.22   Physical Characteristics

To be suitable for human use, water should be free from all impurities which are offensive to the senses of sight, taste, and smell. The physical characteristics which might be offensive include turbidity, color, taste, odor, and temperature.

TURBIDITY: The presence of suspended material in water causes cloudiness which is known as turbidity. Clay, silt, finely divided organic material, plankton, and other inorganic materials give water this appearance. Turbidities in excess of 5 turbidity units are easily visible in a glass of water, and this level is usually objectionable for aesthetic reasons. Turbidity's major danger in drinking water is that it can harbor bacteria as well as exert a high demand on chlorine. Water that has been filtered to remove the turbidity should have considerably less than one turbidity unit. Good treatment plants consistently obtain finished water turbidity levels of 0.05 to 0.3 units.

COLOR: Dissolved organic material from decaying vegetation and certain inorganic matter cause color in water. Occasionally, excessive blooms of algae or the growth of other aquatic microorganisms may also impart color. Iron and manganese may be the cause of consumer complaints (red or black water). While the color itself is not objectionable from the standpoint of health, its presence is aesthetically objectionable and suggests that the water needs better treatment. In some instances, however, a color in the water indicates more than an aesthetic problem. For example, an amber color in the water could indicate the presence of humic substances which could later be formed into trihalomethanes or it could indicate acid waters from mine drainage.

TEMPERATURE: The most desirable drinking waters are consistently cool and do not have temperature fluctuations of more than a few degrees. Groundwater and surface water from mountainous areas generally meet these requirements. Most individuals find that water having a temperature between 50° and 60°F (10° and 15°C) is most pleasing while water over 86°F (30°C) is not acceptable. The temperature of groundwaters varies with the depth of the aquifer. Water from very deep wells (more than 1,000 ft or 300 m) may be quite warm. Temperature also affects sensory perception of tastes and odors.

TASTES: Each area's natural waters have a distinctive taste related to the dissolved mineral characteristics of local geology. Occasionally, algal growths also impart a distinctive taste. However, taste is rarely measured since most water treatment plants cannot alter a water's mineral characteristics.

ODORS: Growths of algae in a water supply can give the water an unpleasant odor. Some groundwaters may contain hydrogen sulfide which will produce a disagreeable rotten egg odor.

2.23   Chemical Characteristics

The nature of the materials that form the earth's crust affects not only the quantity of water that may be recovered, but also its chemical makeup. As surface water infiltrates and percolates downward to the water table, it dissolves some of the minerals contained in soils and rocks. Groundwater, therefore, sometimes contains more dissolved minerals than surface water. The use and disposal of chemicals by society can also affect water quality.

Chemical analysis of a domestic water supply is broken down into three areas:

1. Inorganic chemicals which include the toxic (poisonous) metals—arsenic, barium, cadmium, chromium, lead, mercury, selenium, and silver; and the nonmetals—fluoride and nitrate;

2. Organic chemicals which include the volatile organics such as benzene, methylene chloride, and trichloroethylene and other organic chemicals such as the pesticides (chlorinated hydrocarbons), pentachlorophenol, aldicarb, dibromodichloropropane (DBCP), PCBs, and simazine; and

3. The general mineral constituents which include alkalinity, calcium, chloride, copper, foaming agents (MBAS), iron, magnesium, manganese, pH, sodium, sulfate, zinc, specific conductance, total dissolved solids, and hardness (calcium and magnesium).

Upper limits for the concentrations of the chemicals listed in this section have been established by the Safe Drinking Water Act. For a summary of the Drinking Water Regulations established by the Act, see the poster included with this manual. Chapter 22, "Drinking Water Regulations," *WATER TREATMENT PLANT OPERATION*, Volume II, contains a detailed discussion of the Safe Drinking Water Act.

2.24 Biological Factors

Water for domestic uses must be made free from disease-producing (pathogenic) organisms. These organisms include bacteria, protozoa, spores, viruses, cysts, and helminths (parasitic worms).

Many organisms which cause disease in humans originate with the fecal discharges of infected individuals. To monitor and control the activities of human disease carriers is seldom practical. For this reason, it is necessary to take precautions to prevent contamination of a normally safe water source or to institute treatment methods which will produce a safe water.

Unfortunately, the specific disease-producing organisms present in water are not easily isolated and identified. The techniques for comprehensive bacteriological examination are complex and time-consuming. Therefore, it has been necessary to develop tests which indicate the relative degree of contamination in terms of an easily defined quality. The most widely used test involves estimation of the number of bacteria of the coliform group, which are always present in fecal wastes and vastly outnumber specific disease-producing organisms. Coliform bacteria normally inhabit the intestinal tract of humans, but are also found in most animals and birds, as well as in the soil. The Drinking Water Standards in the Safe Drinking Water Act have established upper limits for the concentration of coliform bacteria in a series of water samples with a goal of zero coliforms in all samples. The Maximum Contaminant Level (MCL) for coliforms for systems analyzing fewer than 40 samples per month is no more than one sample per month may be total coliform positive.

To further ensure protection against the spread of waterborne diseases, the Surface Water Treatment Rule (SWTR), which took effect December 31, 1990, requires that all public water systems disinfect their water. If the source of supply (the raw water) is surface water or groundwater under the influence of surface water, the SWTR also requires water suppliers to install filtration equipment unless the source water meets very high standards for purity.

2.25 Radiological Factors

The development and use of atomic energy as a power source and the mining of radioactive materials have made it necessary to examine the safe limits of exposure for humans. Such limits include concentrations of radioactive material taken into the body in drinking water.

QUESTIONS

Write your answers in a notebook and then compare your answers with those on pages 33 and 34.

2.2D List the common physical characteristics of water.

2.2E What causes turbidity in water?

2.2F Chemical analysis of a domestic water supply measures what three general types of chemical concentrations?

2.2G Why are coliform bacteria used to measure the bacteriological quality of water?

2.2H Why must upper limit concentrations of radioactive materials in drinking water be established?

2.3 THE SAFE DRINKING WATER ACT

Water has many important uses and each requires a certain specific level of water quality. The major concern of the operators of water treatment plants and water supply systems is to produce and deliver to consumers water that is safe and pleasant to drink. The water should be acceptable to domestic and commercial water users and many industries. Some industries, such as food and drug processors and the electronics industry, require higher quality water. Many industries will locate where the local water supply meets their specific needs while other industries may have their own water treatment facilities to produce water suitable for their needs.

On December 16, 1974, the Safe Drinking Water Act (SDWA) was signed into law. The SDWA gave the federal government, through the U.S. Environmental Protection Agency (EPA), the authority to:

● Set national standards regulating the levels of contaminants in drinking water;

● Require public water systems to monitor and report their levels of identified contaminants; and

● Establish uniform guidelines specifying the acceptable treatment technologies for cleansing drinking water of unsafe levels of pollutants.

While the SDWA gave EPA responsibility for developing drinking water regulations, it gave state regulatory agencies the opportunity to assume primary responsibility for enforcing those regulations.

Implementation of the SDWA has greatly improved basic drinking water purity across the nation. However, recent EPA surveys of surface water and groundwater indicate the presence of synthetic organic chemicals in 20 percent of the nation's water sources, with a small percentage at levels of concern. In addition, research studies suggest that some naturally occurring contaminants may pose even greater risks to human health than the synthetic contaminants. Further, there is growing concern about microbial and radon contamination.

In the years following passage of the SDWA, Congress felt that EPA was slow to regulate contaminants and states were lax in enforcing the law. Consequently, in 1986 and again in 1996 Congress enacted amendments designed to strengthen the 1974 SDWA. These amendments set deadlines for the establishment of maximum contaminant levels, placed greater emphasis on enforcement, authorized penalties for tampering with drinking water supplies, mandated the complete elimination of lead from drinking water, and placed considerable emphasis on the protection of underground drinking water sources.

All public water systems must comply with the SDWA regulations. A PUBLIC WATER SYSTEM is any publicly or privately owned water supply system that:

1. Has at least 15 service connections, or

2. Regularly serves an average of at least 25 individuals daily at least 60 days out of the year.

Drinking water regulations also take into account the type of population served by the system and classify water systems as community or noncommunity systems. Therefore, in order to understand what requirements apply to any specific system, it is first necessary to determine whether the system is considered a community system or a noncommunity system. A COMMUNITY WATER SYSTEM is defined as one which:

1. Has at least 15 service connections used by all-year residents, or

2. Regularly serves at least 25 all-year residents.

Any public water system that is not a community water system is classified as a NONCOMMUNITY WATER SYSTEM. Restaurants, campgrounds, and hotels could be considered noncommunity systems for purposes of drinking water regulations.

In addition to distinguishing between community and noncommunity water systems, EPA identifies some small systems as NONTRANSIENT NONCOMMUNITY systems if they regularly serve at least 25 of the same persons over 6 months per year. This classification applies to water systems for facilities such as schools or factories where the consumers served are nearly the same every day but do not actually live at the facility. In general, nontransient noncommunity systems must meet the same requirements as community systems.

A TRANSIENT NONCOMMUNITY water system is a system that does not regularly serve drinking water to at least 25 of the same persons over 6 months per year. This classification is used by EPA only in regulating nitrate levels and total coliform. Examples of a transient noncommunity system might be campgrounds or service stations if those facilities do not meet the definition of a community, noncommunity, or nontransient noncommunity system.

On August 6, 1996, the President signed new Safe Drinking Water Act (SDWA) amendments into law as Public Law (PL) 104-182. These amendments made sweeping changes to the existing SDWA, created several new programs, and included a total authorization of more than $12 billion in federal funds for various drinking water programs and activities from fiscal year (FY) 1997 through FY 2003.

Topics covered in the amendments include arsenic research, assistance for water infrastructure and watersheds, assistance to colonias (low-income communities located along the U.S.-Mexico border), backwash water recycling, bottled water, capacity development (technical, financial, and managerial), consumer awareness, contaminant selection and standard-setting authority, definitions of public water system, community water system and noncommunity water system, disinfectants and disinfection by-products, drinking water studies and research, effective date of regulations, enforcement, environmental finance centers and capacity clearinghouse, estrogenic substances screening program, conditions that could qualify a water system for an exemption, groundwater disinfection, groundwater protection programs, lead plumbing and pipes, monitoring and information gathering, monitoring relief, monitoring for unregulated contaminants, occurrence of contaminants in drinking water database, operator certification,

primacy, public notification, drinking water regulations for radon, review of NPDWRs (National Primary Drinking Water Regulations), risk assessment application to establishing NPDWRs, small systems (technical assistance, treatment technology, variances), source water quality assessment and petition programs, state revolving loan fund, authorization to promulgate an NPDWR for sulfate, Surface Water Treatment Rule (SWTR) compliance, variance treatment technologies, water conservation and waterborne disease study and training. For further information and details see "Overview of the Safe Drinking Water Act Amendments of 1996," by Frederick W. Pontius, *JOURNAL AMERICAN WATER WORKS ASSOCIATION*, October 1996, pages 22-33.

Operators are urged to develop close working relationships with their local regulatory agencies to keep themselves informed of the frequent changes in regulations and requirements.

The maximum contaminant levels (MCLs) and also the sampling and testing requirements established by the Safe Drinking Water Act are summarized on the poster included with this manual. For additional information on the Safe Drinking Water Act, see *WATER TREATMENT PLANT OPERATION*, Volume II, Chapter 22, "Drinking Water Regulations."

QUESTIONS

Write your answers in a notebook and then compare your answers with those on page 34.

2.3A   Which industries require extremely high-quality water?

2.3B   Who must comply with the Safe Drinking Water Act (SDWA) regulations?

2.3C   What is the definition of a public water system?

2.3D   Define "community water system."

2.4   WATER TREATMENT

In the operation of water treatment plants, three basic objectives are controlling:

1. Production of a safe drinking water,

2. Production of an aesthetically pleasing drinking water, and

3. Production of drinking water at a reasonable cost with respect to capital and also operation and maintenance costs.

From a public health perspective, production of a safe drinking water, one that is free of harmful bacteria and toxic materials, is the first priority. But, it is also important to produce a high-quality water which appeals to the consumer. Generally, this means that the water must be clear (free of turbidity), colorless, and free of objectionable tastes and odors. Consumers also show a preference for water supplies that are nonstaining (plumbing fixtures and washing clothes), noncorrosive to plumbing fixtures and piping, and do not leave scale deposits or spot glassware.

Consumer sensitivity to the environment (air quality, water quality, noise) has significantly increased in recent years. With regard to water quality, consumer demands have never been greater. In some instances, consumers have substituted bottled water to meet specific needs, namely, for drinking water and cooking purposes.

Design engineers select water treatment processes on the basis of the type of water source, source water quality, and desired finished water quality established by drinking water regulations and consumer desires. Table 2.1 is a summary of typical water treatment processes as they relate to the source and quality of the raw water.

Operators of water treatment facilities must be very conscientious in order to produce a high-quality finished water. Also they must realize that water can degrade in the distribution or delivery system. The remainder of this manual contains chapters written by operators on how to produce and deliver high-quality drinking water to consumers.

QUESTIONS

Write your answers in a notebook and then compare your answers with those on page 34.

2.4A   What is the first priority for operating a water treatment plant?

2.4B   What type of water is appealing to consumers?

2.5 ARITHMETIC ASSIGNMENT

A good way to learn how to solve arithmetic problems is to work on them a little bit at a time. In this operator training manual we are going to make a short arithmetic assignment at the end of every chapter. If you will work this assignment at the end of every chapter, you can easily learn how to solve waterworks arithmetic problems.

Turn to the Appendix, "How to Solve Water Treatment Plant Arithmetic Problems," at the back of this manual and read the following sections:

1. *OBJECTIVES*,

2. A.0, *HOW TO STUDY THIS APPENDIX*, and

3. A.1, *BASIC ARITHMETIC*.

Solve all of the problems in Sections A.10, Addition; A.11, Subtraction; A.12, Multiplication; A.13, Division; A.14, Rules for Solving Equations; and A.15, Example Problems, on an electronic pocket calculator.

2.6 ADDITIONAL READING

1. *WATER SOURCES*, Third Edition. Obtain from American Water Works Association (AWWA), Bookstore, 6666 West Quincy Avenue, Denver, CO 80235. Order No. 1955. ISBN 1-58321-229-9. Price to members, $73.50; nonmembers, $106.50; price includes cost of shipping and handling.

2. *NEW YORK MANUAL*, Chapter 2,* "Water Sources and Water Uses."

3. *TEXAS MANUAL*, Chapter 2,* "Groundwater Supplies"; Chapter 3,* "Surface Water Supplies"; and Chapter 4,* "Raw Water Quality Management."

* Depends on edition.

Please answer the discussion and review questions next.

DISCUSSION AND REVIEW QUESTIONS

Chapter 2.   WATER SOURCES AND TREATMENT

DO NOT USE IBM ANSWER SHEET. Write the answers to these questions in your notebook. The purpose of these questions is to indicate to you how well you understand the material in the chapter.

1. What has been the impact of drought conditions on the managers of water supplies?

2. Why has the quality of many water supplies deteriorated?

3. How does the water (hydrologic) cycle work?

4. What are water rights?

5. What are the differences between the water quality characteristics of groundwater and surface water supplies?

6. Lakes turn over under what conditions?

7. Algal blooms in reservoirs occur under what general conditions?

8. How can a lake used for a water supply also be used for recreation without endangering water quality?

9. The elevation and slope of water tables and artesian pressure levels can change due to what factors?

10. Who should conduct a sanitary survey? Why?

TABLE 2.1 SOURCES AND TREATMENT OF WATER

GROUNDWATER

WATER QUALITY PROBLEM	TREATMENT [a]
1. Coliforms or Microbial Contamination	1. Disinfection (Chlorination)
2. Sulfide Odors (Rotten Egg)	2a. Aeration 2b. Oxidation (Chlorination) 2c. Desulfuration (Sulfur Dioxide)
3. Excessive Hardness (Calcium and Magnesium)	3a. Ion Exchange Softening 3b. Lime (and Soda) Softening
4. Iron and/or Manganese	4a. Sequestration (Polyphosphates) 4b. Removal by Special Ion Exchange 4c. Permanganate and Greensand 4d. Oxidation by Aeration* 4e. Oxidation With Chlorine* 4f. Oxidation With Permanganate* * Filtration Must Follow Oxidation
5. Dissolved Minerals (High Total Dissolved Solids)	5a. Ion Exchange 5b. Reverse Osmosis
6. Corrosivity (Low pH)	6a. pH Adjustment With Chemicals 6b. Carbon Dioxide Stripping by Aeration 6c. Corrosion Inhibitor Addition (Zinc Phosphate, Silicate)
7. Preventive Treatment (Fluoridation)	7. Add Fluoride Chemicals
8. Sand	8. Sand Separators
9. Nitrate	9a. Anion Exchange 9b. Reverse Osmosis

SURFACE WATER

WATER QUALITY PROBLEM	TREATMENT [a]
1. Coliforms or Microbial Contamination	1a. Disinfection (Chlorination) 1b. Disinfection (Other Oxidants—Ozone, Chlorine Dioxide, Chloramination) 1c. Coagulation, Flocculation, Sedimentation, Filtration, and Disinfection
2. Turbidity, Color	2. Coagulation, Flocculation, Sedimentation, and Filtration
3. Odors (Organic Materials)	3a. Clarification (Coagulation, Flocculation, Sedimentation, and Filtration) 3b. Oxidation (Chlorination or Permanganate) 3c. Special Oxidation (Chlorine Dioxide) 3d. Adsorption (Granular Activated Carbon)
4. Iron and/or Manganese	4a. Sequestration (Polyphosphates) 4b. Removal by Special Ion Exchange 4c. Permanganate and Greensand 4d. Oxidation by Aeration* 4e. Oxidation With Chlorine* 4f. Oxidation With Permanganate* * Filtration Must Follow Oxidation
5. Excessive Hardness (Calcium and Magnesium)	5a. Ion Exchange Softening 5b. Lime (and Soda) Softening
6. Dissolved Minerals (High Total Dissolved Solids)	6a. Ion Exchange 6b. Reverse Osmosis
7. Corrosivity (Low pH)	7a. pH Adjustment With Chemicals 7b. Corrosion Inhibitor Addition (Zinc Phosphate, Silicate)

8. Preventive Treatment
    a. Fluoridation
    b. Trihalomethanes (THMs)
	8a. Add Fluoride Chemicals
	8b. (1) Do not Prechlorinate.
	        Disinfect With Ozone, Chlorine Dioxide, or Chloramination
	    (2) Remove THM Precursors
	    (3) Remove THMs After They Are Formed

[a] For details on the treatment processes, refer to the appropriate chapters in this series of operator training manuals.

SUGGESTED ANSWERS

Chapter 2. WATER SOURCES AND TREATMENT

Answers to questions on page 22.

2.0A Drought conditions have forced water supply managers to recognize the need to conserve and properly budget our water resources. Increasing costs of energy make it expensive to pump water from deeper groundwater levels or to transport water over long distances.

2.0B Energy consumption and water shortages are linked together because (1) with falling groundwater levels energy is required to pump from deeper levels, and (2) water projects that use large amounts of electricity to pump water over long distances are requiring water users to pay increased water rates to cover higher energy costs.

2.0C Groundwater becomes contaminated from (1) seawater intrusion, (2) seepage from septic tank leaching systems and agricultural drainage systems, (3) surface runoff into poorly constructed wells, and (4) seepage from improper disposal of hazardous wastes.

Answers to questions on page 24.

2.1A The water (hydrologic) cycle is the cycle or path water follows from evaporation from oceans, to formation of clouds, to precipitation, to runoff, to evaporation and transpiration back to the atmosphere, and eventually back to the ocean.

2.1B The three basic types of water rights are:

1. Riparian,
2. Appropriative, and
3. Prescriptive.

2.1C In general, surface water supplies are characterized by soft water, suspended solids, turbidity, some color, and microbial contamination.

2.1D In general, groundwater supplies are characterized by higher concentrations of dissolved solids, dissolved gases, low color, relatively high hardness, and freedom from microbial contamination.

2.1E Before selecting a location and constructing a water supply intake in a river or stream, careful consideration must be given to (1) the stream bottom, its degree of scour, and the settling out of silt, and (2) design of the intake to make sure that it can withstand the forces which will act upon it during times of flood, heavy silting, ice conditions, and adverse runoff conditions.

2.1F Filtration and disinfection are considered essential water treatment processes to reliably treat physically and bacteriologically contaminated surface waters for domestic use.

2.1G Lakes and reservoirs used for domestic water supplies can be protected from contamination by (1) proper construction and location of recreation facilities; and (2) careful evaluation of adverse effects of the construction of wastewater collection, treatment, and disposal facilities near the reservoir or its tributaries.

Answers to questions on page 26.

2.1H Groundwater flows within an aquifer because of the pressure differences between the areas of recharge and discharge.

2.1I The "safe yield" of an aquifer can be exceeded if wells extract water from an aquifer over a period of time at a rate such that the aquifer will become depleted or bring about other undesired results, such as seawater intrusion and land subsidence.

2.1J Possible uses of reclaimed wastewater include: (1) crop irrigation, (2) greenbelt irrigation, (3) golf course irrigation, (4) landscape irrigation, (5) industrial reuse, (6) groundwater recharge, (7) landscape impoundments, and (8) wetlands/marsh enhancement.

2.1K Reclaimed water should receive treatment that is appropriate for the intended use of the water.

Answers to questions on page 28.

2.2A The purpose of a sanitary survey is to detect all health hazards and to evaluate their present and future importance.

2.2B For an existing water supply, a sanitary survey should be made frequently enough to control all health hazards and to maintain good, sanitary water quality.

2.2C When conducting a sanitary survey, protective measures that should be investigated include control of sources of waste discharges (municipal wastewater treatment plants, sources of fecal pollution, industrial wastes, oil field brines, acid waters from mines, sanitary landfills, and agricultural drain waters), fishing, boating, landing of airplanes, swimming, wading, ice cutting, and permitting animals on shoreline areas.

Answers to questions on page 29.

2.2D The common physical characteristics of water are observable color, turbidity, temperature, taste, and odor.

2.2E Turbidity in water is caused by the presence of suspended material such as clay, silt, finely divided organic material, plankton, and other inorganic material.

2.2F The three general types of chemicals measured by a chemical analysis of domestic water supplies are inorganic, organic, and general mineral concentrations.

2.2G   Coliforms are used to measure the bacteriological quality of drinking water because the test indicates fecal contamination, whereas specific disease-producing organisms are not easily isolated and identified.

2.2H   The development and use of atomic energy as a power source and the mining of radioactive materials have made it necessary to establish upper limit concentrations of radioactive materials in drinking water.

Answers to questions on page 30.

2.3A   Food and drug processors and also electronics industries require extremely high-quality water.

2.3B   All public water systems must comply with the Safe Drinking Water Act (SDWA) regulations.

2.3C   A public water system is defined as any publicly or privately owned water system that has at least 15 service connections, or that regularly serves at least 25 people at least 60 days out of the year.

2.3D   A community water system is one which:

1. Has at least 15 service connections used by all-year residents, or
2. Regularly serves at least 25 all-year residents.

Answers to questions on page 31.

2.4A   The first priority for operating a water treatment plant is the production of a safe drinking water, one that is free of harmful bacteria and toxic materials.

2.4B   Water that appeals to consumers must be clear (free of turbidity), colorless, and free of objectionable tastes and odors. Consumers also show a preference for water supplies that are nonstaining (plumbing fixtures and washing clothes), noncorrosive to plumbing fixtures and piping, and do not leave scale deposits or spot glassware.

CHAPTER 3

RESERVOIR MANAGEMENT AND INTAKE STRUCTURES

by

Richard H. Barnett

Revised by

Jeanne Ballestero

TABLE OF CONTENTS

Chapter 3.   RESERVOIR MANAGEMENT AND INTAKE STRUCTURES

OBJECTIVES

Chapter 3.   RESERVOIR MANAGEMENT AND INTAKE STRUCTURES

Following completion of Chapter 3, you should be able to:

1. Describe the importance of reservoir management;

2. Identify the causes of reservoir water quality problems;

3. Justify the need for a reservoir management program;

4. Implement the appropriate methods of reservoir management and water quality improvement;

5. Develop a laboratory and monitoring program;

6. Describe the purpose of intake structures;

7. Identify various types of intake structures, gates, and screens;

8. Safely operate, maintain, and troubleshoot intake facilities; and

9. Keep necessary records on the operation and maintenance of reservoir water quality management programs and intake structures.

WORDS

Chapter 3. RESERVOIR MANAGEMENT AND INTAKE STRUCTURES

ACRE-FOOT   ACRE-FOOT

A volume of water that covers one acre to a depth of one foot, or 43,560 cubic feet (1,233.5 cubic meters).

ADSORPTION (add-SORP-shun)   ADSORPTION

The gathering of a gas, liquid, or dissolved substance on the surface or interface zone of another material.

AERATION (air-A-shun)   AERATION

The process of adding air to water. Air can be added to water by either passing air through water or passing water through air.

AEROBIC (AIR-O-bick)   AEROBIC

A condition in which atmospheric or dissolved molecular oxygen is present in the aquatic (water) environment.

ALGAE (AL-gee)   ALGAE

Microscopic plants which contain chlorophyll and live floating or suspended in water. They also may be attached to structures, rocks or other submerged surfaces. Excess algal growths can impart tastes and odors to potable water. Algae produce oxygen during sunlight hours and use oxygen during the night hours. Their biological activities appreciably affect the pH, alkalinity, and dissolved oxygen of the water.

ALGAL (AL-gull) BLOOM   ALGAL BLOOM

Sudden, massive growths of microscopic and macroscopic plant life, such as green or blue-green algae, which can, under the proper conditions, develop in lakes and reservoirs.

ALIPHATIC (AL-uh-FAT-ick) HYDROXY ACIDS   ALIPHATIC HYDROXY ACIDS

Organic acids with carbon atoms arranged in branched or unbranched open chains rather than in rings.

ANAEROBIC (AN-air-O-bick)   ANAEROBIC

A condition in which atmospheric or dissolved molecular oxygen is *NOT* present in the aquatic (water) environment.

ANION (AN-EYE-en)   ANION

A negatively charged ion in an electrolyte solution, attracted to the anode under the influence of a difference in electrical potential. Chloride ion (Cl^-) is an anion.

BIOCHEMICAL OXYGEN DEMAND (BOD)   BIOCHEMICAL OXYGEN DEMAND (BOD)

The rate at which organisms use the oxygen in water while stabilizing decomposable organic matter under aerobic conditions. In decomposition, organic matter serves as food for the bacteria and energy results from its oxidation. BOD measurements are used as a measure of the organic strength of wastes in water.

CATHODIC (ca-THOD-ick) PROTECTION   CATHODIC PROTECTION

An electrical system for prevention of rust, corrosion, and pitting of metal surfaces which are in contact with water or soil. A low-voltage current is made to flow through a liquid (water) or a soil in contact with the metal in such a manner that the external electromotive force renders the metal structure cathodic. This concentrates corrosion on auxiliary anodic parts which are deliberately allowed to corrode instead of letting the structure corrode.

CATION (CAT-EYE-en)   CATION

A positively charged ion in an electrolyte solution, attracted to the cathode under the influence of a difference in electrical potential. Sodium ion (Na^+) is a cation.

CHELATION (key-LAY-shun)   CHELATION

A chemical complexing (forming or joining together) of metallic cations (such as copper) with certain organic compounds, such as EDTA (ethylene diamine tetracetic acid). Chelation is used to prevent the precipitation of metals (copper). Also see SEQUES-TRATION.

COLIFORM (COAL-i-form)   COLIFORM

A group of bacteria found in the intestines of warm-blooded animals (including humans) and also in plants, soil, air and water. Fecal coliforms are a specific class of bacteria which only inhabit the intestines of warm-blooded animals. The presence of coliform bacteria is an indication that the water is polluted and may contain pathogenic (disease-causing) organisms.

COLLOIDS (CALL-loids)   COLLOIDS

Very small, finely divided solids (particles that do not dissolve) that remain dispersed in a liquid for a long time due to their small size and electrical charge. When most of the particles in water have a negative electrical charge, they tend to repel each other. This repulsion prevents the particles from clumping together, becoming heavier, and settling out.

COMPLETE TREATMENT   COMPLETE TREATMENT

A method of treating water which consists of the addition of coagulant chemicals, flash mixing, coagulation-flocculation, sedimentation and filtration. Also called conventional filtration.

CONDUCTIVITY   CONDUCTIVITY

A measure of the ability of a solution (water) to carry an electric current.

DECOMPOSITION, DECAY   DECOMPOSITION, DECAY

The conversion of chemically unstable materials to more stable forms by chemical or biological action. If organic matter decays when there is no oxygen present (anaerobic conditions or putrefaction), undesirable tastes and odors are produced. Decay of organic matter when oxygen is present (aerobic conditions) tends to produce much less objectionable tastes and odors.

DENSITY (DEN-sit-tee)   DENSITY

A measure of how heavy a substance (solid, liquid or gas) is for its size. Density is expressed in terms of weight per unit volume, that is, grams per cubic centimeter or pounds per cubic foot. The density of water (at 4°C or 39°F) is 1.0 gram per cubic centimeter or about 62.4 pounds per cubic foot.

DESTRATIFICATION (de-STRAT-uh-fuh-KAY-shun)   DESTRATIFICATION

The development of vertical mixing within a lake or reservoir to eliminate (either totally or partially) separate layers of temperature, plant, or animal life. This vertical mixing can be caused by mechanical means (pumps) or through the use of forced air diffusers which release air into the lower layers of the reservoir.

DETENTION TIME   DETENTION TIME

(1) The theoretical (calculated) time required for a small amount of water to pass through a tank at a given rate of flow.

(2) The actual time in hours, minutes or seconds that a small amount of water is in a settling basin, flocculating basin or rapid-mix chamber. In storage reservoirs, detention time is the length of time entering water will be held before being drafted for use (several weeks to years, several months being typical).

$$\text{Detention Time, hr} = \frac{(\text{Basin Volume, gal})(24 \text{ hr/day})}{\text{Flow, gal/day}}$$

DIATOMS (DYE-uh-toms)   DIATOMS

Unicellular (single cell), microscopic algae with a rigid (box-like) internal structure consisting mainly of silica.

DIMICTIC (die-MICK-tick)   DIMICTIC

Lakes and reservoirs which freeze over and normally go through two stratification and two mixing cycles within a year.

DIRECT FILTRATION   DIRECT FILTRATION

A method of treating water which consists of the addition of coagulant chemicals, flash mixing, coagulation, minimal flocculation, and filtration. The flocculation facilities may be omitted, but the physical-chemical reactions will occur to some extent. The sedimentation process is omitted.

ELECTROLYTE (ee-LECK-tro-LITE)   ELECTROLYTE

A substance which dissociates (separates) into two or more ions when it is dissolved in water.

EPILIMNION (EP-uh-LIM-knee-on)   EPILIMNION

The upper layer of water in a thermally stratified lake or reservoir. This layer consists of the warmest water and has a fairly uniform (constant) temperature. The layer is readily mixed by wind action.

EUTROPHIC (you-TRO-fick)   EUTROPHIC

Reservoirs and lakes which are rich in nutrients and very productive in terms of aquatic animal and plant life.

EUTROPHICATION (you-TRO-fi-KAY-shun)   EUTROPHICATION

The increase in the nutrient levels of a lake or other body of water; this usually causes an increase in the growth of aquatic animal and plant life.

EVAPOTRANSPIRATION (ee-VAP-o-TRANS-purr-A-shun)   EVAPOTRANSPIRATION

(1)   The process by which water vapor passes into the atmosphere from living plants. Also called transpiration.

(2)   The total water removed from an area by transpiration (plants) and by evaporation from soil, snow and water surfaces.

FLUSHING   FLUSHING

A method used to clean water distribution lines. Hydrants are opened and water with a high velocity flows through the pipes, removes deposits from the pipes, and flows out the hydrants.

HEAD   HEAD

The vertical distance (in feet) equal to the pressure (in psi) at a specific point. The pressure head is equal to the pressure in psi times 2.31 ft/psi.

HYPOLIMNION (HI-poe-LIM-knee-on)   HYPOLIMNION

The lowest layer in a thermally stratified lake or reservoir. This layer consists of colder, more dense water, has a constant temperature and no mixing occurs.

INORGANIC   INORGANIC

Material such as sand, salt, iron, calcium salts and other mineral materials. Inorganic substances are of mineral origin, whereas organic substances are usually of animal or plant origin. Also see ORGANIC.

LITTORAL (LIT-or-al) ZONE   LITTORAL ZONE

(1)   That portion of a body of fresh water extending from the shoreline lakeward to the limit of occupancy of rooted plants.

(2)   The strip of land along the shoreline between the high and low water levels.

MESOTROPHIC (MESS-o-TRO-fick)   MESOTROPHIC

Reservoirs and lakes which contain moderate quantities of nutrients and are moderately productive in terms of aquatic animal and plant life.

METALIMNION (MET-uh-LIM-knee-on)   METALIMNION

The middle layer in a thermally stratified lake or reservoir. In this layer there is a rapid decrease in temperature with depth. Also called the THERMOCLINE.

METHYL ORANGE ALKALINITY   METHYL ORANGE ALKALINITY

A measure of the total alkalinity in a water sample. The alkalinity is measured by the amount of standard sulfuric acid required to lower the pH of the water to a pH level of 4.5, as indicated by the change in color of methyl orange from orange to pink. Methyl orange alkalinity is expressed as milligrams per liter equivalent calcium carbonate.

MILLIGRAMS PER LITER, mg/L   MILLIGRAMS PER LITER, mg/L

A measure of the concentration by weight of a substance per unit volume. For practical purposes, one mg/L of a substance in fresh water is equal to one part per million parts (ppm). Thus a liter of water with a specific gravity of 1.0 weighs one million milligrams. If water contains 10 milligrams of calcium, the concentration is 10 milligrams per million milligrams, or 10 milligrams per liter (10 mg/L), or 10 parts of calcium per million parts of water, or 10 parts per million (10 ppm).

MONOMICTIC (mo-no-MICK-tick)   MONOMICTIC

Lakes and reservoirs which are relatively deep, do not freeze over during the winter months, and undergo a single stratification and mixing cycle during the year. These lakes and reservoirs usually become destratified during the mixing cycle, usually in the fall of the year.

NUTRIENT NUTRIENT

Any substance that is assimilated (taken in) by organisms and promotes growth. Nitrogen and phosphorus are nutrients which promote the growth of algae. There are other essential and trace elements which are also considered nutrients.

OLIGOTROPHIC (AH-lig-o-TRO-fick) OLIGOTROPHIC

Reservoirs and lakes which are nutrient poor and contain little aquatic plant or animal life.

ORGANIC ORGANIC

Substances that come from animal or plant sources. Organic substances always contain carbon. (Inorganic materials are chemical substances of mineral origin.) Also see INORGANIC.

OVERTURN OVERTURN

The almost spontaneous mixing of all layers of water in a reservoir or lake when the water temperature becomes similar from top to bottom. This may occur in the fall/winter when the surface waters cool to the same temperature as the bottom waters and also in the spring when the surface waters warm after the ice melts. This is also called "turnover."

OXIDATION (ox-uh-DAY-shun) OXIDATION

Oxidation is the addition of oxygen, removal of hydrogen, or the removal of electrons from an element or compound. In the environment, organic matter is oxidized to more stable substances. The opposite of REDUCTION.

OXIDATION-REDUCTION POTENTIAL (ORP) OXIDATION-REDUCTION POTENTIAL (ORP)

The electrical potential required to transfer electrons from one compound or element (the oxidant) to another compound or element (the reductant); used as a qualitative measure of the state of oxidation in water treatment systems. ORP is measured in millivolts, with negative values indicating a tendency to reduce compounds or elements and positive values indicating a tendency to oxidize compounds or elements.

PERIPHYTON (pair-e-FI-tawn) PERIPHYTON

Microscopic plants and animals that are firmly attached to solid surfaces under water such as rocks, logs, pilings and other structures.

pH (pronounce as separate letters) pH

pH is an expression of the intensity of the basic or acidic condition of a liquid. Mathematically, pH is the logarithm (base 10) of the reciprocal of the hydrogen ion activity.

$$pH = Log \frac{1}{[H^+]}$$

The pH may range from 0 to 14, where 0 is most acidic, 14 most basic, and 7 neutral. Natural waters usually have a pH between 6.5 and 8.5.

PHOTOSYNTHESIS (foe-toe-SIN-thuh-sis) PHOTOSYNTHESIS

A process in which organisms, with the aid of chlorophyll (green plant enzyme), convert carbon dioxide and inorganic substances into oxygen and additional plant material, using sunlight for energy. All green plants grow by this process.

PHYTOPLANKTON (FI-tow-PLANK-ton) PHYTOPLANKTON

Small, usually microscopic plants (such as algae), found in lakes, reservoirs, and other bodies of water.

POTABLE (POE-tuh-bull) WATER POTABLE WATER

Water that does not contain objectionable pollution, contamination, minerals, or infective agents and is considered satisfactory for drinking.

PRECIPITATE (pre-SIP-uh-TATE) PRECIPITATE

(1) An insoluble, finely divided substance which is a product of a chemical reaction within a liquid.

(2) The separation from solution of an insoluble substance.

PRECURSOR, THM (pre-CURSE-or) PRECURSOR, THM

Natural organic compounds found in all surface and groundwaters. These compounds *MAY* react with halogens (such as chlorine) to form trihalomethanes (tri-HAL-o-METH-hanes) (THMs); they *MUST* be present in order for THMs to form.

REAERATION (RE-air-A-shun) REAERATION

The introduction of air through forced air diffusers into the lower layers of the reservoir. As the air bubbles form and rise through the water, oxygen from the air dissolves into the water and replenishes the dissolved oxygen. The rising bubbles also cause the lower waters to rise to the surface where oxygen from the atmosphere is transferred to the water. This is sometimes called surface reaeration.

REAGENT (re-A-gent) REAGENT

A pure chemical substance that is used to make new products or is used in chemical tests to measure, detect, or examine other substances.

REDUCTION (re-DUCK-shun) REDUCTION

Reduction is the addition of hydrogen, removal of oxygen, or the addition of electrons to an element or compound. Under anaerobic conditions (no dissolved oxygen present), sulfur compounds are reduced to odor-producing hydrogen sulfide (H_2S) and other compounds. The opposite of OXIDATION.

SECCHI (SECK-key) DISC SECCHI DISC

A flat, white disc lowered into the water by a rope until it is just barely visible. At this point, the depth of the disc from the water surface is the recorded Secchi disc transparency.

SEPTIC (SEP-tick) SEPTIC

A condition produced by bacteria when all oxygen supplies are depleted. If severe, the bottom deposits produce hydrogen sulfide, the deposits and water turn black, give off foul odors, and the water has a greatly increased chlorine demand.

SEQUESTRATION (SEE-kwes-TRAY-shun) SEQUESTRATION

A chemical complexing (forming or joining together) of metallic cations (such as iron) with certain inorganic compounds, such as phosphate. Sequestration prevents the precipitation of the metals (iron). Also see CHELATION.

SEWAGE SEWAGE

The used household water and water-carried solids that flow in sewers to a wastewater treatment plant. The preferred term is WASTEWATER.

STRATIFICATION (STRAT-uh-fuh-KAY-shun) STRATIFICATION

The formation of separate layers (of temperature, plant, or animal life) in a lake or reservoir. Each layer has similar characteristics such as all water in the layer has the same temperature. Also see THERMAL STRATIFICATION.

THERMAL STRATIFICATION (STRAT-uh-fuh-KAY-shun) THERMAL STRATIFICATION

The formation of layers of different temperatures in a lake or reservoir. Also see STRATIFICATION.

THERMOCLINE (THUR-moe-KLINE) THERMOCLINE

The middle layer in a thermally stratified lake or reservoir. In this layer there is a rapid decrease in temperature with depth. Also called the METALIMNION.

THRESHOLD ODOR THRESHOLD ODOR

The minimum odor of a water sample that can just be detected after successive dilutions with odorless water. Also called odor threshold.

THRESHOLD ODOR NUMBER (TON) THRESHOLD ODOR NUMBER (TON)

The greatest dilution of a sample with odor-free water that still yields a just-detectable odor.

TRIHALOMETHANES (THMs) (tri-HAL-o-METH-hanes) TRIHALOMETHANES (THMs)

Derivatives of methane, CH_4, in which three halogen atoms (chlorine or bromine) are substituted for three of the hydrogen atoms. Often formed during chlorination by reactions with natural organic materials in the water. The resulting compounds (THMs) are suspected of causing cancer.

TURBIDITY (ter-BID-it-tee) TURBIDITY

The cloudy appearance of water caused by the presence of suspended and colloidal matter. In the waterworks field, a turbidity measurement is used to indicate the clarity of water. Technically, turbidity is an optical property of the water based on the amount of light reflected by suspended particles. Turbidity cannot be directly equated to suspended solids because white particles reflect more light than dark-colored particles and many small particles will reflect more light than an equivalent large particle.

WASTEWATER WASTEWATER

A community's used water and water-carried solids (including used water from industrial processes) that flow to a treatment plant. Storm water, surface water, and groundwater infiltration also may be included in the wastewater that enters a wastewater treatment plant. The term "sewage" usually refers to household wastes, but this word is being replaced by the term "wastewater."

ZOOPLANKTON (ZOE-PLANK-ton) ZOOPLANKTON

Small, usually microscopic animals (such as protozoans), found in lakes and reservoirs.

CHAPTER 3.   RESERVOIR MANAGEMENT AND INTAKE STRUCTURES

(Lesson 1 of 2 Lessons)

3.0   IMPORTANCE OF RESERVOIR WATER QUALITY MANAGEMENT

3.00   Use of Surface Reservoirs (Impoundments) as Domestic Water Supplies

During the past few decades, more and more people in both cities and rural areas have become either partially or wholly dependent on surface reservoirs and lakes as a source for their water supplies. As populations have increased, domestic, municipal, industrial, recreational, and agricultural water usage has also increased, creating demands on water supplies that cannot be met directly by groundwater or surface water diversions from streams and rivers.

These increased demands have been met for the most part by constructing dams and reservoirs which provide carryover storage for excess runoff and provide a dependable water supply during the dry season of the year and during periods of prolonged drought. Particularly in the western United States, a majority of the major cities receive domestic water from surface lakes and reservoirs. In most cases the water is stored in one or more major reservoirs before it is delivered to the consumers. In those areas which do depend directly on local water supplies, there may be several large reservoirs which capture runoff from local watersheds and store it for future use.

The capacities of reservoirs used as domestic water supplies range from less than 100 acre-feet to several million acre-feet. The time water may be stored ranges from weeks or months to several years.

Methods of managing lakes and reservoirs used for domestic water supplies vary widely depending on local situations. In addition to serving domestic water needs, a reservoir may be used for flood control purposes, for hydroelectric power generation, for regulating downstream releases, for recreational purposes, or for providing water for agricultural, municipal, and industrial uses. The amount and type of public use allowed on reservoirs also varies widely according to individual situations. Some allow motor boats, some allow only boats without motors; most do not allow any body-contact water sports but some allow complete body-contact sports such as swimming and water skiing.

Small lakes in remote areas may be open for public use only a few days each year while large lakes and reservoirs located near metropolitan areas may accommodate several million visitors annually. The methods of treating water supplies from reservoirs range from disinfection only, to *DIRECT FILTRATION,*[1] to *COMPLETE TREATMENT,*[2] which may even include softening and activated carbon filtration. Each reservoir should have a water quality management program which is designed to meet the reservoir's individual requirements.

3.01   Factors Affecting Water Quality

Water quality within lakes and reservoirs is influenced and controlled by many factors. Of major concern is the fact that many of the conditions which adversely affect water quality in domestic water supply reservoirs result from our use of the environment. In order to control and maintain water quality, our activities must be controlled. The occurrence of acid rainfall is of concern in many areas of the United States and Europe. Pollution from both motor vehicles and industrial plants has increased the acidity of rain in some areas to the point that when runoff reaches lakes and reservoirs, the biological balance is severely affected. Fish die-offs are obvious and easily detected but other biological upsets or trends may not be so obvious.

The impacts of our activities within a given reservoir's drainage area are also a major concern. Wastewater, agricultural runoff, grazing of livestock, drainage from mining areas, runoff from urban areas, and industrial discharges may all lead to deterioration in physical, chemical, or biological water quality within a reservoir. Increased turbidity and siltation may result from farming practices, fires, construction (soil grading), and logging operations. If not properly controlled, public use of a reservoir may result in reduced water quality.

Natural factors that may affect the quality of water in a given lake or reservoir include the following:

---

[1] *Direct Filtration.   A method of treating water which consists of the addition of coagulant chemicals, flash mixing, coagulation, minimal flocculation, and filtration. The flocculation facilities may be omitted, but the physical-chemical reactions will occur to some extent. The sedimentation process is omitted.*

[2] *Complete Treatment.   A method of treating water which consists of the addition of coagulant chemicals, flash mixing, coagulation-flocculation, sedimentation and filtration. Also called conventional filtration.*

1. Climate: temperature, intensity and direction of wind movements, type, pattern, intensity and duration of precipitation;

2. Watershed and Drainage Areas: geology, topography, type and extent of vegetation, and use by native animals;

3. Wildfires (caused by lightning); and

4. Reservoir Area: geology, land form including depth, area, and bottom topography, and surface vegetation at the time the reservoir is filled.

QUESTIONS

Write your answers in a notebook and then compare your answers with those on page 84.

3.0A List two common sources of water other than lakes or reservoirs.

3.0B What methods are used to treat domestic water delivered from water supply reservoirs?

3.0C How do our activities cause deterioration of water quality in reservoirs?

3.0D What natural factors may result in lowered water quality (degradation) in reservoirs?

3.1 CAUSES OF WATER QUALITY PROBLEMS

3.10 Nutrients[3]

Many water quality problems in domestic water supply reservoirs occur in reservoirs containing moderate or large quantities of nutrients such as phosphate, nitrate, and organic nitrogen compounds. These nutrients may act as a fertilizer in a lake to stimulate the growth of algae just as they stimulate growth on a lawn, garden, or orchard. Reservoirs and lakes that are rich in nutrients and thus very productive in terms of aquatic animal and plant life are commonly referred to as eutrophic (you-TRO-fick). Reservoirs that are nutrient-poor and contain little plant or animal life are classed as oligotrophic (AH-lig-o-TRO-fick). Between these two types of reservoirs are mesotrophic (MESS-o-TRO-fick) reservoirs containing moderate amounts of nutrients able to support moderate levels of plant and animal life.

In productive reservoirs aquatic plants such as pond weeds, water hyacinths, tules, and sedges may be abundant in the shallow water or *LITTORAL ZONES*[4] (Figure 3.1). Productive lakes usually support large populations of phytoplankton (very small plants) and/or zooplankton (very small animals) at various times during the year. A sudden large increase in plankton populations is commonly referred to as a "bloom." Phytoplankton blooms in particular are referred to as "algal blooms" (pronounced AL-gull). The duration and the amount of population growth of an individual algal bloom depends upon various environmental factors including light, temperature, and nutrient conditions. An individual bloom may contain from one to sev-

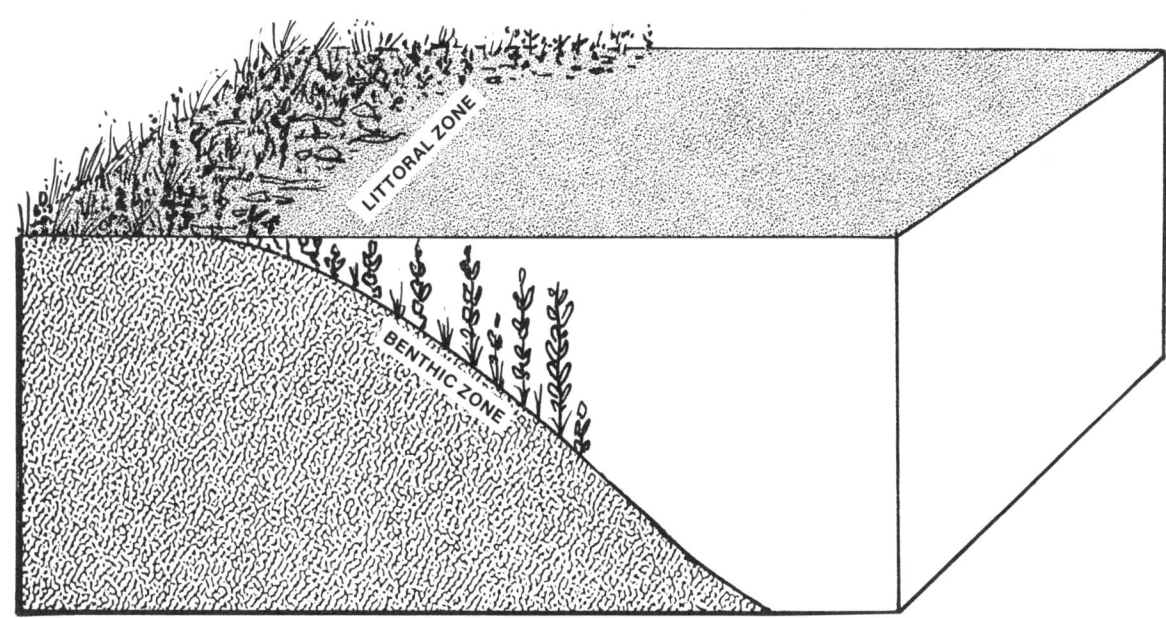

Fig. 3.1 Littoral zone

---

[3] *Nutrient. Any substance that is assimilated (taken in) by organisms and promotes growth. Nitrogen and phosphorus are nutrients which promote the growth of algae. There are other essential and trace elements which are also considered nutrients.*

[4] *Littoral (LIT-or-al) Zone. (1) That portion of a body of fresh water extending from the shoreline lakeward to the limit of occupancy of rooted plants. (2) The strip of land along the shoreline between the high and low water levels.*

eral types of algae and may last from a few days to several weeks or even months.

3.11   Algal Blooms

Several common water quality problems in domestic water supply reservoirs may be related to algal blooms. These problems will be discussed individually but are summarized as follows:

1. Taste and odor problems,

2. Shortened filter runs of complete treatment plants,

3. Increased pH (which reduces chlorination efficiency),

4. Dissolved oxygen depletion, and

5. Organic loading.

QUESTIONS

Write your answers in a notebook and then compare your answers with those on page 84.

3.1A   Large quantities of what nutrients are undesirable in a water supply reservoir?

3.1B   List the three classes of reservoirs based on nutrient content and productivity in terms of aquatic animal and plant life.

3.1C   What is an "algal bloom"?

3.12   Tastes and Odors

Objectionable tastes and odors in the domestic water supply are often related to the occurrence of algal blooms. The nature of these tastes and odors is related to the particular type of algae but may change as the intensity of the algal bloom changes. For example, some algae produce a grassy odor when populations are moderate, but a much more intense odor, described as a septic or pigpen odor, when populations are large or algae are dying and decaying. Among the more common types of tastes and odors produced by algae are the following: fishy, aromatic, grassy, septic, musty, and earthy. Approximately forty types of algae have been identified as taste and odor producers. The extent of taste and odor problems caused by algal blooms ranges from slight consumer objection to total rejection of the supply for domestic uses. Taste complaints often arise when the water is used for drinking or for making coffee or tea because objectionable tastes seem to be more noticeable when the water is at room temperature and above than when the water is cold. Odors are

frequently most noticeable when the hot water supply is in use, particularly when it is used for showers, cooking, and dishwashing.

In many situations, chlorination of the water supply reduces the level of tastes and odors; however, there are some instances in which the tastes and odors appear to be stronger following chlorination. Water treatment costs increase significantly when tastes and odors must be removed. Fortunately many conventional plants are capable of reducing or eliminating tastes and odors when properly operated.

Stratification of reservoirs into thermal layers, as a result of climatic conditions, will be discussed in detail in Section 3.17, "Thermal Stratification." In connection with algae, however, tastes and odors are usually strongest in the thermal layer of the reservoir where the bloom occurs. In most cases, this takes place within the upper layer of water. Figure 3.2[5] illustrates this fact by showing *THRESHOLD ODOR*[6] profiles during three separate algal blooms in Lake Casitas, California. These blooms occurred when thermal stratification existed.

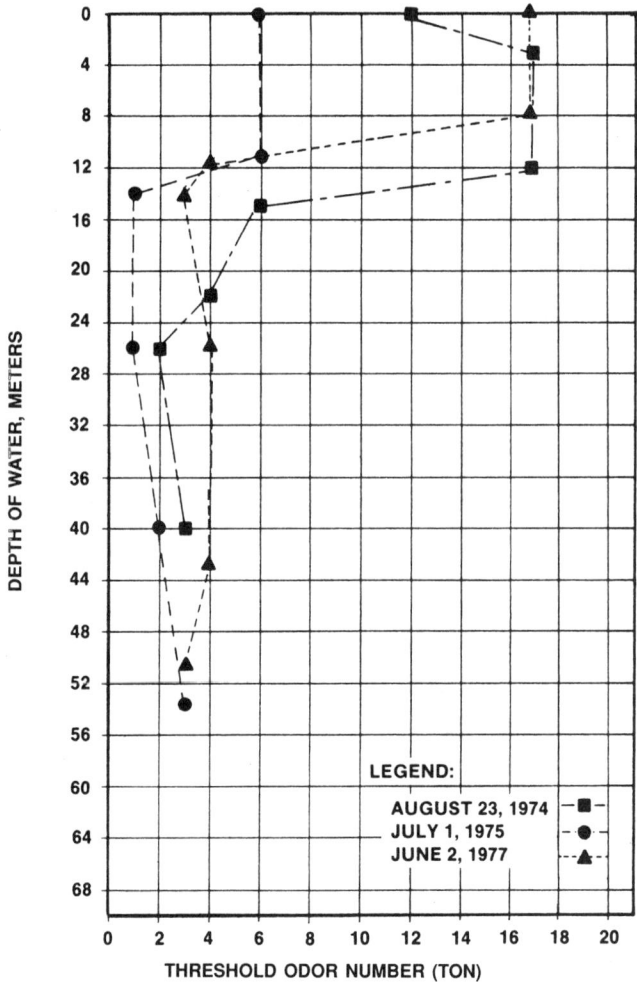

Fig. 3.2   Lake Casitas odor profiles

---

[5] *Figure 3.2 was prepared by taking a boat out onto Lake Casitas on the three dates shown. A water sampler was lowered into the Lake at various depths, a sample collected, and then analyzed to determine the threshold odor number (TON). The odor profile drawing was prepared by marking a plotting point at each depth (on the left side) at the measured TON (across the bottom). The plotting points for each day were connected to produce the "odor profiles."*

[6] *Threshold Odor.   The minimum odor of a water sample that can just be detected after successive dilutions with odorless water. Also called odor threshold.*

The odors (represented by higher TON readings) are concentrated within the upper, warmer layer of water and they have not been mixed into the deeper colder layers to any significant extent.

A problem that has been noted in several Southern California domestic water supply reservoirs is that when the fall *OVERTURN*[7] occurs, obnoxious tastes and odors are brought downward from the upper portion of the reservoir and mixed throughout the entire body of water. Even when multilevel intakes (water inlets) are available, it is not possible to select a depth where taste and odor problems are minimized. In lakes and reservoirs that freeze over during winter months, algal blooms (late algae) have been known to occur underneath the ice, causing taste and odor problems in the deeper waters.

Considerable research has been devoted to identifying taste and odor problems that result from blooms of free-floating (planktonic) algae. To identify algae species and estimate population levels, water samples are collected either with or without the aid of various types of plankton nets (Figure 3.3). Samples are then examined in the laboratory using standard procedures. Research has shown that algae that are growing attached (periphyton) to bottom sediments and structures or submerged plants can be a major contributor to taste and odor problems. These algal growths cannot be sampled and evaluated by conventional methods, so trained scuba divers collect samples and map the extent of algal growths.

One species of blue-green algae in particular, *OSCILLATORIA CURVICEPS*,[8] has been identified as being responsible for certain earthy, musty tastes and odors. Considerable research is also underway throughout the United States to identify and quantify organic compounds that cause taste and odor and to link a particular compound to the species of algae which produces it. Two compounds, Geosmin and Methylisoborneal (MIB), which have been linked to earthy, musty tastes and odors, can generate consumer complaints when they are present in the water supply in concentrations as low as a few parts per trillion (nanograms per liter).

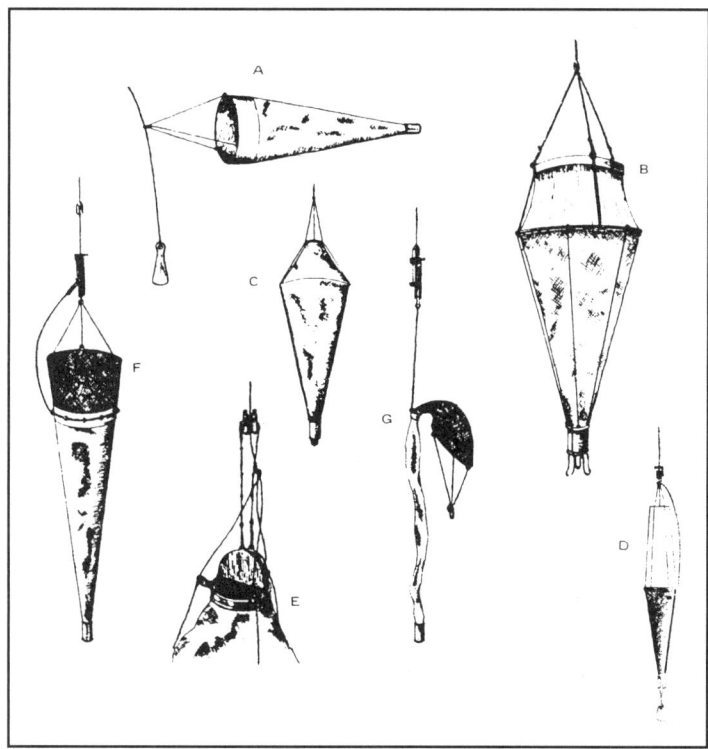

(A) Simple conical tow-net; (B) Hensen net; (C) Apstein net; (D) Juday net; (E) Apstein net with semicircular closing lids; (F) Nansen closing net, open; (G) Nansen closing net, closed.

Fig. 3.3   Plankton sampling nets

(Reprinted from *STANDARD METHODS FOR THE EXAMINATION OF WATER AND WASTEWATER*, 15th Edition, American Public Health Association, Washington, D.C., 1980, by permission of the American Public Health Association)

---

[7] *Overturn.   The almost spontaneous mixing of all layers of water in a reservoir or lake when the water temperature becomes similar from top to bottom. This may occur in the fall/winter when the surface waters cool to the same temperature as the bottom waters and also in the spring when the surface waters warm after the ice melts. This is also called "turnover."*

[8] *See STANDARD METHODS FOR THE EXAMINATION OF WATER AND WASTEWATER, 20th Edition, Color Plates 28 and 29, after page 10-161. Obtain from American Water Works Association (AWWA), Bookstore, 6666 West Quincy Avenue, Denver, CO 80235. Order No. 10079. Price to members, $167.00; nonmembers, $212.00; price includes cost of shipping and handling.*

QUESTIONS

Write your answers in a notebook and then compare your answers with those on page 84.

3.1D   What types of tastes and odors are produced by algae?

3.1E   How can chlorine affect tastes and odors?

3.1F   Where are tastes and odors found in reservoirs?

3.13   Shortened Filter Runs

A second major problem associated with algal blooms is that certain species of algae, specifically diatoms, tend to clog filters at water treatment plants and thereby reduce both filtration rates and the duration of filter runs. Zooplankton, microscopic, planktonic animals can also clog filters when found in large numbers. Filter runs also may be shortened due to gases released by algae. Normal filter runs commonly extend from 30 to 100 hours before cleaning is required, while short filter runs caused by the presence of algae may be less than 10 hours in length. In extreme cases, the clogging may occur so frequently that the amount of water needed to backwash filters for cleaning is greater than the amount of water treated and sent to the distribution system.

Reduced filtration rates and increased frequency of filter backwashing are reflected in an inability to meet system water demands and increased water treatment costs.

3.14   Increased pH[9]

Algal blooms are often associated with marked fluctuations in pH of the water in the upper layer of the reservoir where blooms occur. The pH is frequently raised from a level near 7 to near 9 or above as a result of these blooms. Chlorination efficiency and coagulation are greatly reduced at the higher pH levels. This increases water treatment costs because pH must be adjusted down, or more chlorine must be added. Since algae remove carbon dioxide from solution and convert it into cellular material as they grow, the carbonate equilibrium (balance) is affected. pH values as high as 9.8 can be reached where algae are in high concentrations and under favorable light conditions.

Increases and decreases in pH are caused by *PHOTOSYNTHESIS*[10] during daylight hours and respiration by algae during darkness. During the daylight hours the pH will increase, while the pH is lowered at night. Respiration by algae results in an increase in carbon dioxide in water (lowering of pH), while photosynthesis results in a decrease in carbon dioxide in the water (increasing of pH). These fluctuations in pH can adversely affect both the coagulation and disinfection treatment processes.

3.15   Dissolved Oxygen Depletion

As algal blooms progress, the dissolved oxygen content at depths where the bloom occurs normally increases markedly as a result of photosynthesis. Supersaturation of dissolved oxygen occurs when the dissolved oxygen exceeds the saturation value for the existing water temperature. Dissolved oxygen supersaturation in surface waters is common during major algal blooms.

When algal cells die, however, this abundant oxygen is used up by the bacteria that feed upon (metabolize) the algae cells. Following severe algal blooms, dissolved oxygen in both surface and deeper waters may be reduced to the point that fish kills occur. Fish kills due to low oxygen levels may also occur when there is a combination of extremely heavy algal bloom and a sudden reduction in the amount of sunlight available. Under these conditions, the photosynthetic activity of algae slows down. With less oxygen being produced, the algae use oxygen stored in the water for respiration. If this condition persists for a considerable length of time, the water may lose most of its oxygen, causing both the algae and the fish to die of oxygen starvation. Frequently fish kills from dead algae are the result of the algae clogging the gills of the fish.

Major fish die-offs in domestic water supply reservoirs nearly always generate complaints from the general public, particularly from those who use the water as a drinking supply. The cleanup and disposal of dead fish from a large reservoir can place a large financial and staffing burden on the water utility.

In stratified lakes, oxygen depletion in the colder, deeper waters often occurs following algal blooms. Water quality problems related to this condition will be included in the discussion on stratified lakes in Section 3.17, "Thermal Stratification."

3.16   Organic Loading

As a result of algal blooms, major increases in organic matter naturally occur within water supply reservoirs. The most notable impacts of this increased organic loading are increased color in the water supply and a major increase in chlorine demand. Solution of both of these problems is again reflected in increased water treatment costs.

---

[9] *pH (pronounce as separate letters).   pH is an expression of the intensity of the basic or acidic condition of a liquid. Mathematically, pH is the logarithm (base 10) of the reciprocal of the hydrogen ion activity.*

$$pH = Log \frac{1}{[H^+]}$$

The pH may range from 0 to 14, where 0 is most acidic, 14 most basic, and 7 neutral. Natural waters usually have a pH between 6.5 and 8.5.

[10] *Photosynthesis (foe-toe-SIN-thuh-sis).   A process in which organisms, with the aid of chlorophyll (green plant enzyme), convert carbon dioxide and inorganic substances into oxygen and additional plant material, using sunlight for energy. All green plants grow by this process.*

The organic loading resulting from algal blooms is often associated with high TRIHALOMETHANE[11] levels following free residual chlorination (a water treatment disinfection process). The organic matter contains THM PRECURSORS[12] which react with the chlorine to form trihalomethanes. The U.S. Environmental Protection Agency (EPA) has adopted a maximum contaminant level (MCL) of 80 µg/L (micrograms per liter) for trihalomethanes in domestic water supplies, based upon the average of quarterly samplings within the water distribution system. The EPA trihalomethane standards apply to water systems serving more than 10,000 people and to all surface water systems that meet the criteria to avoid filtration.

Increased trihalomethane (THM) levels resulting from algal blooms may exceed the maximum contaminant levels (MCLs). Either a change in disinfection methods or use of activated carbon filtration will normally be required to lower THM levels to within acceptable limits. If either of these procedures is required, it will result in substantial increases in water treatment costs.

Two helpful references for identifying algae and the problems related to them are ALGAE IN WATER SUPPLIES[13] and STANDARD METHODS.[14]

QUESTIONS

Write your answers in a notebook and then compare your answers with those on page 84.

3.1G What problems do algae cause on filters?

3.1H What is the influence of algal blooms on pH?

3.1I What is the influence of algal blooms on dissolved oxygen?

3.1J Increased organic loadings from algal blooms can cause what kinds of water quality problems?

3.17 Thermal Stratification

Thermal stratification develops in lakes and reservoirs in the spring when the surface waters begin to warm. As summer approaches, the weather warms, the longer days mean longer periods of heating of the water by the sun's rays, and the brisk spring winds subside. Under these conditions the surface waters warm rapidly, expand, and become lighter than the lower waters. Although the wind may continue to blow, its contribution to mixing of the lake waters diminishes because of the resistance to mixing resulting from different water densities caused by the increased water temperatures. The greater the difference in water temperatures, the greater the difference in water densities that create the resistance to mixing. When layers of different temperatures occur within a lake, the lake is considered thermally stratified.

Another common type of water quality problem occurs in productive, thermally stratified reservoirs. Reservoirs and lakes within the United States generally fall into one of two classifications relative to annual thermal stratification cycles. Those relatively deep-water lakes and reservoirs that do not freeze over during winter months undergo a single stratification and destratification (mixing) cycle and are classed as "monomictic" (one mixing). In some areas lakes have one winter turnover that may last from September to mid-May (whenever the wind blows the lake turns over). Lakes and reservoirs that freeze over normally go through two stratification and destratification (mixing) cycles and are classed as "dimictic" (two mixings).

In a monomictic lake, the water temperature during winter months is uniform (the same) from top to bottom; the water density throughout the lake is uniform; and the water is mixed only by wind currents. With continued cooling, the surface

water becomes more dense and sinks to the bottom. As the season progresses into spring, the sun's rays warm the upper portion of the lake or reservoir faster than the deeper portion. The decrease in density of the warmer water on top slows the vertical mixing action within the lake and a barrier is formed between the upper and lower layers. The upper layer, which continues mixing, is known as the EPILIMNION (EP-uh-LIM-knee-on) (Figure 3.4). The middle layer is the zone of rapid temperature decrease with depth and is called the THERMOCLINE (THUR-moe-KLINE) or METALIMNION (MET-uh-LIM-knee-on). The lowest layer of colder, denser water is the HYPOLIMNION (HI-poe-LIM-knee-on). When these conditions exist, the lake is said to be thermally stratified. The lake remains in this stratified or layered condition through the summer and into the fall or early winter when the surface waters become as cool as deeper waters, the density barrier is broken, and destratification or mixing ("turnover") takes place. Figure 3.5 illustrates the thermal stratification cycle in a

---

[11] Trihalomethanes (THMs) (tri-HAL-o-METH-hanes).   Derivatives of methane, CH_4, in which three halogen atoms (chlorine or bromine) are substituted for three of the hydrogen atoms. Often formed during chlorination by reactions with natural organic materials in the water. The resulting compounds (THMs) are suspected of causing cancer.

[12] Precursor, THM (pre-CURSE-or).   Natural organic compounds found in all surface and groundwaters. These compounds MAY react with halogens (such as chlorine) to form trihalomethanes (tri-HAL-o-METH-hanes) (THMs); they MUST be present in order for THMs to form.

[13] ALGAE IN WATER SUPPLIES — AN ILLUSTRATED MANUAL ON THE IDENTIFICATION, SIGNIFICANCE AND CONTROL OF ALGAE IN WATER SUPPLIES, by C. Mervin Palmer, U.S. Public Health Service Publication No. 657, reprinted 1962. No longer in print but may be available in some libraries.

[14] STANDARD METHODS FOR THE EXAMINATION OF WATER AND WASTEWATER, 20th Edition, See Section 10900, Identification of Aquatic Organisms, C. Key for Identification of Freshwater Algae Common in Water Supplies and Polluted Waters (Color Plates 28-35). Obtain from American Water Works Association (AWWA), Bookstore, 6666 West Quincy Avenue, Denver, CO 80235. Order No. 10079. Price to members, $167.00; nonmembers, $212.00; price includes cost of shipping and handling.

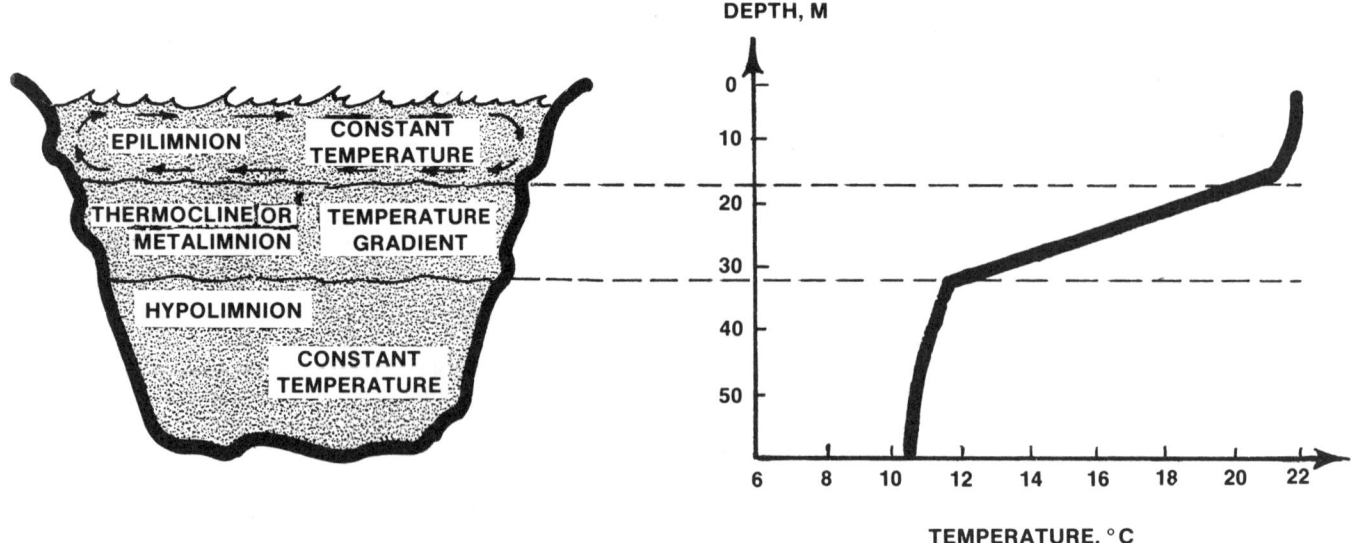

Fig. 3.4   *Thermally stratified lake or reservoir*

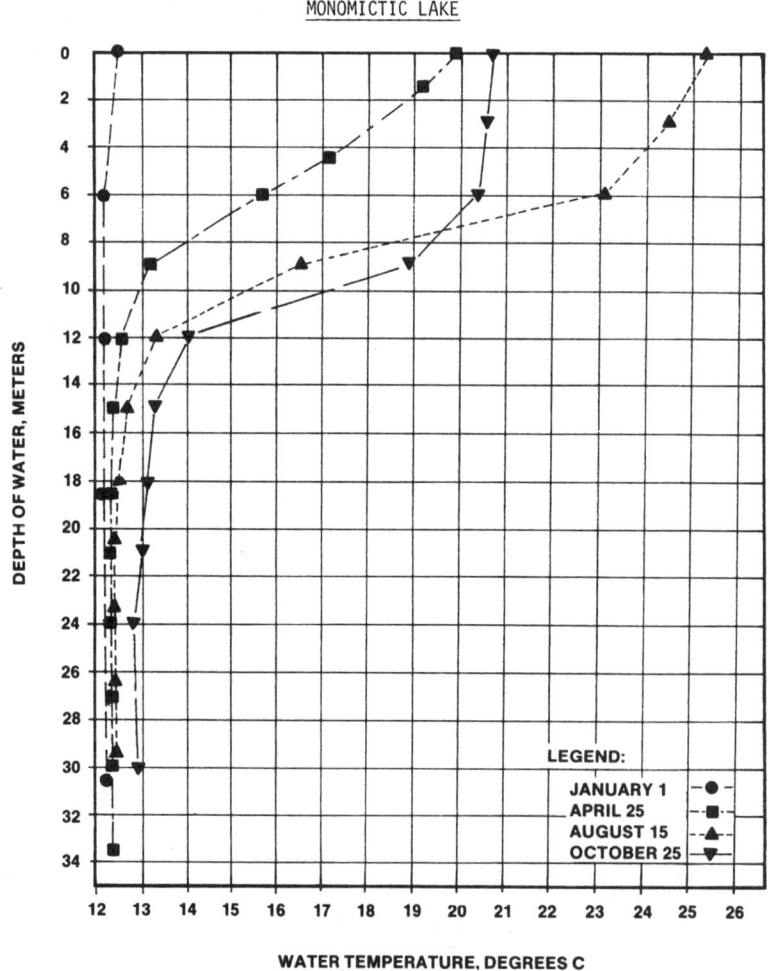

Fig. 3.5   *Average temperature profiles in Lake Casitas on January 1, April 25, August 15, and October 25 for the period 1962 through 1967*

monomictic (single mixing cycle) lake by showing average temperature profiles in Lake Casitas on January 1, April 25, August 15, and October 25 for the period 1962 through 1967.

Once thermal stratification occurs and natural mixing ceases in the metalimnion (middle) and hypolimnion (lower) zones of a productive lake, major changes in water quality begin to take place. *BIOCHEMICAL OXYGEN DEMAND*[15] within the metalimnion and hypolimnion may lead to total dissolved oxygen depletion, resulting in *ANAEROBIC*[16] conditions within these zones. Major contributors to the biochemical oxygen demand are the organisms that decompose dead algal cells as they fall into these zones. Depending upon specific conditions, oxygen depletion may be completed in a few weeks or any time up to several months after thermal stratification begins. When the hypolimnion becomes anaerobic, it usually remains so until the reservoir turnover (destratification) occurs in late fall or early winter. Figure 3.6 illustrates the progress of dissolved oxygen stratification in monomictic Lake Cachuma, California, during 1980. Dissolved oxygen profiles are shown on April 20, June 14, September 20, and December 13. Note that by December 13, reservoir turnover (destratification) had

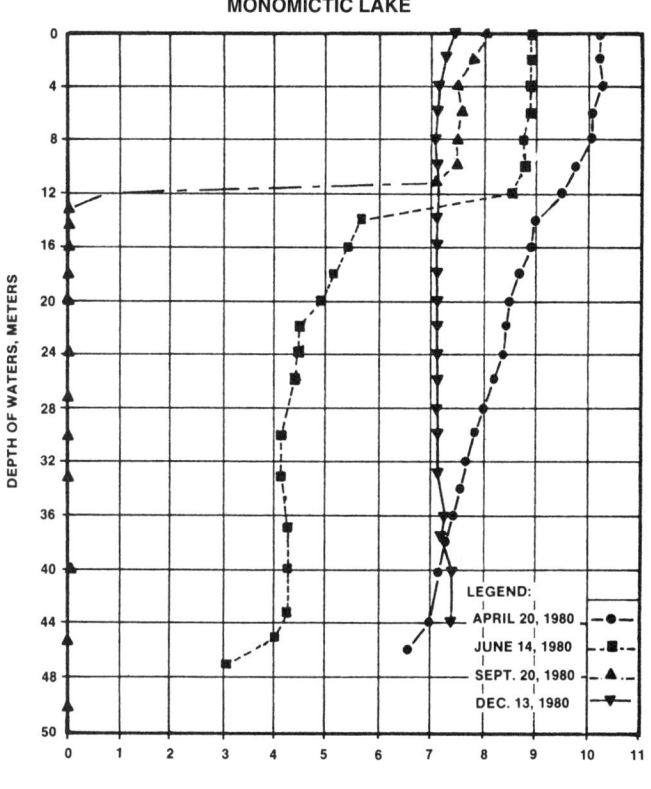

MONOMICTIC LAKE

LEGEND:
APRIL 20, 1980 —●—
JUNE 14, 1980 -■-
SEPT. 20, 1980 -▲-
DEC. 13, 1980 ▼

DISSOLVED OXYGEN CONTENT, mg/*L*

Fig. 3.6 Dissolved oxygen profiles in Lake Cachuma on April 20, June 14, September 20, and December 13, 1980

occurred and dissolved oxygen content was nearly uniform from top to bottom.

An easy way to measure the temperature profile in a thermally stratified lake or reservoir during the summer is to use a maximum-minimum thermometer. Lower the thermometer to various depths from top to bottom. The minimum temperature is the water temperature at each depth.

3.18 Anaerobic Conditions

When anaerobic conditions exist in either the metalimnion or hypolimnion of a stratified lake or reservoir, water quality problems may make the water unappealing for domestic use without costly water treatment procedures. Most of these problems are associated with *REDUCTION*[17] in the stratified waters. The first notable problem may be the presence of a strong rotten egg odor in waters drawn from the anaerobic zone. This odor usually indicates the presence of hydrogen sulfide (H_2S) that occurs as sulfate and is reduced to sulfide within the anaerobic bottom sediments. The reduction to H_2S is brought about by certain anaerobic bacteria. Another group of sulfate-reducing bacteria are capable of producing H_2S by attacking organic matter and liberating (freeing) H_2S from protein material. The presence of hydrogen sulfide in domestic water supplies is generally unacceptable to consumers.

A second major problem in anaerobic water occurs when iron and/or manganese exist in bottom sediments in the reduced state and pass into solution. The presence of either iron or manganese in appreciable quantities within the domestic supply can lead to major "dirty water" problems. In addition to appearing reddish, brown, or just plain dirty, the water may stain clothes during washing and stain porcelain fixtures such

as sinks, bathtubs, and toilet bowls. The color or staining occurs as a result of the iron and/or manganese being changed into the oxidized state *AFTER* it enters the distribution system. *OXIDATION*[18] may occur during disinfection (chlorine is a strong oxidant) or within tanks and reservoirs when the water becomes aerated. Water containing either iron or manganese in quantities exceeding the maximum contaminant levels (MCLs) listed in Federal Secondary Drinking Water Standards is generally unacceptable to consumers

---

[15] *Biochemical Oxygen Demand (BOD). The rate at which organisms use the oxygen in water while stabilizing decomposable organic matter under aerobic conditions. In decomposition, organic matter serves as food for the bacteria and energy results from its oxidation. BOD measurements are used as a measure of the organic strength of wastes in water.*

[16] *Anaerobic (AN-air-O-bick). A condition in which atmospheric or dissolved molecular oxygen is NOT present in the aquatic (water) environment.*

[17] *Reduction (re-DUCK-shun). Reduction is the addition of hydrogen, removal of oxygen, or the addition of electrons to an element or compound. Under anaerobic conditions (no dissolved oxygen present), sulfur compounds are reduced to odor-producing hydrogen sulfide (H_2S) and other compounds. The opposite of OXIDATION.*

[18] *Oxidation (ox-uh-DAY-shun). Oxidation is the addition of oxygen, removal of hydrogen, or the removal of electrons from an element or compound. In the environment, organic matter is oxidized to more stable substances. The opposite of REDUCTION.*

(iron, 0.3 mg/L[19] and manganese, 0.05 mg/L). If iron or manganese accumulates in the distribution system, extensive *FLUSHING*[20] is required to clean out the system.

When significant levels of dissolved oxygen are present, iron and manganese exist in an oxidized state and normally *PRECIPITATE*[21] into the reservoir bottom sediments. Under reducing conditions, however, iron is changed from the oxidized ferric state into the soluble ferrous state, and manganese changes from the oxidized manganic state into the soluble manganous state. Once either or both of these metals pass into solution within the anaerobic zones, they remain there until reservoir turnover occurs and they are oxidized and again precipitate into the bottom sediments. When a lake or reservoir becomes anaerobic, manganese normally passes into solution earlier than iron. Following turnover, manganese usually precipitates out after the iron has precipitated.

In summary, it is often difficult if not impossible to select a level from which to draw acceptable domestic water in a monomictic, productive lake during summer and fall months when thermal stratification exists. Water in the upper levels may contain high quantities of taste- and odor-causing compounds, unacceptably warm water, high organic content, and much higher than desired pH. The deeper, cooler water may not be acceptable due to the presence of hydrogen sulfide gas, iron and manganese, or other problems related to anaerobic conditions.

Water quality problems within dimictic lakes and reservoirs may be similar to those which occur in monomictic lakes and reservoirs, except that instead of occurring primarily during summer and fall months, they may also exist during winter months. Water normally reaches its greatest density at 4°C (39°F). As the freezing point of water is near 0°C (32°F), the ice-covered upper, colder water is less dense than deeper, slightly warmer waters. This condition results in thermal stratification during periods when surface water temperatures fall below 4°C (39°F). In addition to the normal fall turnover which occurs in monomictic lakes, a second turnover takes place following the spring thaw. When the surface waters warm up to 4°C (39°F) they are denser than the deeper, warmer waters and turnover takes place once again. Anaerobic conditions and their related water quality problems may exist in these lakes and reservoirs during most of the winter months as well as during summer and fall months.

3.19 Watershed Conditions

Domestic water supply reservoirs may at times experience major problems with water quality as a result of conditions within the drainage area or watershed in combination with climatic conditions. In many areas of the United States, the major portion of surface runoff into a reservoir occurs during a very short period of time. In semi-arid areas such as the Southwest, 75 percent or more of the annual runoff may occur as a result of only three or four major storms. Most runoff from these storms occurs within a few days during and following the storm. Likewise, in mountainous areas where snowmelt is the major source of runoff, most of the inflow occurs within a relatively short period of time during and following the spring snowmelt. Reservoirs subject to these conditions may experience sudden and dramatic increases in *TURBIDITY*,[22] nutrient loading, and organic loading depending on geological, topographical, and vegetative conditions within the watershed.

Turbidity problems, which occur during and following periods of major runoff, are reflected in reduced rates of flow through the filters and shortened filter runs at water treatment plants. At reservoirs that do not filter water prior to service to consumers, federal and state maximum allowable turbidity levels may be exceeded. The length of time that turbidity will affect water quality and water treatment practices depends on the extent of turbidity loading which occurs, mixing in the reservoir as a result of wind and other currents, and the nature of the particles causing turbidity. Larger suspended particles, such as sand and silt, may settle out within a few days or weeks; colloids, such as fine clays, may cause problems for an extended period of time. When storm waters are colder than reservoir waters, the high-turbidity storm water sometimes flows into the reservoir underneath the warmer reservoir waters. This can cause the greatest turbidity increases to occur within the deeper zones. Within a few days, mixing takes place and turbidity becomes fairly uniform throughout the reservoir. Later, suspended particles begin to settle out and the waters within the upper zones show the least turbidity while the deepest waters have the greatest.

Increased levels of turbidity are a serious concern for water treatment plant operators because increased turbidity has a high chlorine demand. This could result in a decreased chlorine residual and an increasing possibility of bacterial contamination if the operator is not alert. An outbreak of giardiasis *(Giardia lamblia)* occurred in Pennsylvania when a small reservoir "turned over" and high-turbidity conditions developed.

Nutrient loading of a reservoir from its drainage area may result in increased productivity (algal blooms). This tends to occur during wet years when increased runoff raises the nutrient loading in the reservoir.

In watersheds containing large quantities of vegetation such as chaparral, it has been noted that there is an increase in organic loading and in associated THM precursors immediately following periods of major runoff. This condition appears to be related to the large quantities of organic material which are associated with the vegetation. Figure 3.7 illustrates total potential trihalomethane content[23] in Lake Casitas at various depths

---

[19] *Milligrams Per Liter, mg/L.   A measure of the concentration by weight of a substance per unit volume. For practical purposes, one mg/L of a substance in fresh water is equal to one part per million parts (ppm). Thus a liter of water with a specific gravity of 1.0 weighs one million milligrams. If water contains 10 milligrams of calcium, the concentration is 10 milligrams per million milligrams, or 10 milligrams per liter (10 mg/L), or 10 parts of calcium per million parts of water, or 10 parts per million (10 ppm).*

[20] *Flushing.   A method used to clean water distribution lines. Hydrants are opened and water with a high velocity flows through the pipes, removes deposits from the pipes, and flows out the hydrants.*

[21] *Precipitate (pre-SIP-uh-TATE).   (1) An insoluble, finely divided substance which is a product of a chemical reaction within a liquid. (2) The separation from solution of an insoluble substance.*

[22] *Turbidity (ter-BID-it-tee).   The cloudy appearance of water caused by the presence of suspended and colloidal matter. In the waterworks field, a turbidity measurement is used to indicate the clarity of water. Technically, turbidity is an optical property of the water based on the amount of light reflected by suspended particles. Turbidity cannot be directly equated to suspended solids because white particles reflect more light than dark-colored particles and many small particles will reflect more light than an equivalent large particle.*

[23] *Total potential trihalomethane content was determined by attempting to duplicate actual Lake Casitas conditions in the laboratory. Lake samples were collected at various depths, a 3 mg/L chlorine dose was added to each sample, the sample was stored for 100 hours, and then the sample was analyzed for trihalomethanes.*

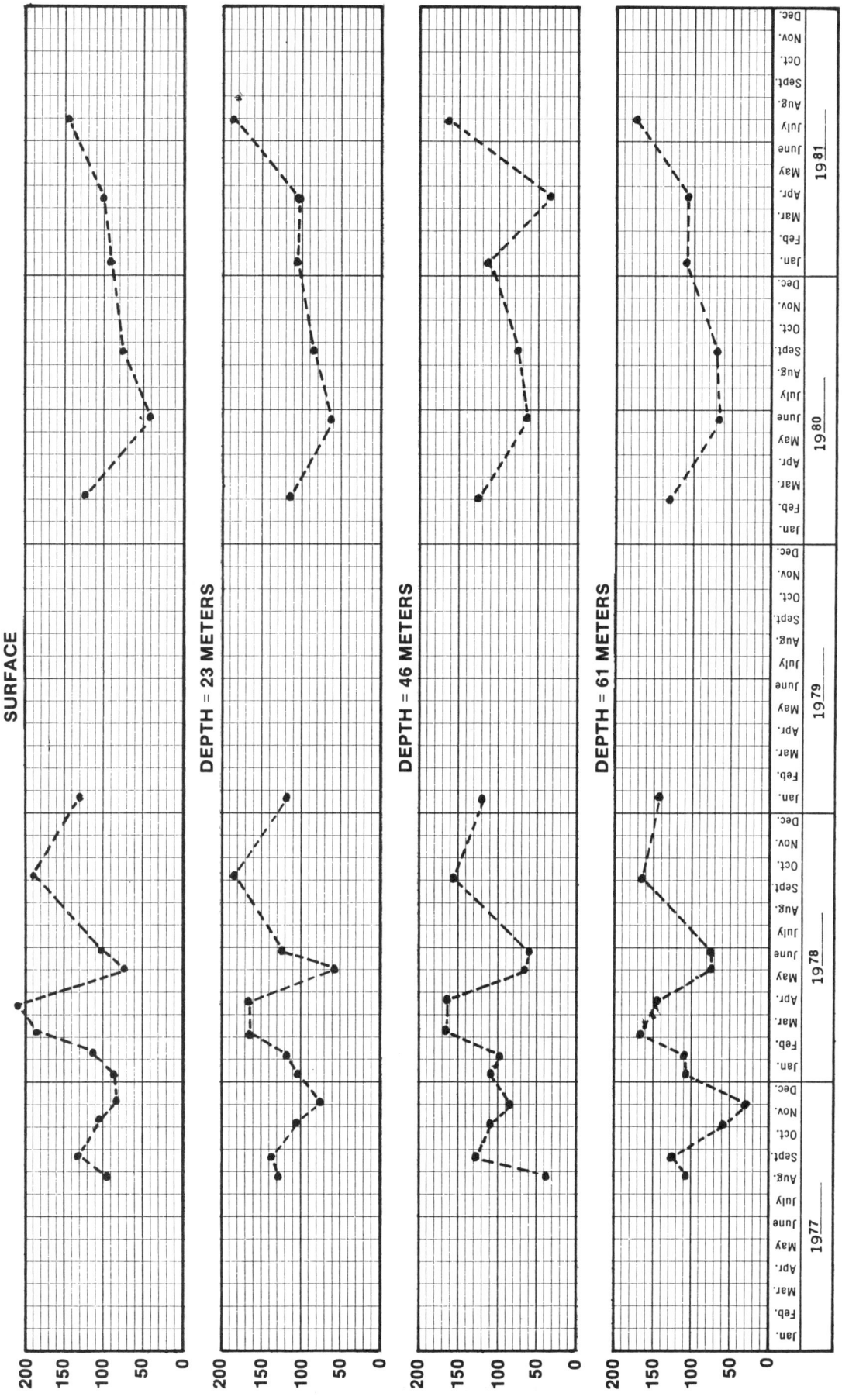

Fig. 3.7   Lake Casitas—Total potential trihalomethane content at various depths

during the periods of August 1977 to January 1979 and March 1980 to August 1981. The dramatic increases in total potential trihalomethane content during the period of January through April, 1978, are associated with the unusually wet winter of 1978. Following the end of the runoff season, the total potential trihalomethane content dropped substantially. The summertime increases in total potential trihalomethane levels are most likely a result of increased organic loading during algal blooms.

Water quality improvement programs should be designed to evaluate water quality problems and their causes within a given lake or reservoir. Since no two reservoirs or lakes are exactly alike, each one will require a water quality improvement program prepared specifically for its individual situation.

QUESTIONS

Write your answers in a notebook and then compare your answers with those on page 84.

3.1K   When a lake warms in the spring or summer, how does the decrease in density of the warmer surface water influence mixing action within the lake?

3.1L   How soon may oxygen depletion be completed after thermal stratification occurs?

3.1M   What problems are caused by anaerobic conditions in reservoirs?

3.1N   What problems can occur in reservoirs that experience large inflows during short periods?

3.2   PURPOSES OF RESERVOIR MANAGEMENT PROGRAMS

3.20   Improvement and Maintenance of Water Quality

Water quality management programs can be a very effective tool for controlling water quality problems in domestic water supply lakes and reservoirs. By evaluating the water quality problems that occur in a given reservoir and the various water quality management and control alternatives available, it may be possible to design a program to eliminate or at least control most problems within acceptable limits. During recent years, great progress has been made toward managing water quality within productive and eutrophic lakes and reservoirs. In many cases it has been possible to control the frequency and intensity of algal blooms and the water quality effects associated with these blooms. Frequently dissolved oxygen depletion can be controlled or eliminated within the metalimnion and hypolimnion, thereby eliminating or controlling iron, manganese, and hydrogen sulfide problems.

Properly designed water quality management programs can also be effective in controlling silt loading, turbidity levels, nutrient loading, and organic loading in many water supply reservoirs. The actual procedures for preparing water quality management programs are discussed in the remainder of this lesson. In order to be feasible, water quality management programs must be prepared for each specific reservoir and be economically as well as technically justifiable.

3.21   Reduction of Water Treatment Costs

Operation, maintenance, and capital costs of water treatment facilities have all increased drastically during recent years. Proper management of water quality within the reser-

voir can prove to be an effective tool in controlling these costs. Cost savings may be realized through increased length of filter runs, reduced chemical costs and alternative treatment methods. For example, if taste and odor problems and organic loading are controlled within the reservoir, the need for activated carbon treatment at the plant may be reduced or eliminated. By controlling algal blooms and silt loading, it may be possible to consider using direct filtration only instead of complete treatment. Iron and manganese control or removal within the treatment plant may not be necessary if levels are controlled within the reservoir. Cost savings must be evaluated on an individual basis, but may range from a few dollars to tens of dollars per million gallons treated. Major savings of hundreds of thousands to millions of dollars may be realized if capital costs for facilities such as sedimentation basins and activated carbon filters can be eliminated.

3.22   Improvement and Maintenance of Fishery, Recreational, and Property Values

In productive reservoirs and lakes where dissolved oxygen depletion occurs within the metalimnion and hypolimnion during summer and fall months, fish are forced to move into the warmer waters of the epilimnion. This not only reduces the size of habitat available to fish, but it may also limit the number of species that exist within a given lake or reservoir. In many areas of the United States, summer temperatures within the epilimnion reach maximums of 75 to 80°F (24 to 27°C). Cold water species of fish, such as trout and salmon, may not survive at these temperatures. If the deeper, colder waters can be prevented from becoming oxygen depleted, both warm and cold water species of fish can be maintained throughout the year.

Proper reservoir management techniques can prevent or minimize fish kills. Reduction in the intensity and severity of algal blooms reduces the hazard of fish kills as a result of rapid oxygen depletion or clogging of their gills with algae. Fish kills that occur in frozen lakes as a result of dissolved oxygen depletion in waters beneath the ice may also be reduced or eliminated by proper management techniques.

Proper lake management naturally results in increased appeal of the reservoir for recreational purposes. Recreational values are also increased when algae problems are reduced. Large mats and scums of algae are unappealing to swimmers, bathers, and water skiers. Objectionable odors associated with these blooms reduce the reservoir area's appeal as a site for camping and picnicking. Property values around the edge of a lake or reservoir may increase significantly when major algae problems are reduced or eliminated.

QUESTIONS

Write your answers in a notebook and then compare your answers with those on pages 84 and 85.

3.2A What reservoir water quality problems can be controlled or eliminated by reservoir management programs?

3.2B How can reservoir management programs reduce water treatment costs?

3.2C What happens to trout and salmon when dissolved oxygen depletion occurs within the metalimnion and hypolimnion during summer and fall months?

3.3 METHODS OF RESERVOIR MANAGEMENT

3.30 Removal of Trees and Brush From Areas To Be Flooded

In areas where new reservoirs are to be formed for the purpose of providing domestic water supplies by construction of dams or other means, it is often advisable to remove trees and brush from the areas to be flooded. The purpose of vegetation removal is to reduce the organic and nutrient loading the reservoir will receive as it fills.

If a large quantity of trees, brush, and other vegetation is left within the reservoir site, organisms will decompose and thus recycle the material after the area is flooded. This will release nutrients and organic matter into reservoir waters. Organisms that decompose the vegetative material will consume dissolved oxygen, thereby increasing the rate of oxygen depletion when thermal stratification of the reservoir takes place. Nutrients that are released during decomposition may lead to more and bigger algal blooms. Organics released during decomposition may contribute to increased color, chlorine demand, and trihalomethane levels following chlorination.

Removal of major vegetation from the reservoir area is best accomplished by mechanical means. Little is accomplished if the material is cut and left on site so that it can decompose by natural processes. When practical, trees can be cut for firewood or timber and the stumps removed mechanically. In some cases, wood is burned on the site and the ashes are removed from the area. In other cases trees, including stumps, are removed by mechanical means.

Brush can be removed by a single tractor or by two or more tractors operating parallel to each other with a large chain stretched between them.

Once vegetation removal is accomplished, regrowth must be controlled by mechanical means until the reservoir is filled. When the filling cycle is completed, reservoir levels will seldom be lowered long enough for major vegetation regrowth to occur.

QUESTIONS

Write your answers in a notebook and then compare your answers with those on page 85.

3.3A Why should trees and brush be removed from areas to be flooded by reservoirs?

3.3B What water quality problems may be caused by organics released during the decomposition of vegetation covered by water in a reservoir?

3.3C How can vegetation be removed from a reservoir site?

3.31 Watershed Management

3.310 Need for Watershed Management

The primary purpose of any watershed management program should be to control, minimize, or eliminate any practices within the watershed area that are harmful to water quality within the domestic water supply reservoir. In order for watershed management programs to be effective, they must be both technically and economically feasible. The general public must be convinced that they are getting their dollars' worth when investing in watershed management programs. If it is much less costly to cure a water quality problem with specific treatment procedures than it is to prevent the problem from occurring through watershed management, it may be difficult, and even unwise, to try to convince the public that watershed controls should be implemented. If factors such as aesthetic or recreational values of the reservoir are also affected by the reduced water quality, the public should be informed so that the value of the water resource from an aesthetic or recreational viewpoint can be evaluated. While the public is very dollar conscious, it is also sensitive to environmental issues and will support well-designed programs for protecting the environment.

Because of the tremendous differences in size, topography, vegetative conditions, and state of development in watersheds, it is impossible to discuss methods of watershed management that would apply in all areas. As with reservoir management programs, each watershed management program should be designed on the basis of potential water quality problems which must be solved or prevented and the options available for solving or preventing them. In most cases, the best tool for managing watersheds is probably the regulatory process. Regulations and ordinances that control or eliminate practices within a watershed that are detrimental to water quality can be adopted as needed on a local, county, state, or even federal level. All types of practices which may have to be controlled within a given watershed are too numerous to discuss in detail, but the most common practices causing water quality problems and means of controlling them are discussed in the following sections.

3.311 Wastewater

Contamination of the domestic water reservoir by wastewater (sewage) can lead to water quality problems ranging from significant to severe. Two major types of problems may result from raw wastewater contamination: nutrient loading of the lake or reservoir and microbial contamination. In some cases, nutrient loading due to prolonged wastewater contamination may be great enough to convert a previously unproductive or moderately productive body of water into a very productive, highly *EUTROPHIC* [24] lake. Examples of lakes that have experienced water quality deterioration due to nutrient loading from wastewater are numerous.

Microbial contamination of a lake or reservoir by wastewater may pose a hazard not only to the domestic water supply but also to persons using the reservoir for recreational purposes such as swimming and water skiing. Diseases caused by protozoans, bacteria, and viruses may all result if the microbial contamination is severe. While conventional water treatment practices do provide protection from microbial contamination to domestic consumers, the source of supply should still be protected. Some states have regulations relating to the maximum allowable *COLIFORM* [25] bacteria content in domestic water supply reservoirs.

The major source of wastewater contamination in domestic water supply reservoirs is usually wastewater disposal systems, such as septic tank leaching systems. There are two major methods by which contamination from septic tank systems can be controlled. The first and most dependable solution is to replace all septic tank leaching systems with well-designed sewer systems that collect all wastewater and transport it to centralized treatment facilities. Wastewater treatment facilities should be located outside of the reservoir watershed or at least located so that the treatment plant effluent does not enter the domestic water supply. Unless very expensive nutrient removal treatment processes are used, the nutrient loading in a reservoir from a treatment plant effluent may have a significant impact on algae productivity.

The second major solution is to adopt ordinances or conditions that regulate the design and installation of septic tank leaching systems to ensure that they function properly and do not contaminate the domestic water supply. Provisions should be made for all systems in the watershed to be brought up to established standards. Periodic inspections should verify that the septic tank leaching systems continue to function properly.

3.312   Fertilization

Fertilization of crops and landscaping with materials containing high concentrations of nitrogen compounds can result in these nutrients being carried away in surface runoff and contributing to productivity in the water supply reservoir if fertilization practices are not controlled. Large quantities of fertilizers containing nitrogen compounds are often used in commercial agricultural production, on golf courses and other park-like areas, and for home lawns, orchards, and gardens. Nitrogen compounds in excess of those required for plant growth may be leached downward into the groundwater following irrigation and/or precipitation. In many instances the groundwater eventually ends up in the surface water supply reservoir through underflow or surface runoff after it is pumped and used for irrigation. Phosphate-base fertilizers usually do not present the problems that nitrogen fertilizers do because of the soil binding characteristics of phosphate compounds. Once phosphate compounds have entered the soil, they tend to remain there unless taken up by plants.

Fertilization practices can be partially controlled by prohibiting use of fertilizers on nonessential crops, but are best controlled through long-term public education programs. People who need to use large quantities of fertilizers can be properly trained to apply only enough for adequate plant growth and to apply fertilizers at times when there is minimal chance of excess fertilizer being washed into the water supply. In some major agricultural areas, it is becoming common practice to have a leaf analysis done on crop plants to determine fertilizer needs and application rates. This practice not only protects the water supply from overfertilization, it is good business practice for the farmer.

3.313   *Industrial Discharges*

Industrial discharges are usually best monitored and controlled by the discharger and a regulatory agency other than the local agency which manages and operates the water supply. In most areas, this responsibility falls under the regulation of a state water pollution control agency. Local water supply agencies can work closely with the pollution control agency in identifying sources of industrial discharges and regulating them so that pollution problems are minimized. Water pollution control programs have been established in each state and are administered under federal legislation which sets controls on industrial discharges.

If mining occurs within the watershed and materials mined present a hazard to the water supply, strict regulations should be adopted to prevent water supply contamination. Cyanide contamination may occur in the runoff from gold mining operations. Erosion from mine tailings and exposed areas should be strictly controlled. Recent investigations have shown very high counts of asbestos fibers in portions of some water supplies. A major source of these asbestos fibers is abandoned mining areas through which surface runoff drains. Asbestos fiber concentrations in domestic water supplies present serious concern regarding links between asbestos and some types of cancer. A limit of 7 million fibers per liter of water has been established.

Oil and gas exploration and drilling must also be regulated so that major contamination of the water supply does not occur. Every agency that operates a surface water supply should prepare and have available an emergency contingency plan which can be implemented immediately if a major spill of hazardous material occurs. County and state health depart-

---

[24] *Eutrophic (you-TRO-fick).   Reservoirs and lakes which are rich in nutrients and very productive in terms of aquatic animal and plant life.*

[25] *Coliform (COAL-i-form).   A group of bacteria found in the intestines of warm-blooded animals (including humans) and also in plants, soil, air and water. Fecal coliforms are a specific class of bacteria which only inhabit the intestines of warm-blooded animals. The presence of coliform bacteria is an indication that the water is polluted and may contain pathogenic (disease-causing) organisms.*

ments, the Environmental Protection Agency, and state pollution control agencies may provide assistance in preparing and implementing hazardous spill emergency plans.

3.314 Soil Grading and Farming Practices

Some watersheds contain soils which, when eroded, may significantly contribute to turbidity in the surface water supply reservoir. It may be necessary to control soil grading and farming practices in sensitive areas. Of particular concern are soils that contain colloidal clays and similar particles which do not settle out in a reasonable period of time within the reservoir. Soil grading and farming practices can be controlled through regulations and ordinances. The major thrust of these rules would not be to prohibit soil disturbances (such as those that occur when farming, logging, or constructing housing subdivisions and industrial parks), but to limit disturbances to those times of the year when danger of erosion from surface runoff is at a minimum. Ordinances can also be used to limit the length of time that soil is left exposed and to specify methods of replanting and replacing ground cover.

QUESTIONS

Write your answers in a notebook and then compare your answers with those on page 85.

3.3D   What should be the primary purpose of a watershed management program?

3.3E   What problems can be caused in reservoirs from raw wastewater contamination?

3.3F   How can problems caused by fertilizers be controlled?

3.3G   How can the adverse impacts of soil disturbances from farming, logging, and construction be minimized?

3.315 Livestock Grazing

Grazing of livestock can be an invaluable tool for controlling vegetative growth and the potential for wildfires in certain watersheds. However, grazing can also contribute to significant deterioration in water quality within surface water supply reservoirs. Overgrazing can expose soils and lead to increased erosion and turbidity problems. Also nutrients concentrated in cattle manure may be washed into the reservoir during periods of high runoff and cause *EUTROPHICATION.*[26] In addition, accumulations of manure may contribute to microbial contamination of the domestic water reservoir during major runoff periods. The most notable microbial pathogen associated with cattle feces is *Cryptosporidium*, a protozoan parasite that causes gastroenteritis in humans. Some of the diseases contracted by humans are carried by animals. In most watersheds, animal grazing can be held within acceptable limits through the implementation of controls and regulations. Grazing can be eliminated or controlled for a limited period of time prior to and during periods of high runoff when fresh manure accumulations would create the greatest hazard to the water supply.

3.316 Pesticides and Herbicides

Unregulated general use of pesticides and herbicides can cause serious problems of contamination in domestic water supply reservoirs. This problem has been minimized on a na-

tional level through federal and local regulations. The federal government monitors and regulates the use of pesticides and herbicides on watershed lands it owns or controls. Certain materials are prohibited from use on these lands, some are restricted in their application, and some are approved for general use. All uses are reported on at least an annual basis. In some areas, agricultural commissioners regulate the use of pesticides and herbicides and will usually cooperate with local agencies in setting up controls for specific problem areas such as watersheds. Operators should be notified when pesticides and herbicides are being applied that may pose a threat to the water supply.

3.317 Wildfires

Wildfires in certain watersheds can create more problems for a domestic water supply reservoir than any other source of contamination or pollution. During the runoff period following the fire, large amounts of debris, nutrients, silt, and other pollutants may enter the reservoir. Turbidity will usually increase tremendously and will have an adverse effect on water treatment plants. Depending on individual situations and the extent of burning within the watershed, water quality may be degraded (reduced) for a relatively short period of time or for up to several years or more. Reservoir storage space lost to debris and silt accumulations may never be recovered without expensive silt removal.

Fire prevention and control programs are an absolute must in watersheds where major fire hazards exist. Public information programs can often be helpful in preventing fires from occurring. Federal and local fire control agencies can often cooperate in firebreak, vegetation control, and other fire control programs. Controlled burning may be a valuable tool in many areas. One fire control measure that is being researched and developed in the western United States is the conversion of chaparral and brushlands to grasslands. This conversion greatly reduces the amount of fuel available to burn. In some instances, the total available water supply can be increased because of increased runoff and reduced *EVAPOTRANSPIRATION*[27] from the grasses.

3.318 Control of Land Use

Most of the Lake Casitas watershed in California lies within a National Forest and presents little hazard to the water quality of the lake except from wildfires and those mining and oil and gas operations permitted. When it was demonstrated that mining operations could significantly reduce the water quality, the Secretary of Interior withdrew permission to mine any U.S. Forest lands within the watershed. The hazards from mining

---

[26] Eutrophication (you-TRO-fi-KAY-shun). The increase in the nutrient levels of a lake or other body of water; this usually causes an increase in the growth of aquatic animal and plant life.

[27] Evapotranspiration (ee-VAP-o-TRANS-purr-A-shun). (1) The process by which water vapor passes into the atmosphere from living plants. Also called transpiration. (2) The total water removed from an area by transpiration (plants) and by evaporation from soil, snow and water surfaces.

operations can be greatly reduced through such efforts of local, state, and federal agencies.

Approximately 3,000 acres of the Casitas watershed around the lake was privately owned and considered prime land for subdivision and other types of development in 1972. Studies indicated that any major development would increase the nutrient loading of the lake and lead to serious deterioration of water quality. Because Lake Casitas is located near the Los Angeles metropolitan area, it was anticipated that development would occur within a few years. As a result of the studies and much effort on the part of local citizens and governmental agencies and representatives, the U.S. Congress approved funds to buy the privately owned lands and preserve them as open space. A local ordinance was passed which limits public use of the open space lands. Residents in the watershed area who wished to remain were granted 25-year or life estates. The remaining residents must comply with conditions of an agreement controlling activities that would significantly deteriorate water quality. In the case of Lake Casitas, watershed management has proven to be a major factor in the overall program to control and maintain water quality within the lake. This sort of control would not have been politically or economically feasible until one single agency held title to all portions of the watershed.

3.319   Highway Storm Water Runoff

Storm water runoff from highways has the potential to cause adverse impacts on reservoir water quality. Automobiles, trucks, and buses can leave pollutants on roadway surfaces. The motoring public can discard trash and litter. Winds can blow pollutants on roadways from adjacent lands, and precipitation can deposit pollutants on highways from great distances. Another source of pollutants is highway maintenance activities.

Pollutants of concern in highway storm water runoff include toxic metals, nutrients, bacteriological constituents, oil and grease, floating materials, trash and litter, pesticides, herbicides, and deicing salts. Pollutants that might enter waterways as a result of accidents or spills are also a concern.

QUESTIONS

Write your answers in a notebook and then compare your answers with those on page 85.

3.3H   How can the use of pesticides and herbicides be controlled in a watershed?

3.3I   What problems can be created as a result of a wildfire?

3.3J   Under what conditions should an agency consider acquiring title to land in a watershed?

3.32   Algae Control by Chemical Methods

3.320   Purpose of Chemical Methods

Chemical control of both planktonic (free-floating) and attached aquatic growths (periphyton) in domestic water supply reservoirs is primarily used to prevent or control taste and odor problems resulting from algal blooms. Many of these algae control programs, however, are expensive and only minimally effective. In some cases it may be much more feasible and economical to limit nutrient availability (and thus prevent major algal blooms) than to correct the problems these blooms cause by the application of chemicals.

A second major purpose of controlling algal blooms by chemical means is to reduce the overall biological productivity. This will reduce the rate of oxygen depletion in the lower parts of the lake. Chemical control programs must be carried out in such a manner that the algal bloom is prevented from becoming intense if the rate of oxygen depletion following die-off is to be reduced. The rate of oxygen depletion in the deeper waters will probably be increased if the bloom is allowed to become intense before chemicals are applied. If algal populations are high when chemical control is started, a large die-off may occur over a very short period of time. This will produce a rapid decomposition of the algal bodies that may cause depletion of dissolved oxygen as well as an increase in tastes and odors. Fish die-offs could result from dead algal bodies clogging fish gills or from low dissolved oxygen levels.

A third reason for using chemicals for control of algae is to maintain acceptable aesthetic conditions in the lake or reservoir. Unsightly algal scums, odors, and lack of water clarity may all be controlled by proper application of chemicals for algae control.

3.321   Chemicals Available

There are very few chemicals that can be economically used for controlling algal growths in domestic water supply lakes and reservoirs. Extreme caution must be exercised in selecting chemicals. Many of the most effective materials are not approved for use in domestic waters due to potential hazards for the health of people, fish, and crops. State and/or local health agencies should be consulted before any chemical algae control program is started. Some states or local areas may require permits from, or consultation with, agencies such as the Department of Fish and Game and Department of Agriculture.

Copper sulfate pentahydrate ($CuSO_4 \cdot 5\ H_2O$) (also called "bluestone"), either by itself or in conjunction with certain other chemicals, is the only algicide in common use in domestic water reservoirs at the present time. (Chlorine, used as a bactericide or oxidizing agent, may also produce the effects of an algicide.) Copper sulfate is toxic to many species of algae at relatively low concentrations. Copper sulfate does not present a health hazard to either the workers applying it or to domestic water users if proper application and safety procedures are followed. However, copper sulfate may be a hazard to trout at levels below those necessary to control some algae. Copper sulfate may cause toxins to be released by algae. Copper sulfate remains relatively inexpensive when compared with other chemicals.

Research by the Metropolitan Water District of Southern California and others, carried out in conjunction with the U.S. Environmental Protection Agency, has pinpointed one major area of concern when copper compounds are regularly added to the water supply source. Water that contains even very low concentrations of copper has been found to cause significant corrosion problems in water distribution systems, particularly those with galvanized piping. The problem appears to be most severe in newer systems where no buildup of calcium or other compounds (which serve to protect the inside of the pipe) has occurred. This problem is more severe in systems supplied with highly mineralized water. Copper residuals in water entering the distribution system must be monitored closely following copper sulfate treatments in order to compile a record of copper concentrations for future reference. The action level for copper at the consumer's tap is 1.3 mg/L.

QUESTIONS

Write your answers in a notebook and then compare your answers with those on page 85.

3.3K   Why are chemicals used in domestic water supply reservoirs to prevent or control attached and floating aquatic growths?

3.3L   What chemical other than copper sulfate ($CuSO_4 \cdot 5 H_2O$) may be used as an algicide?

3.322   Chemical Doses

Alkalinity, suspended matter, and water temperature are the three major water quality indicators that affect the efficiency of using copper sulfate as an algicide. Alkalinity of the water is the principal factor that reduces the effectiveness of copper sulfate. In alkaline waters, the copper ions react with bicarbonate and carbonate ions to form insoluble complexes (joined together) that precipitate from solution and reduce the amount of biologically active copper. Once the copper is removed from the ionized form, it is no longer effective as an algicide. Bartsch[28] emphasizes that the copper sulfate dosage should be dependent upon the alkalinity of the water and states that this combination has resulted in successful treatment in various lakes in the midwestern United States. If the *METHYL ORANGE ALKALINITY*[29] of the water is less than 50 mg/*L*, the copper sulfate is effective at the rate of 0.9 pound of copper sulfate per acre-foot (volume of water) (0.00033 kg/cu m or 0.33 mg/*L*). If the methyl orange alkalinity is greater than 50 mg/*L* the rate should be 5.4 pounds of copper sulfate per acre (surface area of water) (6.06 kg/ha or 0.000605 kg/sq m). In waters with a high alkalinity, the dosage is not dependent upon depth since precipitation of copper would make it ineffective very far below the surface (copper crystals do not dissolve as they fall through the water).

Experience by various researchers and agencies confirms Bartsch's findings that, in most cases, copper sulfate is fully effective as an algicide when the alkalinity is 0 to 50 mg/*L*. These experiences further reveal that at alkalinity concentrations ranging from 50 to 150 mg/*L*, copper sulfate dosages must be increased as the alkalinity increases. When the alkalinity exceeds 150 mg/*L*, the use of copper sulfate by itself as an algicide would not normally be recommended because of its very low effectiveness.

pH of the water is important for two reasons. The effectiveness of copper sulfate as an algicide depends on the pH. Also the pH level influences the precipitation of copper whose presence is essential to control algae.

Suspended matter in the reservoir or lake being treated with copper sulfate can reduce the effectiveness of copper as an algicide. Such suspended matter provides sites or masses other than algae bodies where the copper is *ADSORBED.*[30]

Organic material, both living and dead, adsorbs the copper. Suspended inorganic sediment is also a significant factor influencing the loss of copper available to kill the algae.

Water temperature plays a major role in how well the copper sulfate kills the algae. When the water temperature drops to the 50°F (10°C) level, algae do not respond to treatment as they do at higher temperatures. Higher rates of copper sulfate application will generally be required when water temperature drops below 50°F (10°C). In many reservoirs, major blooms of algae do not occur in colder waters with temperatures below 50°F (10°C). In the majority of cases, the problem blooms take place after surface water temperatures have warmed during spring and summer months.

The amount of copper sulfate required for effective control of algae is also influenced by the species to be treated. Not all algae are alike in their reaction to copper sulfate. Several tiny planktonic green algae, some of the green flagellates, and filamentous blue-green algae are somewhat resistant to the toxic effects of copper sulfate. Most diatoms are quite susceptible to treatment, though they often bloom in large numbers following copper sulfate treatment for other algae. Many of the major taste- and odor-producing algae and filter-clogging algae are controlled effectively with low rates of application.

> In summary, the following factors influence the concentration of copper sulfate needed for effective control of any particular algal bloom: (1) species of algae, (2) amount of algae, (3) alkalinity of the water, (4) pH of the water, (5) water temperature, and (6) quantity of suspended matter and organic material in the water.

The maximum rate of application of copper sulfate in domestic water supply sources is currently influenced by regulations limiting the concentration of copper in *POTABLE WATER.*[31] The Environmental Protection Agency (EPA) has established an action level of 1.3 mg/*L* for copper at the consumer's tap. Another important consideration is the tolerance of fish and other aquatic organisms to copper. Copper sulfate products are currently registered through the EPA for use in controlling algae. To comply with these federal requirements, copper sulfate products properly labeled for the intended use must be selected.

---

[28] Bartsch, A.F., *PRACTICAL METHODS FOR CONTROL OF ALGAE AND WATER WEEDS, Public Health Reports,* 69:749-757, 1954.

[29] *Methyl Orange Alkalinity.   A measure of the total alkalinity in a water sample. The alkalinity is measured by the amount of standard sulfuric acid required to lower the pH of the water to a pH level of 4.5, as indicated by the change in color of methyl orange from orange to pink. Methyl orange alkalinity is expressed as milligrams per liter equivalent calcium carbonate.*

[30] *Adsorption (add-SORP-shun).   The gathering of a gas, liquid, or dissolved substance on the surface or interface zone of another material.*

[31] *Potable (POE-tuh-bull) Water.   Water that does not contain objectionable pollution, contamination, minerals, or infective agents and is considered satisfactory for drinking.*

Investigation and field experiments by various individuals and agencies have indicated that effective use of copper sulfate as an algicide can be accomplished in highly alkaline water when the copper sulfate is combined with *ALIPHATIC HYDROXY ACIDS*.[32] These acids have proven effective for delaying the chemical reaction of copper with the bicarbonate and carbonate ions of the water and thereby preventing the immediate precipitation of copper in alkaline waters. The most commonly used of these acids is citric acid which is normally mixed with the copper sulfate in a ratio of approximately two parts copper sulfate to one part citric acid regardless of the alkalinity. The Casitas Municipal Water District has used a copper sulfate plus citric acid mixture with satisfactory results for over ten years in Lake Casitas. Lake Casitas water has an alkalinity of approximately 150 mg/L. Several commercially prepared *CHELATED*[33] copper compounds for use in alkaline waters are presently available and may prove to be economical for use in many situations.

QUESTIONS

Write your answers in a notebook and then compare your answers with those on page 85.

3.3M   Why is the dose of copper sulfate based on either surface area or volume of reservoir?

3.3N   How does suspended particulate matter in a reservoir reduce the effectiveness of copper as an algicide?

3.3O   What is the major factor limiting the maximum rate of application of copper sulfate in the sources of a domestic water supply?

3.323   Methods of Chemical Application

Methods of using copper sulfate compounds range from very simple to very elaborate depending on the size of the reservoir to be treated, frequency of treatment, and rates of application. Depending on the method of application, copper sulfate may be purchased in dry form with crystal size ranging from snowflake to large diamond size of up to one inch (25 mm) or more in diameter. Some of the most commonly used methods for applying copper sulfate compounds are summarized in the following paragraphs.

1. The simplest method, and one which may be most applicable in very small lakes and reservoirs, is to drag burlap bags containing the copper material through the water using a boat. The reservoir surface is normally crisscrossed in a zigzag fashion so as to cover all of the water surface. Mixing of the material within the water is accomplished by wind, diffusion, and gravity. Boat speed, number of bags used, and the size of the crystals can all be regulated to produce the desired rate of application.

2. Dry copper sulfate crystals may be dumped into a hopper mounted on a boat in such a way that they can be fed into a broadcaster (such as the type used for spreading grass seed) which distributes them onto the lake surface. If the entire surface area of the reservoir is to be treated, the material can be applied in a crisscross pattern. Holes can be drilled or cut in the bottom of the hopper to control how much material is released onto the broadcaster. Boat speed, width of area covered by broadcaster, and rate of copper sulfate fed through the holes can be calculated and adjusted to produce the desired application rate.

3. Perhaps the most efficient and safest method of applying copper sulfate is to mix it into solution and spray it onto the reservoir surface (Figure 3.8) or pump it into the reservoir through a length of pipe (preferably plastic) which contains a number of holes. The pipe may be 15 to 20 feet (4.5 to 6 m) long with the holes about two feet (0.5 m) apart. The pipe can be mounted behind the boat below the surface of the water, and perpendicular to the direction of travel. The primary advantage of this submerged application instead of spraying is that the application of the solution is unaffected by wind. The disadvantages of the submerged application are that it is difficult to tell when the holes are plugged and the delivery system may be damaged or caught in submerged trees or brush, especially in shallow water. Figure 3.9 shows the spreading method of applying copper sulfate from a boat.

Mixing algicide chemicals into solution is a good method when citric acid is combined with the copper sulfate to prevent the precipitation of copper in alkaline waters. Copper sulfate in snowflake-size crystals works best for this method of application. Copper sulfate is commonly supplied in 80-pound (36-kg) bags which are easy to move and store. Citric acid, which is supplied in granular form in bags, is also easy to use.

To get the materials into solution, they are loaded on the boat or barge in bags and fed into a hopper designed so that different sized holes can be fitted into the bottom outlet. The various hole sizes allow adjustment of the rate at which material is released from the hopper. If two chemicals are to be mixed, a divided hopper can be used with different sized holes so that the feed rate of each chemical can be adjusted independently. If the hopper is made of steel, it can be coated with an epoxy material in order to limit corrosion.

The materials are fed from the hopper into a corrosion-resistant tank where they are mixed into solution using water pumped into the tank from the lake or reservoir. Once in solution, the materials are applied to the lake or reservoir by pumping through the spray nozzles or through submerged outlets. By proper regulation of flow through the pumps, the solution level within the tank can be regulated. The desired rate of application of chemicals can be obtained by considering the speed of the boat, the rate at which material is fed from the hopper and pumped into the lake, and the area covered by the spray apparatus or submerged outlet system.

---

[32] *Aliphatic (AL-uh-FAT-ick) Hydroxy Acids.   Organic acids with carbon atoms arranged in branched or unbranched open chains rather than in rings.*

[33] *Chelation (key-LAY-shun).   A chemical complexing (forming or joining together) of metallic cations (such as copper) with certain organic compounds, such as EDTA (ethylene diamine tetracetic acid). Chelation is used to prevent the precipitation of metals (copper).*

Note spray from chemical solution hitting water
on right of boat.

Fig. 3.8   Application of copper sulfate and citric acid

When computing the amount of copper sulfate required to produce the required dosage, it must be remembered that copper sulfate pentahydrate contains approximately 25 percent copper. For a one mg/*L* dose of copper to be obtained in one million gallons of water, 33.4 pounds of copper sulfate should be added.

$$\frac{8.34 \text{ lbs copper/million gallons}}{25\% \text{ available copper}} \times 100\% = \frac{33.4 \text{ lbs copper sulfate}}{\text{million gal water}}$$

Note that 8.34 pounds of a substance in a million gallons of water is equal to one mg/*L*. By knowing the quantity of water to be treated in millions of gallons, and the desired dose, it is relatively simple to compute the amount of copper sulfate required per application.

FORMULAS

In order to calculate the dose of copper sulfate to be applied to a reservoir we need information regarding (1) the reservoir, and (2) the desired dosage in terms of either the reservoir surface area or volume.

1. Reservoir information is usually stated in terms of surface area and volume on the basis of water surface level or elevation. Sometimes these data are presented as depth vs. surface area or depth vs. volume curves (Figure 3.10). If these curves are available, simply observe the depth of water or surface elevation and obtain the reservoir surface area and volume from the curves. The information used to plot these curves may be available in a table also.

If the reservoir volume is given in acre-feet, you may have to convert this number to a volume in million gallons.

Volume, gallons = (Volume, ac-ft)(43,560 sq ft/ac)(7.48 gal/cu ft)

The reservoir volume in acre-feet is multiplied by 43,560 square feet per acre to give us a volume in cubic feet. We multiply this number by 7.48 gallons per cubic foot to obtain a volume in gallons.

Frequently we want the volume in millions of gallons, instead of gallons.

$$\text{Volume, Million Gallons} = \frac{(\text{Volume, Gallons})}{1,000,000/\text{Million}}$$

$$= \frac{(5,000,000 \text{ Gallons})}{1,000,000/\text{Million}}$$

$$= 5.0 \text{ Million Gallons}$$

When we multiply a volume in gallons by 1 Million/1,000,000, we are only changing the units. In the above example we changed a volume of 5,000,000 gallons to 5 million gallons, which are both the same.

2. The copper sulfate required in pounds may be determined on the basis of reservoir volume or reservoir surface area. Another important factor is whether the desired dose is given in terms of a concentration of *COPPER* in milligrams of copper per liter of water or as an application of *COPPER SULFATE* in pounds of copper sulfate per acre of water surface area.

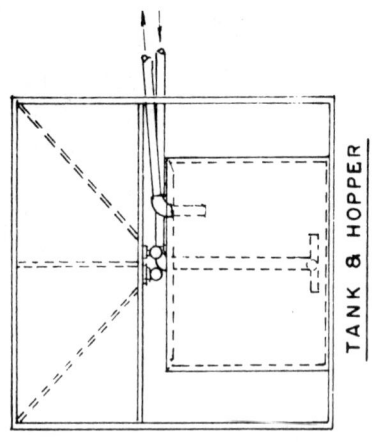

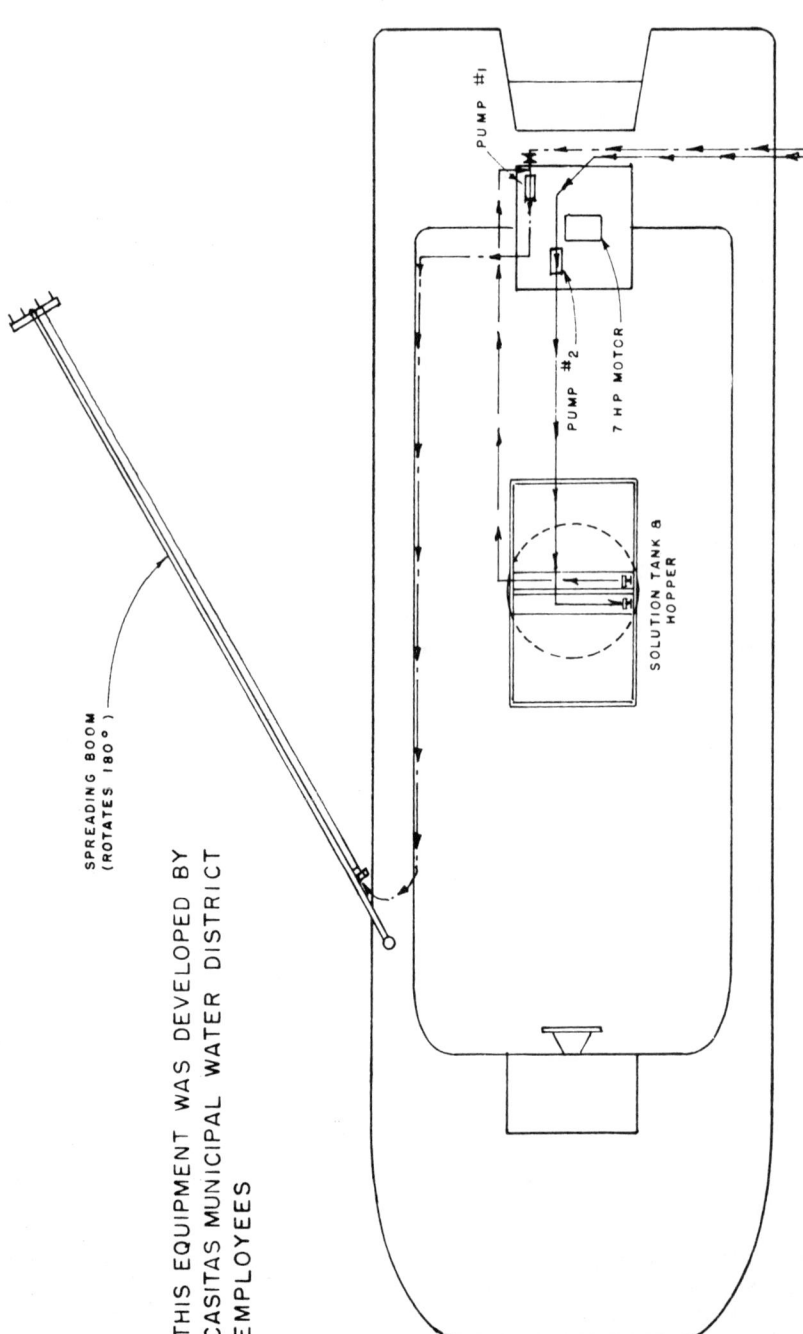

TANK & HOPPER

SPREADING BOOM
(ROTATES 180°)

THIS EQUIPMENT WAS DEVELOPED BY
CASITAS MUNICIPAL WATER DISTRICT
EMPLOYEES

PUMP #1

PUMP #2

7 HP MOTOR

SOLUTION TANK &
HOPPER

MOTOR –   7 HP WISCONSIN WITH
          CLUTCH ASSEMBLY

PUMP #1 – 1¼" MODEL 6400 JABSCO

PUMP #2 – 1" MODEL 777 JABSCO

NOZZLE. – 4 ⅛" PIPE

BOAT   – 21'-4" BOSTON WHALER (1972)

MOTOR – 85 HP OUTBOARD

Fig. 3.9 Copper sulfate spreading equipment

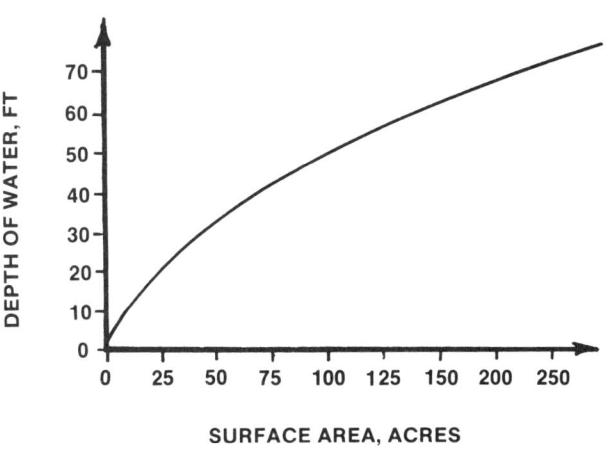
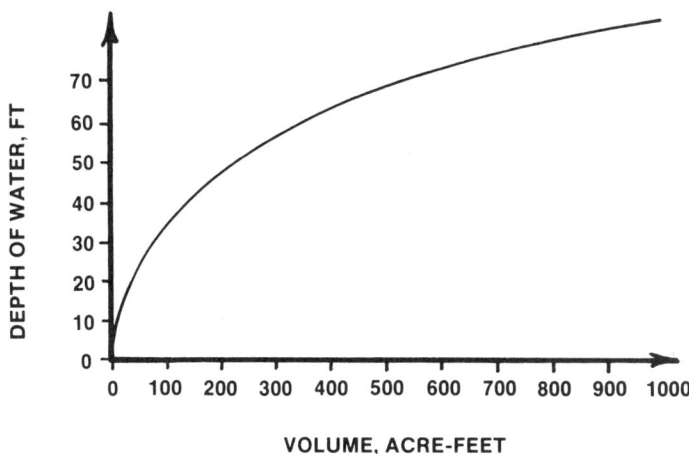

Fig. 3.10   Depth vs. surface area and volume curves

If the desired dose is given as mg/L of copper, calculate the amount of copper needed in pounds as follows:

Copper, lbs = (Volume, M Gal)(Dose, mg/L)(8.34 lbs/gal)

We can check these units by knowing that one liter of water weighs one kilogram. Also one kilogram is equal to 1,000 grams and one gram equals 1,000 milligrams. Therefore one liter weighs one million milligrams. One mg/L is sometimes referred to as one part per million (ppm) or one milligram per million milligrams or one pound per million pounds.

$$\text{Copper, lbs} = \frac{(\text{Volume, M Gal})(\text{Dose, mg})(8.34\ \text{lbs/gal})}{1\ \text{Million mg}}$$

$$= \frac{(\text{Volume, M Gal})(\text{Dose, lbs})(8.34\ \text{lbs/gal})}{1\ \text{Million lbs}}$$

$$= \text{Pounds of Copper}$$

When we add copper sulfate to water, 25 percent of the copper sulfate is copper or for every four pounds of copper sulfate we add to the reservoir, we are actually adding one pound of copper.

$$\text{Copper Sulfate, lbs} = \frac{(\text{Copper, lbs})(100\%)}{25\%}$$

By combining these two formulas, we obtain the formula for calculating the pounds of copper sulfate needed when we have the reservoir volume in million gallons and the copper dose in mg/L.

$$\substack{\text{Copper} \\ \text{Sulfate,} \\ \text{lbs}} = \frac{(\text{Volume, M Gal})(\text{Dose, mg/}L)(8.34\ \text{lbs/gal})(100\%)}{25\%}$$

If the copper sulfate dose is given in pounds of copper sulfate per acre of water surface, use the following formula:

Copper Sulfate, lbs = (Surface Area, ac)(Dose, lbs/ac)

If the alkalinity of the water is greater than 150 mg/L, add one pound of citric acid for every two pounds of copper sulfate.

$$\text{Citric Acid, lbs} = \frac{(\text{Copper Sulfate, lbs})(1\ \text{lb Citric Acid})}{2\ \text{lbs Copper Sulfate}}$$

NOTE: Actual desired copper concentrations in mg/L, copper sulfate doses in pounds per acre, and citric acid doses may vary with each reservoir. Only by monitoring doses and by observing and analyzing results can operators determine the most cost-effective copper sulfate program for their reservoirs.

EXAMPLE 1

A small storage reservoir has a surface area of five acres and contains 80 acre-feet of water. How many pounds of copper sulfate pentahydrate are needed for a 0.5 mg/L dose of copper? Copper sulfate pentahydrate contains 25 percent copper. Assume the alkalinity is 40 mg/L.

Known		**Unknown**
Surface Area, ac	= 5 ac	Copper Sulfate, lbs
Volume, ac-ft	= 80 ac-ft	
Copper Dose, mg/L	= 0.5 mg/L	
Copper,%	= 25%	
Alkalinity, mg/L	= 40 mg/L	

$$\substack{\text{Copper Sulfate,} \\ \text{lbs}} = \frac{(\text{Volume, M Gal})(\text{Dose, mg/}L)(8.34\ \text{lbs/gal})(100\%)}{\text{Copper, \%}}$$

1. Convert acre-feet to million gallons.

Volume, M Gal = (Volume, ac-ft)(43,560 sq ft/ac)(7.48 gal/cu ft)

$$= (80\ \text{ac-ft})(43,560\ \text{sq ft/ac})(7.48\ \text{gal/cu ft})$$

$$= \frac{26,066,304\ \text{gal}}{(1,000,000/\text{Million})}$$

$$= 26.07\ \text{Million Gallons}$$

NOTE: When we multiply or divide an equation by 1,000,000/M or by M/1,000,000 we do not change anything except the units. This is just like multiplying an equation by 12 in/ft or 60 min/hr; all we are doing is changing units.

2. Calculate the pounds of copper sulfate needed.

$$\text{Copper Sulfate, lbs} = \frac{(\text{Volume, M Gal})(\text{Dose, mg/}L)(8.34 \text{ lbs/gal})(100\%)}{\text{Copper, \%}}$$

$$= \frac{(26.07 \text{ MG})(0.5 \text{ mg/}L)(8.34 \text{ lbs/gal})(100\%)}{25\%}$$

$$= 435 \text{ lbs Copper Sulfate}$$

NOTE: In large reservoirs the total volume of water may not need to be treated. From experience the operator may decide to treat only the top 20 feet (6 meters) or down to the thermocline.

EXAMPLE 2

How many pounds of copper sulfate pentahydrate would be required for the reservoir in Example 1 if the alkalinity of the water was 175 mg/L? Since the alkalinity is greater than 150 mg/L, the recommended copper dose is 5.4 lbs of copper sulfate per acre of water surface area.

Known	Unknown
Surface Area, ac = 5 ac	Copper Sulfate, lbs
Volume, ac-ft = 80 ac-ft	
Copper Sulfate Dose, lbs/ac = 5.4 lbs/ac	
Alkalinity, mg/L = 175 mg/L	

Calculate the pounds of copper sulfate pentahydrate needed.

$$\text{Copper Sulfate, lbs} = (\text{Surface Area, ac})(\text{Dose, lbs/ac})$$

$$= (5 \text{ ac})(5.4 \text{ lbs/ac})$$

$$= 27 \text{ lbs Copper Sulfate}$$

NOTE: If the alkalinity is greater than 150 mg/L, citric acid may have to be mixed with the copper sulfate and the copper sulfate may have to be added to the reservoir more frequently. When citric acid is added, copper will remain in solution and the dosage should be based on volume of water, not surface area. The dosage of citric acid is usually one pound of citric acid for every two pounds of copper sulfate, regardless of how much the alkalinity is over 150 mg/L.

EXAMPLE 3

How many pounds of citric acid would be required for the reservoir in Example 2 if the recommended dose of citric acid is one pound of citric acid for every two pounds of copper sulfate applied? The copper sulfate dose in Example 2 was 27 pounds.

Known	Unknown
Copper Sulfate, lbs = 27 lbs	Citric Acid, lbs
Citric Acid Dose $= \dfrac{1 \text{ lb Acid}}{2 \text{ lbs Copper Sulfate}}$	

Calculate the pounds of citric acid needed.

$$\text{Citric Acid, lbs} = \frac{(\text{Copper Sulfate, lbs})(1 \text{ lb Citric Acid})}{2 \text{ lbs Copper Sulfate}}$$

$$= \frac{(27 \text{ lbs Copper Sulfate})(1 \text{ lb Citric Acid})}{2 \text{ lbs Copper Sulfate}}$$

$$= 13.5 \text{ lbs Citric Acid}$$

When applying copper sulfate or any other chemical, you will have to experiment to find the best dosage for your situation. Whether your doses are based on the entire reservoir volume, or only on the surface area, the amount of chemical required will vary with location, time of year, and water quality. Another important variable is the frequency of application of chemicals. To determine the best dosage and frequency of dosing, you will have to develop and analyze results of a reservoir monitoring program.

3.324 Monitoring

In reservoirs where algae are a potential problem, the operator must have a monitoring program capable of anticipating a possible algal bloom. When the data reveal that a bloom is likely, the operator must take the necessary treatment action to prevent the bloom. After a bloom occurs it is more difficult, if not almost impossible, to control the bloom and correct the bad effects on water quality.

Whenever a chemical algae control program is started, monitoring should be carried out before, during, and after use of chemicals. Before, and for several days after the chemical application, data on type of algae, amount of algae, and where they are located should be collected in order to evaluate the effectiveness of treatment at the dosage applied. Careful evaluation must be made to determine if algae die-offs actually occur as a result of the chemical application or if they simply die off due to natural circumstances. This can best be accomplished by monitoring bloom/die-off cycles under natural conditions when no chemical treatment is carried out. Careful monitoring of algicide residual concentrations should be practiced during and following treatment in order to determine if the desired dose is obtained and the extent of the algicide distribution. For example, if the water body is to be treated to a depth of 15 feet (4.5 m), it may be necessary to adjust application methods in order to obtain an effective residual to this depth. Accurate data should be kept on the actual algicide concentration (copper, for example) in the reservoir or water supply in case legal questions regarding causes of fish die-off or system corrosion arise. Monitoring levels of copper in water being released into state-owned streams or rivers should also be conducted and may be a regulation in some states.

3.325 Recordkeeping

Full and accurate recordkeeping is an important part of any chemical algae control program. These records are valuable when evaluating current and historical treatment programs, for designing new or revising existing programs, and for showing compliance with federal, state, and local regulations.

3.326 Safety

Employee and public safety is an important concern in chemical algae control programs. Proper procedures for handling and applying chemicals must be strictly followed. Besides following precautions listed by the chemical manufacturer, anyone applying chemicals should refer to the regulations of federal and state agencies (such as their state OSHA program). Particular caution must be exercised when applying copper sulfate in dry form to protect employees and the public from the dust. Special clothing, gloves, and breathing apparatus should be required. In some cases, it may be necessary to close the lake or reservoir to public use during periods of chemical application in order to protect the public from dust or spray.

If chemicals are to be applied from a boat, operators must observe water safety procedures. Personal flotation devices must be carried on the boat and, in most cases, must be worn while working. Employee training in water safety and first-aid procedures is always a good idea and may be legally required.

3.327 References

Publications containing more information about algae and weed control by chemical methods should be used as references when planning and implementing algae control programs. Information included in these publications ranges from methods of applying chemicals to estimated copper concentrations required to control specific groups and species of algae. Information from the following publications was used extensively in the preparation of this section on algae control.

1. *ALGAE IN WATER SUPPLIES—AN ILLUSTRATED MANUAL ON THE IDENTIFICATION, SIGNIFICANCE AND CONTROL OF ALGAE IN WATER SUPPLIES*, by C. Mervin Palmer, U.S. Public Health Service Publication No. 657, reprinted 1962. No longer in print but may be available in some libraries.

2. *THE USE OF COPPER SULFATE IN CONTROL OF MICROSCOPIC ORGANISMS*, by Frank E. Hale, Ph.D. Presented by Phelps Dodge Refining Corporation.

3. *INVESTIGATIONS OF COPPER SULFATE FOR AQUATIC WEED CONTROL*. A Water Resources Technical Publication, Research Report No. 27, U.S. Department of Interior, Bureau of Reclamation. Obtain from National Technical Information Service (NTIS), 5285 Port Royal Road, Springfield, VA 22161. Order No. PB95-145918. Price, $33.50, plus $5.00 shipping and handling per order.

4. *HOW TO IDENTIFY AND CONTROL WATER WEEDS AND ALGAE*. Created and produced by Applied Biochemists, Inc. Obtain from Cygnet Enterprises West, Inc., 5063 Commercial Circle, Suite C, Concord, CA 94520. Price, $9.95, plus shipping and handling.

QUESTIONS

Write your answers in a notebook and then compare your answers with those on page 85.

3.3P List three methods of applying copper sulfate compounds to a reservoir.

3.3Q How is the effectiveness of a chemical algae control program evaluated?

3.3R What safety precautions should be taken by a person applying copper sulfate in the dry form?

End of Lesson 1 of 2 Lessons on Reservoir Management and Intake Structures

Please answer the discussion and review questions next.

DISCUSSION AND REVIEW QUESTIONS

Chapter 3.   RESERVOIR MANAGEMENT AND INTAKE STRUCTURES

(Lesson 1 of 2 Lessons)

At the end of each lesson in this chapter you will find some discussion and review questions. The purpose of these questions is to indicate to you how well you understand the material in the lesson. Write the answers to these questions in your notebook before continuing.

1. What purposes may a reservoir be used for other than as a source for a domestic water supply?

2. What types of public use may be allowed on water supply reservoirs?

3. Algal blooms can cause what types of problems in domestic water supply reservoirs?

4. Under what circumstances are tastes and odors in a domestic water supply most noticeable?

5. How can a water agency meet acceptable trihalomethane levels?

6. Iron and manganese can cause what kinds of water quality problems in drinking water?

7. Why is it difficult to find the best level from which to withdraw acceptable domestic water from a monomictic, productive lake during the summer when thermal stratification exists?

8. Reservoir water quality management programs can control or eliminate what types of water quality problems?

9. How can the recreational values of a reservoir be improved by proper reservoir management?

10. How can water quality problems caused by septic tank leaching systems be solved?

11. What are the advantages and limitations of grazing livestock on the watershed of a water supply reservoir?

12. Why is copper sulfate used to control algal blooms?

13. How does alkalinity in water reduce the efficiency of using copper sulfate as an algicide?

14. How can the desired rate of copper sulfate application to a reservoir be achieved?

CHAPTER 3. RESERVOIR MANAGEMENT AND INTAKE STRUCTURES

(Lesson 2 of 2 Lessons)

3.33 Reaeration and Artificial Destratification

3.330 Terminology

To help you understand this section, we will first define three important terms.

1. Aeration (air-A-shun). The process of adding air to water. Air can be added to water by either passing air through water or passing water through air.

2. Reaeration (RE-air-A-shun). The introduction of air through forced air diffusers into the lower layers of the reservoir. As the air bubbles form and rise through the water, oxygen from the air dissolves into the water and replenishes the dissolved oxygen. Also the rising bubbles cause the lower waters to rise to the surface where oxygen from the atmosphere is transferred to the water. This is sometimes called surface reaeration.

3. Destratification (de-STRAT-uh-fuh-KAY-shun). The development of vertical mixing within a lake or reservoir to eliminate (either totally or partially) separate layers of temperature, plant, or animal life. This vertical mixing can be caused by mechanical means (pumps) or through the use of forced air diffusers which release air into the lower layers of the reservoir.

In this section when we use the term "reaeration-destratification," we are using air to destratify the reservoir. A relatively small amount of replenishment of the dissolved oxygen in the water actually occurs far below the saturation point. Also the reservoir may be only partially destratified as a result of this procedure. The entire reservoir may not be destratified, but mixing will occur between the upper and lower layers of water.

3.331 Purpose of Reaeration-Destratification Programs

The primary purpose of reaeration-destratification programs in domestic water supply reservoirs is usually to eliminate, control, or minimize the negative effects on domestic water quality that occur during periods of thermal stratification and dissolved oxygen depletion. A secondary purpose may be to increase recreational values of the reservoir through expanded and improved fisheries and improved aesthetic conditions. In reservoirs that freeze over during the winter, reaeration-destratification equipment may be used to reduce winter fish kills in waters that normally become anaerobic, and to prevent portions of the lake or pond from freezing over. Prevention of freezing can be accomplished by circulating water vertically within a lake or reservoir. This practice is particularly helpful in marina areas when open water can be maintained around boat docks and related facilities.

Water quality improvement is obtained by the addition of dissolved oxygen to zones within the lake that would normally become anaerobic during periods of thermal stratification. In reservoirs that have a single outlet gate located at a depth where anaerobic conditions exist during portions of the year, reaeration-destratification systems offer a method of improving water quality for delivery to water treatment-distribution facilities. In reservoirs with multilevel outlets that permit selection of the depth from which water is withdrawn, it is possible to select the depth at which the best water quality exists.

Under certain conditions reaeration-destratification programs may not solve all the problems that they are intended to solve. In large lakes and reservoirs, it is very difficult to design and operate the system so that an adequate amount of oxygen is added to the water at the lowest cost. A monitoring program must be in use prior to and along with reaeration-destratification programs in order to evaluate both positive and negative impacts on water quality.

Researchers and reservoir operators who have used and evaluated reaeration-destratification programs are still debating the impact of these programs on algal blooms. Some investigators have reported significant reductions in the intensity of algal blooms and a shift in the species of algae responsible for the blooms. In some cases taste- and odor-producing blue-green algae have been replaced by certain species of green algae, which do not cause taste and odor problems. Other investigators have noted increases in intensity of algal blooms and taste and odor problems following start-up of reaeration-destratification systems.

Reaeration-destratification effects on algal blooms appear to be related to what happens to nutrient conditions within the reservoir as a result of reaeration-destratification. In lakes and reservoirs that are anaerobic on the bottom and stratified, nutrients (particularly phosphate and nitrate compounds) may be released in large quantities from the bottom sediments into the hypolimnion. Following turnover, these nutrients are mixed into the upper waters where they are available for promoting algal growth under certain environmental conditions. By eliminating anaerobic zones, reaeration-destratification systems may control or eliminate this release of nutrients from the bottom sediments, thereby reducing algal blooms. In some instances, however, reaeration-destratification systems themselves may cause the nutrients within bottom sediments or deeper portions of the reservoir to be mixed upward into the surface waters. When this happens, nutrients become available to algae, thereby increasing algal blooms during certain periods of the year. The number of reservoirs where algal blooms have been decreased by reaeration-destratification appears to be much larger than reservoirs where algal blooms have been increased.

Reservoir operators should collect enough water quality data (temperature, dissolved oxygen, pH, nutrients, alkalinity, suspended matter, turbidity, *SECCHI DISC*[34] transparency) to make their own analysis of the effects of reaeration-destratification programs upon algal and other environmental conditions.

3.332 Methods of Reaeration

There are two basic methods of maintaining or even increasing dissolved oxygen concentrations within zones of reservoirs which would be partially or fully oxygen-depleted when thermal stratification exists. There are many ways (some much more economical than others) in which each of these methods can be used. The first method accomplishes atmospheric reaeration by either altering or totally eliminating thermal stratification and is commonly referred to as destratification. The second method adds dissolved oxygen directly to the hypolimnion without significantly altering the pattern of thermal stratification and is referred to as hypolimnetic reaeration.

QUESTIONS

Write your answers in a notebook and then compare your answers with those on page 86.

3.3S   What is the primary purpose of reaeration-destratification programs in domestic water supply reservoirs?

3.3T   How can water quality be improved by a reaeration-destratification program?

3.3U   List the two basic methods of maintaining or increasing dissolved oxygen concentrations in reservoirs when thermal stratification exists.

3.333 Destratification

Destratification (either total or partial) is accomplished by inducing vertical mixing within the reservoir. This can be done by mechanical means (pumps) or through the use of diffused air which releases air bubbles into the hypolimnion of the res-

ervoir, usually in the deepest portion. Mechanical systems accomplish destratification and mixing either by pumping hypolimnetic waters to the surface or by pumping surface waters downward.

Common practice with diffused air systems using air compressors is to locate them either onshore or on a floating barge or platform as near as possible to the point where the air is to be released into the reservoir. Air is delivered from the compressors into an air supply line which is connected to a system of diffusers. The diffusers release the air near the bottom of the lake or reservoir (Figures 3.11 and 3.12). As the air bubbles rise toward the surface they act like a pump, carrying the colder, denser water upward. Being more dense, the colder water tends to eventually sink downward, causing vertical circulation. With complete destratification, surface waters are cooled and deeper waters are warmed until an equilibrium is reached and temperatures are nearly equal from top to bottom. One major disadvantage of complete destratification is that deeper waters may become warmer than desired for domestic water and for certain species of fish. However, cooler surface temperatures reduce evaporation losses.

Dissolved oxygen is added to deeper waters as they mix with upper waters and make contact with the atmosphere at the surface. Some oxygen enters the water through transfer from the bubbles as they rise toward the surface. Through proper design and operation, destratification systems can be used to adjust temperatures as well as dissolved oxygen levels.

In any given reservoir, the rate of oxygen depletion within the metalimnion and hypolimnion zones may vary considerably from one year to the next depending on algal blooms, die-off, and other biological factors. During years of heavy runoff, higher nutrient inflow may increase biological productivity and the rate of oxygen depletion. In these high runoff years, it may be necessary to circulate water from the deeper zones through more complete mixing.

3.334 Mechanical or Hydraulic Mixing

Destratification of lakes and reservoirs by mechanical or hydraulic mixing has been practiced and evaluated to a much lesser extent than diffused-air mixing. The hydraulic system pumps water from one level of the reservoir and jets it into another area of different *DENSITY*.[35] The pumped water stream induces circulation and mixing. Several different types of mechanical systems have been designed and used, but few, if any, of these systems have remained in operation to control water quality on a routine basis. With many of these systems, power requirements are very high when compared to similar diffused-air mixing systems, making them less efficient and more costly to operate. Dortch[36] has indicated that hydraulic destratification can be an effective means of mixing and may be more efficient than air for mixing large reservoirs and lakes.

Reaeration systems all have one thing in common: they increase the dissolved oxygen content of the hypolimnetic

---

[34] *Secchi (SECK-key) Disc.   A flat, white disc lowered into the water by a rope until it is just barely visible. At this point, the depth of the disc from the water surface is the recorded Secchi disc transparency.*

[35] *Density (DEN-sit-tee).   A measure of how heavy a substance (solid, liquid or gas) is for its size. Density is expressed in terms of weight per unit volume, that is, grams per cubic centimeter or pounds per cubic foot. The density of water (at 4° C or 39° F) is 1.0 gram per cubic centimeter or about 62.4 pounds per cubic foot.*

[36] *Dortch, Mark S., "Method of Total Lake Destratification," published in DESTRATIFICATION OF LAKES AND RESERVOIRS TO IMPROVE WATER QUALITY, Australian Water Resources Council Conference Series No. 2, Canberra.*

Fig. 3.11   Aeration system showing float, hose, and diffuser

Fig. 3.12   Reaeration-destratification system in operation

waters while at the same time they maintain thermal stratification. Hypolimnetic reaeration is accomplished by injecting very small air bubbles or pure oxygen into the hypolimnion of a lake or reservoir or by spraying hypolimnetic water into the air and returning it to the hypolimnion. Hypolimnetic aeration has special applications in ice-covered lakes where it may be necessary to control dissolved oxygen depletion and water quality while at the same time preventing open water conditions from occurring.

In some frozen lakes used by the public, diffused air systems may present a safety hazard by creating open water conditions. In lakes that support cold water fisheries, it may be necessary to practice hypolimnetic aeration to maintain suitable water temperatures. Hypolimnetic aerators that use pure oxygen for reaeration are often not economical for use in large domestic water supplies.

3.335 Development of Reaeration-Destratification Programs

Publications that contain valuable references and information relative to design, operation, advantages, disadvantages, and economies of reaeration-destratification systems should be closely reviewed before deciding to install a system in a particular reservoir. One such publication is:

A GUIDE TO AERATION/CIRCULATION TECHNIQUES FOR LAKE MANAGEMENT, October 1976. Environmental Research Laboratory, U.S. Environmental Protection Agency, Corvallis, OR 97330. Obtain from National Technical Information Service (NTIS), 5285 Port Royal Road, Springfield, VA 22161. Order No. PB-264126/4. Price, $47.50, plus $5.00 shipping and handling per order.

QUESTIONS

Write your answers in a notebook and then compare your answers with those on page 86.

3.3V   How can destratification be accomplished?

3.3W   What factors could cause the rate of dissolved oxygen depletion in a reservoir to vary considerably from one year to the next?

3.34 Managing Frozen Reservoirs
by Dick Krueger

In cold climates, impoundment facilities such as open reservoirs and constructed or natural lakes can have ice formations to varying degrees of severity during winter months. To guarantee trouble-free use and operation of these facilities, certain steps should be taken in preparation for cold winters to

minimize the harmful effects ice formations have on waterworks operations.

3.340 Physical Effects of Ice Formation

3.3400 WATER LEVEL

In anticipation of ice formation on lakes and reservoirs, operators should regulate the water level of the reservoir and maintain an optimum level until the reservoir freezes over. Then the water level should be lowered, sagging the ice cover and reducing the ice pressure on structures and embankments, thus minimizing damages due to ice formation.

3.3401 LAKE LEVEL MEASUREMENT

In order to manage available water resources efficiently, operators have to balance withdrawal from and inflow to the reservoir and determine a withdrawal rate that will guarantee an uninterrupted minimum water supply to the community for domestic and firefighting purposes. This can only be accomplished by continuous monitoring of the water level under the ice.

A float arrangement, though simple and reliable, is impractical when ice formation is encountered. An alternative measuring system should be used, such as a manometer or a bubbler tube. The bubbler tube should be equipped with a heating tape (Pyrotenax Cable) to ensure that no ice will form along the bubbler tube pipe.

3.3402 INTAKE SCREENS

Inspections of intake screens should be carried out frequently to ensure that debris, ice (sheet) buildup, or frazzle (granular) ice will not obstruct the water flow. In case ice or frazzle ice is encountered, a portable steam generator with sufficient length of hose should be available to counter any serious buildup.

3.3403 INTAKES

If the intake design allows for withdrawal of water from different depths of the lake, the top intakes should be closed during winter months and water should be withdrawn from lower elevations. Depending on the depth of the lake, the temperature of water withdrawn from lower elevations will be a few degrees warmer than water withdrawn from close to the surface due to density stratification. Under freezing conditions the heaviest (most dense) water is on the bottom and is warmer than surface water.

Caution should be exercised when withdrawing water from near the bottom of the lake or below the thermocline (transition zone) unless the treatment plant is equipped to routinely handle water from below the thermocline and the increased treatment costs can be justified. Prolonged observation over a period of several years and a thorough understanding of local conditions are mandatory for making ongoing decisions concerning depth of withdrawal. (See Section 3.341, "Effects on Raw Water Quality," for more details.)

3.3404 SILT SURVEY

Silt surveys of most reservoirs should be conducted periodically to get an updated measure of silting as it affects available water storage. These surveys can be done conveniently and with great accuracy by putting a grid on the ice surface. At selected points the water depth can then be measured using sounding equipment or by coring holes through the ice and using a plummet (plumb bob).

3.3405   *RECREATIONAL USE OF RESERVOIR ICE SURFACES*

Ideally, reservoir ice surfaces should not be made accessible for recreational use of any kind, including speed skating, skating, ice sailing, and ice fishing. However, if local authorities approve certain activities on the ice, operators should insist on provision of certain safeguards against possible pollution. These provisions include time limits put on activities on the ice, proper and close supervision, leakproof waste containers, adequate toilet facilities, and proper care of these facilities.

3.341   *Effects on Raw Water Quality*

Lakes may be classified according to depth as follows:

1. First order, >200 feet (50 meters);

2. Second order, 25 to 200 feet (7.5 to 60 meters); and

3. Third order, <25 feet (7.5 meters).

Lakes of the first order exhibit relatively little circulation because the large volume of deep water maintains a stable temperature. Third-order lakes have circulation primarily controlled by wind and wave action and thus are usually well circulated in open water times, thus reducing stagnation effects. These lakes are highly susceptible to stagnation once they are covered by ice. Lakes of the second order may have two circulation or turnover periods, one in the spring and one in the fall. Generally speaking, due to their prevalence, the second-order lakes are more frequently encountered by operators.

In second-order lakes the temperature at the bottom during winter, when the surface is frozen, is not far from that of maximum density (39.2°F or 4°C). The heaviest water is at the bottom, the lightest is at the top, with the intermediate layers arranged in the order of their density. Under these conditions the water is in comparatively stable equilibrium, but is inversely stratified. This is the period of winter stagnation.

The degree of water quality problems is highly affected by the depth of the water inlet. Water from the lowest layer of the lake can become stagnant and the reduction of sulfate to sulfide can occur when the dissolved oxygen is depleted. Such conditions are responsible for odor problems in drinking water. The water from the stagnant zone may also become acidified due to higher concentrations of carbon dioxide which will combine with water to form carbonic acid. If the water becomes sufficiently acidic, reduced forms of iron and manganese may be dissolved from lake bottom materials, resulting in taste and staining problems. After the lake surface is frozen, lake algae might bloom causing subsequent severe odor problems. Prudent treatment with copper sulfate in late autumn, when plankton counts are low, may prevent such occurrence of an algal bloom under the ice.

During periods of stagnation, deposits of organic matter accumulate at the lake bottom, ammonia levels increase, decomposition of organic matter takes place, dissolved oxygen disappears, and nitrate, sulfate, and iron compounds become reduced, free ammonia and nitrite increase, and the content of free carbonic acid increases. A monitoring program that collects samples at various depths to determine dissolved oxygen, metals, and nitrogen values can define the extent of this problem.

To prevent stagnant water at lower elevations from rising to the top during spring circulation (lake turnover), the stagnant water should be discarded whenever possible. This can be accomplished by wasting it through flood gates if the design allows for bottom withdrawal to the flood gates.

3.35   Dam and Reservoir Maintenance

The type and frequency of dam and reservoir maintenance will depend on the size and type of dam. Sometimes these activities are the responsibility of people other than the waterworks operator. The topics listed are presented to give you an idea of some of the items that might be your responsibility. You will have to prepare a program for your facilities depending on what needs to be done.

3.350   *Dam Inspection and Maintenance*

Dams must be inspected regularly to avoid a catastrophic disaster. Some of the most serious dam failures have been small water storage reservoirs located above residential subdivisions. Dams should be inspected after heavy rains. Look for evidence of sink holes (holes in the ground), weep holes (water coming out of holes below the dam), and evidence of burrowing animals.

3.351   *Reservoir Maintenance*

Before draining a reservoir for maintenance, determine when, where, and how you will discharge the water in the reservoir. Lower the water level fairly slowly. If the water level drops too quickly the embankments may slip out and be damaged. If the reservoir is lined and the groundwater is high, the lining could be damaged. Lined reservoirs may have to be drained only during periods of low groundwater levels.

Shoreline vegetation such as weeds and cattails should be controlled. Cattails and other weeds can serve as a breeding area for mosquitoes. Mechanical or manual techniques can be used to remove and control vegetation.

QUESTIONS

Write your answers in a notebook and then compare your answers with those on page 86.

3.3X   How should the water level in a reservoir be regulated after a reservoir freezes over?

3.3Y   How can ice (sheet) buildup or frazzle (granular) ice be prevented from obstructing water flow through intake screens?

3.3Z   What water quality problems could develop in a frozen reservoir?

3.4   LABORATORY AND MONITORING PROGRAMS

3.40   Purpose

Using the water quality laboratory and related monitoring programs, information can be collected that is essential in developing and evaluating methods of managing water quality in domestic water supply reservoirs. Laboratory and monitoring programs are essential from both an operations and a legal standpoint to determine whether physical, chemical, and biological water quality indicators are in compliance with federal, state, and local water quality standards. Water quality data are also an essential tool in optimizing operations of the water treatment plant in relation to treatment costs and techniques, and operation and maintenance programs.

Each agency that operates a surface water supply reservoir must design and operate laboratory and monitoring programs that reflect its own financial and technical resources and need for information. Large municipalities or federal, state, or county agencies sometimes provide water to hundreds of thousands of persons. Generally, these agencies have much greater analytical and technical capability than the laboratory operated by a local agency that serves small populations. The difference in laboratory capability is often due primarily to the difference in financial resources of the two agencies. The large agency's laboratory may serve several functions such as performing analyses for wastewater treatment facilities, wastewater discharge control programs, and domestic water supply treatment plants, as well as collecting data on sources of water supply. The small agency may operate its laboratory and monitoring program primarily to manage the surface water supply reservoir and related facilities. There are many more small water supply systems in the United States than there are large municipal or regional systems. The small system operator must be able to manage a water supply as well as large system operators. With proper planning, training, and forethought, the small agency can do a thorough and efficient job of conducting laboratory and monitoring programs within its own financial capabilities.

3.41   Procedures

In many instances, a small reservoir management agency does not have a large enough volume of samples to justify purchasing equipment and training personnel for certain types of analyses. In these cases, it is often much more practical and economical to contract for the work to be performed by a commercial water laboratory or a consulting firm. Arrangements can sometimes be made with county, state, or federal agencies to have specialized samples collected and analyzed. County, state, and federal agencies can be particularly helpful

when a new or unusual problem develops in the water supply. Examples of types of analyses that may be best performed by an outside or commercial laboratory include trihalomethanes and other organics, *Giardia* and *Cryptosporidium*, general and toxic minerals, radioactivity, pesticides, certain toxic heavy metals (such as lead and mercury), and nutrients.

The agency operating a surface water reservoir should be able to perform analyses that produce data needed on a routine basis for conducting day-to-day operations. Much of the data needed to evaluate physical and chemical conditions in a water supply reservoir can be collected using a single multi-probe instrument (Figure 3.13). This instrument costs approximately $1,000 to over $10,000, depending on the number of probes necessary to measure the water quality indicators and whether the instrument has recording or data transmitting capabilities. Types of data that can be measured at any location and depth within a lake or reservoir with such an instrument include the following: temperature, dissolved oxygen content, CONDUCTIVITY,[37] pH, and *OXIDATION-REDUCTION POTENTIAL*.[38] A thorough survey of all these water quality indicators can be made in a single working day in most reservoirs. In Lake Casitas, for example, the above data can be collected at depth intervals of 5 feet (1.5 meters) to 20 feet (6 meters) at seven separate stations in approximately six hours. The frequency of the data survey on any given lake is usually related to how rapidly water quality is changing within the lake.

If algal blooms are a problem in a particular reservoir, the operating agency should develop laboratory and personnel capabilities for monitoring and identifying algae. The laboratory should be able to develop information on the intensity and extent of algal blooms, the major species of algae involved in any given bloom, and the water quality problems that develop as a result of the bloom. The local agency should be able to

Note the above-water portion of a multi-probe instrument on the top left side of lab bench

Fig. 3.13   Water quality monitoring

---

[37] *Conductivity.   A measure of the ability of a solution (water) to carry an electric current.*

[38] *Oxidation-Reduction Potential (ORP).   The electrical potential required to transfer electrons from one compound or element (the oxidant) to another compound or element (the reductant); used as a qualitative measure of the state of oxidation in water treatment systems. ORP is measured in millivolts, with negative values indicating a tendency to reduce compounds or elements and positive values indicating a tendency to oxidize compounds or elements.*

collect data on taste and odor conditions related to algal blooms. If chemical methods for controlling algal blooms are used, the local agency should be capable of monitoring chemical dosage levels and residuals.

On domestic water supply reservoirs that develop anaerobic zones and their related problems, the local agency may need to monitor iron, manganese, and hydrogen sulfide concentrations. Iron and manganese analyses can be contracted out to a commercial laboratory, but are usually performed "in house" due to the immediate need for the results. The best device for evaluating hydrogen sulfide conditions is often the human nose. If hydrogen sulfide can be smelled in freshly collected unaerated samples, it is present in concentrations objectionable to consumers. However, it is still necessary to measure hydrogen sulfide concentrations to determine improvements. Also, some forms of sulfide may cause odors only after heating.

Information on specific laboratory procedures and training in carrying out these procedures can often be obtained from state health departments. Details on sampling and laboratory procedures are also contained in Chapter 11, "Laboratory Procedures," and also in Volume II, Chapter 21, "Advanced Laboratory Procedures."

3.42 Recordkeeping

One of the most important functions of a well-designed laboratory and monitoring program is recordkeeping. This is true whether analyses are performed by the local agency or by some outside laboratory. Records provide the basic foundation upon which management programs are designed, implemented, and evaluated. Records can be used to evaluate rates of water quality deterioration or improvement and have value as a predictive tool in determining when water quality problems will occur or cease.

In many agencies personnel will come and go, but water quality management programs do not suffer if the person who leaves has kept complete and accurate records relating to water quality problems and management programs. Records are of little value if they are only compiled by laboratory personnel and filed away. They must be regularly reviewed and evaluated by persons responsible for making decisions relative to water quality management. Records pertaining to monitoring and laboratory analyses, which are necessary to indicate compliance with federal and state Primary and Secondary Drinking Water Standards, are legally required. State and federal regulations specify how these records should be kept and for how long. Recordkeeping is also a specific requirement of federal and state laboratory certification programs.

3.43 Safety

Emphasis on knowing and implementing proper safety procedures should be an important part of any laboratory and monitoring program. Specific information on laboratory safety requirements and procedures can be obtained from either state or federal Occupational Safety and Health Act (OSHA) offices. For additional information on laboratory safety, see Chapter 11, "Laboratory Procedures," and Chapter 20, "Safety," in Volume II of this manual.

Anyone responsible for laboratory or monitoring program safety should try to comply with all state and federal safety requirements at all times. If this attitude is emphasized in day-to-day operations, then inspections by safety enforcement officers will never be a cause for concern.

Safety hazards involved in water quality sampling may be even greater than those encountered within the laboratory. The main safety hazard encountered during reservoir sampling is drowning. Most reservoir sampling programs involve the use of a boat, dock, barge, or some similar piece of equipment. When conducting sampling, proper flotation equipment (such as life vests) should be worn. People involved in reservoir and stream sampling should at least know how to swim. In many cases, it is advisable to have two persons involved in sampling at a specific location. If sampling is to be carried out on streams and rivers that are a source of supply to the reservoir, additional safety measures may be required.

QUESTIONS

Write your answers in a notebook and then compare your answers with those on page 86.

3.4A   What are the purposes of water quality laboratory and monitoring programs?

3.4B   What types of laboratory analyses should be performed by operating agencies?

3.4C   If algal blooms are a problem, what type of laboratory capability should be available to the operating agency?

3.4D   What is the main safety hazard encountered during reservoir sampling?

3.5   INTAKE STRUCTURES (Figures 3.14, 3.15, and 3.16)

3.50   Purpose of Intake Structures

Intake structures and related facilities at water supply reservoirs may be more appropriately referred to as "intake-outlet" facilities, as they take in water from the reservoir for outlet downstream. "Intake structure" and "outlet structure" are terms which are often used interchangeably to describe the same facility. In domestic water supply lakes and reservoirs, these facilities may be used to deliver water to water treatment plants, directly to the distribution system, or for returning water to the river or stream downstream of the reservoir. In some cases, a single intake-outlet system is used to provide for downstream releases to the stream or river and for delivery to the treatment plant or distribution system. In other instances, the facilities that provide for release to the stream or river are entirely separate from those providing service to the domestic water system. River or stream intake structures simply serve to provide raw water for the treatment plant.

Intake facilities should always be constructed on the basis of the specific function which they must serve at a given lake, reservoir, stream or river. They must be capable of supplying the maximum rate of flow required for the water treatment plant. Water supply lakes and reservoirs must also release adequate flow for downstream uses. In situations where intake facilities provide service to pressurized systems, they must be

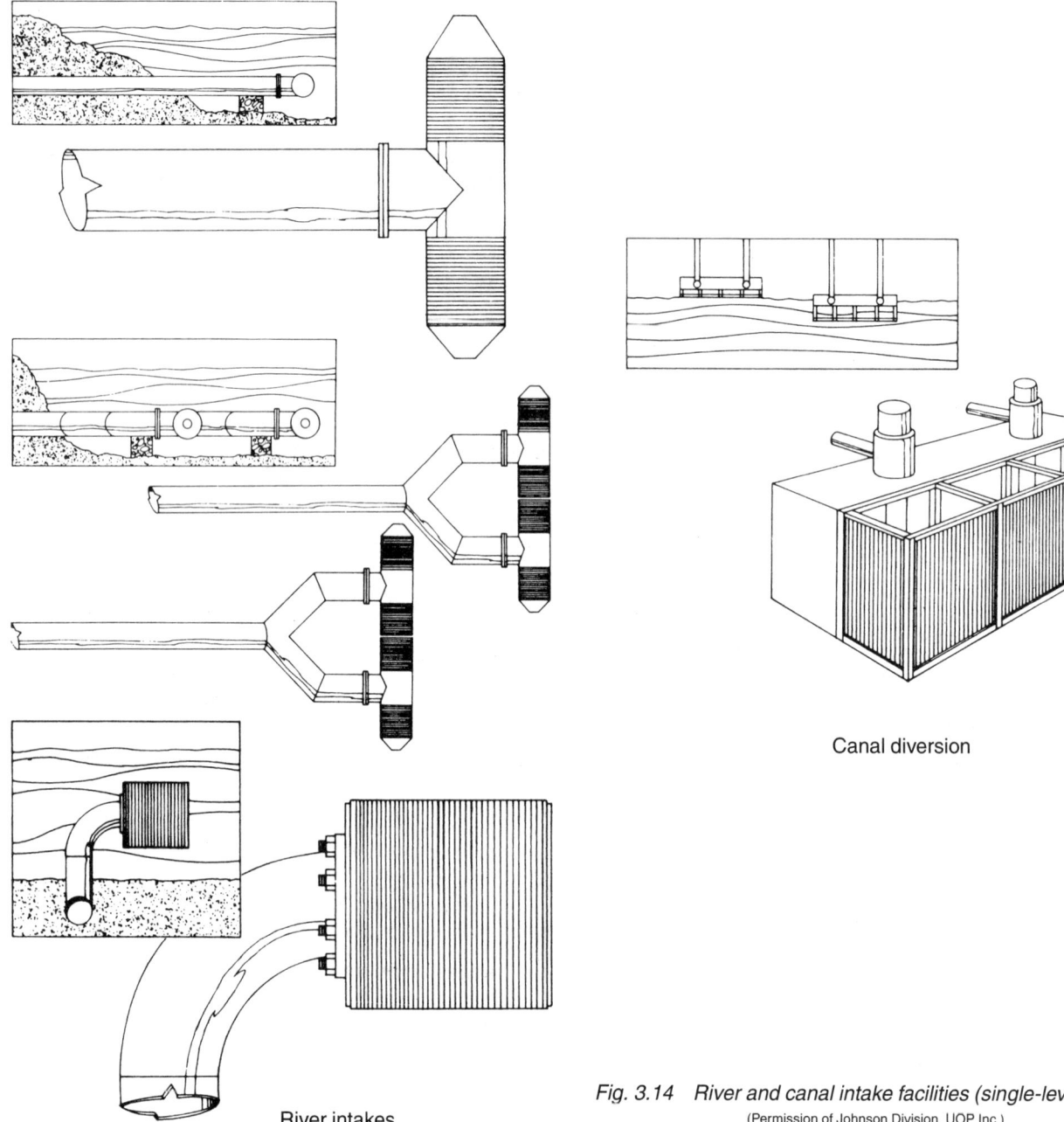

River intakes

Canal diversion

Fig. 3.14   River and canal intake facilities (single-level intakes)
(Permission of Johnson Division, UOP Inc.)

designed so that minimum operating pressures within the system are maintained when the reservoir is drawn down to its minimum operating level. Today greater emphasis is placed on constructing intake structures that permit selection of the depth at which water is drawn from the reservoir, lake, stream, or river.

Intake facilities should be constructed in such a manner that they prevent algal scums, trash, logs, and fish from entering the system. To reduce the danger of silt being drawn into the intake system, the water inlet should not be located at low points where silt buildup is anticipated. Always be very careful when operating the lowest level valves if they have not been in use for some time just in case silt buildup has occurred.

One of the more important considerations in the construction of intake facilities is the ease of operation and maintenance over the expected lifetime of the facility. Every intake structure must be constructed with consideration for operator safety. *CATHODIC PROTECTION*[39] systems, which minimize

---

[39] *Cathodic (ca-THOD-ick) Protection.   An electrical system for prevention of rust, corrosion, and pitting of metal surfaces which are in contact with water or soil. A low-voltage current is made to flow through a liquid (water) or a soil in contact with the metal in such a manner that the external electromotive force renders the metal structure cathodic. This concentrates corrosion on auxiliary anodic parts which are deliberately allowed to corrode instead of letting the structure corrode.*

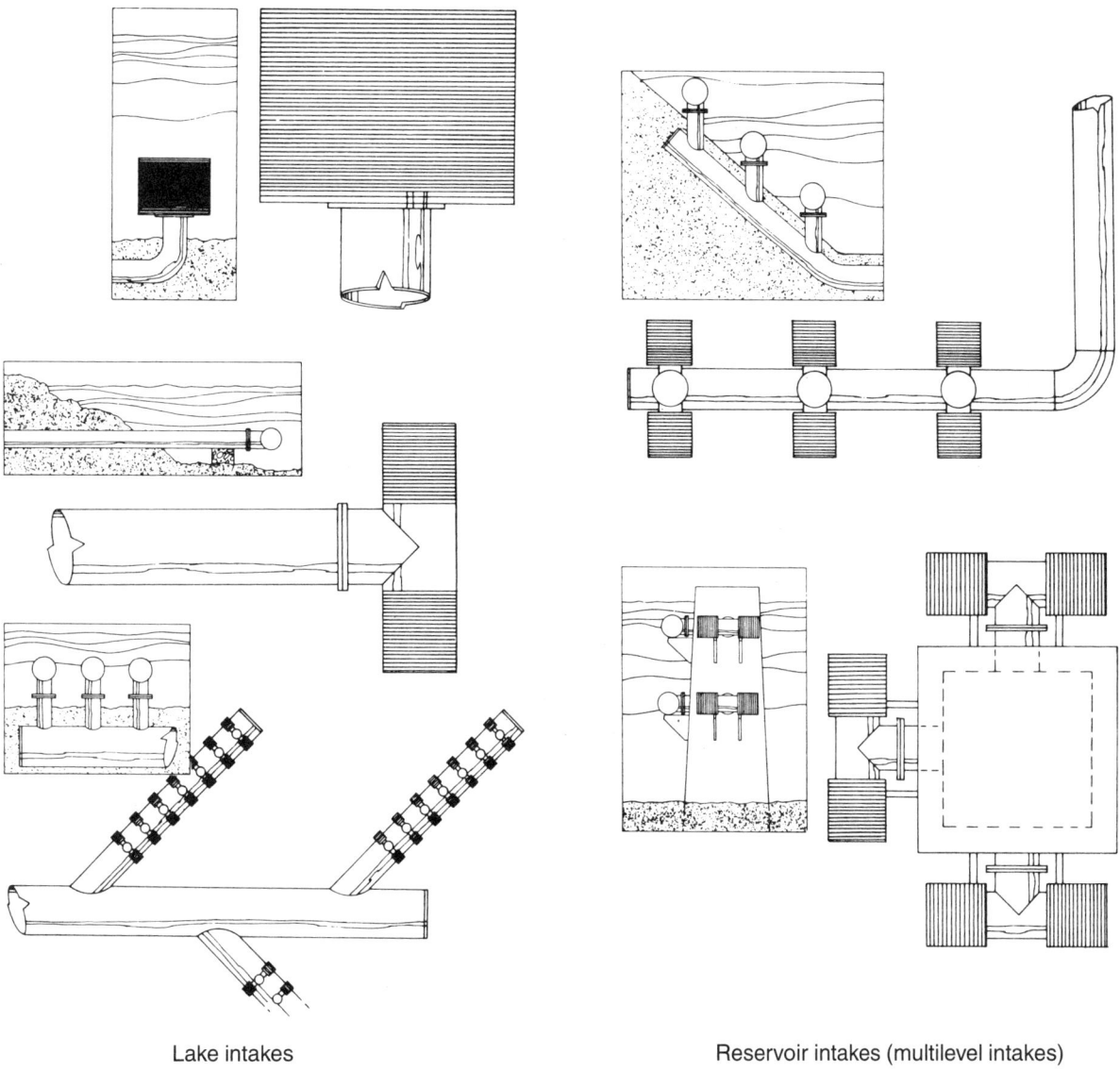

Lake intakes   Reservoir intakes (multilevel intakes)

Fig. 3.15   Lake and reservoir intake structures
(Permission of Johnson Division, UOP Inc.)

the rate of corrosion of metal parts, are a vital part of many intake systems.

3.51   Types of Intake-Outlet Structures

1. Single-Level Intakes (Figure 3.14). Single- or fixed-level intake systems are commonly used in the distribution systems of domestic water supply streams and reservoirs. The single-level intake is usually located in the deepest portion of the stream or reservoir so that water service can still be provided even when the body of water is down to its minimum operating level. In cases of reservoirs where one intake structure supplies both the domestic water system and releases to streams and rivers, the inlet may be located very close to the bottom so that the reservoir can be drained. Single-inlet intake structures are most suitable in relatively shallow lakes and reservoirs that do not stratify significantly and that exhibit fairly uniform water quality

from top to bottom throughout the entire year. These structures may also function well in deeper lakes that are relatively nonproductive and do not experience water quality problems as a result of stratification.

Rivers and streams usually are well mixed and are not stratified. Single-level inlet structures are generally constructed to draw off water from the lowest possible depth in the event of drought conditions.

The advantage of single-inlet intake structures is that they are usually much less complicated and therefore much less costly to construct than multilevel structures. Because of their simplicity, they are both easier and less costly to operate and maintain than multilevel intake structures. In deep reservoirs that often remain nearly full, however, it may be difficult to inspect these facilities and perform necessary repairs and maintenance.

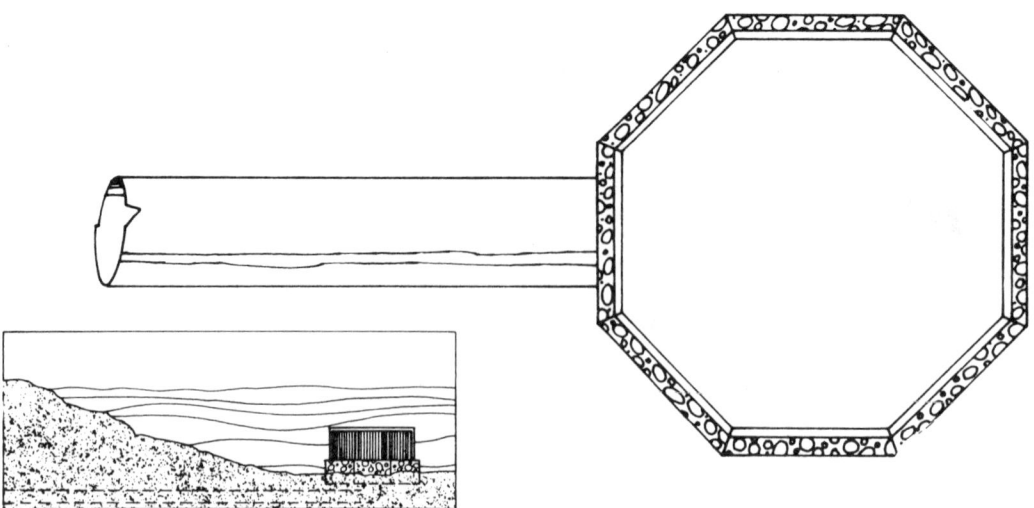

Fig. 3.16   Velocity cap intake structure
(Permission of Johnson Division, UOP Inc.)

Major disadvantages of single fixed-level inlet facilities become apparent when they are used in deeper, productive (eutrophic) lakes and reservoirs. If the inlet is located within the hypolimnion, below the depth at which the thermocline forms, major water quality problems may affect water delivered to the treatment plant or distribution system. Water entering the inlet during spring, summer, and fall months may be anaerobic, may contain high concentrations of iron and/or manganese, and may contain the rotten egg smell caused by the presence of hydrogen sulfide.

Gates or valves that allow water to be taken through the single-level intake may be located either at the point of inlet from the reservoir or stream, in the delivery system at some point downstream, or at both locations.

2. Multilevel Intakes (Figure 3.15). The most satisfactory intake structures in water supply reservoirs are usually those that have inlets to the system at depths ranging from near the surface to the deeper zones. The major advantage of multilevel intake systems in domestic water reservoirs is that they make it possible to serve water from the depth where the best quality of water is located. To obtain good water quality it may be necessary to draw water from different levels during different seasons of the year. When downstream releases to streams and rivers are required, multilevel structures allow for releases from a depth where temperature and dissolved oxygen conditions are acceptable for protecting downstream fish. Fish kills may result from anaerobic water being released downstream from a reservoir.

Multilevel intake structures are most commonly found in a vertical tower located in the deeper portion of the lake and extending above the water surface. Access to the facility may be by bridge, pier, or boat. Inlet gates may be operated from a deck on top of the tower, from some remote control house, or from another location, depending upon equipment design. Inlet gates are commonly located at specific vertical intervals along the face or faces of the tower. Usually, all inlets feed into the pipeline system which extends from the bottom of the tower to the treatment plant, distribution system, or point of release to a river. Each inlet is equipped

with an individually operated gate or valve at the point of inlet. An additional gate or valve is usually located in the pipeline at some point downstream of the intake structure. This arrangement allows for dual control over waters entering the system. If valves or gates at either the inlet or within the pipeline fail, the system can still be shut down.

Some reservoirs contain multilevel intake structures that are inclined rather than vertical. These facilities are commonly located on the inclined face of an earth-fill dam or some similar slope and extend from the maximum water surface level to deeper portions of the reservoir. Inlets with individually operated intake gates or valves are located at intervals along the inclined structure. As with vertical structures, an additional gate or valve is normally located in the pipeline at some point downstream. The inclined intake structure is often a concrete conduit or tunnel with the pipeline located inside and extending from the bottom of the structure downstream to the transmission pipeline. The intake ports extend from the pipeline through the concrete conduit or tunnel and into the reservoir.

For example, the inclined intake structure at Lake Casitas, California, contains nine intake gates located within the reservoir at depth intervals of 24 feet (7.2 m) (see Figure 3.17). When the reservoir is full, water may be drawn from depths ranging from approximately 25 feet (7.5 m) to 217 feet (65 m). Intake gates and related facilities are remotely operated from a control house, which is located on top of Casitas Dam at the entrance to the intake structure.

Fig. 3.17 Casitas intake tower

Economy, topography, and ease of access are major considerations in determining whether vertical or inclined intake structures should be installed at a given reservoir or stream. In reservoirs or streams that freeze, the effects of ice on intake structures must be considered. Structures may be endangered not only from ice pressures from the side, but also from uplift if a reservoir is filling and the ice mass lifts vertically. In some cases, reservoir reaeration may be used to prevent ice from forming around the intake. See Section 3.34, "Managing Frozen Reservoirs," for additional details.

Selection of the level to withdraw water from a lake or reservoir depends on the water quality in the various layers of water. Once the layer has been selected, regular monitoring is required to continually withdraw water with a good quality. Winds and uniform water temperatures from top to bottom can cause water quality at various depths to change very quickly. To be prepared for changing conditions, the water treatment plant operator should:

1. Maintain a log of wind direction and velocity, at least during the summer months; in warm climates this should be done all year;

2. Be alert for onshore winds;

3. Maintain close surveillance of threshold odor test results;

4. Be prepared to make necessary changes in treatment plant operation to combat any sudden increases in taste and odor problems; and

5. Do everything possible to keep any water with a bad taste or odor that gets through the treatment plant from getting out into the distribution system.

QUESTIONS

Write your answers in a notebook and then compare your answers with those on page 86.

3.5A What is the purpose of "intake-outlet" facilities in domestic water supply lakes and reservoirs?

3.5B Why do some intake systems require cathodic protection?

3.5C What may happen to fish if anaerobic water is released downstream from a reservoir?

3.52 Types of Intake Gates

Any one of several types of intake gates or valves may be used at the inlet to the intake structure or as a control valve within the pipeline system downstream of the intake. The most commonly used types include slide gates (steel or cast iron), gate valves, and butterfly valves. Some gates and valves operate only in the fully open or fully closed positions. Others operate as flow-control or regulating valves and gates at any position from fully open to fully closed. Some gates and valves will perform satisfactorily with *HEAD*[40] or pressure on only one side while others are designed to perform best with fairly constant head or pressure on both the upstream and downstream sides.

Gates and valves that regulate releases from small impoundments into small distribution facilities are often designed to be operated manually while those in larger installations frequently use electrical power. Both mechanical gates and hydraulically operated gates and valves are in common use in intake facilities.

3.53 Intake Screens and Trash Racks

Most intake structures in domestic reservoirs, lakes, streams, and rivers are installed in such a manner that inlet gates, fish screens, and trash racks are all combined into a single structure. The primary purpose of fish screens, trash racks, log stops, and other protective facilities is to prevent or minimize the entry of foreign material and fish into the intake system.

Intake screens are installed in panels or in cylindrical forms. Usually they are made of stainless steel. Various types of screens include "vee wire," traveling screens, woven wire, and slotted plates as shown in Figures 3.18, 3.19, and 3.20. In order to do an effective job, these facilities must be designed so that they are easy to service and maintain, have a relatively long life, and do an effective job of protecting the water delivery system. In cases where screens, racks, and related structures are constructed of steel or metal, they should be coated with corrosion-resistant material. Sometimes it is necessary to install a cathodic protection system in order to extend the service life of these facilities.

The type of screen, trash rack, or log stop used in a given intake structure depends on a number of factors including: depth or depths at which inlets are located; location of the intake structure in relation to where debris accumulates in the reservoir or stream; frequency and intensity of algal scum and/or algal mass accumulations; quantity and type of debris encountered; and the size, depth of distribution, and number of fish, crayfish, and other forms of aquatic life. Nonproductive lakes and rivers that contain little aquatic life and receive little or no debris load may only need bar screens. Lakes and rivers with large algal populations might require fine-mesh screens. In lakes, streams, and reservoirs that contain large quantities of debris and trash, log booms, hanging screen curtains, and similar facilities may be effective in protecting the area surrounding the intake structure. Stream intake structures are usually designed to deflect floating debris away from the inlet ports. This procedure minimizes the time required to clean the screens.

---

[40] *Head. The vertical distance (in feet) equal to the pressure (in psi) at a specific point. The pressure head is equal to the pressure in psi times 2.31 ft/psi.*

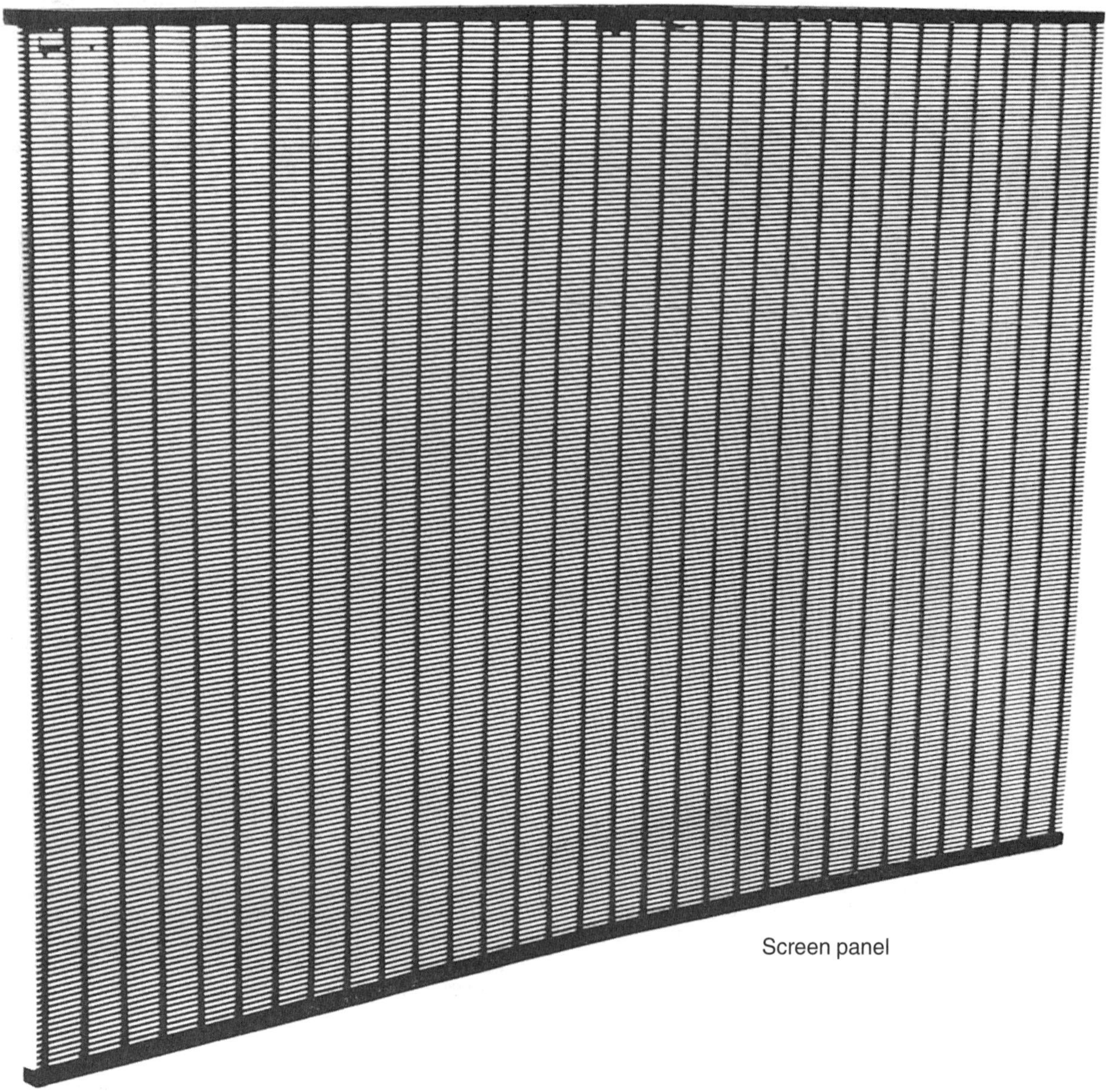

Screen panel

Cylindrical intake screens

Fig. 3.18   Vee wire passive screens
(Permission of Johnson Division, UOP Inc.)

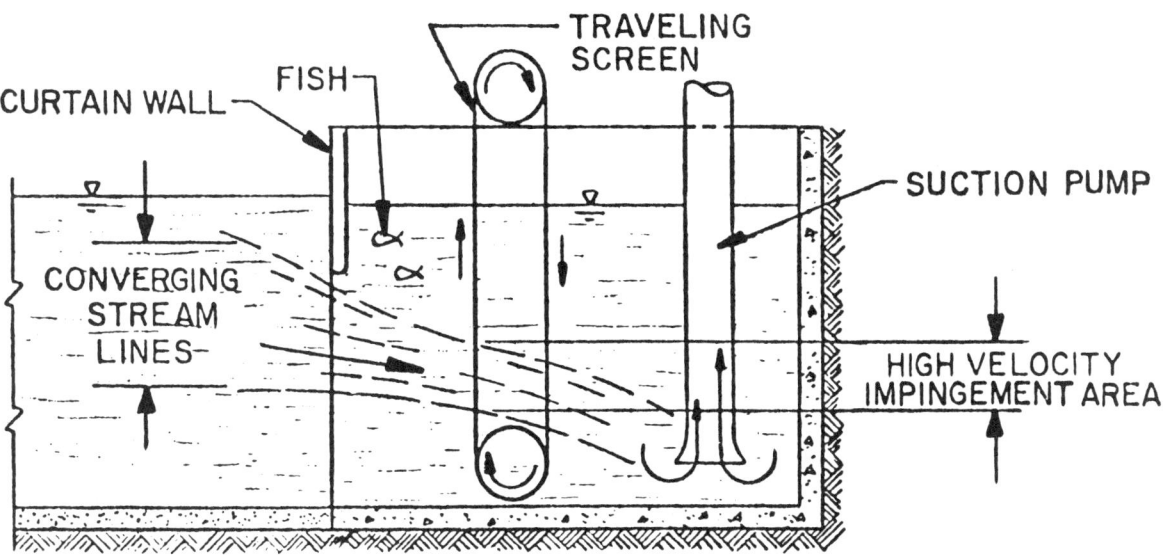

Fig. 3.19   Traveling screen
(Permission of Johnson Division, UOP Inc.)

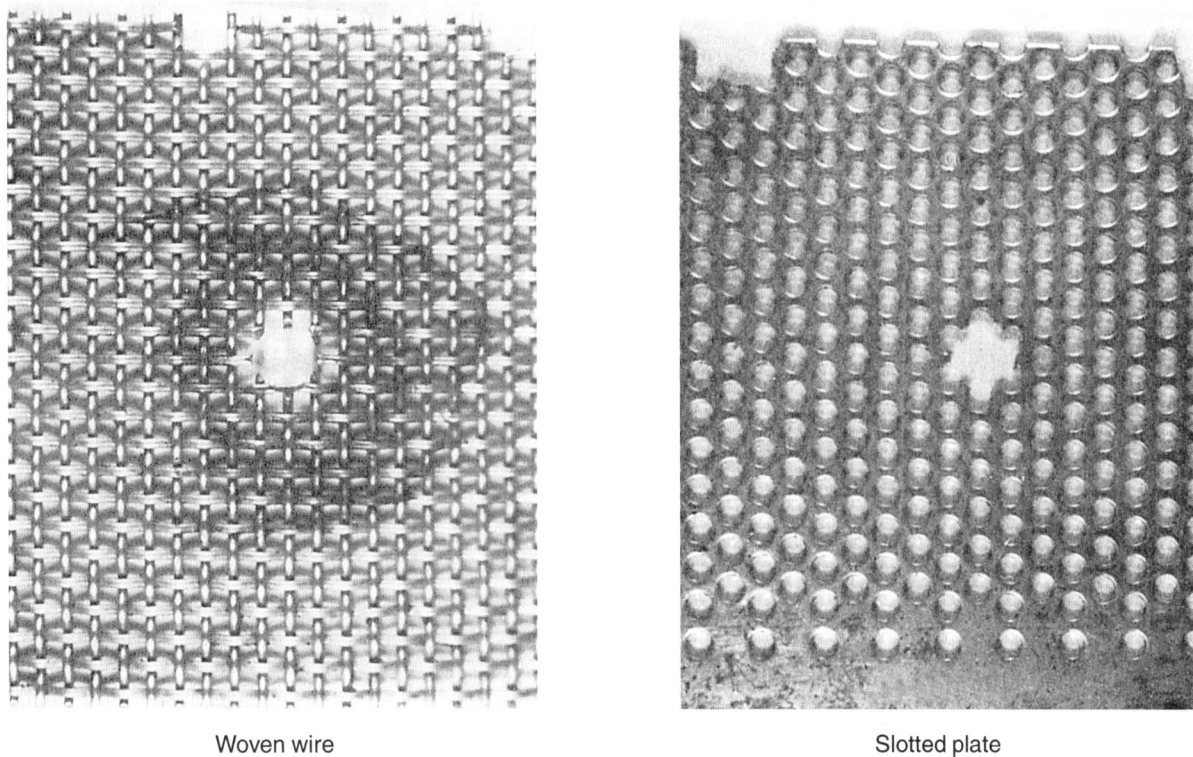

Woven wire   Slotted plate

NOTE: Holes are results of corrosion tests.

Fig. 3.20   Intake screens

(Source: Smith, Lawrence W., CLOGGING, CLEANING, AND CORROSION STUDY OF POSSIBLE FISH SCREENS FOR THE PROPOSED PERIPHERAL CANAL,
California Department of Water Resources, Sacramento, CA, 1982)

QUESTIONS

Write your answers in a notebook and then compare your answers with those on page 86.

3.5D   What are the most common types of intake gates?

3.5E   List the factors that influence the type of screen needed in a specific reservoir.

3.54   Operation and Maintenance Procedures

Well-designed operation and maintenance programs are absolutely necessary if intake structures and related facilities are to perform as intended. Operating criteria, equipment manufacturers' operating instructions, and standard operating procedures should be bound into a manual and used for reference by operators responsible for operating and maintaining intake facilities. If written references containing standard operating and maintenance procedures are not available for a particular facility, they should be prepared with the assistance of knowledgeable operators, design engineers, and equipment manufacturers' representatives.

Screens and trash racks are often designed so that they can be removed for inspection, maintenance, and cleaning. Screens can be drawn to the surface, worked on, and replaced by one person in a short period of time. Screens and trash racks that are not removable should have some provision for cleaning them in place. Mechanical or hydraulic jet cleaning devices may be used, or divers may be employed to clean the screens.

Intake structures and related facilities should be inspected, operated, and tested periodically, preferably at regular intervals. If the reservoir is not drawn down to the level of the deepest intake at periodic intervals, inspections may be conducted by divers (either scuba or hard hat). In recent years, some agencies have used remote controlled video (TV) units to inspect deep-water facilities. The submarine-type unit containing the camera is operated from above the water and can be maneuvered into any position along the intake structure. A video receiver (TV screen) located on a barge, the shoreline, or other above-water structure allows operators to complete a detailed inspection of gates, screens, and other structures. Broken, worn, and corroded facilities are easily identified. One major advantage of the remotely operated video unit is that it allows a number of persons with different areas of expertise to participate in the inspection. When only divers are used, others must depend entirely on what the divers see or photograph to evaluate conditions.

Proper service and lubrication of intake facilities is particularly important. The following factors are major causes of faulty operation of gates and valves:

1. Settlement or shifting of support structure, which could cause binding of gates;

2. Worn, corroded, loose, or broken parts;

3. Lack of use;

4. Lack of lubrication;

5. Vibration;

6. Improper operating procedures;

7. Design errors or deficiencies;

8. Failure of power source or circuit failure; and

9. Vandalism.

Intake screens may be cleaned manually by operators or automatically by mechanical means. Also the screens may be cleaned in place or they may be removed from the inflowing water and cleaned.

If screens are located out in rivers or lakes, the operator may reverse the flow to clean the screen. Some screens have devices that measure the difference in head or head loss between the water surface upstream and downstream from the screen. When a specified head loss is exceeded, a cleaning cycle is started. The cleaning cycle could consist of high-pressure water sprays, which clean the screen in place, or the screen could be lifted out of the water for cleaning.

Manual methods of cleaning screens include the use of rakes, brooms, bristle brushes, and water sprays. Bristle brushes are made of nylon or polypropylene strips of bristles with lengths from 1.5 to 2.5 inches (38 to 64 mm). Screens may be cleaned manually either under water or removed from the flow stream. Woven wire and slotted plate screens require operators to develop a cleaning program and schedule.

Screens that are undersized or improperly designed require operators to spend considerable time cleaning the screens. When this happens the operator should budget funds to improve the screening facilities.

If frazzle (granular) ice plugs a screen, see Section 3.34, "Managing Frozen Reservoirs," for cleaning procedures.

Mechanical cleaning devices require a regular maintenance schedule, including lubrication.

3.55   Records

Records containing a history of operations and maintenance performed on intake facilities are vitally important. By keeping a record of when and under what conditions failures or malfunctions occur, it may be possible to take preventive action. Operators come and go, but if adequate records are maintained, new operators are in a better position to perform their jobs properly.

3.56   Safety

When working around intake structures, proper safety procedures involving use of electrical and mechanical equipment and water safety should always be observed. Proper safety procedures should be documented and included in the manual containing the standard operating procedures. When working in boats or around open bodies of water, two-person crews are recommended.

3.57   Summary

In summary, properly constructed, operated, and maintained intake facilities serve a vital function in the overall domestic reservoir water quality management program. Once good quality water is obtained in the lake, stream, or reservoir, the intake system must be able to deliver water free of debris and trash to treatment or distribution facilities efficiently and satisfactorily.

QUESTIONS

Write your answers in a notebook and then compare your answers with those on page 86.

3.5F What should be done if written standard operating and maintenance procedures are not available for a particular intake structure?

3.5G How can screens and trash racks that are not removable be cleaned?

3.5H How can inlets that are always under water be inspected?

3.5I List the major causes of faulty operation of gates and valves.

3.6 ARITHMETIC ASSIGNMENT

Turn to the Appendix, "How to Solve Water Treatment Plant Arithmetic Problems," at the back of this manual and read the following sections:

1. A.2, *AREAS*,

2. A.3, *VOLUMES*,

3. A.10, *BASIC CONVERSION FACTORS*,

4. A.11, *BASIC FORMULAS*, and

5. A.12, *HOW TO USE THE BASIC FORMULAS*.

Check all of the arithmetic in Sections A.2, *AREAS* (A.20, A.21, A.22, A.23, A.24, A.25, and A.26) and A.3, *VOLUMES* (A.30, A.31, A.32, A.33, and A.34) on an electronic pocket calculator. You should be able to get the same answers.

3.7 ADDITIONAL READING

1. *WATER SOURCES*, Third Edition. Obtain from American Water Works Association (AWWA), Bookstore, 6666 West Quincy Avenue, Denver, CO 80235. Order No. 1955. ISBN 1-58321-229-9. Price to members, $73.50; nonmembers, $106.50; price includes cost of shipping and handling.

2. *WATER TRANSMISSION AND DISTRIBUTION*, Third Edition. Obtain from American Water Works Association (AWWA), Bookstore, 6666 West Quincy Avenue, Denver, CO 80235. Order No. 1957. ISBN 1-58321-231-0. Price to members, $83.50; nonmembers, $125.00; price includes cost of shipping and handling.

3. *NEW YORK MANUAL*, Chapter 6,* "Water Quality."

4. *TEXAS MANUAL*, Chapter 4,* "Raw Water Quality Management," and Chapter 7,* "Pretreatment of Surface Water Supplies."

* Depends on edition.

End of Lesson 2 of 2 Lessons on Reservoir Management and Intake Structures

Please answer the discussion and review questions next.

DISCUSSION AND REVIEW QUESTIONS

Chapter 3. RESERVOIR MANAGEMENT AND INTAKE STRUCTURES

(Lesson 2 of 2 Lessons)

Write the answers to these questions in your notebook. The question numbering continues from Lesson 1.

15. What happens to algal blooms as a result of a reaeration-destratification program?

16. Why might complete destratification of a reservoir be undesirable?

17. Under what conditions might an agency contract out laboratory analyses?

18. How can the entrance of silt into the intake system be minimized?

19. What is the major advantage of multilevel intake systems in domestic water reservoirs? The major limitation?

20. Why should operators keep records on the operation and maintenance of intake facilities?

SUGGESTED ANSWERS

Chapter 3.   RESERVOIR MANAGEMENT AND INTAKE STRUCTURES

ANSWERS TO QUESTIONS IN LESSON 1

Answers to questions on page 47.

3.0A   Two common sources of water other than lakes or reservoirs are groundwater and surface water diversions from streams and rivers.

3.0B   The methods of treating domestic water delivered from reservoirs range from disinfection only, to direct filtration, to complete treatment, which may include softening and activated carbon filtration.

3.0C   Deterioration of water quality due to our activities results from wastewater, agricultural runoff, grazing of livestock, runoff from mining areas, runoff from urban areas, industrial discharges, farming practices, fires, and logging operations.

3.0D   Natural factors that may lower water quality in reservoirs include climate, the structure and uses of watershed and drainage areas, wildfires (lightning), and reservoir geology and vegetation.

Answers to questions on page 48.

3.1A   Large quantities of nutrients such as phosphate, nitrate and organic nitrogen compounds are undesirable in a water supply reservoir.

3.1B   The three classes of reservoirs based on nutrient content and productivity in terms of aquatic animal and plant life are (1) eutrophic, (2) mesotrophic, and (3) oligotrophic.

3.1C   An "algal bloom" is a very large increase in plankton (algae) populations over a very short period of time.

Answers to questions on page 50.

3.1D   Types of tastes and odors produced by algae include fishy, aromatic, grassy, septic, musty, and earthy.

3.1E   In many instances, chlorination of a water supply reduces the level of tastes and odors; however, there are some instances in which tastes and odors are more intense following chlorination.

3.1F   Tastes and odors may be found in the upper layer of a thermally stratified reservoir, throughout a reservoir during periods of overturn, and throughout deeper waters when a reservoir is frozen over.

Answers to questions on page 51.

3.1G   Certain species of algae tend to clog filters and thereby reduce both filtration rates and the duration of filter runs. These conditions cause increased water treatment costs. Under extremely adverse conditions the clogging may occur so frequently that the amount of water used to backwash filters may be greater than the amount of water reaching the distribution system.

3.1H   Algal blooms can raise the pH level from near 7 to 9 or above. Increases and decreases in pH are caused by photosynthesis during daylight hours and respiration by algae during darkness.

3.1I   As an algal bloom progresses, the dissolved oxygen content at depths where the bloom occurs normally increases markedly as a result of photosynthesis. When the algal cells die, this oxygen is used in the decomposition of the cells by organisms, mainly bacteria, that feed upon (metabolize) the algal cells.

3.1J   Increased organic loadings from algal blooms can cause decreased oxygen levels and also increased levels of color and chlorine demand. Also, high trihalomethane levels may occur following free residual chlorination.

Answers to questions on page 56.

3.1K   When a lake surface warms in the spring or summer, the decrease in density of the warmer water reduces the mixing action within the lake and a barrier is formed between the upper and lower layers.

3.1L   Oxygen depletion may be completed any time from a few weeks to several months after thermal stratification begins.

3.1M   Anaerobic conditions in reservoirs cause the release of hydrogen sulfide (rotten egg odors), and cause iron and manganese in bottom sediments to go into solution (into the water).

3.1N   Reservoirs that experience large inflows during short periods may experience sudden and dramatic increases in turbidity, nutrient loading, and organic loading depending on geological, topographical, and vegetative conditions within the watershed.

Answers to questions on page 57.

3.2A   Reservoir water quality problems that can be controlled or eliminated by reservoir management programs include: (1) frequency and intensity of algal blooms and the water quality effects associated with these blooms, (2) dissolved oxygen depletion in the metalimnion and hypolimnion, thereby eliminating or controlling iron, manganese, and hydrogen sulfide problems, (3) silt loading, (4) high turbidity levels, (5) nutrient loading, and (6) organic loading.

3.2B   Reservoir management programs can reduce water treatment costs by controlling tastes and odors and also organic loadings within reservoirs. By controlling algal blooms and silt loadings, chemical costs can be reduced and length of filter runs can be increased. Control of iron and manganese in a reservoir may eliminate the need for control or removal by a treatment plant.

3.2C When dissolved oxygen depletion occurs within the metalimnion and hypolimnion during summer and fall months, trout and salmon are forced up into the warmer epilimnion. The temperatures may be too high for these fish to survive.

Answers to questions on page 57.

3.3A Trees and brush should be removed from areas to be flooded by reservoirs to reduce the organic and nutrient loading the reservoir will receive as it fills.

3.3B Water quality problems caused by organics released during decomposition include increased color, increased chlorine demand, and increased levels of trihalomethanes following chlorination. Nutrients are also released during decomposition which will support an increase in algae production.

3.3C Vegetation can be removed from a reservoir site by mechanical means and hauled from the area for disposal. Also the vegetation can be burned at the site and the ashes hauled out of the area.

Answers to questions on page 59.

3.3D The primary purpose of a watershed management program should be to control, minimize, or eliminate practices within the watershed of a domestic water supply reservoir that would lower water quality.

3.3E Problems caused in reservoirs by raw wastewater contamination include nutrient loading and microbial contamination.

3.3F Fertilizers can be partially controlled by prohibiting their use on nonessential crops, but are best controlled by public education programs. Leaf analysis on crop plants can determine fertilizer needs and application rates.

3.3G Adverse impacts from soil disturbances due to farming, logging, and construction can be minimized by ordinances that limit such activities to those times of the year when the danger of erosion from surface runoff is at a minimum.

Answers to questions on page 60.

3.3H Use of pesticides and herbicides can be controlled in a watershed by rules and regulations.

3.3I During the runoff period following a fire, large quantities of debris, nutrients, silt, and other pollutants may enter a water supply reservoir. Turbidity will usually increase and will have an adverse effect on water treatment plants.

3.3J An agency should consider acquiring title to land in a watershed in order to control or maintain water quality when it is neither politically nor economically feasible to implement controls needed to manage the area properly.

Answers to questions on page 61.

3.3K Chemicals are used to prevent or control attached and floating aquatic growths in domestic water supply reservoirs primarily to prevent or control taste and odor problems resulting from algal blooms.

3.3L Chlorine, used as a bactericide or oxidizing agent, may also produce the effects of an algicide.

Answers to questions on page 62.

3.3M The dose of copper sulfate is based on surface area for waters with a methyl orange alkalinity greater than 50 mg/L because precipitation of copper makes it ineffective against algae below the surface (copper crystals do not dissolve as they fall through the water). For low alkaline waters, the dose is based on the volume of water.

3.3N Suspended particulate matter in a reservoir reduces the effectiveness of copper as an algicide by providing sites or masses other than algae bodies where the copper is adsorbed.

3.3O Regulations limiting the concentration of copper in potable water are the major factor limiting the maximum rate of application of copper sulfate in domestic water supply sources. Another important consideration is the tolerance of fish and other aquatic organisms to copper.

Answers to questions on page 67.

3.3P Three methods of applying copper sulfate compounds to a reservoir include:

1. Dragging burlap bags containing the copper sulfate compounds through the water using a boat,
2. Broadcasting or spreading copper sulfate crystals on the water surface, and
3. Spraying or pumping a copper sulfate solution onto or into the reservoir.

3.3Q The effectiveness of a chemical algae control program is evaluated by monitoring data on type of algae, amount of algae, and where they are located prior to, during, and following application of chemicals. Careful evaluation must be made to determine if algae die-offs actually occur as a result of the chemical application or if they simply die off due to natural circumstances.

3.3R Safety precautions that should be taken by a person applying copper sulfate in the dry form include special clothing, gloves, and breathing apparatus. Also, personal flotation devices must be available on any boat used to spread the chemicals and, in most cases, must be worn while working.

ANSWERS TO QUESTIONS IN LESSON 2

Answers to questions on page 70.

3.3S   The primary purpose of reaeration-destratification programs in domestic water supply reservoirs is usually to eliminate, control, or minimize the negative effects on domestic water quality that occur during periods of thermal stratification and dissolved oxygen depletion.

3.3T   A reaeration-destratification program can improve water quality by adding dissolved oxygen to zones within a lake that would normally become anaerobic during periods of thermal stratification.

3.3U   The two basic methods of maintaining or increasing dissolved oxygen concentrations in reservoirs when thermal stratification exists are (1) destratification, and (2) hypolimnetic reaeration.

Answers to questions on page 72.

3.3V   Destratification is accomplished by inducing vertical mixing within the reservoir. Such mixing can be achieved either by (1) mechanical means, or (2) the use of diffused air which is released near the bottom of the reservoir.

3.3W   The rate of dissolved oxygen depletion in a reservoir could vary considerably from one year to the next depending on algal blooms, die-off, and other biological factors. Amount of runoff and available nutrients are also important.

Answers to questions on page 73.

3.3X   After a reservoir freezes over, the operator should lower the water level. This will sag the ice cover and reduce the ice pressure on structures and embankments, thus minimizing damage due to ice formation.

3.3Y   If ice or frazzle ice builds up on an intake screen, apply steam from a portable steam generator.

3.3Z   Water quality problems that could develop in frozen reservoirs include:

1. The reduction of sulfate to sulfide when the dissolved oxygen is depleted can cause odor problems,
2. Reduced forms of iron and manganese can cause taste and staining problems, and
3. Late algal blooms can cause odor problems.

Answers to questions on page 75.

3.4A   Water quality laboratory and monitoring programs (1) serve as a research tool in managing water quality, (2) help the operator comply with operational and legal guidelines, and (3) help to optimize operations of water treatment plants in relation to treatment costs, techniques, and operation and maintenance programs.

3.4B   The operating agency should perform those laboratory analyses that produce data needed on a routine basis for conducting the day-to-day operations.

3.4C   If algal blooms are a problem, the operating agency should be able to develop specific information on the intensity and extent of algal blooms, the major species of algae involved in any given bloom, and the water quality problems that develop as a result of the bloom. Also the agency should be able to obtain data on taste and odor conditions related to algal blooms. If algae are controlled by chemicals, chemical dosage levels and residuals should be monitored.

3.4D   The main safety hazard encountered during reservoir sampling is drowning.

Answers to questions on page 79.

3.5A   Intake-outlet facilities deliver water from a reservoir to water treatment plants, directly to the distribution system, or return water to the river or stream downstream of the reservoir.

3.5B   Some intake systems require cathodic protection to minimize the rate of corrosion of metal parts.

3.5C   Fish kills may result from anaerobic water being released downstream from a reservoir.

Answers to questions on page 82.

3.5D   The most common types of intake gates are slide gates, gate valves, and butterfly valves.

3.5E   The type of screen needed in a specific reservoir depends on depth or depths at which inlets are located, location of the intake structure in relation to debris accumulation in the reservoir, the frequency and intensity of algal scum and/or algal mass accumulations, the quantity and type of debris encountered, and the size, depth of distribution, and number of fish, crayfish, and other forms of aquatic life.

Answers to questions on page 83.

3.5F   If written standard operating and maintenance procedures are not available for a particular intake structure, the procedures should be prepared with the assistance of knowledgeable operators, design engineers, and equipment manufacturers' representatives.

3.5G   Screens and trash racks that are not removable can be cleaned by mechanical or hydraulic jet cleaning devices, or divers may be employed.

3.5H   Inlets that are always under water can be inspected by divers (scuba or hard hat) or by remote controlled video units.

3.5I   The major causes of faulty operation of gates and valves are: (1) settlement or shifting of support structure; (2) worn, corroded, loose, or broken parts; (3) lack of use; (4) lack of lubrication; (5) vibration; (6) improper operating procedures; (7) design errors or deficiencies; (8) failure of power source or circuit failure; and (9) vandalism.

CHAPTER 4

COAGULATION AND FLOCCULATION

by

Jim Beard

TABLE OF CONTENTS

Chapter 4.   COAGULATION AND FLOCCULATION

OBJECTIVES

Chapter 4.   COAGULATION AND FLOCCULATION

Following completion of Chapter 4, you should be able to:

1. Describe the need for coagulation and flocculation,

2. Perform a jar test,

3. Select the proper coagulant and determine the dosage,

4. Adjust chemical feed rates,

5. Select optimum speeds for flash mixers and flocculators,

6. Collect samples from the coagulation and flocculation basins,

7. Start up and shut down a coagulation-flocculation process, and

8. Operate and maintain coagulation-flocculation processes.

WORDS

Chapter 4.   COAGULATION AND FLOCCULATION

ALKALINITY (AL-ka-LIN-it-tee)   ALKALINITY

The capacity of water to neutralize acids. This capacity is caused by the water's content of carbonate, bicarbonate, hydroxide, and occasionally borate, silicate, and phosphate. Alkalinity is expressed in milligrams per liter of equivalent calcium carbonate. Alkalinity is not the same as pH because water does not have to be strongly basic (high pH) to have a high alkalinity. Alkalinity is a measure of how much acid must be added to a liquid to lower the pH to 4.5.

ANIONIC (AN-eye-ON-ick) POLYMER   ANIONIC POLYMER

A polymer having negatively charged groups of ions; often used as a filter aid and for dewatering sludges.

APPARENT COLOR   APPARENT COLOR

Color of the water that includes not only the color due to substances in the water but suspended matter as well.

BATCH PROCESS   BATCH PROCESS

A treatment process in which a tank or reactor is filled, the water is treated or a chemical solution is prepared, and the tank is emptied. The tank may then be filled and the process repeated.

BUFFER   BUFFER

A solution or liquid whose chemical makeup neutralizes acids or bases without a great change in pH.

CATIONIC POLYMER   CATIONIC POLYMER

A polymer having positively charged groups of ions; often used as a coagulant aid.

CHARGE CHEMISTRY   CHARGE CHEMISTRY

A branch of chemistry in which the destabilization and neutralization reactions occur between stable negatively charged and stable positively charged particles.

COAGULANTS (co-AGG-you-lents)   COAGULANTS

Chemicals that cause very fine particles to clump (floc) together into larger particles. This makes it easier to separate the solids from the water by settling, skimming, draining or filtering.

COAGULATION (co-AGG-you-LAY-shun)   COAGULATION

The clumping together of very fine particles into larger particles (floc) caused by the use of chemicals (coagulants). The chemicals neutralize the electrical charges of the fine particles, allowing them to come closer and form larger clumps. This clumping together makes it easier to separate the solids from the water by settling, skimming, draining or filtering.

COLLOIDS (CALL-loids)   COLLOIDS

Very small, finely divided solids (particles that do not dissolve) that remain dispersed in a liquid for a long time due to their small size and electrical charge. When most of the particles in water have a negative electrical charge, they tend to repel each other. This repulsion prevents the particles from clumping together, becoming heavier, and settling out.

COLOR   COLOR

The substances in water that impart a yellowish-brown color to the water. These substances are the result of iron and manganese ions, humus and peat materials, plankton, aquatic weeds, and industrial waste present in the water.

COMPOSITE (come-PAH-zit) (PROPORTIONAL) SAMPLE   COMPOSITE (PROPORTIONAL) SAMPLE

A composite sample is a collection of individual samples obtained at regular intervals, usually every one or two hours during a 24-hour time span. Each individual sample is combined with the others in proportion to the rate of flow when the sample was collected. The resulting mixture (composite sample) forms a representative sample and is analyzed to determine the average conditions during the sampling period.

CONTINUOUS SAMPLE   CONTINUOUS SAMPLE

A flow of water from a particular place in a plant to the location where samples are collected for testing. This continuous stream may be used to obtain grab or composite samples. Frequently, several taps (faucets) will flow continuously in the laboratory to provide test samples from various places in a water treatment plant.

DETENTION TIME   DETENTION TIME

(1) The theoretical (calculated) time required for a small amount of water to pass through a tank at a given rate of flow.

(2) The actual time in hours, minutes or seconds that a small amount of water is in a settling basin, flocculating basin or rapid-mix chamber. In storage reservoirs, detention time is the length of time entering water will be held before being drafted for use (several weeks to years, several months being typical).

$$\text{Detention Time, hr} = \frac{(\text{Basin Volume, gal})(24 \text{ hr/day})}{\text{Flow, gal/day}}$$

DISINFECTION BY-PRODUCT (DBP)   DISINFECTION BY-PRODUCT (DBP)

A contaminant formed by the reaction of disinfection chemicals (such as chlorine) with other substances in the water being disinfected.

DIVERSION   DIVERSION

Use of part of a stream flow as a water supply.

FLOC   FLOC

Clumps of bacteria and particulate impurities that have come together and formed a cluster. Found in flocculation tanks and settling or sedimentation basins.

FLOCCULATION (FLOCK-you-LAY-shun)   FLOCCULATION

The gathering together of fine particles after coagulation to form larger particles by a process of gentle mixing.

GRAB SAMPLE   GRAB SAMPLE

A single sample of water collected at a particular time and place which represents the composition of the water only at that time and place.

HEAD LOSS   HEAD LOSS

The head, pressure or energy (they are the same) lost by water flowing in a pipe or channel as a result of turbulence caused by the velocity of the flowing water and the roughness of the pipe, channel walls, or restrictions caused by fittings. Water flowing in a pipe loses head, pressure or energy as a result of friction losses. The head loss through a filter is due to friction losses caused by material building up on the surface or in the top part of a filter.

INORGANIC   INORGANIC

Material such as sand, salt, iron, calcium salts and other mineral materials. Inorganic substances are of mineral origin, whereas organic substances are usually of animal or plant origin. Also see ORGANIC.

JAR TEST   JAR TEST

A laboratory procedure that simulates a water treatment plant's coagulation/flocculation units with differing chemical doses and also energy of rapid mix, energy of slow mix, and settling time. The purpose of this procedure is to *ESTIMATE* the minimum or ideal coagulant dose required to achieve certain water quality goals. Samples of water to be treated are commonly placed in six jars. Various amounts of chemicals are added to each jar, stirred and the settling of solids is observed. The dose of chemicals that provides satisfactory settling, removal of turbidity and/or color is the dose used to treat the water being taken into the plant at that time. When evaluating the results of a jar test, the operator should also consider the floc quality in the flocculation area and the floc loading on the filter.

LAUNDERING WEIR (LAWN-der-ing weer)   LAUNDERING WEIR

Sedimentation basin overflow weir. A plate with V-notches along the top to ensure a uniform flow rate and avoid short-circuiting.

MOLECULAR WEIGHT   MOLECULAR WEIGHT

The molecular weight of a compound in grams is the sum of the atomic weights of the elements in the compound. The molecular weight of sulfuric acid (H_2SO_4) in grams is 98.

Element	Atomic Weight	Number of Atoms	Molecular Weight
H	1	2	2
S	32	1	32
O	16	4	64
			98

MONOMER (MON-o-MER) MONOMER

A molecule of low molecular weight capable of reacting with identical or different monomers to form polymers.

NOM (NATURAL ORGANIC MATTER) NOM (NATURAL ORGANIC MATTER)

Humic substances composed of humic and fulvic acids that come from decayed vegetation.

NONIONIC (NON-eye-ON-ick) POLYMER NONIONIC POLYMER

A polymer that has no net electrical charge.

ORGANIC ORGANIC

Substances that come from animal or plant sources. Organic substances always contain carbon. (Inorganic materials are chemical substances of mineral origin.) Also see INORGANIC.

pcu (PLATINUM COBALT UNITS) pcu (PLATINUM COBALT UNITS)

Platinum cobalt units are a measure of color using platinum cobalt standards by visual comparison.

PARTICLE COUNT PARTICLE COUNT

The results of a microscopic examination of treated water with a special particle counter which classifies suspended particles by number and size.

PARTICULATE (par-TICK-you-let) PARTICULATE

A very small solid suspended in water which can vary widely in size, shape, density, and electrical charge. Colloidal and dispersed particulates are artificially gathered together by the processes of coagulation and flocculation.

PLAN VIEW PLAN VIEW

A diagram or photo showing a facility as it would appear when looking down on top of it.

POLYANIONIC (poly-AN-eye-ON-ick) POLYANIONIC

Characterized by many active negative charges especially active on the surface of particles.

POLYELECTROLYTE (POLY-ee-LECK-tro-lite) POLYELECTROLYTE

A high-molecular-weight (relatively heavy) substance having points of positive or negative electrical charges that is formed by either natural or manmade processes. Natural polyelectrolytes may be of biological origin or derived from starch products and cellulose derivatives. Manmade polyelectrolytes consist of simple substances that have been made into complex, high-molecular-weight substances. Used with other chemical coagulants to aid in binding small suspended particles to larger chemical flocs for their removal from water. Often called a POLYMER.

POLYMER (POLY-mer) POLYMER

A long chain molecule formed by the union of many monomers (molecules of lower molecular weight). Polymers are used with other chemical coagulants to aid in binding small suspended particles to larger chemical flocs for their removal from water.

PRECIPITATE (pre-SIP-uh-TATE) PRECIPITATE

(1) An insoluble, finely divided substance which is a product of a chemical reaction within a liquid.

(2) The separation from solution of an insoluble substance.

PROFILE PROFILE

A drawing showing elevation plotted against distance, such as the vertical section or *SIDE* view of a pipeline.

RAW WATER RAW WATER

(1) Water in its natural state, prior to any treatment.

(2) Usually the water entering the first treatment process of a water treatment plant.

REAGENT (re-A-gent) REAGENT

A pure chemical substance that is used to make new products or is used in chemical tests to measure, detect, or examine other substances.

REPRESENTATIVE SAMPLE REPRESENTATIVE SAMPLE

A sample portion of material or water that is as nearly identical in content and consistency as possible to that in the larger body of material or water being sampled.

SHORT-CIRCUITING   SHORT-CIRCUITING

A condition that occurs in tanks or basins when some of the flowing water entering a tank or basin flows along a nearly direct pathway from the inlet to the outlet. This is usually undesirable since it may result in shorter contact, reaction, or settling times in comparison with the theoretical (calculated) or presumed detention times.

SIMULATE   SIMULATE

To reproduce the action of some process, usually on a smaller scale.

SLUDGE (sluj)   SLUDGE

The settleable solids separated from water during processing.

SPECIFIC GRAVITY   SPECIFIC GRAVITY

(1) Weight of a particle, substance, or chemical solution in relation to the weight of an equal volume of water. Water has a specific gravity of 1.000 at 4°C (39°F). Particulates in raw water may have a specific gravity of 1.005 to 2.5.

(2) Weight of a particular gas in relation to the weight of an equal volume of air at the same temperature and pressure (air has a specific gravity of 1.0). Chlorine has a specific gravity of 2.5 as a gas.

SUPERNATANT (sue-per-NAY-tent)   SUPERNATANT

Liquid removed from settled sludge. Supernatant commonly refers to the liquid between the sludge on the bottom and the scum on the water surface of a basin or container.

TOTAL ORGANIC CARBON (TOC)   TOTAL ORGANIC CARBON (TOC)

TOC measures the amount of organic carbon in water.

TRIHALOMETHANES (THMs) (tri-HAL-o-METH-hanes)   TRIHALOMETHANES (THMs)

Derivatives of methane, CH_4, in which three halogen atoms (chlorine or bromine) are substituted for three of the hydrogen atoms. Often formed during chlorination by reactions with natural organic materials in the water. The resulting compounds (THMs) are suspected of causing cancer.

TRUE COLOR   TRUE COLOR

Color of the water from which turbidity has been removed. The turbidity may be removed by double filtering the sample through a Whatman No. 40 filter when using the visual comparison method.

TURBIDIMETER   TURBIDIMETER

An instrument for measuring and comparing the turbidity of liquids by passing light through them and determining how much light is reflected by the particles in the liquid. The normal measuring range is 0 to 100 and is expressed as Nephelometric Turbidity Units (NTUs).

TURBIDITY (ter-BID-it-tee)   TURBIDITY

The cloudy appearance of water caused by the presence of suspended and colloidal matter. In the waterworks field, a turbidity measurement is used to indicate the clarity of water. Technically, turbidity is an optical property of the water based on the amount of light reflected by suspended particles. Turbidity cannot be directly equated to suspended solids because white particles reflect more light than dark-colored particles and many small particles will reflect more light than an equivalent large particle.

TURBIDITY UNITS (TU)   TURBIDITY UNITS (TU)

Turbidity units are a measure of the cloudiness of water. If measured by a nephelometric (deflected light) instrumental procedure, turbidity units are expressed in nephelometric turbidity units (NTU) or simply TU. Those turbidity units obtained by visual methods are expressed in Jackson Turbidity Units (JTU) which are a measure of the cloudiness of water; they are used to indicate the clarity of water. There is no real connection between NTUs and JTUs. The Jackson turbidimeter is a visual method and the nephelometer is an instrumental method based on deflected light.

WEIR (weer)   WEIR

(1)   A wall or plate placed in an open channel and used to measure the flow of water. The depth of the flow over the weir can be used to calculate the flow rate, or a chart or conversion table may be used to convert depth to flow.

(2)   A wall or obstruction used to control flow (from settling tanks and clarifiers) to ensure a uniform flow rate and avoid short-circuiting.

WET CHEMISTRY   WET CHEMISTRY

Laboratory procedures used to analyze a sample of water using liquid chemical solutions (wet) instead of, or in addition to, laboratory instruments.

CHAPTER 4.   COAGULATION AND FLOCCULATION

(Lesson 1 of 3 Lessons)

4.0   NATURE OF PARTICULATE IMPURITIES IN WATER

PARTICULATE[1] impurities in water result from land erosion, pickup of minerals, and the decay of plant material. Additional impurities are added by airborne contamination, industrial discharges, and by animal wastes. Thus, surface water sources, polluted by people and nature, are likely to contain suspended and dissolved organic (plant or animal origin) and inorganic (mineral) material, and biological forms such as bacteria and plankton.

These particulates (commonly called suspended solids) cover a broad size range. Larger sized particles, such as sand and heavy silts, can be removed from water by slowing down the flow to allow for simple gravity settling. These particles are often called *SETTLEABLE SOLIDS*. Settling of larger sized particles occurs naturally when surface water is stored for a sufficient period of time in a reservoir or a lake. Smaller sized particles, such as bacteria and fine clays and silts, do not readily settle and treatment is required to produce larger particles that are settleable. These smaller particles are often called *NONSETTLEABLE SOLIDS* or *COLLOIDAL*[2] *MATTER*.

4.1   NEED FOR COAGULATION AND FLOCCULATION

The purpose of coagulation and flocculation is to remove particulate impurities, especially nonsettleable solids, and color from the water being treated. Nonsettleable particles in water are removed by the use of *COAGULATING* chemicals. These chemicals cause the particles to clump together forming floc. When pieces of floc clump together, they form larger, heavier floc which will settle out.

In the *COAGULATION PROCESS*,[3] chemicals are added that will initially cause the particles to become destabilized and clump together. The particles gather together to form larger particles in the *FLOCCULATION PROCESS*[4] (see Figure 4.1).

With few exceptions, surface waters require treatment to remove particulate impurities and color before distribution of water to the consumer.

QUESTIONS

Write your answers in a notebook and then compare your answers with those on page 133.

4.0A   What is the purpose of coagulation and flocculation?

4.1A   What happens in the coagulation and flocculation processes?

4.2   COAGULATION

4.20   Process Description

The term coagulation describes the effect produced when certain chemicals are added to raw water containing slowly settling or nonsettleable particles. The small particles begin to form larger or heavier floc which will be removed by sedimentation and filtration.

The mixing of the coagulant chemical and the raw water to be treated is commonly referred to as *FLASH MIXING*. The primary purpose of the flash mix process is to rapidly mix and equally distribute the coagulant chemical throughout the water. The entire process occurs in a very short time (several seconds), and the first results are the formation of very small particles.

4.21   Coagulants

In practice, chemical coagulants are referred to either as primary coagulants or as coagulant aids. Primary coagulants neutralize the electrical charges of the particles which causes them to begin to clump together. The purpose of coagulant aids is to add density to slow-settling flocs and add toughness so the floc will not break up in the following processes. In view of this definition, coagulant aids could be called flocculation or sedimentation aids.

Metallic salts (aluminum sulfate (commonly called alum), ferric sulfate, ferrous sulfate) and synthetic (manmade) organic *POLYMERS*[5] (cationic, anionic, nonionic) are commonly used as coagulation chemicals in water treatment because they are effective, relatively low in cost, available, and easy to handle, store, and apply.

---

[1] *Particulate (par-TICK-you-let).   A very small solid suspended in water which can vary widely in size, shape, density, and electrical charge. Colloidal and dispersed particulates are artificially gathered together by the processes of coagulation and flocculation.*

[2] *Colloids (CALL-loids).   Very small, finely divided solids (particles that do not dissolve) that remain dispersed in a liquid for a long time due to their small size and electrical charge. When most of the particles in water have a negative electrical charge, they tend to repel each other. This repulsion prevents the particles from clumping together, becoming heavier, and settling out.*

[3] *Coagulation (co-AGG-you-LAY-shun).   The clumping together of very fine particles into larger particles (floc) caused by the use of chemicals (coagulants). The chemicals neutralize the electrical charges of the fine particles, allowing them to come closer and form larger clumps. This clumping together makes it easier to separate the solids from the water by settling, skimming, draining or filtering.*

[4] *Flocculation (FLOCK-you-LAY-shun).   The gathering together of fine particles after coagulation to form larger particles by a process of gentle mixing.*

[5] *Polymer (POLY-mer).   A long chain molecule formed by the union of many monomers (molecules of lower molecular weight). Polymers are used with other chemical coagulants to aid in binding small suspended particles to larger chemical flocs for their removal from water.*

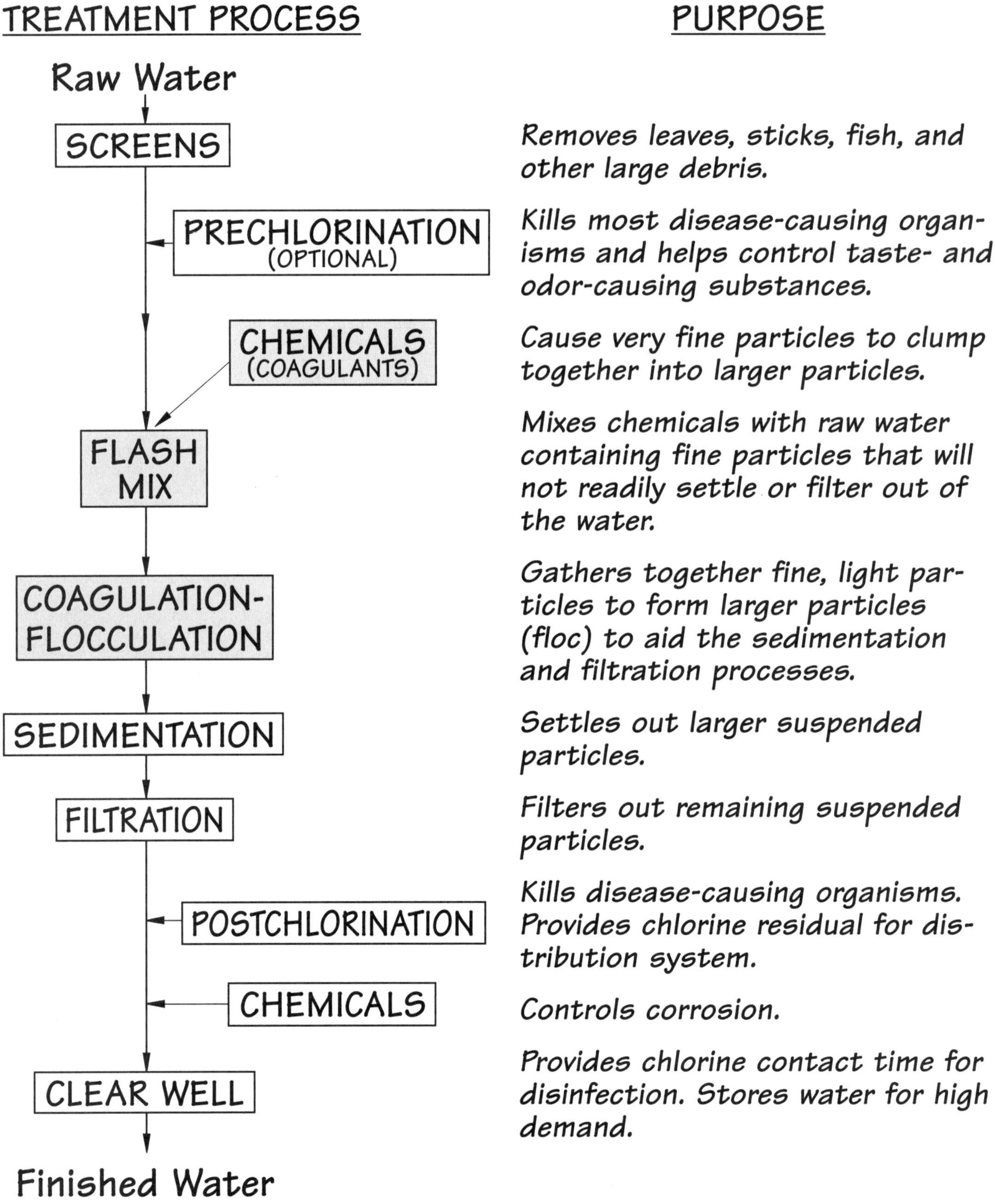

TREATMENT PROCESS

Raw Water
↓
SCREENS
↓
PRECHLORINATION
(OPTIONAL) →
↓
CHEMICALS
(COAGULANTS)
↘
FLASH
MIX
↓
COAGULATION-
FLOCCULATION
↓
SEDIMENTATION
↓
FILTRATION
↓
POSTCHLORINATION →
↓
CHEMICALS →
↓
CLEAR WELL
↓
Finished Water

PURPOSE

Removes leaves, sticks, fish, and other large debris.

Kills most disease-causing organisms and helps control taste- and odor-causing substances.

Cause very fine particles to clump together into larger particles.

Mixes chemicals with raw water containing fine particles that will not readily settle or filter out of the water.

Gathers together fine, light particles to form larger particles (floc) to aid the sedimentation and filtration processes.

Settles out larger suspended particles.

Filters out remaining suspended particles.

Kills disease-causing organisms. Provides chlorine residual for distribution system.

Controls corrosion.

Provides chlorine contact time for disinfection. Stores water for high demand.

Fig. 4.1   Process diagram of typical plant

When metallic salts, such as aluminum sulfate or ferric sulfate, are added to water, a series of reactions occur with the water and with other ions in the water. Sufficient chemical quantities must be added to the water to exceed the solubility limit of the metal hydroxide, resulting in the formation of a precipitate (floc). The resulting floc formed will then adsorb on particles (turbidity) in the water.

The synthetic organic polymers (cationic, anionic, nonionic) used in water treatment consist of a long chain of small subunits or "monomers." The polymer chain can have a linear or branched structure, ranging in length from a fraction of a micron (one micron = 0.001 millimeter) to as much as 10 microns. The total number of monomers in a synthetic polymer can be varied to produce materials of different *MOLECULAR WEIGHTS*,[6] which vary from about 100 to 10,000,000.

The polymers normally used in water treatment contain ionizable groups on the monomeric units (carboxyl, amino, sulfonic groups), and are commonly referred to as "polyelectrolytes." Polymers with positively charged groups on the monomeric units are referred to as "cationic" polyelectrolytes, while polymers with negatively charged groups are called "anionic" polyelectrolytes. Polymers without ionizable groups are referred to as "nonionic" polymers.

Cationic polymers have the ability to adsorb on negatively charged particles (turbidity) and neutralize their charge. They can also form an interparticle bridge that collects (entraps) the particles. Anionic and nonionic polymers also form interparticle bridges, which aid in the collection and removal of particles from water.

While alum is perhaps the most commonly used coagulant chemical, cationic polymers are used in the water treatment field as both a primary coagulant (in place of alum or other metallic salts) and as a coagulant aid (used in conjunction with alum and other metallic salts). Anionic and nonionic polymers have also proven to be effective in certain applications as coagulant aids and filter aids.

One of the problems that will confront the water treatment plant operator in the selection of an appropriate polymer is that there are a tremendous number of polymers available in the marketplace, and no universal evaluation method has been generally adopted for polymer selection. Thus, the operator should use caution in the selection and use of polymers and should take note of the following considerations regarding polymer use:

1. Polymer overdosing will adversely affect coagulation efficiency and when used as a filter aid, overdosing can result in accelerated head loss buildup;

2. Not all water supplies can be treated with equal success;

3. Some polymers lose their effectiveness when used in the presence of a chlorine residual; and

4. Some polymers are dosage limited. The operator should obtain the maximum safe dosage that can be applied from the specific chemical manufacturer.

Since universal standards do not exist for the selection and use of organic polymers, the operator should be careful to select only those products that have been approved by state and federal regulatory agencies for use in potable water treatment. The chemical supplier should be required to provide written evidence of this approval. Many chemical suppliers have considerable experience in dealing with many types of water and may be able to recommend the "best" polymer for your plant.

The National Sanitation Foundation (877-867-3435) publishes a list of polymers approved for potable water usage. This list contains the name of the manufacturer and the maximum concentration recommended. A list of primary coagulants and coagulant aids is shown in Table 4.1.

TABLE 4.1   CHEMICAL COAGULANTS USED IN WATER TREATMENT

Chemical Name	Chemical Formula	Primary Coagulant	Coagulant Aid
ALUMINUM SULFATE	$Al_2(SO_4)_3 \cdot 14\ H_2O$	X	
FERROUS SULFATE	$FeSO_4 \cdot 7\ H_2O$	X	
FERRIC SULFATE	$Fe_2(SO_4)_3 \cdot 9\ H_2O$	X	
FERRIC CHLORIDE	$FeCl_3 \cdot 6\ H_2O$	X	
CATIONIC POLYMER	Various	X	X
CALCIUM HYDROXIDE	$Ca(OH)_2$	X[a]	X
CALCIUM OXIDE	CaO	X[a]	X
SODIUM ALUMINATE	$Na_2Al_2O_4$	X[a]	X
BENTONITE	Clay		X
CALCIUM CARBONATE	$CaCO_3$		X
SODIUM SILICATE	Na_2SiO_3		X
ANIONIC POLYMER	Various		X
NONIONIC POLYMER	Various		X

[a] Used as primary coagulants only in water softening processes.

---

[6] *Molecular Weight.   The molecular weight of a compound in grams is the sum of the atomic weights of the elements in the compound. The molecular weight of sulfuric acid (H_2SO_4) in grams is 98.*

Element	Atomic Weight	Number of Atoms	Molecular Weight
H	*1*	*2*	*2*
S	*32*	*1*	*32*
O	*16*	*4*	*64*
			98

4.22 Basic Coagulant Chemistry

The theory of coagulation is very complex. However, this discussion of coagulation chemistry is presented to help you understand the coagulation process.

Coagulation is a physical and chemical reaction occurring between the *ALKALINITY*[7] of the water and the coagulant added to the water which results in the formation of insoluble flocs (floc that will not dissolve).

For a specific coagulant (such as aluminum sulfate or alum), the pH of the water determines which hydrolysis species (chemical compounds) predominate. Lower pH values tend to favor positively charged species which are desirable for reacting with negatively charged colloids and particulates, forming insoluble flocs and removing impurities from the water.

The best pH for coagulation usually falls in the range of pH 5 to 7. The proper pH range must be maintained because coagulants generally react with the alkalinity in water. Residual alkalinity in the water serves to *BUFFER*[8] (prevent pH from changing) the system and aids in the complete precipitation of the coagulant chemicals. The amount of alkalinity in the source (raw) water is generally not a problem unless the alkalinity is very low. Alkalinity may be increased by the addition of lime or soda ash.

Polymers are generally added in the coagulation process to stimulate or improve the formation of insoluble flocs.

Generally, the operator has no control over the pH and alkalinity of the source water. Hence, evaluation of these water quality indicators may play a major role in selecting the type of chemical coagulants to be used at a particular water treatment plant, or in changing the type of coagulant normally used if significant changes in pH and alkalinity occur in the raw water.

In some instances the natural alkalinity in the raw water may be too low to produce complete precipitation of alum. In these cases lime is often added to ensure complete precipitation. Care must be used to keep the pH within the desired range.

Overdosing as well as underdosing of coagulants may lead to reduced solids removal efficiency. This condition can be corrected by carefully performing *JAR TESTS*[9] and verifying process performance after making any changes in the operation of the coagulation process.

4.23 Process Performance Considerations

4.230 Methods of Mixing

In modern water treatment plants, it is desirable to complete the coagulation reaction (mixing of chemicals into the water) in as short a time as possible—preferably within a period of several seconds since the reaction time is short. To accomplish the mixing of the chemicals with the water to be treated, several methods can be used (see Figure 4.2):

1. Hydraulic mixing using flow energy in the system,

2. Mechanical mixing,

3. Diffusers and grid systems, and

4. Pumped blenders.

4.231 Types of Mixers

In order for complete coagulation and flocculation to take place, the coagulant must make contact with all of the suspended particles. This is accomplished by "flash mixing."

Mixing can be satisfactorily achieved with a number of different types of mixing devices. Hydraulic mixing with baffles or throttling valves works well in systems that have sufficient water velocity (speed) to cause turbulence in the water being treated. The turbulence in the flowing water mixes the chemicals with the water.

Mechanical mixers (paddles, turbines, and propellers) are frequently used in coagulation facilities. Mechanical mixers are versatile and reliable; however, they generally use the greatest amount of electric energy for mixing the coagulant with the water being treated.

Diffusers and grid systems consisting of perforated tubes or nozzles can be used to disperse the coagulant into the water being treated. These systems can provide uniform (equal) distribution of the coagulant over the entire coagulation basin. However, they are generally sensitive to flow changes and may require frequent adjustments to produce the proper amount of mixing.

Pumped blenders (Figure 4.3) have also been used for mixing in coagulation facilities. In this system, the coagulant is added directly to the water being treated through a diffuser in

---

[7] *Alkalinity (AL-ka-LIN-it-tee). The capacity of water to neutralize acids. This capacity is caused by the water's content of carbonate, bicarbonate, hydroxide, and occasionally borate, silicate, and phosphate. Alkalinity is expressed in milligrams per liter of equivalent calcium carbonate. Alkalinity is not the same as pH because water does not have to be strongly basic (high pH) to have a high alkalinity. Alkalinity is a measure of how much acid must be added to a liquid to lower the pH to 4.5.*

[8] *Buffer.   A solution or liquid whose chemical makeup neutralizes acids or bases without a great change in pH.*

[9] *Jar Test.   A laboratory procedure that simulates a water treatment plant's coagulation/flocculation units with differing chemical doses and also energy of rapid mix, energy of slow mix, and settling time. The purpose of this procedure is to ESTIMATE the minimum or ideal coagulant dose required to achieve certain water quality goals. Samples of water to be treated are commonly placed in six jars. Various amounts of chemicals are added to each jar, stirred and the settling of solids is observed. The dose of chemicals that provides satisfactory settling, removal of turbidity and/ or color is the dose used to treat the water being taken into the plant at that time. When evaluating the results of a jar test, the operator should also consider the floc quality in the flocculation area and the floc loading on the filter.*

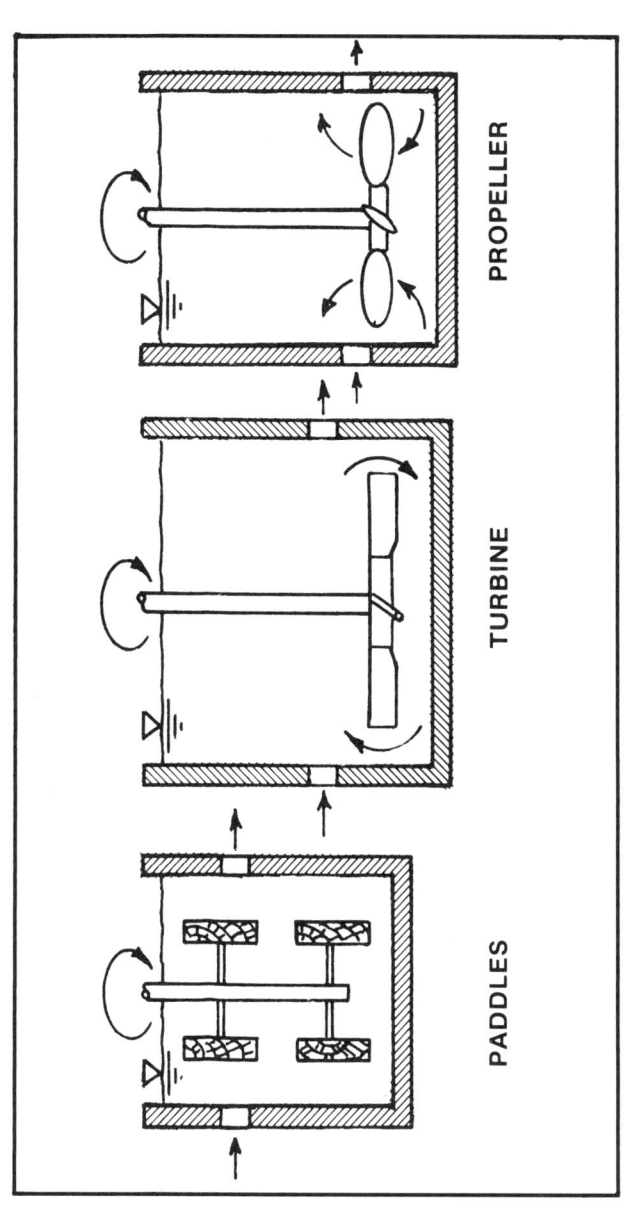

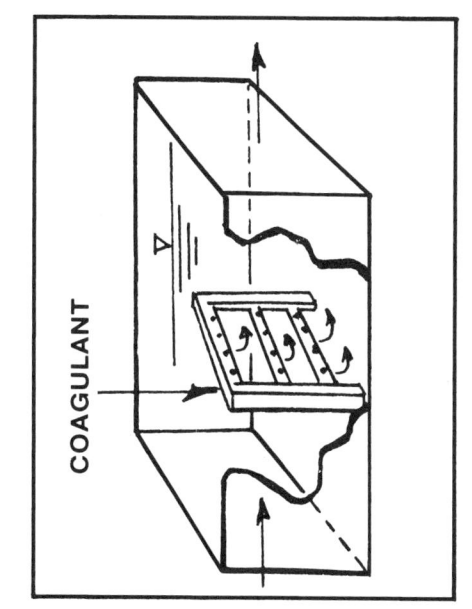

Fig. 4.2   Methods of flash mixing

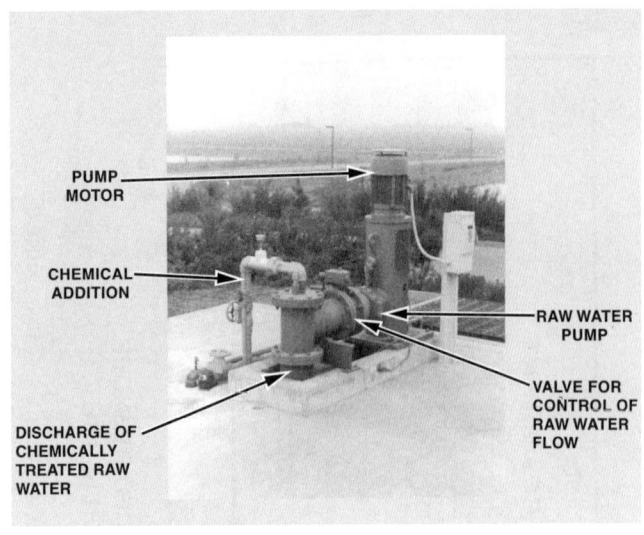

PUMP MOTOR

CHEMICAL ADDITION

RAW WATER PUMP

VALVE FOR CONTROL OF RAW WATER FLOW

DISCHARGE OF CHEMICALLY TREATED RAW WATER

Fig. 4.3   Pump used in pumped blender system

a pipe. This system can provide rapid dispersion of the coagulant and does not create any significant *HEAD LOSS*[10] in the system. Electric energy consumption is considerably less than that of a comparable mechanical mixer.

4.232   Coagulation Basins

Mixing of the chemical coagulant can be satisfactorily accomplished in a special rectangular tank with mixing devices. Mixing may also occur in the influent channel or a pipeline to the flocculation basin if the flow velocity is high enough to produce the necessary turbulence. The shape of the basin is part of the flash-mix system design.

QUESTIONS

Write your answers in a notebook and then compare your answers with those on page 133.

4.2A   What is the primary purpose of flash mixing?

4.2B   Why are both primary coagulants and coagulant aids used in the coagulation process?

4.2C   List four methods of mixing coagulant chemicals into the plant flow.

4.2D   What is a hydraulic mixing device?

4.3   FLOCCULATION

4.30   Process Description

Flocculation is a slow stirring process that causes the gathering together of small, coagulated particles into larger, settleable particles. The flocculation process provides contact between particles to promote their gathering together into floc

for ease of removal by sedimentation and filtration. Generally, these contacts or collisions between particles result from gentle stirring created by a mechanical or hydraulic means of mixing.

4.31   Floc Formation

Floc formation is controlled by the rate at which collisions occur between particles and by the effectiveness of these collisions in promoting attachment between particles. The purpose of flocculation is to create a floc of a good size, density, and toughness for later removal in the sedimentation and filtration processes. The best floc size ranges from 0.1 mm to about 3 mm, depending on the type of removal processes used (direct filtration vs. conventional filtration which is discussed in Chapter 6, "Filtration").

4.32   Process Performance Considerations

An efficient flocculation process involves the selection of the right stirring time (detention time), the proper stirring intensity, a properly shaped basin for uniform mixing, and mechanical equipment or other means of creating the stirring action. Insufficient mixing will result in ineffective collisions and poor floc formation. Excessive mixing may tear apart the flocculated particles after they have clumped together.

4.320   Detention Time

Detention time is usually not a critical factor in the coagulation or flash-mixing process if the chemical coagulants are satisfactorily dispersed into the water being treated and mixed for at least several seconds. Detention time is required for the necessary chemical reactions to take place. Some operators have been able to reduce coagulant dosages by increasing the amount of detention time between the point of addition of the coagulant and the flocculation basins. In the flocculation process, however, stirring (detention) time is *QUITE* important. The minimum detention time recommended for flocculation ranges from about 5 to 20 minutes for direct filtration systems and up to 30 minutes for conventional filtration. (The size and shape of the flocculation facility also influence the detention time needed for optimum floc development.)

4.321   Types of Flocculators (Stirrers)

Two types of mechanical flocculators are commonly installed, horizontal paddle wheel types and vertical flocculators (see Figures 4.4 and 4.5). Both types can provide satisfactory performance; however, the vertical flocculators usually require less maintenance since they eliminate submerged bearings and packings. Vertical flocculators can be of the propeller, paddle, or turbine types.

Some flocculation can also be accomplished by the turbulence resulting from the roughness in conduits or channels, or by the dissipated energy of head losses associated with weirs, baffles, and orifices.[11] Generally, these methods find only limited use owing to disadvantages such as very localized distribution of turbulence, inadequate detention time, and widely variable turbulence resulting from flow fluctuations.

---

[10] *Head Loss.   The head, pressure or energy (they are the same) lost by water flowing in a pipe or channel as a result of turbulence caused by the velocity of the flowing water and the roughness of the pipe, channel walls, or restrictions caused by fittings. Water flowing in a pipe loses head, pressure or energy as a result of friction losses. The head loss through a filter is due to friction losses caused by material building up on the surface or in the top part of a filter.*

[11] *Weirs, baffles, and orifices are flow regulating devices used in flocculation processes to create turbulence which will mix chemicals with the water. Weirs and baffles are boards or plates that water flows over while orifices are holes in walls that water flows through.*

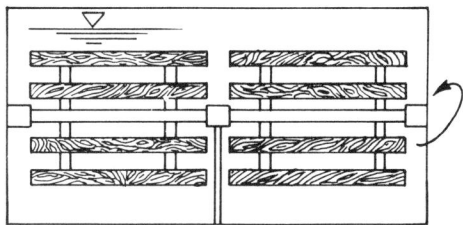

HORIZONTAL PADDLE WHEEL

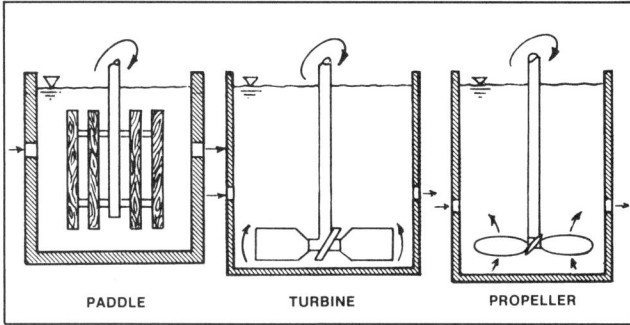

VERTICAL FLOCCULATORS
(INSTALLED IN FLOCCULATION BASINS)

*Fig. 4.4   Types of mechanical flocculators
(there are other types)*

4.322   Flocculation Basins

The actual shape of flocculation basins is determined partially by the flocculator selected, but also for compatibility with adjoining structures (sedimentation basins). Flocculation basins for horizontal flocculators are generally rectangular in shape, while basins for vertical flocculators are nearly square. The depth of flocculation basins is usually about the same as the sedimentation basins.

The best flocculation is usually achieved in a compartmentalized basin. The compartments (most often three) are separated by baffles to prevent *SHORT-CIRCUITING*[12] of the water being treated. The turbulence can be reduced gradually by reducing the speed of the mixers in each succeeding tank (see Figure 4.6) or by reducing the surface area of the paddles. This is called tapered-energy mixing. The reason for reducing the speed of the stirrers is to prevent breaking apart the large floc particles that have already formed. If you break up the floc you have not accomplished anything and will overload the filters.

Fig. 4.5   Horizontal paddle wheel flocculator

The solids-contact process (upflow clarifiers) is used in some water treatment plants to improve the overall solids removal process. These units combine the coagulation, flocculation, and sedimentation processes into a single basin. A detailed discussion of solids-contact units is given in Chapter 5, "Sedimentation," Section 5.24, "Solids-Contact Clarification."

QUESTIONS

Write your answers in a notebook and then compare your answers with those on page 133.

4.3A   What is flocculation?

4.3B   How long is the typical mixing time in the coagulation process?

4.3C   What is the recommended minimum detention time for flocculation?

4.3D   What is an advantage of vertical flocculators over horizontal flocculators?

4.3E   Why are the compartments in flocculation basins separated by baffles?

4.4   INTERACTION WITH OTHER TREATMENT PROCESSES

In previous sections, the need for coagulation and flocculation was discussed. As you've seen, these processes are required to *PRECONDITION* or prepare nonsettleable particles present in the raw water for removal by sedimentation and filtration. Small particles, without proper coagulation and flocculation, are too light to settle out and will not be large enough to be trapped during filtration. In this regard, it is convenient to consider coagulation-flocculation as one treatment process.

Since the purpose of coagulation-flocculation is to promote particulate removal, the effectiveness of the sedimentation and filtration processes, as well as overall plant performance, depends upon successful coagulation-flocculation. Disinfec-

---

[12] *Short-Circuiting.   A condition that occurs in tanks or basins when some of the flowing water entering a tank or basin flows along a nearly direct pathway from the inlet to the outlet. This is usually undesirable since it may result in shorter contact, reaction, or settling times in comparison with the theoretical (calculated) or presumed detention times.*

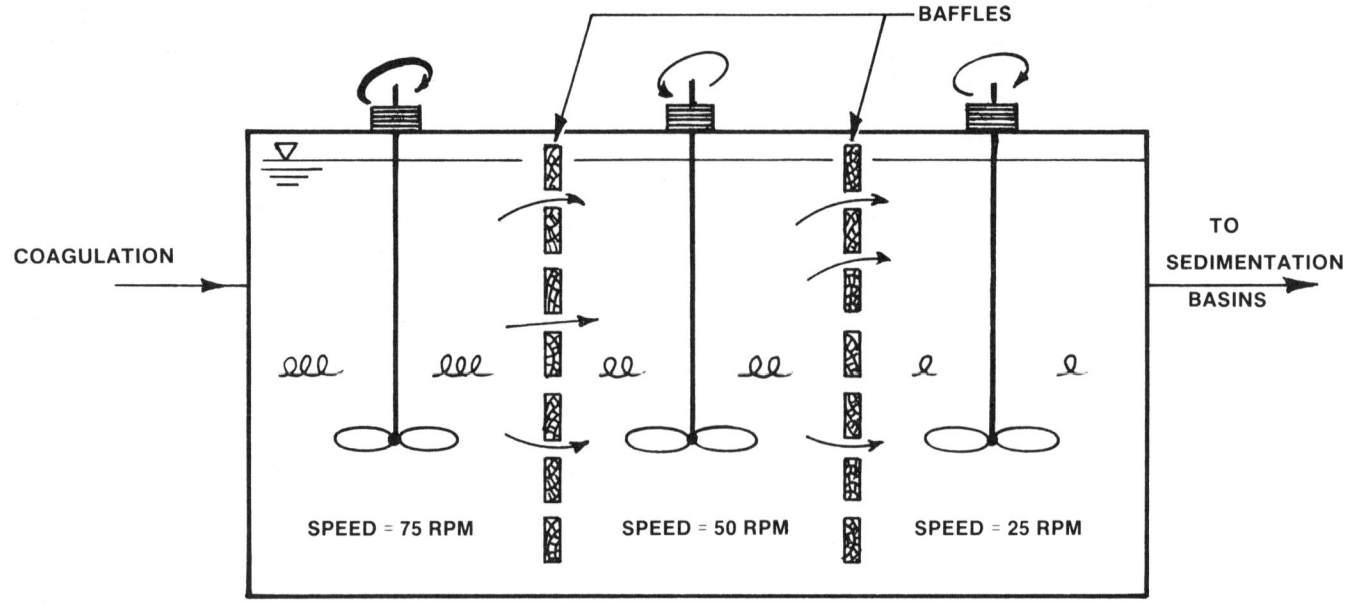

Typical flocculation basin

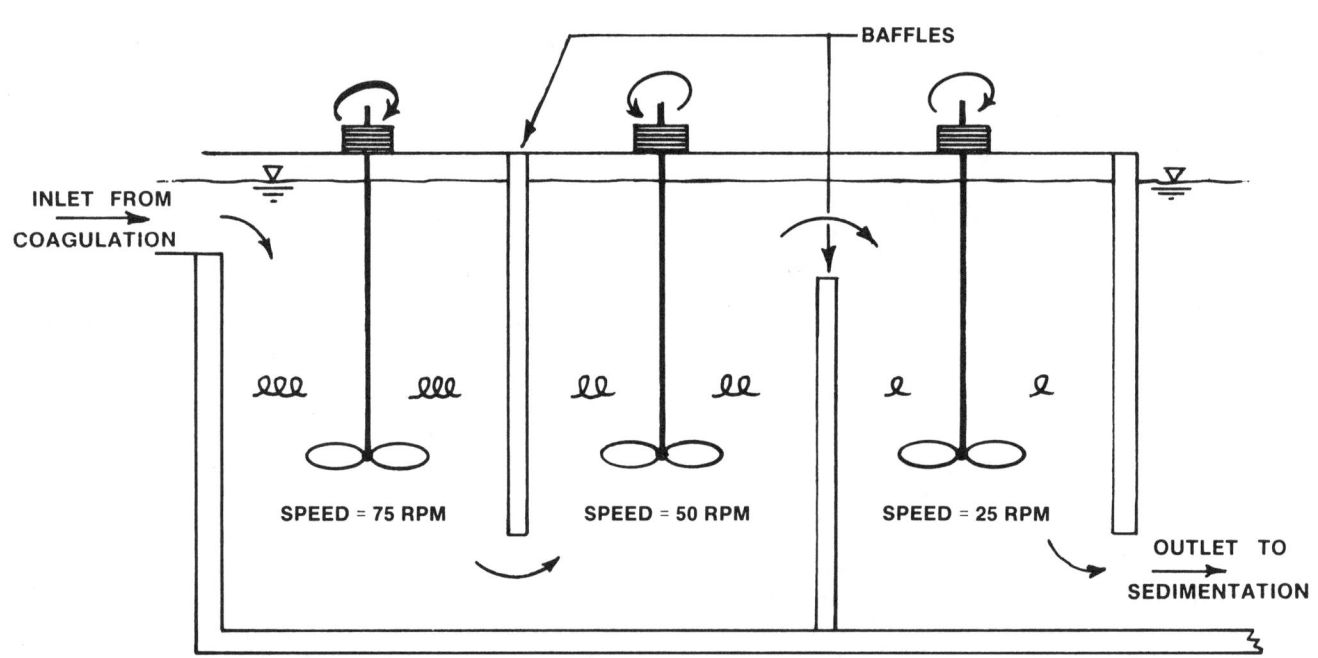

Flocculation basin with alternate baffle arrangement

Fig. 4.6   Typical flocculation basins

tion of the water can also be affected by poor coagulation-flocculation performance. Bacteria and other disease-causing organisms can be bound up in suspended particles and thereby shielded from disinfection if the solids removal processes before final disinfection, especially filtration, are ineffective. Effective coagulation-flocculation promotes the removal of natural organic compounds. Removal of these compounds will reduce the formation of trihalomethanes following the use of chlorine for disinfection.

4.5 PROCESS CONTROL

In theory, the chemical reactions and the formation of floc associated with the coagulation-flocculation process are rather complex. Yet from a practical viewpoint, the operator of a water treatment plant must be able to measure and control the performance of these processes on a day-to-day basis.

The most important consideration in coagulation-flocculation process control is selection of the proper type and amount of coagulant chemical(s) to be added to the water being treated. This determination is commonly made in the laboratory with the aid of a jar testing apparatus (Figure 4.7). When se-

lecting a particular type of coagulant chemical, consideration must be given to the quantity and solids content of the sludge created and the means of ultimate disposal. Jar tests should be run at least daily and more often when the quality of the raw water changes. Changes in the raw water may require changes in the amount of chemical and/or type of chemical.

Procedures for determining proper coagulant dosage, as well as means for measuring and controlling coagulation-flocculation process performance, are described in the following sections.

QUESTIONS

Write your answers in a notebook and then compare your answers with those on pages 133 and 134.

4.4A   Why is coagulation-flocculation important to other treatment processes?

4.5A   What is the most important consideration in coagulation-flocculation process control?

Start of test

End of test (after settling)

Fig. 4.7   Jar test apparatus with mechanical stirrers

End of Lesson 1 of 3 Lessons on Coagulation and Flocculation

Please answer the discussion and review questions next.

DISCUSSION AND REVIEW QUESTIONS

Chapter 4.   COAGULATION AND FLOCCULATION

(Lesson 1 of 3 Lessons)

At the end of each lesson in this chapter you will find some discussion and review questions. The purpose of these questions is to indicate to you how well you understand the material in the lesson. Write the answers to these questions in your notebook before continuing.

1. What is the difference between the coagulation and flocculation processes?

2. What is the difference between primary coagulants and coagulant aids?

3. Why have pumped blenders become more popular for mixing coagulant chemicals in recent years?

4. What is the purpose of the flocculation process?

5. How does the coagulation-flocculation process affect other treatment processes?

6. How are the proper type and amount of coagulant chemical(s) determined?

CHAPTER 4.   COAGULATION AND FLOCCULATION

(Lesson 2 of 3 Lessons)

4.6   OPERATING PROCEDURES ASSOCIATED WITH NORMAL PROCESS CONDITIONS

4.60   Indicators of Normal Operating Conditions

Coagulation-flocculation is a *PRETREATMENT* process for the sedimentation and filtration processes. Most of the suspended solids are removed in the sedimentation basins and filtration is the final step in the solids removal process. Thus, the coagulation-flocculation process should be operated and controlled to improve filtration and thus produce a filtered water that is low in turbidity.

The measurement of filtered water turbidity on either a periodic (*GRAB SAMPLE*[13]) or continuous basis by a *TURBIDIMETER*[14] will give the operator a good indication of overall process performance. However, the operator cannot rely solely on filtered water turbidity for complete process control. The difficulty in relying on a single water quality indicator, such as filtered water turbidity, is that it takes a considerable amount of time to transport the water through the various treatment processes. Depending on the amount of water being processed, the total transit time through the treatment plant can vary from 2 to 6 hours or more. This means that a change in coagulant dosage at the front end of the plant will not be noticed in the final finished water quality for a period of 2 to 6 hours or more, depending on flow conditions. Thus, turbidity, as well as other water quality indicators such as pH, temperature, chlorine demand, and floc quality, must be monitored throughout the water treatment process. Poor process performance can be spotted early and corrective measures can be taken.

Process control guidelines for a specific plant are often developed to assist the operator in making these determinations.

These guidelines are partially based on theory and partially based on experience, but also must be combined with practical knowledge of the source water conditions as well as known performance characteristics of the treatment facilities used for a variety of different treatment conditions.

4.61   Process Actions

In the normal operation of the coagulation-flocculation process, the operator performs a variety of jobs within the water treatment plant. The number and type of functions that each operator will perform vary considerably depending on the size and type of plant and the number of people working in the plant. In smaller plants, the operator is required to control almost all process actions as well as perform most routine maintenance activities. Regardless of the plant size, *ALL* operators should be thoroughly familiar with the routine and special operations and maintenance procedures associated with each treatment process.

---

[13] *Grab Sample.   A single sample of water collected at a particular time and place which represents the composition of the water only at that time and place.*

[14] *Turbidimeter.   An instrument for measuring and comparing the turbidity of liquids by passing light through them and determining how much light is reflected by the particles in the liquid. The normal measuring range is 0 to 100 and is expressed as Nephelometric Turbidity Units (NTUs).*

Typical jobs performed by an operator in the normal operation of the coagulation-flocculation process include the following:

1. Monitor process performance,

2. Evaluate water quality conditions (raw and treated water),

3. Check and adjust process controls and equipment, and

4. Visually inspect facilities.

Monitoring process performance is an ongoing activity. As discussed in Section 4.60, filtered water turbidity levels are controlled to a great extent by the efficiency of the coagulation-flocculation process. Early detection of a pretreatment failure is extremely important because considerable time elapses while the water flows through the coagulation, flocculation, sedimentation, and filtration processes.

Process performance can be monitored with the aid of continuous water quality analyzers which automatically measure a specific water quality indicator such as turbidity. However, reliable and accurate water quality analyzers are expensive and, in certain cases, automated equipment is not readily available for measuring all water quality indicators of concern to the operator. Thus, a combination of techniques must be used by the operator to evaluate process performance including visual observations and periodic laboratory tests to supplement any continuous water quality monitors.

Visual observations and laboratory tests of coagulation-flocculation process performance should be performed on a routine basis. The most common laboratory tests are turbidity, alkalinity, pH, color, temperature, and chlorine demand. The frequency of these observations and tests depends on how much the quality of the source water supply can or does change. In treatment plants where the source water is stored in a large upstream lake or reservoir, the water quality is generally more stable or constant than water taken directly from rivers or streams.

In the case of direct *DIVERSIONS*[15] from a stream or river, water quality conditions will vary seasonally as well as daily. In extreme cases (during heavy runoff periods) even hourly changes in source water quality can be expected. Thus, the appropriate frequency of performing certain tests may be as often as hourly, or perhaps only once per eight-hour shift.

Visual checks of the coagulation-flocculation process generally include an observation of the turbulence of the water in the flash-mixing channel or chamber (you can see improper flow patterns), and close observation of the size and distribution of floc in the flocculation basins. An uneven distribution of floc could be an indication of short-circuiting in the flocculation basin. Floc particles that are too small or too large may not settle properly and could cause trouble during removal in the following processes. These observations are frequently supplemented by laboratory evaluations which are necessary to provide better data. For example, floc settling characteristics require laboratory evaluation based on trial and experience methods using the jar test.

Speed adjustment of flocculators, if such a feature is provided, should take into account the following items:

1. Volume of floc to be formed. If source water turbidity is low, a small pinpoint floc may be best suited for removal on the filters (direct filtration). Lower flocculator speeds are appropriate here. On the other hand, high turbidity source waters generally require near-maximum flocculator speed to produce a readily settleable floc.

2. Visual observations. Short-circuiting may indicate flocculator mixing intensity is not sufficient, while floc breakup is an indication that the mixing turbulence (speed) may be too high for the type of floc formed (large alum floc).

3. Water temperature. Lower water temperature[16] requires higher mixing turbulence, so speed should be increased.

Unfortunately, these concepts are not easily measured. Experience and judgment are needed. One of the real limitations in process control is too much reliance on the settled turbidity value. While turbidity gives an indirect measurement of suspended solids concentration, it does not describe particle size, density, volume, nor the ability of a particular filter to handle the applied waters.

Typical coagulation-flocculation monitoring locations are shown in Figure 4.8. Water quality indicators used to evaluate coagulant dosage and process performance include turbidity, temperature, alkalinity, pH, color, and chlorine demand. These water quality indicators are further discussed in Section 4.10, "Laboratory Tests."

Based on an overall evaluation of process performance, the operator may need to make minor changes in chemical feed dosages, or adjust the speed of the flash mixer or flocculators, if variable-speed units are provided. These are normal actions associated with minor changes in source water quality such as turbidity or temperature fluctuations.

Flash mixers are generally less sensitive to speed adjustments than flocculators since their primary purpose is to disperse the chemicals rapidly into the water being treated. This reaction is almost instantaneous in such small quantities.

Process equipment, such as chemical feeders, should be checked regularly to ensure that they are accurately feeding the desired amount (feed rate) of chemical. Operation and maintenance of process equipment is discussed in more detail in Section 4.11, "Process and Support Equipment Operation and Maintenance."

The operator should routinely perform a visual inspection of the overall coagulation-flocculation physical facilities. This is a part of good housekeeping practice. Leaves, twigs, and other debris can easily build up in the influent channel or in the flocculation basins. If ignored, this material may get into other processes where it can foul meters, water quality monitors, pumps, or other mechanical equipment. In some cases taste and odor problems can develop from microorganisms that can grow in debris and sediment that accumulate in plant facilities. See Chapter 9, "Taste and Odor Control," for additional information.

---

[15] Diversion.   Use of part of a stream flow as a water supply.

[16] The rate of all chemical reactions decreases with temperature. For example, 50°F (10°C) water slows down chemical reaction times to one-half their speed at 70°F (21°C).

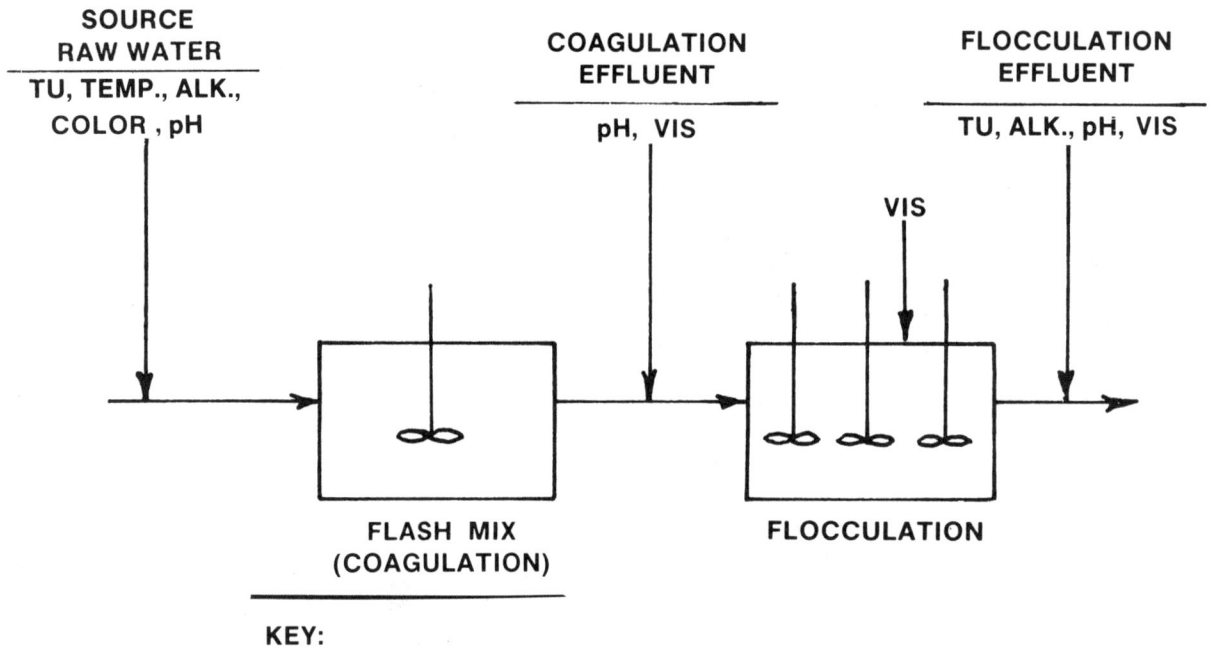

Fig. 4.8   Coagulation-flocculation process monitoring guidelines and sample points

QUESTIONS

Write your answers in a notebook and then compare your answers with those on page 134.

4.6A   Which processes remove suspended solids after the coagulation-flocculation process?

4.6B   How is the effectiveness of the solids removal processes commonly monitored?

4.6C   List the typical functions performed by an operator in the normal operation of the coagulation-flocculation process.

4.6D   Which laboratory tests would you use to monitor the coagulation-flocculation process?

4.6E   What would you look for when visually observing the performance of a coagulation-flocculation process?

4.62   Process Operation

4.620   Need for Experimentation

In order to illustrate how to operate the actual coagulation-flocculation process, the procedures used in a typical plant will be followed on a step-by-step basis. You must realize that there are many plants that are different from our typical plant and that many waters behave differently than the water being treated here. In actual practice, you will have to experiment with your plant and with your water, but these guidelines may be very helpful to you.

4.621   Physical Facilities

Figure 4.9 shows the overall *PLAN VIEW*[17] of the coagulation-flocculation processes in our typical plant. Chemicals are added to the source water in the flash-mix chamber and mixed by the flash mixer. The water flows from the flash mixer through a distribution channel to the flocculation basins. From the flocculation basins the water flows to the sedimentation basins (Chapter 5) and then to the filters (Chapter 6). From the treatment plant design drawings (blueprints), the dimensions (length, width, and depth) of the facilities can be determined.

4.622   Detention Times[18]

With the information obtained from the design drawings, we can calculate the expected detention times in the flash-mix chamber, distribution channel, and flocculation basins. These times are important when determining the optimum (best mini-

---

[17] *Plan View.   A diagram or photo showing a facility as it would appear when looking down on top of it.*
[18] *Detention Time.   (1) The theoretical (calculated) time required for a small amount of water to pass through a tank at a given rate of flow. (2) The actual time in hours, minutes or seconds that a small amount of water is in a settling basin, flocculating basin or rapid-mix chamber. In storage reservoirs, detention time is the length of time entering water will be held before being drafted for use (several weeks to years, several months being typical).*

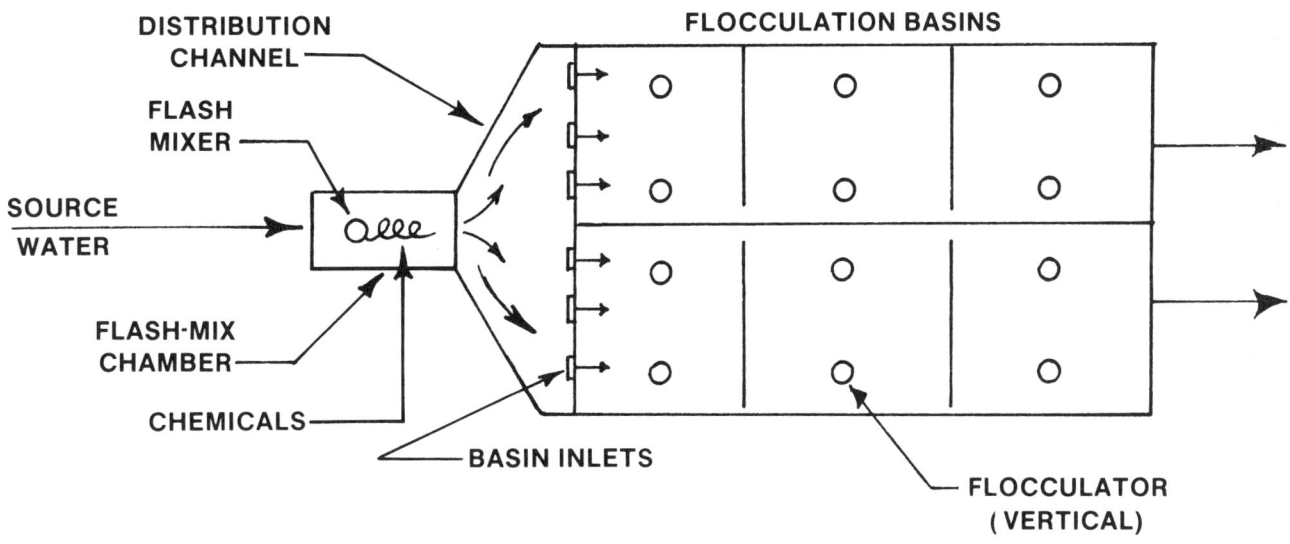

DISTRIBUTION
CHANNEL

FLASH
MIXER

SOURCE
WATER

FLASH-MIX
CHAMBER

CHEMICALS

BASIN INLETS

FLOCCULATION BASINS

FLOCCULATOR
(VERTICAL)

Fig. 4.9   Overall plan view of a typical coagulation-flocculation process

mum) chemical dosage for the water you are treating by using the jar test. Also these times are necessary for the desired chemical reactions to occur.

FORMULAS

In order to calculate the detention times of tanks, basins, or clarifiers, we must know the volume of the container.

1. To calculate the volume of a rectangular tank or basin in cubic feet,

Volume, cu ft = (Length, ft)(Width, ft)(Depth, ft)

2. To calculate the volume of a circular tank or clarifier in cubic feet,

$$\text{Volume, cu ft} = \frac{(\pi)}{4}(\text{Diameter, ft})^2(\text{Depth})$$

The term $\pi/4$ is equal to 0.785, so this term is commonly used.

Volume, cu ft = (0.785)(Diameter, ft)^2(Depth, ft)

3. Frequently, we need the volume in gallons, rather than cubic feet.

Volume, gal = (Volume, cu ft)(7.48 gal/cu ft)

4. To calculate the detention time of any chamber, tank, basin, or clarifier,

$$\frac{\text{Detention Time,}}{\text{minutes}} = \frac{(\text{Volume, gal})(24 \text{ hr/day})(60 \text{ min/hr})}{\text{Flow, gal/day}}$$

or

$$\frac{\text{Detention Time,}}{\text{hours}} = \frac{(\text{Volume, gal})(24 \text{ hr/day})}{\text{Flow, gal/day}}$$

Detention times are determined by dividing the volume in gallons by the flow in gallons per day. This will give us the detention time in days. We multiply by 24 hours per day to obtain the detention time in hours. If we wish to convert the detention time from hours to minutes, we multiply by 60 minutes per hour.

EXAMPLE 1

A water treatment plant treats a flow of 2.4 MGD or 2,400,000 gallons per day. The flash-mix chamber is 2.5 feet square (length and width are both 2.5 feet) and the depth of water is 3 feet. Calculate the detention time in seconds.

Known	Unknown
Length, ft = 2.5 ft	Detention Time, sec
Width, ft = 2.5 ft	
Depth, ft = 3.0 ft	
Flow, GPD = 2,400,000 GPD	

1. Determine the volume of the flash-mix chamber in cubic feet.

Volume, cu ft = (Length, ft)(Width, ft)(Depth, ft)

= (2.5 ft)(2.5 ft)(3.0 ft)

= 18.75 cu ft

2. Convert the volume of the flash-mix chamber from cubic feet to gallons.

Volume, gal = (Volume, cu ft)(7.48 gal/cu ft)

= (18.75 cu ft)(7.48 gal/cu ft)

= 140.25 gallons

3. Calculate the detention time of the flash-mix chamber in seconds.

$$\frac{\text{Detention Time,}}{\text{sec}} = \frac{(\text{Volume, gal})(24 \text{ hr/day})(60 \text{ min/hr})(60 \text{ sec/min})}{\text{Flow, gal/day}}$$

$$= \frac{(140.25 \text{ gal})(24 \text{ hr/day})(60 \text{ min/hr})(60 \text{ sec/min})}{2,400,000 \text{ gal/day}}$$

= 5.0 sec

EXAMPLE 2

A water treatment plant treats a flow of 2.4 MGD or 2,400,000 gallons per day. The flocculation basin is 8 feet deep, 15 feet wide, and 45 feet long. Calculate the detention time in minutes.

Known	Unknown
Known	**Unknown**
Flow, GPD = 2,400,000 GPD	Detention Time, min
Depth, ft = 8 ft	
Width, ft = 15 ft	
Length, ft = 45 ft	

1. Determine the volume of the flocculation basin in cubic feet.

$$\text{Volume, cu ft} = (\text{Length, ft})(\text{Width, ft})(\text{Depth, ft})$$
$$= (45 \text{ ft})(15 \text{ ft})(8 \text{ ft})$$
$$= 5,400 \text{ cu ft}$$

2. Convert the volume of the flocculation basin from cubic feet to gallons.

$$\text{Volume, gal} = (\text{Volume, cu ft})(7.48 \text{ gal/cu ft})$$
$$= (5,400 \text{ cu ft})(7.48 \text{ gal/cu ft})$$
$$= 40,392 \text{ gallons}$$

3. Calculate the detention time of the flocculation basin in minutes.

$$\text{Detention Time, minutes} = \frac{(\text{Volume, gal})(24 \text{ hr/day})(60 \text{ min/hr})}{\text{Flow, gal/day}}$$
$$= \frac{(40,392 \text{ gal})(24 \text{ hr/day})(60 \text{ min/hr})}{2,400,000 \text{ gal/day}}$$
$$= 24 \text{ minutes}$$

Many operators prepare curves of flow vs. detention time for the basins in their plant. These curves allow for easy selection of stirring times when performing jar tests. See Appendix A, "Preparation of Detention Time Curves," for detailed procedures.

4.623 The Jar Test

1. Preparation for Test

For a simple, illustrated, step-by-step procedure on how to run a "standard" jar test, see Chapter 11, "Laboratory Procedures," Section 11.3, Water Laboratory Tests, 6, Jar Test. You must realize that it is almost impossible to exactly duplicate in the jar test the flow-through conditions that are occurring in your treatment plant. The jar test attempts to duplicate in the laboratory what is occurring in the plant in the relation between detention times, mixing conditions, and settling conditions. By watching the jar test floc form and settle, you can get a good idea of what should happen in your plant for the test chemical dose. The jar test should be used as an indication of what you can expect in your water treatment plant. By closely watching the floc form in the flocculators and settle out in the sedimenta-

tion basin of your plant, you can also get a good indication of whether or not you are near the best coagulant dose. Also, by observing the performance of your filters and by looking at the laboratory test results, you will gain additional information that will help you make the necessary adjustments to the actual chemical feed rates. We'll discuss all of these items in more detail in the remainder of this chapter and in the next two chapters (Chapter 5, "Sedimentation," and Chapter 6, "Filtration").

In our typical plant, and thus in the jar test, we are going to use liquid alum as the primary coagulant and a *CATIONIC POLYMER*[19] as a coagulant aid.[20] The size of sample used may be 500 mL, 1,000 mL, or 2,000 mL. We will use one-liter (1,000-mL) water samples since most "gang stirrers" or jar test apparatus (Figure 4.7) are designed for one-liter containers.

Jar test *REAGENTS*[21] can be prepared at several concentrations depending on the desired dosage. Table 4.2 lists the most commonly used chemical concentrations for water treatment.

In this example, we will assume the following reagent concentrations are best for our tests:

1. Alum at a 10-mg/L concentration for each mL of stock solution added to one liter of raw water.

2. Cationic polymer at a 1-mg/L concentration for each mL of stock solution added to one liter of raw water.

NOTES

1. If you wish to avoid mixing your own chemical reagents for your jar tests, contact your suppliers. Find out from your chemical suppliers the concentrations of chemicals being delivered to your treatment plant. Contact your reagent or laboratory chemical supplier and indicate the strengths of the stock solutions (1,000 mg/L or 0.1%) you need to perform the jar tests. There will be a charge for these specially prepared reagents, but the costs are usually considerably less than the value of your time to prepare these reagents. A good practice is to prepare jar test reagents using samples of the chemicals actually used in the plant, rather than reagent grade chemicals. Sometimes trace impurities in industrial chemicals can have significant effects.

Always *use* fresh reagents.

---

[19] *Cationic Polymer. A polymer having positively charged groups of ions; often used as a coagulant aid.*

[20] *For additional information on the selection and use of coagulant chemicals, see Chapter 10, "Plant Operation."*

[21] *Reagent (re-A-gent). A pure chemical substance that is used to make new products or is used in chemical tests to measure, detect, or examine other substances.*

TABLE 4.2 DRY CHEMICAL CONCENTRATIONS USED FOR JAR TESTING[a]

Approx. Dosage Required, mg/L[b]	Grams/Liter to Prepare[c]	1 mL Added to 1 Liter Sample Equals	Stock Solution Conc., mg/L (%)
1-10 mg/L	1 gm/L	1 mg/L	1,000 mg/L (0.1%)
10-50 mg/L	10 gm/L	10 mg/L	10,000 mg/L (1.0%)
50-500 mg/L	100 gm/L	100 mg/L	100,000 mg/L (10.0%)

[a] From *JAR TEST* by E.E. Arasmith, Linn-Benton Community College, Albany, OR.
[b] Use this column, which indicates the approximate dosage required by raw water, to determine the trial dosages to be used in the jar test.
[c] This column indicates the grams of dry chemical that should be used when preparing the stock solution. The stock solution consists of the chemical plus enough water to make a one-liter solution.

2. For details on how to prepare fresh reagents, see Chapter 11, "Laboratory Procedures," Section 11.3, Water Laboratory Tests, 6, Jar Test, Paragraph 1, Stock Coagulant Solution, (a) Dry alum, and (b) Liquid alum.

3. If you have the laboratory capability, a good practice is to sample new chemical supplies when they are delivered to confirm the quality and strength of the chemical provided.

4. Pages 112 and 113 are typical data sheets used with jar tests. To measure filterability, collect a sample of water from above the settled floc. Pass the water through Whatman No. 40 filter paper and measure the turbidity of the filtered water. Another filterability measurement is to record the time required for 100 mL to pass through the filter paper.

QUESTIONS

Write your answers in a notebook and then compare your answers with those on page 134.

4.6F What is the sequence of solids-removal processes following the flash mixer in the typical treatment plant?

4.6G Why do many operators prepare flow vs. detention time curves for basins in their plant?

4.6H Estimate the detention time (seconds) in a flash-mix chamber 6 feet long, 4 feet wide, and 5 feet deep when the rate of flow is 10 MGD.

4.6I How can jar test reagents be obtained?

2. Establish Range of Dosages

Compare your current raw water quality conditions with past records and experiences to determine a range of suitable dosages for the jar tests. Seasonal differences are an important consideration. Laboratory and operations records provide a wealth of information to the operator.

For our typical plant, let's assume that records indicate the proper range of dosages for current water quality conditions are as follows:

Alum, mg/L = 10 to 20 mg/L

Cationic Polymer, mg/L = 0.5 to 3 mg/L

3. Establish Test Sequence

Six tests can be run simultaneously on the standard laboratory jar test apparatus (stirring machine). In setting up a multiple-jar series, remember that comparisons between jars that differ in only *ONE* variable (alum dosage, for example) are the most useful. When only one variable is changed and everything else remains the same, any changes in the final outcome will be due to that single variable you changed.

For our typical plant, let's assume that in test sequence No. 1 you decide to evaluate the effect of increasing the polymer dosage while holding the alum dosage constant at 10 mg/L. You will be increasing polymer dosage from 0.5 mg/L to 3 mg/L.

TEST SEQUENCE NO. 1

Jar No.	1	2	3	4	5	6
Alum, mg/L	10	10	10	10	10	10
Cationic Polymer, mg/L	0.5	1.0	1.5	2.0	2.5	3.0

For test sequence No. 2, let's evaluate the effects of varying the polymer dosage at a higher alum dosage (15 mg/L).

TEST SEQUENCE NO. 2

Jar No.	1	2	3	4	5	6
Alum, mg/L	15	15	15	15	15	15
Cationic Polymer, mg/L	0.5	1.0	1.5	2.0	2.5	3.0

Other combinations of alum and cationic polymer may be evaluated by trial and experience to determine the optimum (best possible) dosages. You may wish to try various levels of alum doses first and then try combinations with polymers.

4. Perform Tests

Collect the sample to be tested from the plant flow when you are ready to perform the jar test. The temperature of the water sample being tested should be approximately the same as the temperature of the water being treated. Clear plastic beakers may be used instead of glass beakers because the temperature of the water will change less during the test.

STOCKTON-EAST WATER TREATMENT PLANT
JAR TEST

DATE ____/____/____   TIME _____   OPERATOR _____

RAW WATER DATA		MIXING SEQUENCE	

SOURCE _____

			RPM	TIME

TEMP _____ °C   pH _____   1. _____   _____

TURBIDITY _____ NTU   COLOR _____   2. _____   _____

ALKALINITY _____ mg/l   HARDNESS _____ mg/l   3. _____   _____

_____   _____   4. _____   _____

CHEMICALS (mg/l)		JAR					
		1	2	3	4	5	6
1. ALUM LIQUID / DRY							
2.							
3.							
4.							
5.							
6.							
FLOC CHARACTERISTICS	AFTER FLASH MIX						
	AFTER RAPID MIX						
	5 MIN. SLOW MIX						
	10 MIN. SLOW MIX						
	15 MIN. SLOW MIX						
FLOC SETTLING	5 MIN.						
	10 MIN.						
	20 MIN.						
	30 MIN.						
SETTLED WATER QUALITY	TURBIDITY						
	pH						
	COLOR						
	FILTERABILITY						

COMMENTS:

Jar Test Data Sheet

	Flash Mix	Slow Mix			Date		
		1st Stage	2nd Stage		**Analyst**		
Minutes					**Subject**		
RPM							

Raw Water Characteristics

| Temperature °F | Turbidity JTU | pH | Total Alkalinity mg/L CaCO₃ |
| | | | |

Floc Characteristics

Chemical Used			Floc First Seen Minutes	Description Size, nature, etc.	Settling Rate
Jar No.					
1					
2					
3					
4					
5					
6					

Settled

| Temp. °F | Total Alkalinity mg/L CaCO₃ | Turbidity TU | pH | Color |
| | | | | |

Filtered

| Turbidity TU | Color |
| | |

Conclusions:

However, be sure the plastic beakers are clean before starting the test.

We will use the following jar test procedure:

1. Record test sequence and water quality data;

2. Collect sample;

3. Fill beakers to one-liter mark with water to be tested; and

4. Add measured volumes of chemicals into each beaker (jar) *AS QUICKLY AS POSSIBLE*. In test sequence No. 1, the following reagent volumes would be added:

Jar No.	1	2	3	4	5	6
Alum Dosage, mg/L	10	10	10	10	10	10
Reagent Volume, mL (1%)	1	1	1	1	1	1
Cat. Polymer Dosage, mg/L	0.5	1	1.5	2	2.5	3
Reagent Volume, mL (0.1%)	0.5	1	1.5	2	2.5	3

5. Quickly lower the stirring paddles into the beakers and start them immediately. Operate the paddles for one minute at 80 RPM.

   NOTE: The one minute was based on the detention times in the flash mix and distribution channels with a flow of 3.0 MGD. If the flows were less than 3.0 MGD, we would use more than one minute. Also, if we felt a higher mixing speed (100 RPM) would produce conditions more like actual hydraulic conditions in our plant, we would use a higher mixing speed. Table 4.3 shows the jar mixing times used by one lab and how they are adjusted as the flows change. This lab uses both a flash mix (100 RPM) and a rapid mix (85 RPM). Refer to Appendix A, "Preparation of Detention Time Curves," for procedures on how to prepare a chart similar to Table 4.3 for your plant.

   Settling times cannot be accurately simulated in a one- or two-liter jar test beaker, so these times are shown for information purposes only.

   Adjustment of flash mixer and flocculator speeds in the plant is partly by trial and experience.

6. Reduce mixer speed to 20 RPM for 20 minutes to simulate flocculation basin conditions. If a different speed or mixing time better simulates actual flocculation basin conditions, then we would use these other conditions.

7. Record the time required for visible floc to form and describe the floc characteristics (pin-head sized floc, flake sized floc) during mixing.

8. Stop stirrers. Allow floc to settle for 30 minutes or for a period similar to your plant conditions. Observe and note how quickly the floc settles, floc appearance, and turbidity of settled water above the floc. A hazy settled water indicates poor coagulation. Properly coagulated water contains floc particles that are well formed and the water between the particles is clear. Describe the results as poor, fair, good, or excellent.

9. Measure the turbidity of the settled water.

10. Evaluate the results of the jar test.

RESULTS OF TEST SEQUENCE NO. 1

Jar No.	1	2	3	4	5	6
Alum, mg/L	10	10	10	10	10	10
Cationic Polymer, mg/L	0.5	1.0	1.5	2.0	2.5	3.0
Settled Water Turbidity, TU	0.8	0.4	0.2	0.3	0.5	0.9

Plot the Settled Water Turbidity, TU, vs. Cationic Polymer, mg/L, as shown in Figure 4.10. The results indicate that the lowest turbidity was produced by an alum dose of 10 mg/L and a cationic polymer dose of 1.5 mg/L. Note that when you overdose, the turbidity will increase as shown in Jar Numbers 4, 5, and 6.

TABLE 4.3 JAR TEST MIXING TIMES FOR A WATER TREATMENT PLANT

Plant Flow Rate	Flash Mix 100 RPM	Rapid Mix 85 RPM	Slow Mix 20 RPM	Settling Time
(MGD)	(min)	(min)	(min)	(hours)
10	1	23	46.5	4.8
12	1	19.4	38.8	4.0
14	1	16.6	33.3	3.5
15	1	15.5	31.0	3.2
16	1	14.5	29.0	3.0
18	1	13.0	26.0	2.7
20	0.5	11.6	23.3	2.4
22	0.5	10.6	21.0	2.2
24	0.5	9.7	19.4	2.0
26	0.5	9.0	18.0	1.9
28	0.5	8.3	16.6	1.7
30	0.5	7.75	15.5	1.6
35	0.5	6.7	13.3	1.4
Rapid-Mix Basin[a]	—	0.16 MG (Total)		
Slow-Mix Basin[a]	—	0.32 MG (Total)		
Sedimentation Basin(s)	—	2.01 MG (Total)		

[a] Flocculation Basins

NOTE: The volumes in million gallons (MG) and plant flow rates (MGD) are used to calculate the mixing or detention times.

5. Evaluation of Test Results

Several factors are important in evaluating jar test results. These factors include:

1. Rate of floc formation,

2. Type of floc particles,

3. Clarity of water between floc particles,

4. Size of floc,

5. Amount of floc formed,

6. Floc settling rate, and

7. Clarity of water (supernatant) above settled floc.

Visible floc formation should begin shortly after the flash-mix portion of the jar test. During flocculation mixing, a

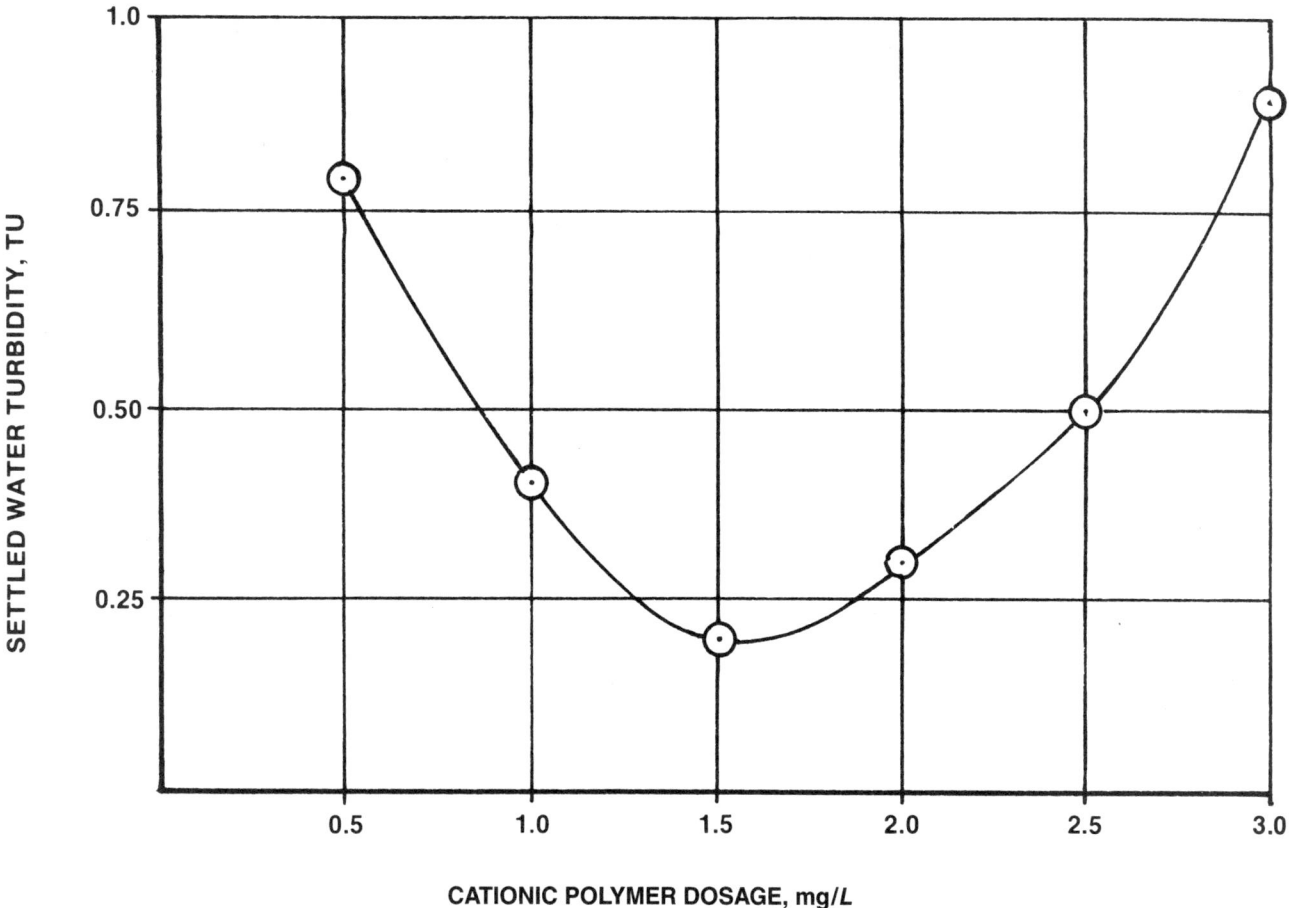

Fig. 4.10   Plot of settled water turbidity vs. cationic polymer dosage

number of small particles will gradually clump together to form larger particles. Floc particles that are discrete (separate) and fairly dense in appearance are usually better than floc particles that have a light, fluffy appearance. Large floc is impressive but it is neither necessary nor always desirable. Large, light floc does not settle as well as smaller, denser floc, and it is more subject to shearing (breaking up by paddles and water turbulence).

The quantity of floc formed is not as critical as floc quality or clarity of the settled water produced. The water between the floc particles should be clear and not hazy or milky in appearance. The best chemical dosage is one which produces a finished water that meets Drinking Water Standards at the lowest cost. Another important consideration is the amount of *SLUDGE*[22] produced. Smaller amounts of sludge are desirable to reduce sludge handling and disposal requirements. Most of the sludge volume consists of precipitates of the added chemicals rather than suspended solids (turbidity) removed.

The rate at which the floc settles after mixing has stopped is another important consideration. The floc should start to settle as soon as the mixer is turned off, and should be almost completely settled (80 to 90 percent) after about 15 minutes. Floc that remains suspended longer than 15 to 20 minutes in the jar test is not likely to settle out in the sedimentation basin, and will increase the load on the filter.

If the floc starts to settle before mixing is completed, or more than about 80 percent of the floc has settled within one or two minutes after mixing has stopped, the floc is too heavy. In your water treatment plant, this can result in the floc settling out in the flocculation basins rather than in the sedimentation basins. (Fortunately this is a rather rare occurrence and indicates that too much chemical has been added.)

6. Frequency of Performing Tests

There is no substitute for experience in evaluating jar test data. Therefore, we recommend that jar tests be performed regularly during periods of high raw water turbidity, even if the plant is producing good quality finished water at the time. This will provide a basis for comparing coagulation-flocculation effectiveness under different conditions and allow "fine-tuning" of the chemical treatment to achieve the best efficiency. Jar tests of flash-mixer water samples (see Section 4.624, "Evaluation of Plant Performance, A., Applying Jar Test Results") should be performed regularly at the start of every shift and more frequently during periods of high turbidity in the raw water. The results of these tests may give an early warning of impending (coming) treatment process problems.

---

[22] *Sludge (sluj).   The settleable solids separated from water during processing. In this case, sediment in the settling basin.*

Always *VERIFY* the effectiveness of a change in treatment based upon jar test results. To verify jar test results with treatment plant performance, obtain a water sample just downstream from the flash mixer. Collect the sample after sufficient time has passed for the treatment change to take effect. This sample should be mixed by the jar test apparatus under the same conditions that the original raw water sample was mixed.

Jar tests are an effective tool for predicting the results of chemical treatment alternatives. However, jar test results are *USELESS* unless applied and verified in your treatment plant.

4.624 Evaluation of Plant Performance

A. APPLYING JAR TEST RESULTS

After evaluation of the jar test results, apply the dosage used to achieve the best jar test results to your water treatment plant operation.

FORMULAS

The settings on chemical feeders will depend on whether you are using a dry chemical feeder or a liquid chemical feeder.

1. Dry chemical feeders are often set on the basis of pounds of chemical fed per day.

$$\text{Feeder Setting, lbs/day} = (\text{Flow, MGD})(\text{Alum, mg/}L)(8.34 \text{ lbs/gal})$$

If you change the alum concentration from mg alum per liter of water to pounds of alum per million pounds of water, nothing changes but the units cancel out OK. This is because one liter weighs one million milligrams.

$$\text{Feeder Setting, lbs/day} = (\text{Flow, } \frac{\text{Mil Gal}}{\text{day}})(\text{Alum, } \frac{\text{mg}}{\text{Mil mg}})(8.34 \frac{\text{lbs}}{\text{gal}})$$

$$= (\frac{\text{Mil Gal}}{\text{day}})(\frac{\text{lbs}}{\text{Mil lbs}})(\frac{\text{lbs}}{\text{Gal}})$$

$$= \text{lbs/day}$$

2. Liquid chemical feeders are often set on the basis of milliliters of chemical solution delivered per minute.

$$\text{Chemical Feeder Setting, m}L\text{/min} = \frac{(\text{Flow, MGD})(\text{Alum Dose, mg/}L)(3.785 \text{ } L\text{/gal})(1,000,000/\text{Million})}{(\text{Liquid Alum, mg/m}L)(24 \text{ hr/day})(60 \text{ min/hr})}$$

We have to multiply flow times dosage to get the amount of chemical needed per unit of time. We multiplied by 3.785 *L*/gal to cancel out the gallons in the flow and the liters in the dose. The concentration of liquid alum in mg/m*L* converted the amount of chemical (alum) needed from one amount in milligrams to a volume in milliliters. Since the flow in MGD gave us an amount of chemical needed on a daily basis, we divided by 24 hr/day and 60 min/hr to convert the amount needed to a per minute basis. The 1,000,000 over 1 Million takes care of the Million in MGD.

Liquid alum may be delivered to plants as 8.0 to 8.5 percent $Al_2(SO_4)_3$ and contains about 5.36 pounds of dry aluminum sulfate (17 percent dry) per gallon (*SPECIFIC GRAVITY*[23] 1.325). This converts to 642,336 mg/*L*, 642.3 gm/*L*, or 642.3 mg/m*L*.

3. To determine a liquid chemical feeder setting in gallons per day (GPD) use the following formula:

$$\text{Chemical Feeder Setting, gal/day} = \frac{(\text{Flow, MGD})(\text{Alum Dose, mg/}L)(8.34 \text{ lbs/gal})}{\text{Liquid Alum, lbs/gal}}$$

From the first formula in this section, we know that when we multiply flow in MGD times dose in mg/*L* times 8.34 lbs/gal, we get pounds of alum per day. By dividing the pounds of alum in a gallon of liquid alum into this number, we'll get the needed gallons of alum per day.

EXAMPLE 3

The best dry alum dose from the jar tests is 10 mg/*L*. Determine the setting on the alum feeder in pounds per day when the flow is 3.0 MGD.

Known	Unknown
Alum Dose, mg/*L* = 10 mg/*L*	Feeder Setting, lbs/day
Flow, MGD = 3.0 MGD	

Calculate the alum feeder setting in pounds per day.

$$\text{Feeder Setting, lbs/day} = (\text{Flow, MGD})(\text{Alum, mg/}L)(8.34 \text{ lbs/gal})$$

$$= (3.0 \text{ MGD})(10 \text{ mg/}L)(8.34 \text{ lbs/gal})$$

$$= 250 \text{ lbs/day}$$

Set the alum feeder to dose alum at a rate of 250 pounds of alum per day.[24] Also set the chemical feeder for the cationic polymer level obtained from the jar test results.

EXAMPLE 4

The optimum liquid alum dose from the jar tests is 10 mg/*L*. Determine the setting on the liquid alum chemical feeder in milliliters per minute when the flow is 3.0 MGD. The liquid alum delivered to the plant contains 642.3 milligrams of alum per milliliter of liquid solution.

---

[23] *Specific Gravity.* (1) Weight of a particle, substance, or chemical solution in relation to the weight of an equal volume of water. Water has a specific gravity of 1.000 at 4°C (39°F). Particulates in raw water may have a specific gravity of 1.005 to 2.5. (2) Weight of a particular gas in relation to the weight of an equal volume of air at the same temperature and pressure (air has a specific gravity of 1.0). Chlorine has a specific gravity of 2.5 as a gas.

[24] From the 250 lbs/day figure, you may easily calculate the lbs/hr equivalent or whatever other weight/time feed rate for which your equipment is calibrated.

$$\text{Feeder Setting, lbs/hr} = \frac{250 \text{ lbs/day}}{24 \text{ hr/day}} = 10.4 \text{ lbs/hr}$$

Known		Unknown
Alum Dose, mg/L	= 10 mg/L	Chemical Feeder Setting, mL/min
Flow, MGD	= 3.0 MGD	
Liquid Alum, mg/mL	= 642.3 mg/mL	

Calculate the liquid alum chemical feeder setting in milliliters per minute.

$$\text{Chemical Feeder Setting, m}L\text{/min} = \frac{(\text{Flow, MGD})(\text{Alum Dose, mg/}L)(3.785\ L\text{/gal})(1,000,000\text{/Million})}{(\text{Liquid Alum, mg/m}L)(24\ \text{hr/day})(60\ \text{min/hr})}$$

$$= \frac{(3.0\ \text{MGD})(10\ \text{mg/}L)(3.785\ L\text{/gal})(1,000,000\text{/Million})}{(642.3\ \text{mg/m}L)(24\ \text{hr/day})(60\ \text{min/hr})}$$

$$= 123\ \text{m}L\text{/min}$$

EXAMPLE 5

The optimum liquid alum dose from the jar tests is 10 mg/L. Determine the setting on the liquid alum chemical feeder in gallons per day when the flow is 3.0 MGD. The liquid alum delivered to the plant contains 5.36 pounds of alum per gallon of liquid solution.

Known		Unknown
Alum Dose, mg/L	= 10 mg/L	Chemical Feeder Setting, GPD
Flow, MGD	= 3.0 MGD	
Liquid Alum, lbs/gal	= 5.36 lbs/gal	

Calculate the liquid alum chemical feeder setting in gallons per day.

$$\text{Chemical Feeder Setting, GPD} = \frac{(\text{Flow, MGD})(\text{Alum Dose, mg/}L)(8.34\ \text{lbs/gal})}{\text{Liquid Alum, lbs/gal}}$$

$$= \frac{(3.0\ \text{MGD})(10\ \text{mg/}L)(8.34\ \text{lbs/gal})}{5.36\ \text{lbs/gal}}$$

$$= 47\ \text{GPD}$$

When the alum and cationic polymer feeders are working properly, collect a sample of well-mixed water from the effluent of the flash-mix chamber to fine-tune process performance. Take the sample to the lab and perform another jar test. First test the alum dosage (is the alum dosage too high or too low?).

1. Add 800 mL of sample water to the first jar and 200 mL of raw water. Since the chemicals have already been mixed in the sample water, floc will start to form. Be sure the sample stays mixed while pouring off the volumes into the six jars.

2. Add 900 mL of sample water to the second jar and 100 mL of raw water.

3. Add 1,000 mL of sample water to the third, fourth, fifth, and sixth jars.

4. Add the alum reagent to the fourth, fifth, and sixth jars in these amounts:

Jar No.	3	4	5	6
1% Alum Reagent Volume, mL	0	0.5	1.0	1.5

5. Mix all six beakers (jars) for one minute at 80 RPM. Adjust time and mixing speed if necessary to simulate plant conditions.

6. Reduce mixer speed to 20 RPM for 20 minutes to simulate flocculation basin conditions.

7. Stop stirrers. Allow floc to settle for 30 minutes. When stirrers are stopped, immediately collect a sample from the flocculation basin effluent. Place 1,000 mL of this sample in a beaker for comparison with the six beakers used in the jar test.

8. Observe how quickly the floc settles, floc appearance, and turbidity of settled water above the floc.

9. Evaluate these jar test results (see Section 4.623, 5, "Evaluation of Test Results"), and make further process adjustments as appropriate.

B. *WALK THROUGH PLANT*

One of the best ways to evaluate the performance of your coagulation-flocculation process is to observe the actual process. When you walk through the treatment plant, take some clear plastic beakers. Dip some water out of each stage of the treatment process. Hold the sample up to a light and look at the clarity of the water between the floc and study the shape and size of the floc. Study the development of the floc from one flocculation chamber to the next and into the sedimentation basin.

1. Observe the floc as it enters the flocculation basins. The floc should be small and well dispersed throughout the flow. If not, the flash mixer may not be providing effective mixing or the chemical dose or feed rate may be too low.

2. Tiny alum floc may be an indication that the chemical coagulant dose is too low. A "popcorn flake" is a desirable floc appearance. If the water has a milky appearance or a bluish tint, the alum dose is probably too high.

3. What does the floc look like as it moves through the flocculation basins? The size of the floc should be increasing. If the floc size increases and then later starts to break up, the mixing intensity of the downstream flocculators may be too high. Try reducing the speed of these flocculators, or increasing the polymer dosage.

4. Does the floc settle out in the sedimentation basin? If a lot of floc is observed flowing over the *LAUNDERING WEIRS*,[25] the floc is too light for the detention time produced by that flow rate. By increasing the chemical coagulant dose or adding a coagulant aid such as a polymer, a heavier, larger floc may be produced. The appearance of fine floc particles washing over the effluent weirs could be an indication of too much alum and the dose should be reduced. *THEREFORE, REGARDLESS OF YOUR PROBLEM, YOU SHOULD MAKE ONLY ONE CHANGE AT A TIME AND EVALUATE THE RESULTS.* This topic will be discussed more in Chapter 5, "Sedimentation."

5. Bring some beakers with samples from various locations back to the laboratory, let them sit for awhile, and then observe the floc settling.

6. How are the filters performing? This topic will be discussed more in Chapter 6, "Filtration."

C. *VARIATIONS OF THE JAR TEST*

In order to improve the effectiveness of their plant performance and reduce operating costs, many operators continuous-

---

[25] *Laundering Weir* (LAWN-der-ing weer). *Sedimentation basin overflow weir. A plate with V-notches along the top to ensure a uniform flow rate and avoid short-circuiting.*

ly experiment to find better ways to do their job. There are many variations and improvements to the "standard" jar test.

1. Some operators have built square jars to use in the jar test because the shape is similar to the actual shape of the flash-mix chamber, flocculation basin, and sedimentation tank. The square jars are built by cementing together sheets of clear acrylic plastic. The jars are 11.5 centimeters square and 21 centimeters deep. A water depth of 15.2 centimeters represents a two-liter sample. Square jars are also available from laboratory supply houses.

2. Use of Whatman No. 40 filter paper. What really counts is the turbidity of the effluent from your plant filters and the performance of those filters. The settled water above the settled floc in the jars is passed through Whatman No. 40 filter paper and the turbidity of the filtered water measured. Tests with such filter papers have shown that the dosage of coagulant needed to achieve a high-quality filtered water may be far below that required to produce a floc that settles well. The lower the chemical dose, the less sludge will be produced. Reducing the chemical dose reduces costs as well.

Another filterability test is to pass 100 mL of water from above the settled floc through the filter paper. Measure the time required for the 100 mL sample to pass through the filter paper and record this value as the filterability.

These tests have also shown that in some cases when a nonsettling floc forms, the flocculators can be shut off without any effect on filtered water turbidity. This could occur when the raw water has a very low turbidity. (For details on these tests, read "Conduct and Uses of Jar Tests," by Herbert E. Hudson, Jr., and E.G. Wagner, in Journal of American Water Works Association, Volume 73, No. 4, pages 218-222, April 1981.)

3. Additional laboratory tests can be performed to better evaluate the use of coagulants and performance of the processes. Tests that may be conducted on samples before and after jar tests or in the plant at the start and end of the processes include turbidity, temperature, pH, alkalinity, and chlorine demand. Compare these results. You might also want to compare starting and final concentrations of aluminum (if using alum) or iron (if using ferric chloride or ferrous sulfate). The difference in residuals is an indication of the amount of chemicals removed by the processes. Large residuals when the tests are completed indicate chemical overdoses or poor sedimentation and/or filtration efficiencies. These residuals may cause floc to form in the water distribution system.

4. Optimum pH. Operators should record the pH during the jar test since some coagulants (especially alum) have an optimum pH range. The pH in the jar after the flash mix that produces the best results becomes a target pH. These operators adjust their chemical feeders to produce the target pH after flash mixing in their treatment plant. This may be an effective means of adjusting chemical feeders for waters

with lower alkalinities where pH changes at least a few tenths (7.5 to 7.0) when alum is added.

Additional jar tests may be run at different pH levels to determine the optimum pH. The chemical dose (alum) could be the same in each jar, but the pH could be changed.

5. Summary. Use the procedures that best suit your needs. Only change one operational variable at a time, and wait to analyze the results before changing another variable. Keep good records. Evaluate the performance of your plant and adjust your procedures as necessary.

4.625 Streaming Current Meters

Streaming current meters are a new device used by operators to optimize coagulant doses. The streaming current meter is a continuous on-line measuring instrument. Properly used, the streaming current meter can function as an on-line jar test.

Most particles in water are anions (negative charge) and most coagulants are cationic (positive charge). The streaming current meter presumes that by bringing the total charge of the water being treated to neutral (zero, 0), the coagulation process has been optimized. Most operators run the charge of their water slightly negative by adjusting the coagulant dose.

QUESTIONS

Write your answers in a notebook and then compare your answers with those on page 134.

4.6J   Why should only one variable at a time be changed when running a jar test?

4.6K   Why should a jar test sample be collected only when you are ready to perform the jar test and not in advance?

4.6L   What factors should be considered when evaluating the results of jar tests?

4.63 Chemical Usage for Small Plants

This section contains arithmetic problems that the operator of a small treatment plant might be expected to solve. Regardless of the size of your water treatment plant, the procedures for solving these problems are similar.

4.630 Calculating Amount of Chemical Required

Operators must be able to estimate the amounts of chemicals used on a monthly basis. Always order enough chemicals in advance so your plant won't run out of needed chemicals.

EXAMPLE 6

A water treatment plant treats an average daily flow of 300,000 GPD. Results from a jar test indicate that the desired polymer dosage for coagulation is 2 mg/L. How many pounds of polymer will be used in 30 days?

Known		Unknown
Flow, GPD	= 300,000 GPD	1. Polymer Used, lbs/day
	= 0.3 MGD	2. Polymer Used, lbs
Polymer Dose, mg/L	= 2 mg/L	
Time, days	= 30 days	

1. Calculate the polymer used in pounds per day.

$$\text{Polymer Used,} \atop \text{lbs/day} = (\text{Flow, MGD})(\text{Dose, mg/}L)(8.34 \text{ lbs/gal})$$

$$= (0.3 \text{ MGD})(2 \text{ mg/}L)(8.34 \text{ lbs/gal})$$

$$= 5 \text{ lbs/day}$$

2. Estimate the polymer used in 30 days.

$$\text{Polymer Used} \atop \text{in 30 Days,} \atop \text{lbs} = (\text{Polymer Used, lbs/day})(\text{Time, days})$$

$$= (5 \text{ lbs/day})(30 \text{ days})$$

$$= 150 \text{ lbs}$$

4.631 Chemical Feeding

The chemical feed rate delivered by chemical feeders must be checked regularly. Jar tests will show the best dosages of chemicals in mg/L. To check on the actual feed rate delivered by a chemical feeder, measure the volume (in gallons for a liquid chemical feeder) or the weight (in pounds for a dry chemical feeder) delivered during a 24-hour period. The flow during this time period also needs to be known and is recorded in gallons per day (GPD) or million gallons per day (MGD).

EXAMPLE 7

During a 24-hour period a plant treated a flow of 1.5 million gallons. Ten pounds of cationic polymer were used in the coagulation process. What is the polymer dosage in mg/L?

Known		Unknown
Flow, MGD	= 1.5 MGD	Polymer Dose, mg/L
Polymer Used, lbs/day	= 10 lbs/day	

Calculate the polymer dose in mg/L. The basic equation is:

$$\text{Polymer Used,} \atop \text{lbs/day} = (\text{Flow, MGD})(\text{Dose, mg/}L)(8.34 \text{ lbs/gal})$$

Rearrange the equation to:

$$\text{Polymer Dose,} \atop \text{mg/}L = \frac{\text{Polymer Used, lbs/day}}{(\text{Flow, MGD})(8.34 \text{ lbs/gal})}$$

$$= \frac{10 \text{ lbs/day}}{(1.5 \text{ MGD})(8.34 \text{ lbs/gal})}$$

$$= 0.8 \text{ mg/}L$$

This actual polymer dose of 0.8 mg/L should be very close to the best dose obtained from the jar test. Also walk through your plant and observe how the water you are treating responds to a polymer dose of 0.8 mg/L. If the floc is not settling properly, the polymer dose may have to be increased or decreased.

EXAMPLE 8

What is the alum dosage in mg/L if a water treatment plant treats a source water with an average flow of 700 GPM and an average turbidity of 10 TU? A total weight of 150 pounds of alum is used during a 24-hour period.

Known		Unknown
Flow, GPM	= 700 GPM	Alum Dose, mg/L
Alum Used, lbs/day	= 150 lbs/day	

1. Convert the flow of 700 GPM to MGD.

$$\text{Flow, MGD} = \frac{(\text{Flow, gal/min})(60 \text{ min/hr})(24 \text{ hr/day})}{1,000,000/\text{M}}$$

$$= \frac{(700 \text{ gal/min})(60 \text{ min/hr})(24 \text{ hr/day})}{1,000,000/\text{M}}$$

$$= 1.0 \text{ MGD}$$

2. Calculate the alum dose in mg/L.

$$\text{Alum Dose,} \atop \text{mg/}L = \frac{\text{Alum Used, lbs/day}}{(\text{Flow, MGD})(8.34 \text{ lbs/gal})}$$

$$= \frac{150 \text{ lbs/day}}{(1.0 \text{ MGD})(8.34 \text{ lbs/gal})}$$

$$= 18.0 \text{ mg/}L$$

NOTE: For an example of how to calculate the lime dosage to provide the alkalinity when using alum, see Chapter 5, Section 5.243, "Arithmetic for Solids-Contact Clarification."

4.632 Preparation of Chemical Solutions

Polymers are frequently used as coagulant aids. Often the operator starts with a dry polymer chemical and wants to prepare a specific solution concentration (as mg/L or as a percent solution). The solution concentration depends on the type of polymer (cationic, nonionic, or anionic) and the polymer's molecular weight (this weight may vary from 100 to ten million). The higher the molecular weight of the polymer, the more difficult it is to mix the polymer with dilution water and to feed the resulting solution to the water being treated; the solution becomes very viscous (thick). Therefore, anionic and nonionic dry polymers used as coagulant aids are often prepared as very dilute (weak) solutions (0.25 to 1.0 percent). Cationic polymers in the dry form can be prepared at higher solution concentrations (say 5 to 10 percent), since their molecular weights are typically small.

High molecular weight polymers are very difficult to prepare. To be effective, polymer solutions must be the same throughout (homogeneous). They must be thin enough to be accurately measured and pumped to the flash-mix chamber.

When mixing a dry polymer with water, sift or spread the polymer evenly over the surface of the water in the mixing chamber. The polymer should be sucked evenly into the hole (vortex) of the stirred water. This will ensure that each particle of polymer is wet individually. This will also ensure even dispersion and prevent the formation of large, sticky balls of polymer which have dry polymer in the middle.

Excessive mixing speeds, mixing time, and the buildup of heat can break down the polymer chain and reduce its effectiveness.

High concentrations of polymer result in very thick, viscous (sticky) solutions. Prepare and use concentrations of polymers that can be metered (measured) easily and pumped accurately to the flash mixer.

The addition of dry polymer to water must be done in a closed system or under an efficient dust collector. Polymer powders on floors and walkways become extremely slippery when wet and are very difficult to remove. In the interest of safety, keep polymers off the floors. Use an inert, absorbent material, such as sand or earth, to clean up spills.

FORMULAS

When mixing either dry or liquid polymers, always follow the directions of the polymer supplier. Polymer solutions are usually prepared in "batches" or as a "batch mixture." Often they are stored after mixing in a "day tank" or an "aging tank" to allow time for all the powder to dissolve and/or the solution to become completely mixed.

DRY POLYMERS

To prepare a specific percent polymer solution use the following formula:

$$\text{Polymer, \%} = \frac{(\text{Dry Polymer, lbs})(100\%)}{(\text{Dry Polymer, lbs} + \text{Water, lbs})}$$

Examination of the above formula reveals three possible unknowns. In addition to wanting to know the percent polymer solution, the operator might wish to know the pounds of dry polymer to use or the amount of water in either pounds or probably gallons to mix with the polymer. In any case, two of the three unknowns must be known to solve the problem. The above basic polymer formula to find the polymer concentration as a percent can be rearranged to solve for the other two unknowns.

$$\text{Dry Polymer, lbs} = \frac{\text{Water, lbs}}{\left(\dfrac{100\%}{\text{Polymer, \%}} - 1.0\right)}$$

$$\text{Water, lbs} = \frac{(\text{Dry Polymer, lbs})(100\%)}{\text{Polymer, \%}} - \text{Dry Polymer, lbs}$$

$$\text{Water, gal} = \frac{\text{Water, lbs}}{8.34 \text{ lbs/gal}}$$

LIQUID POLYMERS

When working with liquid polymers we usually know the percent polymer in the liquid polymer we receive from the supplier. The problem is to determine how much of the supplier's polymer should be mixed with water to give us a tank or barrel with a diluted or lower percent polymer.

The basic liquid polymer formula is as follows:

(Polymer, %)(Volume, gallons) = (Polymer, %)(Volume, gallons)

The volume term could also be expressed in liters. Also the basic liquid polymer formula could be rearranged as follows:

$$\text{Polymer, \%} = \frac{(\text{Polymer, \%})(\text{Volume, gallons})}{\text{Volume, gallons}}$$

or

$$\text{Volume, gallons} = \frac{(\text{Polymer, \%})(\text{Volume, gallons})}{\text{Polymer, \%}}$$

EXAMPLE 9

At the beginning of each shift, or each day, the operator of a small plant mixes a batch of nonionic polymer solution. What is the percent (by weight) polymer solution if the operator adds 60 grams (0.132 pound) of dry polymer to five gallons of water?

Known	**Unknown**
Dry Polymer, lbs = 0.132 lb	Polymer Solution, %
Volume Water, gal = 5 gal	

1. Convert the five gallons of water to pounds.

Water, lbs = (Volume Water, gal)(8.34 lbs/gal)

= (5 gal)(8.34 lbs/gal)

= 41.7 lbs

2. Calculate the polymer solution as a percent.

$$\text{Polymer, \%} = \frac{(\text{Dry Polymer, lbs})(100\%)}{(\text{Dry Polymer, lbs} + \text{Water, lbs})}$$

$$= \frac{(0.132 \text{ lb})(100\%)}{(0.132 \text{ lb} + 41.7 \text{ lbs})}$$

$$= 0.32\%$$

EXAMPLE 10

How many pounds of dry polymer must be added to 50 gallons (417 pounds) of water to produce a 0.5 percent polymer solution?

Known	**Unknown**
Water Volume, gal = 50 gal	Dry Polymer, lbs
Water Weight, lbs = 417 lbs	
Polymer, % = 0.5%	

Calculate the pounds of dry polymer that must be added to 50 gallons of water.

$$\text{Dry Polymer, lbs} = \frac{\text{Water, lbs}}{\left(\dfrac{100\%}{\text{Polymer, \%}} - 1.0\right)}$$

$$= \frac{417 \text{ lbs}}{\left(\dfrac{100\%}{0.5\%} - 1.0\right)}$$

$$= \frac{417 \text{ lbs}}{(200 - 1.0)}$$

$$= 2.1 \text{ lbs}$$

or

$$= (2.1 \text{ lbs})(454 \text{ grams/lb})$$

$$= 951 \text{ grams}$$

EXAMPLE 11

How many gallons of water should be mixed with 1.5 pounds of dry polymer to produce a 0.5 percent polymer solution?

Known	Unknown
Polymer, % = 0.5%	Water, gallons
Dry Polymer, lbs = 1.5 lbs	

1. Calculate the pounds of water needed.

$$\text{Water, lbs} = \frac{(\text{Dry Polymer, lbs})(100\%)}{\text{Polymer, \%}} - \text{Dry Polymer, lbs}$$

$$= \frac{(1.5 \text{ lbs})(100\%)}{0.5\%} - 1.5 \text{ lbs}$$

$$= 300 \text{ lbs} - 1.5 \text{ lbs}$$

$$= 298.5 \text{ lbs}$$

2. Calculate the volume of water in gallons to be mixed with the polymer.

$$\text{Volume Water, gallons} = \frac{\text{Water, lbs}}{8.34 \text{ lbs/gal}}$$

$$= \frac{298.5 \text{ lbs}}{8.34 \text{ lbs/gal}}$$

$$= 35.8 \text{ gallons}$$

EXAMPLE 12

Liquid polymer is supplied to a water treatment plant as a ten percent solution. How many gallons of liquid polymer should be mixed in a barrel with water to produce 50 gallons of 0.5 percent polymer solution?

Known	Unknown
Liquid Polymer, % = 10%	Volume of Liquid Polymer, gal
Polymer Solution, % = 0.5%	
Volume of Polymer Solution, gal = 50 gal	

Calculate the volume of liquid polymer in gallons.

$$\text{Liquid Polymer, gal} = \frac{(\text{Polymer Solution, \%})(\text{Volume of Solution, gal})}{\text{Liquid Polymer, \%}}$$

$$= \frac{(0.5\%)(50 \text{ gal})}{10\%}$$

$$= 2.5 \text{ gallons}$$

NOTE: The volume of the polymer solution is 50 gallons. This means that we have 2.5 gallons of liquid polymer and 47.5 gallons (50 − 2.5 = 47.5) of water mixed together.

4.64 Recordkeeping

One of the most important *ADMINISTRATIVE* functions of the water treatment plant operator is the preparation and maintenance of accurate plant operation records.

In the routine operation of the coagulation-flocculation process, the operator usually makes entries in and maintains records such as daily operations logs and diaries. These records should provide an accurate day-to-day account of actual operating experience.

Accurate records provide the operator, as well as others, with a running account of operations (historical records), and are a great help to the operator in solving current or future process problems. They also provide a factual account of the operation, which is required by law and by regulatory agencies.

In the coagulation-flocculation process, you should keep records of the following items:

1. Source water quality (pH, turbidity, temperature, alkalinity, chlorine demand, and color),

2. Process water quality (pH, turbidity, and alkalinity),

3. Process production inventories (chemicals used, chemical feed rates, amount of water processed, and amount of chemicals in storage), and

4. Process equipment performance (types of equipment in operation, maintenance procedures performed, equipment calibration and adjustments).

Record entries should be neat, legible, and easily found. Keep in mind that many records will be kept and may be used for long periods of time. Entries should reflect the date and time of an event and should be initialed by the operator making the entry for future identification.

Operators should maintain a plot of key process variables. A plot of source water turbidity vs. coagulant dosage should be maintained. If other process variables, such as alkalinity or pH, vary significantly, these should also be plotted.

These graphs should be designed so that the operator can see one year of data at one time. This will give the operator a better understanding of the seasonal variation in these water quality indicators. A sample graph showing a plot of source water turbidity vs. alum dosage is given in Figure 4.11.

4.65 Safety Considerations

In the routine operation of the coagulation-flocculation process, the operator will be exposed to a number of potential hazards such as:

1. Electrical equipment,

2. Rotating mechanical equipment,

3. Water treatment chemicals,

4. Laboratory reagents (chemicals),

5. Slippery surfaces caused by wet polymers,

6. Open-surface, water-filled structures (drowning), and

7. Confined spaces and underground structures, such as valve or pump vaults (toxic and explosive gases, insufficient oxygen).

You must realize that accidents do not just happen, they are caused. Therefore, strict and constant attention to safety procedures cannot be overemphasized.

The operator should be familiar with general first-aid practices such as mouth-to-mouth resuscitation, treatment of common physical injuries, and first aid for chemical exposure

(chlorine). When you come to Chapter 20, "Safety," in Volume II, please read the material carefully and review the chapter from time to time.

4.66 Communications

Good communications are an essential part of the operator's job. Clear and concise written or oral communications are necessary to advise other operators and support personnel of current process conditions and unique or unusual events. In this regard, good recordkeeping is an essential element of the communication process.

QUESTIONS

Write your answers in a notebook and then compare your answers with those on page 134.

4.6M Why should operators keep accurate records?

4.6N What information should be recorded for all entries in a record book?

4.6O List the safety hazards an operator may encounter when operating a coagulation-flocculation process.

4.6P Why are good communications such an essential part of an operator's job?

End of Lesson 2 of 3 Lessons on Coagulation and Flocculation

Please answer the discussion and review questions next.

DISCUSSION AND REVIEW QUESTIONS

Chapter 4. COAGULATION AND FLOCCULATION

(Lesson 2 of 3 Lessons)

Write the answers to these questions in your notebook before continuing. The question numbering continues from Lesson 1.

7. What are the limitations of relying solely on the turbidity of filtered water as an indication of overall process performance?

8. How often should samples be collected and analyzed when operating a coagulation-flocculation process?

9. Why might the jar test guidelines outlined here (mixing speed, time, dose) not be proper for your water treatment plant?

10. What is the basic purpose of the jar test?

11. How frequently should jar tests be performed?

12. Why do some operators use square jars in the jar test?

13. How do accurate records help operators?

14. What types of records should be kept when operating a coagulation-flocculation process?

15. How should an operator communicate with other operators?

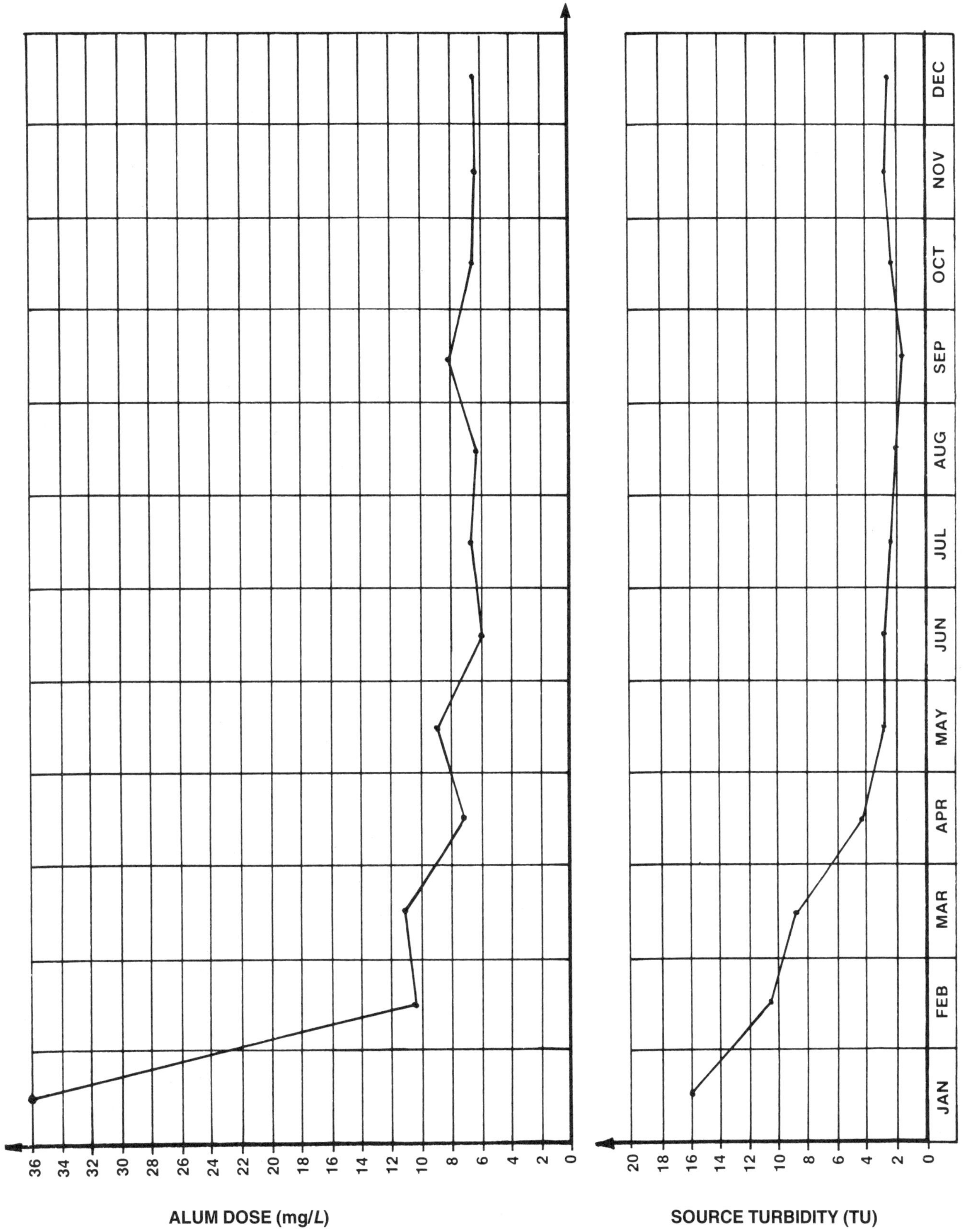

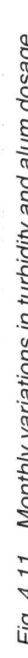

Fig. 4.11 Monthly variations in turbidity and alum dosage

CHAPTER 4. COAGULATION AND FLOCCULATION

(Lesson 3 of 3 Lessons)

4.7 OPERATING PROCEDURES ASSOCIATED WITH ABNORMAL PROCESS CONDITIONS

4.70 Indicators of Abnormal Conditions

Sudden changes in the source water or filtered water turbidity, pH, alkalinity, temperature, or chlorine demand are signals that the operator should immediately review the performance of the coagulation-flocculation process.

4.71 Process Actions

Changes in source water turbidity levels, either increases or decreases, generally require that the operator verify the effectiveness of the coagulant chemicals and dosages being applied at the flash mixer. This is best accomplished by performing a series of jar tests in the laboratory as discussed previously. Remember that decreasing raw water turbidity levels can be just as upsetting to the process and difficult to treat as increasing levels.

Visual observations of flash-mixing intensity as well as the condition of the floc in the flocculation basins may also indicate the need for process changes such as adjustment of mixer speed or coagulant dosage.

Alkalinity, pH, and temperature changes in the source water may have an impact on the clumping together of floc during the coagulation-flocculation process. In addition, water temperature changes may require an adjustment in the level of mixing intensity in flash mixers or flocculators. Temperature changes are usually gradual over time, thus large process adjustments are seldom necessary.

Sudden increases in filtered water turbidity could be caused by poor filter performance (need for backwashing or replacing filter media). However, poor coagulation-flocculation performance is usually the culprit, and the operator must take immediate action to correct the problem, remembering that several hours may pass before changes in the operation of the coagulation-flocculation process are seen in the filter effluent. One quick remedy may be to feed a filter-aid chemical, such as a nonionic polymer, directly to the filter influent. While this may solve the short-term problem, only changes in the coagulation-flocculation process will enhance long-term plant performance. Again, the results of laboratory jar tests should be used as the basis for making process changes. Filter-aid chemicals are discussed further in Chapter 6, "Filtration."

Table 4.4 is a summary of coagulation-flocculation process problems, how to identify the causes of these problems, and also how to go about trying to correct the problems.

QUESTIONS

Write your answers in a notebook and then compare your answers with those on pages 134 and 135.

4.7A What kinds of sudden changes in either raw or filtered water quality are signals that you should immediately review the performance of the coagulation-flocculation process?

4.7B What would you do if you observed a significant change in source water turbidity level?

4.7C How could you verify the effectiveness of the type and dosage of coagulant chemicals being applied at the flash mixer?

4.7D If source water temperature changes suddenly, what changes should an operator consider?

4.72 Recordkeeping

During times of abnormal plant operating conditions, good recordkeeping takes on an added importance, that of documenting the unique conditions or special events and the actions taken by the operator to solve the problem. In addition to the normal log book entry requirements, the operator must also keep accurate notes describing unusual conditions and any steps taken to prevent or resolve problems. Often, a "non-result" following an attempted corrective action is as important as a correct solution to the problem. A "non-result" helps us and others learn from our experience and next time we can try something different. Thus, ALL actions should be carefully documented by the operator.

Detailed notes may be useful in the design of future water treatment plant facilities, or in the modification of existing facilities. In addition, records such as these provide a historical account of actions taken by the operator, and may be helpful from a legal or regulatory agency perspective. Special reports may need to be prepared describing the events and actions taken. The operator's notes provide the facts for these reports.

4.73 Communications

During abnormal periods of operation, you may be required to advise other plant personnel, such as the Senior Operator, Plant Superintendent, Chemist, or Maintenance Mechanic, of the conditions that exist or events that have occurred.

An EMERGENCY RESPONSE procedure should be developed for every water treatment plant so that the proper per-

TABLE 4.4 COAGULATION-FLOCCULATION PROCESS TROUBLESHOOTING

Source Water Quality Changes	Operator Actions	Possible Process Changes
• Turbidity • Temperature • Alkalinity • pH • Color	1. Perform necessary analyses to determine extent of change. 2. Evaluate overall process performance. 3. Perform jar tests if indicated. 4. Make appropriate process changes (see right-hand column, Possible Process Changes). 5. Increase frequency of process monitoring. 6. Verify response to process changes at appropriate time (be sure to allow sufficient time for change to take effect).	1. Change coagulant(s). 2. Adjust coagulant dosage. 3. Adjust flash mixer/flocculator mixing intensity.[a] 4. Add coagulant aid or filter aid. 5. Adjust alkalinity or pH.
Coagulation Process Effluent Quality Changes	**Operator Actions**	**Possible Process Changes**
• Turbidity • Alkalinity • pH	1. Evaluate source water quality. 2. Perform jar tests if indicated. 3. Verify process performance: (a) Coagulant feed rate(s), (b) Flash mixer operation. 4. Make appropriate process changes. 5. Verify response to process changes at appropriate time.	1. Change coagulant(s). 2. Adjust coagulant dosage. 3. Adjust flash mixer intensity (if possible). 4. Adjust alkalinity or pH.
Flocculation Basin Floc Quality Changes	**Operator Actions**	**Possible Process Changes**

• Floc formation	1. Observe floc condition in basin:
	  (a) Dispersion,
	  (b) Size, and
	  (c) Floc strength (breakup).
	2. Evaluate overall process performance.
	3. Perform jar tests if indicated:
	  (a) Evaluate floc size, settling rate, and strength.
	  (b) Evaluate quality of supernatant; clarity (turbidity), pH, and color.
	4. Make appropriate process changes.
	5. Verify response to process changes at appropriate time.
		1. Change coagulant.
		2. Adjust coagulant dosage.
		3. Adjust flash mixer/flocculator mixing intensity.
		4. Add coagulant aid.
		5. Adjust alkalinity or pH.

[a] *NOTE:* Very few plants have provisions for adjusting the flash mixer. However, many plants have variable-speed drives on flocculators to allow for adjustment of mixing intensity.

sonnel can be notified quickly and the emergency resolved. Emergency response procedures should list the *NAMES AND TELEPHONE NUMBERS* (including off-hours, nights, and weekends) of persons to be notified under specific conditions, including health department authorities. Guidelines should be developed to assist the operator in determining when to use these procedures.

Coagulation-flocculation process changes or other conditions which may require notification of others include:

1. Contamination of source water with a chemical spill;

2. Major changes in source or treated water quality (pH, turbidity, alkalinity, and bacteriological quality);

3. Lack of response to changes in coagulant dosage;

4. Equipment failure (chlorinators, chemical feeders, mixers, and pumps); and

5. Power outages.

QUESTIONS

Write your answers in a notebook and then compare your answers with those on page 135.

4.7E   Why is good recordkeeping especially important during abnormal plant operating conditions?

4.7F   Why should an emergency response procedure be developed for every water treatment plant?

4.8   ENHANCED COAGULATION
by Joe Habraken

Enhanced coagulation is a process designed to remove natural organic matter (NOM) from water by adjusting both the coagulant dosage and pH to produce the greatest possible reduction of dissolved and/or suspended organic carbon (*COLOR*[26]), *TOTAL ORGANIC CARBON (TOC)*,[27] *TRIHALOMETHANES (THMs)*,[28] and *DISINFECTION BY-PRODUCTS*

(DBPs).[29] Unlike the "sweep" treatment, where the pH range is achieved by overdosing the coagulant, the optimum pH is arrived at by addition of acid or alkali. The amount of acid will depend on the amount of color and alkalinity present in the raw water. If the color is high and the alkalinity is low, the acid dosage alone may be enough to depress the pH to optimum.

The colloidal and dissolved organics found in some natural waters are the end products of decayed vegetable matter. Generally, natural waters with high organic compounds are low in turbidity (less than 10 NTU). The main constituents of these naturally occurring organic compounds are organic acids called humic substances; they are composed of humic and fulvic acids. Humic acid is a high-molecular-weight, complex macromolecule. Because of its size, the humic acid molecule is less soluble than the smaller fulvic acid molecule. Both acids exhibit *POLYANIONIC*[30] characteristics and impart a yellowish-brown color to the water.

4.80   Chemical Reactions

The fulvic and humic substances found in water are negatively charged anionic polyelectrolytes that owe their stability to their negative charge. To remove these organic acids from water, the first step is to destabilize the molecules. This is accomplished during coagulation when the organic acids react chemically with inorganic coagulant salts. The negative charge of the organic acids is neutralized and destabilized by the positively charged coagulants. In the flocculation process, which follows coagulation, the destabilized particles come together and form larger floc particles that are subsequently removed by settling.

The chemistry that deals with this particular coagulation process is known as charge chemistry. This is a branch of chemistry in which destabilization and neutralization reactions occur between stable negatively charged particles and stable positively charged particles. One of the major factors affecting charge chemistry is pH. The pH range for color removal with aluminum sulfate is 5.5 to 7.0, with the optimum pH often at 5.8. For ferric sulfate, the pH range is 4.0 to 6.2, with the optimum pH often at 4.5. Optimum pH is a function of the specific raw water to be treated, therefore optimum pH may vary. At the lower (optimum) pH, four effects take place that enhance coagulation:

1. The humic and fulvic molecules dissociate (separate) to a lesser degree at lower pH,

2. The coagulant demand decreases correspondingly to the degree of molecular dissociation,

3. Flocculation is improved at lower pH, and

4. Sulfuric acid addition prior to coagulant feed preconditions the organic compounds.

---

[26] *Color.   The substances in water that impart a yellowish-brown color to the water. These substances are the result of iron and manganese ions, humus and peat materials, plankton, aquatic weeds, and industrial waste present in the water.*

[27] *Total Organic Carbon (TOC).   TOC measures the amount of organic carbon in water.*

[28] *Trihalomethanes (THMs) (tri-HAL-o-METH-hanes).   Derivatives of methane, CH_4, in which three halogen atoms (chlorine or bromine) are substituted for three of the hydrogen atoms. Often formed during chlorination by reactions with natural organic materials in the water. The resulting compounds (THMs) are suspected of causing cancer.*

[29] *Disinfection By-Product (DBP).   A contaminant formed by the reaction of disinfection chemicals (such as chlorine) with other substances in the water being disinfected.*

[30] *Polyanionic (poly-AN-eye-ON-ick).   Characterized by many active negative charges especially active on the surface of particles.*

4.81 Process Control

Color determinations are extremely pH dependent and will always increase as the pH of the water increases. Due to this pH dependency, color determinations should be specified with pH values. Color may be determined by visual comparison or by using a spectrophotometer. Regardless of the method used, the sample must be filtered to determine *TRUE COLOR*.[31] Turbidity in any amount will cause the color to be noticeably higher. Color in a sample that has not been filtered is called *APPARENT COLOR*.[32]

The process actions described in Section 4.61 for turbidity removal are also appropriate for color removal. In operating an enhanced coagulation process the operator is attempting to achieve two interrelated objectives at the same time. One objective is to feed the optimal dosage of coagulant to remove the organic color, and the second objective is to adjust pH to the optimal zone for coagulation. Due to the fact that enhanced coagulation is pH dependent, automated pH process control is essential for plant operation. Since acid and/or alkalinity addition will most likely be required to optimize coagulation, a pH backfeed process control loop is required to maintain the proper coagulation pH. The design of the system uses a monitoring pH meter located after the rapid mix zone and feed valve controllers for acid and alkalinity chemical addition. The pH meter sends a constant signal to the chemical feed controllers and will communicate to open or close the feed valve which will increase or decrease chemical feeds respectively.

The cost-effectiveness of enhanced coagulation is directly related to the process control loop system. Generally, acid or alkali is less expensive than the coagulant, so overdosing the coagulant for pH depression (as in sweep-type treatment) is wasteful and costly. In addition to other process problems, overdosing will generate more treatment residues and will increase the concentration of the coagulant residuals in the finished water.

Optimum dosages for acid, alkalinity, and coagulant are determined by performing a series of jar tests as described in Section 4.623. The procedures listed there are used to determine the coagulant dosage. However, to predetermine the optimum acid/alkalinity dosages for pH adjustment and coagulation you must first conduct a series of jar tests following the procedures listed below. The following two jar tests illustrate typical results for determination of optimum coagulant dosage and pH using a low turbidity, high organic source water, and ferric sulfate as the coagulant. The source water characteristics are: color, 150 *pcu*[33]; alkalinity, 90 mg/L; pH, 7.3; and turbidity, 1.2 NTU.

1. Fill six 1-liter beakers with raw source water and add varying coagulant dosages based on an estimated treatment range. Mix at medium speed until coagulant is completely dispersed.

2. Adjust pH for optimal coagulation (pH of 4.5 in *TEST EXAMPLE NO. 1* below) and record milliliters of acid or alkali solution required. To depress the pH, it is best to use sulfuric acid (93 to 96%). To increase the pH, the most common chemicals used are lime and sodium hydroxide. A stock solution should be prepared so that the dose in mg/L may be calculated (see Table 4.2).

TEST EXAMPLE NO. 1

JAR NUMBER	1	2	3	4	5	6
$Fe_2(SO_4)_3$, mg/L (Dry Basis)	40	60	80	100	120	140
Acid, mg/L	60	55	50	45	40	35
Alkali, mg/L	0	0	0	0	0	0
Coagulation pH	4.5	4.5	4.5	4.5	4.5	4.5
Settled/Filtered Color	44	30	20	12	12	11

Jar No. 4 shows the optimal color removal without overdosing the coagulant. This coagulant dose should remain constant for the next jar test to determine the optimal coagulation pH. (*NOTE:* Test Example No. 1 is an attempt to try to identify both optimum pH and coagulant ranges. Some operators prefer to change only one variable at a time, and would therefore conduct separate tests for coagulant dosage and acid/alkali dosage.)

3. Once the ranges for acid and/or alkali dosages are determined, proceed with the jar testing by filling six 1-liter beakers with raw source water.

4. Add the predetermined milliliters of acid or alkali to all six beakers and mix well for precoagulation conditioning.

5. Add coagulant and follow procedure as directed in Section 4.623.

TEST EXAMPLE NO. 2

JAR NUMBER	1	2	3	4	5	6
$Fe_2(SO_4)_3$, mg/L (Dry Basis)	100	100	100	100	100	100
Acid, mg/L	54	48	44	40	35	30
Alkali, mg/L	0	0	0	0	0	0
Coagulation pH	4.0	4.4	4.8	5.2	5.6	6.0
Settled/Filtered Color	16	12	12	14	15	16

Taken together, the results of the two example tests indicate that a coagulant dose of 100 mg/L and an acid dose of 44 to 48 mg/L provide an optimum coagulation pH of 4.4 to 4.8.

6. To pinpoint the optimal pH, run another jar test varying the coagulation pH within the coagulation pH range (pH 4.4 to 4.8) while keeping the coagulant dose constant. The coagulant dose selected should be taken from the prior jar test with the dose that had the best color removal without overdosing.

To further fine-tune the coagulant dose, you could perform another jar test and try variations of coagulant dosage at the optimum pH. In summary, you have to conduct a series of tests to determine *BOTH* optimum pH and coagulant dose.

4.82 Process Actions Under Varying Conditions

Once the optimum pH is determined the operator can use this information to make process changes to fit source water changes. Changes to the source water that will require adjustments are the increase or decrease in color or alkalinity. Considering that the alkalinity demand for coagulation is greater for aluminum salts than for iron salts, there could be the possibility

---

[31] *True Color.* Color of the water from which turbidity has been removed. The turbidity may be removed by double filtering the sample through a Whatman No. 40 filter when using the visual comparison method.

[32] *Apparent Color.* Color of the water that includes not only the color due to substances in the water but suspended matter as well.

[33] *pcu (Platinum Cobalt Units).* Platinum cobalt units are a measure of color using platinum cobalt standards by visual comparison.

that total source water alkalinity will need to be increased to meet the coagulant alkalinity demand prior to coagulant addition optimization.

Table 4.5 summarizes enhanced coagulation process problems and suggests possible corrective actions.

QUESTIONS

Write your answers in a notebook and then compare your answers with those on page 135.

4.8A   What is the main difference between enhanced coagulation and "sweep" type coagulation methods?

4.8B   What is the relationship between pH and color in a water sample?

4.8C   How does turbidity affect the measurement of color in a water sample?

4.8D   How are the optimum dosages of acid, alkalinity, and coagulant determined for enhanced coagulation?

4.9   START-UP AND SHUTDOWN PROCEDURES

4.90   Conditions Requiring Implementation of Start-Up/Shutdown Procedures

Start-up or shutdown of the coagulation-flocculation process is *NOT* a routine operating procedure in most water treatment plants. These procedures generally happen when the plant is shut down for maintenance. In some rare instances, a shutdown may be required due to a major equipment failure.

4.91   Implementation of Start-Up/Shutdown Procedures

Typical actions performed by the operator in the start-up or shutdown of the coagulation-flocculation process are outlined below. These procedures may have to be altered depending on the type of equipment in your plant and on the recommendations of the manufacturers of your equipment.

4.910   Start-Up Procedures

1. Check the condition of all mechanical equipment for proper lubrication and operational status.

2. Make sure all chemical feeders are ready. There should be plenty of chemicals available in the tanks and hoppers and ready to be fed to the raw water.

3. Collect a sample of raw water and immediately run a jar test using fresh chemicals from the supply of chemicals to the feeders.

4. Determine the settings for the chemical feeders and set the feed rates on the equipment.

5. Open the inlet gate or valve to start the raw water flowing.

6. *IMMEDIATELY* start the selected chemical feed systems.

   a. Open valves to start feeding coagulant chemicals and dilution makeup water.

   b. Start chemical feeders.

   c. Adjust chemical feeders as necessary.

7. Turn on the flash mixer at the appropriate time. You may have to wait until the tank or channel is full before turning on the flash mixer. Follow the manufacturer's instructions.

NOTE: *DO NOT ALLOW ANY UNTREATED WATER TO FLOW THROUGH YOUR PLANT.* All raw water must be treated with alum or other appropriate coagulant. Water that has not been treated with a coagulant could flow through your filter without proper treatment (removal of color and particulates) and into your distribution system.

8. Start the sample pumps as soon as there is water at each sampling location. Allow sufficient flushing time before collecting any samples.

TABLE 4.5   ENHANCED COAGULATION PROCESS TROUBLESHOOTING

Treatment Condition Flocculator Effluent	Corrective Action
High coagulation pH with optimum color removal	1. Increase acid feed 2. Decrease alkalinity adjustment in raw water source
High coagulation pH without optimum color removal	1. Increase coagulant 2. Decrease acid feed to maintain optimum pH
Low coagulation pH with optimum color removal	1. Decrease acid feed 2. Increase alkalinity adjustment to raw water source
Low coagulation without optimum color removal	1. Decrease acid if below optimal pH zone 2. Increase coagulant and alkalinity
Loss of acid feed	1. Increase coagulant to achieve optimal pH
Optimal pH without optimized color removal	1. Increase coagulant, decrease acid, or increase alkalinity
Optimal pH and color removal with floc carryover	1. Decrease coagulant 2. Increase polymer 3. Increase removal of settled floc 4. Decrease flow-through velocities in treatment unit
High turbidities and coagulant residuals in settled water	1. Check for floc carryover 2. Adjust polymer feed to enhance settling 3. Jar test to determine optimum acid and coagulant dosage

9. Start the flocculators as soon as the first basin is full of water. Be sure to follow manufacturer's recommendations. If possible and appropriate, make any necessary adjustments in the speed.

10. Inspect mixing chamber and flocculation basins. Observe formation of floc and make any necessary changes.

11. Remove any debris floating on the water surface.

12. Perform water quality analyses and make process adjustments as necessary.

13. Calibrate chemical feeders.

4.911 Shutdown Procedures

1. Close raw water inlet gate or valve to flash-mix chamber or channel.

2. Shut down the chemical feed systems.

   a. Turn off chemical feeders.

   b. Shut off appropriate valves.

   c. Flush or clean chemical feed lines if necessary.

3. Shut down flash mixer and flocculators as water leaves each process. Follow manufacturer's recommendations.

4. Shut down sample pumps before water leaves each sampling location.

5. Waste any water that has not been properly treated.

6. Lock out and tag appropriate electric switches.

7. Dewater basins if necessary. Waste any water that has not been properly treated.

   DO NOT DEWATER BELOW-GROUND BASINS WITHOUT CHECKING GROUNDWATER LEVEL.

   a. Close basin isolation gates or install stop-logs.

   b. Open basin drain valve(s).

   c. Be careful; basin may float or collapse depending on groundwater, soil, or other conditions.

4.92 Recordkeeping

Good records of actions taken during start-up/shutdown operations will assist the operator, as well as other plant operations and maintenance personnel, in conducting future shutdowns. The results of all inspections, equipment adjustments, and any unusual events should be accurately recorded.

4.93 Safety Considerations

Safety procedures are extremely important during start-up and shutdown operations. In the coagulation-flocculation process you will be exposed to a variety of potentially hazardous situations which require the use of extreme caution. The operator may be exposed to electrical hazards, rotating and mechanical equipment, open-surface, water-filled basins (drowning), and empty basins. Also, the bottoms of empty basins are very slippery; a fall could be extremely painful and cause a serious injury. Always use safety devices such as handrails. For example, don't remove handrails to make your job easier.

In some instances, you may be required to make repairs in underground structures such as valve or pump vaults where you may be exposed to toxic and explosive gases or insufficient oxygen. Always make sure that these areas are properly ventilated. Use proper safety equipment, such as hard hats, goggles, rubber boots, gas detectors, and life jackets, when necessary. Never enter a confined area or space alone. Be sure the person watching you has additional help standing by and knows how to evacuate you if you are injured or lose consciousness.

DON'T TAKE ANY CHANCES.

QUESTIONS

Write your answers in a notebook and then compare your answers with those on page 135.

4.9A Under what conditions are coagulation-flocculation processes normally shut down?

4.9B Why is good recordkeeping important during start-up/shutdown operations?

4.9C What safety hazards could be encountered during the start-up or shutdown of a coagulation-flocculation process?

4.10 LABORATORY TESTS

4.100 Process Control Water Quality Indicators

In the operation of the coagulation-flocculation process, the operator will perform a variety of laboratory tests to monitor source water quality and to evaluate process performance. Process control water quality indicators of importance in the operation of the coagulation-flocculation process include turbidity, alkalinity, chlorine demand, color, pH, temperature, odor, and appearance.

4.101 Sampling Procedures

Process water samples will be either grab samples obtained directly from a specific process monitoring location, or continuous samples which are pumped to the laboratory from various locations in the process (raw water, flash mixer effluent, flocculation basin effluent), as shown in Figure 4.8 on page 108. In either case, it is important to emphasize that process samples must be a *REPRESENTATIVE SAMPLE*[34] of actual conditions in the treatment plant. The accuracy and usefulness of laboratory analyses depend on the representative nature of the water samples.

---

[34] *Representative Sample. A sample portion of material or water that is as nearly identical in content and consistency as possible to that in the larger body of material or water being sampled.*

The frequency of sampling for individual process control water quality indicators will vary, depending on the quality of the source water. Certain water quality indicators, such as turbidity, will be routinely monitored, while others, such as alkalinity, are sampled less frequently.

Process grab samples should be collected in clean plastic or glass containers and care should be used to avoid contamination of the sample, especially turbidity and odor samples.

Samples should be analyzed as soon as possible after the sample is collected. Important water quality indicators, such as turbidity, temperature, chlorine demand, color, odor, pH, and alkalinity, can all change while waiting to be analyzed.

4.102   Sample Analysis

Analysis of certain process control water quality indicators, such as turbidity and pH, can be readily performed in the laboratory with the aid of automated analytical instruments, such as turbidimeters and pH meters. Analysis of other process control water quality indicators, such as alkalinity and chlorine demand, may require WET CHEMISTRY[35] procedures which are often performed by a chemist or laboratory technician. A more detailed discussion of sampling and laboratory analytical procedures is contained in Volume II, Chapter 21, "Advanced Laboratory Procedures."

One of the most important laboratory procedures in the operation of the coagulation-flocculation process is the JAR TEST PROCEDURE. As discussed earlier, this procedure is performed to establish the proper type and amount of chemical(s) to be used in the coagulation or flash-mixing process.

4.103   Safety Considerations

Laboratory work may expose the operator to a number of different safety hazards. Care should be exercised in the handling of reagents and glassware. Use protective clothing (safety glasses and aprons) while performing wet chemical analyses, especially when handling dangerous chemicals, such as acid or caustic solutions. Always perform lab tests in a well-ventilated space, and be familiar with the location and use of safety showers and eye wash facilities.

4.104   Recordkeeping

Record all laboratory test results on appropriate data sheets, and document in detail any unusual results.

Upon verification of abnormal test results, be sure to notify the proper personnel of your findings.

QUESTIONS

Write your answers in a notebook and then compare your answers with those on page 135.

4.10A   List the process control water quality indicators of importance in the operation of the coagulation-flocculation process.

4.10B   When should sample analysis be performed to control the coagulation-flocculation process?

4.10C   What safety hazards may be encountered when working in a laboratory?

4.11   PROCESS AND SUPPORT EQUIPMENT OPERATION AND MAINTENANCE

4.110   Types of Equipment

In the operation of the coagulation-flocculation process, the operator will be exposed to a variety of mechanical, electrical, and instrumentation equipment including:

1. Mixers and flocculators,

2. Chemical feeders,

3. Water quality monitors,

4. Pumps,

5. Valves,

6. Flowmeters and gages, and

7. Control systems.

In the coagulation-flocculation process itself, chemical feeders (Figure 4.12) are of particular importance. Chemicals are normally fed at a fixed rate. This can be accomplished by liquid feed (solution) or by dry feed (volumetric or gravimetric). In liquid feed, a diluted solution of known concentration is prepared and fed directly into the water being treated. Liquid chemicals are fed through metering pumps and rotameters. Dry feeders deliver a measured quantity of dry chemical during a specified time interval. VOLUMETRIC FEEDERS deliver a specific volume of chemical during a given time interval, while GRAVIMETRIC FEEDERS deliver a predetermined weight of chemical in a specific unit of time. Generally, volumetric feeders can deliver smaller daily quantities of chemicals than gravimetric feeders, but the performance variables are:

1. Volumetric feeders are simpler and of less expensive construction, and

2. Gravimetric feeders are usually more easily adapted for recording the actual quantities of chemicals fed and for automatic control. For this reason, gravimetric feeders are generally used in large treatment plants.

Water treatment plants should have duplicate chemical feeders. This will permit the operator to maintain full service while a chemical feeder is "off-line" for routine maintenance or major repair. If this feature is not included in your plant, then consider requesting backup equipment as part of the annual operation, maintenance, and repair budget request process.

The ultimate decision of which chemical feeder to use for a given application depends on the type of chemical compound, availability of chemical, chemical form (dry or liquid), and the amount to be fed daily.

4.111   Equipment Operation

Before starting a piece of mechanical equipment, such as a mixer or chemical feeder, be sure that the unit is properly lubricated and its operational status is known. Also be certain that no one is working on the equipment. Be sure all valves are in the proper position before starting chemical feeders.

After start-up, always check for excessive noise and vibration, overheating, and leakage (water, lubricants, and chemi-

---

[35] Wet Chemistry.   Laboratory procedures used to analyze a sample of water using liquid chemical solutions (wet) instead of, or in addition to, laboratory instruments.

Dry chemical makeup and feed system (polymer)

Solution chemical feed system (alum)
(Note graduated cylinder for calibration in right foreground.
The cylinder is about half full.)

Fig. 4.12   Chemical feeders

cals). When in doubt about the performance of a piece of equipment, always refer to operation and maintenance instructions or the manufacturer's technical manual.

Many equipment items, such as valves and mixers, are simple ON/OFF devices with some provision for either speed or position adjustment. Other equipment items, such as pumps and chemical feeders, may require the use of special procedures for priming and calibration. Detailed operating and repair procedures are usually given in the plant operations and maintenance instructions for specific pieces of equipment.

During the course of normal operation, equipment should be periodically inspected for noise and vibration, leakage, overheating, or other signs of abnormal operation. Electric motors should always be kept free of dirt, moisture, and obstructions to their ventilation openings.

QUESTIONS

Write your answers in a notebook and then compare your answers with those on page 135.

4.11A   What types of equipment are used in connection with the coagulation-flocculation process?

4.11B   How do chemical liquid feeders work in the coagulation process?

4.11C   Selection of a chemical feeder for a given application depends on what factors?

4.112   Safety Considerations

When working around electrical equipment, such as motors, mixers, or flocculators, follow the safety procedures listed below to avoid accidents or injury.

ELECTRICAL EQUIPMENT

1. Always shut off power and lock out and attach safety tag before working on electrical equipment, instruments, con-

trols, wiring, and all mechanical equipment driven by electric motors,

2. Avoid electric shock by using protective gloves,

3. Use a multimeter to test for "live" wires and equipment,

4. Check grounds and avoid grounding yourself in water or on pipes,

5. Ground all electric tools, and

6. Use the buddy system.

MECHANICAL EQUIPMENT

1. Use protective guards on rotating equipment,

2. Do not wear loose clothing, worn gloves, or long hair around rotating equipment, and

3. Clean up all lubricant spills (oil and grease).

OPEN WATER SURFACE STRUCTURES

1. Do not avoid or defeat protective devices, such as handrails, by removing them when they are in the way,

2. Close all openings when finished, and

3. Know the location of all life preservers and wear one when necessary.

VALVE AND PUMP VAULTS

1. Be sure all underground structures are free of hazardous atmospheres (toxic and explosive gases or insufficient oxygen) by using gas detectors,

2. Only work in well-ventilated structures, and

3. Use the buddy system.

HANDLING ALUM

Normal precautions should be used to prevent spraying or splashing of liquid alum, especially when the liquid is hot. Face shields can be worn to protect your eyes. A rubber apron and waterproof sleeves may be used to protect your clothing.

Both liquid alum and dry alum dust must be flushed from your eyes immediately. Use great amounts of warm water. Alum must be washed from your skin because prolonged contact can be irritating.

For more details on safety, see Volume II, Chapter 20, "Safety."

4.113 Preventive Maintenance Procedures

Preventive maintenance programs are designed to ensure the satisfactory long-term operation of treatment plant facilities under a variety of different operating conditions. Scheduled or routine maintenance of valves, mixers, pumps, and chemical feeders is an important part of the preventive maintenance program. The operator will be expected to perform routine maintenance functions as part of an overall preventive maintenance program. Typical functions include:

1. Keeping motors free of dirt and moisture,

2. Ensuring good ventilation (air circulation) in equipment work areas,

3. Checking pumps for leaks, unusual noise, vibrations, or overheating,

4. Maintaining proper lubrication and oil levels,

5. Inspecting for alignment of shafts and couplings,

6. Checking bearings for wear, overheating, and proper lubrication,

7. Exercising infrequently used valves on a regular schedule and checking all valves for proper operation, and

8. Calibrating flowmeters and chemical feeders.

Good recordkeeping is the key to a successful preventive maintenance program. These records provide maintenance and operations personnel with clues for determining the causes of equipment breakdowns, and will often show weaknesses in a particular piece of equipment which can be corrected prior to failure.

For more details on maintenance, see Volume II, Chapter 18, "Maintenance."

QUESTIONS

Write your answers in a notebook and then compare your answers with those on page 135.

4.11D   Underground structures could present what types of hazards?

4.11E   What equipment should be part of a preventive maintenance program for a coagulation-flocculation process?

4.12 ARITHMETIC ASSIGNMENT

Turn to the Appendix, "How to Solve Water Treatment Plant Arithmetic Problems," at the back of this manual and read all of Section A.4, *METRIC SYSTEM*.

In Section A.13, *TYPICAL WATER TREATMENT PLANT PROBLEMS*, read and work the problems in the following sections:

1. A.130, Flows,

2. A.131, Chemical Doses,

3. A.132, Reservoir Management and Intake Structures, and

4. A.133, Coagulation and Flocculation.

4.13 ADDITIONAL READING

1. *NEW YORK MANUAL*, Chapter 4,* "Water Chemistry," and Chapter 7,* "Chemical Coagulation."

2. *TEXAS MANUAL*, Chapter 6,* "Water Chemistry," and Chapter 8,* "Coagulation and Sedimentation."

* Depends on edition.

4.14 ACKNOWLEDGMENTS

Many of the concepts and procedures discussed in this chapter are based on material obtained from the sources listed below.

1. Stone, B.G. Notes from "Design of Water Treatment Systems," CE-610, Loyola Marymount University, Los Angeles, CA, 1977.

2. *WATER QUALITY AND TREATMENT*, Fifth Edition. Obtain from American Water Works Association (AWWA), Bookstore, 6666 West Quincy Avenue, Denver, CO 80235. Order No. 10008. ISBN 0-07-001659-3. Price to members, $119.00; nonmembers, $135.00; price includes cost of shipping and handling.

3. *WATER TREATMENT PLANT DESIGN*, Third Edition, prepared jointly by the American Society of Civil Engineers, American Water Works Association, and Conference of State Sanitary Engineers. Obtain from American Water Works Association (AWWA), Bookstore, 6666 West Quincy Avenue, Denver, CO 80235. Order No. 10009. ISBN 0-07-001643-7. Price to members, $103.50; nonmembers, $135.00; price includes cost of shipping and handling.

4. *OPERATION AND MAINTENANCE MANUAL FOR STOCKTON EAST WATER TREATMENT PLANT*, James M. Montgomery, Consulting Engineers, Inc., 501 Lennon Lane, Walnut Creek, CA 94598, 1979. No longer available.

End of Lesson 3 of 3 Lessons
on
Coagulation and Flocculation

Please answer the discussion and review questions next.

DISCUSSION AND REVIEW QUESTIONS

Chapter 4.   COAGULATION AND FLOCCULATION

(Lesson 3 of 3 Lessons)

Write the answers to these questions in your notebook. The question numbering continues from Lesson 2.

16. What would you do if you discovered abrupt (sudden) changes in the source water or filtered water turbidity?

17. What changes in the coagulation-flocculation process would you consider if a change in source water quality caused water quality problems in the filtered water?

18. Which coagulation-flocculation process changes may require notification of others?

19. How does the operator of an enhanced coagulation process adjust the pH of the water to achieve optimal coagulation?

20. Why must representative samples be collected?

SUGGESTED ANSWERS

Chapter 4.   COAGULATION AND FLOCCULATION

ANSWERS TO QUESTIONS IN LESSON 1

Answers to questions on page 97.

4.0A   The purpose of coagulation and flocculation is to remove particulate impurities and color from the water being treated.

4.1A   In the *COAGULATION PROCESS*, chemicals are added that will cause the particles to begin to clump together. The particles gather together to form larger, settleable particles in the *FLOCCULATION PROCESS*.

Answers to questions on page 102.

4.2A   The primary purpose of flash mixing is to rapidly mix and uniformly (evenly) distribute the coagulant chemical throughout the water.

4.2B   *PRIMARY COAGULANTS* are used to neutralize the electrical charge of the particles and cause the particles to clump together. *COAGULANT AIDS* are used to add density to slow-settling flocs and to add toughness so the floc will not break up in subsequent processes.

4.2C   Four methods of mixing coagulant chemicals include:

1. Hydraulic mixing,
2. Mechanical mixing,
3. Diffusers and grid systems, and
4. Pumped blenders.

4.2D   Hydraulic mixing devices rely on the turbulence created by flowing water to mix chemicals with the water. Baffles or throttling valves are hydraulic mixers.

Answers to questions on page 103.

4.3A   Flocculation is a slow stirring process that causes the gathering together of small, coagulated particles into larger, settleable floc particles.

4.3B   The typical mixing time in the coagulation process is several seconds.

4.3C   The minimum detention time recommended for flocculation ranges from about 5 to 20 minutes for direct filtration systems and up to 30 minutes for conventional filtration.

4.3D   An advantage of vertical flocculators over horizontal ones is that vertical flocculators usually require less maintenance since they eliminate submerged bearings and packings.

4.3E   The compartments in flocculation basins are separated by baffles to prevent short-circuiting of the water being treated, and to reduce the level of turbulence in each succeeding compartment by reducing the speed of the stirrers. This is called tapered-energy mixing.

Answers to questions on page 105.

4.4A   Coagulation-flocculation influences the effectiveness of the sedimentation, filtration, and disinfection processes. The removal of particulates by the sedimentation and filtration processes depends on effective coagulation-flocculation. Coagulation-flocculation causes bacteria and other disease-causing organisms to be bound up in suspended solids and floc. If not removed by sedimentation and/or filtration, bacteria may be shielded from disinfection and released into the distribution system.

4.5A   The most important consideration in coagulation-flocculation process control is the selection of the proper type and amount of coagulant chemical(s) to be applied to the water being treated.

ANSWERS TO QUESTIONS IN LESSON 2

Answers to questions on page 108.

4.6A   Sedimentation and filtration are used to remove suspended solids after the coagulation-flocculation process.

4.6B   The effectiveness of the solids removal processes is commonly monitored by measuring the turbidity of filtered water.

4.6C   Typical functions performed by an operator in the normal operation of the coagulation-flocculation process include the following:

1. Monitor process performance,
2. Evaluate water quality conditions,
3. Check and adjust process controls and equipment, and
4. Visually inspect facilities.

4.6D   The most common laboratory tests used to monitor the coagulation-flocculation process are turbidity, alkalinity, temperature, color, pH, and chlorine demand.

4.6E   Visual observations of the coagulation-flocculation process generally include observing the degree of agitation of the water in the flash-mixing channel or chamber, and observing the size and distribution of floc in the flocculation basin. An uneven distribution of floc could be an indication of short-circuiting in the flocculation basin, while floc particles that are too small or too large may be difficult to remove in subsequent processes.

Answers to questions on page 111.

4.6F   Water flowing through the typical treatment plant after the flash mixer flows through the flocculation, sedimentation, and filtration processes.

4.6G   Many operators prepare curves of flow vs. detention time for the basins in their plant. These curves allow for easy selection of stirring times when performing jar tests.

4.6H   **Known**   **Unknown**

Flash Mix Chamber   Detention Time, sec

Length, ft   = 6 ft

Width, ft   = 4 ft

Depth, ft   = 5 ft

Flow, MGD   = 10 MGD

1. Calculate the volume of the flash-mix chamber in cubic feet and then in gallons.

$$\text{Volume, cu ft} = (\text{Length, ft})(\text{Width, ft})(\text{Depth, ft})$$
$$= (6 \text{ ft})(4 \text{ ft})(5 \text{ ft})$$
$$= 120 \text{ cu ft}$$

$$\text{Volume, gal} = (120 \text{ cu ft})(7.48 \text{ gal/cu ft})$$
$$= 898 \text{ gal}$$

2. Calculate the detention time in seconds.

$$\text{Detention Time, sec} = \frac{(\text{Volume, gal})(24 \text{ hr/day})(60 \text{ min/hr})(60 \text{ sec/min})}{\text{Flow, gal/day}}$$
$$= \frac{(898 \text{ gal})(24 \text{ hr/day})(60 \text{ min/hr})(60 \text{ sec/min})}{10,000,000 \text{ gal/day}}$$
$$= 7.7 \text{ sec}$$

4.6I   Jar test reagents can be prepared by using the actual chemical coagulants used to treat the water, or they may be obtained from a chemical supplier in some cases.

Answers to questions on page 118.

4.6J   Only *ONE* variable at a time should be changed when running a jar test so that all observed changes can be attributed to the one variable.

4.6K   A jar test sample should not be collected in advance because the temperature of the water being tested should be approximately the same as the temperature of the water being treated.

4.6L   When evaluating the results of jar tests, the following factors should be considered:

1. Rate of floc formation,
2. Type of floc particles,
3. Clarity of water between floc particles,
4. Size of floc,
5. Amount of floc formed,
6. Floc settling rate, and
7. Clarity of water above settled floc.

Answers to questions on page 122.

4.6M   Operators should keep accurate records in order to obtain a running account of operations (historical records); to have a source of information for assistance in solving current or future process problems; and to comply with regulatory requirements.

4.6N   All record entries should include the date, time of an event, and initials of the operator making the entry.

4.6O   The safety hazards an operator may encounter when operating a coagulation-flocculation process include:

1. Electrical equipment,
2. Rotating mechanical equipment,
3. Water treatment chemicals,
4. Laboratory reagents (chemicals),
5. Slippery surfaces caused by wet polymers,
6. Open-surface, water-filled structures (drowning), and
7. Confined spaces and underground structures, such as valve or pump vaults (toxic and explosive gases, insufficient oxygen).

4.6P   Good communications are essential to advise other operators and support personnel of current process conditions and unusual events.

ANSWERS TO QUESTIONS IN LESSON 3

Answers to questions on page 124.

4.7A   Sudden changes in either raw or filtered water turbidity, pH, alkalinity, temperature, or chlorine demand are signals that the operator should immediately review the performance of the coagulation-flocculation process.

4.7B   Significant changes in source water turbidity levels, either increases or decreases, generally require that the operator verify the effectiveness of the coagulant chemicals and dosages being applied at the flash mixer.

4.7C   The best way to verify the effectiveness of the coagulant chemicals and dosages is by performing a series of jar tests in the laboratory to simulate process performance in the treatment plant.

4.7D   Sudden water temperature changes may require coagulant dosage changes and an adjustment in the level of mixing intensity in the flash mixers or flocculators. However, temperature changes are usually gradual changes over time so large process adjustments are seldom necessary.

Answers to questions on page 126.

4.7E   Good recordkeeping is especially important during abnormal plant operating conditions to document the unique or special event and the actions taken by the operator to solve the problem. These records also may be helpful from a legal or regulatory agency perspective.

4.7F   An emergency response procedure should be developed for every water treatment plant so that notification of the proper personnel can be accomplished and the emergency resolved.

Answers to questions on page 128.

4.8A   In the "sweep" type coagulation process, the optimum pH is arrived at by overdosing with coagulant chemicals. In enhanced coagulation, optimum pH is achieved by the addition of acid or alkali.

4.8B   Color determinations are extremely pH dependent and will always increase as the pH of the water increases.

4.8C   Any turbidity in a water sample will increase the measured level of color.

4.8D   Optimum dosages of acid, alkalinity, and coagulant for enhanced coagulation are determined by running a series of jar tests.

Answers to questions on page 129.

4.9A   Coagulation-flocculation processes are normally shut down when the treatment plant is shut down for periodic maintenance and only rarely due to a major process failure.

4.9B   Good documentation of actions taken during start-up/shutdown operations will assist the operator, as well as other plant operations and maintenance personnel, in conducting future shutdowns.

4.9C   Safety hazards that could be encountered during the start-up or shutdown of a coagulation-flocculation process include electric shocks, rotating mechanical equipment, and open-surface, water-filled basins (drowning). Also the bottoms of empty basins are very slippery and a fall could be extremely painful and cause serious injury. Underground structures, such as valve or pump vaults, could contain toxic or explosive gases or insufficient oxygen.

Answers to questions on page 130.

4.10A   The process control water quality indicators of importance in the operation of the coagulation-flocculation process include turbidity, temperature, alkalinity, chlorine demand, color, pH, odor, and appearance.

4.10B   Sample analysis should be performed immediately following sample collection.

4.10C   Safety hazards encountered in the laboratory include the handling of reagents and glassware.

Answers to questions on page 131.

4.11A   The types of equipment used in connection with the coagulation-flocculation process include mixers and flocculators, chemical feeders, water quality monitors, pumps, valves, flowmeters, gages, and control systems.

4.11B   Chemical liquid feeders feed a solution of known concentration directly into the water being treated.

4.11C   Selection of a chemical feeder for a given application depends on the type of chemical compound, availability of chemical, chemical form (dry or liquid), and the amount to be fed daily.

Answers to questions on page 132.

4.11D   Underground structures could present hazardous atmospheres containing toxic or explosive gases or insufficient oxygen.

4.11E   Equipment that should be part of a preventive maintenance program includes valves, mixers, pumps, and chemical feeders.

APPENDIX

A. Preparation of Detention Time Curves
B. Adjustment and Calibration of Chemical Feeders

A. PREPARATION OF DETENTION TIME CURVES

Example 2 (page 110) was a water treatment plant with a flow of 2.4 MGD. The flocculation basin was 8 feet deep, 15 feet wide, and 45 feet long. We calculated a basin volume of 40,392 gallons and a detention time of 24 minutes. The typical flows for this plant range from 0.8 MGD to 2.4 MGD. To plot a detention time curve, we should have at least four plotting points. This means that we must calculate the detention times for at least four flows. For this example we will calculate the detention times for flows of 0.8, 1.2, 1.6, 2.0, and 2.4 MGD. This will provide us with five plotting points.

Known		**Unknown**
Basin Volume, gal	= 40,392 gal	Detention Time, min,
Flows, MGD	= 0.8, 1.2,	for each flow
	1.6, 2.0, and	
	2.4 MGD	

1. Calculate the detention time in minutes for each flow.

$$\text{Detention Time, min} = \frac{(\text{Volume, gal})(24 \text{ hr/day})(60 \text{ min/hr})}{\text{Flow, gal/day}}$$

$$\text{Detention Time, min for 0.8 MGD} = \frac{(40,392 \text{ gal})(24 \text{ hr/day})(60 \text{ min/hr})}{800,000 \text{ gal/day}}$$

$$\text{or} = \frac{58,164,480^{36}}{800,000 \text{ gal/day}}$$

$$= 73 \text{ min}$$

$$\text{Detention Time, min for 1.2 MGD} = \frac{58,164,480}{1,200,000 \text{ gal/day}}$$

$$= 48 \text{ min}$$

$$\text{Detention Time, min for 1.6 MGD} = \frac{58,164,480}{1,600,000 \text{ gal/day}}$$

$$= 36 \text{ min}$$

$$\text{Detention Time, min for 2.0 MGD} = \frac{58,164,480}{2,000,000 \text{ gal/day}}$$

$$= 29 \text{ min}$$

$$\text{Detention Time, min for 2.4 MGD} = \frac{58,164,480}{2,400,000 \text{ gal/day}}$$

$$= 24 \text{ min}$$

2. Summarize the calculated plotting points.

Flow, MGD	Detention Time, min
0.8	73
1.2	48
1.6	36
2.0	29
2.4	24

3. Plot the flow vs. detention time curve as shown in Figure 4.13. This curve can be used to determine the detention time for any flow. If you know the flow you are treating, find the flow on Figure 4.13, move across to the curve, and then down to the detention time scale.

B. ADJUSTMENT AND CALIBRATION OF CHEMICAL FEEDERS

The capacity rating of solution chemical feeders is usually given in units of gallons per minute (GPM) or gallons per hour (GPH), while dry feeders are often rated by the maximum amount of chemical that can be fed in a 24-hour period (pounds per day).

Adjusting or changing the amount of chemical to be fed is generally accomplished by manually changing the feed rate setting on the chemical feeder. Adjustment is physically performed by turning a knob, adjusting a wheel, or by rotating a hand-crank.

---

[36] 58,164,480 will be a constant for all calculations. You can determine this constant for your plant.

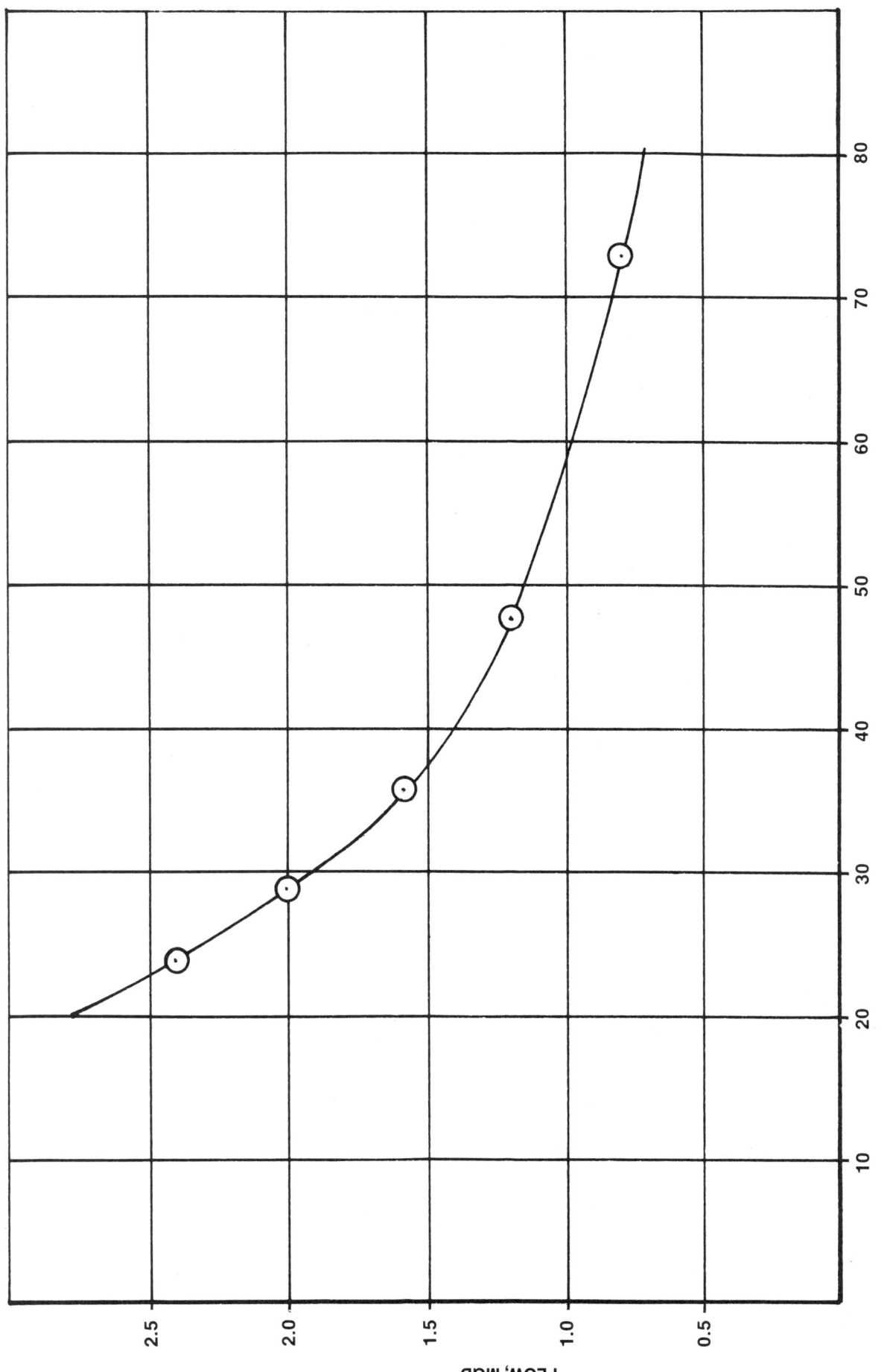

Fig. 4.13   Development of flow vs. detention time curve

Typically, a feed-rate scale is provided on the chemical feeder which is calibrated over a range from zero to 100 percent of maximum feed rate, for both solution and dry chemical feeders. In a solution feed system, if the desired feed rate is 3 GPH and the chemical feeder has a maximum feed rate of 15 GPH, then the feeder would be set at:

$$\text{Scale Setting, \%} = \frac{\text{(Desired Feed Rate, GPH)(100\%)}}{\text{Maximum Feed Rate, GPH}}$$

$$= \frac{\text{(3 GPH)(100\%)}}{\text{15 GPH}}$$

$$= 20\% \text{ of full setting}$$

Likewise, in a dry feed system, if the desired feed rate is 150 pounds per day and the chemical feeder has a maximum feed rate of 300 pounds per day, then the feeder would be set at:

$$\text{Feeder Setting, \%} = \frac{\text{(Desired Feed Rate, lbs/day)(100\%)}}{\text{Maximum Feed Rate, lbs/day}}$$

$$= \frac{\text{(150 lbs/day)(100\%)}}{\text{300 lbs/day}}$$

$$= 50\% \text{ of full scale}$$

Chemical feed systems should be calibrated at least once per shift to verify proper chemical feed rate. Always calibrate chemical feeders against working system pressures to avoid errors. In liquid chemical feed systems, the *VOLUMETRIC METHOD* is probably the most accurate calibration technique. This method involves the use of a calibrated container (usually a graduated cylinder as shown in Figure 4.12, page 131) and a stopwatch to determine the volume of chemical fed during a given time period. Ideally the cylinder and timer are part of the chemical feeder piping system (Figure 4.12). This procedure can, of course, be used to calibrate the chemical feed pump over the full range of feed rates.

To apply the procedure, select an appropriate time period such as 30 to 90 seconds. The time period should be increased when measuring dilute chemical solutions to ensure accurate results. Fill the graduated container with a convenient amount of the chemical solution. Insert one end of a tube into the container with the chemical and attach the other end to the feeder inlet on the suction side of the feeder. Start the feeder and the stopwatch. After a minimum time period (for example, 30 seconds), read the graduation mark on the container that corresponds to the liquid level drawdown. Record the total elapsed time. The following example uses a one-liter graduated cylinder.

EXAMPLE 13

A chemical feeder draws a liquid chemical from a one-liter (1,000-mL) graduated cylinder for 30 seconds. At the end of 30 seconds, the graduated cylinder has 400 mL remaining. What is the chemical feed rate in milliliters per minute and in gallons per minute (GPM)?

Known	**Unknown**
One liter of chemical	Chemical Feed Rate, mL/min and GPM
Starting Level, mL = 1,000 mL	
Final Level, mL = 400 mL	
Feed Time, sec = 30 sec	

1. Determine volume of chemical fed in milliliters.

$$\text{Chemical Fed, mL} = \text{Starting Level, mL} - \text{Final Level, mL}$$
$$= 1,000 \text{ mL} - 400 \text{ mL}$$
$$= 600 \text{ mL}$$

2. Determine chemical feed rate, mL/min.

$$\frac{\text{Chemical Feed}}{\text{Rate, mL/min}} = \frac{\text{Chemical Fed, mL}}{\text{Feed Time, min}}$$

$$= \frac{\text{(600 mL)(60 sec/min)}}{\text{30 sec}}$$

$$= 1,200 \text{ mL/min}$$

3. Calculate chemical feed rate in GPM.

$$\frac{\text{Chemical Feed}}{\text{Rate, GPM}} = \frac{\text{Chemical Feed Rate, mL/min}}{\text{3,785 mL/gal}}$$

$$= \frac{\text{1,200 mL/min}}{\text{3,785 mL/gal}}$$

$$= 0.32 \text{ GPM}$$

Compare your calculated value with the setting on the chemical feeder and any calibration tables you may have. If the values do not agree, recheck your work and then adjust the feeder to deliver the correct amount of chemical. If the chemical feeder has a setting (say, 1, 2, 3), record the chemical feed rate as 0.32 GPM for the actual setting. Adjust the feeder to another setting (higher or lower) and repeat the test to determine the feed rate at the new setting.

Most chemical feed systems are not furnished with volumetric calibration accessories. However, this feature can be readily added to existing systems by purchasing a standard laboratory-grade graduated cylinder (preferably plastic), a stopwatch, and installing the necessary piping and valving on the pump suction. Figure 4.14 is a sketch of a typical installation. Another approach is to open a sample tap and pump directly into a graduated cylinder.

Procedures similar to those described above can be used for dry feed applications by measuring the dry weight of chemical fed during a given time period.

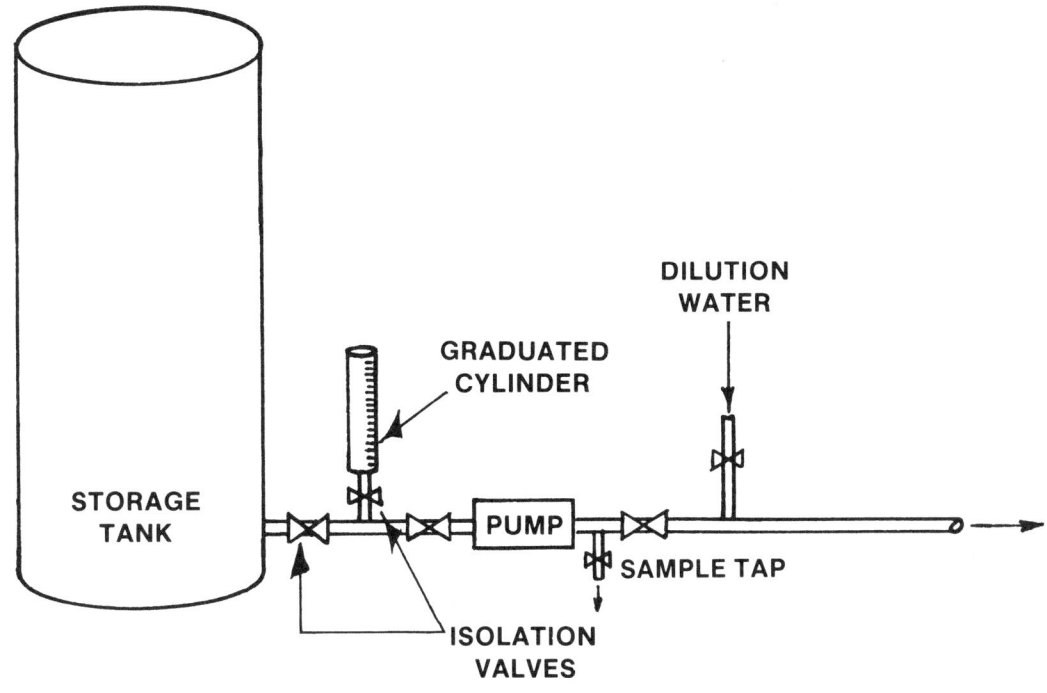

Fig. 4.14   Calibration system

EXAMPLE 14

An empty bucket weighs 0.8 pound. This bucket is placed under a dry chemical feeder. After 24 minutes the bucket weighs 5.6 pounds.[37] Estimate the chemical feed rate in pounds per day.

Known	Unknown
Weight of Empty Bucket, lbs = 0.8 pound	Dry Chemical Feed Rate, lbs/day
Weight of Bucket and Chemical, lbs = 5.6 pounds	
Feed Time, min = 24 minutes	

1. Determine the amount of chemical fed in pounds.

$$\text{Chemical Fed, lbs} = \frac{\text{Weight of Bucket}}{\text{and Chemical, lbs}} - \frac{\text{Weight of Empty}}{\text{Bucket, lbs}}$$

$$= 5.6 \text{ lbs} - 0.8 \text{ lbs}$$

$$= 4.8 \text{ lbs}$$

2. Calculate the dry chemical feed rate in pounds per minute.

$$\frac{\text{Chemical Feed}}{\text{Rate, lbs/min}} = \frac{\text{Chemical Fed, lbs}}{\text{Feed Time, min}}$$

$$= \frac{4.8 \text{ lbs}}{24 \text{ min}}$$

$$= 0.2 \text{ lbs/min}$$

3. Calculate the dry chemical feed rate in pounds per day.

$$\frac{\text{Chemical Feed}}{\text{Rate, lbs/day}} = \left(\frac{\text{Chemical Feed}}{\text{Rate, lbs/min}}\right)(60 \text{ min/hr})(24 \text{ hr/day})$$

$$= (0.2 \text{ lbs/min})(60 \text{ min/hr})(24 \text{ hr/day})$$

$$= 288 \text{ lbs/day}$$

EXAMPLE 15

An empty pie tin weighs 30 grams. This tin is placed under a dry chemical feeder. After 5 minutes the pie tin weighs 450 grams. Estimate the chemical feed rate in grams per minute.

Known	Unknown
Weight of Empty Tin, gm = 30 gm	Dry Chemical Feed Rate, gm/min
Weight of Tin and Chemical, gm = 450 gm	
Feed Time, min = 5 min	

1. Determine the amount of chemical fed in grams.

$$\text{Chemical Fed, gm} = \frac{\text{Weight of Tin}}{\text{and Chemical, gm}} - \frac{\text{Weight of Empty}}{\text{Tin, gm}}$$

$$= 450 \text{ gm} - 30 \text{ gm}$$

$$= 420 \text{ grams}$$

---

[37] If you can't get a bucket this big in your chemical feeder to collect a sample or leave it in the feeder this long, use a pie tin. Some operators will collect three samples of three minutes or five minutes each and use a total amount of chemical fed for nine or fifteen minutes.

2. Calculate the dry chemical feed rate in grams per minute.

$$\frac{\text{Chemical Feed}}{\text{Rate, gm/min}} = \frac{\text{Chemical Fed, gm}}{\text{Feed Time, min}}$$

$$= \frac{420 \text{ grams}}{5 \text{ min}}$$

$$= 84 \text{ grams/min}$$

A check on the amount of liquid chemical used in a given time period, say 24 hours, can be made by measuring the difference (drawdown) in chemical storage tank levels (Figure 4.15). In this case, the operator must compute the volume of chemical used based on the dimensions of the storage tank. Some small chemical storage tanks are mounted on a scale, thus allowing a direct reading of the amount of chemical remaining at the end of each day. The difference between these values is the amount used per day.

EXAMPLE 16

An alum storage tank has an inside diameter (I.D.) of 10 feet and a height of 25 feet. During a 24-hour time period (say 8:00 am Monday to 8:00 am Tuesday), the tank level dropped three inches (3 in/12 in/ft = 0.25 ft). How many gallons of chemical were used? If the chemical feed was constant, what was the chemical feed rate in gallons per minute (GPM)?

Known	Unknown
Tank Diameter, ft = 10 ft	Chemical Used, gal
Tank Height, ft = 25 ft	Chemical Feed Rate, GPM
Tank Drop, ft = 0.25 ft	
Time, hr = 24 hours	

1. Calculate the volume of chemical used in cubic feet.

$$\text{Volume, cu ft} = (\pi/4)(\text{Diameter, ft})^2(\text{Depth Drop, ft})$$

$$= (0.785)(10 \text{ ft})^2(0.25 \text{ ft})$$

$$= 19.63 \text{ cu ft}$$

2. Determine the volume of chemical used in gallons.

$$\text{Chemical Used, gal} = (\text{Volume, cu ft})(7.48 \text{ gal/cu ft})$$

$$= (19.63 \text{ cu ft})(7.48 \text{ gal/cu ft})$$

$$= 147 \text{ gallons}$$

3. Calculate the chemical feed rate in gallons per minute.

$$\frac{\text{Chemical Feed}}{\text{Rate, GPM}} = \frac{\text{Chemical Used, gallons}}{(\text{Time, hours})(60 \text{ min/hr})}$$

$$= \frac{147 \text{ gallons}}{(24 \text{ hours})(60 \text{ min/hr})}$$

$$= 0.10 \text{ GPM}$$

Fig. 4.15   Chemical storage tanks

CHAPTER 5

SEDIMENTATION

by

Jim Beard

TABLE OF CONTENTS

Chapter 5.   SEDIMENTATION

OBJECTIVES

Chapter 5.   SEDIMENTATION

Following completion of Chapter 5, you should be able to:

1. Identify factors affecting the performance of sedimentation basins,

2. Describe various types of sedimentation basins and how they work,

3. Start up and shut down sedimentation basins,

4. Operate and maintain a sedimentation process and basins,

5. Collect samples and analyze results for a sedimentation process,

6. Keep records of a sedimentation process and basins, and

7. Safely perform your duties around a sedimentation basin.

WORDS

Chapter 5.   SEDIMENTATION

ABSORPTION (ab-SORP-shun)   ABSORPTION

The taking in or soaking up of one substance into the body of another by molecular or chemical action (as tree roots absorb dissolved nutrients in the soil).

ADSORPTION (add-SORP-shun)   ADSORPTION

The gathering of a gas, liquid, or dissolved substance on the surface or interface zone of another material.

CATHODIC (ca-THOD-ick) PROTECTION   CATHODIC PROTECTION

An electrical system for prevention of rust, corrosion, and pitting of metal surfaces which are in contact with water or soil. A low-voltage current is made to flow through a liquid (water) or a soil in contact with the metal in such a manner that the external electromotive force renders the metal structure cathodic. This concentrates corrosion on auxiliary anodic parts which are deliberately allowed to corrode instead of letting the structure corrode.

CLARIFIER (KLAIR-uh-fire)   CLARIFIER

A large circular or rectangular tank or basin in which water is held for a period of time during which the heavier suspended solids settle to the bottom. Clarifiers are also called settling basins and sedimentation basins.

COMPLETE TREATMENT   COMPLETE TREATMENT

A method of treating water which consists of the addition of coagulant chemicals, flash mixing, coagulation-flocculation, sedimentation and filtration. Also called conventional filtration.

DENSITY (DEN-sit-tee)   DENSITY

A measure of how heavy a substance (solid, liquid or gas) is for its size. Density is expressed in terms of weight per unit volume, that is, grams per cubic centimeter or pounds per cubic foot. The density of water (at 4°C or 39°F) is 1.0 gram per cubic centimeter or about 62.4 pounds per cubic foot.

DETENTION TIME   DETENTION TIME

(1)   The theoretical (calculated) time required for a small amount of water to pass through a tank at a given rate of flow.

(2)   The actual time in hours, minutes or seconds that a small amount of water is in a settling basin, flocculating basin or rapid-mix chamber. In storage reservoirs, detention time is the length of time entering water will be held before being drafted for use (several weeks to years, several months being typical).

$$\text{Detention Time, hr} = \frac{(\text{Basin Volume, gal})(24 \text{ hr/day})}{\text{Flow, gal/day}}$$

DEWATER   DEWATER

(1)   To remove or separate a portion of the water present in a sludge or slurry. To dry sludge so it can be handled and disposed of.

(2)   To remove or drain the water from a tank or a trench.

DIRECT FILTRATION   DIRECT FILTRATION

A method of treating water which consists of the addition of coagulant chemicals, flash mixing, coagulation, minimal flocculation, and filtration. The flocculation facilities may be omitted, but the physical-chemical reactions will occur to some extent. The sedimentation process is omitted.

EFFLUENT   EFFLUENT

Water or other liquid—raw (untreated), partially or completely treated—flowing *FROM* a reservoir, basin, treatment process, or treatment plant.

INFLUENT   INFLUENT

Water or other liquid—raw (untreated) or partially treated—flowing *INTO* a reservoir, basin, treatment process, or treatment plant.

LAUNDERS (LAWN-ders)   LAUNDERS

Sedimentation basin and filter discharge channels consisting of overflow weir plates (in sedimentation basins) and conveying troughs.

NONPOINT SOURCE   NONPOINT SOURCE

A runoff or discharge from a field or similar source. A point source refers to a discharge that comes out the end of a pipe.

OVERFLOW RATE   OVERFLOW RATE

One of the guidelines for the design of settling tanks and clarifiers in treatment plants. Used by operators to determine if tanks and clarifiers are hydraulically (flow) over- or underloaded. Also called SURFACE LOADING.

$$\text{Overflow Rate, GPD/sq ft} = \frac{\text{Flow, gallons/day}}{\text{Surface Area, sq ft}}$$

PLUG FLOW   PLUG FLOW

A type of flow that occurs in tanks, basins or reactors when a slug of water moves through a tank without ever dispersing or mixing with the rest of the water flowing through the tank.

POINT SOURCE   POINT SOURCE

A discharge that comes out the end of a pipe. A nonpoint source refers to runoff or a discharge from a field or similar source.

PRECIPITATE (pre-SIP-uh-TATE)   PRECIPITATE

(1) An insoluble, finely divided substance which is a product of a chemical reaction within a liquid.

(2) The separation from solution of an insoluble substance.

REPRESENTATIVE SAMPLE   REPRESENTATIVE SAMPLE

A sample portion of material or water that is as nearly identical in content and consistency as possible to that in the larger body of material or water being sampled.

SEDIMENTATION (SED-uh-men-TAY-shun)   SEDIMENTATION

A water treatment process in which solid particles settle out of the water being treated in a large clarifier or sedimentation basin.

SEPTIC (SEP-tick)   SEPTIC

A condition produced by bacteria when all oxygen supplies are depleted. If severe, the bottom deposits produce hydrogen sulfide, the deposits and water turn black, give off foul odors, and the water has a greatly increased chlorine demand.

SHOCK LOAD   SHOCK LOAD

The arrival at a water treatment plant of raw water containing unusual amounts of algae, colloidal matter, color, suspended solids, turbidity, or other pollutants.

SHORT-CIRCUITING   SHORT-CIRCUITING

A condition that occurs in tanks or basins when some of the flowing water entering a tank or basin flows along a nearly direct pathway from the inlet to the outlet. This is usually undesirable since it may result in shorter contact, reaction, or settling times in comparison with the theoretical (calculated) or presumed detention times.

SLURRY (SLUR-e)   SLURRY

A watery mixture or suspension of insoluble (not dissolved) matter; a thin, watery mud or any substance resembling it (such as a grit slurry or a lime slurry).

SUPERNATANT (sue-per-NAY-tent)   SUPERNATANT

Liquid removed from settled sludge. Supernatant commonly refers to the liquid between the sludge on the bottom and the scum on the water surface of a basin or container.

SURFACE LOADING   SURFACE LOADING

One of the guidelines for the design of settling tanks and clarifiers in treatment plants. Used by operators to determine if tanks and clarifiers are hydraulically (flow) over- or underloaded. Also called OVERFLOW RATE.

$$\text{Surface Loading, GPD/sq ft} = \frac{\text{Flow, gallons/day}}{\text{Surface Area, sq ft}}$$

TUBE SETTLER   TUBE SETTLER

A device that uses bundles of small-bore (2 to 3 inches or 50 to 75 mm) tubes installed on an incline as an aid to sedimentation. The tubes may come in a variety of shapes including circular and rectangular. As water rises within the tubes, settling solids fall to the tube surface. As the sludge (from the settled solids) in the tube gains weight, it moves down the tubes and settles to the bottom of the basin for removal by conventional sludge collection means. Tube settlers are sometimes installed in sedimentation basins and clarifiers to improve particle removal.

TURBIDITY UNITS (TU)   TURBIDITY UNITS (TU)

Turbidity units are a measure of the cloudiness of water. If measured by a nephelometric (deflected light) instrumental procedure, turbidity units are expressed in nephelometric turbidity units (NTU) or simply TU. Those turbidity units obtained by visual methods are expressed in Jackson Turbidity Units (JTU) which are a measure of the cloudiness of water; they are used to indicate the clarity of water. There is no real connection between NTUs and JTUs. The Jackson turbidimeter is a visual method and the nephelometer is an instrumental method based on deflected light.

VISCOSITY (vis-KOSS-uh-tee)   VISCOSITY

A property of water, or any other fluid, which resists efforts to change its shape or flow. Syrup is more viscous (has a higher viscosity) than water. The viscosity of water increases significantly as temperatures decrease. Motor oil is rated by how thick (viscous) it is; 20 weight oil is considered relatively thin while 50 weight oil is relatively thick or viscous.

CHAPTER 5. SEDIMENTATION

(Lesson 1 of 2 Lessons)

5.0 PROCESS DESCRIPTION

5.00 Process Definition (Figure 5.1)

The purposes of the sedimentation process are to remove suspended solids (particles) that are denser (heavier) than water and to reduce the load on the filters (Chapter 6). The suspended solids may be in their natural state (such as bacteria, clays, or silts); they may be modified (preconditioned) by prior treatment in the coagulation-flocculation process (to form floc); or may be *PRECIPITATED*[1] impurities (hardness and iron precipitates formed by the addition of chemicals).

Sedimentation is accomplished or helped by decreasing the velocity of the water being treated below the point where it can transport settleable suspended material, thus allowing gravitational forces to remove particles held in suspension. When water is almost still in sedimentation basins, settleable solids will move toward the bottom of the basin.

5.01 Presedimentation

In Chapters 3 and 4, you learned that settling of larger-sized particles occurs naturally when surface water is stored for a sufficient period of time in a reservoir or a natural lake. Gravitational forces acting in the lake accomplish the same purpose as sedimentation in the water treatment plants; larger particles, such as sand and heavy silts, settle to the bottom.

Debris dams, grit basins, or sand traps can also be used to remove some of the heavier particles from the source water. These facilities may be located upstream from the reservoir, diversion works, or treatment plant intake or diversion facilities, and serve to protect the municipal intake pipeline from siltation (settling out of solids). Grit basins may be located between the intake structure and the coagulation-flocculation facilities (Figure 5.2). Thus, presedimentation facilities such as debris dams, impoundments, and grit basins, reduce the solids-removal load at the water treatment plant. At the same time, they provide an equalizing basin which evens out fluctuations in the concentration of suspended solids in the source water.

Presedimentation facilities are often installed in locations where the source water supply is diverted directly from rivers or streams which can be contaminated by overland runoff and *POINT SOURCE*[2] waste discharges. Ideally, surface waters should be stored in a reservoir and transported directly to the water treatment plant in a pipeline. In a reservoir the heavier solids can settle out before they reach the plant. However, geographical, physical, and economic considerations (such as the lack of a suitable dam site) often make this alternative impractical.

QUESTIONS

Write your answers in a notebook and then compare your answers with those on page 186.

5.0A What are the purposes of the sedimentation process?

5.0B How is sedimentation accomplished?

5.0C Presedimentation facilities are installed in what types of locations?

5.1 PROCESS PERFORMANCE CONSIDERATIONS

5.10 Factors Affecting Sedimentation

The size, shape, and weight of the particles to be settled out, as well as physical and environmental conditions in the sedimentation tank, have a significant impact on the type of pretreatment needed and the sedimentation process efficiency.

Factors affecting particle settling include:

1. Particle size and distribution,

2. Shape of particles,

3. Density of particles,

4. Temperature (*VISCOSITY*[3] and *DENSITY*[4]) of water,

---

[1] Precipitate (pre-SIP-uh-TATE).   (1) An insoluble, finely divided substance which is a product of a chemical reaction within a liquid. (2) The separation from solution of an insoluble substance.

[2] Point Source.   A discharge that comes out the end of a pipe. A nonpoint source refers to runoff or a discharge from a field or similar source.

[3] Viscosity (vis-KOSS-uh-tee).   A property of water, or any other fluid, which resists efforts to change its shape or flow. Syrup is more viscous (has a higher viscosity than water. The viscosity of water increases significantly as temperatures decrease. Motor oil is rated by how thick (viscous) it is; 20 weight oil is considered relatively thin while 50 weight oil is relatively thick or viscous.

[4] Density (DEN-sit-tee).   A measure of how heavy a substance (solid, liquid or gas) is for its size. Density is expressed in terms of weight per unit volume, that is, grams per cubic centimeter or pounds per cubic foot. The density of water (at 4°C or 39°F) is 1.0 gram per cubic centimeter or about 62.4 pounds per cubic foot.

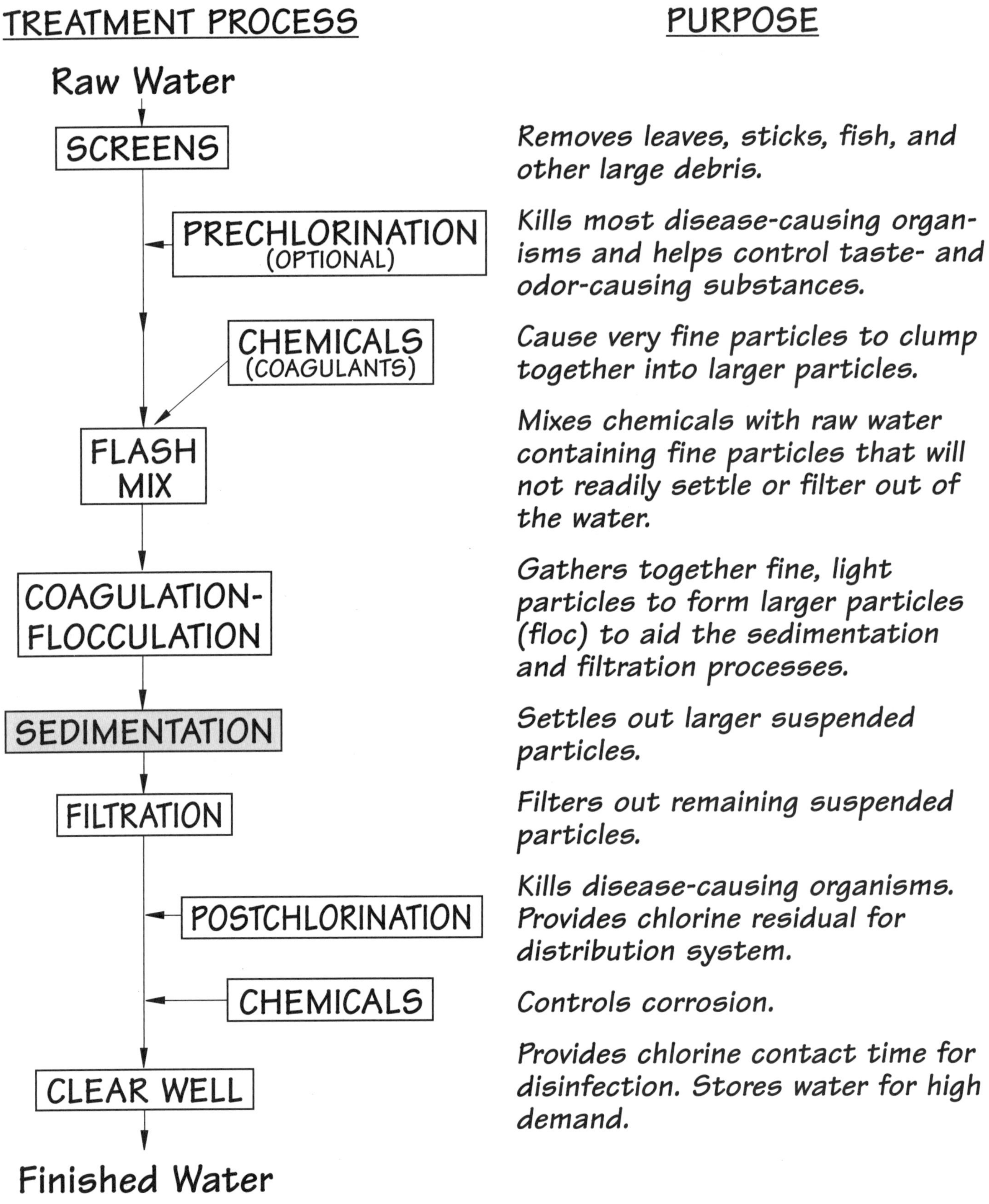

TREATMENT PROCESS

Raw Water
↓
| SCREENS |
↓
← | PRECHLORINATION (OPTIONAL) |
↓
| CHEMICALS (COAGULANTS) |
↓
| FLASH MIX |
↓
| COAGULATION-FLOCCULATION |
↓
| SEDIMENTATION |
↓
| FILTRATION |
↓
← | POSTCHLORINATION |
↓
← | CHEMICALS |
↓
| CLEAR WELL |
↓
Finished Water

PURPOSE

Removes leaves, sticks, fish, and other large debris.

Kills most disease-causing organisms and helps control taste- and odor-causing substances.

Cause very fine particles to clump together into larger particles.

Mixes chemicals with raw water containing fine particles that will not readily settle or filter out of the water.

Gathers together fine, light particles to form larger particles (floc) to aid the sedimentation and filtration processes.

Settles out larger suspended particles.

Filters out remaining suspended particles.

Kills disease-causing organisms. Provides chlorine residual for distribution system.

Controls corrosion.

Provides chlorine contact time for disinfection. Stores water for high demand.

Fig. 5.1   Flow diagram of a typical plant

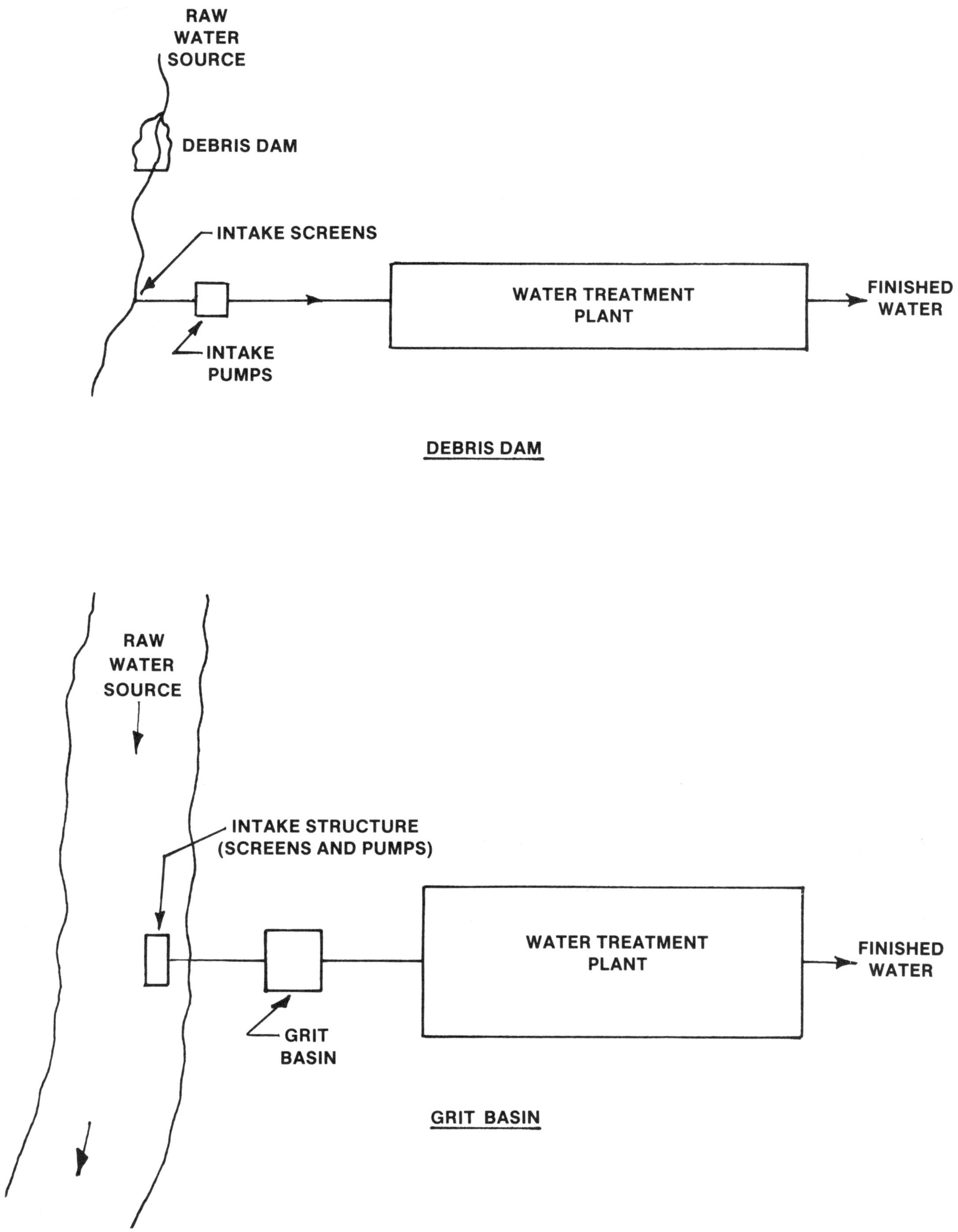

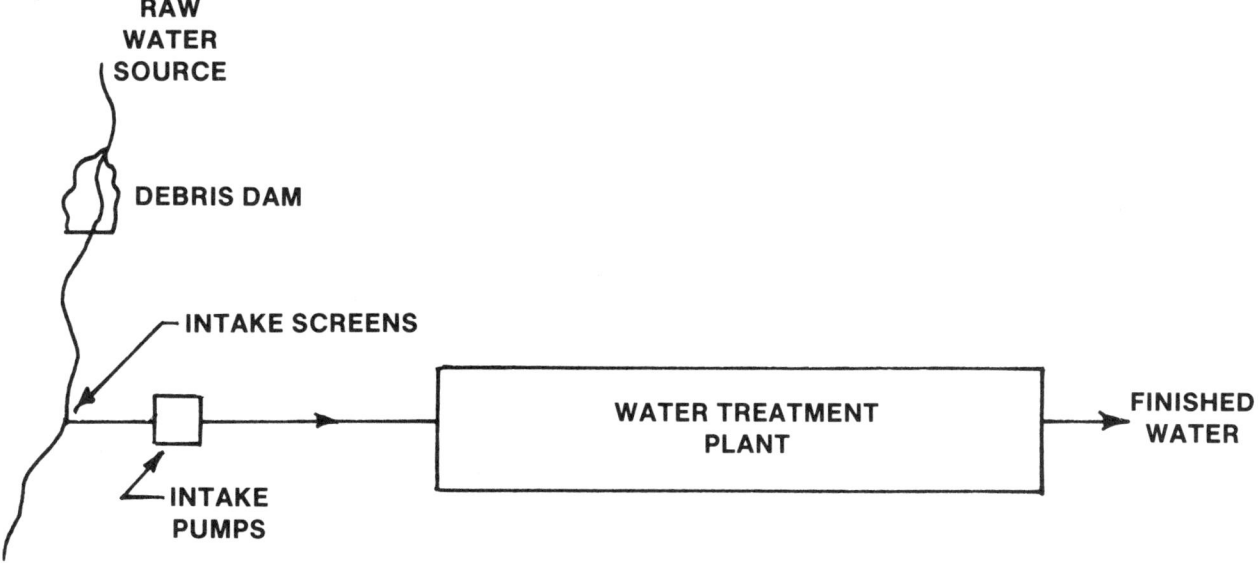

DEBRIS DAM

GRIT BASIN

Fig. 5.2   Presedimentation facilities

5. Electrical charge on particles,

6. Dissolved substances in water,

7. Flocculation characteristics of the suspended material,

8. Environmental conditions (such as wind effects), and

9. Sedimentation basin hydraulic and design characteristics (such as inlet conditions and shape of basin).

5.11   Nature of Particulate Impurities

Because of their size and density, sand and silt particles greater than 10 microns in diameter (1 micron = 0.001 mm) can be removed from water by sedimentation (simple gravitational settling). In contrast, finer particles do not readily settle and treatment is required to produce larger, denser particles (floc) that are settleable (see Table 5.1).

**TABLE 5.1   TYPICAL SIZE OF PARTICLES
IN SURFACE WATERS**

Source	Diameter of Particle (microns)[a]
Coarse Turbidity	1 – 1,000
Algae	3 – 1,000
Silt	10
Bacteria	0.3 – 10
Fine Turbidity	0.1 – 1
Viruses	0.02 – 0.26
Colloids	0.001 – 1

[a] 1 micron = 0.00004 of an inch

The shape of particles also influences particle settling. Smooth circular particles will settle faster than irregular particles with ragged edges.

Most particles have a very slight electrical charge. If all of the particles have a negative charge, they will tend to repel each other and not settle. Since alum consists of aluminum with a positive charge, the negatively charged particles are attracted to the positively charged aluminum ions. This causes the clumping together which helps the particles to settle out.

5.12   Water Temperature

Another consideration in sedimentation is the effect of water temperature changes. The settling rate (settling velocity) of a particle becomes much slower as the temperature drops. The colder the water temperature becomes, the longer particles take to settle out. Water is similar to syrup in this regard. The colder syrup becomes, the longer it would take a marble to settle to the bottom of the container. This means that longer time periods (lower flows) are required for effective settling at colder water temperatures, or that chemical dosages must be adjusted for the slower settling velocities.

5.13   Currents

Several types of currents are found in the typical sedimentation basin:

1. Surface currents caused by winds,

2. Density currents caused by differences in suspended solids concentrations and temperature differences, and

3. Eddy currents produced by the flow of the water coming into and leaving the basin.

Currents in the sedimentation basin are beneficial to the extent that they promote flocculation. Collectively, however, these currents distribute the suspended particles unevenly throughout the basin, thereby reducing the expected performance of the sedimentation basin.

Some of these currents can be substantially reduced in the design of a treatment plant by providing baffled inlets and other hydraulic control features (described in a later section). Others, such as wind-induced currents, can only be eliminated by providing covers or suitable windbreaks for the sedimentation basins. In most instances, basin covers are not economically feasible nor necessarily desirable from an operations and maintenance standpoint.

5.14   Particle Interactions

Suspended particles will continue to clump together (form floc) and other particles will precipitate from solution through flocculation and chemical precipitation in the sedimentation basin. The density and volume of particles will change. As a result, the settling velocities of individual particles will change as larger, denser floc particles are formed when particles of different size and density collide during the sedimentation process. Generally, this results in increased settling velocities.

QUESTIONS

Write your answers in a notebook and then compare your answers with those on page 186.

5.1A   List as many factors as you can recall that affect particle settling in a sedimentation basin.

5.1B   Why is treatment preferred before sedimentation?

5.1C   What types of currents may be found in a typical sedimentation basin?

5.2   SEDIMENTATION BASINS

5.20   Sedimentation Basin Zones

For convenience in discussing sedimentation basins, a typical sedimentation basin can be divided into four zones (see Figure 5.3):

1. Inlet zone,

2. Settling zone,

3. Sludge zone, and

4. Outlet zone.

5.200 Inlet Zone

The inlet to the sedimentation basin should provide a smooth transition from the flocculation basin and should distribute the flocculated water uniformly over the entire cross section of the basin. A properly designed inlet, such as a perforated baffle wall (see Figures 5.4 and 5.5), will significantly reduce SHORT-CIRCUITING[5] of water in the basin. An inlet baffle wall will also minimize the tendency of the water to flow at the inlet velocity straight through the basin, minimize densi-ty currents due to temperature differences, and minimize wind currents, as previously described.

5.201 Settling Zone

The settling zone is the largest portion of the sedimentation basin. This zone provides calm, undisturbed storage of the flocculated water for a sufficient time period (three or more hours) to permit effective settling of the suspended particles in the water being treated.

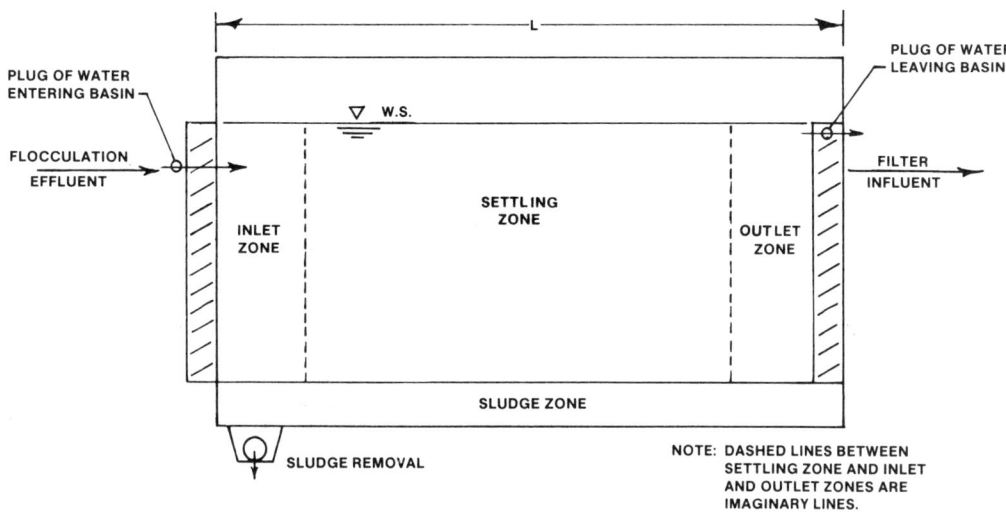

Fig. 5.3 Sedimentation basin zones

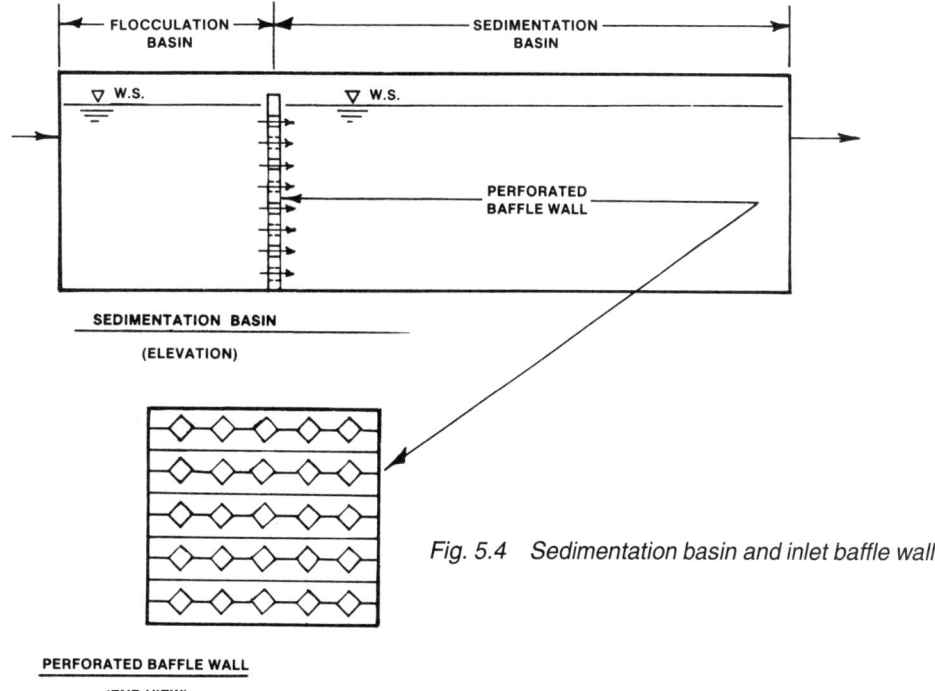

Fig. 5.4 Sedimentation basin and inlet baffle wall

---

[5] Short-Circuiting. A condition that occurs in tanks or basins when some of the flowing water entering a tank or basin flows along a nearly direct pathway from the inlet to the outlet. This is usually undesirable since it may result in shorter contact, reaction, or settling times in comparison with the theoretical (calculated) or presumed detention times.

5.202   Sludge Zone

The sludge zone, located at the bottom of the sedimentation basin, is a temporary storage place for the settled particles (Figure 5.6). Also the sludge zone is used for compression settling of the sludge; as sludge settles onto the zone, its weight further compacts the sludge below it. If the sludge buildup becomes too great, however, the effective depth of the basin will be significantly reduced, causing localized high flow velocities, sludge scouring, and a decrease in process effeciency. Basin inlet structures should be designed to minimize high flow velocities near the bottom of the sedimentation basin which could disturb or scour settled particles in the sludge zone, causing them to become resuspended.

Sludge is removed from the sludge zone by scraper and vacuum devices which move along the bottom of the basin (Figures 5.7 and 5.8) as needed or on a regularly scheduled basis. If the removal devices do not operate over the entire length of the basin, it may have to be drained and flushed to remove the sludge.

5.203   Outlet Zone

The basin outlet should provide a smooth transition from the sedimentation basin to the settled water conduit or channel. The outlet can also control the water level in the basin.

Basin inlet baffles, traveling bridge (top) and vacuum sweep

Sedimentation basin perforated inlet baffle

Fig. 5.5   Sedimentation basin perforated inlet baffle

Fig. 5.6   Sludge buildup in bottom of sedimentation basin

NOTE: Sludge sump is hole in middle.

Fig. 5.7   Sludge scraper

Traveling bridge (top) sludge removal system
(Note covered flocculators in foreground)

Traveling bridge vacuum sweep (sweep at bottom)

Fig. 5.8   Vacuum sweep sludge removal system

Skimming or effluent troughs, commonly referred to as *LAUNDERS* (Figures 5.9, 5.10, 5.11, and 5.12), are frequently used to uniformly collect the settled or clarified water. Adjustable V-notch weirs (Figures 5.11 and 5.12) are generally attached to the launders to enable a uniform draw-off of basin water by controlling the flow. If the water leaving a sedimentation basin flows out unevenly over the weirs or at too high a velocity, floc can be carried over to the filters. The increased loading on the filters may cause shortened filter runs and, therefore, more frequent backwashing.

QUESTIONS

Write your answers in a notebook and then compare your answers with those on page 186.

5.2A   List the four zones into which a typical sedimentation basin can be divided.

5.2B   What is the purpose of the settling zone in a sedimentation basin?

5.2C   What are launders?

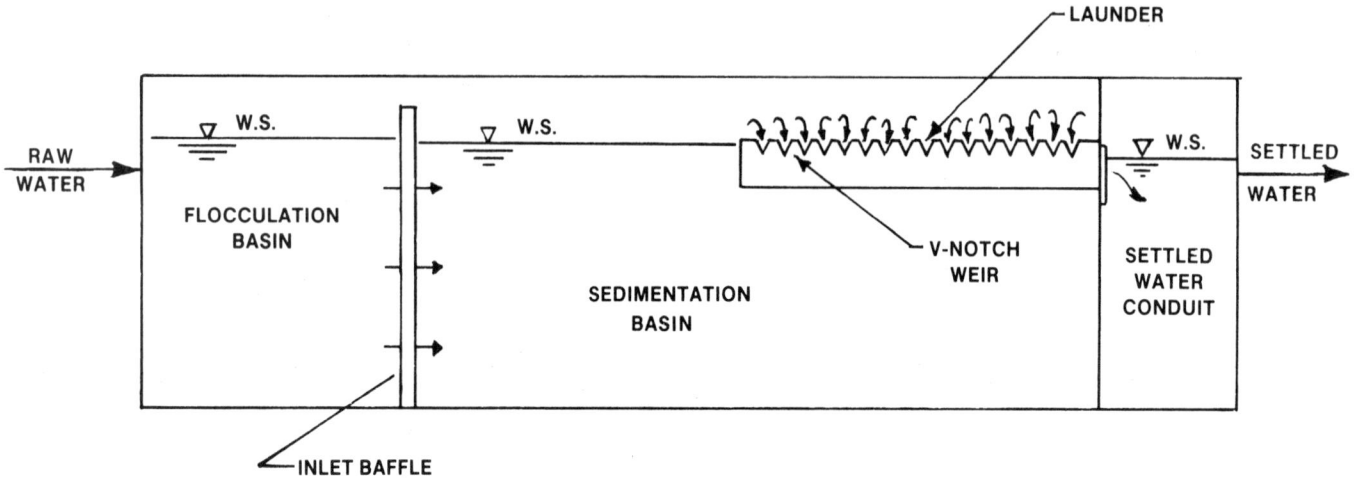

Fig. 5.9   *Typical sedimentation basin outlet (flow over V-notch weirs)*

Fig. 5.10   *Effluent launders in empty sedimentation basin*

Dewatered sedimentation basin
(Note inlet baffle at far end and traveling bridge above baffle)

Sedimentation basin outlet launders and V-notch weirs

Fig. 5.11   Empty sedimentation basin

Water spilling over V-notch weirs into outlet launders

Sedimentation basin outlet launders

Fig. 5.12   Sedimentation basin outlet launders and V-notch weirs

5.21   Basin Types

5.210   Selection of Basin Type

There are a wide variety of basin types and configurations in use today (Figure 5.13). The more common basin types will be described in the following sections to acquaint you with the major characteristics of each style.

5.211   Rectangular Basins

Rectangular sedimentation basins are commonly found in large-scale water treatment plants. Rectangular basins are popular for the following reasons:

1. High tolerance to *SHOCK LOADING*[6] (water quality changes),

2. Predictable performance,

3. Cost-effectiveness,

4. Low maintenance, and

5. Minimal short-circuiting.

5.212   Double-Deck Basins

Double-deck basins (see Figure 5.14) are an adaptation of the rectangular basin design. By stacking one basin on top of another, double-deck basins provide twice the effective sedimentation surface area of a single basin of equivalent land area. Double-deck basins are designed to conserve land area, but are not in common use owing partially to higher operation and maintenance costs. In this design, sludge removal equipment must operate in both decks, and the entire operation may have to be shut down if an equipment problem develops in either deck.

5.213   Circular and Square Basins

Circular or square, horizontal-flow basins, as shown in Figures 5.13, 5.15, and 5.16, are often referred to as *CLARIFIERS*. These basins share some of the performance advantages of rectangular basins; however, they are generally more likely to have short-circuiting and particle removal problems. One of the major problems with square settling basins is the removal of sludge from the corners. This can also be a problem with rectangular basins. Some circular clarifiers are also called solids-contact units or upflow clarifiers. These units are discussed in Section 5.215, "Solids-Contact Units."

---

[6] *Shock Load.   The arrival at a water treatment plant of raw water containing unusual amounts of algae, colloidal matter, color, suspended solids, turbidity, or other pollutants.*

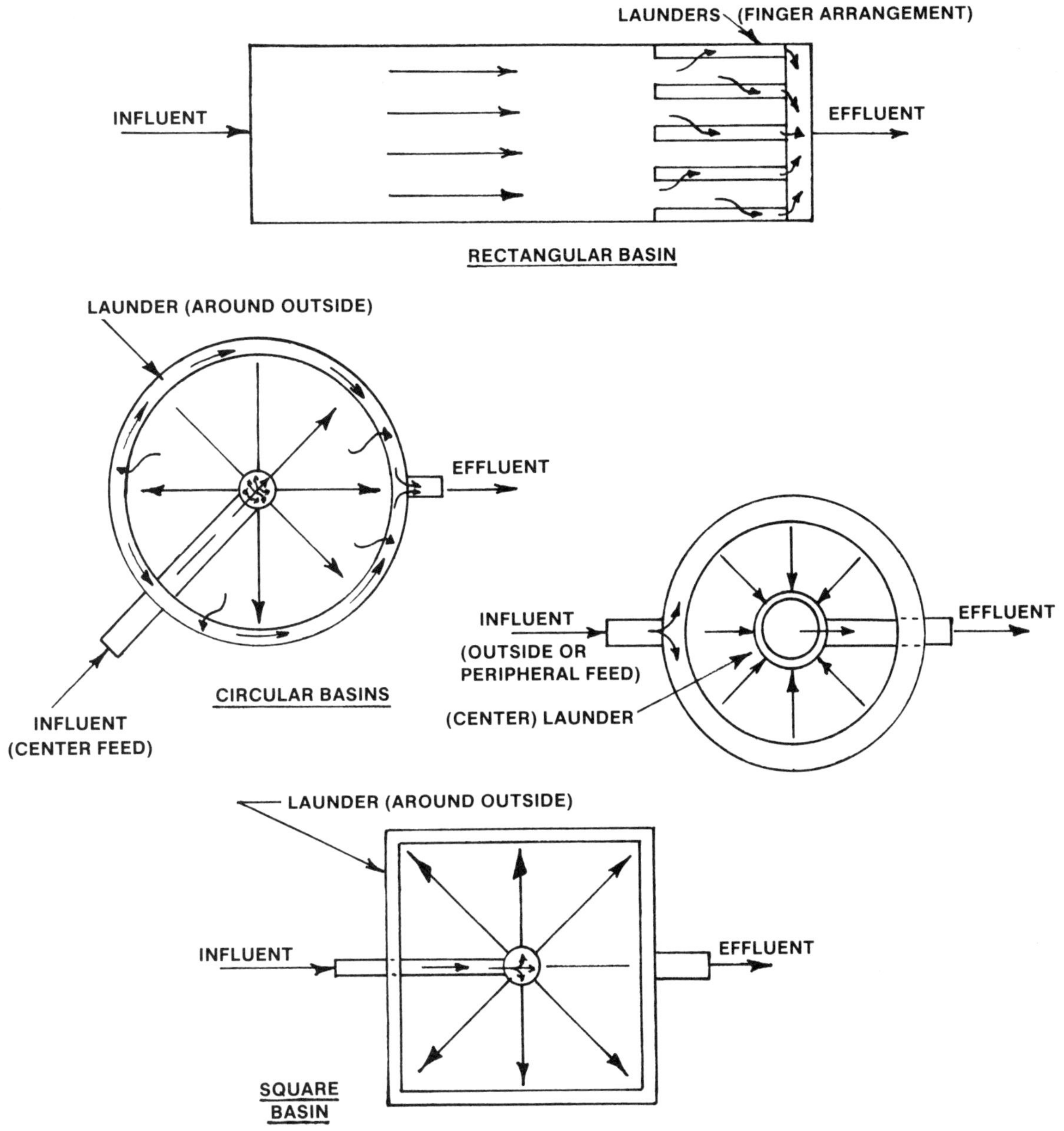

LAUNDERS (FINGER ARRANGEMENT)
INFLUENT
EFFLUENT
RECTANGULAR BASIN
LAUNDER (AROUND OUTSIDE)
EFFLUENT
INFLUENT
(OUTSIDE OR
PERIPHERAL FEED)
EFFLUENT
CIRCULAR BASINS
(CENTER) LAUNDER
INFLUENT
(CENTER FEED)
LAUNDER (AROUND OUTSIDE)
INFLUENT
EFFLUENT
SQUARE
BASIN

Fig. 5.13 Sedimentation basin types (plan views)

OUTLET
W.S.
TOP DECK
INLET
BOTTOM DECK

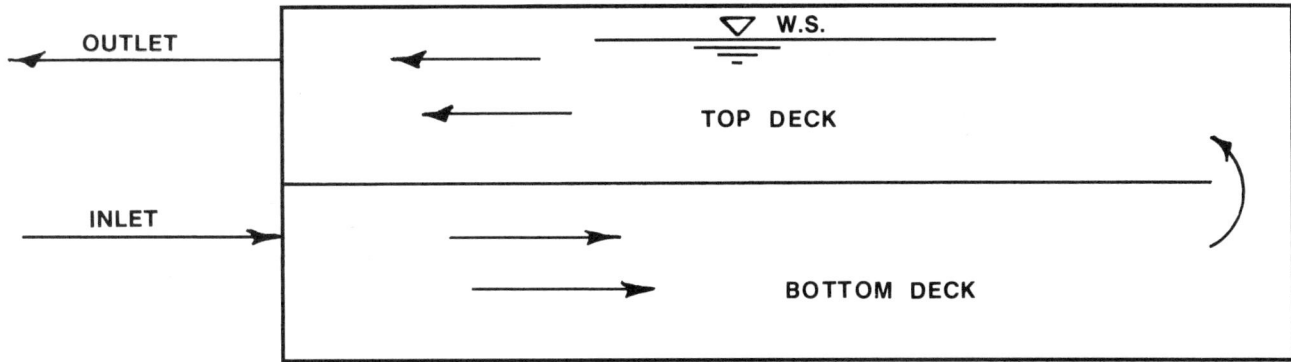

Fig. 5.14 Double-deck sedimentation basin (elevation)

5.214 High-Rate Settlers

High-rate or *TUBE SETTLERS* were developed to increase the settling efficiency of conventional rectangular sedimentation basins. They have been installed in circular basins with successful results.

Water enters the inclined settler tubes and is directed upward through the tubes as shown in Figures 5.17, 5.18, and 5.19. Each tube functions as a shallow settling basin. Together, they provide a high ratio of effective settling surface area per unit volume of water. The settled particles can collect on the inside surfaces of the tubes or settle to the bottom of the sedimentation basin.

Parallel plate or tilted plate settlers can also be used to increase the efficiency of rectangular sedimentation basins, and these function in a manner similar to tube settlers.

High-rate settlers are particularly useful for water treatment applications where site area is limited, in packaged-type water treatment units, and to increase the capacity of existing sedimentation basins. In existing rectangular and circular sedimentation basins, high-rate settler modules can be conveniently installed between the launders. High winds can have an adverse effect on tube settlers.

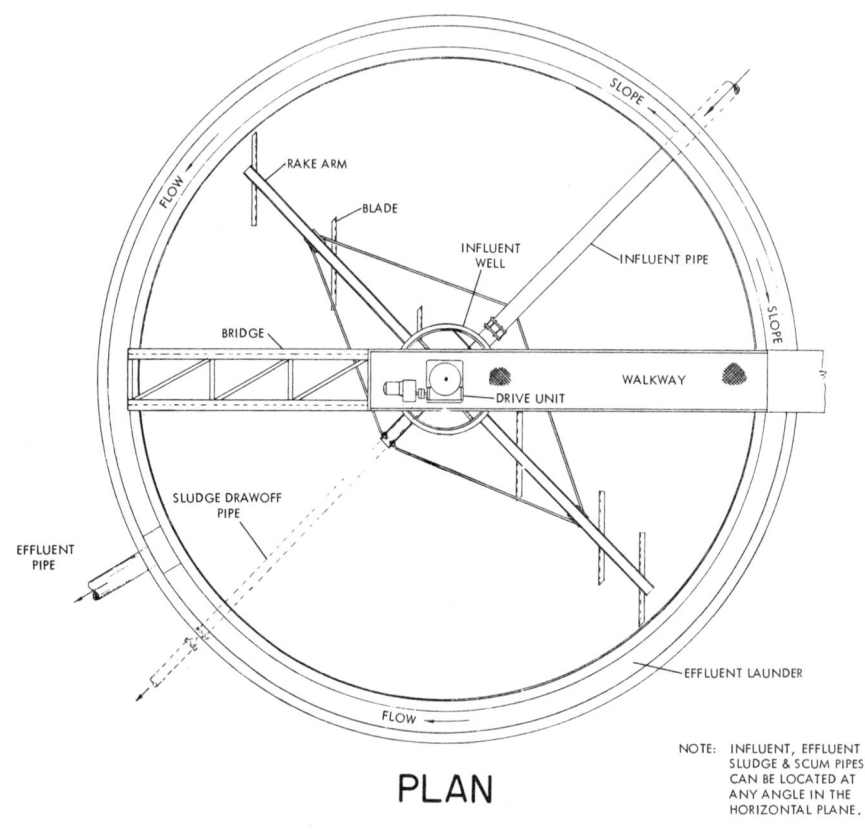

PLAN

NOTE: INFLUENT, EFFLUENT SLUDGE & SCUM PIPES CAN BE LOCATED AT ANY ANGLE IN THE HORIZONTAL PLANE.

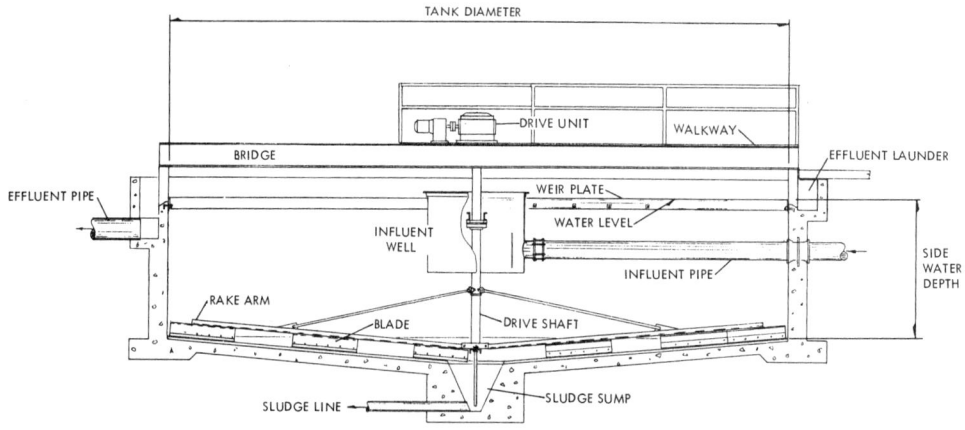

SECTION

Fig. 5.15   Circular clarifier
(Permission of General Filter)

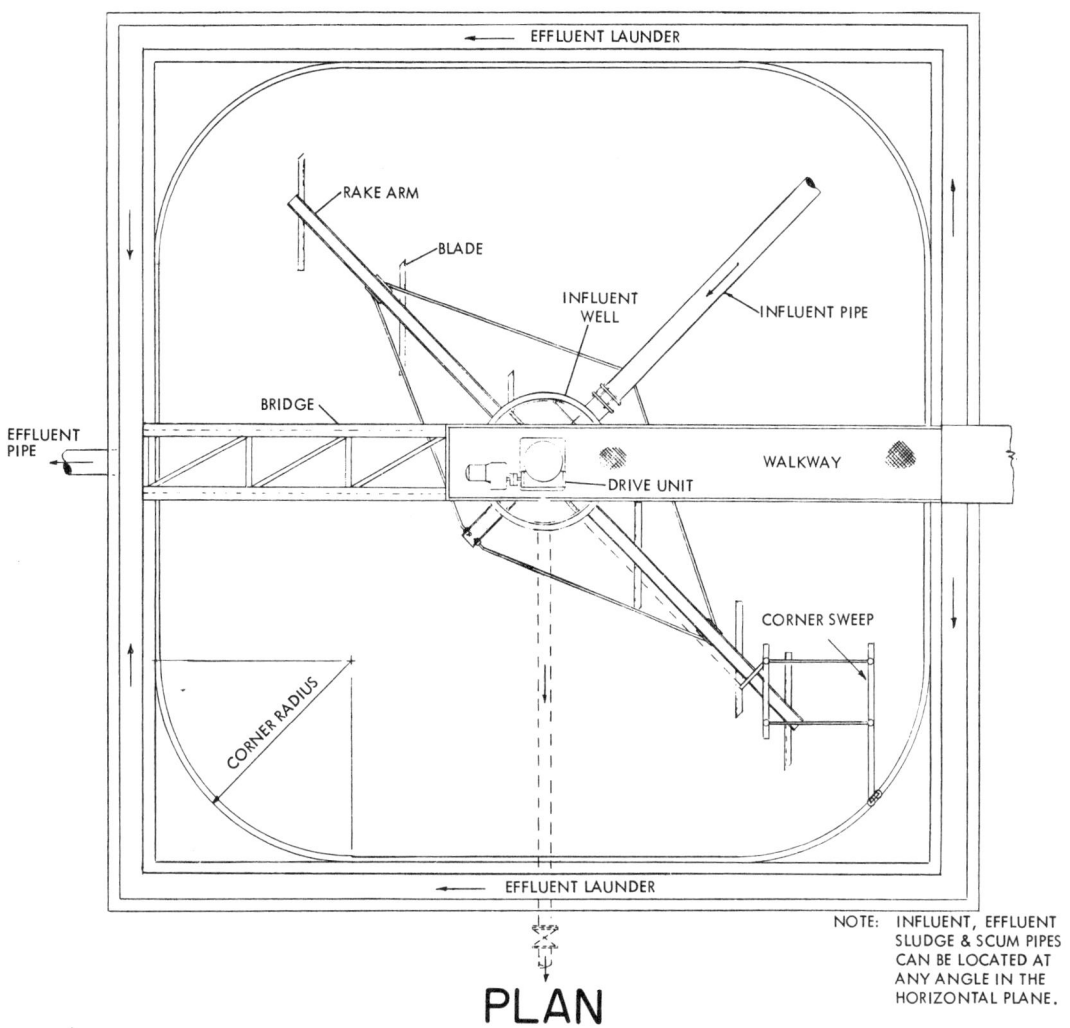

EFFLUENT LAUNDER

RAKE ARM

BLADE

INFLUENT WELL

INFLUENT PIPE

BRIDGE

EFFLUENT PIPE

WALKWAY

DRIVE UNIT

CORNER SWEEP

CORNER RADIUS

EFFLUENT LAUNDER

NOTE: INFLUENT, EFFLUENT
SLUDGE & SCUM PIPES
CAN BE LOCATED AT
ANY ANGLE IN THE
HORIZONTAL PLANE.

PLAN

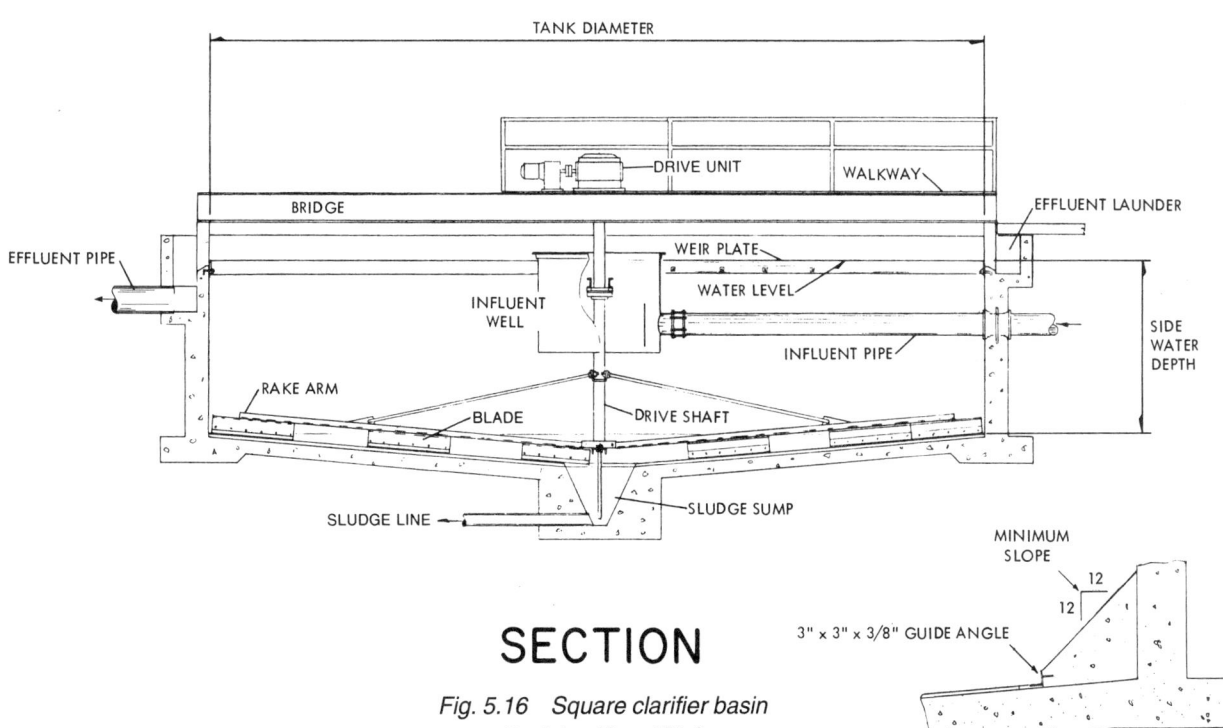

TANK DIAMETER

DRIVE UNIT

WALKWAY

BRIDGE

EFFLUENT LAUNDER

EFFLUENT PIPE

WEIR PLATE

WATER LEVEL

INFLUENT WELL

INFLUENT PIPE

SIDE WATER DEPTH

RAKE ARM

BLADE

DRIVE SHAFT

SLUDGE LINE

SLUDGE SUMP

MINIMUM SLOPE

12

12

3" x 3" x 3/8" GUIDE ANGLE

SECTION

Fig. 5.16   Square clarifier basin
(Permission of General Filter)

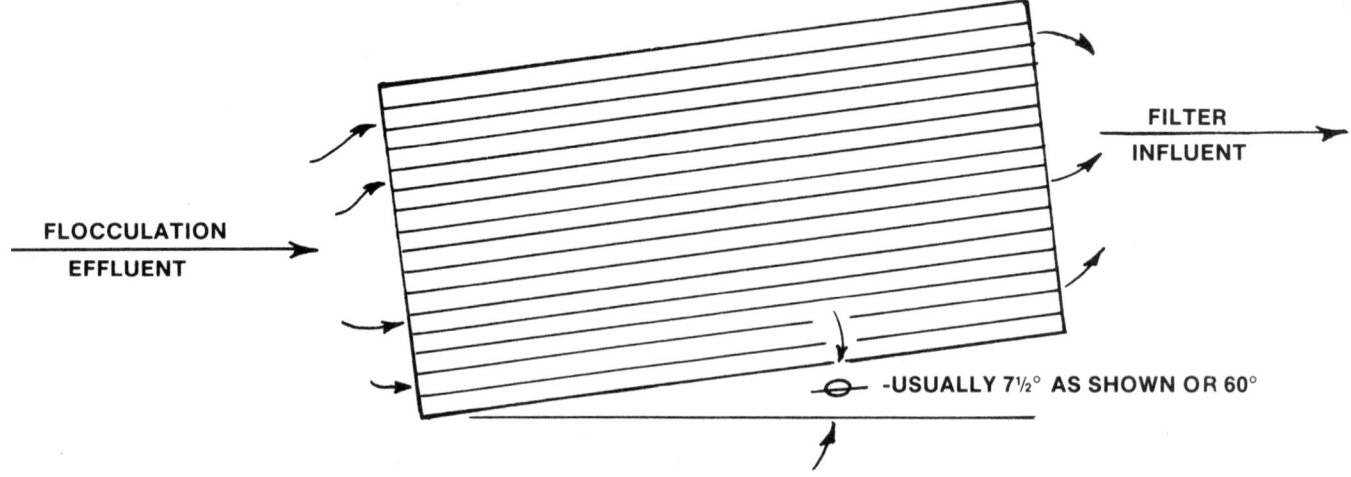

FLOCCULATION EFFLUENT

FILTER INFLUENT

-USUALLY 7½° AS SHOWN OR 60°

Fig. 5.17   Tube settler (installed in a rectangular or circular sedimentation basin)

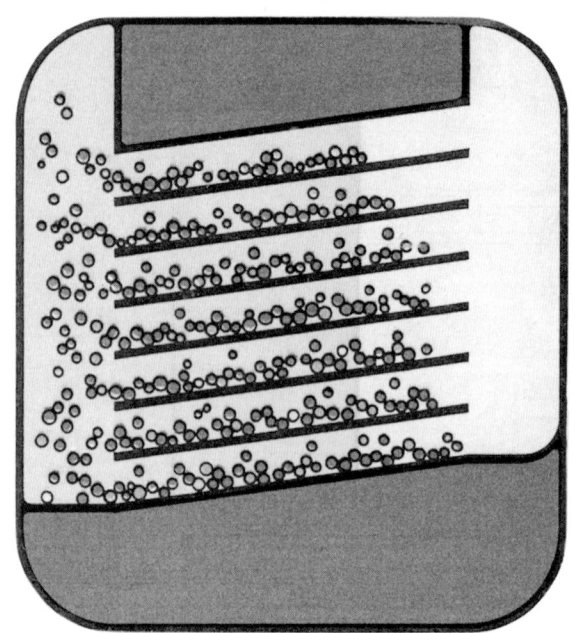

7½° Tube Settlers

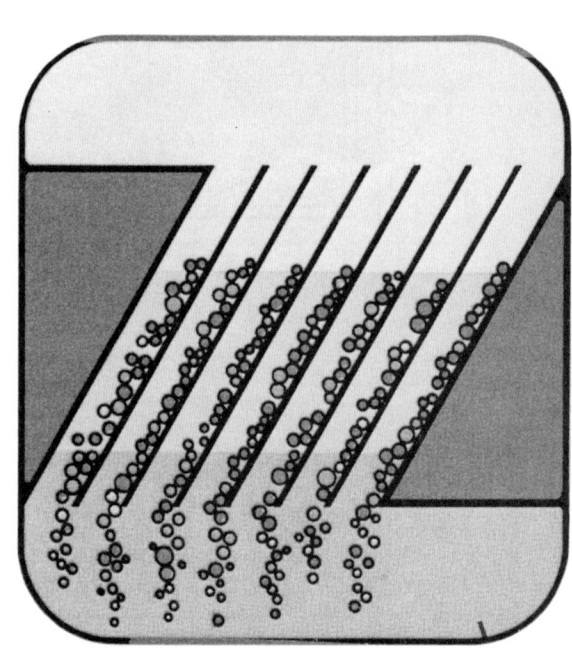

60° Tube Settlers

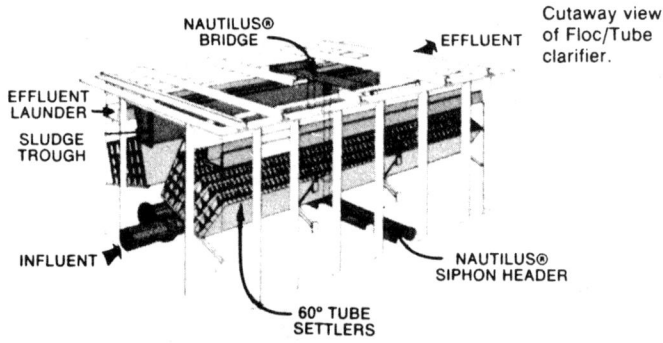

Fig. 5.18   Cutaway view of Floc/Tube clarifier
(Permission of Neptune Microfloc, Inc.) (Now Microfloc Products)

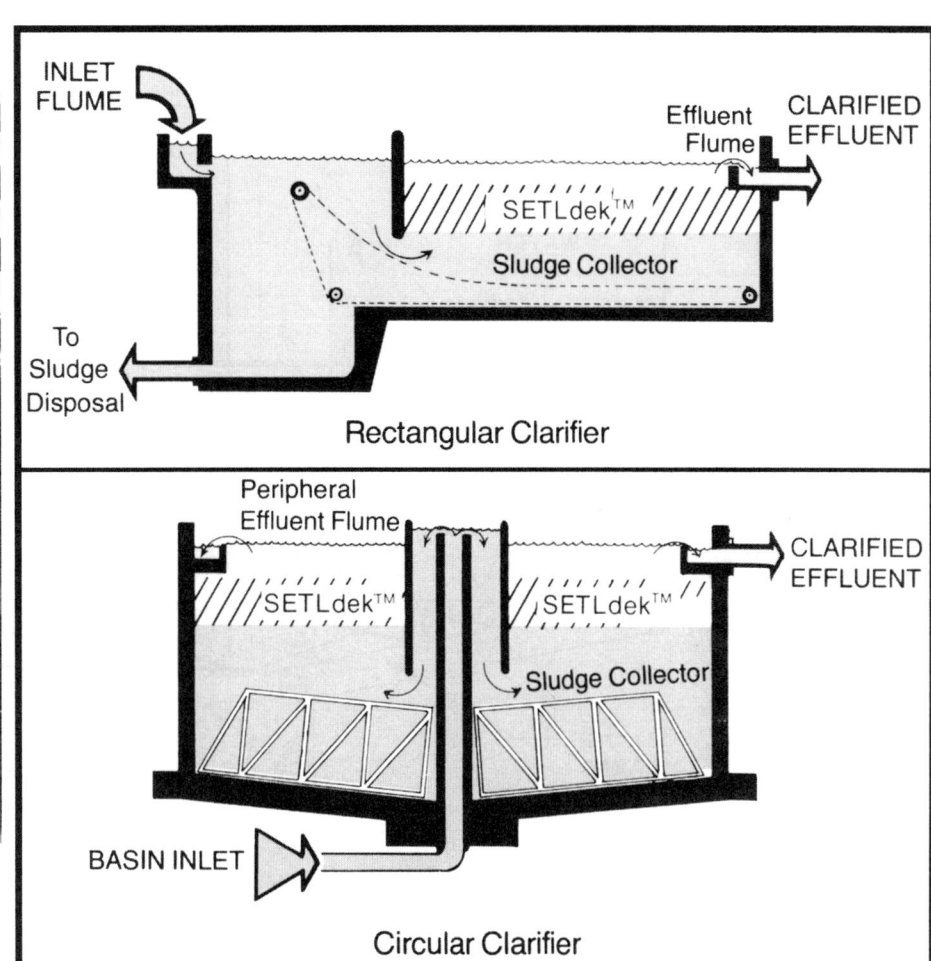

Fig. 5.19   Tube settlers in rectangular and circular clarifiers
(Permission of The Munters Corporation)

5.215   Solids-Contact Units

The solids-contact process, also referred to as "upflow solids-contact clarification" and "upflow sludge-blanket clarification," was developed to improve the overall solids removal process under certain design conditions. These units combine the coagulation, flocculation, and sedimentation processes into a single basin, which may be either rectangular or circular in shape. Flow is generally in an upward direction through a sludge blanket or slurry of flocculated, suspended solids as shown in Figure 5.20.

Solids-contact units generally have provisions for controlled removal of solids so that the concentration of solids retained in the basin can be maintained at some desired level.

Solids-contact units are popular for smaller package-type water treatment plants and also in cold climates where the units have to be inside a building. However, care must be exercised in the operation of these units to ensure that a uniform sludge blanket is formed and is subsequently maintained throughout the solids removal process. The sludge blanket is sensitive to changes in water temperature. Temperature density currents tend to upset the sludge blanket. Loss of the sludge blanket will affect the performance of the filters. Other important operational factors include control of chemical dosage, mixing of chemicals, and control of the sludge blanket.

Under ideal conditions, solids-contact units provide better performance for both turbidity removal and softening processes requiring the precipitation of hardness. With softening processes, chemical requirements are usually lower also. In the case of turbidity removal, coagulant requirements are often higher. In either case, solids-contact units are very sensitive to changes in influent flow or temperature. In these facilities, changes in the rate of flow should be made infrequently, slowly, and with great care.

For additional information about solids-contact units, see Section 5.24, "Solids-Contact Clarification."

QUESTIONS

Write your answers in a notebook and then compare your answers with those on pages 186 and 187.

5.2D   List three possible shapes for sedimentation basins.

5.2E   List the advantages and limitations of double-deck sedimentation basins.

5.2F   Why are rectangular sedimentation basins often preferred over circular basins?

5.2G   During the operation of a solids-contact unit, what items should be of particular concern to the operator?

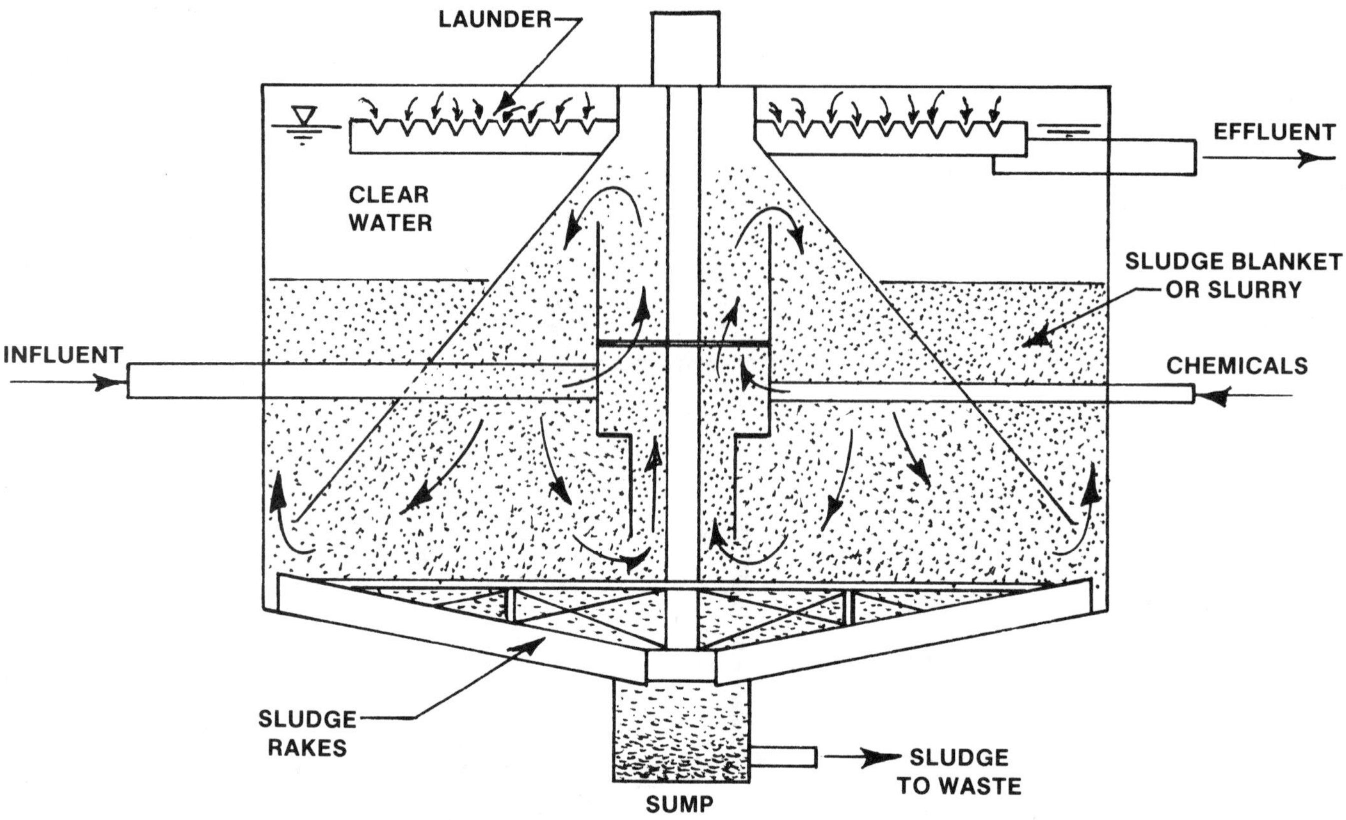

Fig. 5.20   Solids-contact unit

5.22   Basin Layout

A minimum of two sedimentation basins should be provided in all water treatment plants to allow for maintenance, cleaning, and inspection of a basin without requiring a complete plant shutdown.

A typical rectangular-shaped basin layout is shown in Figure 5.21. Note that a chemical application point is provided in the settled water conduit to permit feeding a filter aid chemical, chlorine, or other chemicals prior to filtration.

5.23   Detention Time

There are two definitions for detention time. Detention time (or retention time) is the actual time required for a small amount of water to pass through a sedimentation basin at a given rate of flow. Also, detention time can refer to the theoretical (calculated) time required for a small amount of water to pass through (or be retained within) a basin at a given rate of flow. The detention time is calculated by dividing the volume of the basin by the flow going into the basin. Actual flow-through times for different small amounts of water in the same basin may vary significantly from the calculated detention time due to short-circuiting, effective exchangeable volume (portion of basin through which the water flows), and other hydraulic considerations, such as basin inlet and outlet design. A dye placed at the inlet to a sedimentation basin will produce the curves shown in Figure 5.22 as it leaves the basin. The flatter the curve, the greater the short-circuiting.

FORMULAS

In order to calculate the detention time of a sedimentation basin, the volume of the basin and the flow must be known.

The volume of a basin can be calculated from the dimensions of a basin. The dimensions of a basin can be obtained from the plan drawings for the treatment plant. These drawings will have the length, width, and depth for rectangular sedimentation basins and the diameter and depth of circular clarifiers. The flow can be obtained from a flowmeter or flow records.

To calculate the detention time, divide the flow in gallons per day into the tank volume in gallons. To convert this detention time in days to hours, multiply by 24 hours per day.

For rectangular basins:

$$\text{Basin Volume, gal} = (L, \text{ft})(W, \text{ft})(D, \text{ft})(7.48 \text{ gal/cu ft})$$

For circular basins:

$$\text{Basin Volume, gal} = (0.785)(\text{Diameter, ft})^2(\text{Depth, ft})(7.48 \text{ gal/cu ft})$$

To calculate the theoretical detention time:

$$\text{Detention Time, hr} = \frac{(\text{Basin Volume, gal})(24 \text{ hr/day})}{\text{Flow, gal/day}}$$

If the size of the basin and design detention time for the sedimentation basin are known, the maximum flow for the basin can be calculated by rearranging the detention time formula:

$$\text{Flow, gal/day} = \frac{(\text{Basin Volume, gal})(24 \text{ hr/day})}{\text{Detention Time, hr}}$$

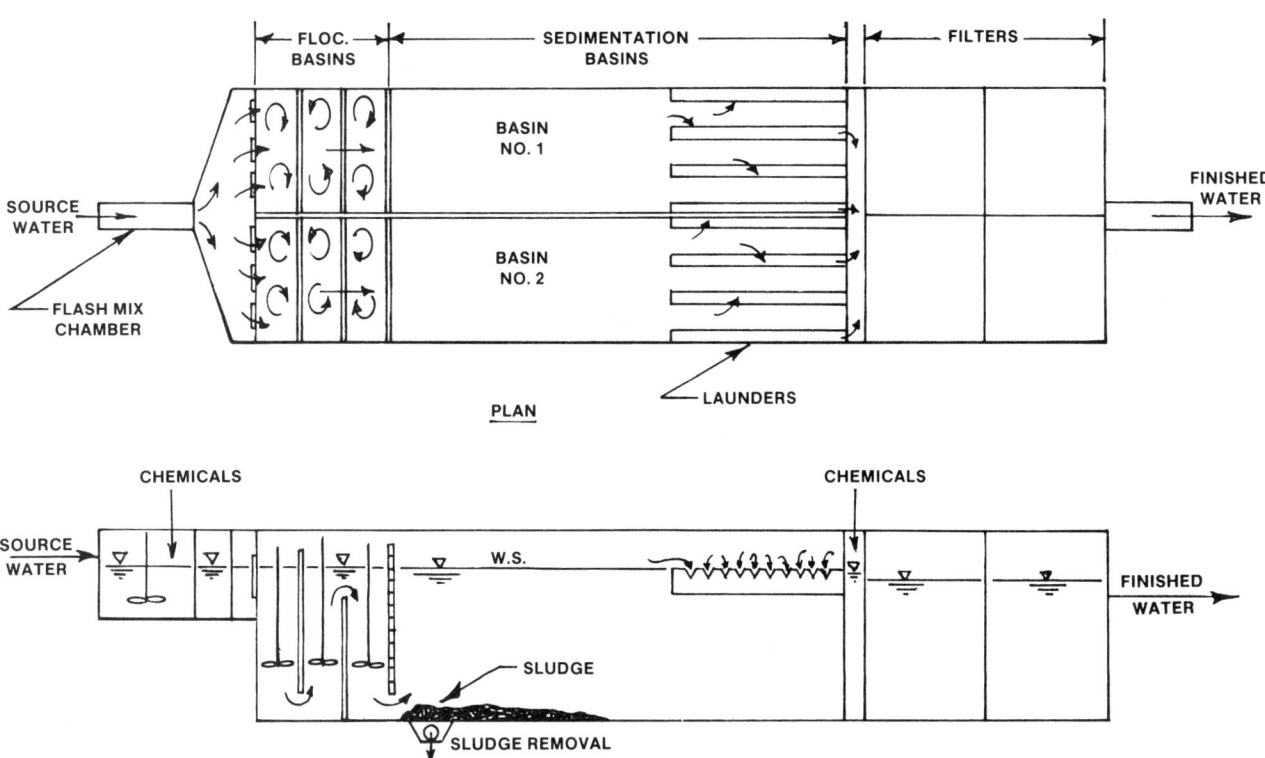

Fig. 5.21   Rectangular sedimentation basin

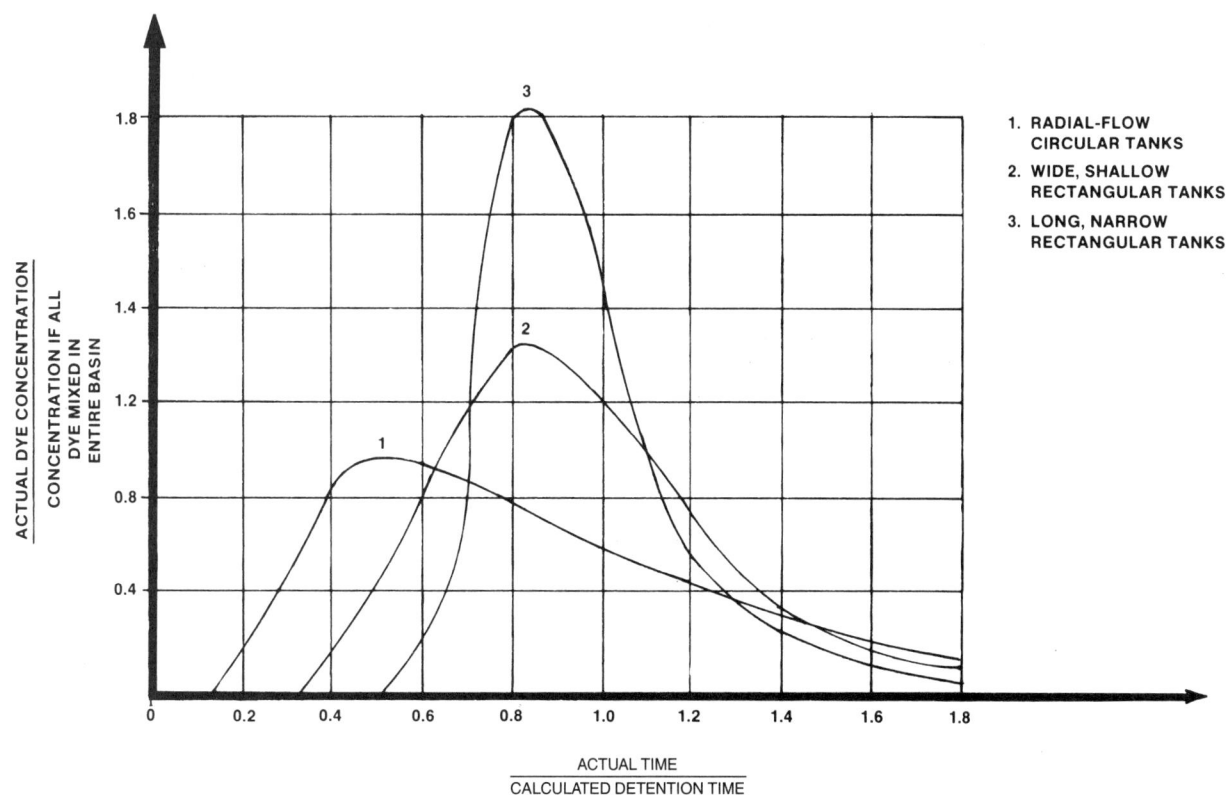

Fig. 5.22   Flow-through times for various types of sedimentation basins

(Adapted from "Studies of Sedimentation Basin Design" by T. R. Camp, *SEWAGE AND INDUSTRIAL WASTES*)

EXAMPLE 1

A water treatment plant treats a flow of 1.5 MGD. An examination of treatment plant design drawings reveals that the rectangular sedimentation basin is 75 feet long, 25 feet wide, and has an effective (water) depth of 12 feet. Calculate the theoretical detention time in hours for the rectangular sedimentation basin.

Known	**Unknown**
Flow, MGD = 1.5 MGD	Detention Time, hr
Length, ft = 75 ft	
Width, ft = 25 ft	
Depth, ft = 12 ft	

1. Calculate the basin volume in gallons.

Basin Volume, gal = (L, ft)(W, ft)(D, ft)(7.48 gal/cu ft)

= (75 ft)(25 ft)(12 ft)(7.48 gal/cu ft)

= 168,300 gallons

2. Determine the theoretical detention time in hours.

$$\text{Detention Time, hr} = \frac{(\text{Basin Volume, gal})(24 \text{ hr/day})}{\text{Flow, gal/day}}$$

$$= \frac{(168,300 \text{ gal})(24 \text{ hr/day})}{1,500,000 \text{ gal/day}}$$

$$= 2.7 \text{ hr}$$

EXAMPLE 2

What is the maximum flow in MGD for the rectangular sedimentation basin in Example 1 if the theoretical detention time is 2 hours?

Known	**Unknown**
Length, ft = 75 ft	Maximum Flow, MGD
Width, ft = 25 ft	
Depth, ft = 12 ft	
Volume, gal = 168,300 gal (from Example 1)	
Detention Time, hr = 2.0 hr	

Calculate the maximum flow in MGD.

$$\text{Flow, gal/day} = \frac{(\text{Basin Volume, gal})(24 \text{ hr/day})}{\text{Detention Time, hr}}$$

$$= \frac{(168,300 \text{ gal})(24 \text{ hr/day})}{2.0 \text{ hr}}$$

$$= 2,019,600 \text{ gal/day}$$

Flow, MGD   = 2.0 MGD

EXAMPLE 3

A water treatment plant has a circular clarifier for a sedimentation basin. The treatment plant design drawings indicate that the clarifier has a diameter of 60 feet and an average water depth of 12 feet. What is the theoretical detention time in hours for the basin when the flow is 2 MGD?

Known	**Unknown**
Diameter, ft = 60 ft	Detention Time, hr
Depth, ft = 12 ft	
Flow, MGD = 2 MGD	

1. Calculate the basin volume in gallons.

Basin Volume, gal = (0.785)(Diameter, ft)^2(Depth, ft)(7.48 gal/cu ft)

= (0.785)(60 ft)^2(12 ft)(7.48 gal/cu ft)

= 253,662 gal

2. Determine the theoretical detention time in hours.

$$\text{Detention Time, hr} = \frac{(\text{Basin Volume, gal})(24 \text{ hr/day})}{\text{Flow, gal/day}}$$

$$= \frac{(253,662 \text{ gal})(24 \text{ hr/day})}{2,000,000 \text{ gal/day}}$$

$$= 3.0 \text{ hr}$$

In this section we have calculated the detention time for a sedimentation basin. From a practical standpoint, however, you can anticipate problems by comparing the actual flow through your water treatment plant with the design flow. Whenever actual flows approach or exceed design flows, problems are likely to develop. You don't have to go through any calculations to know that your plant is hydraulically (flow) overloaded.

Also, when water temperature decreases, be prepared to reduce flows if problems should develop. The colder the water, the longer it takes particles to settle out. By reducing flows you are, of course, increasing the available detention time.

If the demand for water does not allow you to reduce flows, run jar tests with shorter detention times. Adjust the chemical doses as necessary to compensate for the colder water. There is very little an operator can do to control the sedimentation process. Adjusting chemicals and chemical feed rates is the major means by which operators can control water treatment processes.

QUESTIONS

Write your answers in a notebook and then compare your answers with those on page 187.

5.2H   How would you calculate the detention time for a sedimentation basin?

5.2I   A rectangular sedimentation basin 50 feet long, 20 feet wide, and 10 feet deep treats a flow of 0.8 MGD. What is the theoretical detention time?

5.24   Solids-Contact Clarification
by J. T. Monscvitz

5.240   Process Description

Solids-contact units were first used in the Midwest as a means of handling the large amounts of sludge generated by water softening processes. It quickly became clear that this compact, single-unit process could also be used to remove turbidity from drinking water.

Solids-contact clarifiers (Figures 5.23 and 5.24) go by several interchangeable names: solids-contact clarifiers, upflow

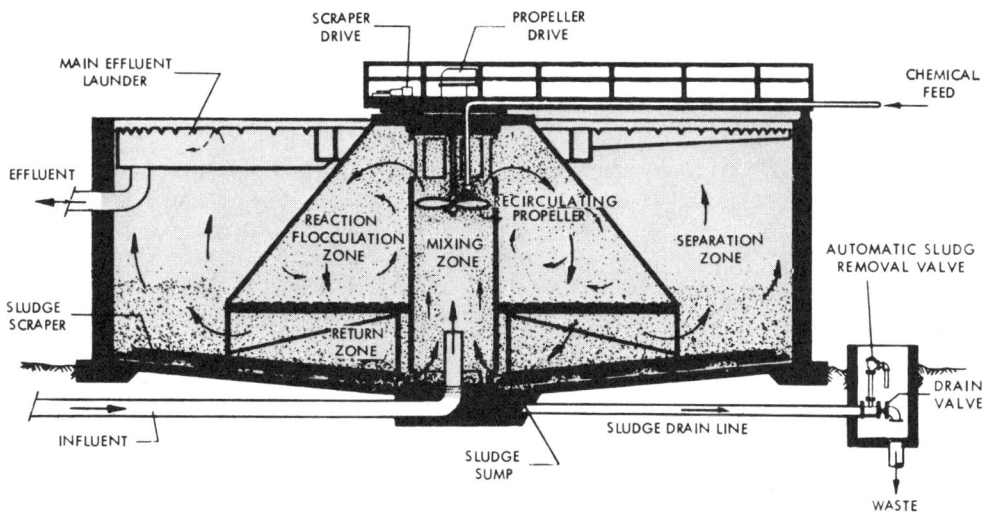

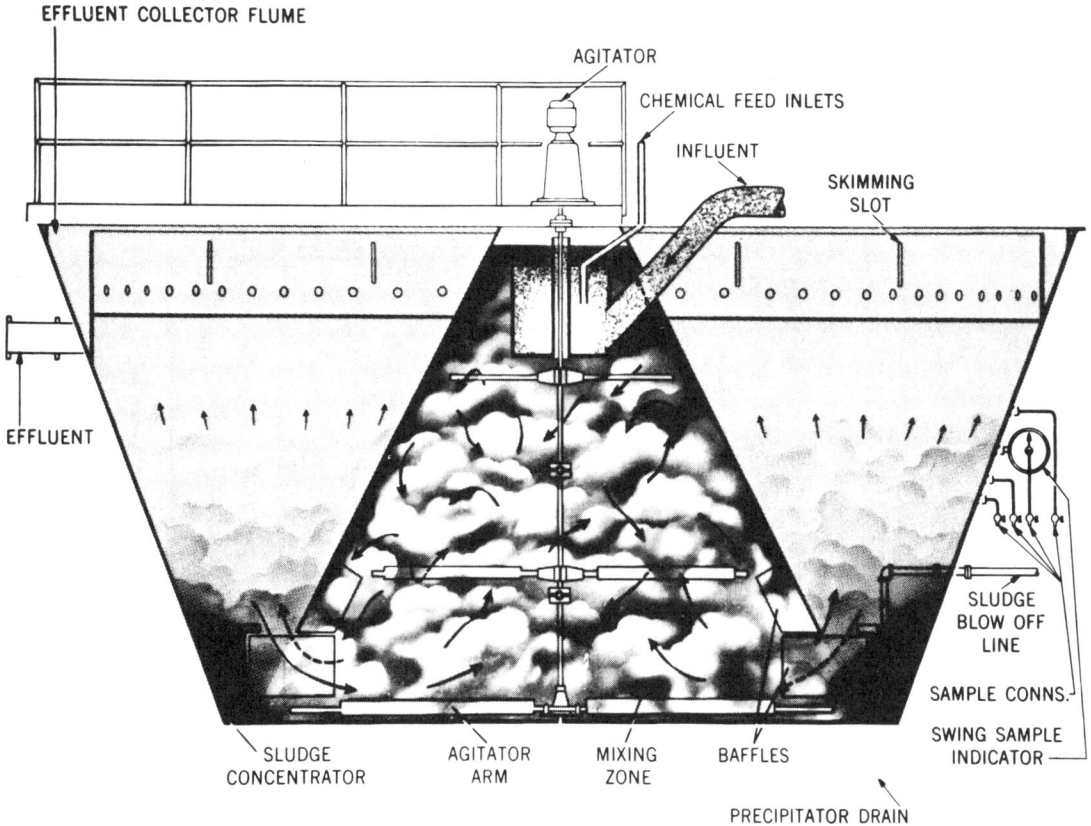

Fig. 5.23   Solids-contact clarifiers

(Permission of Permutit Company)

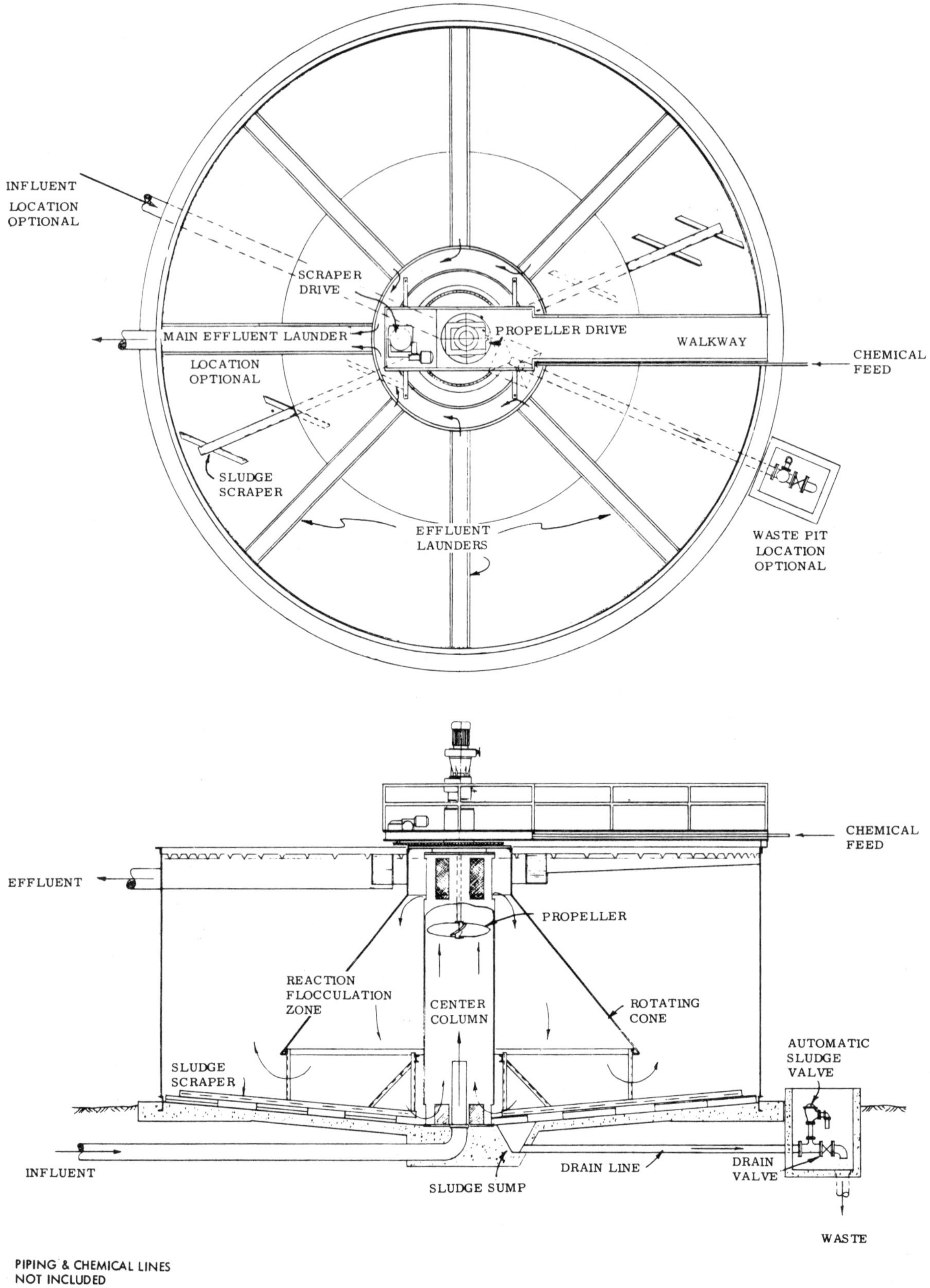

Fig. 5.24   Plan and section views of a solids-contact clarifier
(Permission of General Filter)

clarifiers, reactivators, and precipitators. The basic principles of operation are all the same, even though various manufacturers use different terms to describe how the mechanisms remove solids from water. The settled materials from coagulation or settling are referred to as sludge, and slurry refers to the suspended floc clumps in the clarifier. Sometimes the terms sludge and slurry are used interchangeably.

The internal mechanism consists of three distinct unit processes that function in the same way as any conventional coagulation-flocculation-sedimentation process chain. Sludge produced by the unit is recycled through the process to act as a coagulant aid, thereby increasing the efficiency of the processes of coagulation, flocculation, and sedimentation. This is the same principle that has been used successfully for many years in the operation of separate coagulation, flocculation, and sedimentation processes.

The advantages of using a solids-contact clarifier over operating the same three processes separately are significant, but are sometimes offset by disadvantages. For example, capital and maintenance costs are greatly reduced because the entire chain of processes is accomplished in a single tank. At the same time, though, operation of the single-unit processes requires a higher level of operator knowledge and skill. The operator must have a thorough understanding of how these processes operate and be able to imagine how all of these processes can occur in a small chamber or clarifier at the same time. Operators often have trouble visualizing what is happening in an upflow clarifier and become discouraged or upset when routine problems associated with solids-contact units occur.

A tremendous advantage in the use of the solids-contact units is the ability to adjust the volume of slurry (sludge blanket). By proper operational control, the operator can increase or decrease the volume of slurry in the clarifier as needed to cope with certain problems. During periods of severe taste and odor problems, for example, the operator can increase the sludge level and add activated carbon. The *ADSORPTIVE*[7] characteristics of activated carbon make it highly effective in treating taste and odor problems. Similarly, when coagulation fails because of increased algal activities, the operator can take advantage of the slurry accumulation to carry the plant through the severe periods of the day when the chemicals will fail to react properly because of changes in the pH, alkalinity, carbonate, and dissolved oxygen. In the conventional plant, the operator cannot respond to this type of breakdown in the coagulation process as well as the operator can with a solids-

contact unit. Once algal activities have been determined to be causing the problem (readily checked by pH and DO (dissolved oxygen)) the operator can increase the amount of slurry available during good periods of the day and remove it during periods when the coagulation process is not functioning well. By skillfully making these slurry level adjustments, the operator can maintain a high-quality effluent from the solids-contact unit.

The most serious limitation of solids-contact units is their instability during rapid changes in flow (through-put), turbidity level, and temperature. The solids-contact unit is most unstable during rapid changes in the flow rate. The operator should identify and keep in mind the design flow for which the unit was constructed. For easier use, convert the design flow to an *OVERFLOW RATE.*[8] This rate may be expressed as the rise rate in inches per minute or feet per minute. For each rise rate there will be an optimum (best) slurry level to be maintained within the unit. A rising flow rate will increase the depth of the slurry without increasing its volume or density. Conversely, a decrease in flow rate will reduce the level of slurry without changing its volume.

Sampling taps can be installed to enable the operator to monitor changes in slurry depth or concentration. The level of the slurry can be identified by placing sampling taps at various depths along the wall of the solids-contact reactor. The taps should penetrate the wall and extend into the slurry zone. It is often necessary to modify existing sampling taps or install additional sampling pipes to accomplish this.

By observing and measuring slurry depths at frequent intervals, the operator can easily monitor the rise or fall of the slurry levels; this will enable the operator to promptly make appropriate adjustments in the recirculation device and/or more tightly control the rate of change in the flow rate. This method of operational control is relatively effective in a gravity flow system when the water demands are moderate and the flow rate can be changed slowly. However, operational control in pressure systems is more difficult. Responding to rapid changes in demand or placing pumps into service at full capacity can easily upset an upflow clarifier immediately. In this case, the operator can witness the crisis occurring by observing the sampling taps, but be next to helpless to respond. The slurry will rise very rapidly in the settling zone, approach the overflow weirs, and spill onto the filters with a complete and total breakdown of the plant process.

Solids-contact clarifiers are also sensitive to severe changes in the turbidity of incoming raw water. The operator must be alert to changes in turbidity and must take immediate action. With experience, the operator will learn to accurately forecast when the turbidity may arrive at the reaction zone and will cope with the problem by increasing the chemical dosage prior to the arrival of excess turbidity. Early application of an increased chemical dosage puts the unit in a mode in which the turbidity can be handled successfully. The control of slurry and its influence on operational control during turbidity changes is discussed in greater detail later in this section.

---

[7] Adsorption (add-SORP-shun).   The gathering of a gas, liquid, or dissolved substance on the surface or interface zone of another material.

[8] Overflow Rate.   One of the guidelines for the design of settling tanks and clarifiers in treatment plants. Used by operators to determine if tanks and clarifiers are hydraulically (flow) over- or underloaded. Also called surface loading.

$$\text{Overflow Rate, GPD/sq ft} = \frac{\text{Flow, gallons/day}}{\text{Surface Area, sq ft}}$$

If we divide the overflow rate by 7.48 gallons per cubic foot and also divide by 1,440 minutes per day, we will have converted the overflow rate to a rise rate in feet per minute.

The third factor that exerts a major influence on operation of a solids-contact unit is temperature. Changes in water temperature will cause changes in the density of the water; changes in density influence the particle settling rate. In extremely cold water, consider using polymers, activated silica, powdered calcium carbonate, or some other weighting agent to aid sedimentation without affecting coagulation. Simple heating by the sun on the wall of the tank or on the flocculant particles within the container will cause a certain amount of carryover of solids to occur. Operators who are not familiar with solids-contact units tend to become upset and overreact because of the potential carryover problem. This phenomenon is not a matter of serious concern because as the position of the sun changes, the convection currents change. The clouds of flocculant particles appear and disappear in response to the currents and there is no real need to control this phenomenon if the overall settled turbidity meets your objective. So long as the major portion of the sludge blanket lies in the settling zone, the few clouds of flocculant particles (which look like billowing clouds) really do not significantly harm the operation of the unit or the quality of the water produced.

Dramatic changes in temperatures and flow rates may sometimes make it impossible to control or prevent process upsets. If the slurry rises to the weirs and is carried over onto the filters, reduce the flow rate. If possible, use weighting agents before changing flow rates in cold water. The use of weighting agents may cause problems with the slurry requiring changes in recirculation rates. However, too high a recirculation rate may also cause the slurry to overflow onto the filters. During a change in temperature (cold water), be very careful in changing the flow rate.

QUESTIONS

Write your answers in a notebook and then compare your answers with those on page 187.

5.2J   List two advantages of solids-contact units.

5.2K   How can the level of the slurry or sludge blanket be determined in solids-contact units?

5.2L   What should be done when a rapid change in turbidity is expected?

5.241 Fundamentals of Operation

The operator of a solids-contact unit controls the performance of the unit by adjusting three variables: chemical dosage, recirculation rate, and sludge control.

All three of these variables are interrelated and frequently you may have trouble distinguishing which one is the root of a problem you are encountering. However, if you will use the following analytical techniques, you should be able to separate the three fundamentals into separate groups of symptoms and diagnose the cause of the upset. This is not to say that if you have one problem, it may not co-exist with the other two.

First, let's consider chemical dosage. As in conventional treatment plants, proper consideration must be given to chemical dosage, otherwise the entire system of solids-contact clarification will collapse. There must always be sufficient alkalinity in the raw water to react with the coagulant. Assuming for practical purposes the coagulant used is aluminum sulfate, for every mg/L of alum added, 0.45 mg/L of bicarbonate alkalinity is required to complete the chemical reaction. To drive the chemical reaction sufficiently to the right (that is to say, for precipitation to occur), there should be an excess of 20 mg/L of alkalinity present. You may have to add sodium hydroxide (caustic soda), calcium hydroxide (lime), or sodium carbonate (soda ash) to cause sufficient alkalinity to be present. For example, if there was only 30 mg/L of natural alkalinity present, for every mg/L of alum added you should add 0.35 mg/L of lime if calcium hydroxide was being used.

All of this information can be verified by jar testing, which is fundamental in determining proper coagulation by chemical dosing. You should never attempt to make changes in solids-contact unit operation without first determining the proper chemical dosage through jar testing. For most solids-contact units, use the chemical dosages that produce floc that gives the lowest turbidity within a five-minute settling period after stopping the jar tester. Using the above criteria, the operator now can set the chemical feeders to dose the raw water entering the solids-contact unit.

The next control mechanism is recirculation. Here most often the plant operator is misled by intuitive judgment. The recirculation rate is established by the speed of the impeller, turbine, pumping unit, or by air injection. Any of these devices causes the slurry to recirculate through the coagulation (reaction) zone.

To help you visualize how the slurry should look in the reaction zone, take another look at the lower drawing in Figure 5.23 and note the cloud-like, billowy appearance of the flocculated slurry in this area. Under normal operating conditions, the entire mass of suspended floc clumps billows and flows within the chamber. Its motion is continually being influenced by the mixing of recirculated sludge and incoming raw water. In principle, you are attempting to chemically dose the raw water when it enters the reaction zone and is mixed with the recirculated sludge. Coagulation and flocculation occur in the reaction zone and then the water and sludge pass into the settling zone. Some sludge is recirculated and mixed with incoming raw water and the rest of the sludge settles and is removed from the bottom of the settling area. At the point where water and sludge pass from the reaction zone to the settling zone, approximately one liter of water should rise and one liter of slurry should be returned into the reaction zone.

In order to sort out the effects of chemical dosages, recirculation rate, and sludge control, you should keep a log of the speed (RPM) of the recirculation device. If air is used for mixing, then the cubic feet of air applied per minute should be recorded. There is a direct relationship between the percentage of slurry present and the speed at which the mixing device is traveling.

To control the process, the operator must maintain the correct slurry volume in the reaction zone by exercising control of the rate of recirculation. The percentage of slurry can be determined by performing a volume over volume (V/V) test. The test procedure is as follows: using a 100-mL graduated cylinder, collect a sample from the reaction zone. Let the sample sit for five minutes and then determine the volume of slurry accumulated (mL) using the formula:

$$\text{V/V, \%} = \frac{(\text{Settled Slurry, m}L)(100\%)}{\text{Total Sample Volume, m}L}$$

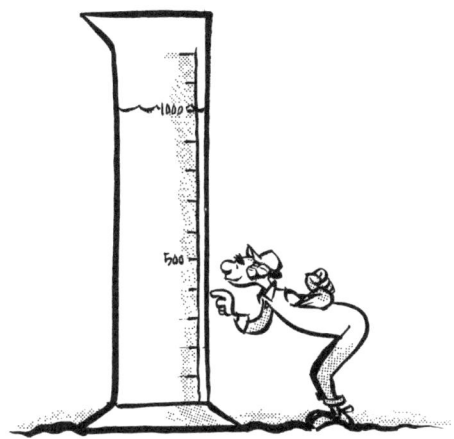

At the same time, observe the clarity of the supernatant (settled water) that remains in the graduated cylinder. The clarity of the water above the slurry (the supernatant) will indicate to you how well the chemical reaction is proceeding. The percentage of accumulated solids by volume will indicate whether a proper amount of slurry is in the reaction zone. Customarily such reactors require 5 to 20 percent solids, or a higher percentage in the graduated cylinder at the end of a five- to ten-minute settling period.

Through recordkeeping and experience, you will find an optimum percentage of solids to maintain. You should perform the above analyses hourly and more frequently when the raw water quality is undergoing change. Accurate records must be kept.

The final step in control of the solids-contact unit is the removal of sludge that has accumulated on the bottom of the clarifier (settling zone). There are several means of sludge collection; some devices are located in areas of clarifiers that hold the sludge and are controlled by opening and closing recirculation gates. Others have scrapers that collect the sludge and move it to a discharge sump. In both cases the sludge is removed by hydraulic means (water pressure) through a control valve. The sludge removal mechanisms are generally on a timer, which operates periodically for a time duration set by the operator. The means of making this judgment is quite simple. Once again, use a graduated cylinder to collect a sample from the sludge discharge line. The sludge being discharged should be 90 to 98 percent solids in a V/V test, as indicated above. A five- to ten-minute period should be sufficient to make this determination. When slurries or sludge weaker than 90 percent is pumped, the operator is discharging a considerable amount of water and not leaving enough sludge to be recirculated into the reaction zone. If the percentage is considerably greater than 90 percent, then too much sludge may be accumulating and the recirculation device could become overloaded with too much return slurry.

If you will visualize the above reactions, you can see that with increased speed of the recirculation device, a larger amount of slurry can be retained in the unit. At the same time, if this amount becomes too great, it may cause the sludge to rise and ultimately spill over the effluent weirs with the treated water. If the recirculation rate is too low, the solids may settle too soon and without sufficient recirculation will not return to the reaction zone. The absence of solids in the reaction zone causes improper coagulation. The net result is a failure of the total solids-contact system.

As you gain experience with your system, you should be able to use the principles described in this section to determine the optimum slurry volume for your solids-contact unit. The goal is to find the correct combination of recirculation rate, chemical dosage, and percentage of solids available for recycling. You should always be aware of the amount of solids in the reaction zone and, based upon practical experience, know approximately the percentage required. Some of the obvious difficulties in this judgment will occur as the raw water turbidity changes. For instance, in muddy streams carrying silt, sedimentation may occur very rapidly. Even with proper chemical dosage, increased circulation rates and higher sludge removal rates will be required to maintain sufficient slurry in the reaction zone. As the raw water turbidity becomes lighter, the increased circulation rate may cause the slurry blanket to rise to an uncontrolled depth in the settling zone. Removing too much sludge will also produce this same effect. All of these problems are readily observed in the V/V test for solids determination in the reaction zone; also, this is cause for increased observations of the V/V during water quality changes.

Another problem may be caused by cold water when the recirculation rate may be too high for the densities of the particles present. A set of recirculation speeds for warm-weather operation may be entirely different from those used during cold weather. As a remedy, the operator may select a nonionic polymer as a weighting agent to increase the settling rate in cold waters. Other alternative chemicals are powdered calcium carbonate or the use of activated silica. A note of caution in chemical dosage determination: the reactions in the jar tester should be reasonably rapid to ensure comparable reactions within the solids-contact unit.

Another important point when determining chemical dosage for a solids-contact unit is that a specific set of jar test guidelines will be needed for each plant. For example, you should determine the volume of the reaction zone and the period of detention of the raw water in that reaction zone. This, along with knowledge of the speed of the recirculation device, should allow you to determine the detention time and flocculator speed in the jar tester.

In the real world, this means if the flow rate of the solids-contact unit is 10 minutes in the reaction zone and the speed is two feet per second (0.6 m/sec), then the jar tester mixer should turn at a speed equal to two feet per second with a coagulation period of 10 minutes. You should duplicate in the jar tester, as nearly as possible, those conditions of chemical dosage, detention period, and mixing speeds that occur in the solids-contact unit. Using these guidelines, you should be able to approach approximate real-world conditions in the laboratory and better optimize chemical dosages.

5.242 Maintenance

Solids-contact units, like all waterworks equipment, require at least a minimum of maintenance. The primary consideration is the recirculating device which needs regular inspection of the belt drive, gear boxes, and lubrication. If the unit has a sludge collector, its drive and gear boxes also require the same attention. Inspect the units daily and lubricate following the manufacturer's recommendations. The contact unit may also need to be drained periodically to inspect the sludge collectors for wear and corrosion.

Sludge collector devices are usually constructed of steel within a concrete container. If a *CATHODIC PROTECTION*[9] system is provided with the unit, weekly readings of the amperes and voltage supply should be recorded. Changes in these readings indicate that a problem may be developing. Periodically inspect the cathodic protection devices. If any defects are detected, correct them as soon as possible.

QUESTIONS

Write your answers in a notebook and then compare your answers with those on page 187.

5.2M   How is the proper chemical dose selected when operating a solids-contact unit?

5.2N   List the devices that may be used to provide recirculation in a solids-contact unit.

5.2O   How is the percentage of slurry present in the reaction zone determined?

5.243   *Arithmetic for Solids-Contact Clarification*

Successful operation of a solids-contact clarification unit requires very little arithmetic. The volume over volume test provides you with an indication of the settleability of the slurry or sludge in the sludge blanket. The detention time in the reaction zone is important to ensure that there is sufficient mixing time and time for the chemical reactions (coagulation) to take place.

If the raw water is low in alkalinity and alum is the coagulant, lime may have to be added to provide enough alkalinity. With the necessary information and by following the step-by-step procedures outlined in this section, the setting on the lime feeder can be easily calculated.

FORMULAS

The volume over volume (V/V) test requires the collection of 100 mL slurry from a solids-contact unit in a 100-mL graduated cylinder. The slurry is allowed to sit for ten minutes and the volume of the settled slurry on the bottom of the graduated cylinder is measured and recorded.

$$V/V, \% = \frac{(\text{Settled Slurry, m}L)(100\%)}{\text{Total Volume, m}L}$$

The detention time in the reaction zone of a solids-contact unit is calculated the same way the detention time in any basin is calculated. The flow is divided into the volume of the reaction zone. Any necessary adjustments are made for units, such as multiplying by 60 minutes per hour to convert a detention time from hours to minutes.

$$\frac{\text{Detention}}{\text{Time, min}} = \frac{(\text{Reaction Zone Volume, gal})(24 \text{ hr/day})(60 \text{ min/hr})}{\text{Flow, gal/day}}$$

To determine the lime dose that must be added to a raw water being treated, we need to know (1) the alkalinity of the raw water, (2) the alkalinity that must be present to ensure complete precipitation of the alum, (3) the amount of alkalinity that reacts with the alum, and (4) the amount of lime that reacts with the alum; that is:

1. Raw Water Alkalinity, mg/L as HCO_3^-,

2. Alkalinity Present for Precipitation, mg/L (at least 30 mg/L),

3. 0.45 mg/L Alkalinity (HCO_3^-) reacts with 1 mg/L Alum, and

4. 0.35 mg/L Lime ($Ca(OH)_2$) reacts with 1 mg/L Alum.

Procedure to Calculate Lime Dose in mg/L

1. Determine the alkalinity available to react with the alum.

$$\frac{\text{Alkalinity Available,}}{\text{mg/}L} = \frac{\text{Raw Water}}{\text{Alkalinity, mg/}L} - \frac{\text{Alkalinity Present}}{\text{for Precipitation, mg/}L}$$

2. Determine the amount of alum that will react with the available alkalinity.

$$\frac{0.45 \text{ mg/}L \text{ Alkalinity}}{1.0 \text{ mg/}L \text{ Alum}} = \frac{\text{Alkalinity Available, mg/}L}{\text{Alum Reacting, mg/}L}$$

or

$$\frac{\text{Alum Reacting,}}{\text{mg/}L} = \frac{(1.0 \text{ mg/}L \text{ Alum})(\text{Alkalinity Available, mg/}L)}{0.45 \text{ mg/}L \text{ Alkalinity}}$$

3. Determine the milligrams per liter of alum that needs additional alkalinity (or is unreacted with). The total alum required is determined by the jar test.

$$\frac{\text{Alum Needing}}{\text{Alkalinity, mg/}L} = \text{Total Alum, mg/}L - \text{Alum Reacting, mg/}L$$

4. Determine the lime dose in milligrams per liter.

$$\frac{1 \text{ mg/}L \text{ Alum}}{0.35 \text{ mg/}L \text{ Lime}} = \frac{\text{Alum Needing Alkalinity, mg/}L}{\text{Lime Dosage, mg/}L}$$

or

$$\frac{\text{Lime Dosage,}}{\text{mg/}L} = \frac{(0.35 \text{ mg/}L \text{ Lime})(\text{Alum Needing Alkalinity, mg/}L)}{1 \text{ mg/}L \text{ Alum}}$$

5. Determine the setting on the lime feeder in pounds per day.

$$\frac{\text{Lime Feed,}}{\text{lbs/day}} = (\text{Flow, MGD})(\text{Lime Dose, mg/}L)(8.34 \text{ lbs/gal})$$

6. Determine the setting on the lime feeder in grams per minute.

$$\frac{\text{Lime Feed,}}{\text{gm/min}} = \frac{(\text{Flow, MGD})(\text{Lime Dose, mg/}L)(3.785 \text{ }L\text{/gal})(1,000,000\text{/Million})}{(24 \text{ hr/day})(60 \text{ min/hr})(1,000 \text{ mg/gm})}$$

We multiplied by 3.785 liters per gallon to cancel out the gallons in MGD and the liters in mg/L. By multiplying 1,000,000/1 Million we took care of the Million units. When we divided by 24 hr/day and 60 min/hr, we converted the feed rate from days to minutes. By dividing by 1,000 mg/gm we changed the amount of lime to be fed from milligrams to grams so we could work with convenient numbers.

---

[9] *Cathodic (ca-THOD-ick) Protection.   An electrical system for prevention of rust, corrosion, and pitting of metal surfaces which are in contact with water or soil. A low-voltage current is made to flow through a liquid (water) or a soil in contact with the metal in such a manner that the external electromotive force renders the metal structure cathodic. This concentrates corrosion on auxiliary anodic parts which are deliberately allowed to corrode instead of letting the structure corrode.*

EXAMPLE 4

A graduated cylinder is filled to the 100-mL level with the slurry from a solids-contact unit. After ten minutes there is 21 mL of slurry on the bottom and 79 mL of clear water remaining in the top part of the cylinder. This is the volume over volume (V/V) test.

Known	**Unknown**
Settled Slurry, mL = 21 mL	V/V,%
Total Volume, mL = 100 mL	

Determine V/V as a percent.

$$V/V, \% = \frac{(\text{Settled Slurry, mL})(100\%)}{\text{Total Volume, mL}}$$

$$= \frac{(21 \text{ mL})(100\%)}{100 \text{ mL}}$$

$$= 21\%$$

EXAMPLE 5

The reaction zone in a solids-contact clarifier is 11 feet in diameter and 4 feet high. Find the detention time in minutes in the reaction zone if the flow is 2 MGD.

Known	**Unknown**
Diameter, ft = 11 ft	Detention Time, min
Height, ft = 4 ft	
Flow, MGD = 2 MGD	
= 2,000,000 GPD	

1. Calculate the volume of the reaction zone in gallons.

$$\text{Volume, gal} = (0.785)(\text{Diameter, ft})^2(\text{Height, ft})(7.48 \text{ gal/cu ft})$$

$$= (0.785)(11 \text{ ft})^2(4 \text{ ft})(7.48 \text{ gal/cu ft})$$

$$= 2,842 \text{ gal}$$

2. Calculate the detention time in the reaction zone in minutes.

$$\text{Detention Time, min} = \frac{(\text{Reaction Zone Volume, gal})(24 \text{ hr/day})(60 \text{ min/hr})}{\text{Flow, gal/day}}$$

$$= \frac{(2,842 \text{ gal})(24 \text{ hr/day})(60 \text{ min/hr})}{2,000,000 \text{ gal/day}}$$

$$= 2.05 \text{ min}$$

EXAMPLE 6

A raw water has an alkalinity of 36 mg/L as bicarbonate (HCO_3^-). A chemical dose of 52 mg/L of alum (from a jar test) is needed to reduce the turbidity from 75 TU down to 1.0 TU. At least 30 mg/L of alkalinity must be present to ensure complete precipitation of the alum added. Find the dose of lime ($Ca(OH)_2$) in mg/L that will be needed to complete this reaction.

Known	**Unknown**
Raw Water Alkalinity, mg/L = 36 mg/L	Lime Dose, mg/L
Total Alum Required, mg/L = 52 mg/L	
Alkalinity Present for Precipitation, mg/L = 30 mg/L	

1. Determine the alkalinity available to react with the alum.

$$\text{Alkalinity Available, mg/L} = \text{Raw Water Alkalinity, mg/L} - \text{Alkalinity Present for Precipitation, mg/L}$$

$$= 36 \text{ mg/L} - 30 \text{ mg/L}$$

$$= 6 \text{ mg/L}$$

2. Determine the amount of alum that will react with the available alkalinity.

$$\text{Alum Reacting, mg/L} = \frac{(1.0 \text{ mg/L Alum})(\text{Alkalinity Available, mg/L})}{0.45 \text{ mg/L Alkalinity}}$$

$$= \frac{(1.0 \text{ mg/L Alum})(6 \text{ mg/L Alkalinity})}{0.45 \text{ mg/L Alkalinity}}$$

$$= 13.3 \text{ mg/L Alum}$$

3. Determine the milligrams per liter of alum that needs additional alkalinity (or is not reacted with).

$$\text{Alum Needing Alkalinity, mg/L} = \text{Total Alum, mg/L} - \text{Alum Reacting, mg/L}$$

$$= 52 \text{ mg/L} - 13.3 \text{ mg/L}$$

$$= 38.7 \text{ mg/L}$$

4. Determine the lime dose in milligrams per liter.

$$\text{Lime Dose, mg/L} = \frac{(0.35 \text{ mg/L Lime})(\text{Alum Needing Alkalinity, mg/L})}{1 \text{ mg/L Alum}}$$

$$= \frac{(0.35 \text{ mg/L Lime})(38.7 \text{ mg/L Alum})}{1 \text{ mg/L Alum}}$$

$$= 13.5 \text{ mg/L}$$

NOTE: This dose may be verified by the use of a jar test by selecting an alum dose and trying different lime doses.

EXAMPLE 7

If the raw water in Example 6 needs a lime dose of 13.5 mg/L, what should be the setting on the lime feeder in (1) pounds per day, and (2) grams per minute when the flow is 2.0 MGD?

Known	**Unknown**
Lime Dose, mg/L = 13.5 mg/L	1. Lime Feed, lbs/day
Flow, MGD = 2.0 MGD	2. Lime Feed, gm/min

1. Determine the setting on the lime feeder in pounds per day.

$$\text{Lime Feed, lbs/day} = (\text{Flow, MGD})(\text{Lime Dose, mg/L})(8.34 \text{ lbs/gal})$$

$$= (2.0 \text{ MGD})(13.5 \text{ mg/L})(8.34 \text{ lbs/gal})$$

$$= 225 \text{ lbs lime/day}$$

2. Determine the setting on the lime feeder in grams per minute.

$$\text{Lime Feed, gm/min} = \frac{\text{(Flow, MGD)(Lime Dose, mg/}L\text{)(3.785 }L\text{/gal)(1,000,000/Million)}}{\text{(24 hr/day)(60 min/hr)(1,000 mg/gm)}}$$

$$= \frac{\text{(2.0 MGD)(13.5 mg/}L\text{)(3.785 }L\text{/gal)(1,000,000/Million)}}{\text{(24 hr/day)(60 min/hr)(1,000 mg/gm)}}$$

$$= 71 \text{ grams lime/minute}$$

5.25 Sludge Handling

5.250 Sludge Characteristics

Water treatment plant sludges are typically alum sludges, with solids concentrations varying from 0.25 to 10 percent when removed from the basin. In gravity flow sludge removal systems, the solids concentration should be limited to about 3 percent. If the sludge is to be pumped, solids concentrations as high as 10 percent can be readily transported.

In horizontal-flow sedimentation basins preceded by coagulation and flocculation, over 50 percent of the floc will settle out in the first third of the basin length. Operationally, this must be considered when establishing the frequency of operation of sludge removal equipment. Also, you must consider the volume or amount of sludge to be removed and the sludge storage volume available in the basin.

5.251 Sludge Removal Systems

Sludge that accumulates on the bottom of sedimentation basins must be removed periodically for the following reasons:

1. To prevent interference with the settling process (such as resuspension of solids due to scouring);

2. To prevent the sludge from becoming *SEPTIC*[10] or providing an environment for the growth of microorganisms that can create taste and odor problems; and

3. To prevent excessive reduction in the cross-sectional area of the basin (reduction of detention time).

In large-scale plants, sludge is normally removed on an intermittent basis with the aid of mechanical sludge removal equipment. However, in smaller plants with low solids loading, manual sludge removal may be more cost-effective.

In manually cleaned basins, the sludge is allowed to accumulate until it reduces settled water quality. High levels of sludge reduce the detention time and floc carries over to the filters. The basin is then dewatered (drained), most of the sludge is removed by stationary or portable pumps, and the remaining sludge is removed with squeegees and hoses. Basin floors are usually sloped toward a drain to help sludge removal. The frequency of shutdown for cleaning will vary from several months to a year or more, depending on source water quality (amount of suspended matter in the water).

In larger plants, a variety of mechanical devices (Figure 5.25) can be used to remove sludge including:

1. Mechanical rakes (Figure 5.7),

2. Drag-chain and flights, and

3. Traveling bridges (Figure 5.8).

Circular or square basins are usually equipped with rotating sludge rakes. Basin floors are sloped toward the center, and the sludge rakes progressively push the sludge toward a center outlet.

In rectangular basins, the simplest sludge removal mechanism is the chain and flight system. An endless chain outfitted with wooden flights (scrapers) pushes the sludge into a sump. The disadvantage of this system and of the rotating rakes previously described is high operation and maintenance costs. Most of the moving (wearing) parts are submerged so the basin has to be dewatered to perform major maintenance.

In an attempt to reduce operation and maintenance costs (as well as capital equipment costs), and to improve sludge removal equipment maintainability, the traveling bridge was developed. This bridge looks like an old highway bridge except it has no deck for cars. The traveling bridge spans the width of the sedimentation basin and travels along the length of the basin walls. Movable sludge sweeps, which are hung from the bridge structure, remove the sludge from the basin floor with suction pumps or by siphon action. There are few submerged parts in this system and these can normally be removed for maintenance without dewatering the basin. Traveling bridge sludge removal systems will operate effectively on the simplest of basin designs.

Sludge may be discharged into sludge basins or ponds for liquid-solids separation. Ultimately the sludge may be disposed of in a landfill. See Chapter 10, "Plant Operation," and Chapter 17 (Volume II), "Handling and Disposal of Process Wastes," for additional details.

QUESTIONS

Write your answers in a notebook and then compare your answers with those on page 187.

5.2P Alum sludge solids concentrations typically vary from _____ to _____ percent when removed from a sedimentation basin.

5.2Q What factors must be considered when determining how frequently you will need to operate sludge removal equipment?

5.2R Why must accumulated sludge be removed periodically from the bottom of sedimentation basins?

---

[10] *Septic (SEP-tick). A condition produced by bacteria when all oxygen supplies are depleted. If severe, the bottom deposits produce hydrogen sulfide, the deposits and water turn black, give off foul odors, and the water has a greatly increased chlorine demand.*

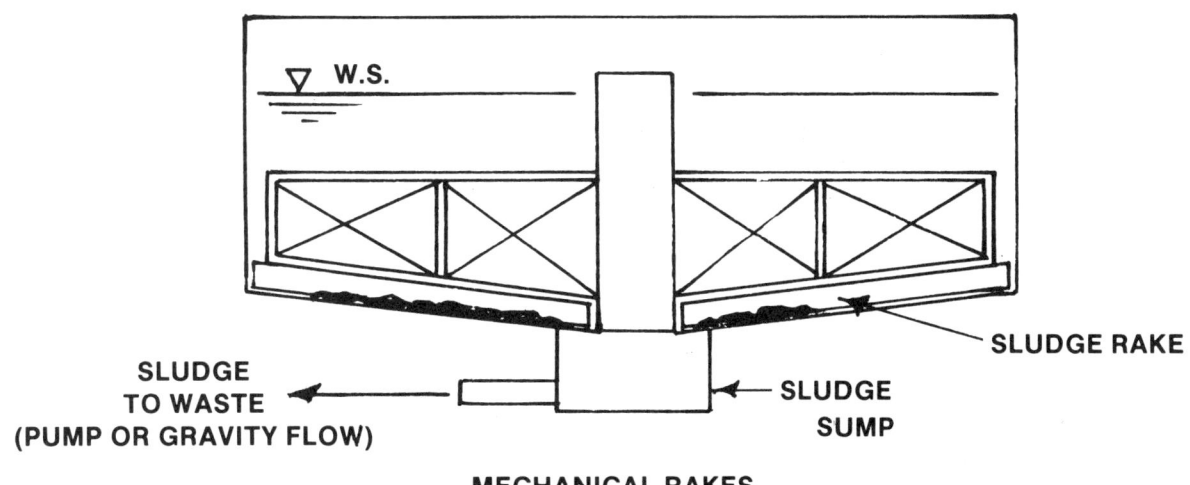

MECHANICAL RAKES
(CIRCULAR OR SQUARE BASINS)

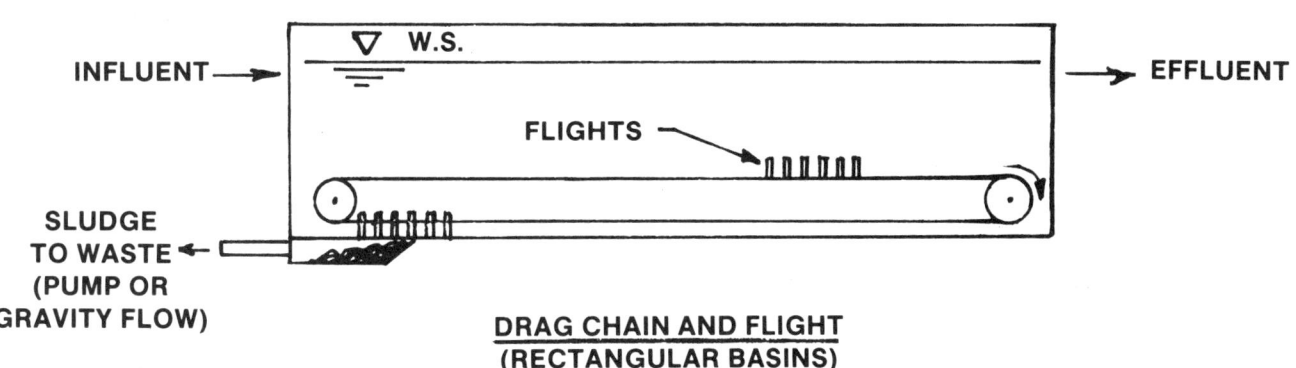

DRAG CHAIN AND FLIGHT
(RECTANGULAR BASINS)

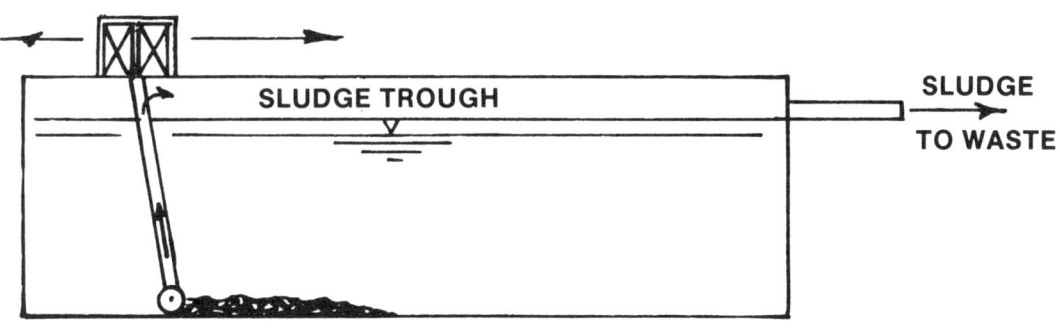

TRAVELING BRIDGE — PUMP OR SIPHON
(RECTANGULAR BASINS)

Fig. 5.25   Mechanical sludge removal systems

5.252   Operation of Sludge Removal Equipment

Accumulated sludge in the basin bottom or sludge sump is periodically removed for further processing (dewatering) and ultimate disposal. The frequency of sludge removal or transfer depends on the rate of sludge buildup and this is directly related to the amount of suspended\material and floc removed in the sedimentation process. Other factors influencing the frequency of sludge removal include the size of the sludge sump and the capacity of the sludge pump. In alum coagulation operations this generally means that sludge removal equipment need only be operated once per shift or perhaps less frequently (daily). If polymers are used as the primary coagulant and the source water suspended solids concentration is low (less than 5 mg/L), sludge removal equipment need only be operated once or twice per week.

In some water treatment plants, the operator measures the depth of the accumulated sludge deposit and uses this information to determine the operating frequency of the sludge removal equipment. This measurement can be made with a sludge blanket sounder, a bubbler tube, an aspirator, or an ultrasonic level indicator.

Perhaps the simplest sludge blanket measuring device is the "sludge blanket sounder." This device consists of a one-quarter inch (6 mm) thick hardware cloth disc about 18 inches (450 mm) in diameter. The disc is suspended from a lightweight chain by a three-point suspension (see Figure 5.26).

In using this sounding device, the disc is slowly lowered into the sedimentation basin. When the disc reaches the top of the sludge blanket it stops its descent, and a depth reading is taken from markings on the chain.

After you have determined an appropriate time interval for sludge equipment operation, the sludge discharge should be periodically checked to determine the concentration of the sludge solids. This is generally done by observation. If the sludge is too thick and bulks, the frequency of sludge removal should be increased. Likewise, if the sludge concentration is too low in solids (soupy), decrease the frequency of sludge removal.

Some water treatment plants are furnished with semiautomatic sludge removal equipment which can be adjusted to change the frequency of sludge removal equipment operation by merely resetting a time clock.

QUESTIONS

Write your answers in a notebook and then compare your answers with those on page 187.

5.2S   How can the depth of sludge in a sedimentation basin be measured?

5.2T   If the sludge being pumped from a sedimentation basin is too low in solids (soupy), what should the operator do?

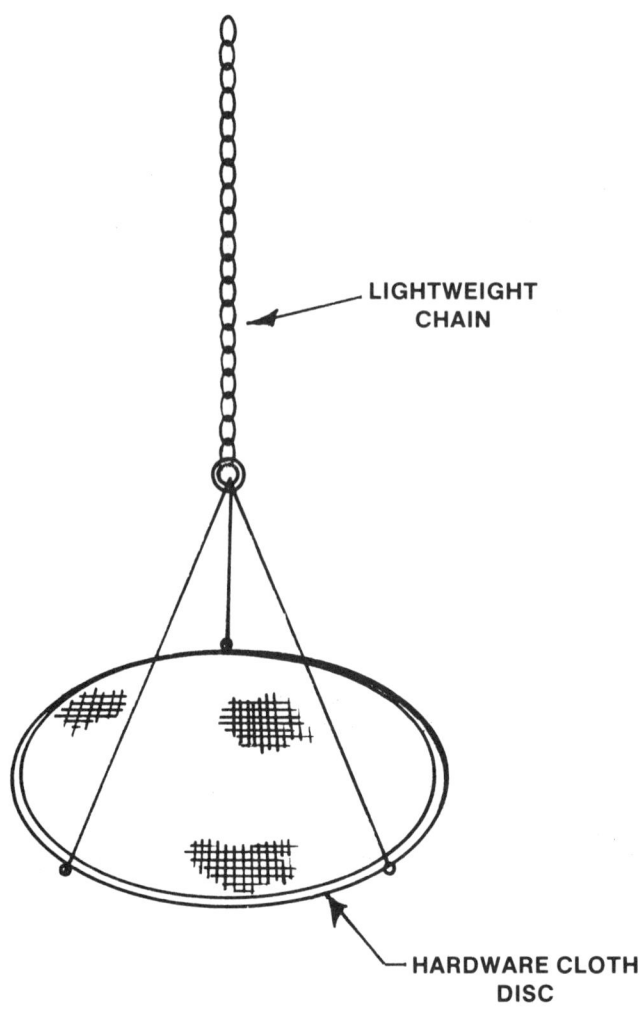

LIGHTWEIGHT CHAIN

HARDWARE CLOTH DISC

Fig. 5.26   Sludge blanket sounder
(Adapted from *MANUAL OF WASTEWATER OPERATIONS*, prepared by the Texas Water Utilities Association)

5.3   INTERACTION WITH OTHER TREATMENT PROCESSES

The purposes of the sedimentation process are to remove suspended solids from the water being treated and to reduce the load on the filters. If adequate detention time and basin surface area are provided in the sedimentation basins, solids removal efficiencies greater than 95 percent can be achieved. However, high sedimentation basin removal efficiencies may not always be the most cost-effective way to remove suspended solids.

In low-turbidity source waters (less than about 10 *TU*[11]), effective coagulation, flocculation, and filtration may produce a satisfactory filtered water without the need for sedimentation.

---

[11] *Turbidity Units (TU).   Turbidity units are a measure of the cloudiness of water. If measured by a nephelometric (deflected light) instrumental procedure, turbidity units are expressed in nephelometric turbidity units (NTU) or simply TU. Those turbidity units obtained by visual methods are expressed in Jackson Turbidity Units (JTU) which are a measure of the cloudiness of water; they are used to indicate the clarity of water. There is no real connection between NTUs and JTUs. The Jackson turbidimeter is a visual method and the nephelometer is an instrumental method based on deflected light.*

In this case, the coagulation-flocculation process is operated to produce a highly filtrable *PINPOINT FLOC*, which does not readily settle due to its small size. Instead, the pinpoint floc is removed by the filters. However, there is a practical limitation in applying this concept to higher turbidity conditions. If the filters become overloaded with suspended solids, they will quickly clog and need frequent backwashing. This can limit plant production and cause a degradation in filtered water quality.

Thus, the sedimentation process should be operated from the standpoint of overall plant efficiency. If the source water turbidity is only 3 TU, and jar tests indicated that 0.5 mg/*L* of cationic polymer is the most effective coagulant dosage, then you cannot expect the sedimentation process to remove a significant fraction of the suspended solids. On the other hand, source water turbidities in excess of 50 TU will probably require a high alum dosage (or other primary coagulant) for efficient solids removal. In this case, the majority of the suspended particles and alum floc should be removed in the sedimentation basin.

5.4 PROCESS CONTROL

The actual performance of sedimentation basins depends on the settling characteristics of the suspended particles and the flow rate through the sedimentation basins. To control the settling characteristics of the suspended particles, adjust the chemical coagulant dose and the coagulation-flocculation process (see Chapter 4). The flow rate through the sedimentation basin controls the efficiency of the process in removing suspended particles. The higher the flow rate, the lower the efficiency (the fewer suspended particles are removed). Once the actual flow rate becomes greater than the design flow rate, you can expect an increase in suspended particles flowing over the V-notch weirs.

From a practical standpoint, you will want to operate sedimentation basins near design flows. However, to achieve the intended removal of suspended particles once design flows are exceeded, suspended particles leaving the sedimentation basin may overload the filters with solids and require additional filter backwashing. Study the settling characteristics of the particles by using laboratory jar tests. Then verify your test re-

sults and make adjustments based on actual performance of the water treatment plant.

During periods of low flows the use of all sedimentation basins may not be necessary. Since the cost to operate a basin is very low, it is common practice to keep all basins in service except during periods of draining for maintenance and repairs.

QUESTIONS

Write your answers in a notebook and then compare your answers with those on page 187.

5.3A Under what circumstances are sedimentation basins needed to treat water?

5.4A The actual performance of sedimentation basins depends on what two major factors?

End of Lesson 1 of 2 Lessons
on
SEDIMENTATION

Please answer the discussion and review questions next.

DISCUSSION AND REVIEW QUESTIONS

Chapter 5. SEDIMENTATION

(Lesson 1 of 2 Lessons)

At the end of each lesson in this chapter you will find some discussion and review questions. The purpose of these questions is to indicate to you how well you understand the material in the lesson. Write the answers to these questions in your notebook before continuing.

1. How would you adjust a sedimentation process to colder water temperatures?

2. What problems are created when the sludge buildup on the bottom of the sedimentation tank becomes too great?

3. Why are rectangular sedimentation basins commonly used to treat water?

4. Why should water treatment plants have at least two sedimentation basins?

5. Sudden changes in what three factors can cause operating problems in solids-contact units?

6. What problems can develop in a solids-contact unit if the recirculation rate is too high or too low?

7. What problems could develop if sludge is allowed to remain too long on the bottom of a sedimentation basin?

8. How often should sludge removal equipment be operated?

CHAPTER 5.   SEDIMENTATION

(Lesson 2 of 2 Lessons)

5.5   OPERATING PROCEDURES ASSOCIATED WITH NORMAL PROCESS CONDITIONS

5.50   Indicators of Normal Operating Conditions

From a water quality standpoint, filter effluent turbidity is a good indication of overall process performance. However, you must still monitor the performance of each of the individual water treatment processes, including sedimentation, in order to anticipate quality or performance changes. Normal operating conditions are considered to be conditions within the operating ranges of your plant, while abnormal conditions are unusual or difficult-to-handle conditions. Changes in raw water quality may be considered a normal condition for many plants and an abnormal condition for other water treatment plants.

In the normal operation of the sedimentation process you will monitor (Figure 5.27):

1. Turbidity of the water entering and leaving the sedimentation basin, and

2. Temperature of the entering water.

Turbidity of the entering water indicates the floc or solids loading on the sedimentation process. Turbidity of the water leaving the basin reveals the effectiveness or efficiency of the sedimentation process. Low levels of turbidity are desirable to minimize the floc loading on the filters.

Temperature of the water entering the sedimentation basin is very important. Usually water temperature changes are gradual, depending on time of the year and the weather. As the water becomes colder, the particles will settle more slowly. To compensate for this change, you should perform jar tests (refer to Chapter 4) and adjust the coagulant dosage to produce a heavier and thus a faster settling floc. Another possibility is that if the demand for water decreases during colder weather, the flow to be treated can be reduced which will produce longer detention times. Longer detention times will allow slower settling particles or floc to be removed in the sedimentation basins.

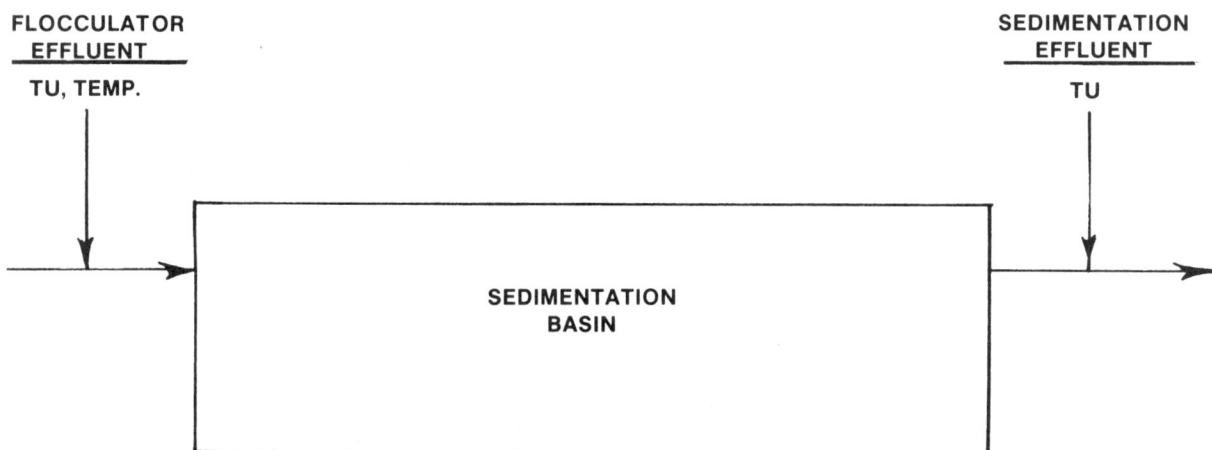

FLOCCULATOR EFFLUENT

TU, TEMP.

SEDIMENTATION EFFLUENT

TU

SEDIMENTATION BASIN

LEGEND:
 TU-TURBIDITY
 TEMP.-TEMPERATURE

Fig. 5.27   Sedimentation process monitoring water quality indicator and sample points

Visual checks of the sedimentation process should include observation of floc settling characteristics, distribution of floc at the basin inlet, and clarity of settled water spilling over the launder weirs. An uneven distribution of floc, or poorly settling floc, may indicate that a raw water quality change has occurred or that operational problems may have developed.

5.51 Process Actions

In rectangular and circular sedimentation basins, you can generally make a judgment about the performance of the sedimentation process by observing how far the floc is visible beyond the basin inlet. When sedimentation is working well, the floc will only be visible for a short distance. When sedimentation is poor, the floc will be visible for a long distance beyond the inlet.

In upflow or solids-contact clarifiers, the depth of the sludge blanket and the density of the blanket are useful monitoring tools. If the sludge blanket is of normal density (measured as milligrams of solids per liter of water) but is very close to the surface, more sludge should be wasted. If the blanket is of unusually light density, the coagulation-flocculation process (chemical dosage) must be adjusted to improve performance.

With any of the sedimentation processes, it is useful to observe the quality of the effluent as it passes over the launder weir. Flocs coming over at the ends of the basin are indicative of density currents, short-circuiting, sludge blankets that are too deep, or high flows. The clarity of the effluent is also a reliable indicator of coagulation-flocculation efficiency.

Process equipment should be checked regularly to ensure adequate performance. Proper operation of sludge removal equipment should be verified each time the equipment is operated, since sludge removal discharge piping systems are subject to clogging. Free-flowing sludge can be readily observed if sight glasses are incorporated in the sludge discharge piping. Otherwise, the outlet of the sludge line should be observed during sludge pumping. Frequent clogging of the sludge discharge line is an indication that the sludge concentration is too high. If this occurs, you should increase the frequency of operation of the sludge removal equipment. This problem can be accurately diagnosed by performing a sludge solids volume analysis[12] in the laboratory, if this capability is available to the operator.

The operator should routinely inspect physical facilities and equipment as part of good housekeeping and maintenance practice. Abnormal equipment conditions should be corrected or reported to maintenance personnel, and basin water surfaces and launders should be kept free of leaves, twigs, and other debris which might jam or foul mechanical equipment, such as valves and meters.

A summary of routine sedimentation process actions is given in Table 5.2. Actual frequency of monitoring should be based on the source of the water supply and variations in the supply.

5.52 Recordkeeping

Recordkeeping is one of the most important *ADMINISTRATIVE* functions of the water treatment plant operator. In the routine daily operation of the sedimentation process, you will maintain a daily operations log of process performance and water quality characteristics. Keep the following records:

1. Influent and effluent turbidity and influent temperature,

2. Process production inventory (amount of water processed and volume of sludge produced), and

3. Process equipment performance (types of equipment in operation, maintenance procedures performed, and equipment calibration).

Entries in logs should be neat and legible, should reflect the date and time of an event, and should be initialed by the operator making the entry.

QUESTIONS

Write your answers in a notebook and then compare your answers with those on pages 187 and 188.

5.5A   What items should an operator monitor during the normal operation of the sedimentation process?

5.5B   How often should visual observations and laboratory evaluation of sedimentation process performance be conducted?

5.5C   How can an operator determine if sludge lines are free flowing?

5.5D   What should be attempted if the sludge line plugs frequently?

5.5E   In the routine operation of the sedimentation process, what types of records should be maintained?

5.6 OPERATING PROCEDURES ASSOCIATED WITH ABNORMAL PROCESS CONDITIONS

5.60 Indicators of Abnormal Conditions

Sudden changes in source or process water quality indicators, such as turbidity, pH, alkalinity, temperature, chlorine demand (source water), and color, are signals that you should immediately review the performance of the coagulation-flocculation process and also the sedimentation process.

---

[12] Collect a sludge sample and pour a known volume into a drying dish. Place the sample dish in a drying oven and evaporate the sample to dryness (usually about one hour) at 103 to 105°C. Weigh the remaining solids.

$$Sludge\ Solids,\ \% = \frac{(Weight\ of\ Sample,\ mg)(1\ mL)(100\%)}{(Volume\ of\ Sample,\ mL)(1,000\ mg)}$$

TABLE 5.2 SUMMARY OF ROUTINE SEDIMENTATION PROCESS ACTIONS

1. Monitor Process Performance and Evaluate Water Quality Conditions	Location	Frequency[a]	Possible Operator Actions
• Turbidity	Influent/Effluent	At least once every 2 hours	1. Increase sampling frequency when process water quality is variable.
• Temperature	Influent	Occasionally	2. Perform jar tests if indicated (see procedure in Chapter 4, "Coagulation and Flocculation").
			3. Make necessary process changes.
			a. Change coagulant.
			b. Adjust coagulant dosage.
			c. Adjust flash mixer/flocculator mixing intensity.
			d. Change frequency of sludge removal.
			4. Verify response at appropriate time.

2. Make Visual Observations	Location	Frequency[a]	Possible Operator Actions
• Floc settling characteristics	First half of basin	At least once per 8-hour shift	1. Perform jar tests if indicated.
• Floc distribution	Inlet	At least once per 8-hour shift	2. Make necessary process changes.
			a. Change coagulant.
			b. Change coagulant dosage.
• Turbidity (clarity) of settled water	Launders or settled water conduit	At least once per 8-hour shift	c. Adjust flash mixer/flocculator mixing intensity.
			d. Change frequency of sludge removal.
			3. Verify response to process changes at appropriate time.

3. Check Process and Sludge Removal Equipment Condition	Location	Frequency[a]	Possible Operator Actions
• Noise	Various	Once per 8-hour shift	1. Correct minor problems.
• Vibration	Various	Once per 8-hour shift	2. Notify others of major problems.
• Leakage	Various	Once per 8-hour shift	
• Overheating	Various	Once per 8-hour shift	

4. Operate Sludge Removal Equipment	Location	Frequency[a]	Possible Operator Actions
• Perform normal operations sequence	Sed. Basin	Depends on process conditions (may vary from once per day to several days or more)	1. Change frequency of operation.
			a. If sludge is too watery, decrease frequency of operation and/or pumping rate.
• Observe conditions of sludge being removed	Sed. Basin	Depends on process conditions (may vary from once per day to several days or more)	b. If sludge is too dense, bulks, or clogs discharge lines, increase frequency of operation and/or pumping rate.
			c. If sludge is septic, increase frequency of operation and/or pumping rate.

5. Inspect Facilities	Location	Frequency[a]	Possible Operator Actions
• Check sedimentation basins	Various	Once every 2 hours	1. Report abnormal conditions.
• Observe basin water levels and depth of water flowing over launder weirs	Various	Once per 8-hour shift	2. Make flow changes (see Chapter 10, "Plant Operation"), or adjust launder weirs.
• Observe basin water surface	Various	Once per 8-hour shift	3. Remove debris from basin water surface.
• Check for algae buildup on basin walls and launders	Various	Occasionally	

[a] Frequency of monitoring should be based on the source of the water supply and variations in the supply.

5.61 Process Actions

Significant changes in source water turbidity levels, either increases or decreases, require that you verify the effectiveness of the sedimentation process in removing suspended solids and floc. Measurement of turbidity levels at the sedimentation basin inlet and outlet will give you a rough idea of process removal efficiency. Grab samples can be used for this determination. Visual observations of floc dispersion and settling characteristics will also help you evaluate process performance.

Increasing source water turbidity levels may be the result of rainfall and runoff into a river, stream, or impoundment serving the treatment plant. If turbidity levels are increasing rapidly, verify the effectiveness of the coagulant chemicals and dosages being applied at the flash mixer. The efficiency of the coagulation-flocculation process directly affects the performance of the sedimentation process. Performance of jar tests (described in Chapter 4, "Coagulation and Flocculation") in the laboratory may be used to simulate process performance in the treatment plant. Use the test results to adjust chemical dosages in the flash mixer.

In the event that higher dosages of alum or other coagulants are required for effective removal of increased suspended solids loads, you may also need to increase the frequency of operation of sludge removal equipment. On the other hand, if source water turbidity levels decrease, less frequent operation of sludge removal equipment may be indicated.

Changes in source water alkalinity and pH caused by storms, waste discharges, or spills have a significant impact on the performance of the sedimentation process as a result of decreased coagulation-flocculation process performance. Again, you need to perform jar tests to assess the impact of source water or process water quality changes.

Water temperature changes may also require a reevaluation of process performance. Decreasing water temperatures lower the rate at which particles settle, while higher water temperatures increase particle settling velocities. Thus, temperature changes may also require that jar tests be performed to establish optimum floc settling rates. Temperature changes are usually gradual over time so sudden changes in temperature are unlikely unless a source water change is made.

Sudden increases in settled water turbidity could spell trouble for the operator in the operation of the filtration process. Floc carryover from the sedimentation basin will cause premature clogging of filters and may result in the degradation of filtered water quality. Filtration process actions are discussed in Chapter 6, "Filtration."

Table 5.3 gives a summary of sedimentation process problems, how to identify the causes of problems, and also how to correct the problems.

QUESTIONS

Write your answers in a notebook and then compare your answers with those on page 188.

5.6A What water quality indicator is used as a rough measure of sedimentation basin process removal efficiency?

5.6B What problems can be created by a sudden increase in settled water turbidity?

5.7 START-UP AND SHUTDOWN PROCEDURES

5.70 Conditions Requiring Implementation of Start-Up and Shutdown Procedures

Start-up or shutdown of the sedimentation process is *NOT* a routine operating procedure in most water treatment plants that are operated on a continuous basis. Implementation of these procedures is generally associated with a complete plant shutdown for periodic maintenance and cleaning, which is generally performed on an annual basis. In some instances, a shutdown may result from a major process failure. Photographs of special features provide a visual record of events or conditions that may be difficult to illustrate when basins are full of water.

5.71 Implementation of Start-Up/Shutdown Procedures

Typical actions performed by the operator in the start-up or shutdown of the sedimentation process are outlined below.

5.710 Start-Up Procedures

1. Check operational status and mode of operation (manual or automatic) of equipment and physical facilities.

   a. Check that basin drain valves are closed.

   b. Check that basin isolation gates or stop logs are removed.

   c. Check that launder weir plates are set at equal elevations.

   d. Check to ensure that all trash, debris, and tools have been removed from basin.

2. Test sludge removal equipment.

   a. Check that mechanical equipment is properly lubricated and ready for operation.

   b. Observe operation of sludge removal equipment.

TABLE 5.3   SEDIMENTATION PROCESS TROUBLESHOOTING

1. Source Water Quality Changes	Operator Actions	Possible Process Changes
• Turbidity • Temperature • Alkalinity • pH • Color	1. Perform necessary analyses to determine extent of change. 2. Evaluate overall process performance. 3. Perform jar tests if indicated. 4. Make appropriate process changes (see right-hand column, Possible Process Changes). 5. Increase frequency of process monitoring. 6. Verify response to process changes at appropriate time (be sure to allow sufficient time for change to take effect).	1. Change coagulant. 2. Adjust coagulant dosage. 3. Adjust flash mixer/flocculator mixing intensity. 4. Change frequency of sludge removal (increase or decrease). 5. Increase alkalinity by adding lime, caustic soda, or soda ash.
2. Flocculation Process Effluent Quality Changes	**Operator Actions**	**Possible Process Changes**
• Turbidity • Alkalinity • pH	1. Evaluate overall process performance. 2. Perform jar tests if indicated. 3. Verify performance of coagulation-flocculation process (see Chapter 4, "Coagulation and Flocculation"). 4. Make appropriate process changes. 5. Verify response to process changes at appropriate time.	1. Change coagulant. 2. Adjust coagulant dosage. 3. Adjust flash mixer/flocculator mixing intensity. 4. Adjust improperly working chemical feeder.
3. Sedimentation Basin Changes	**Operator Actions**	**Possible Process Changes**

• Floc Settling
• Rising or Floating Sludge
	1. Observe floc settling characteristics:
	  a. Dispersion
	  b. Size
	  c. Settling rate
	2. Evaluate overall process performance.
	3. Perform jar tests if indicated:
	  a. Assess floc size and settling rate.
	  b. Assess quality of settled water (clarity and color).
	4. Make appropriate process changes.
	5. Verify response to process changes at appropriate time.
		1. Change coagulant.
		2. Adjust coagulant dosage.
		3. Adjust flash mixer/flocculator mixing intensity.
		4. Change frequency of sludge removal (increase or decrease).
		5. Remove sludge from basin.
		6. Repair broken sludge rakes.

4. Sedimentation Process Effluent Quality Changes	**Operator Actions**	**Possible Process Changes**

• Turbidity
• Color
	1. Evaluate overall process performance.
	2. Perform jar test if indicated.
	3. Verify process performance:
	  a. Coagulation-flocculation process (see Chapter 4, "Coagulation and Flocculation").
	  b. Floc settling characteristics.
	4. Make appropriate process changes.
	5. Verify response to process changes at appropriate time.
		1. Change coagulant.
		2. Adjust coagulant dosage.
		3. Adjust flash mixer/flocculator mixing intensity.
		4. Change frequency of sludge removal (increase or decrease).

5. Upflow Clarifier Process Effluent Quality Changes	**Operator Actions**	**Possible Process Changes**
• Turbidity • Turbidity Caused by Sludge Blanket Coming to Top Due to Rainfall on Watershed	1. See 4 above. 2. Open main drain valve of clarifier.	1. See 4 above. 2. Drop entire water level of clarifier to bring the sludge blanket down.

3. Fill sedimentation basin with water.

   a. Observe proper depth of water in basin.

   b. Remove floating debris from basin water surface.

4. Start sample pumps (allow sufficient flushing time before securing samples).

5. Perform water quality analyses (make process adjustments as necessary).

6. Operate sludge removal equipment. Be sure that all valves are in the proper position (either open or closed).

5.711 Shutdown Procedures

1. Stop flow to sedimentation basin. Install basin isolation gates or stop logs.

2. Turn off sample pumps.

3. Turn off sludge removal equipment.

   a. Shut off mechanical equipment and disconnect where appropriate.

   b. Check that valves are in proper position (either open or closed).

4. Lock out and tag electric switches and equipment.

5. Dewater (drain) basin if necessary.

   a. Be sure water table is not high enough to float the empty basin.

   b. Open basin drain valves.

6. Grease and lubricate all gears, sprockets, and mechanical moving parts that have been submerged *IMMEDIATELY FOLLOWING DEWATERING*. If this is not done, they can freeze up (seize up) in a few hours. Frozen parts will require long hours to repair and can result in equipment breakage.

QUESTIONS

Write your answers in a notebook and then compare your answers with those on page 188.

5.7A   Under what circumstances might a water treatment plant have to be shut down?

5.7B   List the shutdown procedures for a sedimentation basin.

5.7C   Why should photographs be taken during shutdown and start-up procedures?

5.8 LABORATORY TESTS

5.80 Process Control Water Quality Indicators

Process control water quality indicators of importance in the operation of the sedimentation process include turbidity and temperature.

5.81 Sampling Procedures

Process water samples will be either grab samples obtained directly from a specific process monitoring location, or continuous samples which are pumped to the laboratory from various locations in the process (flocculation basin effluent, settled water conduit). Process water samples must be representative of actual conditions in the treatment plant.

Critical water quality indicators such as turbidity will be routinely monitored, perhaps several times per eight-hour shift. Temperature will be evaluated less frequently, especially under stable water quality conditions.

Process grab samples should be collected in clean plastic or glass containers and care should be taken to avoid contamination of the sample.

Sample analysis should be performed immediately following sample collection. When this is not possible, care should be taken to preserve samples by proper storage. See Chapter 11, "Laboratory Procedures," for sample preservation and storage procedures.

5.82 Sample Analysis

Analyses of certain process control water quality indicators (such as turbidity) are easily performed in the laboratory with the aid of automated analytical instruments, such as turbidimeters. The collection of a bad sample or a bad laboratory result is about as useful as no results! To prevent bad results requires constant maintenance and calibration of laboratory equipment and correct lab procedures. Results of lab tests are of no value unless they are used.

QUESTIONS

Write your answers in a notebook and then compare your answers with those on page 188.

5.8A   How are sedimentation process water samples obtained for analysis?

5.8B   How frequently should you monitor the turbidity of the water being treated?

5.9 PROCESS AND SUPPORT EQUIPMENT OPERATION AND MAINTENANCE

5.90 Types of Equipment

The operator of a sedimentation process will need to be familiar with the operation and maintenance of a variety of mechanical, electrical, and electronic equipment including:

1. Sludge removal equipment,

2. Sludge pumps,

3. Sump pumps,

4. Valves,

5. Flowmeters and gages,

6. Water quality monitors, such as turbidimeters, and

7. Control systems.

Sludge removal equipment will constitute the majority of the electro-mechanical equipment used in the sedimentation process. Since a wide variety of systems can be used to remove sludge from the bottom of the sedimentation basins, the operator will need to be thoroughly familiar with the operation and maintenance instructions for each specific equipment item.

5.91 Equipment Operation

Before starting a piece of mechanical equipment, such as a sludge pump, be sure that the unit is properly lubricated and its operational status is known. After start-up and during normal operation, check for excessive noise and vibration, overheating, and leakage (water, lubricants). Check the pump's suction and discharge pressures to be sure the lines are not plugged. When in doubt about the performance of a piece of equipment, always refer to the operation and maintenance instructions.

Many equipment items, such as valves, are simple open or closed devices. Similarly, sump pumps are simple ON/OFF devices. Other equipment, such as sludge pumps, have provisions for flow rate adjustment and may be furnished with sight glasses to visually check sludge flow. Check the flow each time the equipment is operated. Sludge collectors, discharge lines, and troughs should be periodically flushed to maintain a free sludge flow. Calibration of flow rates and measurement of sludge density may require the use of special procedures. Detailed operating, repair, and calibration procedures are usually given in the manufacturer's literature.

5.92 Safety Considerations

To avoid accidents or injury when working around sludge removal equipment, such as pumps and motors, follow the safety procedures listed below.

ELECTRICAL EQUIPMENT

1. Avoid electric shock (use protective gloves),

2. Avoid grounding yourself in water or on pipes,

3. Ground all electric tools,

4. Use the buddy system, and

5. Use a lockout and tag system whenever electrical equipment or electrically driven mechanical equipment is out of service or being worked on.

MECHANICAL EQUIPMENT

1. Keep protective guards on rotating equipment,

2. Do not wear loose clothing around rotating equipment,

3. Keep hands out of valves, pumps, and other pieces of equipment (lock out and tag power switches before cleaning),

4. Clean up all lubricant and sludge spills (slippery surfaces cause falls), and

5. Use a lockout and tag system whenever mechanical equipment is out of service or being worked on.

OPEN-SURFACE WATER-FILLED STRUCTURES

1. Use safety devices, such as handrails and ladders,

2. Close all openings and replace safety gratings when finished working,

3. Know the location of all life preservers, and

4. Use the buddy system.

VALVE AND PUMP VAULTS, SUMPS

1. Be sure all underground or confined structures are free of hazardous atmospheres (use a gas detector to check for toxic or explosive gases, lack of sufficient oxygen),

2. Only work in well-ventilated structures (use ventilation fans),

3. Use the buddy system, and

4. Lock or chain valves when working in an area that could be flooded.

5.93 Corrosion Control

The metallic parts of clarifiers and solids-contact units must be protected from corrosion. A good layer of paint or other protective coating over all metal parts exposed to water is one successful approach. Another approach is the use of a cathodic protection system (Figure 5.28).

Fig. 5.28   Clarifier wired for cathodic protection
(Permission of Wallace and Tiernan Division of Pennwalt Corporation)

Cathodic protection is a process for reducing or eliminating corrosion of metal parts exposed to water, soil, or the atmosphere. In this process the flow of an electric current from the metal surface into water (the cause of corrosion) is stopped by overpowering it by applying a stronger current from another source (the cathodic protection system).

To maintain a high degree of corrosion control, you must inspect and test the operation of the cathodic protection system and its parts regularly. Regular and proper maintenance is critical because the amount of protective current required to prevent corrosion can vary with changes in the conditions of the coatings on the metallic surfaces, in the chemical characteristics of the water being treated, and the operation of the clarifier.

For additional information on how to control corrosion, see Chapter 8, "Corrosion Control."

5.94 Preventive Maintenance Procedures

Preventive maintenance (P/M) programs are designed to ensure the continued satisfactory operation of treatment plant facilities by reducing the frequency of breakdown failures. This is accomplished by performing scheduled or routine maintenance of valves, pumps, and other electrical and mechanical equipment items.

In the normal operation of the sedimentation process, the operator will be expected to perform routine maintenance functions as part of an overall preventive maintenance program. Typical functions include:

1. Keeping electric motors free of dirt and moisture,

2. Ensuring good ventilation (air circulation) in equipment work areas,

3. Checking pumps and motors for leaks, unusual noise and vibrations, overheating, or signs of wear,

4. Maintaining proper lubrication and oil levels,

5. Inspecting for alignment of shafts and couplings,

6. Checking bearings for wear, overheating, and proper lubrication,

7. Checking for proper valve operation (leakage or jamming), and

8. Checking for free flow of sludge in sludge removal collection and discharge systems.

Accurate records are a key element in a successful preventive maintenance program. These records provide maintenance and operations personnel with clues for determining causes for equipment breakdowns, and will assist in spotting similar weaknesses in other equipment items. The frequency of performing periodic maintenance on electrical and mechanical equipment items is often based on preventive maintenance records of prior performance and equipment manufacturers' recommendations.

Good housekeeping practices are an important part of every preventive maintenance program.

For additional details on equipment maintenance, see Volume II, Chapter 18, "Maintenance."

5.10 ARITHMETIC ASSIGNMENT

Turn to the Appendix, "How to Solve Water Treatment Plant Arithmetic Problems," at the back of this manual and read the following sections:

1. A.5, *WEIGHT—VOLUME RELATIONS*, and

2. A.134, Sedimentation.

Check all of the arithmetic in these two sections on an electronic pocket calculator. You should be able to get the same answers.

5.11 ADDITIONAL READING

1. *NEW YORK MANUAL*, Chapter 8,* "Sedimentation."

2. *TEXAS MANUAL*, Chapter 8,* "Coagulation and Sedimentation."

* Depends on edition.

5.12 ACKNOWLEDGMENTS

Many of the concepts and procedures discussed in this chapter are based on material obtained from the sources listed below.

1. Stone, B. G. Notes from "Design of Water Treatment Systems," CE-610, Loyola Marymount University, Los Angeles, CA, 1977.

2. *WATER QUALITY AND TREATMENT*, Fifth Edition. Obtain from American Water Works Association (AWWA), Bookstore, 6666 West Quincy Avenue, Denver, CO 80235. Order No. 10008. ISBN 0-07-001659-3. Price to members, $119.00; nonmembers, $135.00; price includes cost of shipping and handling.

3. *WATER TREATMENT PLANT DESIGN*, Third Edition, prepared jointly by the American Society of Civil Engineers, American Water Works Association, and Conference of State Sanitary Engineers. Obtain from American Water Works Association (AWWA), Bookstore, 6666 West Quincy Avenue, Denver, CO 80235. Order No. 10009. ISBN 0-07-001643-7. Price to members, $103.50; nonmembers, $135.00; price includes cost of shipping and handling.

QUESTIONS

Write your answers in a notebook and then compare your answers with those on page 188.

5.9A What possible items should be checked for after a sludge pump has been started?

5.9B How can a free sludge flow be maintained?

5.9C What types of safety hazards are associated with sludge removal equipment?

End of Lesson 2 of 2 Lessons on SEDIMENTATION

Please answer the discussion and review questions next.

DISCUSSION AND REVIEW QUESTIONS

Chapter 5.   SEDIMENTATION

(Lesson 2 of 2 Lessons)

Write the answers to these questions in your notebook. The question numbering continues from Lesson 1.

9. What would you check during a visual inspection of a sedimentation process?

10. How is the approximate process removal efficiency of a sedimentation basin determined?

11. What process changes would you consider if source water quality abruptly changes?

12. What items should be checked when inspecting the condition of sludge removal equipment?

13. How can the metallic parts of clarifiers be protected from corrosion?

14. What functions should be included in a sedimentation process preventive maintenance program?

SUGGESTED ANSWERS

Chapter 5.   SEDIMENTATION

ANSWERS TO QUESTIONS IN LESSON 1

Answers to questions on page 149.

5.0A   The purposes of the sedimentation process are to remove suspended solids (particles) that are denser (heavier) than the water and to reduce the load on the filters.

5.0B   Sedimentation is accomplished by decreasing the velocity of the water being treated to allow heavier particles to settle out.

5.0C   Presedimentation facilities are installed in locations where the source water supply is diverted directly from rivers or streams. The basins are especially needed if the rivers or streams can be contaminated by overland runoff and point source waste discharges.

Answers to questions on page 152.

5.1A   Factors that affect particle settling in a sedimentation basin include: (1) particle size and distribution, (2) shape of particles, (3) density of particles, (4) temperature (viscosity and density) of water, (5) electrical charge on particles, (6) dissolved substances in water, (7) flocculation characteristics of the suspended material, (8) environmental conditions (wind), and (9) sedimentation basin hydraulic and design characteristics (inlet conditions and shape of basin).

5.1B   Treatment is preferred before sedimentation to convert finer particles that do not readily settle into larger, denser particles (floc) that are settleable.

5.1C   Types of currents that may be found in a typical sedimentation basin include: (1) surface currents induced by winds, (2) density currents caused by differences in suspended solids concentrations and temperature differences, and (3) eddy currents produced by the flow of the water coming into and leaving the basin.

Answers to questions on page 156.

5.2A   The four zones into which a typical sedimentation basin can be divided are: (1) inlet zone, (2) settling zone, (3) sludge zone, and (4) outlet zone.

5.2B   The purpose of the settling zone is to provide a calm, undisturbed storage place for the flocculated water for a sufficient time period to permit effective settling of the suspended particles in the water being treated.

5.2C   Launders are skimming or effluent troughs used to uniformly collect settled water. Adjustable V-notch weirs are generally attached to the launders for controlling the water level in the sedimentation basin.

Answers to questions on page 163.

5.2D   Sedimentation basins are available in circular, rectangular, or square shapes.

5.2E   Double-deck sedimentation basins.

ADVANTAGES

1. Provide twice the surface area of a single basin of equivalent dimensions.
2. Conserve land area.

LIMITATIONS

1. Higher operation and maintenance costs.
2. Entire operation may have to be shut down if an equipment problem develops in either deck.

5.2F Rectangular sedimentation basins are often preferred over circular basins because circular basins are generally more sensitive to short-circuiting and achieve poorer solids removal.

5.2G During the operation of a solids-contact unit, care must be exercised to ensure that a uniform sludge blanket is formed and is subsequently maintained throughout the solids removal process. Other important operational factors include control of chemical dosages, mixing of chemicals, and control of the sludge blanket.

Answers to questions on page 166.

5.2H The detention time is calculated by:

$$\text{Detention Time, hr} = \frac{(\text{Basin Volume, gal})(24 \text{ hr/day})}{\text{Flow, gal/day}}$$

5.2I

Known	**Unknown**
Flow, MGD = 0.8 MGD	Detention Time, hr
Length, ft = 50 ft	
Width, ft = 20 ft	
Depth, ft = 10 ft	

1. Calculate the basin volume in gallons.

$$\text{Basin Volume, gal} = (L, ft)(W, ft)(D, ft)(7.48 \text{ gal/cu ft})$$

$$= (50 \text{ ft})(20 \text{ ft})(10 \text{ ft})(7.48 \text{ gal/cu ft})$$

$$= 74,800 \text{ gallons}$$

2. Determine the theoretical detention time in hours.

$$\text{Detention Time, hr} = \frac{(\text{Basin Volume, gal})(24 \text{ hr/day})}{\text{Flow, gal/day}}$$

$$= \frac{(74,800 \text{ gallons})(24 \text{ hr/day})}{800,000 \text{ gal/day}}$$

$$= 2.2 \text{ hours}$$

Answers to questions on page 170.

5.2J Two advantages of solids-contact units include:

1. Only one reaction unit to contend with,
2. Ability to accumulate slurry during periods of severe taste and odor problems, and
3. Use slurry accumulation to carry plant when coagulation fails because of increased algal activities.

5.2K The level of the slurry or sludge blanket can be determined by sampling taps which are placed at various depths on the wall of the solids-contact reactor.

5.2L Operators should attempt to forecast turbidity changes and adjust chemical dosage prior to the arrival of a turbidity change.

Answers to questions on page 172.

5.2M The jar test is used to determine the proper chemical dose when operating a solids-contact unit. Observations and process monitoring are used to "fine tune" the results from jar tests.

5.2N Recirculation in a solids-contact unit may be provided by impellers, turbines, pumping units, or by air injection.

5.2O The percent slurry in the reaction zone is determined by the volume over volume (V/V) test.

Answers to questions on page 174.

5.2P Alum sludge solids concentrations typically vary from *0.25* to *10* percent when removed from a sedimentation basin.

5.2Q When determining how often to use sludge removal equipment, consider (1) the volume of sludge to be removed, (2) available sludge storage capacity, and (3) the fact that over 50 percent of the floc will settle out in the first third of the basin length. The sludge removal equipment will have to operate frequently enough to prevent excessive sludge buildup over the first third of the basin length.

5.2R Accumulated sludge must be removed periodically from the bottom of sedimentation basins to:

1. Prevent interference with the settling process (such as resuspension of solids due to scouring),
2. Prevent the sludge from becoming septic or providing an environment for the growth of microorganisms that can create taste and odor problems, and
3. Prevent excessive reduction in the cross-sectional area of the basin (reduction of detention time).

Answers to questions on page 176.

5.2S The depth of sludge in a sedimentation basin can be measured with a sludge blanket sounder, a bubbler tube, an aspirator, or an ultrasonic level indicator.

5.2T If the sludge being pumped from a sedimentation basin is too low in solids (soupy), the frequency of sludge removal should be decreased.

Answers to questions on page 177.

5.3A Sedimentation basins are needed to treat water when source water turbidities are high and generally require alum dosage for efficient solids removal.

5.4A The two major factors that influence the performance of sedimentation basins are (1) the settling characteristics of the suspended particles, and (2) the flow rate through the sedimentation basins.

ANSWERS TO QUESTIONS IN LESSON 2

Answers to questions on page 179.

5.5A During the normal operation of the sedimentation process, the operator should monitor the turbidity of the water entering and leaving the basin and the temperature of the water entering the basin.

5.5B Visual observations and laboratory evaluation of sedimentation process performance should be conducted on a routine basis, at least once per eight-hour shift, and more frequently when water quality conditions are fluctuating.

5.5C Free-flowing sludge can be readily observed if sight glasses are incorporated in the sludge discharge piping. Otherwise, the outlet of the sludge line should be observed during sludge pumping.

5.5D Frequent clogging of the sludge discharge line is an indication that the sludge concentration is too high. If this occurs, try increasing the frequency of operation of the sludge removal equipment.

5.5E   In the routine operation of the sedimentation process, the following records should be maintained: (1) influent and effluent turbidity and influent temperature, (2) process production inventory, and (3) process equipment performance.

Answers to questions on page 181.

5.6A   The measurement of turbidity levels at the sedimentation basin inlet and outlet will give a rough idea of process removal efficiency.

5.6B   Sudden increases in settled water turbidity could cause problems in the operation of the filtration process. High settled water turbidity levels resulting from floc carryover from the sedimentation basin will cause premature clogging of filters and may result in the degradation of filtered water quality.

Answers to questions on page 183.

5.7A   A water treatment plant might have to be shut down for periodic maintenance and cleaning, or from a major process failure.

5.7B   Shutdown procedures for a sedimentation basin include:

1. Stop flow to sedimentation basin. Install basin isolation gates or stop logs.
2. Turn off sample pumps.
3. Turn off sludge removal equipment.
    a. Shut off mechanical equipment.
    b. Check that valves are in proper position.
4. Lock out and tag electric switches and equipment.
5. Dewater (drain) basin if necessary.
    a. Be sure water table is not high enough to float the empty basin.
    b. Open basin drain valves.
6. Grease and lubricate all gears, sprockets, and mechanical moving parts that have been submerged *IMMEDIATELY FOLLOWING DEWATERING.*

5.7C   Photographs of special features during start-up and shutdown provide a visual record of events or conditions that may be difficult to illustrate when basins are full of water.

Answers to questions on page 183.

5.8A   Sedimentation process water samples may be either grab samples obtained directly from a specific process monitoring location, or continuous samples which are pumped to the laboratory from various locations in the process.

5.8B   Turbidity should be measured several times during a shift and more frequently when the source water quality is changing rapidly.

Answers to questions on page 185.

5.9A   After a sludge pump has been started, always check for excessive noise and vibration, overheating, and leakage (water, lubricants). Also check the suction and discharge pressures to be sure the lines are not plugged.

5.9B   A free sludge flow can be maintained by periodically flushing sludge collectors, discharge lines, and troughs. Correct operation of the sedimentation basin will usually result in a free sludge flow.

5.9C   Types of safety hazards associated with sludge removal equipment include: (1) electrical equipment, (2) mechanical equipment, (3) open-surface water-filled structures, and (4) valve and pump vaults, sumps.

APPENDIX

A. Design and Operational Guidelines

1. Surface Loading
2. Effective Water Depth
3. Mean Flow Velocity
4. Weir Loading Rate

A. DESIGN AND OPERATIONAL GUIDELINES

Some of the basic guidelines used by engineers to design sedimentation processes are described in this section. Operators should know how these guidelines are obtained and why they are important in order to communicate effectively with design engineers.

By knowing the values used by design engineers and comparing the actual values for your plant, you may be able to identify the cause of operational problems and poor quality finished water.

Detention time is a very important guideline used by designers; however, this topic has already been discussed in Section 5.23, "Detention Time," and will not be repeated again.

1. Surface Loading

Surface loading or overflow rate is one of the most important factors influencing sedimentation. The surface overflow rate translates into a velocity, and it is equal to the settling velocity of the smallest particle the basin will remove.

Surface loading is determined by dividing basin flow rate by the basin surface area as follows:

$$\text{Surface Loading, GPM/sq ft} = \frac{\text{Flow, gallons per minute}}{\text{Surface Area, square feet}}$$

Colder water temperatures require lower basin overflow rates since particle settling velocities also decrease with colder water temperatures. Therefore, your plant can effectively treat only lower flow rates during colder weather. The overflow rate is the same as the theoretical upwelling rate. When this value exceeds the settling rate of a particle, the particle will be unable to settle out and will be carried over to the filters.

EXAMPLE 1

A water treatment plant treats a flow of 1.5 MGD. An examination of treatment plant design drawings reveals that the rectangular sedimentation basin is 75 feet long, 25 feet wide, and has an effective (water) depth of 12 feet. Calculate the surface loading or overflow rate in gallons per minute per square foot of surface area.

Known	Unknown
Flow, MGD = 1.5 MGD	Surface Loading, GPM/sq ft
Length, ft = 75 ft	
Width, ft = 25 ft	
Depth, ft = 12 ft	

1. Convert flow in MGD to GPM.

$$\begin{aligned}\text{Flow, GPM} &= \frac{(\text{Flow, MGD})(1,000,000/\text{Million})}{(24 \text{ hr/day})(60 \text{ min/hr})} \\ &= \frac{(1.5 \text{ MGD})(1,000,000/\text{Million})}{(24 \text{ hr/day})(60 \text{ min/hr})} \\ &= 1,042 \text{ GPM}\end{aligned}$$

2. Determine the surface loading in gallons per minute per square foot of surface area.

$$\begin{aligned}\text{Surface Loading, GPM/sq ft} &= \frac{\text{Flow, GPM}}{\text{Surface Area, sq ft}} \\ &= \frac{1,042 \text{ GPM}}{(75 \text{ ft})(25 \text{ ft})} \\ &= 0.56 \text{ GPM/sq ft}\end{aligned}$$

The surface loading or overflow rate can be converted to a rise rate in feet per minute by dividing the surface loading by 7.48 gallons per cubic foot.

$$\text{Rise Rate, ft/min} = \frac{\text{Surface Loading, GPM/sq ft}}{7.48 \text{ gal/cu ft}}$$

EXAMPLE 2

A solids-contact unit has a surface loading rate of 0.75 GPM/sq ft. What is the rise rate in feet per minute?

Known	Unknown
Surface Loading, GPM/sq ft = 0.75 GPM/sq ft	Rise Rate, ft/min

Calculate the rise rate in feet per minute from the surface loading.

$$\begin{aligned}\text{Rise Rate, ft/min} &= \frac{\text{Surface Loading, GPM/sq ft}}{7.48 \text{ gal/cu ft}} \\ &= \frac{0.75 \text{ GPM/sq ft}}{7.48 \text{ gal/cu ft}} \\ &= 0.1 \text{ ft/min}\end{aligned}$$

2. Effective Water Depth

In theory, the ideal sedimentation basin would have a very shallow depth and a large surface area. However, practical considerations, such as minimum depth required to accommodate mechanical sludge removal equipment, desired flow velocity, and current and wind effects, must all be considered in the selection of an appropriate basin depth.

Sludge volume storage requirements may also be an important consideration in selecting an appropriate basin depth. This is particularly important in manually cleaned basins where sludge removal requires taking the basin off-line for dewatering (draining) to permit removal of accumulated sludge. If the sludge buildup becomes too great, the effective depth of the basin will be significantly reduced. This can cause localized high flow velocities, sludge scouring, and degradation of process efficiency.

3. Mean Flow Velocity

The flow velocity is a function of the basin cross-sectional area and the flow going into the basin. Flow velocity is calculated by dividing the flow rate by the cross-sectional area of the basin.

The flow velocity is not completely uniform or stable throughout the length of the basin due to variations in water density, currents, and reduction in cross-sectional area resulting from sludge accumulation in the bottom of the basin.

The main concern about high velocities is scour. When velocities get too high, the flowing water will pick up some of the settled sludge from the bottom of the basin and carry it on to the filters. This is called "scour," and it reduces sedimentation efficiency.

$$\text{Mean Flow Velocity, ft/min} = \frac{\text{Flow, GPM}}{(\text{Cross-Sectional Area, sq ft})(7.48 \text{ gal/cu ft})}$$

EXAMPLE 3

Determine the mean flow velocity in feet per minute for the rectangular sedimentation basin in Example 1.

Known		Unknown
Flow, MGD	= 1.5 MGD	Mean Flow Velocity, ft/min
	= 1,042 GPM	
Length, ft	= 75 ft	
Width, ft	= 25 ft	
Depth, ft	= 12 ft	

Determine the mean flow velocity in feet per minute.

$$\text{Mean Flow Velocity, ft/min} = \frac{\text{Flow, GPM}}{(\text{Cross-Sectional Area, sq ft})(7.48 \text{ gal/cu ft})}$$

$$= \frac{1,042 \text{ gal/min}}{(25 \text{ ft})(12 \text{ ft})(7.48 \text{ gal/cu ft})}$$

$$= 0.46 \text{ ft/min}$$

4. Weir Loading Rate

Launders outfitted with V-notch weirs are generally installed at the basin outlet to provide uniform collection and distribution of the clear water. While insufficient total weir length may reduce settling efficiency due to current effects, operation over a broad range of weir loading rates will generally not interfere with basin performance. If the weir loading rate becomes too high, floc will be carried out of the sedimentation basins and onto the filters. Generally, weir loadings are more important in shallow basins than in deeper basins.

In cold climates launders use orifices in pipes rather than V-notch weirs on channels. Pipes usually are submerged in two feet (0.6 m) of water to avoid ice problems.

$$\text{Weir Loading, GPM/ft} = \frac{\text{Flow, GPM}}{\text{Weir Length, ft}}$$

EXAMPLE 4

Determine the weir loading rate in gallons per minute per foot of weir of the rectangular sedimentation basin in Example 1. Four effluent launders 12.5 feet long with V-notch weirs on both sides of the launder extend into the basin from the outlet end. Therefore, each launder has 25 feet of weir for a total of 100 feet.

Known		Unknown
Flow, MGD	= 1.5 MGD	Weir Loading Rate, GPM/ft
	= 1,042 GPM	
Length, ft	= 75 ft	
Width, ft	= 25 ft	
Depth, ft	= 12 ft	
Weir Length, ft	= 100 ft	

Calculate the weir loading rate in gallons per minute per foot of weir.

$$\text{Weir Loading Rate, GPM/ft} = \frac{\text{Flow, GPM}}{\text{Weir Length, ft}}$$

$$= \frac{1,042 \text{ GPM}}{100 \text{ ft}}$$

$$= 10.4 \text{ GPM/ft}$$

CHAPTER 6

FILTRATION

by

Jim Beard

NOTICE

Filter performance monitoring requirements were increased by the passage of the Interim Enhanced Surface Water Treatment Rule (IESWTR) in 1998. The Rule imposes more stringent filtered water turbidity limits and requires continuous turbidity monitoring of individual filters. The effective date of the IESWTR filtration provisions was December 2001. The IESWTR applies to all public water systems that use surface water or groundwater under the influence of surface water and serve 10,000 or more people.

Keep in contact with your state drinking water agency to obtain the rules and regulations that currently apply to your water utility. For additional information or answers to specific questions about the new regulations, phone EPA's toll-free Safe Drinking Water Hotline at (800) 426-4791.

TABLE OF CONTENTS

Chapter 6. FILTRATION

OBJECTIVES

Chapter 6.   FILTRATION

Following completion of Chapter 6, you should be able to:

1. Describe the various types of potable water filters and how they work,

2. Explain how other treatment processes affect the performance of the filtration process,

3. Operate and maintain filters under normal and abnormal process conditions,

4. Start up and shut down filtration processes, and

5. Safely perform duties related to the various types of filters.

WORDS

Chapter 6.   FILTRATION

ABSORPTION (ab-SORP-shun)   ABSORPTION

The taking in or soaking up of one substance into the body of another by molecular or chemical action (as tree roots absorb dissolved nutrients in the soil).

ACTIVATED CARBON   ACTIVATED CARBON

Adsorptive particles or granules of carbon usually obtained by heating carbon (such as wood). These particles or granules have a high capacity to selectively remove certain trace and soluble materials from water.

ADSORPTION (add-SORP-shun)   ADSORPTION

The gathering of a gas, liquid, or dissolved substance on the surface or interface zone of another material.

AIR BINDING   AIR BINDING

The clogging of a filter, pipe or pump due to the presence of air released from water. Air entering the filter media is harmful to both the filtration and backwash processes. Air can prevent the passage of water during the filtration process and can cause the loss of filter media during the backwash process.

BACKWASHING   BACKWASHING

The process of reversing the flow of water back through the filter media to remove the entrapped solids.

BASE METAL   BASE METAL

A metal (such as iron) which reacts with dilute hydrochloric acid to form hydrogen. Also see NOBLE METAL.

BREAKTHROUGH   BREAKTHROUGH

A crack or break in a filter bed allowing the passage of floc or particulate matter through a filter. This will cause an increase in filter effluent turbidity. A breakthrough can occur (1) when a filter is first placed in service, (2) when the effluent valve suddenly opens or closes, and (3) during periods of excessive head loss through the filter (including when the filter is exposed to negative heads).

COLLOIDS (CALL-loids)   COLLOIDS

Very small, finely divided solids (particles that do not dissolve) that remain dispersed in a liquid for a long time due to their small size and electrical charge. When most of the particles in water have a negative electrical charge, they tend to repel each other. This repulsion prevents the particles from clumping together, becoming heavier, and settling out.

CONVENTIONAL FILTRATION   CONVENTIONAL FILTRATION

A method of treating water which consists of the addition of coagulant chemicals, flash mixing, coagulation-flocculation, sedimentation and filtration. Also called complete treatment. Also see DIRECT FILTRATION and IN-LINE FILTRATION.

CONVENTIONAL TREATMENT   CONVENTIONAL TREATMENT

See CONVENTIONAL FILTRATION. Also called complete treatment.

DIATOMACEOUS (DYE-uh-toe-MAY-shus) EARTH   DIATOMACEOUS EARTH

A fine, siliceous (made of silica) "earth" composed mainly of the skeletal remains of diatoms.

DIATOMS (DYE-uh-toms)   DIATOMS

Unicellular (single cell), microscopic algae with a rigid (box-like) internal structure consisting mainly of silica.

DIRECT FILTRATION DIRECT FILTRATION

A method of treating water which consists of the addition of coagulant chemicals, flash mixing, coagulation, minimal flocculation, and filtration. The flocculation facilities may be omitted, but the physical-chemical reactions will occur to some extent. The sedimentation process is omitted. Also see CONVENTIONAL FILTRATION and IN-LINE FILTRATION.

EFFECTIVE SIZE (E.S.) EFFECTIVE SIZE (E.S.)

The diameter of the particles in a granular sample (filter media) for which 10 percent of the total grains are smaller and 90 percent larger on a weight basis. Effective size is obtained by passing granular material through sieves with varying dimensions of mesh and weighing the material retained by each sieve. The effective size is also approximately the average size of the grains.

ENTRAIN ENTRAIN

To trap bubbles in water either mechanically through turbulence or chemically through a reaction.

FLUIDIZED (FLEW-id-I-zd) FLUIDIZED

A mass of solid particles that is made to flow like a liquid by injection of water or gas is said to have been fluidized. In water treatment, a bed of filter media is fluidized by backwashing water through the filter.

GARNET GARNET

A group of hard, reddish, glassy, mineral sands made up of silicates of base metals (calcium, magnesium, iron and manganese). Garnet has a higher density than sand.

HEAD LOSS HEAD LOSS

The head, pressure or energy (they are the same) lost by water flowing in a pipe or channel as a result of turbulence caused by the velocity of the flowing water and the roughness of the pipe, channel walls, or restrictions caused by fittings. Water flowing in a pipe loses head, pressure or energy as a result of friction losses. The head loss through a filter is due to friction losses caused by material building up on the surface or in the top part of a filter.

IN-LINE FILTRATION IN-LINE FILTRATION

The addition of chemical coagulants directly to the filter inlet pipe. The chemicals are mixed by the flowing water. Flocculation and sedimentation facilities are eliminated. This pretreatment method is commonly used in pressure filter installations. Also see CONVENTIONAL FILTRATION and DIRECT FILTRATION.

INTERFACE INTERFACE

The common boundary layer between two substances such as water and a solid (metal); or between two fluids such as water and a gas (air); or between a liquid (water) and another liquid (oil).

MICRON (MY-kron) MICRON

μm, Micrometer or Micron. A unit of length. One millionth of a meter or one thousandth of a millimeter. One micron equals 0.00004 of an inch.

NOBLE METAL NOBLE METAL

A chemically inactive metal (such as gold). A metal that does not corrode easily and is much scarcer (and more valuable) than the so-called useful or base metals. Also see BASE METAL.

PARTICLE COUNTER PARTICLE COUNTER

A device which counts and measures the size of individual particles in water. Particles are divided into size ranges and the number of particles is counted in each of these ranges. The results are reported in terms of the number of particles in different particle diameter size ranges per milliliter of water sampled.

PARTICLE COUNTING PARTICLE COUNTING

A procedure for counting and measuring the size of individual particles in water. Particles are divided into size ranges and the number of particles is counted in each of these ranges. The results are reported in terms of the number of particles in different particle diameter size ranges per milliliter of water sampled.

PERMEABILITY (PURR-me-uh-BILL-uh-tee) PERMEABILITY

The property of a material or soil that permits considerable movement of water through it when it is saturated.

PORE PORE

A very small open space in a rock or granular material. Also called an interstice, void, or void space.

SENSITIVITY (PARTICLE COUNTERS)   SENSITIVITY (PARTICLE COUNTERS)

The smallest particle a particle counter will measure and count.

SLURRY (SLUR-e)   SLURRY

A watery mixture or suspension of insoluble (not dissolved) matter; a thin, watery mud or any substance resembling it (such as a grit slurry or a lime slurry).

SPECIFIC GRAVITY   SPECIFIC GRAVITY

(1) Weight of a particle, substance, or chemical solution in relation to the weight of an equal volume of water. Water has a specific gravity of 1.000 at 4°C (39°F). Particulates in raw water may have a specific gravity of 1.005 to 2.5.

(2) Weight of a particular gas in relation to the weight of an equal volume of air at the same temperature and pressure (air has a specific gravity of 1.0). Chlorine has a specific gravity of 2.5 as a gas.

SUBMERGENCE   SUBMERGENCE

The distance between the water surface and the media surface in a filter.

TURBIDIMETER   TURBIDIMETER

An instrument for measuring and comparing the turbidity of liquids by passing light through them and determining how much light is reflected by the particles in the liquid. The normal measuring range is 0 to 100 and is expressed as Nephelometric Turbidity Units (NTUs).

UNIFORMITY COEFFICIENT (U.C.)   UNIFORMITY COEFFICIENT (U.C.)

The ratio of (1) the diameter of a grain (particle) of a size that is barely too large to pass through a sieve that allows 60 percent of the material (by weight) to pass through, to (2) the diameter of a grain (particle) of a size that is barely too large to pass through a sieve that allows 10 percent of the material (by weight) to pass through. The resulting ratio is a measure of the degree of uniformity in a granular material such as filter media.

$$\text{Uniformity Coefficient} = \frac{\text{Particle Diameter}_{60\%}}{\text{Particle Diameter}_{10\%}}$$

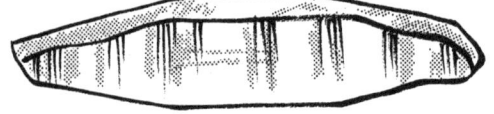

CHAPTER 6. FILTRATION

(Lesson 1 of 2 Lessons)

6.0 PROCESS DESCRIPTION

The purpose of filtration is the removal of particulate impurities and floc from the water being treated. Filtration is the process of passing water through material such as a bed of sand, coal, or other granular substance to remove floc and particulate impurities. These impurities consist of suspended particles (fine silts and clays), COLLOIDS,[1] biological forms (bacteria and plankton), and floc in the water being treated.

Filtration preceded by coagulation, flocculation, and sedimentation is commonly referred to as CONVENTIONAL FILTRATION (Figure 6.1). In the DIRECT FILTRATION process, the sedimentation step is omitted and flocculation facilities are reduced in size or may be omitted. Typical treatment processes for each of these filtration methods are shown in Figure 6.2.

6.1 FILTRATION MECHANISMS

Filtration is a physical and chemical process. The actual removal mechanisms are interrelated and rather complex, but filter removal of turbidity is based on the following factors:

1. Chemical characteristics of the water being treated (particularly source water quality),

2. Nature of suspension (physical and chemical characteristics of particulates suspended in the water),

3. Types and degree of pretreatment (coagulation, flocculation, and sedimentation), and

4. Filter type and operation.

A popular misconception is that particles are removed in the filtration process mainly by physical STRAINING. Straining is a term used to describe the removal of particles from a liquid (water) by passing the liquid through a filter whose PORES[2] are smaller than the particles to be removed. While the straining mechanism does play a role in the overall removal process, especially in the removal of larger particles, it is important to realize that most of the particles removed during filtration are considerably smaller than the pore spaces in the media. This is particularly true at the beginning of the filtration cycle when the pore spaces are clean (not clogged by particulates removed during filtration).

Thus, a number of interrelated removal mechanisms within the filter media itself are relied upon to achieve high removal efficiencies. These removal mechanisms include the following processes:

1. Sedimentation on media (very important),

2. ADSORPTION[3] (very important),

3. Biological action,

4. ABSORPTION[4] (not too important after initial wetting), and

5. Straining.

The relative importance of these removal mechanisms will depend largely on the nature of the water being treated, degree of pretreatment, and filter characteristics.

QUESTIONS

Write your answers in a notebook and then compare your answers with those on page 253.

6.0A What is the major difference between conventional filtration and direct filtration?

6.1A List the particle removal mechanisms involved in the filtration process.

6.2 TYPES OF FILTERS

Over the years a distinction has been made between the older SLOW SAND FILTRATION process (minor application in the United States except for small systems) and modern RAPID SAND FILTRATION (used extensively in the United States). This distinction is made because the removal mechanisms that apply to slow sand filtration are not directly comparable to those used in rapid sand filtration. However, the rapid

---

[1] Colloids (CALL-loids).   Very small, finely divided solids (particles that do not dissolve) that remain dispersed in a liquid for a long time due to their small size and electrical charge. When most of the particles in water have a negative electrical charge, they tend to repel each other. This repulsion prevents the particles from clumping together, becoming heavier, and settling out.

[2] Pore.   A very small open space in a rock or granular material. Also called an interstice, void, or void space.

[3] Adsorption (add-SORP-shun).   The gathering of a gas, liquid, or dissolved substance on the surface or interface zone of another material.

[4] Absorption (ab-SORP-shun).   The taking in or soaking up of one substance into the body of another by molecular or chemical action (as tree roots absorb dissolved nutrients in the soil).

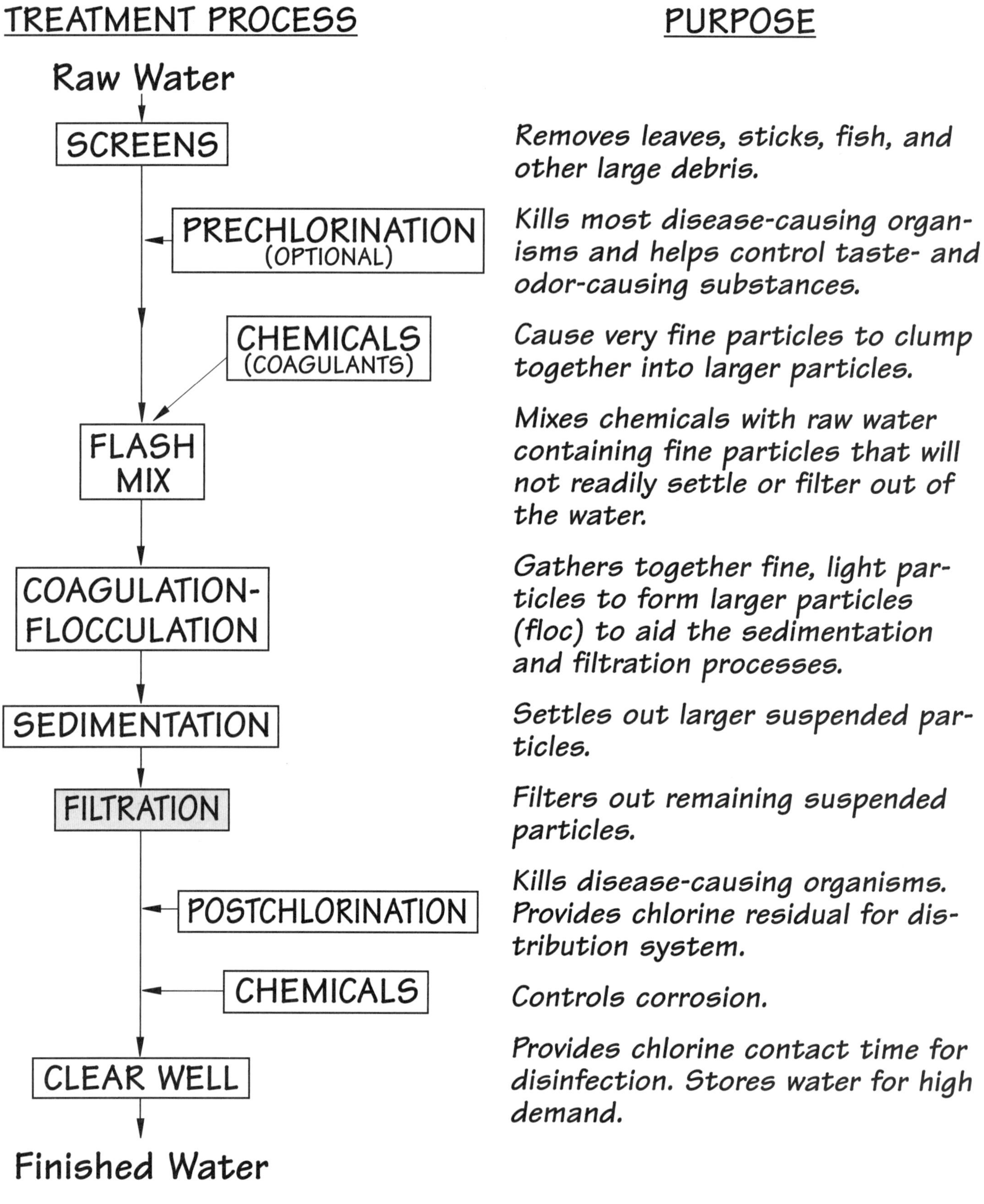

TREATMENT PROCESS

Raw Water

↓

SCREENS

↓

← PRECHLORINATION (OPTIONAL)

↓

CHEMICALS (COAGULANTS)

↓

FLASH MIX

↓

COAGULATION-FLOCCULATION

↓

SEDIMENTATION

↓

FILTRATION

↓

← POSTCHLORINATION

↓

← CHEMICALS

↓

CLEAR WELL

↓

Finished Water

PURPOSE

Removes leaves, sticks, fish, and other large debris.

Kills most disease-causing organisms and helps control taste- and odor-causing substances.

Cause very fine particles to clump together into larger particles.

Mixes chemicals with raw water containing fine particles that will not readily settle or filter out of the water.

Gathers together fine, light particles to form larger particles (floc) to aid the sedimentation and filtration processes.

Settles out larger suspended particles.

Filters out remaining suspended particles.

Kills disease-causing organisms. Provides chlorine residual for distribution system.

Controls corrosion.

Provides chlorine contact time for disinfection. Stores water for high demand.

Fig. 6.1   Typical process flow diagram

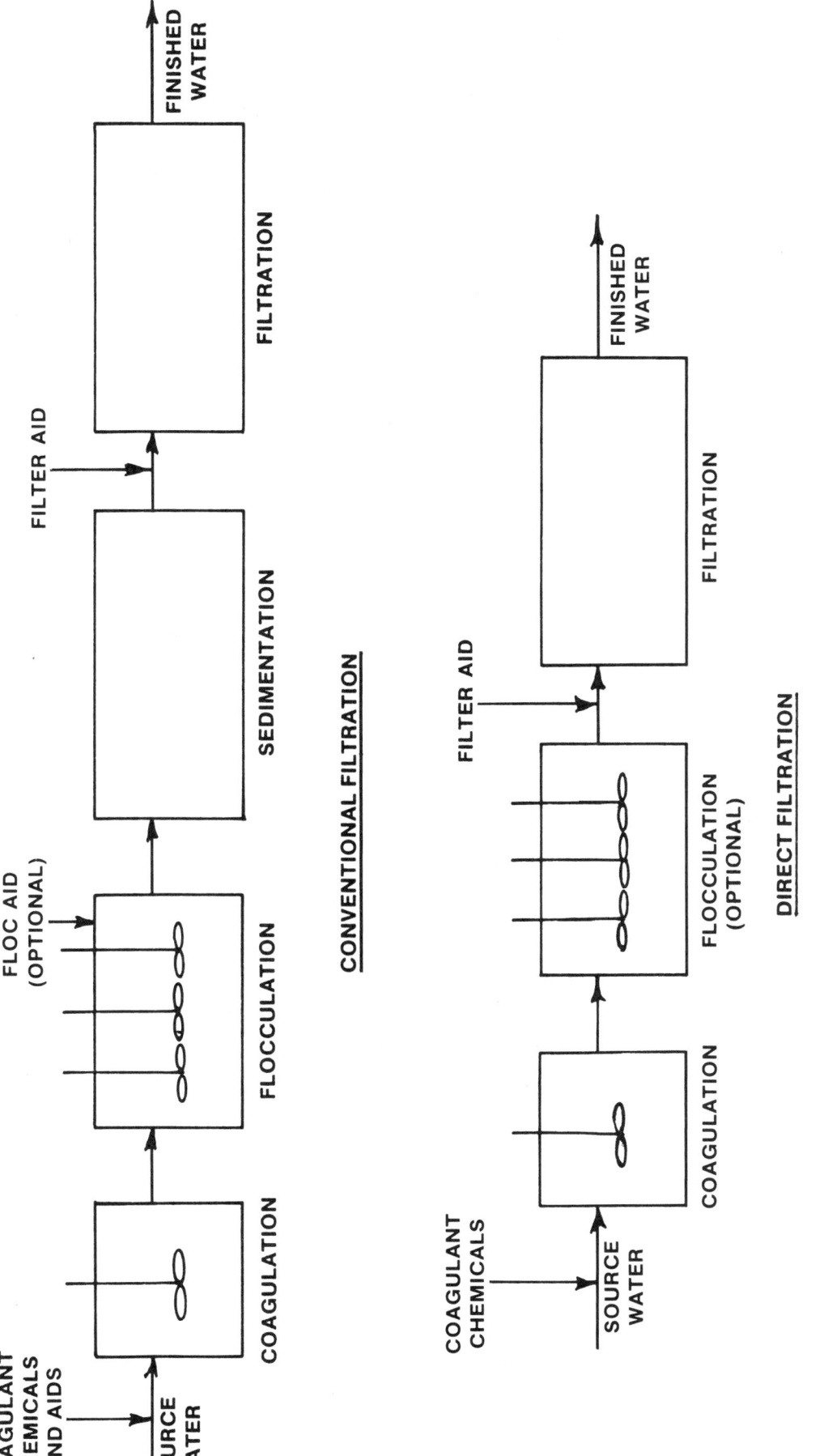

Fig. 6.2 Water treatment filtration processes

sand process has taken on new dimensions in recent years owing to significant improvements in the process. The term "rapid sand filtration" no longer adequately describes the variety of processes used. Thus, a more specific classification system for filter types has developed, and includes the following:

1. Gravity filtration (sand, dual media, and mixed media),

2. Pressure filtration (mixed media),

3. *DIATOMACEOUS EARTH*[5] (precoat) filtration, and

4. Slow sand filtration.

6.20 Gravity Filtration

In all gravity filtration systems the water level or pressure (head) above the media forces the water through the filter media as shown in Figure 6.3. The rate at which water passes through the granular filter media may vary from 2 to about 10 GPM/sq ft (1.36 to 6.79 liters per sec/sq m or 1.36 to 6.79 mm/sec)[6] (commonly referred to as filtration rate). However, many state health authorities limit the maximum filtration rate to 2 or 3 GPM/sq ft (1.36 or 2.04 liters per sec/sq m or 1.36 to 2.04 mm/sec) for gravity filtration. The rate of water flow through the filter is the hydraulic loading or merely the filtration rate. The filtration rate depends on the type of filter media.

Filter media consists of the following substances:

1. Single media (sand),[7]

2. Dual media (sand and anthracite coal), and

3. Multi- or mixed media (sand, anthracite coal, and *GARNET*[8]).

ACTIVATED CARBON[9] can also be used along with the above filter media for removal of tastes, odors, and organic substances.

In gravity filtration the particulate impurities are removed in/on the media, thus causing the filter to clog after a period of filtration time. In spite of this, gravity filtration is very widely used in water treatment plants.

6.21 Pressure Filtration

A pressure filter is similar to a gravity sand filter except that the filter is completely enclosed in a pressure vessel such as a steel tank, and is operated under pressure, as shown in Figure 6.4.

Pressure filters frequently offer lower installation and operation costs in small filtration plants; *HOWEVER*, they are generally somewhat less reliable than gravity filters (depending upon pumped pressure). Some states do not recommend the use of pressure filters for treating surface waters. Maximum filtration rates for pressure filters are in the 2 to 3 GPM/sq ft (1.36 to 2.04 liters per sec/sq m or 1.36 to 2.04 mm/sec) range.

6.22 Diatomaceous Earth Filtration

In diatomaceous earth (precoat) filtration, the filter media is added to the water being treated as a *SLURRY*[10]; it then collects on a septum (a pipe or conduit with porous walls) or other appropriate screening device as shown in Figure 6.5. After the initial precoat application, water is filtered by passing it through the coated screen. The coating thickness may be increased during the filtration process by gradually adding more media—a body feed. In most water treatment applications, diatomaceous earth is used for both the precoat and body-feed operations.

Diatomaceous earth filtration is primarily a straining process, and finds wide application where very high particle removal efficiencies (high clarity water) are required, such as in the beverage and food industries. Precoat filters can be operated as gravity, pressure, or vacuum filters. They are also commonly used in swimming pool installations due to their small size, efficiency, ease of operation, and relatively low cost. They find limited use in larger water treatment plants due to hydraulic (flow), sludge disposal, and other operational considerations.

6.23 Slow Sand Filtration

In slow sand filtration, water is drawn through the filter media (sand) by gravity as it is in the gravity filtration process. However, this is generally where the similarity between these two filtration processes ends.

In the slow sand filtration process, particles are removed by straining, adsorption, and biological action. Filtration rates are extremely low (0.015 to 0.15 GPM/sq ft or 0.01 to 0.1 liter per sec/sq m or 0.01 to 0.1 mm/sec).

The majority of the particulate material is removed in the top several inches of sand, so this entire layer must be physically removed when the filter becomes clogged. This filtration process has found limited application due to the large area required and the need to manually backwash the filters.

Slow sand filtration is becoming popular for small systems because of reliability and minimum operation and maintenance requirements. For additional information, see *SMALL WATER SYSTEM OPERATION AND MAINTENANCE*, Section 4.8, "Slow Sand Filtration," in this series of operator training manuals.

---

[5] Diatomaceous (DYE-oh-toe-MAY-thus) Earth. A fine, siliceous (made of silica) "earth" composed mainly of the skeletal remains of diatoms.

[6] Note that the numbers for liters per second per square meter are the same as for millimeters per second.

$$1 \frac{liter\ per\ sec}{square\ meter} = \frac{(1\ liter/sec)(1\ cu\ meter)(1,000\ mm)}{(square\ meter)(1,000\ liters)(1\ meter)}$$

$$= 1\ mm/sec$$

This occurs because one cubic meter equals 1,000 liters and one meter equals 1,000 millimeters.

[7] Single media filters also may consist of either anthracite coal or granular activated carbon.

[8] Garnet. A group of hard, reddish, glassy, mineral sands made up of silicates of base metals (calcium, magnesium, iron and manganese). Garnet has a higher density than sand.

[9] Activated Carbon. Adsorptive particles or granules of carbon usually obtained by heating carbon (such as wood). These particles or granules have a high capacity to selectively remove certain trace and soluble materials from water.

[10] Slurry (SLUR-e). A watery mixture or suspension of insoluble (not dissolved) matter; a thin, watery mud or any substance resembling it (such as a grit slurry or a lime slurry).

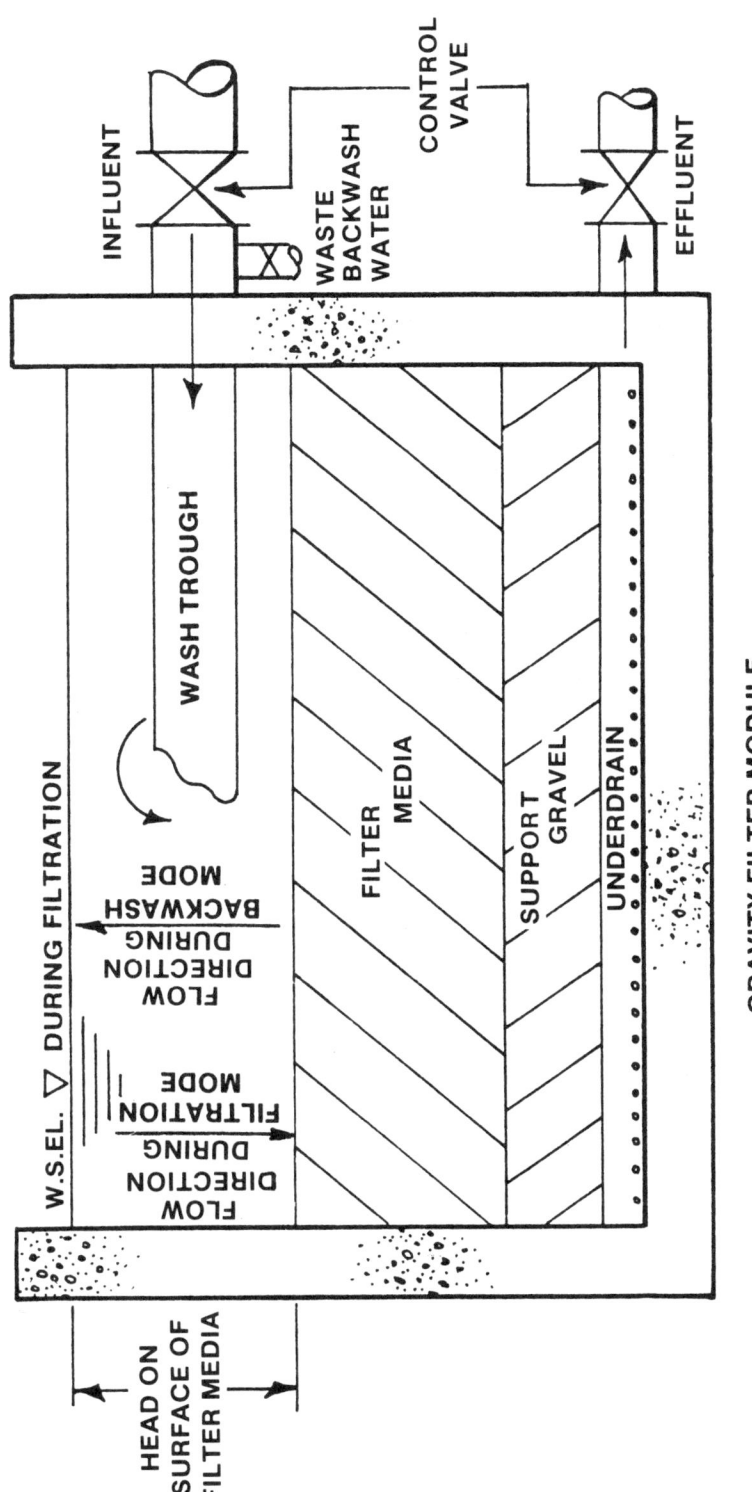

Fig. 6.3 Gravity filter

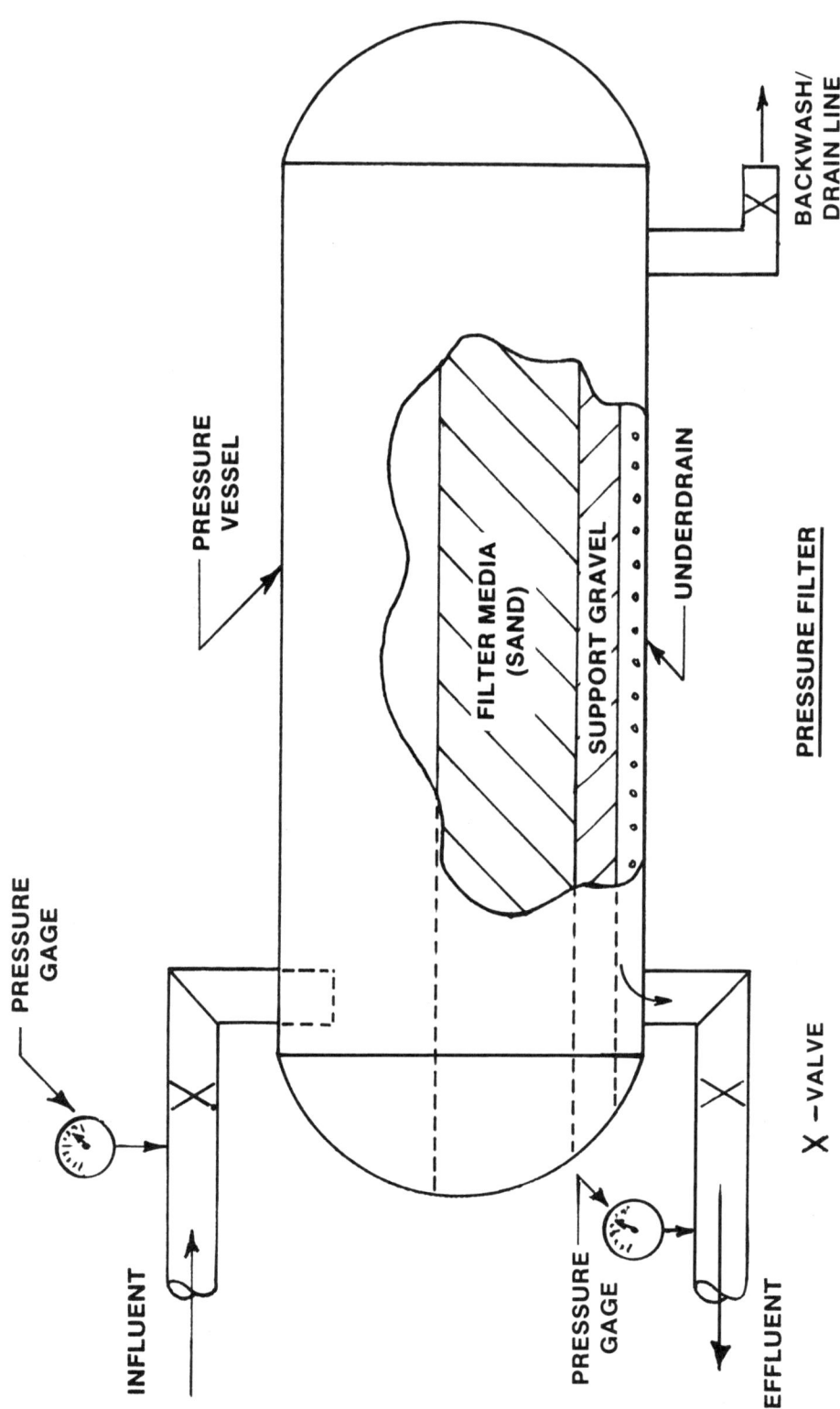

PRESSURE GAGE

INFLUENT

PRESSURE VESSEL

FILTER MEDIA (SAND)

SUPPORT GRAVEL

UNDERDRAIN

BACKWASH/ DRAIN LINE

PRESSURE GAGE

EFFLUENT

X — VALVE

PRESSURE FILTER

Fig. 6.4   Pressure filter

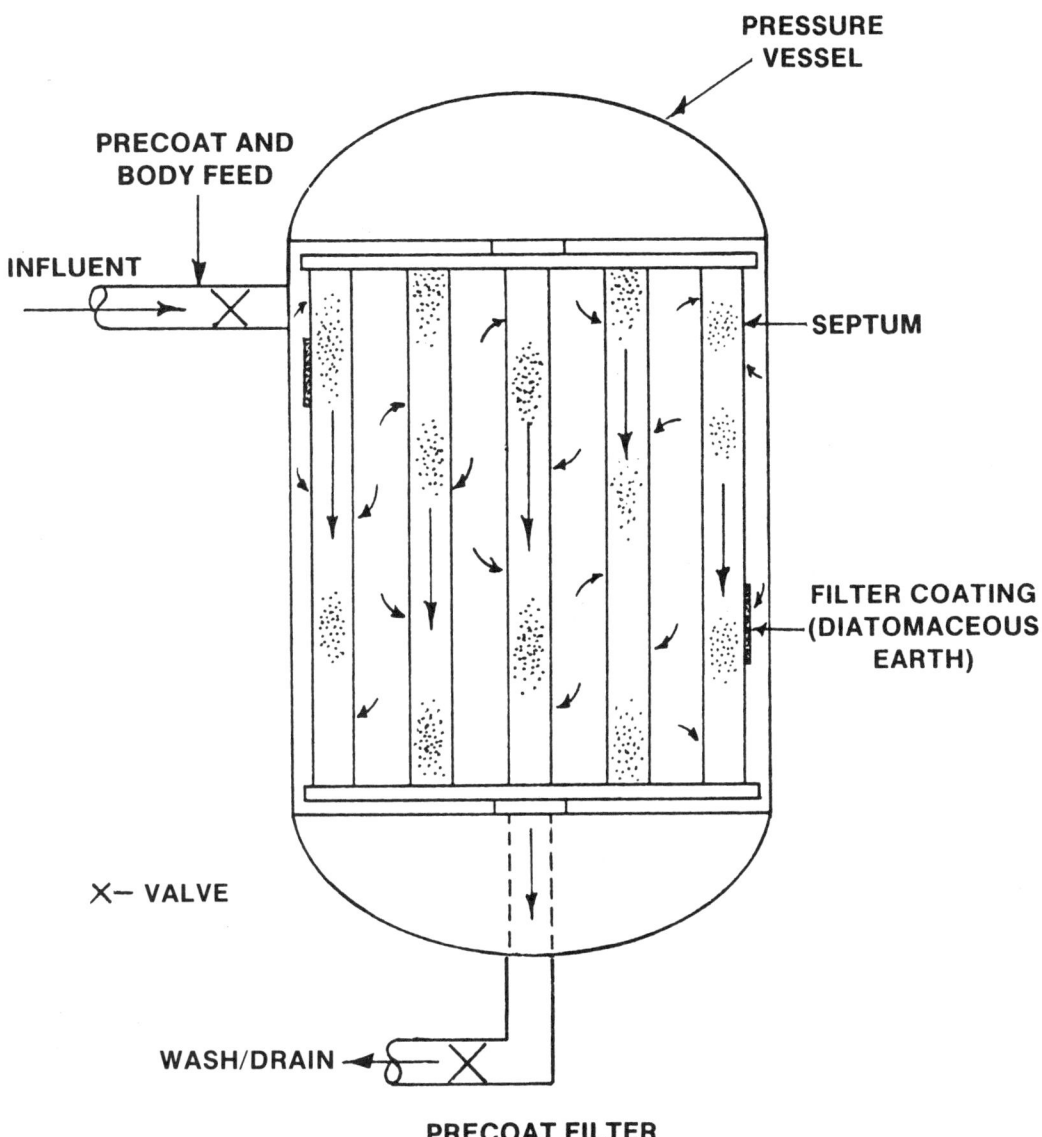

PRESSURE VESSEL

PRECOAT AND BODY FEED

INFLUENT

SEPTUM

FILTER COATING (DIATOMACEOUS EARTH)

X— VALVE

WASH/DRAIN

PRECOAT FILTER

Fig. 6.5 Diatomaceous earth filter

QUESTIONS

Write your answers in a notebook and then compare your answers with those on page 253.

6.2A List the four specific classes of filters.

6.2B What is garnet?

6.2C What material is used for precoat and body-feed operations?

6.3 PROCESS PERFORMANCE CONSIDERATIONS

6.30 Filter Media

Filter media most often selected by designers in gravity filtration consists of sand, anthracite coal, and garnet. However, other inert materials can also be used. Gravel is commonly used to support the filter materials. In the diatomaceous earth filtration process, diatomaceous earth is used due to its high strength (very rigid) and high PERMEABILITY.[11]

Desirable filter media characteristics are as follows:

1. Good hydraulic characteristics (permeable),

2. Does not react with substances in the water (inert and easy to clean),

3. Hard and durable,

4. Free of impurities, and

5. Insoluble in water.

---

[11] Permeability (PURR-me-oh-BILL-oh-tee). The property of a material or soil that permits considerable movement of water through it when it is saturated.

The various filter media are usually classified by the following characteristics:

1. Effective size,

2. Uniformity coefficient,

3. Specific gravity, and

4. Hardness.

The *EFFECTIVE SIZE* (E.S.) refers to the size of a sieve opening that permits 10 percent (by weight) of the particles (sand) to pass through. In other words, 90 percent of the particles (by weight) are larger in diameter than this sieve opening.

Effective Size, mm = Diameter of 10% by weight grains, mm

Some operational problems can be improved by proper selection of media size. To make judgments about media selection, two factors are very important:

1. The time required for turbidity to break through the filter bed, and

2. The time required for the filter to reach limiting (terminal) *HEAD LOSS*.[12]

With a properly selected media, these times are about the same.

If the limiting head loss is frequently a problem and turbidity breakthrough rarely occurs, then a larger media size may be considered. If turbidity breakthrough is frequently a problem and limiting head loss is rarely encountered, then a smaller media size may be considered.

If both head loss and turbidity breakthrough are constantly a problem, a deeper filter bed with a larger media size should be considered. However, increasing the media depth is not always possible without modification of the filter box or tank. Adequate clearance must be allowed between the top of the media and the bottom of the wash water troughs. Otherwise, filter media will be carried out over the wash troughs during backwash.

The relationship between turbidity breakthrough and limiting head loss is also strongly affected by optimum chemical treatment. Poor chemical treatment can often result in either early turbidity breakthrough or rapid head loss buildup.

UNIFORMITY COEFFICIENT (U.C.) refers to the ratio of particle diameters (based on sieve sizes) comprising 60 percent and 10 percent of the media weight, respectively. Media with lower uniformity coefficients are composed of more uniform particles, that is particles that are all closer to the same size. For example, if the 60 percent sieve size grain diameter is 1 mm and the 10 percent sieve size grain diameter is 0.5 mm, then the uniformity coefficient would be:

$$\frac{\text{Uniformity}}{\text{Coefficient}} = \frac{\text{Diameter of 60\% by weight grains, mm}}{\text{Diameter of 10\% by weight grains, mm}}$$

$$= \frac{1.0 \text{ mm}}{0.5 \text{ mm}}$$

$$= 2.0$$

Selection of an appropriate media size and uniformity coefficient depends on the source water quality, filter design, and anticipated filtration rate. Generally the more uniform the media, the slower the head loss buildup. Media with uniformity coefficients of less than 1.5 are readily available. Media with uniformity coefficients of less than 1.3 are only available at a very high cost. Typical filter media characteristics are given in Table 6.1.

TABLE 6.1   TYPICAL FILTER MEDIA CHARACTERISTICS[a]

Material	Size Range (mm)	Specific Gravity	Hardness (MOH scale)
Conventional Sand	0.5 - 0.6	2.6	7
Coarse Sand	0.7 - 3.0	2.6	7
Anthracite Coal	1.0 - 3.0	1.5 - 1.8	3
Garnet	0.2 - 0.4	3.1 - 4.3	6.5 - 7.5
Gravel	1.0 - 50	2.6	7

[a] Adapted from:
 1. Stone, B. G. Notes from "Design of Water Treatment Systems," CE 610, Loyola Marymount University, Los Angeles, CA, 1977.
 2. *WATER QUALITY AND TREATMENT: A HANDBOOK OF COMMUNITY WATER SUPPLIES*, by AWWA, 1990.

6.31   Operational Criteria

6.310   Filter Layout

In gravity filtration, the filter media is usually contained in concrete (also steel or aluminum) filter modules which are all the same size. However, the size will vary widely in surface area from plant to plant. In general, the minimum number of filter modules is four. This allows for one filter to be out of service and still have 75 percent capacity available. Typical gravity filter media sections are shown in Figure 6.6.

6.311   Filter Production and Filtration Rate

Filter production (capacity) rates are measures of the amount of water that can be processed through an individual filter module in a given time period. *FILTER PRODUCTION* is measured in units of million gallons per day (MGD). *FILTRATION RATE* (or hydraulic loading) is commonly used to measure flow of water through a filter and has units of gallons per minute per square foot of filter surface area (GPM/sq ft) or liters per second/square meter or millimeters per second (mm/sec).

Filtration rates used in gravity filtration generally range from about 2 to 10 GPM/sq ft (1.36 to 6.79 liters per sec/sq m or 1.36 to 6.79 mm/sec). The higher filtration rates are found in plants with dual-media filters. In larger sized water treatment plants, filter capacities range from about 5 to 20 MGD (0.2 to 0.8 cu m/sec) for each filter unit.

---

[12] *Head Loss. The head, pressure or energy (they are the same) lost by water flowing in a pipe or channel as a result of turbulence caused by the velocity of the flowing water and the roughness of the pipe, channel walls, or restrictions caused by fittings. Water flowing in a pipe loses head, pressure or energy as a result of friction losses. The head loss through a filter is due to friction losses caused by material building up on the surface or in the top part of a filter.*

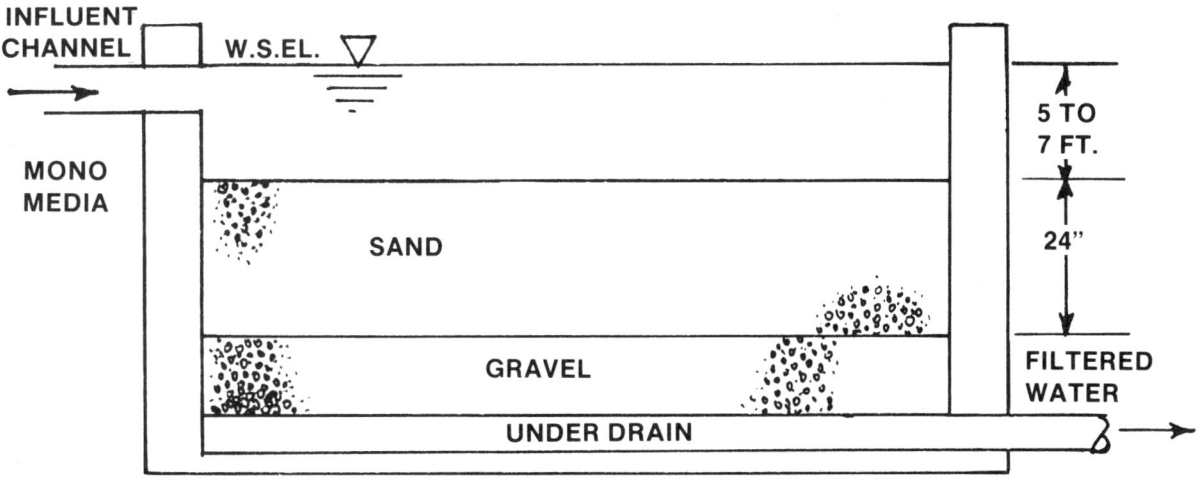

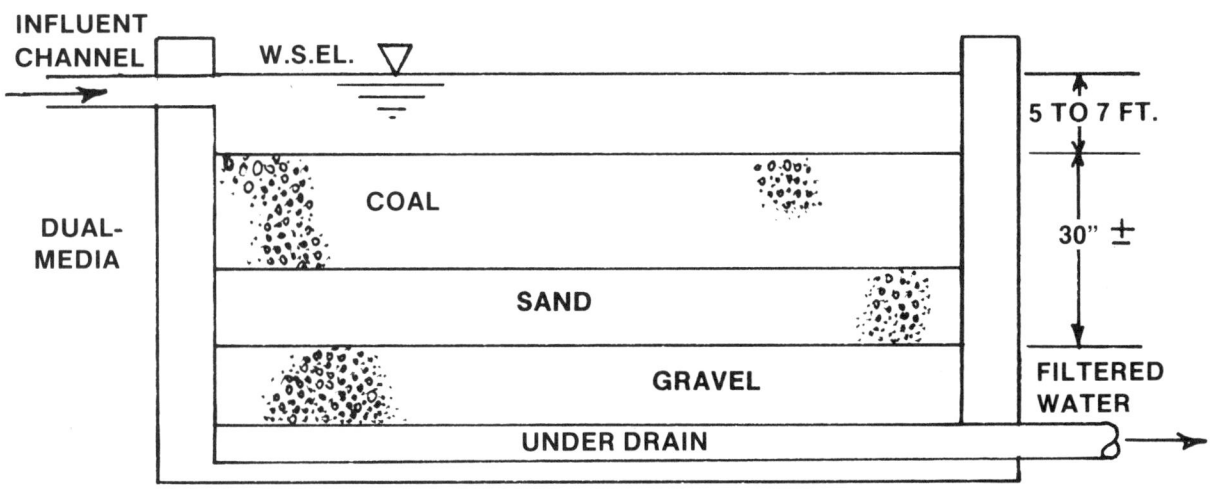

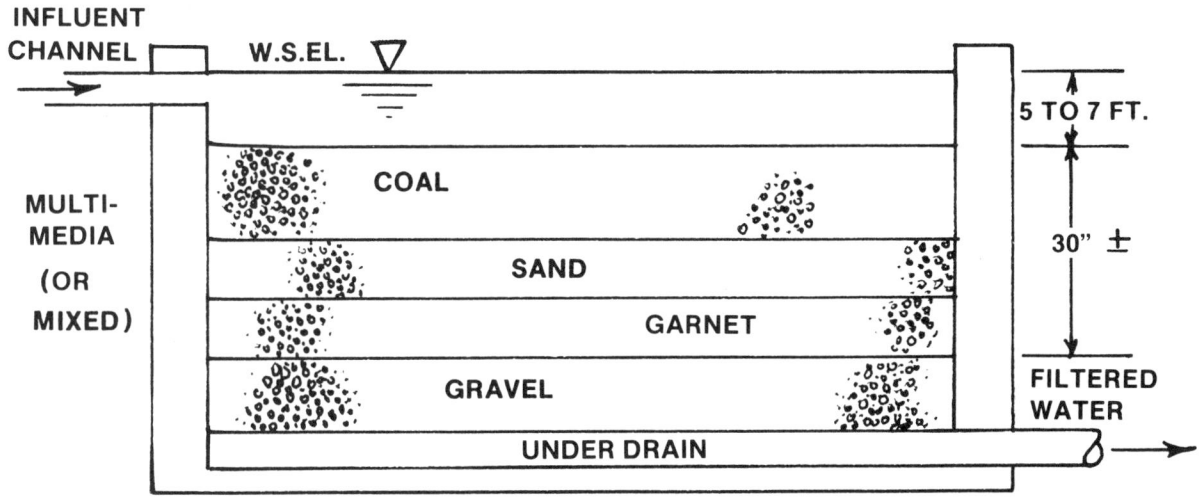

**GRAVITY FILTER MEDIA
CONFIGURATION**

Fig. 6.6   Gravity filter media layouts

6.312 Filtration Efficiency

In the filtration process, filter efficiency is roughly measured by overall plant reduction in turbidity. Reductions of over 99.5 percent can be achieved under optimum conditions, while a poorly operated filter and inadequate pretreatment (coagulation, flocculation, and sedimentation) can result in turbidity removals of less than 50 percent. *THE BEST WAY TO ENSURE HIGH FILTRATION EFFICIENCY IS TO SELECT AN EFFLUENT TURBIDITY GOAL (LEVEL) AND STAY BELOW THE TARGET VALUE (SUCH AS 0.1 TU).*

Filter removal efficiency depends largely on the quality of the water being treated, the effectiveness of the pretreatment processes in conditioning the suspended particles for removal by sedimentation and filtration, and filter operation itself.

Filter unit design and filter media type and thickness also play a role in determining filter removal efficiency, but are less important than water quality and pretreatment considerations. Gravity sand filters usually produce a filtered water turbidity comparable to that of a dual-media filter if the applied water quality is similar. However, the operational differences between sand and dual-media filters are significant. Because of their smaller media grain size (say 0.5 mm), sand filters tend to clog with suspended matter and floc more quickly than dual-media filters. This means frequent backwashing will be required to keep the sand filter operating efficiently. Sand filters have fine, light grains on the top which stop all floc and particulates at the surface of the filter. Dual-media filters have lighter, larger diameter grains in the top layer of the media which stop the larger particles; the smaller particles are usually stopped farther down in the filter. The larger grain size of the anthracite coal layer (up to 1.5 mm) in the top portion of a dual-media filter permits greater depth penetration of solids into the anthracite coal layer and larger solids storage volume in the filter. The sand layer below the anthracite is used as a protective barrier against breakthrough. Chemical dosages are adjusted to keep the solids in the anthracite. Taps at various depths in the filter are used to observe the depth of solids penetration. These characteristics generally produce filter runs that are several times longer and higher filtration rates without a high head loss than those achieved by sand filters. Accordingly, dual-media filters are referred to as "depth filters" while sand filters are known as "surface filters."

Multi-media filters (coal, sand, and garnet) are also used to extend filter run times, and these filters generally perform in a manner similar to gravity sand filters, except that the filter media is enclosed in a pressure vessel. You can get consistently satisfactory filtered water quality with pressure filters if you are careful about the routine operation of this type of filtration system. With the filter media fully enclosed, it is very difficult to assess the media condition by simple visual observation. In addition, excessive pressure in the vessel will force solids as well as water through the filter media. Obviously, this will result in the deterioration of filtered water quality.

In contrast to gravity filtration, diatomaceous earth filtration is essentially a straining process. In many instances, high particulate removal efficiencies can be achieved in diatomaceous earth filtration without any preconditioning processes (coagulation, flocculation, and sedimentation). However, the application of diatomaceous earth filtration in larger water treatment plants is limited by the following considerations:

1. High head losses across the filter,

2. Possible sludge disposal problems, and

3. Potential decreased reliability (in terms of quality of filtered water).

QUESTIONS

Write your answers in a notebook and then compare your answers with those on page 253.

6.3A What material is most often used to support granular filter materials?

6.3B Filtration rate is commonly expressed in what units?

6.3C What is the major operational difference between sand and dual-media filters?

6.32 Filter Operation

6.320 Filtration Mode

In the filtration mode of operation, water containing suspended solids is applied to the surface of the filter media. Depending on the amount of suspended solids in the water being treated and the filtration rate, the filter will exhibit high head loss or "clog" after a given time period (varies from several hours to several days).

CLOGGING may be defined as a buildup of head loss (pressure drop) across the filter media until it reaches some predetermined design limit. Total design head loss in gravity filters generally ranges from about 6 to 10 feet (1.8 to 3.0 m), depending mainly on the depth of the water over the media. Clogging of the filter leads to *BREAKTHROUGH*,[13] a condition in which solids are no longer removed by the already overloaded filter. The solids pass into the filter effluent where they appear as increased turbidity.

A filter is usually operated until just before clogging or breakthrough occurs, or a specified time period has passed, generally within 72 hours. In order to save money, energy, and water by maximizing production before backwashing, some operators run their filters until clogging or breakthrough occurs. This is a poor practice; when breakthrough occurs, there

---

[13] *Breakthrough.   A crack or break in a filter bed allowing the passage of floc or particulate matter through a filter. This will cause an increase in filter effluent turbidity. A breakthrough can occur (1) when a filter is first placed in service, (2) when the effluent valve suddenly opens or closes, and (3) during periods of excessive head loss through the filter (including when the filter is exposed to negative heads).*

will be an increase in filtered water turbidity. See Section 6.91 for detailed procedures on how to check out, start up, and shut down filters.

6.321 Backwashing

After a filter clogs (reaches maximum head loss), or breakthrough occurs, or a specified time period has passed, the filtration process is stopped and the filter is taken out of service for cleaning or "backwashing." BACKWASHING is the process of reversing the flow of water through the filter media to remove the entrapped solids. In order to remove the trapped solids from the filter media in the gravity and pressure filtration processes, the filter media must be expanded or FLUIDIZED[14] by reversing the flow of water. Backwash flow rates ranging from 10 to 25 GPM/sq ft (6.8 to 17 liters per sec/sq m or 6.8 to 17 mm/sec) of filter media surface area are usually required to clean the filter adequately. Insufficient backwash rates may not completely remove trapped solids from the filter media, while too high a backwash rate may cause excessive loss of filter media and media disturbance (mounding). Usually higher backwash rates are required at high temperatures and with larger media. Higher backwash rates are required at higher temperatures to suspend the media because the water is less viscous (like warmer syrup). Backwashing at too high a rate is much more destructive than at too low a backwash rate. See Section 6.911, "Backwash Procedures," for details on how to backwash filters.

In diatomaceous earth filtration, accumulated solids are also removed by reversing the flow through the filter. However, the diatomaceous earth filter coating itself must also be removed; this contributes greatly to the volume of waste sludge produced by this filtration process. An advantage is that less water is required for backwashing.

The pressurized water supply required for backwashing a filter is usually supplied by a backwash pump which uses filtered water as a supply source. An elevated storage tank can be used to store enough water to backwash large filters and multiple filters. A typical backwash pump and control panel are shown in Figure 6.7.

The backwashing process may use about two to four percent of the process water for transporting the wash water to de-watering and final solids disposal processes (see Volume II, Chapter 17, "Handling and Disposal of Process Wastes"). If the wash water can be removed from the water surface in the solids disposal process, a major portion of the water can be recycled through the water treatment plant. In some plants the wash water is recycled directly to the headworks (ahead of flash mixer).

Backwash pump

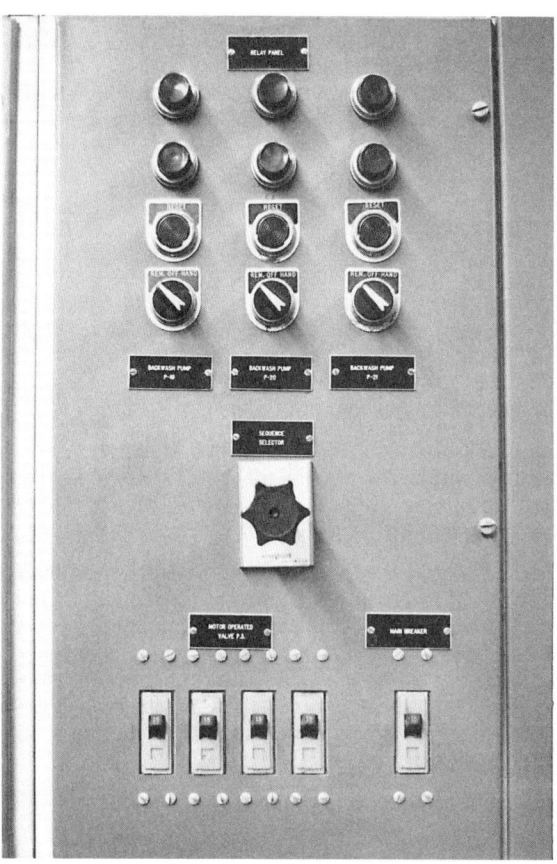

Backwash pump
control panel

Fig. 6.7 Backwash pumping

---

[14] Fluidized (FLEW-id-I-zd). A mass of solid particles that is made to flow like a liquid by injection of water or gas is said to have been fluidized. In water treatment, a bed of filter media is fluidized by backwashing water through the filter.

Wash water troughs are installed in gravity filters to aid in collecting wash water from the filters (see Figures 6.8 and 6.9).

The Filter Backwash Rule requires that recycled filter backwash water, sedimentation basin sludge thickener supernatant, and liquids from sludge dewatering processes be returned upstream of all conventional or direct filtration treatment systems, including coagulation, flocculation, sedimentation (with conventional filtration only), and filtration. Systems may apply to the state for approval to recycle at an alternate location.

The intent of the Filter Backwash Rule is to improve performance at conventional and direct water filtration plants by reducing the opportunity for recycle practices to adversely affect plant performance in a way that would allow microbes such as *Cryptosporidium* to pass through the treatment processes and into the finished drinking water. Surges of recycle flow returned to the treatment processes must also be controlled because surges may adversely affect treatment processes by creating hydraulically overloaded conditions (when plants exceed design capacity) that lower performance of individual treatment processes within a treatment plant resulting in lowered *Cryptosporidium* removal efficiency.

6.322   Surface Wash

In order to produce optimum cleaning of the filter media during backwashing and to prevent mudballs, surface wash (supplemental scouring) is usually required. Surface wash systems provide additional scrubbing action to remove attached floc and other entrapped solids from the filter media. Four types of surface wash systems (Figure 6.10) are:

1. Baylis,
2. Fixed grid,
3. Rotary, and
4. Air scour.

The first three surface wash systems are mechanical or water-powered, while the fourth system uses air to create the washing action.

6.33   Filter Control Systems

Filter control systems (see Figures 6.11 and 6.12) regulate flow rates through the filter by maintaining an adequate head above the media surface. This head (*SUBMERGENCE*[15]) forces water through a gravity filter.

A filter control system also prevents sudden flow increases or surges which could discharge solids trapped on the filter media. If the solids were suddenly dislodged, they would seriously degrade water quality. One way to prevent incoming water from disturbing (scouring) the media is to maintain an adequate depth of water above the media surface. In this way, the force of the incoming water is absorbed before it reaches the media, thus preventing scouring.

Filter control systems commonly used in gravity filtration installations include the following:

1. Rate-of-flow,
2. Split-flow,
3. Declining-rate, and
4. Self-backwash.

In *RATE-OF-FLOW* control systems, each filter effluent control valve is connected to a flowmeter. As the filter run continues and the media begins to clog, the control valve slowly opens to maintain a constant flow of water through the filter. A master controller is required to monitor the overall plant flow and adjust the flow rate of each filter accordingly. With this system, all filters operate at the same flow rate, but the rate is variable, dropping each time a new filter is backwashed. Some filters have influent control valves.

In the *SPLIT-FLOW* control system, the flow to each filter influent is "split" or divided by a weir. With this system equal flow is automatically distributed to each filter. The filter effluent valve position is controlled by the water level in the filter. Each filter operates within a narrow water level range.

In *DECLINING-RATE* filters, flow rate varies with head loss. Each filter operates at the same, but variable, water surface level. This system is relatively simple, but requires an effluent control structure (weir) to provide adequate media submergence. Effluent control structures are usually desirable with the other systems described here as well.

In the *SELF-BACKWASH* (or Streicher design) system, influent flow to each filter is divided by a weir. The water surface level in each filter varies according to head loss, but the flow rate remains constant for each filter. This system reduces the amount of mechanical equipment required for operation and backwashing, such as wash water pumps, and also requires an effluent control structure and a deeper filter box.

QUESTIONS

Write your answers in a notebook and then compare your answers with those on page 253.

6.3D   What two main factors influence the time period before a filter becomes clogged?

6.3E   Under what conditions is the filtration process stopped and the filter taken out of service for cleaning or "backwashing"?

6.3F   List four types of surface wash (supplemental scour) systems for filters.

6.3G   What aspects of the filtration process are actually controlled by the filter control system?

6.4   ACTIVATED CARBON FILTERS

The primary purpose of filtration is to remove suspended particles and floc from the water being treated. Another dimension is added to the filtration process by the use of activated carbon (granular form) as a filter media. The high adsorptive capacity of activated carbon enables it to remove taste- and odor-causing compounds, as well as other trace organics from the water. However, not all organic compounds are removed with the same degree of efficiency. Other methods of removing volatile organics include air stripping.

While activated carbon filtration is very effective in removing taste- and odor-causing compounds, the construction, carbon handling equipment, and operating costs are generally quite high. Still, activated carbon can be added to existing filter beds or can be incorporated as a separate process. Provisions must be made for regeneration or reactivation of "spent" carbon (carbon that has lost its adsorptive capacity). Usually

---

[15] *Submergence.   The distance between the water surface and the media surface in a filter.*

Wash water trough
(Filter water-surface level drawn down prior to backwash)

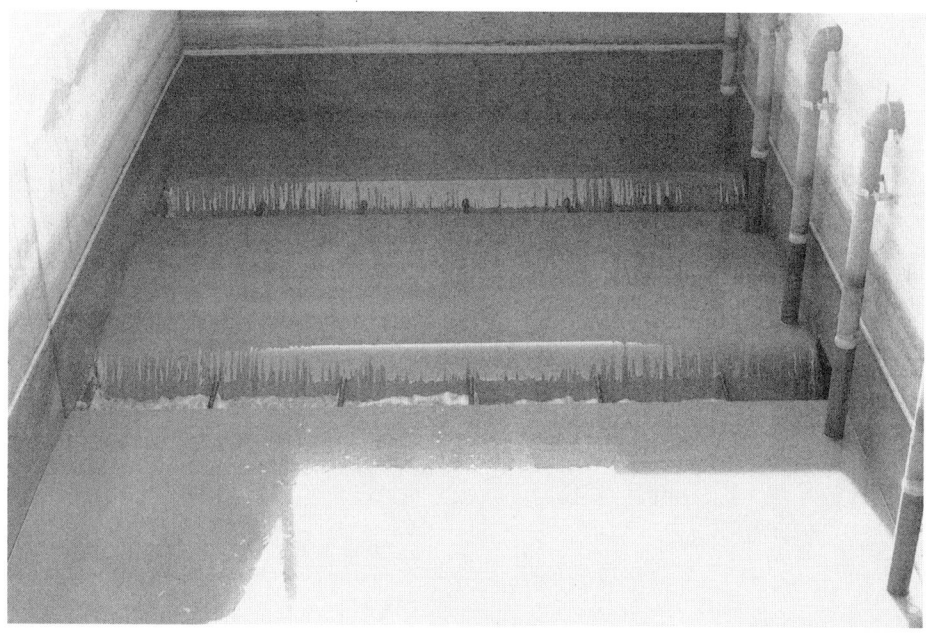

Filter during backwash mode

Fig. 6.8 Backwashing

Removal of solids during backwash

Operator hosing down filter sidewalls during backwash

(*NOTE:* Operator is too close to edge and
could fall in—should be behind guardrails)

Fig. 6.9   Backwashing

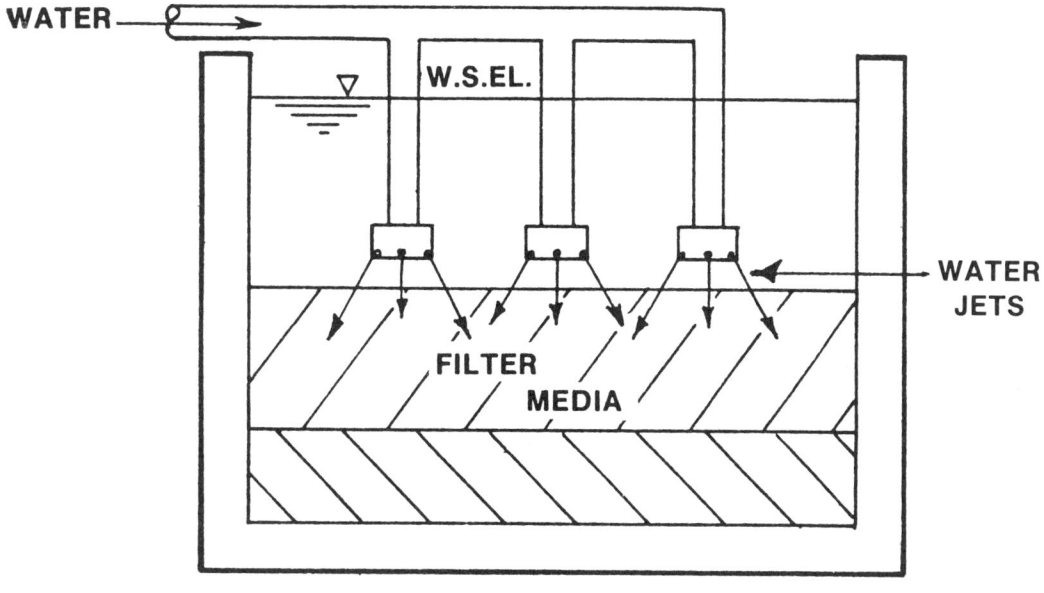

BAYLIS SURFACE WASH

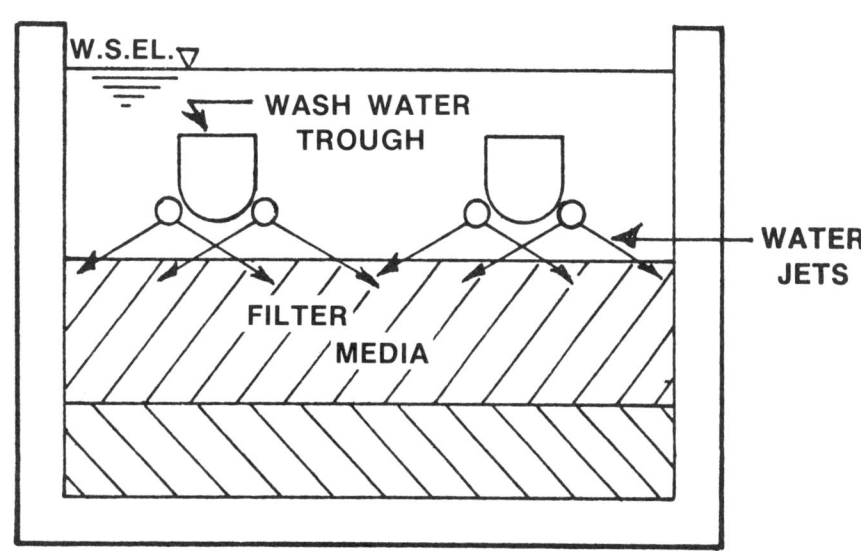

FIXED GRID SURFACE WASH

SUPPLEMENTAL SCOUR SYSTEMS

Fig. 6.10   Surface wash systems

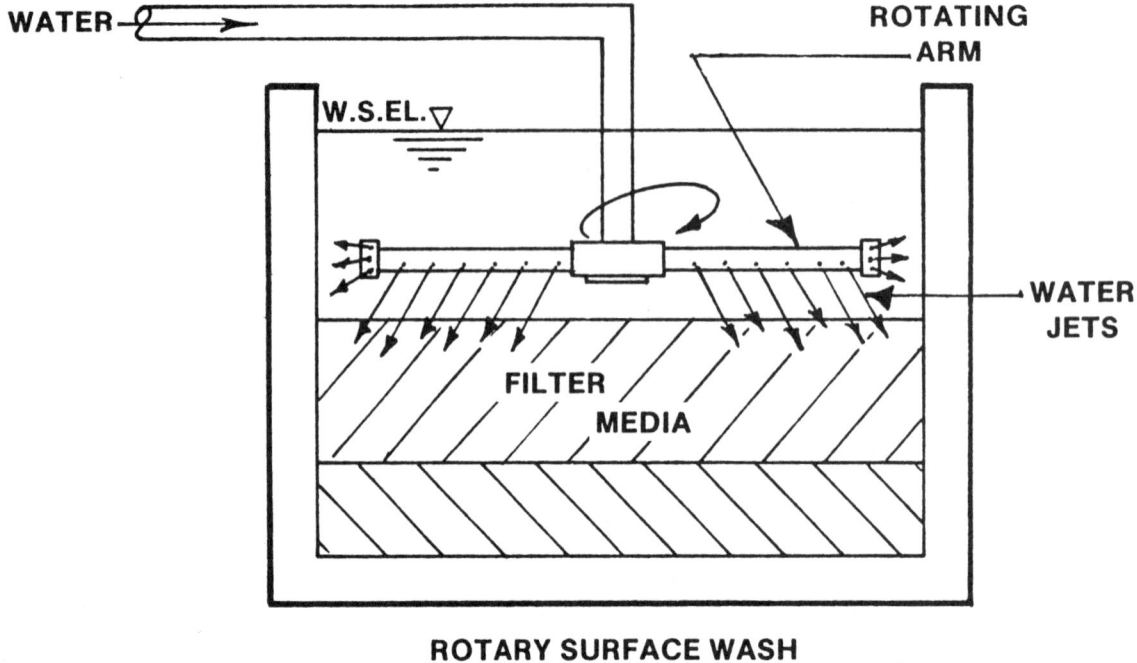

ROTARY SURFACE WASH

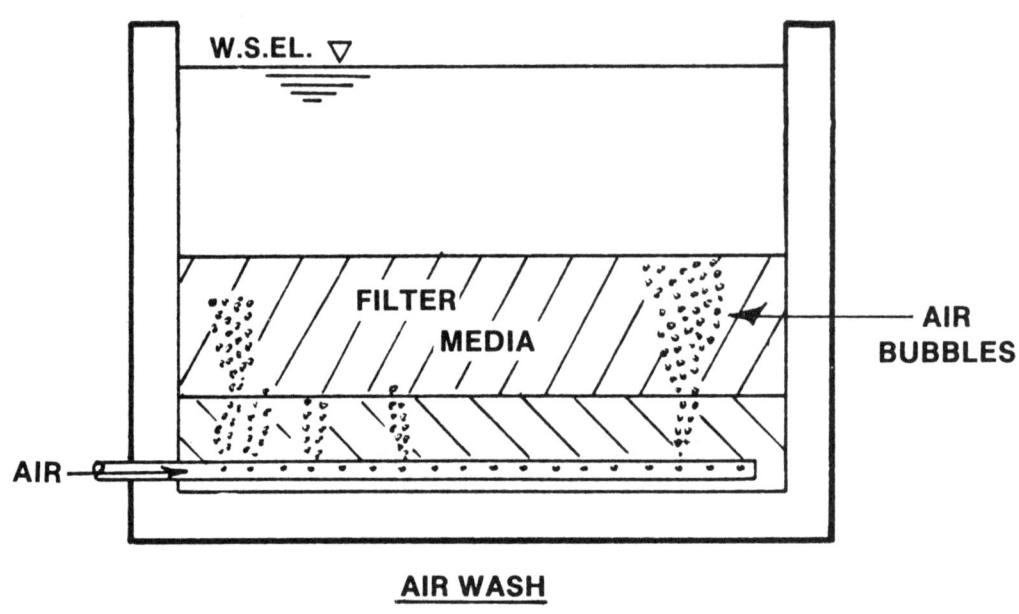

AIR WASH

SUPPLEMENTAL SCOUR SYSTEMS

Fig. 6.10   Surface wash systems (continued)

Filter operating deck

Filter control panels for two filters

Fig. 6.11   Filter controls

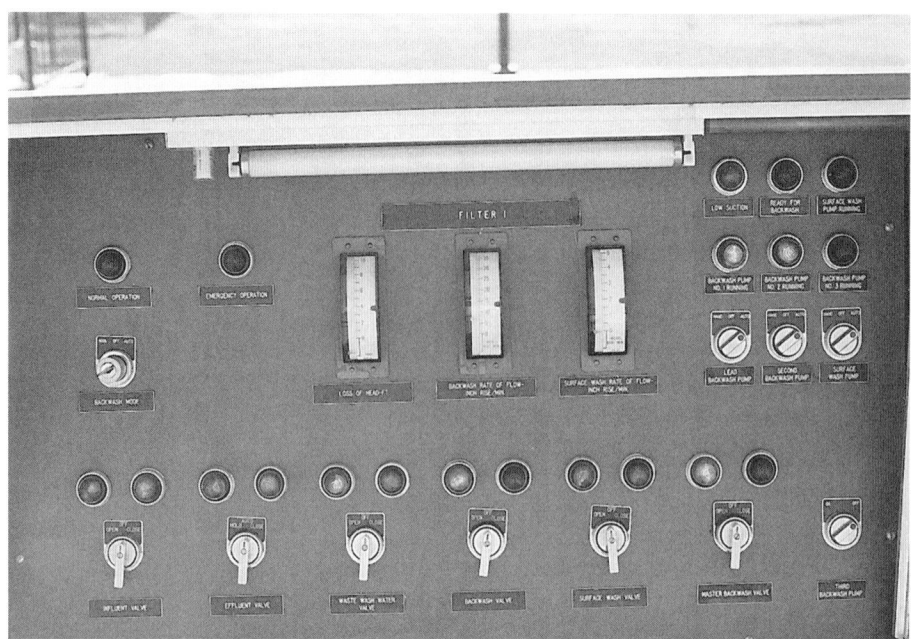

Filter control panel

Filter instrumentation
(Backwash rate, backwash and surface wash water consumption)

Fig. 6.12   Filter controls

arrangements are made with the carbon manufacturer to regenerate or exchange spent carbon.

An alternative to activated carbon filtration for reducing taste and odor levels is the use of powdered activated carbon prior to sedimentation and filtration. This option is usually more cost-effective than activated carbon filtration, but generally will not yield equivalent results, due to operational and physical problems which limit the amount of powdered activated carbon that can be applied to the water being treated. Powdered activated carbon is difficult to handle due to dust problems.

6.5 INTERACTION WITH OTHER TREATMENT PROCESSES

6.50 Importance of Pretreatment

The purpose of filtration is the removal of particulate impurities and floc from the water being treated. In this regard, the filtration process is the final step in the solids removal process which usually includes the pretreatment processes of coagulation, flocculation, and sedimentation.

The degree of pretreatment applied prior to filtration depends on the type of treatment plant (pressure filter, diatomaceous earth filtration, or gravity filtration) and the size of the treatment facility. Most large municipal treatment plants include complete pretreatment facilities and gravity filtration. In any event, the importance of pretreatment prior to filtration cannot be overemphasized.

Floc particles that are carried over into the filter influent must be small enough to penetrate the upper filter media (depth filtration). Floc that is too large will cause the top portion of the filter bed to clog rapidly, thus leading to short filter runs. In addition, large floc (particularly alum floc) is often weak and easily broken up by water turbulence. This can cause degradation of effluent water quality.

Ideally, floc removal is accomplished by contact with the media grains (sand, coal, and garnet) throughout the upper depths of the filter. After the initial coating or conditioning of the media surfaces with floc at the beginning of the filtration cycle, subsequent applications of floc will build up on the material previously deposited on the media surface. This process is often referred to as the "ripening period." Higher filter effluent turbidities may occur during the first few minutes at the beginning of the filter run until the "ripening period" is completed.

At filtration rates below 4 GPM/sq ft (2.72 liters per sec/sq m or 2.72 mm/sec) alum or iron coagulants usually give adequate treatment without assistance.

At filtration rates above 4 or 5 GPM/sq ft (2.72 or 3.40 liters per sec/sq m or 2.72 or 3.40 mm/sec) iron and alum floc will shear in the pores of the filter and short filter runs will result because of turbidity breakthrough. Under these conditions the addition of a polymer coagulant aid or a chemical filter aid is often beneficial.

The polymers most frequently found to be successful in this application are moderate molecular weight cationic polymers (DADMA) and relatively high molecular weight nonionic polymers (polyacrylamides). The cationic polymers are best added ahead of the flocculation process to strengthen the floc formed. The nonionic polymers are generally added in the settled water as it moves toward the filters from the sedimentation basins.

Typical doses for cationic polymers range from 0.25 to 2 mg/L. Typical doses for nonionic polymers range from 0.02 to 0.2 mg/L.

6.51 In-Line Filtration

IN-LINE FILTRATION refers to the addition of coagulant chemicals immediately before the water enters the filtration system (Figure 6.13). This pretreatment method is commonly used in pressure filter installations. Chemical filter aids are added directly to the filter inlet pipe and are mixed by the flowing water. When this is done, separate flocculation and sedimentation facilities are eliminated.

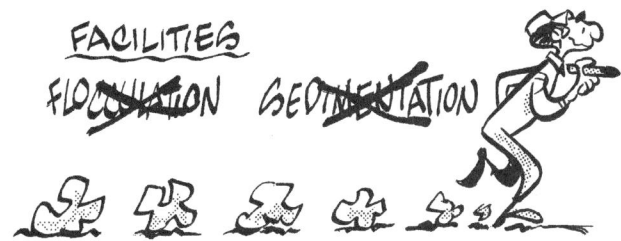

This process is not as efficient in forming floc as conventional or direct filtration when source water quality has variable turbidity and bacterial levels. Problems may develop with the formation of floc in the water after filtration.

6.52 Conventional Filtration (Treatment)

The conventional filtration (treatment) process is used in most municipal treatment plants in the United States. This process includes "complete" pretreatment (coagulation, flocculation, and sedimentation) as shown in Figure 6.13. This system provides a great amount of flexibility and reliability in plant operation, especially when source water quality is highly variable or is high in suspended solids.

A chemical application point just prior to filtration permits the application of a filter aid chemical (such as a nonionic polymer) to assist in the solids removal process, especially during periods of pretreatment process upset, or when operating at high filtration rates.

6.53 Direct Filtration

Direct filtration is considered a feasible alternative to conventional filtration, particularly when source waters are low in turbidity, color, plankton, and coliform organisms. Direct filtration can be defined as a treatment system in which filtration is not preceded by sedimentation, as shown in Figure 6.13. Many direct filtration plants provide rapid mix, short detention without agitation (30 to 60 minutes) followed by filtration. As in conventional filtration, a chemical application point just prior to filtration permits the addition of a filter aid chemical.

6.6 PROCESS CONTROL

In theory, the physical and chemical processes governing solids removal in filtration are rather complex. Yet, from a practical perspective, the treatment plant operator must be provided with the means to measure and control the performance of the filtration process on a day-to-day basis. In this regard, filter influent water quality (turbidity), filter performance (head loss buildup rate and filter run time), and filter effluent water quality (turbidity) are important process control guidelines.

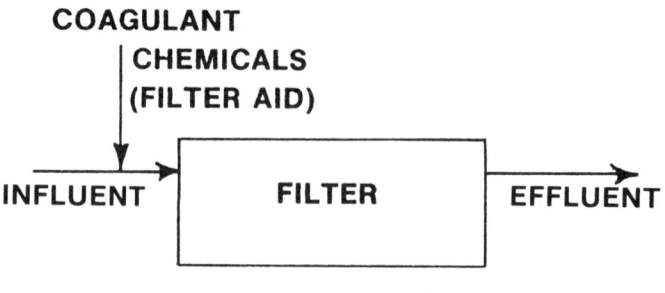

IN-LINE FILTRATION

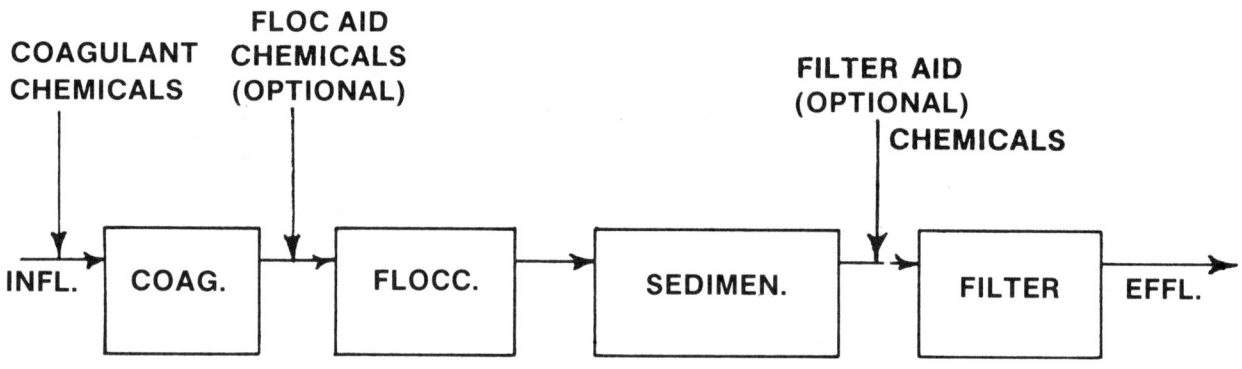

CONVENTIONAL FILTRATION

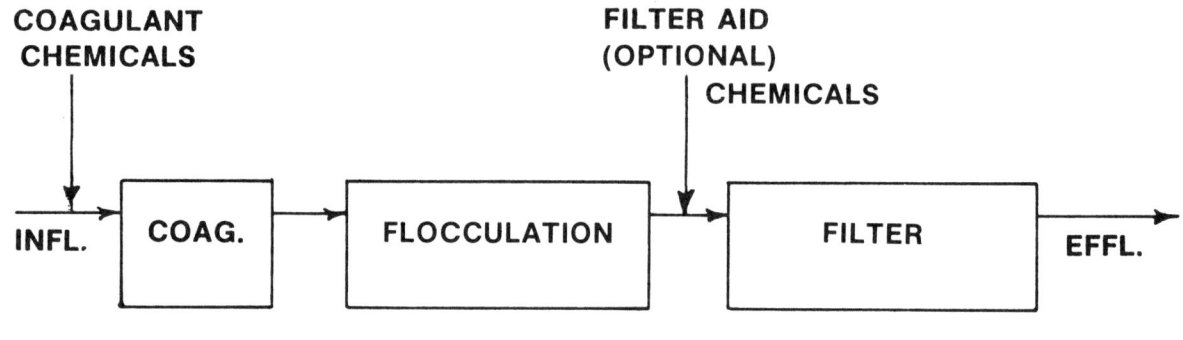

DIRECT FILTRATION

Fig. 6.13   Filtration systems

An efficient means of handling wash waters is to recycle the water to the beginning of the plant (before the flash mixer).

Means for measuring and controlling the filtration process are described in the next lesson.

QUESTIONS

Write your answers in a notebook and then compare your answers with those on pages 253 and 254.

6.4A   What is the primary purpose of using activated carbon (granular form) as filter media?

6.5A   What is "in-line filtration"?

6.5B   When and where are filter aid chemicals used?

6.5C   When is direct filtration used?

6.6A   What factors must an operator measure to control the performance of the filtration process on a day-to-day basis?

End of Lesson 1 of 2 Lessons on FILTRATION

Please answer the discussion and review questions next.

DISCUSSION AND REVIEW QUESTIONS

Chapter 6.   FILTRATION

(Lesson 1 of 2 Lessons)

At the end of each lesson in this chapter you will find some discussion and review questions. The purpose of these questions is to indicate to you how well you understand the material in the lesson. Write the answers to these questions in your notebook before continuing.

1. What is the primary purpose of the filtration process?

2. Why is diatomaceous earth filtration used in some water treatment applications and not in others?

3. What are the desirable characteristics for filter media?

4. If both head loss and turbidity breakthrough are constantly a problem due to improperly sized media, what should be done?

5. What problems can develop if backwash rates are either too high or too low?

CHAPTER 6.   FILTRATION

(Lesson 2 of 2 Lessons)

6.7   OPERATING PROCEDURES ASSOCIATED WITH NORMAL PROCESS CONDITIONS

6.70   Indicators of Normal Operating Conditions

Filtration is the final and most important step in the solids removal process. From a water quality standpoint, filter effluent turbidity will give you a good indication of overall process performance. However, you must also monitor the performance of each of the preceding treatment processes (coagulation, flocculation, and sedimentation) as well as filter effluent water quality, in order to anticipate water quality or process performance changes which might affect filter operation.

In the normal operation of the filtration process, the operator should closely monitor filter influent turbidity as well as filter effluent turbidity levels with a *TURBIDIMETER*[16] (Figure 6.14). Filter influent turbidity levels (settled turbidity) can be checked on a periodic basis (say once every two hours) by securing a grab sample either at the filter or from the laboratory sample tap (if such facilities are provided). However, filter effluent turbidity is best monitored and recorded on a continuous basis by an on-line turbidimeter. If the turbidimeter is provided with an alarm feature, virtually instantaneous response to process failures can be achieved.

Other indicators that can be monitored to determine if the filter is performing normally include head loss buildup and filter effluent color.

A written set of process guidelines should be developed to assist the operator in evaluating normal process conditions and in recognizing abnormal conditions. These guidelines can be developed based on water quality standards, design considerations, water quality conditions, and most importantly, trial and experience.

6.71   Process Actions

In the normal operation of the filtration process, you will perform a variety of functions with emphasis on maintaining a high-quality filtered water. For all practical purposes, the quality of the filter effluent constitutes the final product quality that will be distributed to consumers.

Typical functions you will be expected to perform in the normal operation of the filtration process include the following:

1. Monitor process performance;

2. Evaluate water quality conditions (turbidity) and make appropriate process changes;

3. Check and adjust process equipment (change chemical feed rates);

4. Backwash filters;

5. Evaluate filter media condition (media loss, mudballs, cracking); and

6. Visually inspect facilities.

Monitoring process performance is an ongoing activity. You should look for and attempt to anticipate any treatment process changes or other problems that might affect filtered water quality, such as a chemical feed system failure.

Measurement of head loss buildup (Figure 6.15) in the filter media will give you a good indication of how well the solids removal process is performing. The total designed head loss from the filter influent to the effluent in a gravity filter is usually about 10 feet (3 m). The actual head loss from a point above the filter media to a reference point in the effluent can be monitored as "loss-in-head" (or "loss-of-head"). For example, suppose that a gravity filter is designed for a total potential head loss of 10 feet (3 m). At the beginning of the filtration cycle the actual measured head loss due to clean media and other hydraulic losses is 3 feet (0.9 m). This would permit an additional head loss of 7 feet (2.1 m) due to solids accumulation in the filter. In this example, a practical cutoff point might be established at an additional six feet (1.8 m) of head loss (total of nine feet (2.7 m)) for backwashing purposes.

The rate of head loss buildup is also an important indicator of process performance. Sudden increases in head loss might be an indication of surface sealing of the filter media (lack of depth penetration). Early detection of this condition may permit you to make appropriate process changes such as adjustment of the filter aid chemical feed rate or reduction of filtration rate.

---

[16] *Turbidimeter.   An instrument for measuring and comparing the turbidity of liquids by passing light through them and determining how much light is reflected by the particles in the liquid. The normal measuring range is 0 to 100 and is expressed as Nephelometric Turbidity Units (NTUs).*

FLOW DIAGRAM

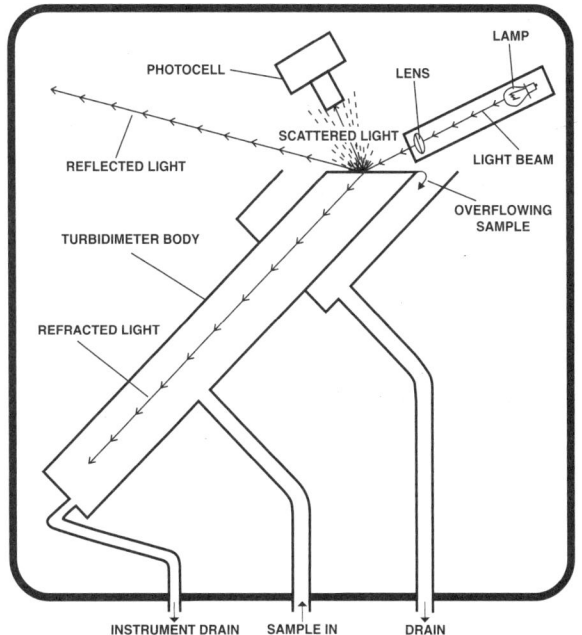

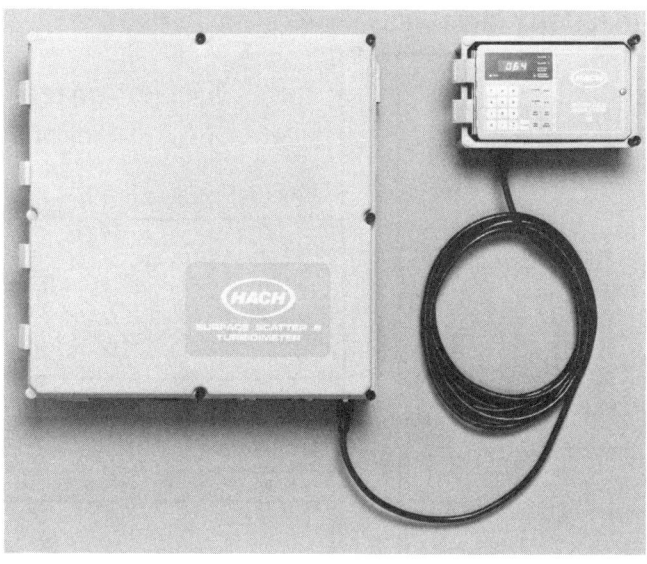

Raw Water On-Line Turbidimeter

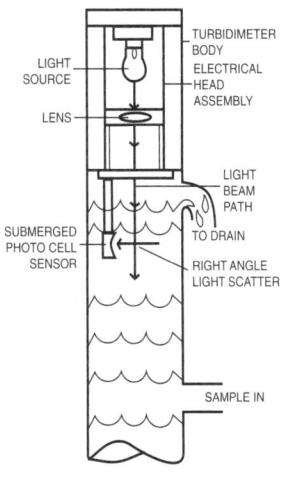

HOW IT WORKS

Hach's Low Range Turbidimeter is a continuous reading nephelometer. A strong light beam is passed through a water sample and measurement of the amount of light scattered by the turbidity particles is made.

If the turbidity of the sample is negligible, no light reaches the photocell sensor. Conversely, the presence of turbidity in the sample results in light being scattered with the corresponding number of NTUs shown on the master indicator.

The most important advantage of the nephelometer is that a very intense beam of light can be passed through the sample. This results in a very high sensitivity which permits accurate measurements of trace turbidity amounts.

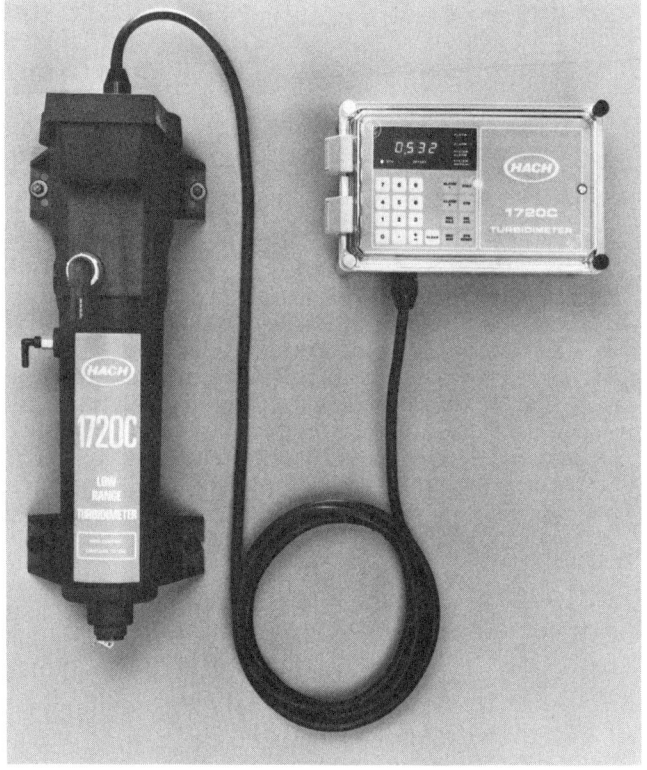

INSTRUMENT BODY   **MASTER INDICATOR**

Fig. 6.14   Continuous flow filter influent and effluent turbidimeter
(Courtesy of the HACH Company)

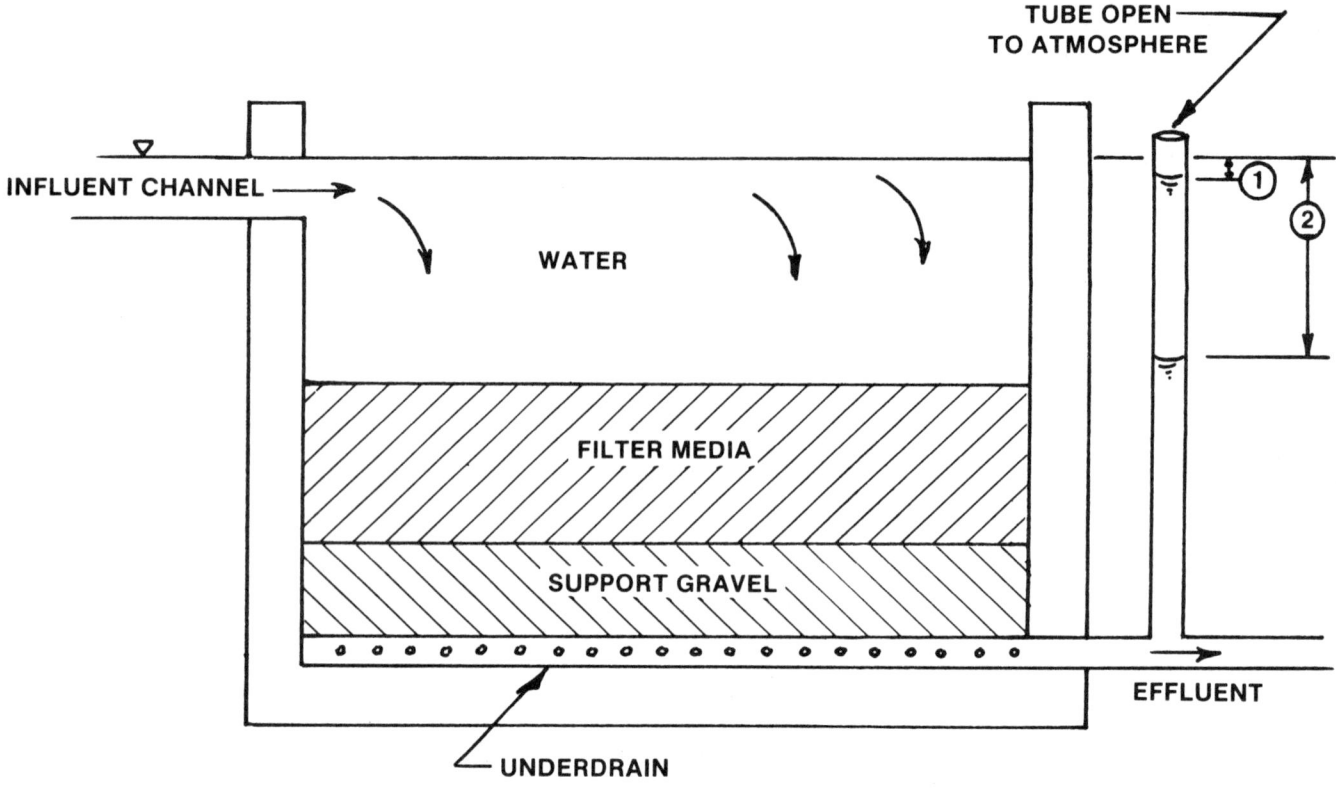

NOTE: IF A TUBE OPEN TO THE ATMOSPHERE WAS INSTALLED
IN THE FILTER EFFLUENT, THEN
① = HEAD LOSS THROUGH FILTER AT START OF RUN, AND
② = HEAD LOSS THROUGH FILTER BEFORE START OF
    BACKWASH CYCLE.

Fig. 6.15   Measurement of head loss buildup

Monitoring of filter effluent turbidity on a continuous basis with an on-line turbidimeter (Figure 6.14) is highly recommended. This will provide you with continuous feedback on the performance of the filtration process. In most instances it is desirable to cut off (terminate) filter operation at a predetermined effluent turbidity level. Preset the filter cutoff control at a point where your experience and tests show breakthrough will soon occur (Figure 6.16).

In the normal operation of the filter process, it is best to calculate when the filtration cycle will be completed on the basis of head loss, effluent turbidity level, and elapsed run time.

A predetermined value is established for each guideline as a cutoff point for filter operation. When any one of these levels is reached, the filter is removed from service and backwashed (see Section 6.9, "Start-Up and Shutdown Procedures," for step-by-step procedures).

Usually plant operators compare filter performance from season to season and from plant to plant by comparing the filter run length in hours. The reason for this is that as the filter run gets shorter, the amount of water wasted in backwash becomes increasingly important when compared to the amount of water produced during the filter run. Percent backwash water statistics are also occasionally collected.

Although of some use, filter run length is not a satisfactory basis for comparing filter runs without considering the filtration rate as well. For example, at a filtration rate of 6 GPM/sq ft (4.1 liters per sec/sq m or 4.1 mm/sec), an 18-hour filter run is quite adequate. Whereas, at a filtration rate of 1.5 GPM/sq ft (1.0 liters per sec/sq m or 1.0 mm/sec), an 18-hour filter run is not satisfactory.

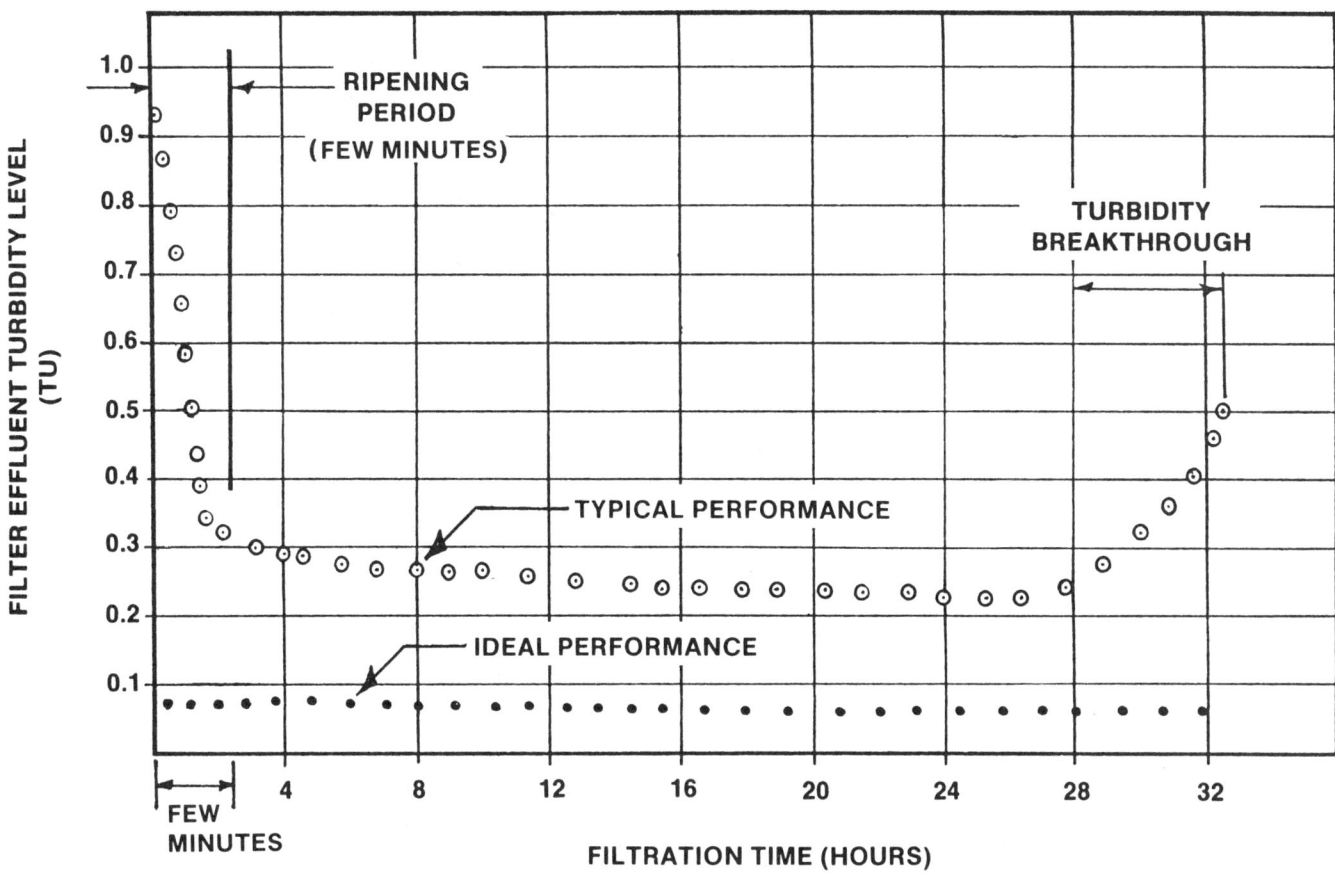

Fig. 6.16   *Typical filter effluent turbidity graph*

The best way to compare filter runs is by using the Unit Filter Run Volume (UFRV) technique. The UFRV is the volume of water produced by the filter during the course of the filter run divided by the surface area of the filter. This is usually expressed in gallons per square foot (liters per square meter). UFRVs of 5,000 gal/sq ft (203,710 liters per square meter) or greater are satisfactory, and UFRVs greater than 10,000 gal/sq ft (407,430 liters per square meter) are desirable. In the examples cited in the previous paragraph, the UFRV for the filter operating at 6 GPM/sq ft (4.1 liters per sec/sq m or 4.1 mm/sec) would be 6,480 gal/sq ft (264,010 liters per sq m), and the filter operating at 1.5 GPM/sq ft (1.0 liters per sec/sq m or 1.0 mm/sec) would be 1,620 gal/sq ft (66,000 liters per sq m).

Water quality indicators used to assess process performance include turbidity and color (Figure 6.17). Based on an assessment of overall process performance, you may need to make changes in the coagulation-flocculation process as described in Chapter 4, or in the sedimentation process as described in Chapter 5.

At least once a year examine the filter media and evaluate its overall condition. Measure the filter media thickness for an indication of media loss during the backwashing process.

Measure mudball accumulation in the filter media to evaluate the effectiveness of the overall backwashing operation (see Appendix A, "Mudball Evaluation Procedure," at the end of this chapter).

Routinely observe the backwash process to qualitatively assess process performance. Watch for media boils (uneven flow distribution) during backwashing, media carryover into the wash water trough, and clarity of the waste wash water near the end of the backwash cycle.

Never "bump" a filter to avoid backwashing. Bumping is the act of opening the backwash valve during the course of a filter run to dislodge the trapped solids and increase the length of the filter run. This is *NOT* a good practice. If your plant has an on-line turbidimeter on the filter effluent, you will get caught.

Upon completion of the backwash cycle, observe the condition of the media surface and check for filter sidewall or media surface cracks. Corrective actions are described in Section 6.8. Routinely inspect physical facilities and equipment as part of good housekeeping and maintenance practice. Correct or report abnormal equipment conditions to the appropriate maintenance personnel. Table 6.2 is a summary of routine filtration process actions.

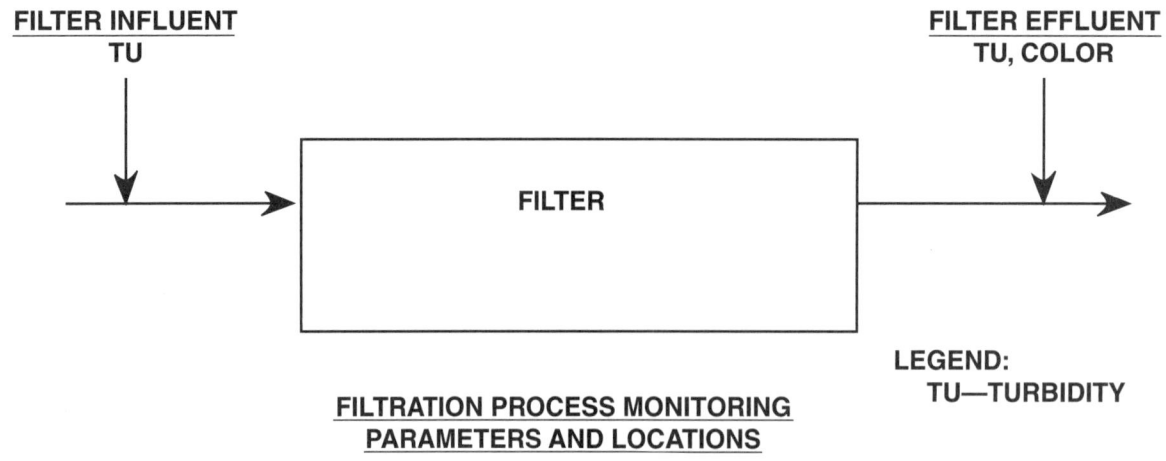

FILTER INFLUENT
TU

FILTER EFFLUENT
TU, COLOR

FILTER

LEGEND:
TU—TURBIDITY

FILTRATION PROCESS MONITORING
PARAMETERS AND LOCATIONS

Fig. 6.17 Filtration process monitoring guidelines and locations

QUESTIONS

Write your answers in a notebook and then compare your answers with those on page 254.

6.7A What is the most important water quality indicator used to monitor the filtration process?

6.7B How can filter effluent turbidity be measured on a continuous basis?

6.7C List some of the typical functions performed by operators in the normal operation of the filtration process.

6.7D What could cause a sudden increase in head loss through a filter?

6.7E How would you change the operation of a filter if there was a sudden increase in head loss through the filter?

6.72 Process Calculations

In the routine operation of the filtration process, you will be expected to perform a variety of process calculations related to filter operation (flow rate, filtration rate), backwashing (backwash rate, surface wash rate), water production, and percent of water production used to backwash filters.

6.720 Filter Efficiency

Normally the efficiency of the filtration process is not calculated as such. The filter effluent turbidity is measured from either a grab sample or an on-line turbidimeter. Whenever the effluent turbidity exceeds 0.5 TU or some other established value, the filter should be backwashed.

6.721 Filtration Rate

Filtration rates are measured in gallons per minute per square foot of filter area (GPM/sq ft), liters per second per square meter of filter area (liters per sec/sq m), or rate of rise in millimeters per second. Filtration rates will vary from 2 to 10 GPM/sq ft (1.36 to 6.79 liters per sec/sq m or 1.36 to 6.79 mm/sec) with most agencies specifying a maximum of 6 GPM/sq ft (4.07 liters per sec/sq m or 4.07 mm/sec) for dual-media units. From a practical standpoint, if your plant has four filters the same size and a plant design flow of 10 MGD, any time more than 2.5 MGD is applied to any filter, you are exceeding the design filtration rate. Problems can develop when design filtration rates are exceeded.

FORMULAS

The filtration rate is described as the flow in gallons per minute (GPM) that is filtered by one square foot of filter surface area (a square one foot wide and one foot long). If the flows are given in million gallons per day (MGD), they must be converted to gallons per minute (GPM).

$$\text{Flow, GPM} = \frac{(\text{Flow, MGD})(1{,}000{,}000/\text{Million})}{(24 \text{ hr/day})(60 \text{ min/hr})}$$

$$\text{Filtration Rate,} \atop \text{GPM/sq ft} = \frac{\text{Flow, GPM}}{\text{Surface Area, sq ft}}$$

The flow rate in gallons per minute can be determined by obtaining the flow from a flowmeter in millions of gallons per day (MGD) and converting this flow to GPM as shown above. Another approach is to turn off the influent valve to a filter and record the time for the water to drop a specified distance. By calculating the velocity of the dropping water and knowing the surface area of the filter, we can calculate the flow rate in gallons per minute (GPM) or any other desired units.

1. To calculate the velocity of the water dropping above the filter, measure the time for the water to drop a specific distance.

$$\text{Velocity, ft/min} = \frac{\text{Water Drop, ft}}{\text{Time, min}}$$

2. To determine the flow of water through the filter in gallons per minute, multiply the velocity (ft/min) times the area (sq ft) times 7.48 gallons per cubic foot (this converts flow from cubic feet per minute to gallons per minute).

Flow, GPM = (Velocity, ft/min)(Surface Area, sq ft)(7.48 gal/cu ft)

TABLE 6.2 SUMMARY OF ROUTINE FILTRATION PROCESS ACTIONS

Monitor Process Performance and Evaluate Water Quality Conditions	Location	Frequency	Possible Operator Actions
Turbidity	Influent/ Effluent	Influent at least once per 8-hour shift. Effluent, monitor continuously.	1. Increase sampling frequency when process water quality is variable. 2. Perform jar tests if indicated (see procedure in Chapter 4, "Coagulation and Flocculation").
Color	Influent/ Effluent	At least once per 8-hour shift.	3. Make necessary process changes. (a) Change coagulant. (b) Adjust coagulant dosage. (c) Adjust flash mixer/flocculator mixing intensity. (d) Change chlorine dosage. (e) Change filtration rate. (f) Backwash filter.
Head Loss		At least three times per 8-hour shift.	4. Verify response to process changes at appropriate time.
Operate Filters and Backwash			
Put filter into service. Change filtration rate. Remove filter from service. Backwash filter. Change backwash rate.	Filter module	Depends on process conditions.	See operating procedures in Sections 6.7, 6.8, 6.9, and 6.10.
Check Filter Media Condition			
Media depth evaluation. Media cleanliness. Cracks or shrinkage.	Filter module	At least monthly.	1. Replace lost filter media. 2. Change backwash procedure. 3. Change chemical coagulants.
Make Visual Observations of Backwash Operation			
Check for media boils. Observe media expansion. Check for media carryover into wash water trough. Observe clarity of wastewater.	Filter module	At least once per day or whenever backwashing occurs when less frequent.	1. Change backwash rate. 2. Change backwash cycle time. 3. Adjust surface wash rate or cycle time. 4. Inspect filter media and support gravel for disturbance.
Check Filtration Process and Backwash Equipment Condition			
Noise Vibration Leakage Overheating	Various	Once per 8-hour shift.	1. Correct minor problems. 2. Notify others of major problems.
Inspect Facilities			
Check physical facilities. Check for algae buildup on filter sidewalls and on wash water troughs.	Various	Once per 8-hour shift.	1. Report abnormal conditions. 2. Remove debris from filter media surfaces. 3. Adjust chlorine dosage to control algae.

The flow rate can also be determined in gallons per minute (GPM) if the total volume of water filtered between filter runs is known. This is done by obtaining the total flow volume in gallons and dividing this value by the length of filter run in hours. If we divide the results by 60 minutes per hour, we will convert the flow from gallons per hour to gallons per minute (GPM).

$$\text{Flow, GPM} = \frac{\text{Total Flow, gallons}}{(\text{Filter Run, hr})(60 \text{ min/hr})}$$

The Unit Filter Run Volume (UFRV) is a measure of filter performance. This value is determined by obtaining the total volume of water filtered between filter runs and dividing this number by the surface area of the filter in square feet.

$$\text{UFRV, gal/sq ft} = \frac{\text{Volume Filtered, gal}}{\text{Filter Surface Area, sq ft}}$$

If the filtration rate in gallons per minute per square foot (GPM/sq ft) is known, this number times the length of filter run in hours times 60 minutes per hour will also give the Unit Filter Run Volume.

$$\text{UFRV,} \atop \text{gal/sq ft} = \left(\text{Filtration Rate,} \atop \text{GPM/sq ft}\right)(\text{Filter Run, hr})(60 \text{ min/hr})$$

EXAMPLE 1

Calculate the filtration rate in GPM/sq ft for a filter with a surface length of 25 feet and a width of 20 feet when the applied flow is 2 MGD (2 million gallons during a 24-hour period).

Known		**Unknown**
Length, ft	= 25 ft	Filtration Rate, GPM/sq ft
Width, ft	= 20 ft	
Flow, MGD	= 2 MGD	

1. Convert the flow from MGD to GPM.

$$\text{Flow, GPM} = \frac{(\text{Flow, MGD})(1{,}000{,}000/\text{Million})}{(24 \text{ hr/day})(60 \text{ min/hr})}$$

$$= \frac{(2 \text{ MGD})(1{,}000{,}000/\text{Million})}{(24 \text{ hr/day})(60 \text{ min/hr})}$$

$$= 1{,}389 \text{ GPM}$$

2. Calculate the surface area of the filter in square feet.

$$\text{Area, sq ft} = (\text{Length, ft})(\text{Width, ft})$$
$$= (25 \text{ ft})(20 \text{ ft})$$
$$= 500 \text{ sq ft}$$

3. Calculate the filtration rate in gallons per minute per square foot.

$$\text{Filtration Rate, GPM/sq ft} = \frac{\text{Flow, GPM}}{\text{Surface Area, sq ft}}$$
$$= \frac{1{,}389 \text{ GPM}}{500 \text{ sq ft}}$$
$$= 2.8 \text{ GPM/sq ft}$$

EXAMPLE 2

Determine the filtration rate in GPM/sq ft for a filter with a surface length of 30 feet and a width of 20 feet. With the influ-ent valve closed, the water above the filter dropped 12 inches in 5 minutes.

Known		**Unknown**
Length, ft	= 30 feet	Filtration Rate, GPM/sq ft
Width, ft	= 20 feet	
Water Drop, in	= 12 in	
Time, min	= 5 min	

1. Calculate the surface area of the filter in square feet.

$$\text{Area, sq ft} = (\text{Length, ft})(\text{Width, ft})$$
$$= (30 \text{ ft})(20 \text{ ft})$$
$$= 600 \text{ sq ft}$$

2. Calculate the velocity of the dropping water in feet per minute.

$$\text{Velocity, ft/min} = \frac{\text{Water Drop, ft}}{\text{Time, min}}$$
$$= \frac{12 \text{ in}}{(12 \text{ in/ft})(5 \text{ min})}$$
$$= 0.2 \text{ ft/min}$$

3. Determine the flow of water through the filter in gallons per minute.

$$\text{Flow, GPM} = (\text{Velocity, ft/min})(\text{Area, sq ft})(7.48 \text{ gal/cu ft})$$
$$= (0.2 \text{ ft/min})(600 \text{ sq ft})(7.48 \text{ gal/cu ft})$$
$$= 898 \text{ GPM}$$

4. Calculate the filtration rate in gallons per minute per square foot.

$$\text{Filtration Rate, GPM/sq ft} = \frac{\text{Flow, GPM}}{\text{Surface Area, sq ft}}$$
$$= \frac{898 \text{ GPM}}{600 \text{ sq ft}}$$
$$= 1.5 \text{ GPM/sq ft}$$

EXAMPLE 3

A filter with media surface dimensions of 42 feet long by 22 feet wide produces a total of 18.5 million gallons during a 73.5-hour long filter run. What is the average filtration rate in GPM/sq ft?

Known		**Unknown**
Length, ft	= 42 ft	Filtration Rate, GPM/sq ft
Width, ft	= 22 ft	
Total Flow, MG	= 18.5 MG	
Filter Run, hr	= 73.5 hr	

1. Find the filter media surface area in square feet.

$$\text{Area, sq ft} = (\text{Length, ft})(\text{Width, ft})$$
$$= (42 \text{ ft})(22 \text{ ft})$$
$$= 924 \text{ sq ft}$$

2. Calculate the average flow rate in gallons per minute.

$$\text{Flow, GPM} = \frac{\text{Total Flow, gallons}}{(\text{Filter Run, hr})(60 \text{ min/hr})}$$

$$= \frac{18{,}500{,}000 \text{ gallons}}{(73.5 \text{ hr})(60 \text{ min/hr})}$$

$$= 4{,}195 \text{ GPM}$$

3. Determine the average filtration rate in gallons per minute per square foot of surface area.

$$\text{Filtration Rate, GPM/sq ft} = \frac{\text{Flow, GPM}}{\text{Area, sq ft}}$$

$$= \frac{4{,}195 \text{ GPM}}{924 \text{ sq ft}}$$

$$= 4.5 \text{ GPM/sq ft}$$

EXAMPLE 4

Determine the Unit Filter Run Volume (UFRV) for the filter in Example 1. The volume of water filtered between backwash cycles was 3.8 million gallons. The filter is 25 feet long and 20 feet wide.

Known		Unknown
Length, ft	= 25 ft	UFRV, gal/sq ft
Width, ft	= 20 ft	
Volume Filtered, gal	= 3,800,000 gal	

Calculate the Unit Filter Run Volume in gallons per square foot of filter surface area.

$$\text{UFRV, gal/sq ft} = \frac{\text{Volume Filtered, gal}}{\text{Filter Surface Area, sq ft}}$$

$$= \frac{3{,}800{,}000 \text{ gal}}{(25 \text{ ft})(20 \text{ ft})}$$

$$= 7{,}600 \text{ gal/sq ft}$$

EXAMPLE 5

Determine the Unit Filter Run Volume (UFRV) for the filter in Example 1. The filtration rate was 2.8 GPM/sq ft during a 46-hour filter run.

Known		Unknown
Filtration Rate, GPM/sq ft	= 2.8 GPM/sq ft	UFRV, gal/sq ft
Filter Run, hr	= 46 hr	

Calculate the Unit Filter Run Volume in gallons per square foot of filter surface area.

$$\text{UFRV, gal/sq ft} = (\text{Filtration Rate, GPM/sq ft})(\text{Filter Run, hr})(60 \text{ min/hr})$$

$$= (2.8 \text{ GPM/sq ft})(46 \text{ hr})(60 \text{ min/hr})$$

$$= 7{,}728 \text{ gal/sq ft}$$

NOTE: The method used to calculate the UFRV for your plant will depend on the information available.

6.722 Backwash Rate

Filter backwash rates are usually given in gallons per minute per square foot (GPM/sq ft) of surface area; or liters per second per square meter (liters per sec/sq m) of surface area; or inches per minute of water "rise rate" (in/min or mm/sec). Backwash rates will vary from 10 to 25 GPM/sq ft (6.8 to 17 liters per sec/sq m or 6.8 to 17 mm/sec). Usually the plant operation and maintenance (O & M) instructions will specify the filter backwash rate in GPM/sq ft or the pumping rate for the backwash pump in GPM. From a practical standpoint, the backwash rate or pumping rate is too low or the length of the backwash cycle too short if the filter is not completely cleaned. This will be obvious if the next filter run is too short or the initial head loss is too high. The backwash pumping or flow rate is too high if excessive amounts of filter media are being lost or disturbed during backwashing. To determine if filter media is being lost, place a burlap bag on the backwash discharge line (if possible) and examine what is caught by the bag. You could also merely observe how much media material remains in the empty wash trough or recovery basin when backwashing is completed.

When filter backwash rates are given in inches per minute (the velocity of the backwash water rising in the filter, such as 24 inches per minute or 10 millimeters per second), you may want to convert this value to other units such as GPM/sq ft. By applying the proper mathematical conversion factors, inches per minute can be converted to gallons per minute per square foot. One GPM per sq ft is approximately equal to 1.6 inches per minute of rise. Most operators and designers use gallons per minute per square foot rather than inches per minute.

FORMULAS

Formulas involving backwash calculations are very similar to the formulas for calculating filtration rates. To calculate the backwash pumping or flow rate in gallons per minute (GPM), multiply the filter surface area in square feet times the desired backwash rate in gallons per minute per square foot.

$$\text{Backwash Pumping Rate, GPM} = (\text{Filter Area, sq ft})(\text{Backwash Rate, GPM/sq ft})$$

To calculate the volume of backwash water needed, multiply the backwash flow in gallons per minute (GPM) times the backwash time in minutes.

$$\text{Backwash Water, gallons} = (\text{Backwash Flow, GPM})(\text{Backwash Time, min})$$

To convert a backwash flow rate from gallons per minute per square foot (GPM/sq ft) to inches per minute of rise, divide by 7.48 gallons per cubic foot to obtain the rise rate in feet per minute. Multiply by 12 inches per foot to obtain the backwash rate in inches per minute of rise.

$$\text{Backwash, in/min} = \frac{(\text{Backwash, GPM/sq ft})(12 \text{ in/ft})}{7.48 \text{ gal/cu ft}}$$

To determine the percent of water used for backwashing, divide the gallons of backwash water by the gallons of water filtered and multiply by 100 percent.

$$\text{Backwash, \%} = \frac{(\text{Backwash Water, gal})(100\%)}{\text{Water Filtered, gal}}$$

EXAMPLE 6

Determine the backwash pumping rate in gallons per minute (GPM) for a filter 30 feet long and 20 feet wide if the desired backwash rate is 20 GPM/sq ft.

Known		Unknown
Length, ft	= 30 ft	Backwash Pumping Rate, GPM
Width, ft	= 20 ft	
Backwash Rate, GPM/sq ft	= 20 GPM/sq ft	

1. Calculate the area of the filter.

$$\text{Filter Area, sq ft} = (\text{Length, ft})(\text{Width, ft})$$
$$= (30\text{ ft})(20\text{ ft})$$
$$= 600\text{ sq ft}$$

2. Determine the backwash pumping rate in GPM.

$$\text{Backwash Pumping Rate, GPM} = (\text{Filter Area, sq ft})(\text{Backwash Rate, GPM/sq ft})$$
$$= (600\text{ sq ft})(20\text{ GPM/sq ft})$$
$$= 12,000\text{ GPM}$$

NOTE: Large filter areas may have backwash tanks instead of pumps because of the large flows required. Variable rate pumps are also used which are programmed to start with a low backwash rate and then to increase the rate.

EXAMPLE 7

Determine the volume or amount of water required to backwash the filter in Example 3 if the filter is backwashed for 8 minutes. The backwash pumping rate calculated in Example 6 is the same as the desired flow rate from the backwash tank.

Known		Unknown
Backwash Flow Rate, GPM	= 12,000 GPM	Backwash Water, gallons
Backwash Time, min	= 8 min	

1. Calculate the volume of backwash water required in gallons.

$$\text{Backwash Water, gallons} = (\text{Backwash Flow, GPM})(\text{Backwash Time, min})$$
$$= (12,000\text{ gal/min})(8\text{ min})$$
$$= 96,000\text{ gallons}$$

EXAMPLE 8

How deep must the water be in the backwash tank in Example 7 to be able to backwash the filter for 8 minutes? The backwash tank is 50 feet in diameter.

Known		Unknown
Backwash Water, gallons	= 96,000 gallons	Water Depth, ft
Tank Diameter, feet	= 50 ft	

1. Convert volume of backwash water from gallons to cubic feet.

$$\text{Backwash Water, cu ft} = \frac{\text{Backwash Water, gallons}}{7.48\text{ gallons/cu ft}}$$
$$= \frac{96,000\text{ gallons}}{7.48\text{ gallons/cu ft}}$$
$$= 12,834\text{ cu ft}$$

2. Calculate the depth of water required in the backwash tank.

$$\text{Water Depth, ft} = \frac{\text{Backwash Water Volume, cu ft}}{\text{Tank Area, sq ft}}$$
$$= \frac{12,834\text{ cu ft}}{(0.785)(50\text{ ft})^2}$$
$$= \frac{12,834\text{ cu ft}}{1,962.5\text{ sq ft}}$$
$$= 6.54\text{ ft}$$

You should have at least 6.6 feet or 6 feet 7 inches of water in the backwash tank.

EXAMPLE 9

Convert a filter backwash rate from 25 gallons per minute per square foot to inches per minute of rise.

Known	Unknown
Backwash, GPM/sq ft = 25 GPM/sq ft	Backwash, in/min

Convert the backwash rate from GPM/sq ft to inches/minute.

$$\text{Backwash, in/min} = \frac{(\text{Backwash, GPM/sq ft})(12\text{ in/ft})}{7.48\text{ gal/cu ft}}$$
$$= \frac{(25\text{ GPM/sq ft})(12\text{ in/ft})}{7.48\text{ gal/cu ft}}$$
$$= 40\text{ in/min}$$

EXAMPLE 10

During a filter run, the total volume of water filtered was 18.5 million gallons. When the filter was backwashed, 96,000 gallons of water were used. Calculate the percent of the product water used for backwashing.

Known

Water Filtered, gal = 18,500,000 gal

Backwash Water, gal = 96,000 gal

Unknown

Backwash,%

Calculate the percent of water used for backwashing.

$$\text{Backwash, \%} = \frac{(\text{Backwash Water, gal})(100\%)}{\text{Water Filtered, gal}}$$

$$= \frac{(96,000 \text{ gal})(100\%)}{18,500,000 \text{ gal}}$$

$$= 0.5\%$$

6.73 Recordkeeping

Recordkeeping is one of the most important *ADMINISTRA-TIVE* jobs required of water treatment plant operators.

You will maintain a daily operations log of process performance data and water quality characteristics. Accurate records of the following items should be maintained:

1. Process water quality (turbidity and color);

2. Process operation (filters in service, filtration rates, loss of head, length of filter runs, frequency of backwash, backwash rates, and UFRV);

3. Process water production (water processed, amount of backwash water used, and chemicals used);

4. Percent of water production used to backwash filters; and

5. Process equipment performance (types of equipment in operation, equipment adjustments, maintenance procedures performed, and equipment calibration).

Entries in logs should be neat and legible, should reflect the date and time of an event, and should be initialed by the operator making the entry.

Figure 6.18 is an example of a typical daily operating record for a water treatment plant with 16 filters. If your plant has four filters or even only one filter, you could develop a similar record sheet for your plant.

6.74 Filter Monitoring Instrumentation

To evaluate filtration process efficiency, you will need to be familiar with the measurement of turbidity. This test can be readily performed in the laboratory with the aid of a turbidimeter. In addition, on-line or continuous water quality monitors, such as turbidimeters (Figure 6.14), will give you an early warning of process failure and will aid in making a rapid assessment of process performance. Section 6.12, "Particle Counters," describes a new technology for monitoring and evaluating filter performance.

You will also need to be familiar with methods used to measure filter media loss and to determine the presence of mud-

balls in the filter media (see Appendix A at the end of this chapter).

QUESTIONS

Write your answers in a notebook and then compare your answers with those on page 254.

6.7F What types of records should be kept when operating a filtration process?

6.7G Calculate the percent of water filtered used for backwashing if a filtration plant uses 0.12 million gallons for backwashing during a period when a total of 5 million gallons of water was filtered.

6.8 OPERATING PROCEDURES ASSOCIATED WITH ABNORMAL PROCESS CONDITIONS

6.80 Indicators of Abnormal Conditions

Abrupt changes in water quality indicators, such as turbidity, pH, alkalinity, threshold odor number (TON), temperature, chlorine demand (source water), chlorine residual (in-process), or color, are signals that the operator should immediately review the performance of the filtration process, as well as pretreatment processes (coagulation, flocculation, and sedimentation).

During a normal filter run, watch for rapid changes in head loss buildup in the filter and turbidity breakthrough. Significant changes in either of these guidelines may indicate an upset or failure in the filtration process or pretreatment processes. Other indicators of abnormal conditions are as follows:

1. Mudballs in filter media,

2. Media cracking or shrinkage,

3. Media boils during backwash,

4. Excessive media loss or visible disturbance,

5. Short filter runs,

6. Filters that will not come clean during backwash, and

7. Algae on walls and media.

6.81 Process Actions

Significant changes in source water turbidity levels, either increases or decreases, require immediate verification of the effectiveness of the filtration process in removing suspended solids and floc. A quick determination of filtration removal efficiency can be made by comparing filter influent and effluent turbidity levels with those of recent record.

In the event that filter turbidity removal efficiency is decreasing, look first at the performance of the coagulation and flocculation processes to determine if the coagulant dosage is correct for current conditions. This may require performing jar tests in the laboratory as described in Chapter 4, "Coagulation and Flocculation," to properly assess treatment conditions.

Increases in source water turbidity and resultant increases in coagulant feed rates may impose a greater load on the filters if the majority of suspended solids and floc are not removed in the settling basins. This condition may require that you decrease filtration rates (put additional filters into service) or backwash filters more frequently.

If pretreatment processes do not readily respond to source water quality changes, it may be necessary to add filter aid chemicals at the filter influent to improve filtration solids remov-

CITY OF SACRAMENTO

SHEET NO. 311
DATE NOV. 7

FILTERS
DAILY OPERATING RECORD

NO.	TIME START	TIME STOP	HOURS OPERATED TODAY	HOURS OPERATED PREVIOUS	HOURS OPERATED TOTAL	HEAD LOSS START	HEAD LOSS STOP	WASH MIN.	WASH M. GALS.	PHYSICAL CONDITION OF FILTERS
1	11/4 08:30	11/7 11:30	11:30	63:30	75:00	0.50	6.00	6	0.14	
2	11/6 21:30			2:30						
3										
4										
5										
6										
7										
8										
9	11/3 14:00	11/7 18:00	18:00	82:00	100.00	0.50	5.50	5	0.14	
10										
11										
12										
13										
14										
15										
16										

NO. OF FILTERS WASHED	2	AVERAGE FILTER RATE—M.G.D. GPM/SQ FT	2.7	
AVERAGE RUN—HOURS	87.5	MAX. HOURLY RATE—M.G.D.	49.0	
TOTAL WASH WATER—M.G.	0.28	TOTAL WATER FILTERED—M.G.	43.6	
PERCENT OF WATER FILTERED	0.64	NO. FILTERS OPERATING	16	
AV. TIME OF WASH—MIN.	5.5	FILTERS OUT PER WASH—MIN.	30	

SHIFT	12 - 8	8 - 4	4 - 12
OPERATOR	NY	JL	LW

W.D. Form 49 Rev.

Fig. 6.18   Filter daily operating record
(Permission of City of Sacramento)

**INSTRUCTIONS FOR COMPLETING
FILTER OPERATING RECORD**

1. Sheet No. 311 indicates that November 7 (today's date) is the 311th day of the year.

2. Filter No. 1 was last backwashed (or started) on November 4 at 8:30 a.m. Today the filter was backwashed (stopped) at 11:30 a.m. on November 7.

3. The filter operated 15 hours, 30 minutes (24:00 − 8:30 = 15:30) on November 4, 24 hours on November 5, and 24 hours on November 6 for a previous total of 63 hours and 30 minutes (15:30 + 24:00 + 24:00 = 63:30). Today (November 7) the filter operated for 11 hours and 30 minutes. Therefore, the total hours operated was 75:00 hours (11:30 + 63:30 = 75:00).

4. The head loss through the filter at the start was 0.50 ft and 6.00 ft at the end of the filter run on November 7.

5. The actual backwash time was 6 minutes and 0.14 million gallons of water were used to backwash the filter.

6. Filter No. 9 was also backwashed today.

7. The data sheet is not completed for the remaining filters.

8. At the bottom of the data sheet, a summary of the performance of all 16 filters is provided. On November 7 two filters were washed.

9. Filter No. 1 ran for 75 hours and Filter No. 9 ran for 100 hours for an average run time per filter of 87.5 hours.

$$\text{Average Run, hr} = \frac{\text{Total Length of Run for Each Filter Washed, hr}}{\text{Total Number of Filters Washed}}$$

$$= \frac{75 \text{ hr} + 100 \text{ hr}}{2}$$

$$= 87.5 \text{ hr}$$

10. Total water used to wash filters is the sum of the water used to wash each filter (0.14 M Gal + 0.14 M Gal = 0.28 M Gal).

11. Percent of water filtered is the Total Wash Water divided by the Total Water Filtered times 100 percent. This value is the percent of water filtered which is used for backwashing.

$$\text{Water Filtered, \%} = \frac{(\text{Total Wash Water, M Gal})(100\%)}{\text{Total Water Filtered, M Gal}}$$

$$= \frac{(0.28 \text{ M Gal})(100\%)}{43.6 \text{ M Gal}}$$

$$= 0.64\%$$

12. Average time of washing filters is the average of the two wash times (6 min + 5 min)/2 = 5.5 min.

13. The average filter rate in gallons per minute per square foot is the total water filtered in gallons per minute divided by the total surface area for all filters in service. The total water filtered in million gallons per day (MGD) must be converted to GPM. There are 16 filters with a surface area of 700 square feet each for a total of 11,200 square feet of filter surface area.

$$\text{Filtration Rate, GPM/sq ft} = \frac{(\text{Total Water Filtered, MG/day})(1,000,000/\text{M})}{(\text{Filter Surface Area, sq ft})(24 \text{ hr/day})(60 \text{ min/hr})}$$

$$= \frac{(43.6 \text{ MG/day})(1,000,000/\text{M})}{(11,200 \text{ sq ft})(24 \text{ hr/day})(60 \text{ min/hr})}$$

$$= 2.7 \text{ GPM/sq ft}$$

14. The maximum hourly rate in MGD is the maximum flow treated by the filters during any time period during the day.

15. The total water filtered is the volume of water filtered during the 24-hour period for November 7. This value can be obtained from the filter effluent flowmeter or other appropriate plant flowmeter.

16. The number of filters operating on November 7 was sixteen.

17. When a filter was out of service for backwashing, the total time for the entire backwash cycle was 30 minutes.

Fig. 6.18   Filter daily operating record (continued)

al efficiency. Filter aid chemicals such as nonionic polymers have proven to be effective in improving filtration performance when fed at low dosage rates (parts per billion). However, care must be exercised in selecting the appropriate feed rate, since overdosing can cause sealing of the filter media resulting in drastically shortened filter runs. Generally, appropriate feed rates are established by trial and experience since this procedure is not easily simulated (duplicated) in the laboratory.

For example, you could begin feeding a polymer as a filter aid at 0.10 mg/L. Note the effectiveness of the results by observing (1) removal of floc, (2) filter effluent turbidity, and (3) length of filter run. Compare these results to results without the polymer. If there was an improvement in filter effluent quality (reduction of turbidity) but a decrease in length of filter run, decrease the polymer feed. Continue to decrease the polymer feed until there is either a lessening of filter effluent quality or filter run time is maximized.

Changes in source water quality such as alkalinity and pH may also affect filtration performance through decreased coagulation-flocculation process performance. This is particularly evident when source water quality changes result from precipitation and runoff, or from algal blooms in a source water reservoir. Again, use of filter aids may improve filtration efficiency until other pretreatment processes are stabilized.

Increases in filter effluent turbidity may also result from floc carryover from the sedimentation process. As described in Chapter 4, "Coagulation and Flocculation," the optimum floc size developed in the flocculation process ranges from about 0.1 to 3.0 mm. In conventional filtration, the optimum floc size is closer to 3.0 mm for settling purposes. However, in the direct filtration process (no sedimentation step) the optimum floc size is closer to 0.1 mm to permit depth penetration of the filter media. When larger floc is not removed in sedimentation (too light), it will be carried over into the filters causing rapid media surface clogging. Hydraulic forces in the filter will shear weak flocs, further contributing to turbidity breakthrough. Reevaluation of coagulation-flocculation and sedimentation performance may be required if floc carryover into the filters reduces filtration efficiency. The size of floc can be estimated by observation, as it is seldom necessary to make an accurate measurement of floc size.

Short filter runs may result from increased solids loading or filter aid overdosing, excessively high filtration rates, excessive mudball formation in the filter media, or clogging of the filter

underdrain system. Possible corrective actions are summarized in Table 6.3.

If you encounter backwash problems, such as media boils, media loss, or failure of the filter to come clean during the backwash process, take immediate corrective actions. Generally, these problems can be solved by adjusting backwash flow rates, surface wash flow rate or duration, or adjusting the time sequence or duration of the backwash cycle. In filters with nozzle-type underdrains, boils are often the result of nozzle failure. In this situation the filter should be taken out of service and the nozzles replaced. Possible operator actions are summarized in Table 6.3.

Problems within the filter itself, such as mudball formation or filter cracks and shrinkage, result from ineffective or improper filter backwashing. Correction of these conditions will require evaluation and modification of backwash procedures.

Table 6.3 gives a summary of filtration process problems, how to identify the causes of problems, and also how to correct the problems.

If filter beds are not thoroughly washed, material filtered from the water is retained on the surface of the filter. This material is sufficiently adhesive to form very small balls. In time these balls of material come together in clumps to form larger masses of mudballs. Usually as time goes on, filter media becomes mixed in and gives the mudballs additional weight. When the mass becomes great enough, it causes the mudballs to sink into the filter bed. These mudballs, if allowed to remain, will cause clogged areas in the filter. Generally, proper surface washing will prevent mudball formation.

6.82 Air Binding[17]

Shortened filter runs can occur because of air-bound filters. This is caused by the release of dissolved air in saturated cold water due to a decrease in pressure. Air is released from the water when passing through the filter bed by differences in pressure produced by friction through the bed. Subsequently the released air is entrapped in the filter bed. Air binding will occur more frequently when large head losses are allowed to develop in the filter. Whenever a filter is operated to a head loss that exceeds the head of water over the media, air will be released. Air-bound filters are objectionable because the air prevents water from passing through the filter and causes shortened filter runs. When an air-bound filter is backwashed, the released air can damage the filter media. When air is released during backwashing, the media becomes suspended in the wash water and is carried out of the filter, thus being no longer available for filtration.

6.83 Excessive Head Loss

If excessive head losses through a filter remain after backwashing, the filter underdrain system and the head loss measurement equipment should be checked. High head losses can be caused by reduction in the size and number of underdrain openings. The underdrain openings can be reduced in size or clogged due to media and also from corrosion or chemical deposits. Excessive or abnormal head loss readings also may be caused by malfunctioning head loss measurement equipment.

---

[17] *Air Binding. The clogging of a filter, pipe or pump due to the presence of air released from water. Air entering the filter media is harmful to both the filtration and backwash processes. Air can prevent the passage of water during the filtration process and can cause the loss of filter media during the backwash process.*

TABLE 6.3   FILTRATION PROCESS TROUBLESHOOTING

Source Water Quality Changes	Operator Actions	Possible Process Changes
Turbidity Temperature Alkalinity pH Color Chlorine demand	1. Perform necessary analysis to determine extent of damage. 2. Assess overall process performance. 3. Perform jar tests if indicated. 4. Make appropriate process changes (see right-hand column, Possible Process Changes). 5. Increase frequency of process monitoring. 6. Verify response to process changes at appropriate time (be sure to allow sufficient time for change to take effect). 7. Add lime or caustic soda if alkalinity is low.	1. Change coagulant. 2. Adjust coagulant dosage. 3. Adjust flash mixer/flocculator mixing intensity. 4. Change frequency of sludge removal (increase or decrease). 5. Change filtration rate (add or delete filters). 6. Start filter aid feed. 7. Adjust backwash cycle (rate, duration).
Sedimentation Process Effluent Quality Changes		
Turbidity or floc carryover	1. Assess overall process performance. 2. Perform jar tests if indicated. 3. Make appropriate process changes. 4. Verify response to process changes at appropriate time.	1. Same as for source water quality changes.
Filtration Process Changes/Problems		
Head loss increase Short filter runs Media surface sealing Mudballs Filter media cracks, shrinkage Filter will not come clean Media boils Media loss Excessive head loss	1. Assess overall process performance. 2. Perform jar tests if indicated. 3. Make appropriate process changes. 4. Verify response to process changes at appropriate time.	1. Change coagulant. 2. Adjust coagulant dosage. 3. Adjust flash mixer/flocculator mixing intensity. 4. Change frequency of sludge removal. 5. Decrease filtration rate (add more filters). 6. Decrease or terminate filter aid feed. 7. Adjust backwash cycle (rate, duration). 8. Manually remove mudballs (hoses or rakes). 9. Replenish lost media. 10. Clear underdrain openings of media, corrosion or chemical deposits; check head loss indicator.
Filter Effluent Quality Changes		
Turbidity breakthrough Color pH Chlorine demand	1. Assess overall process performance. 2. Perform jar tests if indicated. 3. Verify process performance: (a) Coagulation-flocculation process (see Chapter 4, "Coagulation and Flocculation"), (b) Sedimentation process (see Chapter 5, "Sedimentation"), (c) Filtration process (see above suggestions). 4. Make appropriate process changes. 5. Verify response to process changes at appropriate time.	1. Change coagulant. 2. Adjust coagulant dosage. 3. Adjust flash mixer/flocculator mixing intensity. 4. Change frequency of sludge removal. 5. Decrease filtration rate (add more filters). 6. Start filter aid feed. 7. Change chlorine dosage.

QUESTIONS

Write your answers in a notebook and then compare your answers with those on page 254.

6.8A How would you identify an upset or failure in the filtration process or pretreatment processes?

6.8B List the indicators of abnormal filtration process conditions.

6.8C How could you make a quick determination of filtration removal efficiency?

6.8D What problems may be encountered during backwash?

6.8E How does a filter become air bound?

6.9 START-UP AND SHUTDOWN PROCEDURES

6.90 Conditions Requiring Implementation of Start-Up and Shutdown Procedures

Unlike the previously discussed treatment processes (coagulation, flocculation, and sedimentation), start-up or shutdown of the filtration process *IS* a routine operating procedure in most water treatment plants. This is true even if the treatment plant is operated on a continuous basis, since it is common practice for a filter module to be brought into service or taken off line for backwashing. Clean filters may be put into service when a dirty filter is removed for backwashing, when it is necessary to decrease filtration rates, or to increase plant production as a result of increased demand for water. However, most plants keep all filters on line except for backwashing and in service except for maintenance. Filters are routinely taken off line for backwashing when the media becomes clogged with particulates, turbidity breakthrough occurs, or demands for water are reduced.

6.91 Implementation of Start-Up/Shutdown Procedures

Typical actions performed by the operator in the start-up and shutdown of the gravity filtration process are outlined next. These procedures also generally apply to pressure filters. While some of these concepts also apply to diatomaceous earth filtration, the manufacturer's operating procedures will set forth more specific instructions.

Figures 6.19 and 6.20 illustrate sectional views of typical gravity filters. The figures show the valve positions and flow patterns in the filtration and backwash modes of filter operation.

6.910 Filter Check-Out Procedures

Check operational status of filter.

1. Be sure filter media and wash water troughs are clean of all debris such as leaves, twigs, and tools.

2. Check and be sure all access covers and walkway gratings are in place.

3. Make sure process monitoring equipment such as head loss and turbidity systems are operational.

4. Check source of backwash water to ensure it is ready to go. This could be an elevated wash water tank, pumps, or other source.

6.911 Backwash Procedures

1. Filters should be washed before placing them in service.

   a. If filters are to be washed using automatic equipment, check to be sure length of cycle times set for backwash and surface wash cycles are "correct." "Correct" times vary from plant to plant and time of year. These settings should be based on physical observations of actual time required to clean the filter. If filters are usually washed automatically, it is a good idea to occasionally use a manual wash procedure to ensure efficient cleaning of the media during the wash cycle.

   b. The surface wash system should be activated just before the backwash cycle starts to aid in removing and breaking up solids on the filter media and to prevent the development of mudballs. The surface wash system should be stopped before completion of the backwash cycle to permit proper reclassification (settling) of the filter media.

   c. A filter wash should begin slowly (about 5 GPM/square foot, 3.4 liters per sec/sq m or 3.4 mm/sec) for about one minute to permit purging (removing) of any entrapped air from the filter media, and also to provide uniform expansion of the filter bed. After this period the full backwash rate can be applied (15 to 25 GPM/sq ft, 10 to 17 liters per sec/sq m or 10 to 17 mm/sec). Sufficient time should be allowed for cleaning of the filter media. Usually when the backwash water coming up through the filter becomes clear, the media is clean. This generally takes from three to eight minutes. If flooding of wash water troughs or carryover of filter media is a problem, the backwash rate must be reduced. This may be accomplished by adjusting the master backwash valve, thereby throttling the amount of wash water used.

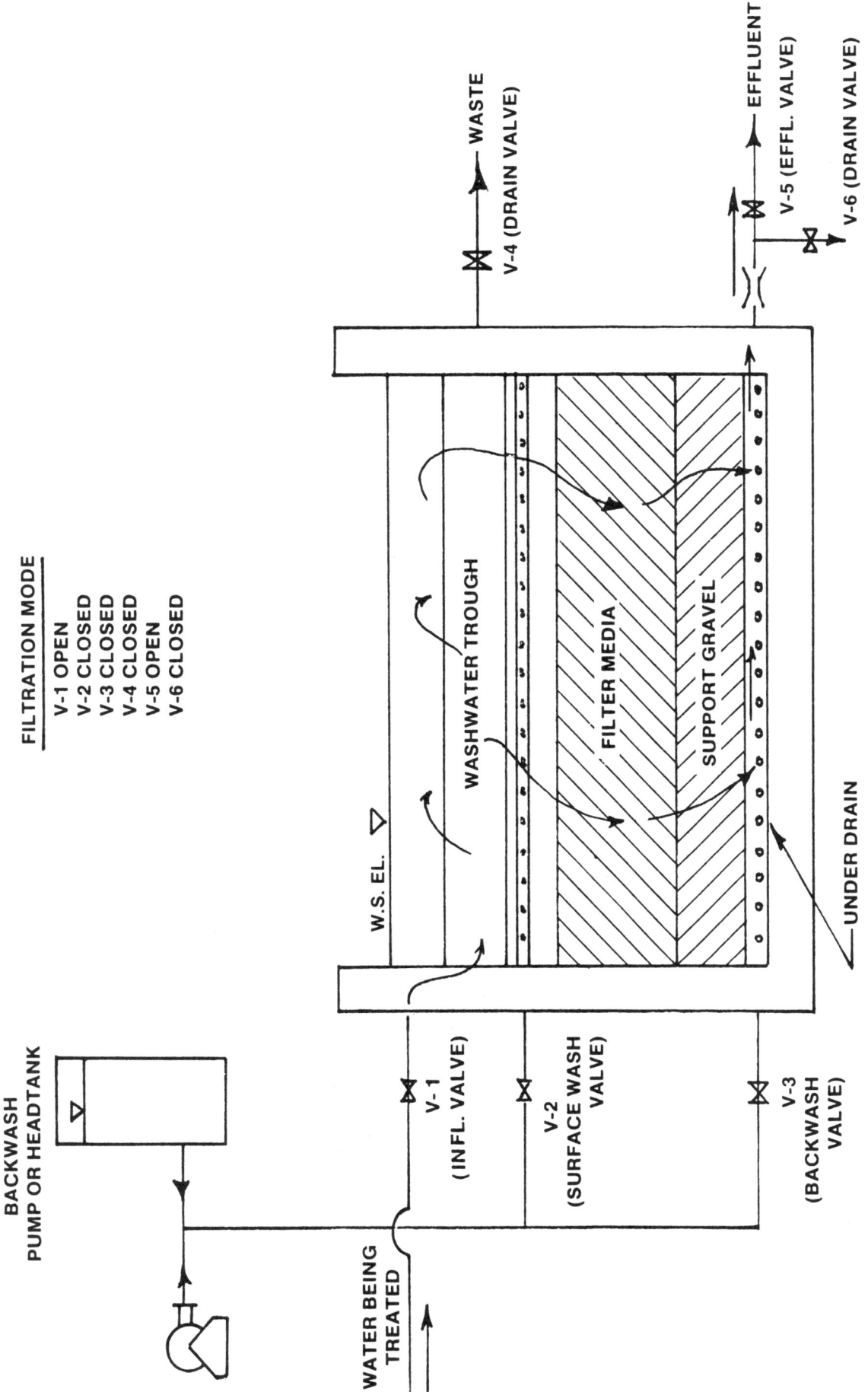

FILTRATION MODE

V-1 OPEN
V-2 CLOSED
V-3 CLOSED
V-4 CLOSED
V-5 OPEN
V-6 CLOSED

W.S. EL.

WASHWATER TROUGH

FILTER MEDIA

SUPPORT GRAVEL

UNDER DRAIN

BACKWASH
PUMP OR HEADTANK

V-1
(INFL. VALVE)

V-2
(SURFACE WASH
VALVE)

V-3
(BACKWASH
VALVE)

WATER BEING
TREATED

Fig. 6.19   Filtration mode of operation

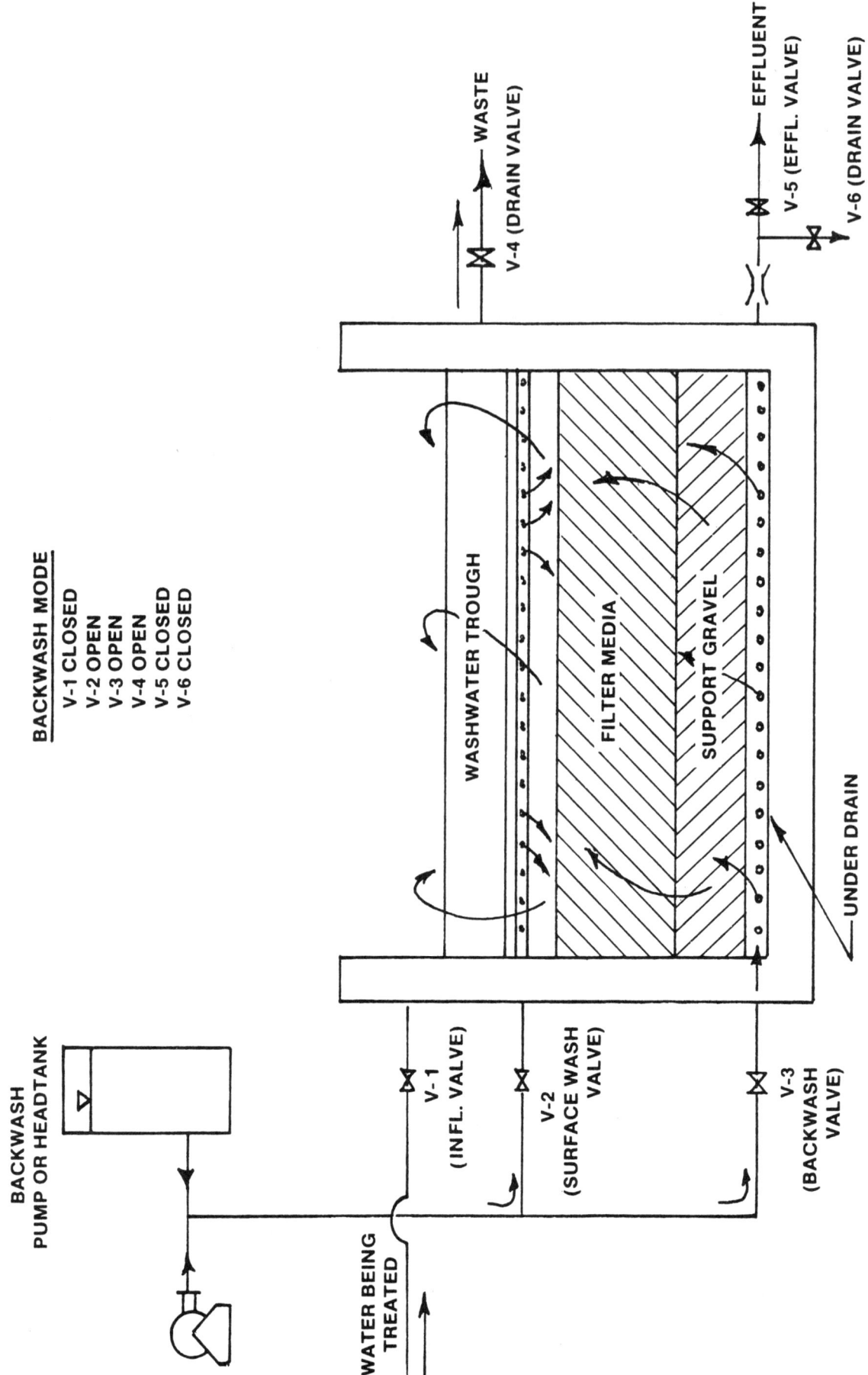

BACKWASH MODE

V-1 CLOSED
V-2 OPEN
V-3 OPEN
V-4 OPEN
V-5 CLOSED
V-6 CLOSED

WASTE
V-4 (DRAIN VALVE)

EFFLUENT
V-5 (EFFL. VALVE)
V-6 (DRAIN VALVE)

WASHWATER TROUGH

FILTER MEDIA

SUPPORT GRAVEL

UNDER DRAIN

BACKWASH
PUMP OR HEADTANK

V-1
(INFL. VALVE)

V-2
(SURFACE WASH
VALVE)

V-3
(BACKWASH
VALVE)

WATER BEING
TREATED

Fig. 6.20   Backwash mode of operation

d. In many water treatment plants waste backwash water is either directly recycled through the plant or is allowed to settle in a tank, pond, or basin and then the supernatant (clear, top portion of water) is pumped back to be recycled through the plant. Usually it is best to gradually add the waste backwash water to the headworks of the treatment plant (before the flash mixer). This is because a large slug of waste backwash water will require changes in chemical dosages due to the additional flow and increased turbidity.

2. Procedures for backwashing a filter are as follows:

a. Log length of filter run since last backwash (Figure 6.18),

b. Close filter influent valve (V-1),

c. Open drain valve (V-4),

d. Close filter effluent valve (V-5),

e. Start surface wash system (open V-2),

f. Slowly start backwash system (open V-3),

g. Observe filter during washing process,

h. When wash water from filter becomes clear (filter media is clean), close surface wash system valve (V-2),

i. Slowly turn off backwash water system (close V-3),

j. Close drain valve (V-4),

k. Log length of wash and number of gallons of water used to clean filter.

6.912 Filter Start-Up Procedures

1. After washing the filters they should be eased on line. With automatic equipment this is generally done by a gradual opening of the filter's effluent valve. Manual operations require a gradual increase of the amount of water treated by the filter. The initial few hours after a filter is placed in service is a time when turbidity breakthrough can pose a problem. For this reason, filters should be eased into service to avoid hydraulic shock loads.

2. Start filter

a. Slowly open filter influent valve (V-1).

b. When proper elevation of water is reached on top of filter, filter effluent valve should be gradually opened (V-5). In many systems the filter effluent valve controls the level of water on the filter. This valve adjusts itself to maintain a constant level of water over the filter media regardless of filtration rate.

c. Some plants have provisions to waste some of the initial filtered water (open V-6). This provision can be very helpful if an initial breakthrough occurs.

d. Perform turbidity or particle counting analyses of filtered water and make process adjustments as necessary.

6.913 Filter Shutdown Procedures

1. Remove filter from service by:

a. Closing influent valve (V-1), and

b. Closing effluent valve (V-5).

2. Backwash filter (see Section 6.911, "Backwash Procedures").

3. If filter is to be out of service for a prolonged period, drain water from filter to avoid algal growth.

4. Note status of filter in operations log.

QUESTIONS

Write your answers in a notebook and then compare your answers with those on page 254.

6.9A Under what conditions may (clean) filters be put back into service?

6.9B When are filters routinely taken off line?

6.9C Why should the surface wash system be activated just before the backwash cycle starts?

6.9D What should be done if a filter will be out of service for a prolonged period?

6.10 PROCESS AND SUPPORT EQUIPMENT OPERATION AND MAINTENANCE

6.100 Types of Equipment

To run a filtration process you must be familiar with the operation and minor (preventive) maintenance of a variety of mechanical, electrical, and electronic equipment including:

1. Filter control valves,

2. Backwash and surface wash pumps,

3. Flowmeters and level/pressure gages,

4. Water quality monitors such as turbidimeters,

5. Process monitors (head loss and water level), and

6. Mechanical and electrical filter control systems.

Since a wide variety of mechanical, electrical, and electronic equipment is used in the filtration process, the operator should be familiar with the operation and maintenance instructions for each specific equipment item or control system.

6.101 Equipment Operation

Before starting a piece of mechanical equipment, such as a backwash pump, be sure that the unit has been serviced on schedule and its operational status is positively known.

After start-up, *ALWAYS* check for excessive noise and vibration, overheating, and leakage (water, lubricants). When in doubt about the performance of a piece of equipment, refer to the manufacturer's instructions.

Much of the equipment used in the filtration process is automated and only requires limited attention by operators during normal operation. However, periodic calibration and maintenance of this equipment is necessary, and this usually involves special procedures. Detailed operating, repair, and calibration procedures are usually described in the manufacturer's literature.

6.102 Preventive Maintenance Procedures

Preventive maintenance (P/M) programs are designed to ensure the continued satisfactory operation of treatment plant facilities by reducing the frequency of breakdown failures. This is accomplished by performing scheduled or routine maintenance on valves, pumps, and other electrical and mechanical equipment items.

In the normal operation of the filtration process, you will be expected to perform routine maintenance functions as part of an overall preventive maintenance program. Typical functions include:

1. Keeping electric motors free of dirt, moisture, and pests (rodents and birds);

2. Ensuring good ventilation (air circulation) in equipment work areas;

3. Checking pumps and motors for leaks, unusual noise and vibrations, or overheating;

4. Maintaining proper lubrication and oil levels;

5. Inspecting for alignment of shafts and couplings;

6. Checking bearings for overheating and proper lubrication;

7. Checking for proper valve operation (leakage or jamming);

8. Checking automatic control systems for proper operation;

9. Checking air/vacuum relief systems for proper functioning, dirt, and moisture;

10. Verifying correct operation of filtration and backwash cycles by observation;

11. Inspecting filter media condition (look for algae and mudballs and examine gravel and media for proper gradation); and

12. Inspecting filter underdrain system (be sure underdrain openings are not becoming clogged due to media, corrosion, or chemical deposits).

Accurate recordkeeping is the most important element of any successful preventive maintenance program. These records provide operation and maintenance personnel with clues for determining the causes of equipment failures. They frequently can be used to forecast impending failures thus preventing costly repairs.

6.103 Safety Considerations

To avoid accidents or injury when working around filtration equipment such as pumps and motors, follow the safety procedures listed below:

ELECTRICAL EQUIPMENT

1. Avoid electric shock (use protective gloves),

2. Avoid grounding yourself in water or on pipes,

3. Ground all electric tools,

4. Lock out and tag electric switches and panels when servicing equipment, and

5. Use the buddy system.

MECHANICAL EQUIPMENT

1. Use protective guards on rotating equipment,

2. Do not wear loose clothing around rotating equipment,

3. Keep hands out of energized valves, pumps, and other pieces of equipment, and

4. Clean up all lubricant and chemical spills (slippery surfaces cause bad falls).

OPEN-SURFACE FILTERS

1. Use safety devices such as handrails and ladders,

2. Close all openings and replace safety gratings when finished working, and

3. Know the location of all life preservers and other safety devices.

VALVE AND PUMP VAULTS, SUMPS, FILTER GALLERIES

1. Be sure all underground or confined structures are free of hazardous atmospheres (toxic or explosive gases, too much or too little oxygen) by checking with gas detectors,

2. Only work in well-ventilated structures (use air circulation fans), and

3. Use the buddy system.

6.11   SURFACE WATER TREATMENT RULE (SWTR)

> #### NOTICE
>
> The filtration guidelines presented in this section are current as this manual is being reprinted in 2004. As a result of passage of the Interim Enhanced Surface Water Treatment Rule (IESWTR) and the Filter Backwash Rule, however, changes in filtration regulations will be phased in over the next few years. By December 2001, for example, water systems using filtration were required to meet more stringent turbidity performance standards and monitor the performance of each individual filter. Operators are urged to contact their state regulatory agency for the regulations that currently apply to their drinking water agency. For additional information or answers to specific questions about the new regulations, phone EPA's toll-free Safe Drinking Water Hotline at (800) 426-4791. Also refer to the poster provided with this manual.

6.110 Description of the SWTR

The Surface Water Treatment Rule (SWTR) is a set of treatment technique requirements that apply to all water systems using surface water and those using groundwater that is under the influence of surface water. The rule requires that these systems properly filter the water unless they can meet certain strict criteria. The rule also requires that all systems using surface water disinfect the water; there are no exemptions from the disinfection requirement.

The SWTR defines surface water as "all water open to the atmosphere and subject to surface runoff." This includes rivers, lakes, streams, and reservoirs, as well as groundwaters that are directly influenced by surface water. The determination as to whether a supply uses surface water is left up to the state. Generally, if a groundwater source has significant and relatively rapid shifts in water quality indicators (such as turbidity, temperature, conductivity, or pH) that closely correlate to climatological or nearby surface water conditions, that source can be considered surface water.

At a minimum, the treatment required for surface water includes disinfection. Very clean and protected source water systems may only be required to disinfect to achieve sufficient removal of coliforms to meet the requirements.

Water supply systems that are required to filter can use a variety of treatment technologies in conjunction with disinfection to meet the expected performance levels. These technologies include conventional filtration, direct filtration, slow sand filtration, and diatomaceous earth filtration.

The general performance criteria to be met by surface water systems are primarily directed toward acute health risks from waterborne microbiological contaminants. The requirements are:

1. At least 99.9 percent (also called three-log for the three nines) removal and/or inactivation of *Giardia* cysts, and

2. At least 99.99 percent (four-log) removal and/or inactivation of enteric (intestinal) viruses.

In general, compliance by the surface water purveyor could be through one of the following alternatives:

1. Meeting the criteria for which filtration is not required and providing disinfection according to the specific requirements in the SWTR, or

2. Providing filtration and meeting disinfection criteria required for those supplies that are filtered.

Mandatory filtration affects the small and medium-sized water systems most severely. A few large surface water systems do not filter their water; more than nine million people drink unfiltered water in Seattle, New York City, and Boston alone. However, most of the unfiltered surface water systems serve communities with fewer than 10,000 residents.

6.111 Filtration Technologies

Conventional filtration. This includes coagulation, flocculation, sedimentation, and filtration. Flows range from about 2 to 6 gallons per minute per square foot of filter surface area. This is a commonly used technology for large systems, and it is fairly complex, with many operational and maintenance requirements.

Direct filtration. This is the same as conventional, except that sedimentation is not included. This category includes in-line filtration, which is the same as direct filtration without the flocculation. Generally, higher quality water is needed for this filtration technology than for conventional treatment.

Slow sand filtration. This process usually does not require chemical pretreatment for most surface waters. Flows are about 0.1 gallon per minute per square foot of filter surface area. This technology is well suited to smaller systems because it has fairly simple operation and maintenance requirements.

Diatomaceous earth filtration. This technology uses a thin layer of diatomaceous earth (a fine, siliceous material) that is deposited on a porous plate to serve as the filter. Chemical pretreatment is usually not necessary. This technology is good for smaller systems because of the relative simplicity of the units and their maintenance requirements.

6.112 Turbidity Requirements

Different turbidity monitoring and turbidity MCLs apply for each type of filtration, as follows:

Type of Filtration	Monitoring Frequency	Turbidity Level
conventional	every 4 hours	<0.5 NTU
direct	every 4 hours	<0.5 NTU
diatomaceous earth	every 4 hours	<1.0 NTU
slow sand	once per day	<1.0 NTU

Continuous turbidity reading may be substituted for the 4-hour sampling if the meter is periodically calibrated. The regulatory agency may reduce monitoring to once per day for systems serving fewer than 500 people.

Conventional or direct filtration. Systems using either conventional or direct filtration must achieve a filtered water turbidity level of less than or equal to 0.5 NTU in 95 percent of the measurements taken for each month. This limit may be increased by the state to 1 NTU if the system proves it can effectively remove *Giardia* cysts at such turbidity levels. At no time can filtered water turbidity exceed 5 NTU.

Slow sand filtration. Systems using slow sand filtration must achieve a filtered water turbidity level of less than or equal to 1 NTU in 95 percent of the measurements for each month. This limit can be increased by the state if there is no interference with disinfection and the turbidity level never exceeds 5 NTU. At no time can filtered water turbidity exceed 5 NTU.

Diatomaceous earth filtration. Systems using diatomaceous earth filtration must achieve a filtered water turbidity level of less than or equal to 1 NTU in 95 percent of the measurements for each month. At no time can the filtered water turbidity exceed 5 NTU.

Other filtration technologies. With the approval of the state, systems may use other filtration technologies provided they meet the same criteria for removal of *Giardia* cysts and viruses as are required for the conventional technologies.

For additional information on monitoring requirements and turbidity requirements of the Surface Water Treatment Rule (SWTR), see Sections 22.2213 and 22.2214 in *WATER TREATMENT PLANT OPERATION*, Volume II.

QUESTIONS

Write your answers in a notebook and then compare your answers with those on page 255.

6.10A   List the types of equipment used in the filtration process.

6.10B   What should be done before starting a piece of mechanical equipment, such as a backwash pump?

6.10C   What safety hazards may be encountered when working around mechanical equipment?

6.11A   What is the SWTR definition of surface water?

6.11B   Under what conditions would a groundwater source be considered surface water?

6.12   PARTICLE COUNTERS[18] (Figure 6.21)

6.120   Need for Particle Counters

The Surface Water Treatment Rule (SWTR) requires that water systems using surface water, or groundwater directly influenced by surface water, achieve a 99.9 percent (three-log) reduction of *Giardia* cysts, and a 99.99 percent (four-log) reduction of viruses through filtration and disinfection. These reductions are typically achieved through a combination of filtration and disinfection. To demonstrate these removals, a water system may use particle counting[19] as a substitute for *Giardia* cyst measurement.

Another threat to public health from drinking water is the exposure to *Cryptosporidium* oocysts. To minimize the potential of public exposure to *Cryptosporidium*, operators try to optimize treatment plant performance. The use of particle counters is one of the best monitoring tools available to optimize plant performance for the removal of particles. At this time there is no consistently reliable analytical technique for measuring *Cryptosporidium* densities in drinking water. However, particle counters can be used as an indirect way to indicate the possible removal of *Cryptosporidium* achieved when plant performance is optimized for particle removal. By monitoring the removal efficiencies of particles in the same size ranges as *Giardia* and *Cryptosporidium*, the operator can estimate the system's *Giardia* and *Cryptosporidium* removal rate.

Operators must realize that a particle counter cannot tell the difference between a particle of clay and a microorganism (protozoan cysts or oocysts). While the SWTR permits a portion of microbiological testing for the presence of cysts or oocysts be substituted for by particle counting, particle counting does not eliminate the need for microbiological testing to establish the presence, absence, or enumeration (counting) of cysts and oocysts. Particle counting must be viewed as a complementary measurement technique to be used in combination with many other measurements—physical, chemical, and microbiological—not as a replacement for any other technique.

6.121   Grab vs. On-Line Particle Counters

Accurate particle count measurements, whether they are based on grab samples or on-line readings, depend on achieving a representative sample. While this is true of all analytical measurements, achieving this is critical for particle counting. The sample lines must be as short as possible and the instrument should be located as close as possible to the measurement point. Some operators argue that a sample cannot even be transported from one room to another without changing the sample. Therefore, it is important to minimize all sources of error in measurement.

Grab Samples

Two types of particle counters are available: dedicated laboratory models that are installed at a fixed location in a laboratory, or portable units. Use of a fixed-location laboratory particle counter complicates achieving a representative sample without much delay in analysis or the use of long sample lines (both are potential sources of significant error). One benefit of the dedicated lab units is that they usually permit the lab technician to use pressure to contain *ENTRAINED*[20] air in solution (thus minimizing error due to the presence of air bubbles in the sample) or to vacuum degas the sample prior to measurement.

A portable unit, on the other hand, can be transported to the sampling location, thus minimizing errors due to sample transport and long sample lines. However, it is more difficult to deal with entrained air when using a portable unit. In addition, if there is insufficient sample pressure to deliver the sample through the instrument, a pump must be used (preferably downstream of the sensor). Pumping can introduce errors and thus should be avoided whenever possible for lab, portable, or on-line measurements. Placing a pump downstream from the sensor complicates interference due to air bubbles because the pressure reduction that results in the cell tends to cause air bubbles to come out of solution. However, placing a pump just ahead of the sensor can change particle size distribution (resulting in a non-representative sample). Each operator must examine the trade-offs and select the option that best minimizes errors.

Portable units normally are self-contained and include a means of measurement, memory, built-in printer, and a means to download data directly to a computer. Portable units are convenient where off-site measurements are desired (mini-

---

[18] *Particle Counter. A device which counts and measures the size of individual particles in water. Particles are divided into size ranges and the number of particles is counted in each of these ranges. The results are reported in terms of the number of particles in different particle diameter size ranges per milliliter of water sampled.*

[19] *Particle Counting. A procedure for counting and measuring the size of individual particles in water. Particles are divided into size ranges and the number of particles is counted in each of these ranges. The results are reported in terms of the number of particles in different particle diameter size ranges per milliliter of water sampled.*

[20] *Entrain. To trap bubbles in water either mechanically through turbulence or chemically through a reaction.*

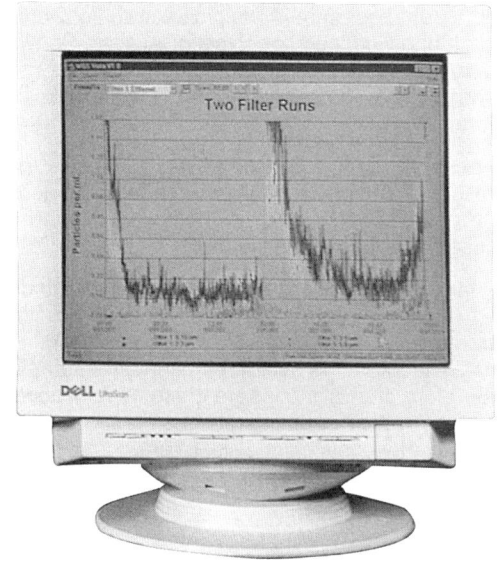

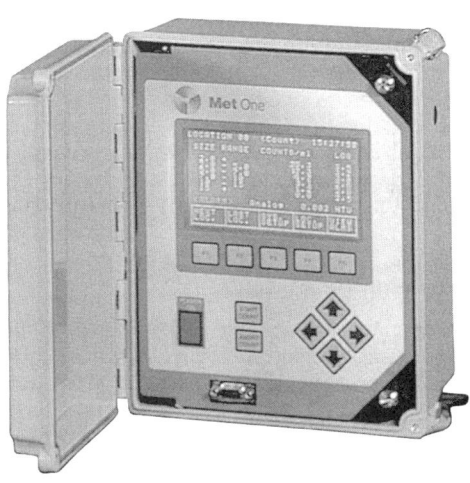

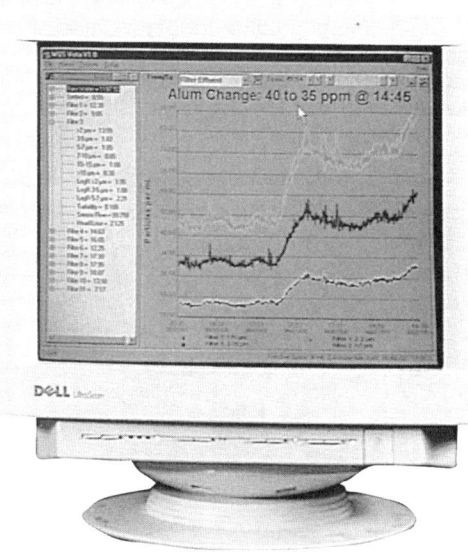

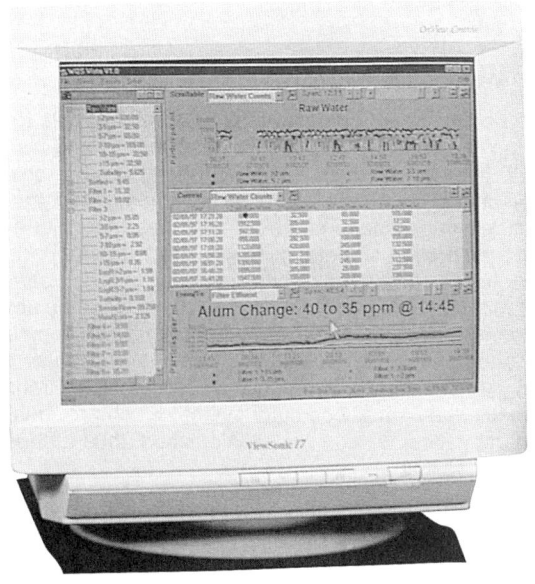

Fig. 6.21   Particle counter, controller, and software

(Reproduced by permission of Met One, Inc., Grants Pass, Oregon)

mizing the need to transport the sample), and they are useful for troubleshooting parts of treatment processes where a permanent on-line instrument is not installed.

Lab and portable units are most useful for QA/QC (quality assurance/quality control) checks and one-time measurements to gather baseline data. The operator must normally spend time and effort to download data to a software package (spreadsheet or database) and then manipulate the data. Persons using either a portable or laboratory unit and downloading data to a computer should be familiar with software packages for data handling, such as Excel or Access. Software for downloading and manipulating data from a particular model of particle counter may be available from the particle counter manufacturer.

Lab or portable particle counting usually involves use of a sample vessel. Any laboratory glassware or equipment used for particle counting must be extremely well cleaned and, if possible, used only for particle counting.

In summary, both fixed-location particle counters and portable models have advantages and limitations. With both types of devices, however, the reliability of the measurements obtained is dependent on the use of representative samples and on the laboratory analyst's consistent use of very good technique. Because particle counting requires rigorous attention to technique, the labor costs can be significant.

On-Line Measurements

On-line units should be located as close as possible to the sample point to provide representative sampling and fast response to process changes. Once a proper sampling location has been chosen and installation has been completed, the on-line particle count measurement is much less technique-dependent than measurements made with laboratory or portable instruments. The laboratory technician only needs to perform periodic maintenance consisting mainly of cleaning the device (maintenance for on-line devices is usually less than maintenance needed for devices installed in the lab or grab sample devices). Most on-line instruments are designed to transmit data directly to a central computer system which provides automated data accumulation and manipulation (thus further reducing labor costs).

Cost per unit is significantly less for on-line instruments than for portable or dedicated laboratory units. In 2004, portable units cost approximately $10,000, and dedicated lab units ranged from $15,000 to more than $20,000. The cost of a single on-line sensor was less than $4,000, as was the software. So, for the cost of a single dedicated lab unit at $20,000, four on-line instruments plus software could be purchased. Or, for the cost of a portable unit ($10,000), two on-line sensors and software could be purchased. Therefore, unless the few benefits of the lab or portable units are required, it is often more economical to install on-line instruments.

6.122 Filter Performance Monitoring

Particulate removal is the main concern of the municipal plant operation. The purpose of the entire treatment process is to prepare the water for particulate removal, first at the sedimentation basin, if used, then ultimately at the filter. Monitoring the efficiency of the filtration process is critically important because if filtration is deficient, particulates, and possibly pathogens, may pass through to the consumer.

Particle counters can help operators optimize plant performance for particle removal by enhancing the performance of the filtration process.

6.1220 Filter Ripening

Particle counts and particle size distribution analyses present an informative picture of effluent quality produced during the filter ripening phase after a filter has been backwashed. The total number of particles will increase during the initial few minutes of the ripening phase. In a properly operated system, the particle counts should then rapidly decrease within a few minutes. Operators should establish operational procedures that make the ripening period as short as possible.

6.1221 Filter Flow Rate

A particle counter can be used to measure the effects of filter flow rate on filter effluent quality. Increases in the flow rate or hydraulic loading on a filter can result in an increase in the particulates detected in the filter effluent. A sudden decrease in filter flow rate can produce a sharp peak of accumulated particulates and then a reduced level of particulates afterward.

6.1222 Filter Run Time

Particle counts show a small spike for a short time immediately after backwash. Particle counts drop (improve) as ripening is achieved. If a filter gradually accumulates particulates and then releases a large number of particulates at periodic intervals, these changes in performance can be recorded. Cumulative particle counts could be used to initiate backwash before backwash is indicated by turbidity data.

Although cumulative particle counts may be at their lowest level at the end of a filter run, the particle size distribution of particulates in the filter effluent could substantially change. With lengthening filter run time, the percentage of large particles passing through the filter often increases while small particles decrease.

Particle counting, when used in conjunction with other measurement techniques, can be a valuable tool to diagnose filtration problems including improperly graded media, insufficient media, disrupted media, short-circuiting, mudballs, defective underdrains, and excessive filter run loading.

6.1223 Filter Media Selection

Particle counters may be used to evaluate and compare the performance of different types of filter media and different arrangements of filter media.

6.1224 Polymer Application

The effectiveness of polymer addition and polymer dose points to improve filter performance can be measured by analyzing the results recorded by particle counters.

6.1225 Other Uses of Particle Counters

Particle count analyses also can be used to optimize other factors affecting filtration efficiency, such as filter start-up rate, filter-to-waste period minimization, and coagulant type and dosage for sedimentation.

6.123 How Particle Counters Work

Two particle counting techniques are used in water treatment plants.

1. Light-scattering instruments pass particles through a light beam and the light scatters in all directions. The scattered light signal measured by the detector depends on the size, shape, and refractive index of the particle, the refractive index of the liquid medium, the polarization state of the light beam, and the relative orientation between the particle, the light beam, and the detector. The light scattering technique is appropriate for measurement of particles less than one *MICRON*[21] (1 μm). While there are sensors available that will measure particles smaller than 0.1 micron, they are not, at present, suitable for use in drinking water applications except in a well-controlled research environment. Usually it is not practical to use a sensor with a *SENSITIVITY*[22] greater than 0.5 micron in water treatment.

   The light-scattering instruments used, if any, are usually 0.5-micron instruments. As sensitivity increases (the greater the sensitivity, the smaller the particle size it can detect) so does the cost, difficulty of maintenance, and "temperamental-ness" of the sensor. Typically, operators are discouraged from using sensors below one micron in water treatment except when they are absolutely the best choice for the given application.

   Many water systems today are consistently achieving filter effluent counts (counts refers to the number of particles counted by the sensor) of less than 10 counts (particles) per milliliter at greater than two microns, and it is not unusual to encounter a plant operating day in and day out near one count per milliliter greater than 2 microns.

2. Light-blocking (sometimes called light obscuration or light interruption) instruments pass particles through a light beam and scatter light in all directions. A photodetector placed directly in the main light beam measures the small drop in the light intensity that occurs when light is scattered away from the main beam by the particle. The amount of light scattered away from the main beam depends on the size, shape, and refractive index of the particle, the refractive index of the liquid medium, and the relative orientation of the particle within the light beam. Light-blocking instruments with good particle sizing characteristics and useful flow rates are available with limiting sensitivities between 1.0 and 2.0 μm.

The amount of light blocked from a detector is the sum of the light absorbed by the particle and the light scattered or reflected by the particle. The size and composition of each particle will determine how much light is scattered and how much is absorbed. Carbon particles will absorb most of the light and scatter very little of it. Organic particles have an index of refraction close to the value of water and tend to refract more light. The result of this is that an organic particle 5 μm (microns) in size will block less light than an inorganic particle of the same size, and the organic particle will appear smaller to the particle counter. For this reason, *Giardia* and *Cryptosporidium* will be counted in size ranges several microns below their actual size. The orientation of the particle as it passes through the light beam will also affect how much light is blocked. These factors make it necessary to count particles over a range of sizes as opposed to an exact size.

The appropriate application for light-blocking sensors is for measuring particles one micron and larger. The most common sensor currently being used has a sensitivity of two microns. This is a good compromise in cost, reliability, and ease of maintenance, and it is sufficient sensitivity for most applications. There is some interest in going to one-micron sensitivity just to "get the numbers up" to show greater numbers of particles removed, however, particle counters with a sensitivity of less than one micron are being used primarily for research purposes. Sensitivity of one micron or greater uses a light-blocking sensor while for sensitivity less than one micron, a light-scattering instrument is used.

A cumulative particle count is the total of all particles greater than the sensor's lower limit. For example, for a sensor with two-micron sensitivity, one may say there are 150 particles/mL greater than two microns. For multiple-channel instruments, one may say there are 150 particles/mL greater than two microns, ten particles/mL greater than five microns, and one particle/mL greater than 20 microns.

A differential particle count expresses the difference between two size ranges. In the previous example, therefore, there would be 140 particles/mL in the range of two to five microns and 149 particles/mL in the range of two to 20 microns. Operators may find it useful to use both cumulative and differential particle count expressions, since conditions that show up in one method may not be readily apparent in the other form.

Some operators prefer to express data in the cumulative form since the differential form can easily be obtained by simple subtraction. It is not always obvious, when expressing data only as differential, how to express the results as cumula-

---

[21] Micron (MY-kron). μm, Micrometer or Micron. A unit of length. One millionth of a meter or one thousandth of a millimeter. One micron equals 0.00004 of an inch.
[22] Sensitivity (Particle Counters). The smallest particle a particle counter will measure and count.

tive counts. See Figure 6.22 for an illustration of cumulative particle counts and differential particle counts.

6.124 Comparison of Particle Counters with Turbidimeters

Almost any application in which turbidity measurement is useful will be an appropriate application for a particle counter. Turbidity is a measure of the relative clarity or cloudiness of a water sample. Turbidity is determined by shining a light into a sample of water and measuring the amount of light scattered by the particles suspended in the sample. The detection device converts light energy into electrical voltage which is scaled to output values in Nephelometric Turbidity Units (NTUs).

Turbidimeters measure the average amount of particulate matter in a slowly changing volume of water, whereas particle counters count and size the particles individually in a constant flow stream. Turbidity readings provide no indication of the number or size of particles. A relatively small amount of large particles can produce exactly the same NTU value as a lot of small particles. For this reason the particle counter is far superior in applications where particle size is used as an indicator of *Giardia-* and *Cryptosporidium*-sized particle removals.

The ability of particle counters to size particles means that trends can be detected much more rapidly; for example, near the end of a filter run, the first particles to break through will be the smaller ones. Since turbidity measurements only indicate the average of all the particles, these smaller particles may not sufficiently affect the overall average for quite some time. A particle counter set to count particles in that range will detect the change immediately. It is common to see indications of filter breakthrough with a particle counter several minutes or even hours before turbidity measurements will change.

Particle counters can detect carbon particles that don't scatter light and cannot be detected by turbidimeters using a single 90-degree light-scatter detector. This is a useful feature when carbon addition is part of the treatment process or when diagnosing problems with carbon filters.

In settled or applied water, particle counters can be used to measure the distribution of particles, and can thereby aid in chemical feed optimization. Polymers have been known to cause strange readings from turbidimeters, resulting in over- or underfeeding. Particle distribution can be correlated with flows to treatment processes to achieve more precise process control.

Particle counters will not replace turbidimeters in water treatment. A turbidity measurement is preferable in some applications, mainly when dealing with high-turbidity waters. Since the particle counter is counting individual particles, there is a physical limit on the particle concentration that can be accurately measured. Operators must realize that particle counters and turbidimeters are tools to help them do a better job, similar to Phillips and slotted screwdrivers—both are important but they do different jobs.

As a general guideline, a turbidity between 5 and 10 NTUs may be over the range of a particle counter. However, actual measurements in the field have identified situations where one NTU was over the range of a particle counter and other situations where 20 NTUs was not over the range of a particle counter. Experience indicates that a typical limit for particle counters is in the range of 5 to 10 NTUs; however, operators will have to test on a site-by-site basis to determine at what NTU point the particle counter will be beyond its range for a particular plant.

Particle counts cannot be correlated with turbidity. Turbidity gives a rough measure of water quality from clear to very turbid (clean to very dirty) water. Particle counts give precise

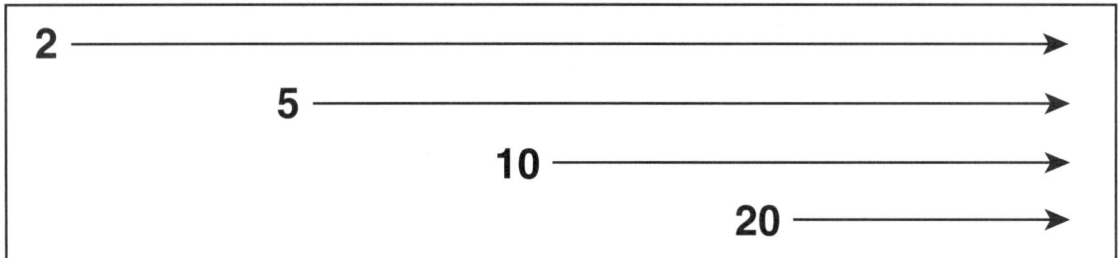

Cumulative Particle Counts: All counts greater than a set threshold. In this example, all counts greater than 2 μm, 5 μm, 10 μm, and 20 μm.

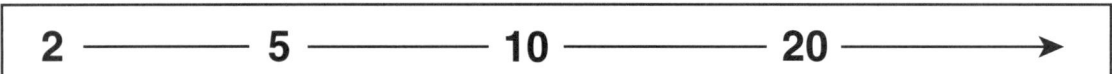

Differential Particle Counts: The particle counts between two thresholds. In this example, counts between 2 and 5 μm, counts between 5 and 10 μm, counts between 10 and 20 μm, then all counts greater than 20 μm.

Fig. 6.22   Illustration of cumulative particle counts and differential particle counts

(Reproduced by permission of Met One, Inc., Grants Pass, Oregon)

information about the particle content in waters ranging from extremely clean to mildly dirty. Also particle counters do not see some very small particles which can be detected by turbidimeters.

6.125 Operation and Maintenance

Sampling

The high sensitivity of the particle counter makes sample handling and delivery critical for proper operation. Every sample collected must be representative of the process stream. Particles can be added to the sample stream if the sample tap is located at the bottom of a pipe where sediment accumulates, or particles can be lost from the sample stream in long sample lines where particles may drop out due to low flow rates and flow velocities. Particle distribution can be altered by sample pumps chopping up larger particles and creating more small particles.

Flow Control

Since all particle count data must be based on sample volume, flow control is critical for accurate and repeatable performance. The simplest and most effective way to achieve constant flow is with an overflow weir (Figures 6.23 and 6.24). As long as enough flow is delivered to maintain some overflow in the weir, a constant flow will be present in the sensor. If this requirement is maintained, flow will only be altered by clogging of the sensor flow cell. The flow rate through a particle counter should be monitored daily to ensure that the flow is within five percent of the calibrated flow rate specified by the manufacturer. Where flow measurement and control are better than five percent of the calibrated flow rate, operators can use electronic flow monitoring to meter the flow and to activate a low flow alarm.

Sample Tubing

Black nylon tubing with a Hytril liner is recommended for sample tubing used with on-line instruments. Hytril has the "slickness" of Teflon, thus particle accumulation in the tubing is minimized. The black outer shell of the tubing discourages biological growths in the tube. Hytril is very expensive, is somewhat stiff, and loses some flexibility with use. Black tubing is definitely preferred when monitoring very clean samples with fewer than 10 particles per milliliter.

When using grab samplers and for some lab applications, operators often use short lengths of Tygon tubing because it is easy to work with, flexible, and inexpensive. Also, operators can see the sample flow to the particle counter.

Bubbles

Bubbles can be introduced into the sample if air is pulled into the sample line or if the temperature of the sample water is allowed to increase while in the sample line. The particle counter will count bubbles if they are large enough, just as a turbidimeter will.

Water weirs must be specifically designed to exhaust gas bubbles in all but the most severe cases. Usually a "debubbling" water weir design is sufficient. Operators may encounter some samples with such severe air entrainment that other measures are necessary. For example, some operators provide high pressure to the sensor and all flow downstream from the sensor, thus minimizing pressure drop across the cell and minimizing the corresponding tendency of gas bubbles to come out of solution in the sensing area. However, this might involve eliminating the water weir altogether. Problems such as these need to be resolved by working with the particle counter supplier.

Initial Start-Up

Be sure that all fittings used in making connections in the sample lines are noncorrosive. Never allow dissimilar metals to be connected. For example, never allow a ductile iron pipe to which a steel coupling is welded into which a galvanized nipple is inserted onto which a brass bushing or reducer is screwed and finally stainless steel tubing is run to serve as a sample line! When looking for just a few particles per mL, the slightest amount of corrosion from the contact of any dissimilar metals can cause a sample error.

Install or place the particle counter as close as practical to the sample point in the flow stream to minimize the length of sample tubing and to ensure rapid response time of the counting of the particles in the sample.

Be sure to flush out all taps and valves before connecting the particle sensor. Sometimes taps or valves have not been used recently and can release a slug of debris that will clog the sensor. Once the installation is complete, all sample lines should be flushed for a short time. If the particle counter output is being recorded, it is easy to determine when the sample lines have been sufficiently flushed because the particle counts per mL will level out or become steady.

Overconcentration

Always monitor the total particle count over the entire monitoring range of the particle counter to ensure that the total particle count is not exceeding the limit of the sensor. If the limit of the sensor is being exceeded, dilution will be necessary. Dilution of samples for on-line particle counters may be difficult. Contact the instrument manufacturer for assistance. An automated on-line dilution system may be available from some manufacturers.

Troubleshooting

The first step is to check and be sure the power is ON. Nearly all difficulties with a particle counter can be solved by cleaning the sample lines. If the particle concentration appears low, check for clogging of the sample line. If the particle concentration appears high, check for contamination, air bubbles, or an increased flow rate through the sensor.

QA/QC (Quality Assurance/Quality Control)

Frequently verify the baseline of the particle counter. When analyzing "particle-free water," the particle counter should indicate less than one particle per mL in the 2 to 150 μm size range. Instrument calibration should be verified by the manufacturer annually. With multiple-sensor systems, verify that the sensors give similar results (within plus or minus ten percent) on one water source. This can be a very time-consuming process until operators gain experience and confidence in particle counters.

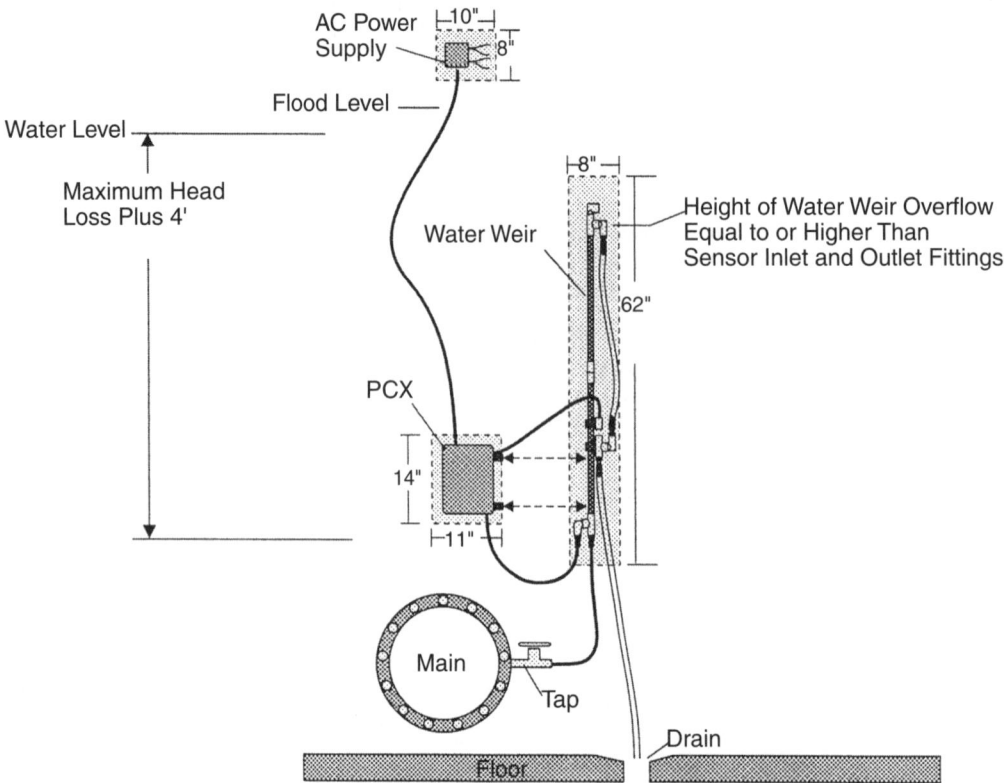

Fig. 6.23   Particle counter mounting diagram
(Reproduced by permission of Met One, Inc., Grants Pass, Oregon)

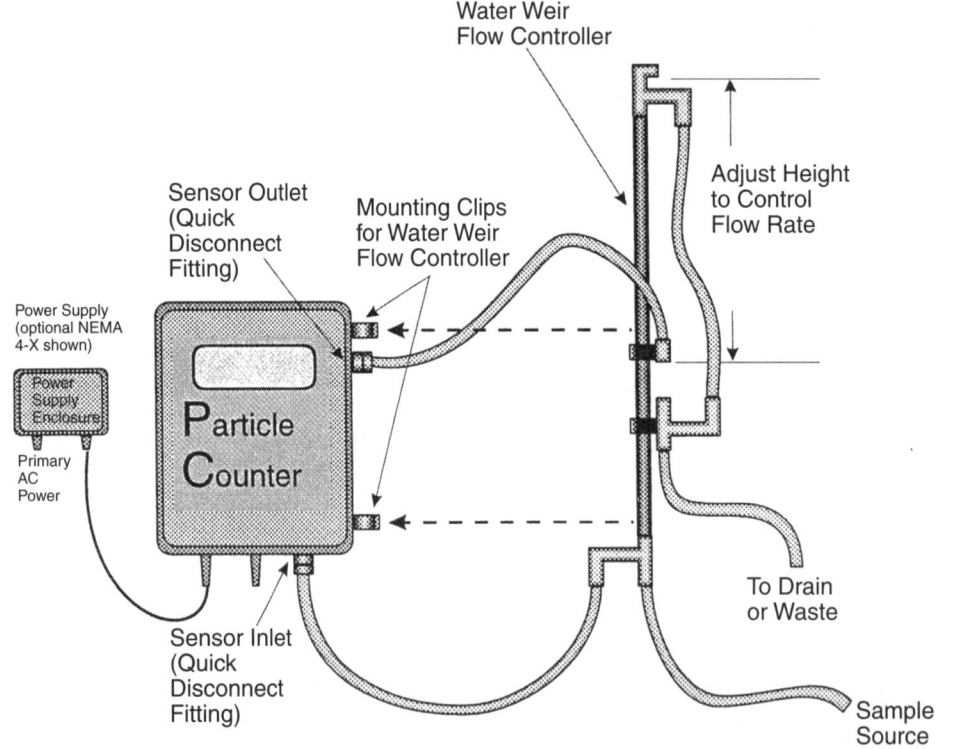

Fig. 6.24   Particle counter connection to flow control weir
(Reproduced by permission of Met One, Inc., Grants Pass, Oregon)

6.126 Actual Plant O & M

This section contains data from pilot plant and treatment plant operation. The data document the effectiveness and usefulness of particle counters.

Figure 6.25 shows a close, but delayed, turbidity response for two pilot filter breakthrough events. Both turbidity readings and particle counts sensed the breakthroughs; however, particle counts warned of the impending turbidity breakthrough by more than 30 minutes. A second pilot filter was switched on line after the first breakthrough (at about 2.5 hours).

Particle Counters Respond Quickly

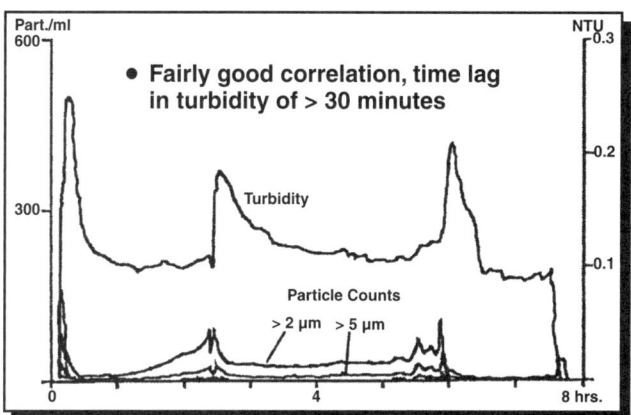

Fig. 6.25   Data from pilot filter at Grants Pass, Oregon
(Reproduced by permission of Met One, Inc., Grants Pass, Oregon)

Notice that the cleanup time for particle counts was much shorter than that for turbidity. Also this second filter did not achieve the low particle counts of the first filter, while its turbidity was very comparable.

Figure 6.26 shows the response of the same pilot filter on a different date. In this case, the turbidity measurement completely missed the filter breakthrough event. Particle counts rose dramatically from 20 particles per mL to over 500 particles per mL, while turbidity increased only 0.05 NTU.

Anticipating Breakthrough

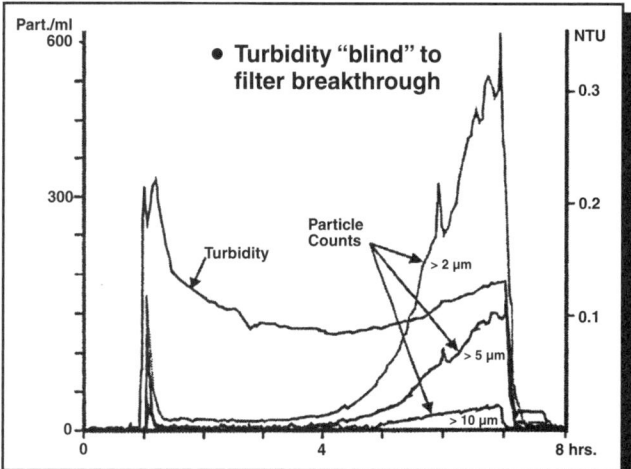

Fig. 6.26   Data from pilot filter at Grants Pass, Oregon
(Reproduced by permission of Met One, Inc., Grants Pass, Oregon)

Figure 6.27 shows the same advanced warning of turbidity breakthrough afforded by particle counting from a full-scale filter at a different plant site. In this case, the particle counts reveal reduced filter performance 5 to 10 hours prior to turbidity.

Anticipating Breakthrough

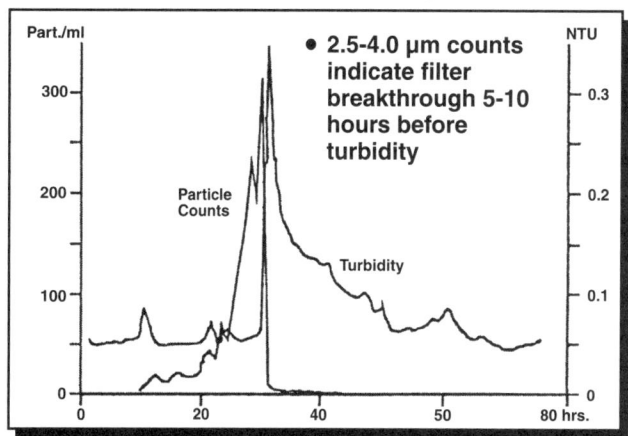

Fig. 6.27   Data from full-scale filter in Utah
(Reproduced by permission of Met One, Inc., Grants Pass, Oregon)

Notice the comparative slopes of the particle counts and turbidity responses. Characteristically, particle counts show a gradual upward trend leading to breakthrough, with a steep drop-off after backwash, whereas turbidity shows a sudden steep increase at breakthrough with a long, gradual drop-off after backwash. This is because turbidimeters actually "see" a lot of very small particles that are not measured by particle counters because the sensitivity of the particle counter selected was purposely limited to 2 microns. A particle counter with a higher sensitivity (sensitivity to less than 2 microns) would be capable of seeing these smaller particles.

Figure 6.28 shows the effects of filter flow rate on effluent particle count. The filter had operated continuously for 92 hours at a flow rate of 2.5 GPM/square foot (102 liters/minute per square meter). The water flow was then stopped for 55 minutes. The filter was restarted at the original flow rate and a particle count was taken; it showed slightly elevated counts. However, with 20 additional minutes, the particle count had again dropped to the original stabilized levels.

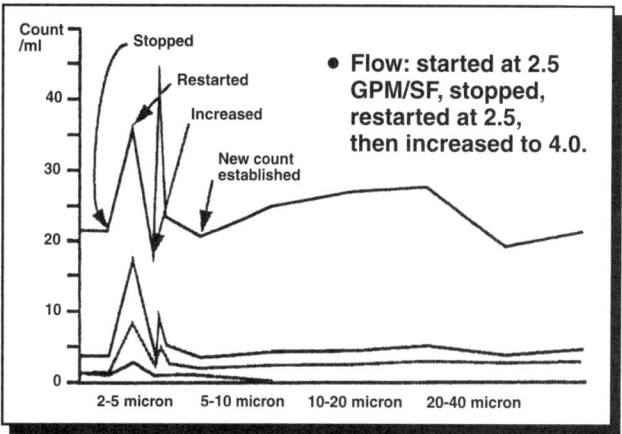

Fig. 6.28   High-rate filter
(Reproduced by permission of Met One, Inc., Grants Pass, Oregon)

The filter flow rate was then increased from 2.5 to 4.0 GPM/ sq ft (102 to 163 liters/minute per square meter). During the transition period, the counts more than doubled in most ranges, reaching a peak eight minutes after the flow change. The counts then decreased steadily over the next 22 minutes to new steady-state levels, only slightly higher than the original stabilized levels. These results indicate that high filter flow rates may be safely achievable on some filter designs. Repeated particle count testing over a variety of seasonal conditions can be used to petition regulatory authorities to authorize increased flow rates through existing filters. Higher flow rates could then avoid, or delay, the cost of constructing additional filters.

Particle counters may be used to assist operators in determining optimum coagulant dosages. However, other tools, such as jar testing, zeta potential, and streaming current, are really better primary tools for determining coagulant dosages. Once these tools have been used, particle counting can assist you in refining the dosages to achieve truly optimal coagulant feed rates. Operators must understand and be able to correctly apply coagulation chemistry to achieve optimal conditions even without the latest sophisticated tools.

Figures 6.29, 6.30, and 6.31 compare the feed and permeate values of turbidity, particle counts, and *Giardia* cysts for both a compromised (damaged) and an intact membrane filter.

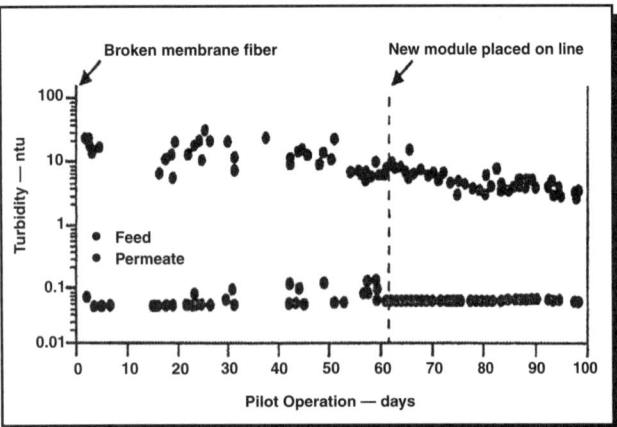

*Fig. 6.29   Loss of membrane integrity
monitored by turbidity*

(Reproduced by permission of Met One, Inc., Grants Pass, Oregon)

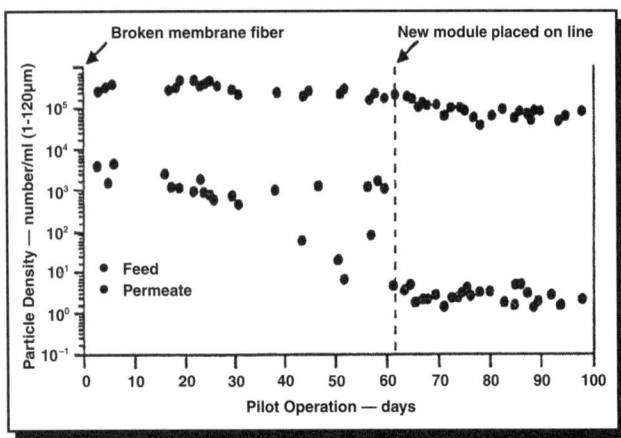

*Fig. 6.30   Loss of membrane integrity
monitored by particle counts*

(Reproduced by permission of Met One, Inc., Grants Pass, Oregon)

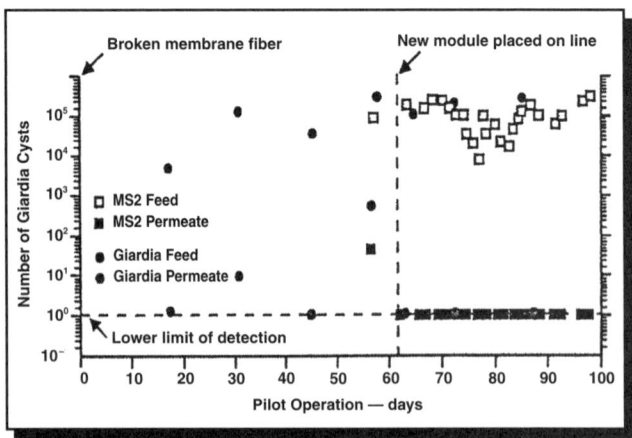

*Fig. 6.31   Loss of membrane integrity
monitored by Giardia cysts*

(Reproduced by permission of Met One, Inc., Grants Pass, Oregon)

During the first 61 days of the test, a UF membrane filter with two broken fibers was used. The effluent turbidity always remained below 0.1 NTU, whereas the permeate particle count read as high as 5,000 particles per mL and *Giardia* cyst recovery measured as high as 400 per sample. After day 61, a new, intact UF membrane was installed with the result that all measurements showed minimal effluent detection levels. These results show that even two broken membrane fibers were revealed by a 3 order of magnitude change in particle counts, while the turbidimeter was unable to conclusively detect the failure.

6.127   Actual Plant Data

Figure 6.32 and Table 6.4 show the actual turbidity NTUs and particle counts for a plant treating a high-quality water and doing an outstanding job of optimizing performance. Data in both the figure and the table cover the same time period (from 10:02:27 PM on August 3, 1998 to 9:26:27 AM on August 4, 1998).

Figure 6.32 shows a computer-generated plot of the continuous filter effluent turbidity NTU measurements for seven filters. Filter effluent turbidity values are all less than 0.066 NTU, with the exception of turbidity spikes (*) due to a particular filter being backwashed. The filter was not on line during the backwash turbidity spike.

At the same water treatment plant, filter effluent particle counts (Table 6.4) for three particle size ranges are recorded every four minutes (only a portion of the August 3 and 4 data is shown in Table 6.4). The spike indicated on August 3 at 10:58:27 PM (the reading of 17 particles in the 2 to 5 μm size range) corresponds with the 0.1 NTU turbidity spike shown in Figure 6.32.

Another illustration of particle count data is shown in Table 6.5, including notations about what was happening with the filter at various points in the operation and backwash cycles.

6.128   References

Material on particle counters in this section is based on information contained in the following references:

- Particle Counter Model PC 2400 D, Operations Manual, Chemtrac® Systems, Inc., Norcross, GA,

- Particle Counting Guidelines, Draft, June 14, 1996, California Department of Health Services, Berkeley, CA, and

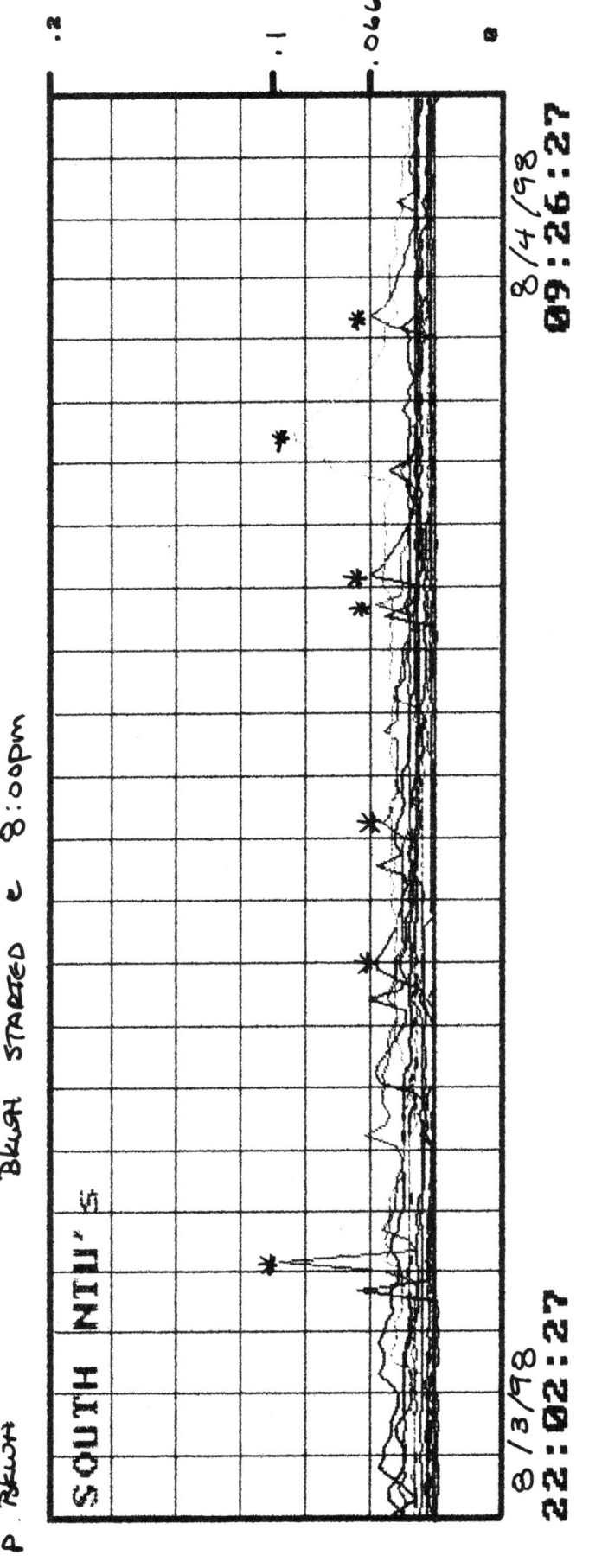

Fig. 6.32   *Plot of continuous effluent turbidity measurements for seven filters*
(Reproduced by permission of San Juan Water District, Granite Bay, California)

* Turbidity spikes due to filter being backwashed. Filter not on line during backwash turbidity spike.

TABLE 6.4   PARTIAL PARTICLE COUNT RECORD:
August 3, 1998 (10:02:27 PM) to August 4, 1998 (9:26:27 AM)

Group	14			
Date	8/4/98		Mils	Sample Time
Run #	4277			Minutes
MG	27			
Hours	20		200	2

Date	Time	2-5μm	5-15μm	15-999μm	
80398	220227	2	0	0	Flow @ 106 mgd
80398	220627	2	0	0	
80398	221027	1	0	0	
80398	221427	8	4	1	
80398	221827	1	0	0	
80398	222227	2	0	0	
80398	222627	3	1	0	
80398	223027	2	0	0	
80398	223427	4	1	0	
80398	223827	2	1	0	
80398	224227	2	0	0	
80398	224627	4	2	0	
80398	225027	1	0	0	
80398	225427	3	1	0	
80398	225827	17	5	0	
80398	230227	9	3	0	
80398	230627	7	2	0	
80398	231027	7	2	0	
80398	231427	6	2	0	
80398	231827	1	0	0	
80498	5427	0	0	0	Flow @ 110 mgd
80498	5827	1	0	0	
80498	10227	1	1	0	
80498	10627	1	0	0	
80498	11027	1	0	0	
80498	84627	2	0	0	
80498	85027	2	0	0	
80498	85427	2	1	0	
80498	85827	2	0	0	
80498	90227	2	1	0	
80498	90627	2	1	0	
80498	91027	2	1	0	
80498	91427	2	1	0	
80498	91827	3	1	0	
80498	92227	2	1	0	
80498	92627	3	1	0	

TABLE 6.5   PARTIAL PARTICLE COUNT RECORD:
February 22, 1998 (5:42:44 AM) to February 26, 1998 (6:02:44 AM)

Group	2					
Date	2/22/98					
Run #	4097		Mils	Sample Time		
MG	110.3		___	Minutes		
Hours	96		200	2		

	Date	Time	2-5μm	5-15μm	15-999μm	
	22298	54244	4	1	0	Flow @ 26 mgd
	22298	54644	57	20	1	
	22298	55044	10	3	0	
	22298	55444	2	0	0	
	22298	55844	1	0	0	
Filter on line ↓	22298	60244	1	0	0	
	22298	60644	1	0	0	
	22298	61044	1	0	0	
	22298	61444	1	0	0	
	22298	61844	1	0	0	
	22298	62244	1	0	0	
	22298	62644	1	0	0	
	22298	63044	1	0	0	
	22298	63444	0	0	0	
	22698	21044	15	2	0	
	22698	21444	16	2	0	
	22698	21844	16	2	0	
	22698	22244	17	2	0	
	22698	22644	17	2	0	
	22698	45444	18	2	0	
Shutting filter off	22698	45844	17	2	0	
	22698	50244	17	2	0	
	22698	50644	16	2	0	
↓	22698	51044	15	2	0	
↓	22698	51444	123	32	1	
Shut one filter down and turn another filter on	22698	51844	145	37	1	
	22698	52244	113	28	1	
	22698	52644	70	15	0	
	22698	53044	52	9	0	
	22698	53444	46	8	0	
	22698	53844	39	6	0	
	22698	54244	39	6	0	
	22698	54644	39	6	0	
	22698	55044	38	6	0	
	22698	55444	46	8	0	
	22698	55844	33	5	0	
	22698	60244	26	3	0	

● Literature provided by Met One, Inc., Grants Pass, Oregon.

For additional information about particle counting, also see "2560. Particle Counting and Size Distribution," *STANDARD METHODS*, 20th Edition, page 2-61.

6.129 Acknowledgment

Material in this section was reviewed by Mike Oblanis, San Juan Water District, California; Terry Engelhardt, Pacific Scientific Instruments, Colorado; and Karl Voight, Contra Costa Water District, California. Their comments and suggestions are greatly appreciated.

QUESTIONS

Write your answers in a notebook and then compare your answers with those on page 255.

6.12A   Particle counters can be used as a substitute for indicating the potential removal of what two microorganisms that are a threat to public health when found in drinking water?

6.12B   Particle counters can be used to enhance the performance of which two water treatment processes?

6.12C   What is the difference between turbidimeter and particle counter measurements?

6.12D   How can improper sampling give incorrect particle counter results?

6.13 ARITHMETIC ASSIGNMENT

Turn to the Appendix, "How to Solve Water Treatment Plant Arithmetic Problems," at the back of this manual and read all of Section A.6, *FORCE, PRESSURE, AND HEAD*.

In Section A.13, *TYPICAL WATER TREATMENT PLANT PROBLEMS*, read and work the problems in Section A.135, Filtration.

6.14 ADDITIONAL READING

1. *NEW YORK MANUAL*, Chapter 9,* "Filtration."

2. *TEXAS MANUAL*, Chapter 9,* "Filtration."

* Depends on edition.

6.15 ACKNOWLEDGMENTS

Many of the concepts and procedures discussed in this chapter are based on material obtained from the sources listed below.

1. Stone, B. G. Notes from "Design of Water Treatment Systems," CE-610, Loyola Marymount University, Los Angeles, CA, 1977.

2. *WATER QUALITY AND TREATMENT*, Fifth Edition. Obtain from American Water Works Association (AWWA), Bookstore, 6666 West Quincy Avenue, Denver, CO 80235. Order No. 10008. ISBN 0-07-001659-3. Price to members, $119.00; nonmembers, $135.00; price includes cost of shipping and handling.

3. *WATER TREATMENT PLANT DESIGN*, Third Edition, prepared jointly by the American Society of Civil Engineers, American Water Works Association, and Conference of State Sanitary Engineers. Obtain from American Water Works Association (AWWA), Bookstore, 6666 West Quincy Avenue, Denver, CO 80235. Order No. 10009. ISBN 0-07-001643-7. Price to members, $103.50; nonmembers, $135.00; price includes cost of shipping and handling.

End of Lesson 2 of 2 Lessons on FILTRATION

Please answer the discussion and review questions next.

DISCUSSION AND REVIEW QUESTIONS

Chapter 6. FILTRATION

(Lesson 2 of 2 Lessons)

Write the answers to these questions in your notebook. The question numbering continues from Lesson 1.

6. How would you monitor a filtration process?

7. How frequently should the performance of the filtration process be evaluated?

8. What would you look for in evaluating the filter media condition after a backwash cycle?

9. Abnormal operating conditions are signaled by sudden changes in what water quality indicators?

10. How could you adjust filter operation if increases in source water turbidity and resultant increases in coagulant feed rates impose a greater load on the plant's active filters?

11. What backwash process adjustments would you make to solve such problems as media boils, media loss, or incomplete cleaning of media?

12. What should you always check for after starting any piece of equipment?

13. What four types of filtration can be used to meet the requirements of the Surface Water Treatment Rule?

14. How can particle counters monitor filter performance?

SUGGESTED ANSWERS

Chapter 6. FILTRATION

ANSWERS TO QUESTIONS IN LESSON 1

Answers to questions on page 199.

6.0A Filtration, preceded by coagulation, flocculation, and sedimentation, is commonly referred to as *CONVENTIONAL FILTRATION*. In the *DIRECT FILTRATION* process, the sedimentation step is omitted. Flocculation facilities are reduced in size or may be omitted.

6.1A The particle removal mechanisms involved in the filtration process include (1) sedimentation on media, (2) adsorption, (3) biological action, (4) absorption, and (5) straining.

Answers to questions on page 205.

6.2A The four specific classes of filters are (1) gravity filtration, (2) pressure filtration, (3) diatomaceous earth filtration, and (4) slow sand filtration.

6.2B Garnet is a group of hard, reddish, glassy, mineral sands made up of silicates of base metals (calcium, magnesium, iron, or manganese).

6.2C In most water treatment applications, diatomaceous earth is used for the precoat and body-feed operations.

Answers to questions on page 208.

6.3A Gravel is used to support filter materials.

6.3B Filtration rate is commonly expressed in units of gallons per minute per square foot of filter surface area or liters per second per square meter of surface area.

6.3C The major operational difference between sand and dual-media filters is that sand filters require more frequent backwashing because of their smaller media grain size. Also dual-media filters permit a higher filtration rate without a high head loss.

Answers to questions on page 210.

6.3D The time period before a filter becomes clogged depends on (1) the amount of suspended solids in the water being treated, and (2) the filtration rate.

6.3E A filter is usually operated until just before clogging or breakthrough occurs, a specified time period has passed, or a specific head loss is reached.

6.3F Four surface wash systems for filters include (1) Baylis, (2) fixed grid, (3) rotary, and (4) air scour.

6.3G Filter control systems are provided (1) to control flow rates through the filter, and (2) to maintain an adequate head above the media surface.

Answers to questions on page 219.

6.4A The primary purpose of using activated carbon (granular form) as filter media is to remove taste- and odor-causing compounds, as well as other trace organics from the water.

6.5A "In-line filtration" refers to the addition of filter aid chemicals immediately prior to filtration. Coagulant chemicals are added directly to the filter inlet pipe and are mixed by the flowing water.

6.5B   Filter aid chemicals are usually added just prior to filtration to aid in the solids removal process during normal operation and during periods of pretreatment process upset, or when operating at high filtration rates.

6.5C   Direct filtration is used when source waters are low in turbidity, color, plankton, and coliform organisms.

6.6A   An operator must be provided with a means to measure (1) filter influent water quality (turbidity), (2) filter performance (head loss buildup rate and filter run time), and (3) filter effluent water quality (turbidity) in order to control the performance of the filtration process on a day-to-day basis.

ANSWERS TO QUESTIONS IN LESSON 2

Answers to questions on page 224.

6.7A   The most important water quality indicator used to monitor the filtration process is the filter influent and effluent turbidity.

6.7B   Filter effluent turbidity is best monitored on a continuous basis by an on-line turbidimeter.

6.7C   Typical functions performed by operators in the normal operation of the filtration process include:

1. Monitor process performance;
2. Evaluate water quality conditions (turbidity) and make appropriate changes;
3. Check and adjust process equipment (change chemical feed rates);
4. Backwash filters;
5. Evaluate filter media condition (media loss, mudballs, cracking); and
6. Visually inspect facilities.

6.7D   A sudden increase in head loss might be an indication of surface sealing of the filter media (lack of depth penetration).

6.7E   If a sudden increase in head loss through a filter occurs, the operation of the filter can be changed by adjusting the filter aid chemical feed rate or by reducing the filtration rate.

Answers to questions on page 229.

6.7F   Types of records that should be kept when operating a filtration process include (1) process water quality, (2) process operation, (3) process water production, (4) percent of water production used to backwash filters, and (5) process equipment performance.

6.7G   Calculate the percent of water filtered used for backwashing if a filtration plant uses 0.12 million gallons for backwashing during a period when a total of 5 million gallons of water was filtered.

Known	Unknown
Wash Water, MG = 0.12 MG	Water Filtered,%
Water Filtered, MG = 5 MG	

Calculate the percent of water filtered used for backwashing.

$$\text{Water Filtered, \%} = \frac{(\text{Wash Water, M Gal})(100\%)}{\text{Water Filtered, M Gal}}$$

$$= \frac{(0.12 \text{ M Gal})(100\%)}{5 \text{ M Gal}}$$

$$= 2.4\%$$

Answers to questions on page 234.

6.8A   Upsets or failures in the filtration process or pretreatment processes may be detected by rapid changes in head loss buildup in the filter and/or turbidity breakthrough.

6.8B   The indicators of abnormal filtration process conditions include:

1. Mudballs in filter media,
2. Media cracking or shrinkage,
3. Media boils during backwash,
4. Excessive media loss or visible disturbance,
5. Short filter runs,
6. Rapid head loss buildup,
7. Turbidity breakthrough,
8. Filters that will not come clean during backwash, and
9. Algae on walls and media.

6.8C   Filtration removal efficiency can be quickly determined by comparing filter influent and effluent turbidity levels with those of recent record.

6.8D   Problems that may be encountered during backwash include media boils, media loss, and failure of the filter to come clean during the backwash process.

6.8E   A filter can become air bound by the release of dissolved air in saturated cold water due to the decrease in pressure. Decreases in pressure occur when water passes through the filter media, especially when the filter operates at a head loss that exceeds the head of water over the media.

Answers to questions on page 237.

6.9A   Clean filters are commonly put into service when a dirty filter is removed for backwashing, when it is necessary to decrease filtration rates, or to increase plant production as a result of increased demand for water. However, most plants keep all filters on line except for backwashing and in service except for maintenance.

6.9B   Filters are routinely taken off line for backwashing when the media becomes clogged with particulates, turbidity breakthrough occurs, or demands for water are reduced.

6.9C   The surface wash system should be activated just before the backwash cycle starts to aid in breaking up solids on the filter media and to prevent the development of mudballs.

6.9D   If a filter will be out of service for a prolonged period, drain water from the filter to avoid algal growth.

Answers to questions on page 240.

6.10A Types of mechanical, electrical, and electronic equipment used in the filtration process include:

1. Filter control valves,
2. Backwash and surface wash pumps,
3. Flowmeters and level/pressure gages,
4. Water quality monitors,
5 Process monitors, and
6. Mechanical and electrical filter controls.

6.10B Before starting a piece of mechanical equipment, such as a backwash pump, be sure that the unit is properly serviced and its operational status is positively known.

6.10C Safety hazards that may be encountered when working around mechanical equipment include (1) rotating equipment, (2) slippery surfaces, and (3) energized valves, pumps, and other equipment.

6.11A The SWTR defines surface water as "all water open to the atmosphere and subject to surface runoff."

6.11B If a groundwater source has significant and relatively rapid shifts in water quality indicators such as turbidity, temperature, conductivity, or pH that closely correlate to climatological or nearby surface water conditions, that source can be considered surface water.

Answers to questions on page 252.

6.12A Particle counters can be used as a substitute for indicating the potential removal of *Giardia* and *Cryptosporidium*.

6.12B Particle counters can be used to enhance the performance of sedimentation and filtration processes.

6.12C Turbidimeters measure the average amount of particulate matter in a slowly changing volume of water, whereas particle counters count and size the particles individually in a constant flow stream.

6.12D Improper sampling can add particles to the flow stream, lose particles in long sample lines, and alter particle distribution when sample pumps chop up larger particles.

APPENDIX

A.   Mudball Evaluation Procedure

MUDBALL EVALUATION PROCEDURE[23]

1. Frequency of mudball evaluation. If mudballs in the top of the filter material are a problem, use this procedure on a monthly basis. If mudballs are not a problem, an annual check is sufficient.

2. Sample for mudballs using mudball sampler shown in Figure 6.33.

3. Backwash the filter to be sampled and drain the filter to at least 12 inches below the surface of the top media layer (or layer of interest).

4. Push the mudball sampler 6 inches into the sand. Tilt the handle until it is nearly level and lift the sampler full of media.

5. Empty the contents of the sampler into a bucket.

6. Repeat steps 4 and 5 four more times from different locations in the filter.

7. Use a 10-mesh sieve for separating the mudballs from the media. Hold the sieve in a bucket or tub of water so the sieve is nearly submerged.

8. Take a handful of media from the bucket containing the samples of media and mudballs and place the material in the sieve.

9. Gently raise and lower the sieve about one-half inch at a time until the sand is washed away from the mudballs.

10. Shift the mudballs to one side of the sieve by tipping the submerged sieve and gently shaking the sieve.

11 Repeat steps 8, 9, and 10 until the entire sample has been washed in the sieve and all of the mudballs have been separated from the sand. If there are so many mudballs that the washing process is hindered, move some of the mudballs to the measuring cylinder described in the next step.

12 Use a 1,000-mL graduated cylinder (a smaller or larger cylinder may be used depending on the volume or amount of mudballs in the sand). Fill the graduated cylinder to the 500-mL mark with water.

13. Allow the water to drain from the mudballs on the sieve. When the draining has stopped (no more dripping), transfer the mudballs to the graduated cylinder.

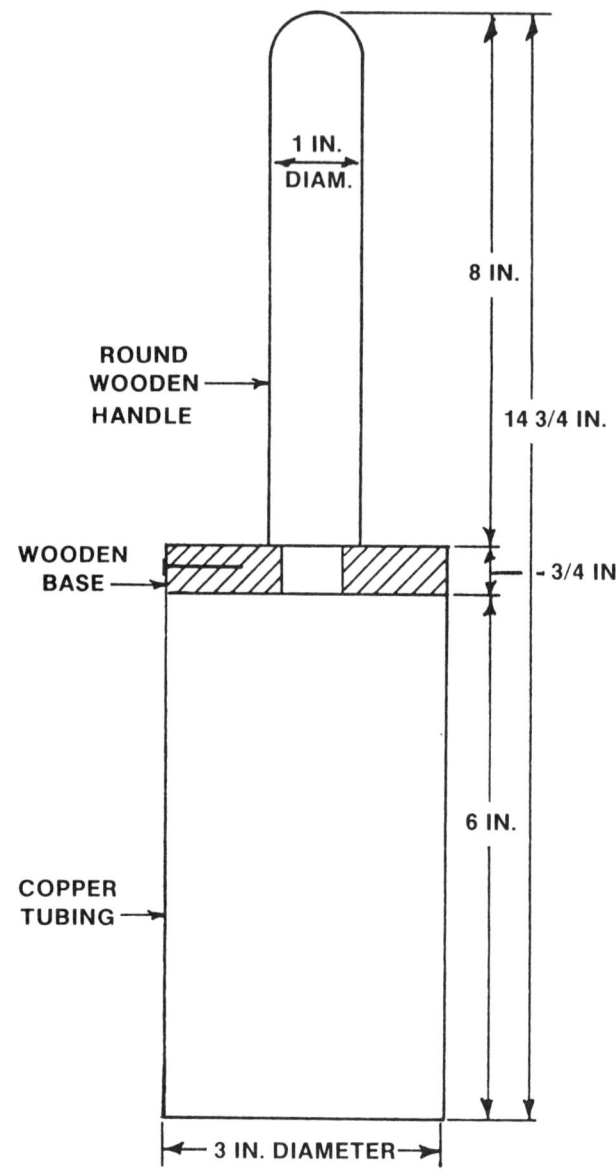

Fig. 6.33   Mudball Sampler

---

[23] Procedures in the Appendix were adapted from procedures in WATER QUALITY AND TREATMENT. Obtain from American Water Works Association (AWWA), Bookstore, 6666 West Quincy Avenue, Denver, CO 80235. Order No. 10008. ISBN 0-07-001659-3. Price to members, $119.00; nonmembers, $135.00; price includes cost of shipping and handling.

14. Record the new level of water in the graduated cylinder.

15. Determine the "Mudball Volume, mL," by subtracting 500 from the new level of water.

16. Calculate the volume of mudballs as a percent. The total volume of sand and mudballs sampled was 3,540 mL if the "mudball sampler" was full.

$$\text{Mudball Volume, \%} = \frac{(\text{Mudball Volume, m}L)(100\%)}{3,540 \text{ m}L}$$

17. Evaluate the condition of the filtering material using Table 6.6.

18. *NOTE:* Mudballs sink more readily in anthracite than sand. Therefore, modify this procedure to collect samples from the bottom six inches of anthracite.

TABLE 6.6   CONDITION OF FILTER

Mudball Volume, %	Condition of Filtering Material
0.0 to 0.1	Excellent
0.1 to 0.2	Very Good
0.2 to 0.5	Good
0.5 to 1.0	Fair
1.0 to 2.5	Fairly Bad
2.5 to 5.0	Bad
Over 5.0	Very Bad

EXAMPLE 11

Evaluate the condition of filtering material on the basis of a mudball evaluation. When the drained mudballs were added to the graduated cylinder, the water rose from the 500-mL mark up to the 583-mL mark. The total volume sampled was 3,540 mL.

Known	Unknown
Initial Cylinder Level, mL = 500 mL	Condition of
Final Cylinder Level, mL = 583 mL	Filtering Material
Total Volume Sample, mL = 3,540 mL	

1. Determine the mudball volume in milliliters.

$$\text{Mudball Volume, m}L = \text{Final Cylinder Level, m}L - \text{Initial Cylinder Level, m}L$$

$$= 583 \text{ m}L - 500 \text{ m}L$$

$$= 83 \text{ m}L$$

2. Calculate the mudball volume as a percent.

$$\text{Mudball Volume, \%} = \frac{(\text{Mudball Volume, m}L)(100\%)}{3,540 \text{ m}L}$$

$$= \frac{(83 \text{ m}L)(100\%)}{3,540 \text{ m}L}$$

$$= 2.3\%$$

3. Determine the condition of the filtering material from Table 6.6.

From Table 6.6, the condition of a sample with a mudball volume of 2.3 percent is *FAIRLY BAD.*

CHAPTER 7

DISINFECTION

by

Tom Ikesaki

NOTICE

Drinking water rules and regulations are continually changing. Two major new laws were signed by the president in December 1998: the Disinfectants/Disinfection By-Products (D/DBP) Rule and the Interim Enhanced Surface Water Treatment Rule (IESWTR). Several other drinking water laws are being developed and are expected to be signed into law over the next two or three years. The regulations described in this chapter are current as of publication of this manual in 2004. Please see Section 7.01, "Safe Drinking Water Laws," and the poster included with this manual for information about recently passed laws and anticipated future regulations.

Keep in contact with your state drinking water agency to obtain the rules and regulations that currently apply to your water utility. For additional information or answers to specific questions about the regulations, phone EPA's toll-free Safe Drinking Water Hotline at (800) 426-4791.

TABLE OF CONTENTS

Chapter 7. DISINFECTION

OBJECTIVES

Chapter 7.   DISINFECTION

Following completion of Chapter 7, you should be able to:

1. Describe the factors that influence disinfection,

2. Explain the process of disinfection using chlorine, hypochlorite, chlorine dioxide, and chloramines,

3. Describe the breakpoint chlorination process,

4. Identify the various points of chlorine application,

5. Operate and maintain chlorination equipment,

6. Handle chlorine safely,

7. Select the proper chlorine dosage,

8. Start up and shut down chlorination equipment,

9. Troubleshoot chlorination systems,

10. Develop and conduct a chlorine safety program, and

11. Operate and maintain disinfection processes other than with chlorine.

WORDS

Chapter 7.   DISINFECTION

AIR GAP   AIR GAP

An open vertical drop, or vertical empty space, that separates a drinking (potable) water supply to be protected from another water system in a water treatment plant or other location. This open gap prevents the contamination of drinking water by backsiphonage or backflow because there is no way raw water or any other water can reach the drinking water supply.

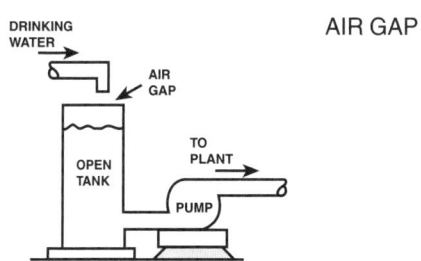

AIR PADDING   AIR PADDING

Pumping dry air (dew point –40°F) into a container to assist with the withdrawal of a liquid or to force a liquified gas such as chlorine out of a container.

AMBIENT (AM-bee-ent) TEMPERATURE   AMBIENT TEMPERATURE

Temperature of the surrounding air (or other medium). For example, temperature of the room where a gas chlorinator is installed.

AMPEROMETRIC (am-PURR-o-MET-rick)   AMPEROMETRIC

A method of measurement that records electric current flowing or generated, rather than recording voltage. Amperometric titration is a means of measuring concentrations of certain substances in water.

AMPEROMETRIC (am-PURR-o-MET-rick) TITRATION   AMPEROMETRIC TITRATION

A means of measuring concentrations of certain substances in water (such as strong oxidizers) based on the electric current that flows during a chemical reaction. Also see TITRATE.

BACTERIA (back-TEAR-e-ah)   BACTERIA

Bacteria are living organisms, microscopic in size, which usually consist of a single cell. Most bacteria use organic matter for their food and produce waste products as a result of their life processes.

BREAKPOINT CHLORINATION   BREAKPOINT CHLORINATION

Addition of chlorine to water until the chlorine demand has been satisfied. At this point, further additions of chlorine will result in a free chlorine residual that is directly proportional to the amount of chlorine added beyond the breakpoint.

BUFFER CAPACITY   BUFFER CAPACITY

A measure of the capacity of a solution or liquid to neutralize acids or bases. This is a measure of the capacity of water for offering a resistance to changes in pH.

CARCINOGEN (CAR-sin-o-JEN)   CARCINOGEN

Any substance which tends to produce cancer in an organism.

CATALYST (CAT-uh-LIST)   CATALYST

A substance that changes the speed or yield of a chemical reaction without being consumed or chemically changed by the chemical reaction.

CHLORAMINATION (KLOR-ah-min-NAY-shun)   CHLORAMINATION

The application of chlorine and ammonia to water to form chloramines for the purpose of disinfection.

CHLORAMINES (KLOR-uh-means)   CHLORAMINES

Compounds formed by the reaction of hypochlorous acid (or aqueous chlorine) with ammonia.

CHLORINATION (KLOR-uh-NAY-shun)   CHLORINATION

The application of chlorine to water, generally for the purpose of disinfection, but frequently for accomplishing other biological or chemical results (aiding coagulation and controlling tastes and odors).

CHLORINE DEMAND   CHLORINE DEMAND

Chlorine demand is the difference between the amount of chlorine added to water and the amount of residual chlorine remaining after a given contact time. Chlorine demand may change with dosage, time, temperature, pH, and nature and amount of the impurities in the water.

   Chlorine Demand, mg/L = Chlorine Applied, mg/L − Chlorine Residual, mg/L

CHLORINE REQUIREMENT   CHLORINE REQUIREMENT

The amount of chlorine which is needed for a particular purpose. Some reasons for adding chlorine are reducing the number of coliform bacteria (Most Probable Number), obtaining a particular chlorine residual, or oxidizing some substance in the water. In each case a definite dosage of chlorine will be necessary. This dosage is the chlorine requirement.

CHLORINE RESIDUAL   CHLORINE RESIDUAL

The concentration of chlorine present in water after the chlorine demand has been satisfied. The concentration is expressed in terms of the total chlorine residual, which includes both the free and combined or chemically bound chlorine residuals.

CHLOROPHENOLIC (klor-o-FEE-NO-lick)   CHLOROPHENOLIC

Chlorophenolic compounds are phenolic compounds (carbolic acid) combined with chlorine.

CHLORORGANIC (klor-or-GAN-ick)   CHLORORGANIC

Organic compounds combined with chlorine. These compounds generally originate from, or are associated with, life processes such as those of algae in water.

COLIFORM (COAL-i-form)   COLIFORM

A group of bacteria found in the intestines of warm-blooded animals (including humans) and also in plants, soil, air and water. Fecal coliforms are a specific class of bacteria which only inhabit the intestines of warm-blooded animals. The presence of coliform bacteria is an indication that the water is polluted and may contain pathogenic (disease-causing) organisms.

COLORIMETRIC MEASUREMENT   COLORIMETRIC MEASUREMENT

A means of measuring unknown chemical concentrations in water by measuring a sample's color intensity. The specific color of the sample, developed by addition of chemical reagents, is measured with a photoelectric colorimeter or is compared with "color standards" using, or corresponding with, known concentrations of the chemical.

COMBINED AVAILABLE CHLORINE   COMBINED AVAILABLE CHLORINE

The total chlorine, present as chloramine or other derivatives, that is present in a water and is still available for disinfection and for oxidation of organic matter. The combined chlorine compounds are more stable than free chlorine forms, but they are somewhat slower in disinfection action.

COMBINED AVAILABLE CHLORINE RESIDUAL   COMBINED AVAILABLE CHLORINE RESIDUAL

The concentration of residual chlorine that is combined with ammonia, organic nitrogen, or both in water as a chloramine (or other chloro derivative) and yet is still available to oxidize organic matter and help kill bacteria.

COMBINED CHLORINE   COMBINED CHLORINE

The sum of the chlorine species composed of free chlorine and ammonia, including monochloramine, dichloramine, and trichloramine (nitrogen trichloride). Dichloramine is the strongest disinfectant of these chlorine species, but it has less oxidative capacity than free chlorine.

COMBINED RESIDUAL CHLORINATION   COMBINED RESIDUAL CHLORINATION

The application of chlorine to water to produce combined available chlorine residual. The residual can be made up of monochloramines, dichloramines, and nitrogen trichloride.

DPD (pronounce as separate letters)   DPD

A method of measuring the chlorine residual in water. The residual may be determined by either titrating or comparing a developed color with color standards. DPD stands for N,N-diethyl-p-phenylene-diamine.

DEW POINT   DEW POINT

The temperature to which air with a given quantity of water vapor must be cooled to cause condensation of the vapor in the air.

DIATOMS (DYE-uh-toms)   DIATOMS

Unicellular (single cell), microscopic algae with a rigid (box-like) internal structure consisting mainly of silica.

DISINFECTION (dis-in-FECT-shun)   DISINFECTION

The process designed to kill or inactivate most microorganisms in water, including essentially all pathogenic (disease-causing) bacteria. There are several ways to disinfect, with chlorination being the most frequently used in water treatment. Compare with STERILIZATION.

EDUCTOR (e-DUCK-ter)   EDUCTOR

A hydraulic device used to create a negative pressure (suction) by forcing a liquid through a restriction, such as a Venturi. An eductor or aspirator (the hydraulic device) may be used in the laboratory in place of a vacuum pump. As an injector, it is used to produce vacuum for chlorinators. Sometimes used instead of a suction pump.

EJECTOR   EJECTOR

A device used to disperse a chemical solution into water being treated.

ELECTRON   ELECTRON

(1) A very small, negatively charged particle which is practically weightless. According to the electron theory, all electrical and electronic effects are caused either by the movement of electrons from place to place or because there is an excess or lack of electrons at a particular place.

(2) The part of an atom that determines its chemical properties.

ENTERIC   ENTERIC

Of intestinal origin, especially applied to wastes or bacteria.

ENZYMES (EN-zimes)   ENZYMES

Organic substances (produced by living organisms) which cause or speed up chemical reactions. Organic catalysts and/or biochemical catalysts.

FREE AVAILABLE RESIDUAL CHLORINE   FREE AVAILABLE RESIDUAL CHLORINE

That portion of the total available residual chlorine composed of dissolved chlorine gas (Cl_2), hypochlorous acid (HOCl), and/or hypochlorite ion (OCl^-) remaining in water after chlorination. This does not include chlorine that has combined with ammonia, nitrogen, or other compounds.

HTH (pronounce as separate letters)   HTH

High **T**est **H**ypochlorite. Calcium hypochlorite or $Ca(OCl)_2$.

HEPATITIS (HEP-uh-TIE-tis)   HEPATITIS

Hepatitis is an inflammation of the liver caused by an acute viral infection. Yellow jaundice is one symptom of hepatitis.

HETEROTROPHIC (HET-er-o-TROF-ick)   HETEROTROPHIC

Describes organisms that use organic matter for energy and growth. Animals, fungi and most bacteria are heterotrophs.

HYDROLYSIS (hi-DROLL-uh-sis)   HYDROLYSIS

(1) A chemical reaction in which a compound is converted into another compound by taking up water.

(2) Usually a chemical degradation of organic matter.

HYPOCHLORINATION (HI-poe-KLOR-uh-NAY-shun)   HYPOCHLORINATION

The application of hypochlorite compounds to water for the purpose of disinfection.

HYPOCHLORITE (HI-poe-KLOR-ite)   HYPOCHLORITE

Chemical compounds containing available chlorine; used for disinfection. They are available as liquids (bleach) or solids (powder, granules, and pellets) in barrels, drums, and cans. Salts of hypochlorous acid.

IDLH   IDLH

Immediately **D**angerous to **L**ife or **H**ealth. The atmospheric concentration of any toxic, corrosive, or asphyxiant substance that poses an immediate threat to life or would cause irreversible or delayed adverse health effects or would interfere with an individual's ability to escape from a dangerous atmosphere.

INJECTOR WATER   INJECTOR WATER

Service water in which chlorine is added (injected) to form a chlorine solution.

MPN (pronounce as separate letters)   MPN

MPN is the **M**ost **P**robable **N**umber of coliform-group organisms per unit volume of sample water. Expressed as a density or population of organisms per 100 mL of sample water.

MOTILE (MO-till)   MOTILE

Capable of self-propelled movement. A term that is sometimes used to distinguish between certain types of organisms found in water.

NEWTON   NEWTON

A force which, when applied to a body having a mass of one kilogram, gives it an acceleration of one meter per second per second.

NITRIFICATION (NYE-truh-fuh-KAY-shun)   NITRIFICATION

An aerobic process in which bacteria reduce the ammonia and organic nitrogen in water into nitrite and then nitrate.

NITROGENOUS (nye-TRAH-jen-us)   NITROGENOUS

A term used to describe chemical compounds (usually organic) containing nitrogen in combined forms. Proteins and nitrates are nitrogenous compounds.

ORTHOTOLIDINE (or-tho-TOL-uh-dine)   ORTHOTOLIDINE

Orthotolidine is a colorimetric indicator of chlorine residual. If chlorine is present, a yellow-colored compound is produced. This reagent is no longer approved for chemical analysis to determine chlorine residual.

OXIDATION (ox-uh-DAY-shun)   OXIDATION

Oxidation is the addition of oxygen, removal of hydrogen, or the removal of electrons from an element or compound. In the environment, organic matter is oxidized to more stable substances. The opposite of REDUCTION.

OXIDIZING AGENT   OXIDIZING AGENT

Any substance, such as oxygen (O_2) or chlorine (Cl_2), that will readily add (take on) electrons. The opposite is a REDUCING AGENT.

PALATABLE (PAL-uh-tuh-bull)   PALATABLE

Water at a desirable temperature that is free from objectionable tastes, odors, colors, and turbidity. Pleasing to the senses.

PASCAL   PASCAL

The pressure or stress of one newton per square meter. Abbreviated Pa.

> 1 psi = 6,895 Pa = 6.895 kN/sq m = 0.0703 kg/sq cm

PATHOGENIC (PATH-o-JEN-ick) ORGANISMS   PATHOGENIC ORGANISMS

Organisms, including bacteria, viruses or cysts, capable of causing diseases (giardiasis, cryptosporidiosis, typhoid, cholera, dysentery) in a host (such as a person). There are many types of organisms which do *NOT* cause disease. These organisms are called non-pathogenic.

PHENOLIC (fee-NO-lick) COMPOUNDS   PHENOLIC COMPOUNDS

Organic compounds that are derivatives of benzene.

POSTCHLORINATION   POSTCHLORINATION

The addition of chlorine to the plant effluent, *FOLLOWING* plant treatment, for disinfection purposes.

POTABLE (POE-tuh-bull) WATER   POTABLE WATER

Water that does not contain objectionable pollution, contamination, minerals, or infective agents and is considered satisfactory for drinking.

PRECHLORINATION   PRECHLORINATION

The addition of chlorine at the headworks of the plant *PRIOR TO* other treatment processes mainly for disinfection and control of tastes, odors, and aquatic growths. Also applied to aid in coagulation and settling.

PRECURSOR, THM (pre-CURSE-or) PRECURSOR, THM

Natural organic compounds found in all surface and groundwaters. These compounds *MAY* react with halogens (such as chlorine) to form trihalomethanes (tri-HAL-o-METH-hanes) (THMs); they *MUST* be present in order for THMs to form.

REAGENT (re-A-gent) REAGENT

A pure chemical substance that is used to make new products or is used in chemical tests to measure, detect, or examine other substances.

REDUCING AGENT REDUCING AGENT

Any substance, such as a base metal (iron) or the sulfide ion (S^{2-}), that will readily donate (give up) electrons. The opposite is an OXIDIZING AGENT.

REDUCTION (re-DUCK-shun) REDUCTION

Reduction is the addition of hydrogen, removal of oxygen, or the addition of electrons to an element or compound. Under anaerobic conditions (no dissolved oxygen present), sulfur compounds are reduced to odor-producing hydrogen sulfide (H_2S) and other compounds. The opposite of OXIDATION.

RELIQUEFACTION (re-LICK-we-FACK-shun) RELIQUEFACTION

The return of a gas to the liquid state; for example, a condensation of chlorine gas to return it to its liquid form by cooling.

RESIDUAL CHLORINE RESIDUAL CHLORINE

The concentration of chlorine present in water after the chlorine demand has been satisfied. The concentration is expressed in terms of the total chlorine residual, which includes both the free and combined or chemically bound chlorine residuals.

ROTAMETER (RODE-uh-ME-ter) ROTAMETER

A device used to measure the flow rate of gases and liquids. The gas or liquid being measured flows vertically up a tapered, calibrated tube. Inside the tube is a small ball or bullet-shaped float (it may rotate) that rises or falls depending on the flow rate. The flow rate may be read on a scale behind or on the tube by looking at the middle of the ball or at the widest part or top of the float.

SAPROPHYTES (SAP-row-FIGHTS) SAPROPHYTES

Organisms living on dead or decaying organic matter. They help natural decomposition of organic matter in water.

STERILIZATION (STARE-uh-luh-ZAY-shun) STERILIZATION

The removal or destruction of all microorganisms, including pathogenic and other bacteria, vegetative forms and spores. Compare with DISINFECTION.

TITRATE (TIE-trate) TITRATE

To *TITRATE* a sample, a chemical solution of known strength is added drop by drop until a certain color change, precipitate, or pH change in the sample is observed (end point). Titration is the process of adding the chemical reagent in small increments (0.1 – 1.0 milliliter) until completion of the reaction, as signaled by the end point.

TOTAL CHLORINE TOTAL CHLORINE

The total concentration of chlorine in water, including the combined chlorine (such as inorganic and organic chloramines) and the free available chlorine.

TOTAL CHLORINE RESIDUAL TOTAL CHLORINE RESIDUAL

The total amount of chlorine residual (value for residual chlorine, including both free chlorine and chemically bound chlorine) present in a water sample after a given contact time.

TRIHALOMETHANES (THMs) (tri-HAL-o-METH-hanes) TRIHALOMETHANES (THMs)

Derivatives of methane, CH_4, in which three halogen atoms (chlorine or bromine) are substituted for three of the hydrogen atoms. Often formed during chlorination by reactions with natural organic materials in the water. The resulting compounds (THMs) are suspected of causing cancer.

TURBIDITY (ter-BID-it-tee) TURBIDITY

The cloudy appearance of water caused by the presence of suspended and colloidal matter. In the waterworks field, a turbidity measurement is used to indicate the clarity of water. Technically, turbidity is an optical property of the water based on the amount of light reflected by suspended particles. Turbidity cannot be directly equated to suspended solids because white particles reflect more light than dark-colored particles and many small particles will reflect more light than an equivalent large particle.

CHAPTER 7.   DISINFECTION

(Lesson 1 of 3 Lessons)

7.0   PURPOSE OF DISINFECTION

7.00   Making Water Safe for Consumption

Our single most important natural resource is water. Without water we could not exist. Unfortunately, safe water is becoming very difficult to find. In the past, safe water could be found in remote areas, but with population growth and related pollution of waters, there are very few natural waters left that are safe to drink without treatment of some kind.

Water is the universal solvent and therefore carries all types of dissolved materials. Water also carries biological life forms which can cause diseases. These waterborne pathogenic organisms are listed in Table 7.1. Most of these organisms and the diseases they transmit are no longer a problem in the United States due to proper water protection, treatment, and monitoring. However, many developing regions of the world still experience serious outbreaks of various waterborne diseases.

TABLE 7.1   PATHOGENIC ORGANISMS (DISEASES) TRANSMITTED BY WATER

Bacteria

Salmonella (salmonellosis)
Shigella (bacillary dysentery)
Bacillus typhosus (typhoid fever)
Salmonella paratyphi (paratyphoid)
Vibrio cholerae (cholera)

Viruses

Enterovirus
Poliovirus
Coxsackie Virus
Echo Virus
Andenovirus
Reovirus
Infectious Hepatitis

Intestinal Parasites

Entamoeba histolytica (amoebic dysentery)
Giardia lamblia (giardiasis)
Ascaris lumbricoides (giant roundworm)
Cryptosporidium (cryptosporidiosis)

One of the cleansing processes in the treatment of safe water is called disinfection. Disinfection is the selective destruction or inactivation of pathogenic organisms. Don't confuse disinfection with sterilization. Sterilization is the complete destruction of all organisms. Sterilization is not necessary in water treatment and is also quite expensive.

7.01   Safe Drinking Water Laws

In the United States, the U.S. Environmental Protection Agency is responsible for setting drinking water standards and for ensuring their enforcement. This agency sets federal regulations which all state and local agencies must enforce. The Safe Drinking Water Act (SDWA) and its amendments contain specific maximum allowable levels of substances known to be hazardous to human health. In addition to describing maximum contaminant levels (MCLs), these federal drinking water regulations also give detailed instructions on what to do when the MCL for a particular substance is exceeded.

The Surface Water Treatment Rule (SWTR) requires disinfection of all surface water supply systems as protection against exposure to viruses, bacteria, and Giardia. Table 7.2, shows an example of the regulations for COLIFORM[1] bacteria which are supposed to be killed by disinfection. See Volume II, Chapter 22, "Drinking Water Regulations," and the poster provided with this manual for details of the SDWA's rules and regulations.

As previously noted (page 259), drinking water regulations are constantly changing. The Interim Enhanced Surface Water Treatment Rule (IESWTR) and the Disinfectant/Disinfection By-Products (D/DBP) Rule were passed in 1998 and further modifications of these rules are being developed. The goal of the IESWTR is to increase public protection from illness caused by the Cryptosporidium organism. The new D/DBP Rule, which applies to water systems using a disinfectant during treatment, limits the amount of certain potentially harmful disinfection by-products that may remain in drinking water after treatment.

Water systems serving 10,000 or more people had three years (from December 16, 1998) to comply with both the D/DBP and the IESWTR regulations. Small systems (serving fewer than 10,000 people) had five years to comply.

QUESTIONS

Write your answers in a notebook and then compare your answers with those on page 356.

7.0A   What are pathogenic organisms?

7.0B   What is disinfection?

7.0C   Drinking water standards are established by what agency of the United States government?

7.0D   MCL stands for what words?

---

[1] Coliform (COAL-i-form).   A group of bacteria found in the intestines of warm-blooded animals (including humans) and also in plants, soil, air and water. Fecal coliforms are a specific class of bacteria which only inhabit the intestines of warm-blooded animals. The presence of coliform bacteria is an indication that the water is polluted and may contain pathogenic (disease-causing) organisms.

TABLE 7.2 MICROBIAL STANDARDS [a]

Contaminant	Maximum Contaminant Level (MCL)	Monitoring Requirement— Surface Only or Combination	Check Sampling, Reporting, and Public Notice
TOTAL COLIFORM	< 40 samples/month, no more than 1 positive ≥ 40 samples/month, no more than than 5% positive	Compliance is based on the presence or absence of total coliforms. All coliform positives must be tested for presence of fecal coliform or *E. coli.* The total number of samples is based on the population served (see Table 7.3).	Repeat samples are required for each coliform-positive sample. All samples must be collected the same day. At least one sample from same tap as original, another from an upstream connection, and one from downstream. All coliform positives must be tested for presence of fecal coliform or *E. coli.* If repeat sample is fecal coliform positive or if the original fecal coliform or *E. coli* positive is followed by a total coliform positive, the state must be notified on the same business day.
GIARDIA LAMBLIA	3-Log (99.9%) removal or inactivation	Based on calculated residual disinfectant CT values.	Failure to meet total percent inactivation on more than two days in a month is a violation. State must be notified within one business day when disinfectant residual is less than 0.2 mg/L.
VIRUSES	4-Log (99.99%) removal or inactivation		If residual is less than 0.2 mg/L, then sampling must be every 4 hours until residual is restored.

[a] See Volume II, Chapter 22, "Drinking Water Regulations," and the poster provided with this manual for more details.

7.1 FACTORS INFLUENCING DISINFECTION

7.10 pH

The pH of water being treated can alter the efficiency of disinfectants. Chlorine, for example, disinfects water much faster at a pH around 7.0 than at a pH over 8.0.

7.11 Temperature

Temperature conditions also influence the effectiveness of the disinfectant. The higher the temperature of the water, the more efficiently it can be treated. Water near 70 to 85°F (21 to 29°C) is easier to disinfect than water at 40 to 60°F (4 to 16°C). Longer contact times are required to disinfect water at lower temperatures. To speed up the process, operators often simply use larger amounts of chemicals. Where water is exposed to the atmosphere, the warmer the water temperature the greater the dissipation rate of chlorine into the atmosphere.

7.12 Turbidity

Under normal operating conditions, the turbidity level of water being treated is very low by the time the water reaches the disinfection process. Excessive turbidity will greatly reduce the efficiency of the disinfecting chemical or process. Studies in water treatment plants have shown that when water is filtered to a turbidity of one unit or less, most of the bacteria have been removed.

The suspended matter itself may also change the chemical nature of the water when the disinfectant is added. Some types of suspended solids can create a continuing demand for the chemical, thus changing the effective germicidal (germ-killing) properties of the disinfectant.

7.120 Organic Matter

Organics found in the water can consume great amounts of disinfectants while forming unwanted compounds. *TRIHALO-METHANES*[2] are an example of undesirable compounds formed by reactions between chlorine and certain organics. Disinfecting chemicals often react with organics and *REDUCING AGENTS*[3] (Section 7.13). Then, if any of the chemical remains available after this initial reaction, it can act as an effective disinfectant. The reactions with organics and reducing agents, however, will have significantly reduced the amount of chemical available for disinfection.

[2] *Trihalomethanes (THMs) (tri-HAL-o-METH-hanes). Derivatives of methane, CH_4, in which three halogen atoms (chlorine or bromine) are substituted for three of the hydrogen atoms. Often formed during chlorination by reactions with natural organic materials in the water. The resulting compounds (THMs) are suspected of causing cancer.*
[3] *Reducing Agent. Any substance, such as a base metal (iron) or the sulfide ion (S^{2-}), that will readily donate (give up) electrons. The opposite is an oxidizing agent.*

**TABLE 7.3 TOTAL COLIFORM SAMPLING REQUIREMENTS
ACCORDING TO POPULATION SERVED**

Population Served	Minimum Number of Routine Samples Per Month[a]	Population Served	Minimum Number of Routine Samples Per Month[a]
25 to 1,000 [b]	1 [c]	59,001 to 70,000	70
1,001 to 2,500	2	70,001 to 83,000	80
2,501 to 3,300	3	83,001 to 96,000	90
3,301 to 4,100	4	96,001 to 130,000	100
4,101 to 4,900	5	130,001 to 220,000	120
4,901 to 5,800	6	220,001 to 320,000	150
5,801 to 6,700	7	320,001 to 450,000	180
6,701 to 7,600	8	450,001 to 600,000	210
7,601 to 8,500	9	600,001 to 780,000	240
8,501 to 12,900	10	780,001 to 970,000	270
12,901 to 17,200	15	970,001 to 1,230,000	300
17,201 to 21,500	20	1,230,001 to 1,520,000	330
21,501 to 25,000	25	1,520,001 to 1,850,000	360
25,001 to 33,000	30	1,850,001 to 2,270,000	390
33,001 to 41,000	40	2,270,001 to 3,020,000	420
41,001 to 50,000	50	3,020,001 to 3,960,000	450
50,001 to 59,000	60	3,960,001 or more	480

[a] A noncommunity water system using groundwater and serving 1,000 persons or fewer may monitor at a lesser frequency specified by the state until a sanitary survey is conducted and the state reviews the results. Thereafter, noncommunity water systems using groundwater and serving 1,000 persons or fewer must monitor in each calendar quarter during which the system provides water to the public, unless the state determines that some other frequency is more appropriate and notifies the system (in writing). In all cases, noncommunity water systems using groundwater and serving 1,000 persons or fewer must monitor at least once/year.

A noncommunity water system using surface water, or groundwater under the direct influence of surface water, regardless of the number of persons served, must monitor at the same frequency as a like-sized community public system. A noncommunity water system using groundwater and serving more than 1,000 persons during any month must monitor at the same frequency as a like-sized community water system, except that the state may reduce the monitoring frequency for any month the system serves 1,000 persons or fewer.

[b] Includes public water systems which have at least 15 service connections, but serve fewer than 25 persons.

[c] For a community water system serving 25 to 1,000 persons, the state may reduce this sampling frequency if a sanitary survey conducted in the last five years indicates that the water system is supplied solely by a protected groundwater source and is free of sanitary defects. However, in no case may the state reduce the sampling frequency to less than once/quarter.

7.121 Inorganic Matter

Inorganic compounds such as ammonia (NH_3) in the water being treated can create special problems. In the presence of ammonia, some oxidizing chemicals form side compounds causing a partial loss of disinfecting power. Silt can also create a chemical demand. It is clear, then, that the chemical properties of the water being treated can seriously interfere with the effectiveness of disinfecting chemicals.

7.13 Reducing Agents

Chlorine combines with a wide variety of materials, especially reducing agents. Most of the reactions are very rapid, while others are much slower. These side reactions complicate the use of chlorine for disinfection. The demand for chlorine by reducing agents must be satisfied before chlorine becomes available to accomplish disinfection. Examples of inorganic reducing agents present in water that will react with chlorine include hydrogen sulfide (H_2S), ferrous ion (Fe^{2+}), manganous ion (Mn^{2+}), ammonia (NH_3), and the nitrite ion (NO_2^-). Organic reducing agents in water also will react with chlorine and form chlorinated organic materials of potential health significance.

7.14 Microorganisms

7.140 Number and Types of Microorganisms

The concentration of microorganisms is important because the higher the number of microorganisms, the greater the demand for a disinfecting chemical. The resistance of microorganisms to specific disinfectants varies greatly. Non-spore-forming bacteria are generally less resistant than spore-forming bacteria. Cysts and viruses can be very resistant to certain types of disinfectants.

7.141 Removal Processes

Pathogenic organisms can be removed from water, killed, or inactivated by various physical and chemical water treatment processes. These processes are:

1. *COAGULATION.* Chemical coagulation followed by sedimentation and filtration will remove 90 to 95 percent of the pathogenic organisms, depending on which chemicals are used. Alum usage can increase virus removals up to 99 percent.

2. *SEDIMENTATION.* Properly designed sedimentation processes can effectively remove 20 to 70 percent of the pathogenic microorganisms. This removal is accomplished by allowing the pathogenic organisms (as well as nonpathogenic organisms) to settle out by gravity, assisted by chemical floc.

3. *FILTRATION.* Filtering water through granular filters is an effective means of removing pathogenic and other organisms from water. The removal rates vary from 20 to 99+ percent depending on the coarseness of the filter media and the type and effectiveness of pretreatment.

4. *DISINFECTION.* Disinfection chemicals such as chlorine are added to water to kill or inactivate pathogenic microorganisms.

In previous chapters, you have already studied the first three processes. The fourth, disinfection, is the subject of this chapter.

QUESTIONS

Write your answers in a notebook and then compare your answers with those on page 356.

7.1A How does pH influence the effectiveness of disinfection?

7.1B How does the temperature of the water influence disinfection?

7.1C What two factors influence the effectiveness of disinfection on microorganisms?

7.2 PROCESS OF DISINFECTION

7.20 Purpose of Process

The purpose of disinfection is to destroy harmful organisms. This can be accomplished either physically or chemically. Physical methods may (1) physically remove the organisms from the water, or (2) introduce motion that will disrupt the cells' biological activity and kill or inactivate them.

Chemical methods alter the cell chemistry causing the microorganism to die. The most widely used disinfectant chemical is chlorine. Chlorine is easily obtained, relatively cheap, and most importantly, leaves a *RESIDUAL CHLORINE*[4] that can be measured. Other disinfectants are also used. There has been increased interest in disinfectants other than chlorine because of the *CARCINOGENIC*[5] compounds that chlorine may form (trihalomethanes or THMs).

This chapter will focus primarily on the use of chlorine as a disinfectant. However, let's take a brief look first at other disinfection methods and chemicals. Some of these are being more widely applied today because of the potential adverse side effects of chlorination.

7.21 Agents of Disinfection

7.210 Physical Means of Disinfection

A. *ULTRAVIOLET RAYS* can be used to destroy pathogenic microorganisms. To be effective, the rays must come in contact with each microorganism. The ultraviolet energy disrupts various organic components of the cell causing a biological change that is fatal to the microorganisms.

This system has not had widespread acceptance because of the lack of measurable residual and the cost of operation. Currently, use of ultraviolet rays is limited to small or local systems and industrial applications. Ocean-going ships have used these systems for their water supply.

Advances in UV technology and concern about disinfection by-products (DBPs) produced by other disinfectants have prompted a renewed interest in UV disinfection. See Section 7.90, "Ultraviolet (UV) Systems," for a more detailed description of this process.

B. *HEAT* has been used for centuries to disinfect water. Boiling water for about 5 minutes will destroy essentially all microorganisms. This method is very energy intensive and thus very expensive. The only practical application is in the event of a disaster when individual local users are required to boil their water.

C. *ULTRASONIC WAVES* have been used to disinfect water on a very limited scale. Sonic waves destroy the microorganism by vibration. This procedure is not yet practical and is very expensive.

7.211 Chemical Disinfectants (Other Than Chlorine)

A. *IODINE* has been used as a disinfectant in water since 1920, but its use has been limited to emergency treatment of water supplies. Although it has long been recognized as a good disinfectant, iodine's high cost and potential physiological effects (pregnant women can suffer serious side effects) have prevented widespread acceptance. The recommended dosage is two drops of iodine (tincture of iodine which is 7 percent available iodine) in a liter of water.

---

[4] *Residual Chlorine.* The concentration of chlorine present in water after the chlorine demand has been satisfied. The concentration is expressed in terms of the total chlorine residual, which includes both the free and combined or chemically bound chlorine residuals.

[5] *Carcinogen (CAR-sin-o-JEN).* Any substance which tends to produce cancer in an organism.

B. *BROMINE* has been used only on a very limited scale for water treatment because of its handling difficulties. Bromine causes skin burns on contact. Because bromine is a very reactive chemical, residuals are hard to obtain. This also limits its use. Bromine can be purchased at swimming pool supply stores.

C. *BASES* such as sodium hydroxide and lime can be effective disinfectants but the high pH leaves a bitter taste in the finished water. Bases can also cause skin burns when left too long in contact with the skin. Bases effectively kill all microorganisms (they sterilize rather than just disinfect water). Although this method has not been used on a large scale, bases have been used to sterilize water pipes.

D. *OZONE* has been used in the water industry since the early 1900s, particularly in France. In the United States it has been used primarily for taste and odor control. The limited use in the United States has been due to its high costs, lack of residual, difficulty in storing, and maintenance requirements.

Although ozone is effective in disinfecting water, its use is limited by its solubility. The temperature and pressure of water being treated regulate the amount of ozone that can be dissolved in the water. These factors tend to limit the disinfectant strength that can be made available to treat the water.

Many scientists claim that ozone destroys all microorganisms. Unfortunately, significant residual ozone does not guarantee that a water is safe to drink. Organic solids may protect organisms from the disinfecting action of ozone and increase the amount of ozone needed for disinfection. In addition, ozone residuals cannot be maintained in metallic conduits for any period of time because of ozone's reactive nature. The inability of ozone to provide a residual in the distribution system is a major drawback to its use. However, recent information about the formation of trihalomethanes by chlorine compounds has resulted in renewed interest in ozone as an alternative means of disinfection. See Section 7.91, "Ozone," for a description of disinfection using ozone.

QUESTIONS

Write your answers in a notebook and then compare your answers with those on page 356.

7.2A   List the physical agents that have been used for disinfection (chlorine is not a physical agent).

7.2B   List the chemical agents that have been used for disinfection other than chlorine.

7.2C   What is a major limitation to the use of ozone?

7.22   Chlorine (Cl$_2$)

7.220   *Properties of Chlorine*

Chlorine is a greenish-yellow gas with a penetrating and distinctive odor. The gas is two-and-a-half times heavier than air. Chlorine has a very high coefficient of expansion. If there is a temperature increase of 50°F (28°C) (from 35°F to 85°F or 2°C to 30°C), the volume will increase by 84 to 89 percent. This expansion could easily rupture a cylinder or a line full of liquid chlorine. For this reason all chlorine containers must not be filled to more than 85 percent of their volume. One liter of liquid chlorine can evaporate and produce 450 liters of chlorine gas.

Chlorine by itself is nonflammable and nonexplosive, but it will support combustion. When the temperature rises, so does the vapor pressure of chlorine. This means that when the temperature increases, the pressure of the chlorine gas inside a chlorine container will increase. This property of chlorine must be considered when:

1. Feeding chlorine gas from a container, and

2. Dealing with a leaking chlorine cylinder.

7.221   *Chlorine Disinfection Action*

The exact mechanism of chlorine disinfection action is not fully known. One theory holds that chlorine exerts a direct action against the bacterial cell, thus destroying it. Another theory is that the toxic character of chlorine inactivates the *ENZYMES*[6] which enable living microorganisms to use their food supply. As a result, the organisms die of starvation. From the point of view of water treatment, the exact mechanism of chlorine disinfection is less important than its demonstrated effects as a disinfectant.

When chlorine is added to water, several chemical reactions take place. Some involve the molecules of the water itself, and some involve organic and inorganic substances suspended in the water. We will discuss these chemical reactions in more detail in the next few sections of this chapter. First, however, there are some terms associated with chlorine disinfection that you should understand.

When chlorine is added to water containing organic and inorganic materials, it will combine with these materials and form chlorine compounds. If you continue to add chlorine, you will eventually reach a point where the reaction with organic and inorganic materials stops. At this point, you have satisfied what is known as the *CHLORINE DEMAND*.

The chemical reactions between chlorine and these organic and inorganic substances produce chlorine compounds. Some of these compounds have disinfecting properties; others do not. In a similar fashion, chlorine reacts with the

---

[6] Enzymes (EN-zimes).   Organic substances (produced by living organisms) which cause or speed up chemical reactions. Organic catalysts and/or biochemical catalysts.

water itself and produces some substances with disinfecting properties. The total of all the compounds with disinfecting properties *PLUS* any remaining free (uncombined) chlorine is known as the *CHLORINE RESIDUAL*. The presence of this measurable chlorine residual is what indicates to the operator that all possible chemical reactions have taken place and that there is still sufficient *AVAILABLE RESIDUAL CHLORINE* to kill the microorganisms present in the water supply.

Now, if you add together the amount of chlorine needed to satisfy the chlorine demand and the amount of chlorine residual needed for disinfection, you will have the *CHLORINE DOSE*. This is the amount of chlorine you will have to add to the water to disinfect it.

Chlorine Dose, mg/L = Chlorine Demand, mg/L + Chlorine Residual, mg/L

where

Chlorine Demand, mg/L = Chlorine Dose, mg/L − Chlorine Residual, mg/L

and

Chlorine Residual, mg/L = Combined Chlorine Forms, mg/L + Free Chlorine, mg/L

7.222   Reaction With Water

Free chlorine combines with water to form hypochlorous and hydrochloric acids:

Chlorine + Water $\leftrightarrows$ Hypochlorous Acid + Hydrochloric Acid

$$Cl_2 + H_2O \leftrightarrows HOCl + HCl$$

Depending on the pH, hypochlorous acid may be present in the water as the hydrogen ion and hypochlorite ion (Figure 7.1).

Hypochlorous Acid $\leftrightarrows$ Hydrogen Ion + Hypochlorite Ion

$$HOCl \leftrightarrows H^+ + OCl^-$$

In solutions that are dilute (low concentration of chlorine) and have a pH above 4, the formation of HOCl (hypochlorous acid) is most complete and leaves little free chlorine (Cl_2). The hypochlorous acid is a weak acid and hence is poorly dissociated (broken up into ions) at pH levels below 6. Thus any free chlorine or hypochlorite (OCl^-) added to water will immediately form either HOCl or OCl^-; the species formed is thereby controlled by the pH value of the water. This is extremely important since HOCl and OCl^- differ in disinfection ability. HOCl has a much greater disinfection potential than OCl^-. Normally in water with a pH of 7.5, approximately 50 percent of the chlorine present will be in the form of HOCl and 50 percent in the form of OCl^-. The higher the pH level, the greater the percent of OCl^-.

7.223   Reaction With Impurities in Water

Most waters that have been processed still contain some impurities. In this section we will discuss some of the more common impurities that react with chlorine and we will examine the effects of these reactions on the disinfection ability of chlorine.

A. Hydrogen sulfide (H_2S) and ammonia (NH_3) are two inorganic substances that may be found in water when it reaches the disinfection stage of treatment. Their presence can complicate the use of chlorine for disinfection purposes.

This is because hydrogen sulfide and ammonia are what is known as *REDUCING AGENTS*. That is, they give up electrons easily. Chlorine reacts rapidly with these particular reducing agents producing some undesirable results.

Hydrogen sulfide produces an odor that smells like rotten eggs. It reacts with chlorine to form sulfuric acid and elemental sulfur (depending on temperature, pH, and hydrogen sulfide concentration). Elemental sulfur is objectionable because it can cause odor problems and will precipitate as finely divided white particles which are sometimes colloidal in nature.

The chemical reactions between hydrogen sulfide and chlorine are as follows:

Hydrogen Sulfide + Chlorine + Oxygen Ion → Elemental Sulfur + Water + Chloride Ions

$$H_2S + Cl_2 + O^{2-} → S\downarrow + H_2O + 2 Cl^-$$

The chlorine required to oxidize hydrogen sulfide to sulfur and water is 2.08 mg/L chlorine to 1 mg/L hydrogen sulfide. The complete oxidation of hydrogen sulfide to the sulfate form is as follows:

Hydrogen Sulfide + Chlorine + Water → Sulfuric Acid + Hydrochloric Acid

$$H_2S + 4 Cl_2 + 4 H_2O → H_2SO_4 + 8 HCl$$

Thus, 8.32 mg/L of chlorine are required to oxidize one mg/L of hydrogen sulfide to the sulfate form. Note that in both reactions the chlorine is converted to the chloride ion (Cl^- or HCl) which has no disinfecting power and produces no chlorine residual. In waterworks practice we always chlorinate to produce a chlorine residual; therefore the second reaction (complete oxidation of hydrogen sulfide) occurs before we have any chlorine residual in the water we are treating.

When chlorine is added to water containing ammonia (NH_3), it reacts rapidly with the ammonia and forms *CHLORAMINES*.[7] This means that less chlorine is available to act as a disinfectant. As the concentration of ammonia increases, the disinfectant power of the chlorine drops off at a rapid rate.

B. When organic materials are present in water being disinfected with chlorine, the chemical reactions that take place may produce suspected carcinogenic compounds (trihalomethanes). The formation of these compounds can be prevented by limiting the amount of prechlorination and by removing the organic materials prior to chlorination of the water.

QUESTIONS

Write your answers in a notebook and then compare your answers with those on page 356.

7.2D   How is the chlorine dosage determined?

7.2E   How is the chlorine demand determined?

7.2F   List two inorganic reducing chemicals with which chlorine reacts rapidly.

7.2G   What does chlorine produce when it reacts with organic matter?

[7] *Chloramines (KLOR-uh-means).   Compounds formed by the reaction of hypochlorous acid (or aqueous chlorine) with ammonia.*

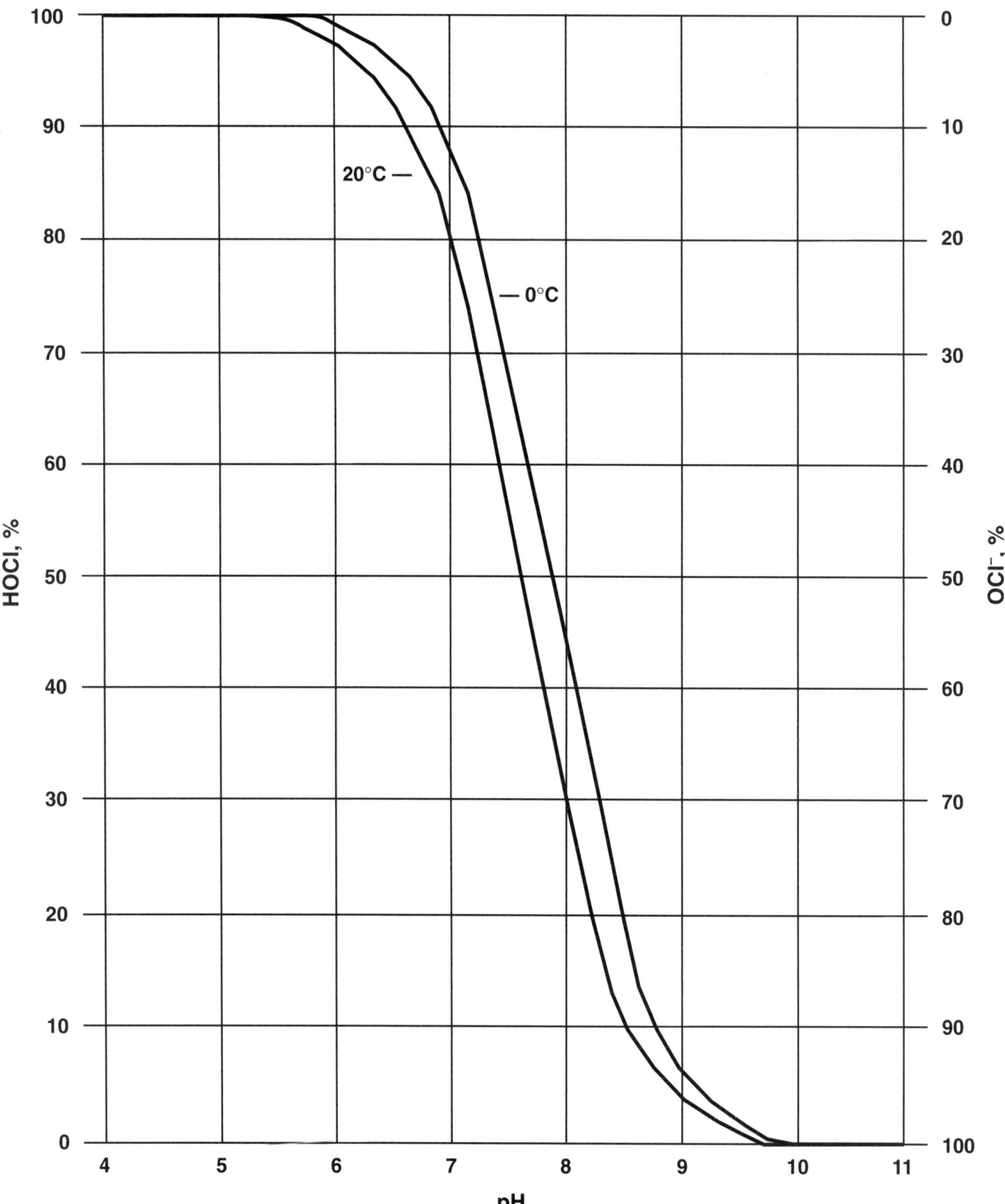

*Fig. 7.1   Relationship between hypochlorous acid (HOCl),
hypochlorite ion (OCl⁻), and pH*

7.23 Hypochlorite (OCl⁻)

7.230 Reactions With Water

The use of hypochlorite to treat potable water achieves the same result as chlorine gas. Hypochlorite may be applied in the form of calcium hypochlorite ($Ca(OCl)_2$) or sodium hypochlorite (NaOCl). The form of calcium hypochlorite most frequently used to disinfect water is known as **H**igh **T**est **H**ypochlorite (HTH). The chemical reactions of hypochlorite in water are similar to those of chlorine gas.

CALCIUM HYPOCHLORITE

$$\text{Calcium Hypochlorite} + \text{Water} \rightarrow \text{Hypochlorous Acid} + \text{Calcium Hydroxide}$$

$$Ca(OCl)_2 + 2\ H_2O \rightarrow 2\ HOCl + Ca(OH)_2$$

SODIUM HYPOCHLORITE

$$\text{Sodium Hypochlorite} + \text{Water} \rightarrow \text{Hypochlorous Acid} + \text{Sodium Hydroxide}$$

$$NaOCl + H_2O \rightarrow HOCl + NaOH$$

Calcium hypochlorite (HTH) is used by a number of small water supply systems. A problem occurs in these systems when sodium fluoride is injected at the same point as the hypochlorite. A severe crust forms when the calcium and fluoride ions combine.

7.231 Differences Between Chlorine Gas and Hypochlorite Compound Reactions

The only difference between the reactions of the hypochlorite compounds and chlorine gas is the "side" reactions of the end products. The reaction of chlorine gas tends to lower the pH (increases the hydrogen ion (H^+) concentration) by the formation of hydrochloric acid which favors the formation of hypochlorous acid (HOCl). The hypochlorite tends to raise the pH with the formation of hydroxyl ions (OH^-) from the calcium or sodium hydroxide. At a high pH of around 8.5 or higher, the hypochlorous acid (HOCl) is almost completely dissociated to the ineffective hypochlorite ion (OCl^-) (Figure 7.1). This reaction also depends on the *BUFFER CAPACITY*[8] (amount of bicarbonate, HCO_3^-, present) of the water.

$$\underset{\text{Acid}}{\text{Hypochlorous}} \rightleftharpoons \underset{\text{Ion}}{\text{Hydrogen}} + \underset{\text{Ion}}{\text{Hypochlorite}}$$

$$HOCl \rightleftharpoons H^+ + OCl^-$$

7.232 On-Site Chlorine Generation

Small water systems are generating chlorine on site for their water treatment processes. On-site generation (OSG) of process chlorine is attractive due to the lower safety hazards and costs involved. On-site generated chlorine systems produce 0.8 percent sodium hypochlorite. This limited solution strength (about $1/15$ the strength of commercial bleach and $1/7$ that of household bleach) is below the lower limit deemed a "hazardous liquid," with obvious economic and safety advantages.

Operators' only duties with on-site generation systems are to observe the control panel daily for proper operating guidelines and to dump bags of salt every few weeks. Since the assemblies (which are quite small) include an ion exchange water softener, mineral deposits forming with the electrolytic cell are minimal, with an acid cleaning being necessary only every few months. Cell voltage is controlled at a low value to maximize electrode life, which is about three years. Process brine strength and cell current determine chlorine production at the anode, with hydrogen gas continually vented from the cathode. The units include provisions for storing the chlorine solution in order to deliver chlorine for several days in the event of a power failure or other problems causing equipment failure.

7.24 Chlorine Dioxide (ClO₂)

7.240 Reaction in Water

Chlorine dioxide may be used as a disinfectant. Chlorine dioxide does not form carcinogenic compounds that may be formed by other chlorine compounds. Also it is not affected by ammonia, and is a very effective disinfectant at higher pH levels. In addition, chlorine dioxide reacts with sulfide compounds, thus helping to remove them and eliminate their characteristic odors. Phenolic tastes and odors can be controlled by using chlorine dioxide.

Chlorine dioxide reacts with water to form chlorate and chlorite ions in the following manner:

$$\underset{\text{Dioxide}}{\text{Chlorine}} + \text{Water} \rightarrow \underset{\text{Ion}}{\text{Chlorate}} + \underset{\text{Ion}}{\text{Chlorite}} + \underset{\text{Ions}}{\text{Hydrogen}}$$

$$2\ ClO_2 + H_2O \rightarrow ClO_3^- + ClO_2^- + 2\ H^+$$

7.241 Reactions With Impurities in Water

A. INORGANIC COMPOUNDS

Chlorine dioxide is an effective *OXIDIZING AGENT*[9] with iron and manganese and does not leave objectionable tastes or odors in the finished water. Because of its oxidizing ability, chlorine dioxide usage must be monitored and the dosage will have to be increased when treating waters with iron and manganese.

B. ORGANIC COMPOUNDS

Chlorine dioxide does not react with organics in water. Therefore, there is little danger of the formation of potentially dangerous trihalomethanes.

---

[8] *Buffer Capacity.* A measure of the capacity of a solution or liquid to neutralize acids or bases. This is a measure of the capacity of water for offering a resistance to changes in pH.

[9] *Oxidizing Agent.* Any substance, such as oxygen (O_2) or chlorine (Cl_2), that will readily add (take on) electrons. The opposite is a reducing agent.

QUESTIONS

Write your answers in a notebook and then compare your answers with those on page 356.

7.2H   How do chlorine gas and hypochlorite influence pH?

7.2I   How does pH influence the relationship between HOCl and OCl⁻?

7.25   Breakpoint Chlorination[10]

In determining how much chlorine you will need for disinfection, remember you will be attempting to produce a certain chlorine residual in the form of *FREE AVAILABLE RESIDUAL CHLORINE*.[11] Chlorine in this form has the highest disinfecting ability. *BREAKPOINT CHLORINATION* is the name of this process of adding chlorine to water until the chlorine demand has been satisfied. Further additions of chlorine will result in a chlorine residual that is directly proportional to the amount of chlorine added beyond the breakpoint. Public water supplies are normally chlorinated *PAST THE BREAKPOINT.*

Take a moment here to look at the breakpoint chlorination curve in Figure 7.2. Assume the water being chlorinated contains some manganese, iron, nitrite, organic matter, and ammonia. Now add a small amount of chlorine. The chlorine reacts with (oxidizes) the manganese, iron, and nitrite. That's all that happens—no disinfection and no chlorine residual (Figure 7.2, points 1 to 2). Add a little more chlorine, enough to react with the organics and ammonia; *CHLORORGANICS*[12] and *CHLORAMINES*[13] will form. The chloramines produce a combined chlorine residual—a chlorine residual combined with other substances so it has lost some of its disinfecting strength. Combined residuals have rather poor disinfecting power and may cause tastes and odors (points 2 to 3).

With just a little more chlorine the chloramines and some of the chlororganics are destroyed (points 3 to 4). Adding just one last amount of chlorine we get *FREE AVAILABLE RESIDUAL CHLORINE* (beyond point 4)—free in the sense that it has not reacted with anything and available in that it *CAN* and *WILL* react if need be. Free available residual chlorine is the best residual for disinfection. It disinfects faster and without

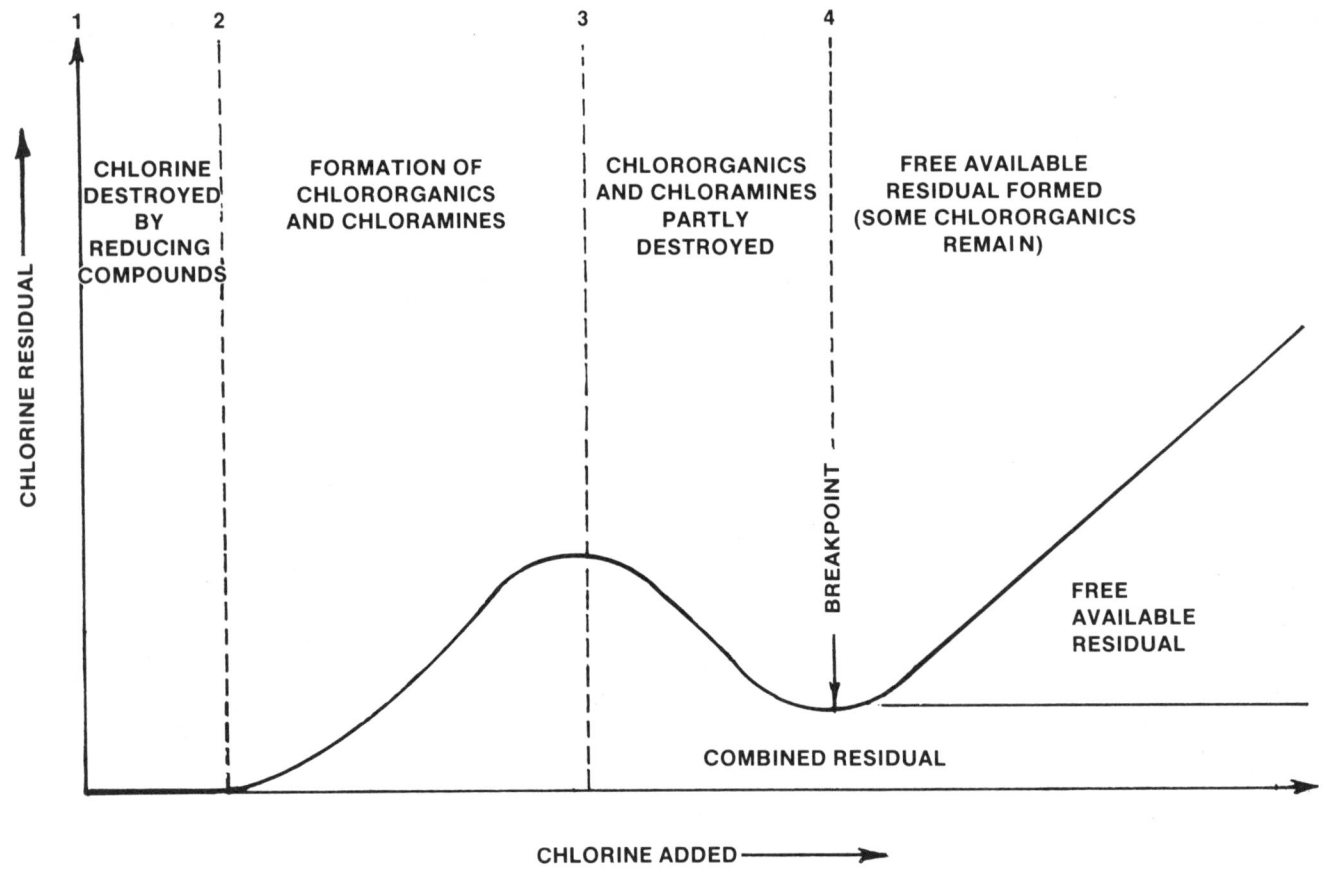

Fig. 7.2   Breakpoint chlorination curve

---

[10] *Breakpoint Chlorination.   Addition of chlorine to water until the chlorine demand has been satisfied. At this point, further additions of chlorine will result in a free chlorine residual that is directly proportional to the amount of chlorine added beyond the breakpoint.*

[11] *Free Available Residual Chlorine.   That portion of the total available residual chlorine composed of dissolved chlorine gas (Cl₂), hypochlorous acid (HOCl), and/or hypochlorite ion (OCl⁻) remaining in water after chlorination. This does not include chlorine that has combined with ammonia, nitrogen, or other compounds.*

[12] *Chlororganic (klor-or-GAN-ick).   Organic compounds combined with chlorine. These compounds generally originate from, or are associated with, life processes such as those of algae in water.*

[13] *Chloramines (KLOR-uh-means).   Compounds formed by the reaction of hypochlorous acid (or aqueous chlorine) with ammonia.*

the "swimming pool" odor of combined residual chlorine. Free available residual chlorine begins to form at the breakpoint; the process is called *BREAKPOINT CHLORINATION*. In water treatment plants today it is common practice to go "past the breakpoint." This means that the treated water will have a low chlorine residual, but the residual will be a very effective disinfectant because it is in the form of *FREE AVAILABLE RESIDUAL CHLORINE*.

CAUTION: Ammonia must be present to produce the breakpoint chlorination curve from the addition of chlorine. Sources of ammonia in raw water include fertilizer in agricultural runoff and discharges from wastewater treatment plants. High-quality raw water without any ammonia will not produce a breakpoint curve. Therefore, if there is no ammonia present in the water, and chlorinated water smells like chlorine, and a chlorine residual is present, DO NOT ADD MORE CHLORINE.

In plants where trihalomethanes (THMs) are not a problem, sufficient chlorine is added to the raw water (prechlorination) to go "past the breakpoint." The chlorine residual will aid coagulation, control algae problems in basins, reduce odor problems in treated water, and provide sufficient chlorine contact time to effectively kill or inactivate pathogenic organisms. Therefore the treated water will have a very low chlorine residual, but the residual will be a very effective disinfectant.

Let's look more closely at some of the chemical reactions that take place during chlorination. When chlorine is added to waters containing ammonia (NH_3), the ammonia reacts with hypochlorous acid (HOCl) to form monochloramine, dichloramine, and trichloramine. The formation of these chloramines depends on the pH of the solution and the initial chlorine-ammonia ratio.

Ammonia + Hypochlorous Acid → Chloramine + Water

$NH_3 + HOCl$	$NH_2Cl + H_2O$	Monochloramine
$NH_2Cl + HOCl$	$NHCl_2 + H_2O$	Dichloramine
$NHCl_2 + HOCl$	$NCl_3 + H_2O$	Trichloramine[14]

As the chlorine to ammonia-nitrogen ratio increases, the ammonia molecule becomes progressively more chlorinated. At Cl_2:NH_3-N weight ratios higher than 7.6:1, all available ammonia is theoretically oxidized to nitrogen gas and chlorine residuals are greatly reduced. The actual Cl_2:NH_3-N ratio for breakpoint for a given source water will usually be greater than 7.6:1 (typically 10:1 for most water), depending on the levels of other substances present in the water (such as nitrite and organic nitrogen). Once this point is reached, additional chlorine dosages yield an equal and proportional increase in free available chlorine.

At the pH levels that are usually found in water (pH 6.5 to 7.5), monochloramine and dichloramine exist together. At pH levels below 5.5, dichloramine exists by itself. Below pH 4.0, trichloramine is the only compound found. The mono- and dichloramine forms have definite disinfection powers and are of interest in the measurement of chlorine residuals. Dichloramine has a more effective disinfecting power than monochloramine. However, dichloramine is not recommended as a disinfectant because of taste and odor problems. Chlorine reacts with *PHENOLIC COMPOUNDS*[15] and salicylic acid (both are leached into water from leaves and blossoms) to form *CHLOROPHENOL*[16] which has an intense medicinal odor. This reaction goes much slower in the presence of monochloramine.

7.26 Chloramination
by David Foust

7.260 Use of Chloramines

Chloramines have been used as an alternative disinfectant by water utilities for over seventy years. An operator's decision to use chloramines depends on several factors, including the quality of the raw water, the ability of the treatment plant to meet various regulations, operational practices, and distribution system characteristics. Chloramines have proven effective in accomplishing the following objectives:

1. Reducing the formation of trihalomethanes (THMs) and other disinfection by-products (DBPs),

2. Maintaining a detectable residual throughout the distribution system,

3. Penetrating the biofilm (the layer of microorganisms on pipeline walls) and reducing the potential for coliform regrowth,

4. Killing or inactivating *HETEROTROPHIC*[17] plate count bacteria, and

5. Reducing taste and odor problems.

7.261 Methods for Producing Chloramines

There are three primary methods by which chloramines are produced: (1) preammoniation followed by later chlorination, (2) addition of chlorine and ammonia at the same time (concurrently), and (3) prechlorination/postammoniation.

1. *PREAMMONIATION FOLLOWED BY LATER CHLORINATION*

   In this method, ammonia is applied at the rapid-mix unit process and chlorine is added downstream at the entrance to the flocculation basins. This approach usually produces lower THM levels than the postammoniation method. Preammoniation to form chloramines (monochloramine) does not produce phenolic tastes and odors, but this method may not be as effective as postammoniation for controlling tastes and odors associated with *DIATOMS*[18] and anaerobic bacteria in source waters.

2. *CONCURRENT ADDITION OF CHLORINE AND AMMONIA*

   In this method, chlorine is applied to the plant influent and, at the same time or immediately thereafter, ammonia is introduced at the rapid-mix unit process. Concurrent

---

[14] More commonly called nitrogen trichloride.
[15] Phenolic (fee-NO-lick) Compounds. Organic compounds that are derivatives of benzene.
[16] Chlorophenolic (klor-o-FEE-NO-lick). Chlorophenolic compounds are phenolic compounds (carbolic acid) combined with chlorine.
[17] Heterotrophic (HET-er-o-TROF-ick). Describes organisms that use organic matter for energy and growth. Animals, fungi and most bacteria are heterotrophs.
[18] Diatoms (DYE-uh-toms). Unicellular (single cell), microscopic algae with a rigid (box-like) internal structure consisting mainly of silica.

chloramination produces the lowest THM levels of the three methods.

3. *PRECHLORINATION/POSTAMMONIATION*

In prechlorination/postammoniation, chlorine is applied at the head of the plant and a free chlorine residual is maintained throughout the plant processes. Ammonia is added at the plant effluent to produce chloramines. Because of the longer free chlorine contact time, this application method will result in the formation of more THMs, but it may be necessary to use this method to meet the disinfection requirements of the Surface Water Treatment Rule (SWTR). A major limitation of using chloramine residuals is that chloramines are less effective as a disinfectant than free chlorine residuals.

7.262 Chlorine to Ammonia-Nitrogen Ratios

After a method of chloramine application has been selected, the best ratio of chlorine to ammonia-nitrogen (by weight) and the desired chloramine residual for each system must be determined. A dosage of three parts of chlorine to one part ammonia (3:1) will form monochloramines. This 3:1 ratio provides an excess of ammonia-nitrogen which will be available to react with any chlorine added in the distribution system to boost the chloramine residual.

Higher chlorine to ammonia-nitrogen weight ratios such as 4:1 and 5:1 also have been used successfully by many water agencies. However, the higher the chlorine to ammonia-nitrogen ratio, the less excess ammonia will be available for rechlorination. Some agencies have found it necessary to limit the amount of excess available ammonia to prevent incomplete *NITRIFICATION.*[19]

Monochloramines form combined residual chlorine (rising part of curve in Figure 7.3) as the chlorine dose is increased in the presence of ammonia. As the chlorine dose increases, the combined residual increases and excess ammonia decreases. The maximum chlorine to ammonia ratio that can be achieved is 5:1. At a chlorine dose above the 5:1 ratio, the combined residual actually decreases and the total ammonia-nitrogen also begins to decrease as it is oxidized by the additional chlorine. Dichloramines form during this oxidation and may cause tastes and odors. As the chlorine dose is further increased, breakpoint chlorination will eventually occur. Trichloramines are formed past the breakpoint and also may form tastes and odors. As with breakpoint chlorination, further additions of chlorine will result in a chlorine residual that is proportional to the amount of chlorine added beyond the breakpoint.

Calculating the chlorine to ammonia-nitrogen ratio on the basis of actual quantity of chemicals applied can lead to incorrect conclusions regarding the finished water quality. In applications in which chlorine is injected before the ammonia, chlorine demand in the water will reduce the amount of chlorine available to form the combined residual. In such applications the *applied* ratio will be greater than the *actual* ratio of chlorine to ammonia-nitrogen leaving the plant.

As an example, assume that an initial dosage of 5.0 mg/*L* results in a free chlorine residual of 3.5 mg/*L* at the ammonia application point; it can be concluded that a chlorine demand

of 1.5 mg/*L* exists. If ammonia-nitrogen is applied at a dose of 1.0 mg/*L*, the applied chlorine to ammonia-nitrogen ratio is 5:1, whereas the actual ratio in water leaving the plant is only 3.5:1.

7.263 Special Water Users

Although chloramines are nontoxic to healthy humans, they can have a weakening effect on individuals with kidney disease who must undergo kidney dialysis. Chloramines must be removed from the water used in the dialysis treatments. Granular activated carbon and ascorbic acid are common substances used to reduce chloramine residuals. All special water users should be notified before chloramines are used as a disinfectant in municipal waters.

Also, like free chlorine, chloramines can be deadly to fish. They can damage gill tissue and enter the red blood cells causing a sudden and severe blood disorder. For this reason, all chloramine compounds must be removed from the water prior to any contact with fish.

7.264 Blending Chloraminated Waters

Care must be taken when blending chloraminated water with water that has been disinfected with free chlorine. Depending on the ratio of the blend, these two different disinfectants can cancel each other out resulting in very low disinfectant residuals. When chlorinated water is blended with chloraminated water, the chloramine residual will decrease after the excess ammonia has been combined (Figure 7.3). Knowing the amount of uncombined ammonia available is important in determining how much chlorinated water can be blended with a particular chloraminated water without significantly affecting the monochloramine residual. Knowing how much uncombined ammonia-nitrogen is available is also important before you make any attempt to boost the chloramine residual by adding chlorine.

7.265 Chloramine Residuals

When measuring combined chlorine residuals (chloramines) in the field, analyze for total chlorine. No free chlorine should be present at chlorine to ammonia-nitrogen ratios of 3:1 to 5:1. Care must be taken when attempting to measure free chlorine with chloraminated water because the chloramine residual will interfere with the DPD method for measuring free chlorine. (See Section 7.7, "Measurement of Chlorine Residual," for more specific information about the methods commonly used to measure chlorine residuals, including the DPD method.)

---

[19] *Nitrification (NYE-truh-fuh-KAY-shun). An aerobic process in which bacteria reduce the ammonia and organic nitrogen in water into nitrite and then nitrate.*

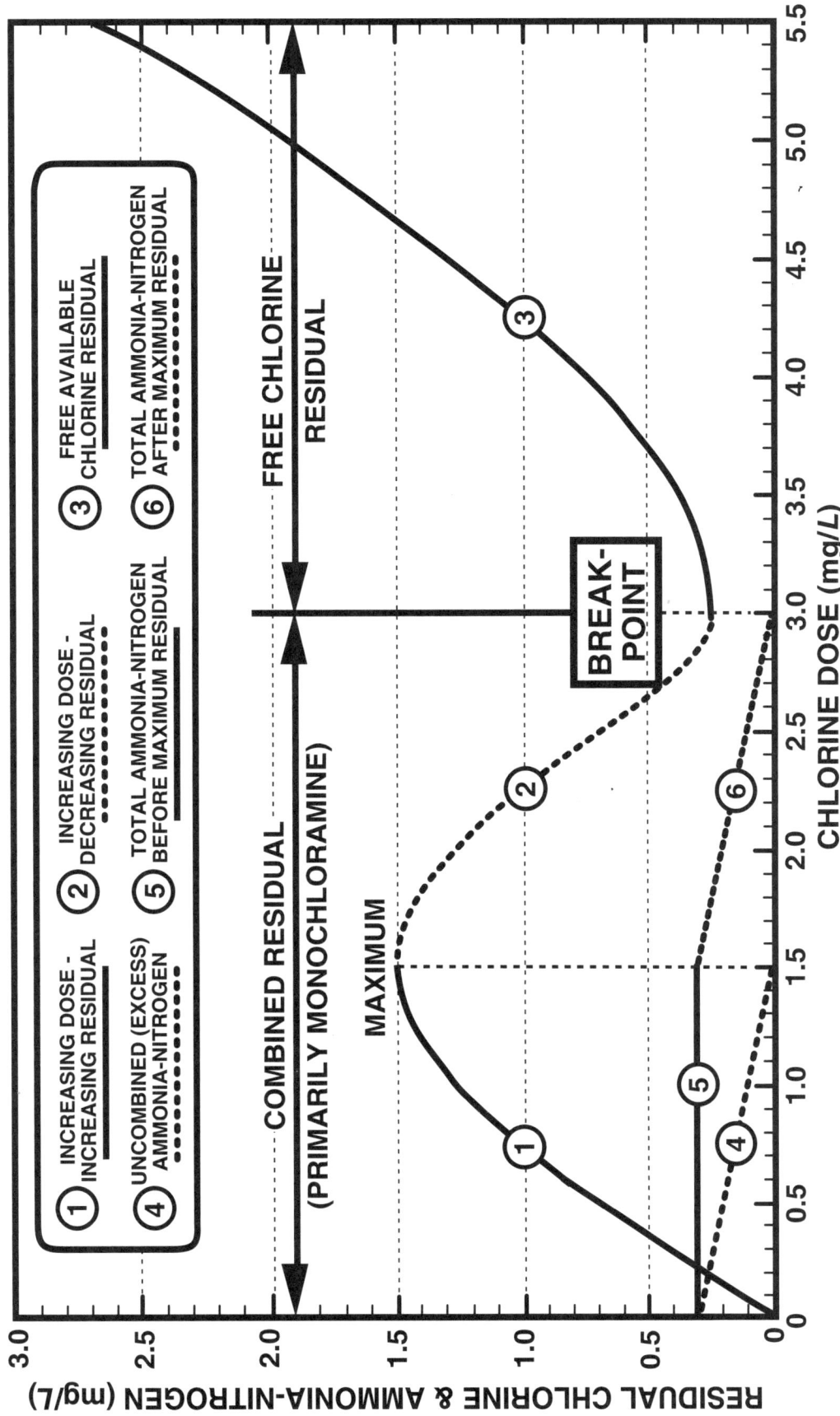

Fig. 7.3 Typical chloramination dose-residual curve

7.266   Nitrification

Nitrification is an important and effective microbial process in the oxidation of ammonia in both land and water environments. Two groups of organisms are involved in the nitrification process: ammonia-oxidizing bacteria (AOB) (see Figure 7.4) and nitrite-oxidizing bacteria. Nitrification has been well recognized as a beneficial treatment for the removal of ammonia in municipal wastewater.

When nitrification occurs in chloraminated drinking water, however, the process may lower the water quality unless the nitrification process reaches completion. Incomplete or partial nitrification causes the production of nitrite from the growth of AOB. This nitrite, in turn, rapidly reduces free chlorine and can interfere with the measurement of free chlorine. The end result may be a loss of total chlorine and ammonia and an increase in the concentration of heterotrophic plate count bacteria.

Factors influencing nitrification include the water temperature, the detention time in the reservoir or distribution system, excess ammonia in the water system, and the chloramine concentration used. The conditions most likely to lead to nitrification when using chloramines are a pH of 7.5 to 8.5, a water temperature of 25 to 30°C, a free ammonia concentration in the water, and a dark environment. The danger in allowing nitrification episodes to occur is that you may be left with very low or no total chlorine residual.

7.267   Nitrification Prevention and Control

When using chloramines for disinfection, an early warning system should be developed to detect the signs that nitrification is beginning to occur so that you can prevent or at least control the nitrification process. The best way to do this is to set up a regularly scheduled monitoring program. The warning signs to watch for include decreases in ammonia level, total chlorine level and pH, increases in nitrite level, and an increase in heterotrophic plate count bacteria. In addition, action response levels should be established for chloraminated distribution systems and reservoirs. Normal background levels of nitrite should be measured and then alert levels should be established so that increasing nitrite levels will not be overlooked.

An inexpensive way to help keep nitrite levels low is to reduce the detention times through the reservoirs and the distribution system, especially during warmer weather. Adding more chlorine to reservoir inlets and increasing the chlorine to ammonia-nitrogen ratio from 3:1 up to 5:1 at the treatment plant effluent will further control nitrification by decreasing the amount of uncombined ammonia in the distribution system. However, at a chlorine and ammonia-nitrogen ratio of 5:1, it is

Nitrification is a biological process caused by naturally occurring ammonia-oxidizing bacteria. These bacteria feed on free ammonia and convert it to nitrite and then nitrate. They thrive in covered reservoirs during warm summer months and are very resistant to chloramine disinfection. The by-products of their biological breakdown can support the growth of coliform bacteria.

THE NITRIFICATION PROCESS

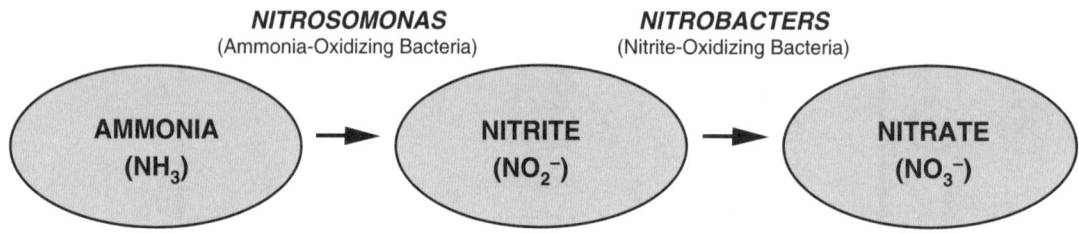

NITROSOMONAS
(Ammonia-Oxidizing Bacteria)

NITROBACTERS
(Nitrite-Oxidizing Bacteria)

AMMONIA (NH_3) → NITRITE (NO_2^-) → NITRATE (NO_3^-)

Factors Influencing
Water Temperature
Detention Time
Excess Ammonia
Chloramine Concentration

Early Warning Signs
Decrease in Ammonia
Decrease in Chlorine
Decrease in pH
Increase in Nitrite
Increase in Plate Count Bacteria

Optimum Conditions
pH   7.5 - 8.5
Temp.   25 - 30 C
Free Ammonia
Dark Environment

Prevention and Control
Decrease Detention Times
Decrease Free Ammonia
Increase Disinfection Dosage Ratio
Breakpoint Chlorinate
Establish a Flushing Program

Fig. 7.4   The nitrification process

critical that the chlorine and ammonia feed systems operate accurately and reliably because an overdose of chlorine can reduce the chloramine residual.

Other strategies for controlling nitrification include establishing a flushing program and increasing the chloramine residual. A uniform flushing program should be a key component of any nitrification control program. Flushing reduces the detention time in low-flow areas, increases the water velocity within pipelines to remove sediments and biofilm that would harbor nitrifying bacteria, and draws higher disinfectant residuals into problem areas. Increasing the chloramine residual in the distribution system to greater than 2.0 mg/*L* is also effective in preventing the onset of nitrification.

QUESTIONS

Write your answers in a notebook and then compare your answers with those on page 356.

7.2J   What is breakpoint chlorination?

7.2K   An operator's decision to use chloramines depends on what factors?

7.2L   What are the three primary methods by which chloramines are produced?

7.2M   Why is the *applied* chlorine to ammonia-nitrogen ratio usually greater than the *actual* chlorine to nitrogen ratio leaving the plant?

7.2N   Incomplete nitrification causes the production of nitrite which produces what problems in disinfection of water?

7.27   Chlorine Residual Testing

7.270   *Importance*

Many small system operators attempt to maintain a chlorine residual throughout the distribution system. Chlorine is very effective in biological control and especially in elimination of coliform bacteria that might reach water in the distribution system through cross connections or leakage into the system. A chlorine residual also helps to control any microorganisms that could produce slimes, tastes, or odors in the water in the distribution system.

Adequate control of coliform "aftergrowth" is usually obtained only when chlorine residuals are carried to the farthest points of the distribution system. To ensure that this is taking place, make daily chlorine residual tests. A chlorine residual of about 0.2 mg/*L* measured at the extreme ends of the distribution system is usually a good indication that a free chlorine residual is present in all other parts of the system. This small residual can destroy a small amount of contamination, so a lack of chlorine residual could indicate the presence of heavy contamination. If routine checks at a given point show measurable residuals, any sudden absence of a residual at that point should alert the operator to the possibility that a potential problem has arisen which needs prompt investigation. Immediate action that can be taken includes retesting for chlorine residual, then checking chlorination equipment, and finally searching for a source of contamination which could cause an increase in the chlorine demand.

7.271   *Chlorine Residual Curve*

The chlorine residual curve procedure is a quick and easy way for an operator to estimate the proper chlorine dose, es-

pecially when surface water conditions are changing rapidly such as during a storm.

Fill a CLEAN five-gallon bucket from a sample tap located at least two 90-degree elbows (or where chlorine is completely mixed with the water in the pipe) AFTER the chlorine has been injected into the pipe. Immediately measure the chlorine residual and record this value on the "time zero" line of your record sheet (Figure 7.5). This is the initial chlorine residual. At 15-minute intervals, vigorously stir the bucket using an up and down motion. (A large plastic spoon works well for this purpose.) Collect a sample from one or two inches below the water surface and measure the chlorine residual. Record this chlorine residual value on the record sheet. For at least one hour, collect a sample every 15 minutes, measure the chlorine residual, and record the results to indicate the "chlorine demand" of the treated water. Plot these recorded values on a chart or graph paper as shown on Figure 7.5. Connect the plotted points to create a chlorine residual curve. If the chlorine residual after one hour is not correct (about 0.2 mg/*L*), increase or decrease the initial chlorine dose so the final chlorine residual will be approximately at the desired ultimate chlorine residual in the water distribution system. Repeat this procedure until the desired TARGET initial chlorine residual will achieve the desired chlorine residual throughout the distribution system.

Precautions that must be taken when performing this test include being sure the five-gallon plastic test bucket is clean and only used for this purpose. A new bucket does not need to be used for every test, but the bucket should be new when the first test is performed. The stirrer should also be clean. DO NOT USE THE STIRRER FOR THE CHLORINE SOLUTION MIXING AND HOLDING TANK. During the test the bucket should be kept cool so that the chlorine gas does not escape from the water and give false chlorine residual values.

The chlorine demand for groundwater changes slowly, or not at all; therefore, the "initial or target" chlorine residual does not have to be checked more frequently than once a month. Always be sure to measure the chlorine residual in the distribution system on a daily basis. This is also a good check that the chlorination equipment is working properly and that the chlorine stock solution is the correct concentration.

The chlorine demand for surface water can change continuously, especially during storms and the snow melt season. Experience has proven that the required "initial or target" chlorine residual at time zero is directly tied to the turbidity of the finished (treated) water. The higher the finished water turbidity,

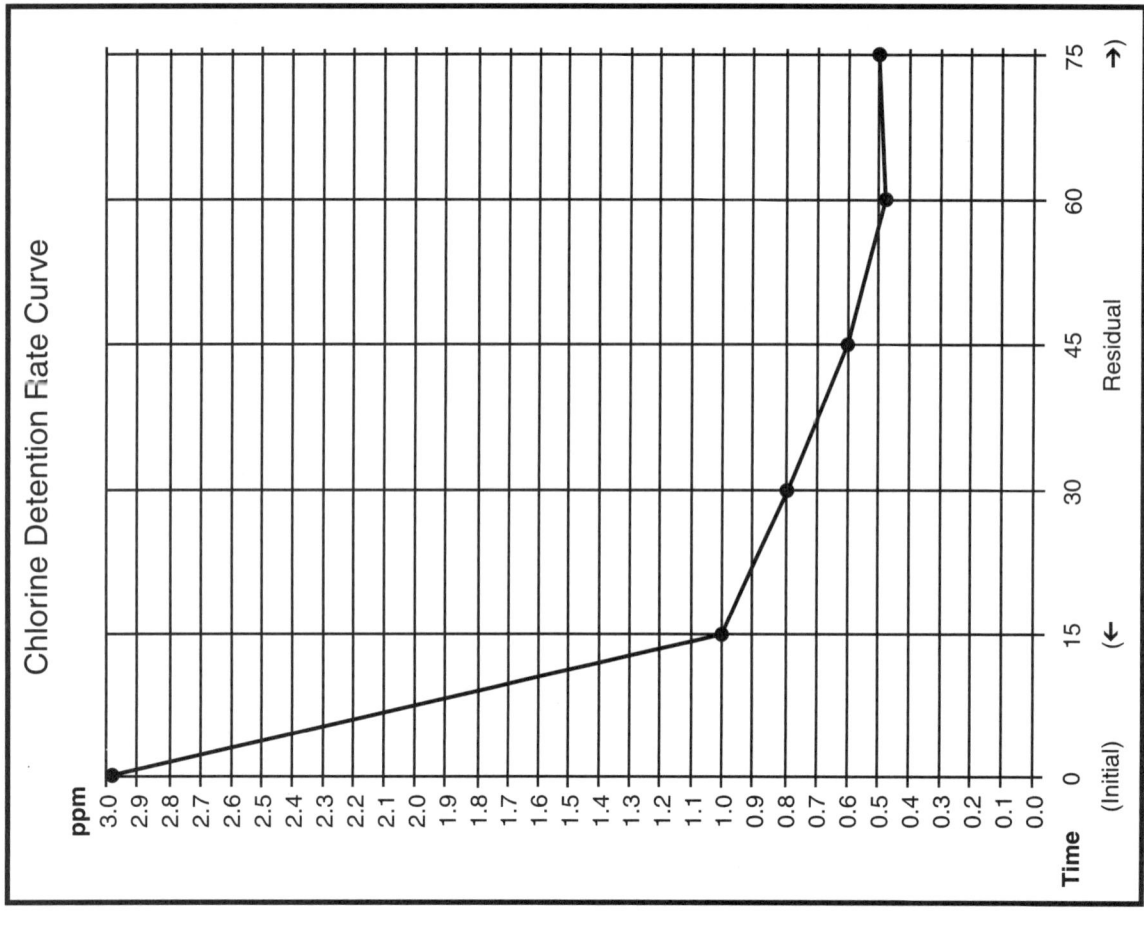

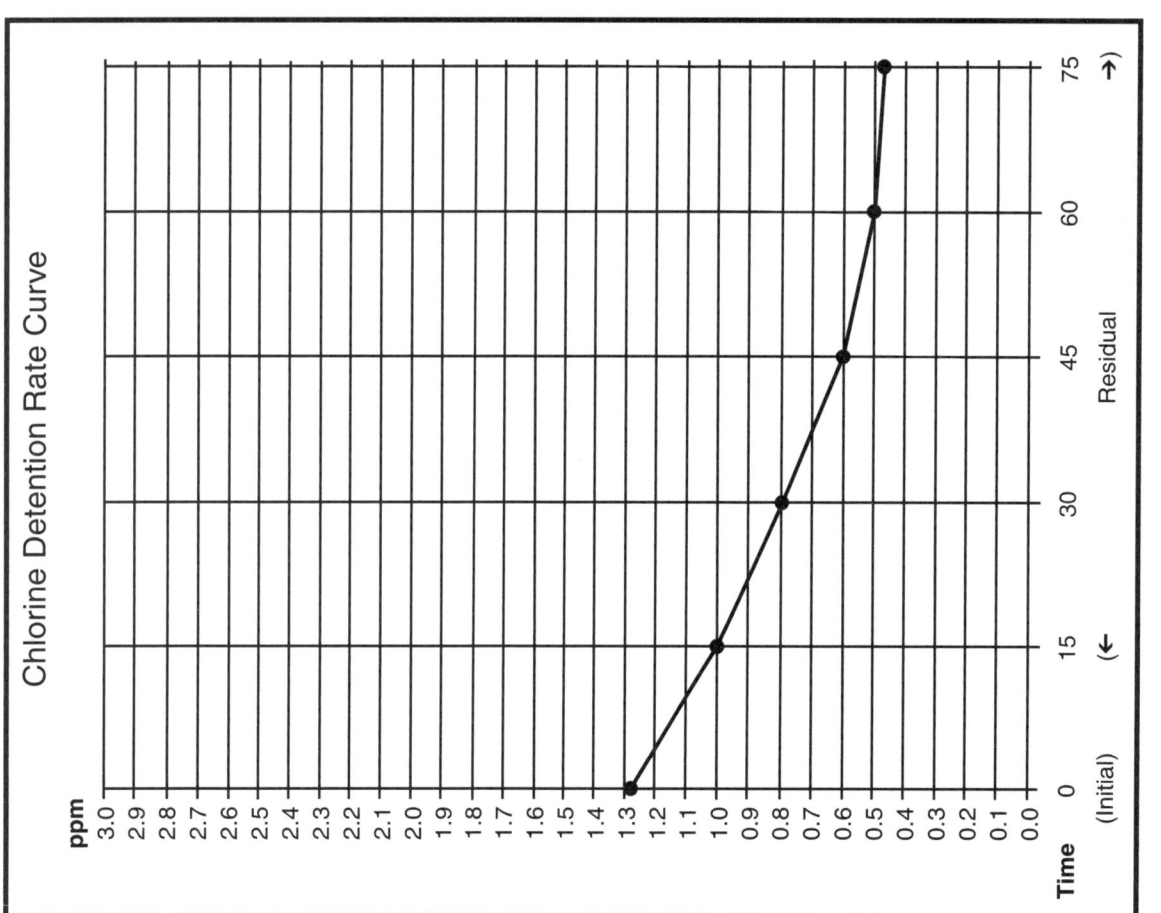

Fig. 7.5   Chlorine detention rate curves

the higher the "initial" chlorine residual value will have to be to ensure the desired chlorine residual in the distribution system. Careful documentation of this information in your records will greatly reduce the lag time in chlorine addition changes to maintain the desired residual in the distribution system and the delivery of safe drinking water to your consumers. Experience and a review of your records will indicate that for a given turbidity value, you can estimate the desired "initial" chlorine residual, which will require a given chlorinator output level for a given water flow rate.

Acknowledgment

The information in Sections 7.270 and 7.271 was developed by Bill Stokes. His suggestions and procedures are greatly appreciated.

7.272 Critical Factors

Both *CHLORINE RESIDUAL* and *CONTACT TIME* are essential for effective killing or inactivation of pathogenic microorganisms. Complete initial mixing is very important. Changes in pH affect the disinfection ability of chlorine and you must reexamine the best combination of contact time and chlorine residual when the pH fluctuates. Critical factors influencing disinfection are summarized as follows:

1. Effectiveness of upstream treatment processes. The lower the turbidity (suspended solids, organic content, reducing agents) of the water, the better the disinfection.

2. Injection point and method of mixing to get disinfectant in contact with water being disinfected. Depends on whether using prechlorination or postchlorination.

3. Temperature. The higher the temperature, the more rapid the rate of disinfection.

4. Dosage and type of chemical. Usually the higher the dosage, the faster the disinfection rate. The form (chloramines or free chlorine residual) and type of chemical also influence the disinfection rate.

5. pH. The lower the pH, the better the disinfection.

6. Contact time. With good initial mixing, the longer the contact time, the better the disinfection.

7. Chlorine residual.

7.28 CT Values

The purpose of the Surface Water Treatment Rule (SWTR) is to ensure that pathogenic organisms are removed and/or inactivated by the treatment process. To meet this goal, all systems are required to disinfect their water supplies. For some water systems using very clean source water and meeting the other criteria to avoid filtration, disinfection alone can achieve the 3-log (99.9%) *Giardia* and 4-log (99.99%) virus inactivation levels required by the SWTR. For extremely clean source waters there may be virtually no *Giardia* or viruses and achieving 3-log or 4-log inactivation levels will be impossible and not necessary.

Several methods of disinfection are in common use, including free chlorination, chloramination, use of chlorine dioxide, and application of ozone. The concentration of chemical needed and the length of contact time needed to ensure disinfection are different for each disinfectant. Therefore, the efficiency of the disinfectant is measured by the time "T" in minutes of the disinfectant's contact in the water and the concentration "C" of the disinfectant residual in mg/L measured at the end of the contact time. The product of these two factors (CxT) provides a measure of the degree of pathogenic

inactivation. The required CT value to achieve inactivation is dependent upon the organism in question, type of disinfectant, pH, and temperature of the water supply.

Time or "T" is measured from point of application to the point where "C" is determined. "T" must be based on peak hour flow rate conditions. In pipelines, "T" is calculated by dividing the volume of the pipeline in gallons by the flow rate in gallons per minute (GPM). In reservoirs and basins, dye tracer tests must be used to determine "T." In this case "T" is the time it takes for 10 percent of the tracer to pass the measuring point.

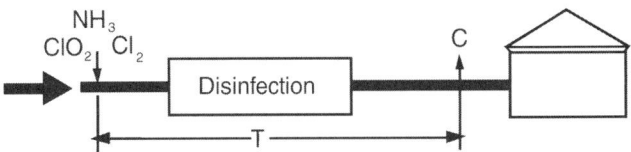

A properly operated filtration system can achieve limited removal or inactivation of microorganisms. Because of this, systems that are required to filter their water are permitted to apply a factor that represents the microorganism removal value of filtration when calculating CT values to meet the disinfection requirements. The factor (removal credit) varies with the type of filtration system. Its purpose is to take into account the combined effects of both disinfection and filtration in meeting the SWTR microbial standards.

Please refer to the Arithmetic Appendix at the end of this manual (Section A.16, "Calculation of CT Values") for instructions on how to perform these calculations for a water treatment plant.

For more detailed information about the requirements and application of the Surface Water Treatment Rule, you may wish to order a copy of the publication, *GUIDANCE MANUAL FOR COMPLIANCE WITH THE FILTRATION AND DISINFECTION REQUIREMENTS FOR PUBLIC WATER SYSTEMS USING SURFACE WATER SOURCES*. It is available from American Water Works Association (AWWA), Bookstore, 6666 West Quincy Avenue, Denver, CO 80235. Order No. 20271. ISBN 0-89867-558-8. Price to members, $56.50; nonmembers, $83.50; price includes cost of shipping and handling.

7.29 Process Calculations

FORMULAS

There are two basic chlorination process calculations:

1. Chlorine dosage, mg/L, and

2. Chlorine demand, mg/L.

To calculate the chlorine dosage (dose) of the water being treated, we need to know the chlorine fed to the water being treated in pounds of chlorine per day and the amount of water treated in million gallons per day. With this information we can calculate the chlorine dosage in milligrams of chlorine per liter of water. Let's recall our basic chemical feed formula:

$$\text{Chemical Feed, lbs/day} = (\text{Flow, MGD})(\text{Dose, mg}/L)(8.34 \text{ lbs/gal})$$

This formula can be rearranged using Davidson's Pie Method (see the Arithmetic Appendix, Section A.12, "How to Use the Basic Formulas").

$$\text{Chlorine Dose, mg}/L = \frac{\text{Chemical Feed, lbs/day}}{(\text{Flow, MGD})(8.34 \text{ lbs/gal})}$$

When we determine the chlorine dose in milligrams of chlorine per liter for the water we are treating, we must add enough chlorine to the water to meet the chlorine demand and the desired chlorine residual.

$$\text{Chlorine Dose, mg}/L = \text{Chlorine Demand, mg}/L + \text{Chlorine Residual, mg}/L$$

If we wish to know the chlorine demand, then we rearrange the terms by subtracting the chlorine residual from the chlorine dose.

$$\text{Chlorine Demand, mg}/L = \text{Chlorine Dose, mg}/L - \text{Chlorine Residual, mg}/L$$

EXAMPLE 1

A chlorinator is set to feed 20 pounds of chlorine in 24 hours to a flow of 0.85 MGD. Find the chlorine dose in mg/L.

Known	Unknown
Chlorinator Setting, lbs/24 hr = 20 lbs Cl/24 hr	Chlorine Dose, mg/L
Flow, MGD = 0.85 MGD	

Calculate the chlorine dose in mg/L.

$$\begin{aligned}\text{Chlorine Dose, mg}/L &= \frac{\text{Chlorine Feed, lbs/day}}{(\text{Water Treated, Million gal/day})(8.34 \text{ lbs/gal})}\\[2mm] &= \frac{20 \text{ lbs Cl/day}}{(0.85 \text{ Million gal/day})(8.34 \text{ lbs/gal})}\\[2mm] &= \frac{20 \text{ lbs Cl/day}}{7.1 \text{ Million lbs water/day}}\\[2mm] &= 2.8 \text{ lbs chlorine/million lbs water}\\[2mm] &= 2.8 \text{ ppm (\textbf{P}arts \textbf{P}er \textbf{M}illion parts)}\\[2mm] &= 2.8 \text{ mg}/L\end{aligned}$$

EXAMPLE 2

Find the chlorine demand in mg/L for the water being treated in Example 1 with a chlorine dose of 2.8 mg/L. The chlorine residual after 30 minutes of contact time is 0.5 mg/L.

Known	Unknown
Chlorine Dose, mg/L = 2.8 mg/L	Chlorine Demand, mg/L
Chlorine Residual, mg/L = 0.5 mg/L	

Find the chlorine demand in mg/L.

$$\begin{aligned}\text{Chlorine Demand, mg}/L &= \text{Chlorine Dose, mg}/L - \text{Chlorine Residual, mg}/L\\[2mm] &= 2.8 \text{ mg}/L - 0.5 \text{ mg}/L\\[2mm] &= 2.3 \text{ mg}/L\end{aligned}$$

QUESTIONS

Write your answers in a notebook and then compare your answers with those on page 356.

7.2O   What actions should an operator take when there is a sudden absence of chlorine residual in the distribution system?

7.2P   How does the length of chlorine contact time affect the disinfection process?

7.2Q   How is the efficiency of a disinfectant measured?

7.3   POINTS OF CHLORINE APPLICATION (Figure 7.6)

7.30   Prechlorination

The two most common points (locations) of chlorination in a water treatment plant are prechlorination and postchlorination. Prechlorination is the application of chlorine ahead of any other treatment processes. While prechlorination may increase the formation of trihalomethanes in raw water containing organic *PRECURSOR*[20] (THM) compounds and tastes and odors when phenolic compounds are present, it provides the following benefits:

1. Control of algae and slime growths,

2. Control of mudball formation,

3. Improved coagulation,

4. Reduction of tastes and odors,

5. Increased chlorine contact time, and

6. Increased safety factor in disinfection of heavily contaminated waters.

7.31   Postchlorination

Postchlorination is the application of chlorine after the water has been treated but before it enters the distribution system. This is the primary point of disinfection and it is normally the last application of any disinfectant.

7.32   Rechlorination

Rechlorination is the practice of adding chlorine in the distribution system. This practice is common when the distribution system is long or complex. The application point could be any place where adequate mixing is available.

---

[20] *Precursor, THM (pre-CURSE-or).   Natural organic compounds found in all surface and groundwaters. These compounds MAY react with halogens (such as chlorine) to form trihalomethanes (tri-HAL-o-METH-hanes) (THMs); they MUST be present in order for THMs to form.*

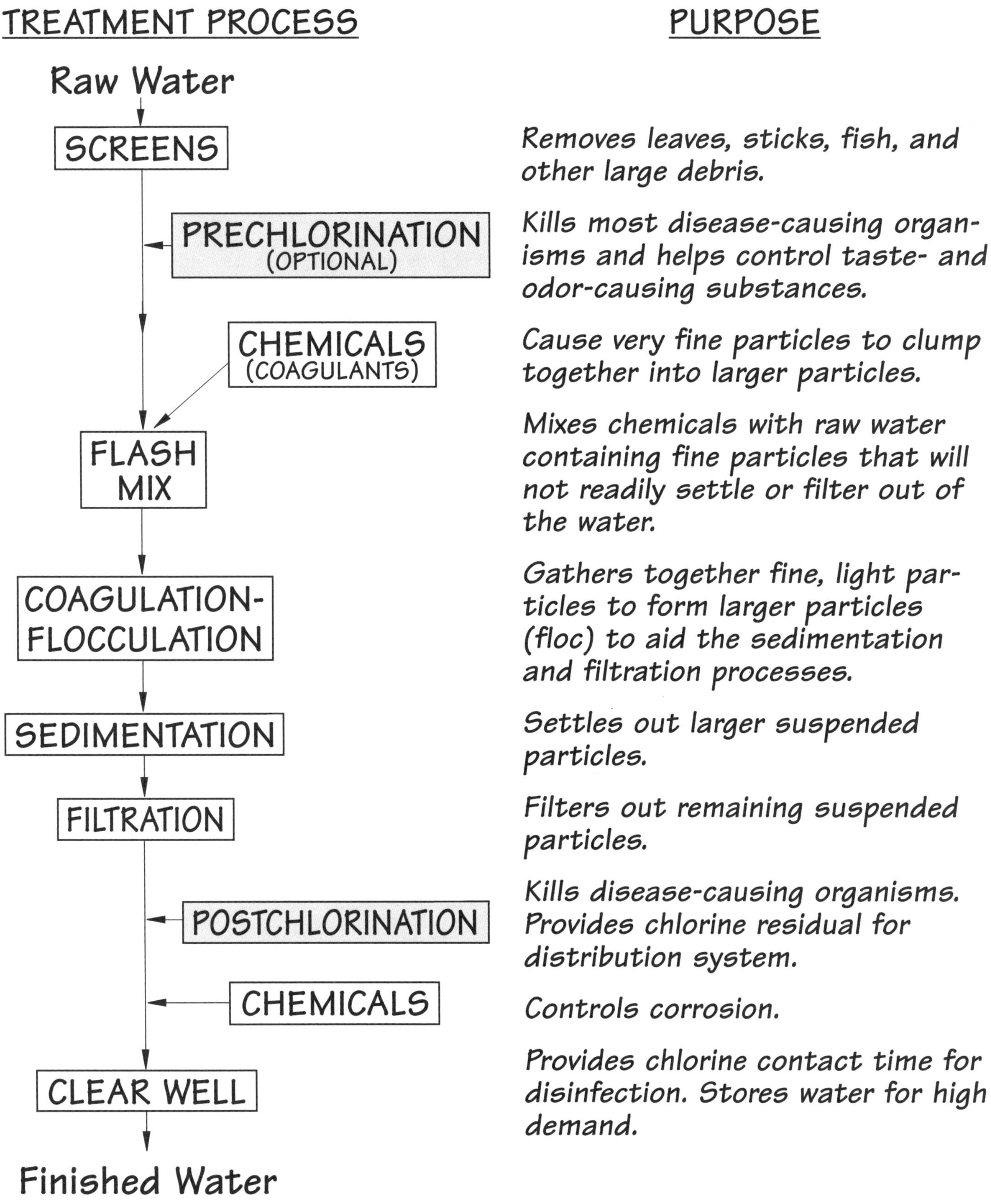

TREATMENT PROCESS

Raw Water

↓

SCREENS

↓

PRECHLORINATION
(OPTIONAL)

↓

CHEMICALS
(COAGULANTS)

↓

FLASH
MIX

↓

COAGULATION-
FLOCCULATION

↓

SEDIMENTATION

↓

FILTRATION

↓

POSTCHLORINATION

↓

CHEMICALS

↓

CLEAR WELL

↓

Finished Water

PURPOSE

Removes leaves, sticks, fish, and other large debris.

Kills most disease-causing organisms and helps control taste- and odor-causing substances.

Cause very fine particles to clump together into larger particles.

Mixes chemicals with raw water containing fine particles that will not readily settle or filter out of the water.

Gathers together fine, light particles to form larger particles (floc) to aid the sedimentation and filtration processes.

Settles out larger suspended particles.

Filters out remaining suspended particles.

Kills disease-causing organisms. Provides chlorine residual for distribution system.

Controls corrosion.

Provides chlorine contact time for disinfection. Stores water for high demand.

Fig. 7.6   Typical flow diagram for a water treatment plant

7.33   Wells

Chlorination of wells is required in some areas and is a good practice whenever wells are used for public water supplies. This is usually accomplished with a small system and can be automated for ease of operation.

7.34   Mains

Mains are usually not a problem and are not chlorinated except in long pipelines and in complex systems. Mains must be chlorinated after initial installation and any repairs.

7.35   Tanks and Reservoirs

Usually tanks and reservoirs are not chlorinated unless specific problems develop which cannot be solved by other means. Tanks must be chlorinated after initial installation, repairs, maintenance, repainting, and cleaning. In other words, any time a tank has been drained or entered, it should be chlorinated.

7.36   Water Supply Systems

See *SMALL WATER SYSTEM OPERATION AND MAINTENANCE* (another manual in this series) for detailed procedures on how to disinfect wells and *WATER DISTRIBUTION SYSTEM OPERATION AND MAINTENANCE* for procedures on how to disinfect mains, tanks, and reservoirs.

7.37   Additional Reading

1. *THE CHLORINATION/CHLORAMINATION HANDBOOK* by Gerald F. Connell. Obtain from American Water Works

Association (AWWA), Bookstore, 6666 West Quincy Avenue, Denver, CO 80235. Order No. 20415. ISBN 0-89867-886-2. Price to members, $68.50; nonmembers, $98.50; price includes cost of shipping and handling.

2. *HANDBOOK OF CHLORINATION AND ALTERNATIVE DISINFECTANTS*, Fourth Edition, by George C. White. Obtain from John Wiley & Sons, Inc., Customer Care Center (Consumer Accounts), 10475 Crosspoint Boulevard, Indianapolis, IN 46256. ISBN 0-471-29207-9. Price, $220.00, plus $5.00 shipping and handling.

QUESTIONS

Write your answers in a notebook and then compare your answers with those on pages 356 and 357.

7.3A   List the two most common points (locations) of chlorination in a water treatment plant.

7.3B   Under what conditions should waters not be prechlorinated?

7.3C   What are the benefits of prechlorination?

End of Lesson 1 of 3 Lessons on DISINFECTION

Please answer the discussion and review questions next.

DISCUSSION AND REVIEW QUESTIONS

Chapter 7.   DISINFECTION

(Lesson 1 of 3 Lessons)

At the end of each lesson in this chapter you will find some discussion and review questions. The purpose of these questions is to indicate to you how well you understand the material in the lesson. Write the answers to these questions in your notebook before continuing.

1. What is the difference between disinfection and sterilization?

2. How does organic matter in water influence disinfection?

3. How do physical and chemical methods accomplish disinfection?

4. Why is chlorine dioxide considered an effective disinfectant?

5. What are the benefits and limitations of prechlorination?

CHAPTER 7.   DISINFECTION

(Lesson 2 of 3 Lessons)

7.4   OPERATION OF CHLORINATION EQUIPMENT

7.40   Hypochlorinators

A hypochlorinator is a piece of equipment used to feed liquid chlorine (bleach) solutions. Hypochlorinators used on small water systems are very simple and relatively easy to install. Typical installations are shown in Figures 7.7 and 7.8. Hypochlorinator systems usually consist of a chemical solution tank for the hypochlorite, diaphragm-type pump (Figure 7.9), power supply, water pump, pressure switch, and water storage tank.

There are two methods of feeding the hypochlorite solution into the water being disinfected. The hypochlorite solution may be pumped directly into the water (Figure 7.10). In the other method, the hypochlorite solution is pumped through an *EJECTOR*[21] (also called an eductor or injector) which draws in additional water for dilution of the hypochlorite solution (Figure 7.11).

TYPICAL INSTALLATION

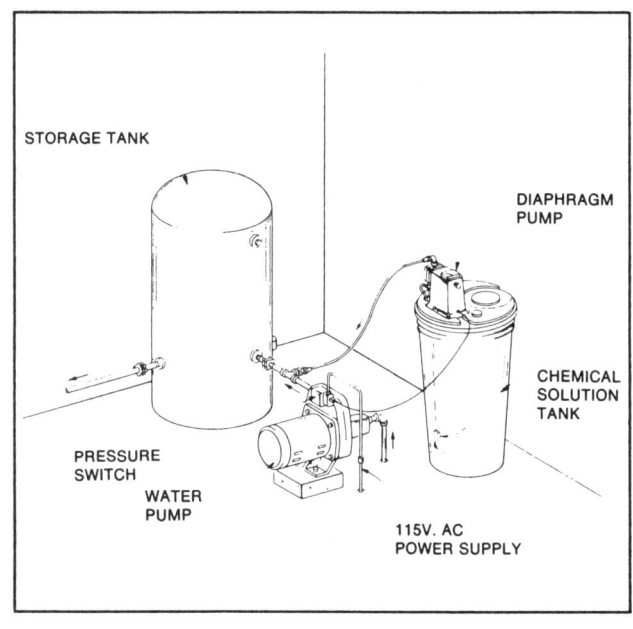

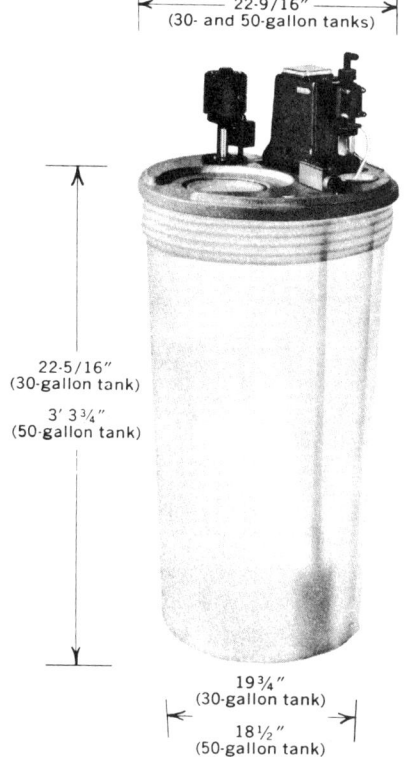

Pump-tank system for chemical mixing and metering. Cover supports pump, impeller-type mixer, and liquid-level switch.

Fig. 7.7   Typical hypochlorinator installation

(Permission of Wallace & Tiernan Division, Pennwalt Corporation)

---

[21] *Ejector.   A device used to disperse a chemical solution into water being treated.*

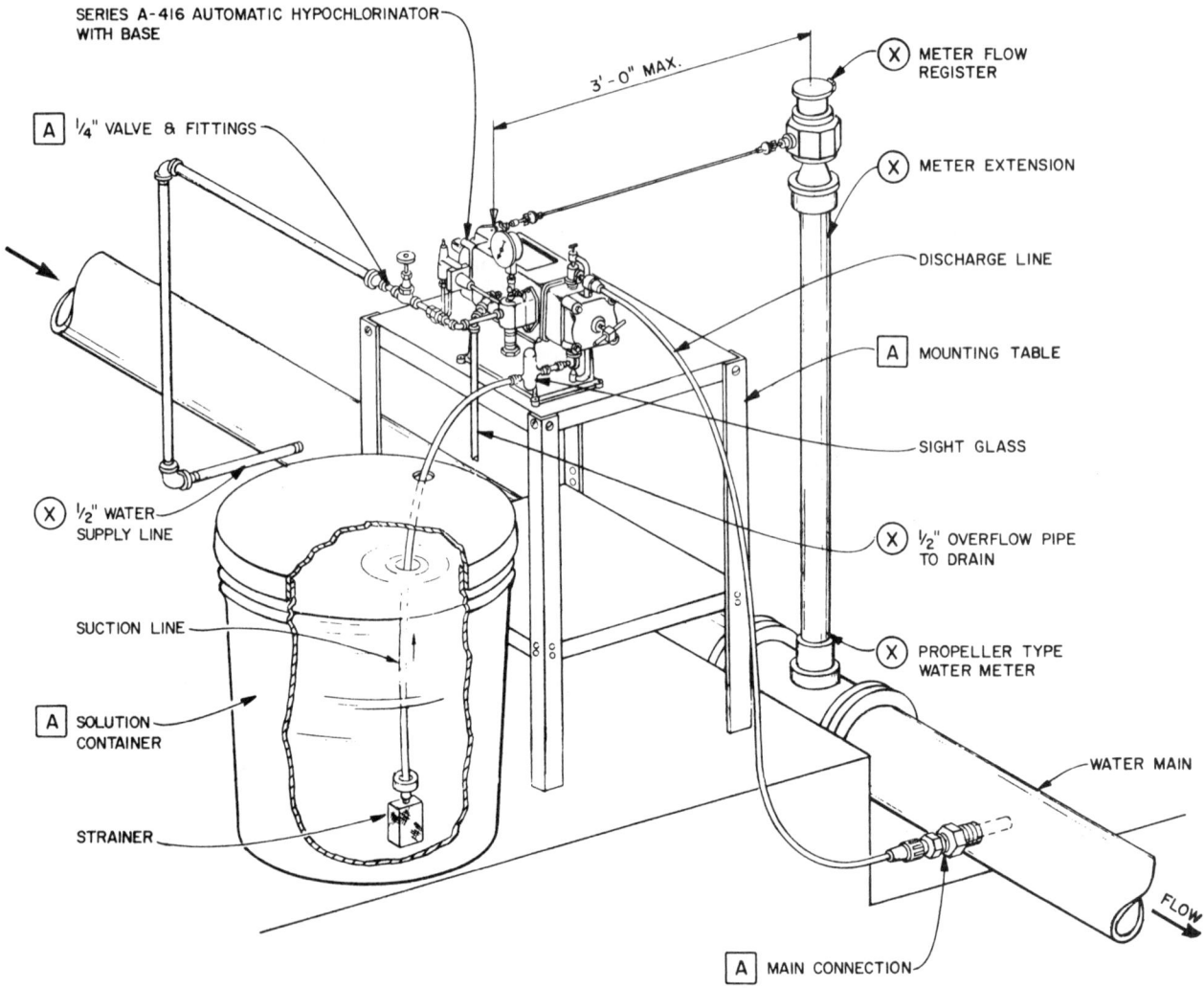

SERIES A-416 AUTOMATIC HYPOCHLORINATOR
WITH BASE

3'-0" MAX.

Ⓧ METER FLOW
REGISTER

Ⓐ ¼" VALVE & FITTINGS

Ⓧ METER EXTENSION

DISCHARGE LINE

Ⓐ MOUNTING TABLE

SIGHT GLASS

Ⓧ ½" WATER
SUPPLY LINE

Ⓧ ½" OVERFLOW PIPE
TO DRAIN

SUCTION LINE

Ⓧ PROPELLER TYPE
WATER METER

Ⓐ SOLUTION
CONTAINER

WATER MAIN

STRAINER

FLOW

Ⓐ MAIN CONNECTION

Ⓧ *NOT FURNISHED BY W & T.*

Ⓐ *ACCESSORY ITEM FURNISHED ONLY IF
SPECIFICALLY LISTED IN QUOTATION
AND AS CHECKED ON THIS DRAWING.*

NOTE: Hypochlorinator paced by a propeller-type water meter.

Fig. 7.8   *Typical hypochlorinator installation*
(Permission of Wallace & Tiernan Division, Pennwalt Corporation)

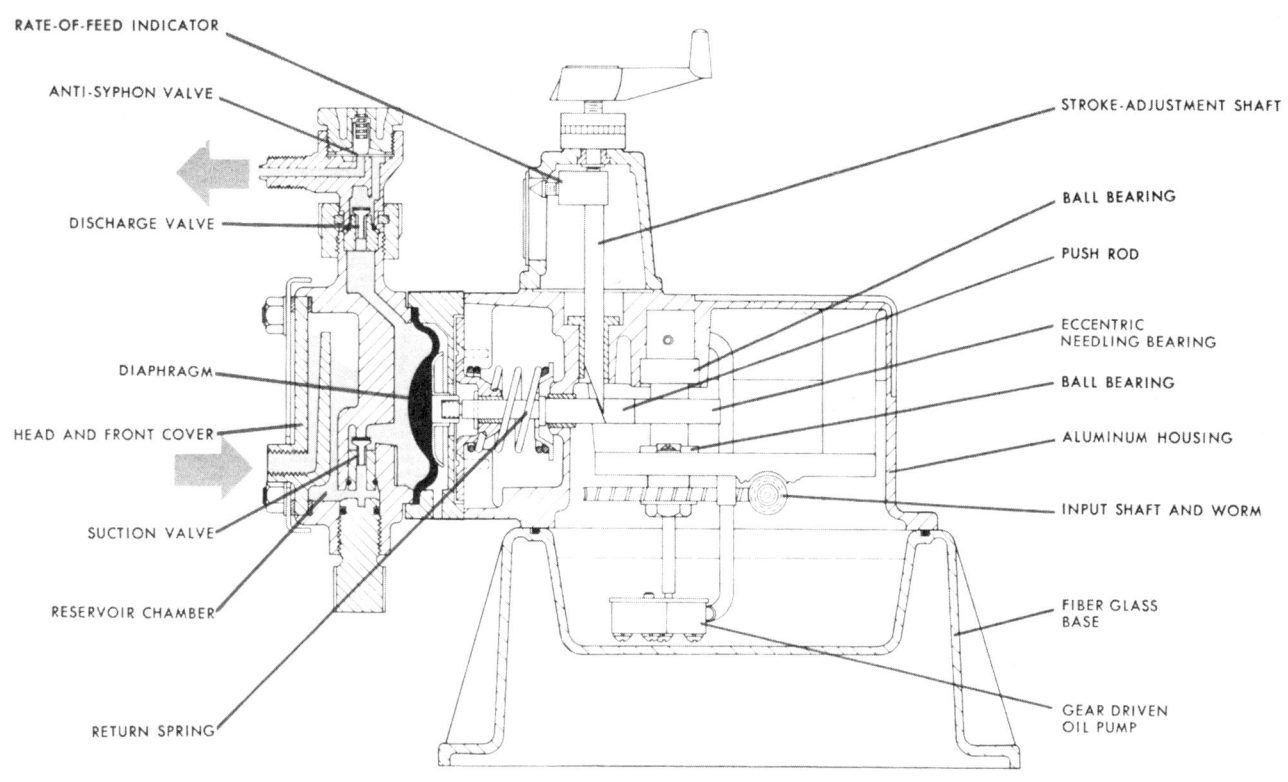

RATE-OF-FEED INDICATOR

ANTI-SYPHON VALVE

DISCHARGE VALVE

DIAPHRAGM

HEAD AND FRONT COVER

SUCTION VALVE

RESERVOIR CHAMBER

RETURN SPRING

STROKE-ADJUSTMENT SHAFT

BALL BEARING

PUSH ROD

ECCENTRIC
NEEDLING BEARING

BALL BEARING

ALUMINUM HOUSING

INPUT SHAFT AND WORM

FIBER GLASS
BASE

GEAR DRIVEN
OIL PUMP

Belt guard
removed
to show
step pulley

Fig. 7.9   Diaphragm-type pump
(Permission of Wallace & Tiernan Division, Pennwalt Corporation)

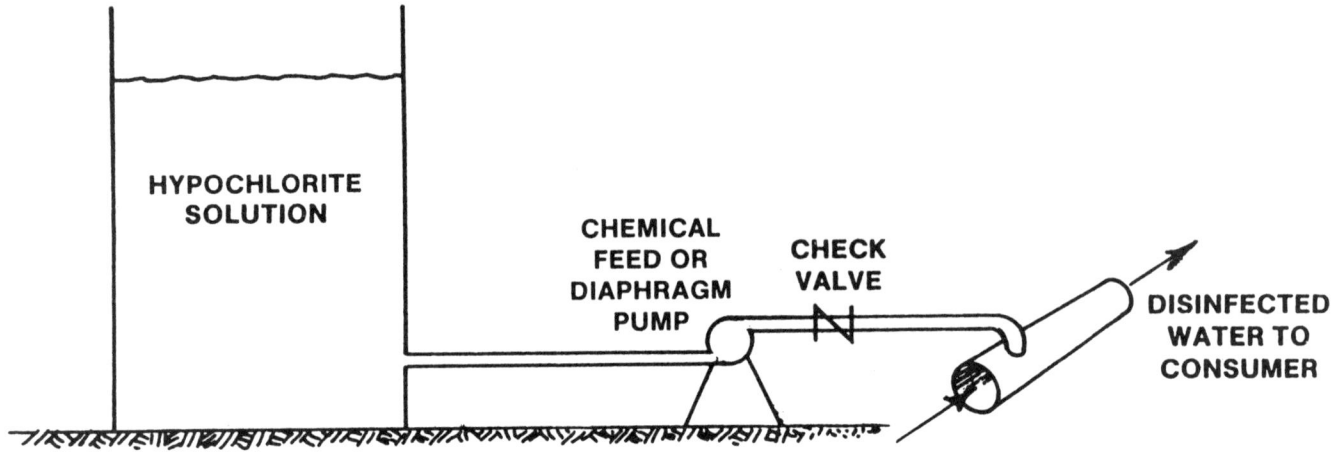

Fig. 7.10   Hypochlorinator direct pumping system

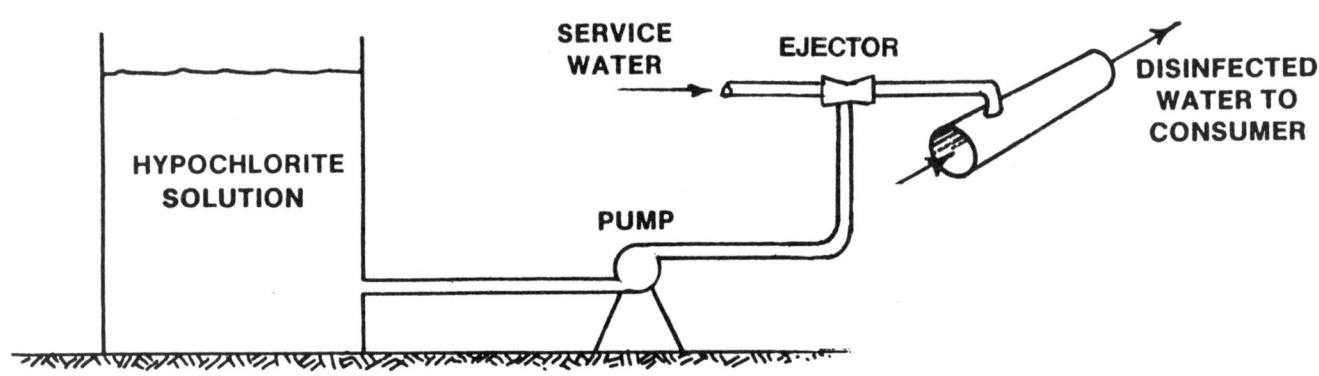

NOTE: Pump is chemical feed or diaphragm pump

Fig. 7.11   Hypochlorinator injector feed system

QUESTIONS

Write your answers in a notebook and then compare your answers with those on page 357.

7.4A List the major parts of a typical hypochlorinator system.

7.4B What are the two common methods of feeding hypochlorite to the water being disinfected?

7.41 Chlorinators

A typical installation for gaseous chlorine is shown in Figure 7.12. The figure shows the exhaust fan installed at floor level. A potential problem for this type of installation is that any chlorine drawn through the fan could corrode the wiring in the fan or the controls and cause a failure of the ventilation system. A better system is one which draws air from the roof and pushes air and any chlorine out the floor vents. (See Section 7.53, "Installation," item 8, for more details about chlorine room ventilation systems and requirements.)

Chlorine gas may be removed from chlorine containers by a valve and piping arrangement to the chlorinators, as shown in Figure 7.12, and delivered by vacuum-controlled solution feed chlorinators (Figures 7.12, 7.13, and 7.14). The chlorine gas is controlled, metered, introduced into a stream of *INJECTOR WATER*,[22] and then is conducted as a solution to the point of application.

The components of a typical vacuum-controlled chlorinator are shown in Figures 7.13 and 7.14. The purpose of each part is summarized in Table 7.4.

In many smaller systems, chlorine gas is withdrawn with equipment installed directly on the cylinder (Figures 7.15, 7.16, and 7.17). A direct-mounted chlorinator meters prescribed (preset or selected) doses of chlorine gas from a chlorine cylinder, conveys it under a vacuum, and injects it into the water supply. Direct cylinder mounting is the safest and simplest way to connect the chlorinator to the chlorine cylinder. The valves on the cylinder and chlorinator inlet are connected by a positive metallic yoke, which is sealed by a single lead or fiber gasket.

CHLORINATOR FLOW PATH

Chlorine gas flows from a chlorine container to the gas inlet (see Figure 7.14). After entering the chlorinator, the gas passes through a spring-loaded pressure regulating valve which maintains the proper operating pressure. A *ROTAMETER*[23] is used to indicate the rate of gas flow. The rate is controlled by an orifice. The gas then moves to the injector where it is dissolved in water. This mixture leaves the chlorinator as a chlorine solution (HOCl) ready for application.

The operating vacuum is provided by a hydraulic injector. The water supplied by this injector absorbs the chlorine gas. The resulting chlorine solution is conveyed to a chlorine diffuser through a corrosion-resistant conduit. A vacuum regulating valve dampens fluctuations and gives smoother operation. A vacuum relief prevents excessive vacuum within the equipment.

The primary advantage of vacuum operation is safety. If a failure or breakage occurs in the vacuum system, the chlorinator either stops the flow of chlorine into the equipment or allows air to enter the vacuum system rather than allowing chlorine to escape into the surrounding atmosphere. In case the chlorine inlet shutoff fails, a vent valve discharges the incoming gas to a chlorine treatment facility outside of the chlorinator building.

CHLORINATOR PARTS AND THEIR PURPOSE

THE EJECTOR: The ejector, fitted with a Venturi nozzle, creates the vacuum that moves the chlorine gas. Water supplied by a pump moves across the Venturi nozzle creating a differential pressure which establishes the vacuum. The gas chlorinator is able to transport the chlorine gas to the water supply by reducing the gas pressure from the chlorine cylinder to less than the atmospheric pressure (vacuum). Figure 7.12 illustrates such an arrangement. The flow diagram in Figure 7.16 is a cutaway view of the ejector and check valve assembly.

In the past it was not uncommon to find the ejector and the vacuum regulator mounted inside some type of cabinet. However, it makes better sense to locate the ejector at the site where the chlorine is to be applied, eliminating the necessity of pumping the chlorine over long distances and the associated problems inherent with gas pressure lines. Also, by placing the ejector at the application point, any tubing break will cause the chlorinator to shut down. This halting of operation stops the flow of gas and any damage that could result from a chlorine solution leak.

CHECK VALVE ASSEMBLY: The vacuum created by the ejector moves through the check valve assembly. This assembly prevents water from back-feeding, that is, entering the vacuum-regulator portion of the chlorinator (Figure 7.16).

RATE VALVE: The rate valve controls the flow rate at which chlorine gas enters the chlorinator. The rate valve controls the vacuum level and thus directly affects the action of the diaphragm assembly in the vacuum regulator. A reduction in vacuum lets the diaphragm close, causing the needle valve to reduce the inlet opening which restricts chlorine gas flow to the chlorinator. An increase in the rate valve setting applies more vacuum to the diaphragm assembly, pulling the needle valve back away from the inlet opening and permitting an increased chlorine gas flow rate.

DIAPHRAGM ASSEMBLY: This assembly connects directly to the inlet valve of the vacuum regulator, as described above. A vacuum (of at least 20 inches (508 mm) of water column) exists on one side of the diaphragm; the other side is open to atmospheric pressure through the vent. This differential in pressure causes the diaphragm to open the chlorine inlet valve allowing the gas to move (under vacuum) through the rotameter, past the rate valve and through the tubing to the check valve assembly, into the ejector nozzle area, and then to the point of application. If for some reason the vacuum is lost, the diaphragm will seat the needle valve on the inlet, stopping chlorine gas flow to the chlorinator.

INTERCONNECTION MANIFOLD: If several gas cylinders provide the chlorine gas, direct cylinder mounting is not possible. An interconnection manifold made of seamless steel pipe and flexible connectors of cadmium-plated copper fitted with isolation valves must be used as the bridge between the chlorinator and the various cylinders.

---

[22] *Injector Water. Service water in which chlorine is added (injected) to form a chlorine solution.*

[23] *Rotameter (RODE-uh-ME-ter). A device used to measure the flow rate of gases and liquids. The gas or liquid being measured flows vertically up a tapered, calibrated tube. Inside the tube is a small ball or bullet-shaped float (it may rotate) that rises or falls depending on the flow rate. The flow rate may be read on a scale behind or on the tube by looking at the middle of the ball or at the widest part or top of the float.*

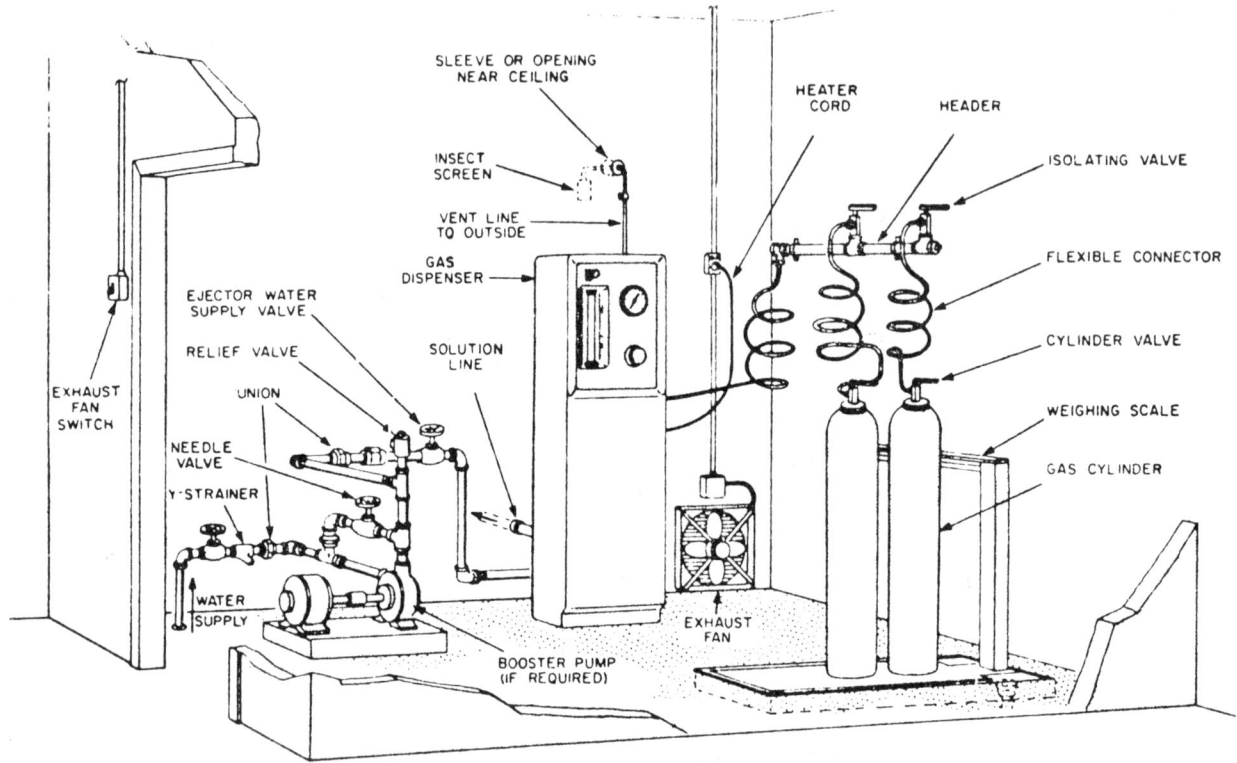

Fig. 7.12   Typical gas dispenser installation

NOTE: If chlorine gas is pulled through exhaust fan as shown, chlorine could corrode the wiring and the fan could fail when needed. A better design is to install the fan elsewhere and use "forced-air" ventilation which will push any
chlorine gas out floor vents.

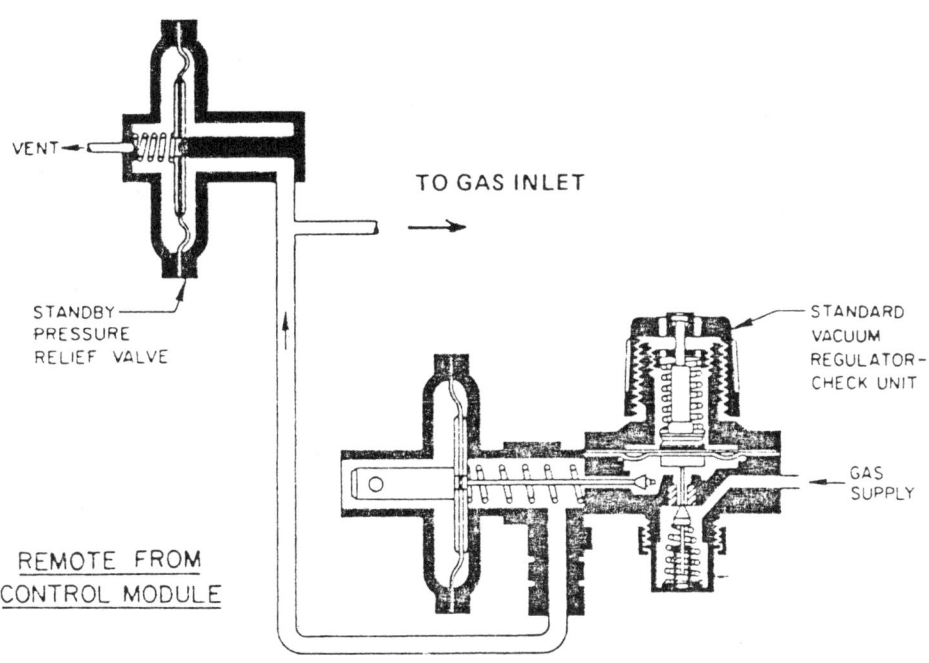

VENT

TO GAS INLET

STANDBY
PRESSURE
RELIEF VALVE

STANDARD
VACUUM
REGULATOR-
CHECK UNIT

GAS
SUPPLY

REMOTE FROM
CONTROL MODULE

Fig. 7.13   Chlorinator gas pressure controls
(Permission of Wallace & Tiernan Division, Pennwalt Corporation)

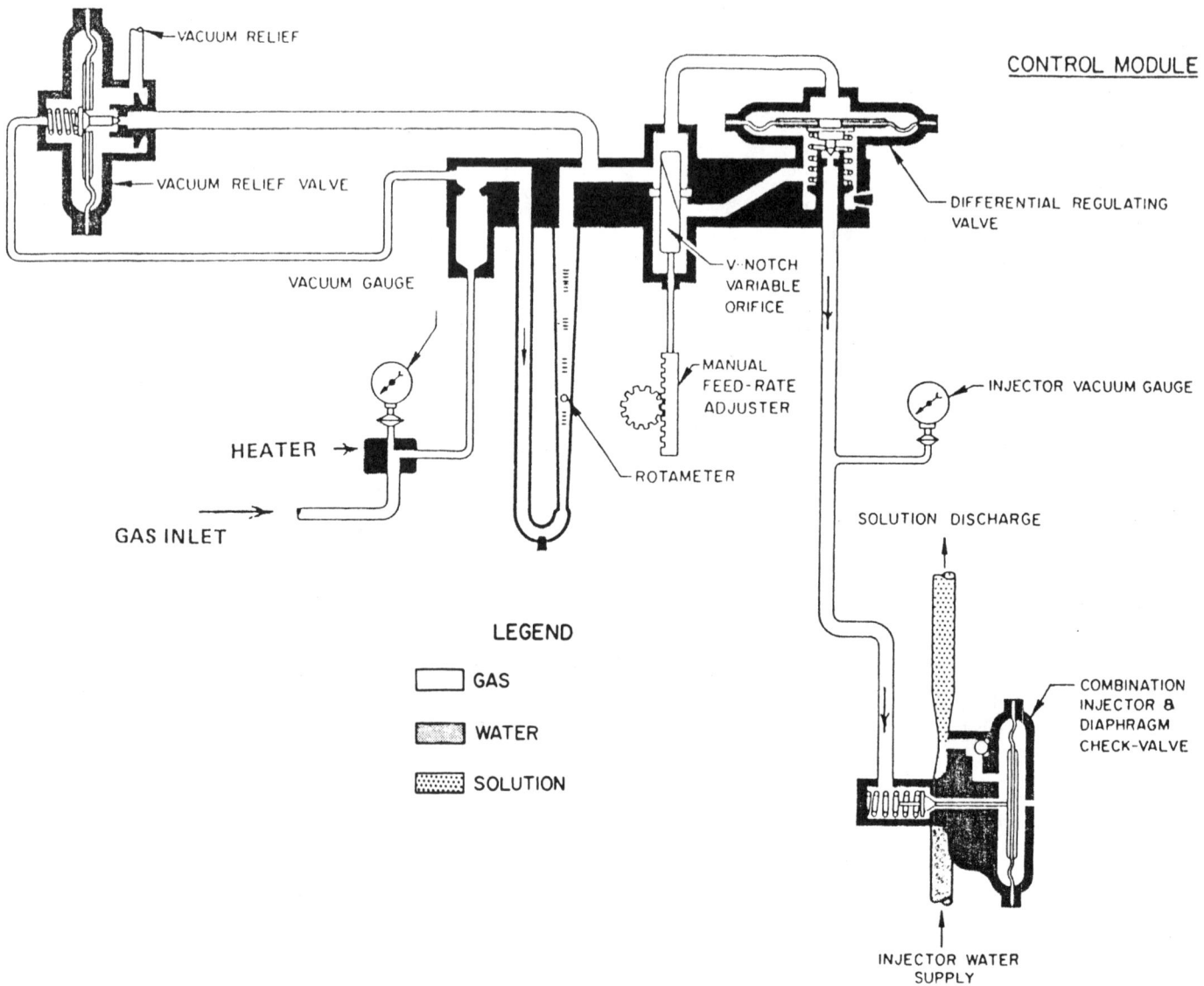

Fig. 7.14   Vacuum solution-feed chlorinator
(Permission of Wallace & Tiernan Division, Pennwalt Corporation)

TABLE 7.4 CHLORINATOR PARTS AND PURPOSES (Figures 7.13 and 7.14)

Part	Purpose
1. Pressure Gage (Not shown on Figure 7.13)	Indicates chlorine gas pressure at chlorinator system from chlorine manifold and supply (20 psi minimum and 40 psi maximum or 137.9 kilo *PASCAL*[24] minimum and 275.8 kPa maximum).
2. Gas Supply	Provides source of chlorine gas from containers to chlorinator system.
3. Vacuum Regulator-Check Unit	Maintains a constant vacuum on chlorinator.
4. Standby Pressure Relief	Relieves excess gas pressure on chlorinator.
5. Vent	Discharges any excess chlorine gas (pressure) to atmosphere outside of chlorination building or to a chlorine treatment facility.
6. Gas Inlet	Allows entrance of chlorine gas to chlorinator. Gas flows from chlorine container through supply line and gas manifold to inlet.
7. Heater	Prevents *RELIQUEFACTION*[25] of chlorine gas.
8. Vacuum Gage	Indicates vacuum on chlorinator system.
9. Rotameter Tube and Float	Indicates chlorinator feed rate. (Read the widest part or top of the float or center of ball for rate marked on tube.)
10. Differential Regulating (Reducing) Valve	Regulates (reduces) chlorinator chlorine gas pressure. Serves to maintain a constant differential across the orifice in order to obtain repeatable chlorine gas flow rates at a given orifice opening regardless of fluctuations in the injector vacuum.
11. Plug and Variable Orifice	Controls chlorine feed rate by regulating flow of chlorine gas.
12. Vacuum Relief Valve	Relieves excess vacuum by allowing air to enter system and reduce vacuum.
13. Vacuum Relief	Provides source of air to reduce excess vacuum.
14. Injector Vacuum Gage	Indicates vacuum at the injector.
15. Diaphragm Check Valve	Regulates chlorinator vacuum which in turn adjusts chlorinator feed rate. Receives signal from chlorine feed rate controls and then adjusts feed rate by regulating vacuum.
16. Manual Feed Rate Adjuster	Regulates chlorine feed rate manually. Most chlorination systems have automatic feed rate controls with a manual override.
17. Injector Water Supply	Provides source of water for chlorine solution. Must provide sufficient pressure and volume to operate injector.
18. Injector	Mixes or injects chlorine gas into water supply. Creates sufficient vacuum to operate chlorinator and to pull metered amount of chlorine gas.
19. Solution Discharge	Discharges solution mixture of chlorine and water.

The steel gas manifold with chlorine valve is mounted to the chlorinator. The flexible connector links the rigid manifold and the chlorine cylinder. The isolation valve between the flexible connector and the cylinder valve provides a way to close off the flexible connector when a new gas cylinder must be attached. This limits the amount of moisture that enters the system. Moisture in the system will combine with the chlorine gas to produce hydrochloric acid and cause corrosion. *CORROSION CAN CAUSE THE MANIFOLD TO FAIL.*

The chlorine is usually injected directly into the water supply pipe and there may not be contact chambers or mixing units. The location of the injection point is important. The injection should never be on the intake side of the pump as it will cause corrosion problems. There should be a check valve and a meter to monitor the chlorine dose.

On most well applications a chlorine booster pump is needed to overcome the higher water pump discharge pressures. If the well produces sand in the water, this pump will wear rapidly and become unreliable. In this situation, the chlorine solution should be introduced down the well through a polyethylene tube.

The polyethylene tube (¹/₂ inch or 12 mm) must be installed in the well so as to discharge a few inches below the suction screen. The chlorinator should operate *ONLY* when the pump is running. The chlorine solution flowing through the polyeth-

---

[24] *Pascal. The pressure or stress of one newton per square meter. Abbreviated Pa.*

    1 psi = 6,895 Pa = 6.895 kN/sq m = 0.0703 kg/sq cm

[25] *Reliquefaction (re-LICK-we-FACK-shun). The return of a gas to the liquid state; for example, a condensation of chlorine gas to return it to its liquid form by cooling.*

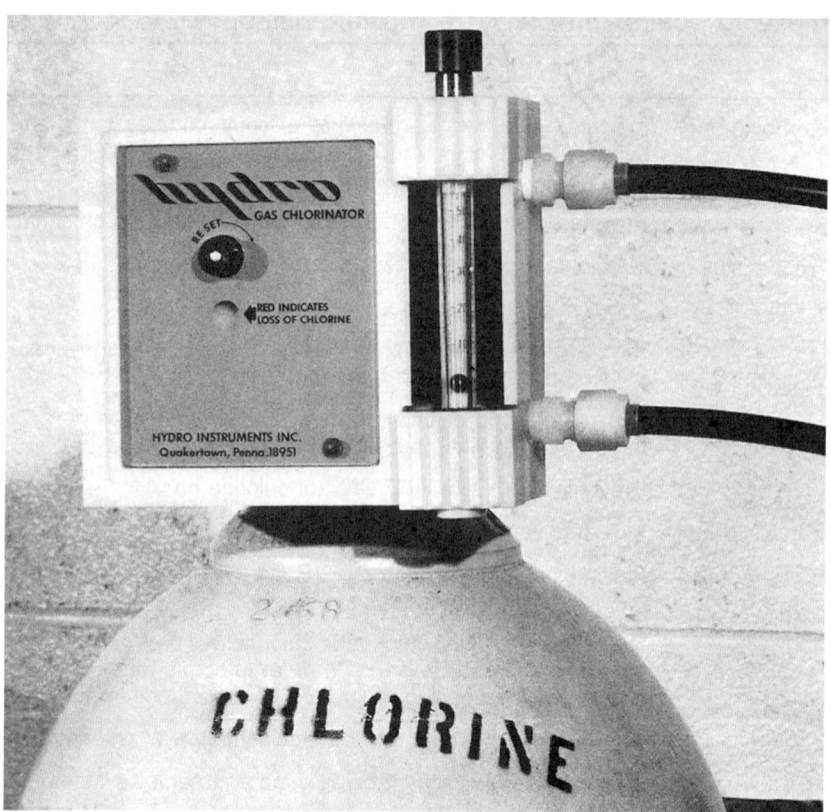

(Permission of Hydro Instruments, Inc.)

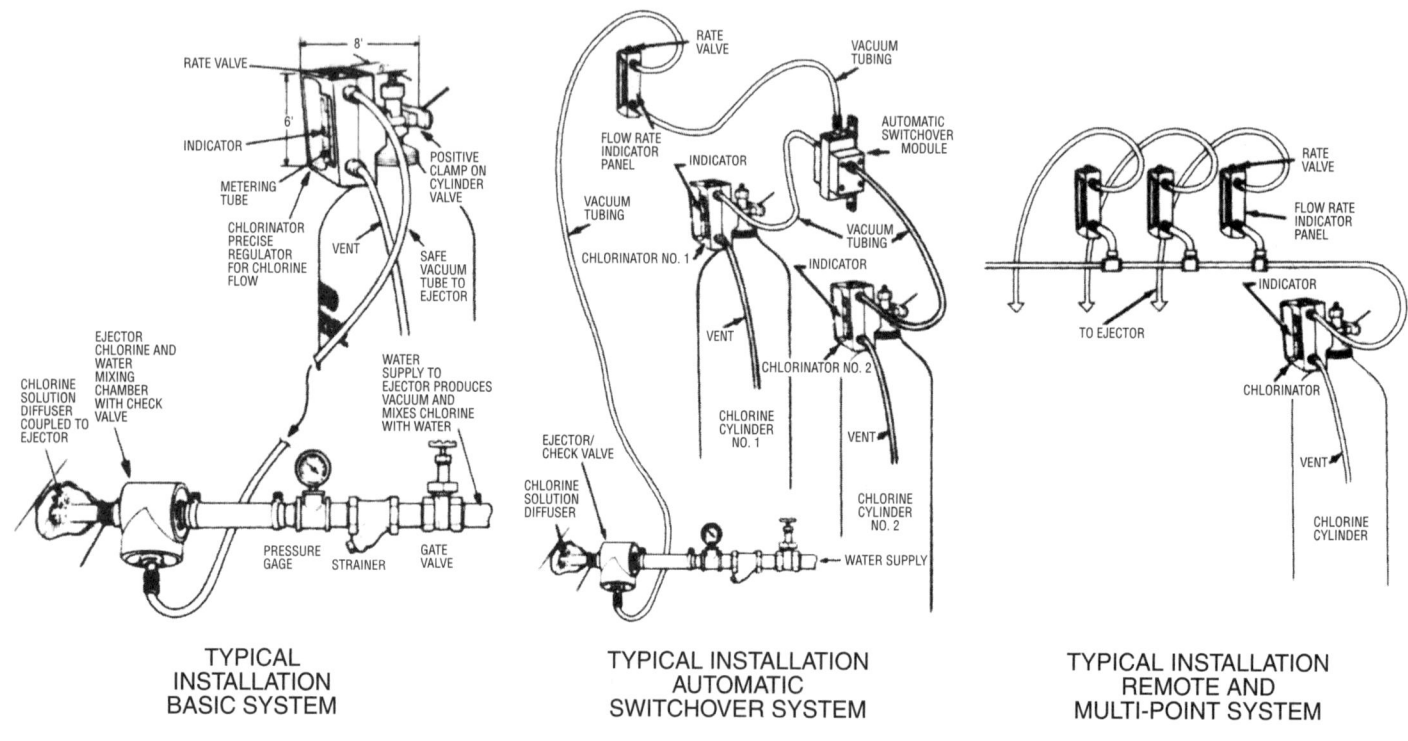

TYPICAL
INSTALLATION
BASIC SYSTEM

TYPICAL INSTALLATION
AUTOMATIC
SWITCHOVER SYSTEM

TYPICAL INSTALLATION
REMOTE AND
MULTI-POINT SYSTEM

(Permission of Chlorinators Incorporated)

Fig. 7.15   Cylinder-mounted gas chlorinators

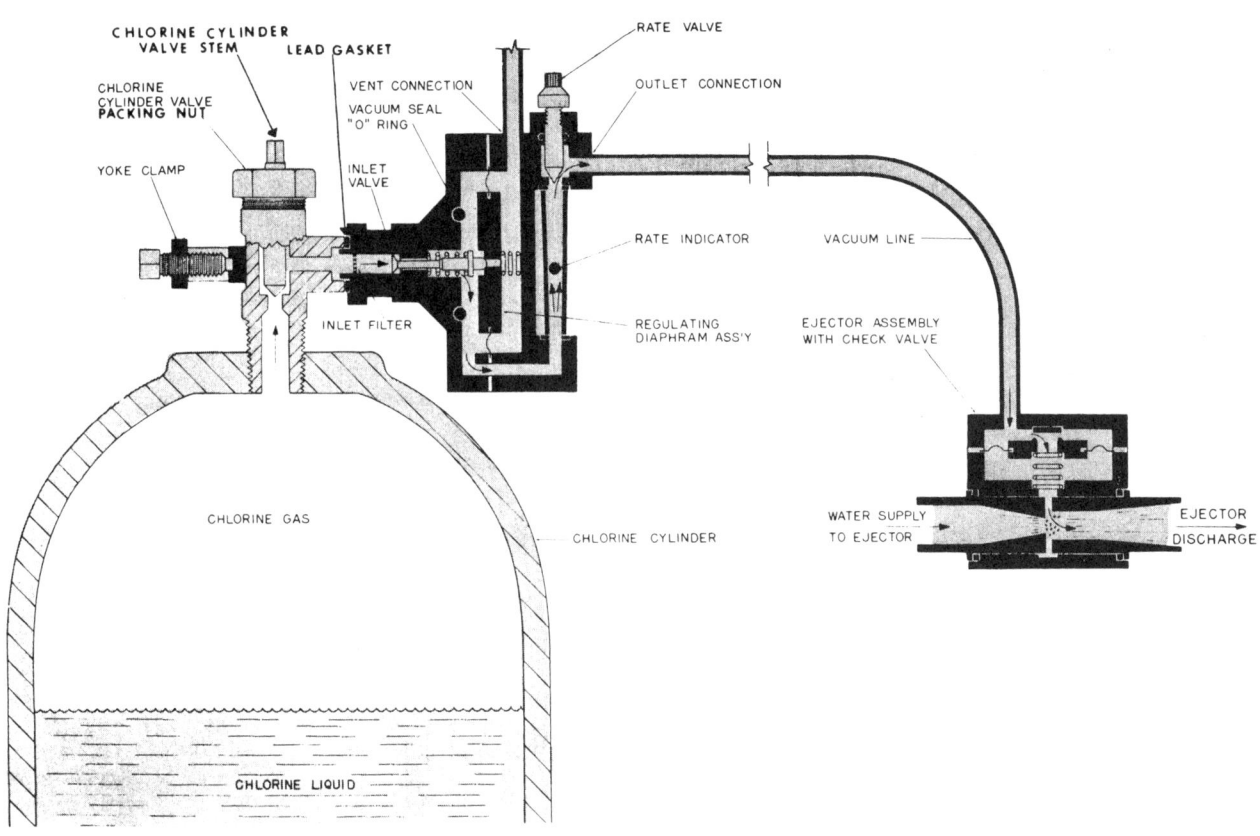

Fig. 7.16   Direct cylinder mounted connection from chlorine gas supply to chlorinator
(Permission of Capital Controls Company, Colmar, PA)

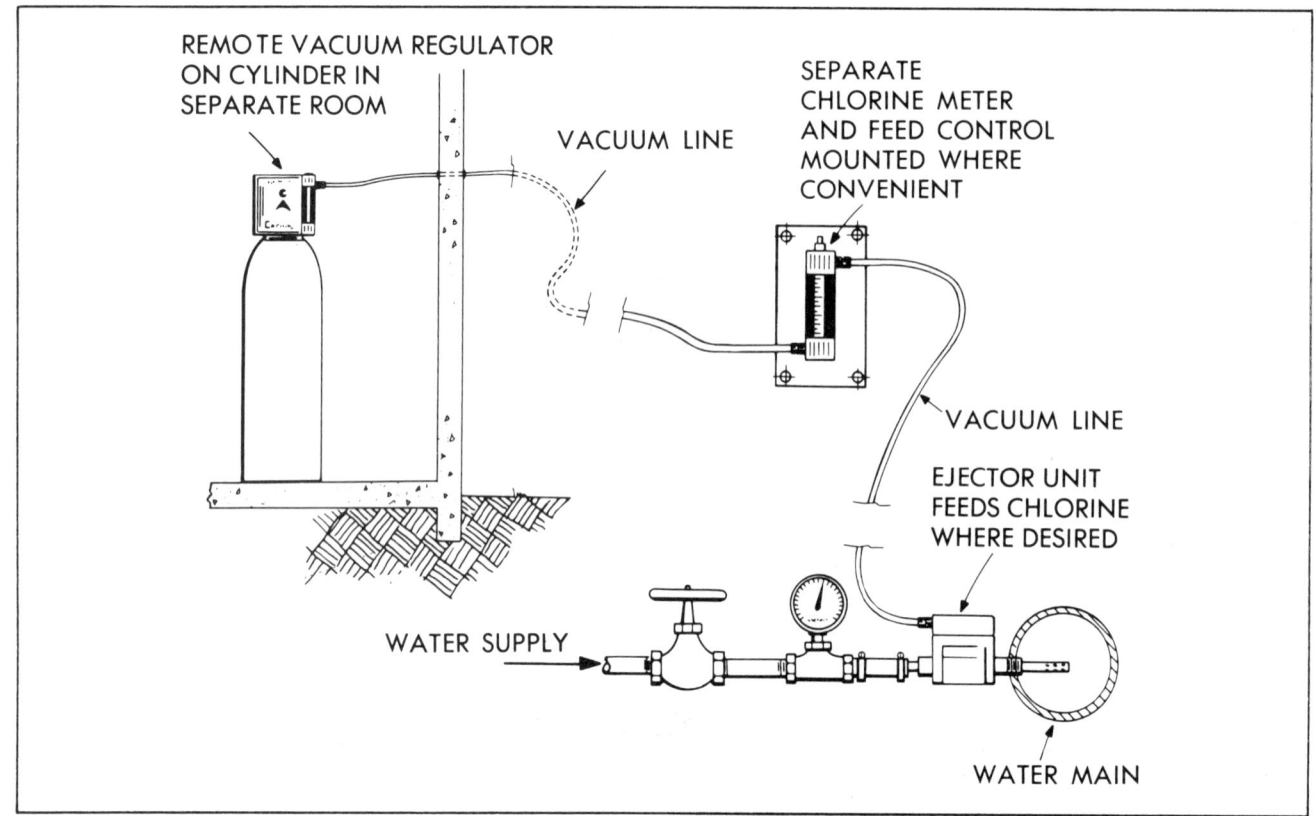

Fig. 7.17   Remote chlorinator with single meter and ejector installed on pipe
(Permission of Capital Controls Company, Colmar, PA)

ylene tube is extremely corrosive. If the tube does not discharge into flowing water, the effect of the solution touching the metal surface can be disastrous. Wells have been destroyed by corrosion from chlorine.

QUESTIONS

Write your answers in a notebook and then compare your answers with those on page 357.

7.4C   How is the rate of gas flow in a chlorinator measured?

7.4D   What is the primary advantage of vacuum system chlorinators?

7.42   Chlorine Containers

7.420   *Plastic (Figures 7.7 and 7.8)*

Plastic containers are commonly used for storage of hypochlorite solution. The container size depends on usage. Plastic containers should be large enough to hold a two or three days' supply of hypochlorite solution. The solution should be prepared every two or three days. If a larger amount of solution is mixed, the solution may lose its strength and thus affect the chlorine feed rate. Normally a week's supply of hypochlorite should be in storage and available for preparing hypochlorite solutions. Store the hypochlorite in a cool, dark place. Sodium hypochlorite can lose from two to four percent of its available chlorine content per month at room temperature. Therefore, manufacturers recommend a maximum shelf life of 60 to 90 days.

7.421   *Steel Cylinders (Figures 7.18 and 7.19)*

Cylinders containing 100 to 150 pounds (45 to 68 kg) of chlorine are convenient for very small treatment plants with capacities of less than 0.5 MGD (1,890 cu m/day). A fusible plug is placed in the valve below the valve seat (Figure 7.20). This plug is a safety device. The fusible metal softens or melts at 158 to 165°F (70 to 74°C) to prevent buildup of excessive pressures and the possibility of rupture due to a fire or high surrounding temperatures. Cylinders will not explode under normal conditions and can be handled safely.

The following are safe procedures for handling and storing chlorine cylinders:

1. Move cylinders with a properly balanced hand truck (Figure 7.21) with clamp supports that fasten about two-thirds of the way up the cylinder.

2. 100- and 150-pound (45- to 68-kg) cylinders can be rolled in a vertical position. Avoid lifting these cylinders except with approved equipment. Use a lifting clamp, cradle, or carrier. Never lift with homemade chain devices, rope slings, or magnetic hoists. Never roll, push, or drop cylinders off the back of trucks or loading docks.

3. Always replace the protective cap when moving a cylinder.

4. Keep cylinders away from direct heat (steam pipes or radiators) and direct sun, especially in warm climates.

5. Transport and store cylinders in an upright position.

6. Firmly secure cylinders to an immovable object (Figure 7.19).

NOTES:

1. Scale for weighing chlorine cylinders and chlorine.

2. Flexible tubing (pigtail).

3. Cylinders chained to wall.

Fig. 7.19 Typical chlorine cylinder station for water treatment
(Courtesy of PPG Industries)

Chlorine Cylinder

Protection
Hood

Valve

Neck Ring

Cylinder
Body

Foot Ring

B

Net Cylinder Contents	Approx. Tare, Lbs.*	Dimensions, Inches	
		A	B
100 Lbs.	73	8¼	54½
150 Lbs.	92	10¼	54½

*Stamped tare weight on cylinder shoulder does not include valve protection hood.

Fig. 7.18 Chlorine cylinder
(Courtesy of PPG Industries)

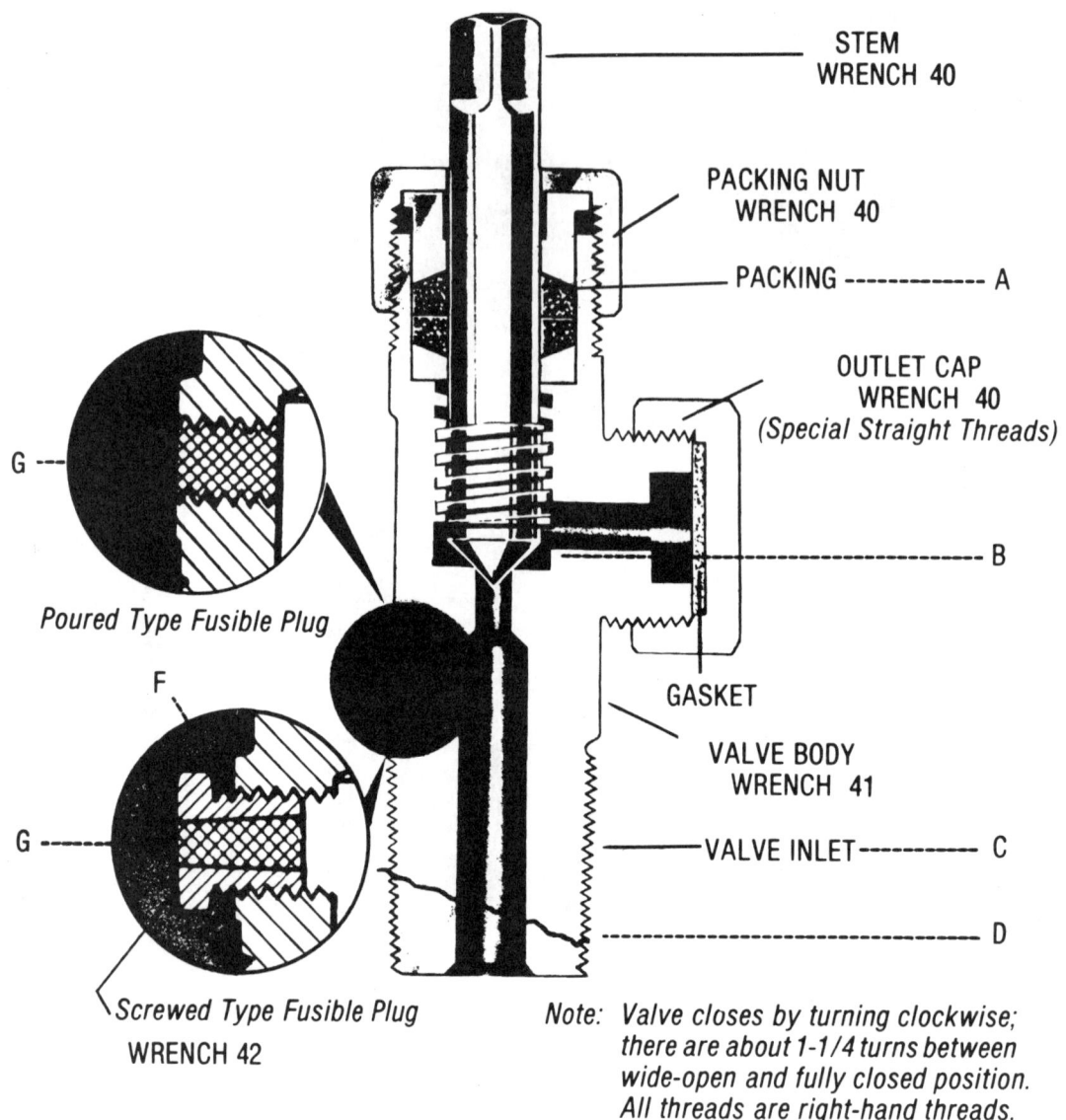

STEM
WRENCH 40

PACKING NUT
WRENCH 40

PACKING ------------ A

OUTLET CAP
WRENCH 40
(Special Straight Threads)

B

Poured Type Fusible Plug

G

F

G

GASKET

VALVE BODY
WRENCH 41

VALVE INLET ---------- C

D

Screwed Type Fusible Plug
WRENCH 42

Note: *Valve closes by turning clockwise; there are about 1-1/4 turns between wide-open and fully closed position. All threads are right-hand threads.*

TYPICAL VALVE LEAKS OCCUR THROUGH . . .

A - VALVE PACKING GLAND   E - VALVE BLOWN OUT*
B - VALVE SEAT   F - FUSIBLE PLUG THREADS
C - VALVE INLET THREADS   G - FUSIBLE METAL OF PLUG
D - BROKEN OFF VALVE   H - VALVE STEM BLOWN OUT*

Not shown on above drawing.

Fig. 7.20   *Standard chlorine cylinder valve*
(Permission of Chlorine Specialties, Inc.)

Fig. 7.21   Hand truck for chlorine cylinder
(Courtesy of PPG Industries)

7. Store empty cylinders separately from full cylinders. All empty chlorine cylinders must be tagged as empty.

    NOTE: Never store chlorine cylinders near turpentine, ether, anhydrous ammonia, finely divided metals, hydrocarbons, or other materials that are flammable in air or will react violently with chlorine.

8. Remove the outlet cap from the cylinder and inspect the threads on the outlet. Cylinders having outlet threads that are corroded, worn, cross-threaded, broken, or missing should be rejected and returned to the supplier.

9. The specifications and regulations of the U.S. Interstate Commerce Commission require that chlorine cylinders be tested at 800 psi (5,516 kPa or 56.24 kg/sq cm) every five years. The date of testing is stamped on the dome of the cylinder. Cylinders that have not been tested within that period of time should be rejected and returned to the supplier.

7.422   Ton Tanks *(Figures 7.22 and 7.23)*

1. Ton tanks are of welded steel construction and have a loaded weight of as much as 3,700 pounds (1,680 kg). They are about 80 inches (200 cm) in length and 30 inches (75 cm) in outside diameter. The ends of the tanks are crimped inward to provide a substantial grip for lifting clamps (Figure 7.24).

2. Most ton tanks have eight openings for fusible plugs and valves (Figures 7.24 and 7.25). Generally, two operating valves are located on one end near the center. There are six or eight fusible metal safety plugs, three or four on each end. These are designed to melt within the same temperature range as the safety plug in the cylinder valve.

> **WARNING**
>
> IT IS **VERY IMPORTANT** THAT FUSIBLE PLUGS SHOULD NOT BE TAMPERED WITH **UNDER ANY CIRCUMSTANCES** AND THAT THE TANK SHOULD NOT BE HEATED. ONCE THIS PLUG OPENS, ALL OF THE CHLORINE IN THE TANK WILL BE RELEASED.

For safe handling of ton tanks, follow these procedures:

1. Ship ton tanks by rail in multi-unit cars. They also may be transported by truck or semi-trailer (Figure 7.26).

2. Handle ton tanks with a suitable lift clamp in conjunction with a hoist or crane of at least two-ton capacity (Figure 7.24).

3. For storage and for use, lay ton tanks on their sides, above the floor or ground, on steel or concrete supports. They should not be stacked more than one high and should be separated by 30 inches (0.75 m) for access in case of leaks.

4. Place ton tanks on trunnions (pivoting mounts) that are equipped with rollers so that the withdrawal valves may be positioned one above the other (Figure 7.27). The upper valve will discharge chlorine gas, and the lower valve will discharge liquid chlorine (see Figure 7.24). The ability to rotate tanks is also a safety feature. In case of a liquid leak, the container can be rolled so that the leaking chlorine escapes as a gas rather than a liquid.

5. Use trunnion rollers (Figure 7.27) that do not exceed 3½ inches (9 cm) in diameter so that the containers will not rotate too easily and be turned out of position.

6. Equip roller shafts with a zerk-type lubrication fitting. Roller bearings are not advised because of the ease with which they rotate.

7. Use locking devices to prevent ton tanks from rolling while connected.

QUESTIONS

Write your answers in a notebook and then compare your answers with those on page 357.

7.4E   What type of container is commonly used to store hypochlorite?

7.4F   How large a supply of hypochlorite should be available?

7.4G   What is the purpose of the fusible plug?

7.4H   What is removed by the upper and lower valves of ton chlorine tanks?

7.43   Removing Chlorine From Containers

Whenever you do any work or maintenance involving the removal of chlorine from containers, a self-contained breathing apparatus (Figure 7.28) should be worn, or at least readily available. This is especially true when searching for chlorine leaks.

The maximum rate of chlorine removal from a 150-pound (68-kg) cylinder is 40 pounds (18 kg) of chlorine per day. If the rate of removal is greater, "freezing" can occur and less chlorine will be delivered.

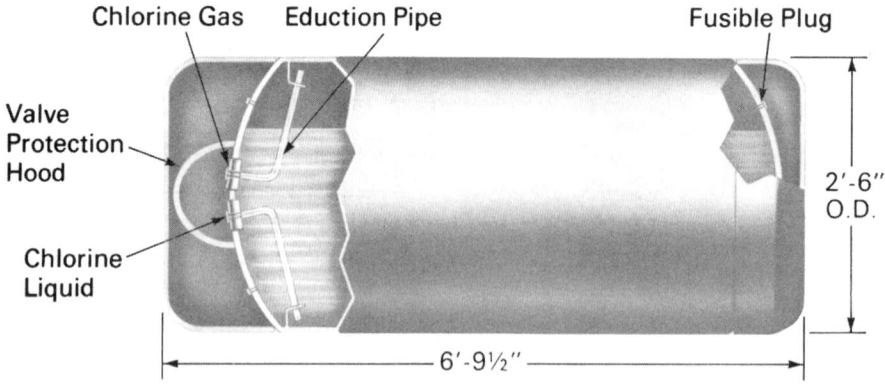

Chlorine Gas   Eduction Pipe

Fusible Plug

Valve
Protection
Hood

Chlorine
Liquid

2'-6"
O.D.

6'-9½"

Fig. 7.22   Chlorine ton container
(Courtesy of PPG Industries)

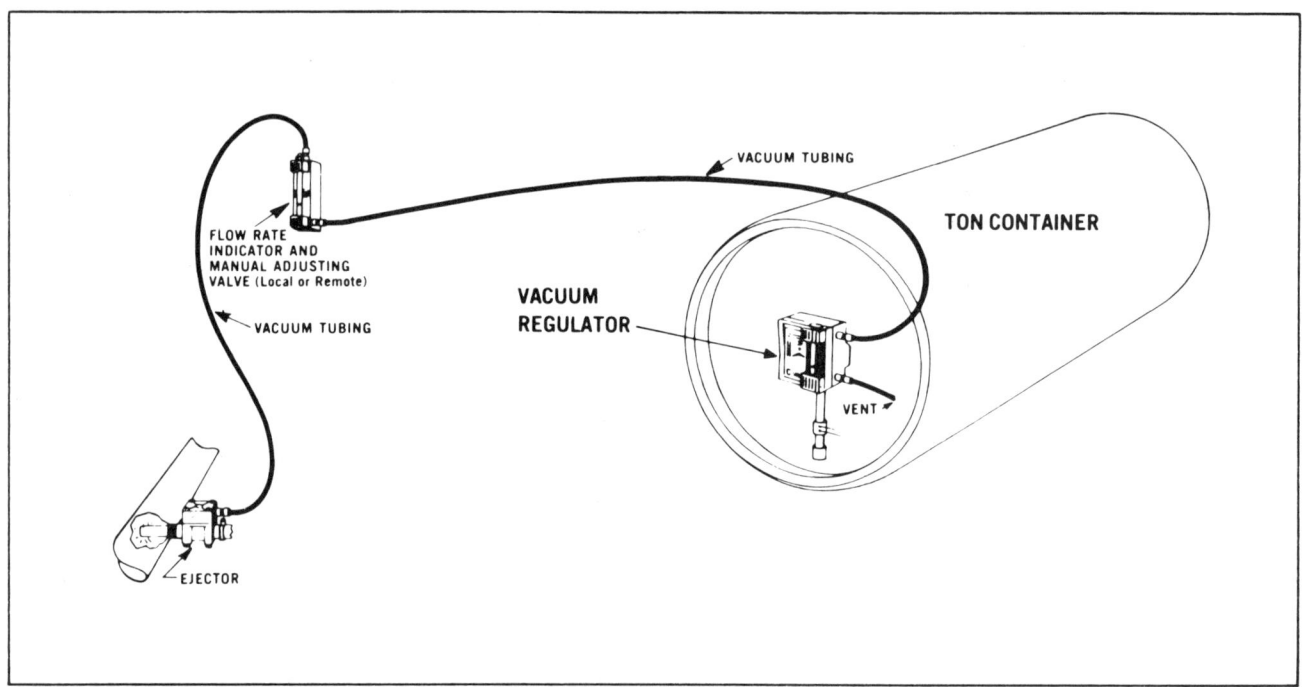

VACUUM TUBING

TON CONTAINER

FLOW RATE
INDICATOR AND
MANUAL ADJUSTING
VALVE (Local or Remote)

VACUUM
REGULATOR

VACUUM TUBING

VENT

EJECTOR

Fig. 7.23   Ton container-mounted remote metering arrangement
(Permission of Capital Controls Company, Colmar, PA)

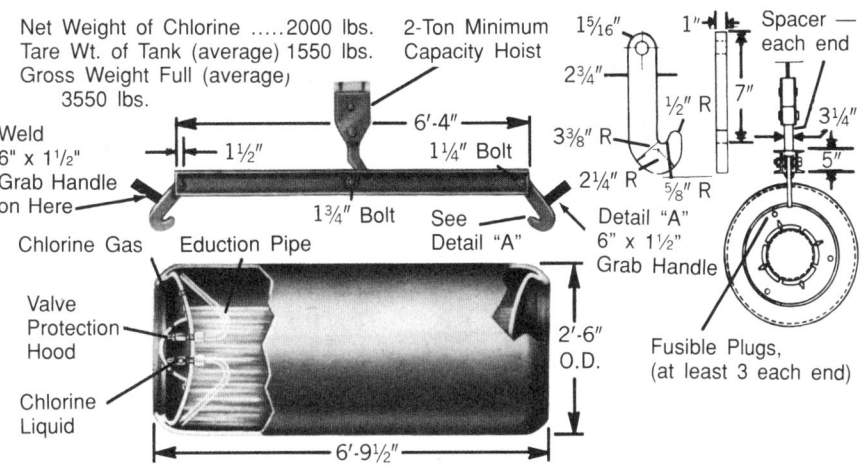

Net Weight of Chlorine .....2000 lbs.
Tare Wt. of Tank (average) 1550 lbs.
Gross Weight Full (average) 3550 lbs.

2-Ton Minimum Capacity Hoist

Weld 6" x 1½" Grab Handle on Here

6'-4"

1½"

1¼" Bolt

1¾" Bolt

See Detail "A"

Chlorine Gas

Eduction Pipe

Valve Protection Hood

Chlorine Liquid

2'-6" O.D.

6'-9½"

1⁵⁄₁₆"

2¾"

3⅜" R

2¼" R

½" R

⅝" R

1"

7"

3¼"

5"

Spacer — each end

Detail "A" 6" x 1½" Grab Handle

Fusible Plugs, (at least 3 each end)

Fig. 7.24 Ton tank lifting beam
(Courtesy of PPG Industries)

NOTE: Grab Handles recommended by Denis Unreiner, Medicine Hat, Alberta, Canada.

Ton Tank Valve   100 and 150 lb Cylinder Valve

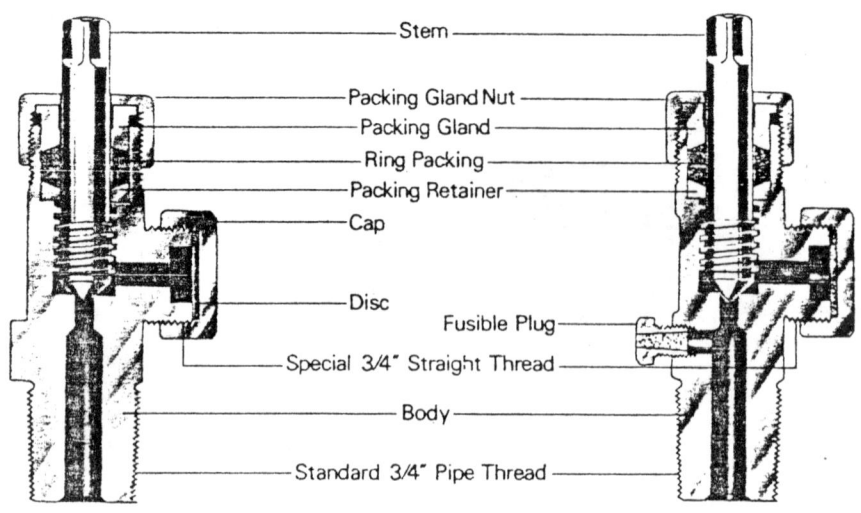

Fig. 7.25   Comparison of ton tank valve with cylinder valve
(Courtesy of PPG Industries)

> **WARNING**
>
> When frost appears on valves and flex connectors, the chlorine gas may condense to liquid (reliquify). The liquid chlorine may plug the chlorine supply lines (sometimes this is referred to as chlorine ice or frozen chlorine). If you disconnect the chlorine supply line to unplug it, *BE VERY CAREFUL.* The liquid chlorine in the line could reevaporate, expand as a gas, build up pressure in the line, and cause liquid chlorine to come shooting out the open end of a disconnected chlorine supply line.

7.430   Connections

The outlet threads on container valves are specialized threads; they are not ordinary tapered pipe threads. Use only the fittings and gaskets furnished by your chlorine supplier or chlorinator equipment manufacturer when making connections to chlorine containers. Do not try to use regular pipe thread fittings. Whenever you make a new connection, always use a new gasket. The outlet threads on container valves should always be inspected before being connected to the chlorine system. Containers with outlet threads that are badly worn, cross threaded, or corroded should be rejected and returned to the supplier. The connecting nut on the chlorine system should also be inspected and replaced if it develops any of these problems. Since the threads on the cylinder connection may become worn, yoke-type connectors (Figure 7.29) are recommended.

Flexible ⅜-inch 2,000-pound (psi) (0.95-cm, 13,790-kPa or 140-kg/sq cm) annealed (toughened) copper tubing (pigtail, or coiled horizontally at least three times, see Figure 7.19) is recommended for connection between chlorine containers and stationary piping. Care should be taken to prevent sharp bends in the tubing because this will weaken it and eventually the tubing will start leaking. Many operators recommend use

of a sling to hold the tubing when disconnecting it from an empty cylinder to prevent the tubing from flopping around and getting kinked or getting dirt inside it. Cap or plug the connectors and/or valves when they are not connected to prevent entry of dirt/debris and moisture.

To simplify changing containers, you will also need a shutoff valve located just beyond the container valve or at the beginning of the stationary piping.

7.431   Valves

Do not use wrenches longer than six inches (15 cm), pipe wrenches, or wrenches with an extension on container valves. If you do, you could exert too much force and break the valve. Use only a square end-open or box wrench (see Figure 7.29) (this wrench can be obtained from your chlorine supplier). To unseat the valve, strike the end of the wrench with the heel of your hand to rotate the valve stem in a counterclockwise direction. Then open slowly. One complete turn permits maximum discharge. Do not force the valve beyond the full open position or you may strip the internal valve stem threads. If the valve is too tight to open in this manner, loosen the packing gland nut (Figure 7.25) slightly to free the stem. Open the valve, then retighten the packing nut. If you are uncertain how to loosen the nut, you should return the container to the supplier. Do not use organic lubricants on valves and lines because further chemical reactions can block the lines.

7.432   Ton Tanks

One-ton tanks (Figure 7.24) must be *PLACED ON THEIR SIDES WITH THE VALVES IN A VERTICAL POSITION* so either chlorine gas or liquid chlorine may be removed. Connect the flexible tubing to the *TOP VALVE* to remove chlorine gas from a tank (Figure 7.30). The *BOTTOM VALVE* is used to remove liquid chlorine and is used only with a chlorine evaporator. The valves are similar to those on the smaller chlorine cylinders (fusible plugs are not located at valves on ton containers) and must be handled with the same care.

Fig. 7.26   Ton container hoist and lifting beam removing ton containers from semi-trailer
(Courtesy of PPG Industries)

Trunnions for rotating ton containers

Close-up of trunnions in photo above

Fig. 7.27   Storage of ton containers
(Courtesy of PPG Industries)

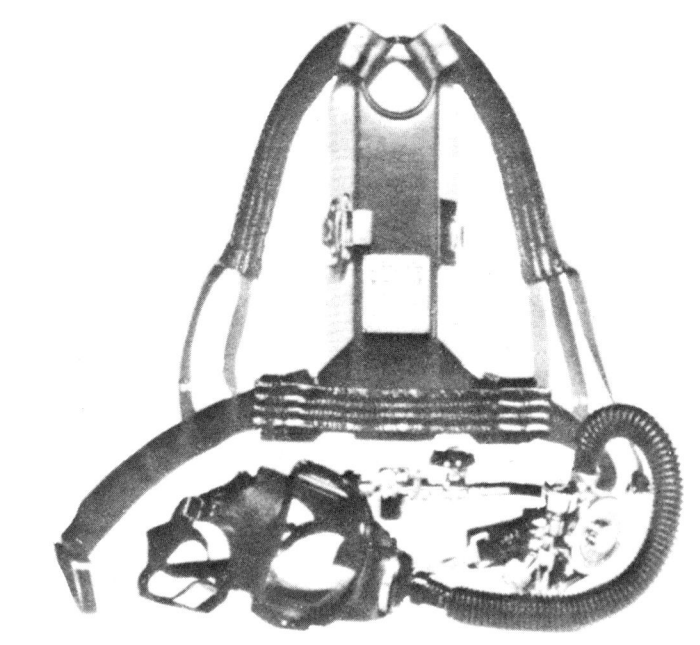

Fig. 7.28 Self-contained breathing apparatus
(Courtesy of PPG Industries)

Manifolded gas valves of ton containers

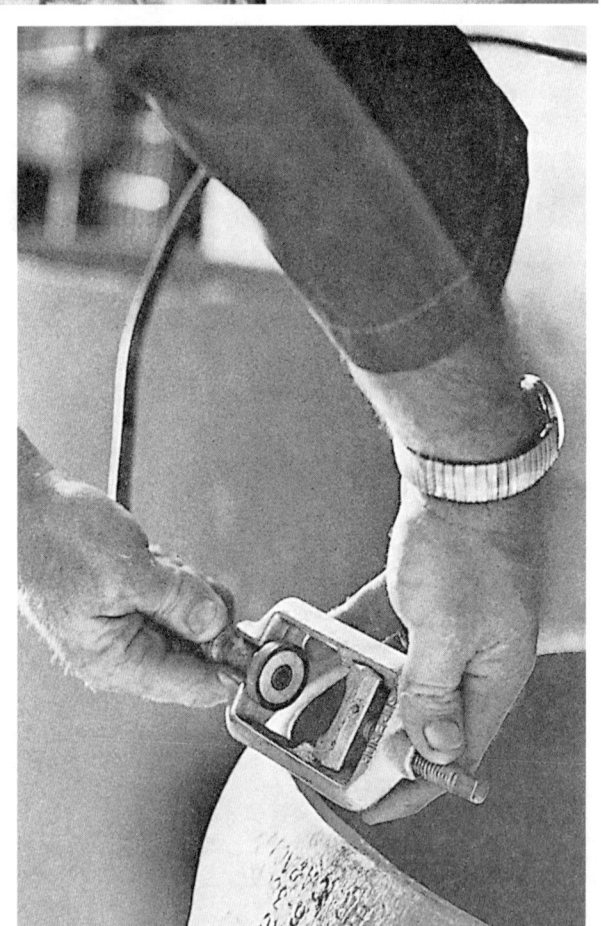

Yoke and adapter for ton container valve connection

Fig. 7.30   Withdrawing chlorine from ton containers
(Courtesy of PPG Industries)

Fig. 7.29   Yoke and adapter-type connection
(Permission of The Chlorine Institute, Inc.)

QUESTIONS

Write your answers in a notebook and then compare your answers with those on page 357.

7.4I How can copper tubing be prevented from getting kinks or dirt inside the tubing?

7.4J Why are one-ton tanks placed on their sides with the valves in a vertical position?

7.44 Performance of Chlorination Units

Before attempting to start or stop any chlorination system, read the manufacturer's literature and your plant's operation and maintenance instructions to become familiar with the equipment. Review the plans and drawings of the facility. Determine what equipment, pipelines, pumps, tanks, and valves are to be placed into service or are in service. The current status of the entire system must be known before starting or stopping any portion of the system. This section provides the procedures for a typical system and will give you ideas for your system.

7.440 Hypochlorinators

1. *START-UP OF HYPOCHLORINATORS*

   a. Prepare the chemical solution. Most agencies buy commercial or industrial hypochlorite at a strength of 12 to 15 percent chlorine. This solution is usually diluted down to a two percent solution. If using commercially prepared solutions, you will need to calculate feed rates.

   b. Lock out the electric circuit and then inspect it. Normally no adjustments are needed. Look for frayed wires. Turn power back on. Leave solution switch off.

   c. Turn on the chemical pump. Make adjustments while the pump is running. Never adjust the pump while it is off because damage to pump will occur.

   d. Calibrate pump to ensure accurate delivery of chlorine solution. See Chapter 4, "Coagulation and Flocculation," Appendix B, "Adjustment and Calibration of Chemical Feeders."

   e. Make sure solution is being fed into the system. Measure chlorine residual just downstream from where solution is being fed into system. You may have a target residual you wish to maintain at the beginning of the system such as 2.0 mg/L.

   f. Check chlorine residual in system. Residual should be measured at the most remote test location within the distribution system and should be at least 0.2 mg/L free residual chlorine. This chlorine residual is necessary to protect the treated water from any recontamination. Adjust chemical feed as needed.

2. *SHUTDOWN OF HYPOCHLORINATORS*

   a. Short Duration

      (1) Turn water supply pump off. You do not want to pump any unchlorinated water and possibly contaminate the rest of the system. *NEVER PUMP ANY UNCHLORINATED WATER INTO YOUR DISTRIBUTION SYSTEM.*

      (2) Turn hypochlorinator off.

      (3) When making repairs, lock out the circuit or pull the plug from an electric socket.

   b. Long Duration

      (1) Obtain and/or place another hypochlorinator in service.

3. *NORMAL OPERATION OF HYPOCHLORINATORS*

   Normal operation of a hypochlorinator requires routine observation and preventive maintenance.

   DAILY

   a. Inspect the building to make sure only authorized personnel have been there.

   b. Read and record the level of the solution tank at the same time every day.

   c. Read the meters and record the amount of water pumped.

   d. Check chlorine residual (at least 0.2 mg/L) in the system and adjust chlorine feed rate as necessary. Try to maintain a chlorine residual of 0.2 mg/L at the most remote point in the distribution system. The suggested free chlorine residual for treated water or well water is 0.5 mg/L at the point of chlorine application provided the 0.2 mg/L residual is maintained throughout the distribution system and coliform test results are negative.

   e. Check chemical feed pump operation. Most hypochlorinators have a dial with a range from 0 to 10 which adjusts the chlorine feed rate. Start with a setting around 6 or 7 on the dial and use a two percent hypochlorite solution. The pump should operate in the upper ranges of the dial so that the strokes or pulses from the pump will be close together. In this way, the chlorine will be fed continuously to the water being treated. Adjust the feed rate after testing residual levels.

   WEEKLY

   a. Clean the building.

   b. Replace the chemicals and wash the chemical storage tank. Try to have a 15- to 30-day supply of chlorine in storage for future needs. When preparing hypochlorite solutions, prepare only enough for a two- or three-day supply.

   MONTHLY

   a. Check the operation of the check valve.

   b. Perform any required preventive maintenance suggested by the manufacturer.

c. Cleaning

Commercial sodium hypochlorite solutions (such as Clorox) contain an excess of caustic (sodium hydroxide or NaOH). When this solution is diluted with water containing calcium and also carbonate alkalinity, the resulting solution becomes supersaturated with calcium carbonate. This calcium carbonate tends to form a coating on the poppet valves in the solution feeder. The coated valves will not seal properly and the feeder will fail to feed properly.

Use the following procedure to remove the carbonate scale:

(1) Fill a one-quart (one-liter) Mason jar half full of tap water.

(2) Place one fluid ounce (20 mL) of 30 to 37 percent hydrochloric acid (swimming pool acid) in the jar. *ALWAYS ADD ACID TO WATER, NEVER THE REVERSE.*

(3) Fill the jar with tap water.

(4) Place the suction hose of the hypochlorinator in the jar and pump the entire contents of the jar through the system.

(5) Return the suction hose to the hypochlorite solution tank and resume normal operation.

You can prevent the formation of the calcium carbonate coatings by obtaining the dilution water from an ordinary home water softener.

NORMAL OPERATION CHECKLIST

a. Check chemical usage. Record solution level and the water pump meter reading or number of hours of pump operation. Calculate the amount of chemical solution used and compare with the desired feed rate. See Example 3.

b. Determine if every piece of equipment is operating.

c. Inspect the lubrication of the equipment.

d. Check the building for any possible problems.

e. Clean up the area.

FORMULAS

When operating a hypochlorinator you should compare the actual chlorine dose applied to the water being treated with the desired chlorine dose in milligrams per liter. The actual dose is calculated by determining the amount of chlorine actually used and the amount of water treated. The amount of chlorine used is found by measuring the amount of hypochlorite solution used and knowing the strength of the hypochlorite solution. The amount of water used is determined from a flowmeter.

To calculate the amount of water treated, determine the amount in gallons from a flowmeter and convert this amount from gallons to pounds.

Water, lbs = (Water Treated, gal)(8.34 lbs/gal)

To calculate the amount of hypochlorite used in gallons, determine the volume of hypochlorite used in gallons.

$$\frac{\text{Hypochlorite,}}{\text{gallons}} = (0.785)(\text{Diameter, ft})^2(\text{Depth, ft})(7.48 \text{ gal/cu ft})$$

To determine the pounds of chlorine used to disinfect the water being treated, we have to convert the hypochlorite used from gallons to pounds of chlorine by considering the strength of the hypochlorite solution.

$$\text{Chlorine, lbs} = (\text{Hypochlorite, gal})(8.34 \text{ lbs/gal})\left(\frac{\text{Hypochlorite, \%}}{100\%}\right)$$

Finally, to estimate the actual chlorine dose in milligrams of chlorine per liter of water treated, we divide the pounds of chlorine used by the millions of pounds of water treated. Pounds of chlorine per million pounds of water is the same as parts per million or milligrams per liter (ppm = mg/L).

$$\text{Chlorine Dose, mg/L} = \frac{\text{Chlorine Used, lbs}}{\text{Water Treated, Million lbs}}$$

EXAMPLE 3

Water pumped from a well is disinfected by a hypochlorinator. A chlorine dose of 1.2 mg/L is necessary to maintain an adequate chlorine residual throughout the system. During a one-week time period, the water meter indicated that 2,289,000 gallons of water were pumped. A two percent sodium hypochlorite solution is stored in a three-foot diameter plastic tank. During this one-week period, the level of hypochlorite in the tank dropped 2 feet 8 inches (2.67 feet). Does the chlorine feed rate appear to be too high, too low, or about OK?

Known	Unknown
Desired Chlorine Dose, mg/L = 1.2 mg/L	1. Actual Chlorine Dose, mg/L
Water Pumped, gal = 2,289,000 gal	2. Is Actual Dose OK?
Hypochlorite, % = 2%	
Chemical Tank Diameter, ft = 3 ft	
Chemical Drop in Tank, ft = 2.67 ft	

1. Calculate the pounds of water disinfected.

Water, lbs = (Water Pumped, gallons)(8.34 lbs/gal)

= (2,289,000 gal)(8.34 lbs/gal)

= 19,090,000 lbs

= 19.09 Million lbs

2. Calculate the volume of two percent sodium hypochlorite used in gallons.

$$\frac{\text{Hypochlorite,}}{\text{gallons}} = (0.785)(\text{Diameter, ft})^2(\text{Depth, ft})(7.48 \text{ gal/cu ft})$$

= (0.785)(3 ft)²(2.67 ft)(7.48 gal/cu ft)

= 141.1 gallons

3. Determine the pounds of chlorine used to disinfect the water.

$$\frac{\text{Chlorine,}}{\text{lbs}} = (\text{Hypochlorite, gal})(8.34 \text{ lbs/gal})\left(\frac{\text{Hypochlorite, \%}}{100\%}\right)$$

= (141.1 gal)(8.34 lbs/gal)$\left(\frac{2\%}{100\%}\right)$

= 23.5 lbs Chlorine

4. Estimate the chlorine dose in mg/L.

$$\text{Chlorine Dose, mg/}L = \frac{\text{Chlorine, lbs}}{\text{Water, Million lbs}}$$

$$= \frac{23.5 \text{ lbs Chlorine}}{19.09 \text{ Million lbs Water}}$$

$$= 1.23 \text{ mg/}L$$

Since the actual estimated chlorine dose (1.23 mg/L) was similar to the desired dose of 1.2 mg/L, the chlorine feed rate appears OK.

4. *ABNORMAL OPERATION OF HYPOCHLORINATORS*

a. Inform your supervisor of the problem.

b. If the hypochlorinator malfunctions, it should be repaired or replaced immediately. See the shutdown operation (Step 2 in this section).

c. Solution tank level.

   (1) If Too Low: Check the adjustment of the pump.
   Check the hour meter on water pump.

   (2) If Too High: Check the chemical pump.
   Check the hour meter on water pump.

d. Determine if the chemical pump is not operating.

TROUBLESHOOTING GUIDELINES

(1) Check the electrical connection.

(2) Check the circuit breaker.

(3) Check for stoppages in the flow lines.

CORRECTIVE MEASURES

(1) Shut off the water pump so that no unchlorinated water is pumped into the system.

(2) Check for a blockage in the solution tank.

(3) Check the operation of the check valve.

(4) Check the electric circuits.

(5) Replace the chemical feed pump with another pump while repairing the defective unit.

e. The solution is not being pumped into the water line.

TROUBLESHOOTING

(1) Check the solution level.

(2) Check for blockages in the solution line.

5. *MAINTENANCE OF HYPOCHLORINATORS*

Hypochlorinators on small systems may be sealed systems and cannot be repaired so replacement is the only solution. Some units are repairable and can be serviced by following manufacturer's instructions. Maintenance requirements are normally minor such as oil changes and lubrication of moving parts. Review the manufacturer's specifications and instructions for maintenance requirements.

QUESTIONS

Write your answers in a notebook and then compare your answers with those on page 357.

7.4K What would you do before attempting to start any chlorination system?

7.4L What should be the chlorine residual in the most remote part of the distribution system?

7.4M Why should a hypochlorite feed pump be operated in the upper end of its range (at 6 or 7 in a range of 0 to 10)?

7.441 Chlorinators

1. *SAFETY EQUIPMENT REQUIRED AND AVAILABLE OUTSIDE THE CHLORINATOR ROOM*

a. Protective clothing

   (1) Gloves

   (2) Rubber suit

b. Self-contained pressure-demand air supply system (Figure 7.28)

c. Chlorine leak detector/warning device should be located outside the chlorine room and should have a battery backup in case of a power failure. The chlorine sensor unit should be in the chlorine room and connected to the leak detector/warning device which is located outside the chlorine storage room.

2. *START-UP OF CHLORINATORS*

Procedures for start-up, operation, shutdown, and troubleshooting are outlined in this section and are intended to be typical procedures for all types of chlorinators. For specific directions, see the manufacturer's literature and operation and maintenance instructions for your plant. During emergencies you must act quickly and may not have time to check out each of the steps outlined below, but you still must follow established procedures. Work in pairs. Never work alone when hooking up a chlorine system.

a. Gas Chlorinators

   Start-up procedures for chlorinators using chlorine gas from containers are outlined in this section.

   (1) Be sure chlorine gas valve at the chlorinator is closed. This valve should already be closed since the chlorinator is out of service.

   (2) All chlorine valves on the supply line should have been closed during shutdown. Be sure they are still closed. If any valves are required to be open for any reason, this exception should be indicated by a tag on the valve.

(3) Inspect all tubing, manifolds, and valve connections for potential leaks and be sure all joints are properly gasketed.

(4) Check chlorine solution distribution lines to be sure that system is properly valved to deliver chlorine solution to desired point of application.

(5) Open the chlorine metering orifice slightly by adjusting chlorine feed-rate control.

(6) Start the injector water supply system. This is usually a potable water supply protected by an *AIR GAP*[26] or air-break system. Injector water is pumped at an appropriate flow rate and the flow through the injector creates sufficient vacuum in the injector to draw chlorine. Chlorine is absorbed and mixed in the water at the injector. This chlorine solution is conveyed to the point of application.

(7) Examine injector water supply system.

   (a) Note reading of injector supply pressure gage. If reading is abnormal (different from usual reading), try to identify cause and correct. Injectors should operate at not less than 50 psi inlet pressure.

   (b) Note reading of injector vacuum gage. If system does not have a vacuum gage, have one installed. If the vacuum reading is less than normal, the machine may function at a lower feed rate, but will be unable to deliver at rated capacity.

(8) Inspect chlorinator vacuum lines for leaks.

(9) Crack open the chlorine container valve and allow gas to enter the line. Inspect all valves and joints for leaks by placing an ammonia-soaked rag[27] near each valve and joint. A polyethylene "squeeze bottle" filled with ammonia water to dispense ammonia vapor may also be used. Care should be taken to avoid spraying ammonia water on any leak or touching the soaked cloth to any metal. The formation of a white cloud of vapor will indicate a chlorine leak. Start with the valve at the chlorine container, move down the line and check all joints between this valve and the next one downstream. *NEVER* apply ammonia solutions directly to any valve because an acid will form which will eat away the valve fittings. If the downstream valve passes the ammonia test, open the valve and continue to the next valve. If there are no leaks to the chlorinator, open the cylinder valve approximately one (1) complete turn to obtain maximum discharge and continue with the start-up procedure.

(10) Inspect the chlorinator.

   (a) Chlorine gas pressure at the chlorinator should be between 20 and 30 psi (137.9 kPa and 206.85 kPa or 1.4 and 2.1 kg/sq cm). However, in the summer on very hot days the pressure in a chlorine cylinder may exceed 150 psi (1,035 kPa or 10.5 kg/sq cm).

   (b) Operate chlorinator at complete range of feed rates.

   (c) Check operation on manual and automatic settings.

(11) Chlorinator is ready for use. Set chlorinator at desired feed rate. Log in the time the system is placed into operation and the application point.

b. Chlorinators With Evaporators[28] (Figure 7.31)

Start-up procedures for chlorinators using liquid chlorine from containers are outlined in this section. In most plants liquid chlorine is delivered in one-ton containers. However, liquid chlorine may be delivered in railroad tank cars to very large plants.

(1) Inspect all joints, valves, manifolds, and tubing connections in chlorination system, including application lines, for proper fit and for leaks. Make sure that all joints have gaskets.

(2) If chlorination system has been broken open or exposed to the atmosphere, verify that the system is dry. Usually, once a system has been dried out, it is never opened again to the atmosphere. However, if moisture enters the system, from the air or by other means, it readily mixes with chlorine and forms hydrochloric acid which will corrode the pipes, valves, joints, and fittings. *CORROSION CAN CAUSE LEAKS AND REQUIRE THAT THE ENTIRE SYSTEM BE REPLACED.*

To verify that the system is dry, determine the *DEW POINT*[29] (must be lower than −40°F or −40°C). If not dry, turn the evaporator on, pass dry air through the evaporator and force this air through the system. If this step is omitted and moisture remains in the system, serious corrosion damage can result and the entire system may have to be repaired.

(3) Start up the evaporators. Fill the water bath and adjust the device according to the manufacturer's directions. Water baths or evaporators should be equipped with low water alarms and automatic shutoffs in case of excessive heat. Meltdown can occur with lack of water and chlorine is explosive at high temperatures.

---

[26] *Air Gap. An open vertical drop, or vertical empty space, that separates a drinking (potable) water supply to be protected from another water system in a water treatment plant or other location. This open gap prevents the contamination of drinking water by backsiphonage or backflow because there is no way raw water or any other water can reach the drinking water supply.*

[27] *Use a concentrated ammonia solution containing 28 to 30 percent ammonia as NH_3 (this is the same as 58 percent ammonium hydroxide, NH_4OH, or commercial 26° Baumé).*

[28] *Only the very largest water treatment plants require chlorinators equipped with evaporators.*

[29] *Dew Point. The temperature to which air with a given quantity of water vapor must be cooled to cause condensation of the vapor in the air. Dew point test equipment and procedures are available from Lectrodryer, 135 Quality Drive, Richmond, KY 40475, and TSI Incorporated, Alnor Products, 500 Cardigan Road, Shoreview, MN 55126-3996.*

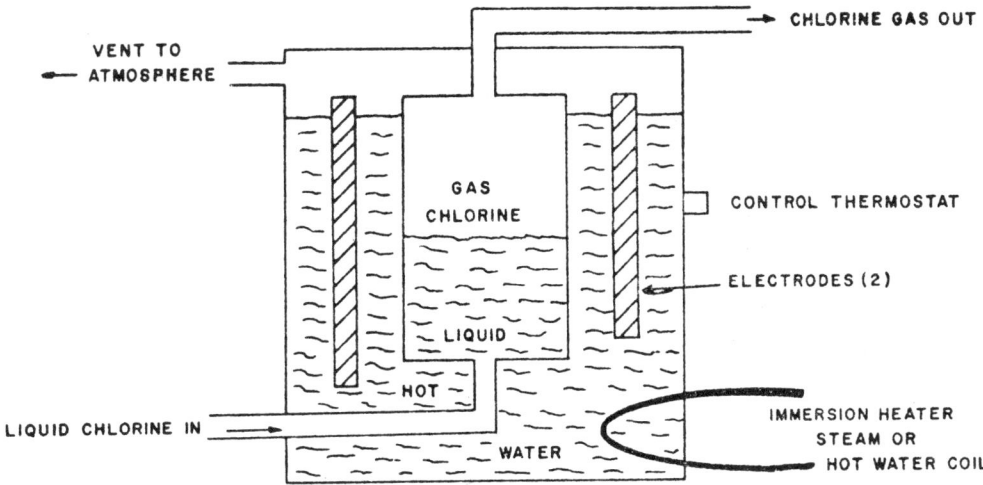

Fig. 7.31   Liquid chlorine evaporator
(Source: *BASIC GAS CHLORINATION WORKSHOP MANUAL*, 1972, published by Training and
Licensing Section, Ministry of Environment, Ontario, Canada)

(4) Turn on the heaters on the evaporators.

(5) Wait until the temperature of the evaporators reaches 180°F (82°C). This may take over an hour on large units.

(6) Inspect and close all valves on the chlorine supply line. One chlorine valve on the evaporator (the inlet valve) should be open to allow for expansion of the chlorine if heated.

(7) Open the chlorine metering orifice slightly by adjusting the chlorine feed rate knob. This is to prevent damage to the rotameter.

(8) Start the injector water supply system.

(9) Examine injector water supply system.

   (a) Note reading of injector water supply pressure gage. If gage reading is abnormal (different than usual reading), try to identify the cause and correct it. Injectors should operate at not less than 50 psi inlet pressure.

   (b) Note reading of injector vacuum gage. If system does not have a vacuum gage, have one installed. If the vacuum reading is less than normal, the machine may function at a lower feed rate, but will be unable to deliver at the maximum rated capacity.

(10) Inspect chlorinator vacuum lines for leaks.

(11) On the system connected to the gas side of the supply container, crack open the CHLORINE GAS LINE at the chlorine container. All liquid chlorine systems should be checked by using gas because of the danger of leaks (one liter of liquid will produce 450 liters of gas). Inspect the joints between this valve and the next one downstream. If this valve passes the ammonia leak test, continue to the next valve down the line. Follow this procedure until the evaporator is reached. Before allowing chlorine to enter the evaporator and the chlorinator, make sure that all valves between the evaporator and the chlorinator are open. Heat in the evaporator will expand the

gas, and if the system is closed, excessive pressure can develop. Chlorine should never be trapped in a line, an evaporator, or a chlorinator. After the gas test, reconnect to liquid side and retest at connection and pigtail valve.

(12) If no problems develop, the gas line can be put in service by opening the valve one complete turn.

(13) Check the operation of the chlorinator.

   (a) Operate over complete range of chlorine feed rates.

   (b) Check operation on manual and automatic settings.

(14) If the chlorinator is operating properly, close the gas valve on the container, wait a few minutes to evacuate pigtail connector, close valve from pigtail to manifold, disconnect pigtail, replace gas valve cap, remove liquid valve cap, clean valve face and threads, install new gasket to connect pigtail, crack valve open and then close. Check for leak at new connection. If no leak, open liquid valve two turns (usually one turn open is sufficient).

(15) After admitting liquid chlorine to the system, wait until the temperature of the evaporator again reaches 180°F (82°C) and full working pressure (100 psi, 690 kPa or 7 kg/sq cm). Inspect the evaporator by looking for leaks around pipe joints, unions, and valves.

(16) The system is ready for normal operation.

3. *SHUTDOWN OF A CHLORINATOR (when operating from container gas connection)*

   Work in pairs. A plan should be used where both people are not exposed to the chlorine at the same time.

   a. Short-Term Shutdown

      The following is a typical procedure for shutting down a chlorinator for a time period of less than one week.

   (1) Have safety equipment available in the event of a chlorine leak.

(2) Close chlorine container gas outlet valve.

(3) Allow chlorine gas to completely evacuate the system through the injector. Chlorine gas pressure gages will fall to zero psi on the manifold and the chlorinator.

(4) Close chlorinator gas discharge valve. The chlorinator may remain in this condition indefinitely and is ready to be placed back in service by reopening the chlorinator discharge valve and the chlorine container gas outlet valve. After these valves have been reopened, inspect for chlorine leaks throughout the chlorination system.

b. Long-Term Shutdown

(1) Perform steps one through four above for short-term shutdown.

(2) Turn off chlorinator power switch, lock out, and tag.

(3) Secure chlorinator gas manifold and chlorinator valve in closed position.

QUESTIONS

Write your answers in a notebook and then compare your answers with those on page 357.

7.4N When starting a gas chlorinator, how is the system checked for chlorine gas leaks?

7.4O List the steps to follow when shutting down a chlorinator for a long time period.

7.45 Normal and Abnormal Operation

Normal operation of the chlorination process requires regular observation of facilities and a regular preventive maintenance program. When something abnormal is observed or discovered, corrective action must be taken. This section outlines normal operation procedures and also possible operator responses to abnormal conditions. Exact procedures will depend on the type of equipment in your plant.

7.450 Container Storage Area

DAILY

1. Inspect building or area for ease of access by authorized personnel to perform routine and emergency duties.

2. Be sure fan and ventilation equipment are operating properly.

3. Read scales, charts, or meters at the same time every day to determine use of chlorine and any other chemicals. Notify plant superintendent when chlorine supply is low.

4. Look at least once per shift for chlorine and chemical leaks.

5. Try to maintain temperature of storage area below temperature of chlorinator room.

6. Determine manifold pressure before and after chlorine pressure regulating valve.

7. Be sure all chlorine containers are properly secured.

WEEKLY

1. Clean building or storage area.

2. Check operation of chlorine leak-detector alarm.

MONTHLY

1. Exercise all valves, including flex connector's auxiliary, manifold, filter bypass, and switchover valves.

2. Inspect all flex connectors and replace any that have been kinked or flattened.

3. Inspect hoisting equipment.

   a. Cables: frayed or cut.

   b. Container hoisting beam and hooks: cracked or bent.

   c. Controls: operate properly; do not stick or respond sluggishly; cords not frayed; safety chains or cables in place.

   d. Hoist travels on monorail easily and smoothly.

4. Examine building ventilation.

   a. Ducts and louvers: clean and operate freely.

   b. Fans and blowers: operate properly; guards in place; equipment properly lubricated.

5. Perform preventive maintenance as scheduled. These duties may include:

   a. Lubricating equipment (monthly to quarterly).

   b. Repacking valves and regulators (6 to 12 months).

   c. Cleaning and replacing valve seats and stems (annually).

   d. Cleaning filters and replacing glass wool (annually). (*CAUTION:* Glass wool soaked with liquid chlorine or chlorine impurities may burn your skin or give off sufficient chlorine gas to be dangerous.)

   e. Painting equipment if needed (annually).

   f. Inspecting condition and parts of all repair kits and safety equipment (monthly).

7.451 Evaporators (Figure 7.31)

Evaporators are used to convert liquid chlorine to gaseous chlorine for use by gas chlorinators.

DAILY

1. Check evaporator water bath to be sure water level is at midpoint of sight glass.

2. In most evaporators the water bath temperature is between 160 and 195°F (71 and 91°C). Low alarm should sound at 160°F (71°C) and high alarm should sound at 200°F (93°C).

3. Determine chlorine inlet pressure to evaporator. Pressure should be same pressure as on supply manifold from containers (20 to 100 psi, 138 to 690 kPa or 1.4 to 7 kg/sq cm).

4. Determine chlorine outlet temperature from evaporator. Typical range is 90 to 105°F (32 to 41°C). High alarm should sound at 110°F (43°C). At low temperatures the chlorine pressure-reducing valve (CPRV) will close due to the low temperature in the water bath.

5. Check chlorine pressure-reducing valve (CPRV) operation.

6. If evaporator is equipped with water bath recirculation pump at back of evaporator, determine if pump is operating properly.

7. Look for leaks and repair any discovered.

ABNORMAL EVAPORATOR CONDITIONS

1. Evaporator water level low. Water level not visible in sight glass.

   Troubleshooting Measures

   a. Determine actual level of water.

   b. Monitor temperature of water.

   c. Check temperature and pressure of chlorine in evaporator system and feed lines back to containers and chlorinators for possible overpressure of system (pressure should not exceed 100 psi, 690 kPa or 7 kg/sq cm).

   Corrective Action

   a. If chlorine pressure on system is near or over 100 psi (690 kPa or 7 kg/sq cm), close supply container valves to stop chlorine addition to the system, increase feed rate of chlorinator to use chlorine in the system, and drop the pressure back down to a safe range. *NOTE:* Alarm system is usually set to sound at 110 psi (758 kPa or 7.7 kg/sq cm). If system pressure was at or over 110 psi (758 kPa or 7.7 kg/sq cm), inspect alarm circuit to determine why the alarm failed.

   b. If temperature of water bath is set at an abnormal level, find the cause. Water bath levels are usually set in the following sequence:

      (1) 160°F or 71°C: low temperature alarm.
      (2) 185 to 195°F or 85 to 91°C: normal operating range.
      (3) 185°F or 85°C: actuates pressure-reducing valve (PRV) to open position.
      (4) 200°F or 93°C: high temperature alarm.

   If temperature is above 200°F (93°C), alarm should have sounded. Open control switch on evaporator heaters to stop current flow to heating elements. If temperature is in normal range, return to correcting the original problem of a low water level.

   c. Evaporator water levels are controlled by a solenoid valve. First check to see if drain valve is fully closed and then use the following steps:

      (1) Override solenoid valve and fill water bath to proper level. *NOTE:* The installation of a manual bypass on the water line would allow continued operation of the evaporator in the event of solenoid failure and would facilitate solenoid valve replacement with the unit on line.

      (2) If water cannot be added, take evaporator out of service:

         (a) Try to remove all liquid chlorine left in line,
         (b) Switch to another evaporator, or
         (c) Switch supply to gas side of containers. May have to connect more containers to supply sufficient chlorine gas.

2. Low water temperature in evaporator.

   a. Check chemical (chlorine) flow-through rate. Rate may have exceeded unit's capacity and may require two evaporators to be on line to handle chemical feed rate.

   b. Inspect immersion heaters for proper operation. First examine control panel for thermal overload on breaker. Most evaporators are equipped with two to three heating elements. An inspection of the electrical system will indicate if the breakers are shorted or open and will locate the problem. Replace any heating elements that have failed.

   c. If no spare evaporators are available, operate from the valves on the chlorine gas supply. If necessary, reduce the chlorine feed rate to keep the chlorination system working properly.

3. No chlorine gas flow to the chlorinator.

   a. Inspect pressure-reducing valve downstream from evaporator and determine if valve is in the open or closed position.

      (1) Valve may be closed due to low water temperature in evaporator (less than 185°F or 85°C).
      (2) Valve may be closed due to loss of vacuum on system or loss of continuity of electrical control circuits which may have been caused by a momentary power drop. Correct problem and reset valve.
      (3) Valve may be out of adjustment and restricting gas flow through the valve due to a low pressure setting.

   b. Inspect supply containers and manifold. Possible sources of lack of chlorine gas flow to chlorinators include:

      (1) Containers empty.
      (2) Container chlorine supply lines incorrectly connected to gas instead of liquid side of containers. High flow rates of gas will remove gas from container faster than it can change from liquid to gas. This will cause a reduced flow of chlorine gas. When this happens a frost may appear on valves and flex connectors. Reduce flow of water being treated and/or connect more chlorine containers.

         WARNING. When frost appears on valves and flex connectors, the chlorine gas may condense to liquid chlorine (reliquify). The liquid chlorine may plug the chlorine supply lines (sometimes this is referred to as chlorine ice or frozen chlorine). If you disconnect the chlorine supply line to unplug it, *BE VERY CAREFUL.* The liquid chlorine in the line could reevaporate, expand as a gas, build up pressure in the line, and cause liquid chlorine to come shooting out the open end of a disconnected chlorine supply line.

(3) Chlorine manifold filters plugged. Check pressure upstream and downstream from filter. Pressure drop should not exceed 10 psi (6.9 kPa or 0.7 kg/sq cm). Frost on manifold may indicate an excessive flow of chlorine gas through the filters.

(4) Inspect manifold and system for closed valves. Most systems operate properly with all chlorine valves at only ONE TURN OPEN position.

MONTHLY

1. Exercise all valves, including inlet, outlet, pressure reducing (PRV), water, drain, and fill valves.

2. Inspect evaporator cathodic protection meter (if so equipped). Cathodic protection protects the metal water tank and piping from corrosion due to electrolysis. Electrolysis is the flow of electric current and is the reverse of metal plating. In electrolysis, the flow of certain compounds away from the metal causes corrosion and holes in a short time. This type of corrosion is controlled by either a sacrificial anode made of magnesium and zinc or by applying small electric currents to suppress or reverse the normal corroding current flow.

3. Check setting of PRV (pressure-reducing valve) in order to maintain desired pressure of chlorine gas to chlorinators.

4. Inspect heating and ventilating equipment in chlorinator area.

5. Perform scheduled routine preventive maintenance.

   a. Drain and flush water bath.

   b. Clean evaporator tank.

   c. Check heater elements.

   d. Repack gasket and reseat pressure-reducing valves (annually).

   e. Replace anodes (annually).

   f. Paint system (annually).

7.452 Chlorinators, Including Injectors

DAILY

1. Check injector water supply pressure. Pressures will range from 40 to 90 psi (276 to 520 kPa or 2.8 to 6.3 kg/sq cm) depending on system.

2. Determine injector vacuum. Values will range from 15 to 25 inches (38 to 64 cm) of mercury.

3. Check chlorinator vacuum. Values will range from 5 to 10 inches (13 to 25 cm) of mercury.

4. Determine chlorinator chlorine supply pressure. Values will range from 20 to 40 psi (138 to 276 kPa or 1.4 to 2.8 kg/sq cm) after the pressure regulating valve.

5. Read chlorinator feed rate on rotameter tube. Is feed rate at required level? Record rotameter reading and time.

6. Calculate the chlorine usage.

7. Examine and record mode of control.

   a. Manual

   b. Automatic (single input)

   c. Automatic (dual input)

8. Measure chlorine residual at application point.

9. Inspect system for chlorine leaks.

10. Inspect auxiliary components.

   a. Flow signal input. Does chlorinator feed rate change when flow changes? Chlorinator response is normally checked by biasing (adjusting) flow signal which may drive dosage control unit on chlorinator to full open or closed position. When switch is released, chlorinator will return to previous feed rate. During this operation the unit should have responded smoothly through the change. If the response was not smooth, look for mechanical problems of binding, lubrication, or vacuum leaks.

   b. If chlorinator also is controlled by a residual analyzer, be sure the analyzer is working properly. Check the following items on the residual analyzer. Be sure to follow the manufacturer's instructions.

      (1) Actual chlorine residual is properly indicated.
      (2) Recorder alarm set point.
      (3) Recorder control set point.
      (4) Sample water flow.
      (5) Sample water flow to cell block after dilution with fresh water.
      (6) Adequate flow of dilution water.
      (7) Filter system and drain.
      (8) Run comparison tests of chlorine residual. Do tests match with analyzer output readings?
      (9) If residual analyzer samples two streams, start other stream flow and compare tested residual of that stream with analyzer output readings. Standardize analyzer output readings against tested residuals. Enter changes and corrections in log.
      (10) Change recorder chart daily or as necessary. If a strip chart is used, date the chart for recordkeeping purposes.
      (11) Check recorder output signal controlling chlorinator for control responses on feed rate. Correct feed rates through ratio controller.

WEEKLY

1. Put chlorinator on manual control. Operate feed rate adjustment through full range from zero to full scale (250, 500, 1,000, 2,000, 4,000, 6,000, 8,000, or 10,000 pounds/day). At each end of scale check:

   a. Chlorinator vacuum.

   b. Injector vacuum.

   c. Solution line pressure.

d. Chlorine pressure at chlorinator.

If any of the readings do not produce normal set points, make proper adjustments.

a. Injector should produce necessary vacuum at chlorinator (5 to 10 inches or 13 to 25 centimeters of mercury).

b. Adjust PRV to obtain sufficient pressure and chemical feed for full feed rate operation of chlorinator. This CPRV is on the main chlorine supply line and is usually located near or in the chlorine container area.

2. If unit performs properly through complete range of feed rates, return unit to automatic control. If any problems develop, locate source and correct.

3. Clean chlorine residual analyzer (see Section 7.7, "Measurement of Chlorine Residual"), including the following items:

a. Clean filters.

b. Clean sample line.

c. Clean hydraulic dilution wells and baffles.

d. Flush discharge hoses and pipes.

e. Clean and flush cell block.

f. Fill buffer reservoirs.

g. Check buffer pump and feed rate.

h. Wipe machine clean and keep it clean.

4. Calculate chlorine usage so that replacement supply containers can be ordered and constant chlorination can be maintained. Try to have a 15- to 30-day supply of chlorine in storage.

MONTHLY

1. Exercise all chlorine valves.

2. Inspect heaters and room ventilation equipment.

3. Check chlorinator vent line to outside of structure for any obstructions that could prevent free access to the atmosphere. Bugs and wasps like vent lines for nests. If this becomes a problem, install a fine wire mesh over the open end of the vent line.

4. Inspect unit for vacuum leaks.

5. Clean rotameter sight glass.

6. Inspect all drain lines and hoses.

7. Perform scheduled routine maintenance.

a. Repack seat and stem of valves.

b. Inspect tubing and fittings for leaks. Wash and dry thoroughly before reassembling.

c. Inspect control system.

(1) Electrical and electronics.
(2) Pneumatics.
(3) Lubrication.
(4) Calibration of total system.

d. Chlorine analyzer.

(1) Lubrication of chart drives, filter drives, and pumps.
(2) Clean and flush all piping and hoses, filters, tubing, cell blocks, and hydraulic chambers.
(3) Clean acid and iodide reservoirs.
(4) Calibrate unit with known standards.
(5) Repaint unit if needed.

e. Inspect safety equipment, including self-contained breathing equipment and repair kits.

ANNUALLY

1. Disassemble, clean, and regasket chlorinator.

POSSIBLE ABNORMAL CONDITIONS

1. Chlorine leak in chlorinator.

Shut off gas flow to chlorinator. Leave injector on line. Place another chlorinator on line, if available, until repairs are accomplished. Allow the chlorinator to operate under a vacuum with zero psi showing on the chlorine pressure gage for three to five minutes to remove chlorine gas. If the system must be opened to repair the leak, the vacuum on the chlorinator may minimize the potential for residual chlorine to be released to the atmosphere. If effective evacuation of the unit cannot be ensured, appropriate respiratory protection must be used.

2. Gas pressure too low, less than 20 psi (138 kPa or 1.4 kg/sq cm). Alarm indicated. Check chlorine supply:

a. Empty containers, switch to standby units.

b. Evaporator shut down. See Section 7.451, "Evaporators."

c. Inspect manifold for closed valves or restricted filters. Correct by switching to another manifold or cleaning/replacing filters, or setting valves and controls to proper position.

3. Injector vacuum too low.

   a. Adjust injector to achieve required vacuum.

   b. Inspect injector water supply system.

   (1) Pump off: start pump.
   (2) Strainers dirty: clean strainers.
   (3) Pump worn out and will not deliver appropriate flow and pressure to injector: use other unit and/or repair or replace pump.
   (4) Inspect valves in system. Place valves in proper position.

   c. Inspect solution line discharge downstream from injector. Check for the following items:

   (1) Valve closed or partially closed.
   (2) Line broken or restriction reducing flow or increasing back pressure.
   (3) Diffuser plugged, thus restricting flow and creating a higher back pressure on discharge line and injector. Clean diffuser and flush pipe.
   (4) *NOTE:* The installation of a pressure gage on the discharge side of the injector or on the solution line would alert the operator to abnormal system back pressure.

4. Low chlorine residual. Alarm indicator is on from chlorine residual analyzer.

   Determine actual chlorine residual and compare with residual reading from chlorine analyzer. If residual analyzer is indicating a different chlorine residual, recalibrate analyzer and readjust. If chlorine residual analyzer is correct and chlorine residual is low, check the following items:

   a. Sample pump

   (1) Operation, flow, and pressure.
   (2) Sample lines clean and free of solids or algae that could create a chlorine demand.
   (3) Strainer dirty and restricting flow, thus preventing adequate pressure (15 to 20 psi, 103 to 138 kPa, or 1.0 to 1.4 kg/sq cm) at analyzer.

   b. Control system if chlorinator is on automatic control. If chlorine feed rate remains too low, switch chlorinator from automatic control to manual control. Set chlorina-

tor to proper feed rate as determined by previous adequate feed rates.

   c. Chlorine demand higher than the amount one chlorinator can supply. Place additional chlorinator on line.

TROUBLESHOOTING DIRECT MOUNT CHLORINATORS

See Table 7.5.

QUESTIONS

Write your answers in a notebook and then compare your answers with those on page 357.

7.4P   Normal operation of a chlorinator includes daily inspection of what facilities or areas?

7.4Q   What is the purpose of evaporators?

7.4R   What abnormal conditions could be encountered when operating an evaporator?

7.4S   What are possible chlorinator abnormal conditions?

7.4T   How can you determine if the chlorine residual analyzer is working properly?

7.453   Summary, Daily Operation

Actual procedures for operating chlorination equipment will vary from plant to plant, region to region, season to season, and water to water. The following procedures are provided to serve as guidelines to help you develop your own procedures.

1. *PRECHLORINATION*

   If trihalomethanes are not a problem, prechlorination is a very cost-effective means of disinfecting water. Prechlorination should be at a chlorine dosage that will produce a free chlorine residual (past the breakpoint) of 0.5 to 1.5 mg/L before the flocculation basins. If trihalomethanes are a problem, see Volume II, Chapter 15, "Trihalomethanes." If phenolic compounds are present, discontinue prechlorination, adsorb phenols with activated carbon and remove by filtration, and disinfect by postchlorination.

2. *POSTCHLORINATION*

   Regardless of whether prechlorination is practiced, a free chlorine residual of at least 0.5 mg/L should be main-

TABLE 7.5   DIRECT MOUNT CHLORINATOR TROUBLESHOOTING GUIDE

Operating Symptom	Probable Cause	Remedy
1. Water in the chlorine metering tube.	Check valve failure, deposits on seat of check valve, or check valve seat distorted by high pressure.	Clean deposits from check ball and seat with dilute muriatic acid. Badly distorted check valve may have to be replaced.
2. Water venting to atmosphere.	Excess water pressure in the vacuum regulator.	Remove vacuum regulator from chlorine cylinder and allow chlorinator to pull air until dry.
3. No indication on flowmeter when vacuum is present.	Vacuum leak due to bad or brittle vacuum tubing, connections, rate valve o-rings, or gasket on top of flowmeter.	Check the vacuum tubing, rate valve o-rings, and flowmeter gasket for vacuum leaks. Replace bad tubing connectors, o-rings, or gasket.
4. Indication on flowmeter but air present, not chlorine gas.	Connection below meter tube gasket leaks.	Check connections and replace damaged elements.

tained in the clear well or distribution reservoir immediately downstream from the point of postchlorination.

3. *DISTRIBUTION SYSTEM*

Postchlorination dosages should be adequate to produce a free chlorine residual of 0.2 mg/*L* at the farthest point in the distribution system at all times.

Very often when consumers complain about chlorine tastes in their drinking water, the chlorine dose has been *INADEQUATE*. When the chlorine dose is inadequate, the measured chlorine residual is frequently combined available chlorine. By increasing the chlorine dose, you chlorinate past the breakpoint and the chlorine residual may be lower, but the residual contains free available chlorine. One way to determine if you have reached the chlorine breakpoint is to increase the chlorine dose rate. If the residual chlorine increases in proportion to the increased dose, then you are chlorinating past the breakpoint. For a discussion of break-point chlorination, see Section 7.25, "Breakpoint Chlorination."

The objective of disinfection is the destruction or inactivation of pathogenic organisms, and the ultimate measure of the effectiveness is the bacteriological result (negative coliform test results). The measurement of chlorine residual does supply a tool for practical control. If the chlorine residual value commonly effective in most water treatment plants does not yield satisfactory bacteriological kills in a particular plant, the residual chlorine that does produce satisfactory results must be determined and used as a control in that plant. In other words, the 0.5 mg/*L* chlorine residual, while generally effective, is not a rigid standard but a guide that may be changed to meet local requirements.

FORMULAS

To determine if the chlorinator setting is high enough to produce a free available residual chlorine past the breakpoint, we can increase the chlorinator feed rate. If all of the chlorine fed at the new or higher setting is converted to chlorine residual in milligrams per liter, we know that the original setting was at or past the breakpoint. The expected increase in chlorine residual can be calculated by using the formula:

$$\text{Expected Increase in Residual, mg/}L = \frac{\text{New Setting, lbs/day} - \text{Old Setting, lbs/day}}{(\text{Flow, MGD})(8.34 \text{ lbs/gal})}$$

To determine the actual increase in free chlorine residual, find the difference in milligrams per liter between the new and old free available residual chlorine.

$$\text{Actual Increase, mg/}L = \text{New Residual, mg/}L - \text{Old Residual, mg/}L$$

EXAMPLE 4

A chlorinator is set to feed chlorine to a treated water at a dose of 27 pounds of chlorine per 24 hours. This dose rate produces a free chlorine residual of 0.4 mg/*L*. When the chlorinator setting is increased to 30 pounds per 24 hours, the free chlorine residual increases to 0.6 mg/*L*. If the average 24-hour flow is 1.6 MGD, is the water being chlorinated past the breakpoint?

Known	**Unknown**
Old Setting, lbs/day = 27 lbs/day	1. Increase in Residual, mg/*L*
New Setting, lbs/day = 30 lbs/day	2. Past Breakpoint?
Old Residual, mg/*L* = 0.4 mg/*L*	
New Residual, mg/*L* = 0.6 mg/*L*	
Flow, MGD = 1.6 MGD	

1. Calculate the expected increase in chlorine residual if chlorination is at the breakpoint.

$$\text{Expected Increase in Residual, mg/}L = \frac{\text{New Setting, lbs/day} - \text{Old Setting, lbs/day}}{(\text{Flow, MGD})(8.34 \text{ lbs/gal})}$$

$$= \frac{(30 \text{ lbs/day} - 27 \text{ lbs/day})}{(1.6 \text{ MGD})(8.34 \text{ lbs/gal})}$$

$$= 0.2 \text{ mg/}L$$

2. Determine actual increase in free chlorine residual in mg/*L*.

$$\text{Actual Increase, mg/}L = \text{New Residual, mg/}L - \text{Old Residual, mg/}L$$

$$= 0.6 \text{ mg/}L - 0.4 \text{ mg/}L$$

$$= 0.2 \text{ mg/}L$$

3. Is the water being chlorinated past the breakpoint? Yes, since the calculated expected increase in chlorine residual (0.2 mg/*L*) is approximately the same as the actual increase in the chlorine residual (0.2 mg/*L*), we can conclude that we are chlorinating past the breakpoint.

7.454 Laboratory Tests

1. Chlorine Residual in System

   a. Daily chlorine residual tests using the *DPD*[30] *METHOD*[31] should be taken at various locations in the system. A remote tap is ideal for one sampling location. Take the test sample from a tap as close to the main as possible. Allow the water to run at least 5 minutes before sampling to ensure a representative sample from the main.

      Operators using the DPD colorimetric method to test water for a free chlorine residual need to be aware of a potential error that may occur. If the DPD test is run on

---

[30] *DPD (pronounce as separate letters). A method of measuring the chlorine residual in water. The residual may be determined by either titrating or comparing a developed color with color standards. DPD stands for N,N-diethyl-p-phenylene-diamine.*

[31] *See Chapter 11, "Laboratory Procedures," for details on how to perform the DPD test for measuring chlorine residual. Also see Section 7.7, "Measurement of Chlorine Residual," later in this chapter.*

water containing a combined chlorine residual, a precipitate may form during the test. The particles of precipitated material will give the sample a turbid appearance or the appearance of having color. This turbidity can produce a positive test result for free chlorine residual when there is actually no chlorine present. Operators call this error a "false positive" chlorine residual reading.

b. Chlorine residual test kits are available for small systems.

2. Bacteriological Analysis (Coliform Tests)

Samples should be taken routinely in accordance with EPA and health department requirements. Take samples according to approved procedures.[32] Be sure to use a sterile plastic or glass bottle. If the sample contains any chlorine residual, sufficient sodium thiosulfate should be added to neutralize all of the chlorine residual. Usually 0.1 milliliter of 10 percent sodium thiosulfate in a 120-mL (4-oz) bottle is sufficient for distribution systems. The "thio" should be added to the sample bottle before sterilization.

7.46 Troubleshooting Gas Chlorinator Systems

Operating Symptom	Probable Cause	Remedy
1. Injector vacuum reading low	1. Hydraulic system 2. Flow restricted 3. Low pressure 4. High pressure 5. Back pressure 6. Low flow of water	1. Check injector water supply system 2. Adjust injector orifice 3. Close throat 4. Open throat 5. Change injector and/or increase water supply to injector 6. Increase pump output
2. Leaking joints	1. Missing gasket	1. Repair joint
3. Chlorinator will not reach maximum point	1. Faulty injector (no vacuum) 2. Restriction in supply 3. Faulty chlorinator 4. Leaks 5. Wrong orifice	1. Repair injector system 2. Find restriction in supply system 3. Check for vacuum leaks 4. Repair leaks 5. Install proper orifice
4. Chlorinator will feed OK at maximum output, but will not control at low rates	1. Vacuum regulating valve 2. If equipped with CPRV (Chlorine Pressure Reducing Valve)	1a. Repair diaphragm 1b. Check valve capsule 2. Clean CPRV cartridge, CPRV diaphragm, and CPRV gaskets
5. Chlorinator does not feed	1. Supply 2. Piping	1. Renew Cl$_2$ supply 2a. Open valve 2b. Clean filter
6. Variable vacuum control, formerly working well, now will not go below 30% feed. Signal OK	1. CPRV	1. Clean CPRV
7. Variable vacuum control reaches full feed, but will not go below 50% feed. CPRV OK	1. Signal vacuum too high	1a. Hole in diaphragm 1b. Clean dirty filter disks 1c. Clean converter nozzle
8. Variable vacuum control won't go to full feed. Gas pressure OK. CPRV OK	1. Plugged restrictor 2. Air leak in signal	1. Clean restrictor 2. Repair air leak
9. Freezing of manometer	1. Rate too high 2. Restriction in manometer orifice	1. Lower rate 2. Clean piping

---

[32] See Chapter 11, "Laboratory Procedures," for proper procedures for collecting and analyzing samples for chlorine residuals and coliform tests.

7.47 Disinfection Troubleshooting

TABLE 7.6 DISINFECTION TROUBLESHOOTING GUIDE

Operating Symptom	Probable Cause	Remedy
1. Increase in coliform level	Low chlorine residual	Raise chlorine dose
2. Drop in chlorine level	a. Increase in chlorine demand	Raise chlorine dose and find out why chlorine demand increased or chlorine feed rate dropped
	b. Drop in chlorine feed rate	

7.48 Chlorination System Failure

IF YOUR CHLORINATION SYSTEM FAILS, DO NOT ALLOW UNCHLORINATED WATER TO ENTER THE DISTRIBUTION SYSTEM. Never allow unchlorinated water to be delivered to your consumers. If your chlorination system fails and cannot be repaired within a reasonable time period, notify your supervisor and officials of the health department. To prevent this problem from occurring, your plant should have backup or standby chlorination facilities.

7.49 Acknowledgment

Some of the material in this section on gas chlorinators was prepared by Joe Habraken, Treatment Supervisor, City of Tampa, Florida. His contribution is greatly appreciated.

QUESTIONS

Write your answers in a notebook and then compare your answers with those on page 357.

7.4U What is the suggested free chlorine residual for treated water (measured at a point just beyond post-chlorination)?

7.4V What is the suggested free chlorine residual for the farthest point in the distribution system?

7.4W How would you determine if you were chlorinating at the breakpoint?

End of Lesson 2 of 3 Lessons on DISINFECTION

Please answer the discussion and review questions next.

DISCUSSION AND REVIEW QUESTIONS

CHAPTER 7. DISINFECTION

(Lesson 2 of 3 Lessons)

Write the answers to these questions in your notebook before continuing. The question numbering continues from Lesson 1.

6. Why should chlorine containers and cylinders be stored where they won't be heated?

7. Why should precautions be taken to avoid sharp bends in copper tubing used to convey chlorine?

8. How should valves on chlorine containers be opened?

9. What should be done to the water supply pump when the hypochlorinator is shut down?

10. Why must liquid chlorine piping systems be dry?

11. If consumers complain about chlorine tastes in their drinking water, what is the most likely cause of this problem?

CHAPTER 7. DISINFECTION

(Lesson 3 of 3 Lessons)

7.5 MAINTENANCE

Necessary maintenance of chlorination equipment requires a thorough understanding of the manufacturer's literature. Generally the daily, weekly, and monthly operating procedures also contain the appropriate maintenance procedures. Do not attempt maintenance tasks that you are not qualified to perform. There are too many possibilities for serious accidents when you are not qualified.

7.50 Hypochlorinators

See Section 7.440, "Hypochlorinators," 3. Normal Operation of Hypochlorinators (page 311), and 5. Maintenance of Hypochlorinators (page 313).

7.51 Chlorinators

See Section 7.441, "Chlorinators" (page 313), and Section 7.45, "Normal and Abnormal Operation" (page 316).

7.52 Chlorine Leaks

Your sense of smell can detect chlorine concentrations as low as 3 ppm. Portable and permanent automatic chlorine detection devices (Figure 7.32) can detect chlorine concentrations of 1 ppm or less. Whenever you must deal with a chlorine leak, always follow the safe procedures outlined in Section 7.8, "Chlorine Safety Program."

CHLORINE LEAKS MUST BE TAKEN CARE OF IMMEDIATELY OR THEY WILL BECOME WORSE. Corrective measures should be undertaken only by trained operators wearing proper safety equipment. All operators should be trained to repair chlorine leaks. Always work in pairs when looking for and repairing leaks. All other persons should leave the danger area during repairs until conditions are safe again.

If the leak is large, all persons in the adjacent areas should be warned and evacuated. Obtain help from your fire department. They have self-contained breathing equipment and can help evacuate people. The police department can help control curious sightseers. Repair crews and drivers of emergency vehicles must realize that vehicle engines will quit operating in the vicinity of a large chlorine leak because of the lack of oxygen. You must always consider your neighbors . . . PEOPLE, animals, and plants.

1. *BEFORE ANY NEW SYSTEM IS PUT INTO SERVICE*, it should be cleaned, dried, and tested for leaks. Clean and dry pipelines by flushing and steaming from the high end to allow condensate and foreign materials to drain out, or by the use of commercially available cleaning solvents compatible with chlorine. After the empty line is heated thoroughly, blow dry air through the line until it is dry. After drying, test the system for tightness with 150 psi (1,034 kPa or 10.5 kg/sq cm) dry air. Apply soapy water to the outside of joints to detect leaks. Small quantities of chlorine gas may now be introduced into the line, the test pressure built up with air, and the system tested for leaks with ammonia. Whenever a new system is tested for leaks, at least one chlorinator should be on the line to withdraw chlorine from the system in case of a leak. The same is true in case of an emergency leak at any installation. If a chlorinator is not running, at least one or more should be started. Preferably, all available chlorinators should be put on line.

2. *TO FIND A CHLORINE LEAK*, tie a rag on a stick, *DIP THE RAG*[33] in a strong ammonia solution,[34] and hold the rag near the suspected leaks. A polyethylene "squeeze bottle" filled with ammonia water to dispense ammonia vapor may also be used. Care should be taken to avoid spraying ammonia water on any leak or touching the cloth to any metal. White fumes will indicate the exact location of the leak. Location of leaks by this method may not be possible for large leaks which diffuse the gas over large areas. Do not use an ammonia spray bottle because the entire room could turn white if it is full of chlorine gas. Also any chlorine deposits will draw water from the ammonia and form an acid which will eat away any material it contacts.

3. *IF THE LEAK IS IN THE EQUIPMENT* in which the chlorine is being used, close the valves on the chlorine container at once. Repairs should not be attempted while the equipment is in service. All chlorine piping and equipment that is to be repaired by welding should be flushed with water or steam. Before returning equipment to use, it *MUST* be cleaned, dried, and tested as previously described.

---

[33] *A one-inch (2.5-cm) paint brush may be used instead of a rag.*

[34] *Use a concentrated ammonia solution containing 28 to 30 percent ammonia as NH_3 (this is the same as 58 percent ammonium hydroxide, NH_4OH, or commercial 26° Baumé).*

Portable Chlorine Leak Detector

(Permission of Leak-Tec Division American Gas & Chemical Co., Ltd.)

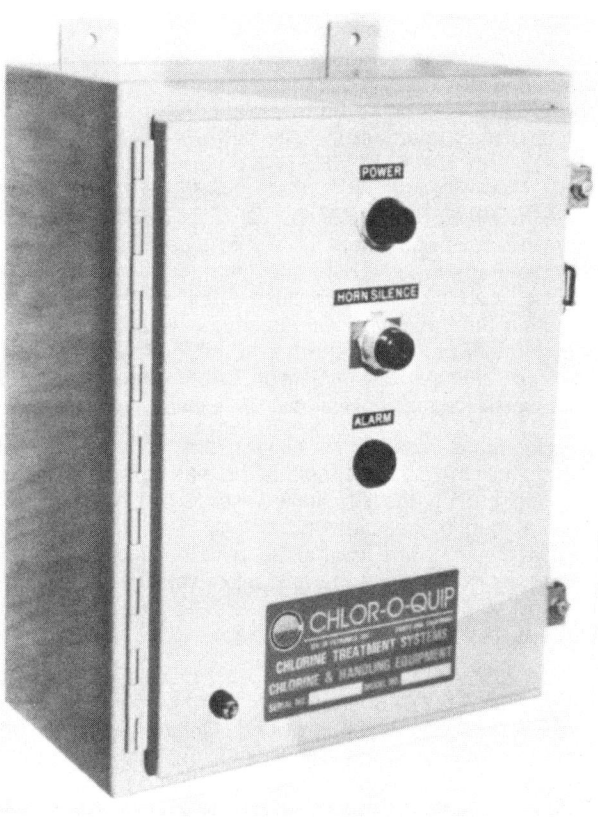

Audio-Visual Chlorine Alarm Panel

(Permission of Chlor-O-Quip, a Divison of Filtronics, Inc.)

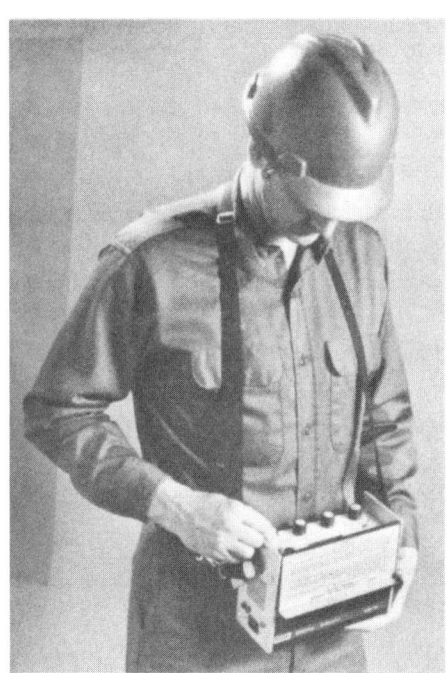

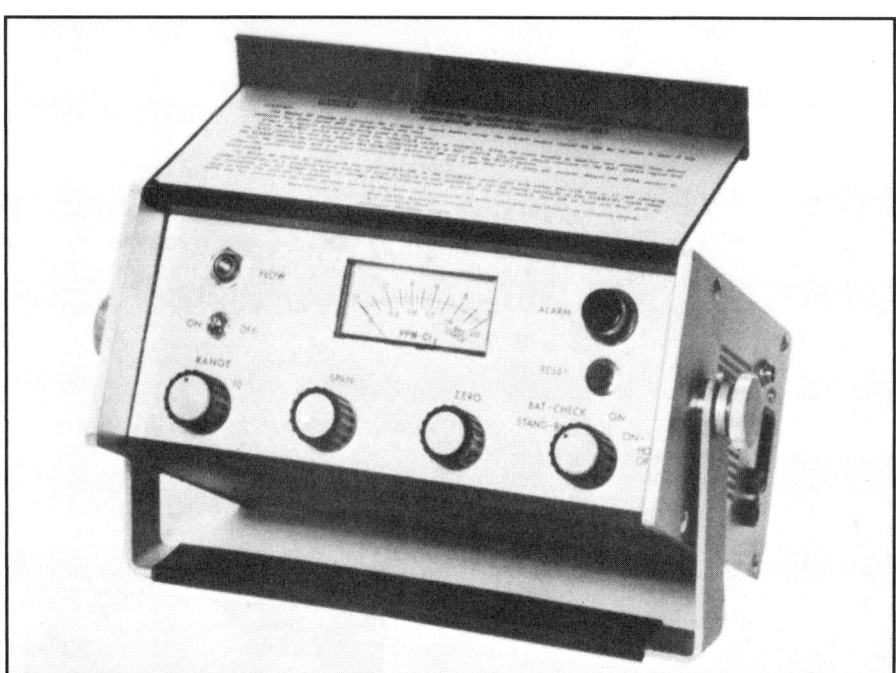

Fig. 7.32   Chlorine leak detectors

(Permission of MSA, Mine Safety Appliances Company)

4. *IF THE LEAK IS IN A CHLORINE CYLINDER OR CONTAINER,* use the emergency repair kit supplied by most chlorine suppliers (Figure 7.33). These kits can be used to stop most leaks in a chlorine cylinder or container and can usually be delivered to a plant within a few hours if one is not already at the site of the leak. *IT IS ADVISABLE TO HAVE EMERGENCY REPAIR KITS AVAILABLE AT YOUR PLANT AT ALL TIMES AND TO TRAIN PERSONNEL FREQUENTLY IN THEIR USE.* Respiratory protective equipment should be located outside chlorine storage areas. The repair kits may be located within chlorine storage areas because during an emergency requiring their use, you will already be wearing approved respiratory protection and hence they will be accessible. Refer to Figure 7.20 in Section 7.421, "Steel Cylinders," for typical locations of cylinder valve leaks.

5. If chlorine is escaping as a liquid from a cylinder or a ton tank, turn the container so that the leaking side is on top. In this position, the chlorine will escape only as a gas, and the amount that escapes will be only $\frac{1}{15}$ as much as if the liquid chlorine were leaking. Keeping the chlorinators running also will reduce the amount of chlorine gas leaking out of a container. Increase the feed rate to cool the supply tanks as much as possible.

6. *FOR SITUATIONS IN WHICH A PROLONGED OR UNSTOPPABLE LEAK* is encountered, emergency disposal of chlorine should be provided. Chlorine may be absorbed in solutions of caustic soda (sodium hydroxide), soda ash (sodium carbonate), or agitated hydrated lime (calcium hydroxide) slurries (Table 7.7). Chlorine should be passed into the solution through an iron pipe or a properly weighted rubber hose to keep it immersed in the absorption solution. The container should not be immersed because the leaks will be aggravated due to the corrosive effect, and the container may float when partially empty. In some cases it may be advisable to move the container to an isolated area. Discuss the details of such precautions with your chlorine supplier.

TABLE 7.7 CHLORINE ABSORPTION SOLUTIONS[a]

Absorption Solution	Container Size (lb net)	Chemical (lb)	Water (gal)
Caustic Soda (100%)	100	125	40
(Sodium Hydroxide)	150	188	60
	2,000	2,500	800
Soda Ash	100	300	100
(Sodium Carbonate)	150	450	150
	2,000	6,000	2,000
Hydrated Lime[b]	100	125	125
(Calcium Hydroxide)	150	188	188
	2,000	2,500	2,500

[a] Source: The Chlorine Institute, Inc.
[b] Hydrated lime solution must be continuously and vigorously agitated while chlorine is to be absorbed.

7. *NEVER PUT WATER ON A CHLORINE LEAK.* A mixture of water and chlorine will increase the rate of corrosion of the container and make the leak larger. Besides, water may warm the chlorine, thus increasing the pressure and forcing the chlorine to escape faster.

8. *LEAKS AROUND VALVE STEMS* can often be stopped by closing the valve or tightening the packing gland nut. Tighten the nut or stem by turning it clockwise.

9. *LEAKS AT THE VALVE DISCHARGE OUTLET* can often be stopped by replacing the gasket or adapter connection.

10. *LEAKS AT FUSIBLE PLUGS AND CYLINDER VALVES* usually require special handling and emergency equipment. Call your chlorine supplier immediately and obtain an emergency repair kit for this purpose if you do not have a kit readily available.

11. *PIN HOLE LEAKS* in the walls of cylinders and ton tanks can be stopped by using a clamping pressure saddle with a turnbuckle available in repair kits. This is only a temporary measure, and the container must be emptied as soon as possible.

If a repair kit is not available, use your ingenuity. One operator stopped a pinhole leak temporarily until a repair kit arrived by placing several folded layers of neoprene packing over a leak, a piece of scrap steel plate over the packing, wrapping a chain around the cylinder and steel plate, and applying leverage pressure with a crowbar.

Dry ice has been applied to chlorine containers to cool the liquid and thus reduce the amount of gas escaping through a pinhole leak.

12. *A LEAKING CONTAINER* must not be shipped. If the container leaks or if the valves do not work properly, keep the container until you receive instructions from your chlorine supplier for returning it. If a chlorine leak develops in transit, keep the vehicle moving until it reaches an open area.

13. Do not accept delivery of containers showing evidence of leaking, stripped threads, or abuse of any kind.

14. If a chlorine container develops a leak, be sure your supplier does not charge you for the unused chlorine.

15. Chlorine leaks may be detected by chlorine gas detection devices (Figure 7.32). Alarm systems may be connected to these devices. Be sure to follow the manufacturer's recommendations regarding frequency of checking and testing detection devices and alarm systems.

QUESTIONS

Write your answers in a notebook and then compare your answers with those on page 358.

7.5A If chlorine is escaping from a cylinder, what would you do?

7.5B How can chlorine leaks around valve stems be stopped?

7.5C How can chlorine leaks at the valve discharge outlet be stopped?

7.53 Installation

The following are some features of importance when working with chlorine facilities. Also examine these items when reviewing plans and specifications.

1. Chlorinators should be located as near the point of application as possible.

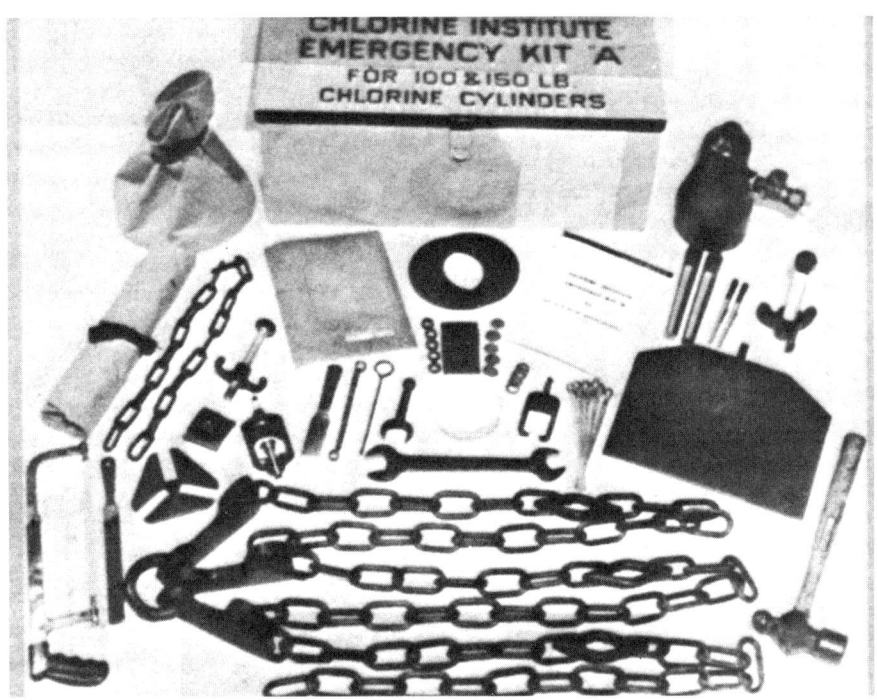

Emergency Kit "A" for Chlorine Cylinders

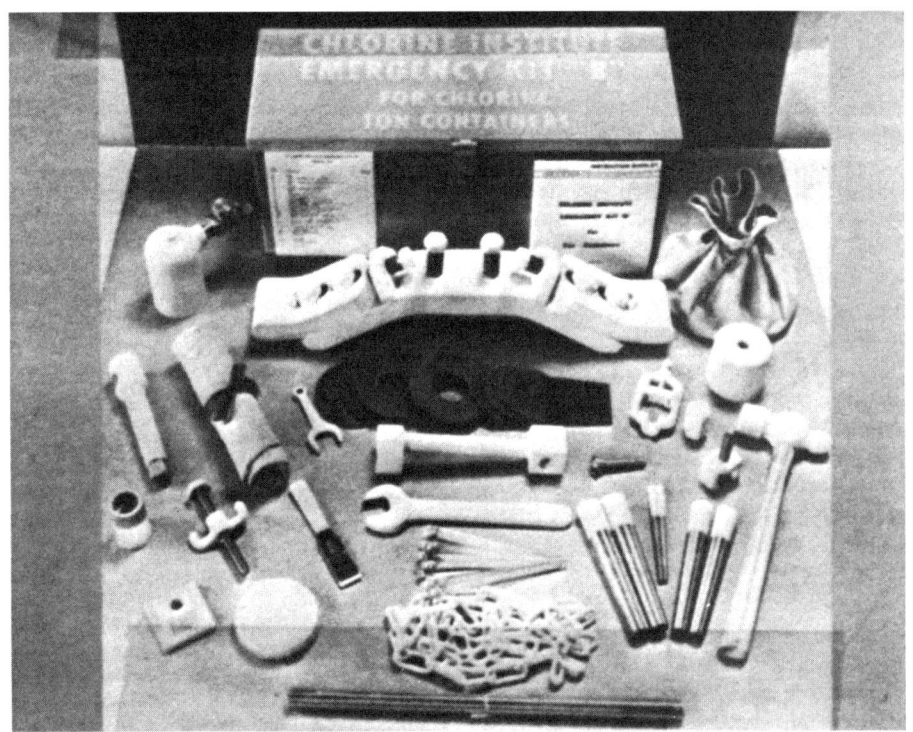

Emergency Kit "B" for Chlorine Ton Containers

Fig. 7.33   Chlorine Institute emergency repair kits

(Permission of The Chlorine Institute, Inc.)

2. There must be a separate room for chlorinators and chlorine container storage (above ground) to prevent chlorine gas leaks from damaging equipment and harming personnel. There should be no access to this room from a room containing equipment or where personnel work.

3. Ample working space around the equipment and storage space for spare parts should be provided.

4. There should be an ample supply of water to operate the chlorinator at required capacity under maximum pressure conditions at the chlorinator injector discharge.

5. The building should be adequately heated. The temperature of the chlorine cylinder and chlorinator should be above 50°F (10°C). Line heaters may be used to keep chlorine piping and chlorinator at higher temperatures to prevent condensing of gas into liquid in the pipelines and chlorinator. In general, a temperature difference of 5 to 10°F (3 to 6°C) above surroundings is recommended. The maximum temperature at which a chlorine cylinder is stored should not exceed 100°F (38°C).

6. It is not advisable to draw more than 40 pounds (18 kg) of chlorine from any one 100- to 150-pound (45- and 68-kg) cylinder in a 24-hour period. At higher chlorine withdrawal rates, the chlorine gas is removed from the cylinder faster than the liquid chlorine is being converted to chlorine gas. When this happens the actual flow of chlorine gas will be reduced and become less than the desired rate. The maximum allowable chlorine withdrawal rate varies with temperature with the maximum withdrawal rate increasing as the temperature increases. With ton containers, the limit of chlorine gas withdrawal is about eight pounds of chlorine per day per °F (6.4 kg/°C) *AMBIENT TEMPERATURE*.[35] When evaporators are provided, these limitations do not apply.[36]

7. There should be adequate light.

8. Adequate ventilation is important in a chlorinator room to remove any leaking chlorine gas that would be hazardous to personnel and damaging to equipment. The 1991 Uniform Fire Code requires proper ventilation of chlorine storage rooms and rooms where chlorine is used. Mechanical exhaust systems must draw air from the room at a point no higher than 12 inches (30.5 cm) above the floor at a rate of not less than one cubic foot of air per minute per square foot (0.00508 cu m/sec/sq m) of floor area in the storage area. (The system should not draw air through the fan itself because chlorine gas can damage the fan motor. See *NOTE* on page 294.) Normally ventilation from chlorine storage rooms is discharged to the atmosphere. When a chlorine leak occurs the ventilated air containing the chlorine should be routed to a treatment system for processing. A caustic scrubbing system can be used to treat air containing chlorine from a leak. The treatment system should be designed to reduce the maximum allowable discharge concentration of chlorine to one-half the IDLH (**I**mmediately **D**angerous to **L**ife or **H**ealth) at the point of discharge to the atmosphere. The IDLH for chlorine is 10 ppm. A secondary standby source of power is required for the chlorine detection, alarm, ventilation, and treatment systems.

9. Adequate measuring and controlling of chlorine dosage are required. Scales and recorders indicating loss in weight are desirable as a continuous check and as a record of the continuity of chlorination. *WEIGH* chlorine containers and *RECORD* the weights at the same time every day. Compare actual weight of chlorine used with calculated use based on chlorinator setting. Also compare results of chlorine residual tests with calculated dosage.

10. There should be continuity of chlorination. When chlorination is practiced for disinfection, it is needed continuously when the plant is operating for the protection of the water consumers. To ensure continuous chlorination, the chlorine gas lines from cylinders should feed to the manifold so that the cylinders can be removed without interrupting the feed of gas. Duplicate units with automatic cylinder switchover should be provided. Hypochlorinators are sometimes used during emergencies.

QUESTIONS

Write your answers in a notebook and then compare your answers with those on page 358.

7.5D   Why should chlorinators be located in a separate room?

7.5E   Why is adequate ventilation important in a chlorinator room?

7.5F   How can chlorination rates be checked against the chlorinator setting?

7.5G   When and how often should the weights of chlorine containers be recorded?

7.6   CHLORINE DIOXIDE FACILITIES
(Figures 7.34, 7.35, and 7.36)

7.60   Use of Chlorine Dioxide

Use of chlorine dioxide as a disinfecting chemical instead of chlorine is of considerable interest to operators. A major reason for this interest is that trihalomethanes are not formed when disinfecting with chlorine dioxide. Other reasons for considering chlorine dioxide include the fact that chlorine dioxide is effective in killing bacteria and viruses. Chlorine dioxide is much more effective than chlorine in killing bacteria in the pH range from 8 to 10. Chlorine dioxide does not combine with ammonia and is more selective in its reaction with many organics than chlorine, therefore less chlorine dioxide is required to achieve equivalent chlorine residuals and bacteria kills in waters containing such contaminants.

Most existing chlorination units may be used to produce chlorine dioxide. In addition to the existing chlorination system, a diaphragm pump, solution tank, mixer, chlorine dioxide generating tower, and electrical controls are needed. The diaphragm pump and piping must be made of corrosion-resistant materials because of the corrosive nature of chlorine dioxide. Usually, PVC or polyethylene pipe is used.

Special precautions must be taken when handling sodium chlorite. Sodium chlorite is usually supplied as a salt and is

---

[35] *Ambient (AM-bee-ent) Temperature.   Temperature of the surrounding air (or other medium). For example, temperature of the room where a gas chlorinator is installed.*

[36] *For procedures on how to calculate maximum withdrawal rates, see "Maximum Withdrawal Rates from Chlorine, Sulfur Dioxide, and Ammonia Cylinders," by Robert J. Baker in the October, 1984, issue of OPFLOW, published by AWWA.*

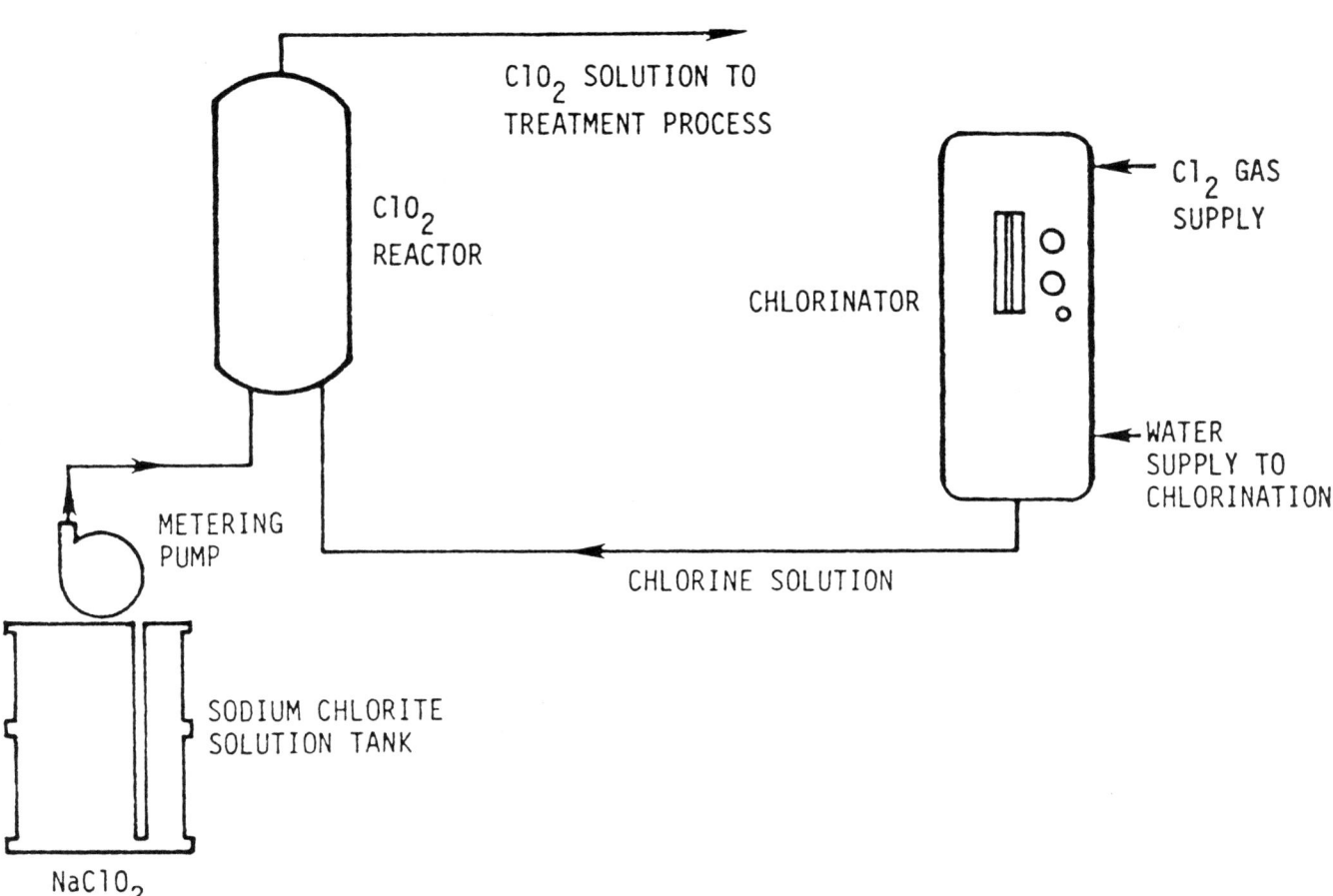

Fig. 7.34   Chlorine dioxide facility

(Source: *AN ASSESSMENT OF OZONE AND CHLORINE DIOXIDE TECHNOLOGIES FOR TREATMENT OF MUNICIPAL WATER SUPPLIES, EXECUTIVE SUMMARY.*
U.S. Environmental Protection Agency, Cincinnati, OH 45268, EPA-600/8-78-018, October 1978)

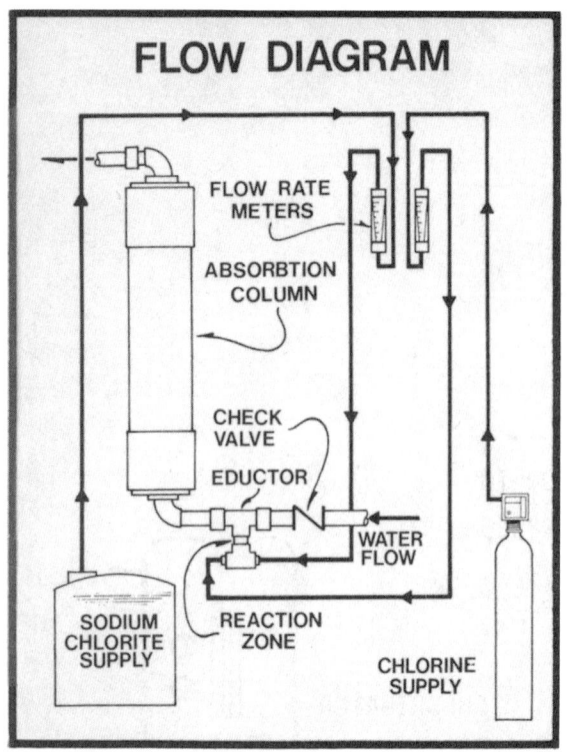
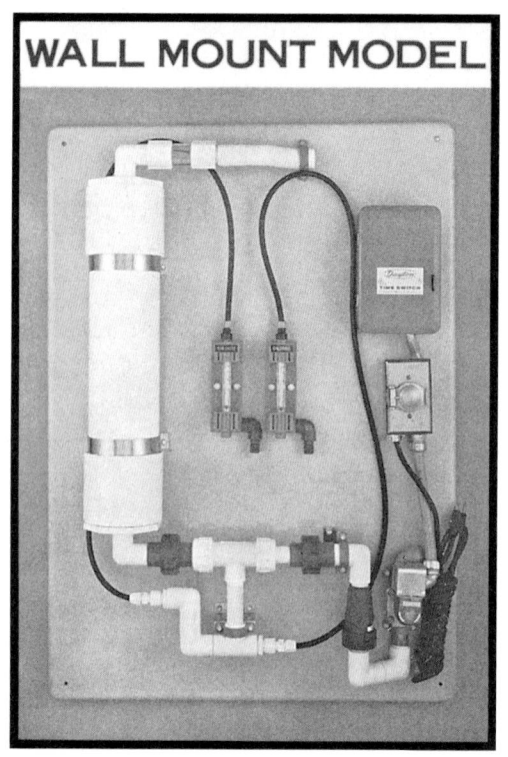

Chlorine-Chlorite Process (see Figure 7.34)

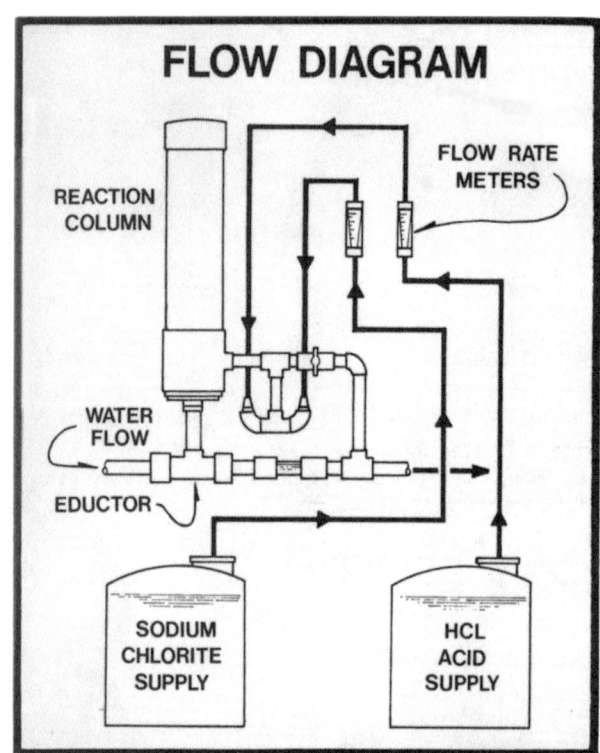
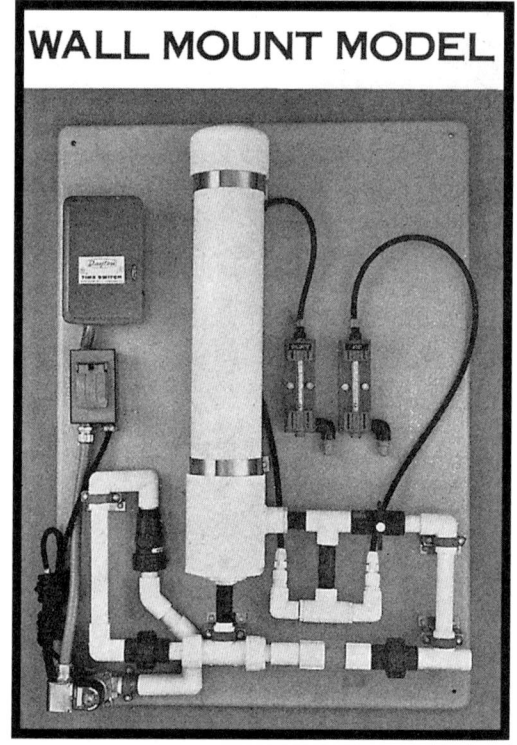

Acid-Chlorite Process

Fig. 7.35   Methods of generating chlorine dioxide
(Permission of Rio Linda Chemical Company)

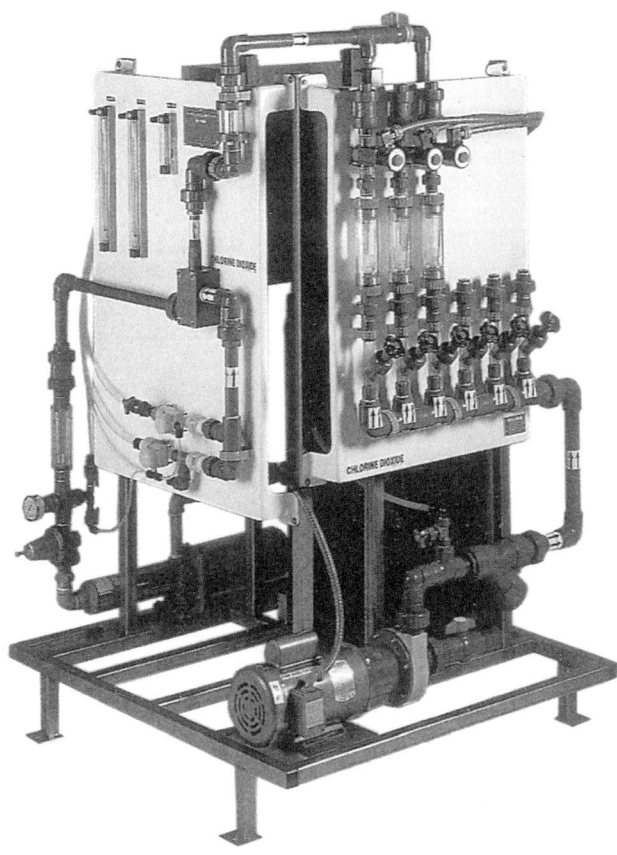

Fig. 7.36   Platform-mounted chlorine dioxide generator
(Permission of Vulcan Chemical Technologies, Inc.)

very combustible around organic compounds. Whenever spills occur, sodium chlorite must be neutralized with anhydrous sodium sulfite. Combustible materials (including combustible gloves) should not be worn when handling sodium chlorite. If sodium chlorite comes in contact with clothing, the clothes should be removed immediately and soaked in water to remove all traces of sodium chlorite or they should be burned immediately. If you follow safe procedures, you can safely handle sodium chlorite. Chlorine dioxide has not been widely used to treat drinking water because costs are higher than other more commonly used disinfection methods.

7.61 Safe Handling of Chemicals

Due to the corrosive nature of the chemicals involved in chlorine dioxide generation (sodium chlorite and chlorine), certain precautions should be taken in the handling of these chemicals.

1. Sodium chlorite solutions have very strong bleaching capabilities. *Always wear protective clothing, face shield or goggles, and chemical-resistant, noncombustible gloves and boots.*

2. An emergency shower and eye wash station should be located in the immediate vicinity. *If sodium chlorite comes in contact with eyes or skin, flush immediately with large amounts of fresh water. Soak contaminated clothing in water or burn it.*

3. Dried sodium chlorite presents a fire hazard, particularly if allowed to come in contact with organic materials, for example, cotton, leather, wood, or oils. *Do not allow the solution to dry. Flush spills with large amounts of fresh water.*

4. Thoroughly rinse rags and mops used for cleanup prior to disposal or storage.

5. Chlorine gas is a respiratory irritant. *If a leak occurs, evacuate the vicinity immediately.*

Refer to the Material Safety Data Sheets (MSDSs) for more information on the safe handling of sodium chlorite, chlorine dioxide solutions, and chlorine gas.

7.62 Operation

7.620 Pre-Start Procedures

The following section provides a comprehensive list of pre-start procedures that must be taken to ensure safe and efficient operation of the chlorine dioxide generator.

Debris such as PVC or metal filings can seriously damage or plug the eductor assembly and rotameters, resulting in poor performance and delays in start-up. If the water and chemical lines have not been properly flushed during installation, follow the procedures outlined in steps 1 through 4.

1. Eductor Water Supply Line (Figure 7.37)

   a. Close the shutoff valve.

   b. Remove the filter from the Y-strainer.

c. Disconnect the water line from the generator inlet.

d. Fully open the shutoff valve and allow the water to flow for two or three minutes.

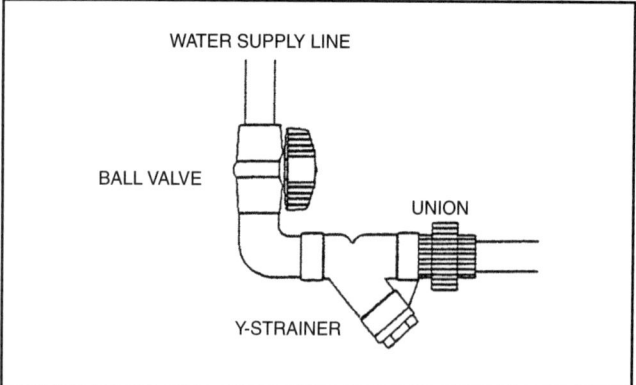

Fig. 7.37   Water supply line
(Permission of Vulcan Chemical Technologies, Inc.)

e. Close the shutoff valve.

f. Replace the Y-strainer filter.

g. Reconnect the water line to the generator.

h. Fully open the shutoff valve.

2. Chlorite Supply Line (Figure 7.38)

a. Disconnect the tubing from the supply line to the chlorite rotameter.

b. Remove the filter element from the chemical filter.

c. Disconnect the upper union of the ball check valve.

d. Connect a water source to one end of the chlorite supply line. Be sure the water supply you choose is clean and has sufficient pressure to properly flush debris from the line.

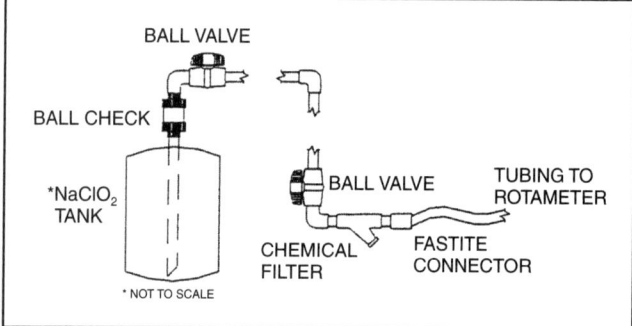

Fig. 7.38   Chlorite supply line
(Permission of Vulcan Chemical Technologies, Inc.)

e. Open both ball valves and flush the line for at least two to three minutes.

f. Turn off the water supply and close both ball valves.

g. Reconnect the chemical line to the ball check valve.

h. Replace the filter element of the chemical filter.

i. Reconnect the tubing from the chlorite supply line to the chlorite rotameter.

j. Fully open both ball valves.

3. Chlorine Supply Line (Figure 7.39)

a. Ensure that the shutoff valve on the chlorine cylinder is fully closed.

b. Disconnect the chlorine supply line from the vacuum demand regulator.

c. Disconnect the tubing from the supply line to the chlorine rotameter.

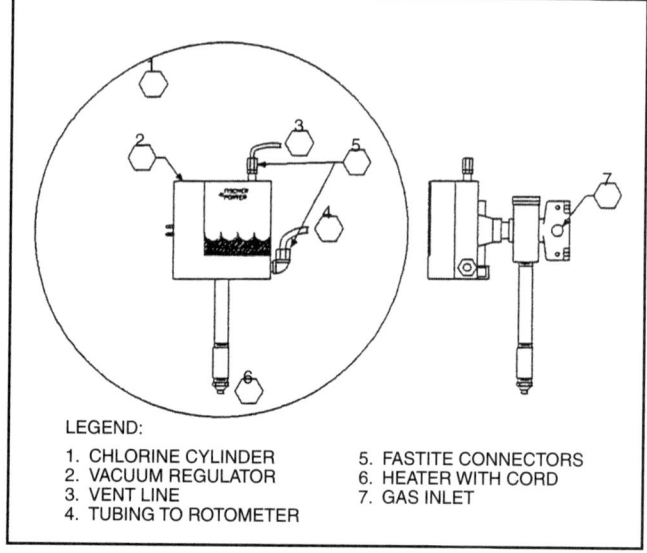

LEGEND:
1. CHLORINE CYLINDER
2. VACUUM REGULATOR
3. VENT LINE
4. TUBING TO ROTOMETER
5. FASTITE CONNECTORS
6. HEATER WITH CORD
7. GAS INLET

Fig. 7.39   Chlorine supply
(Permission of Vulcan Chemical Technologies, Inc.)

d. Connect a source of **DRY** compressed air to one end of the line. To prevent damaging the line and fittings, be sure the air pressure does not exceed 50 psi.

e. Apply several short bursts of air through the supply line.

f. Disconnect the compressed air.

g. Reconnect the tubing from the chlorine supply line to the chlorine rotameter.

NOTE: **Do not reconnect the chlorine supply line to the vacuum demand regulator at this time.**

4. Chlorine Dioxide Distribution Lines

a. Disconnect the chlorine dioxide distribution line from the generator output.

b. Disconnect the distribution line from the injection point.

c. Connect a clean water source to the generator end of the distribution line and flush for several minutes.

d. Disconnect the water source from the distribution line.

e. Reconnect the chlorine dioxide distribution line to the injection point.

f. Reconnect the distribution line to the generator output.

5. Generator Cycle

Water testing is the final pre-start test and will determine the readiness of the system for operation.

a. Ensure the shutoff valves for both the chlorite and chlorine supplies are in the OFF position.

b. Fully open the eductor water supply valve.

c. Ensure any shutoff valves on the distribution line are open.

d. With the generator running and the chemical shutoff valves closed, the generator should pull full vacuum. The meter floats will fall to and remain at zero. **If the floats bounce off the bottom of the meters, there is a leak in the system that must be found before you proceed**. Open the chlorine shutoff valve; the rotameter float will begin to rise indicating air flow.

e. Close the eductor water supply valve.

This completes the system pre-start tests.

7.621   Start-Up

Once the pre-start checks have been completed and any problems that were detected have been resolved, the generator is ready for start-up.

1. Reconnect the chlorine supply line to the vacuum demand regulator.

2. Open all the shutoff valves for both the chlorine and chlorite supply lines.

3. Open both the chlorine and the sodium chlorite rate valves (located at the top of the rotameters). (Refer to Figure 7.40, items 1 and 2.)

CAUTION: **Do not open the rate valves all the way against the stops. This can stress the valve and eventually cause vacuum leaks.**

4. Determine the chlorine dioxide demand. Open the eductor supply valve. The generator should begin to run.

5. Use the chlorine rate valve (2) to adjust the feed rate to correspond to the feed rate chart located on the generator.

6. Adjust the chlorite rate valve (1) until the rotameter reads 100. Use the manual sodium chlorite valve (12) to adjust the feed rate to correspond to the feed rate chart. The production of chlorine dioxide will be evident by a yellow-green solution in the sight glass (5).

It will take 15 to 20 seconds for the system to reach full vacuum and stabilize. No further adjustments are required to the generator unless the production demand changes.

7.622   Shutdown

1. Short Term

A short-term shutdown would be less than 24 hours, as in the case of maintenance or troubleshooting.

a. Open the eductor supply valve, allowing the generator to run.

b. Close the sodium chlorite supply line shutoff valve. This will allow the full system vacuum to be applied to the chlorine line.

c. Close the chlorine shutoff valve at the vacuum demand regulator. Allow the generator to run until the float in the chlorine rotameter falls to zero. Disconnect the feed line from the vacuum regulator. This will allow air to be drawn in and will purge the chemical feed line.

d. Disconnect the Teflon tubing from the chemical filter on the chlorite supply line and immerse the tubing in a container of clean, warm water. This will flush the chlorite line. Allow the generator to flush in this fashion for at least three or four minutes.

e. Close the eductor supply valve.

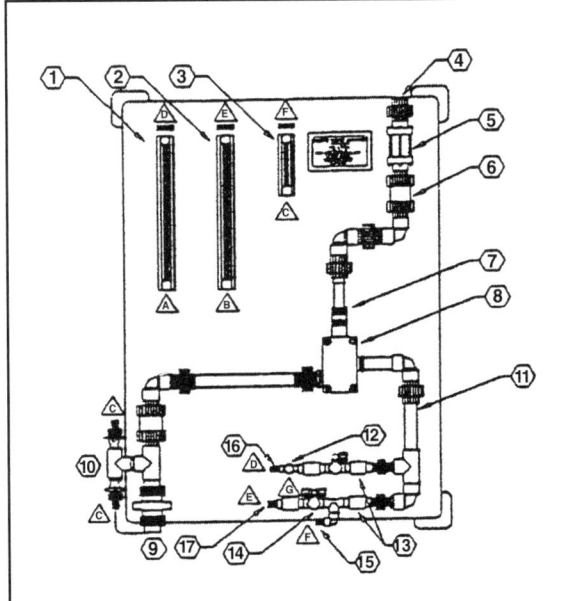

LEGEND:

1. SODIUM CHLORITE ROTAMETER
2. Cl_2 ROTAMETER
3. WATER BLEED ROTAMETER
4. ClO_2 SOLUTION OUTLET
5. SIGHT GLASS
6. ClO_2 SOLUTION CHECK VALVE
7. EJECTOR NOZZLE
8. EJECTOR BLOCK
9. WATER INLET
10. BALL VALVES SUPPLY FOR HYDRAULIC CHECK VALVE AND WATER BLEED
11. ClO_2 REACTION COLUMN
12. MANUAL SODIUM CHLORITE CONTROL VALVE
13. TEFLON CHECK VALVE
14. HYDRAULIC CHECK VALVE
15. WATER BLEED INLET
16. SODIUM CHLORITE INLET
17. Cl_2 INLET

A-G TUBE CONNECTORS/LOCATIONS

Fig. 7.40   Chlorine dioxide generator
(Permission of Vulcan Chemical Technologies, Inc.)

2. Long Term

If the system is to be shut down for more than 24 hours, as in a seasonal application, additional precautions should be taken.

a. Follow all of the steps outlined in item 1, "Short Term."

b. Remove the filter from the Y-strainer and allow the water to drain.

c. Disconnect the Teflon tubing from the water and two chemical inlets (Figure 7.40, points D-F). Loosen the two unions at the bottom of the reaction column. Open the plastic retaining clamps and remove both check valve assemblies. Use low pressure air to dry the assemblies by placing the air nozzle on the *inlet* side of the assemblies.

d. Loosen the union on the water inlet and remove the connecting pipe. Empty the water from this section and dry with compressed air.

e. Loosen the union on the solution outlet (Figure 7.40, item 4) and dry the rest of the generator assembly using compressed air.

f. Open the rate valves on the rotameters (Figure 7.40, items 1 and 2) and allow them to drain.

NOTE: **Do not use pressurized air to dry these components**.

If distribution lines are located in areas subject to freezing, they must also be drained and properly dried.

g. To drain the distribution lines, disconnect the line at the injection point and open all shutoff valves in the distribution line.

Once all of the assemblies are thoroughly dry, the system should be reassembled to prevent the loss of any components. The unit is now ready for storage.

7.63 Maintenance

Use of a periodic maintenance schedule is the best safeguard against production losses of chlorine dioxide and unscheduled downtime. The recommended maintenance schedule shown in Table 7.8 is for systems operating under optimum conditions. Factors such as *dirty* water, *hard* water, or *wet* chlorine can increase the frequency of maintenance that is required. Keeping accurate maintenance logs will help you to establish a schedule specific to your system.

7.64 Troubleshooting

Table 7.9 lists the symptoms of several problems that sometimes develop with chlorine dioxide generators, the probable causes of these symptoms, and some possible solutions.

7.65 Acknowledgment

Material in this section was provided by Vulcan Chemical Technologies (formerly Rio Linda Chemical Company) for their manual wall-mount chlorine dioxide generator. Their cooperation and assistance are sincerely appreciated.

QUESTIONS

Write your answers in a notebook and then compare your answers with those on page 358.

7.6A What additional equipment is necessary to use an existing chlorination unit to produce chlorine dioxide?

7.6B What hazards are associated with the handling of sodium chlorite?

7.6C What factors could increase the maintenance needed on chlorine dioxide generators?

7.7 MEASUREMENT OF CHLORINE RESIDUAL

7.70 Methods of Measuring Chlorine Residual

AMPEROMETRIC TITRATION[37] provides for the most convenient and most repeatable chlorine residual results. However, amperometric titration equipment is more expensive than equipment for other methods. *DPD TESTS*[38] can be used and are less expensive than other methods, but this method requires the operator to match the color of a sample with the colors on a comparator. See Chapter 11, "Laboratory Procedures," for detailed information on these tests. ORP (oxidation-reduction potential) probes can also be used to measure chlorine residuals.

Residual chlorine measurements of treated water should be taken at least three times per day on small systems and once every two hours on large systems. Residuals are measured to ensure that the treated water is being adequately disinfected. A free chlorine residual of at least 0.5 mg/L in the treated water is usually recommended.

ALL surface water systems and ground water systems under the influence of surface water must provide disinfection. Systems are required to monitor the disinfectant residual leaving the plant and at various points in the distribution system. The water leaving the plant must have at least 0.2 mg/L of the disinfectant, and the samples taken in the distribution system must have a detectable residual. Certain guidelines must be followed to ensure that there is enough contact time between the disinfectant and the water so that the microorganisms are inactivated.

If at any time the disinfectant residual leaving the plant is less than 0.2 mg/L, the system is allowed up to four hours to correct the problem. If the problem is corrected within this time, it is not considered a violation but the regulatory agency must be notified. The disinfectant residual must be measured continuously. For systems serving fewer than 3,300 people, this may be reduced to once per day.

The disinfectant in the distribution system must be measured at the same frequency and location as the total coliform samples. Measurements for heterotrophic plate count (HPC) bacteria may be substituted for disinfectant residual measurements. If the HPC is less than 500 colonies per mL, then the sample is considered equivalent to a detectable disinfectant residual. For systems serving fewer than 500 people, the regulatory agency may determine the adequacy of the disinfectant residual in place of monitoring.

---

[37] *Amperometric (am-PURR-o-MET-rick) Titration. A means of measuring concentrations of certain substances in water (such as strong oxidizers) based on the electric current that flows during a chemical reaction.*

[38] *DPD Test. A method of measuring the chlorine residual in water. The residual may be determined by either titrating or comparing a developed color with color standards. DPD stands for N,N-diethyl-p-phenylene-diamine.*

TABLE 7.8 PERIODIC MAINTENANCE SCHEDULE

Service Interval	Bi-Monthly	Monthly	Quarterly	Yearly
Water Filter				
Inspect	X			
Clean	X			
Rebuild			X	
Ball Checks				
Inspect		X		
Rebuild			X	
Generator				
Inspect	X			
Clean			X	
Tubing and Connectors				
Inspect			X	
Replace				X
Rotameters				
Inspect	X			
Rebuild				X
Sightglass				
Inspect	X			
Clean			X	
Rebuild				X
Nozzle and Throat				
Inspect			X	
Clean			X	
Replace				X
Reaction Column				
Inspect	X			
Clean			X	
Check Valve				
Inspect	X			
Clean		X		
Teflon Solenoid				
Inspect	X			
Clean		X		
Rebuild				
Chemical Filter				
Inspect		X		
Clean		X		
Replace			X	

TABLE 7.9 CHLORINE DIOXIDE GENERATOR TROUBLESHOOTING GUIDE

Symptom	Probable Cause	Remedy
1. Low water flow	Input supply restricted	Eliminate restriction
	Y-strainer blocked	Clean filter element
	Nozzle and throat plugged or damaged	Clean or replace
	Output blocked	Fully open outlet shutoff valve or clear blockage
2. Low sodium chlorite flow	Water flow low	Refer to item 1 of this table
	Rotameter setting incorrect	Adjust setting
	Manual rate valve setting incorrect	Adjust setting
	Chemical filter blocked	Clean or replace filter
	Chemical feed check valve failure	Repair or replace
	Rotameter inlet plugged	Disassemble and clean
	Teflon solenoid plugged	Clean or replace
	Reaction column blocked	a. Open water bleed and flush column
		b. Clean reaction column
	Nozzle and throat plugged or damaged	Clean or replace
3. Low chlorine flow	Rotameter setting incorrect	Adjust setting
	Vacuum regulator failure	Repair or replace
	Rotameter inlet plugged	Disassemble and clean
	Chemical check valve failure	Clean or replace
	Teflon solenoid plugged	Clean or replace
	Reaction column blocked	a. Open water bleed and flush column
		b. Clean reaction column
	Nozzle and throat plugged or damaged	Clean or replace
	Vacuum leak	Find and repair leak
4. No sodium chlorite flow	Out of chemical	Replenish supply
	Chemical shutoff valve closed	Open valve
	Rotameter rate valve closed	Open valve
	Manual rate valve closed	Open valve
	Teflon solenoid failure	Clean or replace
5. No chlorine flow	Out of chemical	Replenish supply
	Chemical shutoff valve closed	Open valve
	Rotameter rate valve closed	Open valve
	Teflon solenoid failure	Clean or replace
	Check valve salted out	Clean
	Vacuum leak	Find and repair leak
6. Bubbles in sodium chlorite meter	Vacuum leak	Find and repair leak
	Chemical level low	Replenish supply
	Bottom O-ring in rotameter leaking	a. Rotate meter tube
		b. Replace O-ring
7. Low vacuum	Low water flow	Refer to item 1 in this table
	Low water pressure	Check water supply line pressure
	Water bleed setting too high	Adjust setting
	Reaction column blocked	Clean reaction column
8. Falling meter ball	Vacuum leak	Find and repair leak
	Unstable water supply	Correct condition
	Chemical feed line incorrectly installed	Reinstall feed line
	Chemical feed line incorrectly sized	Re-size feed line
	Feed rate below 10 percent of system capacity	Adjust sodium chlorite feed rate
	Rate valve incorrectly adjusted	Adjust rate valve
9. Chlorite in chemical lines	Chemical check valve failure	Clean or replace
	Teflon solenoid failure	Clean or replace

7.71 Amperometric Titration for Free Residual Chlorine

1. Place a 200-mL sample of water in the titrator.

2. Start the agitator.

3. Add 1 mL of pH 7 buffer.

4. Titrate with 0.00564 N phenylarsene oxide solution.

5. End point is reached when one drop will cause a deflection on the microammeter and the deflection will remain.

6. mL of phenylarsene oxide used in titration is equal to mg/L of free chlorine residual.

7.72 DPD Colorimetric Method for Free Residual Chlorine (Figures 7.41 and 7.42)

This procedure is for the use of prepared powder pillows.

1. Collect a 100-mL sample.

2. Add color reagent.

3. Match color sample with a color on the comparator to obtain the chlorine residual in mg/L.

Operators using the DPD colorimetric method to test water for a free chlorine residual need to be aware of a potential error that may occur. If the DPD test is run on water containing a combined chlorine residual, a precipitate may form during the test. The particles of precipitated material will give the sample a turbid appearance or the appearance of having color. This turbidity can produce a positive test result for free chlorine residual when there is actually no chlorine present. Operators call this error a "false positive" chlorine residual reading.

7.73 ORP Probes

ORP (oxidation-reduction potential) probes are being used to optimize chlorination processes in water treatment plants. ORP (also called the redox potential) is a direct measure of the effectiveness of a chlorine residual in disinfecting the water being treated. Chlorine forms that are toxic to microorganisms (including coliforms) are missing one or more electrons in their molecular structure. They satisfy their need for electrons by taking electrons from any organic substances or microorganisms present in the water being treated. When microorganisms lose electrons they become inactivated and can no longer transmit a disease or reproduce.

The ability of chlorine to take electrons (the electrical attraction or electrical potential) is the ORP and is measurable in millivolts. The strength of the millivoltage (or the redox measurement) is directly proportional to the oxidative disinfection strength of the chlorine in the treatment system. The higher the concentration of chlorine disinfectant, the higher the measured ORP voltage. Conversely, the higher the concentration of organics (chlorine demanding substances), the lower the measured ORP voltage. The redox sensing unit (ORP probe) measures the voltage present in the water being treated and thus provides a direct measure of the disinfecting power of the disinfectant present in the water.

In a typical installation, a High Resolution Redox (HRR) chlorine controller monitors the chlorine residual using a redox (ORP) probe suspended in the chlorine contact chamber approximately 6.5 minutes downstream from the chlorine injection point. The controller converts the redox signal to a 4- to 20-milliamp (mA) signal that automatically adjusts the chlorine feed rate from the chlorinator.

The High Resolution Redox (HRR) units control the chlorination chemical feed rates according to actual demand in the treatment processes. These HRR units automatically treat the water with the chlorine dosages required to maintain chemical residuals in the ideal ranges, regardless of changes in the chemical demand or water flow.

Maintenance for the chlorine ORP probe consists of cleaning the unit's sensor once a month.

QUESTIONS

Write your answers in a notebook and then compare your answers with those on page 358.

7.7A What three methods are used to measure chlorine residual in treated water?

7.7B How often should treated water residual chlorine measurements be made?

7.7C What does an ORP probe measure in a disinfection system?

7.7D What happens to a microorganism when it loses an electron?

7.7E What maintenance is required on ORP probes?

7.8 CHLORINE SAFETY PROGRAM

Every good safety program begins with cooperation between the employee and the employer. The employee must take an active part in the overall program. The employee must be responsible and should take all necessary steps to prevent accidents. This begins with the attitude that as good an effort as possible must be made by everyone. Safety is everyone's priority. The employer also must take an active part by supporting safety programs. There must be funding to purchase equipment and to enforce safety regulations required by OSHA and state industrial safety programs. The following items should be included in all safety programs.

1. Establishment of a formal safety program.

2. Written rules and specific safety procedures.

3. Periodic hands-on training using safety equipment.

   a. Leak-detection equipment

   b. Self-contained breathing apparatus (Figure 7.28)

   c. Atmospheric monitoring devices

4. Establishment of emergency procedures for chlorine leaks and first aid.

5. Establishment of a maintenance and calibration program for safety devices and equipment.

6. Provide police and fire departments with tours of facilities to locate hazardous areas and provide chlorine safety information.

All persons handling chlorine should be thoroughly aware of its hazardous properties. Personnel should know the location and use of the various pieces of protective equipment and be instructed in safety procedures. In addition, an emergency procedure should be established and each individual should

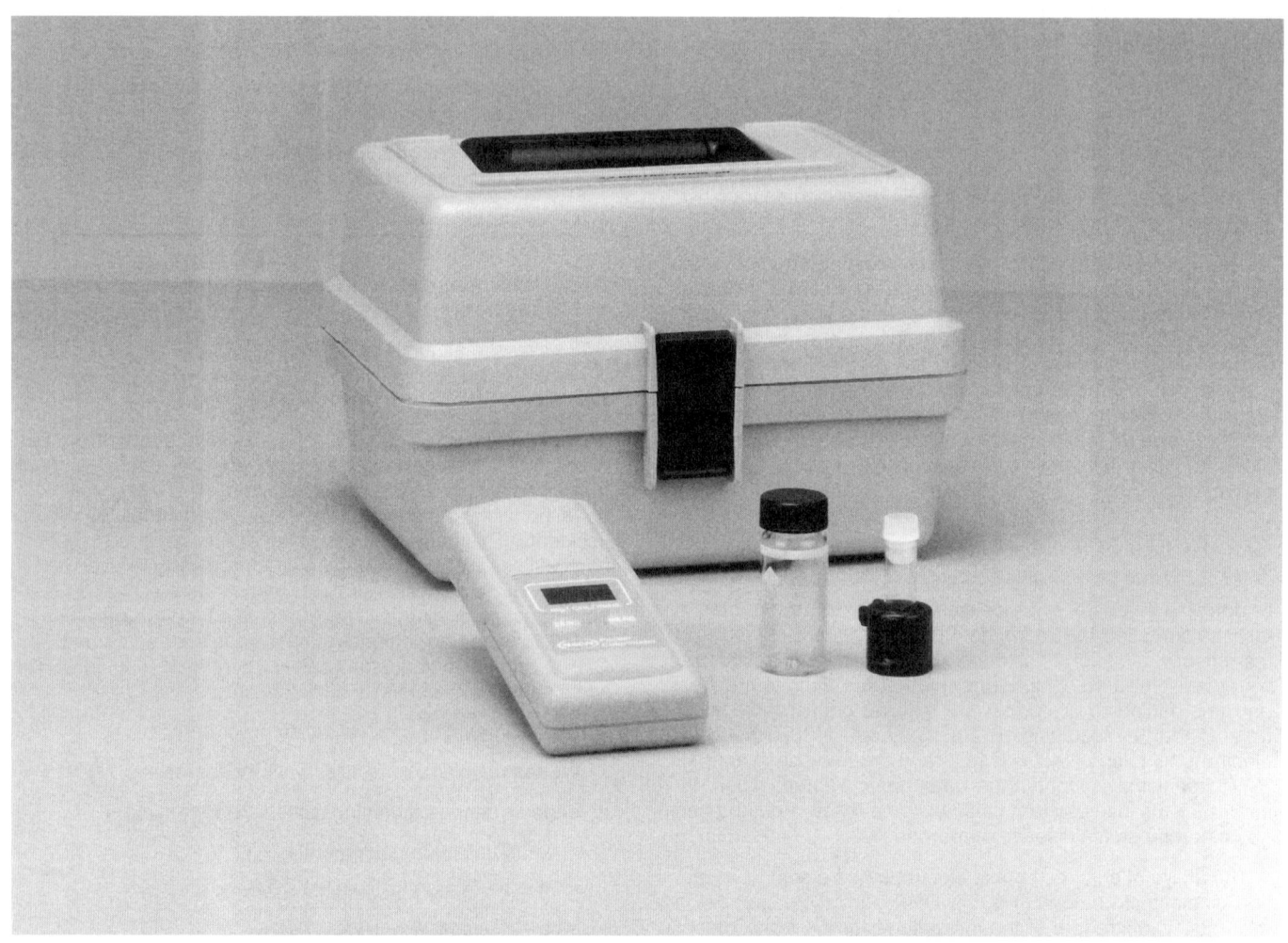

Fig. 7.41   Direct-reading colorimeter for free chlorine residuals
(Courtesy of the HACH Company)

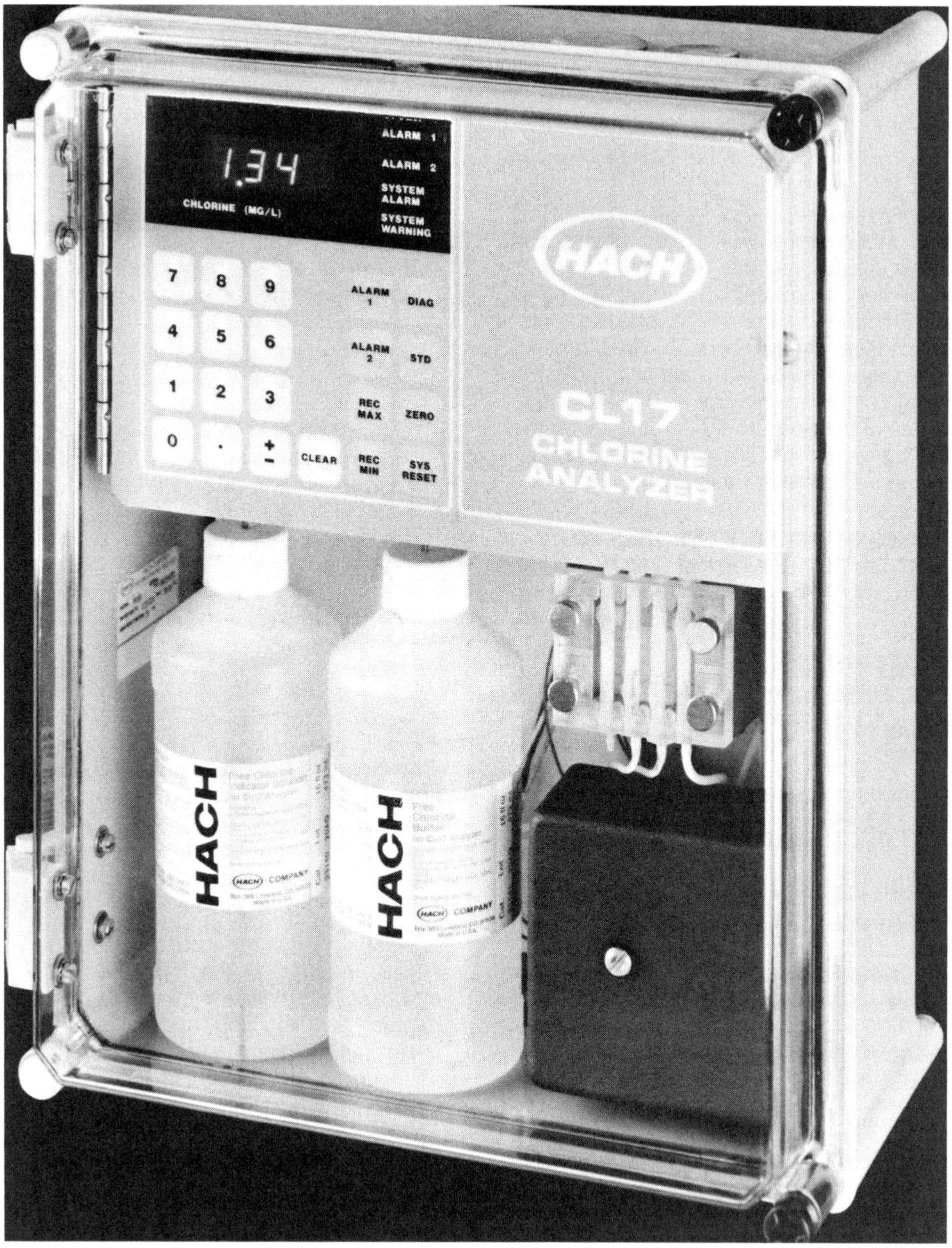

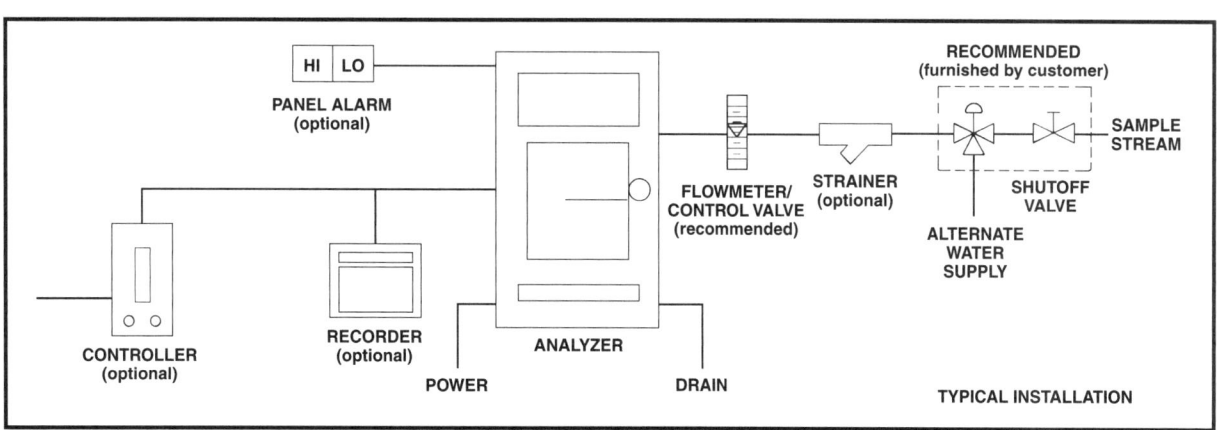

Fig. 7.42   Continuous on-line free chlorine residual analyzer
(Courtesy of the HACH Company)

be instructed how to follow the procedures. An emergency checklist also should be developed and available. For additional information on this topic, see The Chlorine Institute's *CHLORINE MANUAL*, 6th Edition.[39] Also see Volume II, Chapter 20, "Safety."

7.80   Chlorine Hazards

Chlorine is a gas that is 2.5 times heavier than air, extremely toxic, and corrosive in moist atmospheres. Dry chlorine gas can be safely handled in steel containers and piping, but with moisture must be handled in corrosion-resistant materials such as silver, glass, Teflon, and certain other plastics. Chlorine gas at container pressure should never be piped in silver, glass, Teflon, or any other material that cannot handle the pressure. Even in dry atmospheres, the gas is very irritating to the mucous membranes of the nose, to the throat, and to the lungs; a very small percentage in the air causes severe coughing. Heavy exposure can be fatal (see Table 7.10).

TABLE 7.10   PHYSIOLOGICAL RESPONSE TO CONCENTRATIONS OF CHLORINE GAS [a, b]

Effect	Parts of Chlorine Gas Per Million Parts of Air by Volume (ppm)
Slight symptoms after several hours' exposure	1
Detectable odor	0.3 to 3.5
Noxiousness	5
Throat irritation	15
Coughing	30
Dangerous from one-half to one hour	40
Death after a few deep breaths	1,000

[a] Adapted from data in U.S. Bureau of Mines *TECHNICAL PAPER 248* (1955).

[b] The maximum **P**ermissible **E**xposure **L**imit (PEL) is 0.5 ppm (8-hour weighted average). The IDLH level is 10 ppm. IDLH is the **I**mmediately **D**angerous to **L**ife or **H**ealth concentration. The atmospheric concentration of any toxic, corrosive, or asphyxiant substance that poses an immediate threat to life or would cause irreversible or delayed adverse health effects or would interfere with an individual's ability to escape from a dangerous atmosphere.

WARNING

WHEN ENTERING A ROOM THAT MAY CONTAIN CHLORINE GAS, OPEN THE DOOR SLIGHTLY AND CHECK FOR THE SMELL OF CHLORINE. **NEVER** GO INTO A ROOM CONTAINING CHLORINE GAS WITH HARMFUL CONCENTRATIONS IN THE AIR WITHOUT A SELF-CONTAINED AIR SUPPLY, PROTECTIVE CLOTHING, AND HELP STANDING BY. HELP MAY BE OBTAINED FROM YOUR CHLORINE SUPPLIER AND YOUR LOCAL FIRE DEPARTMENT.

Most people can usually detect concentrations of chlorine gas above 0.3 ppm and you should not be exposed to concentrations greater than 1 ppm. However, chlorine gas can deaden your sense of smell and cause a false sense of security. *NEVER* rely on your sense of smell to protect you from chlorine because *YOUR* sense of smell might not be able to detect harmful levels of chlorine.

7.81   Why Chlorine Must Be Handled With Care

You must always remember that chlorine is a hazardous chemical and must be handled with respect. Concentrations of chlorine gas in excess of 1,000 ppm (0.1% by volume in air) may be fatal after a few breaths.

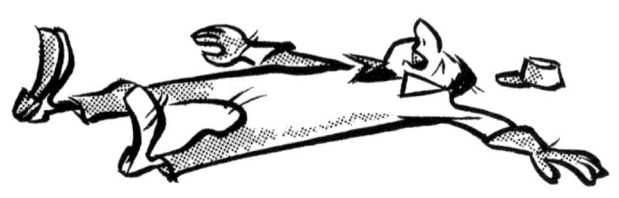

Because the characteristic sharp odor of chlorine is noticeable even when the amount in the air is small, it is usually possible to get out of the gas area before serious harm is suffered. This feature makes chlorine less hazardous than gases such as carbon monoxide, which is odorless, and hydrogen sulfide, which impairs your sense of smell in a short time.

Inhaling chlorine causes general restlessness, panic, severe irritation of the throat, sneezing, and production of much saliva. These symptoms are followed by coughing, retching and vomiting, and difficulty in breathing. Chlorine is particularly irritating to persons suffering from asthma and certain types of chronic bronchitis. Liquid chlorine causes severe irritation and blistering on contact with the skin.

7.82   Protect Yourself From Chlorine

Every person working with chlorine should know the proper ways to handle it, should be trained in the use of self-contained breathing apparatus (SCBA), methods of detecting hazards, and should know what to do in case of emergencies. The clothing of persons exposed to chlorine can be saturated with chlorine which will irritate the skin if exposed to moisture or sweat. These people should not enter confined spaces before their clothing is purged of chlorine (stand out in the open air for awhile). This is particularly applicable to police and fire department personnel who leave the scene of a chlorine leak and ride back to their stations in closed vehicles. Suitable protective clothing for working in an atmosphere containing chlorine includes a disposable rainsuit with hood to protect your body, head, and limbs, and rubber boots to protect your feet.

---

[39] *Write to: The Chlorine Institute, Inc., 1300 Wilson Boulevard, Arlington, VA 22209. Pamphlet 1. Price to members, $15.00; nonmembers, $30.00; plus 10 percent of order total for shipping and handling.*

WARNING

CANISTER TYPE 'GAS MASKS' ARE USUALLY **IN-ADEQUATE** AND **INEFFECTIVE** IN SITUATIONS WHERE CHLORINE LEAKS OCCUR AND ARE THEREFORE NOT RECOMMENDED FOR USE UNDER ANY CIRCUMSTANCES. **SELF-CONTAINED AIR OR OXYGEN SUPPLY TYPE BREATHING APPARATUS ARE RECOMMENDED.** OPERATORS SERVING ON "EMERGENCY CHLORINE TEAMS" MUST BE CAREFULLY SELECTED AND RECEIVE REGULAR APPROVED TRAINING. THEY MUST BE PROVIDED THE PROPER EQUIPMENT WHICH RECEIVES REGULAR MAINTENANCE AND IS READY FOR USE AT ALL TIMES.

Self-contained air supply and positive pressure/demand breathing equipment must fit properly and be used properly. Pressure demand and rebreather kits may be safer. Pressure demand units use more air from the air bottle which reduces the time a person may work on a leak. There are certain physical constraints when using respiratory protection. Contact your local safety regulatory agency to determine these requirements.

Before entering an area containing a chlorine leak, wear protective clothing. A chemical suit will prevent chlorine from contacting the sweat on your body and forming hydrochloric acid. Chemical suits are very cumbersome, but should be worn when the chlorine concentration is high. A great deal of practice is required to perform effectively while wearing a chemical suit.

The best protection that one can have when dealing with chlorine is to respect it. Each individual should practice rules of safe handling and good *PREVENTIVE MAINTENANCE.*

PREVENTION IS THE BEST EMERGENCY TOOL YOU HAVE.

PLAN AHEAD.

1. Have your fire department and other available emergency response agencies tour the area so that they know where the facilities are located. Give them a clearly marked map indicating the location of the chlorine storage area, chlorinators, and emergency equipment.

2. Have regularly scheduled practice sessions in the use of respiratory protective devices, chemical suits, and chlorine repair kits. Involve all personnel who may respond to a chlorine leak.

3. Have a supply of ammonia available to detect chlorine leaks.

4. Write emergency procedures:

   Prepare a CHLORINE EMERGENCY LIST of names of companies and phone numbers of persons to call during an emergency and ensure that all involved personnel are trained in notification procedures. This list should be posted at plant telephones and should include:

   a. Fire department,

   b. Chlorine emergency personnel,

   c. Chlorine supplier, and

   d. Police department.

5. Follow established procedures during all emergencies.

   a. Never work alone during chlorine emergencies.

   b. Obtain help immediately and quickly repair the problem. *PROBLEMS DO NOT GET BETTER.*

   c. Only authorized and properly trained persons with adequate equipment should be allowed in the danger area to correct the problem.

   d. If you are caught in a chlorine atmosphere without appropriate respiratory protection, shallow breathing is safer than breathing deeply. Recovery depends upon the duration and amount of chlorine inhaled, so it is important to keep that amount as small as possible.

   e. If you discover a chlorine leak, leave the area immediately unless it is a very minor leak. Small leaks can be found by using a rag soaked with ammonia. A white gas will form near the leak so it can be located and corrected.

   f. Use approved respiratory protection and wear disposable clothing when repairing a chlorine leak.

   g. Notify your police department that you need help if it becomes necessary to stop traffic on roads and to evacuate persons in the vicinity of the chlorine leak.

6. Develop emergency evacuation procedures for use during a serious chlorine leak. Coordinate these procedures with your police department and other officials. Ensure that all facility personnel are thoroughly trained in any evacuation procedure developed.

7. Post emergency procedures in all operating areas.

8. Inspect equipment and routinely make any necessary repairs.

9. At least twice weekly, inspect area where chlorine is stored and where chlorinators are located. Remove all obstructions from the area.

10. Schedule routine maintenance on *ALL* chlorine equipment at least once every six months or more frequently.

11. Have health appraisal for employees on chlorine emergency duty. All those who have heart and/or respiratory problems should not be allowed on emergency teams. There may be other physical constraints. Contact your local safety regulatory agency for details.

REMEMBER:

Small amounts of chlorine cause large problems. Leaks never get better.

7.83   First-Aid Measures

MILD CASES

Whenever you have a mild case of chlorine exposure (which does happen from time to time around chlorination equipment), you should first leave the contaminated area. Move slowly, breathe lightly without exertion, remain calm, keep warm, and resist coughing. Notify other operators and have them repair the leak immediately.

If clothing has been contaminated, remove as soon as possible. Otherwise the clothing will continue to give off chlorine gas which will irritate the body even after leaving the contaminated area. Immediately wash any area affected by chlorine. Shower and put on clean clothes.

If the victim has slight throat irritation, immediate relief can be accomplished by drinking milk. Drinking spirits of peppermint also will help reduce throat irritation. A mild stimulant such as hot coffee or hot tea is often used for coughing. See a physician.

EXTREME CASES

1. Follow established emergency procedures.

2. Always use proper safety equipment. Do not enter area without a self-contained breathing apparatus.

3. Remove patient from affected area immediately. Call a physician and begin appropriate treatment immediately.

4. First aid:

   a. Remove contaminated clothes to prevent clothing giving off chlorine gas which will irritate the body.

   b. Keep patient warm and cover with blankets if necessary.

   c. Place patient in a comfortable position on back.

   d. If breathing is difficult, administer oxygen if equipment and trained personnel are available.

   e. If breathing seems to have stopped, begin artificial respiration immediately. Mouth-to-mouth resuscitation or any of the approved methods may be used.

   f. EYES!

      If even a small amount of chlorine gets into the eyes, they should be flushed immediately with large amounts of lukewarm water so that all traces of chlorine are flushed from the eyes (at least 15 minutes). Hold the eyelids apart forcibly to ensure complete washing of all eye and lid tissues.

5. See a physician.

7.84   Hypochlorite Safety

Hypochlorite does not present the hazards that gaseous chlorine does and therefore is safer to handle. When spills occur, wash with large volumes of water. The solution is messy to handle. Hypochlorite causes damage to your eyes and skin upon contact. Immediately wash affected areas thoroughly with water. Consult a physician if the area appears burned. Hypochlorite solutions are very corrosive. Hypochlorite compounds are nonflammable; however, they can cause fires when they come in contact with organics or other easily oxidizable substances.

7.85   Chlorine Dioxide Safety

Chlorine dioxide is generated in much the same manner as chlorine and should be handled with the same care. Of special concern is the use of sodium chlorite to generate chlorine dioxide. Sodium chlorite is very combustible around organic compounds. Whenever spills occur, sodium chlorite must be neutralized with anhydrous sodium sulfite. Combustible materials (including combustible gloves) should not be worn when handling sodium chlorite. If sodium chlorite comes in contact with clothing, the clothes should be removed immediately and soaked in water to remove all traces of sodium chlorite or the clothes should be burned immediately.

7.86   Operator Safety Training

Training is a concern to everyone, especially when your safety and perhaps your life is involved. Every utility agency should have an operator chlorine safety training program that introduces new operators to the program and updates previously trained operators. As soon as a training session ends, obsolescence begins. People will forget what they have learned if they don't use and practice their knowledge and skills. Operator turnover can dilute a well-trained staff. New equipment and also new techniques and procedures can dilute the readiness of trained operators. An ongoing training program could include a monthly luncheon seminar, a monthly safety bulletin that is to be read by every operator, and outside speakers who reinforce and refresh specific elements of safety training.

7.87   CHEMTREC (800) 424-9300

Safely handling chemicals used in daily water treatment is an operator's responsibility. However, if the situation ever gets out of hand, there are emergency teams that will respond with help anywhere there is an emergency. If an emergency does develop in your plant and you need assistance, call CHEMTREC (Chemical Transportation Emergency Center) for assistance. CHEMTREC will provide immediate advice for those at the scene of an emergency and then quickly alert experts whose products are involved for more detailed assistance and appropriate follow-up.

CHEMTREC'S EMERGENCY TOLL-FREE PHONE NUMBER IS (800) 424-9300.

QUESTIONS

Write your answers in a notebook and then compare your answers with those on page 358.

7.8A   What properties make chlorine gas so hazardous?

7.8B   What type of breathing apparatus is recommended when repairing chlorine leaks?

7.8C   What first-aid measures should be taken if a person comes in contact with chlorine gas?

7.9 OPERATION OF OTHER DISINFECTION PROCESSES

7.90 Ultraviolet (UV) Systems

Just beyond the visible light spectrum there is a band of electromagnetic radiation which we commonly refer to as ultraviolet (UV) light. When ultraviolet radiation is absorbed by the cells of microorganisms, it damages the genetic material in such a way that the organisms are no longer able to grow or reproduce, thus ultimately killing them. This ability of UV radiation to disinfect water has been understood for almost a century, but technological difficulties and high energy costs prevented widespread use of UV systems for disinfection. Today, however, with growing concern about the safety aspects of handling chlorine and the possible health effects of chlorination by-products, UV disinfection is gaining in popularity. Technological advances are being made and several manufacturers now produce UV disinfection systems for water and wastewater applications. As operating experience with installed systems increases, UV disinfection is expected to become a practical alternative to the use of chlorination at many water treatment plants.

7.900 Types of UV Systems

The usual source of the UV radiation for disinfection systems is from low pressure mercury vapor UV lamps which have been made into multi-lamp assemblies, as shown in Figure 7.43. Each lamp is protected by a quartz sleeve and each has watertight electrical connections. The lamp assemblies are mounted in a rack (or racks) and these racks are immersed in the flowing water. The racks may be mounted either within an enclosed vessel or in an open channel (Figure 7.44). Most of the UV installations in North America are of the open channel configuration.

When UV lamps are installed in open channels, they are typically placed either horizontal and parallel to the flow (Figure 7.45) or vertical and perpendicular to the flow. In the horizontal and parallel-to-flow open channel lamp configuration, the lamps are arranged into horizontal modules of evenly spaced lamps. The number of lamps per module establishes the water depth in the channel. For example, 16 lamps could be stacked 3 inches apart to provide disinfection for water flowing through a 48-inch deep open channel.

Each horizontal lamp module has a stainless-steel frame. Each module is fitted with a waterproof wiring connector to the power distribution center. The connectors allow each module to be disconnected and removed from the channel separately for maintenance. The horizontal lamp modules are arranged in a support rack to form a lamp bank that covers the width of the UV channel and there may be several such lamp banks along the channel. The number of UV banks per channel is determined by the required UV dosage to achieve the target effluent quality.

In the vertical and perpendicular-to-flow lamp configuration, rows of lamps are grouped together into vertical modules. Each vertical lamp module has a stainless-steel support frame and can be removed individually from the channel for cleaning or inspection. The electrical wirings for the lamps are located within the frame above the water level. Each individual lamp can be removed from the top of the frame without removing the entire module from the channel. The length of the UV lamps establishes the depth of water in the channel. One or more vertical modules are installed to cover the width of the channel. As with the horizontal arrangement, the number of

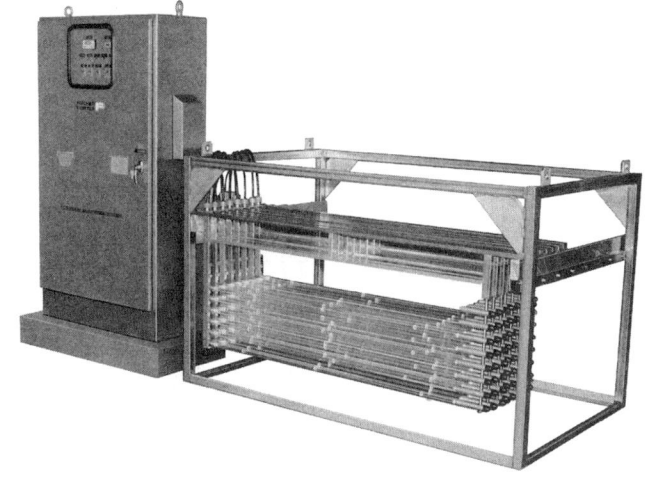

Fig. 7.43 UV lamp assembly
(Reproduced with permission of Fischer & Porter (Canada) Limited)

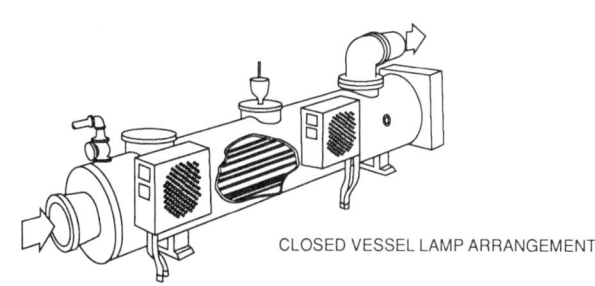

CLOSED VESSEL LAMP ARRANGEMENT

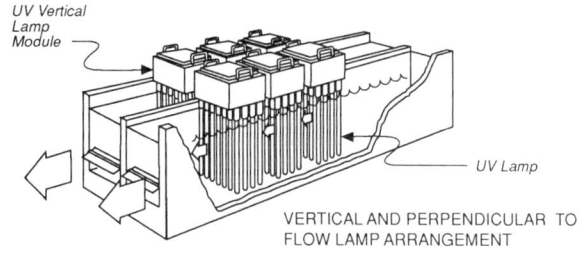

UV Vertical Lamp Module

UV Lamp

VERTICAL AND PERPENDICULAR TO FLOW LAMP ARRANGEMENT

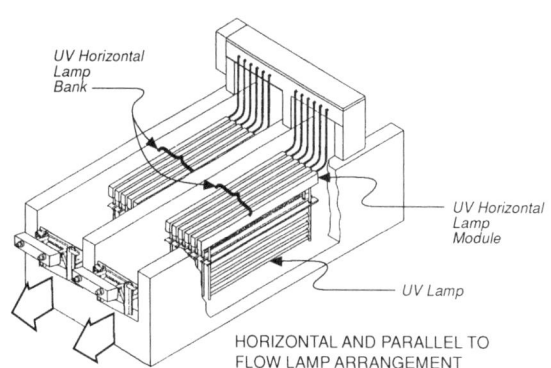

UV Horizontal Lamp Bank

UV Horizontal Lamp Module

UV Lamp

HORIZONTAL AND PARALLEL TO FLOW LAMP ARRANGEMENT

Fig. 7.44 Typical UV lamp configurations
(Source: "Ultraviolet Disinfection," by CH2M Hill, reproduced with permission of CH2M Hill)

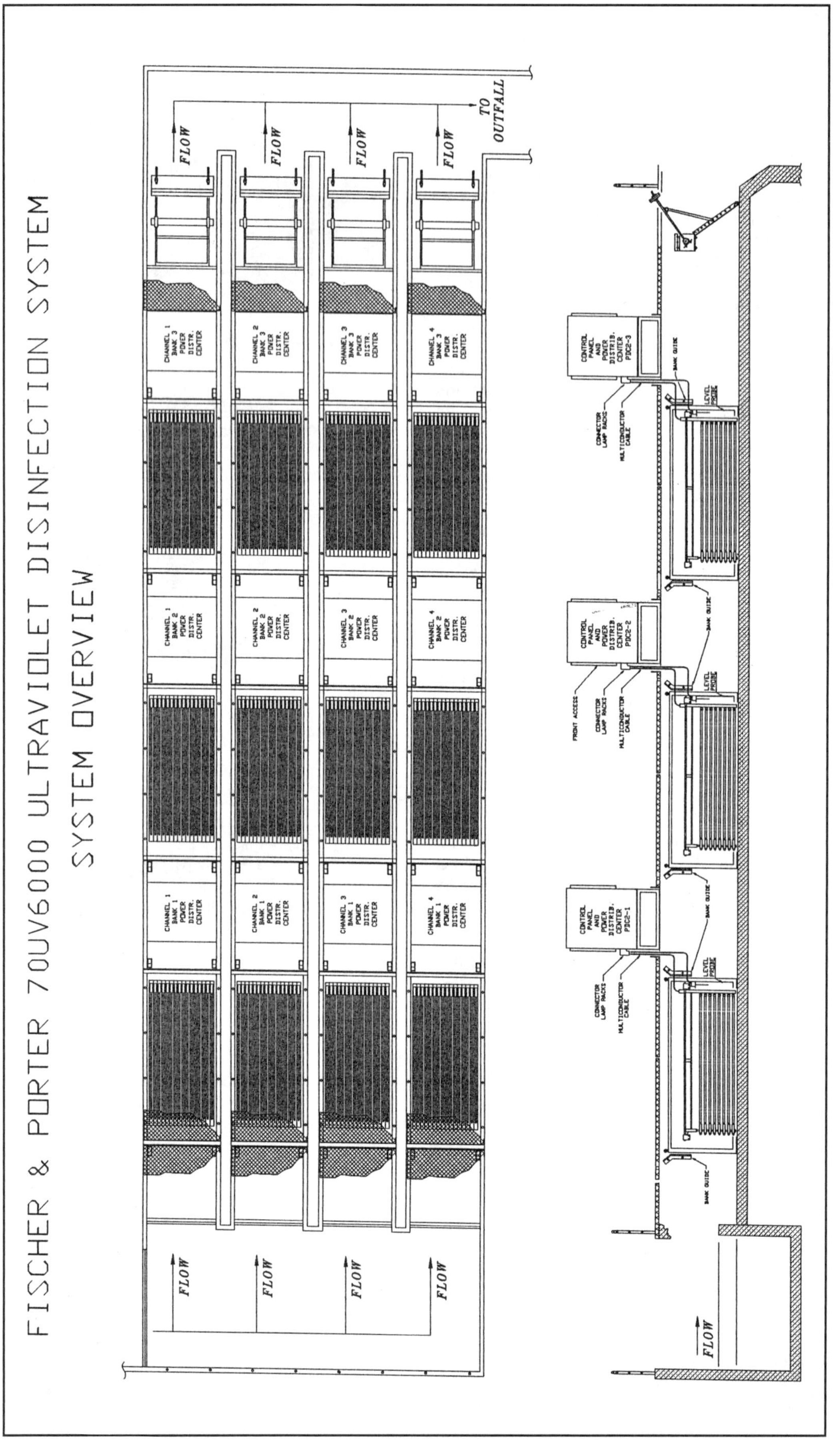

Fig. 7.45 Horizontal, in-channel UV lamp installation
(Reproduced with permission of Fischer & Porter (Canada) Limited)

vertical lamp modules per channel will depend on the UV dosage needed to achieve the desired effluent quality.

When it is necessary to maintain pressure within the water transmission system, UV lamps can be installed in a closed pressure vessel, as shown at the top of Figure 7.44.

Another type of UV system, the thin film type, uses a chamber with many lamps spaced one-quarter inch (6 mm) apart. This system has been used in the wastewater industry for a 9-MGD (0.35-cu m/sec) secondary treatment plant that has been operating for several years.

Operators may also occasionally encounter a Teflon tube UV disinfection system, although this design is not in common use. Water flows in a thin-walled Teflon tube past a series of UV lamps. UV light penetrates the Teflon tube and is absorbed by the fluid. The advantage to this system is that water never comes in contact with the lamps. However, scale does eventually build up on the pipe walls and must be removed, or the Teflon tube must be replaced. This type of system has generally been replaced by the quartz sleeve systems described earlier.

7.901 Safety

WARNING: The light from a UV lamp can cause serious burns to your eyes and skin. *ALWAYS* take precautions to protect them. *NEVER* look into the uncovered parts of the UV chamber without proper protective glasses. Do not plug a UV unit into an electrical outlet or switch a unit on without having the UV lamps properly secured in the UV water chamber and the box closed.

UV lamps contain mercury vapor, a hazardous substance that will be released if a lamp is broken. Handle UV lamps with care and be prepared with the proper equipment to ensure operator safety in case of an accident.

7.902 Operation

The operation of ultraviolet water disinfection systems requires very little operator attention. To prevent short-circuiting and ensure that all microorganisms receive sufficient exposure to the UV radiation, the water level over the lamps must be maintained at the appropriate level. Water levels in channels can be controlled by weirs or automatic control gates.

Lamp output declines with use so the operator must monitor the output intensity and replace lamps that no longer meet design standards, as well as replacing any lamps that simply burn out. Lamp intensity monitors can be installed to assist the operator in monitoring the level of light output. Lamp failure indicators connected to the main UV control panel will alert the operator when a lamp burns out and requires replacement. In addition, computerized systems are available to monitor and record the age (burn time) of each lamp.

Care must be taken not to exceed the maximum design turbidity levels and flow velocities when using this type of equipment. Suspended particles will shield microorganisms from the UV light and thus protect them from its destructive effects. Flows should be somewhat turbulent to ensure complete exposure of all organisms to the UV light, but flow velocity must be controlled so that the water is exposed to UV radiation long enough for the desired level of disinfection to occur.

Since ultraviolet rays leave no chemical residual like chlorine does, bacteriological tests must be made frequently to ensure that adequate disinfection is being achieved by the ultraviolet system. In addition, the lack of residual disinfectant means that no protection is provided against recontamination after the treated water has left the disinfection facility. When the treated water is exposed to visible light, the microorganisms can be reactivated. Microorganisms that have not been killed have the ability to heal themselves when exposed to sunlight. The solution to this problem is to design UV systems with a high efficiency for killing microorganisms.

7.903 Maintenance

A UV system is capable of continuous use if a simple maintenance routine is performed at regular intervals. By checking the following items regularly, the operator of a UV system can determine when maintenance is needed.

1. Check UV monitor for significant reduction in lamp output,

2. Monitor process for major changes in normal flow conditions such as incoming water quality,

3. Check for fouling of quartz sleeves and UV intensity monitor probes,

4. Check indicator light display to ensure that all of the UV lamps are energized,

5. Monitor elapsed time meter, microbiological results, and lamp log sheet to determine when UV lamps require replacement, and

6. Check quartz sleeves for discoloration. This effect of UV radiation on the quartz is called solarization. Excessive solarization is an indication that a sleeve is close to the end of its useful service life. Solarization reduces the ability of the sleeves to transmit the necessary amount of UV radiation to the process.

Maintenance on UV systems requires two tasks: cleaning the quartz sleeves and changing the lamps.

Algae and other attached biological growths may form on the walls and floor of the UV channel. This slime can slough off, potentially hindering the disinfection process. If this condition occurs, the UV channel should be dewatered and hosed out to remove accumulated algae and slimes.

7.9030 QUARTZ SLEEVE FOULING

Fouling of the quartz sleeves occurs when cations such as calcium, iron, or aluminum ions attach to protein and colloidal matter that crystallizes on the quartz sleeves. As this coating builds up on the sleeves, the intensity of the UV light decreases to the point where the buildup has to be removed for the system to remain effective. The rate at which fouling of the quartz sleeves occurs depends on several factors including:

1. Types of treatment processes prior to UV disinfection,

2. Quality of water being treated,

3. Chemicals used in the treatment processes,

4. Length of time that the lamps are submerged, and

5. Velocity of the water through the UV system. Very low or stagnant flows are especially likely to permit the settling of solids and the resulting fouling problems.

7.9031 SLEEVE CLEANING

How often quartz sleeves need to be cleaned will depend on the quality of the water being treated and the water treatment chemicals used prior to disinfection. Dipping the UV modules for five minutes in a suitable cleaning solution will completely remove scale that has deposited on the quartz sleeves. Cleaning is best done using an inorganic acid solution with a pH between 2 and 3. The two most suitable cleaning solutions are nitric acid in strengths to approximately 50 percent concentration and a 5 percent or 10 percent solution of phosphoric acid. To clean the system while still continuing to disinfect normal flows, single modules can be removed from the channel, cleaned, and then reinstalled. The other modules remaining on line while one is being cleaned should still be able to provide for continuous disinfection.

In-channel cleaning of UV lamps is another option, but it has some disadvantages. A back-up channel is required and a much greater volume of acid solution is needed. Also, additional equipment and storage tanks for chemicals are required. Precautions must be taken to prevent damage to concrete channels from the acid cleaning solution. Epoxy coatings normally used to protect concrete from acid attack are not used in UV disinfection systems because the epoxy tends to break down under high UV-light intensities.

The type and complexity of the cleaning system will depend on the size of the system and the required frequency of cleaning. Table 7.11 offers some guidelines.

TABLE 7.11   RECOMMENDED UV LAMP CLEANING METHODS

Peak Flow, MGD	Location of Lamps	Type of Cleaning	Work Hours per MGD
<5	out of channel	manual wipe	1
5-20	out of channel	immersion in cleaning tank	0.5
>20	out of channel	remove bank of lamps	0.25
>20	in channel	isolate a channel	0.25

7.9032   LAMP MAINTENANCE

The lamps are the only components that have to be changed on a regular basis. Their service life can be from 7,500 hours to 20,000 hours. This considerable variation can be attributed to three factors:

1. *LEVEL OF SUSPENDED SOLIDS IN THE WATER TO BE DISINFECTED AND THE FECAL COLIFORM LEVEL TO BE ACHIEVED.* Better-quality effluents or less-stringent fecal coliform standards require smaller UV doses. Since lamps lose intensity with age, the smaller the UV dose required, the greater the drop in lamp output that can be tolerated.

2. *FREQUENCY OF ON/OFF CYCLES.* High cycling rates contribute to lamp electrode failure, the most common cause of lamp failure. Limiting the number of ON/OFF cycles to a maximum of 4 per 24 hours can considerably prolong lamp life.

3. *OPERATING TEMPERATURE OF THE LAMP ELECTRODES.* System temperature usually depends on system conditions. Systems with both lamp electrodes operating at the same temperature (both electrodes submerged in the water) operate up to three times longer than systems where the two electrodes operate at different temperatures. This can occur in systems with lamps protruding through a bulkhead where only one electrode is immersed in the water and the other electrode is surrounded by air. If the air temperature is routinely higher than the water temperature, the service life of the lamp electrodes will be shortened.

The largest drop in lamp output occurs during the first 7,500 hours. This decrease is between 30 and 40 percent. Thereaf-ter the annual decrease in lamp output (usually 5 to 10 percent) is caused by the decreased volume of gases within the lamps and by a compositional change of the quartz (solarization) which makes it more opaque to UV light.

7.9033   DISPOSAL OF USED LAMPS

Contact your appropriate regulatory agency to determine the proper way to dispose of used UV lamps. *DO NOT THROW USED LAMPS IN A GARBAGE CAN TO GET RID OF THEM* because of the hazardous mercury in the lamps.

7.904   Acknowledgments

The authors wish to thank CH2M Hill, Trojan Technologies, Inc., London, Ontario, and Fischer & Porter (Canada) Limited for providing information about UV systems and for permitting the use of illustrations and portions of other proprietary materials.

7.91   Ozone

7.910   Equipment

Ozone is normally prepared on site because it is very unstable. Ozonation equipment (Figures 7.46 and 7.47) consists of four major parts:

1. Gas preparation unit,

2. Electrical power unit,

3. Ozone generator, and

4. Contactor.

Gages, controls, safety equipment, and housing are also needed.

7.911   Gas Preparation

The gas preparation unit to produce dry air usually consists of a commercial air dryer with a dew point monitoring system. This is the most critical part of the system.

7.912   Electrical Supply Unit

This unit is normally a very special electrical control system. The most common electrical supply unit provides low frequency, variable voltage. For large installations, medium frequency, variable voltage is used to reduce power costs and because it allows for higher output of ozone.

7.913   Ozone Generator (Figures 7.47 and 7.48)

This unit consists of a pair of electrodes separated by a gas space and a layer of glass insulation. An oxygen-containing gas (air) is passed through the empty space as a high-voltage alternating current is applied. An electrical discharge occurs across the gas space and ozone is formed when a portion of the oxygen is ionized and then becomes associated with non-ionized oxygen molecules.

$$\text{Oxygen From Air} + \overset{\text{High}}{\underset{\text{Voltage}}{\text{Electrical}}} \rightarrow \text{Ionized Oxygen} + \text{Heat}$$

$$O_2 \qquad \rightarrow \qquad 2(O)$$

and

$$\text{Ionized Oxygen} + \text{Non-Ionized Oxygen} \rightarrow \text{Ozone}$$

$$2(O) \quad + \quad 2(O_2) \quad \rightarrow 2(O_3)$$

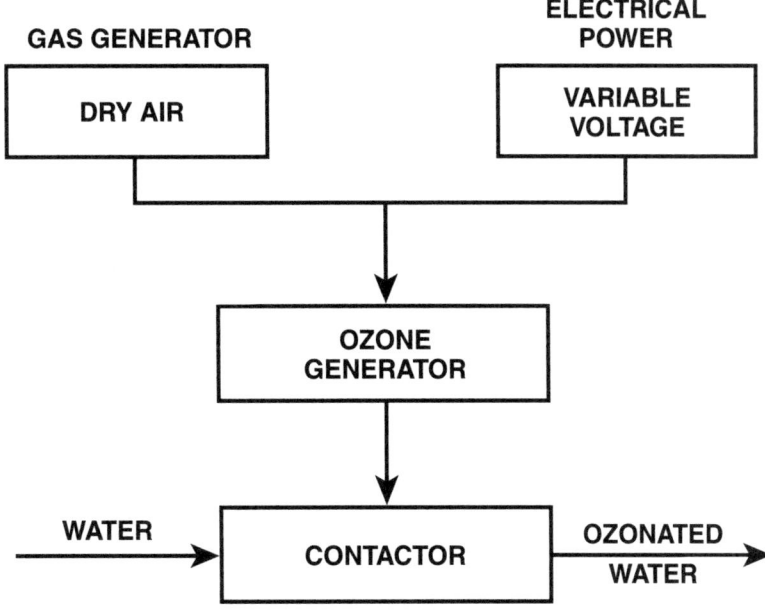

Fig. 7.46 Ozonation equipment

7.914 Ozone Contactor

This unit is a mixing chamber for the ozone-rich material and the process water. The objective is to dissolve enough ozone in the water to achieve disinfection at the lowest possible cost. These units are available in many configurations.

1. Multi-Stage Porous Diffuser

   a. Single application of an ozone-rich stream

   b. Application of ozone to second state

2. Eductor System

   a. Total flow through eductor

   b. Partial plant flow through the eductor

3. Turbine

   a. Positive pressure

   b. Negative pressure

4. Packed bed

   a. Concurrent ozone-rich flow

   b. Countercurrent ozone-rich flow

5. Two-level diffuser

   a. Lower chamber off gases applied to upper chamber

   b. Application of ozone-rich gas to lower chamber

For disinfection purposes, the diffuser-type ozone contact (#5 above) is the most commonly used design. The off gases must be treated prior to release to the atmosphere. The most common method of treatment is the use of activated carbon and dilution.

7.915 Ozone Residuals

Residual ozone is measured by the iodometric method. The procedure is as follows:

1. Collect an 800-mL sample in a 1-liter wash bottle.

2. Pass pure air or nitrogen through sample and then through an absorber containing 400 mL KI solution. Continue for 5 to 10 minutes at a rate of 1.0 liter/minute to purge all ozone from sample.

3. Transfer KI solution.

4. Add 20 mL 1 N H_2SO_4 to reduce pH to 2.

5. Titrate with a 0.005-N sodium thiosulfate solution.

6. Add several drops of starch.

7. End point is reached when the purple color is discharged (solution becomes colorless).

8. Repeat test using blank or distilled water.

9. Calculation

$$\text{mg } O_3/L = \frac{(A \pm B) \times N \times 24{,}000}{mL \text{ of sample}}$$

where:

A = mL of titrant for sample,

B = mL of titrant for blank (positive if turned blue and negative if had to back titrate blank), and

N = normality of sodium thiosulfate

Continuous on-line ozone residual analyzers are available similar to the continuous on-line chlorine residual analyzers (Figure 7.42).

OZONE GENERATION

The basic system for ozone production is presented in the diagram below. The ozone-producing unit, or ozonator, is fed an oxygen-containing gas that has first been cleaned and dried (A). As oxygen molecules (O_2) travel through the discharge gap (B) and are subjected to corona discharge, some are separated into free oxygen atoms (O) which combine with intact molecules to form ozone (O_3). Power (C) is required to generate the corona discharge, and the heat given off during ozonation is dissipated by some form of cooling process (D).

Instrumentation provides monitoring and control of ozone production.

Ozonator dielectric assemblies convert oxygen to ozone by means of a high-voltage corona discharge.

B. The ozonator dielectric assembly consists of two electrode surfaces separated by a glass dielectric and an open gap. Feed gas moves through the gap during ozone production.

Deflectors at the ends of the ozonator tubes prevent discharge arcs, allowing higher-voltage operation for more efficient ozone production, and ensuring longer life for the ozonator dielectric.

C. Electrical power is used to create the corona discharge. Ozone production is controlled by varying the voltage supplied to the ozonator tubes. Feed gas, rich in ozone, leaves the ozonator tubes and is carried to the contactor, where it is mixed with the material to be treated.

Dual-tower regenerative dryers reduce moisture content of the feed gas. Gas entering the ozonator has been dried to a dew point of about –51C° (–60°F).

A. Feed gas (air or oxygen) is filtered to remove particulate matter, run through the compressor package, then filtered again prior to entering the dryers.

D. The corona discharge within the ozonator tubes creates a considerable amount of heat. Ozonators are cooled by a continuous flow of water in the outer shell of the tube-and-shell design.

Special high-voltage fuses protect each dielectric assembly.

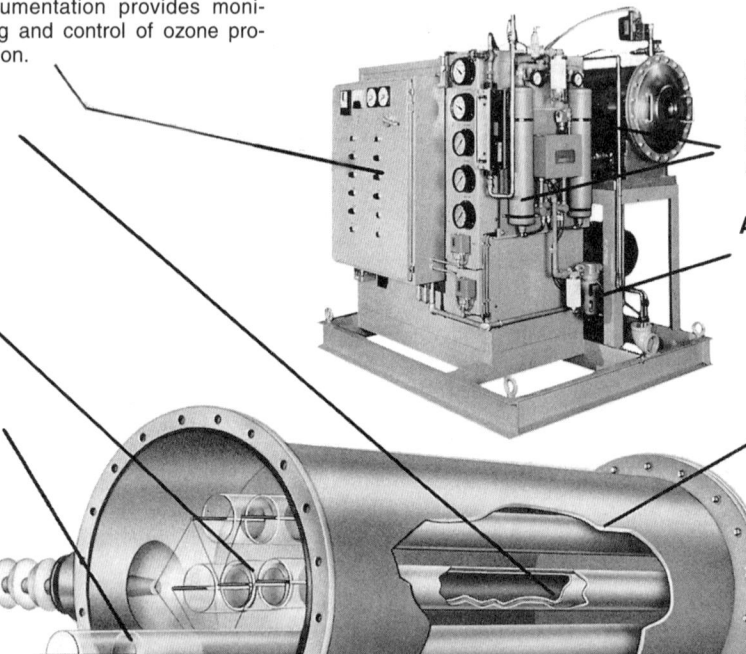

Fig. 7.47   Ozone generation
(Permission of Walsbach Ozone Systems Corporation)

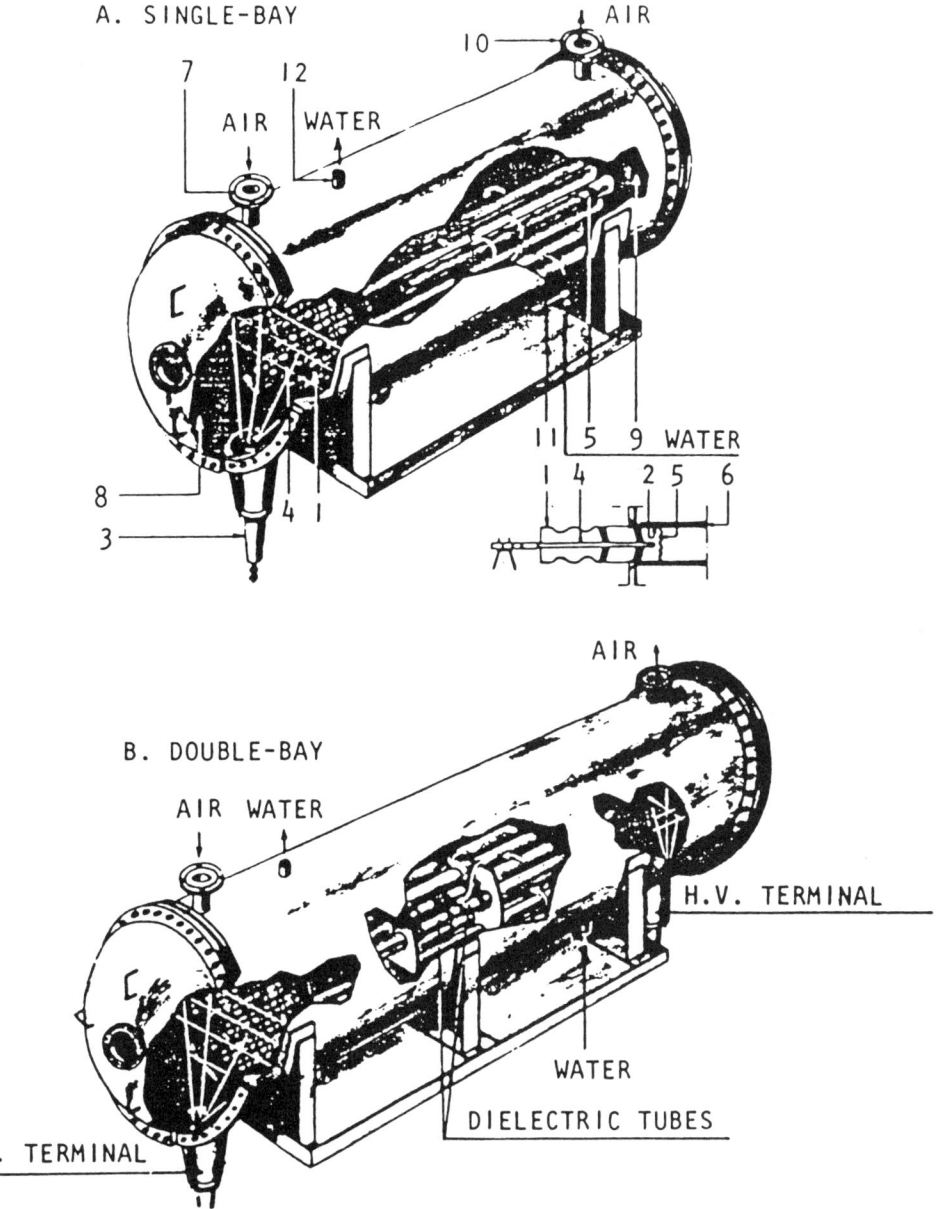

A. SINGLE-BAY

AIR

7 12

AIR WATER

10

8

3 4 1

11 5 9 WATER
1 4 2 5 6

B. DOUBLE-BAY

AIR

AIR WATER

H.V. TERMINAL

WATER

DIELECTRIC TUBES

H.V. TERMINAL

1. DIELECTRIC TUBE
2. METALLIC COATING
3. H.V. TERMINAL
4. CONTACT
5. CENTERING PIECE
6. IONIZATION GAP
7. AIR INLET
8. FRONT CHAMBER
9. REAR CHAMBER
10. AIR OUTLET
11. WATER INLET
12. WATER OUTLET

Fig. 7.48 Typical details of horizontal tube-type ozone generator

7.916   Safety

Ozone is a toxic gas and also is a hazard to plants and animals. When ozone breaks down in the atmosphere as a result of photochemical reactions (reactions taking place in the presence of sunlight), the resulting atmospheric pollutants can be very harmful. However, ozone is less of a hazard than gaseous chlorine since chlorine is normally manufactured and delivered to the plant site. Ozone is produced on the site, it is used in low concentrations, and it is not stored under pressure. Problem leaks can be stopped by turning the unit off. Ozone irritates nasal passages in low concentrations.

Ozone production equipment has various fail-safe protection devices which will automatically shut off the equipment when a potential hazard develops.

7.917   Maintenance

Electrical equipment and pressure vessels should be inspected monthly by trained operators. A yearly preventive maintenance program should be conducted by factory representatives or by an operator trained by the manufacturer. Lubrication of the moving parts should be done according to the manufacturer's recommended schedule.

7.918   Applications of Ozone (Figure 7.49)

In addition to using ozone after filtration for bacterial disinfection and viral inactivation, ozone may also be used for several other purposes in treating drinking waters. Ozone may be used prior to coagulation for treating iron and manganese, helping flocculation, and removing algae. When ozone is applied before filtration, it may be used for oxidizing organics, removing color, or treating tastes and odors.

7.919   Advantages and Limitations of Ozone

ADVANTAGES

1. Better virucide than chlorine.

2. Removes color, odor, and tastes (phenols).

3. Increases dissolved oxygen.

4. Oxidizes iron, manganese, sulfide, and organics.

LIMITATIONS

1. Cost.

2. Maintenance.

3. Safety.

4. Disinfection—unpredictable. Not consistently able to meet MPN of less than 5 percent coliform-positive samples.

5. Operational constraints.

6. No track record.

7.92   Mixed-Oxidants (MIOX) System (Figure 7.50)

Mixed-oxidants disinfection systems are providing good protection from *Cryptosporidium* and *Giardia*. Field studies using

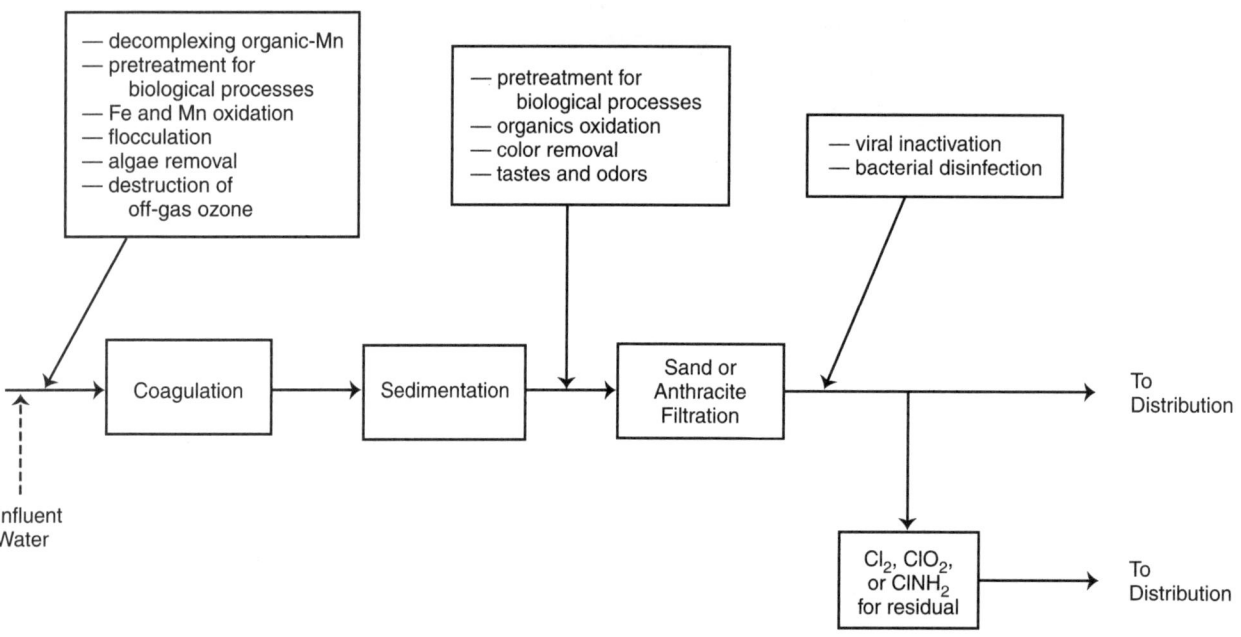

Fig. 7.49   Typical ozone application points in drinking water processes

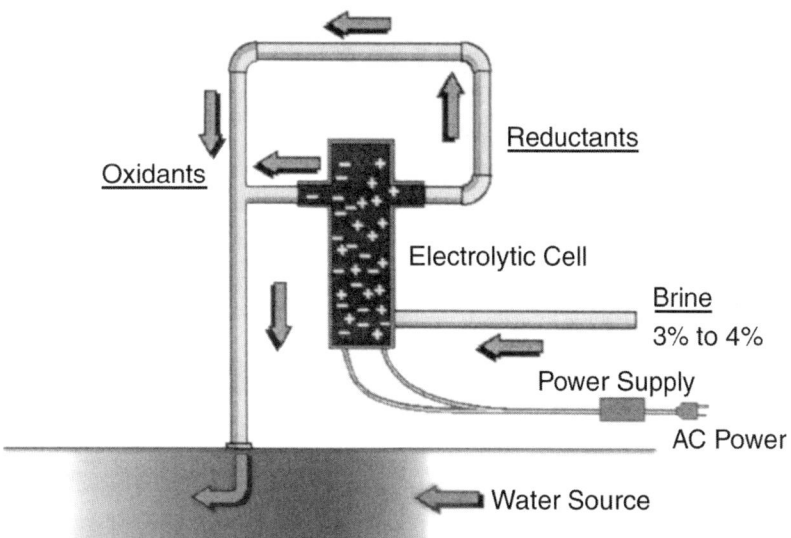

Features		Benefits
Strong Water Disinfection	➤	Kills virus, Giardia, cholera, and Cryptosporidium. Two to six times more effective than chlorine.
Excellent Water Quality	➤	No chemical aftereffects—MIOX-treated water appears, smells, and tastes good.
Meets EPA Standards	➤	Properly dosed water maintains required chlorine residual.
No Harmful Products Used	➤	No toxic gas or liquid chemicals required; uses only salt water.
Low Operating Cost	➤	Operating cost is about 2-4¢/1,000 gallons of treated water.
Wide Range of Capacities	➤	*Standard Cell*—treats up to 6,000 gallons per hour *Large Cell*—treats 20,000 gallons per hour
Simple Operation/Low Maintenance	➤	Operated and maintained by water system operators with minimal training.
Remote Operation	➤	Operates on site anywhere, using any salt, and is adaptable to virtually any power source.

Fig. 7.50   Mixed-oxidants system

(Permission of MIOX Corporation)

mixed oxidants have achieved 99.99 percent reduction of *Giardia* after 30 minutes, and eliminated *E. coli* and cholera after 30 minutes. Mixed-oxidants treatment has reduced total trihalomethanes/disinfection by-products (DBPs) produced when compared with conventional chlorination.

The mixed-oxidants disinfection system works by placing salt and water in a membraneless electrolytic cell (Figure 7.50) to produce a concentrated solution of the oxidants that work together as a disinfectant. The system produces on demand a liquid stream of disinfectant that is safe to handle but can be immediately injected into raw water. No mixing of chemicals or gases is required. The disinfectant is safe to store, produces no toxic fumes, and poses no explosive threat.

The mixed-oxidant cell generates a mixed-oxidant solution electrolytically from a sodium chloride (NaCl) brine. The mixed-oxidant cell separates solutions generated at the anode (oxidants) and at the cathode (reductants) into two fluid streams.

The composition of the mixed-oxidant solution can be varied by changing the brine salt (NaCl) concentration, voltage, and brine flow rate. For example, the cell operating guidelines may be adjusted to produce less chlorine/hypochlorite, more chlorine dioxide and ozone, and a higher oxidation reduction potential by lowering the salt concentration and raising the voltage.

The mixed-oxidant solution is more effective than chlorine/hypochlorite, chlorine dioxide, and ozone used individually. The fast-reacting oxidants (chlorine dioxide and ozone) satisfy the oxidant demand while providing exceptional disinfection and leaving chlorine/hypochlorite as a disinfectant residual. Disinfection occurs within minutes of treatment. These disinfection systems produce no chemical aftereffects and the treated water has no tastes or odors.

QUESTIONS

Write your answers in a notebook and then compare your answers with those on page 358.

7.9A   What precautions must be exercised by the operators of ultraviolet disinfection systems to ensure adequate disinfection?

7.9B   Why is ozone normally prepared on site?

7.9C   Why is ozone less of a hazard than gaseous chlorine?

7.10   TYPICAL CHLORINATION ARITHMETIC PROBLEMS

7.100   Chlorinators

EXAMPLE 5

At 8:00 a.m. on Monday morning a chlorine cylinder weighs 83 pounds. At 8:00 a.m. Tuesday morning the same cylinder weighs 69 pounds. What is the chlorinator feed rate in pounds per 24 hours?

Known	Unknown
Monday Weight, lbs = 83 lbs	Chlorinator Feed Rate, lbs/24 hours
Tuesday Weight, lbs = 69 lbs	
Time Period, hr = 24 hours	

Calculate the chlorinator feed rate in pounds per 24 hours.

$$\text{Chlorinator Feed Rate, lbs/24 hr} = \frac{\text{Initial Weight, lbs} - \text{Final Weight, lbs}}{\text{Time Period, hr}}$$

$$= \frac{83 \text{ lbs} - 69 \text{ lbs}}{24 \text{ hours}}$$

$$= 14 \text{ lbs/24 hours}$$

$$\text{or} = 14 \text{ lbs/day}$$

NOTE: The chlorinator setting should read 14 pounds per 24 hours.

EXAMPLE 6

Estimate the chlorine dose in mg/*L* for the chlorinator in Example 5. The flow totalizer read 12,982,083 gallons at 8:00 a.m. on Monday morning and 13,528,924 at 8:00 a.m. on Tuesday morning. Fourteen pounds of chlorine were fed during the 24-hour period.

Known	Unknown
Monday Flow, gal = 12,982,083 gal	Chlorine Dose, mg/*L*
Tuesday Flow, gal = 13,528,924 gal	
Chlorine Feed, lbs/day = 14 lbs/day	
Time Period, days = 1 day	

Calculate the chlorine dose in mg/*L*.

$$\text{Chlorine Dose, mg/}L = \frac{(\text{Chlorine Feed, lbs/day})(\text{Time, days})}{(\text{Tues Flow, M Gal} - \text{Mon Flow, M Gal})(8.34 \text{ lbs/gal})}$$

$$= \frac{(14 \text{ lbs/day})(1 \text{ day})}{(13.53 \text{ M Gal} - 12.98 \text{ M Gal})(8.34 \text{ lbs/gal})}$$

$$= \frac{14 \text{ lbs Chlorine}}{(0.55 \text{ M Gal})(8.34 \text{ lbs/gal})}$$

$$= \frac{14 \text{ lbs Chlorine}}{4.59 \text{ M lbs Water}}$$

$$= 3.05 \text{ lbs Chlorine/M lbs Water}$$

$$= 3.1 \text{ mg/}L$$

EXAMPLE 7

A centrifugal pump delivers approximately 200 GPM (gallons per minute) against typical operating heads. If the desired chlorine dosage is 3 mg/*L*, what should be the setting on the rotameter for the chlorinator (lbs chlorine per 24 hours)?

Known	Unknown
Pump Flow, GPM = 200 GPM	Rotameter Setting, lbs chlorine/24 hours
Chlorine Dose, mg/*L* = 3 mg/*L*	

1. Convert pump flow to million gallons per day (MGD).

$$\text{Flow, MGD} = \frac{(200 \text{ GPM})(60 \text{ min/hr})(24 \text{ hr/day})}{1,000,000/\text{M}}$$

$$= 0.288 \text{ MGD}$$

NOTE: When we multiply or divide an equation by 1,000,000/M we do not change anything except the units. This is just like multiplying an equation by 12 in/ft or 60 min/hr; all we are doing is changing units.

2. Calculate the rotameter setting in pounds of chlorine per 24 hours.

$$\text{Rotameter Setting, lbs/day} = (\text{Flow, MGD})(\text{Dose, mg/}L)(8.34 \text{ lbs/gal})$$

$$= (0.288 \text{ M Gal/day})(3 \text{ lbs/M lbs})(8.34 \text{ lbs/gal})$$

$$= 7.2 \text{ lbs/day}$$

$$= 7.2 \text{ lbs/24 hours}$$

NOTE: We use the "lbs/day formula" regularly in our work to calculate the setting on a chemical feeder (lbs chlorine per day) and the loading on a treatment process (lbs BOD per day). An explanation of how the units in the formula cancel helps us understand how to use and apply the formula to problems such as this one in Example 7.

Calculate the setting on a chlorinator in pounds of chlorine per day to treat a flow of 0.288 million gallons per day (0.288 MGD) at a chlorine dose of 3 milligrams of chlorine per liter of water (3 mg/*L*). To perform this calculation we need to realize that one liter of water weighs one million milligrams. Therefore,

$$\frac{mg}{L} = \frac{mg}{1 \text{ Million mg}} = \frac{lbs}{1 \text{ Million lbs}}$$

Calculate the chlorinator setting in pounds of chlorine per day.

lbs Cl/day = (Flow, MGD)(Dose, mg/L)(8.34 lbs/gal)

or = (Flow, MGD)(Dose, lbs/M lbs)(8.34 lbs/gal)

$$= \frac{0.288 \text{ M gal H}_2\text{O}}{\text{day}} \times \frac{3 \text{ lbs Cl}}{1 \text{ M lbs H}_2\text{O}} \times \frac{8.34 \text{ lbs H}_2\text{O}}{\text{gal H}_2\text{O}}$$

In this formula the million (*M*) on the top and bottom of the formula cancel, the gallons of water (*gal H₂O*) on top and bottom cancel and the pounds of water (*lbs H₂O*) on top and bottom cancel. This leaves us with pounds of chlorine (*lbs Cl*) on the top and day (*day*) on the bottom. The answer is the chlorinator setting in pounds of chlorine per day (*lbs Cl/day*).

EXAMPLE 8

Using the results from Example 7 (a chlorinator setting of 7.2 lbs/day), how many pounds of chlorine would be used during one week if the pump hour meter showed 100 hours of pump operation? If the chlorine cylinder contained 78 pounds of chlorine at the start of the week, how many pounds of chlorine should be remaining at the end of the week?

Known	**Unknown**
Chlorinator Setting, lbs/day = 7.2 lbs/day	1. Chlorine Used, lbs/week
Chlorine Cylinder, lbs = 78 lbs	2. Chlorine Remaining, lbs

1. Calculate the chlorine used in pounds per week.

$$\text{Chlorine Used, lbs/week} = (\text{Chlorinator Setting, lbs/day})(\text{Time, hr/week})$$

$$= (7.2 \text{ lbs/day})\left(\frac{100 \text{ hr/wk}}{24 \text{ hr/day}}\right)$$

$$= 30 \text{ lbs Chlorine/week}$$

2. Determine the amount of chlorine that should be in the cylinder at the end of the week.

$$\text{Chlorine Remaining, lbs} = \text{Chlorine at Start, lbs} - \text{Chlorine Used, lbs}$$

$$= 78 \text{ lbs} - 30 \text{ lbs}$$

$$= 48 \text{ lbs chlorine remaining at end of week}$$

EXAMPLE 9

Given the pumping and chlorination system in Examples 7 and 8, if 30 pounds of chlorine are used during an average week, how many 150-pound chlorine cylinders will be used per month (assume 30 days per month)?

Known	**Unknown**
Chlorine Used, lbs/week = 30 lbs/week	1. Amount of Chlorine Used per Month, lbs
Chlorine Cyl, lbs/cyl = 150 lbs/cyl	2. Number of 150-lb Cylinders Used per Month

1. Calculate the amount of chlorine used in pounds of chlorine per month.

$$\text{Chlorine Used, lbs/mo} = (\text{Chlorine Used, lbs/week})(\text{Number Weeks/mo})$$

$$= (30 \text{ lbs/week})\left[\frac{(1 \text{ week})(30 \text{ days})}{(7 \text{ days})(1 \text{ mo})}\right]$$

$$= 129 \text{ lbs/mo}$$

2. Determine the number of 150-pound chlorine cylinders used per month.

$$\text{Cylinders Used, number/month} = \frac{\text{Chlorine Used, lbs/mo}}{\text{Chlorine Cylinders, lbs/cylinder}}$$

$$= \frac{129 \text{ lbs/mo}}{150 \text{ lbs/cylinder}}$$

$$= 0.86 \text{ Cylinders/month}$$

This installation requires less than one 150-pound chlorine cylinder per month.

7.101 Hypochlorinators

EXAMPLE 10

Water from a well is being treated by a hypochlorinator. If the hypochlorinator is set at a pumping rate of 50 gallons per day (GPD) and uses a three percent available chlorine solution, what is the chlorine dose in mg/L if the pump delivers 350 GPM?

Known	**Unknown**
Hypochlorinator, GPD = 50 GPD	Chlorine Dose, mg/L
Hypochlorite,% = 3%	
Pump, GPM = 350 GPM	

1. Convert the pumping rate to MGD.

$$\text{Pumping Rate, MGD} = \frac{(350 \text{ GPM})(60 \text{ min/hr})(24 \text{ hr/day})}{1,000,000/\text{M}}$$

$$= 0.50 \text{ MGD}$$

2. Calculate the chlorine feed rate in pounds per day.

$$\text{Chlorine Feed, lbs/day} = \frac{(\text{Flow, gal/day})(\text{Hypochlorite, \%})(8.34 \text{ lbs/gal})}{100\%}$$

$$= \frac{(50 \text{ gal/day})(3\%)(8.34 \text{ lbs/gal})}{100\%}$$

$$= 12.5 \text{ lbs/day}$$

3. Estimate the chlorine dose in mg/L.

$$\text{Chlorine Dose, mg/L} = \frac{\text{Chlorine Feed, lbs/day}}{(\text{Flow, MGD})(8.34 \text{ lbs/gal})}$$

$$= \frac{12.5 \text{ lbs Chlorine/day}}{(0.50 \text{ M gal/day})(8.34 \text{ lbs/gal})}$$

$$= 3 \text{ lbs Chlorine/M lbs Water}$$

$$= 3 \text{ mg/L}$$

EXAMPLE 11

Water pumped from a well is disinfected by a hypochlorinator. During a one-week time period, the water meter indicated that 1,098,000 gallons of water were pumped. A 2.0 percent sodium hypochlorite solution is stored in a 2.5-foot diameter plastic tank. During this one-week time period, the level of hypochlorite in the tank dropped 18 inches (1.50 ft). What was the chlorine dose in mg/*L*?

Known	**Unknown**
Water Treated, M Gal = 1.098 M Gal	Chlorine Dose, mg/*L*
Hypochlorite, % = 2.0%	
Hypochlorite Tank D, ft = 2.5 ft	
Hypochlorite Used, ft = 1.5 ft	

1. Calculate the pounds of water disinfected.

$$\text{Water, lbs} = (\text{Water Treated, M Gal})(8.34 \text{ lbs/gal})$$

$$= (1.098 \text{ M Gal})(8.34 \text{ lbs/gal})$$

$$= 9.16 \text{ M lbs Water}$$

2. Calculate the gallons of hypochlorite solution used.

$$\text{Hypochlorite, gal} = (0.785)(\text{Diameter, ft})^2(\text{Drop, ft})(7.48 \text{ gal/cu ft})$$

$$= (0.785)(2.5 \text{ ft})^2(1.5 \text{ ft})(7.48 \text{ gal/cu ft})$$

$$= 55.0 \text{ gallons}$$

3. Determine the pounds of chlorine used to treat the water.

$$\text{Chlorine, lbs} = (\text{Hypochlorite, gal})\left(\frac{\text{Hypochlorite, \%}}{100\%}\right)(8.34 \text{ lbs/gal})$$

$$= (55.0 \text{ gal})\left(\frac{2.0\%}{100\%}\right)(8.34 \text{ lbs/gal})$$

$$= 9.17 \text{ lbs Chlorine}$$

4. Estimate the chlorine dose in mg/*L*.

$$\text{Chlorine Dose, mg/}L = \frac{\text{Chlorine Used, lbs}}{\text{Water Treated, Million lbs}}$$

$$= \frac{9.17 \text{ lbs Chlorine}}{9.16 \text{ M lbs Water}}$$

$$\text{or} = \frac{1.0 \text{ lbs Chlorine}}{1 \text{ M lbs Water}}$$

$$= 1.0 \text{ mg/}L$$

EXAMPLE 12

Estimate the required concentration of a hypochlorite solution (%) if a pump delivers 600 GPM from a well. The hypochlorinator can deliver a maximum of 120 GPD and the desired chlorine dose is 1.8 mg/*L*.

Known	**Unknown**
Pump Flow, GPM = 600 GPM	Hypochlorite Strength, %
Hypochl Flow, GPD = 120 GPD	
Chlorine Dose, mg/*L* = 1.8 mg/*L*	

1. Calculate the flow of water treated in million gallons per day.

$$\text{Water Treated, M Gal/day} = \frac{(600 \text{ GPM})(60 \text{ min/hr})(24 \text{ hr/day})}{1,000,000/\text{M}}$$

$$= 0.86 \text{ MGD}$$

2. Determine the pounds of chlorine required per day.

$$\text{Chlorine Required, lbs/day} = (\text{Flow, MGD})(\text{Dose, mg/}L)(8.34 \text{ lbs/gal})$$

$$= (0.86 \text{ MGD})(1.8 \text{ mg/}L)(8.34 \text{ lbs/gal})$$

$$= 12.9 \text{ lbs Chlorine/day}$$

3. Calculate the hypochlorite solution strength as a percent.

$$\text{Hypochlorite Strength, \%} = \frac{(\text{Chlorine Required, lbs/day})(100\%)}{(\text{Hypochlorinator Flow, GPD})(8.34 \text{ lbs/gal})}$$

$$= \frac{(12.9 \text{ lbs/day})(100\%)}{(120 \text{ GPD})(8.34 \text{ lbs/gal})}$$

$$= 1.3\%$$

EXAMPLE 13

A hypochlorite solution for a hypochlorinator is being prepared in a 55-gallon drum. If 10 gallons of 5 percent hypochlorite is added to the drum, how much water should be added to the drum to produce a 1.3 percent hypochlorite solution?

Known	**Unknown**
Drum Capacity, gal = 55 gal	Water Added, gal
Hypochlorite, gal = 10 gal	
Actual Hypo, % = 5%	
Desired Hypo, % = 1.3%	

or

$$\text{Desired Hypo, \%} = \frac{(\text{Hypo, gal})(\text{Hypo, \%})}{\text{Hypo, gal} + \text{Water Added, gal}}$$

Rearrange the terms in the equation.

$$(\text{Desired Hypo, \%})(\text{Hypo, gal} + \text{Water Added, gal}) = (\text{Hypo, gal})(\text{Hypo, \%})$$

$$(\text{Desired Hypo, \%})(\text{Hypo, gal}) + (\text{Desired Hypo, \%})(\text{Water Added, gal}) = (\text{Hypo, gal})(\text{Hypo, \%})$$

$$(\text{Desired Hypo, \%})(\text{Water Added, gal}) = (\text{Hypo, gal})(\text{Hypo, \%}) - (\text{Desired Hypo, \%})(\text{Hypo, gal})$$

Calculate the volume of water to be added in gallons.

$$\text{Water Added, gal} = \frac{(\text{Hypo, gal})(\text{Actual Hypo, \%}) - (\text{Desired Hypo, \%})(\text{Hypo, gal})}{\text{Desired Hypo, \%}}$$

$$= \frac{(10 \text{ gal})(5\%) - (1.3\%)(10 \text{ gal})}{1.3\%}$$

$$= \frac{50 - 13}{1.3}$$

$$= 28.5 \text{ gallons of Water}$$

Add 28.5 gallons of water to the 10 gallons of 5 percent hypochlorite in the drum.

7.11 ARITHMETIC ASSIGNMENT

Turn to the Appendix, "How to Solve Water Treatment Plant Arithmetic Problems," at the back of this manual and read all of Section A.7, *VELOCITY AND FLOW RATE*.

In Section A.13, *TYPICAL WATER TREATMENT PLANT PROBLEMS*, read and work the problems in Section A.136, Disinfection.

7.12 ADDITIONAL READING

1. *NEW YORK MANUAL*, Chapter 10,* "Chlorination."

2. *TEXAS MANUAL*, Chapter 10,* "Disinfection of Water."

3. *CHLORINE MANUAL*, Sixth Edition. Obtain from the Chlorine Institute, Inc., 1300 Wilson Boulevard, Arlington, VA 22209. Pamphlet 1. Price to members, $15.00; nonmembers, $30.00; plus 10 percent of order total for shipping and handling.

4. *CHLORINE MANUAL*. Obtain from PPG Industries, Inc., Chemicals Group, One PPG Place, Pittsburgh, PA 15272. Available with the purchase of chlorine.

5. *CHLORINE SAFE HANDLING BOOKLET*. Obtain from PPG Industries, Inc., Chemicals Group, One PPG Place, Pittsburgh, PA 15272. No charge.

* Depends on edition.

QUESTIONS

Write your answers in a notebook and then compare your answers with those on pages 358 and 359.

7.10A Estimate the chlorine demand in milligrams per liter of a water that is dosed at 2.0 mg/*L*. The chlorine residual is 0.2 mg/*L* after a 30-minute contact period.

7.10B What should be the setting on a chlorinator (lbs chlorine per 24 hours) if a pump usually delivers 600 GPM and the desired chlorine dosage is 4.0 mg/*L*?

7.10C Water from a well is being disinfected by a hypochlorinator. If the hypochlorinator is set at a pumping rate of 60 gallons per day (GPD) and uses a 2 percent available chlorine solution, what is the chlorine dose rate in mg/*L*? The pump delivers 400 GPM.

End of Lesson 3 of 3 Lessons on DISINFECTION

Please answer the discussion and review questions next.

DISCUSSION AND REVIEW QUESTIONS

Chapter 7. DISINFECTION

(Lesson 3 of 3 Lessons)

Write the answers to these questions in your notebook. The question numbering continues from Lesson 2.

12. Why is the room temperature important for proper chlorinator operation?

13. How should chlorinator rooms be ventilated?

14. Why must chlorination be continuous?

15. What are the advantages and limitations of the use of chlorine dioxide for disinfection?

16. List the items that should be included in all chlorine safety programs.

17. What precautions must be exercised by the operators of ultraviolet disinfection systems to ensure adequate disinfection?

18. What are the possible uses of ozone for treating drinking water?

SUGGESTED ANSWERS

Chapter 7.   DISINFECTION

ANSWERS TO QUESTIONS IN LESSON 1

Answers to questions on page 270.

7.0A   Pathogenic organisms are disease-producing organisms.

7.0B   Disinfection is the selective destruction or inactivation of pathogenic organisms.

7.0C   The U.S. Environmental Protection Agency establishes drinking water standards.

7.0D   MCL stands for **M**aximum **C**ontaminant **L**evel.

Answers to questions on page 273.

7.1A   Most disinfectants are more effective in water with a pH around 7.0 than at a pH over 8.0.

7.1B   Relatively cold water requires longer disinfection time or greater quantities of disinfectants.

7.1C   The number and type of organisms present in water influence the effectiveness of disinfection on microorganisms.

Answers to questions on page 274.

7.2A   Physical agents that have been used for disinfection include (1) ultraviolet rays, (2) heat, and (3) ultrasonic waves.

7.2B   Chemical agents that have been used for disinfection other than chlorine include (1) iodine, (2) bromine, (3) bases (sodium hydroxide and lime), and (4) ozone.

7.2C   A major limitation to the use of ozone is the inability of ozone to provide a residual in the distribution system.

Answers to questions on page 275.

7.2D   Chlorine Dose, mg/L = Chlorine Demand, mg/L + Chlorine Residual, mg/L

7.2E   Chlorine Demand, mg/L = Chlorine Dose, mg/L – Chlorine Residual, mg/L

7.2F   Hydrogen sulfide and ammonia are two inorganic reducing chemicals with which chlorine reacts rapidly.

7.2G   When chlorine reacts with organic matter, suspected carcinogenic compounds (trihalomethanes) may be formed.

Answers to questions on page 278.

7.2H   Chlorine gas tends to lower the pH while hypochlorite tends to increase the pH.

7.2I   The higher the pH level, the greater the percent of OCl$^-$.

Answers to questions on page 283.

7.2J   Breakpoint chlorination is the addition of chlorine to water until the chlorine demand has been satisfied and further additions of chlorine result in a free available residual chlorine that is directly proportional to the amount of chlorine added beyond the breakpoint.

7.2K   An operator's decision to use chloramines depends on the ability to meet various regulations, the quality of the raw water, operational practices, and distribution system characteristics.

7.2L   The three primary methods by which chloramines are produced include: (1) preammoniation followed by chlorination, (2) concurrent addition of chlorine and ammonia, and (3) prechlorination/postammoniation.

7.2M   The *applied* chlorine to ammonia-nitrogen ratio is usually greater than the *actual* chlorine to nitrogen ratio leaving the plant because of the chlorine demand of the water.

7.2N   Production of nitrite rapidly reduces free chlorine and can interfere with the measurement of free chlorine. The end result of incomplete nitrification may be a loss of total chlorine and ammonia and an increase in the concentration of heterotrophic plate count bacteria.

Answer to questions on page 286.

7.2O   Any sudden absence of chlorine residual in the distribution system indicates that a potential problem has arisen which needs prompt investigation. Immediate action that can be taken includes retesting for chlorine residual, checking chlorination equipment, and searching for a source of contamination which could cause an increase in the chlorine demand.

7.2P   With good initial mixing, the longer the contact time, the better the disinfection.

7.2Q   The efficiency of a disinfectant is measured by the time "T" in minutes of the disinfectant's contact in water and the concentration "C" of the disinfectant residual in mg/L measured at the end of the contact time. The product of these two factors (CxT) provides a measure of the degree of pathogenic inactivation.

Answers to questions on page 288.

7.3A   The two most common points (locations) of chlorination in a water treatment plant are:

1. Prechlorination ahead of any other treatment processes, and
2. Postchlorination after the water has been treated and before it enters the distribution system.

7.3B Waters should not be prechlorinated when the raw waters contain organic compounds. The addition of chlorine will result in the formation of trihalomethanes and tastes and odors (if phenolic compounds are present).

7.3C The benefits of prechlorination include (1) control of algae and slime growths, (2) control of mudball formation, (3) improved coagulation, (4) reduction of tastes and odors, (5) increased chlorine contact time, and (6) increased safety factor in disinfection of heavily contaminated water.

ANSWERS TO QUESTIONS IN LESSON 2

Answers to questions on page 293.

7.4A Typical hypochlorinator systems consist of a chemical solution tank for the hypochlorite, diaphragm-type pump, power supply, water pump, pressure switch, and water storage tank.

7.4B The two common methods of feeding hypochlorite to the water being disinfected include (1) pumping directly into the water, and (2) pumping through an ejector which draws in additional water for dilution of the hypochlorite solution.

Answers to questions on page 300.

7.4C The rate of gas flow in a chlorinator is measured by the use of a rotameter.

7.4D The primary advantage of vacuum system chlorinators is safety. If a failure or breakage occurs in the vacuum system, the chlorinator either stops the flow of chlorine into the equipment or allows air to enter the vacuum system rather than allowing chlorine to escape into the surrounding atmosphere.

Answers to questions on page 303.

7.4E Plastic containers are commonly used to store hypochlorite.

7.4F Normally, a week's supply of hypochlorite should be available.

7.4G The fusible plug is a safety device. The fusible metal softens or melts at 158 to 165°F (70 to 74°C) to prevent buildup of excessive pressures and the possibility of rupture due to fire or high surrounding temperatures.

7.4H The upper valve discharges chlorine gas, and the lower valve discharges liquid chlorine from ton chlorine tanks.

Answers to questions on page 311.

7.4I A sling can be used to hold the tubing to prevent it from flopping around and getting kinked or getting dirt inside.

7.4J One-ton chlorine tanks are placed on their sides with the valves in a vertical position so either chlorine gas or liquid chlorine may be removed.

Answers to questions on page 313.

7.4K Before attempting to start (or stop) any chlorination system, read the operation and maintenance instructions for your plant and the manufacturer's literature to become familiar with the equipment. Review the plans and drawings of the facility. Determine what equipment, pipelines, pumps, tanks, and valves are to be placed into service or are in service. The current status of the entire system must be known before starting or stopping any portion of the system.

7.4L A chlorine residual of 0.2 mg/L should be maintained in the most remote part of the distribution system.

7.4M At the upper end of its operating range, the frequency of the strokes or pulses of a hypochlorite pump will be close together and the chlorine will be fed continuously to the water being treated.

Answers to questions on page 316.

7.4N When starting a gas chlorinator, inspect all valves and joints for leaks by placing an ammonia-soaked rag near each valve and joint. The formation of a white cloud of vapor will indicate a chlorine leak.

7.4O The following steps should be performed when shutting down a chlorinator for a long time period.

1. Have safety equipment available.
2. Close chlorine container gas outlet valve.
3. Allow chlorine gas to completely evacuate the system through the injector.
4. Close chlorinator gas discharge valve.
5. Turn off chlorinator power switch, lock out, and tag.
6. Secure chlorinator gas manifold and chlorinator valve in closed position.

Answers to questions on page 320.

7.4P Normal operation of a chlorinator includes daily inspection of container storage area, evaporators, and chlorinators, including injectors.

7.4Q Evaporators are used to convert liquid chlorine to gaseous chlorine for use by gas chlorinators.

7.4R Abnormal conditions that could be encountered when operating an evaporator include (1) too low a water level, (2) low water temperatures, and (3) no chlorine gas flow to chlorinator.

7.4S Possible chlorinator abnormal conditions include (1) chlorine leaks, (2) chlorine gas pressure too low, (3) injector vacuum too low, and (4) chlorine residual too low.

7.4T The chlorine residual analyzer can be tested by measuring the actual chlorine residual and comparing this result with the residual indicated by the analyzer.

Answers to questions on page 323.

7.4U The suggested free chlorine residual for treated water is 0.5 mg/L at a point just beyond postchlorination.

7.4V The suggested free chlorine residual for the farthest point in the distribution system is 0.2 mg/L.

7.4W To determine if you were chlorinating at the breakpoint, increase the chlorinator feed rate. If the increase in the free chlorine residual is the same as the increased dosage, then you were chlorinating at or past the breakpoint.

ANSWERS TO QUESTIONS IN LESSON 3

Answers to questions on page 326.

7.5A If chlorine is escaping from a cylinder as a liquid, turn the cylinder so that the leak is on top and the chlorine will escape as a gas.

7.5B Chlorine leaks around valve stems can often be stopped by closing the valve or tightening the packing gland nut. Tighten the nut or stem by turning it clockwise.

7.5C Chlorine leaks at the valve discharge outlet can often be stopped by replacing the gasket or adapter connection.

Answers to questions on page 328.

7.5D Chlorinators should be in a separate room because chlorine gas leaks can damage equipment and are hazardous to operators.

7.5E Adequate ventilation is important in a chlorinator room to remove any leaking chlorine gas that would be hazardous to personnel and damaging to equipment.

7.5F Chlorination rates can be checked by the use of scales and recorders to measure weight loss.

7.5G The weights of chlorine containers should be measured and recorded at the same time every day.

Answers to questions on page 334.

7.6A To produce chlorine dioxide in an existing chlorination system, a diaphragm pump, solution tank, mixer, chlorine dioxide generating tower, and electrical controls are also needed.

7.6B Special precautions must be taken when handling sodium chlorite. Spills must be neutralized immediately. Sodium chlorite is very combustible around organic compounds. If sodium chlorite comes in contact with clothing, the clothes should be removed immediately and soaked in water to remove all traces of sodium chlorite or they should be burned immediately.

7.6C Factors such as dirty water, hard water, or wet chlorine can increase the frequency of maintenance required on chlorine dioxide generators.

Answers to questions on page 337.

7.7A Chlorine residual is measured in treated water by the use of (1) amperometric titration, (2) DPD colorimetric method, and (3) ORP probes.

7.7B Residual chlorine measurements of treated water should be taken three times per day on small systems and once every two hours on large systems. The disinfectant in the distribution system must be measured at the same frequency and location as the total coliform samples.

7.7C In a disinfection system, ORP is a direct measure of the effectiveness of a chlorine residual in disinfecting the water being treated.

7.7D When a microorganism loses an electron, it becomes inactivated and can no longer transmit a disease or reproduce.

7.7E Maintenance for the chlorine ORP probe consists of cleaning the unit's sensor once a month.

Answers to questions on page 342.

7.8A Chlorine gas is extremely toxic and corrosive in moist atmospheres.

7.8B A properly fitting self-contained air or oxygen supply type of breathing apparatus, positive pressure/demand breathing equipment, or rebreather kits are used when repairing a chlorine leak.

7.8C First-aid measures depend on the severity of the contact. Move the victim away from the gas area, remove contaminated clothes and keep the victim warm and quiet. Call a doctor and fire department immediately. Keep the patient breathing.

Answers to questions on page 352.

7.9A The operators of ultraviolet disinfection systems must be sure that maximum design turbidity levels and flows are not exceeded. Also bacteriological tests must be made frequently to ensure that adequate disinfection is being achieved.

7.9B Ozone is normally prepared on site because it is very unstable.

7.9C Ozone is less of a hazard than gaseous chlorine since chlorine is normally manufactured elsewhere and delivered to the plant site. Ozone is produced on the site, it is used in low concentrations, and it is not stored under pressure. Problem leaks can be stopped by turning the unit off.

Answers to questions on page 355.

7.10A

Known	**Unknown**
Chlorine Dose, mg/L = 2.0 mg/L	Chlorine Demand, mg/L
Chlorine Residual, mg/L = 0.2 mg/L	

Calculate the chlorine demand in mg/L.

$$\text{Chlorine Demand, mg/}L = \text{Chlorine Dose, mg/}L - \text{Chlorine Residual, mg/}L$$

$$= 2.0 \text{ mg/}L - 0.2 \text{ mg/}L$$

$$= 1.8 \text{ mg/}L$$

7.10B

Known	**Unknown**
Pump Flow, GPM = 600 GPM	Chlorinator Setting, lbs chlorine/24 hr
Chlorine Dose, mg/L = 4.0 mg/L	

1. Convert the pump flow to million gallons per day (MGD).

$$\text{Flow, MGD} = \frac{(600 \text{ GPM})(60 \text{ min/hr})(24 \text{ hr/day})}{1,000,000/\text{M}}$$

$$= 0.864 \text{ MGD}$$

2. Calculate the chlorinator setting in pounds of chlorine per 24 hours.

$$\text{Chlorinator Setting, lbs/24 hr} = (\text{Flow, MGD})(\text{Dose, mg/}L)(8.34 \text{ lbs/gal})$$

$$= (0.864 \text{ MGD})\left(\frac{4.0 \text{ lbs Chlorine}}{1 \text{ M lbs Water}}\right)(8.34 \text{ lbs/gal})$$

$$= 28.8 \text{ lbs Chlorine/day}$$

$$= 28.8 \text{ lbs Chlorine/24 hr}$$

7.10C

Known	**Unknown**
Hypochlorinator, GPD = 60 GPD	Chlorine Dose, mg/L
Hypochlorite, % = 2%	
Pump, GPM = 400 GPM	

1. Convert the pumping rate to MGD.

$$\text{Pumping Rate, MGD} = \frac{(400 \text{ GPM})(60 \text{ min/hr})(24 \text{ hr/day})}{1,000,000/M}$$

$$= 0.576 \text{ MGD}$$

2. Calculate the chlorine feed rate in pounds per day.

$$\text{Chlorine Feed, lbs/day} = \frac{(\text{Flow, gal/day})(\text{Hypochlorite, \%})(8.34 \text{ lbs/gal})}{100\%}$$

$$= \frac{(60 \text{ GPD})(2\%)(8.34 \text{ lbs/gal})}{100\%}$$

$$= 10.0 \text{ lbs Chlorine/day}$$

3. Estimate the chlorine dose in mg/L.

$$\text{Chlorine Dose, mg/L} = \frac{\text{Chlorine Feed, lbs/day}}{(\text{Flow, MGD})(8.34 \text{ lbs/gal})}$$

$$= \frac{10.0 \text{ lbs Chlorine/day}}{(0.576 \text{ MGD})(8.34 \text{ lbs/gal})}$$

$$= 2.1 \text{ lbs Chlorine/M lbs Water}$$

$$= 2.1 \text{ mg/L}$$

CHAPTER 8

CORROSION CONTROL

by

Jack Rossum

Revised by

Mike Pollen

TABLE OF CONTENTS

Chapter 8.   CORROSION CONTROL

OBJECTIVES

Chapter 8.   CORROSION CONTROL

Following completion of Chapter 8, you should be able to:

1. Recognize adverse effects of corrosion,

2. Describe how a pipe corrodes,

3. Determine if corrosion problems exist in your system,

4. Determine if a water is saturated with calcium carbonate,

5. Select the proper chemical to control corrosion,

6. Determine the proper chemical dose to control corrosion,

7. Use cathodic protection to control corrosion,

8. Prevent soil corrosion (external corrosion),

9. Troubleshoot and solve corrosion problems, and

10. Explain and implement the Lead and Copper Rule.

WORDS

Chapter 8. CORROSION CONTROL

ALKALINITY (AL-ka-LIN-it-tee)   ALKALINITY

The capacity of water to neutralize acids. This capacity is caused by the water's content of carbonate, bicarbonate, hydroxide, and occasionally borate, silicate, and phosphate. Alkalinity is expressed in milligrams per liter of equivalent calcium carbonate. Alkalinity is not the same as pH because water does not have to be strongly basic (high pH) to have a high alkalinity. Alkalinity is a measure of how much acid must be added to a liquid to lower the pH to 4.5.

AMPERAGE (AM-purr-age)   AMPERAGE

The strength of an electric current measured in amperes. The amount of electric current flow, similar to the flow of water in gallons per minute.

AMPERE (AM-peer)   AMPERE

The unit used to measure current strength. The current produced by an electromotive force of one volt acting through a resistance of one ohm.

ANAEROBIC (AN-air-O-bick)   ANAEROBIC

A condition in which atmospheric or dissolved molecular oxygen is *NOT* present in the aquatic (water) environment.

ANION (AN-EYE-en)   ANION

A negatively charged ion in an electrolyte solution, attracted to the anode under the influence of a difference in electrical potential. Chloride ion (Cl^-) is an anion.

ANNULAR (AN-you-ler) SPACE   ANNULAR SPACE

A ring-shaped space located between two circular objects, such as two pipes.

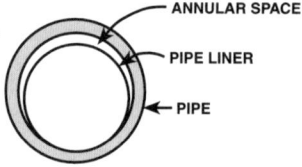

ANODE (an-O-d)   ANODE

The positive pole or electrode of an electrolytic system, such as a battery. The anode attracts negatively charged particles or ions (anions).

ATOM   ATOM

The smallest unit of a chemical element; composed of protons, neutrons and electrons.

BASE METAL   BASE METAL

A metal (such as iron) which reacts with dilute hydrochloric acid to form hydrogen. Also see NOBLE METAL.

BUFFER CAPACITY   BUFFER CAPACITY

A measure of the capacity of a solution or liquid to neutralize acids or bases. This is a measure of the capacity of water for offering a resistance to changes in pH.

C FACTOR   C FACTOR

A factor or value used to indicate the smoothness of the interior of a pipe. The higher the C Factor, the smoother the pipe, the greater the carrying capacity, and the smaller the friction or energy losses from water flowing in the pipe. To calculate the C Factor, measure the flow, pipe diameter, distance between two pressure gages, and the friction or energy loss of the water between the gages.

$$\text{C Factor} = \frac{\text{Flow, GPM}}{193.75(\text{Diameter, ft})^{2.63}(\text{Slope})^{0.54}}$$

CALCIUM CARBONATE ($CaCO_3$) EQUIVALENT

CALCIUM CARBONATE ($CaCO_3$) EQUIVALENT

An expression of the concentration of specified constituents in water in terms of their equivalent value to calcium carbonate. For example, the hardness in water which is caused by calcium, magnesium and other ions is usually described as calcium carbonate equivalent. Alkalinity test results are usually reported as mg/L $CaCO_3$ equivalents. To convert chloride to $CaCO_3$ equivalents, multiply the concentration of chloride ions in mg/L by 1.41, and for sulfate, multiply by 1.04.

CATALYST (CAT-uh-LIST)

CATALYST

A substance that changes the speed or yield of a chemical reaction without being consumed or chemically changed by the chemical reaction.

CATALYZE (CAT-uh-LIZE)

CATALYZE

To act as a catalyst. Or, to speed up a chemical reaction.

CATHODE (KA-thow-d)

CATHODE

The negative pole or electrode of an electrolytic cell or system. The cathode attracts positively charged particles or ions (cations).

CATHODIC (ca-THOD-ick) PROTECTION

CATHODIC PROTECTION

An electrical system for prevention of rust, corrosion, and pitting of metal surfaces which are in contact with water or soil. A low-voltage current is made to flow through a liquid (water) or a soil in contact with the metal in such a manner that the external electromotive force renders the metal structure cathodic. This concentrates corrosion on auxiliary anodic parts which are deliberately allowed to corrode instead of letting the structure corrode.

CATION (CAT-EYE-en)

CATION

A positively charged ion in an electrolyte solution, attracted to the cathode under the influence of a difference in electrical potential. Sodium ion (Na^+) is a cation.

COMPOUND

COMPOUND

A pure substance composed of two or more elements whose composition is constant. For example, table salt (sodium chloride, $NaCl$) is a compound.

CORROSION

CORROSION

The gradual decomposition or destruction of a material by chemical action, often due to an electrochemical reaction. Corrosion may be caused by (1) stray current electrolysis, (2) galvanic corrosion caused by dissimilar metals, or (3) differential-concentration cells. Corrosion starts at the surface of a material and moves inward.

CORROSION INHIBITORS

CORROSION INHIBITORS

Substances that slow the rate of corrosion.

CORROSIVE GASES

CORROSIVE GASES

In water, dissolved oxygen reacts readily with metals at the anode of a corrosion cell, accelerating the rate of corrosion until a film of oxidation products such as rust forms. At the cathode where hydrogen gas may form a coating on the cathode and slow the corrosion rate, oxygen reacts rapidly with hydrogen gas forming water, and again increases the rate of corrosion.

CORROSIVITY

CORROSIVITY

An indication of the corrosiveness of a water. The corrosiveness of a water is described by the water's pH, alkalinity, hardness, temperature, total dissolved solids, dissolved oxygen concentration, and the Langelier Index.

COUPON

COUPON

A steel specimen inserted into water to measure the corrosiveness of water. The rate of corrosion is measured as the loss of weight of the coupon (in milligrams) per surface area (in square decimeters) exposed to the water per day. 10 decimeters = 1 meter = 100 centimeters.

CURRENT

CURRENT

A movement or flow of electricity. Water flowing in a pipe is measured in gallons per second past a certain point, not by the number of water molecules going past a point. Electric current is measured by the number of coulombs per second flowing past a certain point in a conductor. A coulomb is equal to about 6.25 x 10^{18} electrons (6,250,000,000,000,000,000 electrons). A flow of one coulomb per second is called one ampere, the unit of the rate of flow of current.

DEAD END

DEAD END

The end of a water main which is not connected to other parts of the distribution system by means of a connecting loop of pipe.

DEPOLARIZATION DEPOLARIZATION

The removal or depletion of ions in the thin boundary layer adjacent to a membrane or pipe wall.

DIELECTRIC (DIE-ee-LECK-trick) DIELECTRIC

Does not conduct an electric current. An insulator or nonconducting substance.

ELECTROCHEMICAL REACTION ELECTROCHEMICAL REACTION

Chemical changes produced by electricity (electrolysis) or the production of electricity by chemical changes (galvanic action). In corrosion, a chemical reaction is accompanied by the flow of electrons through a metallic path. The electron flow may come from an external source and cause the reaction, such as electrolysis caused by a D.C. (direct current) electric railway or the electron flow may be caused by a chemical reaction as in the galvanic action of a flashlight dry cell.

ELECTROCHEMICAL SERIES ELECTROCHEMICAL SERIES

A list of metals with the standard electrode potentials given in volts. The size and sign of the electrode potential indicates how easily these elements will take on or give up electrons, or corrode. Hydrogen is conventionally assigned a value of zero.

ELECTROLYSIS (ee-leck-TRAWL-uh-sis) ELECTROLYSIS

The decomposition of material by an outside electric current.

ELECTROLYTE (ee-LECK-tro-LITE) ELECTROLYTE

A substance which dissociates (separates) into two or more ions when it is dissolved in water.

ELECTROLYTIC (ee-LECK-tro-LIT-ick) CELL ELECTROLYTIC CELL

A device in which the chemical decomposition of material causes an electric current to flow. Also, a device in which a chemical reaction occurs as a result of the flow of electric current. Chlorine and caustic (NaOH) are made from salt (NaCl) in electrolytic cells.

ELECTROMOTIVE FORCE (E.M.F.) ELECTROMOTIVE FORCE (E.M.F.)

The electrical pressure available to cause a flow of current (amperage) when an electric circuit is closed. Also called VOLTAGE.

ELECTROMOTIVE SERIES ELECTROMOTIVE SERIES

A list of metals and alloys presented in the order of their tendency to corrode (or go into solution). Also called the GALVANIC SERIES. This is a practical application of the theoretical ELECTROCHEMICAL SERIES.

ELECTRON ELECTRON

(1) A very small, negatively charged particle which is practically weightless. According to the electron theory, all electrical and electronic effects are caused either by the movement of electrons from place to place or because there is an excess or lack of electrons at a particular place.

(2) The part of an atom that determines its chemical properties.

ELEMENT ELEMENT

A substance which cannot be separated into its constituent parts and still retain its chemical identity. For example, sodium (Na) is an element.

GALVANIC CELL GALVANIC CELL

An electrolytic cell capable of producing electric energy by electrochemical action. The decomposition of materials in the cell causes an electric (electron) current to flow from cathode to anode.

GALVANIC SERIES GALVANIC SERIES

A list of metals and alloys presented in the order of their tendency to corrode (or go into solution). Also called the ELECTROMOTIVE SERIES. This is a practical application of the theoretical ELECTROCHEMICAL SERIES.

HARDNESS, WATER HARDNESS, WATER

A characteristic of water caused mainly by the salts of calcium and magnesium, such as bicarbonate, carbonate, sulfate, chloride and nitrate. Excessive hardness in water is undesirable because it causes the formation of soap curds, increased use of soap, deposition of scale in boilers, damage in some industrial processes, and sometimes causes objectionable tastes in drinking water.

HYDROLYSIS (hi-DROLL-uh-sis) HYDROLYSIS

(1) A chemical reaction in which a compound is converted into another compound by taking up water.

(2) Usually a chemical degradation of organic matter.

INTERFACE

The common boundary layer between two substances such as water and a solid (metal); or between two fluids such as water and a gas (air); or between a liquid (water) and another liquid (oil).

ION

An electrically charged atom, radical (such as SO_4^{2-}), or molecule formed by the loss or gain of one or more electrons.

LANGELIER INDEX (L.I.)

An index reflecting the equilibrium pH of a water with respect to calcium and alkalinity. This index is used in stabilizing water to control both corrosion and the deposition of scale.

$$\text{Langelier Index} = pH - pH_s$$
$$\text{where } pH = \text{actual pH of the water, and}$$
$$pH_s = \text{pH at which water having the same alkalinity and calcium content is just saturated with calcium carbonate.}$$

LOGARITHM (LOG-a-rith-m)

The exponent that indicates the power to which a number must be raised to produce a given number. For example: if $B^2 = N$, the 2 is the logarithm of N (to the base B), or $10^2 = 100$ and $\log_{10} 100 = 2$. Also abbreviated to "log."

MECHANICAL JOINT

A flexible device that joins pipes or fittings together by the use of lugs and bolts.

MOLECULE (MOLL-uh-KULE)

The smallest division of a compound that still retains or exhibits all the properties of the substance.

NPDES PERMIT

National **P**ollutant **D**ischarge **E**limination **S**ystem permit is the regulatory agency document issued by either a federal or state agency which is designed to control all discharges of potential pollutants from point sources and storm water runoff into U.S. waterways. NPDES permits regulate discharges into navigable waters from all point sources of pollution, including industries, municipal wastewater treatment plants, sanitary landfills, large agricultural feedlots and return irrigation flows.

NOBLE METAL

A chemically inactive metal (such as gold). A metal that does not corrode easily and is much scarcer (and more valuable) than the so-called useful or base metals. Also see BASE METAL.

OHM

The unit of electrical resistance. The resistance of a conductor in which one volt produces a current of one ampere.

OXIDATION (ox-uh-DAY-shun)

Oxidation is the addition of oxygen, removal of hydrogen, or the removal of electrons from an element or compound. In the environment, organic matter is oxidized to more stable substances. The opposite of REDUCTION.

OXIDIZING AGENT

Any substance, such as oxygen (O_2) or chlorine (Cl_2), that will readily add (take on) electrons. The opposite is a REDUCING AGENT.

pH (pronounce as separate letters)

pH is an expression of the intensity of the basic or acidic condition of a liquid. Mathematically, pH is the logarithm (base 10) of the reciprocal of the hydrogen ion activity.

$$pH = \text{Log} \frac{1}{[H^+]}$$

The pH may range from 0 to 14, where 0 is most acidic, 14 most basic, and 7 neutral. Natural waters usually have a pH between 6.5 and 8.5.

POLARIZATION

The concentration of ions in the thin boundary layer adjacent to a membrane or pipe wall.

REDUCING AGENT

Any substance, such as base metal (iron) or the sulfide ion (S^{2-}), that will readily donate (give up) electrons. The opposite is an OXIDIZING AGENT.

REDUCTION (re-DUCK-shun)   REDUCTION

Reduction is the addition of hydrogen, removal of oxygen, or the addition of electrons to an element or compound. Under anaerobic conditions (no dissolved oxygen present), sulfur compounds are reduced to odor-producing hydrogen sulfide (H_2S) and other compounds. The opposite of OXIDATION.

REPRESENTATIVE SAMPLE   REPRESENTATIVE SAMPLE

A sample portion of material or water that is as nearly identical in content and consistency as possible to that in the larger body of material or water being sampled.

SACRIFICIAL ANODE   SACRIFICIAL ANODE

An easily corroded material deliberately installed in a pipe or tank. The intent of such an installation is to give up (sacrifice) this anode to corrosion while the water supply facilities remain relatively corrosion free.

SALINITY   SALINITY

(1) The relative concentration of dissolved salts, usually sodium chloride, in a given water.

(2) A measure of the concentration of dissolved mineral substances in water.

SATURATION   SATURATION

The condition of a liquid (water) when it has taken into solution the maximum possible quantity of a given substance at a given temperature and pressure.

SLAKE   SLAKE

To mix with water so that a true chemical combination (hydration) takes place, such as in the slaking of lime.

SLURRY (SLUR-e)   SLURRY

A watery mixture or suspension of insoluble (not dissolved) matter; a thin, watery mud or any substance resembling it (such as a grit slurry or a lime slurry).

STRAY CURRENT CORROSION   STRAY CURRENT CORROSION

A corrosion activity resulting from stray electric current originating from some source outside the plumbing system such as D.C. grounding on phone systems.

SUPERSATURATED   SUPERSATURATED

An unstable condition of a solution (water) in which the solution contains a substance at a concentration greater than the saturation concentration for the substance.

TITRATE (TIE-trate)   TITRATE

To *TITRATE* a sample, a chemical solution of known strength is added drop by drop until a certain color change, precipitate, or pH change in the sample is observed (end point). Titration is the process of adding the chemical reagent in small increments (0.1 – 1.0 milliliter) until completion of the reaction, as signaled by the end point.

TUBERCLE (TOO-burr-cull)   TUBERCLE

A protective crust of corrosion products (rust) which builds up over a pit caused by the loss of metal due to corrosion.

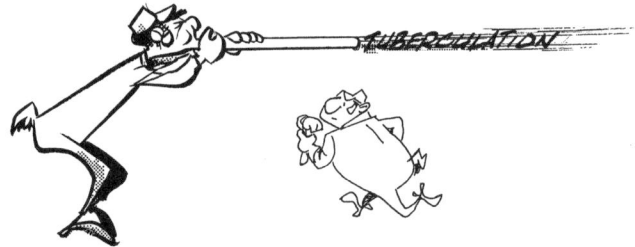

TUBERCULATION (too-BURR-que-LAY-shun)   TUBERCULATION

The development or formation of small mounds of corrosion products (rust) on the inside of iron pipe. These mounds (tubercles) increase the roughness of the inside of the pipe thus increasing resistance to water flow (decreases the C Factor).

VOLTAGE   VOLTAGE

The electrical pressure available to cause a flow of current (amperage) when an electric circuit is closed. Also called ELECTROMOTIVE FORCE (E.M.F.).

CHAPTER 8. CORROSION CONTROL

(Lesson 1 of 2 Lessons)

8.0 ADVERSE EFFECTS OF CORROSION[1]

Corrosive water can cause very serious problems in both water supply facilities and water treatment plants. Many hundreds of millions of dollars in damage occurs each year due to corrosive conditions in water systems. Water main replacement is often required when *TUBERCULATION*[2] reduces the carrying capacity of a main. Tuberculation increases pipe roughness which causes an increase in pump energy costs and may reduce distribution system pressures. Leaks in water mains are usually caused by corrosion and may eventually require the replacement of a water main. Figure 8.1 shows corrosion damage to pipes due to corrosive soils.

Many other serious problems are caused by corrosive water. Corrosive water causes materials to deteriorate and go into solution (be carried by the water). Corrosion of toxic metal pipe materials such as lead can create a serious health hazard. Corrosion of iron may produce a flood of unpleasant telephone calls from consumers complaining about rusty water, stained laundry, and bad tastes. Corrosive drinking water causes internal corrosion (the inside of the pipe corrodes) and corrosive soils and moisture cause external corrosion (the outside of the pipe corrodes).

8.1 PROCESS OF CORROSION

8.10 Definition of Corrosion

This chapter explains what causes corrosion and how to control corrosion. Corrosion is a very complex problem, involving numerous chemical, electrical, physical, and biological factors. Corrosion can occur both on the interior and exterior of metal piping and equipment. Fortunately, there are a few basic concepts that describe the major principles of corrosion. Selecting and implementing an effective corrosion-control program requires an understanding of these basic concepts.

Corrosion in the water treatment industry can be defined as the gradual decomposition or destruction of a material (such as a metal or cement lining) as it reacts with water. The severity and type of corrosion depend on the chemical and physical characteristics of the water and the material.

8.11 Electrochemical Corrosion: The Galvanic Cell[3,4]

In this section we will explain the corrosion reaction so you can (1) understand what causes corrosion, and (2) determine whether a corrosion problem exists in your system.

Metallic (metal) corrosion in potable water is *ALWAYS* the result of an *ELECTROCHEMICAL REACTION.*[5] An electrochemical reaction is a chemical reaction where the flow of electric current[6] itself is an essential part of the reaction. If the electric current is stopped by breaking the circuit, the chemical reaction will stop. Also, if the chemical reaction is stopped by removing one of the reacting chemicals, the flow of electric current will stop. For corrosion to occur, both of the factors—electric current and chemical reaction—must be present.

The deterioration of metal during corrosion is called an electrochemical reaction since both a chemical and an electrical process are occurring. Figure 8.2 shows a simplified electrochemical corrosion reaction with the following components:

- *ANODE*[7]   Point from which metal is lost and electric current begins

- *CATHODE*[8]   Point where electric current leaves the metal and flows to the anode through the electrolyte

- *ELECTROLYTE*[9] Conducting solution (usually water with dissolved salts)

In addition, the anode and the cathode must be joined. In Figure 8.2, the only metal present is iron. At the anode a mol-

---

[1] *Corrosion. The gradual decomposition or destruction of a material by chemical action, often due to an electrochemical reaction. Corrosion may be caused by (1) stray current electrolysis, (2) galvanic corrosion caused by dissimilar metals, or (3) differential-concentration cells. Corrosion starts at the surface of a material and moves inward.*

[2] *Tuberculation (too-BURR-que-LAY-shun). The development or formation of small mounds of corrosion products (rust) on the inside of iron pipe. These mounds (tubercles) increase the roughness of the inside of the pipe thus increasing resistance to water flow (decreases the C Factor).*

[3] *Galvanic Cell. An ELECTROLYTIC CELL[4] capable of producing electric energy by electrochemical action. The decomposition of materials in the cell causes an electric (electron) current to flow from cathode to anode.*

[4] *Electrolytic (ee-LECK-tro-LIT-ick) Cell. A device in which the chemical decomposition of material causes an electric current to flow. Also, a device in which a chemical reaction occurs as a result of the flow of electric current. Chlorine and caustic (NaOH) are made from salt (NaCl) in electrolytic cells.*

[5] *Electrochemical Reaction. Chemical changes produced by electricity (electrolysis) or the production of electricity by chemical changes (galvanic action). In corrosion, a chemical reaction is accompanied by the flow of electrons through a metallic path. The electron flow may come from an external source and cause the reaction, such as electrolysis caused by a D.C. (direct current) electric railway or the electron flow may be caused by a chemical reaction as in the galvanic action of a flashlight dry cell.*

[6] *Also referred to as "electron transfer" from one type of atom to another.*

[7] *Anode (an-O-d). The positive pole or electrode of an electrolytic system, such as a battery. The anode attracts negatively charged particles or ions (anions).*

[8] *Cathode (KA-thow-d). The negative pole or electrode of an electrolytic cell or system. The cathode attracts positively charged particles or ions (cations).*

[9] *Electrolyte (ee-LECK-tro-LITE). A substance which dissociates (separates) into two or more ions when it is dissolved in water.*

CORROSION OF PLAIN CARBON STEEL N
EXPOSED 14 YEARS AT 5 TEST SITES

CORROSION OF PLAIN CAST IRON G
EXPOSED 14 YEARS AT 5 TEST SITES

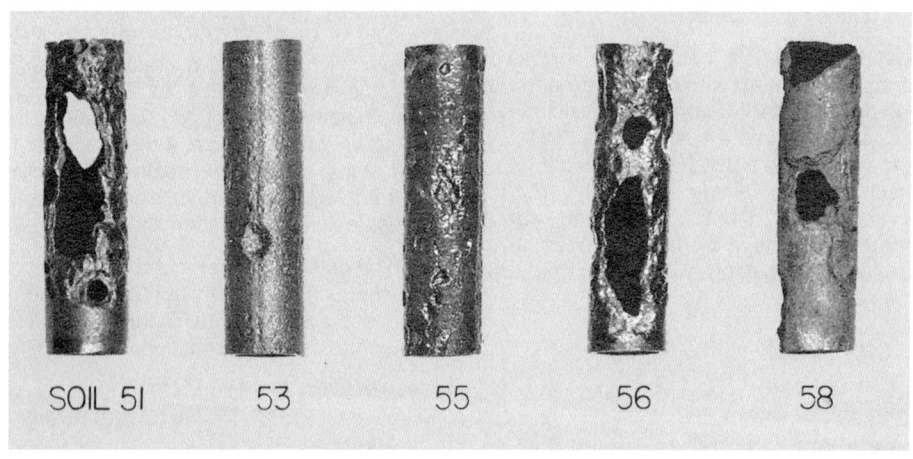

Fig. 8.1   Exterior corrosion due to corrosive soils
(Permission of ARMCO)

ecule of iron dissolves into the water as a ferrous (Fe^{2+}) ion (chemical reaction), and two *ELECTRONS*[10] ($2e^-$) flow to the cathode (electrical reaction). At the cathode, the electrons leave the metal at the point of contact with the electrolyte (water) and react with hydrogen ions (H^+) in the water to form hydrogen gas (H_2). The hydrogen ions are always present in the water from the normal dissociation of water ($H_2O \rightleftarrows H^+ + OH^-$). The electrolyte (water) is in contact with both the anode and the cathode, completing the circuit.

At the anode, the dissolved iron reacts with oxygen and the water, forming a rust film composed initially of ferrous hydroxide [$Fe(OH)_2$] as shown in Figure 8.3. Additional water and oxygen then react with the ferrous hydroxide to form ferric hydroxide [$Fe(OH)_3$], which becomes a second layer over the ferrous hydroxide.

This multilayered rust deposit is known as a *TUBERCLE*.[11] Tubercles can increase in quantity to the point that the carrying capacity of the pipe is significantly reduced. Also, during periods of high flow rates, the tubercles may dislodge, resulting in rusty or red-colored water. The formation of a rust coating on the pipe has another important effect on the rate of corrosion. As the rust film forms, it begins to cover and protect the anode, slowing the rate of corrosion. If the rust film is flushed away, the corrosion reaction accelerates again.

A more complex form of electrochemical corrosion is caused by the joining of dissimilar metals. This type of corrosion is called *GALVANIC* corrosion. Figure 8.4 shows a galvanic corrosion model resulting from the joining of sections of copper and iron pipe. Figure 8.5 contains photos of corrosion damage caused by galvanic corrosion. Like the corrosion cell in Figure

[10] *Electron.   (1) A very small, negatively charged particle which is practically weightless. According to the electron theory, all electrical and electronic effects are caused either by the movement of electrons from place to place or because there is an excess or lack of electrons at a particular place. (2) The part of an atom that determines its chemical properties.*
[11] *Tubercle (TOO-burr-cull).   A protective crust of corrosion products (rust) which builds up over a pit caused by the loss of metal due to corrosion.*

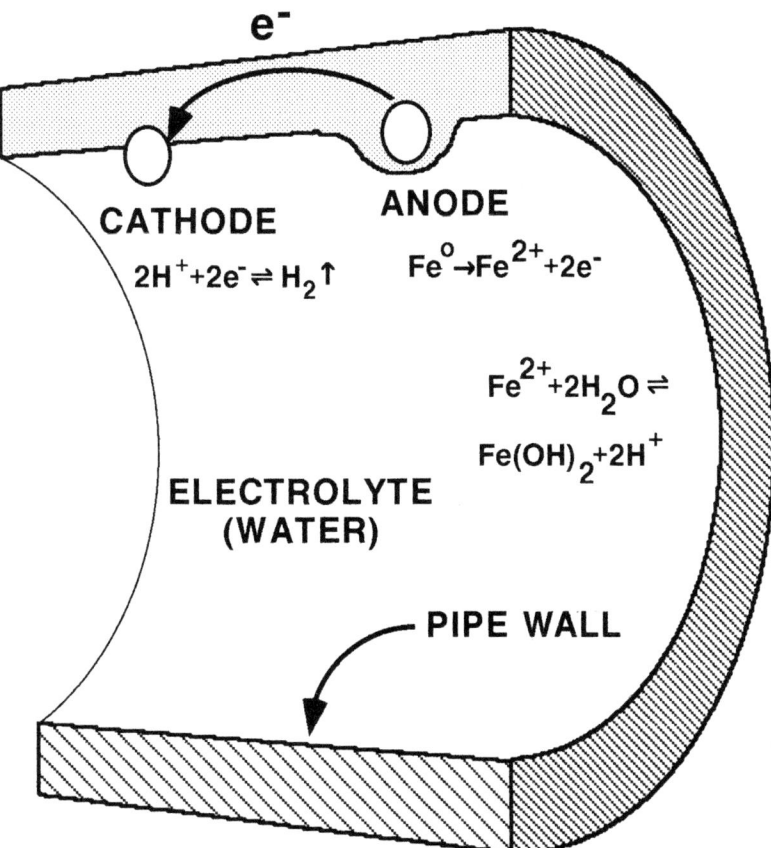

Fig. 8.2 Simplified anode and cathode reactions of iron in contact with water.
Source of H^+ ions is the normal dissociation of water, $H_2O \rightleftarrows H^+ + OH^-$.

8.2, the galvanic corrosion cell in Figure 8.4 also has an anode, a cathode, and electrolyte, as well as a connection between the anode and cathode. This galvanic cell, however, has two dissimilar metals, copper for the cathode and iron for the anode. The electrochemical corrosion reaction is otherwise very similar.

The degree to which a particular metal will become anodic (corrode) in a galvanic reaction is related to its tendency to enter into solution. Another way to view this is the tendency of a metal to revert to its natural ore state (ferric hydroxide, for example) from a refined, or finished, metal state (steel, for example). The relative tendency of various metals to revert to an ore state can be shown by their positions on a *GALVANIC SERIES*[12] in which the most active metals are listed at the top as is shown in Table 8.1.

The higher the level of activity, the greater the tendency for that metal to corrode. Also the farther two metals are apart on the galvanic series, the greater the galvanic corrosion potential. The more active metal of any two in the galvanic series will always become the anode. When iron and copper water pipes are joined, the iron will corrode if water contains dis-

solved oxygen and the copper will be protected. Because of their very active positions on the galvanic series, zinc and magnesium make excellent anodes and are commonly used as *SACRIFICIAL ANODES*[13] in water tanks or for buried pipelines. These very reactive metals, called *BASE METALS*,[14] will corrode preferentially to aluminum and iron, for example.

On the cathodic side of the galvanic series, the least reactive metals are called *NOBLE METALS*.[15] One of the most noble metals is gold, which has been known for millennia to be resistant to corrosion, even when worn as jewelry in continuous contact with oils and acids on the skin. Stainless steel is also cathodic to most other metals, which, although quite expensive, is why it is often used in critical chemical industry applications where corrosion potential is great. Stainless steel is commonly used in high-pressure reverse osmosis desalination systems due to its excellent corrosion resistance in the presence of highly conductive seawater.

If a steel nail is immersed in a solution of a copper salt, such as a solution of bluestone or copper sulfate ($CuSO_4 \cdot 5\ H_2O$), metallic copper will be formed or "plate out" on the surface of the nail. In the same manner, if a reservoir has been treated

---

[12] *Galvanic Series. A list of metals and alloys presented in the order of their tendency to corrode (or go into solution). Also called the electromotive series. This is a practical application of the theoretical electrochemical series.*

[13] *Sacrificial Anode. An easily corroded material deliberately installed in a pipe or tank. The intent of such an installation is to give up (sacrifice) this anode to corrosion while the water supply facilities remain relatively corrosion free.*

[14] *Base Metal. A metal (such as iron) which reacts with dilute hydrochloric acid to form hydrogen.*

[15] *Noble Metal. A chemically inactive metal (such as gold). A metal that does not corrode easily and is much scarcer (and more valuable) than the so-called useful or base metals.*

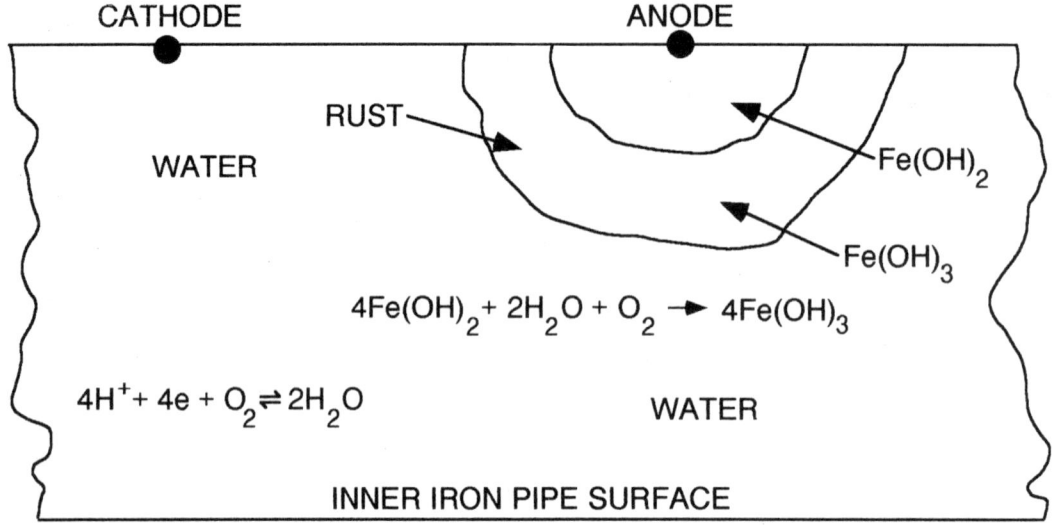

CATHODE

ANODE

RUST

WATER

Fe(OH)$_2$

Fe(OH)$_3$

$$4Fe(OH)_2 + 2H_2O + O_2 \rightarrow 4Fe(OH)_3$$

$$4H^+ + 4e + O_2 \rightleftharpoons 2H_2O$$

WATER

INNER IRON PIPE SURFACE

Fig. 8.3   Role of oxygen in iron corrosion

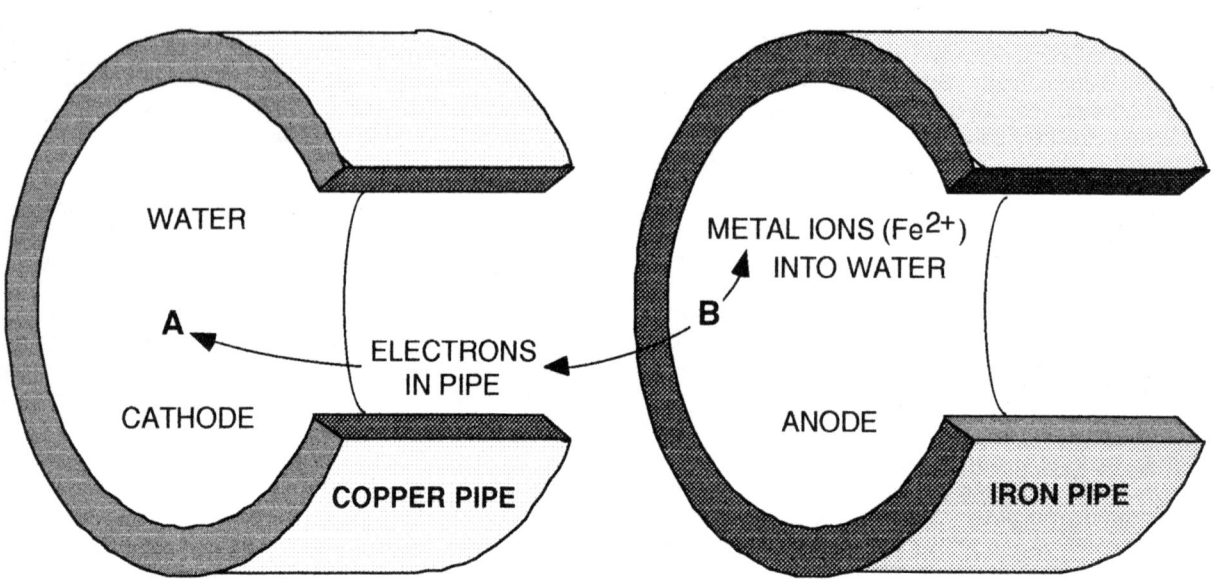

WATER

A

CATHODE

COPPER PIPE

METAL IONS (Fe^{2+}) INTO WATER

B

ELECTRONS IN PIPE

ANODE

IRON PIPE

Note that areas A and B are located on the inner pipe surface.

Fig. 8.4   Simplified galvanic cell

Local galvanic cell

Corrosion at corporation stop hole caused by joining two dissimilar metals

Fig. 8.5   Galvanic cells in water mains
(Permission of ARMCO)

TABLE 8.1 GALVANIC SERIES

		Galvanic Series
ANODE	(Most Active)	Magnesium
		Zinc
		Aluminum (2S)
		Cadmium
		Aluminum (175T)
		Steel or Iron
		Cast Iron
		Lead—Tin Solder
		Lead
		Nickel
		Brasses
		Copper
		Bronzes
		Stainless Steel (304)
		Monel Metal
		Stainless Steel (316)
		Silver
		Graphite
CATHODE	(Least Active)	Gold

with bluestone to control algal growths, there is a possibility that copper will plate out on steel pipes in the distribution system. The metallic copper might then act as a cathode in a galvanic cell and cause corrosion of the steel.

A similar phenomenon occurs in the so-called "dezincification" of brass. Brass is an alloy of zinc and copper. When brass corrodes, the result is a mass of spongy copper having nearly the same shape as the original brass. For a long time people believed that the zinc dissolved and left the copper. Today we know that the alloy itself dissolves and, since the copper is more noble (less reactive) than the brass, the copper plates out in more or less its original position.

As can be seen in the galvanic series, when copper and lead solder are in contact, the lead becomes the anode and will corrode in preference to the copper. The relatively high toxicity of lead and its anodic tendencies are the reasons that lead has been banned from use in potable water distribution systems by the U.S. Government under the 1986 Safe Drinking Water Act Amendments and the Lead and Copper Rule. It is important to note that other factors such as water chemistry, biological films, and physical characteristics (temperature and flow rate) all play a role in the severity of the corrosion reaction. These other factors can be used to help mitigate (lessen) the effects of galvanic and other forms of corrosion. Control of lead corrosion usually involves changes in water chemistry.

An important feature of galvanic corrosion is the relative size of the anode and cathode. The level of galvanic electric CURRENT[16] increases as the area of the cathode increases.

A large cathode will generate a high level of electrical potential. If that current is directed at a small anode, a relatively large amount of metal will dissolve from the available anode area and deep pits will form there. In an extreme case the anodic metal may develop PINHOLES all the way through the wall of a tank or pipe in a relatively short time. An example of this mechanism is a steel water tank which has been protectively lined with a paint coating. If the steel tank is connected to a large copper distribution system line, for example, the corrosion potential will be high; the tank is the anode and the distribution system is the cathode. If the tank coating is imperfect, the entire energy potential of the cathode will be directed to the HOLIDAY, or imperfection, in the coating and severe pitting leading to penetration of the tank wall may result.

Using our understanding of galvanic electrochemistry, corrosion protection for such a tank can be gained by uncoupling the copper distribution system line from the tank. Specially designed DIELECTRIC[17] couplings, which are partially constructed of ceramic, plastic, or other nonconductive materials, are used to separate the copper and steel. Additionally, sacrificial zinc or magnesium anodes can be attached to the tank wall and suspended in the water inside the tank. The sacrificial anodes will dissolve first, protecting the steel if some corrosion potential remains.

CORROSION INHIBITORS[18] that slow the cathodic reaction (in this example, the reaction on the copper electrode) are known as "safe" inhibitors. Inhibitors that act on the anode reaction are called "dangerous" inhibitors because if there is not quite enough inhibitor added, severe pitting will result and corrosion will be worse than if none had been added.

In the ordinary large gate valve used in water distribution systems, brass or bronze parts often make contact with the valve's cast-iron body. The brass or bronze and cast iron create a galvanic cell, but the area of the brass or bronze is so small, compared to the area of the cast iron, that galvanic corrosion is minimal.

We encounter many similar situations in the waterworks field. For example, copper in a valve can be the cathode and the steel or cast iron in the valve the anode. As long as the cathode area is small relative to the anode, corrosion will not be a problem. However, if a copper service line is connected to a steel main, the cathode area is large relative to the anode and corrosion will occur at the anode. The opposite will occur with a brass valve in a steel water line; the cathode area is small and the anode area is large so there will be no problem.

QUESTIONS

Write your answers in a notebook and then compare your answers with those on page 403.

8.0A List the problems that can be created by corrosive waters.

8.1A What is an electrochemical reaction?

---

[16] Current. A movement or flow of electricity. Water flowing in a pipe is measured in gallons per second past a certain point, not by the number of water molecules going past a point. Electric current is measured by the number of coulombs per second flowing past a certain point in a conductor. A coulomb is equal to about 6.25×10^{18} electrons (6,250,000,000,000,000,000 electrons). A flow of one coulomb per second is called one ampere, the unit of the rate of flow of current.

[17] Dielectric (DIE-ee-LECK-trick). Does not conduct an electric current. An insulator or nonconducting substance.

[18] Corrosion Inhibitors. Substances that slow the rate of corrosion.

8.1B What happens to an iron water pipe when the pipe is connected to a copper pipe?

8.1C What is the "dezincification" of brass?

8.1D What happens when copper and lead solder are in contact?

8.1E What is a "dangerous" corrosion inhibitor?

8.1F What will happen if a copper service line is connected to a steel water main?

8.2 FACTORS INFLUENCING CORROSION

As we discussed earlier, corrosion is a complex phenomenon with many possible variables. The essential elements of a corrosion cell, including the specific case of a galvanic cell, have been presented in the previous section. In this section we will discuss how other factors influence the corrosion cell, either increasing or decreasing the rate and severity of the electrochemical reaction. The factors to be considered are physical, chemical, and biological.

8.20 Physical Factors

We will start by considering physical factors. Physical factors that influence corrosion include the type and arrangement of materials used in the system, system pressure, soil moisture, the presence of stray electric currents, temperature, and the water flow velocity.

8.200 System Construction

Some of the effects of the type of materials that make up the anode and cathode were discussed earlier. Also discussed was the influence of the relative size of the anode and cathode on each other, particularly with regard to the corrosion rate and development of pitting. The types of coatings or other protective measures taken on both the inside and outside of pipes and tanks play a major role in corrosion activity.

8.201 System Pressure

If a system is under high pressure, corrosion can be affected greatly. An entire field of corrosion control is concerned with servicing boilers, steam lines, and similar industrial systems under high pressure. This special set of circumstances is not the topic of discussion of this manual, but some of the effects of high pressure on the rate of corrosion are applicable. Higher pressure, for example, increases the maximum concentration of *CORROSIVE GASES*[19] such as carbon dioxide and oxygen that can exist in water.

8.202 Soil Moisture

For buried distribution system lines, contact with moist soil can cause external pipe corrosion. The moisture functions as the electrolyte, the same as water inside the pipe does in the case of internal corrosion.

8.203 Stray Electric Current

Grounding of electric circuits to water pipes is a common practice, but one that is not recommended because it can lead to corrosion of the pipes. *STRAY CURRENT CORROSION*[20] is much more pronounced from direct current (D.C.) grounding than from alternating current (A.C.) grounding. As a rule of thumb, A.C. current effects are only about one percent as great as D.C. current effects. Both types of stray current effects are most pronounced on the outside of pipes and fixtures, since water inside pipes has a much greater resistance to the rate of current flow than does the metal.

Homeowners may notice the effect of stray current corrosion when pitting penetrates fixtures or pipes, resulting in leaks. Unfortunately, due to the nature of stray current corrosion, the anode can occur some distance from the point where the ground contacts the pipe. This may add to the difficulty of identifying this type of corrosion problem.

ELECTROLYSIS[21] is the decomposition of a material by the passage of an exterior source of D.C. electric current. When a D.C. current flows (plus to minus) from a metal into soil, the metal is corroded, except for very noble metals such as platinum. Prior to 1940, electrolysis from stray electrical amperage from streetcar power systems caused a great deal of damage to metallic water mains, but presently there are few such sources of direct current. Modern electric transit systems, however, may be a problem again in the future.

Another form of electrolysis of water mains that can result from stray currents may be generated by *CATHODIC PROTECTION*[22] systems installed by utilities other than the water utility. Cathodic protection systems are discussed in Section 8.4, "Methods of Controlling Corrosion."

8.204 Temperature

The rate of chemical reactions usually increases as temperatures rise. Since chemical reactions are involved in corrosion, temperature generally has the effect of increasing the corrosion rate. However, there are several exceptions to this general rule.

As the temperature of water increases, the amount of calcium carbonate that can remain dissolved in water is reduced. This means that protective calcium carbonate film formation may be improved under this circumstance. The deposition may be so enhanced, however, that severe scaling may result, clogging hot water lines and fittings, and coating hot water heating elements. Heavy scaling of heating elements will increase the amount of energy used, and may result in their failure due to overheating.

---

[19] *Corrosive Gases.* In water, dissolved oxygen reacts readily with metals at the anode of a corrosion cell, accelerating the rate of corrosion until a film of oxidation products such as rust forms. At the cathode where hydrogen gas may form a coating on the cathode and slow the corrosion rate, oxygen reacts rapidly with hydrogen gas forming water, and again increases the rate of corrosion.

[20] *Stray Current Corrosion.* A corrosion activity resulting from stray electric current originating from some source outside the plumbing system such as D.C. grounding on phone systems.

[21] *Electrolysis* (ee-leck-TRAWL-uh-sis). The decomposition of material by an outside electric current.

[22] *Cathodic* (ca-THOD-ick) *Protection.* An electrical system for prevention of rust, corrosion, and pitting of metal surfaces which are in contact with water or soil. A low-voltage current is made to flow through a liquid (water) or a soil in contact with the metal in such a manner that the external electromotive force renders the metal structure cathodic. This concentrates corrosion on auxiliary anodic parts which are deliberately allowed to corrode instead of letting the structure corrode.

Higher temperature can also alter the form of corrosion. In cold water, pitting may be the dominant form of attack on anodic metals, resulting in short service life. Although the rate of corrosion may increase as the temperature rises, the form of corrosion in the same water may change to a uniform or more generalized metal loss, actually resulting in a longer service life for the pipe.

Another special case of temperature effects is the influence on zinc/iron galvanic corrosion. In hot water heaters with temperatures exceeding 140°F (60°C), the normal condition of zinc being anodic to iron may reverse, resulting in the zinc becoming cathodic and the iron corroding. Penetration of the steel tank wall can result.

8.205   Flow Velocity

Flow velocity has several significant influences on corrosion. Moderate flow rates are often beneficial while very high or low flow rates usually increase the rate of corrosion.

● Negative effects: Under stagnant water flow conditions, corrosion is usually more severe, particularly in the form of pitting and tuberculation in iron pipes. However, highly oxygenated water can become even more corrosive under higher flow conditions as the movement of water increases the contact of oxygen with the pipe surface.

In the extreme case of very high velocities, EROSION CORROSION can occur, particularly in copper pipes. At rates exceeding 5 ft/sec (1.5 m/sec), copper tubing will erode rapidly. This is usually noticeable at joints or elbows, and results in structural damage to the pipe. This problem often becomes evident by the occurrence of persistent leaks in even relatively new copper tubing. Circulating hot water systems in large buildings are particularly susceptible to erosion corrosion due to both flow velocities and high temperature effects.

● Beneficial effects: Water that has protective properties, such as the tendency to deposit calcium carbonate films, or to which a corrosion inhibitor has been added, will be less corrosive under moderate flow conditions. Film formation requires deposition of calcium carbonate or the inhibitor (usually a phosphate or silicate compound) on the surface of the metal. In stagnant water, deposition is limited. Under very high flows, it may be scoured off the pipe as it forms, or erosion corrosion may occur faster than the film deposits. If the flow rate is moderate, less than 5 ft/sec (1.5 m/sec), deposition of protective films is enhanced. We will discuss the effect of film formation as a corrosion-control measure in later sections of this chapter.

8.21   Chemical Factors

A very important part of understanding the theory of corrosion is an understanding of the chemistry involved. Various chemical factors influence corrosion, such as pH, alkalinity, chlorine residual, levels of dissolved solids and dissolved gases such as oxygen and carbon dioxide, and the types and concentrations of various minerals present in the water. The major chemical factors are discussed in this section.

8.210   Alkalinity

Alkalinity is a measure of the buffering capacity, or the ability of a particular quality of water to resist a change in pH. Alkalinity is primarily composed of carbonate (CO_3^{2-}) and bicarbonate (HCO_3^-) ions. Acids, compounds which contain free H^+ ions, react with carbonate and bicarbonate:

Carbonate Ion   + Hydrogen Ion (Acid) → Bicarbonate Ion

$$CO_3^{2-} \quad + \quad H^+ \quad \rightarrow \quad HCO_3^-$$

Bicarbonate Ion + Hydrogen Ion (Acid) → Carbonic Acid

$$HCO_3^- \quad + \quad H^+ \quad \rightarrow \quad H_2CO_3$$

Conversely, bases (OH^- ions) can react with bicarbonate to form carbonate:

Bicarbonate Ion + Hydroxide Ion (Base) → Carbonate Ion + Water

$$HCO_3^- \quad + \quad OH^- \quad \rightarrow \quad CO_3^{2-} \quad + H_2O$$

In either case, the acid or the base is neutralized by the carbonate or bicarbonate. The effect of a high concentration of carbonate and bicarbonate in water (high alkalinity) is that the water has a strong tendency to resist a change in pH. Water with low alkalinity will have little BUFFER CAPACITY[23] and may become acidic very easily. Carbonate is also necessary to react with calcium in the water to form protective calcium carbonate ($CaCO_3$) films on the inside of pipes:

Carbonate Ion   +   Calcium Ion   →   Calcium Carbonate

$$CO_3^{2-} \quad + \quad Ca^{2+} \quad \rightarrow \quad CaCO_3$$

The simplest form of corrosion control in many water systems is to simply add more alkalinity in the form of lime, soda ash or caustic soda, or directly as calcium carbonate in the form of crushed limestone. This will be discussed further in Section 8.322, "Calcium Carbonate Saturation."

8.211   pH

The hydrogen ion is extremely active (corrosive) at pH values below 4. Neither chlorine nor hydrogen ions are usually present in sufficient concentrations in potable water to have a significant effect on corrosion. Low pH (<7.0) water tends to be corrosive and high pH (>7.5) water is protective of pipe materials. Very high pH water may have a tendency to deposit excessive amounts of scale. The pH of the water is influenced by the level of alkalinity present. The ability of an acid or a base to change the pH when added to a given amount of water is a direct function of the alkalinity.

Certain pH ranges offer less protection than would otherwise be assumed. For example, at pH values very near 8.3, the transition point between carbonate and bicarbonate, the buffering system is weak, and pH values slightly lower, 7.8 to 8.0 for example, may be more protective.

8.212   Dissolved Oxygen

Oxygen is often considered the most corrosive component of water chemistry. Oxygen plays an important role in corrosion both at the anode and at the cathode. The hydrogen gas released at the cathode (Figure 8.2) coats the cathode area, slowing the rate of corrosion. This phenomenon is called POLARIZATION.[24] However, when dissolved oxygen (O_2) reacts with the hydrogen gas (H_2) to form water (H_2O), the gas is removed from circulation and the rate of corrosion at the cathode accelerates. The removal of hydrogen from the cathode is called DEPOLARIZATION.[25]

---

[23] Buffer Capacity.   A measure of the capacity of a solution or liquid to neutralize acids or bases. This is a measure of the capacity of water for offering a resistance to changes in pH.

[24] Polarization.   The concentration of ions in the thin boundary layer adjacent to a membrane or pipe wall.

[25] Depolarization.   The removal or depletion of ions in the thin boundary layer adjacent to a membrane or pipe wall.

At the anode, dissolved oxygen reacts with iron as it dissolves into the water to form ferric hydroxide (rust). This process is known as *OXIDATION*.[26] It produces red water and, when rust particles attach to the anode area, tubercles are formed. The formation of a rust coating can actually slow the rate of corrosion, until rapid flushing of the water lines clears the coating and again exposes the anode to corrosive oxygen.

8.213 Dissolved Solids

Solids dissolved in water are present as ions, which increase the electrical conductivity of the water. Generally, the higher the dissolved solids, or salt, content of the water, the greater the potential for corrosion to occur due to the increased conductivity. Some dissolved solids are involved in scale formation, possibly slowing the rate of corrosion if a protective film is formed. All scale-forming components such as iron oxide (rust) and calcium carbonate (limestone) are present first as dissolved solids in the water before they deposit on the surface of pipes and fixtures.

8.214 Hardness[27]

The dissolved form of some of the principal scale-forming components in water is referred to as hardness. Hardness is comprised primarily of calcium and magnesium ions, but may also include such ions as iron, manganese, and strontium. All hardness ions have the common property of forming a scale on the inside of pipes or fixtures under conditions of high enough concentration, and at elevated pH and temperature levels.

Planned deposition of calcium carbonate film is one of the most common corrosion-control measures used in water systems. There are several methods of measuring the relative level of calcium carbonate saturation in water. One of the simpler methods is called the Marble Test, which directly measures whether a water sample will increase in hardness and pH when dosed with an excess of calcium carbonate. The Marble Test and a more extensive determination of the saturation level of calcium carbonate called the Langelier Index are discussed in Section 8.322, "Calcium Carbonate Saturation."

If hardness levels are too high (a condition referred to as oversaturation), calcium carbonate deposits can clog pipes and reduce flows. They can also coat hot water heating elements, increasing power consumption. Conversely, water containing very little hardness tends to be more aggressive since it does not deposit a protective calcium carbonate film.

8.215 Chloride (Cl^-) and Sulfate (SO_4^{2-})

Chloride and sulfate ions in water may inhibit the formation of protective scales by keeping hardness ions in solution. The relative amount of alkalinity compared to chloride and sulfate greatly affects this tendency. It is recommended that the alkalinity, expressed as *CALCIUM CARBONATE (CaCO₃) EQUIVALENT*,[28] be five times higher than the sum of chloride and sulfate ions, also expressed as $CaCO_3$ equivalents.

8.216 Phosphate and Silicate

These compounds have a tendency to form protective films in water systems when present in high enough concentrations and when in the correct chemical form for the particular conditions of the water. Phosphate and silicate compounds are frequently added at the water treatment works as a corrosion-control method. The selection and use of phosphate and silicate inhibitors are discussed in Lesson 2 of this chapter.

8.217 Trace Metals

Trace metals of significance in corrosion control include copper, iron, lead, and zinc. These metals are referred to as "trace" since they are normally present in relatively low concentrations. When present at high levels in distribution system water, they are usually indicators of corrosion of the pipes and fittings. Copper and lead usually indicate corrosion of copper pipe and lead solder or service lines. Iron usually results from the corrosion of iron or steel pipe and fittings, and zinc may result from corrosion of galvanized pipe.

Due to their relatively high toxicity to the consumer, copper and lead have been specifically identified by the US EPA in a set of regulations known as the "Lead and Copper Rule," finalized in 1991. These regulations have a very significant impact on all water utilities in the Untied States. Control of lead and copper in water systems involves corrosion monitoring and, if necessary, corrosion control. The Lead and Copper Rule is discussed in Section 8.6 of this chapter.

Iron and zinc may be involved extensively in the formation of protective films, limiting the rate of corrosion. Zinc is also a common corrosion-control additive, usually in a compound formed of zinc and phosphate. When used as an additive for corrosion control, zinc levels may rise to several mg/L in the water. The use of zinc compounds in corrosion control is discussed in Lesson 2.

8.22 Biological Factors

Two types of microorganisms that can play an important role in corrosion of water distribution systems are iron bacteria and sulfate-reducing bacteria. Both can increase the rate of corrosion and the formation of undesirable corrosion by-products. Iron bacteria use dissolved iron as an energy source, and sulfate-reducing bacteria use sulfate for their energy. Since both types of bacteria can grow in dense masses, they can be relatively tolerant to disinfection by chlorine. Both types of bacteria can be particularly troublesome in low-flow areas of distribution systems.

8.220 Iron Bacteria

Iron bacteria are often present in a filamentous (string-like) form, resulting in slimy, reddish or brown-colored masses. These may be present on well screens, inside pipes or water storage tanks, or on the inside of fixtures. Some of the types of iron bacteria common to water systems include *Crenothrix*, *Sphaerotilus*, and *Gallionella*, all of which are filamentous forms.

---

[26] Oxidation (ox-uh-DAY-shun). Oxidation is the addition of oxygen, removal of hydrogen, or the removal of electrons from an element or compound. In the environment, organic matter is oxidized to more stable substances. The opposite of reduction.

[27] Hardness, Water. A characteristic of water caused mainly by the salts of calcium and magnesium, such as bicarbonate, carbonate, sulfate, chloride and nitrate. Excessive hardness in water is undesirable because it causes the formation of soap curds, increased use of soap, deposition of scale in boilers, damage in some industrial processes, and sometimes causes objectionable tastes in drinking water.

[28] Calcium Carbonate (CaCO₃) Equivalent. An expression of the concentration of specified constituents in water in terms of their equivalent value to calcium carbonate. For example, the hardness in water which is caused by calcium, magnesium and other ions is usually described as calcium carbonate equivalent. Alkalinity test results are usually reported as mg/L CaCO₃ equivalents. To convert chloride to CaCO₃ equivalents, multiply the concentration of chloride ions in mg/L by 1.41, and for sulfate, multiply by 1.04.

Iron bacteria convert ferrous (dissolved) iron into ferric hydroxide precipitate, and deposit the rust particles on or in the slime sheaths surrounding the cells. The deposited iron can be released during periods of high water flow velocity, contributing to "red water" conditions. An unpleasant musty odor is often associated with the presence of iron bacteria. Corrosion can be accelerated underneath deposits of iron bacteria on pipes and tanks, resulting in pitting and tuberculation.

8.221 Sulfate-Reducing Bacteria

The presence of sulfate-reducing bacteria can often be readily distinguished by the characteristic rotten egg odor of hydrogen sulfide. Using sulfate as an energy source, sulfate-reducing bacteria produce sulfide as a by-product. Some of the by-product sulfide may be released as gaseous hydrogen sulfide, generating the characteristic rotten egg odor, or it may react with metals producing black metal sulfide deposits on the inside of pipes and fixtures. Free hydrogen sulfide (H_2S) can react with water forming sulfuric acid (H_2SO_4), which is extremely corrosive to metals.

QUESTIONS

Write your answers in a notebook and then compare your answers with those on page 403.

8.2A   How can homeowners notice the effect of stray current corrosion?

8.2B   What causes "erosion corrosion" in copper tubing?

8.2C   List the chemical factors that influence corrosion.

8.2D   Why does water with a higher dissolved solids content have a greater potential for corrosion?

8.2E   What is the impact of calcium carbonate on corrosion?

8.23 Oxygen Concentration Cell

Although galvanic cells are responsible for some corrosion problems, by far the most common corrosion cell is the oxygen concentration cell. In order to understand an oxygen concentration cell, think in terms of the DEAD END[29] created by a six-inch (150-mm) dry barrel fire hydrant installed on an eight-inch (200-mm) water main (Figure 8.6). The fire hydrant assembly will normally consist of an 8" x 8" x 6" (200-mm x 200-mm x 150-mm) tee, six-inch (150-mm) nipple, a six-inch (150-mm) gate valve, 10 to 20 feet (3 to 6 m) of six-inch (150-mm) pipe, and the fire hydrant. When corrosion starts in the six-inch (150-mm) pipe, the following chemical reaction occurs:

Iron → Ferrous Ion + Electrons

Fe   → Fe^{2+} + 2 e^-

This is the anode reaction of the galvanic cell. The ferrous ions (Fe^{2+}) formed by this reaction will in turn react with dissolved oxygen and water as follows:

Ferrous Ions		Dissolved Oxygen		Water		Ferric Hydroxide (Solid)		Hydrogen Ions
4 Fe^{2+}	+	O_2	+ 10 H_2O →			4 Fe (OH)$_3$	+	8 H^+

This is chemical shorthand for saying that four ferrous ions and one molecule of oxygen react with ten water molecules to form four molecules of solid (precipitate) ferric hydroxide and eight hydrogen ions.

Just as they did in the galvanic cell, the electrons from the anode reaction flow through the metallic path (pipe) to the eight-inch (200-mm) main where they react with dissolved oxygen which is continually being replenished by the flowing water.

Oxygen	+	Electrons	+	Water	→	Hydroxide Ions
O_2	+	4 e^-	+	2 H_2O →		4 OH^-

This is the cathode reaction.

The continued production of ferrous ions within the six-inch (150-mm) pipe completely removes dissolved oxygen from the water in the dead end and also scavenges any dissolved oxygen that may diffuse in from the flowing water in the eight-inch (200-mm) pipe. The hydrogen ions produced by this reaction can lower the pH in the water to around 5.2 to 5.8.

Also, the production of positively charged ferrous ions in the six-inch (150-mm) pipe requires that there be an inflow of negatively charged ions to maintain electrical neutrality in the water. The common negative ions present in water are bicarbonate (HCO_3^-), chloride (Cl^-), and sulfate (SO_4^{2-}).

The absence of oxygen and the low pH value in the six-inch (150-mm) pipe make conditions ideal for the growth of ANAEROBIC[30] bacteria. The action of these bacteria on traces of organic matter and on reducing the sulfate ions to sulfide are responsible for the foul odors usually found in the dead ends of water mains.

The typical distribution system contains many dead ends. In addition to those at the end of the system, other dead end conditions include the ANNULAR SPACES[31] (circular cavities) created by mechanical couplings, tapping sleeves (Figure 8.7), and even the bonnets (covers) of gate valves, but the number of dead ends from pits (holes) in the pipe surfaces far exceeds the number of dead ends from all other causes combined.

Pits may be started by anything that will shield the metal surface from dissolved oxygen in the water, such as bits of clay, dirt, sand, or a colony of bacteria. Also, impurities in the metal may cause a local anode to form.

As the iron ions from the anode react with the dissolved oxygen in the water, the resulting ferric hydroxide forms a membrane over the anode. As the reactions continue and the membrane ages, it turns into a crust. The membrane, and later the crust, protect the anode area from dissolved oxygen in the water, thus causing the oxygen concentration cell to intensify itself. As the iron ions leave the metal, a pit is formed that grows deeper and deeper. As this occurs the crust becomes thicker and thicker. In this way, a mound of iron rust is built up over the pit. This mound is called a TUBERCLE.[32]

As the ferric hydroxide ($Fe(OH)_3$) ages, it forms other minerals (such as ferric oxide (Fe_2O_3) or iron rust). Eventually, the crust becomes so thick that negative ions cannot enter the pit,

---

[29] Dead End.   The end of a water main which is not connected to other parts of the distribution system by means of a connecting loop of pipe.

[30] Anaerobic (AN-air-O-bick).   A condition in which atmospheric or dissolved molecular oxygen is NOT present in the aquatic (water) environment.

[31] Annular (AN-you-ler) Space.   A ring-shaped space located between two circular objects, such as two pipes.

[32] Tubercle (TOO-burr-cull).   A protective crust of corrosion products (rust) which builds up over a pit caused by the loss of metal due to corrosion.

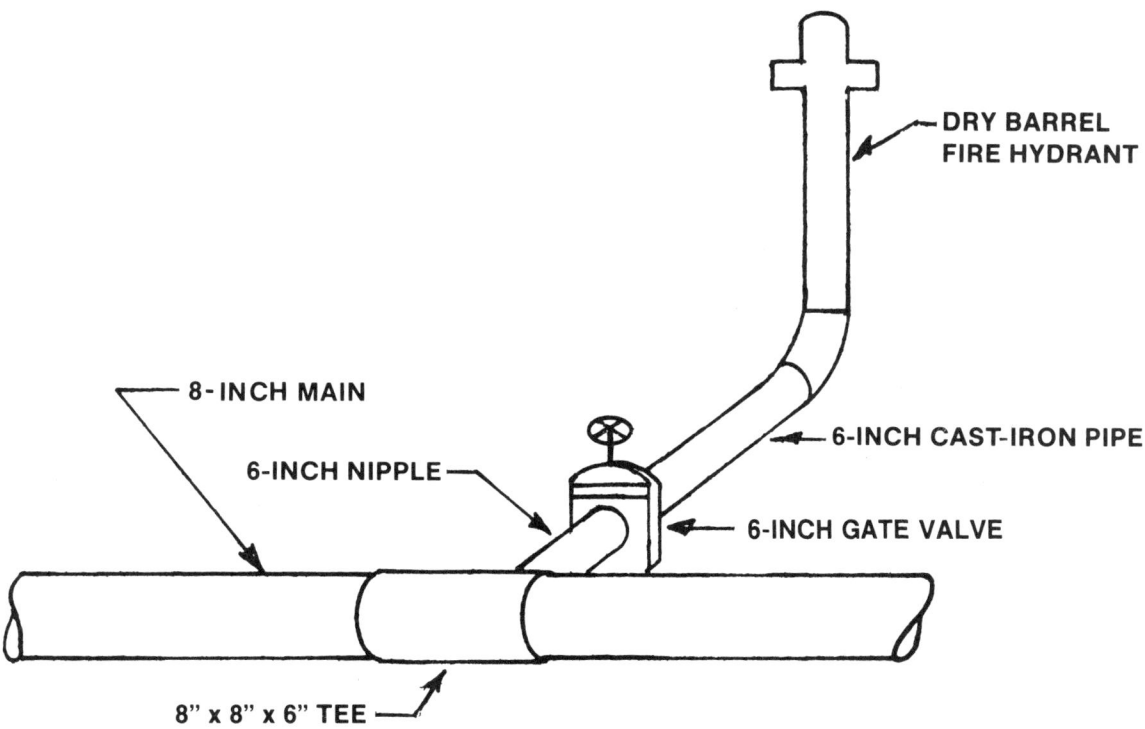

Fig. 8.6   Dead end caused by installation of fire hydrant

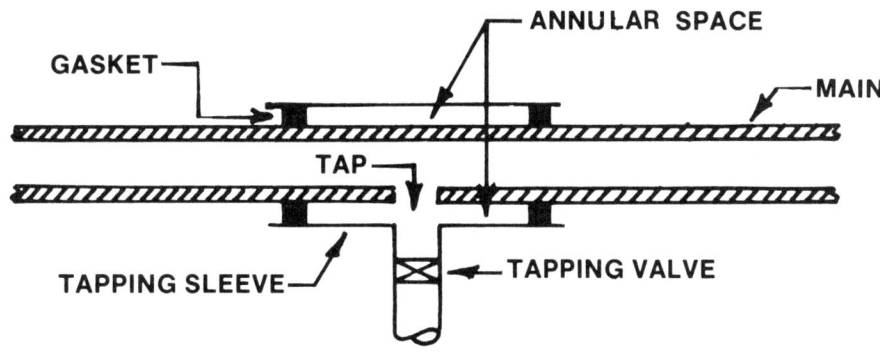

TAPPING SLEEVE (TOP VIEW)

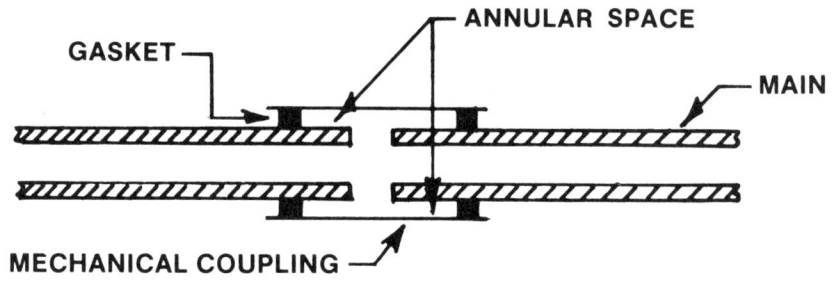

MECHANICAL COUPLING

Fig. 8.7   Creation of annular spaces in distribution systems

nor can iron ions escape, and the corrosion stops. At this point the pit is said to be inactive.

The reaction of dissolved oxygen with ferrous ions is very slow at low pH values. When the pH is less than seven, the reaction is so slow that tubercles do not form and the pits are not self-perpetuating. New pits keep starting in different places, so that corrosion appears to be uniform over the surface of the pipe.

QUESTIONS

Write your answers in a notebook and then compare your answers with those on page 403.

8.2F   What is the most common type of corrosion cell?

8.2G   Where can an oxygen concentration cell be started?

8.2H   How can pits be started on a metallic surface under water?

8.2I   What is a tubercle?

8.3   HOW TO DETERMINE IF CORROSION PROBLEMS EXIST

The Lead and Copper Rule requires water systems to control corrosion to protect the public from the harmful effects of lead, copper, or other toxic metals in drinking water. To determine if the water you are treating is causing corrosion problems, you can examine materials removed from your distribution system for signs of corrosion damage. Chemical tests on the water can be used to indicate the corrosiveness of a water. *IF THE DISTRIBUTION SYSTEM HAS AN INCREASING NUMBER OF LEAKS AND THE CONSUMERS ARE COMPLAINING ABOUT DIRTY OR RED WATER, THESE ARE THE MOST COMMON INDICATORS OF CORROSION PROBLEMS.*

8.30   Examine Materials Removed From Distribution System

Corrosion rates may be measured by inserting special steel specimens called *COUPONS*[33] (Figure 8.8) in the water mains. After a period of time, usually a month or two, the coupons are removed and the loss of weight and nature of corrosion damage are measured. The rate of corrosion is measured by the loss of weight of the coupon between weighing time intervals. Be sure to scrape off all encrustations before weighing the coupons. These tests should be made under the supervision of an experienced corrosion engineer because standard procedures must be used to obtain results that can be compared with other water supply systems.

Much can be learned about the corrosiveness of a water by prowling through the scrap heap and examining sections of water mains that have been taken out of service for various reasons. When examining internal corrosion damage on a pipe, pay particular attention to the maximum depth of the pits. When pit depth equals the wall thickness of a pipe, a leak develops. As a rule of thumb, you can assume (for internal corrosion only) that pit depth increases with the cube root of time ($\sqrt[3]{Time}$). Thus, if a pit depth reaches a certain value

in one year, it will about double this depth in eight years (2 x 2 x 2 = 8).

Leaks are often detected by the observation of wet spots above a pipeline. All reports of leaks should be recorded. If the number of leaks is large, plot the location of the leaks on a map to identify trouble spots.

If effervescence (bubbles) occurs when a drop of dilute hydrochloric (HCl) acid is placed on an obvious cathodic area (such as a brass ring on a gate valve), this indicates the presence of a calcium carbonate ($CaCO_3$) film that may be too thin to see. This indicates that the water is, at worst, only moderately corrosive.

QUESTIONS

Write your answers in a notebook and then compare your answers with those on page 404.

8.3A   Why is corrosion of water system facilities a public health concern?

8.3B   How can corrosion rates be measured?

8.3C   How can leaks in water mains be detected?

8.3D   How can you detect a film of calcium carbonate that is too thin to see?

8.31   Flow Tests

A very simple and useful *CORROSIVITY*[34] test is to measure the change in water flow through a 20-foot (6-meter) length of half-inch (12.5-mm) standard black iron pipe under a constant head of one foot (0.3 m). The initial flow rate will be small enough that the flow rate can be determined by measuring the time required to fill a one-quart (or one-liter) container. If the water is highly corrosive, the flow rate will be reduced by as much as 50 percent in two weeks. In other words, the fill time will be twice as long. If the water is relatively noncorrosive, the flow reduction will be only 10 percent or less after two weeks. This test may be run for longer time periods, but the changes in flow caused by tuberculation are greatest during the first few weeks. Usually there is little change after 26 weeks.

8.32   Chemical Tests on the Water

8.320   Dissolved Oxygen

Certain chemical tests on the water may be helpful to indicate the corrosiveness of a water. As you've learned, corrosion can only occur in the presence of oxygen. Water supplies taken from lakes or streams will contain dissolved oxygen. If your source water is drawn from wells, measure the dissolved oxygen[35] concentration in each well. If no oxygen is present, the water cannot be corrosive.

By measuring the dissolved oxygen at various points in the distribution system, you can calculate how much oxygen is used up as water passes through the system. Loss of dissolved oxygen indicates either that the water contains oxidizable organic matter or that gross corrosion is occurring.

---

[33] *Coupon.   A steel specimen inserted into water to measure the corrosiveness of water. The rate of corrosion is measured as the loss of weight of the coupon (in milligrams) per surface area (in square decimeters) exposed to the water per day. 10 decimeters = 1 meter = 100 centimeters.*

[34] *Corrosivity.   An indication of the corrosiveness of a water. The corrosiveness of a water is described by the water's pH, alkalinity, hardness, temperature, total dissolved solids, dissolved oxygen concentration, and the Langelier Index (see Section 8.322, "Calcium Carbonate Saturation").*

[35] *Refer to Chapter 11, "Laboratory Procedures," and Volume II, Chapter 21, "Advanced Laboratory Procedures."*

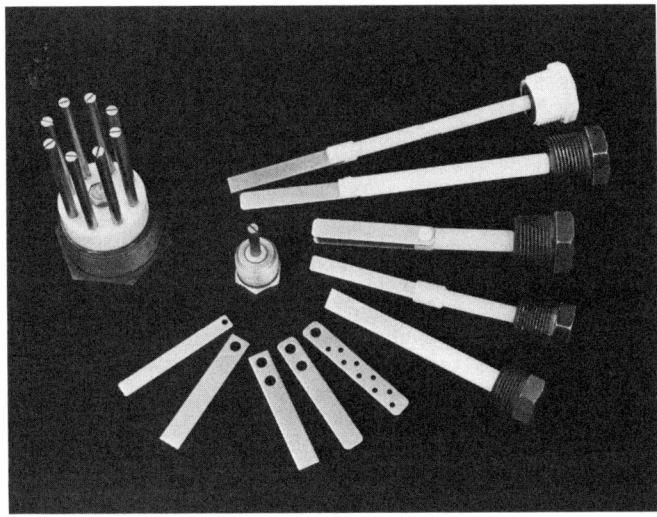

Pipe plug assembly coupons
(Permission of Metal Samples Co., Inc.)

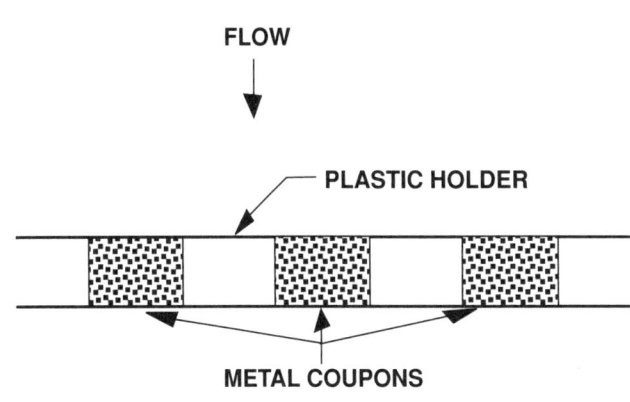

Fig. 8.8 Coupons

When taking samples for dissolved oxygen, you must avoid exposing the sample to air because air contains oxygen. You will need the following equipment and reagents:

1. A four-foot (1.2-m) length of polyethylene tubing with fittings to connect to a hose bib (faucet).

2. BOD bottle. A glass bottle with 300 mL capacity with a tapered, pointed, ground-glass stopper and a flared mouth.

3. Three small (100-mL) plastic bottles with screw-on rubber bulb dispensing pipets. These pipets should deliver approximately 0.5 mL when the bulb is squeezed (an ordinary eye dropper is satisfactory). The bottles are filled with standard reagents available from chemical supply houses.

   a. Manganous sulfate.

   b. Alkaline potassium iodide.

   c. Concentrated sulfuric acid. This is a very dangerous chemical and must be handled with great care.

To obtain a *REPRESENTATIVE SAMPLE*,[36] select a hose bib as close as possible to the meter of a customer on a short water service. Flush the service line at a rate of approximately half a gallon per minute (2 L/min) for five minutes for each 100 feet (30 m) of pipe between the hose bib and the main so that you can be sure you are getting your sample from the main. Do not flush the service line at a high water velocity because this may disturb sediment in the pipe. Therefore, take a long time to flush the line. After the line has been flushed, connect the plastic tube to the hose bib, place the end of the plastic tube at the bottom of the BOD bottle, and turn the hose bib on again at a rate of flow no higher than was used for flushing the service line. Allow the flow to continue until the BOD bottle has

overflowed at least three times its volume. Withdraw the plastic tube taking care not to introduce any air bubbles while removing the tube.

Next, "fix" (stop any chemical reactions involving dissolved oxygen) the sample so it can be transported to the laboratory where the dissolved oxygen can be measured.

1. Gently, but rapidly, add one mL of manganous sulfate reagent (two squirts with dropper) followed by one mL of alkaline potassium iodide. Each time hold the tip of the pipet below the surface of the water so you will not add any dissolved oxygen.

   A heavy brown floc of manganese hydroxide will form at this point if dissolved oxygen is present. A white floc will indicate that there is no dissolved oxygen in the water.

2. Insert the glass stopper without trapping any air bubbles. Mix sample and reagents by rapidly inverting the bottle back and forth.

3. Allow the floc to settle halfway down in the bottle. Invert the bottle and allow the floc to settle halfway again.

4. Carefully remove the stopper and add one mL of concentrated sulfuric acid. Allow the acid to run down the neck of the bottle and into the sample.

5. Mix the sample by inverting again.

The sample now contains an iodine solution that is chemically equivalent to the initial dissolved oxygen concentration in the sample. The final solution is stable and can be transported to the laboratory where the iodine solution can be measured by *TITRATING*[37] with a standard solution of sodium thiosul-

---

[36] *Representative Sample. A sample portion of material or water that is as nearly identical in content and consistency as possible to that in the larger body of material or water being sampled.*

[37] *Titrate (TIE-trate). To TITRATE a sample, a chemical solution of known strength is added drop by drop until a certain color change, precipitate, or pH change in the sample is observed (end point). Titration is the process of adding the chemical reagent in small increments (0.1 – 1.0 milliliter) until completion of the reaction as signaled by the end point.*

fate. See Volume II, Chapter 21, "Advanced Laboratory Procedures," for details.

8.321   Toxic Heavy Metals

Testing for toxic heavy metals in water samples from customers' plumbing has been recommended as a way of determining whether the delivered water is corrosive. However, the results of such samples can be so variable that a great many samples must be taken in order to obtain a meaningful average.

A common source of lead from plumbing is copper tubing with soldered fittings. A water sample taken shortly after a period of heavy use will contain much less lead than a sample taken after standing overnight. Experiments have shown that on new copper tube systems, significant amounts of lead are found in water samples regardless of the water quality. With noncorrosive water, no lead can be detected from copper plumbing one month old, and even with moderately corrosive water, lead is low after approximately two years. With highly corrosive water, excessive lead concentrations may be found after more than ten years.

Copper is a toxic heavy metal that can leach into drinking water from copper and brass materials such as piping, tubing, and fittings. High levels of copper in corrosive waters can produce a bitter or metallic taste in water and cause green stains on plumbing fixtures.

Cadmium is another toxic metal found in samples from plumbing systems; it is found only in very small amounts. Cadmium is a contaminant found in zinc used for galvanizing steel pipes. Cadmium-plated waterworks fittings are not used in the United States. However, they have been used in Europe and have been responsible for serious cases of cadmium poisoning.

Chromium is used for external decorative plating and has virtually no chance to get into the water. Arsenic, antimony, and silver are found in copper and zinc ores, but the quantities found in the refined metals used for plumbing are too small to be significant sources of contamination of drinking water.

QUESTIONS

Write your answers in a notebook and then compare your answers with those on page 404.

8.3E   What does a loss of dissolved oxygen in the water flowing in a distribution system indicate?

8.3F   What toxic metals may enter drinking waters from the customer's plumbing due to corrosive water?

8.322   Calcium Carbonate Saturation[38]

A water is considered stable when it is just saturated with calcium carbonate. In this condition the water will neither dissolve nor deposit calcium carbonate. Water treatment plant operators commonly use two approaches to determine the calcium carbonate saturation level of their water: the Marble Test and the Langelier Index.

MARBLE TEST[39]

To conduct a Marble Test for calcium carbonate saturation, first measure the pH, alkalinity, and hardness of the water sample. Add a pinch of powdered calcium carbonate and then stir the water for five minutes. If the pH, alkalinity, or calcium have increased, the water was undersaturated; if they have decreased, the water was SUPERSATURATED[40] with respect to calcium carbonate.

The water should be stirred in a stoppered flask on a magnetic stirrer in order to avoid the introduction of carbon dioxide from the air. Also, the water being stirred should be at the same temperature as the water in the distribution system. If the Marble Test is made frequently, it is convenient to use an Enslow column. This is a column packed with calcium carbonate granules. The pH, alkalinity, and calcium are measured on a sample stream of water before and after passing through the column. See Section 8.43, "Determination of Chemical Feeder Setting," for a detailed description of an Enslow column.

Results from either the Marble Test or an Enslow column may be as follows:

Initial pH = 8.7 Initial Hardness = 34 mg/L as $CaCO_3$
Final pH = 9.1 Final Hardness = 38 mg/L as $CaCO_3$

The tested water is considered corrosive because the pH and hardness both increased in the test. An increase in these values indicates that the water was undersaturated with calcium carbonate to begin with.

LANGELIER INDEX (L.I.)

As previously mentioned, a water is considered stable when it is just saturated with calcium carbonate; in this condition it will neither dissolve nor deposit calcium carbonate. Thus, in a stable water the calcium carbonate is in equilibrium with the hydrogen ion concentration. If the pH is raised from the equilibrium point (pH_s), the water becomes scale forming and will deposit calcium carbonate. If the pH is lowered from the equilibrium point, the water turns corrosive.

The Langelier Index (L.I.) is the most common index used to indicate how close a water is to the equilibrium point, or the corrosiveness of the water. This index reflects the equilibrium pH of a water with respect to calcium and alkalinity. The Langelier Index can be determined by using equation (1).

$$\text{Langelier Index} = pH - pH_s \qquad (1)$$

where pH = actual pH of the water, and

pH_s = pH at which water having the same alkalinity and calcium content is just saturated with calcium carbonate.

In equation (1) pH_s is defined as the pH value where water of a given calcium content and alkalinity is just saturated with calcium carbonate (at the equilibrium point). For some waters of low calcium content and alkalinity there is no pH value that satisfies this definition; however, for most waters there will be two values for pH_s. These difficulties can be avoided by defining

---

[38] Saturation.   The condition of a liquid (water) when it has taken into solution the maximum possible quantity of a given substance at a given temperature and pressure.

[39] Marble Test.   See Volume II, Chapter 21, "Advanced Laboratory Procedures," for detailed procedures on how to perform the Marble Test.

[40] Supersaturated.   An unstable condition of a solution (water) in which the solution contains a substance at a concentration greater than the saturation concentration for the substance.

pH_s as that pH where a water of given calcium and bicarbonate concentrations is just saturated with calcium carbonate.

T. E. Larson's method[41] for calculating pH_s is a satisfactory approximation when the value of pH_s is calculated using equation (2).

$$pH_s = A + B - \log(Ca^{2+}) - \log(\text{Alkalinity}) \qquad (2)$$

The values for A and B are found in Tables 8.2 and 8.3; (Ca^{2+}) is the calcium hardness as $CaCO_3$; and (Alky) is the alkalinity as $CaCO_3$ (Table 8.4).

This calculation is accurate enough for practical purposes up to a pH_s value of 9.3. Above this value, errors are large.

EXAMPLE 1

Find the pH_s and Langelier Index of a water at 15°C having a TDS of 200 mg/*L*, alkalinity of 100 mg/*L*, and a calcium hardness of 50 mg/*L*. The pH is 8.6.

Known		Unknown
Water Temp, °C	= 15°C	pH_s
TDS, mg/*L*	= 200 mg/*L*	Langelier Index
Alkalinity, mg/*L*	= 100 mg/*L* as $CaCO_3$	
Ca Hardness, mg/*L*	= 50 mg/*L* as $CaCO_3$	
pH	= 8.6	

1. Find the formula values from the tables.

   From Table 8.2 for a water temperature of 15°C,

   A = 2.12

   From Table 8.3 for a TDS of 200 mg/*L*,

   B = 9.80

   From Table 8.4 for Ca of 50 mg/*L* and Alky of 100 mg/*L*,

   $\log(Ca^{2+})$ = 1.70

   $\log(\text{Alky})$ = 2.00

2. Calculate pH_s.

   pH_s = A + B - $\log(Ca^{2+})$ - $\log(\text{Alky})$

   = 2.12 + 9.80 - 1.70 - 2.00

   = 8.22

3. Calculate the Langelier Index.

   Langelier Index = pH - pH_s

   = 8.6 - 8.22

   = 0.38

A positive Langelier Index (pH greater than pH_s) indicates that the water is supersaturated with calcium carbonate ($CaCO_3$) and will tend to form scale. A negative Langelier Index means that the water is corrosive.

Soft, low-alkalinity waters having excessively high pH values are corrosive even though the calculated L.I. may indicate a noncorrosive tendency. In this instance, due to the insufficient amount of calcium ions and alkalinity, no protective calcium carbonate film can form.

The corrosive tendencies of water to particular metals, such as the ones used in distribution systems, are also significantly influenced by the amount of total dissolved solids (TDS). Waters containing TDS exceeding 50 mg/*L* may exhibit corrosive tendencies in spite of a positive Langelier Index. The presence of various ions such as sulfate and chloride ions in water may interfere with the formation and maintenance of a uniform protective calcium carbonate layer on metal surfaces. In addition, the presence of these ions will accelerate the corrosion process.

Because of the various water quality indicators involved, the L.I. should only be used to determine the corrosive tendencies of water within a pH range of 6.5 to 9.5 provided that a sufficient amount of calcium ions and alkalinity over 40 mg/*L* are present in the water.

OTHER CORROSIVITY INDICES

Three other indices for calcium carbonate saturation have been used: (1) Driving Force Index (D.F.I.), (2) Ryznar Index (R.I.), and (3) Aggressive Index (A.I.). These indices will not be described because they are not as widely used as the Langelier Index.

8.33 Complaints

Accurate records should be maintained of all complaints. If more than a minimum number of rusty or red water complaints are received, they should be plotted on a map of the system. The distribution of the plots can tell you where corrosion problems are occurring and indicate how the problem can be corrected. See the section on "Handling Water Quality Complaints" in Chapter 10, "Plant Operation," for more details.

QUESTIONS

Write your answers in a notebook and then compare your answers with those on page 404.

8.3G When is a water considered stable?

8.3H How can water be tested to determine if it is undersaturated or supersaturated with calcium carbonate?

8.3I The Langelier Index is defined by what equation?

8.3J What is the meaning of pH_s?

8.3K Why do some waters not have a meaningful pH_s value?

8.3L Find the pH_s of a water at 10°C having a TDS of 100 mg/*L*, alkalinity of 80 mg/*L* as $CaCO_3$, and a calcium hardness of 40 mg/*L* as $CaCO_3$.

8.3M What do rusty or red water complaints indicate?

End of Lesson 1 of 2 Lessons on CORROSION CONTROL

Please answer the discussion and review questions next.

[41] T. E. Larson and A. M. Buswell, "Calcium Carbonate Saturation Index and Alkalinity Interpretations," JOURNAL AMERICAN WATER WORKS ASSOCIATION, Volume 34, page 1667, 1942.

TABLE 8.2 VALUES OF A FOR VARIOUS TEMPERATURES	
Temperature, °C	A
0	2.34
5	2.27
10	2.20
15	2.12
20	2.05
25	1.98
30	1.91
40	1.76
50	1.62
60	1.47
80	1.18
100	0.88

TABLE 8.3 VALUES OF B FOR VARIOUS LEVELS OF TDS	
TDS, mg/L	B
0	9.63
50	9.72
100	9.75
200	9.80
400	9.86
800	9.94
1,600	10.04

TABLE 8.4 VALUES OF LOG OF CA OR ALKY AS $CaCO_3$ IN mg/L	
Ca or Alky as $CaCO_3$, mg/L	Log_{10}
10	1.00
20	1.30
30	1.48
40	1.60
50	1.70
60	1.78
70	1.84
80	1.90
100	2.00
200	2.30
300	2.48
400	2.60
500	2.70
600	2.78
700	2.84
800	2.90
900	2.95
1,000	3.00

Langelier Index = $pH - pH_s$

where $pH_s = A + B - log(Ca^{2+}) - log(Alky)$

DISCUSSION AND REVIEW QUESTIONS

Chapter 8.   CORROSION CONTROL

(Lesson 1 of 2 Lessons)

At the end of each lesson in this chapter you will find some discussion and review questions. The purpose of these questions is to indicate to you how well you understand the material in the lesson. Write the answers to these questions in your notebook before continuing.

1. What are the adverse effects caused by corrosive waters?

2. What is corrosion?

3. What could happen to unlined steel pipes in a water distribution system if bluestone (copper sulfate) is used to control algal growths in a reservoir?

4. List as many of the physical factors that influence corrosion as you can remember.

5. Why is scaling a problem in hot water systems?

6. How do dissolved solids in water influence corrosion?

7. How can operators determine if the water from their treatment plant is causing corrosion problems?

8. How can a flow test indicate the corrosivity of water?

9. What chemical tests are helpful to determine the corrosiveness of water?

10. What is the meaning of a negative Langelier Index?

CHAPTER 8. CORROSION CONTROL

(Lesson 2 of 2 Lessons)

8.4 METHODS OF CONTROLLING CORROSION

8.40 Calcium Carbonate Saturation

If the water is corrosive, treatment to reduce the corrosivity of the water should be undertaken. Reduction of corrosivity is almost always accomplished by treating the water with chemicals (Figure 8.9) so that the water is saturated or slightly supersaturated with calcium carbonate. Chemicals should be fed AFTER filtration; otherwise a slight excess chemical may result in cementing of filter sands. (Samples for turbidity measurements should be taken after filtration but before chemical feed. Small amounts of turbidity could be introduced with the chemical and might produce misleading results suggesting poor filter performance.) The chemical feed can take place before, after, or along with postchlorination. However, samples should be taken only after postchlorination because the chlorine may react with the chemicals used to reduce corrosivity. For example, chlorine gas will lower the pH of the water, while hypochlorite compounds will raise the pH.

Selection of a chemical to achieve calcium carbonate saturation will depend on the water quality characteristics of the water and the cost of chemicals. Quicklime and hydrated lime should be added to waters that have a low hardness and low alkalinity in order to form calcium carbonate. Caustic soda or soda ash may be added to waters with high levels of hardness and alkalinity because there is already sufficient calcium alkalinity to form calcium carbonate, but the pH must be increased to reach saturation conditions.

Quicklime is calcium oxide (CaO); it is the least expensive of the four chemicals. However, lime requires expensive equipment to SLAKE[42] or hydrate it. This procedure is cost effective only for very large water treatment plants.

Hydrated lime is calcium hydroxide ($Ca(OH)_2$); it is slightly more costly than unslaked lime. Hydrated lime is only slightly soluble in water so it cannot be fed as a true solution. Lime SLURRY[43] reacts with carbon dioxide to form limestone (calcium carbonate or $CaCO_3$) so pipes, pumps, and solution feeders tend to become plugged with scale or deposits very rapidly. For this reason, hydrated lime is best fed using a dry feeder. Use of either form of lime will add hardness to the water. This extra hardness may be a slight disadvantage where the water already contains too much calcium or hardness. However, the lime is advantageous for waters that contain so little calcium that they cannot otherwise be saturated with calcium carbonate.

Lime should be fed to the water after it has passed through the filters, but before it enters the clear well. Provisions should be made to collect turbidity samples or measure turbidity be-tween the filters and the location of the lime feed because lime will cause an increase in the turbidity of the filtered water.

Lime can be very difficult to handle. If you are having problems with calcium carbonate forming in the pipes that are delivering lime to the point of application, consider the construction of open channels so the deposits of lime can be cleaned from the channel with a hoe. Other possibilities include the use of flexible pipelines (rather than rigid pipe) such as plastic hose for ease in breaking loose the calcium carbonate deposits. The outlet of any pipe or hose should not be submerged. To reduce the problem of cleaning either pipes or open channels, minimize the length of the solution lines by placing the chemical feeder as close as possible to the point of application.

Caustic soda (NaOH or sodium hydroxide) is more expensive than lime, but is available in a 50 percent solution that can be fed directly with less expensive solution feeders. There are some waters, however, with initial calcium and alkalinity levels so low that calcium carbonate ($CaCO_3$) saturation cannot be reached by feeding caustic soda. When considering the use of caustic soda, problems associated with the feeding of caustic soda must be considered. Caustic soda crystallization (freezing) occurring at temperatures below around 50°F (10°C) can be a problem. At locations where caustic soda is fed to the water being treated, the caustic should not be fed in a closed conduit or pipe because this will encourage clogging.

Caustic soda must be handled with great care because caustic soda (1) dissolves human skin, (2) produces heat when mixed with water, and (3) reacts with amphoteric metals (such as aluminum) generating hydrogen gas which is flammable and may explode if ignited. When handling caustic soda you should control the mists with good ventilation. You must protect your nose and throat with an approved respiratory system. For eye protection you must wear chemical worker's gog-

---

[42] Slake. To mix with water so that a true chemical combination (hydration) takes place, such as in the slaking of lime.

[43] Slurry (SLUR-e). A watery mixture or suspension of insoluble (not dissolved) matter; a thin, watery mud or any substance resembling it (such as a grit slurry or a lime slurry).

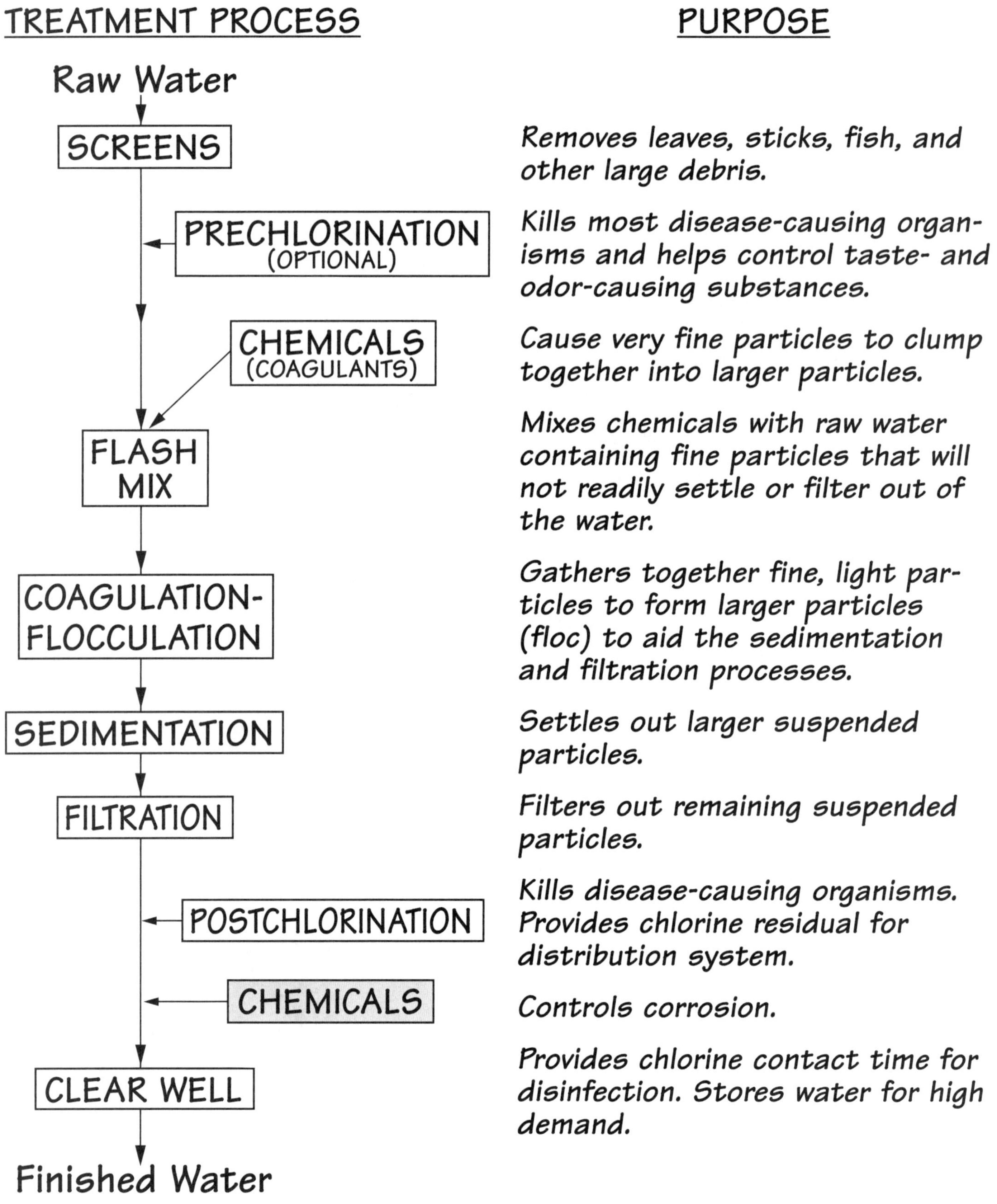

TREATMENT PROCESS

PURPOSE

Raw Water

SCREENS

Removes leaves, sticks, fish, and other large debris.

PRECHLORINATION
(OPTIONAL)

Kills most disease-causing organisms and helps control taste- and odor-causing substances.

CHEMICALS
(COAGULANTS)

Cause very fine particles to clump together into larger particles.

FLASH
MIX

Mixes chemicals with raw water containing fine particles that will not readily settle or filter out of the water.

COAGULATION-
FLOCCULATION

Gathers together fine, light particles to form larger particles (floc) to aid the sedimentation and filtration processes.

SEDIMENTATION

Settles out larger suspended particles.

FILTRATION

Filters out remaining suspended particles.

POSTCHLORINATION

Kills disease-causing organisms. Provides chlorine residual for distribution system.

CHEMICALS

Controls corrosion.

CLEAR WELL

Provides chlorine contact time for disinfection. Stores water for high demand.

Finished Water

Fig. 8.9   Typical process flow diagram

gles and/or a full face shield to protect your eyes. You should protect your body by being fully clothed, and by using impervious gloves, boots, apron, and face shield.

Soda ash is sodium carbonate (Na_2CO_3). Sodium carbonate will dissolve in water up to approximately $1\frac{1}{3}$ pounds per gallon (0.16 kg/L) and can thus be fed with solution feeders. Soda ash can be used in waters of low alkalinity (where caustic cannot) provided the calcium hardness is greater than about 30 mg/L (as $CaCO_3$). Since such waters are rare, it is not a common method of treatment. When comparing the applications of soda ash and caustic soda, consider the cost of increasing the alkalinity by one mg/L. This will require consideration of the cost per pound of chemical, the percent purity of the chemical, and the change in alkalinity or pH resulting from the application of the chemical.

Utilities should exercise caution when applying compounds containing sodium. Use of corrosion-control chemicals containing sodium should be carefully reviewed if the added sodium will increase the level in the water to more than 20 mg/L. Public health officials are concerned because evidence indicates that sodium is one important factor in the development of high blood pressure in susceptible individuals.

QUESTIONS

Write your answers in a notebook and then compare your answers with those on page 404.

8.4A   How can the corrosivity of a water be reduced?

8.4B   What chemicals may be added to waters to reduce the corrosivity?

8.41   Selection of Corrosion-Control Chemicals

Once you have analyzed your water and determined that you have a potential corrosion problem, the next step is to select the appropriate chemical or chemicals. Selection of chemicals depends on the characteristics of the water, where the chemicals can be applied, how they can be applied and mixed with the water, and the costs of the chemicals. You want to solve your corrosion problem by the most cost-effective means.

When you multiply the calcium hardness by the alkalinity (both in mg/L as $CaCO_3$) and the product is less than 1,000, then the treatment required may be complicated. For example,

both lime and carbon dioxide may be required. A qualified expert's advice is desirable to determine the proper chemical doses.

If the calcium hardness times (multiplied by) the alkalinity is between 1,000 and 5,000, either lime or soda ash (Na_2CO_3) will be satisfactory. The decision regarding which chemical to use will depend on the cost of equipment and the cost of chemicals.

If the calcium hardness times the alkalinity is greater than 5,000, either lime or caustic (NaOH) may be used. Soda ash will be ruled out because of expense.

8.42   Determination of Chemical Dose

The chemical dose required to saturate the water with calcium carbonate may be determined graphically[44] or by a trial and experiment calculation that is practical only with the help of a computer.

In any event, the calculations are not exact and the chemical dose must be checked by the Marble Test.[45] Only slightly more time is required to find the proper chemical dose by experiment than by the use of the graphical approach or a computer. Results obtained by these methods should be verified by the Marble Test anyway.

To determine the chemical dose experimentally, first calculate the value of pH_s using Larson's formula (Equation 2, page 385). Prepare a solution containing 1.000 gram per liter of the chemical to be used (one mL will then contain one mg of chemical). Be sure the dilution water does not contain any carbon dioxide. Carbon dioxide may be removed from distilled water by boiling. Deionized water is usually satisfactory. Treat a one-liter sample of the water to be tested with one-mL portions of the chemical solution (mixing well) until the calculated value of pH_s is reached. As one-mL portions of the chemical solution are added, the pH of the sample will gradually increase until the pH_s value is reached. The degree of saturation is then measured by the Marble Test. Compare the total hardness (mg/L as $CaCO_3$) of the water before and after the Marble Test. If the total hardness (as $CaCO_3$) is reduced by more than 10 mg/L (the water is supersaturated with $CaCO_3$), try a smaller chemical dose and repeat the procedure until the total hardness is decreased by between 0 and 10 mg/L as $CaCO_3$.

To conduct the Marble Test, first measure the pH, alkalinity, and hardness. Stir a one-half liter sample of the water being tested with approximately 0.5 gram of pulverized marble ($CaCO_3$) for five minutes. Filter the water and again measure the pH, alkalinity, and hardness. A decrease in all three values means the water was supersaturated; an increase in all three values means the water was undersaturated; and no change

[44] See D. T. Merril and R. L. Sanks, "Corrosion Control by Deposition of $CaCO_3$," JOURNAL AMERICAN WATER WORKS ASSOCIATION, Part 1, page 592, November 1977; Part 2, page 634, December 1977; and Part 3, page 12, January 1978.
[45] Marble Test.   See Volume II, Chapter 21, "Advanced Laboratory Procedures," for detailed procedures on how to perform the Marble Test.

indicates the water was just saturated with $CaCO_3$. When stirring the sample use a magnetic stirrer and a nearly full, glass-stoppered bottle to prevent a loss or gain of carbon dioxide from the air. Stir fresh water samples as rapidly as possible to be sure that the temperature of the water stays nearly constant during the test.

Water that is just saturated will form a $CaCO_3$ scale only on the cathodic corrosion areas, but water that is well supersaturated will form a scale on all surfaces exposed to flowing water. The thickest scale will form on the surfaces where the water velocity is the highest (up to 5 ft/sec or 1.5 m/sec) because these have the greatest contact of calcium carbonate with the surface. The thickness of scale in water mains seldom exceeds one-eighth inch (3 mm). However, ridges of scale may form transverse (perpendicular) to the flow and very seriously reduce the carrying capacity of the water main. When the velocities are higher (greater than 5 ft/sec or 1.5 m/sec), the scale can be washed or eroded away.

Furthermore, calcium carbonate, unlike most salts, is less soluble in hot water than in cold water. Therefore, even if a water is just saturated as it enters the distribution system, it may become highly supersaturated in hot water systems. Scale has a strong tendency to form on heat transfer surfaces in hot water heaters. Because calcium carbonate does not conduct heat as well as steel, this results in a lowered heating efficiency and in overheating of the heat-transferring metal. Hot water pipes are sometimes found almost completely plugged with a calcium carbonate ($CaCO_3$) deposit. Scale formation can be inhibited by feeding a one- to two-mg/L solution of sodium trimetaphosphate.[46] This chemical eventually reverts to the less effective chemical called orthophosphate in the distribution system. The chemical change occurs slowly in cold waters, but quite rapidly in hot water systems.

8.43   Determination of Chemical Feeder Setting

The desirable feeder setting must be established by analysis of the results of your corrosion-control program. If the chemical composition of the water is fairly constant, periodic Marble Tests may be used. If the chemical composition of the water is variable, an Enslow column[47] (Figure 8.10) is more practical.

Unfortunately there are no standards for constructing an Enslow column, nor are any ready-made columns commercially available. To prepare your own Enslow column, refer to Figure 8.10. Use two pieces of $2^{1}/_{2}$-inch (65-mm) PVC pipe or glass columns approximately 12 inches (30 cm) long. Fill the first tube (A) with powdered chalk ($CaCO_3$). Fill the second tube (B) with marble or limestone ($CaCO_3$) chips, or coarse silica sand. (The second tube traps any chalk powder swept from the first tube.) The limestone used in either or both tubes should be about the same coarseness as filter sand. This limestone sand can be obtained by breaking up limestone, available from a local building materials supplier, or by breaking up chicken grit obtained from a farm supply store. Use plugs (C) of compacted glass wool or other suitable material for supporting the straining media. Stopcocks or pinchcocks (D) are used to regulate the flow. The flask (E) contains the effluent from the Enslow column. Allow this flask to overflow continuously with the calcium carbonate saturated water.

A sample stream of finished water from the plant of approximately $^{1}/_{8}$ GPM (0.5 liters/min) is passed through the two columns. This flow should be adjusted to allow for a contact time of two hours, although one hour could be sufficient. A longer contact period should not cause any problems. If the pH increases as the water flows through the columns, the chemical feed should be increased. If the pH decreases, the chemical feed should be decreased. When there is no pH change, the water is just saturated with calcium carbonate ($CaCO_3$). A slight decrease in pH is acceptable, but any increase in pH should be avoided by increasing the chemical dosage. If the alkalinity is measured before and after the water passes through the columns, the amount of increase in alkalinity would indicate the amount of increase required for the chemical feeder setting.

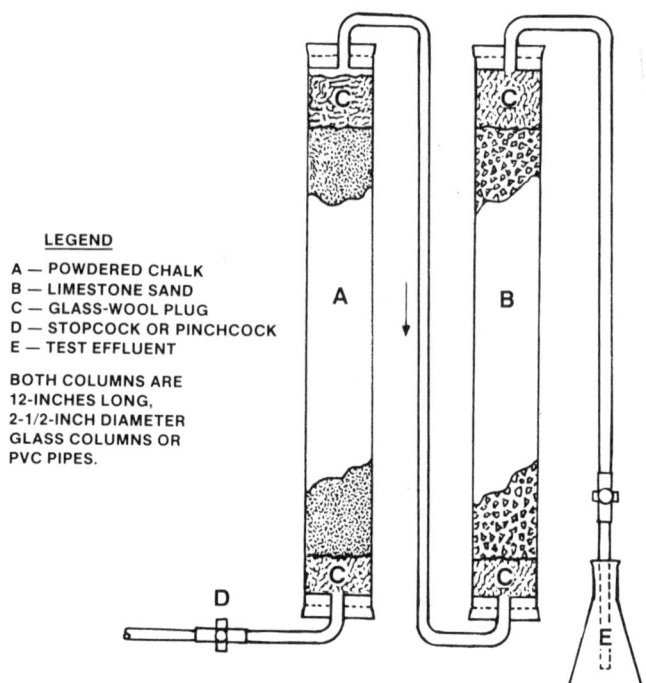

LEGEND

A — POWDERED CHALK
B — LIMESTONE SAND
C — GLASS-WOOL PLUG
D — STOPCOCK OR PINCHCOCK
E — TEST EFFLUENT

BOTH COLUMNS ARE
12-INCHES LONG,
2-1/2-INCH DIAMETER
GLASS COLUMNS OR
PVC PIPES.

Fig. 8.10   Enslow column

(From L. H. Enslow, "The Continuous Stability Indicator,"
WATER WORKS AND SEWERAGE, page 108, March 1939)

---

[46] *Commercial sodium trimetaphosphate is called sodium hexametaphosphate.*

[47] *L. H. Enslow, "The Continuous Stability Indicator," WATER WORKS AND SEWERAGE, pages 107-108, March 1939, and pages 283-284, July, 1939.*

QUESTIONS

Write your answers in a notebook and then compare your answers with those on page 404.

8.4C   What chemicals may be required for corrosion control if the product of calcium hardness multiplied by the alkalinity (both in mg/L as $CaCO_3$) is less than 1,000?

8.4D   What chemicals will be required for corrosion control if the product of calcium hardness multiplied by the alkalinity (both in mg/L as $CaCO_3$) is greater than 5,000?

8.4E   How can the proper chemical dose be determined to produce water that is just saturated with calcium carbonate ($CaCO_3$)?

8.44   Zinc, Silica, and Polyphosphate Compounds

Certain zinc compounds such as zinc phosphate are capable of forming effective cathodic films that will control corrosion. These zinc compounds are largely proprietary (can only be bought from the owner) and the companies that market these compounds usually supply technical advice and assistance without charge. The zinc compound treatments are generally more expensive than treatment with lime or caustic, but they have the advantage that scale is less apt to be a problem. Do not use zinc phosphate compounds to control corrosion caused by water that will be stored in an open reservoir. The phosphate may cause algal blooms. Residual chlorine lasts longer in distribution systems using zinc orthophosphate (ZOP). This has been attributed to decreased chlorine demand as a result of the reduction of iron sediments in distribution piping.

Sodium silicate has been used to treat corrosive waters. A solution of sodium silicate ($Na_2O : 3 SiO_2$) fed at a rate of approximately 12 mg/L as silica is used for the first month, after which the rate is reduced to 8 mg/L. This method of treatment is used by individual customers, such as apartment houses and large office buildings, but is not commonly used by water utilities.

Sodium polyphosphates, usually either tetrasodium pyrophosphate ($Na_4P_2O_7$) or sodium hexametaphosphate, have been used for corrosion control. Solutions of these compounds may form protective films, but because they react with calcium, they reduce the effective calcium concentration and thereby actually increase corrosion rates. The major use of these chemicals in water treatment is to control scale formation in waters that are supersaturated with calcium carbonate.

The deterioration of asbestos-cement pipe may be prevented by maintaining calcium carbonate saturation. There is evidence that the zinc treatments are also effective for this purpose as are treatments using traces of iron, manganese, or silica in the water. Any deterioration of asbestos-cement (AC) pipe will cause an increase in pH and calcium content of water as it passes through the pipe. Tests for pH and calcium should be performed after the pipe has been in service for two months or longer because all AC pipe contains at least traces of "free lime" which will result in an initial increase in water pH when the pipe is placed in service.

8.45   Cathodic Protection

8.450   Need for Cathodic Protection

Mixers, tanks, flocculators, clarifiers, and filter troughs are frequently constructed of steel and require some sort of corrosion protection. Cathodic protection systems are available in both the manual and automatic types. Automatic cathodic protection systems are preferred because the conductivity (total dissolved solids or TDS) in water can change. Cathodic protection systems are very costly to install. However, it is also very costly to shut down a water treatment process, drain the facility, sandblast metal surfaces, apply paint or a protective coating, and put the facility back on line.

8.451   How the Protection System Works

Cathodic protection is a process used to reduce or inhibit corrosion of metal exposed to water or soil. This process consists of the deliberate act of reversing the electrochemical force to check the destruction that naturally occurs to metals whenever they are buried.

The technique introduces into the natural corrosion cycle an external D.C. (direct current) electric current sufficiently strong to offset and cancel out the corrosion-producing action. Key to the system is the use of an auxiliary anode of expendable metal which is immersed or buried in the soil or water (called the electrolyte) a predetermined distance from the metal to be protected. Electric current flowing from the anode to the structure (pipeline, flocculator, or clarifier) in precisely the proper flow can exactly counteract corrosion losses.

The application of cathodic protection is an involved process. Factors affecting a corrosion problem include soil conductivity (which varies considerably even within a limited area), soil moisture content, soil and water characteristics, dissolved oxygen content, temperature, seasonal variations of environment (weather), protective coatings, dissimilar metals to be protected, position of other metallic structures, and stray currents already present in the ground or water.

All cathodic protection systems pass current through the soil or water from anodes connected to the structure that is to be protected. Two basic methods are used. Sacrificial anode material, such as magnesium or zinc, is used to create a galvanic cell. Such anodes are self-energized and are connected directly to the structure to be protected. These anodes are commonly used where it is desirable to apply small amounts of current at many locations.

The other basic method uses anodes energized by a direct current power supply such as a rectifier (Figure 8.11). This method, commonly referred to as "impressed current," uses relatively inert anodes (usually graphite or high-silicon cast iron) connected directly to the positive terminal of a D.C. (direct current) power supply or rectifier, with the pipe or structure being protected connected to the negative terminal. Such systems are generally used where large amounts of current are required at relatively few locations.

8.452   Equipment

The equipment to be used in any given cathodic protection system is very important. A wide selection of rectifiers is available, including air-cooled, oil-immersed, and automatic units. The range of voltage and amperage output is almost infinite. Selection should be based on the particular requirements of the pipe or facilities being protected from corrosion.

Anodes, which serve to distribute the direct current into the earth or water, are manufactured from various metals. Graphite, carbon, high-silicon cast iron, platinum, magnesium, aluminum, and zinc alloys are commonly used. Each has its own particular application. Proper usage is a determining factor in the success or failure of a cathodic protection system.

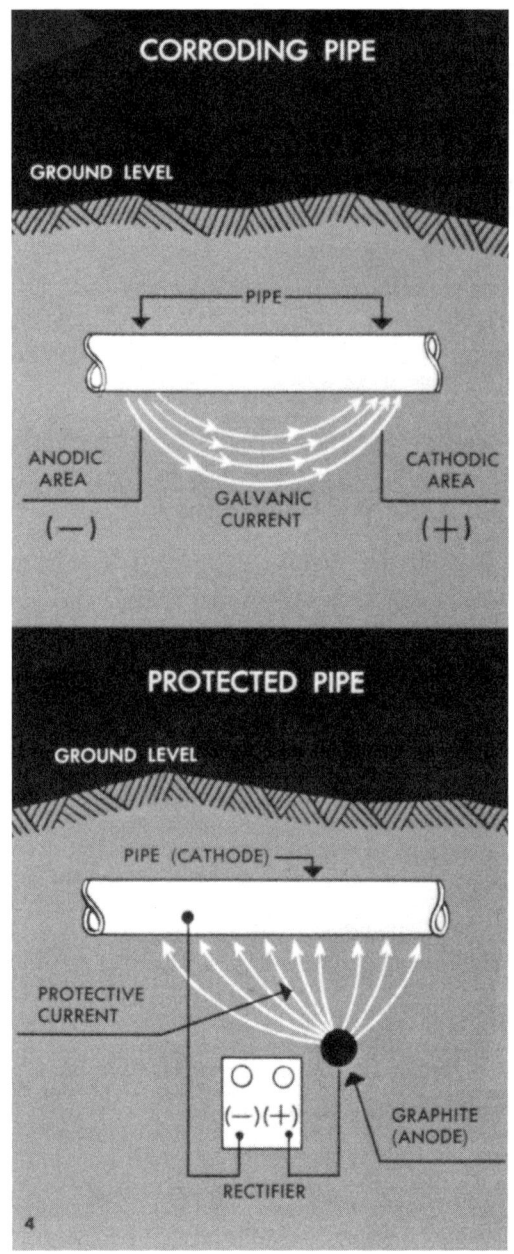

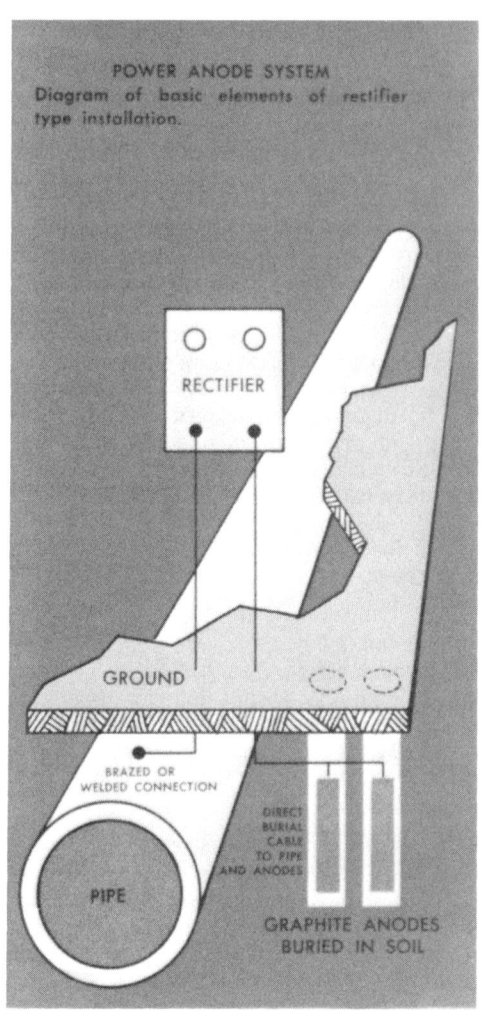

Fig. 8.11   Cathodic protection for buried pipe

(Courtesy of Harco Corporation, Medina, Ohio)

8.453 Protection of Flocculators, Clarifiers, and Filters

Cathodic protection of flocculators, clarifiers (Figure 8.12), and filters is a very effective means of controlling corrosion since maintenance and repair of protective coatings on these facilities is very difficult. Automatic cathodic protection control devices can provide precise corrosion control under nearly all conditions. Anodes with a ten-year design life are usually installed below the low-water level.

8.454 Maintenance

To achieve and maintain a high degree of corrosion control, you must regularly inspect and test the operation of the cathodic protection system and its parts. Regular and proper maintenance is critical because the amount of protective current required to prevent corrosion can vary with changes in the condition of the coatings on the metallic surfaces, in the chemical characteristics of the water being treated, and the operation of the facilities.

An annual maintenance checkout should include a visual inspection of all anodes, wiring, electrical splices and connections, power units, meters, and reference cells. In addition, a complete potential profile should be taken inside the structure to determine the proper automatic controller setting to ensure that corrosion control will be maintained automatically on all submerged surfaces throughout the year.

8.46 Removal of Oxygen

Other methods of water treatment for corrosion control are not practical for domestic water systems. Removal of oxygen is used in boilers and in other water heating systems.

8.47 External Corrosion

8.470 Soil Corrosion

Although corrosion on the outside of the water mains is not the responsibility of the water treatment plant operator, there are many smaller systems where the operator is the person who comes closest to being the "corrosion engineer." For this reason some of the factors influencing soil corrosion will be discussed very briefly. The best measure of the corrosivity of soil is the soil resistivity, which is easily measured using one of the several soil resistance meters on the market. The four-point type is the most useful because it can measure average resistivity down to the depth of the pipeline. Some water systems use soil resistivity as an indication of the kind of pipe to install. If the soil resistivity is greater than 5,000 ohms/cm, serious corrosion is unlikely. Steel pipe, with its superior strength and flexibility, may be used under these conditions. If the soil resistivity is below 500 ohms/cm, nonmetallic pipe such as asbestos-cement or PVC is used. Cast-iron pipe or pipe that is lined and coated with cement mortar is used in the intermediate ranges of soil resistivity.

The chemical reactions involved in external corrosion are electrochemical in nature just like those in internal corrosion. The current paths (flows) are not confined to the inside surfaces of the pipe; therefore, galvanic corrosion and electrolysis are relatively more important and cathodic protection is usually practical.

8.471 Corrosion of Steel Imbedded in Concrete

Galvanic corrosion of ferrous (iron) materials under homes with concrete slab floors has resulted in millions of dollars in damage before the cause was identified. When steel is imbedded in concrete, it assumes the characteristics of a noble metal. Concrete slabs are always poured over a steel mesh or other reinforcing iron. The electrical resistance of the steel is very small compared to that of concrete, so the entire slab behaves as if it were a solid sheet of a noble metal. Since the area of the slabs in a subdivision of slab floor homes is much larger than the area of the pipe, the flow of electrons from the pipes in the area may be fairly large, thus the pipes may be seriously corroded. This problem may be avoided by the use of insulating fittings at the service cock or by the use of plastic service pipe. After the concrete has cured for a few years, the electrical resistance becomes so high that corrosion ceases.

A corrosion problem can develop where a steel water transmission main many miles long runs alongside a second main that is coated with cement. If the two mains are connected at pump stations, corrosion problems can develop. At these pump stations a galvanic current of more than 1.0 ampere might be measured. This instance of corrosion may sometimes be controlled by the installation of insulating fittings in order to interrupt the electric circuit.

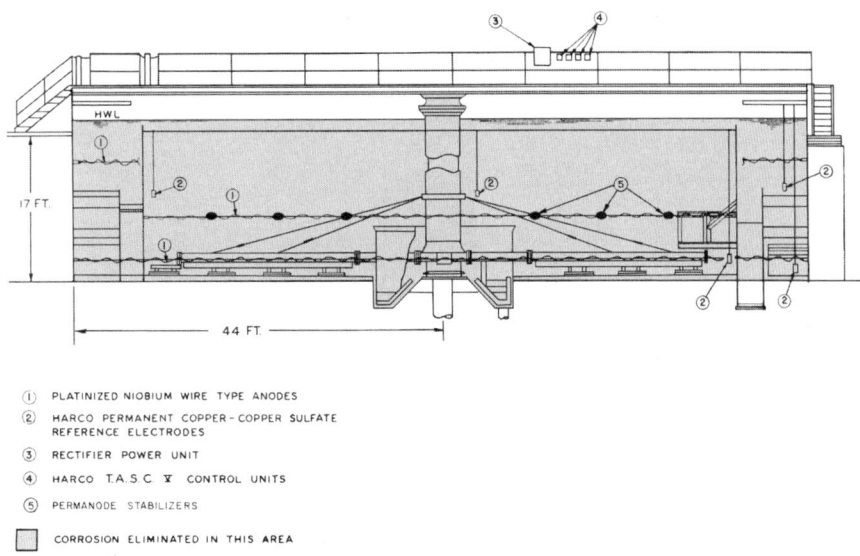

① PLATINIZED NIOBIUM WIRE TYPE ANODES

② HARCO PERMANENT COPPER-COPPER SULFATE REFERENCE ELECTRODES

③ RECTIFIER POWER UNIT

④ HARCO T.A.S.C. Ⅴ CONTROL UNITS

⑤ PERMANODE STABILIZERS

▢ CORROSION ELIMINATED IN THIS AREA

Fig. 8.12 Cathodic protection for a clarifier

(Courtesy of Harco Corporation, Medina, Ohio)

Aluminum should never be imbedded directly in concrete without protection, such as a zinc chromate primer. Aluminum and, to a lesser extent, zinc may be oxidized by hydroxyl ions:

$$2\,Al + 2\,OH^- + 4\,H_2O \;\rightarrow\; 2\,H_2AlO_3^- + 3\,H_2$$

For this reason aluminum is unsatisfactory in highly alkaline aquatic (water) environments.

8.472 Stray Electric Currents

As explained in Lesson 1, electrolysis is the decomposition of a substance by the passage of an exterior source of D.C. (direct current) electric current (Figure 8.13). Internal corrosion caused by electrolysis is practically impossible in a system with properly made joints because the resistance of the water in the pipe is about a billion times as great as the resistance of the pipe itself so that corrosive electric currents must be extremely small by comparison.

Alternating current electrolysis will also corrode metals, but A.C.-caused electrolysis is a rare occurrence and seldom a serious problem.

As in the past, electric transit systems may pose a serious threat again in the future (Figure 8.14). Electrolysis of water mains can result from stray currents generated by cathodic protection installed by the gas company or other utilities, but engineers who install cathodic protection systems are aware of this possibility and can avoid these problems (Figure 8.15). Electrolysis caused by defective grounding of a customer's piping may be avoided by the use of insulating fittings or using plastic service pipe.

QUESTIONS

Write your answers in a notebook and then compare your answers with those on page 404.

8.4F   List one advantage and one limitation of using zinc compounds instead of lime or caustic for corrosion control.

8.4G   Where is the application of cathodic protection practical in water treatment plants?

8.4H   What is the best measure of the corrosivity of soil?

8.4I   How can electrolysis of water mains be caused by other utilities?

8.5 TROUBLESHOOTING

8.50 Internal Pipe Corrosion

Internal corrosion can be detected by rusty water complaints and by examining the insides of pipes for pitting, tubercles, and other evidence of corrosion. To control internal corrosion, treat the water to achieve calcium carbonate saturation.

Select a target pH_s. Dose to this pH. Run a Marble Test on a sample of the treated water. If the pH does not change by more than ±0.2 pH before and after the Marble Test, the target pH_s is satisfactory. If the pH increases during the Marble Test, the target pH_s should be increased by the amount of the pH increase during the Marble Test. Reset the chemical feeder to dose to the new pH. Repeat this procedure until satisfactory results are obtained.

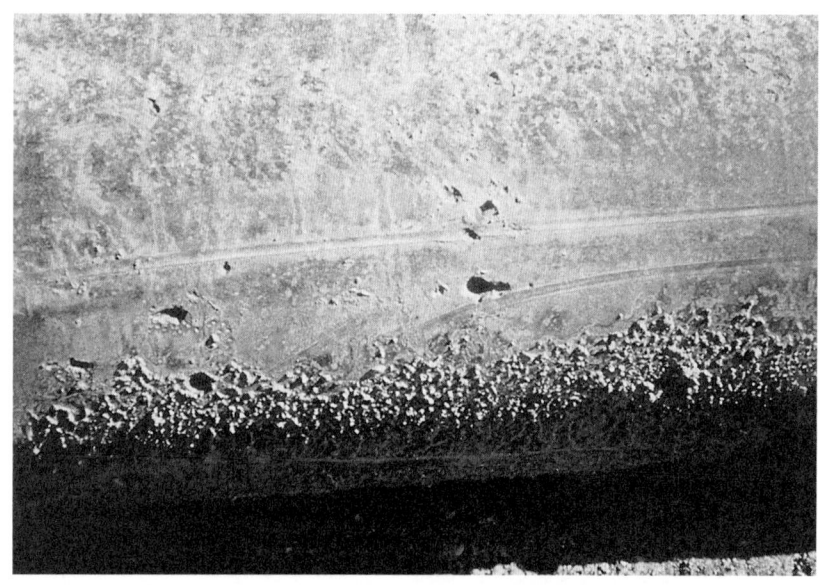

Fig. 8.13   Stray-current electrolysis
(Permission of ARMCO)

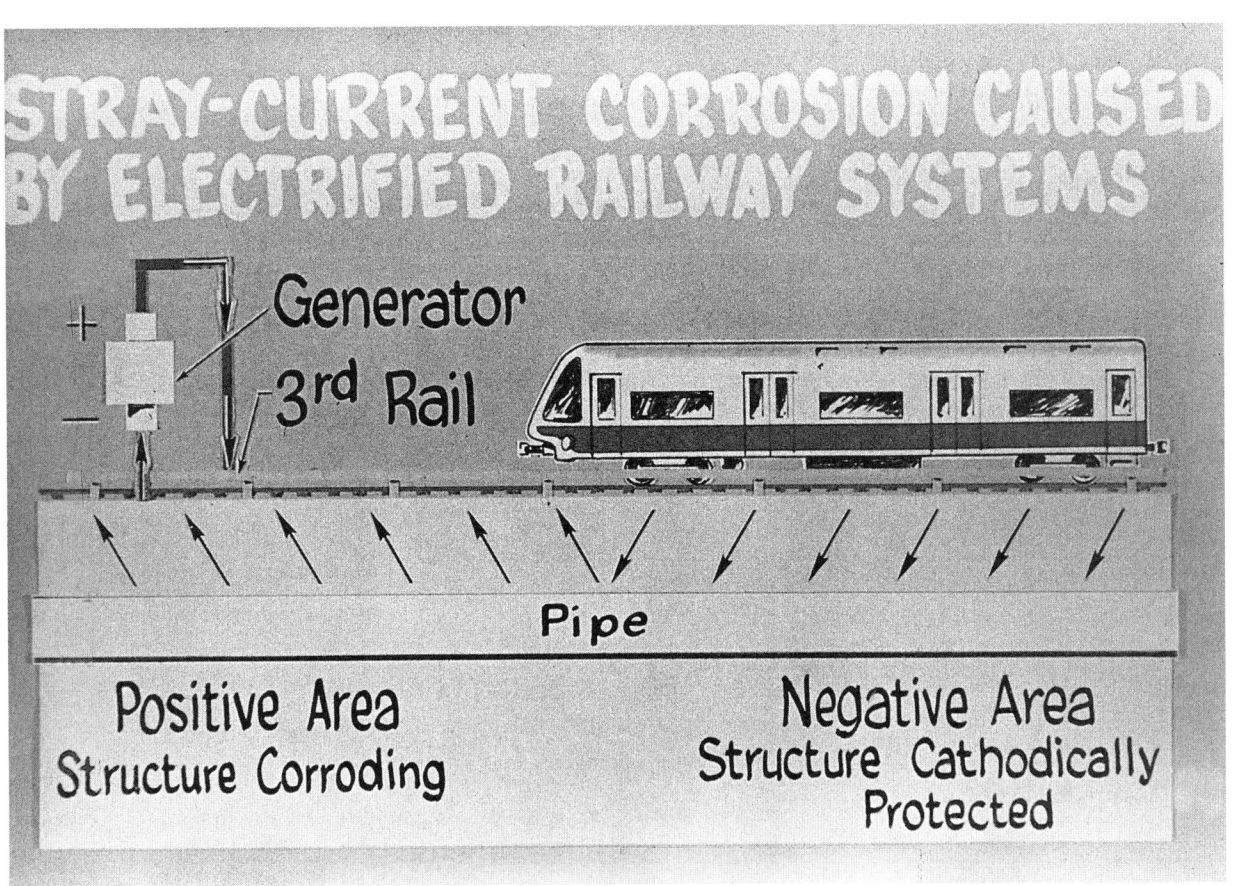

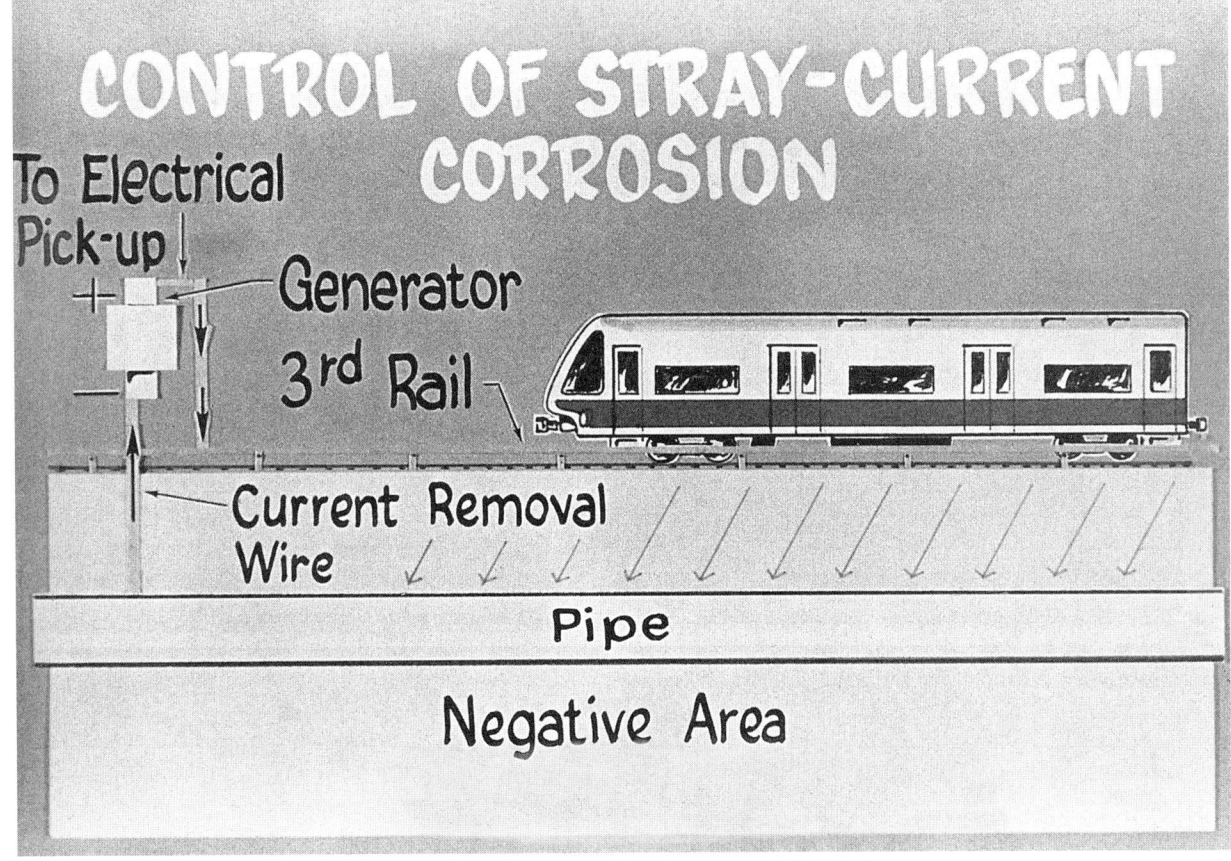

Fig. 8.14   *Stray-current corrosion from electric transit systems*

(Permission of ARMCO)

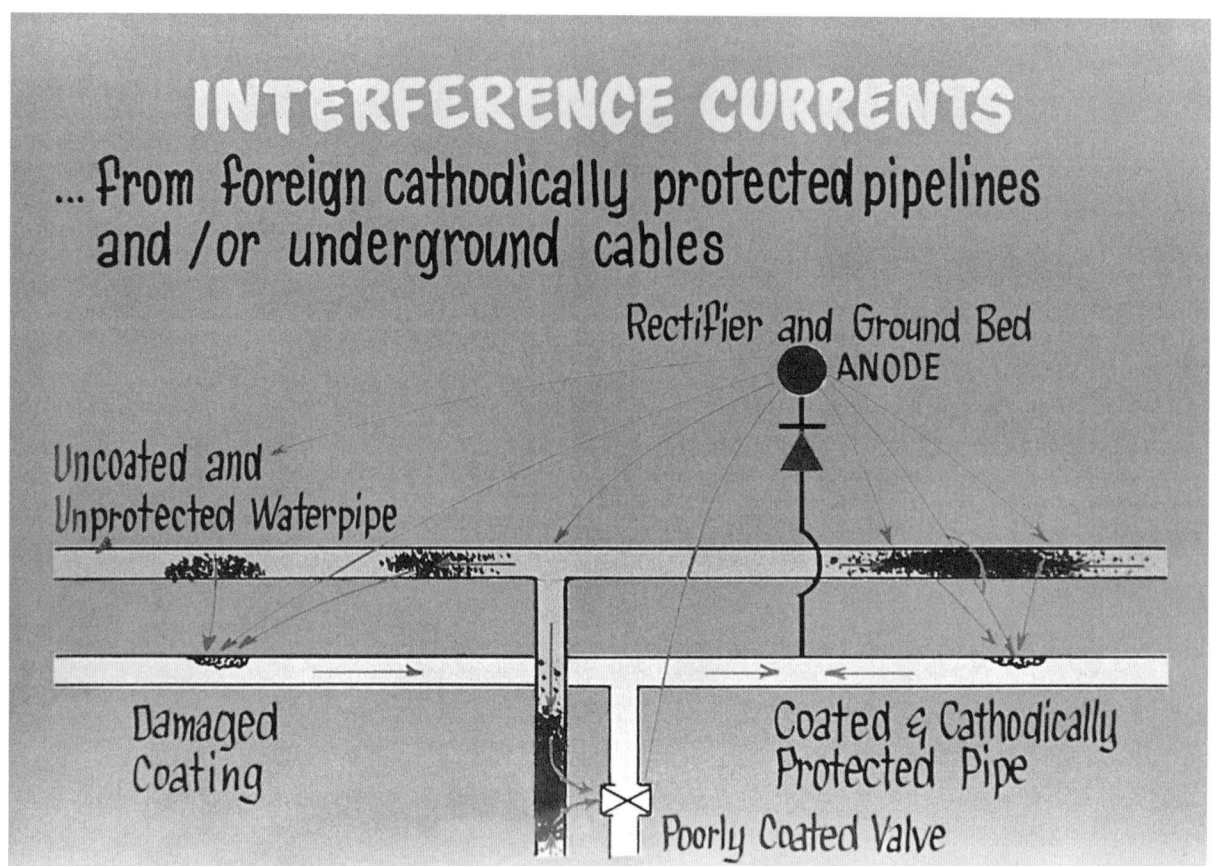

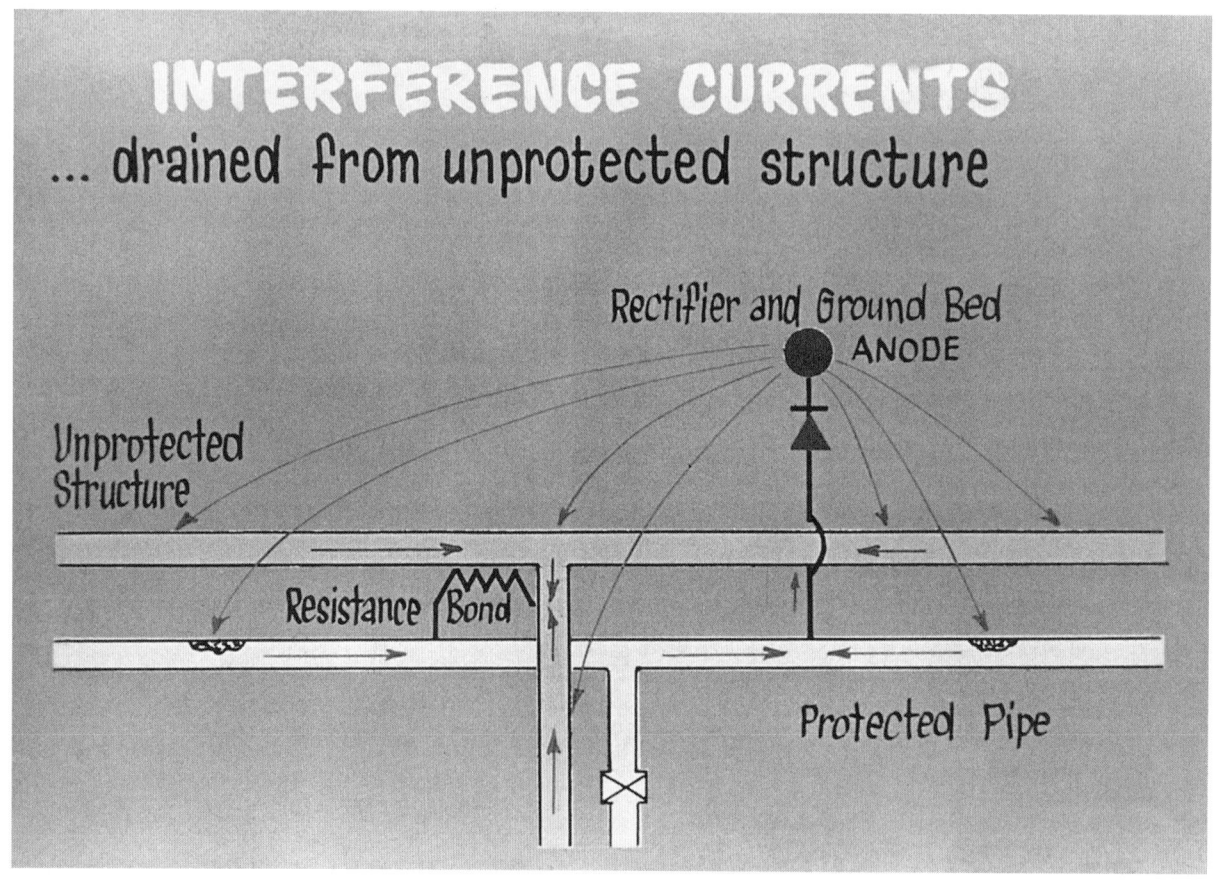

Fig. 8.15 Electrolysis caused by interference currents
(Permission of ARMCO)

8.51 External Pipe Corrosion

External corrosion is detected by observing pinhole leaks or rust on the outside of pipes. If the rusting pipes were installed with the proper bonds and insulating fittings, cathodic protection can be effective. If cathodic protection is not feasible, the pipe will have to be replaced with PVC, plastic pipe, or a cement-coated line.

QUESTIONS

Write your answers in a notebook and then compare your answers with those on pages 404 and 405.

8.5A   How can internal pipe corrosion be detected?

8.5B   How can internal pipe corrosion be controlled?

8.5C   How can external pipe corrosion be detected?

8.5D   How can external pipe corrosion be controlled?

8.6 THE LEAD AND COPPER RULE

8.60 Health Concerns

The health concerns about exposure to lead are best described by the EPA:

Lead is a common, natural and often useful metal found throughout the environment in lead-based paint, air, soil, household dust, food, certain types of pottery, porcelain and pewter, and water. Lead can pose a significant risk to your health if too much of it enters your body. Lead builds up in the body over many years and can cause damage to the brain, red blood cells and kidneys. The greatest risk is to young children and pregnant women. Amounts of lead that won't hurt adults can slow down normal mental and physical development of growing bodies. In addition, a child at play often comes into contact with sources of lead contamination—like dirt and dust—that rarely affect an adult.

Lead in drinking water, although rarely the sole cause of lead poisoning, can significantly increase a person's total lead exposure, particularly the exposure of infants who drink baby formulas and concentrated juices that are mixed with water. The EPA estimates that drinking water can make up 20 percent or more of a person's total exposure to lead.

Lead is unusual among drinking water contaminants in that it seldom occurs naturally in water supplies like rivers and lakes. Lead enters drinking water primarily as a result of the corrosion, or wearing away, of materials containing lead in the water distribution system and household plumbing. (See Figure 8.16.) These materials include lead-based solder used to join copper pipe, brass and chrome-plated brass faucets, and in some cases, pipes made of lead that connect your house to the water main (service lines).

When water stands in lead pipes or plumbing systems containing lead for several hours or more, the lead may dissolve into your drinking water. This means the first water drawn from the tap in the morning, or later in the afternoon after returning from work or school, can contain fairly high levels of lead.

The health effects of copper include stomach and intestinal distress. Prolonged doses result in liver damage. Excess intake of copper or the inability to metabolize copper is called Wilson's disease.

8.61 Regulations

As part of the Safe Drinking Water Act Amendments of 1986, the U.S. Congress directed the Environmental Protection Agency to develop regulations for monitoring and control of lead and copper in drinking water. On June 7, 1991, EPA published the final Lead and Copper Rule. The rule and its implications are both complex and significant. Major features of the rule are explained in this section. More detailed information can be obtained from the EPA or your state regulatory agency, or from the American Water Works Association (AWWA), 6666 West Quincy Avenue, Denver, CO 80235, phone (303) 794-7711.

The issue of lead contamination in water supplies and in other environmental media (air, paint, food) continues to be a subject of considerable activity on the part of individual state governments, Congress, and the EPA. Future amendments to the rules or regulations governing lead contamination in water, monitoring requirements, and mitigation measures may affect how the June 7, 1991, Lead and Copper Rule is implemented. You are encouraged to seek the most current status of lead contamination regulations from the regulatory agencies in your state before starting a monitoring or mitigation program to ensure that your program meets all applicable requirements. The information provided in this section applies to the final Lead and Copper Rule of June 7, 1991, published in the FEDERAL REGISTER.

The 1991 Lead and Copper Rule includes the following elements:

● Maximum Contaminant Level Goals (MCLGs), which are described as "nonenforceable health-based targets," and action levels are established for lead and copper;

● Monitoring requirements for lead, copper, and other corrosion analysis constituents, analytical methods, and laboratory certification requirements;

● Treatment techniques for lead and copper, required if action levels are exceeded during monitoring, including optimal corrosion-control treatment, source water treatment, and lead service line replacement;

● Public notification and public education program requirements;

● Utility system recordkeeping and reporting requirements; and

● Variances and exemptions from the regulations and compliance schedules based on size of the population served by the utility system.

8.62 Monitoring Requirements

The first requirement of the Lead and Copper Rule is for monitoring to determine if either of these metals exceeds the levels at which further action must be taken. An unusual provision of the Lead and Copper Rule is that the monitoring samples must be collected at the consumers' taps, rather than only at the water treatment plant or in the distribution system. The samples must be taken from locations identified by the utility as "high-risk," including:

● Homes with lead solder installed after 1982,

● Homes with lead pipes, and

● Homes with lead service lines.

Contaminant	Low Level Health Effects	Sources in Drinking Water
Lead	**Children:** Altered physical and mental development; interference with growth; deficits in IQ, attention span, and hearing. **Women:** Increased blood pressure; shorter gestational period **Men:** Increased blood pressure	Corrosion of: Lead solder and brass faucets and fixtures Lead service lines (20% of public water systems) Source water (1% of systems)
Copper	Stomach and intestinal distress; Wilson's Disease	Corrosion by-products: Interior household and building pipes Source water (1% of systems)

Public Water System (PWS) and Homeowner Plumbing

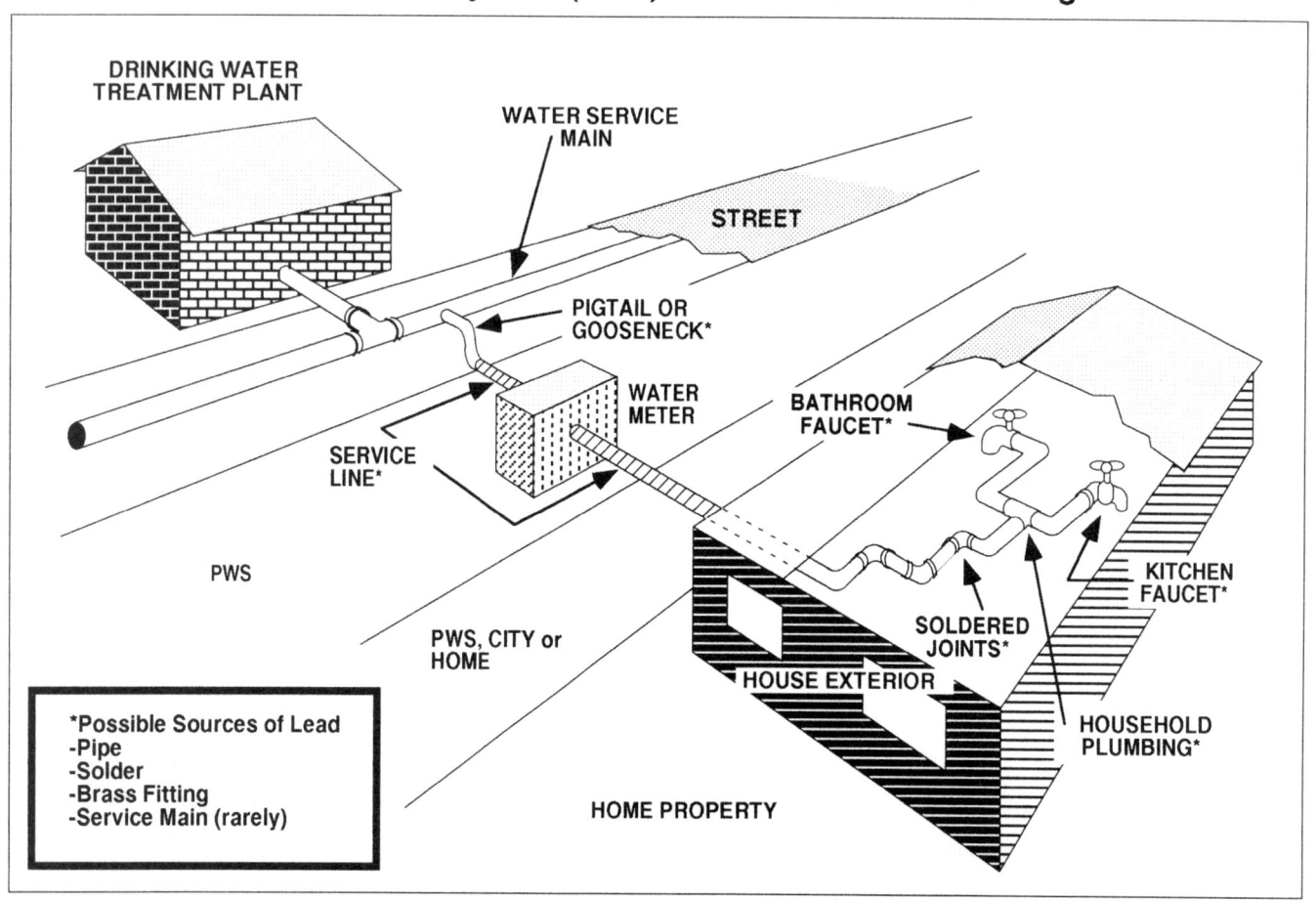

Fig. 8.16   Health effects and sources of lead and copper

8.620 Monitoring Frequency

The number of samples to be collected for lead and copper analysis is based on the size of the distribution system; the sampling frequency is every six months for the initial monitoring program. There are two monitoring periods each year, January to June, and July to December. If the system is in compliance, either as demonstrated by monitoring or after installation of corrosion control, a reduced monitoring frequency can be initiated. The number of sampling sites required based on system size for initial and reduced monitoring are listed in Table 8.5.

8.621 Sampling Procedure

The samples are to be collected as "first-draw" samples from the cold water tap in either the kitchen or bathroom, or from a tap routinely used for consumption of water if in a building other than a home. A first-draw sample is defined as the first liter of water collected from a tap that has not been used for at least 6 hours, but preferably unused no more than twelve hours. Faucet aerators should be removed prior to sample collection. The Lead and Copper Rule allows homeowners to collect samples for the utility as long as the proper sample collection instructions have been provided. The EPA specifically prohibits the utility from disputing the accuracy of a sample collected by a resident.

8.622 Maximum Contaminant Level Goals (MCLGs)

Under the 1977 National Interim Primary Drinking Water Regulations, a maximum contaminant level (MCL) was established for lead, and in 1979, under the National Secondary Drinking Water Regulations, a secondary maximum contaminant level (SMCL) was set for copper. The Lead and Copper Rule replaces those with MCLGs, and with treatment technique requirements when "action levels" are exceeded during system monitoring. The MCLGs and action levels for lead and copper, compared to the older interim limits, are:

	Interim Limits	1991 MCLG	1991 Action Level
LEAD:	0.05 mg/L	Zero	0.015 mg/L
COPPER:	1 mg/L	1.3 mg/L	1.3 mg/L

Action levels are defined in this rule as the value measured in the 90th percentile at the consumer's tap. This means that when all of the samples measured at consumer taps are arranged from lowest to highest value in a list, the value at the point $^9/_{10}$ (90%) of the way up from the lowest number is the value used to determine compliance with the action level. For example, in the following series of lead and copper test results, the 90th percentile is shown:

	Lead	Copper
	0.031	0.950
90th Percentile:	**0.025**	**0.806**
	0.021	0.758
	0.015	0.653
	0.014	0.534
	0.012	0.465
	0.010	0.430
	0.008	0.411
	0.002	0.396
	0.002	0.387

In the previous example, the lead value at the 90th percentile exceeds the 0.015 mg/L action level, requiring the utility to begin implementation of a corrosion-control program. All utilities with populations larger than 50,000 are required to initiate corrosion-control studies unless they can prove through the routine monitoring program that any increase between the source water lead level and the 90th percentile tap samples is less than 0.005 mg/L for two consecutive six-month monitoring periods. For small systems with only 5 sampling sites, the 90th percentile is defined as the average of the 4th and 5th highest values in the series.

8.623 Monitoring Deadlines

The regulatory deadlines to begin the initial monitoring programs varied with system size:

SYSTEM SIZE	POPULATION	MONITORING START DATE
• **Large systems**	>50,000	January, 1992
• **Medium systems**	>3,300 to ≤50,000	July, 1992
• **Small systems**	≤3,300	July, 1993

TABLE 8.5 SAMPLING SITES REQUIRED FOR LEAD AND COPPER ANALYSIS		
System Size (Population)	Sampling Sites Required (Base Monitoring)	Sampling Sites Required (Reduced Monitoring)
>100,000	100	50
10,001 - 100,000	60	30
3,301 - 10,000	40	20
501 - 3,300	20	10
101 - 500	10	5
≤100	5	5

Reduced Monitoring:

- All public water systems that meet the lead and copper action levels or maintain optimal corrosion-control treatment for two consecutive six-month monitoring periods may reduce the number of tap water sampling sites as shown in Table 8.5 and their collection frequency to once per year.

- All public water systems that meet the lead and copper action levels or maintain optimal corrosion-control treatment for three consecutive years may reduce the number of tap water sampling sites as shown in Table 8.5 and their collection frequency to once every three years.

8.624   Other Water Quality Monitoring

In addition to the required lead and copper monitoring program, all large systems and any medium or small systems that exceed the action levels must also monitor for other corrosion-related water quality indicators. These indicators are:

- Alkalinity
- Conductivity
- pH
- Temperature
- Corrosion Inhibitor Concentration (Calcium, Orthophosphate, or Silica)
- Corrosion Inhibitor Dosage Rate (if added)

Specified limits for these water quality indicators are to be set by each state. The monitoring locations include both the point(s) of entry to the distribution system and consumer taps. The sampling program requirements are:

- All large systems, and those medium and small systems that exceed the lead or copper action levels, must collect two samples for each applicable water quality indicator at each lead and copper tap water sample site every six months, and one sample for each applicable water quality indicator at each entry point to the distribution system every two weeks.

- All large systems, and those medium and small systems that exceed the lead or copper action levels, even after installing optimal corrosion control treatment, must continue to collect two samples for each applicable water quality indicator at each lead and copper tap water sample site every six months, and one sample for each applicable water quality indicator at each entry into the distribution system every two weeks.

- The same number of reduced monitoring stations as are allowed with the lead and copper monitoring are allowed for systems in compliance or after successful implementation of corrosion-control treatment.

8.625   Analytical Methods and Certification Requirements

US EPA-approved methods must be used for laboratory analysis of all samples tested in Lead and Copper Rule monitoring programs. Methods are listed as either EPA-approved methods or in *STANDARD METHODS FOR THE EXAMINATION OF WATER AND WASTEWATER*, 20th Edition. Laboratories are not required to be certified to test for alkalinity, calcium, conductivity, orthophosphate, pH, silica, or temperature, but they must be certified to test for copper and lead.

8.63   Treatment Requirements

8.630   Corrosion Treatment Studies

All large water systems (>50,000 population) that do not meet the action levels for copper or lead, or that cannot show that the difference between source water and 90th percentile tap water samples is less than 0.005 mg/*L*, must conduct corrosion-control studies. These studies should include comparison of the following accepted potable water corrosion-control treatments:

- pH and alkalinity adjustment (to reduce acidity),

- Calcium adjustment (to form protective calcium carbonate films inside plumbing), and

- Phosphate- or silica-based inhibitor addition (to form protective films inside plumbing).

The corrosion studies can be conducted using pipe rig/loop tests, metal coupon tests, partial-system tests, or documented analogous treatment. Pipe rig/loop and metal coupon tests directly evaluate the performance of corrosion inhibitors on the corrosion rate of materials similar to those used in the distribution system. Partial-system tests determine the effectiveness of inhibitors on an isolated portion of a distribution system as compared to other sections with other treatments or with no treatment. Documented analogous treatment refers to using data from other similar water systems that have already conducted corrosion studies.

Small or medium-sized systems that do not meet the action levels are also required to install corrosion-control treatment. States have some flexibility in providing smaller systems with standard recommended treatment practices for corrosion control, and in allowing such systems to recommend treatments without extensive field studies. Some states may require full studies of any size system, however. In either case, all water systems are required to implement corrosion treatment within 24 months once approved by the state, and to begin collecting follow-up samples within one year of completion of installation.

8.631   Source Water Treatment

A small number of systems may not be in compliance with the lead and copper action levels due to the occurrence of these metals in the source water. If it is discovered during the routine monitoring that a lead or copper problem exists in the source water feeding the distribution system, the utility must specify a treatment method to remove that contaminant from the source water. One of the following treatment methods or an alternative that is equally effective must be selected:

- Ion exchange
- Reverse osmosis
- Lime softening
- Coagulation/filtration

As with corrosion treatment, once the state has approved the recommended treatment, installation must be accomplished within 24 months, and collection of follow-up samples must begin within one year of completion of installation. See *WATER TREATMENT PLANT OPERATION*, Volume II, for procedures on how to operate and maintain the first three source water treatment processes listed above.

8.632 Lead Service Line Replacement

Special requirements also apply to distribution systems with lead service lines into the homes. If corrosion-control and source water treatment programs fail to achieve compliance with lead action levels for such systems, the following requirements take effect:

- All public water systems that continue to exceed the lead action level after installing optimal corrosion-control treatment and source water treatment must replace lead service lines that contribute in excess of 0.015 mg/L of lead to the tap water.

- Utilities must replace seven percent of their lead service lines per year, or demonstrate that the lines not replaced contribute less than 0.015 mg/L of lead to the tap water. Samples to determine lead service line contribution levels can either be taken directly from a tap in the service line, or from the consumer's tap after the water has been run sufficiently long to ensure that the water being sampled is drawing directly from the sample line. This may be indicated by a change in temperature at the tap.

- A system must replace the entire lead service line unless it can demonstrate that it does not control the entire line. Water systems must offer to replace the owner's portion of the service line (at the owner's expense).

- A system that exceeds the lead action level after installing optimal corrosion-control treatment and source water treatment has a maximum of 15 years to complete the lead service line replacement.

8.633 Treatment for Control of Lead and Copper

A water's pH/alkalinity combination determines that water's tendency to dissolve lead and copper. Corrosion-control treatment often involves adjusting pH and alkalinity to make water less corrosive.

The inhibitors used to control lead and copper act by forming a protective coating over the site of corrosion activity, thus "inhibiting" corrosion. The success of inhibitor addition depends on the ability of the inhibitor to provide a continuous coating throughout the distribution system.

Phosphates are by far the most common inhibitors used in water treatment for corrosion control. Silicates have a more limited application and may be most suitable for small systems with iron and manganese problems. The control of lead using a phosphate inhibitor occurs when a lead-phosphate compound is formed.

Orthophosphates are primarily used to control lead, not copper. Under optimal conditions, orthophosphate treatment is usually more effective in reducing lead than pH/alkalinity adjustment. The optimal pH for orthophosphate treatment is between 7.2 and 7.8. The orthophosphate residual must be maintained continuously throughout the distribution system. Systems that have a raw water naturally in the optimal pH range from 7.2 to 7.8 can treat with orthophosphate alone, provided the water has sufficient alkalinity for a stable pH.

Optimal corrosion-control treatment minimizes lead and copper concentrations at users' taps while ensuring that the treatment does not cause the water system to violate any drinking water regulations. When applying corrosion-control treatment, a substantial amount of time may elapse between the time treatment changes are made and the reduction of lead and copper are detected by the analysis of tap water samples. Be very cautious because by changing the chemistry of water; conditions may get worse before they get better.

When using corrosion inhibitors, operators must be aware of both the positive and negative side effects that may result from the use of inhibitors. Phosphate-based inhibitors may stimulate biofilms in the distribution system. These biofilms may deplete (reduce) disinfection residuals within the distribution system. Consumer complaints regarding red water, dirty water, color, and sediment may result from the action of the inhibitor on existing corrosion by-products within the distribution system. The use of zinc orthophosphate may present problems for wastewater facilities with zinc or phosphorus limits in their *NPDES PERMITS*.[48] Customers with specific water quality needs, such as health care facilities, should be advised of any treatment changes at the plant. The use of sodium-based chemicals will increase total sodium levels in the finished water. The use of silicates may reduce the useful life of domestic hot water heaters due to "glassification" because silicates precipitate rapidly at higher water temperatures.

For systems using alum or other products containing aluminum, the aluminum will bind with the orthophosphate at a ratio of 1:4 and will interfere with the formation of an effective inhibitor film. For example, if the target orthophosphate residual is 0.5 mg/L and the water has 0.1 mg/L of aluminum, only 0.1 mg/L of the orthophosphate will be available to coat the pipes while 0.4 mg/L will be bound to the aluminum. To solve this problem, measure the aluminum level at the entry to the distribution system, multiply the amount by four, and add the result to the desired operating orthophosphate residual. Calculating the orthophosphate dose in this manner will give you a sufficient residual of orthophosphate for effective corrosion control in the distribution system.

8.64 Public Education and Reporting Requirements

8.640 Public Education

If a utility fails to meet the lead action level, an EPA-developed public education program must be initiated. The program for community water supplies includes mail-out notices in the utility bills, announcements in major local daily and weekly newspapers, distribution of pamphlets to schools, health care, and day care facilities, and submittal of public service announcements to at least five of the largest radio and television stations in the area. Each of these actions must be repeated every twelve months as long as the lead action level is exceeded, with the exception of the radio and television announcements, which must be distributed every six months. For noncommunity systems that exceed the lead action level, in-

---

[48] *NPDES Permit. National Pollutant Discharge Elimination System permit is the regulatory agency document issued by either a federal or state agency which is designed to control all discharges of potential pollutants from point sources and storm water runoff into U.S. waterways. NPDES permits regulate discharges into navigable waters from all point sources of pollution, including industries, municipal wastewater treatment plants, sanitary landfills, large agricultural feedlots and return irrigation flows.*

formation posters must be placed in public locations and brochures must be distributed to each person served by the system.

In any routine six-month monitoring period that the lead action levels are complied with, no public education activity is required. The public education programs are designed to provide the consumer with information on how to minimize their exposure to lead in the water through such activities as:

- Flushing the taps before use,
- Cooking with cold rather than hot tap water,
- Checking new plumbing for lead solder, and
- Testing their water for lead.

Utilities that do not meet the lead action level must offer to sample the tap water of any consumer who requests it, although the consumer can be charged for the cost of the sample collection and analysis. Mandatory health effects language (a portion of which was quoted in Section 8.60) is provided by the EPA and required to be used in the public education program. Copies of the public education materials can be obtained from the EPA, the state regulatory agency, or from the American Water Works Association.

8.641 Reporting and Recordkeeping Requirements

Under the provisions of the Lead and Copper Rule, utilities must provide separate reports for tap water monitoring programs, source water monitoring, corrosion-control treatment, source water treatment, lead service line replacement activities, and public education programs. All utilities are required to keep complete, original records of sampling data, test results, reports, surveys, letters, evaluations, schedules, state determinations, and any other information required to comply with the Lead and Copper Rule for at least 12 years.

8.7 ARITHMETIC ASSIGNMENT

Turn to the Appendix, "How to Solve Water Treatment Plant Arithmetic Problems," at the back of this manual and read all of Section A.8, *PUMPS*. Also work the example problems and check the arithmetic using your calculator.

In Section A.13, *TYPICAL WATER TREATMENT PLANT PROBLEMS*, read and work the problems in Section A.137, Corrosion Control.

8.8 ADDITIONAL READING

1. *NEW YORK MANUAL*, Chapter 15,* "Corrosion and Corrosion Control."

2. *TEXAS MANUAL*, Chapter 11,* "Special Water Treatment."

3. *EXTERNAL CORROSION: INTRODUCTION TO CHEMISTRY AND CONTROL* (M27). Obtain from American Water Works Association (AWWA), Bookstore, 6666 West Quincy Avenue, Denver, CO 80235. Order No. 30027. ISBN 0-89867-366-6. Price to members, $46.50; nonmembers, $68.50; price includes cost of shipping and handling.

4. "Corrosion Control by Deposition of $CaCO_3$ Films, A Practical Approach for Plant Operators," by Douglas T. Merrill and Robert L. Sanks. *JOURNAL AMERICAN WATER WORKS ASSOCIATION*; Part 1, November 1977, pages 592-599; Part 2, December 1977, pages 634-640; and Part 3, January 1978, pages 12-18.

* Depends on edition.

Another excellent source of information about all aspects of corrosion control is the National Association of Corrosion Engineers (NACE). Write to NACE International, 1440 South Creek Drive, Houston, TX 77084-4906. NACE's website (www.nace.org) offers a variety of information including technical papers on corrosion-related topics, testing and materials standards, abstracts of publications, and training and certification opportunities.

8.9 SUMMARY

Now that you've made it through this chapter on corrosion control—CONGRATULATIONS. You have just finished the toughest chapter in all of our manuals. Corrosion is a very complex topic. We have tried to make this subject as understandable as possible, without oversimplifying all of the factors that influence corrosion. Likewise there are many possible solutions to corrosion problems and a combination of solutions may be necessary to solve any corrosion problems in your water treatment and water distribution facilities.

QUESTIONS

Write your answers in a notebook and then compare your answers with those on page 405.

8.6A List the important elements of the 1991 Lead and Copper Rule.

8.6B What is the definition of an action level in the Lead and Copper Rule?

8.6C What are the accepted potable water corrosion-control treatments?

8.6D How do chemical inhibitors control lead and copper in water distribution systems?

8.6E How must a utility attempt to educate the public if it fails to meet the lead action levels?

End of Lesson 2 of 2 Lessons on CORROSION CONTROL

Please answer the discussion and review questions next.

DISCUSSION AND REVIEW QUESTIONS

Chapter 8. CORROSION CONTROL

(Lesson 2 of 2 Lessons)

Write the answers to these questions in your notebook. The question numbering continues from Lesson 1.

11. Why must caustic soda be handled with great care?

12. How should corrosion-control chemicals be selected?

13. How does the cathodic protection system to control corrosion work?

14. Why can concrete slab floors cause corrosion problems?

15. Where must monitoring samples be collected for the Lead and Copper Rule?

16. What must a utility do if the lead content in drinking water exceeds the action level?

17. What activities could a utility suggest in its public education program to minimize the consumers' exposure to lead in their drinking water?

SUGGESTED ANSWERS

Chapter 8. CORROSION CONTROL

ANSWERS TO QUESTIONS IN LESSON 1

Answers to questions on pages 376 and 377.

8.0A Problems that can be created by corrosive waters include:

1. Economic losses resulting from corrosion damage,
2. Replacement of water mains,
3. Reduced carrying capacity of mains,
4. Reduction of distribution system pressures,
5. Increases in pump energy costs,
6. Corrosion of lead that may create a serious health hazard, and
7. Customer complaints resulting from rusty water, stained laundry, and bad tastes.

8.1A An electrochemical reaction occurs when chemical changes are produced by electricity (electrolysis) or electricity is produced by chemical changes (galvanic action).

8.1B The iron pipe will corrode into the water when it is connected to a copper pipe if the water contains dissolved oxygen.

8.1C When brass corrodes (dezincification), the copper plates out as a mass of spongy copper having nearly the same shape as the original brass.

8.1D When copper and lead solder are in contact, the lead becomes the anode and will corrode in preference to the copper.

8.1E A "dangerous" corrosion inhibitor is one that acts on the anode reaction. If not quite enough inhibitor is added, severe pitting will result and corrosion will be worse than if no inhibitor had been added.

8.1F If a copper service line is connected to a steel water main, the cathode area is large relative to the anode and corrosion will occur at the anode.

Answers to questions on page 380.

8.2A Homeowners may notice the effect of stray current corrosion when pitting penetrates fixtures or pipes, resulting in leaks.

8.2B "Erosion corrosion" in copper tubing is caused by water velocities over five feet per second.

8.2C Chemical factors that influence corrosion include pH, alkalinity, chlorine residual, levels of dissolved solids and dissolved gases such as oxygen and carbon dioxide, and the types and concentrations of various minerals present in water.

8.2D Water with a higher dissolved solids content has a greater potential for corrosion due to the increased conductivity of the water.

8.2E A thin film or coating of calcium carbonate can drastically inhibit corrosion.

Answers to questions on page 382.

8.2F The oxygen concentration cell is the most common type of corrosion cell.

8.2G An oxygen concentration cell can be started in the dead end of a water main.

8.2H Pits can be started on a metallic surface under water by anything that will shield the metal surface from dissolved oxygen in the water, such as bits of clay, dirt, sand, or a colony of bacteria. Also, impurities in the metal may cause a local anode to form.

8.2I A tubercle is a small protective crust of rust which builds up over a pit caused by the loss of metal from corrosion.

Answers to questions on page 382.

8.3A Corrosion of water system facilities is a public health concern because the corrosion process could release lead, copper, or other toxic metals into the water.

8.3B Corrosion rates may be measured by inserting steel specimens called "coupons" in water mains. After a period of time, usually a month or two, the coupons are removed and the loss of weight is measured.

8.3C Leaks in water mains can often be detected by the observation of wet spots above the water main.

8.3D A film of calcium carbonate too thin to see can be detected by placing a drop of dilute hydrochloric (HCl) acid on an obvious cathodic area and observing the area for effervescence (bubbles).

Answers to questions on page 384.

8.3E Loss of dissolved oxygen in the water flowing in a distribution system indicates either that the water contains organic matter or that corrosion is occurring.

8.3F Toxic metals that may enter drinking waters from the customer's plumbing due to corrosive water include:

1. Lead from soldered fittings used with copper tubing,
2. Copper from copper pipes, tubing, and fittings, and
3. Cadmium found in zinc used for galvanizing steel pipes.

Answers to questions on page 385.

8.3G A water is considered stable when it is just saturated with calcium carbonate.

8.3H The Marble Test and Langelier Index are used to determine if water is undersaturated or supersaturated with calcium carbonate.

8.3I The Langelier Index is defined by the equation

$$L.I. = pH - pH_s.$$

8.3J pH_s is that pH where water is just saturated with calcium carbonate.

8.3K Some waters do not have a true pH_s value because they contain insufficient calcium and/or alkalinity to become saturated regardless of their pH.

8.3L

	Known		**Unknown**
Water Temp, °C	= 10°C		pH_s
TDS, mg/L	= 100 mg/L		
Alky, mg/L	= 80 mg/L		
Ca Hardness, mg/L	= 40 mg/L		

1. Find the formula values from the tables.

From Table 8.2 for a water temperature of 10°C,

A = 2.20

From Table 8.3 for a TDS of 100 mg/L,

B = 9.75

From Table 8.4 for Ca of 40 mg/L and Alky of 80 mg/L,

$log(Ca^{2+})$ = 1.60

$log(Alky)$ = 1.90

2. Calculate pH_s.

$$\begin{aligned} pH_s &= A + B - log(Ca^{2+}) - log(Alky) \\ &= 2.20 + 9.75 - 1.60 - 1.90 \\ &= 8.45 \end{aligned}$$

8.3M Rusty or red water complaints indicate that the water is corrosive.

ANSWERS TO QUESTIONS IN LESSON 2

Answers to questions on page 389.

8.4A The corrosivity of a water is often reduced by treating the water so that the water is saturated or slightly supersaturated with calcium carbonate.

8.4B The corrosivity of water may be reduced by adding quicklime (CaO), hydrated lime (Ca(OH)$_2$), caustic soda (NaOH), or soda ash (Na$_2$CO$_3$).

Answers to questions on page 391.

8.4C If the product of calcium hardness multiplied by the alkalinity (both in mg/L as CaCO$_3$) is less than 1,000, both lime and carbon dioxide may be required.

8.4D If the product of calcium hardness multiplied by the alkalinity (both in mg/L as CaCO$_3$) is greater than 5,000, either lime or caustic (NaOH) may be used. Soda ash (Na$_2$CO$_3$) should not be used because of the expense.

8.4E The proper chemical dose to produce water that is just saturated with calcium carbonate (CaCO$_3$) can be determined by (1) graphical methods, (2) trial and experiment calculations with a computer, and (3) use of the Marble Test or an Enslow column.

Answers to questions on page 394.

8.4F List one advantage and one limitation of using zinc compounds over lime or caustic for corrosion control.

ADVANTAGE

1. Scale is less apt to be a problem with zinc compounds.
2. Zinc treatments prevent deterioration of asbestos-cement pipe.

LIMITATION

Zinc compounds are generally more expensive than treatment with lime or caustic.

8.4G Cathodic protection is practical in water treatment plants to protect mixers, tanks, flocculators, clarifiers, and filter troughs.

8.4H The best measure of the corrosivity of soil is the soil resistivity, which is measured by a soil resistance meter.

8.4I Electrolysis of water mains can result from currents generated by cathodic protection installed by the gas company or other utilities.

Answers to questions on page 397.

8.5A Internal pipe corrosion can be detected (1) by rusty water complaints, and (2) by examining the insides of pipes for pitting, tubercles, and other evidence of corrosion.

8.5B Internal pipe corrosion can be controlled by treating the water to achieve calcium carbonate saturation.

8.5C External pipe corrosion is detected by observing pinhole leaks or rust on the outside of pipes.

8.5D External pipe corrosion can be controlled by the installation of the proper bonds and insulating fittings, and by the use of cathodic protection.

Answers to questions on page 402.

8.6A The important elements of the 1991 Lead and Copper Rule include:

1. Maximum Contaminant Level Goals (MCLGs) and action levels for lead and copper;

2. Monitoring requirements for lead, copper, and other corrosion analysis constituents, analytical methods, and laboratory certification requirements;

3. Treatment techniques for lead and copper, required if action levels are exceeded during monitoring, including optimal corrosion-control treatment, source water treatment, and lead service line replacement;

4. Public notification and public education program requirements;

5. Utility system recordkeeping and reporting requirements; and

6. Variances and exemptions from the regulations and compliance schedules based on size of the population served by the utility agency.

8.6B Action levels in the Lead and Copper Rule are defined as the value measured in the 90th percentile at the consumer's tap.

8.6C The accepted potable water corrosion-control treatments include:

1. pH and alkalinity adjustment (to reduce acidity),
2. Calcium adjustment (to form protective calcium carbonate films inside plumbing), and
3. Phosphate- or silica-based inhibitor addition (to form protective films inside plumbing).

8.6D The inhibitors used to control lead and copper act by forming a protective coating over the site of corrosion activity, thus "inhibiting" corrosion.

8.6E If a utility fails to meet the lead action level, the program for educating the public includes mail-out notices in the utility bills, announcements in major local daily and weekly newspapers, distribution of pamphlets to schools, health care, and day care facilities, and submittal of public service announcements to at least five of the largest radio and television stations in the area.

CHAPTER 9

TASTE AND ODOR CONTROL

by

Russ Bowen

Revised by

Jeanne Ballestero

TABLE OF CONTENTS

Chapter 9.   TASTE AND ODOR CONTROL

OBJECTIVES

Chapter 9.   TASTE AND ODOR CONTROL

Following completion of Chapter 9, you should be able to:

1. Explain the importance of taste and odor control,

2. Identify causes of tastes and odors,

3. Locate sources of tastes and odors,

4. Prevent development of tastes and odors,

5. Treat or eliminate tastes and odors,

6. Develop a taste and odor monitoring program, and

7. Develop a taste and odor control strategy.

WORDS

Chapter 9.   TASTE AND ODOR CONTROL

ABSORPTION (ab-SORP-shun) ABSORPTION

The taking in or soaking up of one substance into the body of another by molecular or chemical action (as tree roots absorb dissolved nutrients in the soil).

ADSORBATE (add-SORE-bait) ADSORBATE

The material being removed by the adsorption process.

ADSORBENT (add-SORE-bent) ADSORBENT

The material (activated carbon) that is responsible for removing the undesirable substance in the adsorption process.

ADSORPTION (add-SORP-shun) ADSORPTION

The gathering of a gas, liquid, or dissolved substance on the surface or interface zone of another material.

AERATION (air-A-shun) AERATION

The process of adding air to water. Air can be added to water by either passing air through water or passing water through air.

AEROBIC (AIR-O-bick) AEROBIC

A condition in which atmospheric or dissolved molecular oxygen is present in the aquatic (water) environment.

AIR STRIPPING AIR STRIPPING

A treatment process used to remove dissolved gases and volatile substances from water. Large volumes of air are bubbled through the water being treated to remove (strip out) the dissolved gases and volatile substances.

ALGAE (AL-gee) ALGAE

Microscopic plants which contain chlorophyll and live floating or suspended in water. They also may be attached to structures, rocks or other submerged surfaces. Excess algal growths can impart tastes and odors to potable water. Algae produce oxygen during sunlight hours and use oxygen during the night hours. Their biological activities appreciably affect the pH, alkalinity, and dissolved oxygen of the water.

AMBIENT (AM-bee-ent) TEMPERATURE AMBIENT TEMPERATURE

Temperature of the surrounding air (or other medium). For example, temperature of the room where a gas chlorinator is installed.

ANAEROBIC (AN-air-O-bick) ANAEROBIC

A condition in which atmospheric or dissolved molecular oxygen is *NOT* present in the aquatic (water) environment.

BACTERIA (back-TEAR-e-ah) BACTERIA

Bacteria are living organisms, microscopic in size, which usually consist of a single cell. Most bacteria use organic matter for their food and produce waste products as a result of their life processes.

BIOLOGICAL GROWTH BIOLOGICAL GROWTH

The activity and growth of any and all living organisms.

CARCINOGEN (CAR-sin-o-JEN) CARCINOGEN

Any substance which tends to produce cancer in an organism.

CHLOROPHENOLIC (klor-o-FEE-NO-lick) CHLOROPHENOLIC

Chlorophenolic compounds are phenolic compounds (carbolic acid) combined with chlorine.

CHLORORGANIC (klor-or-GAN-ick)   CHLORORGANIC

Organic compounds combined with chlorine. These compounds generally originate from, or are associated with, life processes such as those of algae in water.

CROSS CONNECTION   CROSS CONNECTION

A connection between a drinking (potable) water system and an unapproved water supply. For example, if you have a pump moving nonpotable water and hook into the drinking water system to supply water for the pump seal, a cross connection or mixing between the two water systems can occur. This mixing may lead to contamination of the drinking water.

DECANT WATER   DECANT WATER

Water that has separated from sludge and is removed from the layer of water above the sludge.

DECHLORINATION (dee-KLOR-uh-NAY-shun)   DECHLORINATION

The deliberate removal of chlorine from water. The partial or complete reduction of residual chlorine by any chemical or physical process.

DECOMPOSITION, DECAY   DECOMPOSITION, DECAY

The conversion of chemically unstable materials to more stable forms by chemical or biological action. If organic matter decays when there is no oxygen present (anaerobic conditions or putrefaction), undesirable tastes and odors are produced. Decay of organic matter when oxygen is present (aerobic conditions) tends to produce much less objectionable tastes and odors.

DEGASIFICATION (DEE-GAS-if-uh-KAY-shun)   DEGASIFICATION

A water treatment process which removes dissolved gases from the water. The gases may be removed by either mechanical or chemical treatment methods or a combination of both.

DIATOMS (DYE-uh-toms)   DIATOMS

Unicellular (single cell), microscopic algae with a rigid (box-like) internal structure consisting mainly of silica.

ENZYMES (EN-zimes)   ENZYMES

Organic substances (produced by living organisms) which cause or speed up chemical reactions. Organic catalysts and/or bio-chemical catalysts.

EUTROPHICATION (you-TRO-fi-KAY-shun)   EUTROPHICATION

The increase in the nutrient levels of a lake or other body of water; this usually causes an increase in the growth of aquatic animal and plant life.

FLAGELLATES (FLAJ-el-LATES)   FLAGELLATES

Microorganisms that move by the action of tail-like projections.

FUNGI (FUN-ji)   FUNGI

Mushrooms, molds, mildews, rusts, and smuts that are small non-chlorophyll-bearing plants lacking roots, stems and leaves. They occur in natural waters and grow best in the absence of light. Their decomposition may cause objectionable tastes and odors in water.

HYDROPHILIC (HI-dro-FILL-ick)   HYDROPHILIC

Having a strong affinity (liking) for water. The opposite of HYDROPHOBIC.

HYDROPHOBIC (HI-dro-FOE-bick)   HYDROPHOBIC

Having a strong aversion (dislike) for water. The opposite of HYDROPHILIC.

IMHOFF CONE   IMHOFF CONE

A clear, cone-shaped container marked with graduations. The cone is used to measure the volume of settleable solids in a specific volume (usually one liter) of water.

INORGANIC   INORGANIC

Material such as sand, salt, iron, calcium salts and other mineral materials. Inorganic substances are of mineral origin, whereas organic substances are usually of animal or plant origin. Also see ORGANIC.

INTERFACE

The common boundary layer between two substances such as water and a solid (metal); or between two fluids such as water and a gas (air); or between a liquid (water) and another liquid (oil).

KJELDAHL (KELL-doll) NITROGEN

Nitrogen in the form of organic proteins or their decomposition product ammonia, as measured by the Kjeldahl Method.

METABOLISM (meh-TAB-uh-LIZ-um)

(1)   The biochemical processes in which food is used and wastes are formed by living organisms.

(2)   All biochemical reactions involved in cell formation and growth.

MICROBIAL (my-KROW-bee-ul) GROWTH

The activity and growth of microorganisms, such as bacteria, algae, diatoms, plankton and fungi.

MICROORGANISMS (MY-crow-OR-gan-IS-zums)

Living organisms that can be seen individually only with the aid of a microscope.

NPDES PERMIT

National **P**ollutant **D**ischarge **E**limination **S**ystem permit is the regulatory agency document issued by either a federal or state agency which is designed to control all discharges of potential pollutants from point sources and storm water runoff into U.S. waterways. NPDES permits regulate discharges into navigable waters from all point sources of pollution, including industries, municipal wastewater treatment plants, sanitary landfills, large agricultural feedlots and return irrigation flows.

NEPHELOMETRIC (NEFF-el-o-MET-rick)

A means of measuring turbidity in a sample by using an instrument called a nephelometer. A nephelometer passes light through a sample and the amount of light deflected (usually at a 90-degree angle) is then measured.

OLFACTORY (ol-FAK-tore-ee) FATIGUE

A condition in which a person's nose, after exposure to certain odors, is no longer able to detect the odor.

ORGANIC

Substances that come from animal or plant sources. Organic substances always contain carbon. (Inorganic materials are chemical substances of mineral origin.) Also see INORGANIC.

OXIDATION (ox-uh-DAY-shun)

Oxidation is the addition of oxygen, removal of hydrogen, or the removal of electrons from an element or compound. In the environment, organic matter is oxidized to more stable substances. The opposite of REDUCTION.

OZONATION (O-zoe-NAY-shun)

The application of ozone to water for disinfection or for taste and odor control.

PERCOLATION (PURR-co-LAY-shun)

The slow passage of water through a filter medium; or, the gradual penetration of soil and rocks by water.

PHENOLIC (fee-NO-lick) COMPOUNDS

Organic compounds that are derivatives of benzene.

PHOTOSYNTHESIS (foe-toe-SIN-thuh-sis)

A process in which organisms, with the aid of chlorophyll (green plant enzyme), convert carbon dioxide and inorganic substances into oxygen and additional plant material, using sunlight for energy. All green plants grow by this process.

PLANKTON

(1)   Small, usually microscopic, plants (phytoplankton) and animals (zooplankton) in aquatic systems.

(2)   All of the smaller floating, suspended or self-propelled organisms in a body of water.

PUTREFACTION (PEW-truh-FACK-shun)

Biological decomposition of organic matter, with the production of foul-smelling and -tasting products, associated with anaerobic (no oxygen present) conditions.

REDUCTION (re-DUCK-shun)

Reduction is the addition of hydrogen, removal of oxygen, or the addition of electrons to an element or compound. Under anaerobic conditions (no dissolved oxygen present), sulfur compounds are reduced to odor-producing hydrogen sulfide (H_2S) and other compounds. The opposite of OXIDATION.

RESPIRATION

The process in which an organism uses oxygen for its life processes and gives off carbon dioxide.

SATURATION

The condition of a liquid (water) when it has taken into solution the maximum possible quantity of a given substance at a given temperature and pressure.

SEPTIC (SEP-tick)

A condition produced by bacteria when all oxygen supplies are depleted. If severe, the bottom deposits produce hydrogen sulfide, the deposits and water turn black, give off foul odors, and the water has a greatly increased chlorine demand.

SLURRY (SLUR-e)

A watery mixture or suspension of insoluble (not dissolved) matter; a thin, watery mud or any substance resembling it (such as a grit slurry or a lime slurry).

SUPERCHLORINATION (SUE-per-KLOR-uh-NAY-shun)

Chlorination with doses that are deliberately selected to produce free or combined residuals so large as to require dechlorination.

SUPERNATANT (sue-per-NAY-tent)

Liquid removed from settled sludge. Supernatant commonly refers to the liquid between the sludge on the bottom and the scum on the water surface of a basin or container.

SUPERSATURATED

An unstable condition of a solution (water) in which the solution contains a substance at a concentration greater than the saturation concentration for the substance.

THRESHOLD ODOR NUMBER (TON)

The greatest dilution of a sample with odor-free water that still yields a just-detectable odor.

TOXIC (TOX-ick)

A substance which is poisonous to a living organism.

VOLATILE (VOL-uh-tull)

(1) A volatile substance is one that is capable of being evaporated or changed to a vapor at relatively low temperatures. Volatile substances also can be partially removed by air stripping.

(2) In terms of solids analysis, volatile refers to materials lost (including most organic matter) upon ignition in a muffle furnace for 60 minutes at 550°C. Natural volatile materials are chemical substances usually of animal or plant origin. Manufactured or synthetic volatile materials such as ether, acetone, and carbon tetrachloride are highly volatile and not of plant or animal origin.

REDUCTION

RESPIRATION

SATURATION

SEPTIC

SLURRY

SUPERCHLORINATION

SUPERNATANT

SUPERSATURATED

THRESHOLD ODOR NUMBER (TON)

TOXIC

VOLATILE

CHAPTER 9. TASTE AND ODOR CONTROL

(Lesson 1 of 3 Lessons)

9.0 IMPORTANCE OF TASTE AND ODOR CONTROL

Tastes and odors in drinking water are among the most common and difficult problems that confront waterworks operators. In nationwide surveys, 20 percent of the people served by municipal water systems rate their water as having an objectionable taste and/or odor. In a 1989 American Water Works Association (AWWA) survey, chlorinous and earthy were found to be the primary objectionable odors. Fishy and medicinal odors were also reported by over 25 percent of the respondents to this survey. Of water utilities surveyed, 80 percent responded that treatment for tastes and odors was necessary at least occasionally. Taste and odor problems may occur locally on a persistent, seasonal, occasional, or infrequent basis. Regardless of the frequency or type of taste and odor problem that a utility may face, these survey results indicate that taste and odor problems are widespread. Most water treatment plant operators will have to deal with a taste and odor problem at some time during their career.

Taste and odor, along with "colored water" complaints, are the most common types of water quality complaints received by a water utility. This is because the average consumer uses three senses to evaluate water: sight, smell, and taste. If the water looks dirty or colored, smells bad, or has an objectionable taste, the consumer will rate it as poor-quality water. This is true whether or not any health-related problem exists.

Taste and odor (T&O) problems are likely to have significant effects on a water utility when they occur. First, numerous complaints must be handled. This requires a great deal of staff time, creates a sense of frustration for both the consumer and for office personnel, and may well require the time of operators who could better be used to solve the problem rather than deal with the effects of the problem.

Water that has an objectionable taste or odor is not desirable to the general public. This may cause a number of consumers to begin purchasing bottled water for drinking purposes. Other consumers may switch to alternative water supplies, such as old, poorly maintained private wells which may not be as safe as the public supply. People could unknowingly use water that has a more pleasing taste or smell than the public water supply, not realizing that this water may also be hazardous to their health.

Perhaps the most damaging effect of a taste and odor episode is the loss of public confidence in the water utility's ability to provide a safe, high-quality water. Serious loss of confidence may later result in funding restrictions and increased public relations problems for the utility.

Because of the widespread occurrence of taste and odor problems in public drinking water supplies and the serious effects of these types of problems to the utility, taste and odor control is an area in which training of all water treatment operators is vitally necessary.

The secret to successful taste and odor control is to *PREVENT TASTES AND ODORS FROM EVER DEVELOPING*. This means control of algae and other microorganisms in source waters and treatment plants through preventive treatment. You must treat for tastes and odors *BEFORE* the problem occurs.

QUESTIONS

Write your answers in a notebook and then compare your answers with those on page 443.

9.0A How frequently may taste and odor problems occur at a water treatment plant?

9.0B What are the most common types of water quality complaints received by a water utility?

9.0C What is the most damaging effect of taste and odor problems for a water utility?

9.1 CAUSES OF TASTES AND ODORS

9.10 Specific Taste and Odor Compounds

Considerable headway has been made in the identification of the taste and odor compounds that routinely affect the aesthetic quality of treated water. The following are the most common taste and odor compounds:

● Geosmin—Geosmin is a natural chemical by-product of various species of actinomycetes and blue-green algae. *Oscillatoria*, *Aphanizomenon*, and *Phormidium* have been identified as potent producers of geosmin. This compound is associated with earthy odors in the water.

● 2-Methylisoborneol (MIB)—MIB is a natural chemical compound produced by various species of actinomycetes and blue-green algae. Species of *Oscillatoria*, *Synechococcus*, and *Phormidium* have been identified as producers of MIB. This compound imparts a musty odor to the water.

● Chlorine—Chlorine is the most common disinfectant used in the water industry. It is also a common source of T&O complaints that range from "bleach" to "chlorinous" and "medicinal."

● Chloramines—When ammonia is added to the treatment process, the formation of odoriferous inorganic chloramines is a potential problem. The three main chloramine compounds that can be formed prior to breakpoint chlorination are monochloramine, dichloramine, and trichloramine. Monochloramines rarely cause taste and odor problems unless levels exceed 5 mg/L. However, dichloramine, which has a "swimming pool" or "bleach" odor, has a lower detection level and becomes a problem at levels of 0.9 to 1.3 mg/L. Trichloramine produces a geranium odor at levels of 0.02 mg/L or above.

- Aldehydes—The most common odor problem associated with aldehydes is the fruity odor that develops in ozonated waters. Aldehydes are formed as a result of the oxidation of amino acids and nitriles during the treatment process.

- Phenols and Chlorophenols—Phenols in the water supply react with chlorine to form chlorophenols. Pharmaceutical and medicinal tastes and odors are most often associated with chlorophenols in drinking water.

9.11   Types of Causes

Taste and odor problems may arise from diverse causes and may be the result of a combination of factors. Nonetheless, a general understanding of the conditions that can contribute to taste and odor problems is useful in trying to prevent and treat water of objectionable quality. According to a 1989 AWWA survey, the major causes of taste and odor problems are algal blooms in source waters, disinfectants used, and water distribution system. Determining the cause of a taste and odor episode may be extremely difficult and, in many cases, no definite answers are ever found.

Tastes and odors can be the result of natural conditions or human activities anywhere within the total water supply system. Raw water sources, conveyance facilities, treatment plants, chlorination stations, finished storage reservoirs, distribution systems, and consumer plumbing have all been identified as potential sources of tastes and odors. Each water system must be evaluated individually since each will have unique characteristics that may significantly affect water quality.

QUESTIONS

Write your answers in a notebook and then compare your answers with those on page 443.

9.1A   What two compounds can produce earthy and musty odors in the water?

9.1B   Which inorganic chloramine causes the most odor problems in treated water?

9.1C   List the major causes of tastes and odors in a water system.

9.1D   Where could taste and odor conditions develop in a water supply system?

9.12   Natural Causes

9.120   Biological Growth[1] in Source Waters

Microscopic organisms that grow in water or in the sediments of lakes, reservoirs, and rivers are major contributors to the tastes and odors experienced by water utilities. Various types of algae, actinomycetes, and bacteria have all been reported as the cause of these problems at waterworks across the nation.

Actinomycetes are a group of filamentous bacteria that grow in sediments, water, and aquatic plant life. They have been identified as one of the sources of earthy-musty tastes and odors in both source waters and distribution systems. Geosmin, an earthy-smelling odor compound, and 2-methylisoborneol (MIB), a musty-smelling odor compound, were first isolated from actinomycetes cultures. Correlation between actinomycetes and taste and odor (T&O) events may be difficult

to prove because actinomycetes are not detected with routine microbiological water quality tests.

Numerous types of algae are known to produce taste and odor compounds. The major T&O-producing algal groups include: blue-green algae (*Cyanophyta*), yellow-green algae (*Chrysophyceae*), diatoms (*Bacillariophyceae*) and dinoflagellates (*Pyrrophyta*). The most common odor producers belong to the blue-green algae class. *Anabeana*, *Aphanizomenon*, *Oscillatoria*, and *Microcystis* are the most notorious blue-green odor producers. The odors produced by these organisms range from earthy-musty to septic and are dependent on species, density, and physiological state of the population. Blue-green algae usually develop into "blooms" which float on the top of the water and are visually noticeable. The yellow-green algae, most notably *Dinobryon*, *Mallomonas*, and *Synura*, produce compounds with odors ranging from cucumbery to fishy. These algae do not need to be present in large numbers and are often overlooked as the source of T&O events. Diatoms, in large quantities, can also produce noticeable odors. *Asterionella*, *Cyclotella*, and *Tabellaria* are known to produce fishy odors while *Melosira* and *Fragellaria* produce musty odors. The most common dinoflagellates are *Ceratium* and *Peridinium* which can produce rotten, septic, or fishy odors, respectively, when found in large quantities.

Although planktonic algae are usually associated with T&O events, attached algae (periphyton) can also be a significant source of taste and odor compounds. Attached beds of *Oscillatoria* have been identified in several California lakes as the source of both geosmin and MIB.

Other microorganisms can also produce tastes and odors. Most notably, sulfate-reducing bacteria, a group of anaerobic bacteria, reduces sulfate to hydrogen sulfide which produces the "rotten egg" odor associated with highly polluted waters. These bacteria can also create T&O problems in thermally stratified lakes and reservoirs when the bottom layer becomes anaerobic and sulfate in the sediments and decaying organic matter is reduced by these bacteria to hydrogen sulfide.

Microbial populations can contribute to unpleasant tastes and odors in water in two general ways. As microorganisms grow and multiply, they produce metabolic by-products. These by-products are released into the water and some may lead to the deterioration of taste and odor quality. The concentration of compounds produced by microorganisms is generally very low and may be measured in nanograms per liter (parts per trillion parts). However, even such extremely low levels of these materials may result in taste and odor complaints.

Cellular material of common aquatic microorganisms present in a drinking water supply can also be responsible for taste and odor complaints. As the microorganisms grow, organic matter accumulates within the cells. As long as the organisms are healthy, these cellular components are retained and usually do not affect the taste or odor of a water. When the population begins to die off, either as a result of natural processes or treatment, the cells rupture and the cellular materials are released into the water. This is one reason why, in some cases, water has better taste and odor qualities before being treated than after treatment.

This phenomenon (die-off causing tastes and odors) must be considered in developing plans for copper sulfate treatment of reservoirs. Even relatively high *PLANKTON*[2] counts do not

---

[1] *Biological Growth.   The activity and growth of any and all living organisms. Microbial growth is the same as biological growth.*
[2] *Plankton.   (1) Small, usually microscopic, plants (phytoplankton) and animals (zooplankton) in aquatic systems. (2) All of the smaller floating, suspended or self-propelled organisms in a body of water.*

necessarily indicate that treatment is either required or that it will be beneficial. A large total plankton count may be the result of low numbers of many different types of organisms. This can indicate a well-balanced, diversified plankton community that does not pose a threat to taste and odor quality. Treating with copper sulfate may disrupt this natural balance, allowing an objectionable organism to gain dominance, and may create a taste and odor problem by causing the release of cellular components from killed organisms that otherwise would not have had a detrimental effect on water quality.

Microbial decomposition of organic matter in a water supply may also create offensive tastes and odors through a combination of both metabolic by-product formation and release of cellular materials. Following an algal bloom, significant natural die-off of the predominant organism may result in the release of cellular materials with objectionable qualities. The resulting tastes and odors can be further worsened by the growth of other microorganisms feeding on the dying algal mass. This secondary growth may lead to the production of obnoxious metabolic by-products.

While *DECOMPOSITION*[3] is a critical step in the continuous cycling of nutrients through nature, it can have serious consequences on water quality. For this reason, raw water treatment programs are only useful if operated to prevent massive algal blooms. If treatments are begun only after large populations have developed, the effect may be to accelerate the decomposition process and worsen a taste and odor outbreak. Frequent monitoring of both the type and density of plankton populations in source waters will provide the early warning needed for preventive treatment such as the application of algicide. The application of an algicide to source water should be considered when: (1) the dominant algae is identified as a taste and odor producer, (2) algae is not a dominant organism in the population but is known for its ability to produce potent T&O compounds (for example, *Synura*), and (3) sensory or chemical analysis indicates an increase in the odor of the water.

QUESTIONS

Write your answers in a notebook and then compare your answers with those on page 443.

9.1E   Where and when can an operator expect biological growths to occur which will lead to objectionable tastes and odors?

9.1F   What are the two general ways in which microbial populations can contribute to unpleasant tastes and odors in water?

9.1G   Why does treatment that takes place after a large microbial population growth has occurred sometimes cause a taste and odor problem to get worse?

9.1H   Name two groups of bacteria that are known to produce T&O compounds in water?

9.1I   What common blue-green algae are associated with earthy-musty odors?

9.1J   When should the application of an algicide be considered?

9.121   Environmental Conditions

The effects of lake and reservoir stratification on water quality were discussed in Chapter 3. The depletion of oxygen in the bottom layers of reservoirs provides suitable conditions for the growth of microorganisms capable of producing compounds, such as hydrogen sulfide, which are very objectionable to consumers. The information in Chapter 3 should be reviewed as it relates to the causes of taste and odor problems in drinking water supplies drawn from large reservoirs.

The inflow of many types of pollutants, particularly of organic matter and compounds containing nitrogen and phosphorus, may indirectly cause oxygen-poor conditions in reservoirs, ponds, rivers, and canals. Increased nutrient levels (called *EUTROPHICATION*[4]) may result from either natural conditions or from human activity within the watershed.

Pollutants themselves normally do not cause the oxygen concentration in the water to decrease. Microorganisms capable of growing on the organic materials (pollutants) are responsible for depletion of the oxygen following nutrient enrichment from runoff. Microbial populations are relatively low in unpolluted waters, and the rate of oxygen transfer from the atmosphere to the water is sufficient to prevent oxygen-poor conditions from developing. When available nutrient concentrations increase, rapid microbial growth consumes dissolved oxygen at a rate faster than it can be replaced from the air. Oxygen-poor conditions in water following increased nutrient loading are the result of this rapid microbial growth.

Blooms of *PHOTOSYNTHETIC*[5] algae resulting from increased nutrient concentrations, suitable water temperatures, and favorable sunlight can cause both oxygen depletion and oxygen supersaturation in water during a 24-hour period. The process of photosynthesis occurs during daylight hours and results in oxygen being released into the water. Large populations of algae can produce oxygen faster than it can escape to the atmosphere, leading to afternoon dissolved oxygen levels higher than would normally occur in the absence of major algal activity. When dissolved oxygen levels in water exceed normal saturation levels the condition is known as supersaturation.

At night, algal photosynthesis stops and respiration begins. Respiration is the metabolic process that consumes oxygen and releases carbon dioxide. In the dark hours, the algae use the available oxygen at a rate faster than it can be replenished from the air. By the early morning hours almost all of the dissolved oxygen may have been consumed. This pattern of supersaturation and depletion is known as the diurnal dissolved oxygen cycle and is shown in Figure 9.1. (Notice that the graph in Figure 9.1 shows the *PERCENT* of oxygen saturation during the 24-hour period. This is because oxygen saturation

---

[3] *Decomposition, Decay.   The conversion of chemically unstable materials to more stable forms by chemical or biological action. If organic matter decays when there is no oxygen present (anaerobic conditions or putrefaction), undesirable tastes and odors are produced. Decay of organic matter when oxygen is present (aerobic conditions) tends to produce much less objectionable tastes and odors.*

[4] *Eutrophication (you-TRO-fi-KAY-shun).   The increase in the nutrient levels of a lake or other body of water; this usually causes an increase in the growth of aquatic animal and plant life.*

[5] *Photosynthesis (foe-toe-SIN-thuh-sis).   A process in which organisms, with the aid of chlorophyll (green plant enzyme), convert carbon dioxide and inorganic substances into oxygen and additional plant material, using sunlight for energy. All green plants grow by this process.*

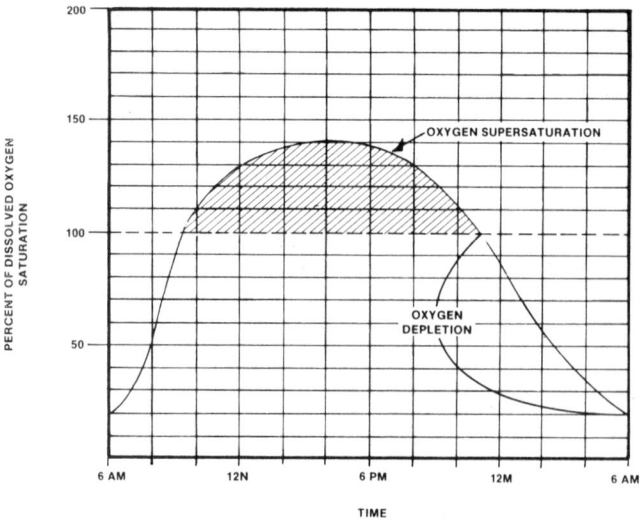

*Fig. 9.1   Diurnal variation in dissolved
oxygen concentrations*

changes with both temperature and elevation. At sea level, for example, oxygen saturation at 10°C is approximately 11 mg/*L* while at 20°C it is only about 9 mg/*L*.)

The importance of diurnal oxygen fluctuations in water is that significantly different conditions can and do exist from day to night. These differences may have a major effect on taste and odor quality. Oxygen depletion during the night may result in fish kills and die-off of aquatic organisms and vegetation which will produce foul tastes and odors in water. Oxygen-poor conditions during darkness may also allow anaerobic organisms to become established and contribute to a general degradation of the aesthetic qualities of the water.

From an operations viewpoint, significant dissolved oxygen fluctuations caused by algae in raw water will also be accompanied by changes in the pH. When algae produce oxygen, carbon dioxide (CO_2) is removed from the water and the pH will increase during the daylight hours. At night during the respiration process, algae will consume oxygen and release carbon dioxide which will lower the pH. These changes in pH caused by increases and decreases of carbon dioxide will influence the chemical doses required to effectively treat the water (coagulation-flocculation, disinfection, corrosion control).

A special case of nutrient enrichment can occur at water treatment plants designed to recycle water used in filter backwashing and settling basin sludge removal. Typically, some type of settling basin or lagoon is provided to allow the sludge to settle before the water is decanted off (separated) and returned to the plant influent. These sludge lagoons produce a water supply in which nutrients previously removed by treatment are concentrated. Microbial populations can flourish in these facilities, and unless careful management is practiced, severe taste and odor problems can originate from sludge settling lagoons. This is especially true when the *DECANT WATER*[6] passes through the plant as a high percentage of total plant flow over extended periods. Frequent treatment with copper sulfate and rotation of sludge lagoons where multiple units are available, or removal of sludge to separate drying facilities are appropriate methods for controlling the taste and odor problems associated with these water recovery systems.

Natural runoff may also lead to tastes and odors by substantially increasing flow velocities in rivers and canals. Sediments deposited on the bottom of channels during low-flow periods may be resuspended by scouring action and transported into the treatment plant. When present in the plant influent, bottom sediments may impart septic, musty, or earthy tastes and odors to the drinking water supply. Problems associated with high-flow conditions may subside as runoff decreases, or may persist for several weeks until natural settling redeposits the suspended material on the channel bottom.

QUESTIONS

Write your answers in a notebook and then compare your answers with those on page 443.

9.1K   What types of materials in runoff waters can lead to oxygen depletion?

9.1L   How can the nutrient levels in a water supply be increased?

9.1M   What causes the diurnal dissolved oxygen cycle?

9.13   Human Causes of Tastes and Odors

9.130   Types of Sources

Many of our activities in the environment lead to objectionable tastes and odors in public water systems. Increased awareness of environmental degradation has led to many new pollution control regulations at both the state and federal levels. These regulations have significantly improved conditions in many of the nation's waterways, but discharges upstream of municipal intake facilities or into raw water storage reservoirs continue to be contributing factors in many taste and odor experiences.

Types of pollution that may enter a municipal water supply upstream of the water treatment plant and result in water quality degradation include inadequately treated municipal wastewaters, domestic wastes from individual homes, industrial discharges, urban runoff, chemical spills, agricultural wastes (manure), and irrigation runoff. Tastes and odors may be created by these discharges directly, or may develop because of microbial activities associated with the pollution.

9.131   Municipal Wastewaters

Inadequately treated municipal wastewaters may enter a water supply due to a process failure at the wastewater treatment plant, mechanical breakdowns, or overloading of the facility. Some older municipal wastewater treatment plants discharge inadequately treated effluent during periods of heavy precipitation because of high flows (inflow and infiltration or I & I) from combined sanitary sewer and storm drain systems. Other wastewater works simply do not have the extensive facilities required to maintain the effluent quality necessary to prevent degradation of the receiving water.

Individual wastewater disposal systems in rural areas of the watershed also contribute pollutants to the water supply. Septic tanks and leach fields may be located too close to rivers for adequate protection of the water supply. Improper siting of individual systems in soils with inadequate *PERCOLATION*[7] rates can create a situation in which nearly untreated wastewater reaches a municipal source. Poor maintenance and substandard installation are also problems of septic systems that may

---

[6] *Decant Water.   Water that has separated from sludge and is removed from the layer of water above the sludge.*
[7] *Percolation (PURR-co-LAY-shun).   The slow passage of water through a filter medium; or, the gradual penetration of soil and rocks by water.*

lead to tastes and odors in community water supplies. Contamination of both surface and groundwater is often associated with individual wastewater disposal system deficiencies in relatively high-density resort and vacation developments.

Municipal and individual wastewater discharges can contribute to tastes and odors in two ways: (1) directly adding odoriferous compounds, such as phenols and aromatic hydrocarbons, to the water and (2) adding nutrients that result in T&O-causing algal blooms. Wastewater discharges have been identified as one of the major sources of phosphorus in surface waters. It is also a significant source of nitrogen. Both nutrients contribute to the eutrophication of a water body and subsequent taste and odor episodes produced by algae blooms.

9.132 Industrial Wastes

Industrial discharges sometimes present significant taste and odor problems for downstream municipal water suppliers. Despite implementation of the National Pollutant Discharge Elimination System (*NPDES PERMIT*[8]), 100 percent control of industrial discharges has not been achieved, nor is it likely to be achieved in the near future. While each individual discharger is required to limit concentrations of chemicals in the effluent according to levels specified by the NPDES permit, water treatment plants downstream of heavily industrialized areas may encounter problems arising from the total effects of all facilities that discharge into the supply. Furthermore, industrial wastewater treatment works are just as likely to experience process and mechanical failures as municipal treatment works are.

9.133 Chemical Spills

Chemical spills into municipal raw water sources can have detrimental effects on taste and odor quality, but the primary concern of water utility operators in such cases must be for the health-related effects due to the toxicity of spilled chemicals. Most spills are the result of accidents at industrial plants, chemical storage facilities, or during transportation. Because chemical spills are unpredictable events, which can lead to a large amount of contamination reaching the water treatment plant in a short time, every water utility should have an Emergency Response Plan[9] to deal with this problem.

QUESTIONS

Write your answers in a notebook and then compare your answers with those on page 443.

9.1N What factors can cause contamination of a water supply by septic tank and leach field systems in rural areas and in resort and vacation developments?

9.1O How can industrial waste discharges cause taste and odor problems?

9.1P What are the sources of most chemical spills?

9.134 Urban Runoff

Urban runoff contributes to tastes and odors, especially in areas where precipitation occurs only during a limited portion of the year. During dry periods, oil, grease, gasoline, and other residues accumulate on paved surfaces. When storms begin, this material is washed into the local receiving water from roadway storm drainage systems. Urban runoff also contains animal droppings from pets and fertilizers used for landscaping. Taste and odor complaints may be received from systems served by water taken downstream of this urban runoff. Usually the complaints slow down after the storm passes, and subsequent storms during the same wet season will typically not create the same degree of problem as the first storm. Nitrate concentrations in runoff can indirectly lead to tastes and odors by increasing nutrient levels in the receiving water bodies, which in turn leads to T&O-producing algal blooms.

9.135 Agricultural Wastes

The contribution to municipal drinking water taste and odor problems by agricultural runoff depends on the nature and extent of farming in the watershed, precipitation patterns, and local irrigation practices. Many municipal water intake facilities are located upstream of major agricultural areas to avoid possible contamination of the water supply by fertilizers, microbial contaminants (*Cryptosporidium* and *Giardia*), pesticides, and herbicides. This siting of intake facilities above heavily cultivated agricultural lands also helps protect against tastes and odors.

Grazing lands are not usually a major source of tastes and odors in municipal supplies. Significant amounts of waste material that could reach raw water sources cannot accumulate in areas where precipitation occurs regularly throughout the year. Those regions that experience limited seasons of precipitation cannot support high herd densities on grazing lands, so the annual quantity of waste material generated is less per acre. Furthermore, long dry periods allow for the drying out of animal wastes, rendering them far less offensive. When storms do occur, the dried material is generally diluted by the heavy precipitation and wet season river flows.

High-density animal feeding and dairy operations can cause problems if located near surface supplies. The high concentration of animal wastes in a confined area can contribute to significant nutrient loading if runoff is allowed to drain into reservoirs or rivers during a storm. Feedlots and dairies are required to control their discharges under the NPDES program, and utilities should work with the health department and water pollution control agency to prevent serious water quality degradation due to runoff from such operations. Water systems faced with concentrated animal waste runoff into the water source need to practice careful reservoir management and river monitoring programs in order to prevent massive taste and odor complaints.

---

[8] *NPDES Permit.* **N**ational **P**ollutant **D**ischarge **E**limination **S**ystem permit is the regulatory agency document issued by either a federal or state agency which is designed to control all discharges of potential pollutants from point sources and storm water runoff into U.S. waterways. NPDES permits regulate discharges into navigable waters from all point sources of pollution, including industries, municipal wastewater treatment plants, sanitary landfills, large agricultural feedlots and return irrigation flows.

[9] *See Volume II, Chapter 23, Section 23.10, "Emergency Response," for detailed information about planning for emergencies and handling the threat of contaminated water supplies.*

Runoff from cultivated fields can contribute both nutrients and objectionable materials to water supplies. Modern, high-intensity farming requires the use of a wide variety of chemicals to achieve maximum crop production. Precipitation or irrigation in excess of the water-holding capacity of the field will lead to runoff that may contain residues of previously applied fertilizers, pesticides, herbicides, and the spreading agent used to apply them.

Often, it will take days for irrigation water to return to the stream or canal. During this time microbial activity may create very high concentrations of objectionable by-products. Even if these return flows represent only a small portion of the total supply, the presence of these microbial by-products in the finished water can lead to consumer complaints of tastes and odors.

9.136   Treatment Plant and Distribution System Housekeeping

Inadequate or incomplete maintenance of water treatment plants and distribution systems will result in water quality deterioration no matter how clean the raw water supply may be. Debris and sediments transported to the plant accumulate during the year in areas such as influent conduits and flocculator basins that are not equipped with sludge removal systems. Sludge removal from settling basins with pumping equipment is never 100 percent complete, and deposits will build up over a period of time. Good housekeeping in and around water treatment plants is required in order to keep the plant in a clean and sanitary condition.

Microorganisms will grow in plant debris and sludges even in the presence of a strong chlorine residual. The conditions that lead to foul, septic, musty, or other types of tastes and odors in raw water supplies may be duplicated on a smaller scale in treatment plants that are not kept clean. In treatment plants with little or no oxidant at the headworks, such as in plants using biological activated carbon filters (BAC), algal growth on the sedimentation basin and filter walls and weirs can cause severe odor problems within the plant. Filamentous algae, such as *Oscillatoria* and *Phormidium*, can colonize these areas and release both geosmin and MIB into the waters. When odors are detected in the treated waters and not in the source waters, a sensory profile of the plant will usually determine the location of the colonized area. Periodic inspection of plant facilities will also help to identify colonized areas within the plant and is a vital part of good water treatment practice and a necessary part of an effective taste and odor prevention program.

Distribution system maintenance is also an important part of taste and odor prevention. Debris, which accumulates in distribution system mains and laterals, provides an environment for bacterial growth. Especially susceptible are low-flow zones and dead ends in which no chlorine residual is maintained. These areas allow for abundant bacterial regrowth in distribution lines which results in stagnant, septic, or foul tastes and odors. Comprehensive flushing programs should be used by utilities as part of a system water quality maintenance effort.

9.137   Household Plumbing

Sometimes a taste and odor problem is traceable directly to the consumer's plumbing system. The age and types of plumbing materials in older homes may contribute to unpleasant-tasting water. The plastic household plumbing in new housing subdivisions may require several days, or longer, before the "plastic" taste of the water disappears. Low flows in some homes, or inadequate flushing of lines and cleaning of strainers and aerators may also contribute to water quality degradation in the consumer's plumbing. However, it is poor practice to attribute widespread complaints about the taste and odor of the water to conditions in consumers' plumbing. If a large num-ber of complaints are received from throughout the system, the chances are very good the problem is with the water supply and not the result of a large number of individual problems.

QUESTIONS

Write your answers in a notebook and then compare your answers with those on pages 443 and 444.

9.1Q   What parts of the plant are likely locations for algal colonization?

9.1R   Why are many municipal water intake facilities located upstream of major agricultural areas?

9.1S   List some sources of agricultural wastes that may cause taste and odor problems in a water supply.

9.1T   How do debris and sludge cause tastes and odors?

9.2   LOCATING TASTE AND ODOR SOURCES

9.20   Potential Sources

A wide variety of conditions that can occur in any portion of a water supply system may cause objectionable tastes and odors at the consumer's tap. When evaluating a taste and odor complaint, no segment of the system, from the raw water supply to the consumer's household plumbing, should be ignored as a potential source of the problem.

The first step in determining the cause of tastes and odors should be to locate where in the overall system the problem is originating. Once the point of origin is known, it is usually easier to determine an underlying cause and to develop plans for correcting the situation. Locating a taste and odor source is often a time-consuming process of elimination that may not yield any conclusive information. Nonetheless, the benefits of successfully identifying sources and causes of taste and odor problems are well worth the effort required.

9.21   Raw Water Sources

The most commonly reported problem faced by water facilities is the development of tastes and odors in the raw water supply (a lake, reservoir, river, or canal) or in the raw water transmission facilities which deliver water from the source to the treatment plant. Any parts of the system that are used to store, transport, or regulate untreated water may provide a suitable habitat for organisms that produce objectionable tastes and odors in the drinking water due to an absence of chlorine residual.

When investigating a taste and odor problem, mentally divide the system into its major component parts based on each component's primary function such as storage, open conveyance channel, and transmission pipelines. Also consider each component's accessibility for sampling, time required for sampling, and the number of samples that can be reasonably analyzed in a timely manner. You must choose sampling locations where the water is representative of the water consumers will receive. Collecting and evaluating surface samples in a strongly stratified reservoir is useless if the water being treated is released from the lower layers.

Some examples of sampling points that would allow you to test major components of raw water supplies include: outlet works of major reservoirs and regulating basins, inlets, and outlets of transmission channels and pipelines, and the plant influent upstream of any chemical additions. Analyze samples from these locations for plankton levels and predominant type, turbidity, pH, Flavor Profile Analysis (FPA) (or, if FPA analysis is not available, threshold odor number (TON)[10]), geosmin, and MIB. Major changes in any of these water quality indicators between sample locations may be the result of conditions contributing to the taste and odor problem. Resampling and inspection (if possible) of that portion of the system between sample points showing remarkably different characteristics should be conducted as soon as possible.

Once the origin of the problem is identified, a sanitary survey of the source water and its watershed should be conducted. The purpose of this survey is to locate the biological or industrial sources of the offending taste or odor or the nutrients responsible for a taste- and odor-producing algal bloom. When resampling and inspecting the segment of the system suspected of being the origin of the taste and odor problem, look for new or expanded residential, commercial, or industrial activity, as well as new or altered tributary streams which could contribute poor quality inflow to rivers and canals. Also look for runoff from agricultural areas if the problem occurred after a heavy rain. Examine pipelines for unauthorized or unintentional *CROSS CONNECTIONS*[11] that could provide a route for contamination to enter the supply. Analyses for various nutrients such as total phosphorus, total *KJELDAHL NITROGEN*,[12] ammonia, nitrate, and nitrite may be useful in determining the extent of nutrient inflow into the water body and the potential for future algal blooms. This information is valuable in developing a source water management plan.

9.22 Treatment Plant

Accumulated debris and sludge in treatment plant facilities will lead to taste and odor deterioration as the water is processed. Algal growth due to poor housekeeping practices is both unsightly and a potential source of tastes and odors in the finished water. Routine inspection and cleaning of all facilities are necessary elements of treatment plant operation.

Collecting samples from various points throughout the treatment plant for laboratory analysis is usually not as productive in locating taste and odor sources as it is in the raw water system. Treatment chemicals, especially chlorine and powdered activated carbon, tend to mask any changes in taste and odor quality that may be occurring within the plant. Comparing taste and odor quality of raw and finished water may be useful in indicating a problem of plant origin, but there is a natural tendency to always rate treated water as of better quality than untreated. Additionally, the presence of strong, easily treated odors in the raw water may invalidate any direct comparison with finished water. However, if a Flavor Profile Analysis Panel is available to the plant, a sensory profile of the treatment plant may reveal the source of a plant-related problem. Plant profiles of geosmin and MIB levels can also be helpful with musty, earthy odors.

Visually inspect basin and filter walls, channels, and weirs for algal or slime (bacterial) growth. Plants without an oxidant at the headworks will often develop problems with attached algae on the sedimentation and filter walls and weirs. As noted earlier, *Oscillatoria* and *Phormidium* are two blue-green algae that readily colonize these areas and can produce both geosmin and MIB. Remove any slime material and send samples to the laboratory for identification. Future growths can sometimes be prevented by regularly washing affected areas with a high-pressure hose. Shock treatment with chlorine through the plant or direct application of a strong chlorine solution to points that prove to be especially difficult to keep clean can also help.

Conduct an evaluation of plant facilities, looking for potential zones of debris accumulation in areas such as influent conduits and flocculator and settling basins. Review records of previous plant inspections and cleanings to see if some areas have a history of particularly heavy sludge buildup. Seldom is it practical to dewater a treatment plant during a taste and odor episode, but plans should be made for regular (yearly) dewatering operations to allow inspection and cleaning of suspected or potential problem facilities.

Windows in structures over filter galleries may allow sunlight to encourage algae and slime growths. Some plants have corrected this problem by covering the windows.

9.23 Distribution System

Conditions within a municipal water supply distribution system can significantly affect water quality received by the public. Just as adequate maintenance and housekeeping are required at the treatment plant, they are a necessary part of distribution system operation. The main causes of tastes and odors in the distribution system are: microbiological activity, disinfection residuals and their by-products, organic or mineral compounds from system materials, and external contaminants from cross-connections.

Taste and odor complaints originating within the distribution system are usually confined to limited areas or zones. Dead ends, low-flow zones, and areas subject to wide flow variations or changes in supply source may experience higher than normal numbers of taste and odor complaints. Records of complaints should be reviewed so such areas can be identified and preventive measures, such as more frequent flushing, can be implemented. You should recognize, however, that any change in your source water may result in taste or odor complaints from an area, even in the absence of any actual quality problems.

Cross connections in the distribution system are potentially very hazardous and can be a source of taste and odor complaints. A variety of contaminants have been introduced to drinking water through cross connections, and often the first warning the supplier has of such a condition are the complaints about a "chemical," "gasoline," or "pesticide" taste or odor in the water. Because of the potential public health hazard, complaints of this nature should receive prompt and careful attention.

Backflow prevention devices that have been improperly installed or bypassed at industrial plants have resulted in the contamination of numerous water systems. Cross connections are also made by contractors, landscape workers, and, with disturbing frequency, municipal employees. Careless or unthink-

---

[10] *See Chapter 11, "Laboratory Procedures," for procedures on how to perform these tests.*

[11] *Cross Connection.   A connection between a drinking (potable) water system and an unapproved water supply. For example, if you have a pump moving nonpotable water and hook into the drinking water system to supply water for the pump seal, a cross connection or mixing between the two water systems can occur. This mixing may lead to contamination of the drinking water.*

[12] *Kjeldahl (KELL-doll) Nitrogen.   Nitrogen in the form of organic proteins or their decomposition product ammonia, as measured by the Kjeldahl Method.*

ing homeowners and amateur repair persons have contaminated systems by creating cross connections. In many cases the problem was located because of the number, location, and type of taste and odor complaints received by the water utility.

Distribution-related taste and odor problems can be reduced by making sure that the water exiting the plant is stable and that the disinfectant used does not deteriorate quickly or produce odoriferous disinfection by-products.

QUESTIONS

Write your answers in a notebook and then compare your answers with those on page 444.

9.2A   What are the most likely sources for the development of tastes and odors?

9.2B   What kind of survey is used to identify the sources of tastes and odors in raw water?

9.2C   How do the chemicals used in a water treatment plant interfere with a search for the source of tastes and odors?

9.2D   Where are the potential sources of tastes and odors within a water treatment plant?

9.2E   What are the main causes of tastes and odors in the distribution system?

9.2F   What types of complaints would alert you to the potential of a cross connection?

9.3   PREVENTION OF TASTES AND ODORS

9.30   Need for Prevention and the Development of a Taste and Odor Monitoring Program

AN IMPORTANT ASPECT OF ANY TASTE AND ODOR CONTROL PROGRAM IS PREVENTION. When dealing with a taste and odor problem, or for that matter any water quality problem, an ounce of prevention is worth far more than a pound of treatment. Preventing problems, at least to the extent prevention is feasible, is usually both more economical and more effective than trying to treat for tastes and odors at the plant. Development of an effective Taste and Odor Monitoring Program is a valuable asset, especially to those systems that experience frequent taste and odor episodes. Such a program is important in predicting the onset of taste and odor episodes, linking odors to specific algal populations, and assessing the effectiveness of control measures.

A Taste and Odor Monitoring Program should include, at a minimum, the following components: (1) routine counting and identification of source water algae populations and attached algae in the plant, and (2) sensory analysis, such as Flavor Profile Analysis (FPA), of source and distribution system wa-

ters, and plant profiles when needed. If appropriate, chemical analysis (closed-loop stripping analysis) of the odor-producing compound can also be a valuable tool in the investigation of taste and odor problems. The information developed through such a program can be incorporated into a more extensive source water management program.

Flavor Profile Analysis is a method that uses a trained panel of individuals to identify, describe, and judge the intensities of taste and odor qualities in the water. This is the only method that provides a description of the type of taste or odor. This method is being used in addition to the more widely used Threshold Odor Number and has been found to be a better tool for monitoring and evaluating water treatment processes and distribution waters.

9.31   Raw Water Management

As explained in Chapter 3, the role of raw water management in providing high-quality water to the consumer cannot be ignored. Once a supply has deteriorated, it usually will take a change of seasons before conditions return to a desirable level. If no alternative water supplies are available to a community, deterioration of the raw water source may mean an extended period of poor-quality water, or an extended period of significantly increased treatment costs, or both.

The major techniques and considerations in developing and implementing a raw water management system were described in detail in Chapter 3. If the water supply is a river rather than a reservoir, many of the treatment approaches described for reservoirs will obviously not work. However, monitoring rivers for changing raw water quality and for sources of pollution that may adversely affect a community water supply are both activities that should be conducted as part of a river management strategy.

9.32   Plant Maintenance

Settled sludges and other debris, which may be transported to the plant by the raw water, need to be removed on a regular basis. These materials provide local environments in which organisms grow and multiply. If left for an extended time, these areas will become septic and impart a foul taste and odor to the water. Depending on the type of material, the quality of the water, and the source and nature of debris that accumulates, rotten egg (hydrogen sulfide), bitter, musty, earthy, swampy, fishy, or grassy tastes and odors will result.

Many treatment plants schedule annual shutdowns to allow inspection and thorough cleaning of all facilities, especially those that are normally submerged. Other, generally larger, plants are designed so that one part of the plant may be taken out of service for inspection and cleaning while another is still operating. This routine cleaning is commonly done during the winter months when flows are lower and full plant capacity is not needed to meet the community's water demands.

Another aspect of plant maintenance that relates to the prevention of taste and odor episodes at the consumer's tap is the use of ongoing programs that will ensure that those facilities that may be required to treat a taste and odor problem are functioning properly at all times. No matter how well managed a surface supply is, or how well the treatment plant is maintained in sanitary condition, surface water conditions will change from time to time and will result in water quality deterioration. If facilities such as powdered activated carbon feed systems are not in functioning condition when they are needed, they are of no value. Periodic inspection and testing of such equipment is necessary to ensure that it will work properly when needed and that emergency repairs will not have to be made just at the time when the equipment is needed most.

9.33 Distribution System Maintenance[13]

The quality of water delivered to the consumer is the result of both adequate treatment and maintenance of that quality through the distribution system. Without procedures to ensure that the distribution system is capable of maintaining water quality, the consumer will not receive high-quality water no matter how well the treatment plant is operated.

Many systems have specified stations within the distribution system which are routinely flushed throughout the year to prevent the development of problems. The locations of these stations and the frequency at which they are flushed is determined from records of complaints and water quality tests. This indicates the importance of good recordkeeping as a tool for water quality management. Records can be used to evaluate the effectiveness of spot flushing, the frequency of flushing, and the need to add or rotate stations during the year.

Annual flushing programs also play a part in maintenance of water quality in the distribution system. Again, records should be used to guide the program. In many cases it is both impossible and unnecessary to flush all parts of the system every year. Review of records may indicate that most zones of a system only need complete flushing every three to five years. Such a program, in combination with routine flushing of problem zones, is more economical and as effective as complete flushing every year.

Flushing alone does not provide an adequate level of protection against the development of tastes and odors in a municipal distribution system. Routine collection of samples for taste and odor analysis using a sensory method such as FPA, especially in systems subject to seasonal outbreaks, can provide an early warning of quality deterioration. These samples should be collected when coliform bacteria samples in the distribution system are collected. This is especially important since tests at the treatment plant may not accurately indicate water quality in the distribution system. The higher chlorine residual and shorter chlorine contact time at the plant tend to give the water at the plant better taste and odor properties than may be encountered in the distribution system.

QUESTIONS

Write your answers in a notebook and then compare your answers with those on page 444.

9.3A What two components should be incorporated into a Taste and Odor Monitoring Program?

9.3B What happens when settled sludges and other debris are allowed to accumulate in the bottoms of channels and tanks in a water treatment plant?

9.3C When are portions of water treatment plants usually taken out of service for inspection and cleaning while the remainder of the plant continues to operate?

9.3D Why should taste and odor treatment equipment be capable of operating properly at all times?

9.3E What sensory method uses a trained panel to identify tastes and odors in both the treatment process and distribution system?

End of Lesson 1 of 3 Lessons on TASTE and ODOR CONTROL

Please answer the discussion and review questions next.

DISCUSSION AND REVIEW QUESTIONS

Chapter 9. TASTE AND ODOR CONTROL

(Lesson 1 of 3 Lessons)

At the end of each lesson in this chapter you will find some discussion and review questions. The purpose of these questions is to indicate to you how well you understand the material in the lesson. Write the answers to these questions in your notebook before continuing.

1. Taste and odor problems are likely to have what types of effects on a water utility when they occur?

2. What will consumers do if they are supplied water with an objectionable taste or odor?

3. How can the use of an algicide in a reservoir cause a taste and odor problem?

4. How can oxygen-poor conditions during the night produce tastes and odors in water?

5. How can high levels of natural runoff lead to tastes and odors?

6. How does pollution cause tastes and odors?

7. What factors influence the contribution of agricultural runoff to taste and odor problems?

8. How can tastes and odors develop within a water treatment plant?

9. What is the first step in determining the cause of a taste and odor problem?

10. How would you attempt to locate the source of a taste and odor problem?

11. Where in a distribution system would you look for the sources of tastes and odors?

12. How is the frequency of flushing mains in a water distribution system determined?

---

[13] See WATER DISTRIBUTION SYSTEM OPERATION AND MAINTENANCE in this series of operator training manuals for additional information on the operation and maintenance of distribution systems.

CHAPTER 9.   TASTE AND ODOR CONTROL

(Lesson 2 of 3 Lessons)

9.4   TASTE AND ODOR TREATMENT

9.40   Methods of Treatment

No single treatment will be applicable to all taste and odor problems because objectionable tastes and odors are the result of so many different causes and because each water system has its own unique characteristics. Each utility must develop procedures to deal with its problems on an individual basis. Successful problem solving depends on understanding some of the important general properties of the most commonly used taste and odor treatment methods.

Taste and odor treatment methods can be divided into two broad categories: removal and destruction. Often both techniques are used at the same time. Multiple treatments, perhaps three or four, may be necessary to produce water of acceptable quality. Removal techniques include optimum coagulation/flocculation/sedimentation, *DEGASIFICATION*,[14] and adsorption. Destruction of tastes and odors is accomplished by various methods of oxidation. The following sections describe the commonly used methods of treatment.

9.41   Improved Coagulation/Flocculation/Sedimentation

Depending on the type of taste and odor, and on the raw water quality, improving sedimentation and associated processes may produce a better tasting water. This is especially true if the taste and odor quality has deteriorated during a period when changes in raw water turbidity, color, or pH have suddenly occurred. Such changes might occur during spring or fall turnover of a lake or reservoir, or during high flows in rivers and canals due to storm runoff. Increases in color and pH may also be the results of an algal bloom, in which case algae levels in the raw water can be expected to increase.

As was discussed in Chapter 4, turbidity, color, and pH can all have a significant effect on coagulation. In addition to removing suspended particulates, inorganic salts such as alum have been shown to reduce the organic content of water,

though turbidity and organic removal may occur at different dosages. In order to determine if increased coagulant dosages will reduce objectionable tastes and odors, the standard jar test procedure can be used followed by tests for both settled water turbidity and threshold odor number.

In cases where tastes and odors increase because of increased algal populations, successful attempts to improve coagulation and sedimentation may produce longer filter runs. In plants where chlorine is added at the headworks, an increase in odor levels may occur due to the destruction of algal cells by the chlorine. The reaction of chlorine on algal cells results in the release of objectionable cellular materials into the water. Further reaction between these cellular products and chlorine may produce *CHLORORGANIC*[15] compounds which impart an even more objectionable taste to the water. When possible, suspension of chlorine at this location and removal of algae in the sedimentation basin will reduce the overall odor levels going on to the filters. Removal of the algae in the sedimentation basin rather than by filtration will also increase filter runs. In plants where chlorine is applied to the water just upstream of the filters, removal of the organisms by sedimentation may improve taste and odor quality by reducing the action of chlorine on the algal cells that would otherwise be trapped on the filters.

The degree of success that may be obtained by improving coagulation may be difficult to assess, especially if the problem is of short duration. However, if it solves the problem, improving coagulation may be the simplest and most economical approach available to the water treatment plant operator for controlling tastes and odors.

QUESTIONS

Write your answers in a notebook and then compare your answers with those on page 444.

9.4A   List the two broad categories of taste and odor treatment methods.

9.4B   How can tastes and odors caused by algae be removed most economically in a water treatment plant?

9.4C   How does the use of chlorine influence taste and odors when water contains algae?

9.42   Aeration Processes and Systems

9.420   Description of Processes

Aeration is the process of mixing air and water together through various means. *THE USE OF AERATION IN TASTE*

---

[14] Degasification (DEE-GAS-if-uh-KAY-shun).   A water treatment process which removes dissolved gases from the water. The gases may be removed by either mechanical or chemical treatment methods or a combination of both.

[15] Chlororganic (klor-or-GAN-ick).   Organic compounds combined with chlorine. These compounds generally originate from, or are associated with, life processes such as those of algae in water.

AND ODOR CONTROL IS EFFECTIVE ONLY IN REMOV-ING GASES AND ORGANIC COMPOUNDS WHICH ARE RELATIVELY VOLATILE.[16] In general, volatile compounds will be noticed as objectionable odors while less volatile compounds are more often associated with objectionable tastes. Aeration is somewhat more successful in treating an odor problem than in treating water with objectionable tastes.

Removal of odor-producing substances that are volatile, as well as other volatile compounds, is known as degasification. Because the compounds being removed exist at a lower concentration in the air than in the water, they will tend to leave the water and move into the air. The more air that is circulated through the water, the greater the amount of objectionable volatile compounds that will be removed from the water. By increasing the aeration rates, the concentration of the objectionable compound present in the water may be reduced to a level at which it no longer causes a problem. As stated above, this process only works if the compound is sufficiently volatile.

Aeration can also destroy some compounds by *OXIDA-TION.*[17] While this can be quite effective for treating *REDUCED*[18] inorganic compounds such as ferrous iron or manganous manganese, it usually is not very effective in the treatment of tastes and odors resulting from the presence of nonvolatile organic compounds. Aeration normally does not provide enough oxidation to attack the taste- and odor-producing organic compounds.

Aeration systems are designed to operate in one of two ways: some systems pass air through the water; other systems pass the water through the air. That is to say, in one type of system air is pumped into the flow of water by some type of air pump, while in the other the water is distributed through the air by nozzles or cascades. A process called air stripping combines elements of both techniques by flowing water over columns of support medium while air is introduced into the water through openings at many points within the support system.

When chemicals are used to treat taste and odor problems, additional treatment by aeration may reduce the chemical dosage needed. The additional benefit to improve taste and odor quality from aeration is usually a minor consideration.

9.421 Air Blowers

Air blowers are basically compressors that supply air under pressure to the water. Large volumes of air are pumped into the water, generally through diffusers along the bottom of a trough or channel, and the air is allowed to rise to the surface. Along the way gases are exchanged between the air bubbles and the surrounding water.

Efficient degasification requires bubbles of very small size to achieve maximum gas transfer. The small bubble size requires that air diffuser orifices be very small. For a specific volume of air, the small orifice size means that the compressor must supply high delivery pressures. In order to make air blowers effective as direct aeration units, a large unit may be needed.

Air blowers can serve as very effective mixing devices since small bubble size is not an important consideration for mixing. Air diffusers located on the bottom of a channel can produce significant turbulence. This turbulence can be used to achieve some degree of degasification by exchanging water between the bottom of the channel and the water surface where some volatile compounds escape to the atmosphere. For odor problems caused by highly volatile substances such as hydrogen sulfide, this type of aerator may provide adequate control. However, general application of air blowers for taste and odor control is usually not a very effective technique and it is not widely used today.

9.422 Cascades and Spray Aerators

Cascades and spray aerators are termed waterfall devices since they aerate water in a manner similar to waterfalls in rivers. These systems pass the water through the air, as opposed to blower devices that introduce air into the water. Both systems are limited to the removal of readily oxidizable or highly volatile compounds just as are air blowers.

Cascade systems are essentially a series of small waterfalls. The water flows down over a series of tiers which may have some type of medium to increase turbulence and improve aeration efficiency. A simple cascade system design consists of a series of concrete steps over which the water flows. When the water reaches the bottom, it flows into a collection basin and is routed on through the treatment plant. A simplified diagram of a cascade system is shown in Figure 9.2.

---

[16] *Volatile (VOL-uh-tull).* (1) A volatile substance is one that is capable of being evaporated or changed to a vapor at relatively low temperatures. Volatile substances also can be partially removed by air stripping. (2) In terms of solids analysis, volatile refers to materials lost (including most organic matter) upon ignition in a muffle furnace for 60 minutes at 550°C. Natural volatile materials are chemical substances usually of animal or plant origin. Manufactured or synthetic volatile materials such as ether, acetone, and carbon tetrachloride are highly volatile and not of plant or animal origin.

[17] *Oxidation (ox-uh-DAY-shun).* Oxidation is the addition of oxygen, removal of hydrogen, or the removal of electrons from an element or compound. In the environment, organic matter is oxidized to more stable substances. The opposite of REDUCTION.

[18] *Reduction (re-DUCK-shun).* Reduction is the addition of hydrogen, removal of oxygen, or the addition of electrons to an element or compound. Under anaerobic conditions (no dissolved oxygen present), sulfur compounds are reduced to odor-producing hydrogen sulfide (H_2S) and other compounds. The opposite of OXIDATION.

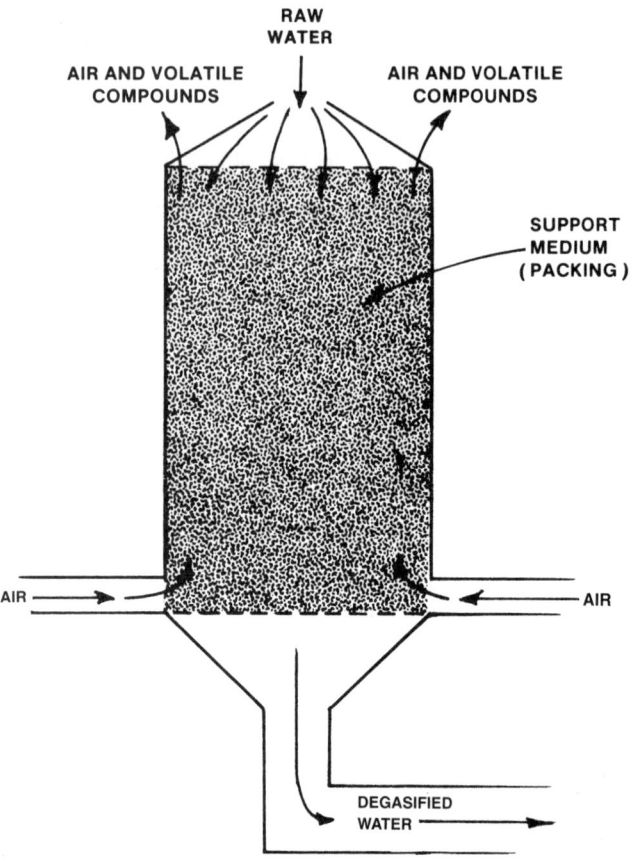

Fig. 9.2   Simple cascade aerator system

Spray aerators, as the name implies, spray water through the air to achieve aeration. In order to be effective, the water must be dispersed into fine droplets. Again, efficiency depends on extensive exposure of the water to air for gas transfer. The number of spray nozzles necessary will depend on plant flow rates, available head, and space limitations. Evaporation losses are a disadvantage of these systems.

Waterfall devices can provide a visual appeal similar to public fountains which they resemble. However, they may encourage biological growths that may contribute to taste and odor problems. Continuous copper sulfate treatment has been applied at some facilities using these open-water systems to control such growth.

9.423   Air Stripping *(Figure 9.3)*

Air stripping is a process that combines elements of both air blowers and waterfall devices to achieve aeration. While not

Fig. 9.3   Updraft air stripping tower

commonly used at conventional water treatment plants, this method may find future application at facilities that must treat both tastes and odors and other trace organic contamination in drinking water supplies.

Air stripping is achieved by flowing water over a support medium, or packing, contained in a tower while pumping air through the packing in the opposite direction. This arrangement of countercurrent (opposite direction) flow provides increased aeration, and therefore improved removal of volatile substances. Water flow is downward through the support medium. As with other aeration devices, air stripping will only be effective in removing compounds that are highly volatile.

QUESTIONS

Write your answers in a notebook and then compare your answers with those on page 444.

9.4D   Aeration is best suited for treating what type of taste and odor problems?

9.4E   What are the two basic ways that aeration systems can be designed?

9.43   Oxidative Processes

9.430   *Types of Processes*

Chemical oxidation is a destructive technique used to control tastes and odors. By application of a strong oxidant, objectionable compounds are chemically modified or broken down into less objectionable by-products. This method is perhaps the most common taste and odor control process because of the widespread use of chlorine, a strong oxidant, for disinfection in water treatment.

In addition to chlorine, there are three other chemicals that deserve some attention as oxidants for the destruction of taste and odor compounds. These are potassium permanganate ($KMnO_4$), ozone (O_3), and chlorine dioxide (ClO_2). All are receiving increased attention in the water industry as alternatives to chlorine at plants that must deal with levels of trihalomethanes in excess of the maximum contaminant level.

9.431   *Chlorine*

Chlorine has been used in water treatment in this country for most of this century, primarily because it is such an effective and relatively inexpensive disinfectant. Because facilities for chlorination are already in place at most water treatment plants, adjustments of chlorine dosages are often used to improve taste and odor quality. In normal cases, where chlorine is applied in a disinfection dosage range, it will often be necessary to increase the normal chlorine dosage to obtain optimal taste and odor treatment.

Many odors in raw water are readily treated by the use of chlorine. Easily detectable odors such as fishy, grassy, or flowery odors can often be decreased significantly with normal chlorination dosages or slight overdoses. Iron and sulfide can also be treated effectively by chlorination. In plants where prechlorination is practiced, it may not even be necessary to change routine treatment to produce a high-quality water. In cases where prechlorination is not routinely applied, some tastes and odors may be eliminated by adding chlorine early in the treatment processes. The success of chlorination as a taste and odor treatment will depend on the type of odor, the seriousness of the problem, the dose applied, and the contact time between chlorine and the water prior to delivery to the consumer.

Increased chlorine doses, including *SUPERCHLORINA-TION*,[19] have been used successfully to treat some difficult taste and odor problems. Higher doses are most beneficial when applied to the plant influent. This allows the greatest contact time and the use of higher doses without adversely affecting consumers located near the treatment plant. Some plants routinely use superchlorination as the initial step in taste and odor control. This process, which uses chlorine doses far in excess of the doses required to meet the chlorine demand, increases the likelihood of adequate oxidation of the target compound. Superchlorination is followed by dechlorination as part of the treatment for tastes and odors. The use of high chlorine doses in pretreatment permits later use of powdered or granular activated carbon to remove excess chlorine.

Despite widespread use and success, chlorine may sometimes be the wrong choice of treatment for a taste and odor problem. Some compounds may become more objectionable after chlorination than before treatment. This is usually the result of incomplete oxidation or substitution of chlorine atoms onto a taste- and odor-causing molecule. For example, runoff from streets and highways contains gasoline residues which are *PHENOLIC COMPOUNDS*[20] (phenols). Chlorination of water containing phenols often results in the production of highly objectionable *CHLOROPHENOLIC*[21] compounds. Chlorophenolic compounds cause noticeable tastes and odors even in concentrations 40 to 200 times *LESS* than the original phenol. In such cases, further laboratory testing is needed to determine if the chlorine dosage should be increased or decreased. The standard jar test procedure followed by tests for odor quality can be used to indicate the direction to proceed (usually the dosage can be reduced).

9.432 Potassium Permanganate

Potassium permanganate ($KMnO_4$) has been used for a number of years in both water and wastewater treatment. Permanganate is a strong chemical oxidizer which can be used to destroy many organic compounds, both natural and manufactured, present in water supplies. Permanganate is also used to oxidize iron, manganese, and sulfide compounds and is often used in conjunction with aeration for the control of these and other taste- and odor-producing substances.

A critical aspect of potassium permanganate treatment is color control. This material produces an intense purple color when mixed with water. As the permanganate ion is reduced during its reaction with compounds that it oxidizes, the color changes from purple to yellow or brown. The final product formed is manganese dioxide (MnO_2), an insoluble precipitate that can be removed by sedimentation and filtration. All of the $KMnO_4$ applied must be converted to the MnO_2 form prior to filtration. If the purple-to-pink color reaches the filters, it will pass into the clear well or distribution system. This may result in the consumer finding pink tap water, or the reaction may continue in the system and the same conditions as exist with naturally occurring manganese will result (staining of plumbing fixtures).

In order to prevent this situation it is necessary to determine the maximum $KMnO_4$ dose that can be safely applied in the plant *BEFORE* starting a permanganate treatment program. To do this, measure the time required for the last trace of pink to disappear from a series of jar tests. Mix the test samples for a time similar to the detention time in the sedimentation basin. You should allow for a margin of error in this determination; for example, use a time equal to one-half to three-quarters of the actual detention time when running these laboratory trials. In actual plant operation, many operators control permanganate dose by adjusting the feed rate so that the pink or purple color is visible only to a specific point in the basin. This involves a simple process of observation, and is important when treating water that exhibits major fluctuations in permanganate demand. At night, however, visual observations are more difficult and you may want to use a reduced daytime dosage. Some plants use on-line permanganate analyzers to assist with permanganate dose control (Figure 9.4).

If raw water is transported to the plant through a pipeline or canal, application of the potassium permanganate at the inlet is a desirable practice. This allows an extended contact time for the reaction to produce manganese dioxide and allows somewhat higher permanganate doses (if required for adequate taste and odor control). This also provides the opportunity for monitoring the plant influent for color or soluble manganese, giving the operator an early warning system for a permanganate overdose.

If a permanganate overdose does occur, powdered activated carbon can be used to control the problem until the per-

---

[19] *Superchlorination (SUE-per-KLOR-uh-NAY-shun). Chlorination with doses that are deliberately selected to produce free or combined residuals so large as to require dechlorination.*

[20] *Phenolic (fee-NO-lick) Compounds. Organic compounds that are derivatives of benzene.*

[21] *Chlorophenolic (klor-o-FEE-NO-lick). Chlorophenolic compounds are phenolic compounds (carbolic acid) combined with chlorine.*

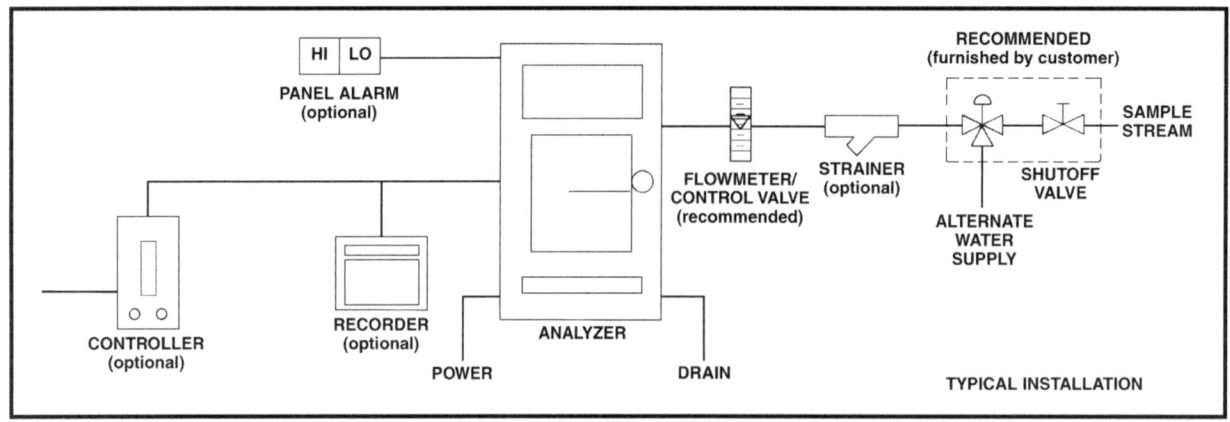

Fig. 9.4   Permanganate on-line analyzer
(Courtesy of the HACH Company)

manganate dosage rate has been adjusted to the correct level. In some extreme cases it may be necessary to increase the pH of the water for a short time to increase the rate at which the manganese is precipitated. While it is better to maintain a permanganate dose that will not cause a problem, or to respond by the addition of powdered activated carbon, pH adjustment to precipitate manganese in the clear well is still preferable to manganese deposition in the distribution system or in the consumers' sinks.

Experience at various water treatment plants has shown that typical permanganate application rates for taste and odor control are about 0.3 to 0.5 mg/L, though the range reported is from 0.1 to 5 mg/L. The cost of this comparatively expensive chemical must be considered when you are trying to set cost-effective dosage ranges.

At one water treatment plant, $KMnO_4$ in the range of 0.3 to 0.5 mg/L has been applied at a river intake structure approximately 13 miles upstream from the plant for taste and odor control. Permanganate treatment has resulted in a 65 percent reduction in complaints at this facility, which has a seasonal problem with a musty-earthy taste in the water. Detention time between $KMnO_4$ application and the treatment plant is normally more than six hours. This allows plenty of time for complete conversion of the permanganate to the manganese dioxide form before the water enters the sedimentation basin.

9.433   Handling of Potassium Permanganate

Potassium permanganate is a dry, crystalline product that is best delivered from a dry feeder into a special mixer immediately prior to application. A portable feeder is shown in Figure 9.5. Permanganate storage facilities must be dry and well ventilated as moisture in the atmosphere can cause caking of the material. This will cause the feeder to clog and prevent accurate delivery of the permanganate.

Ventilation in a permanganate feeding and storage area is important for both operator safety and equipment protection. Potassium permanganate will produce a very fine dust during loading and handling. This dust is irritating to the eyes, mucous membranes, and even to the skin of some sensitive individuals. When handling permanganate, wear goggles, a dust mask, and gloves. Protective outer clothing (such as rain

gear) should also be worn to prevent discoloration[22] of skin and clothes.

Adequate dust control and ventilation are also important for equipment protection. Permanganate and moisture create a very corrosive mixture which will attack metal, including electrical connections. By ventilating the area during loading and maintaining as dry an environment as possible, the service life and reliability of permanganate feed equipment will be improved.

Fig. 9.5   Portable $KMnO_4$ dry feeder used
at intake structure

---

[22] To remove permanganate stains from the skin, a strong solution of sulfite compound (sodium sulfite/meta-bisulfite) works very well.

Never store permanganate in the same area where activated carbon is stored because they are both highly flammable.

QUESTIONS

Write your answers in a notebook and then compare your answers with those on page 444.

9.4F   What types of odors can often be decreased significantly by chlorination?

9.4G   Under what circumstances might the use of chlorine be the wrong treatment for a taste and odor problem?

9.4H   How would you respond to a permanganate overdose?

9.4I   Why must permanganate storage facilities be dry and well ventilated?

9.4J   Why are adequate permanganate dust control and ventilation important for equipment protection?

9.434   Ozone

Ozone (O_3) is an unstable form of oxygen that is an extremely powerful oxidant. The compound in its pure state is a bluish gas, which has a pungent, penetrating odor; it can sometimes be detected around large electric motors. Ozone is produced by passing dry air, or oxygen, through a high energy ionizing unit known as an ozonator. Because it is unstable, ozone must be generated on site for use in water treatment.

Ozone is a stronger oxidant than chlorine and, therefore, destroys a wider range of organic compounds. Ozone has a distinct advantage over chlorine treatment for taste and odor in that objectionable by-products of the reaction do not normally form. This is of particular advantage when the taste and odor is of industrial origin, since the combination of chlorine and some industrial pollutants can lead to more intense tastes and odors than those caused by the original compounds.

Ozone has received much attention in recent years as an agent for oxidizing organic contaminants present in some water supplies and as a means of disinfecting water without producing trihalomethanes (THMs). While not commonly used in this country, ozone treatment has been applied extensively in water treatment in Europe.

Production of ozone on site for water treatment requires specialized equipment. The basic elements of an *OZONATION*[23] system include a source of dry air or oxygen, a condenser and dryer to remove traces of moisture in the feed gas, the ozonator, an enclosed contactor unit where the ozone is mixed with the water, and a means for venting or recycling waste gas from the contactor. These units are shown schematically (skee-MAT-ick-lee) in Figure 9.6.

Because of the specialized equipment requirements and costs, laboratory and pilot-scale testing of the effectiveness of ozonation as a taste and odor control process or for elimination of THMs should be conducted before full-scale installation. Utilities considering purchase of an ozonation system should contact manufacturers or consulting engineering firms for assistance in the evaluation and design of ozone systems.

9.435   Chlorine Dioxide *(also see Chapter 7, Section 7.24, "Chlorine Dioxide (ClO$_2$)," and Section 7.6, "Chlorine Dioxide Facilities")*

Chlorine dioxide (ClO_2) is another chemical oxidant that has received increased attention in recent years as a result of concern about the formation of trihalomethanes in water following chlorination. Chlorine dioxide, like ozone, is a strongly oxidizing, unstable compound. Chlorine dioxide is formed by reacting sodium chlorite and chlorine in a special ClO_2 generator. Because it is unstable, it too must be generated on site at the time it is to be applied to the treatment process. Chlorine dioxide has been used to treat tastes and odors caused by industrial pollution, especially in cases where chlorine has intensified the problem.

While chlorine dioxide has been reported to reduce taste and odor complaints in some cases, other agencies have experienced more severe problems following ClO_2 treatment. The Louisville, Kentucky, Water Company experienced sharp

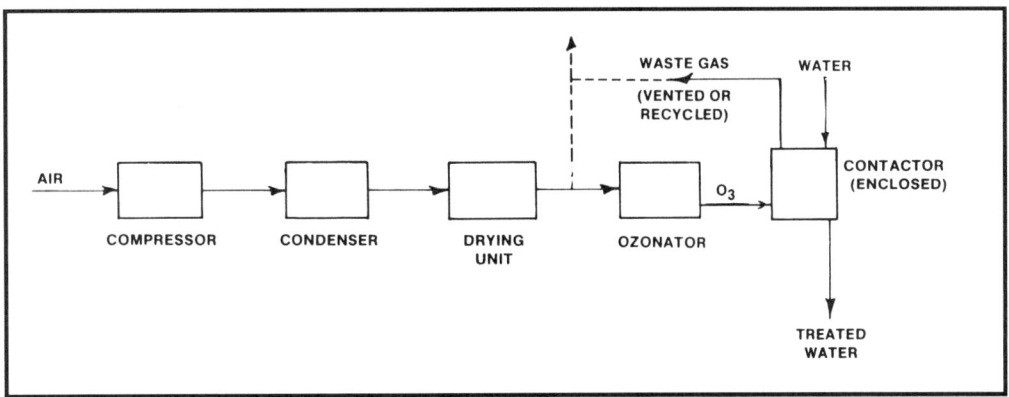

Fig. 9.6   Processes in ozonation

---

[23] *Ozonation (O-zoe-NAY-shun).   The application of ozone to water for disinfection or for taste and odor control.*

increases in taste and odor complaints during initial trials with ClO_2 as a method for lowering distribution system trihalomethane (THM) levels. The problem was resolved by using ClO_2 at the plant influent followed by combined chlorine treatment of water flowing into the clear well.

A further consideration in the use of chlorine dioxide is the formation of unwanted chlorite and chlorate ions. There is concern over the possible health-related effects of these reaction products. Limitations for allowable concentrations have been imposed which may prevent the use of chlorine dioxide at doses adequate to control tastes and odors.

QUESTIONS

Write your answers in a notebook and then compare your answers with those on page 444.

9.4K Why should laboratory and pilot-scale testing of the effectiveness of ozonation as a taste and odor control process be conducted *BEFORE* full-scale installation?

9.4L Why must chlorine dioxide be generated on site?

9.4M What types of taste- and odor-producing wastes have been treated successfully by using chlorine dioxide?

End of Lesson 2 of 3 Lessons on TASTE and ODOR CONTROL

Please answer the discussion and review questions next.

DISCUSSION AND REVIEW QUESTIONS

Chapter 9. TASTE AND ODOR CONTROL

(Lesson 2 of 3 Lessons)

Write the answers to these questions in your notebook before continuing. The question numbering continues from Lesson 1.

13. How would you use jar tests in an attempt to solve a taste and odor problem?

14. What problems may be created by the use of cascades and spray aerators and how are these problems solved?

15. How would you prevent color from potassium permanganate reaching a consumer's tap?

16. What safety precautions should you take when handling potassium permanganate?

17. What is the major advantage of ozone over chlorine treatment for taste and odor control?

CHAPTER 9.   TASTE AND ODOR CONTROL

(Lesson 3 of 3 Lessons)

9.44   Adsorption[24] Processes

9.440   Types of Processes

Adsorption is the process of removing materials from water by adding a material to the water to which the taste- and odor-producing compounds will attach themselves. In water treatment this is accomplished in one of two ways: (1) the addition of powdered activated carbon (PAC) to the treatment process, usually at the influent; or (2) the use of granular activated carbon (GAC) as a filter medium. The material being removed by adsorption is known as the adsorbate, and the material responsible for the removal is known as the adsorbent.

The primary adsorbents for water treatment are the two types of activated carbon mentioned above: powdered and granular. These materials are activated by a process involving high temperature and high-pressure steam treatment. The original source of the carbon may be wood, coal, coconut shells, or even bones. The purpose of the activation process is to significantly increase the surface area of the particles so that more adsorption can take place per pound of carbon. The surface area is increased during activation by the formation of holes and crevices in the carbon resulting in particles that have a very porous structure. The surface of activated carbon may range from 400 to 2,000 square meters per gram (2 to 9 million square feet per pound). This large surface area is responsible for the high degree of effectiveness that can be achieved with this remarkable substance.

An important consideration in evaluating carbon treatment is that the nature of the porous structure will exert a significant effect on the success of the treatment. If the pores of the carbon are too small, the compounds that are being treated will not be able to enter the structure and only a small portion of the available surface area will be used. Activation processes produce carbons with different surface areas. Since adsorption is a surface phenomenon, carbons with greater surface area generally provide greater adsorptive capacity.

Activated carbons are typically rated on the basis of a "phenol number" or an "iodine number." The higher the value, the greater the adsorption capacity of the carbon for phenol or iodine. This is an excellent approach for evaluating the effectiveness of a carbon for removal of phenol or iodine, but there may be no direct relationship between the compounds causing taste and odor and the phenol or iodine of the test. Only by testing various carbons for effective removal of the objectionable taste or odor can a good comparison be made for a particular application.

9.441   Powdered Activated Carbon

9.4410   DESCRIPTION OF PROCESS

Powdered activated carbon (PAC) adsorption is the most common technique used specifically for taste and odor control at water treatment plants in this country. This widespread use is due largely to its nonspecific action over a broad range of taste- and odor-causing compounds. While useful in treating many taste and odor problems, PAC treatment does have limitations, and its effectiveness and required dose rate vary widely from plant to plant.

Powdered activated carbon may be applied to the water at any point in the process prior to filtration. Because carbon must contact the material to be removed in order for adsorption to occur, it is advantageous to apply PAC at plant mixing facilities. Powdered activated carbon is often less effective at removing compounds after chlorination, so application upstream of chlorine treatment is desirable. Chlorine will react with carbon and neutralize the effects of both. From an economic standpoint, therefore, it is not good practice to apply chlorine and PAC near the same location.

Powdered activated carbon is often applied at the plant flash mixer. This location provides high-rate initial mixing and the greatest contact time through the plant. Again, both thorough mixing and long contact time improve the effectiveness of PAC. Another common location for PAC application is the filter influent. While contact time and mixing are drastically reduced compared to flash-mix application, this method ensures that all of the water passes through a PAC layer prior to release into the distribution system. This procedure is sometimes used with $KMnO_4$ treatment to prevent colored water problems in the distribution system.

When PAC is used, dosages may range from 1 to 15 mg/L. Some reports in the literature have indicated that as much as 100 mg/L have been required to adequately treat some serious taste and odor problems. At very high doses, treatment costs become prohibitive, and consideration of granular activated carbon treatment in its own contactor (filter) is warranted.

9.4411   POWDERED ACTIVATED CARBON FEED SYSTEMS

Powdered activated carbon feed systems may be either dry-type feeders or slurry feeders. For small-scale applications the dry-type feed system may be satisfactory, especially if it is only used for short-term, occasional incidents. More frequent PAC

---

[24] *Adsorption (add-SORP-shun).   The gathering of a gas, liquid, or dissolved substance on the surface or INTERFACE[25] zone of another material.*
[25] *Interface.   The common boundary layer between two substances such as water and a solid (metal); or between two fluids such as water and a gas (air); or between a liquid (water) and another liquid (oil).*

treatment requirements, or larger applications are usually best accomplished with *SLURRY* [26] systems.

Carbon slurry tanks require continuous mixing to prevent settling and caking of the carbon. Typical installations provide a vertical bi-level (two-blade) paddle mixer and vertical wall-mounted baffles for mixing. One mixer blade is located near the bottom of the tank and the other is situated about mid-depth. Carbon tanks should have drain sumps, but the suction side of the carbon intake should not extend into the sump. The reason is that during loading operations all types of foreign material can find its way into the tank through the dump chute or in the material itself. At one water treatment plant carbon tanks have been found to contain sand, wood chips, plastic, and paper after having been thoroughly cleaned prior to re-loading.

Because PAC has a tendency to cake, even when continuously mixed, and because of the variety of miscellaneous materials found in carbon tanks, standard practice at some plants is to clean each carbon slurry tank prior to receipt of a new shipment of carbon. The cleaning procedure allows a visual inspection of the tank, including the mixer blades and baffles.

Powdered activated carbon is a somewhat hydrophobic (HI-dro-FOE-bick) material and does not mix readily with water. When carbon is being loaded into slurry tanks, an overhead spray system in the tank should be used. This spray accomplishes some degree of dust control as well as an initial wetting of the carbon. To load a PAC tank, fill it about one-quarter to one-third full of water before starting, and operate the mixers during loading.

Proceed slowly enough to allow complete mixing and wetting of the carbon as it is introduced into the tank. If the mixing capability of the tank is exceeded, the PAC will form a cake on the surface of the water. The caked material will prevent additional carbon from being properly mixed, and a thick, tough layer of floating carbon will form in the tank. Once formed, caked PAC is difficult to disperse into a useful slurry. If PAC does not mix rapidly during loading, stop the operation, check mixers and sprayers, and break up any cake that has formed. The cake can be broken up using breaker bars (crowbars) or wooden paddles (2x4s). If all equipment is operating properly, the loading rate must be reduced to the level at which the system can adequately mix the carbon into a slurry.

Figure 9.7 shows typical PAC feeders and storage tanks.

QUESTIONS

Write your answers in a notebook and then compare your answers with those on page 444.

9.4N   What two forms of activated carbon are used in water treatment?

9.4O   What terms are used to describe the adsorptive capacities or ratings of activated carbons?

9.4P   Why should powdered activated carbon be applied at the plant flash-mixing facilities?

9.4Q   What would you do if a caked layer of carbon starts to form on the surface of the water in the slurry tank?

Feed pumps are located on deck above underground PAC slurry tanks.

PAC tank mixer motor with right-angle drive.

Fig. 9.7   Powdered activated carbon (PAC) feeders and storage tanks

9.4412   POWDERED ACTIVATED CARBON DOSE DETERMINATION

The appropriate dose of PAC for any particular problem will vary depending upon the nature of the problem, the concentration of the material to be removed, the mixing available, the contact time, and the location of application points. Jar tests should be used to determine the necessary PAC doses required to treat the specific taste and odor problem that exists. By simulating mixing speeds and detention times of the various locations at which the carbon might be applied, and by varying the doses of carbon, the jar test can be used to indicate the most effective range of carbon treatment.

Accurate determination of the threshold odor number (TON) of a sample is difficult to make when several jar tests are run all at once. A person's sense of smell becomes rapidly fatigued and after just two or three TON tests individuals tend to become desensitized to the odors present. Because only a limited number of operators are usually available for odor testing, a single operator will often have to do all of the tests. Mis-

---

[26] *Slurry (SLUR-e).   A watery mixture or suspension of insoluble (not dissolved) matter; a thin, watery mud or any substance resembling it (such as a grit slurry or a lime slurry).*

PAC feed pumps. Piping (PVC) in front is slurry suction line from tanks.

Dust collector for bagged carbon loading. Carbon is loaded through lower unit while dust collector is in upper unit.

Close-up of slurry suction line. Hose bib allows backflushing of suction line in case of plugging.

Four-inch "quick-connect" fitting to allow bulk loading of PAC from truck.

Fig. 9.7   Powdered activated carbon (PAC) feeders and storage tanks (continued)

leading results may be obtained if more than a few TON determinations are made by the same person in a short time. Also, if the air around the plant contains the odor of concern, none of the people at the plant site will be able to perform the TON test. When your nose smells the same odor for a long time, eventually you will be unable to detect that odor due to *OL-FACTORY FATIGUE.*[27]

A more practical approach, especially during initial testing, is to compare the *RELATIVE* (subjective) odor of jars dosed at different PAC levels. (This same method can be applied to the preliminary evaluation of all taste and odor treatments.) After mixing for an appropriate time, collect undiluted samples from each jar. Starting with the sample from the jar test treated at the highest level, test (smell) each sample in order of *DE-SCENDING* treatment dose and note your initial reaction to any odor present such as "none," "slightly fishy," "foul," or other descriptive term. After having tested all samples, rate the various treatments from best to worst in terms of odor removal. An example of the results from this type of comparative test are shown in Figure 9.8.

Notice that the test shown in Figure 9.8 included alum and polymer treatment. This test was conducted to evaluate PAC application at the flash mixer where the coagulants are added. If chlorine is added near the PAC delivery point, the jar samples should be treated to reflect this. As in all jar testing, the laboratory trials should simulate plant conditions as closely as possible.

This technique may indicate that PAC treatment alone will not be enough to control the taste and odor problem, which is certainly valuable information. This method is most useful, however, for deciding whether PAC treatment above a certain level provides any additional benefit to water quality. In the example, all jars treated at 6 mg/L and higher gave similar results, indicating that a higher PAC dose would not be useful. Results such as these provide guidelines for establishing the most effective treatment at the most economical costs.

One difficulty often overlooked in the use of PAC is the problem of measuring the actual concentration (lbs/gal) of carbon in the slurry tank. For convenience in calculating feed rates, powdered activated carbon slurry tanks are normally loaded at a rate of one pound of PAC per gallon of water. Unfortunately, off-loading from a bulk delivery into multiple tanks usually results in an unequal division of the carbon among the tanks. The following simple laboratory procedures can be used to determine the actual carbon content of the PAC slurry.

Collect a one-liter sample of the carbon slurry from a recently loaded tank after several hours of mixing. Allow this sample to settle for approximately 2 hours in an *IMHOFF CONE*[28] and record the amount of settled carbon as mL carbon/liter of *SUPERNATANT.*[29] Then dry portions of the settled carbon in a laboratory drying oven and weigh them. In this manner, you can calculate the dry weight of carbon per milliliter of settled carbon. From this information it is easy to calculate the amount of carbon per gallon in the slurry. (See Example 2 on page 436.) Results from several years of testing at the Stockton East Water Treatment Plant indicate that different carbon manufacturers' carbons have their own characteristic values on a "gram per settled mL" basis. Generally, multiplying the mL/L of settled carbon by a factor, which ranges from about

0.0022 to 0.0028 at the Stockton plant, gives the weight of carbon in the tank, expressed as pounds per gallon.

Powdered activated carbon is an abrasive material. Equipment used to feed PAC often requires more frequent inspection, cleaning, and maintenance than other chemical feed equipment at a water treatment plant to ensure proper dosage delivery. Pumping equipment should be routinely inspected for signs of wear. Check valves and slurry suction lines can become clogged with carbon that has caked in the tank. Frequent inspection during actual carbon delivery is necessary when using this equipment.

FORMULAS

To determine the dosage of PAC in either pounds per gallon or pounds per million gallons, we have a choice of two approaches. With the first choice, we convert the milligrams of PAC per liter of water dose from the jar test to pounds per gallon using conversion factors.

$$\text{Desired PAC, lbs/gal} = \frac{(\text{PAC Conc, mg/}L)(3.785\ L/\text{gal})}{(1{,}000\ \text{mg/gm})(454\ \text{gm/lb})}$$

We multiply by 3.785 liters per gallon to convert the liters to gallons. We divide by 1,000 milligrams per gram and 454 grams per pound to convert milligrams to pounds. This gives us an answer in pounds per gallon. Since the pounds of PAC in a gallon of water is very small and we work with millions of gallons of water, we multiply by (1,000,000/1 Million) to get pounds of PAC per million gallons.

The second approach is to refer to our basic loading or chemical feed equation.

$$\text{Feed, lbs/day} = (\text{Flow, MGD})(\text{Dose, mg/}L)(8.34\ \text{lbs/gal})$$

If we changed the flow from million gallons per day to one million gallons, then this would give us a feed of PAC in pounds per one million gallons.

$$\begin{aligned}
\text{Desired PAC,}\atop\text{lbs/M Gal} &= (\text{Volume, M Gal})(\text{Dose, mg/}L)(8.34\ \text{lbs/gal}) \\
&= (1\ \text{M Gal})\frac{(\text{Dose, lbs PAC})}{(1\ \text{M lbs})}\frac{(8.34\ \text{lbs})}{(\text{Gal})} \\
&= \frac{\text{lbs PAC}}{1\ \text{Million Gal}}
\end{aligned}$$

---

[27] *Olfactory (ol-FAK-tore-ee) Fatigue.   A condition in which a person's nose, after exposure to certain odors, is no longer able to detect the odor.*

[28] *Imhoff Cone.   A clear, cone-shaped container marked with graduations. The cone is used to measure the volume of settleable solids in a specific volume (usually one liter) of water.*

[29] *Supernatant (sue-per-NAY-tent).   Liquid removed from settled sludge. Supernatant commonly refers to the liquid between the sludge on the bottom and the scum on the water surface of a basin or container.*

DATE _5-30-02_   TIME _0945_   OPERATOR _JY/JG_

RAW WATER DATA				MIXING SEQUENCE	
SOURCE	PIPELINE			RPM	TIME
TEMP	20 °C	pH	8.2	1. MAX	2 min.
TURBIDITY	1.0 NTU	COLOR	15	2. 85	8 min.
ALKALINITY	65 Mg/L	HARDNESS	66 Mg/L	3. 40	16 min.
				4.	

		JAR					
CHEMICAL (mg/L)		1	2	3	4	5	6
1.	ALUM LIQUID/DRY	6.0	6.0	6.0	6.0	6.0	6.0
2.	CAT POLYMER	1.0	1.0	1.0	1.0	1.0	1.0
3.	PAC	0	2	4	6	8	10
4.							
5.							
6.							
FLOC CHARACTERISTICS	AFTER FLASH MIX						
	AFTER RAPID MIX						
	5 MIN. SLOW MIX						
	10 MIN. SLOW MIX						
	15 MIN SLOW MIX						
FLOC SETTLING	5 MIN						
	10 MIN						
	20 MIN						
	30 MIN						
SETTLED WATER QUALITY	TURBIDITY						
	pH						
	COLOR						
COMMENTS: ODOR :		GRASSY/STALE·DIRTY	GRASS·DIRTY DIRTY	SLIGHTLY DIRTY	NONE	NONE	NONE
SUBJECTIVE RATING : (SCALE OF 1 TO 6)		6 (WORST)	5	4	1	1 (ALL SIMILAR)	1

Fig. 9.8   Jar test

EXAMPLE 1

Results of jar tests indicate that 5 mg/L of powdered activated carbon is the most effective dosage for treating a taste and odor problem. What is the desired concentration in pounds per million gallons?

Known	Unknown
PAC Conc, mg/L = 5 mg/L	Desired PAC, lbs/M Gal

Convert the powdered activated carbon concentration from mg/L to pounds/gallon and pounds/million gallons.

$$\text{Desired PAC, lbs/M Gal} = \frac{(\text{PAC Conc, mg/}L)(3.785\ L/\text{gal})}{(1,000\ \text{mg/gm})(454\ \text{gm/lb})}$$

$$= \frac{(5\ \text{mg/}L)(3.785\ L/\text{gal})}{(1,000\ \text{mg/gm})(454\ \text{gm/lb})}$$

$$\text{or} = (0.000042\ \text{lbs/gal})(1,000,000/\text{Million})$$

$$= 42\ \text{lbs/Million Gal}$$

ALTERNATE SOLUTION TO EXAMPLE 1

$$\text{Desired PAC, lbs/M Gal} = (\text{Volume Water, M Gal})(\text{Dose, mg/}L)(8.34\ \text{lbs/gal})$$

$$= (1\ \text{M Gal})(5\ \text{mg/}L)(8.34\ \text{lbs/gal})$$

$$= 41.7\ \text{lbs/Million Gal}$$

EXAMPLE 2

A sample of PAC slurry was collected from a slurry tank. The one-liter sample was allowed to settle for 24 hours. The amount of settled carbon was recorded as 50 mL carbon per liter of supernatant. Ten mL of the settled carbon was dried and found to weigh 20 mg. Calculate the amount of carbon in the slurry in pounds of dry carbon per gallon of water.

Known		Unknown
Sample Vol, L	= 1 L	PAC Slurry, lbs/gal
Carbon Settled, mL	= 50 mL	
Portion Dried, mL	= 10 mL	
Dried Sample Weight, mg	= 20 mg	

1. Calculate the amount of settled carbon as mg/mL.

$$\text{Settled Carbon, mg/mL} = \frac{\text{Dried Sample Weight, mg}}{\text{Portion Dried, mL}}$$

$$= \frac{20\ \text{mg}}{10\ \text{mL}}$$

$$= 2\ \text{mg/mL}$$

2. Calculate the PAC concentration in the slurry as mg/L.

$$\text{PAC Slurry, mg/}L = \frac{(\text{Carbon Settled, mL})(\text{Settled Carbon, mg/mL})}{\text{Sample Volume, }L}$$

$$= \frac{(50\ \text{mL})(2\ \text{mg/mL})}{1\ L}$$

$$= 100\ \text{mg/}L$$

3. Convert PAC Slurry from mg/L to pounds PAC per gallon.

$$\text{PAC, lbs/gal} = \frac{(\text{PAC, mg/}L)(3.785\ L/\text{gal})}{(1,000\ \text{mg/gm})(454\ \text{gm/lb})}$$

$$= \frac{(100\ \text{mg/}L)(3.785\ L/\text{gal})}{(1,000\ \text{mg/gm})(454\ \text{gm/lb})}$$

$$= 0.000834\ \text{lbs/gal}$$

$$\text{or} = 834\ \text{lbs/Million Gal}$$

9.4413   FILTRATION CONSIDERATIONS WITH POWDERED ACTIVATED CARBON

Use of powdered activated carbon for taste and odor removal may interfere with filter performance at a water treatment plant. Caking of PAC on the surface of filters may cause substantially shorter filter runs than otherwise expected. If this occurs, adjustments to improve PAC removal in the settling process may increase the effective length of filtration. An added advantage of optimizing the settling process is the physical removal of taste and odor components with the settled sludge.

A second and more difficult problem to detect can occur when applying PAC. Because the particle size of powdered activated carbon is nearly microscopic, PAC can penetrate through filters before either head loss or turbidity breakthrough indicate the need for backwashing. Since the PAC particles are black, they absorb light. Therefore, standard *NEPHELO-METRIC*[30] turbidity measurements will not warn of the carbon passing through filters. Penetration of PAC through filters can cause "dirty water" complaints in the distribution system and is particularly apparent to those consumers who have installed home filtration units on their faucets.

Studies conducted at the Stockton East Water Treatment Plant indicate that carbon penetration through filters is related to both the PAC dose and the hydraulic loading on the filters. At higher flow rates and increased carbon loading on the filters, PAC penetrates through the filter beds much earlier in the filter run than at lower flows or carbon doses.

---

[30] *Nephelometric (NEFF-el-o-MET-rick). A means of measuring turbidity in a sample by using an instrument called a nephelometer. A nephelometer passes light through a sample and the amount of light deflected (usually at a 90-degree angle) is then measured.*

Powdered activated carbon penetration through a filter bed occurs in approximately the same manner as turbidity breakthrough. A freshly backwashed filter may show the presence of PAC during the recovery period. Following recovery, PAC filtration is excellent, though the length of time this performance is maintained will vary. Depending on the severity of the PAC penetration problem, special treatment may be necessary. At the Stockton East Plant, filter-aid dosage of a nonionic polymer in the range of 0.005 to 0.010 mg/L has been used to control PAC penetration during recovery. Continued application of this level of filter aid throughout the run, however, increased the rate of head loss, so filters required backwashing at about the same time that PAC began to appear before in the effluent.

One simple method of determining carbon penetration through a filter is to collect a one-liter sample of filter effluent and filter it through a 0.45-μm membrane filter. The presence of PAC in the effluent can be determined from a darkening of the white membrane filter surface. An example of increasing PAC penetration during a filter run as determined by this technique is shown in Figure 9.9.

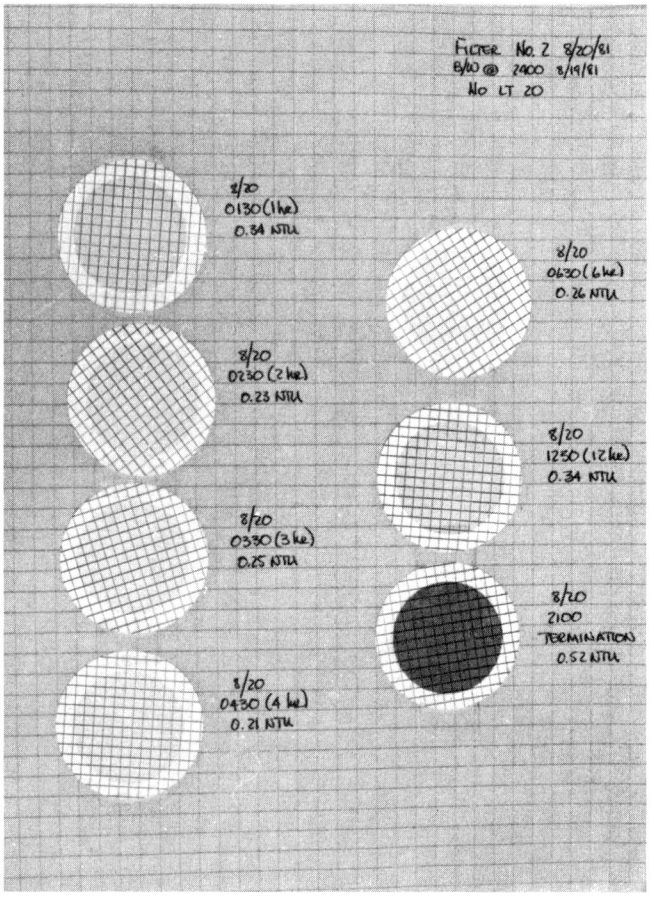

PAC dose was 10 mg/L and filter hydraulic loading was 7.6 GPM/sq ft during test.

Fig. 9.9 PAC penetration through a filter

9.4414 POWDERED ACTIVATED CARBON HANDLING

Powdered activated carbon may be purchased either in individual bags (usually 50 pounds (23 kg)) or by bulk truck load (up to 40,000 lbs (18,180 kg) per load).[31] Regardless of loading arrangement, by bulk or individual container, powdered activated carbon is usually an unpleasant material to work with at a water treatment plant. Hand loading of a carbon storage tank from bags is a time-consuming, dirty job. Off-loading of bulk loads is less time-consuming but not necessarily any less dirty.

Various types of equipment have been used to control the dust generated during PAC handling. The success of any dust collection system depends on how well it meets the particular needs of the facility for carbon handling and on adequate maintenance. Several different approaches to dust control have been used, with more elaborate (and occasionally more successful) designs usually found at larger treatment plants. No matter what type of equipment is used, dust collectors must be clean and functional at the start of loading operations and further cleaning may be required during the operation to maintain dust collection effectiveness.

Operators assigned to carbon loading, especially when bags are used, must wear protective clothing. At the very least, goggles, dust masks, and gloves are required. Protective outer clothing is also highly recommended, and showering facilities should be provided for cleaning up after unloading the carbon.

Because powdered carbon will scavenge (actively remove) oxygen from the air, safety procedures for working in confined spaces must be carefully followed when cleaning tanks. Open the tank fully, set up some means of circulating air in the tank, and work in teams—*NEVER WORK ALONE IN A POWDERED CARBON TANK.* Arrange for a spotter (operator) overhead (outside) to watch the cleaning operation closely and be prepared to assist if the need arises. The operator in the tank should wear a safety harness and the oxygen level in the tank must be monitored continuously.

If bags of PAC are to be stored, it is desirable to have them elevated off the floor to prevent caking due to spilled or sprayed water. Wooden pallets work well for this purpose. Carbon is a combustible material, but tends to smolder rather than burn fiercely. A smoldering pallet may be removed by dragging it out of the area, thereby saving the rest of the material from water and fire damage. For this reason, bags should not be stacked more than three or four high on an individual pallet.

Because carbon is combustible, it should be stored away from other materials, especially flammable materials like potassium permanganate ($KMnO_4$) and HTH (High Test Hypochlorite) compounds. If at all possible, an isolated storage facility is highly desirable. Electrical wiring and switches in a carbon storage area should be of special design for flammable storage area use. Whenever there is dust in the air, a spark could cause an explosion. The area should be prominently marked with "No Smoking" signs and this regulation should be strictly enforced.

---

[31] *Large, one-ton (910-kg) "bladders" or carboys of PAC are available in some areas. Special facilities are required for the use of these containers.*

9.442   Granular Activated Carbon

Granular activated carbon (GAC) is activated in much the same way as powdered activated carbon, although the raw material is usually a lignite coal. Granular carbon, as the name implies, is made up of larger particles than PAC. Typical surface area for GAC is in the range of 400 to 800 square meters per gram (2 to 4 million square feet per pound). The basic adsorption process with granular carbon is the same as with powdered carbon, except that the granular carbon is placed in a stationary bed through which the water flows.

Granular carbon filters are capable of filtering water in the same manner as rapid-sand or dual-media filters, and can produce low-turbidity water when operated under similar conditions. GAC filters, which are also referred to as biologically activated carbon filters because of the high levels of bacteria that will develop within the filter bed, have received attention as a means of treating trace organic contamination of drinking water. Generally, the effective life of a GAC filter being used for taste and odor control will be much greater than one used for removal of potentially toxic or carcinogenic trace organics.

Operational procedures for GAC filters closely resemble procedures used with other typical water treatment filters. Two unique considerations for GAC filtration are empty bed contact time (EBCT) and regeneration interval of the carbon. Empty bed contact time is the time that the water is actually in the filter bed (use calculated volume *NOT* including carbon when determining contact time). Successful application of GAC filtration for taste and odor control usually requires a minimum of five minutes, although some cases have required contact times greater than 15 minutes. The EBCT is an important design consideration in GAC units to be used for taste and odor control since the filters must be large enough to provide adequate contact time as the water is filtered.

Periodic regeneration of GAC is necessary as the capacity of the filter to adsorb and retain organic compounds decreases with time. The time between regenerations will vary with the type of material being removed and the volume of water treated. Regeneration is accomplished essentially in the same manner as initial activation, and most utilities replace spent carbon with fresh rather than attempting to regenerate. Regeneration intervals of three to seven years have been reported by utilities using GAC filters for the control of tastes and odors.

Retrofitting of existing filters with GAC is an expensive modification. Initial installation costs are high and long-term expense for regeneration or replacement must be considered. However, in situations where high doses of PAC must be used, or where taste and odor control is required during a substantial portion of the year, evaluation of GAC filtration may indicate that it is a more economical long-term alternative. Because of the expense, thorough investigation in the laboratory and in pilot-scale tests should be made prior to plant installation.

QUESTIONS

Write your answers in a notebook and then compare your answers with those on page 445.

9.4R   What precautions must operators take because powdered activated carbon is so abrasive?

9.4S   What adjustments would you make if PAC tends to cake on the surface of the filters?

9.4T   How can carbon penetration through a filter be detected?

9.4U   What is empty bed contact time in a granular activated carbon filter?

9.5   IDENTIFYING A TASTE AND ODOR PROBLEM

Before an effective control strategy for a taste and odor problem can be implemented, the fact that a problem exists must be recognized. This may seem obvious, but delayed response (or no response at all) to the onset of a taste and odor problem in water systems frequently occurs when the problem is not identified at the time it develops. This is especially true in organizations that have had no previous experience with taste and odor problems, or in agencies where complaints from consumers are received at one location while operation and water quality activities are conducted at another.

Every water utility must have an effective communications network to deal with all types of consumer complaints. No utility can expect to respond in a timely manner to legitimate consumer dissatisfaction with the water unless it operates under adequate complaint recording and notification procedures.

A good practice is to supply any personnel who deal with the public, either by telephone or in person, with a standard water quality complaint form (Figure 9.10). The form should include, at a minimum, a place for recording the date and time of the complaint, the location (address), and as complete a description of the problem given by the consumer as is possible to obtain.

Dissatisfied consumers who contact the water supplier should be treated courteously and respectfully, and should be considered a valuable resource for the utility in its efforts to provide a high-quality water for public consumption. If the problem has already been identified by the utility, the personnel in positions that will involve public contact should be provided with a clear, easily understood explanation of the problem and the actions being taken to correct the situation. If the

WATER QUALITY COMPLAINT FORM		
DATE CALL RECEIVED	CALL TAKEN BY	TIME
TYPE OF COMPLAINT (TASTE, ODOR, COLOR, TURBIDITY, PLANKTON, BACTERIA)		
NAME AND ADDRESS OF CONSUMER MAKING COMPLAINT		
LOCATION OF COMPLAINT		
DESCRIP-TION OF COMPLAINT		
COMMENTS OF FIELD INVESTI-GATOR		
RESPONSE (INCLUDE DATE, YOUR NAME, AND NAME OF PERSON TO WHOM YOU TALKED)		
CONCLUSIONS		

See reverse for results of analyses. Send completed copy to Water Quality Manager.

Fig. 9.10   Typical water quality complaint form

(Permission of The Metropolitan Water District of Southern California)

problem is likely to be corrected in the near future, a date at which the consumer can expect an improvement in the water should be provided.

Notification of the proper personnel within an organization when a complaint has been received is vital to the utility's ability to respond to the problem. When taste and odor complaints are received, it is important that the operators and those responsible for water quality control be provided with the information. Unless the people who are in a position to take action are aware of a problem, no solution is likely to be found. A standard notification procedure for complaints should be established and followed for providing the necessary information to the appropriate units within the organization.

In many cases, it is not necessary to rely on consumer complaints to identify a taste and odor problem or to anticipate the onset of a taste and odor episode. Seasonal tastes and odors in the source water plague many utilities. Many utilities are plagued by seasonal tastes and odors in the source water. By keeping complete and accurate records of taste and odor outbreaks it is often possible to predict the start of problems and take corrective action before the consumers are affected. Records must be maintained from one year to the next and organized to allow easy retrieval of information.

Routine testing of the raw and finished water for taste and odor quality is an important, ongoing activity of every water supplier. A Taste and Odor Monitoring Program as outlined in Section 9.30 should be implemented. Even if no previous record of taste and odor problems exists, testing should be carried out as part of a preventive program. If plant operators identify a problem before it reaches the distribution system, action to control objectionable tastes and odors can be taken without exposing the consuming public to unpalatable water. Taste tests of the finished water and threshold odor number determinations of the raw and finished water should be conducted on a regular basis. The frequency of these tests should be increased during periods of the year when tastes and odors have occurred in the past and during any time that taste and odor treatments are being used.

QUESTIONS

Write your answers in a notebook and then compare your answers with those on page 445.

9.5A   What information should be recorded on a water quality complaint form?

9.5B   How should operators be kept informed of taste and odor problems?

9.5C   What tests can operators conduct to identify taste and odor problems before they reach the consumer?

9.6   DEVELOPING A TASTE AND ODOR CONTROL STRATEGY

The first step in developing a taste and odor control strategy is to recognize that a problem exists and determine the extent of the problem. If only one complaint has been received, the problem may exist within the consumer's home plumbing and no action at the treatment plant can be expected to improve the situation. If several complaints have been received from one area of the distribution system but the rest of the system has not experienced problems, the problem is probably the result of local conditions in the distribution system. Factors that may affect one area of a system include chlorine residuals above or below normal, low flows due to low demands, or a cross connection. However, because other, more serious water quality

problems may be associated with a localized taste and odor problem, prompt response to such complaints is required.

In systems that use multiple water sources, taste and odor complaints may arise when a new source of water is provided to an area. The complaints do not necessarily mean one water is better than the other—only that the two are different. This frequently occurs in systems that use both a surface supply and groundwater, although use of two or more different surface supplies may also lead to complaints. When the utility switches back to the original source, more complaints can be expected. In cases where water from two different sources is occasionally introduced into the distribution system at opposite ends of the system, the point in the system where these waters meet can also be a source of odor complaints. Utility personnel who handle complaints must be made aware of the situation so they can explain to consumers that a change in supply does not necessarily mean a change in water quality.

If the entire system is being affected by a taste and odor problem, it is important to try to determine just where the problem originates. If complaints can be associated with the use of a particular raw water supply, changing to an alternative source may resolve the problem. If an alternative water supply is not available, then an evaluation of conditions that might be responsible for the deterioration of water quality is necessary.

Reviewing water quality test results from different locations within the supply system may indicate that a problem has developed at a particular point. Conduct a physical inspection of the suspected area and correct any conditions that appear to be adversely affecting water quality. The raw water source, storage facilities, raw water transmission systems, and facilities at the treatment plant all should be evaluated as potential sources of the problem. Finished water storage facilities within systems have also been sources of tastes and odors for many water facilities.

Collect samples at a number of locations from both the source and distribution system for analysis in the laboratory. Examine raw water samples for algal composition and density and, using a sensory analysis method such as FPA, determine whether there has been an increase in source water odors. If the offensive odor is identified in the source water but no taste-and odor-producing planktonic algae is detected, an examination of attached algae may be necessary, especially if the odor is earthy or musty. Sensory analysis should also be conducted on distribution system samples to determine the extent of the problem.

Investigations to determine the cause of a taste and odor episode may take several weeks before providing any useful information. While the laboratory is conducting tests to try to determine the cause, the plant operators are faced with the question of how to treat the problem. Evaluation of different treatment techniques should be done in cooperation with the laboratory, but testing needs to be done in such a manner that the results are useful to plant operation, especially during the initial stages of a taste and odor problem.

Do not waste time evaluating treatment programs that are not available for use at the treatment plant. In order to concentrate your initial efforts in treatment evaluation in those areas that the plant can most readily use, write up an inventory of available treatment techniques. (Ideally this information should be collected and updated before a taste and odor problem begins.) Review the operational status of installed equipment during this inventory process. For example, powdered activated carbon feed systems that are not operable because of repair or maintenance work will not be much help in controlling current tastes and odors.

In addition to looking at readily available treatment processes, identify alternative methods for taste and odor control at other locations. An example would be the capacity to superchlorinate the plant influent, or possible locations upstream of the plant at which potassium permanganate could be applied. Collecting this information in the early stages of a taste and odor outbreak can be useful in developing control programs that are both effective and realistic for a particular facility.

Once you have identified the options available for treating the taste and odor, determine what treatment or combination of treatments is effective in reducing or eliminating the objectionable qualities of the water. Test each available treatment in the laboratory for its ability to improve water quality; combinations of treatments should also be investigated. Most treatment plants faced with taste and odor problems use a combination of two or three treatments to produce finished water of the best possible quality. In one particular situation, for example, it may be possible to use $KMnO_4$ at the raw water intake, powdered activated carbon at the flash mixer, improved coagulation, flocculation, and sedimentation, and above-normal chlorination before filtration to reduce the taste and odor properties of the water to a level that the consumers find acceptable. The most important aspect of a taste and odor control evaluation program is to test as many possible combinations as you reasonably can to determine the most effective approach to solving the problem.

Each potential treatment scheme may require changes in the various steps involved. Dose control at each step of the process is critical so that the individual processes work in harmony and do not create additional problems while trying to control the original source of the consumer complaints. In most cases, for example, it is both ineffective and uneconomical to superchlorinate and add powdered activated carbon at the same point in the process stream. But at the same time, it may be unwise to

completely shut down prechlorination just because carbon is being fed at a point close to the chlorine diffuser. Without prechlorination, algae may rapidly develop in downstream facilities at the treatment plant. The savings associated with reduced chlorine use may well be offset by the increase in problems associated with the algal growth in the treatment works.

In developing a taste and odor control treatment program, it is useful to remember that many of the available methods require some minimum dose level in order to be effective. Application of amounts greater than this minimum dose may provide no further benefit to water quality and, in some cases, may actually impair water quality. For example, tests show that a water supply required 10 mg/L PAC for taste and odor control. To save money, you use only 3 mg/L PAC. However, at 3 mg/L there is almost no improvement in the water quality so you may actually be wasting money at this dosage level. The water requires *AT LEAST* 10 mg/L PAC to show any significant improvement. At the same time, treating the water with 15 mg/L PAC does not improve the quality over that which would be achieved by the 10 mg/L treatment. Similarly, treatment with 0.5 mg/L of $KMnO_4$ may reduce the objectionable qualities of the water while 0.3 mg/L may do no good and 1.0 mg/L may well create colored-water problems in addition to the existing taste and odor problem.

Development of a Taste and Odor Monitoring Program that includes both routine counting and identification of algae populations in source waters, and sensory analysis of raw, treated, and distribution system waters is essential. Routine monitoring of the algae populations in source waters, in conjunction with the use of sensory analysis in the form of an FPA panel, can provide a robust early warning system. These analyses will provide valuable information for evaluating the source and cause of a taste and odor problem and the effectiveness of various treatment techniques in controlling the problem.

Sophisticated laboratory analytical methods using a gas chromatography-mass spectrometry instrument have been developed to measure and identify taste- and odor-causing compounds at the part-per-trillion (ppt) level. The Metropolitan Water District of Southern California applied specialized analytical methods to the detection of geosmin and methylisoborneol, compounds known to cause earthy-musty odors and tastes in water. Other laboratories may be expected to develop similar capabilities in the future. However, for laboratories that cannot afford the costs of maintaining this type of instrumentation, an odor panel, experienced in flavor profile analysis, should be the first step in dealing with taste and odor events.

Increased public awareness of potential health hazards associated with drinking water supplies and the legitimate concern of consumers for the quality of the water that they receive have emphasized the need for prompt and effective action during taste and odor incidents. Constant testing of both raw and treated water and adequate training of operators responsible for water treatment are required to provide the public with high-quality, safe, and pleasing supplies of drinking water.

PREVENTION OF TASTE AND ODOR IN PUBLIC WATER SUPPLIES IS USUALLY MORE EFFECTIVE AND MORE ECONOMICAL THAN TREATMENT. To prevent the development of taste and odor problems, operators should (1) keep a clean and well-maintained water treatment plant and distribution system, and (2) study plant records in order to anticipate changes that might lead to taste and odor problems. However, every water system is subject to unusual and uncontrollable changes in source quality. Familiarity with available treatment alternatives and training in the operation and application of these alternatives is an important aspect of operator development and improvement.

QUESTIONS

Write your answers in a notebook and then compare your answers with those on page 445.

9.6A What factors could cause taste and odor complaints in a local area of a distribution system?

9.6B Which locations in a water supply system might contribute to a taste and odor problem?

9.6C Why is it important to write out and update a list of options available for treating taste and odor problems?

9.6D What two analyses can be used as an early warning system for taste and odor events?

9.7 ARITHMETIC ASSIGNMENT

Turn to the Appendix, "How to Solve Water Treatment Plant Arithmetic Problems," at the back of this manual and read Section A.9, *STEPS IN SOLVING PROBLEMS.* Check all of the arithmetic in this section using an electronic pocket calculator. You should be able to get the same answers.

9.8 ADDITIONAL READING

1. *NEW YORK MANUAL,* Chapter 14,* "Taste and Odor Control."

2. *TEXAS MANUAL,* Chapter 5,* "Tastes and Odors in Surface Water Supplies (Interference, Nuisance, and Taste-and Odor-Producing Organisms)."

* Depends on edition.

End of Lesson 3 of 3 Lessons on TASTE and ODOR CONTROL

Please answer the discussion and review questions next.

DISCUSSION AND REVIEW QUESTIONS

Chapter 9. TASTE AND ODOR CONTROL

(Lesson 3 of 3 Lessons)

Write the answers to these questions in your notebook. The question numbering continues from Lesson 2.

18. How would you determine whether a particular activated carbon would be effective in removing specific tastes and odors?

19. Why should an overhead spray system be used when powdered activated carbon is being loaded into slurry tanks?

20. How does powdered activated carbon influence the performance of filters?

21. What safety precautions should be exercised when cleaning a tank that stores powdered activated carbon?

22. How should a water supplier treat dissatisfied consumers who contact them with a complaint?

23. What is the first step in developing a taste and odor control strategy?

24. Why should a taste and odor control evaluation program test as many combinations of treatment schemes as possible?

25. What are the most important steps an operator can take to prevent the development of taste and odor problems?

SUGGESTED ANSWERS

Chapter 9. TASTE AND ODOR CONTROL

ANSWERS TO QUESTIONS IN LESSON 1

Answers to questions on page 415.

9.0A Taste and odor problems may occur at a water treatment plant on a persistent, seasonal, occasional, or infrequent basis.

9.0B Taste and odor, along with colored water complaints, are the most common types of water quality complaints received by a water utility.

9.0C Loss of public confidence in the water utility's ability to provide safe, high-quality water may be the most damaging effect of taste and odor problems.

Answers to questions on page 416.

9.1A Geosmin and MIB are two compounds that can produce taste and odors in water.

9.1B Dichloramine is an inorganic chloramine that can cause odor problems in water.

9.1C Tastes and odors may be caused by either natural or manmade conditions.

9.1D Taste and odor conditions could develop in raw water sources, conveyance facilities, treatment plants, chlorination stations, finished water storage reservoirs, distribution systems, and consumer plumbing.

Answers to questions on page 417.

9.1E Biological growths leading to objectionable tastes and odors can occur during any time of the year and in any area of a water system.

9.1F Microbial populations can contribute to unpleasant tastes and odors in water by (1) the production of metabolic by-products, and (2) the release of cellular materials.

9.1G The conditions that develop after a large microbial growth may increase taste and odor problems if the decomposition process is accelerated. Decomposition may cause both metabolic by-product formation and release of cellular materials.

9.1H Actinomycetes and sulfur-reducing bacteria are two groups of bacteria that can produce tastes and odors in the water.

9.1I *Oscillatoria, Anabaena, Aphanizomenon,* and *Microcystis* are common blue-green algae associated with earthy-musty odors.

9.1J An algicide should be considered when the dominant algae is an identified taste and odor or an algae is identified as a potent T&O producer and sensory or chemical analysis indicates an increase in odor levels in source waters.

Answers to questions on page 418.

9.1K Types of materials in runoff waters that can lead to oxygen depletion are nitrogen- and phosphorus-containing compounds and organic matter.

9.1L Nutrient levels in a water supply can be increased by either natural conditions or human activity in the watershed.

9.1M The diurnal dissolved oxygen cycle is caused by algae that produce oxygen (photosynthesis) during the day, which can create supersaturated conditions. During the night algae remove dissolved oxygen from the water (respiration) and cause the dissolved oxygen level to approach zero.

Answers to questions on page 419.

9.1N Contamination from septic tank and leach field systems in rural areas and resort or vacation developments may be caused by high densities of homes and people, locations too close to rivers, soils with inadequate percolation rates, poor maintenance, and substandard installation.

9.1O Industrial waste dischargers can cause taste and odor problems from the combined effects of all facilities that discharge into the water supply and also from process and mechanical failures of the industrial wastewater treatment processes.

9.1P The sources of chemical spills include accidents at industrial plants, chemical storage facilities, or during transportation.

Answers to questions on page 420.

9.1Q Plant sedimentation walls and weirs and filter walls are likely locations for algal colonization.

9.1R Many municipal water intake facilities have been located upstream of major agricultural areas to avoid possible contamination of the water supply by fertilizers, pesticides, and herbicides. Siting of intake facilities above heavily cultivated agricultural lands also helps protect against tastes and odors.

9.1S Sources of agricultural wastes that may cause taste and odor problems in a water supply include high-density animal feeding and dairy operations and runoff from cultivated fields.

9.1T   Microorganisms will grow in plant debris and sludges even in the presence of a strong chlorine residual. Microbial growths in these materials can lead to foul, septic, musty, or other types of tastes and odors.

Answers to questions on page 422.

9.2A   Taste and odor problems are most likely to develop in any parts of the system that are used to store, transport, or regulate untreated water. These parts may provide a suitable habitat for organisms that produce objectionable tastes and odors in drinking water due to an absence of chlorine residual.

9.2B   A sanitary survey is used to identify the sources of tastes and odors in raw waters.

9.2C   Chemicals such as chlorine and activated carbon tend to mask any changes in taste and odor quality that may be occurring.

9.2D   Potential sources of tastes and odors in a water treatment plant include areas where debris and sludge can accumulate, and the walls, channels, and weirs where algal and slime (bacterial) growth can occur.

9.2E   The main causes of tastes and odors in the distribution system are microbial activity, disinfection residuals and their by-products, organic or mineral compounds from system materials, and external contaminants from cross connections.

9.2F   Complaints regarding "chemical," "gasoline," or "pesticide" tastes or odors should alert you to the possibility of a cross connection.

Answers to questions on page 423.

9.3A   Algae counting and identification, and sensory analysis are two components that should be included in a Taste and Odor Monitoring Program.

9.3B   If settled sludges and other debris are allowed to accumulate for an extended time, they will become septic and impart a foul taste and odor to the water.

9.3C   Portions of a water treatment plant are usually taken out of service for inspection and cleaning during the winter months when flows are lower and full plant capacity is not needed to meet the community's water demands.

9.3D   Taste and odor treatment equipment should be capable of operating properly at all times so the equipment will work whenever a problem occurs.

9.3E   FPA uses a trained panel to identify tastes and odors in water.

ANSWERS TO QUESTIONS IN LESSON 2

Answers to questions on page 424.

9.4A   The two broad categories of taste and odor treatment methods are (1) removal, and (2) destruction.

9.4B   Tastes and odors caused by algae can be removed most economically in a water treatment plant by improving coagulation and sedimentation. This procedure may lengthen filter runs by achieving removal of the algae in the sedimentation basin rather than by filtration.

9.4C   The reaction of chlorine on algal cells results in the release of objectionable cellular materials into the water. This may increase taste and odor problems.

Answers to questions on page 426.

9.4D   Aeration is best suited for treating odor problems caused by volatile compounds. Aeration is less successful in treating taste problems because these are more often associated with less volatile compounds.

9.4E   Aeration systems can be designed to either pass the air through the water, or to pass water through the air.

Answers to questions on page 429.

9.4F   Odors that can often be significantly reduced by chlorination include fishy, grassy, or flowery odors.

9.4G   The use of chlorine in waters containing phenols could result in the production of highly objectionable chlorophenolic compounds. These compounds cause noticeable tastes and odors even in concentrations 40 to 200 times *LESS* than the original phenol.

9.4H   If a permanganate overdose does occur, treat with powdered activated carbon if the facilities are available. A second alternative is to increase the pH of the water for a short time to increase the rate at which the manganese is precipitated.

9.4I   Permanganate storage facilities must be dry and well ventilated because moisture in the atmosphere can cause caking of the material. This will cause the feeder to clog and prevent accurate delivery of the permanganate.

9.4J   Adequate permanganate dust control and ventilation are important because permanganate and moisture create a very corrosive mixture which will attack metal, including electrical connections.

Answers to questions on page 430.

9.4K   Laboratory and pilot-scale testing of the effectiveness of ozonation as a taste and odor control process should be conducted before full-scale installation because of the specialized equipment requirements and costs.

9.4L   Chlorine dioxide must be generated on site because it is unstable and cannot be stored.

9.4M   Chlorine dioxide has been used to treat tastes and odors caused by industrial pollution, especially in cases where chlorine has intensified the problem.

ANSWERS TO QUESTIONS IN LESSON 3

Answers to questions on page 432.

9.4N   The two forms of activated carbon used in water treatment are powdered and granular (PAC and GAC).

9.4O   Activated carbons are rated on the basis of a "phenol number" or an "iodine number."

9.4P   Powdered activated carbon should be applied at plant mixing facilities because carbon must contact the material to be removed in order for adsorption to occur.

9.4Q   If a caked layer of floating carbon starts to form on the surface of the water in the slurry tank, stop the operation, check mixers and sprayers, and break up any cake that has formed. If all equipment is operating properly, the loading rate must be reduced to the level at which the system can adequately mix the carbon into a slurry.

Answers to questions on page 438.

9.4R Because powdered activated carbon is such an abrasive material, equipment used to feed PAC often requires more frequent inspection, cleaning, and maintenance than other chemical feed equipment. Pumping equipment should be routinely inspected for signs of wear. Frequent inspection during actual carbon delivery is necessary when using this equipment.

9.4S If PAC tends to cake on the surface of filters, adjust the settling process to remove more PAC before it reaches the filters.

9.4T Carbon penetration through a filter can be detected by passing one liter of filter effluent through a 0.45-μm membrane filter. The presence of PAC in the effluent can be determined from a darkening of the white membrane filter surface.

9.4U Empty bed contact time is the time that the water is actually in the filter bed.

Answers to questions on page 440.

9.5A Information that should be recorded on a water quality complaint form includes date and time of complaint, location (address), and a complete description of the problem.

9.5B Operators should be informed of taste and odor problems by personnel who receive and log consumer complaints. Also, if operators keep good records they can anticipate when seasonal taste and odor problems will develop and be prepared to respond at the appropriate time.

9.5C Operators can conduct taste tests of the finished water, and threshold odor number determinations of the raw and finished water on a regular basis.

Answers to questions on page 442.

9.6A Factors that could cause taste and odor complaints in a local area of a distribution system include chlorine residuals above or below normal, low flows due to low demands, or a cross connection.

9.6B Locations that might contribute to a taste and odor problem include the raw water source, storage facilities, raw water transmission systems, facilities at the treatment plant, and finished water storage facilities within the system.

9.6C A list of available options for treating taste and odor problems is essential so that when a problem develops, the lab can immediately run tests on the effectiveness of available solutions. Such a list would eliminate time wasted investigating unrealistic solutions.

9.6D Algae monitoring and FPA analysis can be used as an early warning system for taste and odor events.

CHAPTER 10

PLANT OPERATION

by

Jim Beard

TABLE OF CONTENTS

Chapter 10.   PLANT OPERATION

OBJECTIVES

Chapter 10.   PLANT OPERATION

Following completion of Chapter 10, you should be able to:

1. Monitor and control water treatment processes, and

2. Safely operate and maintain a water treatment plant.

Chapter 10 will help you accomplish these objectives by instructing you how to:

1. Regulate flows,

2. Apply chemicals and adjust dosages,

3. Prepare operating reports and records,

4. Maintain equipment and facilities,

5. Develop daily operating procedures for your plant,

6. Respond to emergency conditions,

7. Handle consumer complaints, and

8. Implement energy conservation measures.

WORDS

Chapter 10.   PLANT OPERATION

BOD (pronounce as separate letters)   BOD

Biochemical **O**xygen **D**emand. The rate at which organisms use the oxygen in water while stabilizing decomposable organic matter under aerobic conditions. In decomposition, organic matter serves as food for the bacteria and energy results from its oxidation. BOD measurements are used as a measure of the organic strength of wastes in water.

COMPOSITE (come-PAH-zit) (PROPORTIONAL) SAMPLE   COMPOSITE (PROPORTIONAL) SAMPLE

A composite sample is a collection of individual samples obtained at regular intervals, usually every one or two hours during a 24-hour time span. Each individual sample is combined with the others in proportion to the rate of flow when the sample was collected. The resulting mixture (composite sample) forms a representative sample and is analyzed to determine the average conditions during the sampling period.

CONTINUOUS SAMPLE   CONTINUOUS SAMPLE

A flow of water from a particular place in a plant to the location where samples are collected for testing. This continuous stream may be used to obtain grab or composite samples. Frequently, several taps (faucets) will flow continuously in the laboratory to provide test samples from various places in a water treatment plant.

GRAB SAMPLE   GRAB SAMPLE

A single sample of water collected at a particular time and place which represents the composition of the water only at that time and place.

MATERIAL SAFETY DATA SHEET (MSDS)   MATERIAL SAFETY DATA SHEET (MSDS)

A document which provides pertinent information and a profile of a particular hazardous substance or mixture. An MSDS is normally developed by the manufacturer or formulator of the hazardous substance or mixture. The MSDS is required to be made available to employees and operators whenever there is the likelihood of the hazardous substance or mixture being introduced into the workplace. Some manufacturers are preparing MSDSs for products that are not considered to be hazardous to show that the product or substance is *NOT* hazardous.

PICO   PICO

A prefix used in the metric system and other scientific systems of measurement which means 10^{-12} or 0.000 000 000 001.

PICOCURIE   PICOCURIE

A measure of radioactivity. One picoCurie of radioactivity is equivalent to 0.037 nuclear disintegrations per second.

POLYELECTROLYTE (POLY-ee-LECK-tro-lite)   POLYELECTROLYTE

A high-molecular-weight (relatively heavy) substance having points of positive or negative electrical charges that is formed by either natural or manmade processes. Natural polyelectrolytes may be of biological origin or derived from starch products and cellulose derivatives. Manmade polyelectrolytes consist of simple substances that have been made into complex, high-molecular-weight substances. Used with other chemical coagulants to aid in binding small suspended particles to larger chemical flocs for their removal from water. Often called a POLYMER.

POLYMER (POLY-mer)   POLYMER

A long chain molecule formed by the union of many monomers (molecules of lower molecular weight). Polymers are used with other chemical coagulants to aid in binding small suspended particles to larger chemical flocs for their removal from water.

POWER FACTOR   POWER FACTOR

The ratio of the true power passing through an electric circuit to the product of the voltage and amperage in the circuit. This is a measure of the lag or lead of the current with respect to the voltage. In alternating current the voltage and amperes are not always in phase; therefore, the true power may be slightly less than that determined by the direct product.

REPRESENTATIVE SAMPLE

REPRESENTATIVE SAMPLE

A sample portion of material or water that is as nearly identical in content and consistency as possible to that in the larger body of material or water being sampled.

SUPERNATANT (sue-per-NAY-tent)

SUPERNATANT

Liquid removed from settled sludge. Supernatant commonly refers to the liquid between the sludge on the bottom and the scum on the water surface of a basin or container.

CHAPTER 10.   PLANT OPERATION

(Lesson 1 of 2 Lessons)

10.0 GOALS OF PLANT OPERATION

In the operation of water treatment plants, competent and responsible operators try to achieve three basic objectives:

1. Production of a safe drinking water,

2. Production of an aesthetically pleasing drinking water, and

3. Production of drinking water at a reasonable cost.

From a public health perspective, production of a safe drinking water, one that is free of harmful bacteria and toxic materials, is the first priority. Federal, state, and local regulations (drinking water standards) control all aspects of treatment of public water supplies to ensure the delivery of safe water to consumers.

In addition to providing safe water, it is also important to produce a high-quality water that appeals to the consumer. Generally, this means that the water must be clear (free of turbidity), colorless, and free of objectionable tastes and odors. Consumers also want water supplies that do not stain plumbing fixtures and clothes, do not corrode plumbing fixtures and piping, and do not leave scale deposits or spot glassware.

Not only do consumers want safe and pleasing water, but they want it at a reasonable cost. In this era of rapidly increasing energy costs, treatment plant operators must examine all plant processes to trim costs and maintain a cost-effective operation.

Consumer sensitivity to the environment (air quality, water quality, noise) has significantly increased in recent years. With regard to water quality, consumer demands have never been greater. In some instances, consumers have substituted bottled water for drinking water and cooking purposes.

Again, maintaining a standard of sanitary excellence and meeting the increasing demands of consumers for high-quality drinking water requires conscientious operators to produce and maintain the quality of the water from the plant to the consumer's tap.

QUESTIONS

Write your answers in a notebook and then compare your answers with those on page 488.

10.0A   What is the first priority for operating a water treatment plant?

10.0B   What type of water is appealing to consumers?

10.1 DRINKING WATER CONSIDERATIONS

10.10 Drinking Water Regulations

For centuries people have judged water quality on the basis of taste, smell, and sight. To a large degree, this is still true. However, many consumers today are demanding water that is not only free of objectionable tastes and odors, but also free of harmful bacteria, organic and inorganic chemical contaminants, turbidity, and color.

Current water quality standards include both federal and state regulations. Following passage of the Safe Drinking Water Act (PL 93-523) in 1974, the U.S. Environmental Protection Agency (EPA) was charged with the responsibility of developing and implementing national drinking water regulations. A summary of the maximum contaminant levels (MCLs) established by these regulations and later amendments is shown in Tables 10.1 and 10.2. Primary regulations establish MCLs based on the health significance of the contaminants while the secondary standards are established based on aesthetic considerations and are a state option.

Under the Safe Drinking Water Act, each state was given the option of assuming primary enforcement responsibility for public water systems on the condition that the state would adopt regulations at least as stringent as the EPA regulations and would implement adequate monitoring and enforcement procedures. Most states have agreed to these terms and have been delegated primary responsibility by the EPA.

The poster provided with this manual summarizes the essential elements of current drinking water regulations as well as the major new regulations under development. For each type of regulated contaminant, this poster lists the maximum contaminant level (MCL) or required treatment technique, the known health effects of the contaminant, monitoring requirements, and other important information about the regulations.

Operators are urged to stay in close contact with their state regulatory agencies and become thoroughly familiar with their state requirements. Although this poster is revised frequently and every effort is made to ensure its accuracy at the time of

TABLE 10.1 PRIMARY DRINKING WATER REGULATIONS

Constituent	Maximum Contaminant Level	Constituent	Maximum Contaminant Level
Inorganic Chemicals		*PESTICIDES AND SYNTHETIC ORGANICS* (continued)	
Antimony	0.006 mg/L	Carbofuran	0.04 mg/L
Arsenic	0.05 mg/L	Chlordane	0.002 mg/L
Asbestos	7 million fibers/L	Dalapon	0.2 mg/L
Barium	2.0 mg/L	Dibromochloropropane (DBCP)	0.0002 mg/L
Beryllium	0.004 mg/L	Di(2-ethylhexyl)adipate	0.4 mg/L
Cadmium	0.005 mg/L	Di(2-ethylhexyl)phthalate	0.006 mg/L
Chromium	0.1 mg/L	Dinoseb	0.007 mg/L
Copper	1.3 mg/L [a] (at tap)	Diquat	0.02 mg/L
Cyanide	0.2 mg/L	Endothall	0.1 mg/L
Fluoride	4.0 mg/L	Endrin	0.002 mg/L
Lead	0.015 mg/L [b] (at tap)	Epichlorhydrin	treatment technique
Mercury	0.002 mg/L	Ethylene Dibromide (EDB)	0.00005 mg/L
Nitrate (as N)	10.0 mg/L	Glyphosate	0.7 mg/L
Nitrite (as N) [b]	1.0 mg/L	Heptachlor	0.0004 mg/L
Selenium	0.05 mg/L	Heptachlor Epoxide	0.0002 mg/L
Thallium	0.002 mg/L	Hexachlorobenzene	0.001 mg/L
Organic Chemicals		Hexachlorocyclopentadiene	0.05 mg/L
VOLATILE ORGANICS		Lindane	0.0002 mg/L
Benzene	0.005 mg/L	Methoxychlor	0.04 mg/L
Carbon Tetrachloride	0.005 mg/L	Oxamyl (Vydate)	0.2 mg/L
o-Dichlorobenzene	0.6 mg/L	PCBs	0.0005 mg/L
p-Dichlorobenzene	0.075 mg/L	Pentachlorophenol	0.001 mg/L
1,2-Dichloroethane	0.005 mg/L	Picloram	0.5 mg/L
1,1-Dichloroethylene	0.007 mg/L	Simazine	0.004 mg/L
cis-1,2-Dichloroethylene	0.07 mg/L	Toxaphene	0.003 mg/L
trans-1,2-Dichloroethylene	0.1 mg/L	2,4-D	0.07 mg/L
Dichloromethane	0.005 mg/L	2,3,7,8-TCDD (Dioxin)	0.00000003 mg/L
1,2-Dichloropropane	0.005 mg/L	2,4,5-TP (Silvex)	0.05 mg/L
Ethylbenzene	0.7 mg/L	**Microbial**	
Monochlorobenzene	0.1 mg/L	*Total Coliform	1 per 100 mL
Styrene	0.1 mg/L		<40 samples/mo - no more than 1 positive
Tetrachloroethylene	0.005 mg/L		>40 samples/mo - no more than 5% positive
Toluene	1.0 mg/L	*Giardia lamblia*	3-log (99.9%) removal [c]
1,2,4-Trichlorobenzene	0.07 mg/L	*Legionella*	treatment technique [c]
1,1,1-Trichloroethane	0.2 mg/L	Enteric viruses	4-log (99.99%) removal [c]
1,1,2-Trichloroethane	0.005 mg/L	Heterotrophic bacteria	treatment technique [c]
Trichloroethylene (TCE)	0.005 mg/L	**Physical**	
Vinyl Chloride	0.002 mg/L	*Turbidity [c]	0.5 to 5 NTU
Xylenes (total)	10.0 mg/L	**Radionuclides**	
PESTICIDES AND SYNTHETIC ORGANICS		Gross alpha particles	15 pCi/L
Acrylamide	treatment technique	Gross beta particles [d]	4 mrem/yr
Alachlor	0.002 mg/L	Radium 226 & 228	5 pCi/L
Atrazine	0.003 mg/L	**Disinfection By-Products**	
Benzo(a)pyrene	0.0002 mg/L	TTHMs [e]	0.080 mg/L

[a] Action level for treatment.
[b] Applies to community, nontransient noncommunity, and transient noncommunity water systems.
[c] Applies to systems using surface water or groundwater under the influence of surface water.
[d] Applies to surface water systems serving more than 100,000 persons and any system determined by the state to be vulnerable.
[e] Applies to systems serving more than 10,000 persons and also to all surface water systems that meet the criteria for avoiding filtration.

*NOTE: Only coliforms and turbidity are more or less under the control of the operator; all other items are not influenced significantly by plant treatment processes.

NOTE: YOUR REGULATORY AGENCY MAY HAVE STRICTER REGULATIONS. CONTACT APPROPRIATE OFFICIALS TO DETERMINE THE REGULATIONS THAT APPLY TO YOUR PLANT.

TABLE 10.2   ENVIRONMENTAL PROTECTION AGENCY SECONDARY DRINKING WATER REGULATIONS

Constituent	Maximum Contaminant Level[a]
Aluminum	0.05 - 0.2
Chloride	250
Color*	15 Color Units
Copper	1.0
Corrosivity*	Noncorrosive
Fluoride	2
Foaming Agents (MBAS)	0.5
Iron*	0.3
Manganese*	0.05
Odor*	3 Threshold Odor Number
pH*	6.5 - 8.5
Silver	0.10
Sulfate	250
Total Dissolved Solids (TDS)	500
Zinc	5

[a] mg/L unless noted.

NOTE: All items marked * are more or less under the comtrol of the operator; all other items are not influenced significantly by plant treatment processes.

NOTE: *YOUR REGULATORY AGENCY MAY HAVE STRICTER REGULATIONS. CONTACT APPROPRIATE OFFICIALS TO DETERMINE THE REGULATIONS THAT APPLY TO YOUR PLANT.*

publication, it is extremely difficult to keep pace with changing regulations. An excellent source of up-to-the-minute information about drinking water regulations is EPA's toll-free Safe Drinking Water Hotline at (800) 426-4791. Also see *WATER TREATMENT PLANT OPERATION*, Volume II, Chapter 22, "Drinking Water Regulations," for more information about primary and secondary contaminants and monitoring, sampling, and reporting requirements.

10.11   Monitoring Program

Operators perform a variety of laboratory tests on source water samples, process water samples, and finished water samples to monitor overall water quality and to evaluate process performance. The sampling location and type of water sample used for a particular analysis will vary depending on the purpose of the analysis. Compliance with the maximum contaminant levels (MCLs) is usually measured at the point where water enters the distribution system. However, operators must realize that water can degrade in the distribution or delivery system. In all cases, it is important to stress that the water sample must be representative of actual conditions of the entire flow being sampled.

GRAB SAMPLES[1] are usually adequate for making periodic measurements of water quality indicators. They are especially useful when measuring indicators that may change quickly after collection, such as coliforms, pH, and temperature. When variations in water quality are expected, *COMPOSITE SAMPLES*[2] may be more appropriate. Ideally, process control measurements should be made on a continuous basis by special instrumentation.

A summary of process monitoring guidelines and sample points is shown in Figure 10.1. The frequency of sampling for individual process control water quality indicators will vary from hourly to perhaps once per day, depending on the quality of the source water and the importance of the indicator being evaluated. Thus, certain water quality indicators such as turbidity may be routinely monitored (every four hours or continuously), while others such as alkalinity are sampled less frequently (once per shift).

10.12   Turbidity Removal

Most municipal plants built to treat surface water are designed to remove turbidity, and turbidity is the single water quality indicator over which water treatment plant designers and operators have the greatest control.

In a well-designed and operated treatment plant, very high turbidity removals can be achieved under optimum conditions, while poorly operated treatment plants (inadequate pretreatment and filtration) may only achieve relatively low turbidity removals (filtered water turbidities greater than 1.0 TU). This range of turbidity removal effectiveness is extremely important when one considers the relationship between turbidity and bacteria and other pathogenic organisms. There is considerable evidence which shows that filtered waters with high levels of turbidity cannot be effectively disinfected. Turbidity must be removed in order for disinfection to be effective in killing or inactivating disease-causing organisms.

Thus, the goal of all water treatment plants should be to produce a filtered water with the *LOWEST PRACTICAL LEVEL* of turbidity. Many water treatment plants in the United States treating surface water routinely produce filtered water turbidities in the range of 0.05 to 0.3 TU. A recommended target turbidity level is 0.1 TU.

QUESTIONS

Write your answers in a notebook and then compare your answers with those on page 488.

10.1A   Grab samples are used when measuring what types of water quality indicators?

10.1B   Why are high levels of turbidity removal important?

---

[1] *Grab Sample.   A single sample of water collected at a particular time and place which represents the composition of the water only at that time and place.*

[2] *Composite (come-PAH-zit) (Proportional) Sample.   A composite sample is a collection of individual samples obtained at regular intervals, usually every one or two hours during a 24-hour time span. Each individual sample is combined with the others in proportion to the rate of flow when the sample was collected. The resulting mixture (composite sample) forms a representative sample and is analyzed to determine the average conditions during the sampling period.*

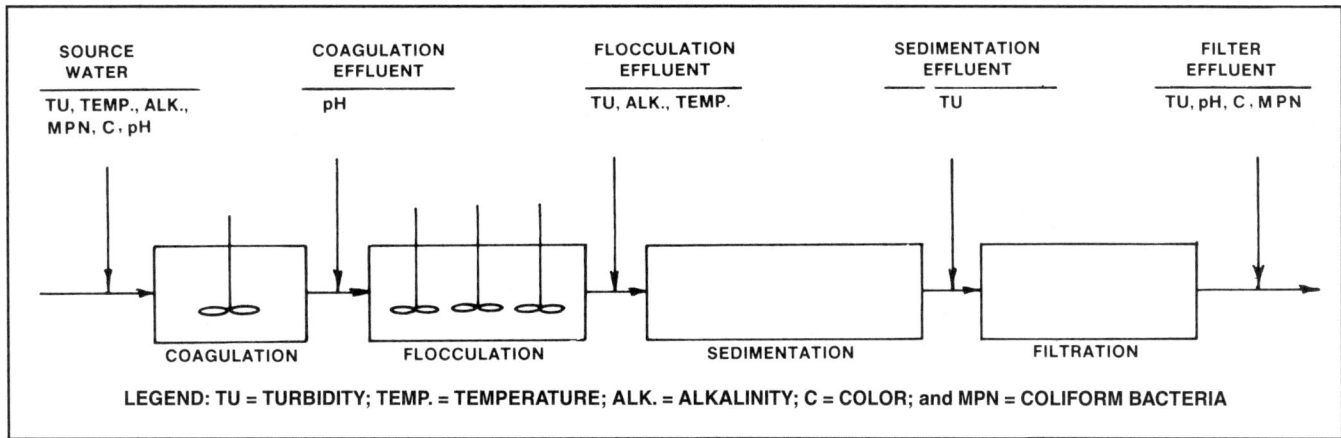

Fig. 10.1 Summary of process monitoring guidelines and sample points

10.2 DAILY OPERATING PROCEDURES

10.20 Prepare a List for Your Plant

Throughout this manual we have explained the step-by-step procedures to effectively operate a water treatment plant. In this section we will outline a checklist of daily activities for the operators of a "typical" plant. *YOU SHOULD PREPARE A SIMILAR LIST FOR YOUR PLANT.* Consider our list and all of the information in this manual. Take into account the size of your plant, type and condition of your plant, raw water characteristics, your distribution system, and the number of operators available to help and their skills. A sample checklist is shown in Table 10.3. (In the boxes an "N" means normal, an "A" means abnormal, and an "O" means out of service.) By using this example and the information contained in this manual, you should be able to prepare a checklist for your plant that reflects your specific needs.

10.21 Daily Tasks

1. Review what has happened during the last shift or since you left the plant and respond to any problems.

   a. An important part of reviewing plant operations is exchanging information with the operator on duty. Shift assignments should be arranged to provide an overlap at shift changes to permit a brief period (perhaps 15 to 30 minutes) for operators to exchange important information on the status of the plant and special problems that need attention or monitoring.

   b. Your review should include a check of raw and finished water quality for any changes. Especially important are the turbidity and chlorine residual of the finished water in the clear well.

   c. Review system pressures.

   d. Check clear well storage and levels of other reservoirs in the distribution system.

   e. Examine records of raw water and distribution system pumping. Inspect records or charts of flowmeters and/or total hours run by each pump.

   f. Check the status of each filter. Do any of them need to be backwashed?

2. Prepare for the day. You have two major concerns: to provide your consumers with drinking water of suitable *QUALITY* and sufficient *QUANTITY*.

   a. *QUALITY*

      Water quality is controlled mainly by the proper application of chemicals. Be sure you have sufficient chemicals to meet today's demands and future demands.

      (1) Alum. Check daily use and amount in storage.

      (2) Polymers. Record daily use, level of day tank, and amount in storage.

      (3) Chlorine. Record daily use, weigh containers, and determine amount in storage.

   b. *QUANTITY*

      Ideally we would like to select an influent pumping rate that will allow our plant to operate at a constant rate for the next 24 hours. This rate should allow the plant to meet all demands and maintain adequate reserves in the clear well and service storage reservoirs. The items listed below should be considered when selecting an influent pumping rate.

      (1) Examine the previous day's circumstances.

         (a) How much water was treated and pumped to the distribution system?

         (b) What were the weather conditions? Was it hot? Overcast? Lawns being watered? Freezing conditions?

      (2) Consider the previous day's average flow based on raw water pumping and records of water pumped to the distribution system.

      (3) Consider storage in clear well. Ideally, the clear well will be nearly full at the start of each day.

      (4) After considering items (1), (2), and (3), select a raw water pumping rate that will allow the plant to operate at a constant rate during the next 24 hours. The clear well will be drawn down during the high demand period during the day and filled at night. See Section 10.4, "Regulation of Flows," for additional details.

TABLE 10.3 ROUTINE PLANT CHECKLIST

DATE_____ OPERATOR_____

       1   2   3

SHIFT   ☐ Midnight-8 a.m.   ☒ 8 a.m.-4 p.m.   ☐ 4 p.m.-Midnight

1. INTAKE STRUCTURE
a. Status of Bar Screens   1 [N] 2 [A]
b. Status of Intake Pumps   1 [N] 2 [N] 3 [O]
COMMENTS: Bar screen in Bay No. 2 partially clogged with debris. Will clean this shift.

2. CHLORINATION SYSTEM
a. Status of Evaporators   1 [N] 2 [N]
b. Status of Chlorinators   1 [N] 2 [N]
c. Status of Chlorine Residual Analyzer   [N]
d. Status of Booster Pumps   1 [O] 2 [O]
e. Chlorine Feed Points Pre- [N]   Intermediate [N]   Post [N]
COMMENTS:_____

3. COAGULANT FEED SYSTEMS
a. Status of Alum Feed Pumps   1 [N] 2 [O]
b. Status of Polymer Feed Pumps   1 [N] 2 [O]
COMMENTS:_____

4. FLASH MIXER/FLOCCULATION SYSTEMS
a. Status of Flash Mixer   [N]
b. Status of Flocculators-Basin No. 1   1 [N] 2 [N] 3 [A] 4 [N]
                   -Basin No. 2   1 [O] 2 [O] 3 [O] 4 [O]
COMMENTS: Flocculator No. 3 in Basin No. 1 starting to develop a whine. Will monitor and advise maintenance.

5. SEDIMENTATION BASINS
a. Basins in Service   1 [N] 2 [O]
b. Status of Sludge Pumps   1 [N] 2 [O]
c. Sludge Lagoons in Service   1 [O] 2 [N] 3 [O]
COMMENTS:_____

6. FILTER STATUS
a. Filters in Service   1 [N] 2 [N] 3 [N] 4 [O]
b. Backwash Cycle: Auto [N]   Manual [O]
c. Filter Run Termination: Loss of Head [N]   Time [O]
d. Status of Filter Aid System [O]
COMMENTS:_____

7. FINISHED WATER CLEAR WELL AND PUMP STATION
a. Clear Well Level   [A]
b. Status of Pumps   1 [N] 2 [N] 3 [O]
c. Status of NaOH Feed System   [N]
COMMENTS: Reservoir level at 9:30 a.m. is low. Will evaluate system demand and adjust plant flow rate.

USE OF CHECKLIST

1. An entry in a box indicates that the unit is in service or out of service and its status (condition).

2. The status of an item is indicated by inserting an "N" (meaning normal), an "A" (meaning abnormal), or an "O" (meaning out of service) in the proper box.

3. Any "A" entry should be briefly explained in the comments section for that system.

4. This checklist is designed to supplement other plant records such as the daily operations log which requires more detailed information and entries. The checklist should not be used in lieu of these other records.

3. Walk through your plant. Start at the intake or headworks and follow the flow of water through your plant to the clear well.

   a. *LOOK* for anything unusual or different in the appearance of the water as it goes through each process. Inspect each piece of equipment.

   b. *LISTEN* for any unusual or different noises from the equipment.

   c. *FEEL* the equipment for excessive temperature and vibrations.

   d. *SMELL* for any signs of developing odors in your finished water. Also be aware of any signs of equipment burning or overheating.

   e. *TASTE* and smell your finished water for any changes or undesirable characteristics.

   f. *SAFETY.* Whenever you walk through your plant, be alert for safety hazards. If you observe anyone using unsafe procedures or if you observe an unsafe condition, correct the situation immediately.

4. Respond to minor problems not taken care of previously.

5. Collect samples for quality control and analyze samples.

   a. Check turbidity, chemical doses (jar tests), chlorine residual, pH, alkalinity, and coliforms.

   b. Check calibrations on finished water recording turbidimeter and chlorine residual analyzer.

6. Perform all scheduled preventive maintenance.

7. Record all necessary data and be sure all records are up to date.

8. Order supplies, including chemicals.

9. Review safety program.

10.22 Tasks To Be Done During the Day

1. Backwash filters (if automatic, everything OK?).

2. System pressures must be observed regularly. If system pressures are monitored and transmitted to a control room, then the operator on duty will be responsible for constant surveillance. If system pressures fluctuate beyond established ranges, adjust pumping rates or number of pumps on line as necessary.

3. Monitor the main control board. Charts provide important information as to present plant status. Alarm panel lights also provide important status information. Beware of lights "ON," particularly "RED" lights as these usually require immediate attention.

4. Storage level in clear well. If clear well level is monitored in a control room, then operator on duty will be responsible for surveillance and any necessary adjustments of influent pumping rates.

5. Pumping rates

   a. Pumping rates into distribution system should be adjusted to maintain system pressures and demands. Pressures should be monitored throughout the system and whenever the pressures start to drop, increase the pump speed or place an additional pump in service.

   b. Raw water pumping rates should be adjusted as necessary to maintain desired levels in clear well throughout the day.

6. Quality control checks

   a. Collect samples for quality control and analyze.

   b. Check turbidity, chemical doses (jar tests), chlorine residual, pH, and alkalinity.

   c. Inspect chlorination system. Is it working properly? Are there any leaks?

7. Repeat these tasks as often as necessary.

10.23 At the End of the Day

Perform the following tasks at the end of your shift or before leaving your plant at night:

1. Repeat the tasks listed under Section 10.22, "Tasks To Be Done During the Day."

2. Anticipate raw water and finished water pumping requirements during the night.

3. Be sure all chemical dosage facilities are prepared to operate until the next operator comes on duty.

4. Secure the plant for the night by checking outside lighting and security systems and locking the plant.

QUESTIONS

Write your answers in a notebook and then compare your answers with those on page 488.

10.2A What should be an operator's first task upon arrival at a water treatment plant?

10.2B How does an operator control finished water quality?

10.2C What tasks should an operator do during the day?

10.2D What tasks should an operator do at the end of the day?

10.3 PROCESS INSTRUMENTATION AND CONTROLS

10.30 Monitored Functions

To assist you in providing consumers with a safe and palatable supply of drinking water, instruments and controls are used to indicate, measure, monitor, record, control, and signal the breakdown of many of the process functions on a continuous or intermittent basis. Regardless of how simple or complex the instrumentation and control systems are for your plant, *YOU* are responsible for controlling plant operation. A summary of typically monitored functions is listed in Table 10.4. In smaller water treatment plants, the number of functions monitored may be limited to the essential functions only. The most common methods of sensing these process functions are listed in Table 10.5.

Instrumentation and controls are communication devices that transmit information (data) from measuring locations in the water treatment plant to a central data collection point (usually a control room). In most modern water treatment plants, this central data collection point is the main control panel (see Figure 10.2). This is the "nerve center" of the plant and is located in the operations control room. From this single

TABLE 10.4   SUMMARY OF COMMONLY MONITORED FUNCTIONS[a]

Function	Monitored Location
Flow	• Raw Water • Service Water • Chemical Solution • Filters • Wash Water • Sludge • Finished Water
Level	• Chemical Tanks and Hoppers • Filters • Wash Water Tank • Clear Well • Recovery Basins
Chlorine Residual	• Each Unit Process • Finished Water
Turbidity	• Raw Water • Each Unit Process • Individual Filters • Finished Water
pH	• Raw Water • Each Unit Process • Finished Water
Weight	• Chlorine Cylinders • Chemical Feeders (Loss-in-Weight Type)
Pressure	• Service Water • Plant Air Supply • Water Level (Bubblers) • Effluent Pump Station • General Piping
Loss-in-Head (Differential Pressure)	• Individual Filters
Sludge Density	• Sedimentation Basin • Solids Disposal Piping
Conductivity	• Raw Water • Finished Water
Temperature	• Raw Water

[a] Adapted from Stone, Reference 1, Section 10.16, "Acknowledgments."

TABLE 10.5   SUMMARY OF COMMON SENSING METHODS[a]

Function	Sensing Method
Flow	• Venturi Meters (Differential Pressure) • Propeller Meter • Magnetic Meter • Sonic Meter • Rotameter
Level	• Float • Bubbler • Probe • Pressure Cell • Sonic
Chlorine	• Amperometric (Electrode) • Colorimetric
Turbidity	• Surface Scatter (High Turbidity) • Nephelometric (Low Turbidity)
pH	• Amperometric (Electrode)
Weight	• Scales (Mechanical) • Load Cells (Electronic or Pressure)
Pressure	• Bourdon Tube (Mechanical) • Electronic • Differential Pressure (D.P. Cell)
Loss-in-Head	• Differential Pressure
Conductivity	• Electronic
Temperature	• Electronic

[a] Adapted from Stone, Reference 1, Section 10.16, "Acknowledgments."

location, the operator can monitor and control most of the major process functions. In many modern water treatment plants, sophisticated electronic methods provide virtually automatic control over major process functions. However, manual controls must be available to back up critical functions in water treatment plants so operators can override the automatic system when necessary.

10.31   Signal Transmission Methods

10.310   Methods Available

There are numerous methods of transmitting data from sensing or measuring locations in the water treatment plant to a central control point. Methods used to transmit data include mechanical, pneumatic, hydraulic, electronic, and electrical. However, electronic systems (millivolt or milliamp) have become the popular choice. When certain data must be transmitted over long distances (greater than about 1,500 feet or 450 m), other systems such as telephone tone, microwave, or radio transmission may be used.

10.311   Mechanical

The most basic type of instrumentation is purely mechanical. These devices are easily understood, reliable, and are generally the most economical. Examples of mechanical devices include:

1. Float valves,

2. Pressure relief valves,

3. Pressure gages,

4. Indicators (valve position and fluid level),

5. Switches (high/low level, pressure), and

6. Scales (chlorine and other chemical weighing devices).

These mechanical devices have been used for many years and are still in use at many of the older plants. Many newer types of sensing and transmitting devices are based on the principles of these mechanical devices.

Fig. 10.2   Main control panel in control room

10.312 Pneumatic

Pneumatic (using air pressure or vacuum) measuring and transmission devices are particularly safe and reliable, and service can be maintained during short-term power outages. In these systems, data are usually transmitted through small-diameter (1/8- to 1/4-inch or 3- to 6-mm) copper tubing. Usually, the air supply for pneumatic instrumentation must be dried to prevent condensation in equipment signal lines, and it must be filtered to remove particulate contaminants that could clog orifices. Most pneumatic transmitters produce a signal in the range of 3 to 15 psi (0.2 to 1.0 kg/sq cm or 1.4 kPa to 6.9 kPa), and the signal is usually linear within that range (that is, 3 psi = 0%, 9 psi = 50%, and 15 psi = 100% of scale).

One drawback of pneumatic systems is that the air volume in the signal lines causes some dampening of the signal. This means that on longer transmission lines (say 300 to 500 feet or 90 to 150 m), delays of 5 to 30 seconds may be experienced. Normally, the maximum transmission distance for pneumatic signals is about 1,000 feet (300 m). Therefore, pneumatic devices frequently provide a direct readout at the measuring location. For remote readings the pneumatic signal is converted to an electronic signal and then transmitted to a receiver in a control room.

10.313 Hydraulic

Hydraulic (using liquid pressure) instrumentation, unlike pneumatic signals, operates over a broad range of pressures. Water, water and oil, and glycerol are frequently used system fluids. The fluid is normally conveyed through small-diameter pipe (1/4 to 1 inch or 6 to 25 mm) constructed of copper or steel.

Hydraulic signals are not dampened like pneumatic signals so signal transmission is virtually instantaneous. These systems can be used to transmit weight (load cells) and pressure data, and are also commonly used to operate (power) mechanical equipment such as valves.

Hydraulic systems are probably the least used signal transmission method in water treatment plants.

10.314 Electronic

Electronic signal transmission has become the most commonly used data gathering and transmission method.

Water treatment plant process data are usually transmitted within a range of 4 to 20 millivolt signals (often expressed as milliamp signals). A major advantage of electronic signal transmission is that extremely low potential (voltage) electronic signals from field devices (such as electrodes in water) can be amplified to millivolt signals. Another advantage of electronics is that converting signals (transducing) from one form to another is quite easy. Solid-state circuitry can be used in this regard to convert variable sensing data to linear signals.

Electronic signals are used to activate common electrical power and control circuitry (120- and 240-volt) to start and stop or otherwise control equipment, and to activate audible and visual alarms. The following list is a summary of the functions commonly performed using electronic instrumentation:

1. Indication (gages (analogs), digital indicators, and cathode ray tube displays),

2. Recording (strip or circular charts),

3. Data logging (magnetic tape or disks),

4. Alarm (audible or visual), and

5. Control (computer, solid-state circuitry, or relay logic systems).

10.315 Electrical

Electrical signals (other than electronic) are also commonly used to perform a variety of signaling, telemetry, and control functions. Ordinary alternating current (A.C.), either 120-volt or 240-volt, is frequently used. However, other voltages, such as 6-, 12-, and 24-volt (A.C. or D.C.) are commonly used for control when a shock hazard may present a problem.

Common functions performed with electrical signals include the following:

1. ON/OFF control (make or break by means of switches),

2. Time-impulse control (signals are transmitted for only a portion of a cycle, for example, 15 seconds), and

3. Pulse rate control (pulses or short transmissions of power).

QUESTIONS

Write your answers in a notebook and then compare your answers with those on page 488.

10.3A   How are data transmitted over distances greater than about 1,500 feet?

10.3B   What precautions must be exercised when using air for pneumatic data transmission?

10.3C   What types of transmission fluids are commonly used in hydraulic systems?

10.3D   List the functions commonly performed by electronic signals.

10.32 Control Methods

Many of the instrumentation systems used in water treatment plants are combinations of the previously described systems, and are commonly referred to as *HYBRID* systems.

Perhaps the most widely used method of automatically controlling pumps, valves, chemical feeders, and other devices is known as *RELAY LOGIC*. Relay logic is a method of switching electrical power on and off in accordance with a predetermined sequence (logic) by means of relays, process switches and contacts, timers, and manual switches. Virtually any sequence of operational control can be achieved using relay logic. This approach is very flexible because the system can be operated by manual controls if the need arises.

10.33 Computers

In many of the newer water treatment plants, computers (microprocessors) are being used to monitor and record data on process functions and status. However, critical functions are wired for manual operation if the computer fails. A cathode ray tube (CRT) is commonly used to visually display selected data when requested by an operator.

Increasing use of computers will minimize the number of visual indicators (such as strip-chart recorders) required to monitor, record, and display water treatment process data.

Computers are being used to optimize process performance as well as monitor and record data. This means increased safety and reliability of plant operation. The expanding use of computers adds an entirely new dimension to water treatment plant operation, and provides challenging new opportunities for operators with the foresight to prepare themselves. For additional information and details, see Volume II, Chapter 19, "Instrumentation."

YOU must realize that plant instrumentation does not relieve you of your *RESPONSIBILITY* to make operational decisions and to exercise operational control. In many "automated" plants the operators yield their human authority to an electronic black box, thereby making themselves slaves to automation. However, this automation was meant to be the slave to human control.

QUESTIONS

Write your answers in a notebook and then compare your answers with those on page 488.

10.3E   What is a hybrid instrumentation system?

10.3F   What is relay logic?

10.4 REGULATION OF FLOWS

10.40 Need for Flow Regulation

Water treatment plant flow rates fluctuate with total system demand. Thus, distribution system or consumer demands generally control water treatment plant operations. Demands for water vary depending on the following considerations:

1. Time of day,

2. Day of week,

3. Season of the year,

4. Prevailing weather conditions,

5. Manufacturing demands (canneries), and

6. Unusual events (fire, main breaks).

To meet these variable system demands, adequate source capacity and treated water storage volume are essential. Usually, a minimum distribution system operating pressure is established for all water service connections, and this value establishes the basic guideline for system operation. When the pressure falls below a predetermined value, additional flow must be provided. This demand establishes the finished water demand. Thus, most water treatment plants are operated on a "demand-feedback" basis, whereby finished water flow requirements (consumer demands) establish raw water and in-process flow requirements. In small water treatment systems the plant may only need to be operated for a portion of the day (one eight-hour shift) in order to produce sufficient water for reservoir storage to meet distribution system demands during a 24-hour period.

To estimate how much water needs to be treated each day, you need to analyze current storage levels in clear wells (or plant storage reservoirs) and in distribution system service storage reservoirs. Also, the expected consumer demand for the day must be estimated. This is where your historical flow records can be very helpful. Important items to consider include trends during the past few days. Is the weather changing and causing an increase or decrease in demand for water? In the late spring or early summer, increasing temperatures can cause an increase in demand for water to irrigate gardens and lawns. High winds and freezing conditions can also create shifts in demand. In many communities the lifestyle of the people can create predictable demands for water on certain days of the week, for example, when people wash their clothes or cars, or do their yard irrigation. Another factor that may influence demand for water is whether or not children are in school or at home "playing with water." In summary, to estimate the flow rate or amount of water to be treated each day by your water treatment plant, review:

1. Clear well and distribution storage needs,

2. Yesterday's and historical consumer demands, and

3. Weather forecasts.

Actual regulation of the flow to water treatment plants depends on the method used to deliver water to the plant. If the raw water is transmitted by gravity from a reservoir through a pipe under pressure, then the flow can be changed by adjusting a butterfly or plug-type valve. If pumping is involved, then a change in the pumping rate and/or the number of pumps in service will be required.

Raw water pumps are either constant speed or variable speed units. In constant rate pumping units using multiple pumps, adjustment of flow is accomplished by adding or removing pumps in service to produce the desired flow. Variable speed pumps can be adjusted by changing the drive speed of the pumps to produce the desired flow. Electricity costs for variable speed pumps are much higher than for constant speed pumping units. Pumps used to pump finished water into the distribution system operate in a similar manner. Many pumps are designed to operate most efficiently in a fairly narrow range of flow and pressure conditions. Therefore, the individual operating characteristics of each pump should be considered so that the most efficient pumping mode can be selected.

The following example shows how to adjust the raw water flow rate to meet system demands. Let's assume that a clear well has a maximum storage capacity of five million gallons, with three million gallons currently in storage, and the plant flow rate is 1.5 MGD.

If the plant is being operated on a continuous basis, the raw water flow rate should be adjusted to approximately three MGD depending on weather conditions and the day of the week.

EXAMPLE

The three MGD will meet the demand of 1.5 MGD and allow us to add 1.5 MG during the 24-hour period to storage. We did not try to make up the entire two million gallon deficit in storage, since this could possibly cause overfilling of the reservoir if distribution system demands suddenly drop.

In this same example, if the plant only operated for eight hours per day, then the flow rate would have to be increased by a factor of three [(3)(3 MGD) = 9 MGD] to achieve the same results.

Let's further assume that the plant has an operating capacity of five MGD and the raw water pumping station has one 1,500-GPM constant speed pump and two 1,000-GPM constant speed pumps. (A flow of one MGD is approximately equal to a flow of 700 GPM for 24 hours.) At the initial flow rate of 1.5 MGD only one of the 1,000-GPM pumps would be needed to provide this flow (1,000 GPM for 24 hours is almost equal to 1.5 MGD). Both 1,000-GPM units in service would almost provide the desired 3 MGD flow (actual flow would be 2.9 MGD). Constant rate distribution system pumps would be adjusted in a manner similar to that just described.

After you have made the flow rate change, you should verify the actual flow rate by reading the raw water flow measuring device. You should periodically check the storage levels (elevations) in the clear well and service storage reservoirs to determine if they are maintaining the desired storage volumes. Plant flows should be adjusted whenever major changes in consumer demand occur.

10.41   Clear Wells

Clear wells (or plant storage reservoirs) are an important part of the water treatment plant; they provide necessary operational storage to average out high and low flow demands. The reservoir is filled when demands are low to compensate for peak periods, which draw the level down (see Figure 10.3). This reservoir also acts as a buffer that prevents frequent ON/OFF cycling of finished water pumps and permits planned changes in treatment plant operation.

10.42   Treatment Process Changes

To maintain adequate clear well and distribution system water storage levels, raw or source water flow changes may be required (either increases or decreases). Raw water flow changes should take into account the travel or detention time between the source of supply (river or lake) and the treatment plant.

When storage demands change and require adjustments in the flow of water through a plant, you may also be required to perform the following functions:

1. Adjust chemical feed rates,

2. Change filtration rates,

3. Perform jar tests,

4. Observe floc formation and floc settling characteristics,

5. Monitor process performance,

6. Collect process water quality samples, and

7. Visually inspect overall process conditions.

Some of these changes may occur automatically if your plant has flow-paced chemical feeders. Changes in chemical feed rates (coagulants and chlorine) are required when using manually operated chemical feeders because they are generally set to feed a specific amount of chemical, and this amount is dependent on the rate of flow. Adjustment and calibration of chemical feeders is discussed in Chapter 4, "Coagulation and Flocculation."

Filters are usually operated at a constant production rate, as described in Chapter 6, "Filtration." Automatic control systems typically maintain uniform flow rates, but the number of filters in service can be changed by starting or stopping individual filter units to meet changing needs.

In addition to the above considerations, each of the other unit treatment processes (coagulation, flocculation, and sedimentation) is designed to operate over a broad range of flow rates. However, in some instances, major flow changes may require either adding or removing facilities from service.

QUESTIONS

Write your answers in a notebook and then compare your answers with those on page 488.

10.4A   What factors influence the amount of water that must be treated each day?

10.4B   How are clear wells operated during peak system demands for water?

10.4C   How are filters operated when changes in demand occur?

10.5   CHEMICAL USE AND HANDLING

10.50   Need for Chemicals

A wide variety of chemicals are used in the water treatment plant in the production of a safe and palatable drinking water supply. They play a crucial role in controlling process perform-

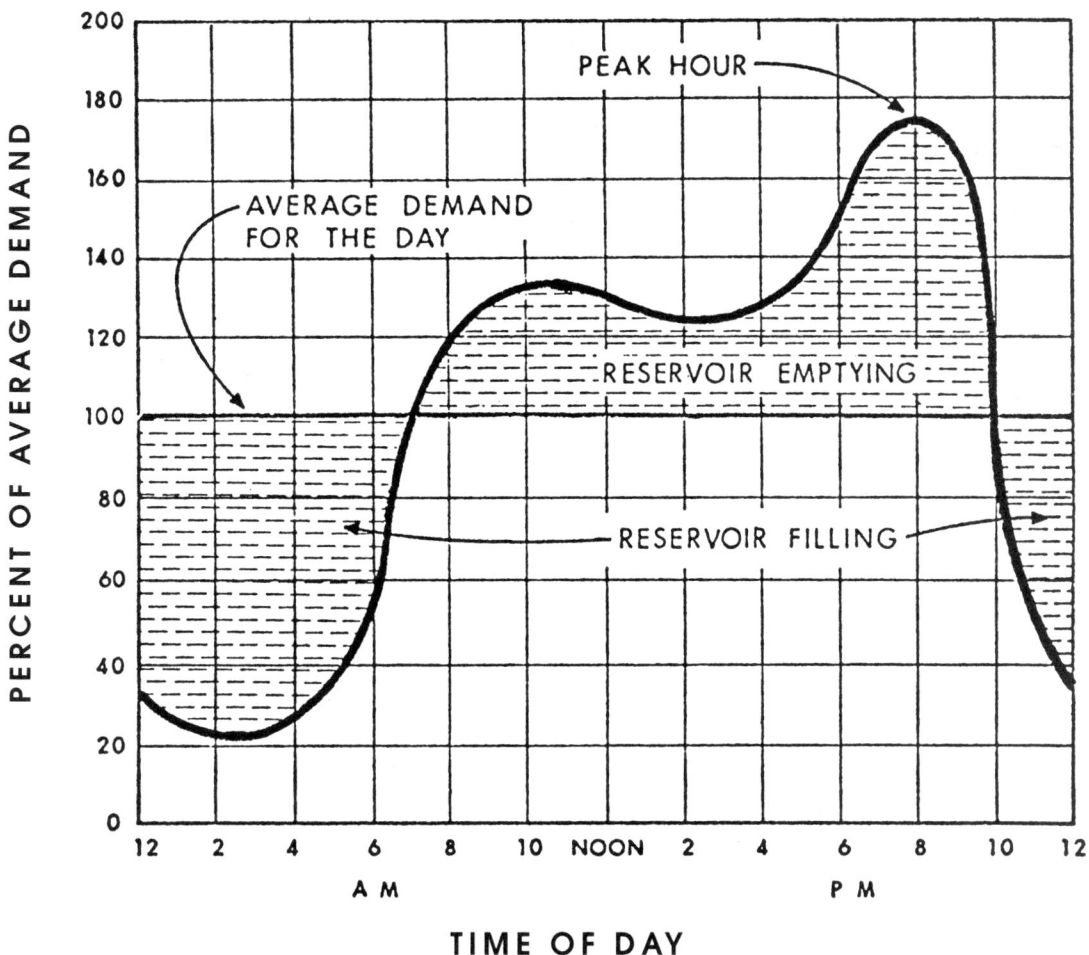

Fig. 10.3 Typical daily variation of system demand
(From WATER DISTRIBUTION OPERATOR TRAINING MANUAL,
By California-Nevada Section, AWWA)

ance and producing a high-quality water. Chemicals are used in the following aspects of water treatment:

1. Clarification (turbidity reduction),

2. Disinfection,

3. Taste and odor control,

4. Algae control,

5. Corrosion/scaling control,

6. Water softening, and

7. Fluoridation.

Operators should be thoroughly familiar with the types of chemicals used in water treatment, specific chemical selection and applications, evaluation methods for determining performance, and safe storage and handling techniques. *ALL* chemicals are potentially dangerous and all necessary precautions must be taken *BEFORE* handling any chemical. All containers, no matter what the use, should bear some form of precautionary labeling.

A MATERIAL SAFETY DATA SHEET (MSDS)[3] is your best source of information about dangerous chemicals. Ask your chemical supplier to furnish you with the MSDS for each chemical you purchase. No chemical should be received, stored, or handled without essential safety information being

---

[3] *Material Safety Data Sheet (MSDS). A document which provides pertinent information and a profile of a particular hazardous substance or mixture. An MSDS is normally developed by the manufacturer or formulator of the hazardous substance or mixture. The MSDS is required to be made available to employees and operators whenever there is the likelihood of the hazardous substance or mixture being introduced into the workplace. Some manufacturers are preparing MSDSs for products that are not considered to be hazardous to show that the product or substance is NOT hazardous.*

provided to those who come into contact with the substance. The MSDS will provide at least the following information:

1. Identification of composition, formula, and common and scientific names;

2. Specific gravity, boiling/freezing points, solubility, and vapor pressure;

3. Incompatible substances and decomposition products;

4. Health hazards;

5. Environmental impacts;

6. Personal protective measures and engineering/administrative controls; and

7. Safe handling, storage, disposal, and cleanup procedures.

REMEMBER, don't work with a chemical unless you understand the hazards involved and are using the protective equipment necessary to protect yourself. Contact your local safety regulatory agency about specific chemicals you may deal with if there is any doubt in your mind about safe procedures.

10.51   Types of Chemicals

The most commonly used chemicals for water treatment are described in Table 10.6. The American Water Works Association has developed standards for many of these chemicals which help to ensure that only quality chemicals are used in water treatment. These standards should be referred to when ordering treatment chemicals (see AWWA Standards, Section B—Treatment).

The choice of specific chemicals to use in a given water treatment plant will vary depending on source water quality, type of treatment to be performed, availability of chemicals, and to some degree on economic considerations.

The initial selection of specific chemicals and anticipated feed rates for a given application are frequently determined by pilot-plant testing of the specific source water. Pilot-plant tests are usually performed by the water treatment plant designer by constructing a "small-scale" treatment plant at or near the source of supply. Pilot tests provide the designer, as well as the treatment plant operator, with valuable information on the "treatability" of a given source of supply over a broad range of water quality conditions.

10.52   Storage of Chemicals

Water treatment chemicals can be stored in a number of ways including:

1. Solid (dry) form (bags, cartons, drums),

2. Liquid form (drums, tanks, cylinders), or

3. Gaseous form (cylinders).

Chemicals should be stored in accordance with the manufacturer's written recommendations and requirements established by regulatory agencies. Regardless of the storage method, always anticipate future chemical requirements so that an ample supply will be on hand when needed. A good practice is to maintain at least a 30-day supply of all commonly used treatment chemicals. Keep a running account of chemical use and storage inventory on a daily basis. Minimum storage quantities should be established for each type of chemical, and this information will indicate when chemicals should be ordered. Minimum storage quantities for some chemicals

such as chlorine may be established by regulatory agencies. *NEVER ALLOW THE SUPPLY OF CHLORINE TO DROP BELOW A 15-DAY SUPPLY AT THE PLANT SITE.*

EXAMPLE 1

Shown below is the amount of chlorine used by a small water treatment plant during one week.

Day of Week	Mon	Tues	Wed	Thurs	Fri	Sat	Sun
Chlorine Used, lbs	43	39	34	38	39	37	29

What was the daily average use of chlorine in pounds per day?

Known	Unknown
Chlorine Used Each Day of Week	Daily Average Use of Chlorine, lbs/day

Calculate the daily average use of chlorine in pounds per day.

$$\text{Avg Use, lbs/day} = \frac{\text{Sum of Chlorine Used Each Day, lbs}}{\text{Total Number of Days, days}}$$

$$= \frac{43 \text{ lbs} + 39 \text{ lbs} + 34 \text{ lbs} + 38 \text{ lbs} + 39 \text{ lbs} + 37 \text{ lbs} + 29 \text{ lbs}}{7 \text{ days}}$$

$$= \frac{259 \text{ lbs}}{7 \text{ days}}$$

$$= 37 \text{ lbs/day}$$

NOTE: We used a seven-day average so each day of the week would be considered.

EXAMPLE 2

The chlorine cylinder on line has less than one day's chlorine supply remaining. Four 150-pound chlorine cylinders are in storage. The plant uses an average of 37 pounds of chlorine per day. How many days' supply of chlorine is available?

Known	Unknown
Chlorine Cylinders = 4 Cylinders	Supply of Chlorine, days
Cylinder Wt, lbs/cyl = 150 lbs/cyl	
Avg Use, lbs/day = 37 lbs/day	

Calculate the available supply of chlorine in days.

$$\text{Supply of Chlorine, days} = \frac{(\text{Cylinder Wt, lbs/cyl})(\text{No. of Cylinders})}{\text{Avg Use, lbs/day}}$$

$$= \frac{(150 \text{ lbs/cyl})(4 \text{ Cylinders})}{37 \text{ lbs/day}}$$

$$= \frac{600 \text{ lbs Chlorine}}{37 \text{ lbs Chlorine/day}}$$

$$= 16 \text{ days}$$

TABLE 10.6   CHEMICAL TYPES AND CHARACTERISTICS[a]

Chemical Name	Chemical Formula	Commercial Concentration	Comments
COAGULANTS			
Aluminum Sulfate (Alum, granular)	$Al_2(SO_4)_3 \cdot 14\ H_2O$	47-50% $(Al_2(SO_4)_3)$	Acidic
Ferric Chloride	$FeCl_3 \cdot 6\ H_2O$	59-61% $FeCl_3$	Acidic
Ferric Sulfate	$Fe_2(SO_4)_3 \cdot 9\ H_2O$	90-94% $Fe_2(SO_4)_3$	Acidic, Staining
Ferrous Sulfate	$FeSO_4 \cdot 7\ H_2O$	55% $(FeSO_4)$	Cakes Dry
Cationic Polymer	—	Varies	Positively Charged
Anionic Polymer	—	Varies	Negatively Charged
Nonionic Polymer	—	Varies	
DISINFECTION			
Sodium Hypochlorite	$NaOCl$	12-15% (Cl_2)	Solution
Calcium Hypo- chlorite (HTH)	$Ca(OCl)_2 \cdot 4\ H_2O$	65-70% (Cl_2)	Powder
Chlorine	Cl_2	99.8% (Cl_2)	Gas/Liquid
Anhydrous Ammonia	NH_3	99-100% (NH_3)	Gas/Liquid
Ammonium Hydroxide	NH_4OH	29.4% (NH_3)	Solution
Ammonium Sulfate	$(NH_4)_2SO_4$	6.3% (NH_3)	Cakes Dry
Chlorine Dioxide	ClO_2	26.3% (Cl_2)	Generated On Site
Ozone	O_3	—	Generated On Site
TASTE AND ODOR			
Activated Carbon	C	—	Insoluble
Potassium Permanganate	$KMnO_4$	100%	Very Soluble
ALGAE CONTROL			
Copper Sulfate	$CuSO_4 \cdot 5\ H_2O$	99% $(CuSO_4)$	
CORROSION CONTROL			
Calcium Hydroxide (Hydrated Lime)	$Ca(OH)_2$	75-99% (CaO)	Basic
Sodium Hydroxide (Caustic Soda)	$NaOH$	98.9% $(NaOH)$	Very Basic
SOFTENING			
Calcium Oxide (Quicklime)	CaO	75-99% (CaO)	
Sodium Carbonate (Soda Ash)	Na_2CO_3	99.4% (Na_2CO_3)	
FLUORIDATION			
Sodium Silicofluoride	Na_2SiF_6	59.8% (F)	Powder
Sodium Fluoride	NaF	43.6% (F)	Powder or Crystal
Fluosilicic Acid	H_2SiF_6	23.8% (F)	Solution

[a] Adapted from AWWA, Section 10.13, "Reference Books," Reference 2.

10.53 Safe Handling of Chemicals

In the routine operation of a water treatment plant, you will come in contact with a variety of potentially dangerous chemicals. While some chemicals are inactive (inert), it is good practice to consider *ALL* chemicals as a potential hazard.

When unloading or transferring chemicals, be especially careful. Know the locations of all safety showers and eye wash fountains. Be familiar with their use and test them periodically to be sure that they function properly.

Wear protective clothing when working with chemicals. Goggles and face shields will protect your eyes and face. Protect other exposed portions of the body by wearing rubber or neoprene gloves, aprons, or other protective clothing. Chemical dust can irritate the eyes and respiratory system. Use respirators when appropriate, and always use dust collectors when such equipment is provided. Promptly wash down or clean up all chemical spills to prevent falls and/or physical contact with the chemical.

A few treatment chemicals such as *CAUSTIC SODA* and *CHLORINE* can be very hazardous to the operator, and extreme care should be taken in the handling of these chemicals. *CAUSTIC SODA* is one of the most dangerous of the common alkalies, and direct contact will cause severe burns.

Caustic can quickly and permanently cloud vision if not immediately flushed out of the eyes. Determine the location of safety showers and eye wash stations *BEFORE* starting to work with caustic soda. Wash down caustic soda spills immediately. When handling caustic soda, wear safety goggles and a face shield; cover your head with a wide brim hat; wear rubber or neoprene gloves, apron, and boots (a full-body protective suit is preferable); and do not tuck pant legs inside boots or shoes. If dust or mist is encountered, use a respirator.

CHLORINE is a strong respiratory irritant, and either prolonged exposure to chlorine gas or high concentrations of chlorine gas could be *FATAL*. Wherever chlorine liquid or gas is stored or used, the following safety equipment should be provided:

1. Shower and eye wash facility,

2. Emergency breathing apparatus (air pack),

3. Chlorine gas detector,

4. Floor-level vents, and

5. Fans that maintain a positive air pressure in the storage facility. *NOTE:* If you pull air with chlorine through a fan, eventually any wiring or controls in the fan can become corroded and fail.

All water treatment plant operators should be fully trained in chlorine safety and leak detection procedures. Whenever you enter a chlorine facility, make sure that the fan is operating. If a chlorine leak is suspected or the chlorine gas concentration in the room is not known, wear a self-contained air pack and use the buddy system. Rubber- or plastic-coated gloves should be worn when handling chlorine containers. When in doubt— always use the "buddy system" and have another operator standing by with an air pack. (See Chapter 7, Section 7.8, "Chlorine Safety Program," for additional information about chlorine safety.)

10.54 First-Aid Procedures

Every operator should be familiar with the following first-aid procedures:

EYE BURNS (GENERAL)

1. Apply a steady flow of water to eyes for at least 15 minutes.

2. Call a physician immediately.

3. *DO NOT* remove burned tissue from the eyes or eyelids.

4. *DO NOT* apply medication (except as directed by a physician).

5. *DO NOT* use compresses.

SKIN BURNS (GENERAL)

1. Remove contaminated clothing immediately (preferably in a shower).

2. Flush affected areas with generous amounts of water.

3. Call a physician immediately.

4. *DO NOT* apply medication (except as directed by a physician).

SWALLOWING OR INHALATION (GENERAL)

1. Call a physician immediately.

2. Read antidote on label of any chemical swallowed. For some chemicals vomiting should be induced, while for other chemicals vomiting should not be induced.

CHLORINE GAS CONTACT

1. If victim is breathing, place on back with head and back in a slightly elevated position. Keep victim warm and comfortable. *CALL A PHYSICIAN IMMEDIATELY.*

2. To check for breathing, tilt the head back (tilting the head back opens the airway and may in itself restore breathing), put your ear over the victim's mouth and nose and listen and feel for air. Look at the victim's chest and see if it is rising and falling. Watch for breathing for three to five seconds. If there is no breathing, perform mouth-to-mouth resuscitation.

   a. Tilt victim's head back and lift the chin. Be sure victim's mouth/throat airway is open.

   b. Gently pinch the victim's nose shut with the thumb and index finger, take a deep breath, seal your lips around the outside of the victim's mouth, create an airtight seal and give victim two full breaths at the rate of 1 to 1½ seconds per breath. Watch for the chest to rise while you breathe into the victim. If you feel resistance when you breathe into the victim, and air will not go in, the most likely cause is that you may not have tilted the head back far enough and the tongue may be blocking the airway. Re-tilt the head and give two full breaths.

   c. Put your ear over the victim's mouth and nose and listen and feel for air. Check for pulse for five to ten seconds.

   d. If victim is not breathing and there is no pulse, give victim 15 compressions (CPR or cardiopulmonary resuscitation) and then two breaths. (Check the pulse at the side of the neck. This pulse is called the carotid pulse.) Feel for the carotid pulse for at least five seconds, but no more than ten seconds.

   e. Repeat step (d) four times and then check for breathing and pulse. Do this after giving the two breaths at the end of the fourth cycle of 15 compressions and two breaths. Tilt the victim's head back and check the carotid pulse for five seconds.

   If you do not find a pulse, then check for breathing for three to five seconds. If breathing is present, keep the airway open and monitor breathing and pulse closely. This means that you should look, listen, and feel for breathing while you keep checking the pulse. If there is no breathing, perform rescue breathing and keep checking the pulse.

f. Continue to give CPR until one of the following things happens:

- The heart starts beating again and the victim begins breathing.

- A second rescuer trained in CPR takes over for you.

- EMS (Emergency Medical Services) personnel arrive and take over.

- You are too exhausted to continue.

g. *DO NOT ATTEMPT TO PERFORM CPR UNLESS YOU ARE QUALIFIED.*

3. Eye irritation caused by chlorine gas should be treated by flushing the eyes with generous amounts of water for not less than 15 minutes. Hold eyelids apart to ensure maximum flushing of exposed areas. *DO NOT* attempt to neutralize with chemicals. *DO NOT* apply any medication (except as directed by a physician).

4. Minor throat irritation can be relieved by drinking milk. *DO NOT* give the victim any drugs (except as directed by a physician).

LIQUID CHLORINE CONTACT

1. Flush the affected area with water. Remove contaminated clothing while flushing (preferably in a shower). Wash affected skin surfaces with soap and water while continuing to flush. *DO NOT* attempt to neutralize with chemicals. Call a physician. *DO NOT* apply medication (except as directed by a physician).

2. If liquid chlorine has been swallowed, immediately give victim large amounts of water or milk, followed with milk of magnesia, vegetable oil, or beaten eggs. *DO NOT* give sodium bicarbonate. *NEVER* give anything by mouth to an unconscious victim. Call emergency response (911) and/or a physician immediately.

QUESTIONS

Write your answers in a notebook and then compare your answers with those on page 488.

10.5A   What is a pilot plant?

10.5B   How are water treatment chemicals stored?

10.5C   What protective clothing should be worn when working with chemicals?

10.5D   What safety equipment should be provided wherever chlorine liquid or gas is stored or used?

10.6   OPERATING RECORDS AND REPORTS

10.60   Written Documents

As mentioned in previous chapters, one of the most important *ADMINISTRATIVE* functions of the water treatment plant operator is the preparation and maintenance of accurate operational records.

Operating records can be separated into two major categories:

1. Physical records, and

2. Performance records.

PHYSICAL records describe the water treatment plant physical facilities and equipment. These records include:

1. Plant design criteria,

2. Construction plans and contract specifications,

3. "As-built" (record) drawings,

4. Equipment fabrication drawings and specifications,

5. Manufacturers' operation and repair manuals for all equipment items,

6. Detailed piping plans and electrical wiring diagrams,

7. Equipment records including manufacturer's name, model number, rated capacity, and date of purchase,

8. Maintenance records on each equipment item,

9. Hydraulic profiles showing pertinent operating water surface elevations throughout the water treatment plant, and

10. Cost records for all major equipment item purchases and repairs.

PERFORMANCE records describe the operation of the water treatment plant and provide the operator, as well as others, with a running account of plant operations (historical records). These records are a valuable resource for the operator trying to solve current process problems and anticipate future needs. Performance records also provide a factual account of the operation which is required to meet legal and regulatory agency requirements.

Typical performance records include the following:

1. Daily operations records (process production inventory, process changes, process equipment performance—see Figure 10.5);

2. Water quality records (source water, process water, finished water—see Figures 10.5 and 10.6);

3. Equipment failure records;

4. Accident records;

5. Consumer complaint records (include follow-up investigations and corrective actions taken);

6. Chemical inventory records (include storage amounts, safe storage levels, procurement records);

7. Charts produced by process records (strip-chart recorders); and

8. Visitor information.

RECORDS MANAGEMENT is a very important part of the overall recordkeeping process. Records should be filed and

Daily Operation Record & Chemical Inventory

Stockton East Water Treatment Plant

Date: Time: Rainfall

PARSHALL FLUME		SRWR
Final		
Start		
Total		
Bellota Intake		
Final		
Start		
Total		
Raw Water		
Final		
Start		
Total		
Plant Effluent		
Final		
Start		
Total		
Potable Water		
Final		
Start		
Total		
Reservoirs		
Raw Water Reservoir Levels	N.	S.
F.W. Reservoir Level	Start Ft.	Final Ft.
Water Source		
NRWR	P-1	
SRWR	P-2	
	P-3	

Column No.			1	2	3	4	5	6	7	8	9	10	11	12	13	14	15	16
Operator			Night				Day				Swing					Total	Average lbs./M.G.	Dose M.G./L.
			0000	0200	0400	0600	0800	1000	1200	1400	1600	1800	2000	2200	2400			
RWR Flow																		
% RWR Flow																		
Pipe Flow																		
District		7401																
Wells		7402																
Chlorine	PRE	1																
	INTER	2																
	POST	3																
Chemical Application	Cationic Polymer Pumps	FD-4																
		FD-5																
	N-1 Poly Pump	FD-6																
	S.C. Pumps	FD-7																
	Alum Pumps	FD-9																
		FD-10																
	NaOH Pumps	FD-11																
		FD-12																
	Carbon Pumps	FD-15																
		FD-16																

	CATIONIC POLY.			NON-IONIC POLY.		
	Tank 15					
	Level	Gal.	Dry Lbs.	Level	Gal.	Dry Lbs.
Start						
Added						
Used						
Finish						

	Alum					
	Tank 1			Tank - 2		
	Level	Gal.	Dry Lbs.	Level	Gal.	Dry Lbs.
Start						
Added						
Used						
Finish						

	NaOH					
	Tank - 4			Tank - 5		
	Level	Gal.	Dry Lbs.	Level	Gal.	Dry Lbs.
Start						
Added						
Used						
Finish						

	Carbon					
	Tank - 11			Tank - 12		
	Level	Gal.	Dry Lbs.	Level	Gal.	Dry Lbs.
Start						
Added						
Used						
Finish						

Filters	Time Start	Time Stop	Loss of Head	Hrs. Prior	Hrs. Today	Hrs. To Date	Hrs. of Completed Run	M.G. Used
Filter No. 1								
Filter No. 2								
Filter No. 3								
Filter No. 4								
Summary:	Ave							

	Chemical Inventory				
	Chlorine		Spare Chemicals		
	North Rack	South Rack	Level	Gal.	Dry Lbs.
Start					
Added					
Used					
Finish					

Notes:

Diesel Fuel	
Start	
Added	
Used	
Finish	

		Surface wash (1000 g.)		Backwash (1000 g.)		Average Wash Water	
Quantity filtered in MG							
No. filters in service		Final Reading		Final Reading		1000 Gals.	
Average filtering Rate of G.P.M./Sq. Ft.		Initial Reading		Initial Reading			
No. of Filters Washed		Total		Total			

Copper Sulfate Feed			Bypass Water Meter Reading	
NRWR			Final	
SRWR			Start	
Backwash			Total	

Bypass Water Meter Reading	
Final	
Start	
Total	

Fig. 10.5 Daily operation record and chemical inventory

Stockton-East Water Treatment Plant
DAILY LABORATORY RECORD

Operator:			Shift One				Shift Two				Shift Three				Date		
Sample Site	Test	Units					Time								Daily Mean	Range	
			0000	0200	0400	0600	0800	1000	1200	1400	1600	1800	2000	2200		Low	High
Plant Influent (SA-1)	Turb.	NTU															
	R.Cl	mg/l															
	pH																
	Temp.	°C															
	Color	C.U.															
	Odor	TON															
	Alk.	mg/l															
	Hard	mg/l															
Settled Water (SA-3)	Turb.	NTU															
	R.Cl	mg/l															
	pH																
	Temp.	°C															
Filtered Water (SA-4)	Turb.	NTU															
	R.Cl	mg/l															
	pH																
	Temp.	°C															
Plant Effluent (SA-5)	Turb.	NTU															
	R.CL	mg/l															
	pH																
	Temp.	°C															
	Color	CU.															
	Odor	TON															
	Alk.	mg/l															
	Hard.	mg/l															
	Flow	MGD															

Plant Flow			Polymer		Alum		NaOH		Chlorine						PAC		Sp. Chem	
									Pre		Inter		Post					
Time	MGD	Source	Time	mg/l	Time	mg/l	Time	mg/l	Time	lb/D	Time	lb/D	Time	lb/D	Time	mg/l	Time	mg/l

Remarks:

Fig. 10.5   Daily laboratory record

STOCKTON EAST WATER DISTRICT
DAILY BACTERIOLOGY RECORD

DATE:			Routine Samples						Raw Water			Special	
Sample Location		SA-5	SA-5	SA-5				Blank					
Sample Date													
Sample Time													
Residual Cl$_2$													
Collected By													
Total Coliform													
Presumptive MPN	Vol.												
Medium: / Positives / 24h													
Date:													
Control: / 48h													
Confirmed MPN	Vol.												
Medium: / Positives / 24h													
Date:													
Control: / 48h													
MPN Index/100 ml													
Membrane Filter	Vol.												
Medium: / Total Count													
Date:													
Control / Coliform Count													
Coliforms/100 ml													
Fecal Coliform													
MPN	Vol.												
Medium: / Positives / 24h													
Date:													
Control: / 48h													
Fecal MPN Index/100 ml													
Membrane Filter	Vol.												
Medium: / Total Count													
Date:													
Control / Fecal Count													
Fecal Coliforms/100 ml													
Std. Plate Count	Vol.												
Medium: / Count													
Date: / SPC / Ml													
Control:													

Set up by:	Temp (°C)	AM	PM	Notes:
Time:	Incubator			
Read D-1:	Water Bath			
Day 2	Laboratory			
Day 3	Time			
Day 4	By			

Fig. 10.6 Daily bacteriology record

cataloged or indexed for future reference. Regulatory agencies may require you to keep certain water quality analyses (bacteriological test results) and customer complaint records on file for specified time periods (ten years for chemical analyses and bacteriological tests). Other records that may have historical value to the operator (source water quality changes and resultant process changes) should be kept as long as they are useful. These records should be maintained in appropriate files properly labeled for easy reference. In addition, a daily diary or pocket notebook should be used by each operator to record unique or unusual events. A typical entry that might appear in an operator's diary or daily operating log is as follows:

8:30 a.m.

Raw water pump No. 1 starting to develop a whine during low flows.

JQO (operator should initial entry)

10.61 Oral Communications

In the previous section we described the types of written documents that the operator uses to assist in the operation of a water treatment plant and to account for the daily operation. To perform effectively you will also need to communicate orally with other operators, supervisors, and other staff members.

Most organizations have a "chain of command" which describes the individuals that you must communicate with and establishes levels and lines of authority.

You must realize that communications are always "two-way," regardless of the size of the organization. There are more people to communicate with in larger organizations. The successful operation of a water treatment plant depends to a large degree on good oral communications. Regardless of plant size, everyone involved must realize that plant operation is not a "one-operator" show.

QUESTIONS

Write your answers in a notebook and then compare your answers with those on page 489.

10.6A   What are the two major categories of operating records?

10.6B   What are performance records?

10.6C   What records may have great historical value to the operator?

10.6D   Why should each operator use a pocket notebook?

End of Lesson 1 of 2 Lessons on PLANT OPERATION

Please answer the discussion and review questions next.

DISCUSSION AND REVIEW QUESTIONS

Chapter 10.   PLANT OPERATION

(Lesson 1 of 2 Lessons)

At the end of each lesson in this chapter you will find some discussion and review questions. The purpose of these questions is to indicate to you how well you understand the material in the lesson. Write the answers to these questions in your notebook before continuing.

1. What are the three basic objectives of operating a water treatment plant?

2. Where are water samples usually collected to measure compliance with water quality standards?

3. How frequently should samples be collected for process control?

4. Prepare a list of daily operating procedures for your plant.

5. How would you determine the raw water pumping rate to a water treatment plant?

6. What is the purpose of a clear well?

7. How are the initial selection of chemical types and anticipated feed rates for a given application in a water treatment plant usually determined?

8. What precautions would you take when handling caustic soda?

9. Why are positive air pressure fans preferred over floor-level exhaust fans in chlorine storage facilities?

10. How long should records be kept?

CHAPTER 10.   PLANT OPERATION

(Lesson 2 of 2 Lessons)

10.7   PLANT MAINTENANCE

10.70   Maintenance Program

The water treatment plant maintenance program is an important part of plant operations. The overall maintenance program should be designed to ensure continuing satisfactory operation of the treatment plant facilities under a variety of different operating conditions. Such a program would include routine and preventive maintenance as well as provisions for effectively handling emergency breakdowns.

All plants should have written instructions on how to operate and maintain the equipment. If your plant does not have written instructions, prepare them now or budget the necessary funds to have someone prepare them. These instructions are very helpful when someone retires or leaves and a new operator is assigned these duties.

The major elements of a good preventive maintenance (PM) program include the following:

1. Planning and scheduling,

2. Records management

3. Spare parts management,

4. Cost and budget control,

5. Emergency repair procedures, and

6. Training program.

10.700   Planning and Scheduling

Planning and scheduling are the foundation of the maintenance program. An important source of information for planning routine maintenance is equipment manufacturers' operating and maintenance instructions. These instructions are usually furnished with the equipment at the time of purchase, or can be obtained directly from the manufacturer. Another important reference source is the plant Operation and Maintenance Manual.

The planning and scheduling effort should define the specific maintenance tasks to be done and the time intervals. The following items are important considerations in developing a good maintenance plan:

1. Routine procedures,

2. Special procedures (equipment overhaul),

3. Skills needed,

4. Special tools and equipment requirements, and

5. Parts availability.

ROUTINE PROCEDURES

In the routine operation of the water treatment plant, the operator will inspect various mechanical equipment items (valves, pumps) and electrical equipment items (motors) to check for proper operation, and will perform a number of maintenance functions as follows:

1. Keep motors free of dirt and moisture,

2. Ensure good ventilation (air circulation) in equipment work areas,

3. Check motors and pumps for leaks, unusual noise, vibrations, or overheating,

4. Maintain proper lubrication and oil levels,

5. Check for alignment of shafts and couplings,

6. Check bearings for overheating and proper lubrication,

7. Check for proper valve and pump operation, and

8. Check calibration of chemical feeders.

These routine tasks should generally be performed on a daily basis.

For additional details on how to develop a routine maintenance program, see Volume II, Chapter 18, "Maintenance."

TOOLS

To be effective in the routine maintenance of plant equipment, you will need to know how to properly use common hand tools to protect equipment and for your own safety. Remember that you cannot perform maintenance procedures safely and properly if you do not have the proper tools.

The exact tools each operator should be familiar with is difficult to specify since the maintenance performed by a plant operator will vary considerably depending on the size of the treatment plant and the number of operators. In larger plants, maintenance personnel will perform virtually all of the routine as well as specialized maintenance functions. In smaller plants the operator will be expected to perform most of the routine maintenance functions. Operators should at least be familiar with the following types of common hand tools:

1. Screwdrivers (slotted and Phillips),

2. Pipe wrenches,

3. Crescent wrenches,

4. Socket wrenches,

5. Allen wrenches,

6. Open-end and box wrenches,

7. Hammers (claw, sledge, mallet),

8. Pliers (vise-grips),

9. Files,

10. Wire brushes, and

11. Putty knives.

Tool loss and replacement can be a problem at any water treatment plant unless procedures are implemented to effectively manage the inventory of tools and equipment. A commonly used procedure is the simple "check-out list." An individual desiring to use a special tool merely signs out for the tool, listing the type of tool, date, and name or initials. When the tool is returned the name is simply struck off the list. Simple 3" x 5" or 5" x 7" cards can be used for this purpose. A color coded tag system may also prove useful for keeping track of tools.

In the event that a tool is lost or damaged, the description of the tool and any special circumstances or information should be noted so that it can be repaired or replaced.

For an additional list of tools used by operators, see *SMALL WATER SYSTEM OPERATION AND MAINTENANCE*, Chapter 4, "Small Water Treatment Plants," in this series of operator training manuals.

SPECIAL PROCEDURES

These procedures are very important for an effective maintenance program.

1. Plan equipment shutdowns to minimize adverse impacts on plant operations.

2. Use equipment repair records to plan and schedule maintenance.

3. Prepare step-by-step procedures or refer to manufacturer's instructions for performing equipment overhauls or other special maintenance tasks.

10.701 Records Management

Good records management is an important administrative feature of the maintenance program and one that is often overlooked. A comprehensive records management system provides the basis for daily task assignments, provides a permanent record of work performed, and becomes a historical reference source for reviewing equipment performance.

The following records should be maintained as a part of a good records management system:

1. Equipment inventory cards,

2. Preventive maintenance schedules,

3. Spare parts lists and reorder information, and

4. Records of work performed.

10.702 Spare Parts Management

Certain parts of mechanical equipment items, such as shaft bearings, require periodic replacement because they have a useful life which is considerably shorter than the predicted overall equipment life. This requires that an adequate stock of spare parts be kept on hand at the treatment plant to facilitate planned replacement.

Spare parts should be stocked on the basis of:

1. The importance of the part to operation,

2. Availability,

3. The effect on operation if the part is defective, and

4. Storage space.

If a part is readily available from a supplier, let the supplier stock it for you. It costs money to stock and warehouse unnecessary spare parts. The actual performance history of a particular piece of equipment may indicate the type and number of spare parts that must be stocked. Spare parts should be promptly reordered whenever they are used.

10.703 Cost and Budget Control

Accurate records of labor and equipment expenditures are an important part of the overall budget and cost control program. Operation and maintenance budgets are usually prepared on an annual basis. A thorough and up-to-date written performance history of equipment operations, repair, and replacement costs will significantly improve the budget planning process. The work order system is a way of keeping track of how much time and money is spent doing various types of work. This information becomes a planning tool and a reporting system to indicate what is being done and how much it costs. This procedure records the parts used in the repairs and the amount of labor required to perform the repair or other maintenance procedures, as well as other kinds of work. These records will also provide valuable information for deciding when to replace a given piece of equipment due to excessive repair costs.

10.704 Emergency Repair Procedures

Identify those pieces of equipment that are critical for your facility to meet the demands for safe drinking water of your consumers. Critical pieces of equipment include raw water pumps, chlorinators, and pumps that deliver finished water to the distribution system. All of these items must have standby or backup equipment. Also, your plant should have standby generators in case of a power outage. If any of these facilities fail, you must have emergency repair procedures to follow, which will enable you to put your facility back in service as soon as possible.

10.705 Training Program

Perhaps as important as any other single element of the operation and maintenance program is training. Training should be an ongoing feature of the operation program and operators should be encouraged to participate. Such training can increase the expertise of maintenance and operations personnel in the general repair of equipment, in specialized procedures required to calibrate and repair selected equipment items, and in their ability to quickly and properly respond to changes in raw and finished water.

Major equipment manufacturers periodically conduct training programs designed to provide operations and maintenance personnel with a "hands-on" familiarity with common mechanical and electrical equipment items.

The certification requirements in most states require successful completion of some form of education or training to qualify for taking a certification examination. In many states some type of education requirement must be met before a certificate can be renewed. In order to do a good job, people need an opportunity to improve their knowledge and skills. The best way for people to improve themselves is with a well-planned training program.

QUESTIONS

Write your answers in a notebook and then compare your answers with those on page 489.

10.7A   List the major elements of a good preventive maintenance program.

10.7B   Why is a records management system important?

10.7C   What items should be included in a cost and budget control program?

10.8   PLANT SAFETY AND SECURITY

10.80   Safety Considerations

In the routine operation of the water treatment plant, the operator will be exposed to many potential hazards including:

1. Electrical equipment (shocks),

2. Rotating mechanical equipment,

3. Open-surface, water-filled structures (drowning),

4. Underground structures (toxic and explosive gases, lack of oxygen or too much oxygen),

5. Water treatment chemicals (acids, alkalies, chlorine gas),

6. Laboratory reagents (chemicals), and

7. Pump stations (high noise levels).

Ample safety devices are generally provided at each water treatment plant to protect the operator, as well as others, from accidents and exposure to chemicals, dust, and other hazardous environments. However, these safety devices are of limited value unless you pay strict attention to safety procedures. If an object appears too heavy to lift, do not try to lift it. Get help or use a lifting device such as a forklift.

When working around mechanical and electrical equipment, plant structures, or chemicals, follow the safety procedures listed below to avoid accidents or injury.

ELECTRICAL EQUIPMENT

1. Avoid electric shock by using protective gloves.

2. Avoid grounding yourself in water or on pipes.

3. Ground all electric tools.

4. Lock out electric circuits and tag out remote controls when working on equipment.

5. Always assume all electrical wires are "live."

6. Never use metal ladders around electrical equipment.

7. When in doubt about a procedure or repair, ask for help.

8. Use the buddy system and be sure your buddy knows how to rescue you when you need help.

MECHANICAL EQUIPMENT

1. Do not remove protective guards on rotating equipment.

2. Do not wear loose clothing around rotating equipment.

3. Secure and lock out drive motors before working on equipment. Tag out remote controls.

4. Clean up all lubricant spills (oil and grease).

OPEN-SURFACE, WATER-FILLED STRUCTURES

1. Do not avoid or defeat protective devices such as handrails by removing them when they are in the way.

2. Close all openings when finished working.

3. Know the location of all life preservers.

UNDERGROUND STRUCTURES AND CONFINED SPACES

1. Know the condition of the environment before entering. Determine if there are any toxic gases present, explosive conditions, or an excess or lack of oxygen. Use detection devices that are capable of monitoring the atmosphere continuously.

2. Use portable ventilation fans to ensure good air circulation.

3. Use the buddy system. Also, be certain your buddy is trained and knows what to do in the event that you get into trouble.

CHEMICALS

1. Wear protective clothing when handling or unloading chemicals.

2. Wear goggles and face shields around all potentially hazardous chemicals.

3. Know the location of all safety showers and eye wash facilities (be sure they work).

4. Be familiar with the care and use of air packs.

5. Know chlorine leak detection and safe handling procedures.

6. Promptly clean up all chemical spills.

PUMP STATIONS

1. Using hearing protection devices.

2. Observe the precautions previously listed for working around electrical and mechanical equipment.

If a hazardous situation exists or if a particular procedure is unsafe, do not proceed—call for help.

Always be sure to report any injury, no matter how slight, to your immediate supervisor. This procedure protects you as well as your employer.

For details on how to develop a safety program and safety procedures, see Volume II, Chapter 20, "Safety."

10.81 Security Considerations

Public access to water treatment plant facilities and grounds can result in vandalism or injury of trespassers or other unwanted visitors. In addition, never rule out the potential for sabotage of a facility such as a water treatment plant. The public water supply affects the entire community.

Fences and gated accesses help to discourage trespassers and other unwanted visitors from entering treatment plant grounds and facilities. Gates should be securely locked during "non-routine" working hours and, in some instances, automatic remotely controlled gates may be required to limit access during all hours of operation. Routinely inspect the plant facilities (at least once per shift) and report any unauthorized persons or unusual events to the proper authorities.

QUESTIONS

Write your answers in a notebook and then compare your answers with those on page 489.

10.8A List the potential hazards an operator could be exposed to during the routine operation of a water treatment plant.

10.8B How often should an operator routinely inspect plant facilities for any evidence of unauthorized persons or unusual events?

10.9 EMERGENCY CONDITIONS AND PROCEDURES

10.90 Emergency Conditions

In the operation of any water treatment plant, abnormal or emergency conditions will occasionally arise which require calm, quick action on the part of the operator. Emergency conditions you may encounter include:

1. Treatment process failures,

2. Process equipment failures,

3. Power failures,

4. Fires, and

5. Floods, earthquakes, or other natural disasters.

You must be able to distinguish between an abnormal condition and a "red-alert" emergency condition. *A "RED ALERT" MEANS THAT YOU MUST IMMEDIATELY SEEK OUTSIDE HELP.* Typical red-alert emergencies include sabotage, raw water contamination, chemical spills, fires, serious injury, and a chlorine leak.

10.91 Treatment Process Failures

10.910 Changes in Raw Water Quality

Treatment process failures generally result from an abrupt or unexpected change in source water quality. A typical example of this condition occurs when the source water suspended solids concentration abruptly increases (high turbidity) as a result of precipitation and runoff into the source water supply.

Other, less common, examples are accidental wastewater contamination or chemical spills in the source water system.

Turbidity fluctuations resulting from precipitation and runoff should be anticipated by the operator, and every effort should be made to obtain samples of the water for jar testing as soon as possible. This will allow the operator to make planned adjustments to the treatment process and avoid major process upsets.

Contamination of the source water system by wastewater or chemical spills is nearly impossible to anticipate, so planned adjustments to the treatment process to correct for these problems is unlikely. However, an early warning of wastewater contamination of the source water may be a sudden drop in chlorine residual in the treatment process or a sudden increase in the chlorine demand of the water being treated. Immediate adjustment of the chlorine dosage should be made and additional bacteriological tests should be performed to define the extent of the problem. Don't wait for the results from bacteriological tests to tell you that you have a problem because by then it is too late.

Accidental chemical spills are perhaps the most hazardous situation to deal with since normal treatment process monitoring techniques may not detect the problem. In most cases, the operator must rely on outside notification of this event. Special sampling may be required to define the extent of the problem.

In cases where a treatment process upset results in the failure to meet a specific water quality standard, the operator must promptly notify supervisory personnel and the appropriate local health authorities. In extreme cases, complete process shutdown and/or public notification may be required. During these periods, the operator must work closely with health authorities.

10.911 Operator Error

Occasionally all of us make a mistake. If we are on top of everything and operating our plant as intended, an error in one process may be eliminated or reduced by another process.

Proper chemical doses can be difficult to maintain. If you discover the chemical dose is too high or too low, immediately make the proper adjustment. Try to monitor the doses more frequently.

If you discover insufficient or no chlorine residual in your clear well, immediately increase the chlorine dosage to the finished water. Review your records. If everything is working properly, all of your quality control tests are looking good, and the turbidity level in the clear well is low, you probably will experience no serious problems.

However, if you discover no chlorine residual in your clear well and your plant is having operating difficulties, you are in trouble. If the turbidity is high in the clear well, try to add chlorine to the clear well or to the finished water pump discharge to achieve the desired chlorine residuals in the clear well and through the entire distribution system. Review your records and the operation of your chlorination system. Determine why there is no chlorine residual in the clear well and correct the situation.

Unfortunately errors can and do happen. After an error has occurred, try to develop procedures that will prevent the error from occurring again. Share your experience with other operators so they won't make the same error. Working together is important and can help everyone.

10.92 Process Equipment Failures

Process equipment failures may also result in treatment process upsets in the event that chemical feeders, chlorinators, or other primary process equipment items fail to operate satisfactorily. The best safeguard against premature process equipment failures is a good preventive maintenance program. The operator plays a vital role in the preventive maintenance program by performing daily inspections of process equipment and making minor adjustments and repairs when necessary.

In certain essential processes, such as in chlorination systems, extra equipment is usually built into the system to provide backup in case an individual system part fails. This feature should also be included in other primary process systems such as chemical feed systems.

In the event of a process failure that results in the failure to meet a water quality standard, promptly notify supervisory personnel and the appropriate local health authorities. In extreme cases, process shutdown and/or public notification may be necessary. Let's examine possible equipment failures and how you might respond to them. We are assuming that you don't have standby facilities or they have failed too. If you consider what would happen at your plant if these failures occur, you may be able to justify the installation of essential standby equipment.

1. Intake screens

    If intake screens become plugged or broken, shut the plant down and unplug or repair the screens. Standby or alternate screens obviously are essential. A bypass system may allow continuous operation and avoid the need to shut down the plant.

2. Grit basin

    Mechanical collector fails and can't be corrected or adjusted by above-water repair procedures. Whenever facilities must be dewatered for emergency repairs, try to fill up all water storage facilities by early evening. Dewater facility and repair at night when demands are low.

3. Prechlorination facility

    a. Shut down facilities and repair immediately. Try to avoid allowing unchlorinated water to pass through your plant and having to rely solely on postchlorination.

    or

    b. If postchlorination facilities are adequate, you may wish to rely strictly on postchlorination. Under these conditions, increase surveillance of chlorine residuals.

4. Alum or polymer feeder

    Shut down influent pumps. Repair chemical feeder. Do not allow water to flow past point of chemical application without alum or polymer. Otherwise, turbidity will pass through filters and may exceed EPA Primary Drinking Water Standards.

5. Rapid mix or flash mix

    Consider moving point of chemical application to a location where water turbulence can help to achieve hydraulic mixing.

6. Flocculators

    a. Underwater. Wait until scheduled dewatering of facility and then repair.

    b. Mechanical. Repair as soon as possible.

7. Sedimentation tank

    Mechanical sludge collector fails and can't be repaired or adjusted by above-water procedures. Dewater facility and repair at night when demands are low.

8. Filters

    Valve or backwash system fails. Take failed portion (bank) out of service and repair.

9. Postchlorination facility

    a. Increase prechlorination doses, if possible.

    or

    b. If you are using postchlorination only, shut down and repair immediately. Notify supervisors and proper authorities.

10. Corrosion-control chemical feeder

    Repair as soon as possible.

QUESTIONS

Write your answers in a notebook and then compare your answers with those on page 489.

10.9A   What three factors usually cause treatment process failures?

10.9B   How would you safeguard against premature process equipment failures?

10.9C   What would you do if the prechlorination facility in your plant failed?

10.93 Power Failures

A backup electrical power source is usually provided at water treatment plants for use in the event of commercial power failure. Engine-generator sets powered by diesel fuel, natural gas, or liquid petroleum gas provide the standby capability to furnish a limited amount of electrical power to keep the water treatment plant in service during periods of commercial power failure. In most cases, it is not practical to provide emergency power to meet all treatment plant demands. Therefore, only critical process functions (such as chemical feeders, mixers, flocculators, and process pumps) are included on the emergency power bus. This power bus is usually connected to primary process equipment items by a transfer switch that automatically transfers power to the backup or standby source during failures.

At the onset of a commercial power failure, take the following actions:

1. Notify the commercial power supplier of the outage;

2. If the power failure originated at the treatment plant, notify electrical maintenance personnel immediately;

3. Restart process equipment that shut off during the power failure (prepare a sequence for your plant so that only one piece of equipment at a time is restarted to avoid overload);

4. Check chlorination equipment and safety devices for proper operation;

5. Check the engine-generator set for proper operation;

6. Notify supervisory personnel of the condition; and

7. Visually inspect all process equipment and check the performance of unit treatment processes.

During brief periods of power outage, most primary plant process functions can continue to operate. However, for extended periods of power outage, it may be necessary to reduce plant production since filter backwashing systems are usually not connected to the emergency power bus due to the high energy demand. As filters stop working due to head loss buildup or turbidity breakthrough, they should simply be removed from service. If insufficient clean filters are available to replace them, the plant flow rate will have to be reduced.

When commercial power is restored, take the following actions:

1. Restart process equipment that shut off during the transfer or was off line during the outage (one at a time),

2. Backwash dirty filters and return them to service,

3. Increase plant flow rate as appropriate,

4. Visually inspect all process equipment and performance of unit treatment processes,

5. Verify process and treated water quality, and

6. Notify supervisory personnel of conditions.

10.94 Fires

If a fire occurs at the water treatment plant, immediately notify the local fire department and then determine the source and severity of the fire. Depending on the type of fire (structure, chemical, electrical), use the appropriate fire safety equipment at the plant in an attempt to extinguish the fire. *DO NOT* try to be a hero! If the fire is too involved, wait for the fire department to arrive (response time is usually short).

After calling the fire department, notify plant supervisory personnel promptly of the emergency condition at the plant.

If you have not already done so, make yourself thoroughly familiar with the care and use of fire safety equipment and learn the special procedures to be observed in dealing with chemical and electrical fires. You won't have time to study the equipment after a fire starts.

10.95 Natural Disasters

Fortunately, natural disasters such as floods and earthquakes are relatively rare events. Most water treatment plants are designed with these events in mind and adequate safety features are usually built into the plant to minimize damage caused by floods or earthquakes. Water treatment plants are normally located on sites that are above the standard flood plain, or special measures are taken to prevent facilities from flooding during a heavy rainstorm.

Emergency preparedness and earthquake safety in the design of new structures are very important. Since there is not much that operators, or anyone else, can do during catastrophic events, only additional planning and emergency preparation will help protect water supplies.

Following any major flood, earthquake, or other natural disaster, take the following actions:

1. Inspect accessibility of all facilities,

2. Check condition and function of all process equipment,

3. Check structures and chemical storage tanks for structural or other damage,

4. Check the plant piping system for leaks or other visible signs of damage,

5. Prepare a preliminary damage report, and

6. Report conditions to plant supervisory personnel.

10.96 Communications

In the event of an emergency, you will be required to advise other plant personnel of the existing conditions or events that have occurred.

An *EMERGENCY RESPONSE PROCEDURE* should be developed for every water treatment plant so that notification of the proper personnel can be readily accomplished and the emergency resolved. Emergency response procedures should list the *NAMES AND TELEPHONE NUMBERS* of persons to be notified under specified conditions, including health department authorities. Guidelines should be developed to assist the operator in determining when to implement these procedures. Alternate communication methods must be considered because telephone service may be lost during an emergency. Be sure to review the emergency response procedures at least once a year and verify the accuracy of all names and telephone numbers.

QUESTIONS

Write your answers in a notebook and then compare your answers with those on page 489.

10.9D What happens at most water treatment plants when commercial power fails?

10.9E What would you do if a fire occurred at your water treatment plant?

10.9F After a major flood, what action should be taken by a water treatment plant operator?

10.10 SLUDGE HANDLING AND DISPOSAL

10.100 Discharge Standards

The problem of water treatment plant sludge disposal is very important. Federal laws include sludge from a water treatment plant as an industrial waste and require proper handling and disposal. Under the National Pollutant Discharge Elimination System (NPDES) provision of the federal laws, a permit must be obtained for wastewater discharges (process sludge) from a water treatment plant into a surface water or groundwater source. This permit sets discharge limits on water quality characteristics such as pH, total suspended solids, settleable solids, flow, and *BOD*.[4]

10.101 Sludge Sources

Suspended solids in the source water represent the major source of sludge solids to be disposed of as a result of water treatment. The treatment chemicals themselves, especially alum, constitute a secondary source of sludge solids. Another major source of sludge is the precipitate from the lime-soda ash softening process.

In most water treatment plants, over 99 percent of the suspended solids in the source water are removed by the sedimentation and/or filtration processes. These processes concentrate the source water solids and treatment chemicals, which are then collected and processed to reduce their volume before final disposal.

10.102 Sludge Processing and Disposal

For procedures on how to process and dispose of sludge, see Chapter 17, "Handling and Disposal of Process Wastes," in Volume II. Another helpful reference is *ADVANCED WASTE TREATMENT*, Chapter 3, "Residual Solids Management." This publication is available from Office of Water Programs, California State University, Sacramento, 6000 J Street, Sacramento, CA 95819-6025. Price, $45.00.

QUESTIONS

Write your answers in a notebook and then compare your answers with those on page 489.

10.10A What do the letters NPDES stand for?

10.10B How are most suspended solids in the source water removed at water treatment plants?

10.11 HANDLING WATER QUALITY COMPLAINTS

10.110 Guidelines for Handling Complaints

No discussion of water treatment plant operation would be complete without discussing consumer complaints, investigation procedures, and possible causes for complaints. The following guidelines and procedures are taken from "Procedural Manual for Handling Water Quality Complaints," prepared by the System Water Quality Committee, California-Nevada Section, American Water Works Association.

Some basic guidelines that should be followed in dealing with consumer complaints are listed below:

1. Be friendly and courteous to the consumer at all times,

2. Assure the consumer that you are pleased that they have taken the trouble to call about their problem,

3. Ask the consumer to describe the problem,

4. Listen carefully and calmly to the consumer's explanation,

5. Review with the consumer the explanation of the problem and ask questions as required to make certain you understand the problem,

6. Do not argue with the consumer,

7. Make every effort to give the consumer an immediate, clear, and accurate answer to the problem,

8. If it is necessary to contact the consumer at their place of business or residence, assure them that it will be scheduled as soon as possible,

9. Do your best to assure the consumer that the problem has been or will be resolved, and

10. If the consumer cannot be satisfied, offer to refer the person to someone in management.

Remember, it is often consumer complaints that alert you to developing problems of water quality or service.

10.111 Investigating Complaints

In some cases, it will be possible to resolve the consumer's problem on the telephone. More often, however, it is best to send operators into the field to determine the cause of the condition. Tables 10.7 through 10.14 describe the more common types of consumer complaints, investigation procedures, and possible causes for complaints.

---

[4] *BOD (pronounce as separate letters).* **B**iochemical **O**xygen **D**emand. *The rate at which organisms use the oxygen in water while stabilizing decomposable organic matter under aerobic conditions. In decomposition, organic matter serves as food for the bacteria and energy results from its oxidation. BOD measurements are used as a measure of the organic strength of wastes in water.*

TABLE 10.7   AIR IN WATER OR MILKY WATER

Complaint Investigation	Possible Causes For Complaint
1. *INFORMATION NEEDED* What is location of premises? Determine pressure zone of consumer's premises. Are the premises new, or has some new galvanized pipe been installed recently? When was air or milkiness first noticed? Has the water to the premises been shut off recently? Is the air or milkiness in both the hot and cold water? 2. *FIELD INVESTIGATION* Check water at consumer's premises. Eliminate air in water by: (a) Flushing house lines, if necessary. (b) Flushing hydrants or blowoffs, if necessary. Take sample to laboratory, if necessary. Report results of laboratory tests to consumer.	1. *DISTRIBUTION SYSTEM* Shutdown of mains Low pressure in mains Leaking pump glands Temperature changes in water Cross connections Miscellaneous causes 2. *PRIVATE PLUMBING SYSTEM* Overheating of hot water systems Warming up of cold water lines Zinc from galvanized pipe Cross connections Miscellaneous causes

TABLE 10.8   DIRTY, COLORED, OR CONTAINS FOREIGN PARTICLES

Complaint Investigation	Possible Causes For Complaint
1. *INFORMATION NEEDED* What is location of premises? Determine pressure zone of premises. When was dirty water first detected? What does the water look like? Does the water have a color? Are both the hot and cold water dirty? Is the water dirty at all faucets? 2. *FIELD INVESTIGATION* Check water at consumer's premises. Eliminate dirty water by: (a) Flushing hydrants or blowoffs. (b) Flushing house line, if necessary. Take sample to laboratory, if necessary. Report results of laboratory tests to customer.	1. *DISTRIBUTION SYSTEM* Water treatment plant problems Breaks in mains Dead ends Cross connections New, recoated, or repainted water mains, tanks, and reservoirs Fires Flushing of fire hydrants Disturbance of consumer's service line Changes in pressure zones Pipe coatings and sand 2. *PRIVATE PLUMBING SYSTEM* Hot water systems Cross connections House piping Plumbing repairs

TABLE 10.9   HARD WATER, SCALE, SPOTS ON GLASSWARE

Complaint Investigation	Possible Causes For Complaint
1. *INFORMATION NEEDED* What is location of customer's premises? Determine source of water supplied to customer. When was hard water or scale first detected? What was the means of measurement of water being harder than usual? 2. *FIELD INVESTIGATION* Check water at customer's premises. Take samples to laboratory, if necessary. Report results of laboratory tests to customer.	1. *DISTRIBUTION SYSTEM* Change of supply Cross connection Other causes 2. *PRIVATE PLUMBING SYSTEM* Cross connection Other causes 3. *GENERAL CONSUMER CONCERNS* Spots on bottles, glassware, boiler scale Soft water use, such as steam irons, batteries Soap rings in tubs, washing machines Types of soap vs. hardness Hardness determination Water softening

TABLE 10.10 SICKNESS OR SKIN IRRITATION

Complaint Investigation	Possible Causes For Complaint
1. *INFORMATION NEEDED* What is location of customer's premises? Determine source of water supplied to customer. When did the sickness first occur? Why is it thought that the sickness is due to water? Have all members of the family been affected? Have the affected members of the family been out of town recently? Has a doctor been consulted? 2. *FIELD INVESTIGATION* Check taste, odor, color, and turbidity of water at customer's premises. Check for cross connections. Take sample to laboratory for bacteriological and partial chemical test. Report results of laboratory tests to consumer.	1. *DISTRIBUTION SYSTEM* Change in supply Cross connection Other causes 2. *PRIVATE PLUMBING SYSTEM* Cross connection Other causes 3. *GENERAL CONSUMER CONCERNS* Consumer's senses affected by illness, medication, and/or diet

TABLE 10.11 TASTES AND ODORS

Complaint Investigation	Possible Causes For Complaint
1. *INFORMATION NEEDED* What is location of customer's premises? Determine source of water supplied to customer's premises. When was taste or odor first detected? Is the taste or odor in both the hot and cold water? Does consumer have a pressurized hose (gun-type nozzle)? 2. *FIELD INVESTIGATION* Check water at customer's premises. Suggest that consumer flush house lines. Flush hydrants or blowoffs, if necessary. Take sample to laboratory, if necessary. Report results of laboratory tests to customer.	1. *DISTRIBUTION SYSTEM* Raw water Water treatment plant Disinfection of new mains, tanks, or reservoirs Dead ends Cross connections Water from a different source Miscellaneous causes 2. *PRIVATE PLUMBING SYSTEM* Hot water tanks Cross connections Old piping Exposed water lines Compounds added by customer to control corrosion or to protect boilers Kitchen sink odors Miscellaneous causes 3. *GENERAL CONSUMER CONCERNS* Consumer's senses affected by tastes and odors

TABLE 10.12 WORMS OR BUGS

Complaint Investigation	Possible Causes For Complaint
1. *INFORMATION NEEDED* What is location of customer's premises? Determine source of water supplied to customer's premises. Where were the organisms first found? How would you describe the organisms? 2. *FIELD INVESTIGATION* Check water and sample of organisms at customer's premises. Flush house lines, if necessary. Flush hydrants or blowoffs, if necessary. Report results of laboratory analysis to customer.	1. *DISTRIBUTION SYSTEM* Distribution reservoirs Cross connections Dead ends Main breaks, fires Water treatment plants Miscellaneous sources 2. *PRIVATE PLUMBING SYSTEM* Cross connections Organisms in bathtubs, bowls, wash basins Organisms from miscellaneous sources

TABLE 10.13   AQUARIUM FISH PROBLEMS

Complaint Investigation	Possible Causes For Complaint
1. *INFORMATION NEEDED* What is location of customer's premises? When did the fish first start to die? When was water last added to the aquarium? Were new fish added? Have any new foods or plants been added? Were any sprays used near the aquarium? Is the aquarium new, or have new materials been used? 2. *FIELD INVESTIGATION* Check water at customer's premises. Take sample to laboratory, if necessary. Report results of laboratory tests to customer.	1. *DISTRIBUTION SYSTEM* pH of water Chlorine residual Copper content Cross connections 2. *PRIVATE PLUMBING SYSTEM* pH of water Chlorine residual Copper content Cross connections 3. *GENERAL CONSUMER CONCERNS* Change in temperature Dissolved oxygen Plant or insect sprays Toxic materials used in construction of aquarium Overloading of aquarium Overfeeding or underfeeding of fish Fish diseases Chemicals used to prevent diseases Miscellaneous causes

TABLE 10.14   GARDEN DAMAGE

Complaint Investigation	Possible Causes For Complaint
1. *INFORMATION NEEDED* What is location of customer's premises? Source of supply? When was plant damage first noticed? Use of fertilizers or garden sprays? Possible animal (dogs, cats) damage. Possible damage by gophers, moles. Plant sensitivity to sun, water, soil? Frequency of watering? 2. *FIELD INVESTIGATION* Test water at customer's premises for pH and chlorine residual. Take sample to laboratory, if necessary. Report results of laboratory tests to customer.	1. *DISTRIBUTION SYSTEM* pH of water Chlorine residual Copper residual Cross connections 2. *PRIVATE PLUMBING SYSTEM* pH of water Chlorine residual Cross connections 3. *GENERAL CONSUMER CONCERNS* Improper care of acid-loving plants Over- or underfertilizing Over- or underwatering of plants Plant diseases or insects Damage by garden sprays or powders Damage by dogs or cats Damage by gophers, moles, deer, or other animals Miscellaneous causes

QUESTIONS

Write your answers in a notebook and then compare your answers with those on page 489.

10.11A   List the common types of consumer complaints.

10.11B   What could be done to solve a consumer's complaint that there is air in the water?

10.11C   What would you do if a consumer complained that the drinking water was causing a sickness?

10.12   ENERGY CONSERVATION

10.120   Energy Considerations

In the operation of a water treatment plant, a considerable amount of energy may be consumed for lighting, heating and air conditioning, and powering numerous electric motors located throughout the plant. Operators can have a positive impact on overall treatment plant operating costs if energy conservation procedures are followed on a routine basis.

10.121   Energy Conservation Procedures

LIGHTING usually represents less than five percent of the electric energy use in a water treatment plant. Even though this is a small percentage, positive measures can be taken to reduce these costs as follows:

1. Turn lights off when leaving a room or work area,

2. Turn lights off in unoccupied areas,

3. Limit yard lighting to essential areas,

4. Replace existing lamps with high-efficiency lamps as the old ones burn out or require replacing, and

5. Convert mercury vapor lamps to more efficient high-pressure sodium vapor lamps.

HEATING, VENTILATING, AND AIR CONDITIONING equipment consume substantial amounts of fuel and electric energy. Energy savings can be gained here by simply adjusting thermostats to more efficient settings. Thermostats should be set at 78°F (26°C) for cooling, and at 68°F (20°C) for heating.

ELECTRIC MOTORS consume the greatest amount of electricity in most water treatment plants, over 90 percent of total electric energy consumed. Considerable savings can be achieved by replacing old electric motors with high-efficiency motors as the old ones burn out.

Consideration should also be given to installing capacitors at the treatment plant to correct a low plant *POWER FACTOR.*[5] Capacitors offset the reaction power used by inductive devices (electric motors) and improve the overall power factor of the plant. This can result in considerable energy savings.

10.122   Power Management

The most sophisticated approach to energy conservation, and perhaps the most beneficial, is through a power management program. A power management program starts by identifying each source of energy use in the water treatment plant. These sources are then tabulated, ranked by size of load, and prioritized to define the importance of each load source in the overall program. This basic evaluation will frequently point to areas where immediate savings can be achieved through simple changes in routine operations (for example, avoid backwashing a filter when another high-load source is in operation to reduce peak power charges).

Some utility suppliers have a "time-of-use" billing schedule, which provides the user with significant price breaks during "off-peak" demand periods. If possible, filter backwashing or other discretionary functions can be performed during off-peak periods to take advantage of these lower rates.

In its most sophisticated form, power management can be used to control all of the electrical loads in the water treatment plant. With the aid of a simple computer, each load can be monitored and controlled to provide the most cost-effective operating mode. In some instances, noncritical loads can be turned off (shed) for short time periods while higher priority loads are on line. This can result in significant operational cost savings without compromising the safety and reliability of plant operations.

10.123   Power Cost Analysis

In order to properly assess the impact of power costs on your overall plant operating budget it is helpful to perform a monthly power cost analysis similar to the one shown in Figure 10.7, and discussed in the following paragraphs.

Basic to any power cost analysis is reading and recording electric energy use on a monthly basis. This is accomplished by reading the plant utility meter or meters. Using the proper meter multiplier factor (1,000 in this example, column C), you can then determine the total amount of electric energy used in kWh (kilowatt hours) during the month (shown in column E).

The kW (kilowatt) demand can also be read on the utility meter. This value represents the greatest single energy demand during the month (or prior months), and is generally billed as a separate cost component. Overall energy costs can generally be reduced by keeping demand to a minimum. Notice in column D that the demand normally varies with plant flow. This results from the higher demand for electrical equipment (pumps and motors) to produce the higher flow rates and operating pressures required during the warmer months. If you are careful, you can avoid creating high peak demands

---

[5] *Power Factor.   The ratio of the true power passing through an electric circuit to the product of the voltage and amperage in the circuit. This is a measure of the lag or lead of the current with respect to the voltage. In alternating current the voltage and amperes are not always in phase; therefore, the true power may be slightly less than that determined by the direct product.*

ELECTRIC UTILITY _____   UNIT _____

UTILITY ACCT. No. _____   LOCATION _____

CITY OF PHOENIX, ARIZONA
WATER & SEWERS DEPARTMENT
DIVISION OF WATER PRODUCTION
POWER COST ANALYSIS
FISCAL YEAR 19 81 to 19 82

A	B	C	D	E	F	G	H	I	J	K	L	M	N	P	REMARKS
READING DATE	READING	MULTI-PLIER	KW DEMAND	KWH	KWH per KW DEMAND ($\frac{E}{D}$)	TOTAL POWER COST	WATER METER READING x1000	GALLONS WATER PUMPED M.G.	POWER COST per M.G. ($\frac{G}{I}$)	COST of CHEMICALS	CHEMICAL COST per M.G. ($\frac{K}{I}$)	WATER COST per M.G.	KWH per M.G. ($\frac{E}{I}$)	*TOTAL POWER CHEMICAL & WATER COST per M.G. (J+L+M)	REMARKS
6-30-81	425.1	1000					7,402,853								
7-31-81	969.9	1000	1072	544,800	508.2	32,661.82	8,167,061	764.2	42.74	27,079.58	35.43	11.80	713	89.97	Fed Activated Carbon for Taste & Odor Control
8-31-81	1,525.9	1000	1064	556,300	522.6	33,283.36	8,988,688	821.6	40.51	25,290.26	30.78	11.80	677	83.09	"
9-30-81	2,069.9	1000	1072	564,000	526.1	33,572.11	9,793,313	804.6	41.73	27,457.87	34.13	11.80	701	87.66	"
10-31-81	2,518.7	1000	1080	428,800	397.0	24,179.31	10,552,931	759.6	31.83	20,801.13	27.38	11.80	565	71.01	"
11-30-81	2,909.9	1000	1016	391,200	385.0	22,191.75	11,204,146	651.2	34.08	19,004.29	29.18	11.80	601	75.06	"
12-31-81	3,210.7	1000	904	300,800	332.7	17,435.19	11,735,615	531.5	32.80	9,190.50	17.29	11.80	566	61.89	
1-31-82	3,522.7	1000	920	312,000	339.1	17,801.75	12,163,703	428.1	41.58	8,571.69	20.02	14.50	729	76.10	
2-28-82	3,839.5	1000	945	316,800	335.2	18,069.16	12,588,965	425.3	42.49	9,601.35	22.58	14.50	745	79.57	
3-31-82	4,157.9	1000	952	318,400	334.5	18,034.57	12,975,611	386.7	46.64	22,305.77	57.68	14.50	823	118.82	High coagulant dose due to high turbidity
4-30-82	4,543.5	1000	1120	385,600	344.3	21,510.10	13,393,901	418.2	51.58	11,395.31	27.25	14.50	922	93.33	
5-31-82	5,021.9	1000	1300	478,400	368.0	26,560.71	14,049,165	655.3	40.53	10,689.01	16.31	14.50	730	71.34	
6-30-82	5,649.1	1000	1215	627,200	516.2	34,246.07	14,810,375	761.2	44.99	10,052.36	13.21	14.50	824	72.70	
ANNUAL TOTALS or AVERAGES			12,660 / 1055	5,224,000 / 435,333	4,908.9 / 409.1	299,604.90 / 24,967.08		7,407.5 / 617.3	491.50 / 40.96	201,439.02 / 16,786.59	331.24 / 27.60	157.80 / 13.15	8,596 / 716	980.54 / 81.71	

* For comparison purposes only. Labor and fixed charges not included.

Fig. 10.7   Power cost analysis

during the warmer months by shifting nonessential operations requiring electrical energy to off-peak hours. Some examples, such as backwashing filters during non-peak hours, have been discussed in the preceding sections.

Total power cost, as shown in column G, can be obtained from the monthly energy bill. Power costs can be calculated in convenient units such as $/MG as shown in column J. In this example, column J was obtained by dividing the total power cost shown in column G ($) by the total gallons of water pumped as shown in column I (MG). Monthly power costs are useful for checking the current budget allocation for energy and for preparing the following year's budget.

Keeping track of other treatment costs such as chemical costs (shown in columns K and L) and water cost (shown in column M) can also be useful to the operator as well as to supervisory personnel. If large variations in treatment costs appear in any given month, this analysis form will provide you with important clues to help explain or solve the problem.

The remarks column is provided for making comments on any unusual conditions which caused higher energy or chemical demands. These remarks will be helpful in budgeting for the next year, as well as reporting on the current year's performance (annual operating report).

10.13   REFERENCE BOOKS

As discussed in previous chapters, reference books and manuals are a valuable technical resource. A wide variety of reference material is available which describes how to perform routine laboratory tests, routine and complex process performance considerations, as well as other technical aspects of water treatment plant operations.

The following is a suggested list of reference books which will help you understand the more complex aspects of water treatment. This list is not intended to be a critical or complete reference source on the topic, as there are a number of other good reference books available.

SUGGESTED REFERENCES

1. *CHEMISTRY FOR ENVIRONMENTAL ENGINEERING AND SCIENCE*, Fifth Edition, 2003. Obtain from the McGraw-Hill Companies, Order Services, PO Box 182604, Columbus, OH 43272-3031. ISBN 0-07-248066-1. Price, $109.37, plus nine percent of order total for shipping and handling.

2. *STANDARD METHODS FOR THE EXAMINATION OF WATER AND WASTEWATER*, 20th Edition. A joint publication of the American Public Health Association (APHA), American Water Works Association (AWWA), and the Water Environment Federation (WEF) which outlines the accepted laboratory procedures used to analyze the impurities in water and wastewater. Available from American Water Works Association, Bookstore, 6666 West Quincy Avenue, Denver, CO 80235. Order No. 10079. Price to members, $167.00; nonmembers, $212.00; price includes cost of shipping and handling.

3. *WATER QUALITY AND TREATMENT*, Fifth Edition. Obtain from American Water Works Association (AWWA), Bookstore, 6666 West Quincy Avenue, Denver, CO 80235. Order No. 10008. ISBN 0-07-001659-3. Price to members, $119.00; nonmembers, $135.00; price includes cost of shipping and handling.

4. *WATER TREATMENT PLANT DESIGN*, Third Edition, prepared jointly by the American Society of Civil Engineers, American Water Works Association, and Conference of State Sanitary Engineers. Obtain from American Water Works Association (AWWA), Bookstore, 6666 West Quincy Avenue, Denver, CO 80235. Order No. 10009. ISBN 0-07-001643-7. Price to members, $103.50; nonmembers, $135.00; price includes cost of shipping and handling.

10.14   ARITHMETIC ASSIGNMENT

Turn to the Appendix, "How to Solve Water Treatment Plant Arithmetic Problems," at the back of this manual and read Section A.138, Plant Operation. Check all of the arithmetic in this section using an electronic pocket calculator. You should be able to get the same answers.

10.15   ADDITIONAL READING

1. *NEW YORK MANUAL*, Chapter 17,* "Protection of Treated Water," Chapter 18,* "Records and Reports," and Chapter 19,* "Treatment Plant Maintenance and Accident Prevention."

2. *TEXAS MANUAL*, Chapter 16,* "Storage of Potable Water," Chapter 18,* "Effective Public Relations in Water Works Operations," and Chapter 23,* "Emergency Operations."

* Depends on edition.

10.16   ACKNOWLEDGMENTS

Many of the concepts and procedures discussed in this chapter are based on material obtained from the sources listed below.

1. Stone, B. G. Notes from "Design of Water Treatment Systems," CE-610, Loyola Marymount University, Los Angeles, CA, 1977.

2. *WATER QUALITY AND TREATMENT: A HANDBOOK OF PUBLIC WATER SUPPLIES*, American Water Works Association, Third Edition, McGraw-Hill, 1971.

3. *WATER TREATMENT PLANT DESIGN*, prepared jointly by the American Water Works Association, and Conference of State Sanitary Engineers, AWWA, 1989.

4. *OPERATION AND MAINTENANCE MANUAL FOR STOCKTON EAST WATER TREATMENT PLANT*, James M. Montgomery, Consulting Engineers, Inc., 501 Lennon Lane, Walnut Creek, CA 94598, 1979.

QUESTIONS

Write your answers in a notebook and then compare your answers with those on page 489.

10.12A   List the major sources of energy consumption in the operation of a water treatment plant.

10.12B   How can the consumption of electric energy at a water treatment plant be reduced?

End of Lesson 2 of 2 Lessons on PLANT OPERATION

Please answer the discussion and review questions next.

DISCUSSION AND REVIEW QUESTIONS

Chapter 10.   PLANT OPERATION

(Lesson 2 of 2 Lessons)

Write the answers to these questions in your notebook. The question numbering continues from Lesson 1.

11. What records should be maintained as part of a good records management system?

12. What is a work order system?

13. What would you do if the entire chemical feed system (coagulants and polymers) for your plant failed?

14. If a water treatment plant must be shut down for emergency repairs, how would you try to prepare for this event?

15. Outline a procedure for handling complaints.

16. How can energy requirements for lighting be reduced in a water treatment plant?

SUGGESTED ANSWERS

Chapter 10.   PLANT OPERATION

ANSWERS TO QUESTIONS IN LESSON 1

Answers to questions on page 454.

10.0A   The first priority for operating a water treatment plant is the production of a safe drinking water, one that is free of harmful bacteria and toxic materials.

10.0B   Water that appeals to consumers must be clear (free of turbidity), colorless, and free of objectionable tastes and odors. Consumers also show a preference for water supplies that do not stain plumbing fixtures and clothes, do not corrode plumbing fixtures and piping, and do not leave scale deposits or spot glassware.

Answers to questions on page 456.

10.1A   Grab samples are used when measuring water quality indicators that can change after collection, such as coliforms, pH, and temperature.

10.1B   Turbidity must be removed in order for disinfection to be effective in killing or inactivating disease-causing organisms.

Answers to questions on page 459.

10.2A   Your first task upon arrival at a water treatment plant should be to review what has happened during the last shift or since you left the plant.

10.2B   Operators control finished water quality mainly by the proper application of chemicals.

10.2C   Tasks that an operator should do during the day include: (1) backwash or check filters, (2) monitor system pressures, (3) monitor the main control board, (4) check storage in clear well, (5) monitor and adjust pumping rates, and (6) perform quality control checks.

10.2D   Tasks that an operator should do at the end of the day are the same as those listed under 10.2C plus anticipate raw water and finished water pumping rates, check chemical dosage facilities, and secure plant for the night.

Answers to questions on page 462.

10.3A   Data can be transmitted over distances greater than 1,500 feet by telephone tone, microwave, or radio transmission.

10.3B   The air supply for pneumatic data transmission devices must be dried to prevent condensation in equipment signal lines, and filtered to remove particle contaminants.

10.3C   Water, water and oil, and glycerol are frequently used transmission fluids in hydraulic systems.

10.3D   Functions commonly performed by electronic signals include (1) indication, (2) recording, (3) data logging, (4) alarm, and (5) control.

Answers to questions on page 463.

10.3E   A hybrid instrumentation system is a combination of many different types of instrumentation systems.

10.3F   Relay logic is a method of switching electrical power on and off in accordance with a predetermined sequence (logic) by means of relays, process switches and contacts, timers, and manual switches.

Answers to questions on page 464.

10.4A   Factors that influence the amount of water that must be treated each day include current storage levels in clear wells and distribution system service storage reservoirs. Other important factors include expected consumer demand based on historical and current trends as well as weather conditions.

10.4B   During periods of peak demand for water, clear wells are drawn down to provide water to meet the demand.

10.4C   Filters are usually operated at a constant rate. When large flow changes occur, either more filters are put in service or some filters are taken out of service.

Answers to questions on page 469.

10.5A   A pilot plant is a "small-scale" water treatment plant built at or near the source of supply. Pilot tests provide the designer, as well as the treatment plant operator, with valuable information on the "treatability" of a given source of supply over a broad range of water quality conditions.

10.5B   Water treatment chemicals can be stored in a number of ways including (1) solid (dry) form (bags, cartons, drums), (2) liquid form (drums, tanks, cylinders), or (3) gaseous form (cylinders).

10.5C   Protective clothing should be worn when working with chemicals. Goggles and face shields will protect your eyes and face. Other exposed portions of the body should be protected by wearing rubber or neoprene gloves, aprons, or other protective clothing. Use respirators when working with chemical dust, and always use dust collectors when such equipment is provided.

10.5D   Wherever chlorine liquid or gas is stored or used, the following safety equipment should be provided:

1. Shower and eye wash facilities,
2. Emergency breathing apparatus (air pack),
3. Chlorine gas detector,
4. Floor-level vents, and
5. Fans that maintain a positive air pressure.

Answers to questions on page 473.

10.6A The two major categories of operating records are (1) physical records, and (2) performance records.

10.6B Performance records describe the operation of the water treatment plant.

10.6C Records that may have historical value to the operator include source water quality changes and resultant process changes.

10.6D Each operator should use a pocket notebook to record unique or unusual events.

ANSWERS TO QUESTIONS IN LESSON 2

Answers to questions on page 476.

10.7A The major elements of a good preventive maintenance (PM) program include: (1) planning and scheduling, (2) records management, (3) spare parts management, (4) cost and budget control, (5) emergency repair procedures, and (6) operator training program. Also the maintenance program must be capable of dealing effectively with emergency conditions.

10.7B A comprehensive records management system provides the basis for daily task assignments, provides a permanent record of work performed, and becomes a historical reference source for reviewing equipment performance.

10.7C Items that should be included in a cost and budget control program include accurate records of labor and equipment expenditures as well as costs of equipment operations, repair, and replacement. A work order system provides much of this information.

Answers to questions on page 477.

10.8A Potential hazards an operator could be exposed to during the routine operation of a water treatment plant include: (1) electrical equipment (shocks), (2) rotating mechanical equipment, (3) open-surface, water-filled structures (drowning), (4) underground structures (toxic and explosive gases, excess or lack of oxygen), (5) water treatment chemicals, (6) laboratory reagents (chemicals), and (7) pumping stations (high noise levels).

10.8B An operator should routinely inspect plant facilities at least once per shift for any evidence of unauthorized persons or unusual events.

Answers to questions on page 478.

10.9A Treatment process failures are usually caused by (1) changes in raw water quality, (2) operator error, or (3) equipment failure.

10.9B The best safeguard against premature process equipment failures is a good preventive maintenance program.

10.9C If the prechlorination facility failed, (1) shut down immediately and repair, or (2) rely strictly on post-chlorination.

Answers to questions on page 479.

10.9D Emergency power is usually provided at water treatment plants as a backup electrical power source for use in the event of commercial power failure.

10.9E If a fire occurs at your water treatment plant, immediately notify the local fire department and then determine the source and severity of the fire. Depending on the type of fire, use the appropriate fire safety equipment at the plant in an attempt to extinguish the fire.

10.9F After a major flood the operator should try to get the plant back on line and functioning properly. Items that should be checked include: (1) inspecting all facilities for accessibility, (2) checking condition and function of all process equipment, (3) checking structures and chemical storage tanks for structural or other damage, (4) checking the plant piping system for leaks and other visible signs of damage, (5) preparing a preliminary damage report, and (6) reporting conditions to plant supervisory personnel.

Answers to questions on page 480.

10.10A NPDES stands for **N**ational **P**ollutant **D**ischarge **E**limination **S**ystem.

10.10B Most suspended solids are removed by the sedimentation and filtration processes at water treatment plants.

Answers to questions on page 484.

10.11A Common types of consumer complaints include:

1. Air in water or milky water,
2. Dirty, colored, or contains foreign particles,
3. Hard water, scale, spots on glassware,
4. Sickness or skin irritation,
5. Tastes and odors,
6. Worms or bugs,
7. Aquarium fish problems, and
8. Garden damage.

10.11B Air in water can be eliminated by (1) flushing house lines, if necessary, or (2) flushing hydrants or blow-offs, if necessary.

10.11C If a consumer complained regarding water causing a sickness,

1. Check taste, odor, color, and turbidity of water at consumer's premises,
2. Look for cross connections,
3. Take sample to laboratory for bacteriological and partial chemical test, and
4. Report results of laboratory test to consumer.

Answers to questions on page 486.

10.12A In the operation of a water treatment plant, energy may be consumed for lighting, heating and air conditioning, and powering numerous electric motors throughout the plant.

10.12B Consumption of electric energy may be reduced by replacing old electric motors with high-efficiency motors as the old ones burn out. Consideration should also be given to installing capacitors at the treatment plant to correct a low power factor.

CHAPTER 11

LABORATORY PROCEDURES

by

Jim Sequeira

TABLE OF CONTENTS

Chapter 11.   LABORATORY PROCEDURES

OBJECTIVES

Chapter 11.   LABORATORY PROCEDURES

Following completion of Chapter 11, you should be able to:

1. Work safely in a laboratory,

2. Operate laboratory equipment,

3. Collect representative samples and also preserve and transport the samples,

4. Prepare samples for analysis,

5. Describe the limitations of lab tests,

6. Recognize precautions to be taken for lab tests,

7. Record laboratory test results, and

8. Perform the following field or laboratory tests—alkalinity, chlorine residual, chlorine demand, coliform, hardness, jar test, pH, temperature, and turbidity.

WORDS

Chapter 11. LABORATORY PROCEDURES

ACIDIC (uh-SID-ick)   ACIDIC

The condition of water or soil which contains a sufficient amount of acid substances to lower the pH below 7.0.

ALIQUOT (AL-li-kwot)   ALIQUOT

Representative portion of a sample. Often an equally divided portion of a sample.

ALKALI (AL-ka-lie)   ALKALI

Any of certain soluble salts, principally of sodium, potassium, magnesium, and calcium, that have the property of combining with acids to form neutral salts and may be used in chemical water treatment processes.

ALKALINE (AL-ka-LINE)   ALKALINE

The condition of water or soil which contains a sufficient amount of alkali substances to raise the pH above 7.0.

AMBIENT (AM-bee-ent) TEMPERATURE   AMBIENT TEMPERATURE

Temperature of the surrounding air (or other medium). For example, temperature of the room where a gas chlorinator is installed.

AMPEROMETRIC (am-PURR-o-MET-rick)   AMPEROMETRIC

A method of measurement that records electric current flowing or generated, rather than recording voltage. Amperometric titration is a means of measuring concentrations of certain substances in water.

AMPEROMETRIC (am-PURR-o-MET-rick) TITRATION   AMPEROMETRIC TITRATION

A means of measuring concentrations of certain substances in water (such as strong oxidizers) based on the electric current that flows during a chemical reaction. Also see TITRATE.

ASEPTIC (a-SEP-tick)   ASEPTIC

Free from the living germs of disease, fermentation, or putrefaction. Sterile.

BACTERIA (back-TEAR-e-ah)   BACTERIA

Bacteria are living organisms, microscopic in size, which usually consist of a single cell. Most bacteria use organic matter for their food and produce waste products as a result of their life processes.

BLANK   BLANK

A bottle containing only dilution water or distilled water; the sample being tested is not added. Tests are frequently run on a *SAMPLE* and a *BLANK* and the differences are compared. The procedure helps to eliminate or reduce test result errors that could be caused when the dilution water or distilled water used is contaminated.

BUFFER   BUFFER

A solution or liquid whose chemical makeup neutralizes acids or bases without a great change in pH.

BUFFER CAPACITY   BUFFER CAPACITY

A measure of the capacity of a solution or liquid to neutralize acids or bases. This is a measure of the capacity of water for offering a resistance to changes in pH.

CALCIUM CARBONATE ($CaCO_3$) EQUIVALENT   CALCIUM CARBONATE ($CaCO_3$) EQUIVALENT

An expression of the concentration of specified constituents in water in terms of their equivalent value to calcium carbonate. For example, the hardness in water which is caused by calcium, magnesium and other ions is usually described as calcium carbonate equivalent. Alkalinity test results are usually reported as mg/L $CaCO_3$ equivalents. To convert chloride to $CaCO_3$ equivalents, multiply the concentration of chloride ions in mg/L by 1.41, and for sulfate, multiply by 1.04.

CARCINOGEN (CAR-sin-o-JEN)   CARCINOGEN

Any substance which tends to produce cancer in an organism.

CHLORORGANIC (klor-or-GAN-ick)   CHLORORGANIC

Organic compounds combined with chlorine. These compounds generally originate from, or are associated with, life processes such as those of algae in water.

COLORIMETRIC MEASUREMENT   COLORIMETRIC MEASUREMENT

A means of measuring unknown chemical concentrations in water by measuring a sample's color intensity. The specific color of the sample, developed by addition of chemical reagents, is measured with a photoelectric colorimeter or is compared with "color standards" using, or corresponding with, known concentrations of the chemical.

COMPOSITE (come-PAH-zit) (PROPORTIONAL) SAMPLE   COMPOSITE (PROPORTIONAL) SAMPLE

A composite sample is a collection of individual samples obtained at regular intervals, usually every one or two hours during a 24-hour time span. Each individual sample is combined with the others in proportion to the rate of flow when the sample was collected. The resulting mixture (composite sample) forms a representative sample and is analyzed to determine the average conditions during the sampling period.

COMPOUND   COMPOUND

A pure substance composed of two or more elements whose composition is constant. For example, table salt (sodium chloride, NaCl) is a compound.

DPD (pronounce as separate letters)   DPD

A method of measuring the chlorine residual in water. The residual may be determined by either titrating or comparing a developed color with color standards. DPD stands for N,N-diethyl-p-phenylene-diamine.

DESICCATOR (DESS-uh-KAY-tor)   DESICCATOR

A closed container into which heated weighing or drying dishes are placed to cool in a dry environment in preparation for weighing. The dishes may be empty or they may contain a sample. Desiccators contain a substance, such as anhydrous calcium chloride, which absorbs moisture and keeps the relative humidity near zero so that the dish or sample will not gain weight from absorbed moisture.

DISINFECTION (dis-in-FECT-shun)   DISINFECTION

The process designed to kill or inactivate most microorganisms in water, including essentially all pathogenic (disease-causing) bacteria. There are several ways to disinfect, with chlorination being the most frequently used in water treatment. Compare with STERILIZATION.

ELEMENT   ELEMENT

A substance which cannot be separated into its constituent parts and still retain its chemical identity. For example, sodium (Na) is an element.

END POINT   END POINT

Samples of water or wastewater are titrated to the end point. This means that a chemical is added, drop by drop, to a sample until a certain color change (blue to clear, for example) occurs. This is called the *END POINT* of the titration. In addition to a color change, an end point may be reached by the formation of a precipitate or the reaching of a specified pH. An end point may be detected by the use of an electronic device such as a pH meter. The completion of a desired chemical reaction.

FACULTATIVE (FACK-ul-TAY-tive)   FACULTATIVE

Facultative bacteria can use either dissolved molecular oxygen or oxygen obtained from food materials such as sulfate or nitrate ions. In other words, facultative bacteria can live under aerobic or anaerobic conditions.

FLAME POLISHED   FLAME POLISHED

Melted by a flame to smooth out irregularities. Sharp or broken edges of glass (such as the end of a glass tube) are rotated in a flame until the edge melts slightly and becomes smooth.

GRAB SAMPLE   GRAB SAMPLE

A single sample of water collected at a particular time and place which represents the composition of the water only at that time and place.

GRAVIMETRIC   GRAVIMETRIC

A means of measuring unknown concentrations of water quality indicators in a sample by *WEIGHING* a precipitate or residue of the sample.

INDICATOR (CHEMICAL)   INDICATOR (CHEMICAL)

A substance that gives a visible change, usually of color, at a desired point in a chemical reaction, generally at a specified end point.

INORGANIC   INORGANIC

Material such as sand, salt, iron, calcium salts and other mineral materials. Inorganic substances are of mineral origin, whereas organic substances are usually of animal or plant origin. Also see ORGANIC.

INORGANIC WASTE   INORGANIC WASTE

Waste material such as sand, salt, iron, calcium, and other mineral materials which are only slightly affected by the action of organisms. Inorganic wastes are chemical substances of mineral origin; whereas organic wastes are chemical substances of an animal or plant origin.

M or MOLAR   *M* or MOLAR

A molar solution consists of one gram molecular weight of a compound dissolved in enough water to make one liter of solution. A gram molecular weight is the molecular weight of a compound in grams. For example, the molecular weight of sulfuric acid (H_2SO_4) is 98. A one *M* solution of sulfuric acid would consist of 98 grams of H_2SO_4 dissolved in enough distilled water to make one liter of solution.

MPN (pronounce as separate letters)   MPN

MPN is the **M**ost **P**robable **N**umber of coliform-group organisms per unit volume of sample water. Expressed as a density or population of organisms per 100 m*L* of sample water.

MENISCUS (meh-NIS-cuss)   MENISCUS

The curved surface of a column of liquid (water, oil, mercury) in a small tube. When the liquid wets the sides of the container (as with water), the curve forms a valley. When the confining sides are not wetted (as with mercury), the curve forms a hill or upward bulge. When a meniscus forms in a measuring device, the top of the liquid level of the sample is determined by the bottom of the meniscus.

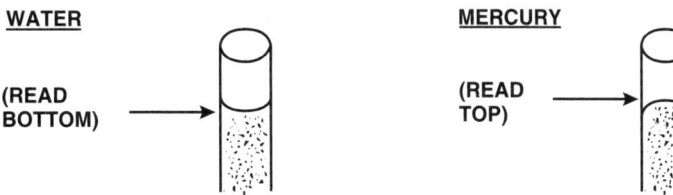

MILLIGRAMS PER LITER, mg/*L*   MILLIGRAMS PER LITER, mg/*L*

A measure of the concentration by weight of a substance per unit volume. For practical purposes, one mg/*L* of a substance in fresh water is equal to one part per million parts (ppm). Thus a liter of water with a specific gravity of 1.0 weighs one million milligrams. If water contains 10 milligrams of calcium, the concentration is 10 milligrams per million milligrams, or 10 milligrams per liter (10 mg/*L*), or 10 parts of calcium per million parts of water, or 10 parts per million (10 ppm).

MOLE   MOLE

The molecular weight of a substance, usually expressed in grams.

MOLECULAR WEIGHT   MOLECULAR WEIGHT

The molecular weight of a compound in grams is the sum of the atomic weights of the elements in the compound. The molecular weight of sulfuric acid (H_2SO_4) in grams is 98.

Element	Atomic Weight	Number of Atoms	Molecular Weight
H	1	2	2
S	32	1	32
O	16	4	64
			98

MOLECULE (MOLL-uh-KULE)   MOLECULE

The smallest division of a compound that still retains or exhibits all of the properties of the substance.

N or NORMAL   *N* or NORMAL

A normal solution contains one gram equivalent weight of reactant (compound) per liter of solution. The equivalent weight of an acid is that weight which contains one gram atom of ionizable hydrogen or its chemical equivalent. For example, the equivalent weight of sulfuric acid (H_2SO_4) is 49 (98 divided by 2 because there are two replaceable hydrogen ions). A one *N* solution of sulfuric acid would consist of 49 grams of H_2SO_4 dissolved in enough water to make one liter of solution.

NEPHELOMETRIC (NEFF-el-o-MET-rick)   NEPHELOMETRIC

A means of measuring turbidity in a sample by using an instrument called a nephelometer. A nephelometer passes light through a sample and the amount of light deflected (usually at a 90-degree angle) is then measured.

NONVOLATILE MATTER   NONVOLATILE MATTER

Material such as sand, salt, iron, calcium, and other mineral materials which are only slightly affected by the actions of organisms and are not lost on ignition of the dry solids at 550°C. Volatile materials are chemical substances usually of animal or plant origin. Also see INORGANIC WASTE and VOLATILE MATTER or VOLATILE SOLIDS.

OSHA (O-shuh)   OSHA

The Williams-Steiger **O**ccupational **S**afety and **H**ealth **A**ct of 1970 (OSHA) is a federal law designed to protect the health and safety of industrial workers and also the operators of water supply systems and treatment plants. The Act regulates the design, construction, operation and maintenance of water supply systems and water treatment plants. OSHA also refers to the federal and state agencies which administer the OSHA regulations.

ORGANIC   ORGANIC

Substances that come from animal or plant sources. Organic substances always contain carbon. (Inorganic materials are chemical substances of mineral origin.) Also see INORGANIC.

ORGANISM   ORGANISM

Any form of animal or plant life. Also see BACTERIA.

OXIDATION (ox-uh-DAY-shun)   OXIDATION

Oxidation is the addition of oxygen, removal of hydrogen, or the removal of electrons from an element or compound. In the environment, organic matter is oxidized to more stable substances. The opposite of REDUCTION.

OXIDATION-REDUCTION POTENTIAL (ORP)   OXIDATION-REDUCTION POTENTIAL (ORP)

The electrical potential required to transfer electrons from one compound or element (the oxidant) to another compound or element (the reductant); used as a qualitative measure of the state of oxidation in water treatment systems. ORP is measured in millivolts, with negative values indicating a tendency to reduce compounds or elements and positive values indicating a tendency to oxidize compounds or elements.

PARTS PER MILLION (PPM)   PARTS PER MILLION (PPM)

Parts per million parts, a measurement of concentration on a weight or volume basis. This term is equivalent to milligrams per liter (mg/L) which is the preferred term.

PATHOGENIC (PATH-o-JEN-ick) ORGANISMS   PATHOGENIC ORGANISMS

Organisms, including bacteria, viruses or cysts, capable of causing diseases (giardiasis, cryptosporidiosis, typhoid, cholera, dysentery) in a host (such as a person). There are many types of organisms which do *NOT* cause disease. These organisms are called non-pathogenic.

PATHOGENS (PATH-o-jens)   PATHOGENS

Pathogenic or disease-causing organisms.

PERCENT SATURATION   PERCENT SATURATION

The amount of a substance that is dissolved in a solution compared with the amount dissolved in the solution at saturation, expressed as a percent.

$$\text{Percent Saturation, \%} = \frac{\text{Amount of Substance That Is Dissolved x 100\%}}{\text{Amount Dissolved in Solution at Saturation}}$$

pH (pronounce as separate letters)   pH

pH is an expression of the intensity of the basic or acidic condition of a liquid. Mathematically, pH is the logarithm (base 10) of the reciprocal of the hydrogen ion activity.

$$pH = \text{Log} \frac{1}{[H^+]}$$

The pH may range from 0 to 14, where 0 is most acidic, 14 most basic, and 7 neutral. Natural waters usually have a pH between 6.5 and 8.5.

POTABLE (POE-tuh-bull) WATER   POTABLE WATER

Water that does not contain objectionable pollution, contamination, minerals, or infective agents and is considered satisfactory for drinking.

PRECIPITATE (pre-SIP-uh-TATE)   PRECIPITATE

(1) An insoluble, finely divided substance which is a product of a chemical reaction within a liquid.

(2) The separation from solution of an insoluble substance.

REAGENT (re-A-gent)   REAGENT

A pure chemical substance that is used to make new products or is used in chemical tests to measure, detect, or examine other substances.

REDUCTION (re-DUCK-shun)   REDUCTION

Reduction is the addition of hydrogen, removal of oxygen, or the addition of electrons to an element or compound. Under anaerobic conditions (no dissolved oxygen present), sulfur compounds are reduced to odor-producing hydrogen sulfide (H_2S) and other compounds. The opposite of OXIDATION.

REPRESENTATIVE SAMPLE   REPRESENTATIVE SAMPLE

A sample portion of material or water that is as nearly identical in content and consistency as possible to that in the larger body of material or water being sampled.

SOLUTION   SOLUTION

A liquid mixture of dissolved substances. In a solution it is impossible to see all the separate parts.

STANDARD METHODS   STANDARD METHODS

STANDARD METHODS FOR THE EXAMINATION OF WATER AND WASTEWATER, 20th Edition. A joint publication of the American Public Health Association (APHA), American Water Works Association (AWWA), and the Water Environment Federation (WEF) which outlines the accepted laboratory procedures used to analyze the impurities in water and wastewater. Available from American Water Works Association, Bookstore, 6666 West Quincy Avenue, Denver, CO 80235. Order No. 10079. Price to members, $167.00; nonmembers, $212.00; price includes cost of shipping and handling.

STANDARD SOLUTION   STANDARD SOLUTION

A solution in which the exact concentration of a chemical or compound is known.

STANDARDIZE   STANDARDIZE

To compare with a standard.

(1) In wet chemistry, to find out the exact strength of a solution by comparing it with a standard of known strength. This information is used to adjust the strength by adding more water or more of the substance dissolved.

(2) To set up an instrument or device to read a standard. This allows you to adjust the instrument so that it reads accurately, or enables you to apply a correction factor to the readings.

STERILIZATION (STARE-uh-luh-ZAY-shun)   STERILIZATION

The removal or destruction of all microorganisms, including pathogenic and other bacteria, vegetative forms and spores. Compare with DISINFECTION.

SUPERNATANT (sue-per-NAY-tent)   SUPERNATANT

Liquid removed from settled sludge. Supernatant commonly refers to the liquid between the sludge on the bottom and the scum on the water surface of a basin or container.

SURFACTANT (sir-FAC-tent)   SURFACTANT

Abbreviation for surface-active agent. The active agent in detergents that possesses a high cleaning ability.

TITRATE (TIE-trate)   TITRATE

To *TITRATE* a sample, a chemical solution of known strength is added drop by drop until a certain color change, precipitate, or pH change in the sample is observed (end point). Titration is the process of adding the chemical reagent in small increments (0.1 – 1.0 milliliter) until completion of the reaction, as signaled by the end point.

TURBIDITY UNITS (TU)   TURBIDITY UNITS (TU)

Turbidity units are a measure of the cloudiness of water. If measured by a nephelometric (deflected light) instrumental procedure, turbidity units are expressed in nephelometric turbidity units (NTU) or simply TU. Those turbidity units obtained by visual methods are expressed in Jackson Turbidity Units (JTU) which are a measure of the cloudiness of water; they are used to indicate the clarity of water. There is no real connection between NTUs and JTUs. The Jackson turbidimeter is a visual method and the nephelometer is an instrumental method based on deflected light.

VOLATILE (VOL-uh-tull)   VOLATILE

(1) A volatile substance is one that is capable of being evaporated or changed to a vapor at relatively low temperatures. Volatile substances also can be partially removed by air stripping.

(2) In terms of solids analysis, volatile refers to materials lost (including most organic matter) upon ignition in a muffle furnace for 60 minutes at 550°C. Natural volatile materials are chemical substances usually of animal or plant origin. Manufactured or synthetic volatile materials such as ether, acetone, and carbon tetrachloride are highly volatile and not of plant or animal origin. Also see NONVOLATILE MATTER.

VOLATILE ACIDS   VOLATILE ACIDS

Fatty acids produced during digestion which are soluble in water and can be steam-distilled at atmospheric pressure. Also called organic acids. Volatile acids are commonly reported as equivalent to acetic acid.

VOLATILE LIQUIDS   VOLATILE LIQUIDS

Liquids which easily vaporize or evaporate at room temperature.

VOLATILE MATTER   VOLATILE MATTER

Matter in water, wastewater, or other liquids that is lost on ignition of the dry solids at 550°C.

VOLATILE SOLIDS   VOLATILE SOLIDS

Those solids in water or other liquids that are lost on ignition of the dry solids at 550°C.

VOLUMETRIC   VOLUMETRIC

A measurement based on the volume of some factor. Volumetric titration is a means of measuring unknown concentrations of water quality indicators in a sample *BY DETERMINING THE VOLUME* of titrant or liquid reagent needed to complete particular reactions.

CHAPTER 11. LABORATORY PROCEDURES

(Lesson 1 of 5 Lessons)

11.0 BASIC WATER LABORATORY PROCEDURES

11.00 Importance of Laboratory Procedures

Water treatment processes cannot be controlled effectively unless the operator has some means to check and evaluate the quality of water being treated and produced. Laboratory quality control tests provide the necessary information to monitor the treatment processes and to ensure a safe and pleasant-tasting drinking water for all who use it. By relating laboratory results to treatment operations, the water treatment or supply system operator can first select the most effective operational procedures, then determine the efficiency of the treatment processes, and identify potential problems before they affect finished water quality. For these reasons, a clear understanding of laboratory procedures is a must for every waterworks operator.

```
NOTICE
THE COLLECTION OF A BAD SAMPLE OR A BAD
LABORATORY RESULT IS ABOUT AS USEFUL AS
NO RESULTS. TO PREVENT BAD RESULTS RE-
QUIRES (1) CONSTANT MAINTENANCE AND CALI-
BRATION OF LABORATORY EQUIPMENT, AND
(2) USE OF CORRECT LAB PROCEDURES. ALSO
RESULTS OF LAB TESTS ARE OF NO VALUE TO
ANYONE UNLESS THEY ARE USED.
```

11.01 Metric System

The metric system is used in the laboratory to express units of length, volume, weight (mass), concentration, and temperature. The metric system is based on the decimal system. All units of length, volume, and weight use factors of 10 to express larger or smaller quantities of these units. Below is a summary of metric and English unit names and their abbreviations.

Type of Measurement	English System	Metric Name	Metric Abbreviation
Length	inch foot yard	meter	m
Temperature	Fahrenheit	Celsius	°C
Volume	quart gallon	liter	L
Weight	ounce pound	gram	gm
Concentration	lbs/gal strength, %	milligrams per liter	mg/L

Many times in the water laboratory we use smaller amounts than a meter, a liter, or a gram. To express these smaller amounts, prefixes are added to the names of the base metric unit. There are many prefixes in use; however, we commonly use two or three prefixes more than any others in the laboratory.

Prefix	Abbreviation	Meaning
centi-	c	1/100 of; or 0.01 times
milli-	m	1/1,000 of; or 0.001 times
micro-	μ	1/1,000,000 of; or 0.000001 times

One centimeter (cm) is 1/100 (one hundredth) of a meter, one milliliter (mL) is 1/1,000 (one thousandth) of a liter, and likewise, one microgram (μgm) is 1/1,000,000 (one millionth) of a gram.

EXAMPLES:

(1) Convert 3 grams into milligrams.

$$1 \text{ milligram } = 1 \text{ mg} = 1/1,000 \text{ grams}$$

$$\text{therefore, 1 gram} = 1,000 \text{ milligrams}$$

$$(3 \text{ grams})(1,000 \text{ mg/gram}) = 3,000 \text{ mg}$$

(2) Convert 750 milliliters (mL) to liters.

$$1 \text{ m}L = 1/1,000 \text{ liter}$$

$$\text{therefore, 1 liter} = 1,000 \text{ m}L$$

$$(750 \text{ m}L)(1 \text{ liter}/1,000 \text{ m}L) = 0.750 \text{ liters}$$

(3) Convert 50 micrograms (μgm) to grams.

$$1 \text{ μgm} = 1/1,000,000 \text{ gram}$$

$$\text{therefore, 1 gram} = 1,000,000 \text{ μgm}$$

$$50 \text{ μgm} \times 1 \text{ gram}/1,000,000 \text{ μgm} = 0.00005 \text{ grams}$$

Larger amounts than a meter, liter, or gram can be expressed using such prefixes as kilo- meaning 1,000. A kilogram is 1,000 grams.

The Celsius (or centigrade) temperature scale is used in the water laboratory rather than the more familiar Fahrenheit scale.

	Fahrenheit (°F)	Celsius (°C)
Freezing point of water	32	0
Boiling point of water	212	100

To convert Fahrenheit to Celsius, you can use the following formula:

$$\text{Temperature, °C} = 5/9(\text{°F} - 32\text{°F})$$

EXAMPLE: Convert 68°F to °C

$$\text{Temperature, °C} = 5/9(°F - 32°F)$$
$$= 5/9(68°F - 32°F)$$
$$= 5/9(36)$$
$$= 20°C$$

To convert Celsius to Fahrenheit, the following formula can be used:

$$\text{Temperature, °F} = 9/5(°C) + 32°F$$

EXAMPLE: Convert 35°C to °F

$$\text{Temperature, °F} = 9/5(°C) + 32°F$$
$$= 9/5(35°C) + 32°F$$
$$= 63 + 32$$
$$= 95°F$$

11.02 Chemical Names and Formulas

In the laboratory, chemical symbols are used as "shorthand" for the names of the elements. The names and symbols for some of these elements are listed below.

Chemical Name	Symbol
Calcium	Ca
Carbon	C
Chlorine	Cl
Copper	Cu
Fluorine	F
Hydrogen	H
Iron	Fe
Lead	Pb
Magnesium	Mg
Manganese	Mn
Nitrogen	N
Oxygen	O
Sodium	Na
Sulfur	S

A compound is a substance composed of two or more different elements and whose composition (proportion of elements) is constant. Generally, all chemical compounds can be divided into two main groups, organic and inorganic. Organic compounds are those which contain the element carbon (C). There are, however, a few simple substances containing carbon which are considered to belong to the realm of inorganic chemistry. These include carbon dioxide (CO_2), carbon monoxide (CO), bicarbonate (HCO_3^-) and carbonate (CO_3^{2-}) as in calcium carbonate ($CaCO_3$).

Many different compounds can be made from the same two or three elements. Therefore, you must carefully read the formula and name to prevent errors and accidents. A chemical formula is a "shorthand" or abbreviated way to write the name of a chemical compound. For example, the name sodium chloride (common table salt) can be written "NaCl." Table 11.1 lists commonly used chemical compounds found in the water laboratory.

TABLE 11.1 NAMES AND FORMULAS OF CHEMICALS COMMONLY USED IN WATER ANALYSES

Chemical Name	Chemical Formula
Acetic Acid	CH_3COOH
Aluminum Sulfate (alum)	$Al_2(SO_4)_3 \cdot 14.3\ H_2O$ [a]
Ammonium Hydroxide	NH_4OH
Calcium Carbonate	$CaCO_3$
Copper Sulfate	$CuSO_4$
Ferric Chloride	$FeCl_3$
Nitric Acid	HNO_3
Phenylarsine Oxide	C_6H_5AsO
Potassium Iodide	KI
Sodium Bicarbonate	$NaHCO_3$
Sodium Hydroxide	$NaOH$
Sulfuric Acid	H_2SO_4

[a] 14.3 H_2O. Alum in the dry form based on 17% Al_2O_3.

Poor results and safety hazards are often caused by using a chemical from the shelf that is *NOT* exactly the same chemical called for in a particular procedure. The mistake usually occurs when the chemicals are not properly labeled or have similar names or formulas. This problem can be eliminated if you use *BOTH* the chemical name and formula as a double check. The spellings of many chemical names are quite similar. These slight differences are critical because the chemicals do not behave alike. For example, the chemicals potassium nitr**a**te (KNO_3) and potassium nitr**i**te (KNO_2) are just as different in meaning chemically as the words f**a**t and f**i**t are to your doctor.

11.03 Helpful References

1. *METHODS FOR CHEMICAL ANALYSIS OF WATER AND WASTES*, U.S. Environmental Protection Agency. Obtain from National Technical Information Service (NTIS), 5285 Port Royal Road, Springfield, VA 22161. Order No. PB84-128677. EPA No. 600-4-79-020. Price, $117.00, plus $5.00 shipping and handling per order.

2. *SIMPLIFIED PROCEDURES FOR WATER EXAMINATION* (M12). Obtain from American Water Works Association (AWWA), Bookstore, 6666 West Quincy Avenue, Denver, CO 80235. Order No. 30012. ISBN 1-58321-182-9. Price to members, $73.50; nonmembers, $103.50; price includes cost of shipping and handling.

3. *STANDARD METHODS FOR THE EXAMINATION OF WATER AND WASTEWATER*, 20th Edition. Obtain from American Water Works Association (AWWA), Bookstore, 6666 West Quincy Avenue, Denver, CO 80235. Order No. 10079. Price to members, $167.00; nonmembers, $212.00; price includes cost of shipping and handling.

4. *HANDBOOK FOR ANALYTICAL QUALITY CONTROL IN WATER AND WASTEWATER LABORATORIES*, U.S. Environmental Protection Agency. Obtain from National Technical Information Service (NTIS), 5285 Port Royal Road, Springfield, VA 22161. Order No. PB-297451/7. EPA No. 600-4-79-019. Price, $52.00, plus $5.00 shipping and handling per order.

5. *MICROBIOLOGICAL METHODS FOR MONITORING THE ENVIRONMENT—WATER AND WASTES*, U.S. Environmental Protection Agency, December 1978. Obtain from National Technical Information Service (NTIS), 5285 Port Royal Road, Springfield, VA 22161. Order No. PB-290329/2. EPA No. 600-8-78-017. Price, $86.50, plus $5.00 shipping and handling per order.

6. *WATER QUALITY.* Obtain from American Water Works Association (AWWA), Bookstore, 6666 West Quincy Avenue, Denver, CO 80235. Order No. 1958. ISBN 1-58321-232-9. Price to members, $73.50; nonmembers, $106.50; price includes cost of shipping and handling.

7. *WATER ANALYSIS HANDBOOK*, Fourth Edition. Obtain from HACH Company, PO Box 389, Loveland, CO 80539-0389. Order No. 2319600. No charge.

QUESTIONS

Write your answers in a notebook and then compare your answers with those on page 568.

11.0A   Why are laboratory quality control tests important?

11.0B   What does the prefix milli- mean?

11.0C   What's the proper name of the chemical compound, $CaCO_3$?

11.1   LABORATORY EQUIPMENT AND TECHNIQUES

11.10   Water Laboratory Equipment

The items of equipment in a water laboratory are the operator's "tools-of-the-trade." In any laboratory there are certain basic pieces of equipment that are used routinely to perform water analysis tests. The following is a brief description of several of the more common items of glassware and pieces of equipment used in the analysis of water.

Volumetric glassware (graduated cylinders and pipets) is calibrated either "to contain" (TC) or "to deliver" (TD). Glassware designed "to deliver" will do so accurately only when the inner surface is so scrupulously clean that water wets the surface immediately and forms a uniform film on the surface upon emptying.

BEAKERS. Beakers are the most common pieces of laboratory equipment. They come in sizes from 1 mL to 4,000 mL. They are used mainly for mixing chemicals and to measure approximate volumes.

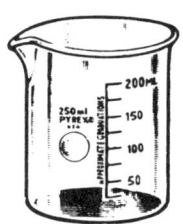

Beaker

GRADUATED CYLINDERS. Graduated cylinders also are basic to any laboratory and come in sizes from 5 mL to 4,000 mL. They are used to measure volumes more accurately than beakers.

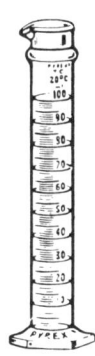

Cylinder, Graduated

PIPETS. Pipets are used to deliver accurate volumes and range in size from 0.1 mL to 100 mL.

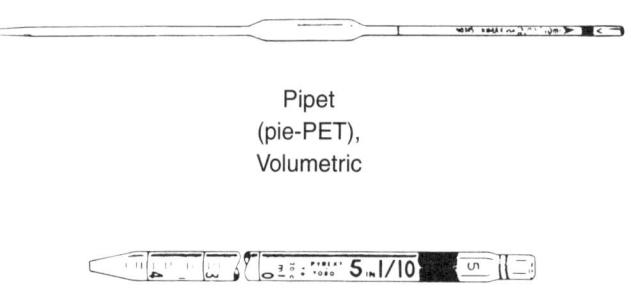

Pipet (pie-PET), Volumetric

Pipet, Serological

BURETS. Burets are also used to deliver accurate volumes. They are especially useful in a procedure called "titration." Burets come in sizes from 10 to 1,000 mL.

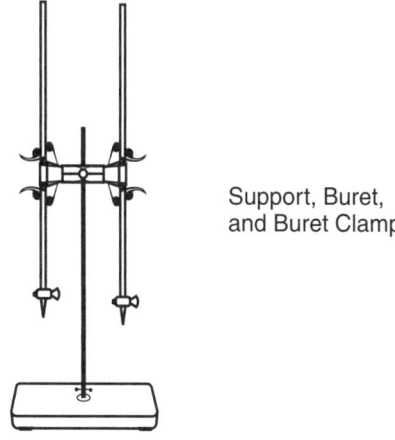

Support, Buret, and Buret Clamp

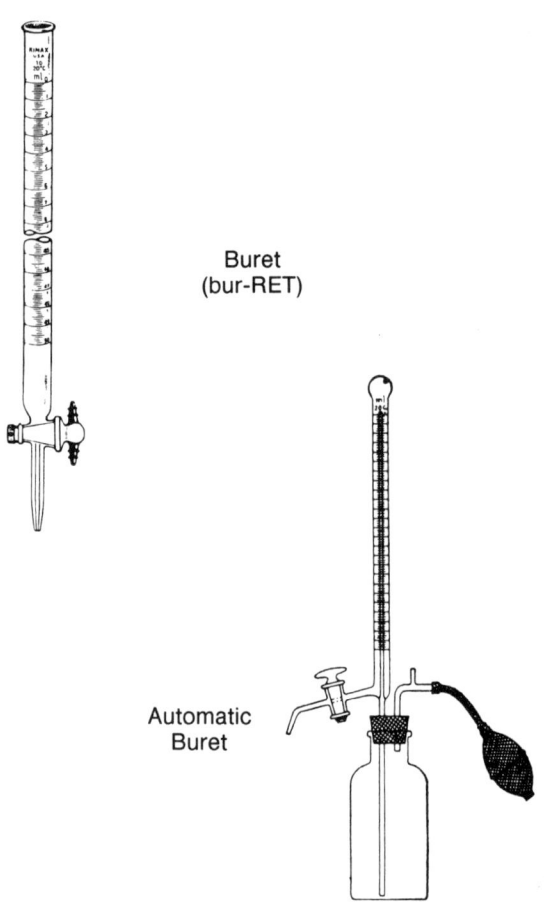

Buret
(bur-RET)

Automatic
Buret

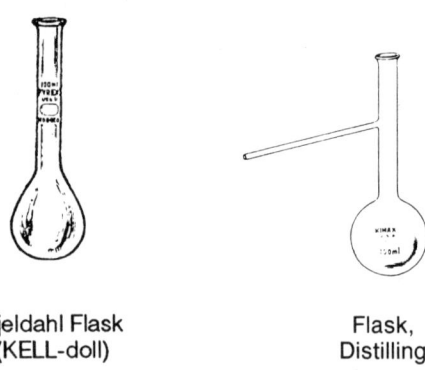

Kjeldahl Flask
(KELL-doll)

Flask,
Distilling

BOTTLES. Bottles are used to store chemicals. to collect samples for testing purposes, and to dispense liquids.

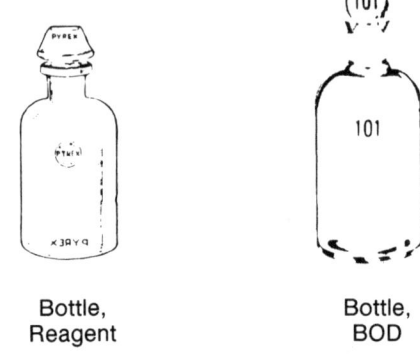

Bottle,
Reagent

Bottle,
BOD

FLASKS. Flasks are used for containing and mixing chemicals. There are many different sizes and shapes.

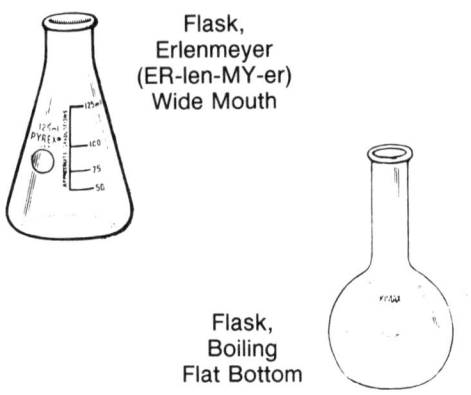

Flask,
Erlenmeyer
(ER-len-MY-er)
Wide Mouth

Flask,
Boiling
Flat Bottom

FUNNELS. A funnel is used for pouring solutions or transferring solid chemicals. This funnel can also be used with filter paper to remove solids from a solution.

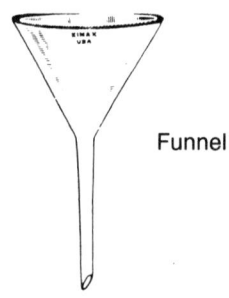

Funnel

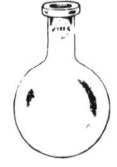

Flask,
Boiling
Round Bottom
Short Neck

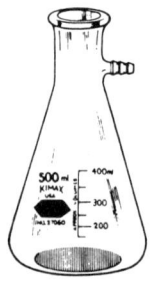

Flask,
Filtering

A Buchner funnel is used to separate solids from a mixture. It is used with a filter flask and a vacuum.

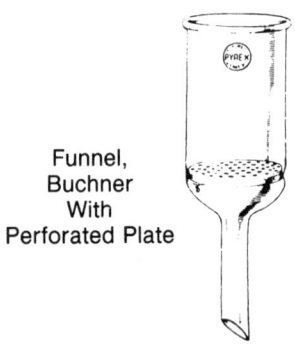

Funnel,
Buchner
With
Perforated Plate

Separatory funnels are used to separate one chemical mixture from another. The separated chemical usually is dissolved in one or two layers of liquid.

OTHER LABWARE AND EQUIPMENT.

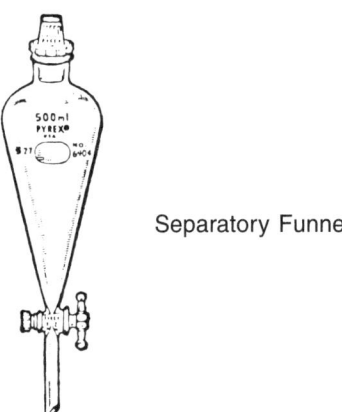

Separatory Funnel

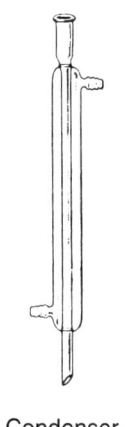

Condenser

TUBES. Test tubes are used for mixing small quantities of chemicals. They are also used as containers for bacterial testing (culture tubes).

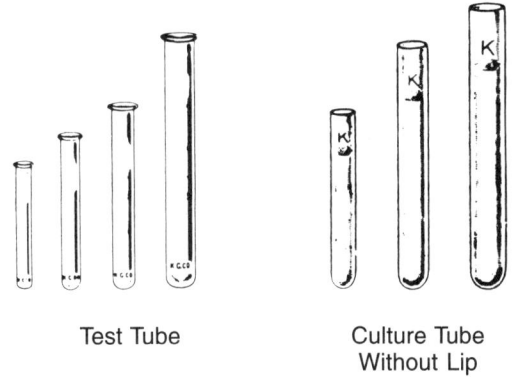

Test Tube

Culture Tube
Without Lip

Dish, Petri

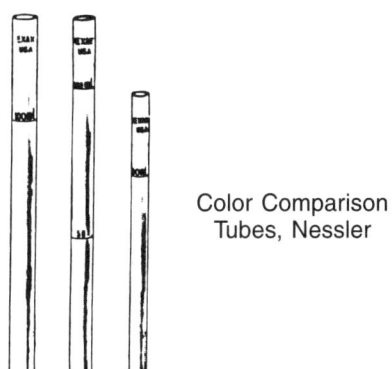

Color Comparison
Tubes, Nessler

Desiccator
(DESS-uh-KAY-tor)

Thermometer, Dial

Oven, Mechanical Convection

Hot Plate

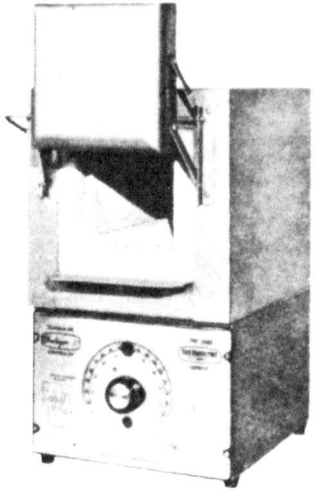

Muffle Furnace, Electric

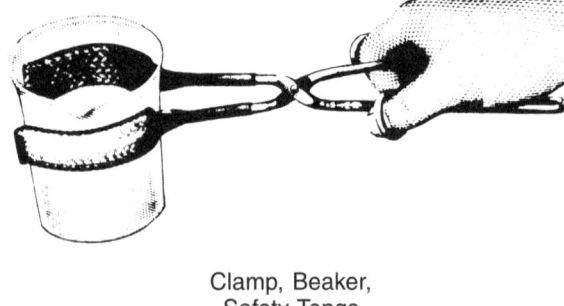

Clamp, Beaker,
Safety Tongs

Clamp, Dish,
Safety Tongs

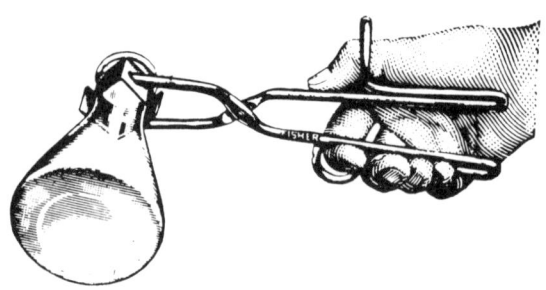

Clamp, Flask,
Safety Tongs

Clamp, Test Tube

Clamp Holder

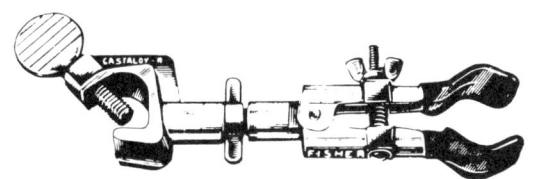

Clamp, Utility

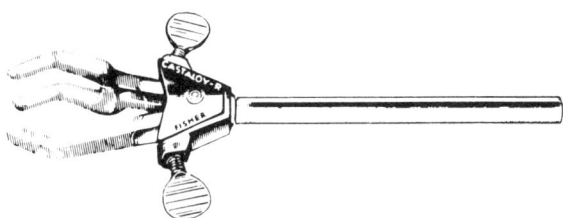

Clamp

Tripod, Concentric
Ring

Burner, Bunsen

Triangle,
Fused

Fume Hood

Portable Dissolved Oxygen Meter
(with computer docking station)
(Courtesy of HACH Company)

Portable pH Meter
(Courtesy of HACH Company)

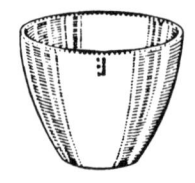

Crucible (CREW-suh-bull),
Porcelain

Pump, Air Pressure and Vacuum

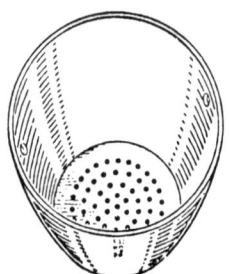

Crucible,
Gooch
(GOO-ch)
Porcelain

Pipet Bulb

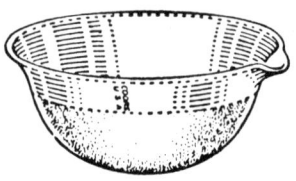

Dish,
Evaporating

Test Paper, pH 1-11

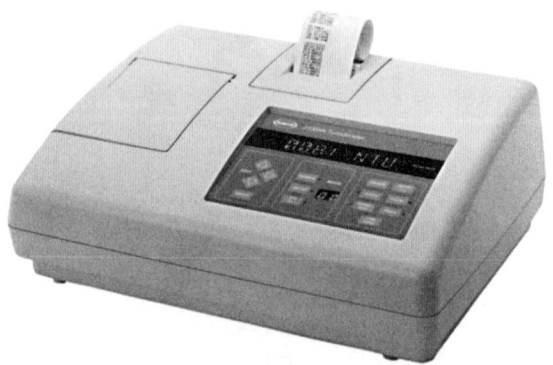

Laboratory Turbidimeter
(ratio mode—ON/OFF—is keypad selectable)
(Courtesy of HACH Company)

Magnetic Stirrer
(Permission of Thermolyne)

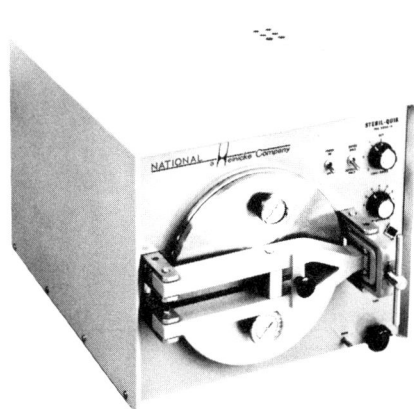

Autoclave
(Permission of Napco)

Incubator
(Permission of Blue M Electric)

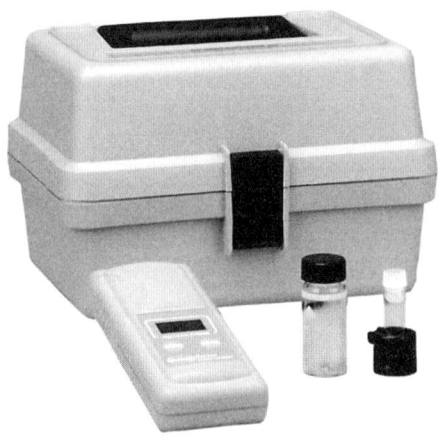

Direct-Reading Colorimeter
(free chlorine residual)
(Courtesy of HACH Company)

Portable Spectrophotometer
(Courtesy of HACH Company)

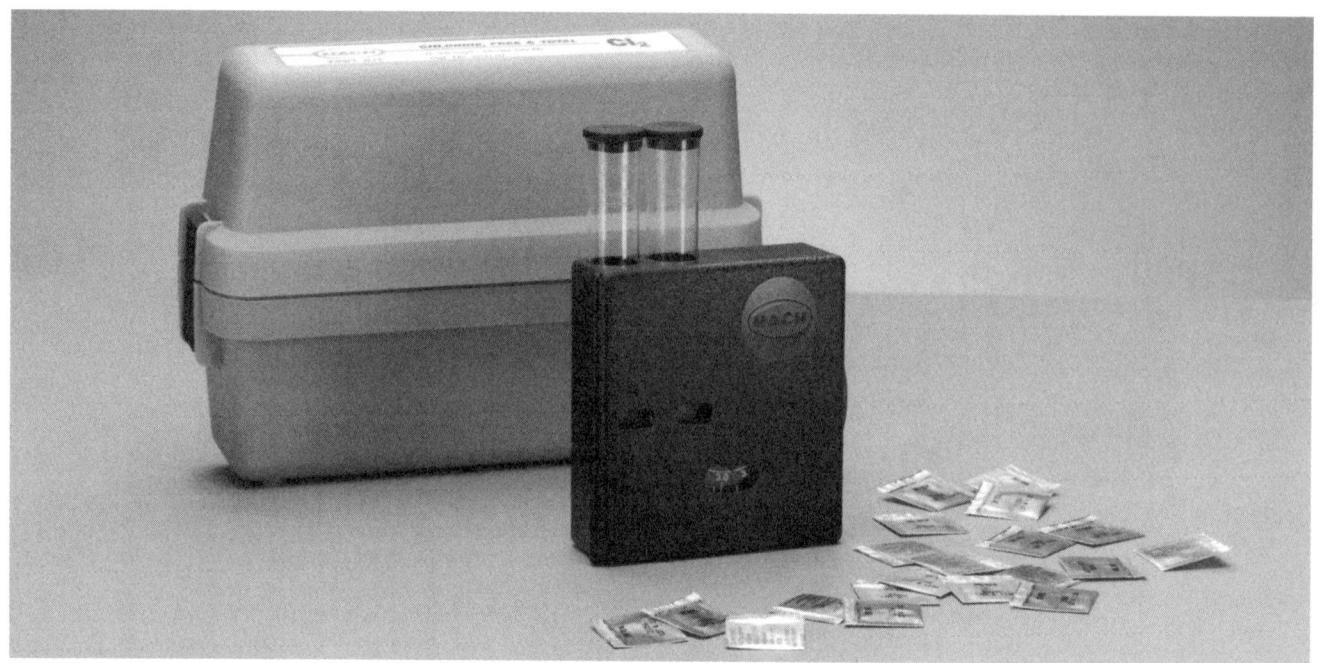

Chlorine Residual Test Kit
(free and total chlorine)
(Courtesy of HACH Company)

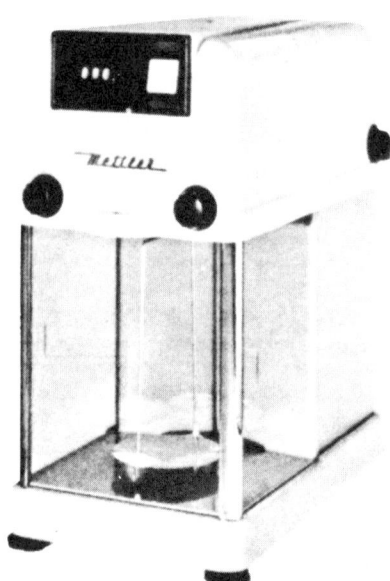

Weight = 95.5580 gm.

Balance, Analytical
(Permission of Mettler)

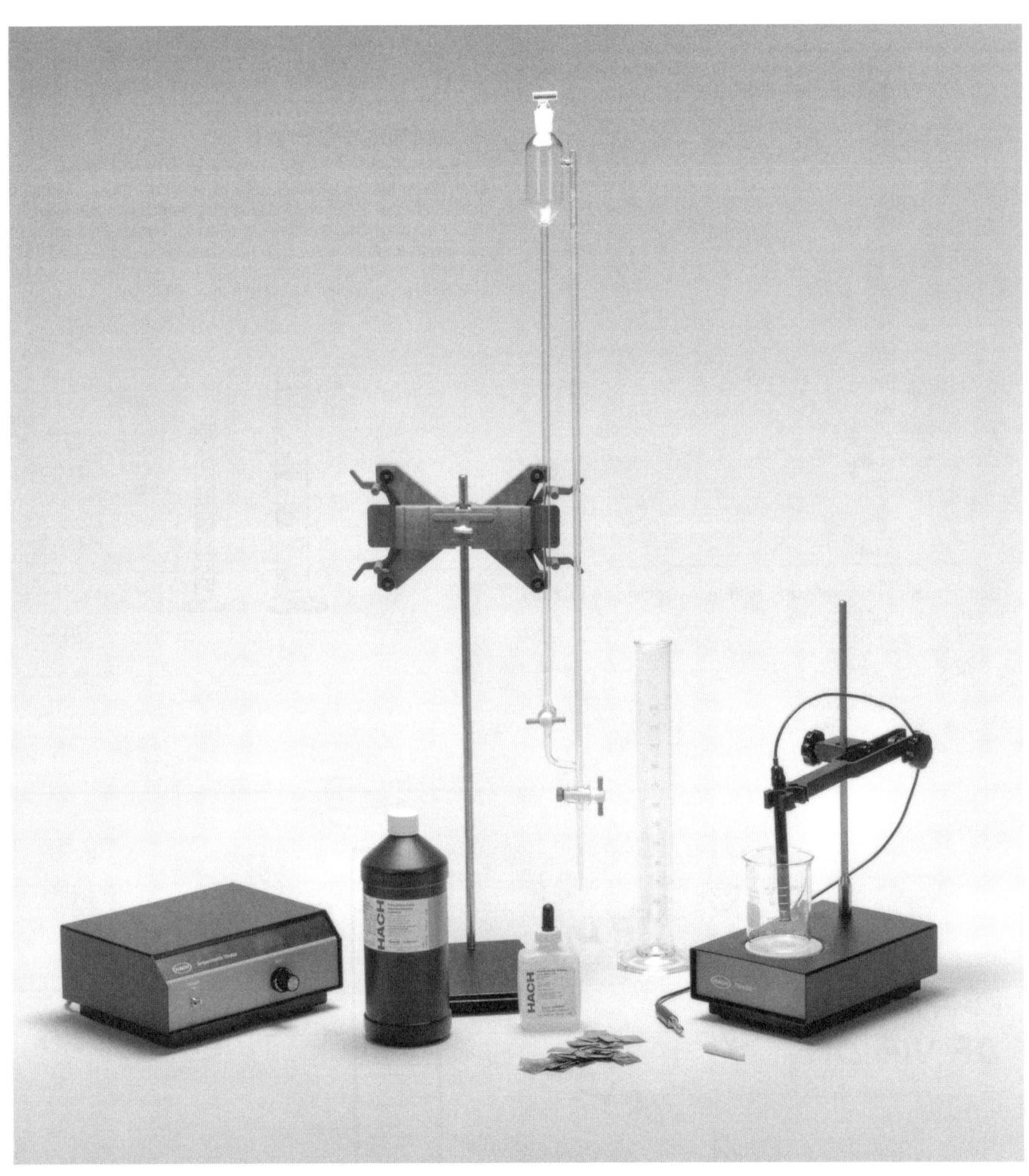

Amperometric Titrator
(Courtesy of HACH Company)

11.11   Use of Laboratory Glassware

BURETS

A buret is used to give accurate measurements of liquid volumes. The stopcock controls the amount of liquid that will flow from the buret. A glass stopcock must be lubricated (stopcock grease) and should not be used with alkaline solutions. A Teflon stopcock never needs to be lubricated.

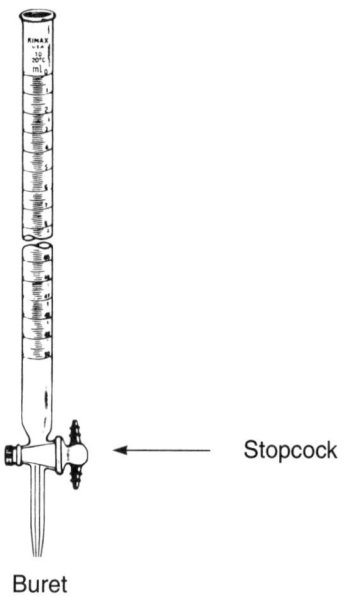

← Stopcock

Buret

Burets come in several sizes, with those holding 10 to 25 milliliters used most frequently.

When a buret is filled with liquid, the surface of the liquid is curved. This curve of the surface is called the meniscus (meh-NIS-cuss). Depending on the liquid, the curve forms a valley, as with water, or forms a hill, as with mercury. Since most solutions used in the laboratory are water based, always read the bottom of the meniscus with your eye at the same level (Figure 11.1). If you have the meniscus at eye level, the closest marks that go all the way around the buret will appear as straight lines, not circles.

GRADUATED CYLINDERS

The graduated cylinder or "graduate" is one of the most often used pieces of laboratory equipment. This cylinder is made either of glass or of plastic and ranges in sizes from 10 m*L* to 4 liters. The graduate is used to measure volumes of liquid with an accuracy *LESS* than burets but *GREATER* than beakers or flasks. Graduated cylinders should never be heated in an open flame because they will break.

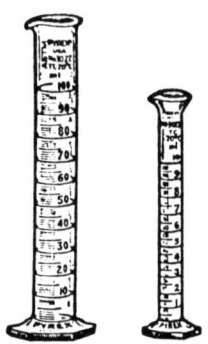

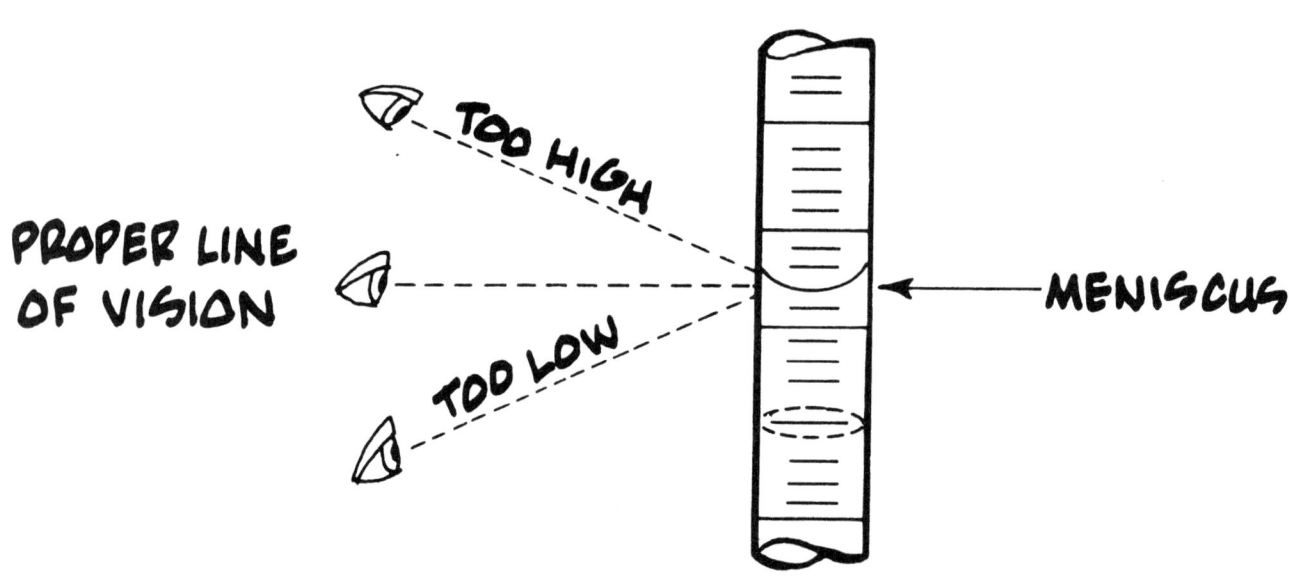

Fig. 11.1   How to read meniscus

FLASKS AND BEAKERS

Beakers and flasks are used for mixing, heating, and weighing chemicals. Most beakers and flasks are *NOT* calibrated with exact volume lines; however, they are sometimes marked with approximate volumes and can be used to estimate volumes.

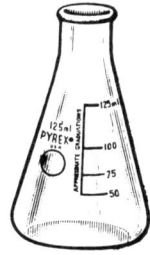

Flask   Beaker

VOLUMETRIC FLASKS

Volumetric flasks are used to prepare solutions; they come in sizes from 10 to 2,000 mL. Volumetric flasks should *NEVER* be heated. Rather than store liquid chemicals in volumetric flasks, the chemicals should be transferred to a storage bottle. Volumetric flasks are more accurate than graduated cylinders.

PIPETS

Pipets are used for accurate volume measurements and transfer. There are three types of pipets commonly used in the laboratory—volumetric pipets, graduated (measuring) or Mohr pipets, and serological pipets.

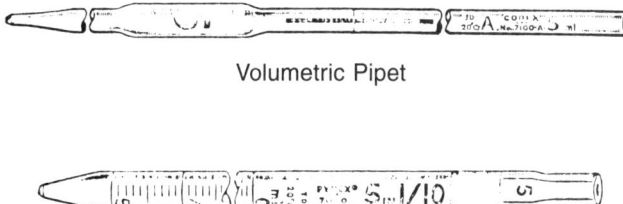

Volumetric Pipet

Graduated or Measuring Pipet

Serological Pipet

Volumetric pipets are available in sizes such as 1, 10, 25, 50, and 100 mL. They are used to deliver a single volume. Measuring and serological pipets, however, will deliver fractions of the total volume indicated on the pipet.

Volumetric pipets should be held in a vertical position when emptying and the outflow should be unrestricted. The tip should be touched to the wet surface of the receiving vessel and kept in contact with it until the emptying is complete. Under no circumstance should the small amount remaining in the tip be blown out.

Measuring and serological pipets should be held in the vertical position. After outflow has stopped, the tip should be touched to the wet surface of the receiving vessel. No drainage period is allowed. When the small amount remaining in the tip is to be blown out and added, this will be indicated by a frosted band near the top of the pipet.

Use of a pipet filler or pipet bulb (page 508) is recommended to draw the sample into a pipet. Never pipet chemical reagent solutions or unknown water samples by mouth. Use the following techniques for best results.

1. Draw liquid up into the pipet past the calibration mark.

2. Quickly remove the bulb and place dry fingertip over the top end of the pipet.

3. Wipe excess liquid from the tip of the pipet using laboratory tissue paper.

4. Lift finger and allow desired amount, or all, of liquid to drain. Pipets can be drained without removing the pipet bulb.

NOTE: There are pipet bulbs with valves that can control the flow of liquid from the pipet without removing the bulb.

ACKNOWLEDGMENTS

Pictures of laboratory glassware and equipment in this manual are reproduced with the permission of VWR Scientific, San Francisco, California, with exceptions noted.

QUESTIONS

Write your answers in a notebook and then compare your answers with those on page 568.

11.1A   For each type of glassware listed below, describe the item and its use or purpose.

1. Beaker
2. Graduated cylinders
3. Pipets
4. Burets

11.1B   What is a meniscus?

11.1C   Why should graduated cylinders never be heated in an open flame?

11.12   Chemical Solutions

Many laboratory procedures do not give the concentrations of standard solutions in grams/liter or milligrams/liter. Instead, the concentrations are usually given as *NORMALITY (N)*,[1] which is the standard designation for solution strengths in chemistry.

EXAMPLES:

0.025 *N* H$_2$SO$_4$ means a 0.025 normal solution of sulfuric acid

2 *N* NaOH   means that the normality of a sodium hydroxide solution is 2

The *LARGER* the number in front of the *N*, the *MORE* concentrated the solution. For example, 1 *N* NaOH solution is more concentrated than a 0.2 *N* NaOH solution.

Another method of specifying the concentration of solutions uses the "a + b system." This means that "a" volumes of concentrated reagent are diluted with "b" volumes of distilled water to form the required solution.

EXAMPLES:

1 + 1 HCl   means 1 volume of concentrated HCl is diluted with 1 volume of distilled water

1 + 5 H$_2$SO$_4$ means 1 volume of concentrated sulfuric acid is diluted with 5 volumes of distilled water

When the exact concentration of a prepared chemical solution is known, it is referred to as a "standard solution." Many times standard solutions can be ordered already prepared from chemical supply companies. Once a standard has been prepared, it can then be used to standardize other laboratory solutions. To standardize a solution means to determine and adjust its concentration accurately, thereby making it a standard solution. "Standardization" is the process of using one solution of known concentration to determine the concentration of another solution. This action often involves a procedure called a "titration."

When preparing standard solutions or reagents, the directions may say to weigh out 7.6992 grams of a chemical and dilute to one liter with distilled water. To weigh out 7.6992 grams of a chemical, determine the weight of a weighing dish and add this weight to the weight of the chemical. Place the weighing dish on the weighing platform of an analytical balance. (See page 510. Some balances have the weighing platform on top for the weighing dish.) Gently add the chemical to the weighing dish until you are slightly below the desired weight. The weighing mechanism should be off while the chemical is being added and then turned on to determine the exact weight. When you get close to the exact weight, place some of the chemical on a spatula. Gently tap the spatula to add very small amounts of chemical to the weighing dish. Continue this procedure until you've reached the exact weight. If you add too much chemical, remove some of the chemical with the spatula and again repeat the procedure until you reach the exact weight.

Another procedure is to place approximately the desired weight in the weighing dish. Weigh this amount exactly. Then add a proportionate amount of distilled water.

EXAMPLE 1

The directions for preparing a standard reagent indicate that you should weigh out 7.6992 grams and dilute to one liter. You weigh out 7.5371 grams. How much water should be added to produce the desired concentration or normality of the standard reagent?

Known	Unknown
Desired Weight, gm = 7.6992 gm	Water, mL
Actual Weight, gm = 7.5371 gm	

The chemical should be diluted to how many milliliters?

$$\text{Dilute to mL} = \frac{(\text{Actual Weight, gm})(1{,}000\ mL)}{\text{Desired Weight, gm}}$$

$$= \frac{(7.5371\ gm)(1{,}000\ mL)}{7.6992\ gm}$$

$$= 979\ mL$$

The 7.5371 grams of chemical should be diluted to 979 mL.

11.13   Titrations

A titration involves the measured addition of a standardized solution, which is usually in a buret, to another solution in a flask or beaker. The solution in the buret is referred to as the

---

[1] ***N*** *or Normal.   A normal solution contains one gram equivalent weight of reactant (compound) per liter of solution. The equivalent weight of an acid is that weight which contains one gram atom of ionizable hydrogen or its chemical equivalent. For example, the equivalent weight of sulfuric acid (H$_2$SO$_4$) is 49 (98 divided by 2 because there are two replaceable hydrogen ions). A one N solution of sulfuric acid would consist of 49 grams of H$_2$SO$_4$ dissolved in enough water to make one liter of solution.*

"titrant" and is added to the other solution until there is a measurable change in the test solution in the flask or beaker. This change is frequently a color change as a result of the addition of another chemical called an "indicator" to the solution in the flask before the titration begins. The solution in the buret is added slowly to the flask until the change, which is called the "end point," is reached. The entire process is the "titration." Figure 11.2 illustrates the four general steps used during a chemical titration.

11.14 Data Recording and Recordkeeping

The use of a laboratory notebook and worksheets are a must for laboratory analysts, water supply system and treatment plant operators. Notebooks and worksheets help you record data in an orderly manner. Too often, hours of work are wasted when test results and other data (such as a sample volume) are written down on a scrap of paper only to be misplaced or thrown away by mistake. Notebooks and worksheets help prevent errors and provide a record of your work. The routine use of laboratory worksheets and notebooks is the only way an operator or a lab person can be sure that all important information is properly recorded.

There is no standard laboratory form. Most operators usually develop their own data sheets for recording test results and other important data. These data sheets should be prepared in a manner that makes it easy for you to record results, review them, and recover these results when it is necessary. Each treatment plant will have different needs for collecting and recording data and may require several different data or worksheets. Figures 11.3 and 11.4 illustrate two typical laboratory worksheets.

11.15 Laboratory Quality Control

Having good equipment and using the correct methods are not enough to ensure correct analytical results. Each operator must be constantly alert to factors in the water treatment process which can lead to poor quality of data. Such factors include sloppy laboratory technique, deteriorated reagents and standards, poorly operating instruments, and calculation mistakes. One of the best ways to ensure quality control in your laboratory is to analyze reference-type samples to provide independent checks on your analysis. These reference-type samples are available from the U.S. Environmental Protection Agency and from commercial sources. From time to time, it is also a good idea to split a sample with one of your fellow operators or another laboratory and compare analytical results. In addition, frequent self-appraisal and evaluation—from sampling to reporting results—can help you gain full confidence in your results.

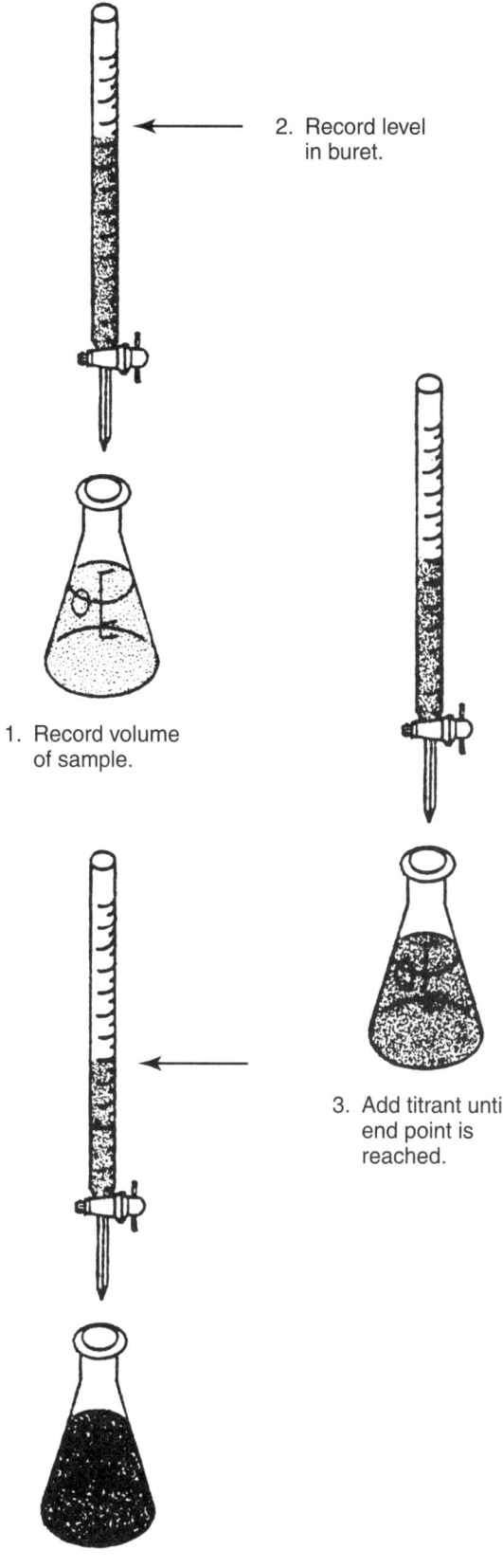

2. Record level in buret.

1. Record volume of sample.

3. Add titrant until end point is reached.

4. Record level of titrant at end point.

Fig 11.2   Titration Steps

WATER ANALYSES REPORT SHEET

(All results in mg/L except as noted)

SAMPLE SOURCE:			
DATE SAMPLED:	BY:		
ANALYST:	DATE COMPLETED:		
A. PHYSICAL PARAMETERS	ADDITIONAL INFORMATION		
color (units)	Water Tr., Vol. II, pg. 471		
odor (units)	Water Tr., Vol. II, pg. 492		
turbidity (NTU)	This Manual, pg. 565		
B. GENERAL MINERAL	ADDITIONAL INFORMATION		
pH (units)	This Manual, pg. 563		
total alkalinity ($CaCO_3$)	This Manual, pg. 529		
specific conductance, μmhos/cm	Water Tr., Vol. II, pg. 489		
total dissolved solids	Water Tr., Vol. II, pg. 497		
total hardness ($CaCO_3$)	This Manual, pg. 559		
calcium	Water Tr., Vol. II, pg. 468		
fluoride	Water Tr., Vol. II, pg. 475		
iron	Water Tr., Vol. II, pg. 479		
manganese	Water Tr., Vol. II, pg. 481		
chloride	Water Tr., Vol. II, pg. 469		
sulfate	Water Tr., Vol. II, pg. 490		

Fig. 11.3   Typical laboratory worksheet

QUESTIONS

Write your answers in a notebook and then compare your answers with those on page 568.

11.1D   What is a "standard solution"?

11.1E   What is the primary purpose of laboratory notebooks and worksheets?

11.1F   List three sources or causes of poor quality of analytical data.

11.16   Laboratory Safety

Safety is just as important in the laboratory as it is outside the lab. State laws and the Occupational Safety and Health Act (OSHA) demand that proper safety procedures be exercised in the laboratory at all times. OSHA specifically deals with "safety at the place of work." The Act requires that "each employer has the general duty to furnish all employees with employment free from recognized hazards causing, or likely to cause, death or serious physical harm."

Personnel working in the water industry must realize that a number of hazardous materials and conditions can exist. Always be alert and careful. Be aware of potential dangers at all times. Safe practice in the laboratory and any time while working around chemicals requires hardly any more effort than unsafe practices, with the important benefits from prevention of injury to you and your fellow operators.

On specific questions of safety, consult your state's General Industrial Safety Orders or OSHA regulations.

11.160   Laboratory Hazards

Working with chemicals and other materials in the water laboratory can be dangerous. Laboratory hazards include:

1. Hazardous materials,
2. Explosions,
3. Cuts and bruises,
4. Electric shock,
5. Fire, and
6. Burns (heat and chemical).

Hazardous materials include (1) corrosive, (2) toxic, and (3) explosive or flammable materials.

Sheet No. ______

Day ______

Date ______

AMERICAN RIVER
WATER TREATMENT PLANT
WATER QUALITY

Washwater used M.G.	
Gals. Alum used	
Lbs. Cl_2 used	
Lbs. Lime used	

Time	Turbidity		Hydrogen Ion Conc.(pH)			Alkalinity			Cl_2 Residual			Chlorine		Chemical Dose			TEMP. °C	
											TAP	Chlorine						
	RAW	Filter Effluent	RAW	Res. Influent	TAP Res.	RAW T-Alk	TAP T-Alk	TAP P-alk	Pre-Chlor	Res. Influent	Res	Pre	Post	Alum	Lime		Raw / Tap	
0001																		
100																		
200																		
300																		
400																		
500																		
600																		
700																		
800																		
900																		
1000																		
1100																		
1200																		
1300																		
1400																		
1500																		
1600																		
1700																		
1800																		
1900																		
2000																		
2100																		
2200																		
2300																		
2400																		
Average																		

Fig. 11.4 Typical operational worksheet

1. Corrosive Materials

   ACIDS

   a. Examples: Hydrochloric or muriatic (HCl), hydrofluoric (HF), glacial acetic (CH_3COOH), nitric (NHO_3), and sulfuric (H_2SO_4).

   b. Acids can be extremely corrosive and hazardous to human tissue, metals, clothing, cement, stone, and concrete.

   c. Commercially available spill cleanup kits should be kept on hand to neutralize the acid in the event of an accidental spill. Baking soda (bicarbonate, *NOT* laundry soda) effectively neutralizes acids. Baking soda can be used on lab and human surfaces without worrying about toxicity.

   BASES (Caustics)

   a. Examples: Sodium hydroxide (caustic soda or lye, NaOH), quicklime (CaO), hydrated lime ($Ca(OH)_2$), and alkaline iodine-azide solution (used in dissolved oxygen test).

   b. Bases are extremely corrosive to skin, clothing, and leather. Caustics can quickly and permanently cloud vision if not immediately flushed out of eyes. Determine location of safety showers and eyewash station *BEFORE* starting to work with dangerous chemicals.

   c. Commercially available spill cleanup materials should be kept on hand for use in the event of an accidental spill. A jug of ordinary vinegar can be kept on hand to neutralize bases and it will not harm your skin.

   MISCELLANEOUS CHEMICALS

   a. Examples: Alum, chlorine, ferric salts (ferric chloride), and other strong oxidants.

2. Toxic Materials

   Examples:

   a. Solids: Cyanide compounds, chromium, orthotolidine, cadmium, mercury, and other heavy metals.

   b. Liquids: Chloroform and other organic solvents.

   c. Gases: Chlorine, ammonia, sulfur dioxide, and chlorine dioxide.

3. Explosive or Flammable Materials

   Examples:

   a. Liquids: Acetone, ethers, and gasoline.

   b. Gases: Propane and hydrogen.

11.161 *Personal Safety and Hygiene*

Laboratory work can be quite dangerous if proper precautions are not taken. *ALWAYS* follow these basic rules:

1. *NEVER* work alone in the laboratory. Someone should always be available to help in case you should have an accident which blinds you, leaves you unconscious, or starts a fire you can't handle. If necessary, have someone check on you regularly to be sure you are OK.

2. Wear protective goggles or eyeglasses at all times in the laboratory. Contact lenses should not be worn, even under safety goggles, because fumes can seep between the lens and the eyeball and irritate the eye.

Safety Glasses

DON'T PIPET HAZARDOUS LIQUIDS BY MOUTH.

3. Never pipet hazardous materials by your mouth.

4. Always wear a lab coat or apron in the laboratory to protect your skin and clothes.

5. Wear insulated gloves when handling hot objects. If there is a danger of hot liquid erupting from a container, wear a face shield, too.

6. Don't keep food in a refrigerator that is used for chemical and/or sample storage.

7. Good housekeeping is an effective way to prevent accidents.

11.162 *Prevention of Laboratory Accidents*

11.1620 *CHEMICAL STORAGE*

An adequate chemical storeroom is essential for safety in the water laboratory. The storeroom should be properly ventilated and lighted and laid out to segregate incompatible chemicals. Order and cleanliness must be maintained. Clearly label and date all chemicals and bottles of reagents.

Store heavy items on or as near to the floor as possible. *VOLATILE LIQUIDS*[2] which may escape as a gas, such as ether, must be kept away from heat sources, sunlight, and electric switches.

Cap and secure cylinders of gas in storage to prevent rolling or tipping. They should also be placed away from any possible sources of heat or open flames.

CLAMPS, RAISED SHELF EDGES, AND PROPER ARRANGEMENT PREVENT STOCKROOM FALLOUT.

Follow usual common sense rules of storage. Good housekeeping is a most significant contribution toward an active safety campaign.

11.1621 MOVEMENT OF CHEMICALS

The next area of concern is the transfer of chemicals, apparatus, gases, or other hazardous materials from the storeroom to the laboratory for use. Use cradles or tilters to facilitate handling carboys or other large chemical vessels.

Drum Tilter

In transporting cylinders of compressed gases, use a trussed hand truck. Never roll a cylinder by its valve. Immediately after they are positioned for use, cylinders should be clamped securely into place to prevent shifting or toppling.

Carry flammable liquids in safety cans or, in the case of reagent-grade chemicals, protect the bottle with a carrier. Always wear protective gloves, safety shoes, and rubber aprons in case of accidental spilling of chemical containers.

11.1622 PROPER LABORATORY TECHNIQUES

Faulty technique is one of the chief causes of accidents and, because it involves the human element, is one of the most difficult to correct.

Because of their nature and prevalence in the laboratory, acids and other corrosive materials constitute a series of hazards ranging from poisoning, burning, and gassing through explosion. Always flush the outsides of acid bottles with water before opening them. Don't lay the stopper down on the countertop where a person might lay a hand or rest an arm on it. Keep all acids tightly stoppered when not in use and make sure no spilled acid remains on the floor, table, or bottle after use. To avoid splashing of acid, don't pour water into acid; *ALWAYS POUR ACID INTO WATER.*

Mercury requires special care. Even a small amount in the bottom of a drawer can poison the atmosphere in a room. After an accident involving mercury, go over the entire area carefully until there are no globules remaining. Keep all mercury containers tightly stoppered.

11.1623 ACCIDENT PREVENTION

ELECTRIC SHOCK. Wherever there are electrical outlets, plugs, and wiring connections, there is a danger of electric shock. The usual "do's" and "don'ts" of protection against shock in the home are equally applicable in the laboratory. Don't use worn or frayed wires. Replace connections when there is any sign of thinning insulation. Ground all apparatus using three-prong plugs. Don't continue to run a motor after liquid has spilled on it. Turn it off immediately and clean and dry the inside thoroughly before attempting to use it again.

Electrical units which are operated in an area exposed to flammable vapors should be explosion-proof. All permanent wiring should be installed by an electrician with proper conduit or BX cable to eliminate any danger of circuit overloading.

CUTS. Some of the pieces of glass used in the laboratory, such as glass tubing, thermometers, and funnels, must be inserted through rubber stoppers. If the glass is forced through the hole in the stopper by applying a lot of pressure, the glass usually breaks. This is one of the most common sources of cuts in the laboratory.

Use care in making rubber-to-glass connections. Lengths of glass tubing should be supported while they are being inserted into rubber. The ends of the glass should be *FLAME POLISHED*[3] and either wetted or covered with a lubricating jelly for ease in joining connections. Never use oil or grease. Wear gloves when making such connections, and hold the tubing as close to the end being inserted as possible to prevent bending or breaking. Also, never try to force rubber tubing or stoppers from glassware. Cut off the rubber or materials.

A FIRST-AID kit must be available in the laboratory.

BURNS. All glassware and porcelain look cold after the red from heating has disappeared. The red is gone in seconds but the glass is hot enough to burn for several minutes. After heating a piece of glass, put it out of the way until cold.

---

[2] *Volatile Liquids.   Liquids which easily vaporize or evaporate at room temperature.*
[3] *Flame Polished.   Melted by a flame to smooth out irregularities. Sharp or broken edges of glass (such as the end of a glass tube) are rotated in a flame until the edge melts slightly and becomes smooth.*

Spattering from acids, caustic materials, and strong oxidizing solutions should be washed off immediately with large quantities of water. Every worker in the water laboratory should have access to a sink and an emergency deluge shower. Keep vinegar and soda handy to neutralize acids and bases (caustic materials).

Many safeguards against burns are available. Gloves, safety tongs, aprons, and emergency deluge showers are but a few examples. Never decide it is too much trouble to put on a pair of gloves or use a pair of tongs to handle a dish or flask that has been heated.

USE TONGS—DON'T JUGGLE HOT CONTAINERS.

Perhaps the most harmful and painful chemical burn occurs when small objects, chemicals, or fumes get into your eyes. Immediately flood your eyes with water or a special "eyewash" solution from a safety kit or from an eye wash station or fountain.

TOXIC FUMES. Use a ventilated laboratory fume hood for routine reagent preparation. Select a hood that has adequate air displacement and expels harmful vapors and gases at their source. An annual check should be made of the entire laboratory building. Sometimes noxious fumes are spread by the heating and cooling system of the building.

When working with chlorine and other toxic substances, always wear a self-contained breathing apparatus. If possible, try to clear the atmosphere with adequate ventilation *BEFORE* entry.

WASTE DISPOSAL. A good safety program requires constant care in disposal of laboratory waste. Corrosive materials should never be poured down an ordinary sink or drain. These substances can corrode away the drain pipe and/or trap. Corrosive acids should be neutralized and poured down corrosion-resistant sinks and sewers using large quantities of water to dilute and flush the acid.

To protect maintenance personnel, use separate covered containers to dispose of broken glass.

DON'T POUR VOLATILE LIQUIDS INTO THE SINK.

FIRE. The laboratory should be equipped with a fire blanket. The fire blanket is used to smother clothing fires. Small fires which occur in an evaporating dish or beaker may be put out by covering the container with a glass plate, wet towel, or wet blanket. For larger fires, or ones which may spread rapidly, promptly use a fire extinguisher. Do not use a fire extinguisher on small beaker fires because the force of the spray will knock over the beaker and spread the fire. Take time to become familiar with the operation and use of your fire extinguishers.

The use of the proper type of extinguisher for each class of fire will give the best control of the situation and avoid compounding the problem. For example, water should not be poured on grease, electrical fires, or metal fires because water could increase the hazards, such as splattering of the fire and electric shock. The classes of fires given here are based on the type of material being consumed.

Fire classifications are important for determining the type of fire extinguisher needed to control the fire. Classifications also aid in recordkeeping. Fires are classified as A, B, C, or D fires based on the type of material being consumed: A, ordinary combustibles; B, flammable liquids and vapors; C, energized electrical equipment; and D, combustible metals. Fire extinguishers are also classified as A, B, C, or D to correspond with the class of fire each will extinguish.

Class A fires: ordinary combustibles such as wood, paper, cloth, rubber, many plastics, dried grass, hay, and stubble. Use foam, water, soda-acid, carbon dioxide gas, or almost any type of extinguisher.

Class B fires: flammable and combustible liquids such as gasoline, oil, grease, tar, oil-based paint, lacquer, and solvents, and also flammable gases. Use foam, carbon dioxide, or dry chemical extinguishers.

Class C fires: energized electrical equipment such as starters, breakers, and motors. Use carbon dioxide or dry chemical extinguishers to smother the fire; both types are nonconductors of electricity.

Class D fires: combustible metals such as magnesium, sodium, zinc, and potassium. Operators rarely encounter this type of fire. Use a Class D extinguisher or use fine dry soda ash, sand, or graphite to smother the fire. Consult with your local fire department about the best methods to use for specific hazards that exist at your facility.

Multipurpose extinguishers are also available, such as a Class BC carbon dioxide extinguisher that can be used to smother Class B and Class C fires. A multipurpose ABC carbon dioxide extinguisher will handle most laboratory fire situations. (When using carbon dioxide extinguishers, remember that the carbon dioxide can displace oxygen—take appropriate precautions.)

There is no single type of fire extinguisher that is effective for all fires so it is important that you understand the class of fire you are trying to control. You must be trained in the use of the different types of extinguishers, and the proper type should be located near the area where that class of fire may occur.

11.163 Acknowledgments

Portions of this section were taken from material written by A. E. Greenberg, "Safety and Hygiene," which appeared in the California Water Pollution Control Association's *OPERATORS' LABORATORY MANUAL*. Some of the ideas and material also came from the *FISHER SAFETY MANUAL*.

11.164 Additional Reading

1. *FISHER SAFETY CATALOG*. Obtain from Fisher Scientific Company, Safety Division, 9999 Veterans Memorial Drive, Houston, TX 77038.

2. *GENERAL INDUSTRY, OSHA. SAFETY AND HEALTH STANDARDS* (CFR, Title 29, Labor Pt. 1900-1910 (most recent edition)). Obtain from the U.S. Government Printing Office, Superintendent of Documents, PO Box 371954, Pittsburgh, PA 15250-7954. Order No. 869-050-00106-3. Price, $61.00.

3. See *SAFETY*, page 1-38, *STANDARD METHODS*, 20th Edition, 1998.

QUESTIONS

Write your answers in a notebook and then compare your answers with those on page 568.

11.1G List five laboratory hazards.

11.1H Why should you not work alone in a laboratory?

11.1I True or False? You may *ADD ACID TO WATER*, but never water to acid.

11.1J How would you dispose of a corrosive acid?

End of Lesson 1 of 5 Lessons on LABORATORY PROCEDURES

Please answer the discussion and review questions next.

DISCUSSION AND REVIEW QUESTIONS

Chapter 11.   LABORATORY PROCEDURES

(Lesson 1 of 5 Lessons)

At the end of each lesson in this chapter you will find some discussion and review questions. The purpose of these questions is to indicate to you how well you understand the material in the lesson. Write the answers to these questions in your notebook before continuing.

1. How do the operators of water treatment plants use the results from laboratory tests?

2. Why must chemicals be properly labeled?

3. How are pipets emptied or drained?

4. How would you titrate a test solution?

5. List as many of the seven basic rules for working in a laboratory as you can remember.

6. Why should work with certain chemicals be conducted under a ventilated laboratory fume hood?

7. Why should water not be poured on certain types of fires?

CHAPTER 11.   LABORATORY PROCEDURES

(Lesson 2 of 5 Lessons)

11.2   SAMPLING

11.20   Importance of Sampling

Sampling is a vital part of studying the quality of water in a water treatment process, distribution system, or source of water supply. The major source of error in the whole process of obtaining water quality information often occurs during sampling. Proper sampling procedures are essential in order to obtain an accurate description of the material or water being sampled and tested. This fact is not well enough recognized and cannot be overemphasized.

In any type of testing program where only small samples (a liter or two) are withdrawn from perhaps millions of gallons of water under examination, there is potential uncertainty because of possible sampling errors. Water treatment decisions based upon incorrect data may be made if sampling is performed in a careless and thoughtless manner. Obtaining good results will depend to a great extent upon the following factors:

1. Ensuring that the sample taken is truly representative of the water under consideration,

2. Using proper sampling techniques, and

3. Protecting and preserving the samples until they are analyzed.

The greatest errors in laboratory tests are usually caused by improper sampling, poor preservation, or lack of enough mixing during testing. The accuracy of your analysis is only as good as the care that was taken in obtaining a representative sample.[4]

11.21   Representative Sampling

11.210   Importance of Representative Sampling

A representative sample must be collected in order for test results to have any significant meaning. Without a representative sample, the test results will not reflect actual water conditions.

The sampling of a tank or a lake that is completely mixed is a simple matter. Unfortunately, most bodies of water are not well mixed and obtaining samples that are truly representative of the whole body depends to a great degree upon sampling technique. A sample that is properly mixed (integrated) by taking small portions of the water at points distributed over the whole body represents the material better than a sample collected from a single point. The more portions taken, the more nearly the sample represents the original. The sample error would reach zero when the size of the sample became equal to the original volume of material being sampled, but for obvious reasons this method of decreasing sample error is not practical. The size of sample depends on which water quality indicators are being tested and how many. Every precaution must be taken to ensure that the sample collected is as representative of the water source or process being examined as is feasible.

11.211   Source Water Sampling

RIVERS. To adequately determine the composition of a flowing stream, each sample (or set of samples taken at the same time) must be representative of the entire flow at the sampling point at that instant. Furthermore, the sampling process must be repeated frequently enough to show significant changes of water quality that may occur over time in the water passing the sampling point.

On small or medium-sized streams, it is usually possible to find a sampling point at which the composition of the water is presumably uniform at all depths and across the stream. Obtaining representative samples in these streams is relatively simple. For larger streams, more than one sample may be required. A portable conductivity meter is very useful in selecting good sample sites.

RESERVOIRS AND LAKES. Water stored in reservoirs and lakes is usually poorly mixed. Thermal stratification and associated depth changes in water composition (such as dissolved oxygen) are among the most frequently observed effects. Single samples can therefore be assumed to represent only the spot of water from which the sample came. Therefore, a number of samples must be collected at different depths and from different areas of the impoundment to accurately sample reservoirs and lakes.

GROUNDWATER. Most of the physical factors which promote mixing in surface waters are absent or much less effective in groundwater systems. Wells usually draw water from a considerable thickness of saturated rock and often from sever-

---

[4] *Representative Sample.   A sample portion of material or water that is as nearly identical in content and consistency as possible to that in the larger body of material or water being sampled.*

al different strata. These water components are mixed by the turbulent flow of water in the well before they reach the surface and become available for sampling. Most techniques for well sampling and exploration are usable only in unfinished or non-operating wells. Usually the only means of sampling the water tapped by a well is the collection of a pumped sample. The operator is cautioned to remember that well pumps and casings can contribute to sample contamination. If a pump has not run for an extended period of time prior to sampling, the water collected may not be representative of the normal water quality.

11.212   In-Plant Sampling

Collection of representative samples within the water treatment plants is really no different from sample collection in a stream or river. The operator simply wants to be sure the water sampled is representative of the water passing that sample point. In many water plants, money is spent to purchase sample pumps and piping only to sample from a point that is not representative of the passing water. A sample tap in a dead area of a reservoir or on the floor of a process basin serves no purpose in helping the plant operator with control of water quality. The operator is urged to find each and every sample point and ensure it is located to provide a useful and representative sample. If the sampling point is not properly located, plan to move the piping to a better location.

11.213   Distribution System Sampling

Representative sampling in the distribution system is a true indication of system water quality. Results of sampling should show if there are quality changes in the entire, or parts of, the system and may point to the source of a problem (such as tastes and/or odors). Sampling points should be selected, in part, to trace the course from the finished water source (at the well or plant), through the transmission mains, and then through the major and minor arteries of the system. A sampling point on a major artery, or on an active main directly connected to it, would be representative of the water quality being furnished to a subdivision of this network. Generally, these primary points are used as "official" sample points in evaluating prevailing water quality.

Obtaining a representative sample from the distribution system is not as easy as it might seem. One would think almost any faucet would do, but experience has shown otherwise. Local conditions at the tap and in its connection to the main can easily make the point unrepresentative of water being furnished to your consumers.

The truest evaluation of water in a distribution system can be obtained from samples drawn directly from the main. You might think that samples taken from a fire hydrant would prove satisfactory, but this is usually not the case. The problem with fire hydrants as sampling points is that they give erratic (uneven) results due to the way they are constructed and their

lack of use. In general, an ideal sample station is one that has a short, direct connection with the main and is made of corrosion-resistant material.

In most smaller water systems, special sample taps are not available. Therefore, customer's faucets must be used to collect samples. The best sample points are front yard faucets on homes supplied by short service lines (homes with short service lines are located on the same side of the street as the water main).

If the customer is home, you should contact the person in the home and obtain permission to collect the sample. Disconnect the hose from the faucet if one is attached and don't forget to reconnect the hose when finished collecting the sample. Open the faucet to a convenient flow for sampling (usually about half a gallon per minute). Allow the water to flow until the water in the service line has been replaced twice. Since 50 feet (15 m) of three-quarter inch (18-mm) pipe contains over one gallon (3.8 liters), four or five minutes will be required to replace the water in the line twice. Collect the sample. Be sure the sample container does not touch the faucet.

Do not try to save time by turning the faucet handle to wide open to flush the service line. This will disturb sediment and incrustations in the line which must be flushed out before the sample can be collected.

FORMULAS

To estimate the flow from a faucet, use a gallon jug and a watch. If you want a flow of half a gallon per minute, then the jug should be half full in one minute or completely full in two minutes.

$$\text{Flow, GPM} = \frac{\text{Volume, gallons}}{\text{Time, minutes}}$$

To calculate the volume of a service line, multiply the area of the pipe in square feet times the length of the pipe in feet to obtain cubic feet. The diameter of a pipe is given in inches, so this value must be divided by 12 inches per foot to obtain a volume in cubic feet. Multiply cubic feet by 7.48 gallons per cubic foot to obtain the volume in gallons.

$$\text{Pipe Volume, cu ft} = \frac{(\text{Pipe Area, sq in})(\text{Pipe Length, ft})}{144 \text{ sq in/sq ft}}$$

$$\text{Pipe Volume, gal} = (\text{Pipe Volume, cu ft})(7.48 \text{ gal/cu ft})$$

To determine the time to allow water to flow from a faucet to flush a service line twice, divide the pipe volume in gallons by the flow in gallons per minute. Then multiply the result by two so the line will be flushed twice.

$$\text{Flushing Time, min} = \frac{(\text{Pipe Volume, gal})(2)}{\text{Flow, gal/min}}$$

EXAMPLE 2

How long should a three-quarter inch service line 80 feet long be flushed if the flow is 0.5 GPM?

Known	Unknown
Diameter, in = ³/₄ in	Flushing Time, min
= 0.75 in	
Length, ft = 80 ft	
Flow, GPM = 0.5 GPM	

Calculate the pipe volume in cubic feet and then in gallons.

$$\text{Pipe Volume, cu ft} = \frac{(\text{Pipe Area, sq in})(\text{Pipe Length, ft})}{144 \text{ sq in/sq ft}}$$

$$= \frac{(0.785)(0.75 \text{ in})^2(80 \text{ ft})}{144 \text{ sq in/sq ft}}$$

$$= 0.245 \text{ cu ft}$$

$$\text{Pipe Volume, gal} = (\text{Pipe Volume, cu ft})(7.48 \text{ gal/cu ft})$$

$$= (0.245 \text{ cu ft})(7.48 \text{ gal/cu ft})$$

$$= 1.833 \text{ gal}$$

Calculate the flushing time for the service line in minutes.

$$\text{Flushing Time, min} = \frac{(\text{Pipe Volume, gal})(2)}{\text{Flow, gal/min}}$$

$$= \frac{(1.833 \text{ gal})(2)}{0.5 \text{ GPM}}$$

$$= 7.3 \text{ min}$$

$$\text{or} = 7 \text{ min} + (0.3 \text{ min})(60 \text{ sec/min})$$

$$= 7 \text{ min and 18 sec}$$

QUESTIONS

Write your answers in a notebook and then compare your answers with those on page 568.

11.2A   What are frequently the greatest causes of errors in laboratory tests?

11.2B   Why must a representative sample be collected?

11.2C   How are sampling points selected in a distribution system?

11.22   Types of Samples

There are generally two types of samples collected by waterworks operators, and either type may be obtained manually or automatically. The two types are grab samples and composite samples.

11.220   *Grab Samples*

A grab sample is a single water sample collected at no specific time. Grab samples will show the water characteristics at the time the sample was taken. A grab sample may be preferred over a composite sample when:

1. The water to be sampled does not flow continuously,

2. The water's characteristics are relatively constant, and

3. The water is to be analyzed for water quality indicators that may change with time, such as dissolved gases, coliform bacteria, residual chlorine, temperature, and pH.

11.221   *Composite Samples*

In many processes, the water quality is changing from moment to moment or hour to hour. A continuous sampler-analyzer would give the most accurate results in these cases. However, since operators themselves are often the sampler-analyzer, continuous analysis would leave little time for anything but sampling and testing. Except for tests which cannot wait due to rapid physical, chemical, or biological changes of the sample (such as tests for dissolved oxygen, pH, and temperature) a fair compromise may be reached by taking samples throughout the day at hourly or two-hour intervals. Each sample should be refrigerated immediately after it is collected. At the end of 24 hours, a portion of each sample is mixed with the other samples. The size of the portion is in direct proportion to the flow when the sample was collected and the total size of sample needed for testing. For example, if hourly samples were collected when the flow was 1.2 MGD, use a 12-mL portion sample, and when the flow was 1.5 MGD, use a 15-mL portion sample. The resulting mixture of portions of samples is called a *COMPOSITE SAMPLE*. In no case, however, should a composite sample be collected for bacteriological examination.

When the samples are taken, they can either be set aside to be combined later or combined as they are collected. In both cases, they should be stored at a temperature of 4°C until they are analyzed.

11.23   Sampling Devices

Automatic sampling devices are wonderful time-savers but are expensive. As with anything automatic, problems do arise and the operator should be on the lookout for potential difficulties.

Manual sampling equipment includes dippers, weighted bottles, hand-operated pumps, and cross-section samplers. Dippers consist of wide-mouth, corrosion-resistant containers (such as cans or jars) on long handles that collect a sample for testing. A weighted bottle is a collection container which is lowered to a desired depth. At this depth a cord or wire removes the bottle stopper so the bottle can be filled (see Figure 11.5).

Some water treatment facilities use sample pumps to collect the sample and transport it to a central location. The pump and its associated piping should be corrosion-resistant and sized to deliver the sample at a high enough velocity to prevent sedimentation in the sample line.

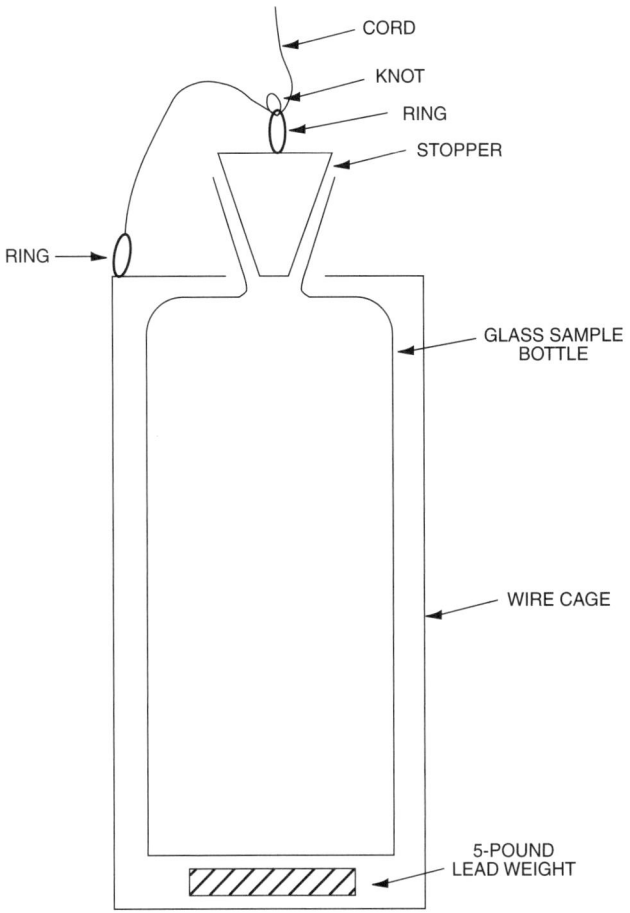

Fig. 11.5 *Sectional view of homemade depth sampler*

Fig. 11.6 *Distribution system sampling station*

Many water agencies have designed and installed special sampling stations throughout their distribution systems (see Figure 11.6). These stations provide an excellent location to sample the actual quality of water in your distribution system.

11.24 Sampling Techniques

11.240 Surface Sampling

A surface sample is obtained by grasping the sample container at the base with one hand and plunging the bottle mouth down into the water to avoid introducing any material floating on the surface. Position the mouth of the bottle into the current and away from the hand of the collector (see Figure 11.7). If the water is not flowing, then an artificial current can be created by moving the bottle horizontally in the direction it is pointed and away from the sampler. Tip the bottle slightly upward to allow air to exit so the bottle can fill. Tightly stopper and label the bottle.

Place the bottle in a weighted frame that holds the bottle securely when sampling from a walkway or other structure above a body of water. Remove the stopper or lid and lower the device to the water surface. A nylon rope which does not absorb water and will not rot is recommended. Face the bottle mouth upstream by swinging the sampling device downstream and then allow it to drop into the water, without slack in the rope. Pull the sample device rapidly upstream and out of the water simulating

(imitating) the scooping motion of the hand-sampling described previously. Take care not to dislodge dirt or other debris that might fall into the open sample container from above. Be sure to label the container when sampling is completed.

11.241 Depth Sampling

Several additional pieces of equipment are needed for collection of depth samples from basins, tanks, lakes, and reservoirs. These depth samplers require lowering the sample device and container to the desired water depth, then opening, filling, and closing the container, and returning the device to the surface. Although depth measurements are best made with a pre-marked steel cable, the sample depths can be determined by pre-measuring and marking a nylon rope at intervals with a non-smearing ink, paint, or fingernail polish. One of the most common commercial devices is called a Kemmerer Sampler (see Figure 11.8). This type of depth sampler consists of a cylindrical tube that contains a rubber stopper or valve at each end. The device is lowered into the water in the open position and the water sample is trapped in the cylinder when the valves are closed by the dropped messenger.

Figures 11.5 and 11.9 show typical depth samplers. These samplers are lowered to the desired depth. A jerk on the cord will remove the stopper and allow the bottle in the depth sampler to fill. Good samples can be collected in depths of water up to 40 feet (12 m).

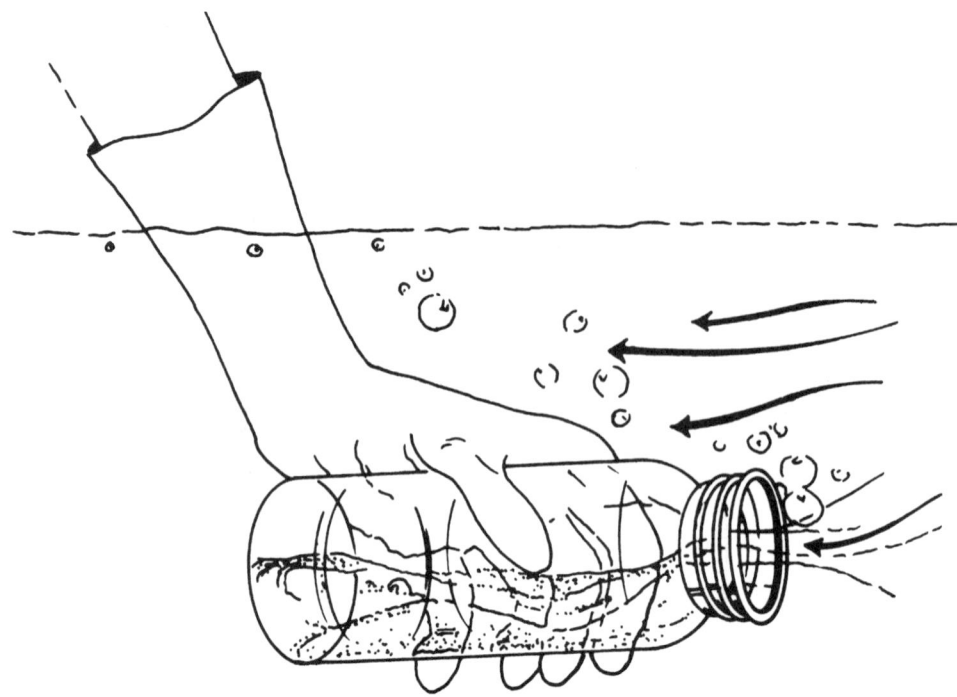

Fig. 11.7   Demonstration of technique used in grab sampling of surface waters
(Source: US EPA "Microbiological Methods For Monitoring the Environment," December 1978)

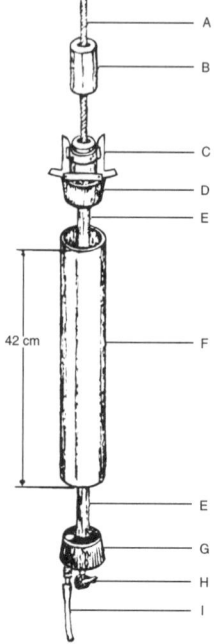

Fig. 11.8   Kemmerer depth sampler. (A) nylon line, (B) messenger, (C) catch set so that the sampler is open, (D) top rubber valve, (E) connecting rod between the valves, (F) tube body, (G) bottom rubber valve, (H) knot at the bottom of the suspension line, and (I) rubber tubing attached to the spring-loaded check valve.

(Source: US EPA "Microbiological Methods For Monitoring the Environment," December 1978)

Fig. 11.9   Depth sampler
(Permission of HACH Company)

11.242 Water Tap Sampling

To collect samples from water main connections, first flush the service line for a brief period of time. Samples should not be taken from drinking fountains, restrooms, or taps that have aerators. Aerators can change water quality indicators such as pH and dissolved oxygen, and can harbor bacteria under some conditions. Do not sample from taps surrounded by excessive foliage (leaves, flowers) or from taps that are dirty, corroded, or are leaking. Never collect a sample from a hose or any other attachment fastened to a faucet. Care must be taken to be sure that the sample collector does not come in contact with the faucet.

11.243 First-Draw Samples

The Lead and Copper Rule calls for first-draw or first-flush samples. These are water samples taken at the customer's tap after the water stands motionless in the plumbing pipes for at least six hours. This usually means taking a sample early in the day before water is used in the kitchen or bathroom.

11.25 Sampling Containers and Preservation of Samples

The shorter the time that elapses between the actual collection of the sample and the analysis, the more reliable your results will be. Samples should be preserved if they are not going to be analyzed immediately due to remoteness of the laboratory or workload. Preservation of some types of samples is essential to prevent deterioration of the sample. Some water quality indicators, such as residual chlorine and temperature, require immediate analysis, while others can be preserved and transported to the laboratory. A summary of acceptable sample containers, preservative, and maximum time between sampling and analysis is shown on Table 11.2.

Whatever type of container you use, clearly identify the sample location, date and time of collection, name of collector, and any other pertinent information.

11.26 Reporting

The water system owner (water utility agency) is responsible for reporting lab results at regular frequencies to the regulatory agency as required by the Safe Drinking Water Act.

11.27 Additional Reading

1. See page 1-27, *COLLECTION AND PRESERVATION OF SAMPLES, STANDARD METHODS*, 20th Edition.

QUESTIONS

Write your answers in a notebook and then compare your answers with those on pages 568 and 569.

11.2D What are the two general types of samples collected by water treatment personnel?

11.2E List three water quality indicators that are usually measured with a grab sample.

11.2F How would you collect a depth sample from a lake?

11.2G Samples should not be collected from water taps under what conditions?

11.2H What information should be recorded when a sample is collected?

End of Lesson 2 of 5 Lessons on LABORATORY PROCEDURES

Please answer the discussion and review questions next.

DISCUSSION AND REVIEW QUESTIONS

Chapter 11. LABORATORY PROCEDURES

(Lesson 2 of 5 Lessons)

Write the answers to these questions in your notebook before continuing. The question numbering continues from Lesson 1.

8. Why are proper sampling procedures important?

9. What is meant by a representative sample?

10. Generally speaking, how would you obtain a representative sample?

11. Under what conditions and why would you preserve a sample?

TABLE 11.2 RECOMMENDATION FOR SAMPLING AND PRESERVATION OF SAMPLES ACCORDING TO MEASUREMENT [a]

Measurement	Vol. Req. (mL)	Container [b]	Preservative	Max. Holding Time [c]
PHYSICAL PROPERTIES				
Color	500	P,G	Cool, 4°C	48 hours
Conductance	500	P,G	Cool, 4°C	28 days
Hardness [d]	100	P,G	HNO_3 to pH <2, H_2SO_4 to pH <2	6 months
Odor	200	G only	Cool, 4°C	6 hours
pH [d]	25	P,G	Det. on site	Immediately
Residue, Filterable	100	P,G	Cool, 4°C	7 days
Temperature	1,000	P,G	Det. on site	Immediately
Turbidity	100	P,G	Cool, 4°C	48 hours
METALS (Fe, Mn)				
Dissolved or Suspended	200	P,G	Filter on site, HNO_3 to pH <2	6 months
Total	100	P,G	Filter on site, HNO_3 to pH <2	6 months
INORGANICS, NONMETALLICS				
Acidity	100	P,G	Cool, 4°C	14 days
Alkalinity	200	P,G	Cool, 4°C	14 days
Bromide	100	P,G	None Req.	28 days
Chloride	50	P,G	None Req.	28 days
Chlorine, Total Residual	500	P,G	Det. on site	Immediately
Cyanide, Total and Amenable to Chlorination	500	P,G	Cool, 4°C, NaOH to pH >12, 0.6 gm ascorbic acid [e]	14 days
Fluoride	300	P	None Req.	28 days
Iodide	100	P,G	Cool, 4°C	24 hours
Nitrogen				
Ammonia	500	P,G	Cool, 4°C, H_2SO_4 to pH <2	28 days
Kjeldahl and Organic	500	P,G	Cool, 4°C, H_2SO_4 to pH <2	28 days
Nitrate-Nitrite	200	P,G	Cool, 4°C, H_2SO_4 to pH <2	28 days
Nitrate	100	P,G	Cool, 4°C	48 hours
Nitrite	100	P,G	Cool, 4°C	48 hours
Dissolved Oxygen				
Probe	300	G with top	Det. on site	Immediately
Winkler	300	G with top	Fix on site, store in dark	8 hours
Phosphorus				
Orthophosphate	50	P,G	Filter on site, Cool, 4°C	48 hours
Elemental	50	G	Cool, 4°C	48 hours
Total	50	P,G	Cool, 4°C, H_2SO_4 to pH <2	28 days
Silica	50	P	Cool, 4°C	28 days
Sulfate	100	P,G	Cool, 4°C	28 days
Sulfide	100	P,G	Cool, 4°C, add zinc acetate plus H_2SO_4 to pH >9	7 days
Sulfite	50	P,G	Det. on site	Immediately

a "Required Containers, Preservation Techniques, and Holding Times." *CODE OF FEDERAL REGULATIONS*, Protection of the Environment, 40, Parts 136-149, 2003. This publication is available from the U.S. Government Printing Office, Superintendent of Documents, PO Box 371954, Pittsburgh, PA 15250-7954. Order No. 869-050-00155-1. Price, $61.00.

b Polyethylene (P) or Glass (G). For metals, polyethylene with a polypropylene cap (no liner) is preferred.

c Holding times listed above are recommended for properly preserved samples based on currently available data. It is recognized that for some sample types, extension of these times may be possible while for other types, these times may be too long. Where shipping regulations prevent the use of the proper preservation technique or the holding time is exceeded, such as the case of a 24-hour composite, the final reported data for these samples should indicate the specific variance.

d Hardness and pH are usually considered chemical properties of water rather than physical properties.

e Use ascorbic acid only if residual chlorine is present.

SPECIAL NOTE: Whenever you collect a sample for a bacteriological test (coliforms), be sure to use a sterile plastic or glass bottle. If the sample contains any chlorine residual, sufficient sodium thiosulfate should be added to neutralize all of the chlorine residual. Usually two drops (0.1 mL) of ten percent sodium thiosulfate for every 100 mL of sample is sufficient, unless you are disinfecting mains or storage tanks.

CHAPTER 11. LABORATORY PROCEDURES

(Lesson 3 of 5 Lessons)

11.3 WATER LABORATORY TESTS

1. Alkalinity

A. Discussion

The alkalinity of a water sample is a measure of the water's capacity to neutralize acids. In natural and treated waters, alkalinity is the result of bicarbonates, carbonates, and hydroxides of the metals of calcium, magnesium, and sodium.

Many of the chemicals used in water treatment, such as alum, chlorine, or lime, cause changes in alkalinity. The alkalinity determination is needed when calculating chemical dosages used in coagulation and water softening. Alkalinity must also be known to calculate corrosivity and to estimate the carbonate hardness of water. Alkalinity is usually expressed in terms of *CALCIUM CARBONATE (CaCO₃) EQUIVALENT.*[5]

There are five alkalinity conditions possible in a water sample: (1) bicarbonate alone, (2) bicarbonate and carbonate, (3) carbonate alone, (4) carbonate and hydroxide, and (5) hydroxide alone. These five conditions may be distinguished and quantities determined from the results of acid titrations by the method given below.

B. What Is Tested

Sample	Common Range, mg/L
Raw and Treated Surface Water	20 - 300
Well Water	80 - 500

C. Apparatus Required

pH meter	graduated cylinder (100 mL)
reference electrode	buret (25 mL)
glass electrode	buret support
magnetic stirrer	beaker (250 mL)
magnetic stir-bar	wash bottle
analytical balance	desiccator

D Reagents

NOTE: Standardized solutions are commercially available for most reagents. Refer to *STANDARD METHODS* if you wish to prepare your own reagents.

1. Sodium carbonate (Na_2CO_3) solution, approximately 0.05 N.

2. Sulfuric acid (H_2SO_4), 0.1 N.

3. Standard sulfuric acid, 0.02 N: Dilute 200 mL 0.10 N standard acid to 1 liter using a volumetric flask. To determine the volume to be diluted, use the following formula:

$$\text{Volume Diluted, mL} = \frac{(\text{Standard, 0.02 } N) \times (1{,}000 \text{ mL})}{(\text{Calculated Normality, 0.10 } N)}$$

$$= \frac{(0.02 \ N) \times (1{,}000 \text{ mL})}{(0.10 \ N)}$$

$$= 200 \text{ mL}$$

E. Procedure

Total and Phenolphthalein Alkalinity

1. Take a clean beaker and add 100 mL of sample (or other sample volume that will give a titration volume of less than 50 mL of acid titrant).

2. Place electrodes of pH meter into beaker containing sample.

3. Stir sample slowly (with a magnetic stirrer if possible).

4. Check pH of sample. If pH is 8.3 or below, then there is no phenolphthalein alkalinity present and you can go to step 6.

5. If the pH is greater than 8.3, titrate very carefully to a pH of 8.3 with 0.02 N H_2SO_4. Record the amount of acid used to this point.

6. Continue to titrate to pH 4.5 with 0.02 N H_2SO_4.[6] Record the total amount of acid used from starting point to finish.

7. Calculate Total and Phenolphthalein (if present) Alkalinities.

F. Example

Results from alkalinity titrations on a finished water sample were as follows:

$$\text{Sample size} = 100 \text{ mL}$$

$$\text{mL titrant used to pH 8.3, A} = 0.5 \text{ mL}$$

$$\text{total mL of titrant used, B} = 6.8 \text{ mL}$$

$$\text{Acid Normality, } N = 0.02 \ N \ H_2SO_4$$

---

[5] *Calcium Carbonate (CaCO₃) Equivalent. An expression of the concentration of specified constituents in water in terms of their equivalent value to calcium carbonate. For example, the hardness in water which is caused by calcium, magnesium and other ions is usually described as calcium carbonate equivalent. Alkalinity test results are usually reported as mg/L CaCO₃ equivalents. To convert chloride to CaCO₃ equivalents, multiply the concentration of chloride ions in mg/L by 1.41, and for sulfate, multiply by 1.04.*

[6] *STANDARD METHODS, 20th Edition, page 2-27, recommends titrating to a pH of 4.5 for routine or automated analyses. However, other pH levels are suggested for various levels of alkalinity.*

(Alkalinity)

OUTLINE OF PROCEDURE FOR ALKALINITY

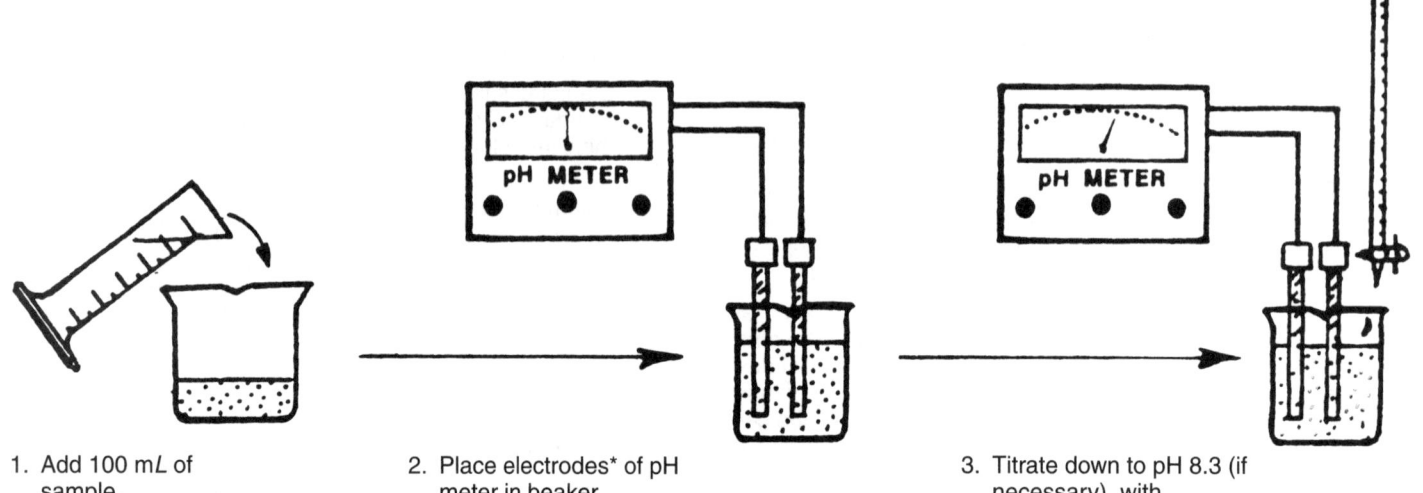

1. Add 100 mL of
   sample.

2. Place electrodes* of pH
   meter in beaker.

3. Titrate down to pH 8.3 (if
   necessary), with
   0.02 N H_2SO_4.

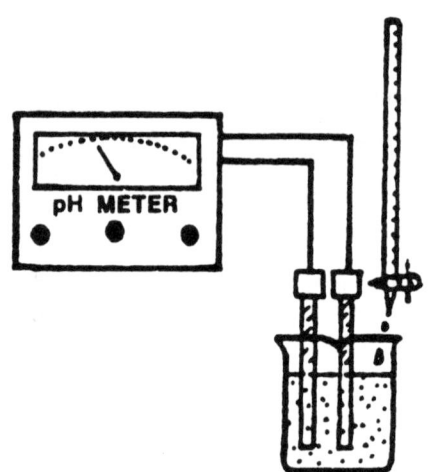

4. Continue to titrate
   to pH 4.5.

* Some pH meters have a single "combination" electrode.

G. Calculations

Phenolphthalein Alkalinity, mg/L as $CaCO_3$

$$= \frac{A \times N \times 50{,}000}{mL \text{ of sample}}$$

$$= \frac{(0.5 \text{ m}L) \times (0.02 \text{ } N) \times 50{,}000}{100 \text{ m}L}$$

$$= 5 \text{ mg}/L$$

Total Alkalinity, mg/L as $CaCO_3$

$$= \frac{B \times N \times 50{,}000}{mL \text{ of sample}}$$

$$= \frac{(6.8 \text{ m}L) \times (0.02 \text{ } N) \times 50{,}000}{100 \text{ m}L}$$

$$= 68 \text{ mg}/L$$

H. Interpretation of Results

From the test results and the information given below, the different types of alkalinity contained in a water sample can be determined.

Alkalinity, mg/L as $CaCO_3$

Titration Result	Bicarbonate	Carbonate	Hydroxide
P = 0	T	0	0
P is less than ¹/₂ T	T − 2P	2P	0
P = ¹/₂ T	0	2P	0
P is greater than ¹/₂ T	0	2T − 2P	2P − T
P = T	0	0	T

where P = phenolphthalein alkalinity
T = total alkalinity

Example: The example in "G" above gave the following results:

phenolphthalein alkalinity = 5 mg/L

total alkalinity = 68 mg/L

Since the phenolphthalein alkalinity (5 mg/L) is less than one half of the total alkalinity (68 mg/L) from the table, then there is bicarbonate and carbonate alkalinity in the water.

The bicarbonate alkalinity in this case is equal to T − 2P or 68 mg/L − (2 × 5 mg/L) = 58 mg/L as $CaCO_3$.

The carbonate alkalinity is equal to 2P or 2 × 5 mg/L = 10 mg/L as $CaCO_3$.

I. Precautions

1. The sample should be analyzed as soon as possible, at least within a few hours after collection.

2. The sample should not be agitated, warmed, filtered, diluted, concentrated, or altered in any way.

J. Reference

1. See page 2-26, STANDARD METHODS, 20th Edition.

(Chlorine Residual)

2. See page 310-1-1 METHODS FOR CHEMICAL ANALYSIS OF WATER AND WASTES, March 1979.

QUESTIONS

Write your answers in a notebook and then compare your answers with those on page 569.

11.3A What chemicals used in water treatment will cause changes in alkalinity?

11.3B Why is it important to know the alkalinity of a water sample?

2. Chlorine Residual

A. Discussion

Chlorine is not only an excellent disinfectant but also serves to react with iron, manganese, protein substances, sulfide, and many taste- and odor-producing compounds to help improve the quality of treated water. In addition, chlorine helps to control microorganisms that might interfere with coagulation and flocculation, keeps filter media free of slime growths, and helps bleach out undesirable color.

There are two general types of residual chlorine produced in chlorinated water. They are (1) free residual chlorine and (2) combined residual chlorine. Free residual chlorine refers to chlorine (Cl_2), hypochlorus acid (HOCl), and the hypochlorite ion (OCl^-). Combined residual chlorine generally refers to the chlorine-ammonia compounds of monochloramine (NH_2Cl), dichloramine ($NHCl_2$), and trichloramine (NCl_3 or nitrogen trichloride). Both types of residuals act as disinfectants, but differ in their capacity to produce a germ-free water supply during the same contact time.

In addition to all the positive aspects of chlorination, there may be some adverse effects. Potentially carcinogenic chlororganic compounds such as chloroform and other THMs may be formed during the chlorination process. To minimize any adverse effects, the operator should be familiar with the concentrations of free and combined residual chlorine produced in a water supply following chlorination. Both residuals are extremely important in producing a potable water that is not only safe to drink but is also free of objectionable tastes and odors.

B. What Is Tested

Source	Common Range Residual Chlorine, mg/L	
	Free	Total
Chlorinated Raw Surface Water (Prechlorination)	0.3 to 3	0.5 to 5
Chlorinated Finished Surface Water (Post-chlorination)	0.2 to 1	0.3 to 1.5
Well Water	0.2 to 1	0.2 to 1

C. Methods

There are eight methods listed for measuring residual chlorine in the 20th Edition of STANDARD METHODS. Selection of the most practical and appropriate procedure in any particular instance generally depends upon the characteristics of the water being examined. The AMPEROMETRIC[7] titration method is a

---

[7] Amperometric (am-PURR-o-MET-rick). A method of measurement that records electric current flowing or generated, rather than recording voltage. Amperometric titration is a means of measuring concentrations of certain substances in water.

(Chlorine Residual)

standard of comparison for determining free or combined chlorine residual. This method is relatively free of interferences but does require greater operator skill to obtain good results. In addition, the titration instrument is expensive.

The *DPD*[8] methods are simpler to perform than amperometric titration but are subject to interferences due to manganese. Field comparator kits are available from several suppliers such as Orbeco-Hellige, Wallace & Tiernan, and HACH.

D.   Apparatus Required

1.   Amperometric Titration Method

See page 4-58, *STANDARD METHODS*, 20th Edition and amperometric titrator's instruction manual.

2.   DPD Colorimetric Method (Field Comparator Kit)

Field Comparator
Sample cells

3.   DPD Titrimetric Method[9]

Graduated cylinder (100 mL)
Pipets (1 and 10 mL)
Flask, Erlenmeyer (250 mL)
Buret (10 mL)
Magnetic stirrer
Magnetic stir-bar
Balance, analytical

E.   Reagents

NOTE: Prepared reagents may be purchased from laboratory chemical supply houses.

Amperometric Titration Method

1.   Standard phenylarsine oxide (PAO) solution, 0.00564 N
CAUTION: Toxic—avoid ingestion.

2.   Acetate buffer solution, pH 4.

3.   Phosphate buffer solution, pH 7.

4.   Potassium iodide solution.

Store in brown glass, stoppered bottle, preferably in the refrigerator. Discard when solution becomes yellow.

DPD Colorimetric Method (Field Comparator Kit)

Use reagents supplied by kit manufacturer.

DPD Titrimetric Method

1.   1 + 3 H_2SO_4. CAREFULLY add 10 mL concentrated sulfuric acid to 30 mL distilled water. Cool.

2.   Phosphate Buffer Solution.

3.   DPD Indicator Solution.

4.   Standard Ferrous Ammonium Sulfate (FAS) Titrant, 0.00282 N.

5.   Potassium iodide, KI, crystals.

F.   Procedure

Amperometric Titration Method

Follow manufacturer's instructions.

DPD Colorimetric Method (Field Comparator Kit)

To measure chlorine residuals you should follow the directions provided by the manufacturer of the equipment or instrument you are using.

If you are disinfecting clear wells, distribution reservoirs, or mains and very high chlorine residuals must be measured, a drop-dilution technique can be used to estimate the chlorine residual. The procedure is as follows:

1.   Add 10 mL of distilled water and one powder pillow of DPD reagent (or 0.5 mL of DPD solution) to the sample tube of the test kit.

2.   Add a sample of the water being tested drop by drop to the sample tube until a color is produced.

3.   Record the number of drops added to the sample tube. Assume one drop equals 0.05 mL.

4.   Determine the chlorine residual in the sample as a result of the color produced and record the residual in milligrams per liter.

EXAMPLE 3

The recorded chlorine residual is 0.3 mg/L. Two drops of sample produced a chlorine residual of 0.3 mg/L in 10 mL of distilled water. Assume 0.05 mL per drop.

Known	Unknown
Chlorine Residual, mg/L = 0.3 mg/L	Actual Chlorine Residual, mg/L
Sample Volume, drops = 2 drops	
Distilled Water, mL = 10 mL	

Calculate the actual residual in milligrams per liter.

$$\text{Actual Chlorine Residual, mg/L} = \frac{(\text{Chlorine Residual, mg/L})(\text{Distilled Water, mL})}{(\text{Sample Volume, drops})(0.05 \text{ mL/drop})}$$

$$= \frac{(0.3 \text{ mg/L})(10 \text{ mL})}{(2 \text{ drops})(0.05 \text{ mL/drop})}$$

$$= 30 \text{ mg/L}$$

---

[8] *DPD (pronounce as separate letters).   A method of measuring the chlorine residual in water. The residual may be determined by either titrating or comparing a developed color with color standards. DPD stands for N,N-diethyl-p-phenylene-diamine.*

[9] *Some regulatory agencies require the use of the DPD Titrimetric Method if chlorine residual testing is used in place of some of the coliform tests.*

DPD Titrimetric Method

FOR FREE RESIDUAL CHLORINE

1. Place 5 m*L* each of buffer reagent and DPD indicator in a 250-m*L* flask and mix.

2. Add 100 m*L* of sample and mix.

3. Titrate rapidly with standard FAS titrant until red color is discharged (disappears).

4. Record amount of FAS used (Reading A). If combined residual chlorine fractions are desired, continue to next step.

FOR COMBINED RESIDUAL CHLORINE

Monochloramine

5. Add one very small crystal of KI and mix. If monochloramine is present, the red color will reappear.

6. Continue titrating carefully until red color again disappears.

7. Record reading of FAS in the buret used to this point (this includes amount used above). This is Reading B.

Dichloramine

8. Add several crystals KI (about 1 gm) and mix until dissolved.

9. Let stand 2 minutes.

10. Continue titrating until red color again disappears. Record amount of FAS used to this point. This includes amounts used in two previous titrations for free and monochloramine. This is Reading C.

NOTE: Manufacturers of laboratory equipment are continually developing faster and more accurate ways to measure chlorine residual. If you have new equipment, follow the manufacturer's procedures.

G. Precautions

1. For accurate results, careful pH control is essential. The pH of the sample, buffer, and DPD indicator together should be between 6.2 and 6.5.

2. If the sample contains oxidized manganese, an inhibitor must be used.

3. Samples should be analyzed as soon as possible after collection.

4. If nitrogen trichloride or chlorine dioxide are present, special procedures are necessary.

5. Operators using the DPD colorimetric method to test water for a free chlorine residual need to be aware of a potential error that may occur. If the DPD test is run on water containing a combined chlorine residual, a precipitate may form during the test. The particles of precipitated material will give the sample a turbid appearance or the appearance of having color. This turbidity can produce a positive test result for free chlorine residual when there is actually no chlorine present. Operators call this error a "false positive" chlorine residual reading.

H. Example

A sample taken after prechlorination at a filtration plant was tested for residual chlorine using the DPD Titrimetric Method.

m*L* of sample = 100 m*L*
Reading A, m*L* = 1.4 m*L*
Reading B, m*L* = 1.6 m*L*
Reading C, m*L* = 2.7 m*L*

I. Calculation

READING	CHLORINE RESIDUAL
A	= mg/*L* free residual chlorine
B – A	= mg/*L* monochloramine
C – B	= mg/*L* dichloramine
C – A	= mg/*L* combined available chlorine
C	= mg/*L* total residual chlorine

Example: The concentrations of the different types of residual chlorine present can be calculated from the information given in (H).

READING		CHLORINE RESIDUAL
A	= 1.4 m*L* = 1.4 mg/*L* free residual chlorine
B – A	= 1.6 – 1.4 = 0.2 mg/*L* monochloramine
C – B	= 2.7 – 1.6 = 1.1 mg/*L* dichloramine
C – A	= 2.7 – 1.4 = 1.3 mg/*L* combined available chlorine
C	= 2.7 m*L* = 2.7 mg/*L* total residual chlorine

J. Interpretation of Results

1. Any "chlorine" taste and odor that may result from chlorination would generally be from the dichloramine or nitrogen trichloride nuisance residuals.

2. Free residual chlorination produces the best results when the free residual makes up more than 80 percent of the total residual. However, this will not be the case if ammonia is added to the water being treated to form chloramines in order to prevent the formation of THMs.

K. Reference

See page 4-53, *STANDARD METHODS*, 20th Edition.

QUESTIONS

Write your answers in a notebook and then compare your answers with those on page 569.

11.3C List some of the important benefits of chlorinating water.

11.3D What is a potential adverse effect from chlorination?

(Chlorine Residual)

OUTLINE OF PROCEDURE FOR CHLORINE RESIDUAL

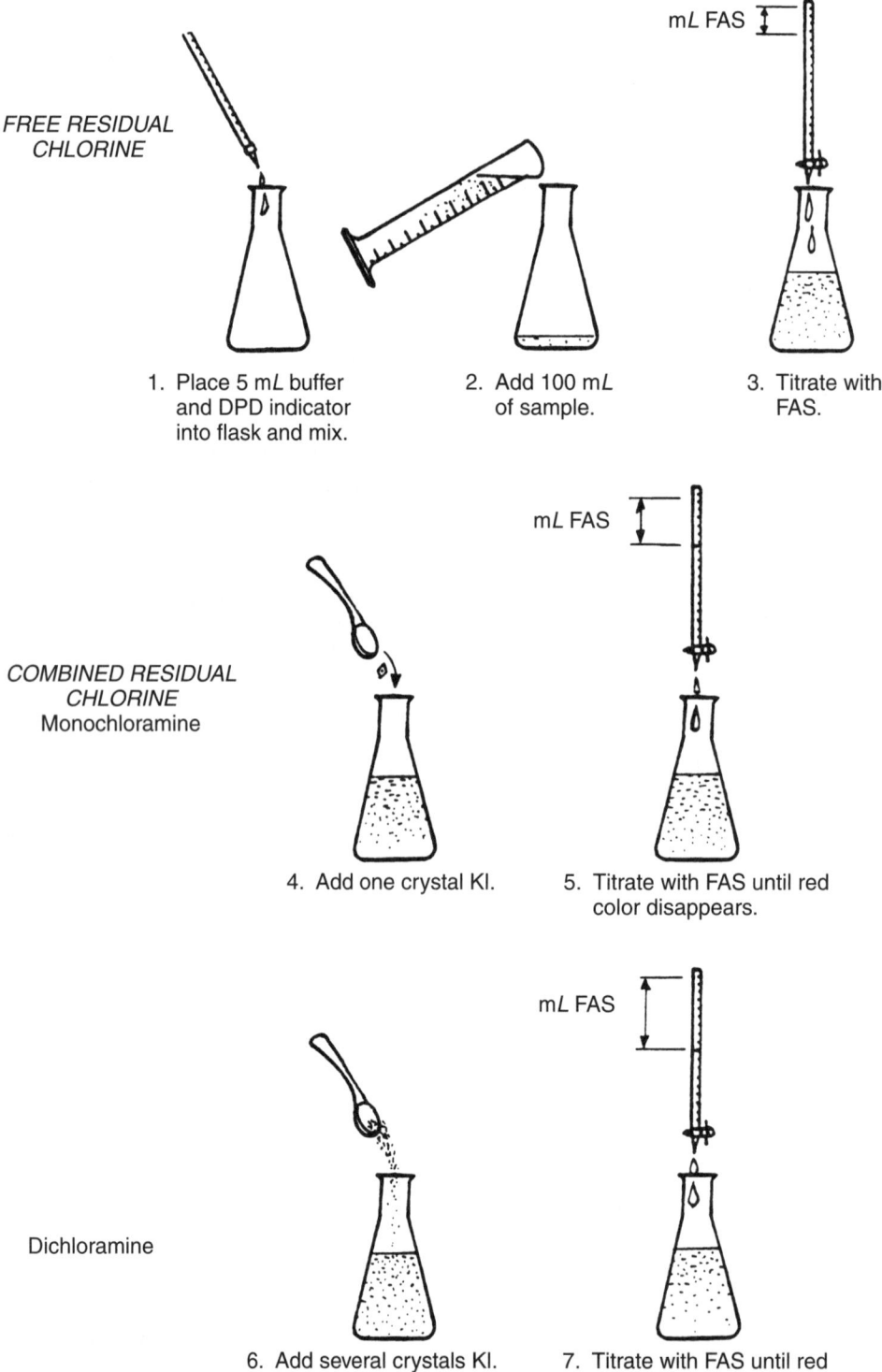

FREE RESIDUAL CHLORINE

1. Place 5 mL buffer and DPD indicator into flask and mix.

2. Add 100 mL of sample.

mL FAS

3. Titrate with FAS.

COMBINED RESIDUAL CHLORINE
Monochloramine

4. Add one crystal KI.

mL FAS

5. Titrate with FAS until red color disappears.

Dichloramine

6. Add several crystals KI.

mL FAS

7. Titrate with FAS until red color disappears.

3. Chlorine Demand

A. Discussion

The chlorine demand of water is the difference between the amount of chlorine applied (or dosed) and the amount of free, combined, or total residual chlorine remaining at the end of a specific contact period. The chlorine demand varies with the amount of chlorine applied, length of contact time, pH, and temperature. Also the presence of organics and reducing agents in water will influence the chlorine demand. The chlorine demand test should be conducted with chlorine gas or with granular hypochlorite, depending upon which form you usually use for chlorination.

The chlorine demand test can be used to determine the best chlorine dosage to achieve specific chlorination objectives. The measurement of chlorine demand is performed by treating a series of water samples in question with known but varying amounts of chlorine or hypochlorite. After the desired contact time, calculation of residual chlorine in the samples will demonstrate which dosage satisfied the requirements of the chlorine demand in terms of the residual desired.

B. What Is Tested

	Common Range Chlorine Demand, mg/L
Surface Water	0.5 to 5
Well Water	0.1 to 1.3

C. Apparatus

In addition to the apparatus described under one of the methods for chlorine residual, the following items are required:

Flasks, Erlenmeyer (1,000 mL)

Pipets (5 and 10 mL)

Graduated cylinder (500 mL)

Flask, volumetric (1,000 mL)

Flask, Erlenmeyer (250 mL)

Buret (25 mL)

D. Reagents

In addition to the reagents described under the method you will use to determine chlorine residual, the following items are also needed:

1. Stock chlorine solution. Obtain a suitable solution from a chlorinator solution line or by purchasing a bottle of household bleach ("Clorox" or similar product). Store in a dark, cool place to maintain chemical strength. Household bleach products usually contain approximately five percent available chlorine which is about 50,000 mg/L.

2. Chlorine dosing solution. If using household bleach, carefully pipet about 10 mL of the bleach into a 1,000 mL volumetric flask. Fill to the mark with chlorine demand-free water and standardize. If using a stock chlorine solution obtained from a chlorinator solution line, simply standardize this solution directly.

Standardization:

a. Place 2 mL acetic acid and 20 mL chlorine demand-free water in a 250-mL flask.

b. Add about 1 gm KI crystals.

c. Measure into the flask a suitable volume of chlorine dosing solution. If using household bleach as your stock solution, add 25 mL of the dosing solution. Note: In measuring the volume of the dosing solution, notice that 1 mL of the 0.025 N thiosulfate titrant is equal to 0.9 mg chlorine.

d. Titrate with standardized 0.025 N thiosulfate titrant until the yellow iodine color almost disappears.

e. Add 1 to 2 mL starch indicator solution.

f. Continue to titrate until blue color disappears.

$$\text{mg/}L \text{ Cl as Cl}_2\text{/m}L = \frac{(\text{m}L \text{ thiosulfate used}) \times N \times 35.45}{\text{m}L \text{ of dosing solution}}$$

where N = normality of thiosulfate titrant

3. Acetic acid, concentrated (glacial).

4. Potassium Iodide, KI, crystals.

5. Standard sodium thiosulfate 0.025 N.

6. Chlorine demand-free water. Prepare chlorine demand-free water from good quality distilled or deionized water by adding sufficient chlorine to give 5 mg/L free chlorine residual. After standing 2 days, this solution should contain at least 2 mg/L free residual chlorine. If not, discard and obtain better quality water. Remove remaining free chlorine by placing the solution in the direct sunlight. After several hours, measure total chlorine residual. Do not use until last trace of chlorine has been removed.

7. Starch indicator.

E. Procedure

1. Measure a 500-mL sample into each of five to ten 1,000-mL flasks or bottles.

2. To the first flask, add an amount of chlorine that leaves no residual at the end of the contact time. Mix.

3. Add increasing amounts of chlorine to successive portions of the sample and mix.

4. Measure residual chlorine after the specific contact time. Record results.

5. On graph paper, plot the residual chlorine (or the amount of chlorine consumed) versus chlorine dosage.

F. Precautions

1. Dose sample portions at time intervals that will leave you enough time for chlorine residual testing at predetermined contact times.

2. Conduct test over the desired contact time. If test objective is to duplicate your plant contact time, then match plant detention time as closely as possible.

3. Keep samples in the dark, protected from sunlight.

4. Keep temperature as constant as possible.

5. Sterilize all glassware if test has a bacteriologic objective.

G. Example

A raw water sample was collected from a river to determine chlorine demand.

Contact time = 30 minutes (plant detention time)

pH = 7.6

Temperature = 15°C

(Chlorine Demand)

OUTLINE OF PROCEDURE FOR CHLORINE DEMAND

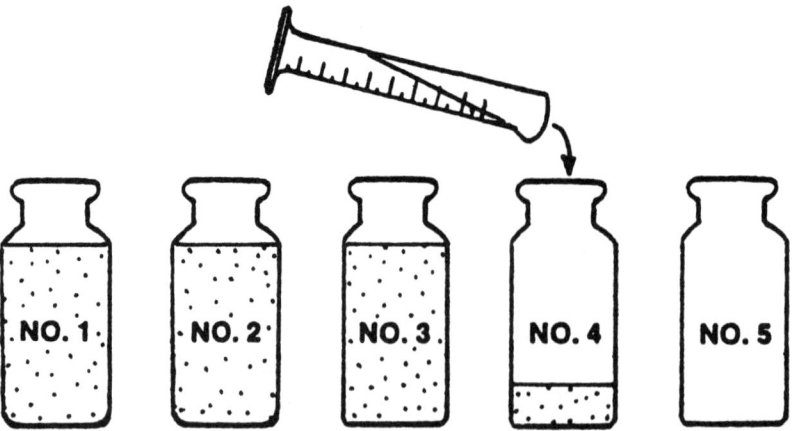

1. Measure 500 mL water sample into each container.

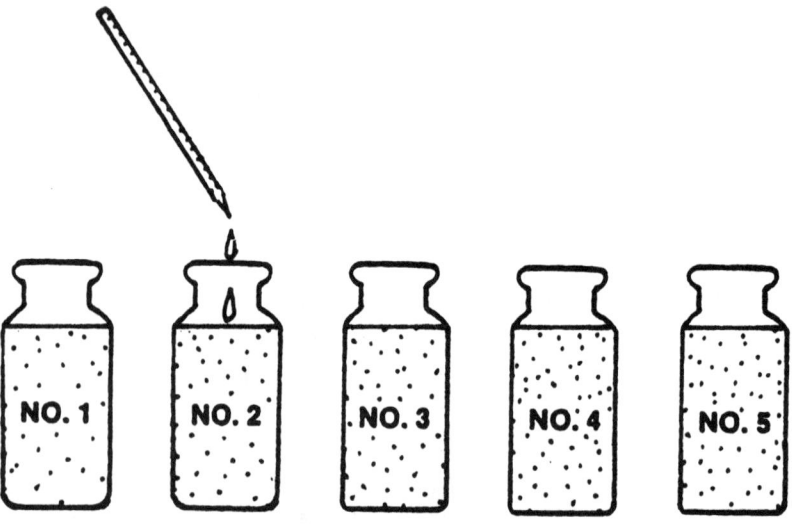

2. Add desired dosage of chlorine to each container.

3. After end of desired contact time, measure chlorine residual. Plot results on graph paper.

Results from the chlorine demand test were as follows:

Flask No.	Chlorine Added, mg/L	Total Residual Chlorine after 30 min, mg/L
1	0.5	0.36
2	1.0	0.82
3	1.5	1.14
4	2.0	0.60
5	2.5	0.75
6	3.0	1.25

H. Calculation

Calculate the chlorine demand by using the formula

$$\text{Chlorine Demand, mg/}L = \text{Chlorine Added, mg/}L - \text{Total Residual Chlorine, mg/}L$$

Flask No.	Chlorine Added, mg/L	$-$	Total Residual Chlorine, mg/L	$=$	Chlorine Demand, mg/L
1	0.5	$-$	0.36	$=$	0.14
2	1.0	$-$	0.82	$=$	0.18
3	1.5	$-$	1.14	$=$	0.36
4	2.0	$-$	0.60	$=$	1.40
5	2.5	$-$	0.75	$=$	1.75
6	3.0	$-$	1.25	$=$	1.75

Valuable knowledge can be gained by plotting the data from test results. Figure 11.10 is a plot of the chlorine added vs. the free chlorine residual after 30 minutes. By drawing a smooth line between the plotting points, a typical breakpoint chlorination curve is produced.

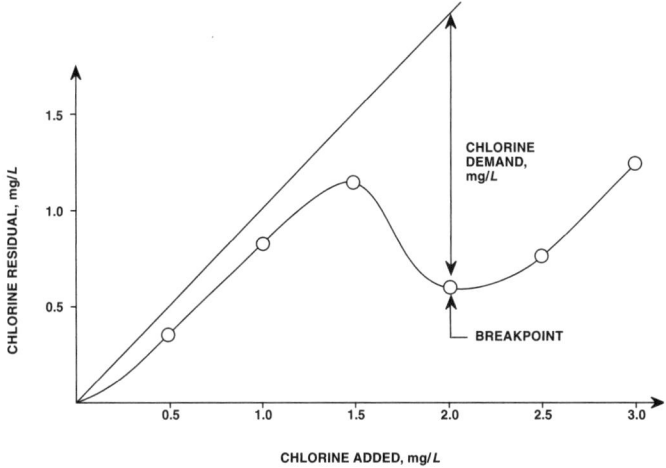

Fig. 11.10   Breakpoint chlorination curve

(Chlorine Demand)

Figure 11.11 is a plot of the chlorine demand curve. Note that the far right end of the curve is flat which indicates that the chlorine demand has been satisfied. The curve reveals that the chlorine demand will increase as chlorine is added to water until you have gone past the breakpoint.

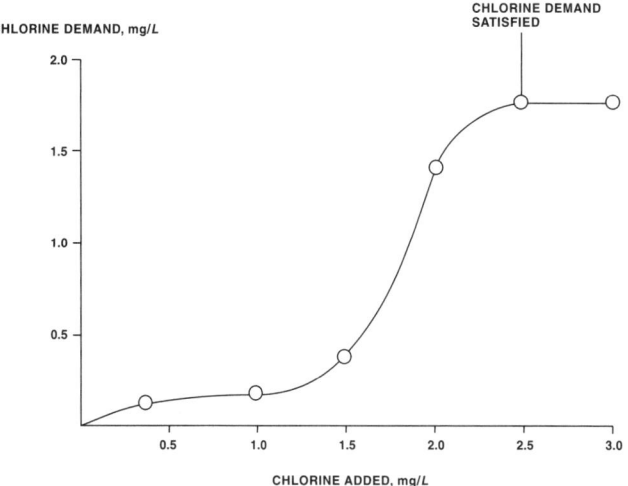

Fig. 11.11   Plot of chlorine demand on graph paper

I. Reference

See page 2-40, STANDARD METHODS, 20th Edition.

QUESTIONS

Write your answers in a notebook and then compare your answers with those on page 569.

11.3E   What conditions can cause variations in the chlorine demand?

11.3F   How does the operator use the results from the chlorine demand test?

11.3G   Calculate the chlorine demand for a raw water sample if chlorine is added at 2.0 mg/L and the free residual chlorine after 30 minutes was 0.4 mg/L.

Please answer the discussion and review questions next.

DISCUSSION AND REVIEW QUESTIONS

Chapter 11.   LABORATORY PROCEDURES

(Lesson 3 of 5 Lessons)

Write the answers to these questions in your notebook before continuing. The question numbering continues from Lesson 2.

12. What precautions must be taken with the sample when conducting the alkalinity test?

13. What could be the causes of tastes and odors that might result from chlorination?

14. What precautions should be exercised when performing a chlorine demand test?

CHAPTER 11.   LABORATORY PROCEDURES

(Lesson 4 of 5 Lessons)

4.   Coliform Bacteria

A.   Discussion

An improperly treated or unprotected water supply may contain microorganisms that are pathogenic, that is, capable of producing disease. Testing for specific pathogenic microorganisms (pathogens which cause diseases such as typhoid, dysentery, cryptosporidiosis, or giardiasis) is very time-consuming and requires special techniques and equipment. So instead of testing for specific pathogens, water is generally analyzed for the presence of an "indicator organism," the coliform group of bacteria.

The presence or absence of coliform bacteria is a good index of the degree of bacteriologic safety of a water supply. In general, coliform bacteria can be divided into fecal or non-fecal groups. Fecal coliform bacteria occur normally in the intestines of humans and other warm-blooded animals. They are discharged in great numbers in human and animal wastes. Coliforms are generally more hardy than true pathogenic bacteria and their absence from water is thus a good indication that the water is bacteriologically safe for human consumption. The presence of coliforms, however, indicates the potential presence of pathogenic organisms that may have entered the water with them, and suggests that the water is not safe to drink.

The coliform group of bacteria includes all the aerobic and *FACULTATIVE*[10] anaerobic gram-negative, nonspore-forming, rod-shaped bacteria that ferment lactose (a sugar) within 48 hours at 35°C (human body temperature). The fecal coliform can grow at a higher temperature (45°C) than the non-fecal coliform.

The bacteriological quality of water supplies is subject to control by federal, state, and local agencies, all of which are governed by the rules and regulations contained in the Safe Drinking Water Act. This law stipulates the methods to be used, the number of samples required, and the maximum levels allowed for coliform organisms in drinking water supplies. The number of samples required is generally based on population served by the water system.

The primary standards (MCLs) for coliform bacteria have been established to indicate the likely presence of disease-causing bacteria. The Total Coliform Rule uses a presence-absence approach (rather than an estimation of how many coliforms are present) to determine compliance with the standards. The Maximum Contaminant Level Goal (MCLG) for coliform is zero.

Compliance under the Total Coliform Rule is determined on a monthly basis. In general, coliform must be absent in at least 95 percent of the samples for those larger systems that collect more than 40 samples per month. Smaller systems that collect fewer than 40 samples cannot have coliform-positive results in more than one sample per month.

Whenever a routine coliform sample is coliform-positive, the regulation calls for determination of the presence of fecal coliform (*E. coli*) and for repeat sampling. Whenever fecal coliform (*E. coli*) are present in the routine sample and the repeat samples are coliform-positive, there is a violation of the MCL, additional repeat sampling is required, and notification of both the state and the public is required.

Also see *WATER TREATMENT PLANT OPERATION*, Volume II, Sections 22.22, "Microbial Standards," and 22.220, "Total Coliform Rule."

B.   Test Methods

Approved water testing procedures for total coliform bacteria are: (1) the Multiple Tube Fermentation Method (sometimes called the Most Probable Number or MPN procedure), (2) the Membrane Filter (MF) Method, (3) Presence-Absence Method (P-A), (4) Colilert™ Method, and (5) Colisure Method. Contact your state regulatory agency to determine if additional testing procedures have been approved since the publication of this manual.

1.   Multiple Tube Fermentation Method (MPN)

The multiple tube coliform test has been a standard method for determining the coliform group of bacteria since 1936. In this procedure tubes of lauryl tryptose broth are inoculated with a water sample. Lauryl tryptose broth contains lactose which is the source of carbohydrates (sugar). The coliform density is then calculated from statistical probability formulas that predict the most probable number (MPN) of coliforms in a 100-mL sample necessary to produce certain combinations of "gas-positive" (gas forming) and gas-negative tubes in the series of inoculated tubes.

There are three distinct test states for coliform testing using the Multiple Tube Fermentation Method—the Presumptive Test, the Confirmed Test, and the Completed Test. Each test makes the coliform test more valid and specific. These tests are described in detail in the following paragraphs.

2.   Membrane Filter (MF) Method

This method was introduced as a tentative method in 1955 and has been an approved test for coliform bacteria since 1960. The basic procedure involves filtering a known volume of water through a membrane filter of optimum pore size for full coliform bacteria retention. As the water passes through the microscopic pores, bacteria are entrapped on the upper surface of the filter. The membrane filter is then placed in contact with either a paper pad saturated with liquid medium or di-

---

[10] *Facultative (FACK-ul-TAY-tive).   Facultative bacteria can use either dissolved molecular oxygen or oxygen obtained from food materials such as sulfate or nitrate ions. In other words, facultative bacteria can live under aerobic or anaerobic conditions.*

rectly over an agar (gelatin-like) medium to provide proper nutrients for bacterial growth. Following incubation under prescribed conditions of time, temperature, and humidity, the cultures are examined for coliform colonies that are counted directly and recorded as a density of coliforms per 100 mL of water sample.

There are certain important limitations to membrane filter methods. Some types of samples cannot be filtered because of excessive turbidity, high non-coliform bacterial densities, or heavy metal (bactericidal) compounds. In addition, coliforms contained in chlorinated supplies sometimes do not give characteristic reactions on the media and hence special procedures must then be used.

3.   Presence-Absence (P-A) Method

The presence-absence (P-A) method for total coliform is a simple modification of the multiple tube (MPN) method. Both lactose broth and lauryl tryptose broth are the essential ingredients of the P-A media. In addition, bromocresol purple is added as a pH indicator.

The procedure involves the addition of 100 mL of sample to 50 mL of sterile P-A broth in a 250-mL milk dilution bottle. This inoculated media is then incubated at 35°C and inspected after 24 and 48 hours for an acid reaction. A distinct yellow color forms in the media when acid conditions exist following lactose fermentation. If gas is also produced, gently shaking the bottle will result in a foaming reaction. Any amount of gas and/or acid (yellow color) constitutes a positive presumptive test requiring confirmation in the same manner as the MPN method.

4.   Colilert™ Method

Colilert, a registered trademark of IDEXX Laboratories, provides simultaneous detection, specific identification, and confirmation of total coliforms and *E. coli* in water. Colilert is a specifically designed reagent formulation that is specific only to coliform bacteria. It provides a specific indicator for the target microbes, total coliforms, and *E. coli*. As the media's nutrients are metabolized, yellow color and fluorescence are released, confirming the presence of total coliforms and *E. coli* respectively.

5.   Colisure Method

The Colisure test is a presence-absence test for coliform bacteria. A sample of water is added to a dehydrated medium and, after 28 to 48 hours, the medium is examined for the presence of coliforms. (The Colisure test is available from IDEXX Laboratories, 1 IDEXX Drive, Westbrook, ME 04092.)

C.   What Is Tested

	Common Range Total Coliforms per 100 mL
Surface waters	50 to 1,000,000
Treated water supplies	0
Well water	0 to 50

D.   Materials Required

1.   Sampling Bottles

Bottles of glass or other material which are watertight, resistant to the solvent action of water, and capable of being sterilized may be used for bacteriologic sampling. Plastic bot-

tles made of nontoxic materials are also satisfactory and eliminate the possibility of breakage during transport. The bottles should hold a sufficient volume of sample for all tests, permit proper washing, and maintain the samples uncontaminated until examinations are complete.

Before sterilization by autoclave, add 0.1 mL 10 percent sodium thiosulfate per 4-ounce bottle (120 mL). This will neutralize a sample containing about 15 mg/L residual chlorine. If the residual chlorine is not neutralized, it would continue to be toxic to the coliform organisms remaining in the sample and give false results.

When filling bottles with sample, do not flush out the sodium thiosulfate or contaminate the bottle or sample. Fill bottles approximately three-quarters full and start the test in the laboratory within six hours. If the samples cannot be processed within one hour, they should be held below 10°C for not longer than six hours.

2.   Glassware

All glassware must be thoroughly cleansed using a suitable detergent and hot water (160°F or 71°C), rinsed with hot water (180°F or 82°C) to remove all traces of residual detergent, and finally rinsed with distilled or deionized water.

3.   Water

Only distilled water or demineralized water which has been tested and found free from traces of dissolved metals and bactericidal and inhibitory compounds may be used for preparation of culture media.

4.   Buffered[11] Dilution Water

Prepare a stock solution by dissolving 34 grams of KH_2PO_4 in 500 mL distilled water, adjusting the pH to 7.2 with 1 N NaOH and dilute to one liter. Prepare dilution water by adding 1.25 mL of the stock solution and 5.0 mL magnesium sulfate (50 grams $MgSO_4 \cdot 7 H_2O$ dissolved in one liter of water) to 1 liter distilled water. This solution can be dispersed into various size dilution blanks or used as a sterile rinse for the membrane filter test.

5.   Media Preparation

Careful media preparation is necessary for meaningful bacteriological testing. Attention must be given to the quality, mixing, and sterilization of the ingredients. The purpose of this care is to ensure that if the bacteria being tested for are indeed present in the sample, every opportunity is presented for the development and ultimate identification. Bacteriological identification is often done by noting changes in the medium; consequently, the composition of the medium must be standardized. Much of the tedium of media preparation can be avoided by purchase of dehydrated media (Difco, BBL, or equivalent) from local scientific supply houses. The operator is advised to make use of these products, and, if only a limited amount of testing is to be done, consider using tubed, prepared media.

MEDIA—MPN (TOTAL COLIFORM)

a.   Lauryl Tryptose Broth

For the Presumptive Coliform Test, dissolve the recommended amount of the dehydrated media in distilled water. Dispense solution into fermentation tubes

---

[11] *Buffer.   A solution or liquid whose chemical makeup neutralizes acids or bases without a great change in pH.*

(Coliform)

containing an inverted glass vial (see illustration of the tube with vial on page 543). For 10-mL water portions from samples, double-strength media is required while all other inoculations require single strength. Directions for preparation are given on the media bottle label.

b. Brilliant Green Bile (BGB) Broth

For the Confirmed Coliform Test, dissolve 40 grams of the dehydrated media in one liter of distilled water. Dispense and sterilize as with Lauryl Tryptose Broth.

c. EC Broth

Prepare this media following the instructions on the bottle.

MEDIA—MEMBRANE FILTER METHOD (TOTAL COLIFORM)

a. M-Endo Broth

Prepare this media by dissolving 48 grams of the dehydrated product in one liter of distilled water which contains 20 mL of ethyl alcohol[12] per liter. Heat solution to boiling only—*DO NOT AUTOCLAVE.* Promptly remove solution from heat and cool. Prepared media should be stored in a refrigerator and used within 96 hours.

b. LES Endo Agar

Prepare this media, used for the two-step procedure, following the instructions on the bottle.

c. M-FC Media

Rehydrate in distilled water containing 10 mL of 1 percent rosolic acid in 0.2 N NaOH. Heat just to boiling then cool to below 45°C. *DO NOT AUTOCLAVE.* Prepared media stored in the refrigerator should be used within 96 hours.

The rosolic acid solution should be stored in the dark and discarded after two weeks or sooner if its color changes from dark red to muddy brown.

MEDIA—PRESENCE-ABSENCE (P-A) METHOD

a. P-A Broth

Prepare this media following the instructions on the media bottle label.

b. Brilliant Green Bile Broth

Same as for MPN method.

c. EC Broth

Same as for MPN method.

MEDIA—COLILERT™ METHOD

a. Available from: IDEXX Laboratories
1 IDEXX Drive
Westbrook, ME 04092
1-800-321-0207

6.   Media Storage

Culture media should be prepared in batches of such size that the entire batch will be used in less than one week.

7.   Autoclaving

Steam autoclaves are used for the sterilization of the liquid media and associated apparatus. They sterilize (kill all organisms) at a relatively low temperature of 121°C within 15 minutes using moist heat.

Components of the media, particularly sugars such as lactose, may decompose at higher temperatures or longer heating times. For this reason adherence to time and temperature schedules is vital. The maximum elapsed time for exposure of the media to any heat (from the time the autoclave door is closed to unloading) is 45 minutes. Preheating the autoclave can reduce total heating time.

Autoclaves operate in a manner similar to the familiar kitchen pressure cooker:

a. Water is heated in a boiler to produce steam.

b. The steam is vented to drive out air.

c. The steam vent is closed when the air is gone.

d. Continued heat raises the pressure to 15 lbs/sq in (103.4 kPa or 1.05 kg/sq cm); (at this pressure, pure steam has a temperature of 121°C at sea level only).

e. The heat and pressure are maintained for 15 minutes.

f. The heat is turned off.

g. The steam vent is opened slowly to vent steam until atmospheric pressure is reached. (Fast venting will cause the liquids to boil and overflow tubes.)

h. Sterile material is removed to cool.

In autoclaving fermentation tubes, a vacuum is formed in the inner tubes. As the tubes cool, the inner tubes are filled with sterile medium. Capture of gas in this inner tube from the culture of bacteria is the evidence of fermentation and is recorded as a *POSITIVE TEST.*

---

[12] *In some states the ethyl alcohol used in bacteriological media preparation CANNOT be the specially denatured alcohol sold by supply houses to people without an alcohol permit. One way to obtain suitable ethanol under these circumstances is to buy a brand of ethanol sold for human consumption.*

E. Procedure for Testing Total Coliform Bacteria

Multiple Tube Fermentation Method

1. General Discussion

The test for coliform bacteria is used to measure the suitability of a water for human use. The test is not only useful in determining the bacterial quality of a finished water, but it can be used by the operator in the treatment plant as a guide to achieving a desired degree of treatment.

Coliform bacteria are detected in water by placing portions of a sample of the water in lauryl tryptose broth. Lauryl tryptose broth is a standard bacteriological media containing lactose (milk) sugar in tryptose broth. The coliform bacteria are those which can grow in this media at 35°C temperature and are able to ferment and produce gas from the lactose within 48 hours. Thus, to detect these bacteria the operator need only inspect the fermentation tubes for gas. A schematic of the coliform test procedure is shown in Figure 11.12.

In states where drinking water coliform samples must be 100 mL in volume, the five-tube multiple tube fermentation method is not appropriate. The test uses ten tubes with a 10-mL sample in each tube.

2. Materials Needed

FOR UNTREATED WATER SAMPLES

a. Fifteen sterile tubes containing 10 mL of lauryl tryptose broth are needed for each sample. Use five tubes for each dilution.

b. Dilution tubes or blanks containing 99 mL of sterile buffered distilled water.

c. A quantity of 1-mL and 10-mL serological pipets. The 1-mL pipets should be graduated in 0.1-mL increments.

d. Incubator set at 35° ± 0.5°C.

e. Thermometer verified to be accurate by comparison with a National Bureau of Standards (NBS) certified thermometer.

FOR DRINKING WATER SAMPLES

a. Ten sterile tubes of 10-mL, double-strength lauryl tryptose broth are needed if 10 mL of sample is added to each tube. Ten mL of lauryl tryptose broth is required in all tubes containing one mL or less of sample.

b. Sterile 10-mL pipet for each sample.

c. Incubator set at 35° ± 0.5°C.

d. Water bath set at 44.5° ± 0.2°C.

e. Thermometer verified to be accurate by comparison with a National Bureau of Standards (NBS) certified thermometer, or equivalent.

3. Technique for Inoculation of Sample (Figures 11.13 and 11.14)

All inoculations and dilutions of water samples must be accurate and made so that no contaminants from the air, equipment, clothes, or fingers reach the sample, either directly or by way of a contaminated pipet. Clean, sterile pipets must be used for each separate sample.

FOR UNTREATED WATER SAMPLES

a. Shake the sample bottle vigorously 20 times before removing sample volumes.

b. Pipet 10 mL of sample directly into each of the first five tubes. Each tube must contain 10 mL lauryl tryptose broth (double strength).

NOTE: You must realize that the sample volume applied to the first five tubes will depend upon the type of water being tested. The sample volume applied to each tube can vary from 10 mL for high-quality waters to as low as 0.00001 mL (applied as 1 mL of a diluted sample) for highly polluted raw water samples.

NOTE: When delivering the sample into the culture medium, deliver sample portions of 1 mL or less down into the culture tube near the surface of the medium. *DO NOT* deliver small sample volumes at the top of the tube and allow them to run down inside the tube; too much of the sample will fail to reach the culture medium.

NOTE: Use 10-mL pipets for 10-mL sample portions, and 1-mL pipets for portions of 1 mL or less. Handle sterile pipet only near the mouthpiece, and protect the delivery end from external contamination.

c. Pipet 1 mL of water sample into each of the next five lauryl tryptose broth (single strength) tubes.

d. Pipet 1/10 mL (0.1 mL) of water sample into each of the next five lauryl tryptose broth (single strength) tubes. This makes a 0.1 (1 to 10) dilution.

At this point you have 15 tubes inoculated and can place these three sets of tubes in the incubator; however, your sample specimen may show gas production in all fermentation tubes. This means your sample was not diluted enough and you have no usable results.

e. To make a 1/100 (0.01) dilution, add 1 mL of well-mixed water sample to 99 mL of sterile buffered dilution water. Mix thoroughly by shaking. Add 1 mL from this bottle directly into each of five more lauryl tryptose broth tubes.

f. To make a 1/1,000 (0.001) dilution, add 0.1 mL from the 1/100 dilution bottle directly into each of five more lauryl tryptose broth tubes.

The first time a sample is analyzed, 25 tubes of lauryl tryptose broth should be prepared. Once you find out what dilutions give usable results for determining the MPN index, you will only need to prepare 15 tubes to analyze subsequent samples from the same source.

(Coliform)

OUTLINE OF PROCEDURE FOR TOTAL COLIFORM

1. Presumptive Test

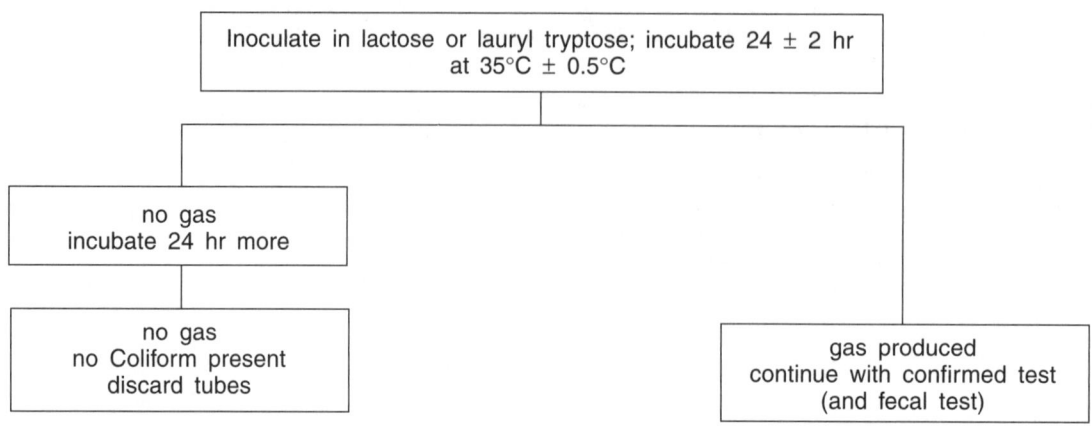

2. Confirmed Test

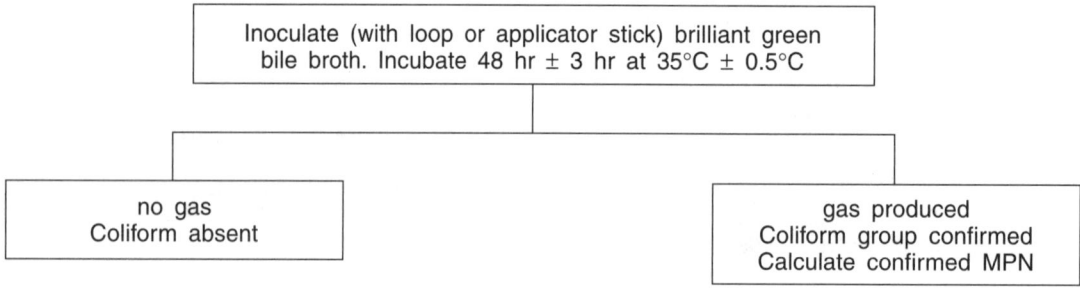

FECAL COLIFORM

1. Presumptive Test
(same as for TOTAL COLIFORM)

2. Fecal Coliform Test

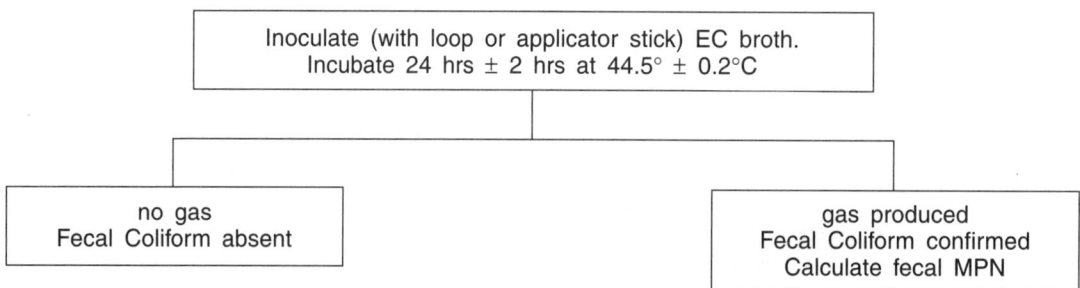

Fig. 11.12   Schematic outline of test procedure for Total Coliform—Multiple Tube Fermentation Method

(Coliform)

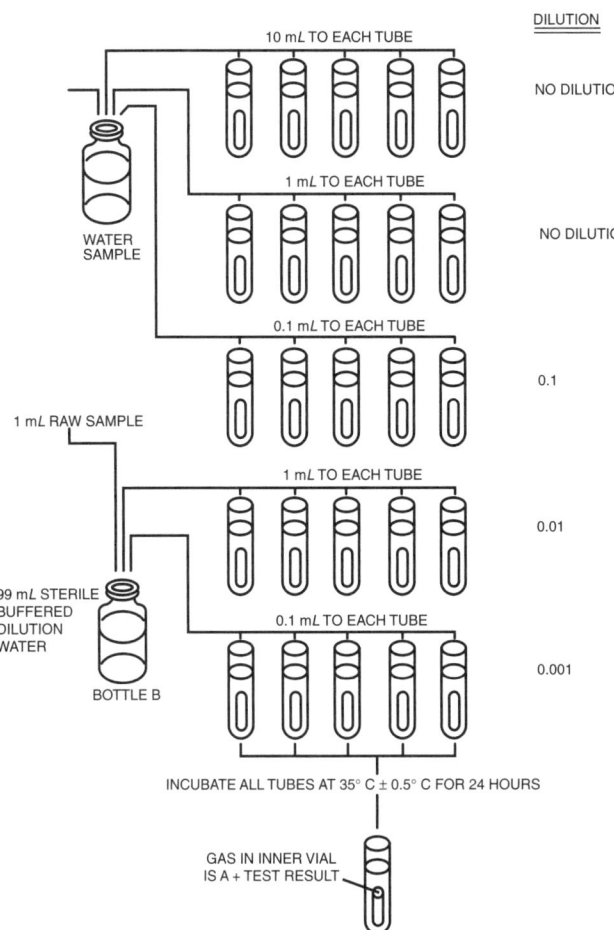

Fig. 11.13   Coliform bacteria test—raw water

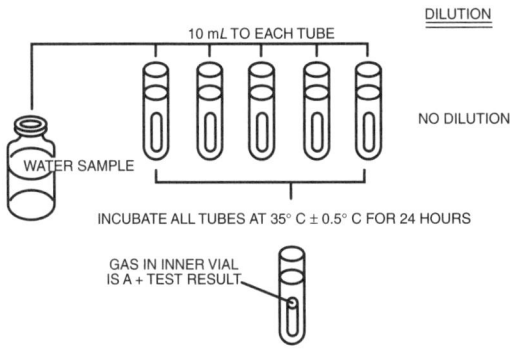

Fig. 11.14   Coliform bacteria test—drinking water

FOR DRINKING WATER SAMPLES

a. Shake the sample bottle vigorously 20 times before removing a sample volume.

b. Pipet 10 mL of sample directly into each of ten tubes containing 10 mL of double-strength lauryl tryptose broth.

4. Incubation (Total Coliform)

a. 24-Hour Lauryl Tryptose (LT) Broth Presumptive Test

Place all inoculated LT broth tubes in 35°C ± 0.5°C incubator. After 24 ± 2 hours have elapsed, examine each tube for gas formation in inverted vial (inner tube). Mark plus (+) on report form such as shown in Figure 11.15 for all tubes that show presence of gas. Mark minus (–) for all tubes showing no gas formation. Immediately perform Confirmation Test on all positive (+) tubes (see paragraph c. below, "24-Hour Brilliant Green Bile (BGB) Confirmation Test"). The negative (–) tubes must be reincubated for an additional 24 hours.

b. 48-Hour Lauryl Tryptose Broth Presumptive Test

Record both positive and negative tubes at the end of 48 ± 3 hours. Immediately perform Total Coliform Confirmation Test on all new positive tubes. For drinking water samples, all positive tubes should also be tested for fecal coliforms (*E. coli*).

c. 24-Hour Brilliant Green Bile (BGB) Confirmation Test

Confirm all presumptive tubes that show gas at 24 or 48 hours. Transfer, with the aid of a sterile 3-mm platinum wire loop (sterile wood applicator or disposable loops may be used also), one loop-full of the broth from the lauryl tryptose broth tubes showing gas, and inoculate a corre-

FOR UNTREATED WATER

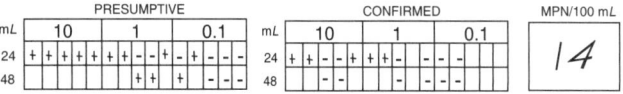

mL = mL OF SAMPLE: 10 mL, 1 mL, AND 0.1 mL
24 = RESULTS AFTER 24 HOURS OF INCUBATION
48 = RESULTS AFTER 48 HOURS OF INCUBATION

FOR DRINKING WATER

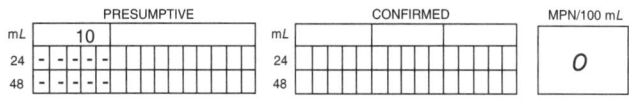

Fig. 11.15   Recorded coliform test results

(Coliform)

sponding tube of BGB broth by mixing the loop of broth in the BGB broth. Discard all positive lauryl tryptose broth tubes after transferring is completed.

Always sterilize inoculation loops and needles in flame immediately before transfer of culture; do not lay loop down or touch it to any nonsterile object before making the transfer. After sterilization in a flame, allow sufficient time for cooling, in the air, to prevent the heat of the loop from killing the bacterial cells being transferred. Sterile wood applicator sticks also are used to transfer cultures, especially in the field where a flame is not available for sterilization. If using hardwood applicators, sterilize by autoclaving before use and discard after each transfer.

After 24 hours have elapsed, inspect each of the BGB tubes for gas formation. Those with any amount of gas are considered positive and are so recorded on the data sheet. Negative BGB tubes are reincubated for an additional 24 hours.

d. 48-Hour Brilliant Green Bile Confirmation Test

(1) Examine tubes for gas at the end of the 48 ± 3 hour period. Record both positive and negative tubes.

(2) Complete reports by determining MPN Index and recording MPN on worksheets.

5. Fecal Coliform (FC) Test

For drinking water samples all presumptive tubes that show gas at 24 or 48 hours must be tested for fecal coliform. This test more reliably indicates the potential presence of pathogenic organisms than do tests for total coliform group of organisms. The procedure described is an *ELEVATED TEMPERATURE TEST FOR FECAL COLIFORM BACTERIA.*

a. Materials Needed

Equipment required for the tests are the same as those required for the 24-Hour Lactose Broth Presumptive Test, plus a water bath set at 44.5 ± 0.2°C, EC Broth media, and a thermometer certified against an NBS thermometer.

b. Procedure

(1) Run lactose broth or lauryl tryptose broth presumptive test.

(2) After 24 hours temporarily retain all gas-positive tubes.

(3) Label a tube of EC broth to correspond with each gas-positive tube of broth from presumptive test.

(4) Shake or mix positive presumptive tubes by rotating them. Transfer one loop-full of culture from each gas-positive culture in presumptive test to the correspondingly labeled tube of EC broth.

(5) Incubate EC broth tubes 24 ± 2 hours at 44.5°C ± 0.2°C in a water bath with water depth sufficient to come up at least as high as the top of the culture medium in the tubes. Place in water bath *AS SOON AS POSSIBLE* after inoculation and always within 30 minutes after inoculation.

(6) After 24 hours remove the rack of EC cultures from the water bath, shake gently, and record gas production for each tube. Gas in any quantity is a positive test.

(7) As soon as results are recorded, discard all tubes. This is a 24-hour test for EC broth inoculations and not a 48-hour test.

(8) Transfer any additional 48-hour gas-positive tubes from the presumptive test to correspondingly labeled tubes of EC broth. Incubate for 24 ± 2 hours at 44.5°C ± 0.2°C and record results on data sheet.

(9) Codify results using the same procedure as for total coliforms and determine MPN of fecal coliforms per 100 mL of sample (Figure 11.15).

EXAMPLE:

mL portion	10	1	0.1
Readings	5	2	1

Read MPN as   70 per 100 mL from Table 11.3

Report results as MPN of fecal coliforms = 70 per 100 mL

6. Recording Results

Results should be recorded on data sheets prepared especially for this test. Examples are shown in Figure 11.15.

7. Method of Calculating the Most Probable Number (MPN)

Select the highest dilution or inoculation with all positive tubes, before a negative tube occurs, plus the next two dilutions.

EXAMPLE 1 (UNTREATED WATER)—Select the underlined inoculations.

mL of sample	10	1	0.1	0.01	0.001
dilutions	0	0	−1	−2	−3
positive tubes	5	1	0	0	0

Read MPN as 30 per 100 mL from Table 11.3
Report MPN as 30 per 100 mL

EXAMPLE 2 (UNTREATED WATER)—Select the underlined inoculations (see Figure 11.16).

mL of sample	10	1	0.1	0.01	0.001
dilutions	0	0	−1	−2	−3
positive tubes	5	5	2	0	0

Read MPN as 50 per 100 mL from Table 11.3
Report results as 500/100 mL

We added one zero to 50 because we started with the 1 mL sample and Table 11.3 begins with one dilution column to the left.

EXAMPLE 3 (DRINKING WATER SAMPLE)—(see Figure 11.17).

mL of sample	10
positive tubes	1

Read MPN as 1.1 per 100 mL from Table 11.4

(Coliform)

TABLE 11.3   MPN INDEX FOR VARIOUS COMBINATIONS OF POSITIVE AND NEGATIVE RESULTS IN A PLANTING SERIES OF FIVE 10-mL, FIVE 1-mL, AND FIVE 0.1-mL PORTIONS OF SAMPLE

Number of tubes giving positive reaction out of			MPN Index
Five 10-mL portions	Five 1-mL portions	Five 0.1-mL portions	(organisms) per 100 mL
0	0	0	<2
0	0	1	2
0	0	2	4
0	1	0	2
0	1	1	4
0	1	2	5
0	2	0	4
0	2	1	6
0	3	0	6
1	0	0	2
1	0	1	4
1	0	2	6
1	0	3	8
1	1	0	4
1	1	1	6
1	1	2	8
1	2	0	6
1	2	1	8
1	2	2	10
1	3	0	8
1	3	1	10
1	4	0	11
2	0	0	4
2	0	1	7
2	0	2	9
2	0	3	12
2	1	0	7
2	1	1	9
2	1	2	12
2	2	0	9
2	2	1	12
2	2	2	14
2	3	0	12
2	3	1	14
2	4	0	15
3	0	0	8
3	0	1	11
3	0	2	13
3	1	0	11
3	1	1	14
3	1	2	17
3	1	3	20
3	2	0	14
3	2	1	17
3	2	2	20

(Coliform)

TABLE 11.3   MPN INDEX FOR VARIOUS COMBINATIONS OF POSITIVE AND NEGATIVE RESULTS IN A PLANTING SERIES OF FIVE 10-mL, FIVE 1-mL, AND FIVE 0.1-mL PORTIONS OF SAMPLE (continued)

Number of tubes giving positive reaction out of			MPN Index
Five 10-mL portions	Five 1-mL portions	Five 0.1-mL portions	(organisms) per 100 mL
3	3	0	17
3	3	1	21
3	4	0	21
3	4	1	24
3	5	0	25
4	0	0	13
4	0	1	17
4	0	2	21
4	0	3	25
4	1	0	17
4	1	1	21
4	1	2	26
4	2	0	22
4	2	1	26
4	2	2	32
4	3	0	27
4	3	1	33
4	3	2	39
4	4	0	34
4	4	1	40
4	5	0	41
4	5	1	48
5	0	0	23
5	0	1	30
5	0	2	40
5	0	3	60
5	0	4	80
5	1	0	30
5	1	1	50
5	1	2	60
5	1	3	80
5	2	0	50
5	2	1	70
5	2	2	90
5	2	3	130
5	2	4	150
5	2	5	180
5	3	0	80
5	3	1	110
5	3	2	140
5	3	3	170
5	3	4	210
5	3	5	250
5	4	0	130
5	4	1	170
5	4	2	220
5	4	3	280

(Coliform)

**TABLE 11.3   MPN INDEX FOR VARIOUS COMBINATIONS OF POSITIVE AND NEGATIVE RESULTS
IN A PLANTING SERIES OF FIVE 10-mL, FIVE 1-mL, AND
FIVE 0.1-mL PORTIONS OF SAMPLE** (continued)

Number of tubes giving positive reaction out of			MPN Index
Five 10-mL portions	Five 1-mL portions	Five 0.1-mL portions	(organisms) per 100 mL
5	4	4	350
5	4	5	430
5	5	0	240
5	5	1	300
5	5	2	500
5	5	3	900
5	5	4	1,600
5	5	5	≥1,600

For Example No. 2

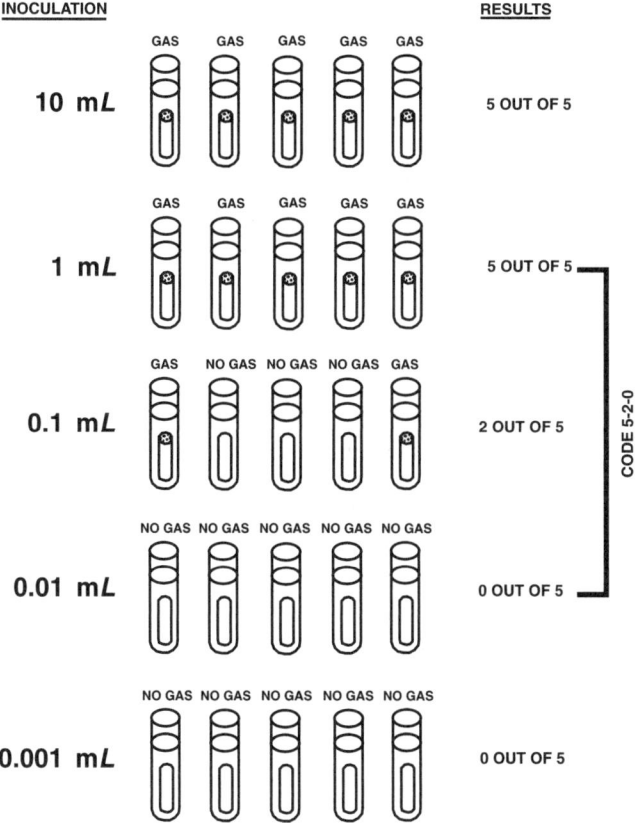

Fig. 11.16   Results of coliform test—untreated water

For Example No. 3

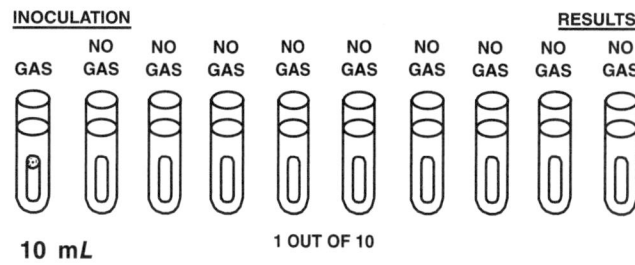

Fig. 11.17   Results of coliform test—drinking water

TABLE 11.4   MPN INDEX FOR VARIOUS COMBINA-TIONS OF POSITIVE AND NEGATIVE RESULTS WHEN TEN 10-mL PORTIONS ARE USED

Number of tubes giving positive reaction out of ten 10-mL portions	MPN Index per 100 mL
0	<1.1
1	1.1
2	2.2
3	3.6
4	5.1
5	6.9
6	9.2
7	12.0
8	16.1
9	23.0
10	>23.0

(Coliform)

When you wish to summarize with a single MPN value the results from a series of samples, the arithmetic mean, the median, or the geometric mean may be used (see following sections for *FORMULAS* and *EXAMPLES*).

FORMULAS

The MPN for combinations not appearing in the table given or for other combinations of tubes or dilutions may be estimated by the formula:

$$\text{MPN/100 mL} = \frac{(\text{No. of positive tubes})(100)}{\sqrt{\left(\begin{array}{c}\text{mL sample in}\\\text{negative tubes}\end{array}\right) \times \left(\begin{array}{c}\text{mL sample in}\\\text{all tubes}\end{array}\right)}}$$

Coliform results may be summarized with a single MPN value by using the arithmetic mean, median, or geometric mean. The arithmetic mean is often called the average.

$$\text{Mean} = \frac{\text{Sum of Items or Values}}{\text{Number of Items or Values}}$$

The median is the middle value in a set or group of data. There are just as many values larger than the median as there are smaller than the median. To determine the median, the data should be written in ascending (increasing) or descending (decreasing) order and the middle value identified.

$$\text{Median} = \text{Middle Value of a Group of Data}$$

The median and geometric mean are used when a sample contains a few very large values. This frequently happens when measuring the MPN of raw water. If **X** is a measurement (MPN) and **n** is the number of measurements, then

$$\text{Geometric Mean} = [(X_1)(X_2)(X_3) \ldots\ldots\ldots\ldots (X_n)]^{1/n}$$

This calculation can be easily performed on many electronic calculators using the power function, y^x.

EXAMPLE 4

The results from an MPN test using five fermentation tubes for each dilution on a sample of water were as follows:

Sample Size, mL	10	1.0	0.1
Number of Positive Tubes Out of Five Tubes	3+	1+	0+

Determine the information necessary to solve the formula:

$$\text{MPN/100 mL} = \frac{(\text{No. of positive tubes})(100)}{\sqrt{\left(\begin{array}{c}\text{mL sample in}\\\text{negative tubes}\end{array}\right) \times \left(\begin{array}{c}\text{mL sample in}\\\text{all tubes}\end{array}\right)}}$$

1   Number of positive tubes. There were 3 positive tubes with 10 mL of sample and 1 positive tube with 1 mL of sample. Therefore, 3 + 1 = 4 positive tubes.

2.   Determine the mL of sample in negative tubes.

Sample Size, mL	Number of Negative Tubes	mL of Sample in Negative Tubes		
10 mL	2–	(10 mL)(2)	=	20 mL
1.0 mL	4–	(1.0 mL)(4)	=	4 mL
0.1 mL	5–	(0.1 mL)(5)	=	0.5 mL
		Total	=	24.5 mL

3.   Determine the mL of sample in all tubes.

Sample Size, mL	Number of Tubes	mL of Sample in All Tubes		
10 mL	5	(10 mL)(5)	=	50 mL
1.0 mL	5	(1.0 mL)(5)	=	5 mL
0.1 mL	5	(0.1 mL)(5)	=	0.5 mL
		Total	=	55.5 mL

4.   Estimate the MPN/100 mL.

$$\text{MPN/100 mL} = \frac{(\text{No. of positive tubes})(100)}{\sqrt{\left(\begin{array}{c}\text{mL sample in}\\\text{negative tubes}\end{array}\right) \times \left(\begin{array}{c}\text{mL sample in}\\\text{all tubes}\end{array}\right)}}$$

$$= \frac{(4)(100)}{\sqrt{(24.5 \text{ mL})(55.5 \text{ mL})}}$$

$$= \frac{400}{36.87}$$

$$= 10.8 \text{ MPN Coliforms/100 mL}$$

$$= 11 \text{ MPN Coliforms/100 mL}$$

EXAMPLE 5

Results from MPN tests during one week were as follows:

Day	S	M	T	W	TH	F	S
MPN/100 mL	2	8	14	6	10	26	4

Estimate the (1) mean, (2) median, and (3) geometric mean of the data in MPN/100 mL.

1.   Calculate the mean.

$$\text{Mean, MPN/100 mL} = \frac{\text{Sum of All MPNs}}{\text{Number of MPNs}}$$

$$= \frac{2 + 8 + 14 + 6 + 10 + 26 + 4}{7}$$

$$= \frac{70}{7}$$

$$= 10 \text{ MPN/100 mL}$$

2.   To determine the median, rearrange the data in ascending (increasing) order and select the middle value (three will be smaller and three will be larger in this example).

Order	1	2	3	4	5	6	7
MPN/100 mL	2	4	6	8	10	14	26
				↑			

$$\text{Median, MPN/100 mL} = \text{Middle value of a group of data}$$
$$= 8 \text{ MPN/100 mL}$$

3.   Calculate the geometric mean for the given data.

$$\text{Geometric Mean, MPN/100 mL} = [(X_1)(X_2)(X_3)(X_4)(X_5)(X_6)(X_7)]^{1/7}$$

$$= [(2)(8)(14)(6)(10)(26)(4)]^{1/7}$$

$$= [1{,}397{,}760]^{1/7}$$

$$= 7.5 \text{ MPN/100 mL}$$

4. Summary

  1. Mean   = 10 MPN/100 mL

  2. Median   = 8 MPN/100 mL

  3. Geometric Mean = 7.5 MPN/100 mL

As you can see from the summary, the geometric mean more nearly describes most of the MPNs. This is the reason why the geometric mean is sometimes used to describe the results of MPN tests when there are a few very large values.

8. Reference

See page 9-47, *STANDARD METHODS*, 20th Edition.

Membrane Filter Method

1. General Discussion

In addition to the fermentation tube test for coliform bacteria, another test is used for these same bacteria in water analysis. This test uses a cellulose ester filter, called a membrane filter, the pore size of which can be manufactured to close tolerances. Not only can the pore size be made to selectively trap bacteria from water filtered through the membrane, but nutrients can be diffused (from an enriched pad) through the membrane to grow these bacteria into colonies. These colonies are recognizable as coliform because the nutrients include fuchsin dye which peculiarly colors the colony. Knowing the number of colonies and the volume of water filtered, the operator can then compare the water tested with water quality standards.

A two-step pre-enrichment technique for chlorinated samples is included at the end of this section. Chlorinated bacteria that are still living have had their enzyme systems damaged and require a 2-hour enrichment media before contact with the selective M-Endo Media.

2. Materials Needed

  a. One sterile membrane filter having a 0.45μ pore size.

  b. One sterile 47-mm petri dish with lid.

  c. One sterile funnel and support stand.

  d. Two sterile pads.

  e. One receiving flask (side-arm, 1,000 mL).

  f. Vacuum pump, trap, suction or vacuum gage, connection sections of plastic tubing, glass "T" hose clamp to adjust pressure bypass.

  g. Forceps (round-tipped tweezers), alcohol, Bunsen burner, grease pencil.

  h. Sterile, buffered, distilled water for rinsing.

  i. M-Endo Media.

  j. Sterile pipets—two 5-mL graduated pipets and one 1-mL pipet for sample or one 10-mL pipet for larger sample. Quantity of 1-mL pipets if dilution of sample is necessary. Also, quantity of dilution water blanks if dilution of sample is necessary.

  k. One moist incubator at 35°C; auxiliary incubator dish with cover.

  l. Enrichment media—lauryl tryptose broth (for pre-enrichment technique).

  m. A binocular, wide-field, dissecting microscope is recommended for counting. The light source should be a cool white, fluorescent lamp.

3. Selection of Sample Size

The size of the sample or *ALIQUOT*[13] will be governed by the expected bacterial density. An ideal quantity will result in the growth of 20 to 80 coliform colonies, but not more than 200 bacterial colonies of all types. The table below lists suggested sample volumes for MF total coliform testing.

| | Quantities Filtered (mL) | | | | |
	100	10	1	0.1	0.01
Well Water	x				
Drinking Water	x				
Lakes	x	x	x		
Rivers		x	x	x	x

When less than 20 mL of sample is to be filtered, a small amount of sterile dilution water should be added to the funnel before filtration. This increase in water volume aids in uniform dispersion of the sample over the membrane filter.

4. Preparation of Petri Dish for Membrane Filter

  a. Sterilize forceps by dipping in alcohol and passing quickly through Bunsen burner flame to burn off the alcohol. An alcohol burner may be used also.

  b. Place sterile absorbent pad into sterile petri dish.

  c. Add 1.8 to 2.0 mL M-Endo Medium to absorbent pad using a sterile pipet. Remove excess media.

5. Procedure for Filtration of Unchlorinated Samples

All filtrations and dilutions of water samples must be accurate; no contaminants from the air, equipment, clothes, or fingers should reach the specimen either directly or by way of the contaminated pipet.

  a. Secure tubing from pump and bypass to receiving flask. Place palm of hand on flask opening and start pump. Adjust suction to 1/4 atmosphere with hose clamp on pressure bypass. Turn pump switch to OFF.

  b. Set sterile filter support stand and funnel on receiving flask. Loosen wrapper. Rotate funnel counterclockwise to disengage pin. Secure wrapper.

  c. Place petri dish on bench with lid up. Write identification on lid with grease pencil.

  d. Unwrap sterile pad container. Light Bunsen burner.

  e. Unwrap membrane filter container.

  f. Sterilize forceps by dipping in alcohol and passing quickly through Bunsen burner to burn off the alcohol.

[13] Aliquot (AL-li-kwot).   Representative portion of a sample. Often an equally divided portion of a sample.

(Coliform)

g. Center membrane filter on filter stand with forceps after lifting funnel. Membrane filter with printed grid should show grid uppermost (Fig. I, next page).

h. Replace funnel and lock against pin (Fig. II).

i. Shake sample or diluted sample. Measure proper aliquot with sterile pipet and add to funnel.

j. Add a small amount of the sterile dilution water to funnel. This will help check for leakage and also aid in dispersing small volumes (Fig. III).

k. Now start vacuum pump.

l. Rinse filter with three 20- to 30-mL portions of sterile dilution water.

m. When membrane filter appears barely moist, switch pump to OFF.

n. Sterilize forceps as before.

o. Remove membrane filter with forceps after first removing funnel as before (Fig. I).

p. Center membrane filter on pad containing M-Endo Media with a rolling motion to ensure water seal. Inspect membrane to ensure no captured air bubbles are present (Fig. IV).

q. Place INVERTED petri dish in incubator for 22 ± 2 hours. Incubate at 35°C.

6. Procedure for Counting Membrane Filter Colonies

a. Remove petri dish from incubator.

b. Remove lid from petri dish.

c. Turn so that your back is to window.

d. Tilt membrane filter in base of petri dish so that green and yellow-green colonies are most apparent. Direct sunlight has too much red to facilitate counting.

e. Count individual colonies using an overhead fluorescent light. Use a low-power (10 to 15 magnifications), binocular, wide-field, dissecting microscope or other similar optical device. The typical colony has a pink to dark red color with a metallic surface sheen. The sheen area may vary from a small pin-head size to complete coverage of the colony surface. Only those showing this sheen should be counted.

f. Report total number of "coliform colonies" on worksheet. Results are often reported as "colony formation units" or CFUs. Use the membranes that show from 20 to 90 colonies and do not have more than 200 colonies of all types (including non-sheen or, in other words, non-coliforms).

EXAMPLE:

A total of 42 colonies grew after filtering a 10-mL sample.

$$\text{Bacteria/100 mL} = \frac{\text{No. of Colonies Counted x 100 mL}}{\text{Sample Volume Filtered, mL x 100 mL}}$$

$$= \frac{(42 \text{ colonies})(100 \text{ mL})}{(10 \text{ mL})(100 \text{ mL})}$$

$$= \frac{(4.2)(100 \text{ mL})}{100 \text{ mL}}$$

$$= 420 \text{ per 100 mL}$$

SPECIAL NOTE:

Inexperienced persons often have great difficulty with connected colonies, with mirror reflections of fluorescent tubes (which are confused with metallic sheen), and with water condensate and particulate matter which are occasionally mistaken for colonies. Thus there is a tendency for inexperienced persons to make errors on the high side in MF counts. At least five apparent coliform colonies should be transferred to lauryl tryptose broth tubes for verification as coliform organisms.

7. Procedure for Filtration of Chlorinated Samples Using Enrichment Technique

a. Place a sterile absorbent pad in the upper half of a sterile petri dish and pipet 1.8 to 2.0 mL sterile lauryl tryptose broth. Carefully remove any surplus liquid.

b. ASEPTICALLY[14] place the membrane filter through which the sample has been passed on the pad.

c. Incubate the filter, without inverting the dish, for 1½ to 2 hours at 35°C in an atmosphere of 90 percent humidity (damp paper towels added to a plastic container with a snap-on lid can be used to produce the humidity).

d. Remove the enrichment culture from the incubator. Place a fresh, sterile, absorbent pad in the bottom half of the petri dish and saturate with 1.8 to 2.0 mL M-Endo Broth.

e. Transfer the membrane filter to the new pad. The used pad of lauryl tryptose may be discarded.

f. Invert the dish and incubate for 20 to 22 hours at 35° ± 0.5°C.

g. Count colonies as in previous method.

8. Procedure for Fecal Coliform

For drinking water samples, all positive total coliform colonies must be tested for fecal coliform. When the membrane filter method is used, growth from the positive total coliform colony should be transferred to a tube of EC media to determine the presence of fecal coliform.

a. Transfer, with the aid of a sterile 3-mm platinum wire loop, growth from each positive total coliform colony to a corresponding tube of EC broth.

b. Incubate inoculated EC broth tube in a water bath at 44.5° ± 0.2°C for 24 hours.

c. Gas production in an EC broth culture within 24 hours or less is considered a positive fecal coliform reaction.

d. Calculate total number of "fecal coliform colonies."

EXAMPLE:

For 100 mL of sample, 2 colonies gave a positive total coliform result. Only 1 of the 2 gave positive results when growth was transferred to EC broth.

Fecal Coliform/100 mL = 1 Coliform/100 mL.

---

[14] Aseptic (a-SEP-tick).   Free from the living germs of disease, fermentation, or putrefaction. Sterile.

OUTLINE OF PROCEDURE FOR INOCULATION OF MEMBRANE FILTER

Fig. I

1. Center membrane filter on filter holder. Handle membrane only on outer 3/16 inch with forceps sterilized before use in ethyl or methyl alcohol and passed lightly through a flame.

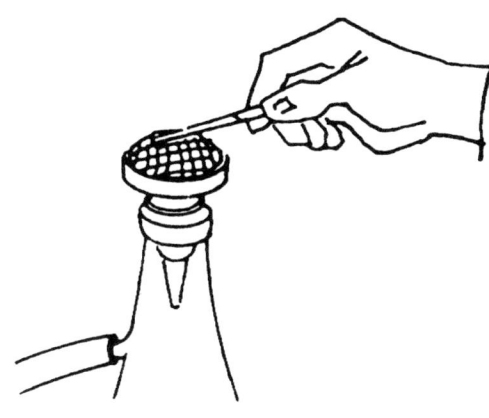

Fig. II

2. Place funnel onto filter holder.

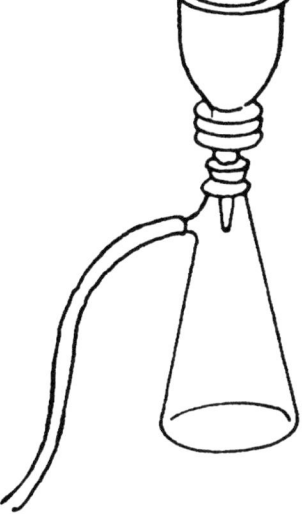

Fig. III

3. Pour or pipet sample aliquot into funnel. Avoid spattering. After suction is applied, rinse four times with sterile, buffered, distilled water.

Fig. IV

4. Remove membrane filter from filter holder with sterile forceps. Place membrane on pad. Cover with petri top.

Fig. V

5. Incubate in <u>inverted</u> position for 22 ± 2 hours.

6. Count coliform-appearing colonies on membrane.

(Coliform)

9. Alternate Membrane Filter Procedure for Fecal Coliform

The following method is a way to test for fecal coliform directly without first running the total coliform test as a separate procedure. This membrane filter procedure for fecal coliform uses an enriched lactose medium (M-FC Broth) that depends on an incubation temperature of $44.5 \pm 0.2°C$ for its selectivity. Since the temperature is *CRITICAL*, incubation takes place in a waterbath using watertight plastic bags.

a. Materials Required

(1) M-FC media.

(2) Culture dishes should be tight-fitting.

(3) Membrane filters.

(4) Watertight plastic bags.

(5) Waterbath set at $44.5 \pm 0.2°C$. The thermometer must be checked against a National Bureau of Standards (NBS) certified thermometer to ensure the accuracy of the water bath temperature.

(6) Use a sample size of 100 mL or whatever size is recommended by your regulatory agency.

b. Preparation of Culture Dish

Place a sterile absorbent pad in each culture dish and pipet (sterile) approximately 2 mL of M-FC medium to saturate the pad. Carefully remove any surplus liquid.

c. Filtration of Sample

Observe the procedure as prescribed for total coliform using membrane filters.

d. Incubation

Place the prepared culture dishes in waterproof plastic bags and immerse in waterbath set at $44.5 \pm 0.2°C$ for 24 hours. All culture dishes should be placed in the water bath within 30 minutes after filtration.

e. Counting

Colonies produced by fecal coliform bacteria are blue. The non-fecal coliform colonies are gray to cream colored. Normally, few non-fecal coliform colonies will be observed due to the selective action of the elevated temperature and the addition of the rosolic acid to the M-FC media.

Examine the cultures under a low-power magnification. Count and calculate fecal coliform density per 100 mL.

$$\frac{\text{Fecal Coliforms}}{\text{per 100 mL}} = \frac{\text{Fecal Colonies Counted x 100 mL}}{\text{Sample Volume Filtered, mL x 100 mL}}$$

EXAMPLE:

A total of 2 colonies grew after filtering a 100-mL sample.

$$\frac{\text{Fecal Coliforms}}{\text{per 100 mL}} = \frac{\text{Fecal Colonies Counted x 100 mL}}{\text{Sample Volume Filtered, mL x 100 mL}}$$

$$= \frac{(2 \text{ Colonies})(100 \text{ mL})}{(100 \text{ mL})(100 \text{ mL})}$$

$$= \frac{(0.02)(100)}{100 \text{ mL}}$$

$$= 2 \text{ per 100 mL}$$

10. Reference

See page 9-56, *STANDARD METHODS*, 20th Edition.

Presence-Absence Method

1. General Discussion

The presence-absence (P-A) test for the coliform group in drinking water is a simple modification of the multiple tube procedure described above. One large test portion (100 mL) in a single culture bottle is used to obtain qualitative information on the presence or absence of coliforms. The media used is a mixture of lactose and lauryl tryptose broths with bromocresol purple added to indicate pH. Following incubation, gas and/or acid (yellow color of media) is produced if coliforms are present. The P-A test procedure is outlined in Figure 11.18.

2. Materials Needed

a. P-A broth. Prepare according to the instructions on the bottle.

b. 250-mL screw-cap, milk dilution bottle with 50 mL sterile triple-strength P-A broth.

c. Autoclave for sterilization of media and glassware.

d. Incubator set at $35°C \pm 0.5°C$.

e. Sterile, 100-mL graduated cylinder.

f. Thermometer verified to be accurate by comparison with a National Bureau of Standards (NBS) certified thermometer.

g. Brilliant Green Bile Broth. Prepare as instructed under MPN method (page 540).

h. Water bath set at $44.5°C \pm 0.2°C$.

i. EC Broth. Prepare according to instructions on bottle.

FOR DRINKING WATER SAMPLES

a. Shake sample approximately 25 times.

b. Inoculate 100 mL into P-A culture bottle and mix thoroughly.

c. Incubate at $35°C \pm 0.5°C$ and inspect after 24 and 48 hours for acid reactions (yellow color). Record both positive and negative culture bottles.

d. A distinct yellow color forms in the culture when acid conditions exist. If gas is also being produced, gentle shaking of the bottle will result in a foaming action. Any amount of gas and/or acid constitutes a positive presumptive test requiring confirmation for total coliform in addition to further testing for fecal coliform.

OUTLINE OF PROCEDURE FOR TOTAL COLIFORM

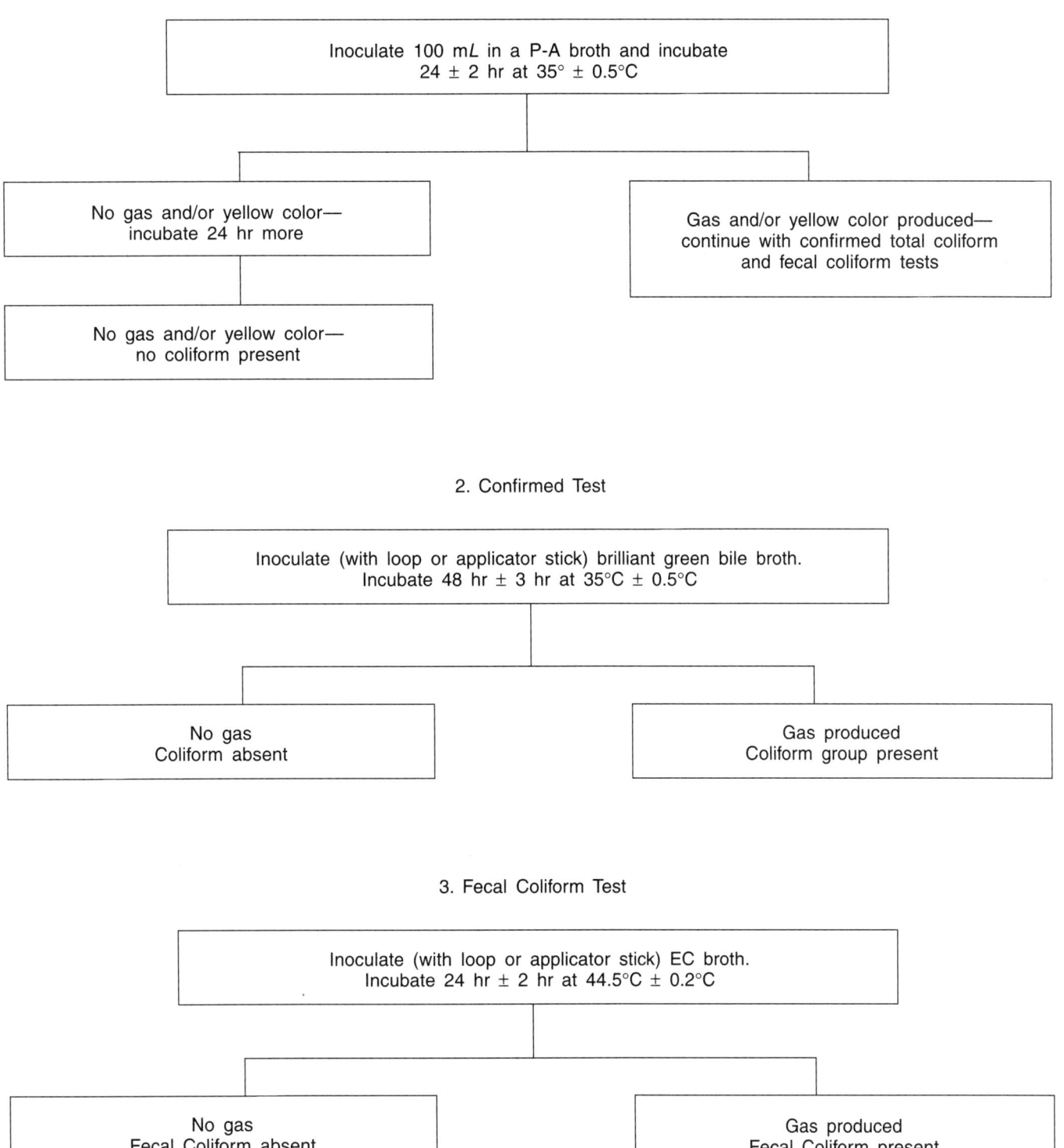

1. Presumptive Test

Inoculate 100 mL in a P-A broth and incubate
24 ± 2 hr at 35° ± 0.5°C

No gas and/or yellow color—
incubate 24 hr more

Gas and/or yellow color produced—
continue with confirmed total coliform
and fecal coliform tests

No gas and/or yellow color—
no coliform present

2. Confirmed Test

Inoculate (with loop or applicator stick) brilliant green bile broth.
Incubate 48 hr ± 3 hr at 35°C ± 0.5°C

No gas
Coliform absent

Gas produced
Coliform group present

3. Fecal Coliform Test

Inoculate (with loop or applicator stick) EC broth.
Incubate 24 hr ± 2 hr at 44.5°C ± 0.2°C

No gas
Fecal Coliform absent

Gas produced
Fecal Coliform present

Fig. 11.18   Schematic outline of test procedure for total coliform—Presence-Absence Method

(Coliform)

e. 24-Hour Brilliant Green Bile Confirmation Test

Confirm all presumptive culture bottles that show gas and/or acid at 24 or 48 hours. Transfer liquid, with aid of a sterile platinum wire loop or sterile wood applicator, from the P-A broth showing gas and/or acid to a tube of brilliant green bile (BGB) broth.

After 24 hours have elapsed, inspect the tube for gas formation. Any amount of gas formation is considered positive and so recorded on the data sheet. Negative BGB tubes are reincubated for an additional 24 hours.

f. 48-Hour Brilliant Green Bile Confirmation Test

Examine tubes for gas at the end of the 48 ± 3 hour period. Record both positive and negative tubes.

g. Recording Results

Gas production in the BGB broth culture within 48 hours confirms the presence of coliform bacteria. Report result as presence-absence test positive or negative for total coliforms in 100 mL of sample.

EXAMPLE 6 (DRINKING WATER SAMPLE)—(see Figure 11.19).

mL of sample, 100 mL

Presence-Absence Method

Colilert™ Method

1. General Discussion

Colilert provides a media that contains specific indicator nutrients for total coliform and E. coli. As these nutrients are metabolized, yellow color and fluorescence are released confirming the presence of total coliform and E. coli respectively. Non-coliform bacteria are chemically suppressed and cannot metabolize the indicator nutri-

ents. Consequently, they do not interfere with the identification of the target microbes. Total coliforms and E. coli are specifically and simultaneously detected and identified in 24 hours or less.

Although the Colilert method can yield Presence/Absence results within 24 hours and is less cumbersome to perform than the Membrane Filtration (MF) method, operators should be aware of the limitations of these tests in evaluating samples for regulatory purposes. The results for the Colilert and MF tests are not always comparable, which may be due to:

- Interferences in the sample that may suppress or mask bacterial growth,

- Greater sensitivity of the Colilert media,

- Added stress to organisms related to filtering, or

- The fact that different media may obtain better growth for some bacteria.

Both test methods are approved by the U.S. Environmental Protection Agency for reporting under the Safe Drinking Water Program. When these tests produce conflicting bacteriological results, however, the safest course of action is to increase monitoring and treatment efforts until the results for both tests are negative.

2. Materials Needed

a. 10 culture tubes each containing Colilert reagent for 10 mL of sample (available from IDEXX Laboratories, 1 IDEXX Drive, Westbrook, ME 04092. Phone 1-800-321-0207).

b. Sterile 10-mL pipets.

c. Incubator at 35°C.

d. Long wavelength ultraviolet lamp.

e. Color and fluorescence comparator.

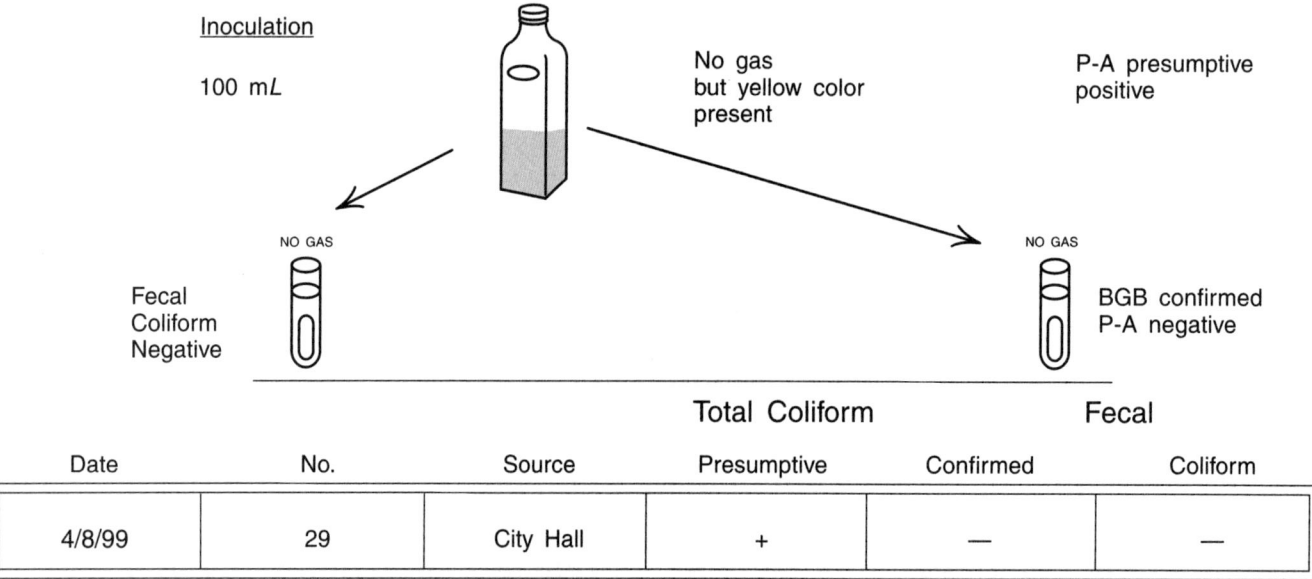

Date	No.	Source	Total Coliform Presumptive	Total Coliform Confirmed	Fecal Coliform
4/8/99	29	City Hall	+	—	—

Fig. 11.19   Record Presence-Absence coliform test results

(Coliform)

3. Test Procedure

    a. Remove the front panel of the Colilert tube box along the perforation line to allow easy access and removal of tube carriers.

    b. Remove a 10-tube carrier from the kit and label each 10-tube MPN test as indicated on the carrier.

    c. Crack the carrier along the front perforation and bend the carrier along the back crease line. Stand the carrier on the table top.

    d. Aseptically fill each Colilert tube with sample water to the level of the back of the tube carrier (10 mL).

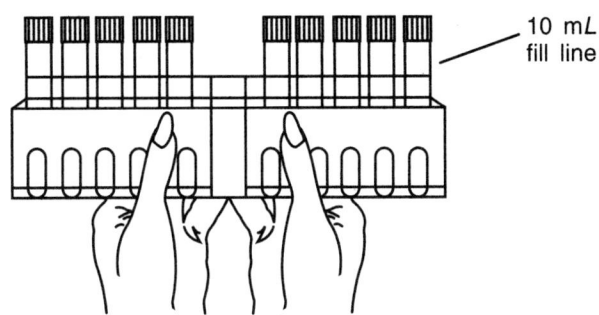

10 mL
fill line

    e. Cap the tubes tightly.

    f. Mix vigorously to dissolve the reagent by repeated inversion of carrier.

    NOTE: High calcium salt concentrations in certain waters may cause precipitate. This will not affect test results.

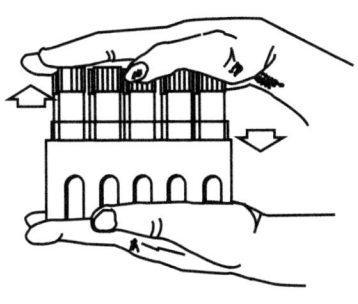

    g. Incubate inoculated reagent tubes in the carrier at 35°C for 24 hours. Incubation should begin within 30 minutes of inoculation.

    NOTE: Incubation exceeding 28 hours should be avoided.

    h. Read tubes in the carrier at 24 hours. If yellow color is seen, check for fluorescence. Color should be uniform throughout the tube. If not, mix by inversion before reading.

4. Test Results and Interpretation

    a. Compare each tube against the color comparator. If the inoculated reagent has a yellow color greater or equal to the comparator, the presence of total coliforms is confirmed.

    b. If any of the tubes are yellow in color, observe each tube for fluorescence by placing carrier 2 to 5 inches from the long wavelength, ultraviolet lamp. If the fluorescence of the tube(s) is greater or equal to the comparator, the presence of *E. coli* is specifically confirmed.

    c. Samples are negative for total coliforms if no color is observed at 24 hours. Should a sample be so lightly yellow after 24 hours' incubation that you cannot definitively read it relative to the positive comparator tube, you may incubate it up to an additional 4 hours. If the sample is coliform positive, the color will intensify. If it does not intensify, consider the sample negative.

    d. To find the concentration of total coliforms or *E. coli* per 100 mL, compare the number of positive reaction tubes per sample set (10 tubes) to the standard MPN (Most Probable Number) chart shown on Table 11.5 or in *STANDARD METHODS.*

TABLE 11.5   MPN INDEX FOR VARIOUS COMBINATIONS OF POSITIVE AND NEGATIVE RESULTS WHEN TEN 10-mL PORTIONS ARE USED

Number of tubes giving positive reaction out of ten 10-mL portions	MPN Index per 100 mL
0	<1.1
1	1.1
2	2.2
3	3.6
4	5.1
5	6.9
6	9.2
7	12.0
8	16.1
9	23.0
10	>23.0

5. Recording Results

    Results should be recorded on data sheets prepared especially for this test. An example is shown in Figure 11.20.

6. Method of Calculating the Most Probable Number (MPN)

EXAMPLE 7 (DRINKING WATER SAMPLE)—(see Figure 11.21).

mL of sample	100 mL
positive tubes	2
yellow color	yes
fluorescence	no

Read MPN as 2.2 per 100 mL for total coliform and <1.1 for *E. coli* from Table 11.5.

(Coliform)

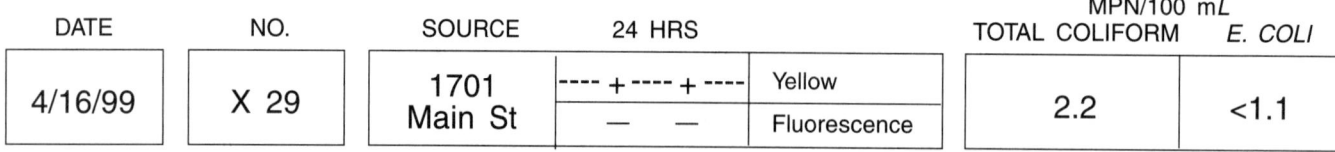

DATE	NO.	SOURCE	24 HRS		MPN/100 mL TOTAL COLIFORM	E. COLI
4/16/99	X 29	1701 Main St	---- + ---- + ----	Yellow	2.2	<1.1
			— —	Fluorescence		

Fig. 11.20   Record coliform test results using Colilert Method

Inoculation

10 mL

Results

2 of 10 - yellow

0 of 10 - fluorescence

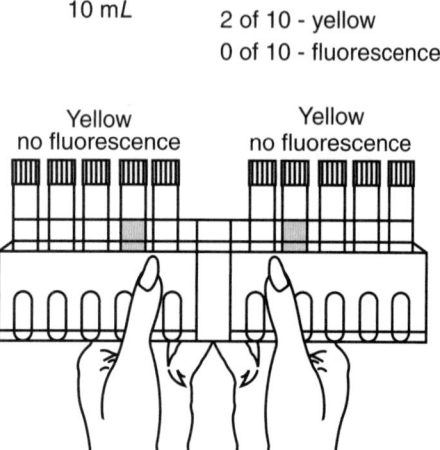

Yellow
no fluorescence   Yellow
no fluorescence

Fig. 11.21   Results of Colilert test—drinking water

Colisure Method

The Colisure test is a presence-absence test for coliform bacteria. A sample of water is added to a dehydrated medium and, after 28 to 48 hours, the medium is examined for the presence of coliforms. (The Colisure test is available from IDEXX Laboratories, 1 IDEXX Drive, Westbrook, ME 04092.)

Membrane Filter Method—Escherichia coli (EPA Method)

1.   General Discussion

This method describes a membrane filter (MF) procedure for identifying and counting *Escherichia coli*. Because the bacterium is a natural inhabitant only of the intestinal tract of warm-blooded animals, its presence in water samples is an indication of fecal pollution and the possible presence of enteric (intestinal) pathogens.

This MF method provides a direct count of bacteria in water based on the development of colonies on the surface of the membrane filter. A water sample is filtered through the membrane that retains the bacteria. After filtration, the membrane containing the bacterial cells is placed on a specially formulated medium, mTEC, incubated at 35°C for 2 hours to revive injured or stressed bacteria, and then incubated at 44.5°C for 22 hours. Following incubation, the filter is transferred to a filter pad saturated with urea substrate. After 15 minutes, yellow or yellow-brown colonies are counted with the aid of a fluorescent lamp and magnifying lens.

2.   Materials and Reagents Required

a.   See page 552, Section 9, "Alternate Membrane Filter Procedure for Fecal Coliform," item a, "Materials Required."

b.   mTEC Agar (Difco 0334-16-0)

Preparation: Add 45.26 grams of dehydrated mTEC medium to 1 L of reagent water in a flask and heat to boiling, until ingredients dissolve. Autoclave at 121°C (15 lb pressure) for 15 minutes and cool in a 44 to 46°C water bath. Pour the medium into each 50- x 10-mm culture dish to a 4- to 5-mm depth (approximately 4 to 6 mL) and allow to solidify. Final pH should be 7.3 ± 0.2. Store in a refrigerator.

c.   Hardness Substrate Medium

Preparation: Add dry ingredients to 100 mL reagent water in a flask. Stir to dissolve and adjust to pH 5.0 with a few drops of 1N HCl. The substrate solution should be a straw yellow color at pH 5.0.

d.   Nutrient Agar (Difco 0001-02, BBL 11471)

Preparation: Add 23 grams of dehydrated nutrient sugar to 1 L of reagent water and mix well. Heat in a boiling water bath to dissolve the agar completely. Dispense in screw-cap tubes and autoclave at 121°C (15 lb pressure) for 15 minutes. Remove tubes and slant. The final pH should be 6.8 ± 0.2.

3.   Procedure

a.   Prepare the mTEC agar and urea substrate as directed in 2.b. and 2.c.

b.   Mark the petri dishes and report forms with sample identification and sample volumes.

c.   Place a sterile membrane filter on the filter base, grid side up, and attach the funnel to the base; the membrane filter is now held between the funnel and the base.

d.   Shake the sample bottle vigorously about 25 times to distribute the bacteria uniformly and measure the desired volume of sample or dilution into the funnel.

e.   For ambient surface waters and wastewaters, select sample volumes based on previous knowledge of pollution level, to produce 20 to 80 *E. coli* colonies on the membranes. Sample volumes of 1 to 100 mL are normally tested at half-log intervals, for example 100, 30, 10, 3 mL.

f. Smaller sample sizes or sample dilutions can be used to minimize the interference of turbidity or high bacterial densities. Multiple volumes of the same sample or sample dilution may be filtered and the results combined.

g. Filter the sample and rinse the sides of the funnel at least twice with 20 to 30 mL of sterile buffered rinse water. Turn off the vacuum and remove the funnel from the filter base.

h. Use sterile forceps to aseptically remove the membrane filter from the filter base and roll it onto the mTEC agar to avoid the formation of bubbles between the membrane and the agar surface. Reseat the membrane if bubbles occur. Close the dish, invert, and incubate at 35°C for 2 hours.

i. After 2 hours incubation at 35°C transfer the plates to Whirl-Pak bags, seal, and place inverted in a 44.5°C water bath for 22 to 24 hours.

j. After 22 to 24 hours, remove the dishes from the water bath. Place absorbent pads in new petri dishes or the lids of the same petri dishes and saturate with urea broth. Aseptically transfer the membranes to absorbent pads saturated with urea substrate and hold at room temperature.

k. After 15 to 20 minutes' incubation on the urea substrate at room temperature, count and record the number of yellow or yellow-brown colonies on those membrane filters containing (ideally) 20 to 80 colonies.

4. Calculation of Results

Select the membrane filter with the number of colonies within the acceptable range (20 to 80) and calculate the count per 100 mL according to the general formula:

$$E.\ coli/100\ mL = \frac{\text{No. } E.\ coli \text{ colonies counted x 100 mL}}{\text{Volume in mL of sample filtered x 100 mL}}$$

5. Reporting Results

Report the results as *E. coli* per 100 mL of sample.

6. Example

A total of 40 *E. coli* colonies were counted after filtering a 50-mL sample.

$$E.\ coli/100\ mL = \frac{\text{No. } E.\ coli \text{ colonies counted x 100 mL}}{\text{Sample volume filtered, mL x 100 mL}}$$

$$= \frac{(40\ E.\ coli)(100\ mL)}{(50\ mL)(100\ mL)}$$

$$= \frac{(0.8\ E.\ coli/mL)(100\ mL)}{100\ mL}$$

$$= 80\ E.\ coli/100\ mL$$

7. References

EPA Test Method: *Escherichia coli* in Water by the Membrane Filter Procedure, Method 1003.1, 1985 (MF-mTEC Method).

See pages 9-47 and 9-56, *STANDARD METHODS*, 20th Edition.

F. Comparison of Coliform Tests

The total coliform bacteria test includes both *Escherichia* and *Aerobacter* coliform bacteria groups. *Aerobacter* and some *Escherichia* can grow in soil. Therefore, not all coliforms found in the total coliform test come from human wastes. *Escherichia coli* (*E. coli*) apparently are all of fecal origin. However, it is difficult to determine *E. coli* without measuring soil coliforms too. The fecal coliform and *Escherichia coli* tests are used in an attempt to more specifically determine the extent of human wastes in water.

QUESTIONS

Write your answers in a notebook and then compare your answers with those on page 569.

11.3H Why are drinking waters not tested for specific pathogens (such as *Cryptosporidium* or *Giardia*)?

11.3I To perform coliform tests, how would you decide how many samples to test?

11.3J Why should sodium thiosulfate be added to coliform sample bottles?

11.3K Steam autoclaves sterilize (kill all organisms) at a pressure of _____ psi and temperature of _____ °C during a _____ minute time period (at sea level).

11.3L Estimate the Most Probable Number (MPN) of coliform group bacteria for a raw water sample from the following test results:

mL of sample	1	0.1	0.01	0.001
Dilutions	0	−1	−2	−3
Readings (+ tubes)	5	5	3	1

11.3M How is the number of coliforms estimated by the membrane filter method?

11.3N What are the incubation conditions for chlorinated samples when using the membrane filter method (enrichment technique)?

11.3O Why is the total coliform test not an adequate measure of the extent of human wastes in water?

End of Lesson 4 of 5 Lessons on LABORATORY PROCEDURES

Please answer the discussion and review questions next.

(Coliform)

COMPARISON OF *E. COLI* AND FECAL COLIFORM TEST PROCEDURES

Item	*E. coli*	Fecal Coliform
Media	mTEC	M-FC
Colony color	Yellow	Blue
Incubation steps	3 steps	1 step

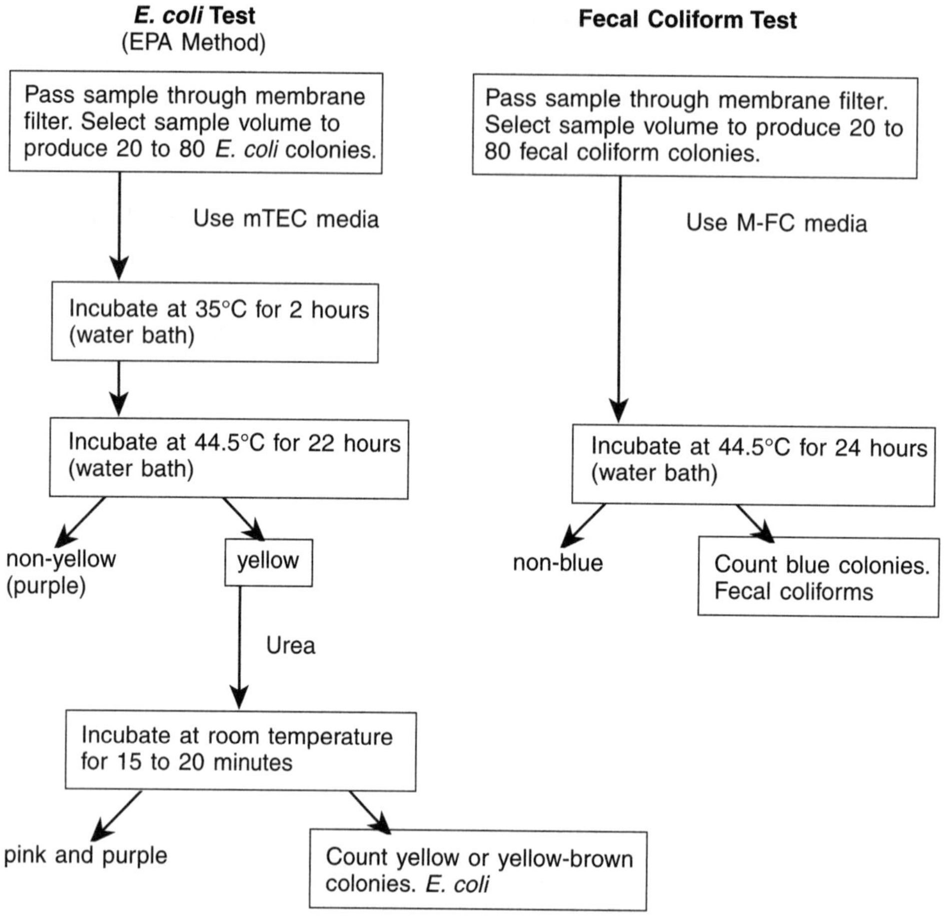

E. coli Test
(EPA Method)

Fecal Coliform Test

Pass sample through membrane filter. Select sample volume to produce 20 to 80 *E. coli* colonies.

Use mTEC media

Incubate at 35°C for 2 hours (water bath)

Incubate at 44.5°C for 22 hours (water bath)

non-yellow (purple)   yellow

Urea

Incubate at room temperature for 15 to 20 minutes

pink and purple   Count yellow or yellow-brown colonies. *E. coli*

Pass sample through membrane filter. Select sample volume to produce 20 to 80 fecal coliform colonies.

Use M-FC media

Incubate at 44.5°C for 24 hours (water bath)

non-blue   Count blue colonies. Fecal coliforms

DISCUSSION AND REVIEW QUESTIONS

Chapter 11.   LABORATORY PROCEDURES

(Lesson 4 of 5 Lessons)

Write the answers to these questions in your notebook before continuing. The question numbering continues from Lesson 3.

15. What is the purpose of the coliform group bacteria test?

16. How would you determine the number of dilutions for an MPN test?

17. What factors can cause errors when counting colonies on membrane filters?

18. How can questionable colonies on membrane filters be verified as coliform colonies?

CHAPTER 11.   LABORATORY PROCEDURES

(Lesson 5 of 5 Lessons)

5.   Hardness

A.   Discussion

Hardness is caused principally by the calcium and magnesium ions commonly present in water. Hardness may also be caused by iron, manganese, aluminum, strontium, and zinc if present in significant amounts. Because only calcium and magnesium are present in significant concentrations in most waters, hardness can be defined as the total concentration of calcium and magnesium ions expressed as the calcium carbonate ($CaCO_3$) equivalent.

There are two types or classifications of water hardness: carbonate and noncarbonate. Carbonate hardness is due to calcium/magnesium bicarbonate or carbonate. Hardness that is due to calcium/magnesium sulfate, chloride, or nitrate is called noncarbonate hardness.

Hard water can cause incrustations (scale) when the water evaporates, or when heated in household hot water heaters and piping. Hardness-producing substances in water also combine with soap to form insoluble precipitates. The common method of minimizing these and other problems due to hardness is water supply softening. This procedure is discussed in Chapter 14, "Softening," in Volume II of *WATER TREATMENT PLANT OPERATION*.

B.   What Is Tested

	Common Range mg/L as $CaCO_3$
Surface Water	30 to 500*
Well Water	80 to 500*

* Levels of hardness depend on local conditions.

C.   Apparatus

Buret (25 mL)
Buret support
Graduated cylinder (100 mL)

Beaker (250 mL)
Magnetic stirrer
Magnetic stir-bar
Flask, Erlenmeyer (500 mL)

Funnel
Hot plate
Flask, volumetric (1,000 mL)

D.   Reagents

NOTE: Standardized solutions are available already prepared from laboratory chemical supply companies.

1.   Buffer solution.

2.   Standard EDTA or CDTA titrant. EDTA is disodium ethylene-diaminetetraacetate dihydrate, also called (ethylenedinitrilo)-tetraacetic acid disodium salt. CDTA is disodium-CDTA (1,2 cyclohexanediaminetetraacetic acid).

3.   Indicator solution.

OUTLINE OF PROCEDURE

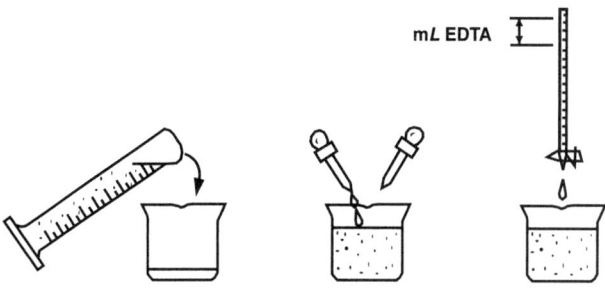

mL EDTA

1.   Add 50 mL of sample to clean beaker.
2.   Add 2 mL of buffer and 2 drops of indicator.
3.   Titrate with EDTA to blue end point.

E.   Procedure

1.   Take a clean beaker and add 50 mL of sample.

2.   Add 2 mL of buffer solution.

3.   Add 2 drops indicator solution.

4.   Titrate with standard EDTA solution until the last reddish tinge disappears from the solution. The solution is pure blue when the end point is reached.

5.   Calculate total hardness.

F.   Example

Results from water hardness testing of a well water sample were as follows:

Sample size   = 50 mL

mL of EDTA titrant used, A = 10.5 mL

(Jar Test)

G. Calculation

$$\text{Hardness, mg/}L \text{ as CaCO}_3 = \frac{A \times 1,000}{mL \text{ of sample}}$$

$$= \frac{10.5 \ mL \times 1,000}{50 \ mL \text{ of sample}}$$

$$= 210 \text{ mg/}L$$

H. Precautions

1. Some metal ions interfere with this procedure by causing fading or indistinct end points. In these cases, an inhibitor reagent should be used. You may titrate with either standard CDTA solution or EDTA solution.

2. The titration should be completed within five minutes to minimize $CaCO_3$ precipitation.

3. A sample volume should be selected that requires less than 15 mL of EDTA titrant to be used.

4. For titrations of samples containing low hardness concentrations (less than 150 mg/L, as $CaCO_3$) a larger sample volume should be used.

I. Reference

See page 2-36, *STANDARD METHODS*, 20th Edition.

QUESTIONS

Write your answers in a notebook and then compare your answers with those on page 569.

11.3P What are the principal hardness-causing ions in water?

11.3Q What problems are caused by hardness in water?

6. Jar Test

A. Discussion

Jar tests are tests designed to show the effectiveness of chemical treatment in a water treatment facility. Many of the chemicals we add to water can be evaluated on a small laboratory scale by the use of a jar test. The most important of these chemicals are those used for coagulation, such as alum and polymers. Using the jar test, the operator can approximate the correct coagulant dosage for plant use when varying amounts of turbidity, color, or other factors indicate raw water quality changes. The jar test is also a very useful tool in evaluating new coagulants or polymers being considered for use on a plant scale.

B. What Is Tested

Raw water, for optimum coagulant dose, which varies depending on coagulant(s) used and water quality.

C. Apparatus

1. A stirring machine with six paddles capable of variable speeds from 0 to 100 revolutions per minute (RPM).

2. An illuminator located underneath the stirring mechanism (optional).

3. Beakers (1,000 mL)

4. Pipets (10 mL)

5. Flask, volumetric (1,000 mL)

6. Balance, analytical

D. Reagents

1. Stock Coagulant Solution

a. Dry alum, $Al_2(SO_4)_3 \cdot 14.3 \ H_2O.$[15] Dissolve 10.0 gm dry alum (17 percent) in 600 mL distilled water contained in a 1,000 mL volumetric flask. Fill to mark. This solution contains 10,000 mg/L or 10 mg/mL.

b. Liquid alum, $Al_2(SO_4)_3 \cdot 49.6 \ H_2O.$[15] The operator should verify the strength of the alum with a hydrometer. Liquid alum is usually shipped as 8.0 to 8.5 percent Al_2O_3 and contains about 5.36 pounds of dry aluminum sulfate (17 percent dry) per gallon (specific gravity 1.325). This converts to 642,350 mg/L. Therefore, add 15.6 mL liquid alum to a 1,000-mL volumetric flask and fill to mark. This solution contains 10,000 mg/L or 10 mg/mL.

c. Table 11.6 indicates the strengths of stock solutions for various dosages.

TABLE 11.6 DRY CHEMICAL CONCENTRATIONS USED FOR JAR TESTING [a]

Approx. Dosage Required, mg/L [b]	Grams/Liter to Prepare [c]	1 mL Added to 1 Liter Sample Equals	Stock Solution Conc., mg/L (%)
1-10 mg/L	1 gm/L	1 mg/L	1,000 mg/L (0.1%)
10-50 mg/L	10 gm/L	10 mg/L	10,000 mg/L (1.0%)
50-500 mg/L	100 gm/L	100 mg/L	100,000 mg/L (10.0%)

[a] From *JAR TEST* by E. E. Arasmith, Linn-Benton Community College, Albany, OR.
[b] Use this column which indicates the approximate dosage required by raw water to determine the trial dosages to be used in the jar test.
[c] This column indicates the grams of dry chemical that should be used when preparing the stock solution. The stock solution consists of the chemical plus enough water to make a one-liter solution.

E. Procedure

1. Collect a two-gallon (8-liter) sample of the water to be tested.

2. Immediately measure six 1,000-mL quantities and place into each of six 1,000-mL beakers.

3. Place all six beakers on stirring apparatus.

4. With a measuring pipet, add increasing dosages of coagulant solution to the beakers as rapidly as possible. Select a series of dosages so that the first beaker will represent an underdose and the last an overdose.

5. With stirring paddles lowered into the beakers, start stirring apparatus and operate it for one minute at a speed of 80 RPM.[16]

---

[15] *Values of 14.3 H_2O and 49.6 H_2O were obtained from ALUMINUM SULFATE published by Stauffer Chemicals. Actual values for commercial alum purchased by your water treatment plant may vary slightly.*
[16] *Use stirring speeds and times which are similar to actual conditions in your water treatment plant.*

(Jar Test)

6. Reduce the stirring speed for the next 30 minutes to 20 RPM.[16]

OUTLINE OF PROCEDURE

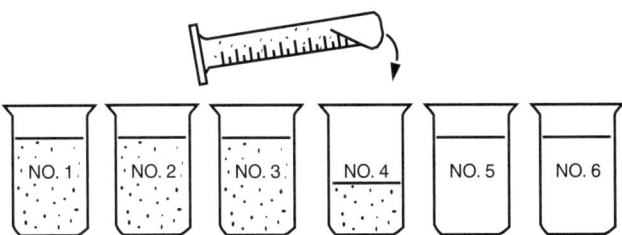

1. Add 1,000 mL to each of 6 beakers.

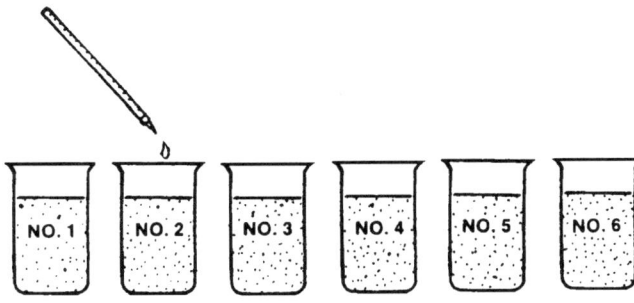

2. Add increasing dosages of coagulant.

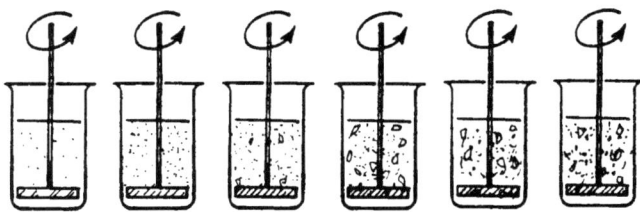

3. Stir for appropriate time period. Evaluate floc quality.

7. Observe and evaluate each beaker as to that specific dosage's floc quality. Record results.

8. Stop the stirring apparatus and allow samples in beakers to settle for 30 minutes.[16] Observe the floc settling characteristics. A hazy sample indicates poor coagulation. Properly coagulated water contains floc particles that are well-formed and dense, with the liquid between the particles clear. Describe results as poor, fair, good, or excellent.

F. Precautions

Without going to extreme measures, it is very difficult to duplicate in the jar test exactly what is happening on a plant scale. The jar test, therefore, should be used as an indication of what is to be expected on a larger scale in a water treatment plant.

There are a number of tests that can be performed to improve the jar test and the interpretation of the results. These tests include:

1. Alkalinity (before and after),

2. pH (before and after),

3. Turbidity of SUPERNATANT[17] (before and after), and

4. Filtered turbidity of supernatant.

See Figure 11.22 for a helpful jar test data sheet.

After estimating the optimum coagulant dosage, run another jar test with the optimum coagulant dosage constant, but vary the pH. These results will give you the optimum pH.

Alkalinity must be monitored very carefully before and after the jar test. Alkalinity must always be AT LEAST half of the coagulant dose. For example, if the optimum coagulant dose is 50 mg/L, the total alkalinity must be at least 25 mg/L. If the natural alkalinity is less than 25 mg/L, adjust the total alkalinity up to 25 mg/L by adding lime. For additional information on how to calculate the amount of lime needed to increase the alkalinity, see page 172, Section 5.243, "Arithmetic for Solids-Contact Clarification," and page 173, Example 6.

G. Example

A sample of river water was collected for jar test analysis to determine the "optimum" alum dosage for effective coagulation.

The following series of alum dosages were added:

Beaker No.	mL Alum Solution Added	Alum Added, mg/L
1	1.0	10
2	1.2	12
3	1.4	14
4	1.6	16
5	1.8	18
6	2.0	20

H. Interpretation of Results

Results of the above jar testing were recorded as follows:

Beaker No.	Alum Dose, mg/L	Floc Quality
1	10	poor
2	12	fair
3	14	good
4	16	excellent
5	18	excellent
6	20	good

The above results seem to indicate that a dose of 16 or 18 mg/L would be optimum. The operator should, however, verify this result with visual observation of what is actually happening in the flocculation basin.

---

[16] Use stirring speeds and times which are similar to actual conditions in your water treatment plant.

[17] Supernatant (sue-per-NAY-tent). Liquid removed from settled sludge. Supernatant commonly refers to the liquid between the sludge on the bottom and the scum on the water surface of a basin or container.

JAR TEST DATA SHEET

JAR NUMBER	SAMPLE SIZE	TEMP START / FINISH	TURB START / FINISH	pH START / FINISH	ALKA START / FINISH	FILTERED TURBIDITY	COAG DOSE	COAG-AID DOSE	ADJUSTED pH	1ST APPEAR	5 MIN	10 MIN	15 MIN	20 MIN	25 MIN	30 MIN	45 MIN	HARD-NESS START / FINISH	COMMENTS
1																			
2																			
3																			
4																			
5																			
6																			

DESCRIPTION OF FLOC IN JARS

CODE: VS = VERY SMALL; S = SMALL; M = MEDIUM
D = DENSE; VD = VERY DENSE; SP = SPARCE
L = LARGE; VL = VERY LARGE

SOURCE:
TIME:
DATE:
COAGULANT AID:
COAGULANT:
pH ADJUST:
ANALYST:

Fig. 11.22 Typical jar test data sheet

(pH)

QUESTIONS

Write your answers in a notebook and then compare your answers with those on page 569.

11.3R   What is the purpose of the jar test?

11.3S   What stirring speeds are used during the jar tests for optimum alum dosage?

7. pH

A. Discussion

The pH of a water indicates the intensity of its acidic or basic strength. The pH scale runs from 0 to 14. Water having a pH of 7 is at the midpoint of the scale and is considered neutral. Such a water is neither acidic nor basic. A pH of greater than 7 indicates basic water. The stronger the basic intensity, the greater the pH. The opposite is true of the acidity. The stronger the intensity of the acidity, the lower will be the pH.

pH SCALE

0 ← INCREASING ACID – – – – 7 – – – – INCREASING BASE → 14

1 ← 2 ← 3 ← 4 ← 5 ← 6   /\   8 → 9 → 10 → 11 → 12 → 13

Neutral

Mathematically, pH is the logarithm (base 10) of the reciprocal of the hydrogen ion activity, or the negative logarithm of the hydrogen ion activity.

$$pH = \text{Log} \frac{1}{[H^+]} = -\text{Log } [H^+]$$

For example, if a water has a pH of 1, then the hydrogen ion activity $[H^+] = 10^{-1} = 0.1$. If the pH is 7, then $[H^+] = 10^{-7} = 0.0000001$. A change in the pH of one unit is caused by changing the hydrogen ion level by a factor of 10 (10 times).

In a solution, both hydrogen ions $[H^+]$ and the hydroxyl ions $[OH^-]$ are always present. At a pH of 7, the activity of both hydrogen and hydroxyl ions equals 10^{-7} *MOLES*[18] per liter. When the pH is less than 7, the activity of hydrogen ions is greater than the hydroxyl ions.

pH plays an important role in the water treatment processes such as disinfection, coagulation, softening, and corrosion control. The pH test also indicates changes in raw and finished water quality.

B. What Is Tested

	Common Range
Surface water	6.5 to 8.5
Finished water	7.5 to 9.0
Well water	6.5 to 8.0

C. Apparatus

1. pH meter

2. Glass electrode

3. Reference electrode

4. Magnetic stirrer

5. Magnetic stir-bar

D. Reagents

1. Buffer tablets for various pH value solutions (available from laboratory chemical supply houses).

2. Distilled water.

E. Procedure

1. Due to the difference between the various makes and models of pH meters commercially available, specific instructions cannot be provided for the correct operation of all instruments. In each case, follow the manufacturer's instructions for preparing the electrodes and operating the instrument.

2. *STANDARDIZE THE INSTRUMENT (pH METER) AGAINST A BUFFER SOLUTION WITH A pH CLOSE TO THAT OF THE SAMPLE.*

3. Rinse electrodes thoroughly with distilled water after removal from buffer solution.

4. Place electrodes in sample and measure pH.

5. Remove electrodes from sample, rinse thoroughly with distilled water.

6. Immerse electrode ends in beaker of pH 7 buffer storage solution.

7. Turn meter to "standby" (or OFF).

F. Precautions

1. To avoid faulty instrument calibration, prepare fresh buffer solutions as needed, and at least once per week (from commercially available buffer tablets).

2. pH meter, buffer solution, and samples should all be near the same temperature because temperature variations will give somewhat erroneous results. Allow a few minutes for the probes to adjust to the buffers before calibrating a pH meter to ensure accurate pH readings.

3. Watch for erratic results arising from electrodes, faulty connections, or fouling of electrodes with oily precipitated matter. Films may be removed from electrodes by placing isopropanol on a tissue or a Q-tip and cleaning the probe.

---

[18] *Mole.   The molecular weight of a substance, usually expressed in grams.*

(Temperature)

4. The temperature compensator on the pH meter adjusts the meter for changes in electrode response with temperature. However, the pH of water also changes with temperature and the pH meter cannot compensate for this change.

5. We recommend standardizing the pH meter using a pH buffer solution close to the pH of the sample. However, if you use another buffer solution with a different pH to determine the calibration of the pH meter for a range of pH values and the pH meter does not give the pH of the second buffer, follow the manufacturer's directions to adjust the pH meter. This procedure is called adjusting the "slope" of the pH meter.

6. If you're measuring pH in colored samples or samples with high solids content, or if you're taking measurements that need to be reported to the US EPA, you should use a pH electrode and meter instead of a colorimetric method or test papers. pH meters are capable of providing ± 0.1 pH accuracy in most applications. In contrast, colorimetric tests provide ± 0.1 pH accuracy only in a limited range. pH papers provide even less accuracy.

G. Interpretation of Results

The pH of water has a very important influence on the effectiveness of chlorine disinfection. Chlorination is a chemical reaction in which chlorine is an oxidizing agent. Chlorine is a more effective oxidant at lower pH values. Simply stated, a chlorine residual of 0.2 mg/*L* at a pH of 7 is just as effective as 1 mg/*L* at a pH of 10. Therefore, five times as much chlorine is required to do the same disinfecting job at pH 10 as it does at pH 7.

The finished water of some treatment facilities is adjusted with lime or caustic soda to a slightly basic pH for the purpose of minimizing corrosion in the distribution system.

H. Reference

See page 4-86, *STANDARD METHODS*, 20th Edition.

QUESTIONS

Write your answers in a notebook and then compare your answers with those on page 569.

11.3T   What does the pH of a water indicate?

11.3U   What precautions should be exercised when using a pH meter?

8. Temperature

A. Discussion

Temperature is one of the most frequently taken tests in the water industry. Accurate water temperature readings are important not only for historical purposes but also because of its influence on chemical reaction rates, biological growth, dissolved gas concentrations, and water stability with respect to calcium carbonate, in addition to its acceptability by consumers for drinking.

B. What Is Tested

	Common Range, °C
Raw and Treated Surface Water	5 to 25
Well Water	10 to 20

C. Apparatus

1. One NBS (National Bureau of Standards) thermometer for calibration of the other thermometers.

2. One Fahrenheit, mercury-filled, 1° subdivided thermometer.

3. One Celsius (formerly called Centigrade), mercury-filled, 1° subdivided thermometer.

4. One metal case to fit each thermometer.

NOTE: There are three types of thermometers and two scales.

SCALES

1. Fahrenheit, marked °F.

2. Celsius, marked °C (formerly Centigrade).

THERMOMETER STYLES

1. Total immersion. This type of thermometer must be totally immersed when read. Readings with this type of thermometer will change most rapidly when removed from the liquid to be recorded.

2. Partial immersion. This type thermometer will have a solid line (water-level indicator) around the stem below the point where the scale starts.

3. Dial. This type has a dial that can be easily read while the thermometer is still immersed. Dial thermometer readings should be checked (calibrated) against the NBS thermometer. Some dial thermometers can be recalibrated (adjusted) to read at a set temperature against the NBS thermometer.

D. Reagents

None required.

E. Procedure

1. Collect as large a volume of sample as is practical. The temperature will have less chance to change in a large volume than in a small container.

2. Immerse the thermometer to the proper depth. Do not touch the bottom or sides of the sample container with the thermometer.

3. Record temperature to the nearest fraction of a degree which can be estimated from the thermometer available.

4. When measuring the temperature of well water samples, allow the water to continuously overflow a small container (a polystyrene coffee cup is ideal). Place the thermometer in the cup. After there has been no change in the temperature reading for one minute, record the temperature. The temperature of water samples collected from a distribution system mainly depends on the soil temperature at the depth of the water main.

(Turbidity)

F. Precautions

To avoid breaking or damaging a glass thermometer, store it in a shielded metal case. Check your thermometer's accuracy against the NBS-certified thermometer by measuring the temperature of a sample with both thermometers simultaneously. Some of the poorer quality thermometers are substantially inaccurate (off as much as 6°F or 3°C).

G. Example

To measure the temperature of treated water, a sample was obtained in a gallon bottle, the thermometer immediately immersed, and a temperature of 15°C recorded after the reading became constant.

H. Calculation

Normally we measure and record temperatures using a thermometer with the proper scale. We could, however, measure a temperature in °C and convert to °F. The following formulas are used to convert temperatures from one scale to the other.

1. If we measure in °F and want °C,

$$°C = \frac{5}{9} \ (°F - 32°F)$$

2. If we measure in °C and want °F,

$$°F = \frac{9}{5} \ (°C) + 32°F$$

3. Sample Calculation

The measured treated water temperature was 15°C. What is the temperature in °F?

$$°F = 9/5(°C) + 32°F$$
$$= 9/5(15°C) + 32°F$$
$$= 27 + 32$$
$$= 59°F$$

I. Reference

See page 2-60, *STANDARD METHODS*, 20th Edition.

QUESTIONS

Write your answers in a notebook and then compare your answers with those on page 570.

11.3V   Why are temperature readings important?

11.3W   Why should the thermometer remain immersed in the liquid while being read?

11.3X   Why should thermometers be calibrated against an accurate NBS-certified thermometer?

9. Turbidity

A. Discussion

The term turbidity is simply an expression of the physical cloudiness of water. Turbidity is caused by the presence of suspended matter such as silt, finely divided organic and inorganic matter, and microscopic organisms such as algae.

The accepted method used to measure turbidity is called the nephelometric method. The nephelometric turbidimeter or nephelometer (Figure 11.23) is designed to measure particle-reflected light at an angle of 90 degrees to the source beam. The greater the intensity of scattered light, the higher the turbidity.

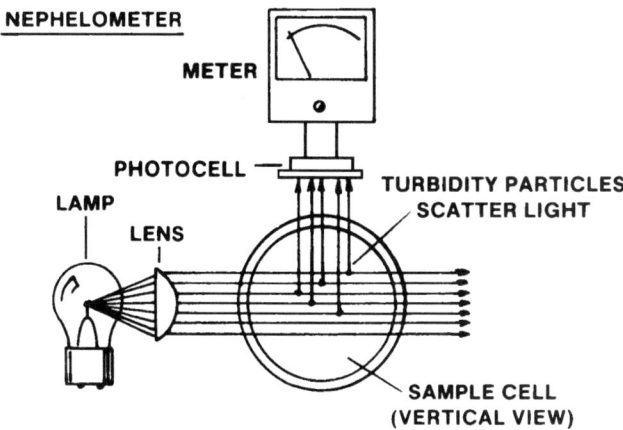

Fig. 11.23   Nephelometer
(Permission of HACH Company)

The *TURBIDITY UNIT (NTU)* [19] is an empirical quantity which is based on the amount of light that is scattered by particles of a polymer reference standard called formazin which produces particles that scatter light in a reproducible manner. Formazin, the primary turbidity standard, is an aqueous (water-based) suspension of an insoluble polymer formed by the condensation reaction between hydrazine sulfate and hexamethylenetetramine.

Secondary turbidity standards are suspensions of various materials formulated to match the primary formazin solutions. These secondary standards are generally used because of their convenience and the instability of dilute formazin primary standard solutions. Examples of these secondary standards include "standards" that are supplied by the turbidimeter manufacturer with the instrument. Periodic checks of these secondary standards against the primary formazin standard are a must and will provide assurance of measurement accuracy.

The turbidity measurement is one of the most important tests the plant operator performs. The Safe Drinking Water Act stipulates specific monitoring requirements for turbidity. Turbidity of treated surface water must be measured every four hours or continuously and the results reported to the appropriate authority. Turbidity testing is the most critical tool in recognizing changes in raw water quality, detecting problems in coagulation and sedimentation, and troubleshooting filtration problems.

The maximum contaminant level (MCL) for turbidity is one TU with five TU allowed under certain circumstances. Water

---

[19] *Turbidity Units (TU).   Turbidity units are a measure of the cloudiness of water. If measured by a nephelometric (deflected light) instrumental procedure, turbidity units are expressed in nephelometric turbidity units (NTU) or simply TU. Those turbidity units obtained by visual methods are expressed in Jackson Turbidity Units (JTU) which are a measure of the cloudiness of water; they are used to indicate the clarity of water. There is no real connection between NTUs and JTUs. The Jackson turbidimeter is a visual method and the nephelometer is an instrumental method based on deflected light.*

(Turbidity)

treatment plant operators should strive to produce a finished water with a turbidity of 0.1 TU or less.

For additional information on monitoring requirements and turbidity requirements of the Surface Water Treatment Rule (SWTR), see Section 22.221, Volume II, and the poster included with this manual.

B. What Is Tested

	Common Range, NTU
Untreated Surface Water	1 to 300
Filtered Water	0.03 to 0.50
Well Water	0.05 to 1.0+

C. Apparatus

1. Turbidimeter: To minimize differences in turbidity measurements, rigorous specifications for turbidimeters are necessary. The turbidimeter should have the following important characteristics:

    a. The turbidimeter should consist of a nephelometer with a light source illuminating the sample, and one or more photoelectric detectors to indicate the intensity of scattered light at a 90° angle with a "readout" device.

    b. The light source should be an intense tungsten filament lamp.

    c. The total distance traveled by the light through the sample should be less than about 5 centimeters.

    d. The instrument should have several measurement ranges. The instrument should be able to measure from 0 to 100 turbidity units, with sufficient sensitivity (on the lowest scale) to detect differences of 0.02 or less in filtered waters having turbidities of less than one unit.

2. Sample tubes. These are usually provided with the instrument.

D. Reagents

1. Turbidity-free water: Pass distilled water through a membrane filter having a pore size no greater than 0.2 microns (available from laboratory supply houses). Discard the first 200 mL collected. If filtration does not reduce turbidity of the distilled water, use unfiltered distilled water.

2. Stock formazin turbidity suspension:[20]

    a. Solution I—Dissolve 1.000 gm hydrazine sulfate in distilled water and dilute to 100 mL in a volumetric flask.

    b. Solution II—Dissolve 10.00 gm hexamethylenetetramine in distilled water and dilute to 100 mL in a volumetric flask.

    c. In a 100-mL volumetric flask, add (using 5-mL volumetric pipets) 5.0 mL Solution I and 5.0 mL of Solution II. Mix and allow to stand 24 hours at 25°C. Then dilute

to the mark and mix. The turbidity of this suspension is considered 400 NTU exactly.

    d. Prepare solutions and suspensions monthly.

3. Standard turbidity suspensions: Dilute 10.00 mL stock turbidity suspension to 100 mL with turbidity-free water. Prepare weekly. The turbidity of this suspension is defined as 40 NTU.

4. Dilute turbidity standards: Dilute portions of the standard turbidity suspension with turbidity-free water as required. Prepare weekly.

E. Procedure

1. Turbidimeter calibration: The manufacturer's operating instructions should be followed. Measure your standard solutions on the turbidimeter covering the range of interest. If the instrument is already calibrated in standard turbidity units, this procedure will check the accuracy of the calibration scales. At least one standard should be run in each instrument range to be used. Some instruments permit adjustments of sensitivity so that scale values will correspond to turbidities. Reliance on an instrument manufacturer's scattering standards for calibrating the instrument is not an acceptable practice unless they are in acceptably close agreement with prepared standards. If a precalibrated scale is not supplied, then calibration curves should be prepared for each usable range of the instrument.

2. Turbidities less than 40 units: Shake the sample to thoroughly disperse the suspended solids. Wait until air bubbles disappear, then pour the sample into the turbidimeter tube. Read the turbidity directly from the instrument scale or from the appropriate calibration curve.

3. Turbidities exceeding 40 units: Dilute the sample with one or more volumes of turbidity-free water until the turbidity falls below 40 units. The turbidity of the original sample is then computed from the turbidity of the diluted sample and the dilution factor.

F. Example

A reservoir sample was collected and the turbidity was found to be greater than 40 units when first checked. 30 mL of this sample was then diluted with 60 mL of turbidity-free water. This diluted sample showed a turbidity of 25 units.

NTU found in diluted sample, A	= 25 NTU
mL of dilution water used, B	= 60 mL
mL of sample volume taken for dilution, C	= 30 mL

G. Calculation

$$\text{Nephelometric Turbidity Units (NTU)} = \frac{(A) \times (B+C)}{C}$$

$$= \frac{(25 \text{ NTU}) \times (60 \text{ mL} + 30 \text{ mL})}{30 \text{ mL}}$$

$$= 75 \text{ NTU}$$

---

[20] *Stock secondary standard turbidity suspensions that require no preparation are available from commercial suppliers and approved for use.*

(Turbidity)

H. Interpretation of Results

Report turbidity results as follows:

Turbidity Reading	Record to Nearest
0.0 to 1.0	0.05
1 to 10	0.1
10 to 40	1
40 to 100	5
100 to 1,000	10
>1,000	100

I. Notes and Precautions

1. Sample tubes must be kept scrupulously clean, both inside and out. Discard them when they become scratched or etched. Never handle them where the light strikes them.

2. Fill the tubes with samples and standards that have been agitated thoroughly, and allow sufficient time for bubbles to escape.

3. Turbidity should be measured in a sample as soon as possible to obtain accurate results. The turbidity of a sample can change after the sample is collected. Shaking the sample will not re-create the original turbidity.

J. Reference

See page 2-8, *STANDARD METHODS*, 20th Edition.

QUESTIONS

Write your answers in a notebook and then compare your answers with those on page 570.

11.3Y   What are the causes of turbidity in water?

11.3Z   How is turbidity measured?

11.4   ARITHMETIC ASSIGNMENT

Turn to the Appendix, "How to Solve Water Treatment Plant Arithmetic Problems," at the back of this manual and read Section A.139, Laboratory Procedures. Check all of the arithmetic in this section using an electronic pocket calculator. You should be able to get the same answers.

11.5   ADDITIONAL READING

1. *NEW YORK MANUAL*, Chapter 4,* "Water Chemistry," Chapter 5,* "Microbiology," and Chapter 21,* "Laboratory Examinations of Water."

2. *TEXAS MANUAL*, Chapter 6,* "Water Chemistry," and Chapter 12,* "Laboratory Examinations."

* Depends on edition.

11.6   WATER LABORATORY TESTS IN *WATER TREATMENT PLANT OPERATION*, Volume II

Laboratory procedures for the following tests are provided in *WATER TREATMENT PLANT OPERATION*, Volume II, Chapter 21, "Advanced Laboratory Procedures."

1. Algae Counts
2. Calcium
3. Chloride
4. Color
5. Dissolved Oxygen
6. Fluoride
7. Iron
8. Manganese
9. Marble Test
10. Metals
11. Nitrate
12. pH
13. Specific Conductance
14. Sulfate
15. Taste and Odor
16. Trihalomethanes
17. Total Dissolved Solids

End of Lesson 5 of 5 Lessons on LABORATORY PROCEDURES

Please answer the discussion and review questions next.

DISCUSSION AND REVIEW QUESTIONS

Chapter 11.   LABORATORY PROCEDURES

(Lesson 5 of 5 Lessons)

Write the answers to these questions in your notebook. The question numbering continues from Lesson 4.

19. What precautions should be considered when performing the hardness determination on a water sample?

20. How could you estimate the most effective dose of alum or a polymer in a water treatment process?

21. What precautions should be exercised when taking temperature measurements?

22. Why should turbidity be measured in a sample as soon as possible?

SUGGESTED ANSWERS

Chapter 11.   LABORATORY PROCEDURES

ANSWERS TO QUESTIONS IN LESSON 1

Answers to questions on page 503.

11.0A   Laboratory quality control tests are important because they provide the necessary information to monitor the treatment processes and ensure a safe and pleasant-tasting drinking water.

11.0B   The prefix milli- means 1/1,000 (0.001) of the unit following.

11.0C   The proper name of the chemical compound $CaCO_3$ is calcium carbonate.

Answers to questions on page 514.

11.1A   Descriptions of laboratory glassware and their use or purpose.

Item	Description	Use or Purpose
1. Beakers	Short, wide cylinders in sizes from 1 mL to 4,000 mL	Mixing chemicals
2. Graduated cylinders	Long, narrow cylinders in sizes from 5 mL to 4,000 mL	Measuring volumes
3. Pipets	Small-diameter, graduated tubes, with a pointed tip, in sizes from 1.0 mL to 100 mL	Delivering accurate volumes
4. Burets	Long tubes with graduated walls and a stopcock in sizes from 10 mL to 1,000 mL	Delivering and measuring accurate volumes used in "titrations"

11.1B   A meniscus is the curve of the surface of a liquid in a small tube.

11.1C   Never heat graduated cylinders in an open flame because they may break.

Answers to questions on page 516.

11.1D   A "standard solution" is a solution in which the exact concentration of a chemical or compound in a solution is known.

11.1E   Laboratory notebooks and worksheets help record data in an orderly manner.

11.1F   Four sources or causes of poor quality of analytical data are:

1. Sloppy laboratory technique,
2. Deteriorated reagents and standards,
3. Poorly operating instruments, and
4. Calculation mistakes.

Answers to questions on page 521.

11.1G   Six laboratory hazards are:

1. Hazardous materials,
2. Explosions,
3. Cuts and bruises,
4. Electric shock,
5. Fire, and
6. Burns (heat and chemical).

11.1H   NEVER work alone in the laboratory. Someone should always be available to help in case you should have an accident which blinds you, leaves you unconscious, or starts a fire you can't handle. If necessary, have someone check on you regularly to be sure you are OK.

11.1I   True. You may ADD ACID TO WATER, but never the reverse.

11.1J   Dispose of small amounts of corrosive acids by pouring the neutralized acid down corrosion-resistant sinks and sewers using large quantities of water to dilute and flush the acid.

ANSWERS TO QUESTIONS IN LESSON 2

Answers to questions on page 524.

11.2A   The greatest errors in laboratory tests are frequently caused by (1) improper sampling, (2) poor sample preservation, and (3) lack of enough mixing during testing.

11.2B   A representative sample must be collected in order for test results to have any significant meaning. Without a representative sample, the test results will not reflect actual water conditions.

11.2C   Sampling points in a distribution system should be selected in order to trace the course from the finished water source (at the well or plant), through the transmission mains, and then through the major and minor arteries of the system.

Answers to questions on page 527.

11.2D   The two general types of samples collected by water treatment personnel are (1) grab samples, and (2) composite samples.

11.2E   Water quality indicators that are usually measured with a grab sample include (1) dissolved gases, (2) coliform bacteria, (3) residual chlorine, (4) temperature, and (5) pH.

11.2F   Depth samples are collected by the use of a Kemmerer Sampler or similar device. The sampling device and container are lowered to the desired depth, then opened, filled, closed, and returned to the surface.

11.2G Samples should not be collected from taps surrounded by excessive foliage (leaves, flowers) or from taps that are dirty, corroded, or are leaking.

11.2H When collecting a sample, record the sample location, date and time of collection, name of collector, and any other pertinent information.

ANSWERS TO QUESTIONS IN LESSON 3

Answers to questions on page 531.

11.3A Chemicals used in water treatment that will cause changes in alkalinity include alum, chlorine, and lime.

11.3B Alkalinity determination is needed when calculating chemical dosages for coagulation and water softening. Alkalinity must also be known to calculate corrosivity and to estimate carbonate hardness of water.

Answers to questions on page 533.

11.3C The benefits of chlorinating water include:

1. Disinfection;
2. Improved quality of water due to chlorine reacting with iron, manganese, protein substances, sulfide, and many taste- and odor-producing compounds;
3. Control of microorganisms that might interfere with coagulation and flocculation;
4. Keeps filter media free of slime growths; and
5. Helps bleach out undesirable color.

11.3D A potential adverse effect from chlorination is the possibility of the formation of carcinogenic chlororganic compounds such as chloroform and other THMs.

Answers to questions on page 537.

11.3E Conditions that can cause variations in the chlorine demand of water include (1) the amount of chlorine applied, (2) length of contact time, (3) pH, (4) temperature, (5) organics, and (6) reducing agents.

11.3F The operator uses the chlorine demand test to determine the best chlorine dosage to achieve specific chlorination objectives.

11.3G Chlorine
Demand, = Chlorine Added, mg/L – Free Residual Chlorine, mg/L
mg/L

= 2.0 mg/L – 0.4 mg/L

= 1.6 mg/L

ANSWERS TO QUESTIONS IN LESSON 4

Answers to questions on page 557.

11.3H Drinking waters are not tested for specific pathogens because the tests are very time-consuming and require special techniques and equipment.

11.3I The number of samples required for coliform tests is generally based on the population served by the water system.

11.3J Sodium thiosulfate should be added to sample bottles for coliform tests to neutralize any residual chlorine in the sample. If residual chlorine is not neutralized, it will continue to be toxic to the coliform organisms remaining in the sample and give false (negative) results.

11.3K Steam autoclaves sterilize (kill all organisms) at a pressure of 15 psi and temperature of 121°C during a 15-minute time period (at sea level).

11.3L

mL of sample	0.1	0.01	0.001
Dilutions	–1	–2	–3
Readings (+ tubes)	5	3	1

MPN = 11,000/100 mL

11.3M The number of coliforms is determined by counting the number of coliform-appearing colonies grown on the membrane filter.

11.3N Incubate the filter, without inverting the dish, for 1½ to 2 hours at 35°C in an atmosphere of 90 percent humidity. Transfer the membrane filter to a new pad enriched with M-Endo Broth. Invert the dish and incubate for 20 to 22 hours at 35°± 0.5°C.

11.3O The total coliform test is not an adequate measure of the extent of human wastes in water because it measures both fecal coliforms (*E. coli*) and soil coliforms. Additional tests such as the MF-mTEC Method are needed to specifically identify and count fecal coliforms.

ANSWERS TO QUESTIONS IN LESSON 5

Answers to questions on page 560.

11.3P The principal hardness-causing ions in water are calcium and magnesium.

11.3Q Problems caused by hardness in water include (1) incrustations when water evaporates or scale when heated, and (2) formation of precipitates when combined with soap.

Answers to questions on page 563.

11.3R The jar test is used to (1) determine the correct coagulant dosage, and (2) evaluate new coagulants or polymers.

11.3S Speeds used during the jar test are as follows:

1. 80 RPM for the first minute,
2. 20 RPM for the next 30 minutes, and
3. Stop stirring (0 RPM) for the next 30 minutes.

These speeds and times should be adjusted if necessary to be similar to actual conditions in the water treatment plant.

Answers to questions on page 564.

11.3T The pH of a water indicates the intensity of its basic or acidic strength.

11.3U Precautions to be exercised when using a pH meter include:

1. Prepare fresh buffer solution weekly for calibration purposes;
2. Have pH meter, samples, and buffer solutions all near the same temperature;
3. Watch for erratic results arising from electrodes, faulty connections, or fouling of electrodes with interfering matter;
4. The pH of water changes with temperature; and
5. If testing waters with a range of pH values, the "slope" of the pH meter may require adjusting.

Answers to questions on page 565.

11.3V Temperature readings are important because temperature influences chemical reaction rates, biological growth, dissolved gas concentrations, and water stability with respect to calcium carbonate. Also consumers are sensitive to the temperature of the water they drink.

11.3W The thermometer should remain immersed in the liquid while being read for accurate results. When removed from the liquid, the reading will change.

11.3X All thermometers should be calibrated against an accurate National Bureau of Standards thermometer because some poorer quality thermometers are substantially inaccurate (off as much as 6°F or 3°C).

Answers to questions on page 567.

11.3Y Turbidity in water can be caused by the presence of suspended matter such as silt, finely divided organic and inorganic matter, and microscopic organisms such as algae.

11.3Z Turbidity is measured by the nephelometric method.

APPENDIX

WATER TREATMENT PLANT OPERATION

(VOLUME I)

Final Examination and Suggested Answers

How to Solve Water Treatment Plant Arithmetic Problems

Water Abbreviations

Water Words

Subject Index

FINAL EXAMINATION

VOLUME I

This final examination was prepared *TO HELP YOU REVIEW* the material in this manual. The questions are divided into five types:

1. True-False,

2. Best Answer,

3. Multiple Choice,

4. Short Answer, and

5. Problems.

To work this examination:

1. Write the answer to each question in your notebook,

2. After you have worked a group of questions (you decide how many), check your answers with the suggested answers at the end of this exam, and

3. If you missed a question and don't understand why, reread the material in the manual.

You may wish to use this examination for review purposes when preparing for civil service and certification examinations.

Since you have already completed this course, you do not have to send your answers to California State University, Sacramento.

True-False

1. Many sources of water are directly suitable for drinking purposes without treatment.

   1. True
   2. False

2. Precipitation has significant microbial content.

   1. True
   2. False

3. Many organisms that cause disease in humans originate with the fecal discharges of infected humans.

   1. True
   2. False

4. Algal blooms rarely take place in the upper layer of water in a reservoir.

   1. True
   2. False

5. When measuring turbidity, many small particles will reflect less light than an equivalent large particle.

   1. True
   2. False

6. Operators should allow as much time as possible for the coagulation reaction.

   1. True
   2. False

7. A "popcorn flake" is a desirable floc appearance.

   1. True
   2. False

8. Polymer powders on floors become extremely slippery when wet.

   1. True
   2. False

9. Solids-contact process units combine the coagulation, flocculation, and sedimentation processes in a single basin.

   1. True
   2. False

10. Sludge produced by a solids-contact clarification unit is recycled through the process to act as a coagulant aid.

    1. True
    2. False

11. Low-turbidity source waters require sedimentation processes.

    1. True
    2. False

12. Process water samples must be representative of actual conditions in the treatment plant.

    1. True
    2. False

13. Poor chemical treatment can often result in either early turbidity breakthrough or rapid head loss buildup.

    1. True
    2. False

14. Backwashing at too high a rate is much more destructive than at too low a backwash rate.

    1. True
    2. False

15. Particle counters can be used to indicate the probable removal of *Cryptosporidium* and *Giardia* achieved through optimized plant performance for particle removal.

    1. True
    2. False

16. Organics found in the water can consume great amounts of chlorine disinfectants while forming unwanted compounds.
    1. True
    2. False

17. A chlorine residual in the form of free available residual chlorine has the highest disinfection ability.
    1. True
    2. False

18. Use regular pipe fittings when making connections with chlorine containers.
    1. True
    2. False

19. Always add acid to water, never the reverse.
    1. True
    2. False

20. For corrosion to occur, both electric current and chemical reaction must be present.
    1. True
    2. False

21. When iron and copper are joined, the iron will be protected and the copper will corrode.
    1. True
    2. False

22. Water containing high levels of hardness tend to be more aggressive.
    1. True
    2. False

23. Increased nutrient levels in source water may result from either natural conditions or from human activity within the watershed.
    1. True
    2. False

24. Cross connections in the distribution system are potentially very hazardous and a source of taste and odor complaints.
    1. True
    2. False

25. Superchlorination is followed by dechlorination as part of the treatment for tastes and odors.
    1. True
    2. False

26. Ideally process control measurements should be made on a continuous basis by special instrumentation.
    1. True
    2. False

27. A material safety data sheet (MSDS) is your best source of information about dangerous chemicals.
    1. True
    2. False

28. Federal laws include sludge from a water treatment plant as an industrial waste and require proper handling and disposal.
    1. True
    2. False

29. The collection of a bad sample or a bad laboratory result is about as useful as no results.
    1. True
    2. False

30. Completely empty volumetric pipets by blowing out the small amount remaining in the tip.
    1. True
    2. False

31. Never pipet chemical reagent solutions or unknown water samples by mouth.
    1. True
    2. False

Best Answer (Select only the closest or best answer.)

1. What is the responsibility of water treatment plant operators?
    1. Avoid boil-water orders
    2. Minimize costs of producing drinking water
    3. Produce safe and pleasant drinking water
    4. Successfully pass operator certification examinations

2. What is the cause of stratification in lakes and reservoirs?
    1. Breezes starting the circulation of surface water
    2. Formation of separate layers of temperature, plant life, or animal life
    3. Lake "turnover"
    4. Uniform water temperature profile in lake or reservoir

3. How extensive should the treatment be for reclaimed water?
    1. All reclaimed water should receive the same degree of treatment
    2. Depends on potential exposure to the public
    3. Depends on the funds available for treatment
    4. Local regulations determine the degree of treatment

4. How can iron, manganese, and hydrogen sulfide problems be controlled in a reservoir?
    1. By controlling algal productivity
    2. By improving the fishery habitat
    3. By preventing dissolved oxygen depletion
    4. By regulating recreational activities

5. What is the best tool for managing watersheds?
    1. Political process
    2. Public hearings
    3. Regulatory process
    4. Voluntary compliance

6. What is destratification of a lake or reservoir?
    1. Elimination of separate layers by algal blooms
    2. Natural turnover during seasonal changes
    3. Vertical mixing to eliminate separate layers of temperature, plant life, or animal life
    4. Wind blowing across water surface and causing mixing

7. What is the purpose of coagulation and flocculation?
    1. To control corrosion
    2. To kill disease-causing organisms
    3. To remove leaves, sticks, fish, and other debris
    4. To remove particulate impurities, especially nonsettleable solids and color

8. After a jar test is completed, what does a hazy settled water indicate?

1. Poor coagulation
2. Poor flocculation
3. Poor mixing
4. Poor turbidity

9. How can an operator verify the effectiveness of the coagulant chemicals and dosages being applied at the flash mixer?

1. Determine finished water chlorine demand
2. Measure raw and finished water temperatures
3. Measure source water turbidity levels
4. Perform a series of jar tests

10. What happens to the settling rate (settling velocity) of particles in water when the water temperature drops?

1. Decreases
2. Depends on weather
3. Increases
4. Stays constant

11. What is the major means by which operators can control water treatment processes?

1. Adjusting chemicals and chemical feed rates
2. Controlling sedimentation process
3. Controlling water temperatures
4. Modifying demands for water

12. Which task should be performed immediately following dewatering of a sedimentation basin?

1. Grease and lubricate all gears, sprockets, and mechanical moving parts that have been submerged
2. Perform necessary laboratory analyses
3. Record all gage and meter readings
4. Stop flow to sedimentation basin

13. How are filter production (capacity) rates measured?

1. GPM
2. GPM/sq ft
3. MGD
4. MGD/sq ft

14. Why is surface wash used during backwashing?

1. To prevent mudballs
2. To provide a backup backwash system
3. To reduce backwash water requirements
4. To wash walls of filter

15. What do significant changes in source water turbidity levels, either increases or decreases, require?

1. Adjustment of filter run time to accommodate change
2. Immediate verification of filtration process effectiveness
3. Observation of head loss to reveal need for process adjustment
4. Performance of jar test to determine optimum chemical dose

16. Why should a filter be drained if it is going to be out of service for a prolonged period?

1. To allow media to dry out
2. To avoid algal growth
3. To prevent filter tank from floating on groundwater
4. To save water

17. Why has there been an increased interest in disinfectants other than chlorine?

1. Chlorine may form carcinogenic compounds (trihalomethanes or THMs)
2. Other disinfectants are cheaper
3. Other disinfectants are more readily available
4. Other disinfectants leave a better disinfecting residual

18. What are the components of a chlorine residual?

1. Chlorine demand plus chlorine dose
2. Combined chlorine forms plus free chlorine
3. Free chlorine plus chlorine dose
4. Hypochlorous acid plus hydrochloric acid

19. Where should the minimum chlorine residual be measured in the distribution system?

1. At the beginning of the distribution system
2. At the most remote location in the system
3. In the distribution system service storage tank
4. Just downstream from where disinfectant is being fed into system

20. How are ORP probes used in the disinfection process?

1. They accurately measure chlorine dosage
2. They measure the liquid chlorine remaining in a chlorine container
3. They provide a direct measure of the disinfecting power of the disinfectant
4. They quickly measure and record chlorine residuals

21. Why are zinc and magnesium commonly used as sacrificial anodes in water tanks or for buried pipes?

1. They will corrode preferentially to aluminum and iron
2. They will form a protective coating over other metals
3. They will last longer than other metals in corrosive waters
4. They will *NOT* corrode when electrons flow in a tank or pipe

22. Why should samples for corrosivity tests be taken only after postchlorination?

1. Because disinfectant chemicals may react with corrosion-control chemicals to either increase or decrease corrosivity
2. Because samples should be collected in areas where corrosion is likely to occur
3. To allow time for mixing of all chemicals
4. To ensure collection of representative samples

23. Where are monitoring samples for the Lead and Copper Rule collected?

1. At the consumers' taps
2. At the entrance to the clear well
3. At the most distant locations in the distribution system
4. At the raw water source

24. What is the secret to successful taste and odor control?

1. Implement an effective public relations program
2. Prevent tastes and odors from ever developing
3. Provide consumers with treatment option information
4. Treat tastes and odors when first complaint is received

25. Taste and odor treatment methods can be divided into which categories?

1. Coagulation/flocculation
2. Degasification and adsorption
3. Oxidation and reduction
4. Removal and destruction

26. Where is powdered activated carbon often applied?

    1. After chlorination
    2. After filtration
    3. At flash mixer
    4. At floc tanks

27. Over which water quality indicator do operators have the greatest control?

    1. Alkalinity
    2. pH
    3. Temperature
    4. Turbidity

28. Which chemical is one of the most dangerous of the common alkalies?

    1. Calcium hydroxide
    2. Caustic soda
    3. Chlorine
    4. Sodium bicarbonate

29. What should an operator do first when a treatment process upset results in the failure to meet a specific drinking water quality standard?

    1. Fix process upset
    2. Notify appropriate local health authorities
    3. Stop operating process
    4. Switch sources of raw water

30. Which piece of laboratory equipment is the most accurate way to measure a volume of liquid?

    1. Beaker
    2. Flask
    3. Graduated cylinder
    4. Pipet

31. Which piece of laboratory equipment is used to titrate a chemical reagent?

    1. Buchner funnel
    2. Buret
    3. Graduated cylinder
    4. Pipet

32. Which style of thermometer will change most rapidly when removed from the liquid sample?

    1. Dial
    2. Digital
    3. Partial immersion
    4. Total immersion

33. What is the accepted method used to measure turbidity?

    1. Candle
    2. Jackson
    3. Light
    4. Nephelometric

Multiple Choice (Select all correct answers.)

1. The size of a water treatment plant as well as the number and specific types of processes it uses depend on which of the following factors?

    1. Cost considerations
    2. Demand for water by population served
    3. Fire protection needs
    4. Impurities in raw water
    5. Water quality (purity) standards

2. Which of the following tasks are typical duties of a water treatment plant operator?

    1. Collect representative water samples
    2. Conduct safety inspections
    3. Maintain plant records
    4. Make periodic operating checks of plant equipment
    5. Perform routine preventive maintenance

3. What are the safety responsibilities of an operator?

    1. Be aware of safety hazards around the plant
    2. Be part of an active safety program
    3. Being sure everyone follows safe procedures
    4. Ensuring that operators take safety seriously
    5. Understanding why safe procedures must be followed at all times

4. Possible sources of groundwater contamination include which of the following items?

    1. Agricultural drainage systems
    2. Entry of surface runoff into wells
    3. Evapotranspiration systems
    4. Seawater intrusion
    5. Seepage from septic tank leaching systems

5. What conditions are favorable for the rapid growth of algae (blooms)?

    1. Dissolved oxygen
    2. Hardness
    3. Nutrients
    4. pH
    5. Temperature

6. What are the basic objectives that control operation of water treatment plants?

    1. Production of a safe drinking water
    2. Production of an aesthetically pleasing drinking water
    3. Production of drinking water at a reasonable cost
    4. Providing a safe working environment
    5. Providing opportunities for professional advancement

7. Which of the following water quality problems may be related to algal blooms?

    1. Dissolved oxygen depletion
    2. Reduced chlorination efficiency
    3. Reduced water temperature
    4. Shortened filter runs of complete treatment plants
    5. Taste and odor problems

8. What is the impact from increased organic loadings on water treatment plants resulting from algal blooms?

    1. High trihalomethane levels
    2. Increased chlorine demand
    3. Increased color
    4. Increased public confidence in water supply
    5. Increased water treatment costs

9. On domestic water supply reservoirs that develop anaerobic zones, what water quality indicators should be monitored?

    1. Biochemical oxygen demand (BOD)
    2. Hardness
    3. Hydrogen sulfide
    4. Iron
    5. Manganese

10. Which of the following are commonly used coagulation chemicals?

1. Alum
2. Hypochlorites
3. Metallic salts
4. Polymers
5. Sodium chloride

11. Which of the following factors are important when evaluating jar test results?

1. Amount of floc formed
2. Clarity of water above settled floc
3. Clarity of water between floc particles
4. Floc settling rate
5. Rate of floc formation

12. Operators of the coagulation-flocculation process will be exposed to which of the following potential safety hazards?

1. Confined spaces
2. Electrical equipment
3. Rotating mechanical equipment
4. Slippery surfaces
5. Water treatment chemicals

13. Which items are adjusted in the enhanced coagulation process?

1. Basic contents
2. Coagulant dosage
3. Detention time
4. Mixing
5. pH

14. What is the purpose of presedimentation facilities?

1. Avoid need for sedimentation basins
2. Even out fluctuations in the concentration of suspended solids
3. Prevent solids settling out before sedimentation
4. Provide an equalizing basin
5. Reduce the solids-removal load at the water treatment plant

15. Why is short-circuiting undesirable in a sedimentation basin?

1. Causes shorter contact times
2. Causes shorter reaction times
3. Causes shorter settling times
4. Minimizes density currents
5. Minimizes temperature differences

16. What kinds of changes cause instability in solids-contact units?

1. DO
2. Flow
3. pH
4. Temperature
5. Turbidity

17. Which water quality indicators should be monitored in the normal operation of the sedimentation process?

1. Chlorine residual
2. Coliforms
3. Suspended solids
4. Temperature
5. Turbidity

18. Filter removal of turbidity depends on which of the following factors?

1. Chemical characteristics of water being treated
2. Filter type
3. Nature of suspended particulates
4. Operation of filter
5. Types and degree of pretreatment

19. Which of the following characteristics are used to classify filter media?

1. Effective size
2. Filterability
3. Hardness
4. Specific gravity
5. Uniformity coefficient

20. How can an operator determine if a filter is NOT completely clean after backwashing?

1. Backwash rate is too low
2. Initial head loss is too high
3. Length of backwash cycle is too short
4. Next filter run is too short
5. Pumping rate is too low

21. Which of the following items are indicators of abnormal filter conditions?

1. Algae on walls and media
2. Media boils during backwash
3. Media cracking or shrinkage
4. Mudballs in filter media
5. Short filter runs

22. Which of the following factors influence chlorine disinfection?

1. Microorganisms
2. pH
3. Reducing agents
4. Temperature
5. Turbidity

23. Chloramines have proven effective in accomplishing which of the following objectives?

1. Killing or inactivating heterotrophic plate count bacteria
2. Maintaining a detectable residual throughout the distribution system
3. Penetrating the biofilm and reducing the potential for coliform regrowth
4. Reducing taste and odor problems
5. Reducing the formation of trihalomethanes (THMs) and other disinfection by-products (DBPs)

24. What are the most probable causes of a drop in chlorine level in treated water?

1. Change in alkalinity
2. Coliform regrowth
3. Drop in chlorine feed rate
4. Increase in chlorine demand
5. Increase in coliform level

25. Which of the following items should be included in all chlorine safety programs?

1. Emergency procedures for chlorine leaks and first aid
2. Maintenance and calibration program for safety devices and equipment
3. Periodic hands-on training using safety equipment
4. Plant tours for police and fire departments
5. Written rules and specific safety procedures

26. Which of the following physical factors influence the rate of corrosion?

    1. Flow velocity
    2. Soil moisture
    3. Stray electric current
    4. Temperature of water
    5. Type of coatings

27. How can operators determine if corrosion problems exist?

    1. Chemical tests on the water
    2. Consumer complaints
    3. Examine materials removed from distribution system
    4. Flow tests
    5. Increasing number of leaks

28. Which of the following are monitored as corrosion-related water quality indicators?

    1. Alkalinity
    2. Conductivity
    3. Hardness
    4. pH
    5. Temperature

29. What are some of the damaging effects for a water utility resulting from a taste and odor episode?

    1. Additional regulatory surveillance
    2. Funding restrictions
    3. Increased public relations problems
    4. Increased sales of bottled water
    5. Loss of public confidence

30. Which of the following are locations that have been identified as possible sources of tastes and odors?

    1. Consumers' plumbing
    2. Conveyance facilities
    3. Finished water storage reservoirs
    4. Raw water sources
    5. Treatment plants

31. The success of chlorination as a taste and odor treatment will depend on which factors?

    1. Contact time
    2. Dose applied
    3. Reservoir stratification
    4. Seriousness of problem
    5. Type of odor

32. What precautions should be taken when handling powdered activated carbon (PAC)?

    1. Follow confined space safety procedures
    2. Store PAC bags on wooden pallets
    3. Store PAC away from flammable materials
    4. Use dust control equipment
    5. Wear protective clothing

33. Where is turbidity commonly monitored in a water treatment plant?

    1. Backwash water
    2. Each unit process
    3. Finished water
    4. Individual filters
    5. Raw water

34. What are the purposes of clear wells (or plant storage reservoirs)?

    1. Act as a buffer that prevents frequent ON/OFF cycling of finished water pumps
    2. Allow clarification of water by providing time for particulates to settle out
    3. Allow filling when demands are low to compensate for peak periods which draw the level down
    4. Permit planned changes in treatment plant operation
    5. Provide operational storage to average out high and low flow demands

35. Which factors can lead to poor quality laboratory data?

    1. Calculation mistakes
    2. Deteriorated reagents
    3. Neatly recorded results in lab notebooks
    4. Poorly operating instruments
    5. Sloppy laboratory technique

36. Which of the following hazards could be encountered in a laboratory?

    1. Burns
    2. Drowning
    3. Electric shock
    4. Explosions
    5. Hazardous materials

37. Grab samples should be used to measure which of the following water quality indicators?

    1. Chlorine residual
    2. Coliform bacteria
    3. Dissolved gases
    4. pH
    5. Temperature

38. An alkalinity determination is needed when calculating chemical dosages used in what water treatment processes?

    1. Coagulation
    2. Filtration
    3. Flocculation
    4. Settling
    5. Softening

Short Answer

1. Why do intake structures at water treatment plants have screens?

2. Why should a water treatment plant operator be present when a new plant is being constructed?

3. What is the reason for keeping adequate, reliable records?

4. How does the water (hydrologic) cycle work?

5. What items should be considered before selecting a location and constructing a water supply intake in a river or stream?

6. Who should conduct a sanitary survey? Why?

7. Who must comply with the Safe Drinking Water Act (SDWA) regulations?

8. What methods are used to treat domestic water delivered from water supply reservoirs?

9. What is the influence of algal blooms on dissolved oxygen?

10. How can the desired rate of copper sulfate application to a reservoir be achieved?

11. What is the main safety hazard encountered during reservoir sampling?

12. Define the following terms:
   a. Grab Sample
   b. Reagent
   c. Short-Circuiting

13. What is the difference between the coagulation and flocculation processes?

14. What is the purpose of the flocculation process?

15. What is the basic purpose of the jar test?

16. What would you do if you discovered abrupt (sudden) changes in the source water or filtered water turbidity?

17. What are the purposes of the sedimentation process?

18. List the four zones into which a typical sedimentation basin can be divided.

19. What problems are created when the sludge buildup on the bottom of the sedimentation tank becomes too great?

20. Why should water treatment plants have at least two sedimentation basins?

21. How is the approximate process removal efficiency of a sedimentation basin determined?

22. What is the primary purpose of the filtration process?

23. What is the primary purpose of using activated carbon (granular form) as filter media?

24. How frequently should the performance of the filtration process be evaluated?

25. What four types of filtration can be used to meet the requirements of the Surface Water Treatment Rule?

26. What is the difference between disinfection and sterilization?

27. How does the temperature of the water influence disinfection?

28. How is the chlorine demand determined?

29. How should valves on chlorine containers be opened?

30. What is the purpose of evaporators?

31. If chlorine is escaping from a cylinder, what would you do?

32. What properties make chlorine gas so hazardous?

33. Why is ozone normally prepared on site?

34. Define the following terms:
   a. Breakpoint Chlorination
   b. Pathogenic Organisms

35. What is corrosion?

36. How do dissolved solids in water influence corrosion?

37. What is the impact of calcium carbonate on corrosion?

38. What is a tubercle?

39. How can water be tested to determine if it is undersaturated or supersaturated with calcium carbonate?

40. How can the corrosivity of a water be reduced?

41. What are the most common types of water quality complaints received by a water utility?

42. How can the use of copper sulfate in a reservoir cause a taste and odor problem?

43. How is the frequency of flushing mains in a water distribution system determined?

44. List the two broad categories of taste and odor treatment methods.

45. What types of taste- and odor-producing wastes have been treated successfully by using chlorine dioxide?

46. What safety precautions should be exercised when cleaning a tank that stores powdered activated carbon?

47. Where are water samples usually collected to measure compliance with water quality standards?

48. What precautions must be exercised when using air for pneumatic data transmission?

49. Why are positive air pressure fans preferred over floor-level exhaust fans in chlorine storage facilities?

50. What three factors usually cause treatment process failures?

51. What is a meniscus?

52. How would you titrate a test solution?

53. True or False? You may *ADD ACID TO WATER*, but never water to acid.

54. Why should water not be poured on certain types of fires?

55. Why are proper sampling procedures important?

56. What is meant by a representative sample?

57. What is a potential adverse effect from chlorination?

58. What are the principal hardness-causing ions in water?

59. Why should turbidity be measured in a sample as soon as possible?

Problems

1. What is the surface area of a rectangular settling basin 50 feet long and 12 feet wide?

   1. 60 square feet
   2. 400 square feet
   3. 500 square feet
   4. 600 square feet
   5. 900 square feet

2. What is the surface area of a circular clarifier 50 feet in diameter?

   1. 962 square feet
   2. 1,256 square feet
   3. 1,963 square feet
   4. 2,512 square feet
   5. 5,024 square feet

3. Determine the setting on a dry alum feeder in pounds per day when the flow is 1.2 MGD. Jar tests indicate that the best alum dose is 9 mg/L.

   1. 70 lbs/day
   2. 75 lbs/day
   3. 90 lbs/day
   4. 100 lbs/day
   5. 130 lbs/day

4. A water treatment plant used 24 pounds of cationic polymer to treat 1.4 million gallons of water during a 24-hour period. What is the polymer dosage in mg/L? Select the closest answer.

   1. 2 mg/L
   2. 5 mg/L
   3. 18 mg/L
   4. 36 mg/L
   5. 48 mg/L

5. Calculate the theoretical detention time for a rectangular sedimentation basin. The basin is 80 feet long, 30 feet wide, 10 feet deep, and treats a flow of 1.8 MGD.

   1. 2.4 hours
   2. 2.7 hours
   3. 2.9 hours
   4. 3.2 hours
   5. 3.5 hours

6. A tank with a capacity of 20,000 cubic feet can hold how many gallons of water?

   1. 2,400 gallons
   2. 2,670 gallons
   3. 37,400 gallons
   4. 149,600 gallons
   5. 166,800 gallons

7. Determine the upward force in pounds on the bottom of an empty sedimentation basin caused by a groundwater depth of 5 feet above the tank bottom. The basin is 10 feet wide and 30 feet long.

   1. 93,600 lbs
   2. 112,320 lbs
   3. 112,500 lbs
   4. 140,400 lbs
   5. 168,480 lbs

8. During a 24-hour period a water treatment plant filters 1.6 million gallons. The surface area of the filters is 500 square feet. Determine the filtration rate in gallons per minute per square foot.

   1. 1.8 GPM/sq ft
   2. 2.2 GPM/sq ft
   3. 2.5 GPM/sq ft
   4. 3.0 GPM/sq ft
   5. 3.1 GPM/sq ft

9. What should be the setting on a chlorinator (lbs chlorine per 24 hours) if a pump usually delivers 400 GPM and the desired chlorine dosage is 4.0 mg/L? Select the closest answer.

   1. 2 lbs chlorine/24 hours
   2. 8 lbs chlorine/24 hours
   3. 14 lbs chlorine/24 hours
   4. 19 lbs chlorine/24 hours
   5. 24 lbs chlorine/24 hours

10. Estimate the chlorine dose in mg/L if a chlorinator feeds at a rate of 18 lbs per 24 hours and the flow is 0.55 MGD.

    1. 3.0 mg/L
    2. 3.3 mg/L
    3. 3.6 mg/L
    4. 3.9 mg/L
    5. 4.2 mg/L

11. Estimate the water velocity in a channel in feet per second if a float travels 20 feet in 10 seconds.

    1. 0.2 ft/sec
    2. 0.25 ft/sec
    3. 2 ft/sec
    4. 4 ft/sec
    5. 5 ft/sec

12. Estimate the water flow rate in an 18-inch diameter pipe in gallons per minute (GPM) when the flow velocity is two feet per second. Select the closest answer.

    1. 1,590 GPM
    2. 2,380 GPM
    3. 2,820 GPM
    4. 3,870 GPM
    5. 4,230 GPM

13. What horsepower must a pump deliver to water that must be lifted 90 feet? The flow is 40 GPM.

    1. 0.9 HP
    2. 1.0 HP
    3. 50 HP
    4. 60 HP
    5. 76 HP

14. How many kilowatt-hours per day are required by a pump with a motor horsepower of 50 horsepower when the pump operates 24 hours per day?

    1. 716 kW-hr/day
    2. 895 kW-hr/day
    3. 960 kW-hr/day
    4. 1,075 kW-hr/day
    5. 1,287 kW-hr/day

15. Calculate the volume of a rectangular sedimentation basin 8 feet deep, 20 feet wide, and 55 feet long.

    1. 1,100 cu ft
    2. 8,800 cu ft
    3. 10,100 cu ft
    4. 75,900 cu ft
    5. 84,900 cu ft

16. Results from a coliform bacteria multiple fermentation tube test of raw water are as follows:

mL of sample	10	1	0.1	0.01	0.001
Dilutions	0	0	−1	−2	−3
Positive tubes	5	5	5	3	1

    What is the most probable number of coliforms per 100 mL in the sample? Use Table 11.3 on page 545.

    1. 110 coliforms per 100 mL
    2. 900 coliforms per 100 mL
    3. 1,100 coliforms per 100 mL
    4. 9,000 coliforms per 100 mL
    5. 11,000 coliforms per 100 mL

Data to answer questions 17 and 18.

Results from MPN tests during one week were as follows:

Day	S	M	T	W	T	F	S
MPN/100 mL	4	7	14	8	9	16	5

17. What is the mean MPN per 100 mL from the data?

1. 6 MPN/100 mL
2. 7 MPN/100 mL
3. 8 MPN/100 mL
4. 9 MPN/100 mL
5. 10 MPN/100 mL

18. What is the median MPN per 100 mL from the data?

1. 6 MPN/100 mL
2. 7 MPN/100 mL
3. 8 MPN/100 mL
4. 9 MPN/100 mL
5. 10 MPN/100 mL

SUGGESTED ANSWERS FOR FINAL EXAMINATION

VOLUME I

True-False

1. False   Many sources of water are *NOT* suitable for drinking without treatment.

2. False   Precipitation has virtually no microbial content.

3. True   Many organisms that cause disease originate from fecal discharges of infected humans.

4. False   In most cases, algal blooms take place in the upper layer of water.

5. False   Many small particles reflect *MORE* light than an equivalent large particle.

6. False   Operators should allow as *SHORT* a time as possible for coagulation reaction.

7. True   A "popcorn flake" is a desirable floc appearance.

8. True   Polymer powders on floors become extremely slippery when wet.

9. True   Solids-contact units combine coagulation, flocculation, and sedimentation in one basin.

10. True   Sludge produced is recycled to act as a coagulant aid.

11. False   Low-turbidity source waters do *NOT* require sedimentation processes.

12. True   Process water samples must be representative of actual plant conditions.

13. True   Poor chemical treatment can result in early turbidity breakthrough or rapid head loss buildup.

14. True   Backwashing at too high a rate is much more destructive than at too low a rate.

15. True   Particle counters can be used to indicate probable removal of *Cryptosporidium* and *Giardia*.

16. True   Organics can consume great amounts of disinfectants and form unwanted compounds.

17. True   Free available residual chlorine has the highest disinfection ability.

18. False   Do *NOT* use regular pipe fittings when making connections with containers.

19. True   Always add acid to water, never the reverse.

20. True   Electric current and chemical reaction must be present for corrosion to occur.

21. False   When iron and copper are joined, iron will corrode and copper is protected.

22. False   Water containing very little hardness tends to be more aggressive.

23. True   Increased nutrient levels may result from natural conditions or human activity.

24. True   Cross connections are potentially very hazardous and a source of complaints.

25. True   Superchlorination is followed by dechlorination for treatment of tastes and odors.

26. True   Process control measurements should be made on a continuous basis.

27. True   An MSDS is your best source of information about dangerous chemicals.

28. True   Federal laws include sludge as an industrial waste and require proper disposal.

29. True   Bad samples or bad lab results are as useful as no results.

30. False   Under no circumstance should the small amount remaining in tip be blown out.

31. True   Never pipet chemical reagent solutions or unknown water samples by mouth.

Best Answer

1. 3   The main responsibility of water treatment plant operators is to produce safe and pleasant drinking water.

2. 2   Stratification is the formation of separate layers of temperature, plant life, or animal life.

3. 2   The extent of treatment for reclaimed water depends on its potential to reach the public.

4. 3   Problems with iron, manganese, and hydrogen sulfide in reservoirs can be controlled by preventing the depletion of dissolved oxygen (DO).

5. 3   The best tool for managing watersheds is the regulatory process.

6. 3 Destratification of a lake or reservoir is a vertical mixing to eliminate separate layers of temperature, plant life, or animal life.

7. 4 Coagulation and flocculation remove particulate impurities, especially nonsettleable solids and color.

8. 1 A hazy settled water after a jar test is completed indicates poor coagulation.

9. 4 Operators can verify the effectiveness of coagulant chemicals and dosages being applied at the flash mixer by performing jar tests.

10. 1 The settling rate of particles in water decreases as the water temperature drops.

11. 1 The major means by which operators can control water treatment processes is by adjusting chemicals and chemical feed rates.

12. 1 Immediately following dewatering of a sedimentation basin, grease and lubricate all gears, sprockets, and mechanical moving parts that have been submerged.

13. 3 Filter production (capacity) rates are measured in MGD.

14. 1 Surface wash is used during backwashing to prevent mudballs.

15. 2 When there are significant changes in source water turbidity levels, immediately verify the effectiveness of the filtration process.

16. 2 When a filter is going to be out of service for a prolonged period, it should be drained to avoid algal growth.

17. 1 There has been an increased interest in disinfectants other than chlorine due to chlorine's potential to form carcinogenic compounds (trihalomethanes or THMs).

18. 2 The components of a chlorine residual are combined chlorine forms plus free chlorine.

19. 2 Measure the minimum chlorine residual at the most remote location in the system.

20. 3 ORP probes provide a direct measure of the disinfecting power of the disinfectant.

21. 1 Zinc and magnesium are used as sacrificial anodes in water tanks or for buried pipes because they will corrode preferentially to aluminum and iron.

22. 1 Samples for corrosivity tests should be taken only after postchlorination because disinfection chemicals may increase or decrease corrosivity.

23. 1 Monitoring samples for the Lead and Copper Rule are collected at the consumers' taps.

24. 2 The secret to successful taste and odor control is to prevent tastes and odors from ever developing.

25. 4 The categories of treatment methods for taste and odor control are removal and destruction.

26. 3 Powdered activated carbon is often applied at the flash mixer.

27. 4 Operators have the greatest control over turbidity.

28. 2 Caustic soda is one of the most dangerous common alkalies.

29. 2 Operators should first notify appropriate local health authorities when a treatment process upset results in the failure to meet a specific drinking water quality standard.

30. 4 Pipets are the most accurate way to measure a volume of liquid.

31. 2 Burets are used to titrate chemical reagents.

32. 4 Total immersion thermometers will change most rapidly when removed from a liquid sample.

33. 4 The nephelometric method is the accepted method used to measure turbidity.

Multiple Choice

1. 1, 2, 3, 4, 5 The size of a water treatment plant as well as the number and specific types of processes it uses depend on cost considerations, demand for water, fire protection needs, impurities in raw water, and water quality (purity) standards.

2. 1, 2, 3, 4, 5 The typical duties of a water treatment plant operator include collecting representative water samples, conducting safety inspections, maintaining plant records, making periodic operating checks of plant equipment, and performing routine preventive maintenance.

3. 1, 2, 3, 4, 5 The safety responsibilities of an operator include being aware of safety hazards, being part of an active safety program, ensuring everyone follows safe procedures, ensuring that operators take safety seriously, and understanding why safe procedures must be followed at all times.

4. 1, 2, 4, 5 Groundwater can become contaminated by agricultural drainage systems, the entry of surface runoff into wells, seawater intrusion, and seepage from septic tank leaching systems.

5. 3, 5 Nutrients and temperature can promote rapid algal growth.

6. 1, 2, 3 The basic objectives that control the operation of water treatment plants include the production of a safe and aesthetically pleasing drinking water at a reasonable cost.

7. 1, 2, 4, 5 Water quality problems that may be related to algal blooms include dissolved oxygen depletion, reduced chlorination efficiency, shortened filter runs, and taste and odor problems.

8. 1, 2, 3, 5 Increased organic loadings on water treatment plants caused by algal blooms are responsible for high trihalomethane levels, increased chlorine demand, increased color, and increased water treatment costs.

9. 3, 4, 5 Monitor hydrogen sulfide, iron, and manganese when domestic water supply reservoirs develop anaerobic zones.

10. 1, 3, 4 Chemicals that are commonly used for coagulation include alum, metallic salts, and polymers.

11. 1, 2, 3, 4, 5   The following factors are important when evaluating jar test results: the amount of floc formed, the clarity of the water above the settled floc and between the floc particles, the floc settling rate, and the rate of floc formation.

12. 1, 2, 3, 4, 5   Operators will be exposed to potential safety hazards in the coagulation-flocculation process that include confined spaces, electrical equipment, rotating mechanical equipment, slippery surfaces, and water treatment chemicals.

13. 2, 5   Coagulant dosage and pH are adjusted in the enhanced coagulation process.

14. 2, 4, 5   Presedimentation facilities serve to even out fluctuations in suspended solids, provide an equalizing basin, and reduce the solids-removal load at the water treatment plant.

15. 1, 2, 3   Short-circuiting in a sedimentation basin causes undesirable shorter contact times, shorter reaction times, and shorter settling times.

16. 2, 4, 5   Rapid changes in flow, temperature, or turbidity can cause instability in solids-contact units.

17. 4, 5   Temperature and turbidity are monitored in the normal operation of the sedimentation process.

18. 1, 2, 3, 4, 5   Filter removal of turbidity depends on several factors including the type of filter, the operation of the filter, the types and degree of pretreatment, the chemical characteristics of the water being treated, and the nature of the suspended particulates.

19. 1, 3, 4, 5   The following characteristics are used to classify filter media: effective size, hardness, specific gravity, and uniformity coefficient.

20. 2, 4   A filter is *NOT* completely clean if the initial head loss is too high or the next filter run is too short.

21. 1, 2, 3, 4, 5   Abnormal filter conditions are indicated by algae on the walls and media, media boils during backwash, media cracking or shrinkage, mudballs in the media, and short filter runs.

22. 1, 2, 3, 4, 5   All of the factors listed influence chlorine disinfection: the numbers and types of microorganisms, pH, reducing agents, temperature, and turbidity.

23. 1, 2, 3, 4, 5   Chloramines have proven effective in killing or inactivating heterotrophic plate count bacteria, in maintaining a detectable residual throughout the distribution system, in penetrating the biofilm and reducing the potential for coliform regrowth, in reducing taste and odor problems, and in reducing the formation of trihalomethanes (THMs) and other disinfection by-products (DBPs).

24. 3, 4   A decrease in the chlorine feed rate or an increase in the chlorine demand will cause the chlorine level in treated water to drop.

25. 1, 2, 3, 4, 5   All chlorine safety programs should include a list of emergency procedures for chlorine leaks and first aid, a maintenance and calibration program for safety devices and equipment, periodic hands-on training in the use of safety equipment, plant tours for police and fire departments, and written rules and specific safety procedures.

26. 1, 2, 3, 4, 5   Physical factors that influence the rate of corrosion include flow velocity, soil moisture, stray electric current, temperature of the water, and the type of coatings.

27. 1, 2, 3, 4, 5   Operators can determine if corrosion problems exist by performing chemical tests on the water, by conducting flow tests, by examining materials removed from the distribution system, by noticing an increasing number of leaks, and by responding to consumer complaints.

28. 1, 2, 4, 5   Alkalinity, conductivity, pH, and temperature are monitored as indicators of corrosion-related water quality.

29. 2, 3, 5   As a result of a taste and odor episode, a water utility could suffer damaging effects such as funding restrictions being imposed, an increase in public relations problems, and a loss of public confidence.

30. 1, 2, 3, 4, 5   Locations that have been identified as possible sources of tastes and odors include consumers' plumbing, conveyance facilities, finished water storage reservoirs, raw water sources, and treatment plants.

31. 1, 2, 4, 5   The following factors will determine the success of chlorination as a taste and odor treatment method: contact time, dose applied, seriousness of the problem, and the type of odor.

32. 1, 2, 3, 4, 5   All of the precautions listed should be taken when handling powdered activated carbon (PAC): follow confined space safety procedures, store PAC bags on wooden pallets away from flammable materials, and use dust control equipment and wear protective clothing when handling PAC.

33. 2, 3, 4, 5   Turbidity is commonly monitored in a water treatment plant at each unit process and in water from individual filters. Finished water turbidity and raw water turbidity also are commonly monitored.

34. 1, 3, 4, 5   Clear wells (or plant storage reservoirs) have several purposes. Clear wells act as a buffer that prevents ON/OFF cycling of finished water pumps. They allow filling when demands are low to compensate for peak periods which draw the level down. Clear wells also permit planned changes in treatment plant operation and they provide operational storage to average out high and low flow demands.

35. 1, 2, 4, 5   Some of the factors that can lead to poor quality laboratory data include errors in calculation, deteriorated reagents, poorly operating instruments, and sloppy laboratory technique.

36. 1, 3, 4, 5   Hazards that could be encountered in a laboratory include burns, electric shock, explosions, and hazardous materials.

37. 1, 2, 3, 4, 5   Grab samples are used to measure chlorine residual, coliform bacteria, dissolved gases, pH, and temperature.

38. 1, 5   An alkalinity determination is needed when calculating chemical dosages used in the water treatment processes of coagulation and softening.

Short Answer

1. Intake structures at water treatment plants have screens to prevent leaves and other debris from clogging or damaging pumps, pipes, and other pieces of equipment in the treatment plant.

2. A water treatment plant operator should be present when a new plant is being constructed in order to be completely familiar with the entire plant layout, including the piping, equipment, and machinery and their intended operation.

3. Adequate, reliable records are important to document (record) the effectiveness of your operation and are required by regulatory agencies.

4. The water (hydrologic) cycle is the process involved in the transfer of moisture from the sea to the land and back to the sea again. Water evaporates from the sea, chills and forms clouds, and falls as precipitation over land and water. The falling water may re-evaporate and repeat the cycle. Precipitation reaching the land may be stored as snow, run off as surface runoff, or percolate down to groundwater. Runoff may be stored in lakes or reservoirs before it returns to the ocean. Water may evaporate from the land or water surfaces or may be converted to water vapor by transpiration from plants. Water that reaches the ocean evaporates and the cycle begins all over again.

5. Before selecting a location and constructing a water supply intake in a river or stream, careful consideration must be given to (1) the stream bottom, its degree of scour, and the settling out of silt, and (2) design of the intake to make sure that it can withstand the forces which will act upon it during times of flood, heavy silting, ice conditions, and adverse runoff conditions.

6. The purpose of a sanitary survey is to detect all health hazards and assess their present and future importance. Persons trained and competent in public health engineering and the epidemiology of waterborne diseases should conduct the sanitary survey. Detection of health hazards will require a thorough field search and investigation. Assessment will require an evaluation of each source and control measures practiced at the source.

7. All public water systems must comply with the Safe Drinking Water Act (SDWA) regulations.

8. The methods of treating domestic water delivered from reservoirs range from disinfection only, to direct filtration, to complete treatment, which may include softening and activated carbon filtration.

9. As an algal bloom progresses, the dissolved oxygen content at depths where the bloom occurs normally increases markedly as a result of photosynthesis. When the algal cells die, this oxygen is used in the decomposition of the cells by organisms, mainly bacteria, that feed upon (metabolize) the algal cells.

10. The desired rate of chemical application can be achieved by considering the speed of the boat, the rate at which the material is being fed from the hopper and pumped into the lake, and the area covered by the spray apparatus or the submerged outlet system.

11. The main safety hazard encountered during reservoir sampling is drowning.

12. Define the following terms:

    a. Grab Sample. A single sample of water collected at a particular time and place which represents the composition of the water only at that time and place.

    b. Reagent. A pure chemical substance that is used to make new products or is used in chemical tests to measure, detect, or examine other substances.

    c. Short-Circuiting. A condition that occurs in tanks or basins when some of the flowing water entering a tank or basin flows along a nearly direct pathway from the inlet to the outlet. This is usually undesirable since it may result in shorter contact, reaction, or settling times in comparison with the theoretical (calculated) or presumed detention times.

13. In the coagulation process, chemicals are added to the water being treated that will cause the suspended particles to clump together. The chemicals are mixed by a flash or rapid mixer. In the flocculation process, the particles gather together to form larger particles by gentle stirring or mixing.

14. The purpose of the flocculation process is to provide particle contacts which will promote the gathering together of floc for ease of removal in the sedimentation and filtration processes.

15. The jar test is used as a rough indication of what is happening in your water treatment plant. The test can be very helpful in determining the best coagulant dosages, but actual plant performance must be observed very closely.

16. Abrupt changes in the source water or filtered water turbidity are signals that you should immediately review the performance of the coagulation-flocculation process.

17. The purposes of the sedimentation process are to remove suspended solids (particles) that are denser (heavier) than the water and to reduce the load on the filters.

18. The four zones into which a typical sedimentation basin can be divided are: (1) inlet zone, (2) settling zone, (3) sludge zone, and (4) outlet zone.

19. If the sludge buildup becomes too great, the effective depth of the basin will be significantly reduced, causing localized high flow velocities, sludge scouring, and a decrease in process efficiency.

20. All water treatment plants should have at least two sedimentation basins to allow for maintenance, cleaning, and inspection of basins without requiring a plant shutdown.

21. The process removal efficiency of a sedimentation basin is determined by measurement of turbidity levels at the sedimentation basin inlet and outlet. This will give a rough idea of process removal efficiency.

22. The primary purpose of the filtration process is the removal of particulate impurities consisting of suspended particles (fine silts and clays), colloids, biological forms (bacteria and plankton), and floc from the water being treated.

23. The primary purpose of using activated carbon (granular form) as filter media is to remove taste- and odor-causing compounds, as well as other trace organics, from the water.

24. Evaluation of overall filtration process performance should be conducted on a routine basis, at least once per 8-hour shift, and more frequently when water quality is fluctuating. Turbidity levels should be checked every two hours.

25. The four types of filtration that can be used to meet the requirements of the Surface Water Treatment Rule are (1) conventional, (2) direct, (3) slow sand, and (4) diatomaceous earth filtration.

26. Disinfection is the selective destruction or inactivation of pathogenic organisms while sterilization is the complete destruction of all organisms.

27. Relatively cold water requires longer disinfection time or greater quantities of disinfectants.

28. Chlorine Demand, mg/L = Chlorine Dose, mg/L – Chlorine Residual, mg/L

29. When opening valves on chlorine containers, do not use wrenches longer than six inches, pipe wrenches, or wrenches with an extension. To unseat, strike the end of the wrench with the heel of your hand to rotate the valve stem in a counterclockwise direction. Then open slowly. One complete turn permits maximum discharge. Do not force valve beyond this point.

30. Evaporators are used to convert liquid chlorine to gaseous chlorine for use by gas chlorinators.

31. If chlorine is escaping from a cylinder as a liquid, turn the cylinder so that the leak is on top and the chlorine will escape as a gas.

32. Chlorine gas is extremely toxic and corrosive in moist atmospheres.

33. Ozone is normally prepared on site because it is very unstable.

34. Define the following terms:

    a. Breakpoint Chlorination. Addition of chlorine to water until the chlorine demand has been satisfied. At this point, further additions of chlorine will result in a free chlorine residual that is directly proportional to the amount of chlorine added beyond the breakpoint.

    b. Pathogenic Organisms. Organisms, including bacteria, viruses or cysts, capable of causing diseases (giardiasis, cryptosporidiosis, typhoid, cholera, dysentery) in a host (such as a person).

35. Corrosion is the gradual decomposition or destruction of a material (such as a metal or a cement lining) as it reacts with water.

36. Dissolved solids increase corrosion by increasing the electrical conductivity of the water.

37. A thin film or coating of calcium carbonate can drastically inhibit corrosion.

38. A tubercle is a small protective crust of rust which builds up over a pit caused by the loss of metal from corrosion.

39. The Marble Test and Langelier Index are used to determine if water is undersaturated or supersaturated with calcium carbonate.

40. The corrosivity of a water is often reduced by treating the water so that the water is saturated or slightly supersaturated with calcium carbonate.

41. Taste and odor, along with colored water complaints, are the most common types of water quality complaints received by a water utility.

42. Treating a reservoir with copper sulfate may disrupt the natural balance of a well-balanced, diversified plankton community and allow an objectionable organism to predominate which could create a taste and odor problem.

43. The frequency of flushing mains should be determined from records of complaints and water quality tests.

44. The two broad categories of taste and odor treatment methods are (1) removal, and (2) destruction.

45. Chlorine dioxide has been used to treat tastes and odors caused by industrial pollution, especially in cases where chlorine has intensified the problem.

46. When cleaning a tank that stores powdered activated carbon, the tank must be fully open, a circulation of air provided, and operators working in the tank should never do so alone.

47. Samples used to measure compliance with water quality standards are usually taken at the entry point to the distribution system.

48. The air supply for pneumatic data transmission devices must be dried to prevent condensation in equipment signal lines, and filtered to remove particle contaminants.

49. Positive pressure fans are preferred because if you pull air with chlorine through a fan, eventually any wiring or controls in the fan will become corroded and fail.

50. Treatment process failures are usually caused by (1) changes in raw water quality, (2) operator error, or (3) equipment failure.

51. A meniscus is the curve of the surface of a liquid in a small tube.

52. Titration involves the addition of a standardized solution, which is usually in a buret, to another solution in a flask or beaker. Slowly add the solution in the buret to the test solution in the flask until the change, which is called the "end point," is reached.

53. True. You may *ADD ACID TO WATER*, but never the reverse.

54. Water should not be poured on grease, electrical fires, or metal fires because water could increase the hazards, such as splattering of the fire and electric shock.

55. Proper sampling procedures are essential in order to obtain an accurate description of the material or water being sampled and tested.

56. A representative sample is a sample portion of material or water that is as nearly identical in content and consistency as possible to that in the larger body of material or water being sampled.

57. A potential adverse effect from chlorination is the possibility of the formation of carcinogenic chlororganic compounds such as chloroform and other THMs.

58. The principal hardness-causing ions in water are calcium and magnesium.

59. Turbidity should be measured in a sample as soon as possible to obtain accurate results. The turbidity of a sample can change after the sample is collected.

Problems

1. What is the surface area of a rectangular settling basin 50 feet long and 12 feet wide?

Known	Unknown
Length, ft = 50 ft	Surface Area, sq ft
Width, ft = 12 ft	

Calculate the surface area of the basin in square feet.

$$\text{Surface Area, sq ft} = (\text{Length, ft})(\text{Width, ft})$$
$$= (50 \text{ ft})(12 \text{ ft})$$
$$= 600 \text{ sq ft}$$

2. What is the surface area of a circular clarifier 50 feet in diameter?

Known	Unknown
Diameter, ft = 50 ft	Surface Area, sq ft

Calculate the surface area of the clarifier in square feet.

$$\text{Surface Area, sq ft} = 0.785(\text{Diameter, ft})^2$$
$$= 0.785(50 \text{ ft})^2$$
$$= 1,963 \text{ sq ft}$$

3. Determine the setting on a dry alum feeder in pounds per day when the flow is 1.2 MGD. Jar tests indicate that the best alum dose is 9 mg/L.

Known	Unknown
Flow, MGD = 1.2 MGD	Feeder Setting, lbs/day
Alum Dose, mg/L = 9 mg/L	

Calculate the alum feeder setting in pounds per day.

$$\text{Feeder Setting, lbs/day} = (\text{Flow, MGD})(\text{Alum, mg/}L)(8.34 \text{ lbs/gal})$$
$$= (1.2 \text{ MGD})(9 \text{ mg/}L)(8.34 \text{ lbs/gal})$$
$$= 90 \text{ lbs/day}$$

4. A water treatment plant used 24 pounds of cationic polymer to treat 1.4 million gallons of water during a 24-hour period. What is the polymer dosage in mg/L? Select the closest answer.

Known	Unknown
Flow, MGD = 1.4 MGD	Polymer Dose, mg/L
Polymer Used, lbs/day = 24 lbs/day	

Calculate the polymer dose in mg/L.

$$\text{Polymer Dose, mg/}L = \frac{\text{Polymer Used, lbs/day}}{(\text{Flow, MGD})(8.34 \text{ lbs/gal})}$$
$$= \frac{24 \text{ lbs/day}}{(1.4 \text{ MGD})(8.34 \text{ lbs/gal})}$$
$$= 2.1 \text{ mg/}L$$

5. Calculate the theoretical detention time for a rectangular sedimentation basin. The basin is 80 feet long, 30 feet wide, 10 feet deep, and treats a flow of 1.8 MGD.

Known	Unknown
Flow, MGD = 1.8 MGD	Detention Time, hr
Length, ft = 80 ft	
Width, ft = 30 ft	
Depth, ft = 10 ft	

1. Calculate the basin volume in gallons.

$$\text{Basin Volume, gal} = (\text{L, ft})(\text{W, ft})(\text{D, ft})(7.48 \text{ gal/cu ft})$$
$$= (80 \text{ ft})(30 \text{ ft})(10 \text{ ft})(7.48 \text{ gal/cu ft})$$
$$= 179,520 \text{ gallons}$$

2. Determine the theoretical detention time in hours.

$$\text{Detention Time, hr} = \frac{(\text{Basin Volume, gal})(24 \text{ hr/day})}{\text{Flow, gal/day}}$$
$$= \frac{(179,520 \text{ gallons})(24 \text{ hr/day})}{1,800,000 \text{ gal/day}}$$
$$= 2.4 \text{ hours}$$

6. A tank with a capacity of 20,000 cubic feet can hold how many gallons of water?

Known	Unknown
Volume, cu ft = 20,000 cu ft	Volume, gal

Convert the volume in cubic feet to volume in gallons.

$$\text{Volume, gal} = (\text{Volume, cu ft})(7.48 \text{ gal/cu ft})$$
$$= (20,000 \text{ cu ft})(7.48 \text{ gal/cu ft})$$
$$= 149,600 \text{ gal}$$

7. Determine the upward force in pounds on the bottom of an empty sedimentation basin caused by a groundwater depth of 5 feet above the tank bottom. The basin is 10 feet wide and 30 feet long.

Known	Unknown
Head, ft = 5 ft	Force, lbs
Width, ft = 10 ft	
Length, ft = 30 ft	

1. Calculate the pressure on the tank bottom in pounds per square foot.

Pressure, lbs/sq ft = (62.4 lbs/cu ft)(Head, ft)

$$= (62.4 \text{ lbs/cu ft})(5 \text{ ft})$$

$$= 312 \text{ lbs/sq ft}$$

2. Calculate the force on the bottom of the tank in pounds.

Force, lbs = (Pressure, lbs/sq ft)(Area, sq ft)

$$= (312 \text{ lbs/sq ft})(10 \text{ ft})(30 \text{ ft})$$

$$= 93,600 \text{ lbs}$$

8. During a 24-hour period a water treatment plant filters 1.6 million gallons. The surface area of the filters is 500 square feet. Determine the filtration rate in gallons per minute per square foot.

Known	Unknown
Water Filtered, MGD = 1.6 MGD	Filtration Rate, GPM/sq ft
Surface Area, sq ft = 500 sq ft	

Determine the filtration rate in gallons per minute per square foot.

$$\text{Filtration Rate, GPM/sq ft} = \frac{\text{(Total Water Filtered, MGD)}(1,000,000/M)}{\text{(Filter Surface Area, sq ft)}(24 \text{ hr/day})(60 \text{ min/hr})}$$

$$= \frac{(1.6 \text{ MGD})(1,000,000/M)}{(500 \text{ sq ft})(24 \text{ hr/day})(60 \text{ min/hr})}$$

$$= 2.2 \text{ GPM/sq ft}$$

9. What should be the setting on a chlorinator (lbs chlorine per 24 hours) if a pump usually delivers 400 GPM and the desired chorine dosage is 4.0 mg/L? Select the closest answer.

Known	Unknown
Pump Flow, GPM = 400 GPM	Chlorinator Setting, lbs Chlorine/24 hours
Chlorine Dose, mg/L = 4.0 mg/L	

1. Convert pump flow from GPM to million gallons per day (MGD).

$$\text{Flow, MGD} = \frac{(400 \text{ gal/min})(60 \text{ min/hr})(24 \text{ hr/day})}{1,000,000/M}$$

$$= 0.576 \text{ MGD}$$

2. Calculate the chlorinator setting in pounds of chlorine per 24 hours.

Chlorinator, Setting lbs/24 hr = (Flow, MGD)(Dose, mg/L)(8.34 lbs/gal)

$$= (0.576 \text{ MGD})(4.0 \text{ mg/L})(8.34 \text{ lbs/gal})$$

$$= \left(\frac{0.576 \text{ M Gal}}{\text{day}}\right)\left(\frac{4.0 \text{ lbs Chlorine}}{1 \text{ M lbs Water}}\right)(8.34 \text{ lbs/gal})$$

$$= 19.2 \text{ lbs Chlorine/day}$$

$$= 19.2 \text{ lbs Chlorine/24 hours}$$

10. Estimate the chlorine dose in mg/L if a chlorinator feeds at a rate of 18 lbs per 24 hours and the flow is 0.55 MGD.

Known	Unknown
Chlorine Feed, lbs/day = 18 lbs/day	Chlorine Dose, mg/L
Flow, MGD = 0.55 MGD	

Calculate the chlorine dose in mg/L.

$$\text{Chlorine Dose, mg/L} = \frac{\text{Chlorine Feed, lbs/day}}{\text{(Flow, MGD)}(8.34 \text{ lbs/gal})}$$

$$= \frac{18 \text{ lbs/day}}{(0.55 \text{ MGD})(8.34 \text{ lbs/gal})}$$

$$= \frac{18 \text{ lbs Chlorine}}{4.59 \text{ M lbs Water}}$$

$$= 3.9 \text{ lbs Chlorine/M lbs Water}$$

$$= 3.9 \text{ mg/L}$$

11. Estimate the water velocity in a channel in feet per second if a float travels 20 feet in 10 seconds.

Known	Unknown
Distance, ft = 20 ft	Velocity, ft/sec
Time, sec = 10 sec	

Calculate the water velocity in feet per second.

$$\text{Velocity, ft/sec} = \frac{\text{Distance, ft}}{\text{Time, sec}}$$

$$= \frac{20 \text{ ft}}{10 \text{ sec}}$$

$$= 2 \text{ ft/sec}$$

12. Estimate the water flow rate in an 18-inch diameter pipe in gallons per minute (GPM) when the flow velocity is two feet per second. Select the closest answer.

Known	Unknown
Diameter, in = 18 in	Flow Rate, gal/min
Diameter, ft = 1.5 ft	
Velocity, ft/sec = 2 ft/sec	

1. Determine the area of the pipe in square feet.

   Area, sq ft $= (0.785)(\text{Diameter, ft})^2$

   $= (0.785)(1.5 \text{ ft})^2$

   $= 1.77 \text{ sq ft}$

2. Determine the flow rate in cubic feet per second.

   Flow Rate, cu ft/sec $= (\text{Velocity, ft/sec})(\text{Area, sq ft})$

   $= (2 \text{ ft/sec})(1.77 \text{ sq ft})$

   $= 3.54 \text{ cu ft/sec}$

3. Convert the flow rate from cubic feet per second to gallons per minute.

   Flow Rate, GPM $= (\text{Flow Rate, cu ft/sec})(7.48 \text{ gal/cu ft})(60 \text{ sec/min})$

   $= (3.54 \text{ cu ft/sec})(7.48 \text{ gal/cu ft})(60 \text{ sec/min})$

   $= 1,589 \text{ GPM}$

13. What horsepower must a pump deliver to water that must be lifted 90 feet? The flow is 40 GPM.

Known	Unknown
Lift, ft = 90 ft	Water, HP
Flow, GPM = 40 GPM	

   Calculate the horsepower the pump must deliver to the water.

   Water, HP $= \dfrac{(\text{Flow, GPM})(\text{Lift, ft})(8.34 \text{ lbs/gal})}{33,000 \text{ ft-lb/min-HP}}$

   $= \dfrac{(40 \text{ GPM})(90 \text{ ft})(8.34 \text{ lbs/gal})}{33,000 \text{ ft-lb/min-HP}}$

   $= 0.9 \text{ HP}$

14. How many kilowatt-hours per day are required by a pump with a motor horsepower of 50 horsepower when the pump operates 24 hours per day?

Known	Unknown
Motor, HP = 50 HP	Power, kW-hr/day

   Calculate the power required in kilowatt-hours per day.

   Power, kW-hr/day $= (\text{Motor, HP})(24 \text{ hr/day})(0.746 \text{ kW/HP})$

   $= (50 \text{ HP})(24 \text{ hr/day})(0.746 \text{ kW/HP})$

   $= 895 \text{ kW-hr/day}$

15. Calculate the volume of a rectangular sedimentation basin 8 feet deep, 20 feet wide, and 55 feet long.

Known	Unknown
Depth, ft = 8 ft	Volume, cu ft
Width, ft = 20 ft	
Length, ft = 55 ft	

   Calculate the volume in cubic feet.

   Volume, cu ft $= (\text{Length, ft})(\text{Width, ft})(\text{Depth, ft})$

   $= (55 \text{ ft})(20 \text{ ft})(8 \text{ ft})$

   $= 8,800 \text{ cu ft}$

16. Results from a coliform bacteria multiple fermentation tube test of raw water are as follows:

mL of sample	10	1	0.1	0.01	0.001
Dilutions	0	0	−1	−2	−3
Positive tubes	5	5	5	3	1

   What is the most probable number of coliforms per 100 mL in the sample? Use Table 11.3 on page 545.

   Select the underlined inoculations.

mL of sample	10	1	0.1	0.01	0.001
Dilutions	0	0	−1	−2	−3
Positive tubes	5	5	5	3	1

   Read MPN as 110 from Table 11.3.
   Report results as 11,000 coliforms per 100 mL.

   We added two zeros to 110 because we started with the 0.1 mL sample and Table 11.3 begins with two dilution columns to the left (from 0.1 mL to 1 mL to 10 mL or two dilution columns to the left).

Data to answer questions 17 and 18.

   Results from MPN tests during one week were as follows:

Day	S	M	T	W	T	F	S
MPN/100 mL	4	7	14	8	9	16	5

17. What is the mean MPN per 100 mL from the data?

   Calculate the mean MPN/100 mL from the data.

   Mean, MPN/100 mL $= \dfrac{\text{Sum of All MPNs}}{\text{Number of MPNs}}$

   $= \dfrac{4 + 7 + 14 + 8 + 9 + 16 + 5}{7}$

   $= \dfrac{63}{7}$

   $= 9 \text{ MPN/100 mL}$

18. What is the median MPN per 100 mL from the data?

   Rearrange the data in ascending (increasing) order and select the middle value (three will be smaller and three will be larger).

Order	1	2	3	4	5	6	7
MPN/100 mL	4	5	7	8	9	14	16

   Median, MPN/100 mL = Middle value of a group of data

   $= 8 \text{ MPN/100 mL}$

APPENDIX

HOW TO SOLVE WATER TREATMENT PLANT
ARITHMETIC PROBLEMS

(VOLUME I)

by

Ken Kerri

TABLE OF CONTENTS

HOW TO SOLVE WATER TREATMENT PLANT ARITHMETIC PROBLEMS

OBJECTIVES

HOW TO SOLVE WATER TREATMENT
PLANT ARITHMETIC PROBLEMS

Following completion of this Appendix, you should be able to:

1. Add, subtract, multiply, and divide;

2. List from memory basic conversion factors and formulas; and

3. Solve water treatment plant arithmetic problems.

APPENDIX

HOW TO SOLVE WATER TREATMENT PLANT ARITHMETIC PROBLEMS

A.0   HOW TO STUDY THIS APPENDIX

This appendix may be worked early in your training program to help you gain the greatest benefit from your efforts. Whether to start this appendix early or wait until later is your decision. The chapters in this manual were written in a manner requiring very little background in arithmetic. You may wish to concentrate your efforts on the chapters and refer to this appendix when you need help. Some operators prefer to complete this appendix early so they will not have to worry about how to do the arithmetic when they are studying the chapters. You may try to work this appendix early or refer to it while studying the other chapters.

The intent of this appendix is to provide you with a quick review of the addition, subtraction, multiplication, and division needed to work the arithmetic problems in this manual. This appendix is not intended to be a math textbook. There are no fractions because you don't need fractions to work the problems in this manual. Some operators will be able to skip over the review of addition, subtraction, multiplication, and division. Others may need more help in these areas. If you need help in solving problems, read Section A.9, "Steps in Solving Problems." Basic arithmetic textbooks are available at every local library or bookstore and should be referred to if needed. Most instructional or operating manuals for pocket electronic calculators contain sufficient information on how to add, subtract, multiply, and divide.

After you have worked a problem involving your job, you should check your calculations, examine your answer to see if it appears reasonable, and if possible have another operator check your work before making any decisions or changes.

A.1   BASIC ARITHMETIC

In this section we provide you with basic arithmetic problems involving addition, subtraction, multiplication, and division. You may work the problems "by hand" if you wish, but we recommend you use an electronic pocket calculator. The operating or instructional manual for your calculator should outline the step-by-step procedures to follow. All calculators use similar procedures, but most of them are slightly different from others.

We will start with very basic, simple problems. Try working the problems and then comparing your answers with the given answers. If you can work these problems, you should be able to work the more difficult problems in the text of this training manual by using the same procedures.

A.10   Addition

2	6.2	16.7	6.12	43
3	8.5	38.9	38.39	39
5	14.7	55.6	44.51	34
				38
				39
2.12	0.12	63	120	37
9.80	2.0	32	60	29
11.92	2.12	95	180	259

				70
4	23	16.2	45.98	50
7	79	43.5	28.09	40
2	31	67.8	114.00	80
13	133	127.5	188.07	240

A.11   Subtraction

7	12	25	78	83
− 5	− 3	− 5	− 30	− 69
2	9	20	48	14

61	485	4.3	3.5	123
− 37	− 296	− 0.8	− 0.7	− 109
24	189	3.5	2.8	14

8.6	11.92	27.32	3.574	75.132
− 8.22	− 3.70	− 12.96	− 0.042	− 49.876
0.38	8.22	14.36	3.532	25.256

A.12   Multiplication

(3)(2)*	= 6		(4)(7)	= 28
(10)(5)	= 50		(10)(1.3)	= 13
(2)(22.99)	= 45.98		(6)(19.5)	= 117
(16)(17.1)	= 273.6		(50)(20,000)	= 1,000,000
(40)(2.31)	= 92.4		(80)(0.433)	= 34.64

(40)(20)(6)	= 4,800
(4,800)(7.48)	= 35,904
(1.6)(2.3)(8.34)	= 30.6912
(0.001)(200)(8.34)	= 1.668
(0.785)(7.48)(60)	= 352.308
(12,000)(500)(60)(24)	= 8,640,000,000 or 8.64 x 10^9
(4)(1,000)(1,000)(454)	= 1,816,000,000 or 1.816 x 10^9

NOTE: The term, x 10^9, means that the number is multiplied by 10^9 or 1,000,000,000. Therefore 8.64 x 10^9 = 8.64 x 1,000,000,000 = 8,640,000,000.

---

* (3)(2) is the same as 3 x 2 = 6.

A.13 Division

$$\frac{6}{3} = 2 \qquad\qquad \frac{48}{12} = 4$$

$$\frac{50}{25} = 2 \qquad\qquad \frac{300}{20} = 15$$

$$\frac{20}{7.1} = 2.8 \qquad\qquad \frac{11,400}{188} = 60.6$$

$$\frac{1,000,000}{17.5} = 57,143 \qquad\qquad \frac{861,429}{30,000} = 28.7$$

$$\frac{4,000,000}{74,880} = 53.4 \qquad\qquad \frac{1.67}{8.34} = 0.20$$

$$\frac{80}{2.31} = 34.6 \qquad\qquad \frac{62}{454} = 0.137$$

$$\frac{250}{17.1} = 14.6 \qquad\qquad \frac{4,000,000}{14.6} = 273,973$$

NOTE: When we divide $1/3 = 0.3333$, we get a long row of 3s. Instead of the row of 3s, we "round off" our answer so $1/3 = 0.33$. For a discussion of rounding off numbers, see Section A.95, "Significant Figures."

A.14 Rules for Solving Equations

Most of the arithmetic problems we work in the water treatment field require us to plug numbers into formulas and calculate the answer. There are a few basic rules that apply to solving formulas. These rules are:

1. Work from left to right.

2. Do all the multiplication and division above the line (in the numerator) and below the line (in the denominator); then do the addition and subtraction above and below the line.

3. Perform the division (divide the numerator by the denominator).

Parentheses ( ) are used in formulas to identify separate parts of a problem. A fourth rule tells us how to handle numbers within parentheses.

4. Work the arithmetic within the parentheses before working outside the parentheses. Use the same order stated in rules 1, 2, and 3: work left to right, above and below the line, then divide the top number by the bottom number.

Let's look at an example problem to see how these rules apply. This year one of the responsibilities of the operators at our plant is to paint both sides of the wooden fence across the front of the facility. The fence is 145 feet long and 9 feet high. The steel access gate, which does not need painting, measures 14 feet wide by 9 feet high. Each gallon of paint will cover 150 square feet of surface area. How many gallons of paint should be purchased?

STEP 1:   Identify the correct formula.

$$\text{Paint Req, gal} = \frac{\text{Total Area, sq ft}}{\text{Coverage, sq ft/gal}}$$

or

$$\underset{\text{gal}}{\text{Paint Req,}} = \frac{(\text{Fence L, ft} \times \text{H, ft} \times \text{No. Sides}) - (\text{Gate L, ft} \times \text{H, ft} \times \text{No. Sides})}{\text{Coverage, sq ft/gal}}$$

STEP 2:   Plug numbers into the formula.

$$\text{Paint Req, gal} = \frac{(145 \text{ ft} \times 9 \text{ ft} \times 2) - (14 \text{ ft} \times 9 \text{ ft} \times 2)}{150 \text{ sq ft/gal}}$$

STEP 3:   Work the multiplication within parentheses.

$$\text{Paint Req, gal} = \frac{(2,610 \text{ sq ft}) - (252 \text{ sq ft})}{150 \text{ sq ft/gal}}$$

STEP 4:   Work the subtraction above the line.

$$\text{Paint Req, gal} = \frac{2,358 \text{ sq ft}}{150 \text{ sq ft/gal}}$$

STEP 5:   Divide the numerator by the denominator.

$$\text{Paint Req, gal} = 15.72 \text{ gal}$$
or 16 gallons of paint will be needed.

Instructions for your electronic calculator can provide you with the detailed procedures for working the practice problems below.

$$\frac{(3)(4)}{2} = 6 \qquad\qquad \frac{64}{(8)(4)} = 2$$

$$\frac{(2+3)(4)}{5} = 4 \qquad\qquad \frac{54}{(4+2)(3)} = 3$$

$$\frac{(7-2)(8)}{4} = 10 \qquad\qquad \frac{48}{(8-3)(4)} = 2.4$$

$$\frac{(0.1)(60)(24)}{3} = 48$$

$$\frac{(12,000)(500)(60)(24)}{(4)(1,000)(1,000)(454)} = 4.76$$

$$\frac{12}{(0.432)(8.34)} = 3.3$$

$$\frac{(274,000)(24)}{200,000} = 32.88$$

A.15 Example Problems

Let's look at the last four problems in the previous Section A.14, "Rules for Solving Equations," as they might be encountered by an operator.

1. To determine the actual chemical feed rate from an alum feeder, an operator collects the alum from the feeder in a bucket for three minutes. The alum in the bucket weighs 0.1 pound.

Known	Unknown
Weight of Alum, lbs = 0.1 lb	Actual Alum Feed, lbs/day
Time, min = 3 min	

Calculate the actual alum feed rate in pounds per day.

$$\underset{\text{lbs/day}}{\substack{\text{Actual Alum}\\\text{Feed Rate,}}} = \frac{(\text{Alum Wt, lbs})(60 \text{ min/hr})(24 \text{ hr/day})}{\text{Time Alum Collected, min}}$$

$$= \frac{(0.1 \text{ lb})(60 \text{ min/hr})(24 \text{ hr/day})}{3 \text{ min}}$$

$$= 48 \text{ lbs/day}$$

2. A solution chemical feeder is calibrated by measuring the time to feed 500 milliliters of chemical solution. The test calibration run required four minutes. The chemical concentration in the solution is 12,000 mg/L or 1.2%. Determine the chemical feed in pounds per day.

Known	**Unknown**
Volume Pumped, mL = 500 mL	Chemical Feed, lbs/day
Time Pumped, min = 4 min	
Chemical Conc, mg/L = 12,000 mg/L	

Estimate the chemical feed rate in pounds per day.

$$\text{Chemical Feed, lbs/day} = \frac{(\text{Chem Conc, mg}/L)(\text{Vol Pumped, m}L)(60 \text{ min/hr})(24 \text{ hr/day})}{(\text{Time Pumped, min})(1{,}000 \text{ m}L/L)(1{,}000 \text{ mg/gm})(454 \text{ gm/lb})}$$

$$= \frac{(12{,}000 \text{ mg}/L)(500 \text{ m}L)(60 \text{ min/hr})(24 \text{ hr/day})}{(4 \text{ min})(1{,}000 \text{ m}L/L)(1{,}000 \text{ mg/gm})(454 \text{ gm/lb})}$$

$$= 4.76 \text{ lbs/day}$$

3. A chlorinator is set to feed 12 pounds of chlorine per day to a flow of 300 gallons per minute (0.432 million gallons per day). What is the chlorine dose in milligrams per liter?

Known	**Unknown**
Chlorinator Feed, lbs/day = 12 lbs/day	Chlorine Dose, mg/L
Flow, MGD = 0.432 MGD	

Determine the chlorine dose in milligrams per liter.

$$\text{Chlorine Dose, mg}/L = \frac{\text{Chlorinator Feed Rate, lbs/day}}{(\text{Flow, MGD})(8.34 \text{ lbs/gal})}$$

$$= \frac{12 \text{ lbs/day}}{(0.432 \text{ MGD})(8.34 \text{ lbs/gal})}$$

$$= 3.3 \text{ mg}/L$$

4. Estimate the operating time of a water softening ion exchange unit before the unit needs regeneration. The unit can treat 274,000 gallons of water before the exchange capacity is exhausted. The average daily flow is 200,000 gallons per day.

Known	**Unknown**
Water Treated, gal = 274,000 gal	Operating Time, hr
Avg Daily Flow, gal/day = 200,000 gal/day	

Estimate the operating time of the ion exchange unit in hours.

$$\text{Operating Time, hr} = \frac{(\text{Water Treated, gal})(24 \text{ hr/day})}{\text{Avg Daily Flow, gal/day}}$$

$$= \frac{(274{,}000 \text{ gal})(24 \text{ hr/day})}{200{,}000 \text{ gal/day}}$$

$$= 32.9^* \text{ hours}$$

---

* We rounded off 32.88 hours to 32.9 hours.

A.2   AREAS

A.20   Units

Areas are measured in two dimensions or in square units. In the English system of measurement the most common units are square inches, square feet, square yards, and square miles. In the Metric system the units are square millimeters, square centimeters, square meters, and square kilometers.

A.21   Rectangle

The area of a rectangle is equal to its length (L) multiplied by its width (W).

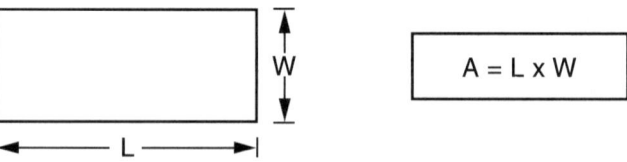

$$A = L \times W$$

EXAMPLE: Find the area of a rectangle if the length is 5 feet and the width is 3.5 feet.

Area, sq ft = Length, ft x Width, ft

= 5 ft x 3.5 ft

= 17.5 ft^2

or   = 17.5 sq ft

EXAMPLE: The surface area of a settling basin is 330 square feet. One side measures 15 feet. How long is the other side?

$$A = L \times W$$

330 sq ft = L, ft x 15 ft

$$\frac{\text{L, ft} \times 15 \text{ ft}}{15 \text{ ft}} = \frac{330 \text{ sq ft}}{15 \text{ ft}} \quad \text{Divide both sides of equation by 15 ft.}$$

$$\text{L, ft} = \frac{330 \text{ sq ft}}{15 \text{ ft}}$$

= 22 ft

A.22   Triangle

The area of a triangle is equal to one-half the base multiplied by the height. This is true for any triangle.

$$A = \tfrac{1}{2} B \times H$$

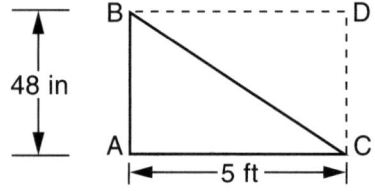

NOTE: The area of ANY triangle is equal to $\frac{1}{2}$ the area of the rectangle that can be drawn around it. The area of the rectangle is B x H. The area of the triangle is $\frac{1}{2}$ B x H.

EXAMPLE: Find the area of triangle ABC.

The first step in the solution is to make all the units the same. In this case, it is easier to change inches to feet.

$$48 \text{ in} = 48 \text{ in} \times \frac{1 \text{ ft}}{12 \text{ in}} = \frac{48}{12} \text{ ft} = 4 \text{ ft}$$

NOTE: All conversions should be calculated in the above manner. Since 1 ft/12 in is equal to unity, or one, multiplying by this factor changes the form of the answer but not its value.

Area, sq ft = ¹/₂(Base, ft)(Height, ft)

$$= \frac{1}{2} \times 5 \text{ ft} \times 4 \text{ ft}$$

$$= \frac{20}{2} \text{ sq ft}$$

$$= 10 \text{ sq ft}$$

NOTE: Triangle ABC is one-half the area of rectangle ABCD. The triangle is a special form called a *RIGHT TRIANGLE* since it contains a 90° angle at point A.

A.23   Circle

A square with sides of 2R can be drawn around a circle with a radius of R.

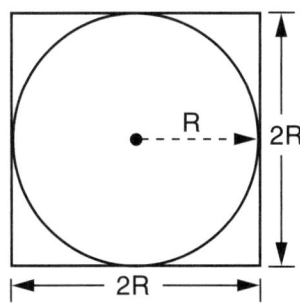

The area of the square is: A = 2R x 2R = $4R^2$.

It has been found that the area of any circle drawn within a square is slightly more than ³/₄ of the area of the square. More precisely, the area of the preceding circle is:

A circle = $3\frac{1}{7} R^2 = 3.14 R^2$

The formula for the area of a circle is usually written:

$$\boxed{A = \pi R^2}$$

The Greek letter π (pronounced pie) merely substitutes for the value 3.1416.

Since the diameter of any circle is equal to twice the radius, the formula for the area of a circle can be rewritten as follows:

$$A = \pi R^2 = \pi \times R \times R = \pi \times \frac{D}{2} \times \frac{D}{2} = \frac{\pi D^2}{4} = \frac{3.14}{4} D^2 = \boxed{0.785 \, D^2}$$

The type of problem and the magnitude (size) of the numbers in a problem will determine which of the two formulas will provide a simpler solution. All of these formulas will give the same results if you use the same number of digits to the right of the decimal point.

EXAMPLE: What is the area of a circle with a diameter of 20 centimeters?

In this case, the formula using a radius is more convenient since it takes advantage of multiplying by 10.

Area, sq cm = $\pi(R, cm)^2$

$$= 3.14 \times 10 \text{ cm} \times 10 \text{ cm}$$

$$= 314 \text{ sq cm}$$

EXAMPLE: What is the area of a clarifier with a 50-foot radius?

In this case, the formula using diameter is more convenient.

Area, sq ft = $(0.785)(\text{Diameter, ft})^2$

$$= 0.785 \times 100 \text{ ft} \times 100 \text{ ft}$$

$$= 7,850 \text{ sq ft}$$

Occasionally the operator may be confronted with a problem giving the area and requesting the radius or diameter. This presents the special problem of finding the square root of the number.

EXAMPLE: The surface area of a circular clarifier is approximately 5,000 square feet. What is the diameter?

A = $0.785 \, D^2$, or

Area, sq ft = $(0.785)(\text{Diameter, ft})^2$

5,000 sq ft = $0.785 \, D^2$ To solve, substitute given values in equation.

$$\frac{0.785 \, D^2}{0.785} = \frac{5,000 \text{ sq ft}}{0.785}$$ Divide both sides by 0.785 to find D^2.

$$D^2 = \frac{5,000 \text{ sq ft}}{0.785}$$

$$= 6,369 \text{ sq ft. Therefore,}$$

D = square root of 6,369 sq ft, or

Diameter, ft = $\sqrt{6,369 \text{ sq ft}}$

Press the √ sign on your calculator and get

D, ft = 79.8 ft

Sometimes a trial-and-error method can be used to find square roots. Since 80 x 80 = 6,400, we know the answer is close to 80 feet.

Try 79 x 79   = 6,241

Try 79.5 x 79.5 = 6,320.25

Try 79.8 x 79.8 = 6,368.04

The diameter is 79.8 ft, or approximately 80 feet.

A.24   Cylinder

With the formulas presented thus far, it would be a simple matter to find the number of square feet in a room that was to be painted. The length of each wall would be added together and then multiplied by the height of the wall. This would give the surface area of the walls (minus any area for doors and windows). The ceiling area would be found by multiplying length times width and the result added to the wall area gives the total area.

The surface area of a circular cylinder, however, has not been discussed. If we wanted to know how many square feet of surface area are in a tank with a diameter of 60 feet and a height of 20 feet, we could start with the top and bottom.

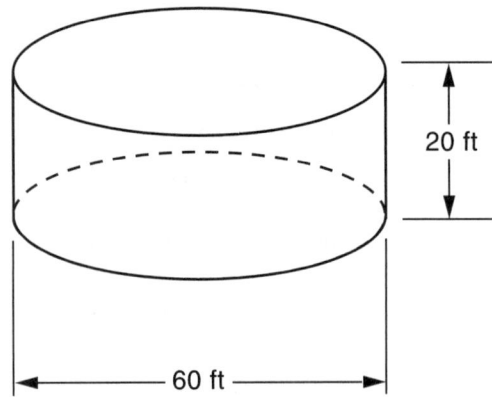

The area of the top and bottom ends are both $\pi \times R^2$.

Area, sq ft = (2 ends)(π)(Radius, ft)^2

= $2 \times \pi \times (30\ ft)^2$

= 5,652 sq ft

The surface area of the wall must now be calculated. If we made a vertical cut in the wall and unrolled it, the straightened wall would be the same length as the circumference of the floor and ceiling.

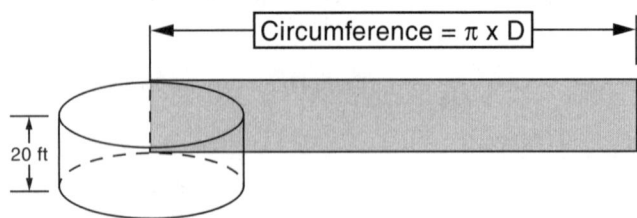

This length has been found to always be $\pi \times D$. In the case of the tank, the length of the wall would be:

Length, ft = (π)(Diameter, ft)

= $3.14 \times 60\ ft$

= 188.4 ft

Area would be:

A_W, sq ft = Length, ft x Height, ft

= 188.4 ft x 20 ft

= 3,768 sq ft

Outside Surface Area to Paint, sq ft = Area of Top and Bottom, sq ft + Area of Wall, sq ft

= 5,652 sq ft + 3,768 sq ft

= 9,420 sq ft

A container has inside and outside surfaces and you may need to paint both of them.

A.25 Cone

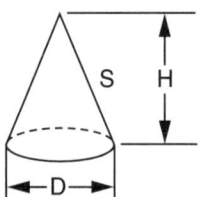

The lateral area of a cone is equal to $1/2$ of the slant height (S) multiplied by the circumference of the base.

$$A_L = 1/2\ S \times \pi \times D = \pi \times S \times R$$

In this case the slant height is not given; it may be calculated by:

$$S = \sqrt{R^2 + H^2}$$

EXAMPLE: Find the entire outside area of a cone with a diameter of 30 inches and a height of 20 inches.

Slant Height, in = $\sqrt{(Radius, in)^2 + (Height, in)^2}$

= $\sqrt{(15\ in)^2 + (20\ in)^2}$

= $\sqrt{225\ sq\ in + 400\ sq\ in}$

= $\sqrt{625\ sq\ in}$

= 25 in

Lateral Area of Cone, sq in = π(Slant Height, in)(Radius, in)

= 3.14 x 25 in x 15 in

= 1,177.5 sq in

Since the entire area was asked for, the area of the base must be added.

Area, sq in = 0.785(Diameter, in)^2

= 0.785 x 30 in x 30 in

= 706.5 sq in

Total Area, sq in = Area of Cone, sq in + Area of Bottom, sq in

= 1,177.5 sq in + 706.5 sq in

= 1,884 sq in

A.26 Sphere

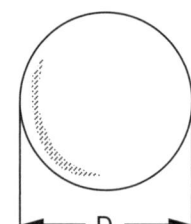

The surface area of a sphere or ball is equal to π multiplied by the diameter squared which is four times the cross-sectional area.

$$A_S = \pi D^2$$

If the radius is used, the formula becomes:

$$A_S = \pi D^2 = \pi \times 2R \times 2R = 4\pi R^2$$

EXAMPLE: What is the surface area of a sphere-shaped water tank 20 feet in diameter?

$$\text{Area, sq ft} = \pi(\text{Diameter, ft})^2$$

$$= 3.14 \times 20 \text{ ft} \times 20 \text{ ft}$$

$$= 1{,}256 \text{ sq ft}$$

A.3 VOLUMES

A.30 Rectangle

Volumes are measured in three dimensions or in cubic units. To calculate the volume of a rectangle, the area of the base is calculated in square units and then multiplied by the height. The formula then becomes:

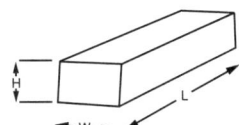

$$\boxed{V = L \times W \times H}$$

EXAMPLE: The length of a box is 2 feet, the width is 15 inches, and the height is 18 inches. Find its volume.

$$\text{Volume, cu ft} = \text{Length, ft} \times \text{Width, ft} \times \text{Height, ft}$$

$$= 2 \text{ ft} \times \frac{15 \text{ in}}{12 \text{ in/ft}} \times \frac{18 \text{ in}}{12 \text{ in/ft}}$$

$$= 2 \text{ ft} \times 1.25 \text{ ft} \times 1.5 \text{ ft}$$

$$= 3.75 \text{ cu ft}$$

A.31 Prism

The same general rule that applies to the volume of a rectangle also applies to a prism.

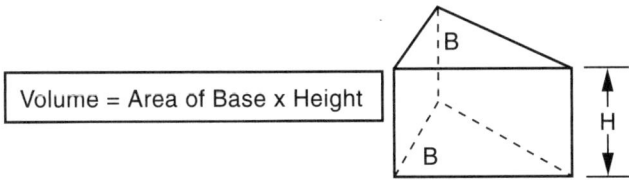

$$\boxed{\text{Volume} = \text{Area of Base} \times \text{Height}}$$

EXAMPLE: Find the volume of a prism with a base area of 10 square feet and a height of 5 feet. (Note that the base of a prism is triangular in shape.

$$\text{Volume, cu ft} = \text{Area of Base, sq ft} \times \text{Height, ft}$$

$$= 10 \text{ sq ft} \times 5 \text{ ft}$$

$$= 50 \text{ cu ft}$$

A.32 Cylinder

The volume of a cylinder is equal to the area of the base multiplied by the height.

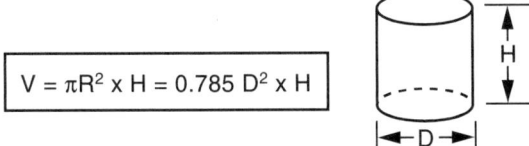

$$\boxed{V = \pi R^2 \times H = 0.785\, D^2 \times H}$$

EXAMPLE: A tank has a diameter of 100 feet and a depth of 12 feet. Find the volume.

$$\text{Volume, cu ft} = 0.785 \times (\text{Diameter, ft})^2 \times \text{Height, ft}$$

$$= 0.785 \times 100 \text{ ft} \times 100 \text{ ft} \times 12 \text{ ft}$$

$$= 94{,}200 \text{ cu ft}$$

A.33 Cone

The volume of a cone is equal to $\frac{1}{3}$ the volume of a circular cylinder of the same height and diameter.

$$\boxed{V = \frac{\pi}{3} R^2 \times H}$$

EXAMPLE: Calculate the volume of a cone if the height at the center is 4 feet and the diameter is 100 feet (radius is 50 feet).

$$\text{Volume, cu ft} = \frac{\pi}{3} \times (\text{Radius})^2 \times \text{Height, ft}$$

$$= \frac{\pi}{3} \times 50 \text{ ft} \times 50 \text{ ft} \times 4 \text{ ft}$$

$$= 10{,}500 \text{ cu ft}$$

A.34 Sphere

The volume of a sphere is equal to $\pi/6$ times the diameter cubed.

$$\boxed{V = \frac{\pi}{6} \times D^3}$$

EXAMPLE: How much gas can be stored in a sphere with a diameter of 12 feet? (Assume atmospheric pressure.)

$$\text{Volume, cu ft} = \frac{\pi}{6} \times (\text{Diameter, ft})^3$$

$$= \frac{\pi}{6} \times \overset{2}{\cancel{12}} \text{ ft} \times 12 \text{ ft} \times 12 \text{ ft}$$

$$= 904.32 \text{ cubic feet}$$

A.4 METRIC SYSTEM

The two most common systems of weights and measures are the English system and the metric system (Le Système International d' Unités, SI). Of these two, the metric system is more popular with most of the nations of the world. The reason for this is that the metric system is based on a system of tens and is therefore easier to remember and easier to use than the English system. Even though the basic system in the United States is the English system, the scientific community uses the metric system almost exclusively. Many organizations have urged, for good reason, that the United States switch to the metric system. Today the metric system is gradually becoming the standard system of measurement in the United States.

As the United States changes from the English to the metric system, some confusion and controversy has developed. For example, which is the correct spelling of the following words:

1. Liter or litre?

2. Meter or metre?

The U.S. National Bureau of Standards, the Water Environment Federation, and the American Water Works Association use litre and metre. The U.S. Government uses liter and meter and accepts no deviations. Some people argue that METRE should be used to measure LENGTH and that METER should be used to measure FLOW RATES (like a water or electric meter). Liter and meter are used in this manual because this is most consistent with spelling in the United States.

One of the most frequent arguments heard against the U.S. switching to the metric system was that the costs of switching manufacturing processes would be excessive. Pipe manufacturers have agreed upon the use of a "soft" metric conversion system during the conversion to the metric system. Past practice in the U.S. has identified some types of pipe by external (outside) diameter while other types are classified by nominal (existing only in name, not real or actual) bore. This means that a six-inch pipe does not have a six-inch inside diameter. With the strict or "hard" metric system, a six-inch pipe would be a 152.4-mm (6 in x 25.4 mm/in) pipe. In the "soft" metric system a six-inch pipe is a 150-mm (6 in x 25 mm/in) pipe. Typical customary and "soft" metric pipe-size designations are shown below:

PIPE-SIZE DESIGNATIONS

Customary, in	2	4	6	8	10	12	15	18
"Soft" Metric, mm	50	100	150	200	250	300	375	450

Customary, in	24	30	36	42	48	60	72	84
"Soft" Metric, mm	600	750	900	1050	1200	1500	1800	2100

In order to study the metric system, you must know the meanings of the terminology used. Following is a list of Greek and Latin prefixes used in the metric system.

PREFIXES USED IN THE METRIC SYSTEM

Prefix	Symbol	Meaning
Micro	μ	1/1 000 000 or 0.000 001
Milli	m	1/1000 or 0.001
Centi	c	1/100 or 0.01
Deci	d	1/10 or 0.1
Unit		1
Deka	da	10
Hecto	h	100
Kilo	k	1000
Mega	M	1 000 000

A.40 Measures of Length

The basic measure of length is the meter.

1 kilometer (km) = 1,000 meters (m)

1 meter (m) = 100 centimeters (cm)

1 centimeter (cm) = 10 millimeters (mm)

Kilometers are usually used in place of miles, meters are used in place of feet and yards, centimeters are used in place

of inches and millimeters are used for inches and fractions of an inch.

LENGTH EQUIVALENTS

1 kilometer	= 0.621 mile	1 mile	= 1.61 kilometers
1 meter	= 3.28 feet	1 foot	= 0.305 meter
1 meter	= 39.37 inches	1 inch	= 0.0254 meter
1 centimeter	= 0.3937 inch	1 inch	= 2.54 centimeters
1 millimeter	= 0.0394 inch	1 inch	= 25.4 millimeters

NOTE: The above equivalents are reciprocals. If one equivalent is given, the reverse can be obtained by division. For instance, if one meter equals 3.28 feet, one foot equals 1/3.28 meter, or 0.305 meter.

A.41 Measures of Capacity or Volume

The basic measure of capacity in the metric system is the liter. For measurement of large quantities the cubic meter is sometimes used.

1 kiloliter (kL) = 1,000 liters (L) = 1 cu meter (cu m)

1 liter (L) = 1,000 milliliters (mL)

Kiloliters, or cubic meters, are used to measure capacity of large storage tanks or reservoirs in place of cubic feet or gallons. Liters are used in place of gallons or quarts. Milliliters are used in place of quarts, pints, or ounces.

CAPACITY EQUIVALENTS

1 kiloliter	= 264.2 gallons	1 gallon	= 0.003785 kiloliter
1 liter	= 1.057 quarts	1 quart	= 0.946 liter
1 liter	= 0.2642 gallon	1 gallon	= 3.785 liters
1 milliliter	= 0.0353 ounce	1 ounce	= 29.57 milliliters

A.42 Measures of Weight

The basic unit of weight in the metric system is the gram. One cubic centimeter of water at maximum density weighs one gram, and thus there is a direct, simple relation between volume of water and weight in the metric system.

1 kilogram (kg) = 1,000 grams (gm)

1 gram (gm) = 1,000 milligrams (mg)

1 milligram (mg) = 1,000 micrograms (μg)

Grams are usually used in place of ounces, and kilograms are used in place of pounds.

WEIGHT EQUIVALENTS

1 kilogram	= 2.205 pounds	1 pound	= 0.4536 kilogram
1 gram	= 0.0022 pound	1 pound	= 453.6 grams
1 gram	= 0.0353 ounce	1 ounce	= 28.35 grams
1 gram	= 15.43 grains	1 grain	= 0.0648 gram

A.43 Temperature

Just as you should become familiar with the metric system, it is also important to become familiar with the centigrade (Celsius) scale for measuring temperature. There is nothing magical about the centigrade scale—it is simply a different size than the Fahrenheit scale. The two scales compare as follows:

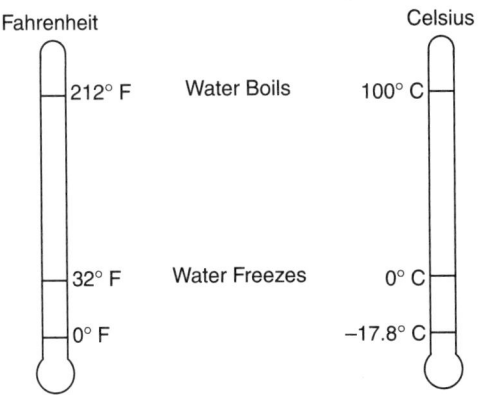

Fahrenheit

212° F   Water Boils   100° C

Celsius

32° F   Water Freezes   0° C

0° F   −17.8° C

The two scales are related in the following manner:

$$\text{Fahrenheit} = (°C \times 9/5) + 32°$$
$$\text{Celsius} = (°F − 32°) \times 5/9$$

EXAMPLE: Convert 20° Celsius to degrees Fahrenheit.

$$°F = (°C \times 9/5) + 32°$$

$$°F = (20° \times 9/5) + 32°$$

$$°F = \frac{180°}{5} + 32°$$

$$= 36° + 32°$$

$$= 68°F$$

EXAMPLE: Convert −10°C to °F

$$°F = (−10° \times 9/5) + 32°$$

$$°F = −90°/5 + 32°$$

$$= −18° + 32°$$

$$= 14°F$$

EXAMPLE: Convert −13°F to °C

$$°C = (°F − 32°) \times \frac{5}{9}$$

$$°C = (−13° − 32°) \times \frac{5}{9}$$

$$= −45° \times \frac{5}{9}$$

$$= −5° \times 5$$

$$= −25°C$$

A.44   Milligrams Per Liter

Milligrams per liter (mg/L) is a unit of measurement used in laboratory and scientific work to indicate very small concentrations of dilutions. Since water contains small concentrations of dissolved substances and solids, and since small amounts of chemical compounds are sometimes used in water treatment processes, the term milligrams per liter is also common in treatment plants. It is a weight/volume relationship.

As previously discussed:

1,000 liters = 1 cubic meter = 1,000,000 cubic centimeters.

Therefore,

1 liter = 1,000 cubic centimeters.

Since one cubic centimeter of water weighs one gram,

1 liter of water = 1,000 grams or 1,000,000 milligrams.

$$\frac{1 \text{ milligram}}{\text{liter}} = \frac{1 \text{ milligram}}{1,000,000 \text{ milligrams}} = \frac{1 \text{ part}}{\text{million parts}} = \frac{1 \text{ part per}}{\text{million (ppm)}}$$

Milligrams per liter and parts per million (parts) may be used interchangeably as long as the liquid density is 1.0 gm/cu cm or 62.43 lb/cu ft. A concentration of 1 milligram/liter (mg/L) or 1 ppm means that there is 1 part of substance by weight for every 1 million parts of water. A concentration of 10 mg/L would mean 10 parts of substance per million parts of water.

To get an idea of how small 1 mg/L is, divide the numerator and denominator of the fraction by 10,000. This, of course, does not change its value since 10,000 ÷ 10,000 is equal to one.

$$1\frac{mg}{L} = \frac{1 \text{ mg}}{1,000,000 \text{ mg}} = \frac{1/10,000 \text{ mg}}{1,000,000/10,000 \text{ mg}} = \frac{0.0001 \text{ mg}}{100 \text{ mg}} = 0.0001\%$$

Therefore, 1 mg/L is equal to one ten-thousandth of a percent, or

1% is equal to 10,000 mg/L.

To convert mg/L to %, move the decimal point four places or numbers to the left.

Working problems using milligrams per liter or parts per million is a part of everyday operation in most water treatment plants.

A.45   Example Problems

EXAMPLE: Raw water flowing into a plant at a rate of five million pounds per day is prechlorinated at 5 mg/L. How many pounds of chlorine are used per day?

$$5 \text{ mg/L} = \frac{5 \text{ lbs chlorine}}{\text{million lbs water}}$$

Chlorine Feed, lbs/day = Conc, lbs/M lbs x Flow, lbs/day

$$= \frac{5 \text{ lbs}}{\text{million lbs}} \times \frac{5 \text{ million lbs}}{\text{day}}$$

$$= 25 \text{ lbs/day}$$

There is one thing that is unusual about the above problem and that is the flow is reported in pounds per day. In most treatment plants, flow is reported in terms of gallons per minute or gallons per day. To convert these flow figures to weight, an additional conversion factor is needed. One gallon of water weighs 8.34 pounds. Using this factor, it is possible to convert flow in gallons per day to flow in pounds per day.

EXAMPLE: A well pump with a flow of 3.5 million gallons per day (MGD) chlorinates the water with 2.0 mg/L chlorine. How many pounds of chlorine are used per day?

$$\text{Flow, lbs/day} = \text{Flow,}\ \frac{\text{M gal}}{\text{day}}\ \text{x}\ \frac{8.34\ \text{lbs}}{\text{gal}}$$

$$= \frac{3.5\ \text{million gal}}{\text{day}}\ \text{x}\ \frac{8.34\ \text{lbs}}{\text{gal}}$$

$$= 29.19\ \text{million lbs/day}$$

$$\begin{array}{c}\text{Chlorine}\\\text{Feed,}\\\text{lbs/day}\end{array} = \text{Level, mg/}L\ \text{x Flow, M lbs/day}$$

$$= \frac{2.0\ \text{mg*}}{\text{million mg}}\ \text{x}\ \frac{29.19\ \text{million lbs}}{\text{day}}$$

$$= 58.38\ \text{lbs/day}$$

---

* Remember that $\dfrac{1\ \text{mg}}{\text{M mg}} = \dfrac{1\ \text{lb}}{\text{M lb}}$. They are identical ratios.

In solving the previous problem, a relation was used that is most important to understand and commit to memory.

> Feed, lbs/day = Flow, MGD x Dose, mg/L x 8.34 lbs/gal

EXAMPLE: A chlorinator is set to feed 50 pounds of chlorine per day to a flow of 0.8 MGD. What is the chlorine dose in mg/L?

$$\begin{array}{c}\text{Conc or Dose,}\\\text{mg/}L\end{array} = \frac{\text{lbs/day}}{\text{MGD x 8.34 lbs/gal}}$$

$$= \frac{50\ \text{lbs/day}}{0.80\ \text{MG/day x 8.34 lbs/gal}}$$

$$= \frac{50\ \text{lbs}}{6.672\ \text{M lbs}}$$

$$= 7.5\ \text{mg/}L,\ \text{or 7.5 ppm}$$

EXAMPLE: A pump delivers 500 gallons per minute to a water treatment plant. Alum is added at 10 mg/L. How much alum is used in pounds per day?

$$\text{Flow, MGD} = \text{Flow, GPM x 60 min/hr x 24 hr/day}$$

$$= \frac{500\ \text{gal}}{\text{min}}\ \text{x}\ \frac{60\ \text{min}}{\text{hr}}\ \text{x}\ \frac{24\ \text{hr}}{\text{day}}$$

$$= 720,000\ \text{gal/day}$$

$$= 0.72\ \text{MGD}$$

$$\begin{array}{c}\text{Alum Feed,}\\\text{lbs/day}\end{array} = \text{Flow, MGD x Dose, mg/}L\ \text{x 8.34 lbs/gal}$$

$$= \frac{0.72\ \text{M gal}}{\text{day}}\ \text{x}\ \frac{10\ \text{mg}}{\text{M mg}}\ \text{x}\ \frac{8.34\ \text{lbs}}{\text{gal}}$$

$$= 60.048\ \text{lbs/day or about 60 lbs/day}$$

A.5   WEIGHT—VOLUME RELATIONS

Another factor for the operator to remember, in addition to the weight of a gallon of water, is the weight of a cubic foot of water. One cubic foot of water weighs 62.4 lbs. If these two weights are divided, it is possible to determine the number of gallons in a cubic foot.

$$\frac{62.4\ \text{pounds/cu ft}}{8.34\ \text{pounds/gal}} = 7.48\ \text{gal/cu ft}$$

Thus we have another very important relationship to commit to memory.

> 8.34 lbs/gal x 7.48 gal/cu ft = 62.4 lbs/cu ft

It is only necessary to remember two of the above items since the third may be found by calculation. For most problems, 8⅓ lbs/gal and 7½ gal/cu ft will provide sufficient accuracy.

EXAMPLE: Change 1,000 cu ft of water to gallons.

$$1,000\ \text{cu ft x 7.48 gal/cu ft} = 7,480\ \text{gallons}$$

EXAMPLE: What is the weight of three cubic feet of water?

$$62.4\ \text{lbs/cu ft x 3 cu ft} + 187.2\ \text{lbs}$$

EXAMPLE: The net weight of a tank of water is 750 lbs. How many gallons does it contain?

$$\frac{750\ \text{lbs}}{8.34\ \text{lbs/gal}} = 90\ \text{gal}$$

A.6   FORCE, PRESSURE, AND HEAD

In order to study the forces and pressures involved in fluid flow, it is first necessary to define the terms used.

FORCE: The push exerted by water on any surface being used to confine it. Force is usually expressed in pounds, tons, grams, or kilograms.

PRESSURE: The force per unit area. Pressure can be expressed in many ways, but the most common term is pounds per square inch (psi).

HEAD: Vertical distance from the water surface to a reference point below the surface. Usually expressed in feet or meters.

An *EXAMPLE* should serve to illustrate these terms.

If water were poured into a one-foot cubical container, the *FORCE* acting on the bottom of the container would be 62.4 pounds.

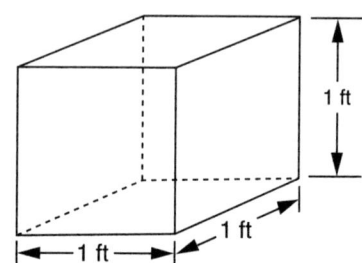

The *PRESSURE* acting on the bottom would be 62.4 pounds per square foot. The area of the bottom is also 12 in x 12 in = 144 sq in. Therefore, the pressure may also be expressed as:

$$\text{Pressure, psi} = \frac{62.4\ \text{lbs}}{\text{sq ft}} = \frac{62.4\ \text{lbs/sq ft}}{144\ \text{sq in/sq ft}}$$

$$= 0.433\ \text{lb/sq in}$$

$$= 0.433\ \text{psi}$$

Since the height of the container is one foot, the *HEAD* would be one foot.

The pressure in any vessel at one foot of depth or one foot of head is 0.433 psi acting in any direction.

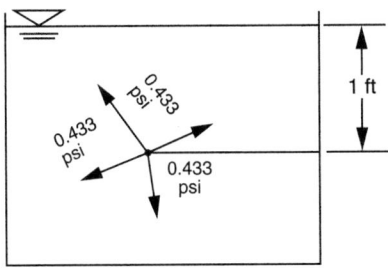

If the depth of water in the previous example were increased to two feet, the pressure would be:

$$p = \frac{2(62.4 \text{ lbs})}{144 \text{ sq in}} = \frac{124.8 \text{ lbs}}{144 \text{ sq in}} = 0.866 \text{ psi}$$

Therefore we can see that for every foot of head, the pressure increases by 0.433 psi. Thus, the general formula for pressure becomes:

$\boxed{p, \text{psi} = 0.433(H, \text{ft})}$	H = feet of head p = pounds per square *INCH* of pressure
	H = feet of head
$\boxed{P, \text{lbs/sq ft} = 62.4(H, \text{ft})}$	P = pounds per square *FOOT* of pressure

We can now draw a diagram of the pressure acting on the side of a tank. Assume a 4-foot deep tank. The pressures shown on the tank are gage pressures. These pressures do not include the atmospheric pressure acting on the surface of the water.

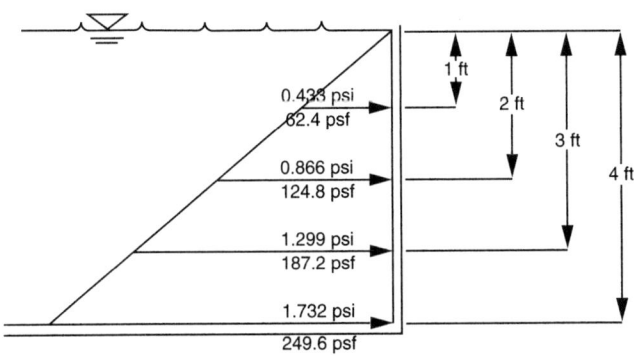

$p_0 = 0.433 \times 0 = 0.0 \text{ psi}$ $P_0 = 62.4 \times 0 = 0.0 \text{ lb/sq ft}$
$p_1 = 0.433 \times 1 = 0.433 \text{ psi}$ $P_1 = 62.4 \times 1 = 62.4 \text{ lbs/sq ft}$
$p_2 = 0.433 \times 2 = 0.866 \text{ psi}$ $P_2 = 62.4 \times 2 = 124.8 \text{ lbs/sq ft}$
$p_3 = 0.433 \times 3 = 1.299 \text{ psi}$ $P_3 = 62.4 \times 3 = 187.2 \text{ lbs/sq ft}$
$p_4 = 0.433 \times 4 = 1.732 \text{ psi}$ $P_4 = 62.4 \times 4 = 249.6 \text{ lbs/sq ft}$

The average *PRESSURE* acting on the tank wall is 1.732 psi/2 = 0.866 psi, or 249.6 psf/2 = 124.8 psf. We divided by two to obtain the average pressure because there is zero pressure at the top and 1.732 psi pressure on the bottom of the wall.

If the wall were 5 feet long, the pressure would be acting over the entire 20-square-foot (5 ft x 4 ft) area of the wall. The total force acting to push the wall would be:

Force, lbs = (Pressure, lbs/sq ft)(Area, sq ft)

= 124.8 lbs/sq ft x 20 sq ft

= 2,496 lbs

If the pressure in psi were used, the problem would be similar:

Force, lbs = (Pressure, lbs/sq in)(Area, sq in)

= 0.866 psi x 48 in x 60 in

= 2,494 lbs*

---

* Difference in answer due to rounding off of decimal points.

The general formula, then, for finding the total force acting on a side wall of a tank is:

F = force in pounds

H = head in feet

L = length of wall in feet

$\boxed{F = 31.2 \times H^2 \times L}$ 31.2 = constant with units of lbs/cu ft and considers the fact that the force results from H/2 or half the depth of the water which is the average depth. The force is exerted at H/3 from the bottom.

EXAMPLE: Find the force acting on a five-foot long wall in a four-foot deep tank.

Force, lbs = 31.2(Head, ft)²(Length, ft)

= 31.2 lbs/cu ft x (4 ft)² x 5 ft

= 2,496 lbs

Occasionally an operator is warned: *NEVER EMPTY A TANK DURING PERIODS OF HIGH GROUNDWATER.* Why? The pressure on the bottom of the tank caused by the water surrounding the tank will tend to float the tank like a cork if the upward force of the water is greater than the weight of the tank.

F = upward force in pounds

H = head of water on tank bottom in feet

$\boxed{F = 62.4 \times H \times A}$ A = area of bottom of tank in square feet

62.4 = a constant with units of lbs/cu ft

This formula is approximately true if the tank doesn't crack, leak, or start to float.

EXAMPLE: Find the upward force on the bottom of an empty tank caused by groundwater depth of 8 feet above the tank bottom. The tank is 20 ft wide and 40 ft long.

$$\text{Force, lbs} = 62.4(\text{Head, ft})(\text{Area, sq ft})$$

$$= 62.4 \text{ lbs/cu ft} \times 8 \text{ ft} \times 20 \text{ ft} \times 40 \text{ ft}$$

$$= 399,400 \text{ lbs}$$

A.7 VELOCITY AND FLOW RATE

A.70 Velocity

The velocity of a particle or substance is the speed at which it is moving. It is expressed by indicating the length of travel and how long it takes to cover the distance. Velocity can be expressed in almost any distance and time units. For instance, a car may be traveling at a rate of 280 miles per five hours. However, it is normal to express the distance traveled per unit time. The above example would then become:

$$\text{Velocity, mi/hr} = \frac{280 \text{ miles}}{5 \text{ hours}}$$

$$= 56 \text{ miles/hour}$$

The velocity of water in a channel, pipe, or other conduit can be expressed in the same way. If the particle of water travels 600 feet in five minutes, the velocity is:

$$\text{Velocity, ft/min} = \frac{\text{Distance, ft}}{\text{Time, minutes}}$$

$$= \frac{600 \text{ ft}}{5 \text{ min}}$$

$$= 120 \text{ ft/min}$$

If you wish to express the velocity in feet per second, multiply by 1 min/60 seconds.

NOTE: Multiplying by $\dfrac{1 \text{ minute}}{60 \text{ seconds}}$ is like multiplying by $\dfrac{1}{1}$; it does not change the relative value of the answer. It only changes the form of the answer.

$$\text{Velocity, ft/sec} = (\text{Velocity, ft/min})(1 \text{ min/60 sec})$$

$$= \frac{120 \text{ ft}}{\text{min}} \times \frac{1 \text{ min}}{60 \text{ sec}}$$

$$= \frac{120 \text{ ft}}{60 \text{ sec}}$$

$$= 2 \text{ ft/sec}$$

A.71 Flow Rate

If water in a one-foot wide channel is one foot deep, then the cross-sectional area of the channel is 1 ft x 1 ft = 1 sq ft.

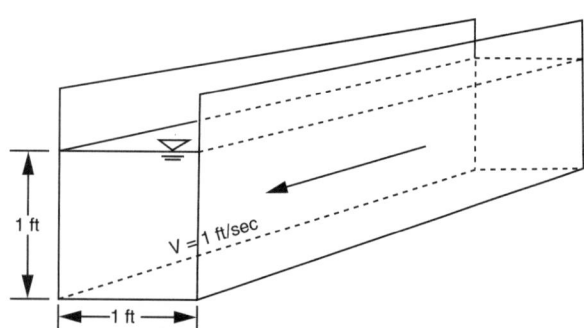

If the velocity in this channel is 1 ft per second, then each second a body of water 1 sq ft in area and 1 ft long will pass a

given point. The volume of this body of water would be 1 cubic foot. Since one cubic foot of water would pass by every second, the flow rate would be equal to 1 cubic foot per second, or 1 CFS.

To obtain the flow rate in the above example the velocity was multiplied by the cross-sectional area. This is another important general formula.

$$\boxed{Q = V \times A}$$

$$Q = \text{flow rate, CFS or cu ft/sec}$$
$$V = \text{velocity, ft/sec}$$
$$A = \text{area, sq ft}$$

EXAMPLE: A rectangular channel 3 feet wide contains water 2 feet deep and flowing at a velocity of 1.5 feet per second. What is the flow rate in CFS?

$$Q = V \times A$$

$$\text{Flow Rate, CFS} = \text{Velocity, ft/sec} \times \text{Area, sq ft}$$

$$= 1.5 \text{ ft/sec} \times 3 \text{ ft} \times 2 \text{ ft}$$

$$= 9 \text{ cu ft/sec}$$

EXAMPLE: Flow in a 2.5-foot wide channel is 1.4 feet deep and measures 11.2 CFS. What is the average velocity?

In this problem we want to find the velocity. Therefore, we must rearrange the general formula to solve for velocity.

$$V = \frac{Q}{A}$$

$$\text{Velocity, ft/sec} = \frac{\text{Flow Rate, cu ft/sec}}{\text{Area, sq ft}}$$

$$= \frac{11.2 \text{ cu ft/sec}}{2.5 \text{ ft} \times 1.4 \text{ ft}}$$

$$= \frac{11.2 \text{ cu ft/sec}}{3.5 \text{ sq ft}}$$

$$= 3.2 \text{ ft/sec}$$

EXAMPLE: Flow in an 8-inch pipe is 500 GPM. What is the average velocity?

$$\text{Area, sq ft} = 0.785(\text{Diameter, ft})^2$$

$$= 0.785(8/12 \text{ ft})^2$$

$$= 0.785(0.67 \text{ ft})^2$$

$$= 0.785(0.67 \text{ ft})(0.67 \text{ ft})$$

$$= 0.785(0.45 \text{ sq ft})$$

$$= 0.35 \text{ sq ft}$$

$$\text{Flow, CFS} = \text{Flow, gal/min} \times \frac{\text{cu ft}}{7.48 \text{ gal}} \times \frac{1 \text{ min}}{60 \text{ sec}}$$

$$= \frac{500 \text{ gal}}{\text{min}} \times \frac{\text{cu ft}}{7.48 \text{ gal}} \times \frac{1 \text{ min}}{60 \text{ sec}}$$

$$= \frac{500 \text{ cu ft}}{448.8 \text{ sec}}$$

$$= 1.114 \text{ CFS}$$

$$\text{Velocity, ft/sec} = \frac{\text{Flow, cu ft/sec}}{\text{Area, sq ft}}$$

$$= \frac{1.114 \text{ cu ft/sec}}{0.35 \text{ sq ft}}$$

$$= 3.18 \text{ ft/sec}$$

A.8   PUMPS

A.80   Pressure

Atmospheric pressure at sea level is approximately 14.7 psi. This pressure acts in all directions and on all objects. If a tube is placed upside down in a basin of water and a 1 psi partial vacuum is drawn on the tube, the water in the tube will rise 2.31 feet.

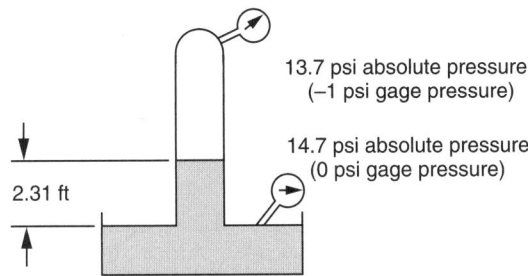

13.7 psi absolute pressure
(–1 psi gage pressure)

14.7 psi absolute pressure
(0 psi gage pressure)

2.31 ft

NOTE: 1 ft of water = 0.433 psi; therefore,

$$1 \text{ psi} = \frac{1}{0.433} \text{ ft} = 2.31 \text{ ft of water}$$

The action of the partial vacuum is what gets water out of a sump or well and up to a pump. It is not sucked up, but it is pushed up by atmospheric pressure on the water surface in the sump. If a complete vacuum could be drawn, the water would rise 2.31 x 14.7 = 33.9 feet; but this is impossible to achieve. The practical limit of the suction lift of a positive displacement pump is about 22 feet, and that of a centrifugal pump is 15 feet.

A.81   Work

Work can be expressed as lifting a weight a certain vertical distance. It is usually defined in terms of foot-pounds.

EXAMPLE: A 165-pound man runs up a flight of stairs 20 feet high. How much work did he do?

$$\text{Work, ft-lbs} = \text{Weight, lbs} \times \text{Height, ft}$$

$$= 165 \text{ lbs} \times 20 \text{ ft}$$

$$= 3,300 \text{ ft-lbs}$$

A.82   Power

Power is a rate of doing work and is usually expressed in foot-pounds per minute.

EXAMPLE: If the man in the above example runs up the stairs in three seconds, how much power has he exerted?

$$\text{Power, ft-lbs/sec} = \frac{\text{Work, ft-lbs}}{\text{Time, sec}}$$

$$= \frac{3,300 \text{ ft-lbs}}{3 \text{ sec}} \times \frac{60 \text{ sec}}{\text{minute}}$$

$$= 66,000 \text{ ft-lbs/min}$$

A.83   Horsepower

Horsepower is also a unit of power. One horsepower is defined as 33,000 ft-lbs per minute or 746 watts.

EXAMPLE: How much horsepower has the man in the previous example exerted as he climbs the stairs?

$$\text{Horsepower, HP} = (\text{Power, ft-lbs/min})\left(\frac{\text{HP}}{33,000 \text{ ft-lbs/min}}\right)$$

$$= 66,000 \text{ ft-lbs/min} \times \frac{\text{Horsepower}}{33,000 \text{ ft-lbs/min}}$$

$$= 2 \text{ HP}$$

Work is also done by lifting water. If the flow from a pump is converted to a weight of water and multiplied by the vertical distance it is lifted, the amount of work or power can be obtained.

$$\text{Horsepower, HP} = \frac{\text{Flow, gal}}{\text{min}} \times \text{Lift, ft} \times \frac{8.34 \text{ lbs}}{\text{gal}} \times \frac{\text{Horsepower}}{33,000 \text{ ft-lbs/min}}$$

Solving the above relation, the amount of horsepower necessary to lift the water is obtained. This is called water horsepower.

$$\text{Water, HP} = \frac{(\text{Flow, GPM})(\text{H, ft})}{3,960^*}$$

$$^* \frac{8.34 \text{ lbs}}{\text{gal}} \times \frac{\text{HP}}{33,000 \text{ ft-lbs/min}} = \frac{1}{3,960}$$

1 gallon weighs 8.34 pounds and 1 horsepower is the same as 33,000 ft-lbs/min.

H or Head in feet is the same as Lift in feet.

However, since pumps are not 100% efficient (they cannot transmit all the power put into them), the horsepower supplied to a pump is greater than the water horsepower. Horsepower supplied to the pump is called brake horsepower.

$$\boxed{\text{Brake, HP} = \frac{\text{Flow, GPM} \times \text{H, ft}}{3,960 \times E_p}}$$

E_p = Efficiency of Pump (Usual range 50-85%, depending on type and size of pump)

Motors are also not 100% efficient; therefore, the power supplied to the motor is greater than the motor transmits.

$$\boxed{\text{Motor, HP} = \frac{\text{Flow, GPM} \times \text{H, ft}}{3,960 \times E_p \times E_m}}$$

E_m = Efficiency of Motor (Usual range 80-95%, depending on type and size of motor)

The above formulas have been developed for the pumping of water and wastewater which have a specific gravity of 1.0. If other liquids are to be pumped, the formulas must be multiplied by the specific gravity of the liquid.

EXAMPLE: A flow of 500 GPM of water is to be pumped against a total head of 100 feet by a pump with an efficiency of 70%. What is the pump horsepower?

$$\text{Brake, HP} = \frac{\text{Flow, GPM} \times \text{H, ft}}{3,960 \times E_p}$$

$$= \frac{500 \times 100}{3,960 \times 0.70}$$

$$= 18 \text{ HP}$$

EXAMPLE: Find the horsepower required to pump gasoline (specific gravity = 0.75) in the above problem.

$$\text{Brake, HP} = \frac{500 \times 100 \times 0.75}{3,960 \times 0.70}$$

$$= 13.5 \text{ HP (gasoline is lighter and requires less horsepower)}$$

A.84 Head

Basically, the head that a pump must work against is determined by measuring the vertical distance between the two water surfaces, or the distance the water must be lifted. This is called the static head. Two typical conditions for lifting water are shown below.

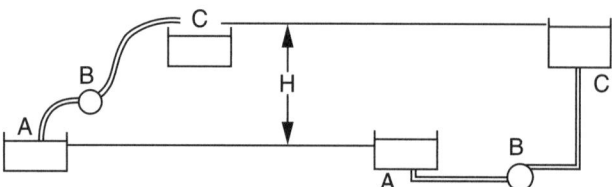

If a pump were designed in the above examples to pump only against head H, the water would never reach the intended point. The reason for this is that the water encounters friction in the pipelines. Friction depends on the roughness and length of pipe, the pipe diameter, and the flow velocity. The turbulence caused at the pipe entrance (point A); the pump (point B); the pipe exit (point C); and at each elbow, bend, or transition also adds to these friction losses. Tables and charts are available in Section A.88 for calculation of these friction losses so they may be added to the measured or static head to obtain the total head. For short runs of pipe which do not have high velocities, the friction losses are generally less than 10 percent of the static head.

EXAMPLE: A pump is to be located 8 feet above a wet well and must lift 1.8 MGD another 50 feet to a storage reservoir. If the pump has an efficiency of 75% and the motor an efficiency of 90%, what is the cost of the power consumed if one kilowatt hour cost 4 cents?

Since we are not given the length or size of pipe and the number of elbows or bends, we will assume friction to be 10% of static head.

Static Head, ft = Suction Lift + Discharge Head, ft

$$= 8 \text{ ft} + 50 \text{ ft}$$

$$= 58 \text{ ft}$$

Friction Losses, ft = 0.1 (Static Head, ft)

$$= 0.1 (58 \text{ ft})$$

$$= 5.8 \text{ ft}$$

Total Dynamic Head, ft = Static Head, ft + Friction Losses, ft

$$= 58 \text{ ft} + 5.8 \text{ ft}$$

$$= 63.8 \text{ ft}$$

$$\text{Flow, GPM} = \frac{1,800,000 \text{ gal}}{\text{day}} \times \frac{\text{day}}{24 \text{ hr}} \times \frac{1 \text{ hr}}{60 \text{ min}}$$

$$= 1,250 \text{ GPM (assuming pump runs 24 hours per day)}$$

$$\text{Motor, HP} = \frac{\text{Flow, GPM} \times H, \text{ ft}}{3,960 \times E_p \times E_m}$$

$$= \frac{1,250 \times 63.8}{3,960 \times 0.75 \times 0.9}$$

$$= 30 \text{ HP}$$

Kilowatt-hr = 30 HP $\times$ 24 hr/day $\times$ 0.746 kW/HP*

$$= 537 \text{ kilowatt-hr/day}$$

Cost = kWh $\times$ \$0.04/kWh

$$= 537 \times 0.04$$

$$= \$21.48/\text{day}$$

---

* See Section A.10, "Basic Conversion Factors," *POWER*, page 614.

A.85 Pump Characteristics

The discharge of a centrifugal pump, unlike a positive displacement pump, can be made to vary from zero to a maximum capacity which depends on the speed, head, power, and specific impeller design. The interrelation of capacity, efficiency, head, and power is known as the characteristics of the pump.

The first relation normally looked at when selecting a pump is the head vs. capacity. The head of a centrifugal pump normally rises as the capacity is reduced. If the values are plotted on a graph they appear as follows:

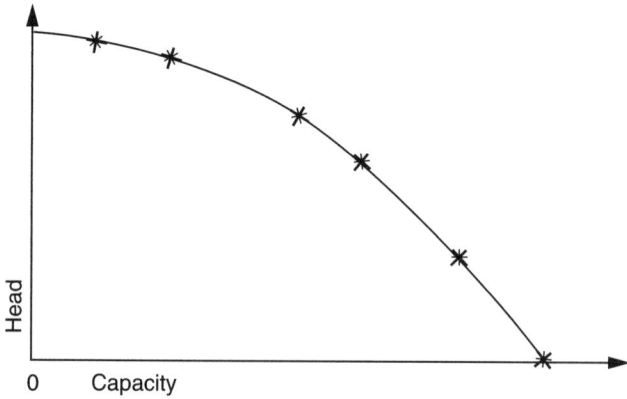

Another important characteristic is the pump efficiency. It begins from zero at no discharge, increases to a maximum, and then drops as the capacity is increased. Following is a graph of efficiency vs. capacity:

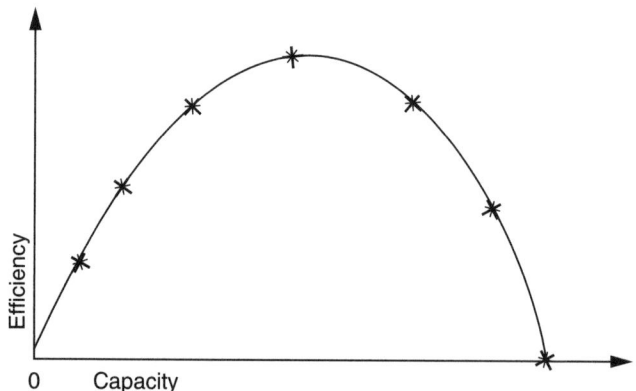

The last important characteristic is the brake horsepower or the power input to the pump. The brake horsepower usually increases with increasing capacity until it reaches a maximum, then it normally reduces slightly.

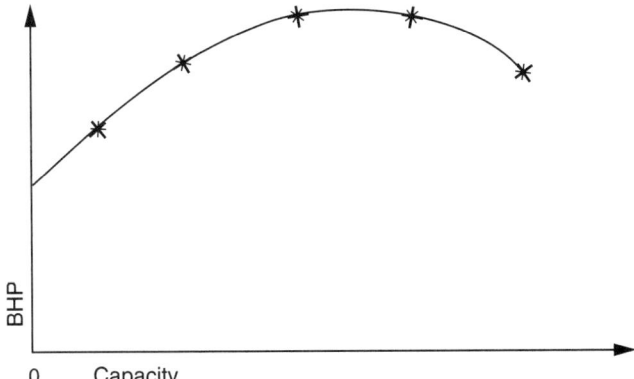

These pump characteristic curves are quite important. Pump sizes are normally picked from these curves rather than calculations. For ease of reading, the three characteristic curves are normally plotted together. A typical graph of pump characteristics is shown as follows:

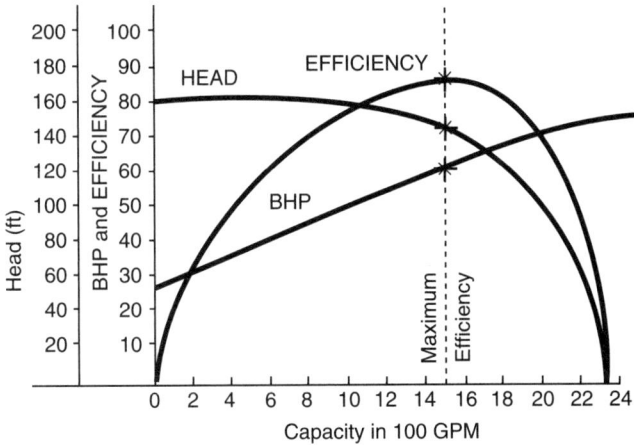

The curves show that the maximum efficiency for the particular pump in question occurs at approximately 1,475 GPM, a head of 132 feet, and a brake horsepower of 58. Operating at this point the pump has an efficiency of approximately 85%. This can be verified by calculation:

$$BHP = \frac{Flow,\ GPM \times H,\ ft}{3,960 \times E}$$

As previously explained, a number can be written over one without changing its value:

$$\frac{BHP}{1} = \frac{GPM \times H}{3,960 \times E}$$

Since the formula is now in ratio form, it can be cross multiplied.

$$BHP \times 3,960 \times E = GPM \times H \times 1$$

Solving for E,

$$E = \frac{GPM \times H}{3,960 \times BHP}$$

$$E = \frac{1,475\ GPM \times 132\ ft}{3,960 \times 58\ HP}$$

$$= 0.85\ or\ 85\%\ (Check)$$

The preceding is only a brief description of pumps to familiarize the operator with their characteristics. The operator does not normally specify the type and size of pump needed at a plant. If a pump is needed, the operator should be able to supply the information necessary for a pump supplier to provide the best possible pump for the lowest cost. Some of the information needed includes:

1. Flow range desired;

2. Head conditions:

    a. Suction head or lift,
    b. Pipe and fitting friction head, and
    c. Discharge head;

3. Type of fluid pumped and temperature; and

4. Pump location.

A.86 Evaluation of Pump Performance

1. Capacity

Sometimes it is necessary to determine the capacity of a pump. This can be accomplished by determining the time it takes a pump to fill or empty a portion of a storage tank or diversion box when all inflow is blocked off.

EXAMPLE:

a. Measure the size of the wet well.

    Length = 10 ft

    Width = 10 ft

    Depth = 5 ft (We will measure the time it takes to lower the well a distance of five feet.)

    Volume, cu ft = L, ft x W, ft x D, ft

               = 10 ft x 10 ft x 5 ft

               = 500 cu ft

b. Record time for water to drop five feet in wet well.

    Time = 10 minutes 30 seconds

         = 10.5 minutes

c. Calculate pumping rate or capacity.

$$Pumping\ Rate,\ GPM = \frac{Volume,\ gallons}{Time,\ minutes}$$

$$= \frac{(500\ cu\ ft)(7.5\ gal/cu\ ft)}{10.5\ min}$$

$$= \frac{3,750}{10.5}$$

$$= 357\ GPM$$

If you know the total dynamic head and have the pump's performance curves, you can determine if the pump is delivering at design capacity. If not, try to determine the cause (see Chapter 18, "Maintenance," in Volume II). After a pump overhaul, the pump's actual performance (flow, head, power, and efficiency) should be compared with the pump manufacturer's performance curves. This procedure for calculating the rate of filling or emptying of a wet well or diversion box can be used to calibrate flowmeters.

2. Efficiency

To estimate the efficiency of the pump in the previous example, the total head must be known. This head may be estimated by measuring the suction and discharge pressures. Assume these were measured as follows:

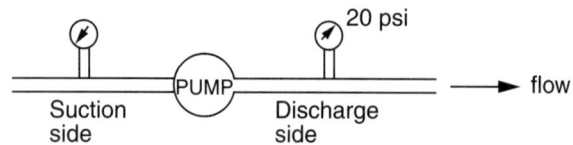

No additional information is necessary if we assume the pressure gages are at the same height and the pipe diameters are the same. Both pressure readings must be converted to feet.

$$\text{Suction Lift, ft} = 2 \text{ in Mercury} \times \frac{1.133 \text{ ft water*}}{1 \text{ in Mercury}}$$

$$= 2.27 \text{ ft}$$

$$\text{Discharge Head, ft} = 20 \text{ psi} \times 2.31 \text{ ft/psi*}$$

$$= 46.20 \text{ ft}$$

$$\text{Total Head, ft} = \text{Suction Lift, ft} + \text{Discharge Head, ft}$$

$$= 2.27 \text{ ft} + 46.20 \text{ ft}$$

$$= 48.47 \text{ ft}$$

---

* See Section A.10, "Basic Conversion Factors," *PRESSURE*, page 614.

Calculate the power output of the pump or water horsepower:

$$\text{Water Horsepower, HP} = \frac{(\text{Flow, GPM})(\text{Head, ft})}{3,960}$$

$$= \frac{(357 \text{ GPM})(48.47 \text{ ft})}{3,960}$$

$$= 4.4 \text{ HP}$$

To estimate the efficiency of the pump, measure the kilowatts drawn by the pump motor. Assume the meter indicates 8,000 watts or 8 kilowatts. The manufacturer claims the electric motor is 80% efficient.

$$\text{Brake Horsepower, HP} = (\text{Power to Elec Motor})(\text{Motor Eff})$$

$$= \frac{(8 \text{ kW})(0.80)}{0.746 \text{ kW/HP}}$$

$$= 8.6 \text{ HP}$$

$$\text{Pump Efficiency, \%} = \frac{\text{Water Horsepower, HP} \times 100\%}{\text{Brake Horsepower, HP}}$$

$$= \frac{4.4 \text{ HP} \times 100\%}{8.6 \text{ HP}}$$

$$= 51\%$$

The following diagram may clarify the previous problem:

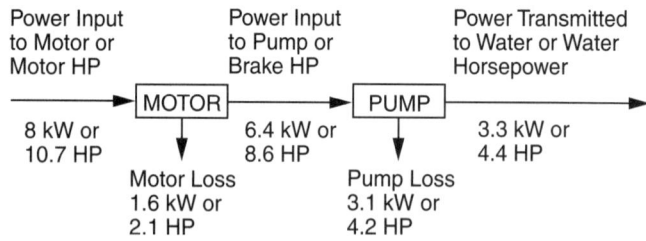

The wire-to-water efficiency is the efficiency of the power input to produce water horsepower.

$$\text{Wire-to-Water Efficiency, \%} = \frac{\text{Water Horsepower, HP}}{\text{Power Input, HP}} \times 100\%$$

$$= \frac{4.4 \text{ HP}}{10.7 \text{ HP}} \times 100\%$$

$$= 41\%$$

The wire-to-water efficiency of a pumping system (pump and electric motor) can be calculated by using the following formula:

$$\text{Efficiency, \%} = \frac{(\text{Flow, GPM})(\text{TDH, ft})(100\%)}{(\text{Voltage, volts})(\text{Current, amps})(5.308)}$$

$$= \frac{(375 \text{ GPM})(48.47 \text{ ft})(100\%)}{(220 \text{ volts})(36 \text{ amps})(5.308)}$$

$$= 41\%$$

The horsepower required by a pump motor is the water horsepower divided by the pump efficiency and the motor efficiency as decimals.

$$\text{Motor, HP} = \frac{\text{Water, HP}}{(E_p)(E_m)}$$

A.87 Pump Speed—Performance Relationships

Changing the velocity of a centrifugal pump will change its operating characteristics. If the speed of a pump is changed, the flow, head developed, and power requirements will change. The operating characteristics of the pump will change with speed approximately as follows:

$$\text{Flow, } Q_n = \left[\frac{N_n}{N_r}\right] Q_r$$

$$\text{Head, } H_n = \left[\frac{N_n}{N_r}\right]^2 H_r$$

$$\text{Power, } P_n = \left[\frac{N_n}{N_r}\right]^3 P_r$$

r = rated

n = now

N = pump speed

Actually, pump efficiency does vary with speed; therefore, these formulas are not quite correct. If speeds do not vary by more than a factor of two (if the speeds are not doubled or cut in half), the results are close enough. Other factors contributing to changes in pump characteristic curves include impeller wear and roughness in pipes.

EXAMPLE: To illustrate these relationships, assume a pump has a rated capacity of 600 GPM, develops 100 ft of head, and has a power requirement of 15 HP when operating at 1,500 RPM. If the efficiency remains constant, what will be the operating characteristics if the speed drops to 1,200 RPM?

Calculate new flow rate or capacity:

$$\text{Flow, } Q_n = \left[\frac{N_n}{N_r}\right]Q_r$$
$$= \left[\frac{1,200 \text{ RPM}}{1,500 \text{ RPM}}\right](600 \text{ GPM})$$
$$= \left(\frac{4}{5}\right)(600 \text{ GPM})$$
$$= (4)(120 \text{ GPM})$$
$$= 480 \text{ GPM}$$

Calculate new head:

$$\text{Head, } H_n = \left[\frac{N_n}{N_r}\right]^2 H_r$$
$$= \left[\frac{1,200 \text{ RPM}}{1,500 \text{ RPM}}\right]^2(100 \text{ ft})$$
$$= \left(\frac{4}{5}\right)^2(100 \text{ ft})$$
$$= \left(\frac{16}{25}\right)(100 \text{ ft})$$
$$= (16)(4 \text{ ft})$$
$$= 64 \text{ ft}$$

Calculate new power requirement:

$$\text{Power, } P_n = \left[\frac{N_n}{N_r}\right]^3 P_r$$
$$= \left[\frac{1,200 \text{ RPM}}{1,500 \text{ RPM}}\right]^3(15 \text{ HP})$$
$$= \left(\frac{4}{5}\right)^3(15 \text{ HP})$$
$$= \left(\frac{64}{125}\right)(15 \text{ HP})$$
$$= \left(\frac{64}{25}\right)(3 \text{ HP})$$
$$= 7.7 \text{ HP}$$

A.88 Friction or Energy Losses

Whenever water flows through pipes, valves, and fittings, energy is lost due to pipe friction (resistance), friction in valves and fittings, and the turbulence resulting from the flowing water changing its direction. Figure A.1 can be used to convert the friction losses through valves and fittings to lengths of straight pipe that would produce the same amount of friction losses. To estimate the friction or energy losses resulting from water flowing in a pipe system, we need to know:

1. Water flow rate,
2. Pipe size or diameter and length, and
3. Number, size, and type of valve fittings.

An easy way to estimate friction or energy losses is to follow these steps:

1. Determine the flow rate,
2. Determine the diameter and length of pipe,
3. Convert all valves and fittings to equivalent lengths of straight pipe (see Figure A.1),
4. Add up total length of equivalent straight pipe, and
5. Estimate friction or energy losses by using Figure A.2. With the flow in GPM and diameter of pipe, find the friction loss per 100 feet of pipe. Multiply this value by equivalent length of straight pipe.

The procedure for using Figure A.1 is very easy. Locate the type of valve or fitting you wish to convert to an equivalent pipe length; find its diameter on the right-hand scale; and draw a straight line between these two points to locate the equivalent length of straight pipe.

EXAMPLE: Estimate the friction losses in the piping system of a pump station when the flow is 1,000 GPM. The 8-inch suction line is 10 feet long and contains a 90-degree bend (long sweep elbow), a gate valve and an 8-inch by 6-inch reducer at the inlet to the pump. The 6-inch discharge line is 30 feet long and contains a check valve, a gate valve, and three 90-degree bends (medium sweep elbows):

SUCTION LINE (8-inch diameter)

Item	Equivalent Length, ft
1. Length of pipe	10
2. 90-degree bend	14
3. Gate valve	4
4. 8-inch by 6-inch reducer	3
5. Ordinary entrance	12
Total equivalent length	43 feet

Friction loss (Figure A.2) = 1.76 ft/100 ft of pipe

DISCHARGE LINE (6-inch diameter)

Item	Equivalent Length, ft
1. Length of pipe	30
2. Check valve	38
3. Gate valve	4
4. Three 90-degree bends (3)(14)	42
Total equivalent length	114 feet

Friction loss (Figure A.2) = 7.73 ft/100 ft of pipe

Estimate the total friction losses in pumping system for a flow of 1,000 GPM.

SUCTION

Loss = (1.76 ft/100 ft)(43 ft) = 0.8 ft

DISCHARGE

Loss = (7.73 ft/100 ft)(114 ft) = 8.8 ft

Total friction losses, ft = 9.6 ft

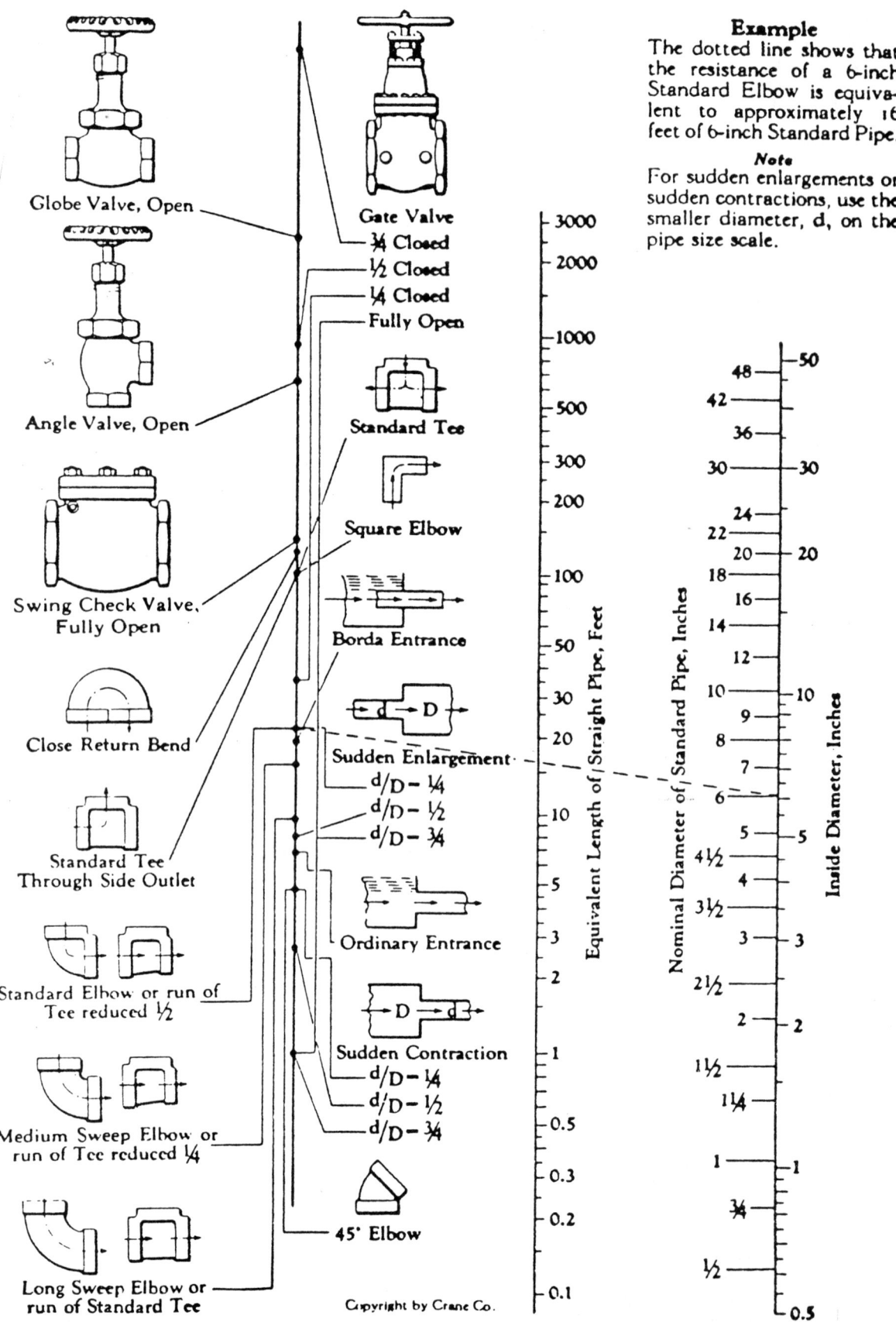

Fig. A.1   Resistance of valves and fittings to flow of water

(Reprinted by permission of Crane Co.)

U.S. GPM	0.5 in. Vel.	Frict.	0.75 in. Vel.	Frict.	1 in. Vel.	Frict.	1.25 in. Vel.	Frict.	1.5 in. Vel.	Frict.	2 in. Vel.	Frict.	2.5 in. Vel.	Frict.
10	10.56	95.9	6.02	23.0	3.71	6.86	2.15	1.77	1.58	.83	.96	.25	.67	.11
20	....	....	12.0	86.1	7.42	25.1	4.29	6.34	3.15	2.94	1.91	.87	1.34	.36
30	....	....	....	....	11.1	54.6	6.44	13.6	4.73	6.26	2.87	1.82	2.01	.75
40	....	....	....	....	14.8	95.0	8.58	23.5	6.30	10.79	3.82	3.10	2.68	1.28
50	....	....	....	....	....	....	10.7	36.0	7.88	16.4	4.78	4.67	3.35	1.94
60	....	....	....	....	....	....	12.9	51.0	9.46	23.2	5.74	6.59	4.02	2.72
70	....	....	....	....	....	....	15.0	68.8	11.03	31.3	6.69	8.86	4.69	3.63
80	....	....	....	....	....	....	17.2	89.2	12.6	40.5	7.65	11.4	5.36	4.66
90	....	....	....	....	....	....	....	....	14.2	51.0	8.60	14.2	6.03	5.82
100	....	....	....	....	....	....	....	....	15.8	62.2	9.56	17.4	6.70	7.11
120	....	....	....	....	....	....	....	....	18.9	88.3	11.5	24.7	8.04	10.0
140	....	....	....	....	....	....	....	....	....	....	13.4	33.2	9.38	13.5
160	....	....	....	....	....	....	....	....	....	....	15.3	43.0	10.7	17.4
180	....	....	....	....	....	....	....	....	....	....	17.2	54.1	12.1	21.9
200	....	....	....	....	....	....	....	....	....	....	19.1	66.3	13.4	26.7
220	....	....	....	....	....	....	....	....	....	....	21.0	80.0	14.7	32.2
240	....	....	....	....	....	....	....	....	....	....	22.9	95.0	16.1	38.1
260	....	....	....	....	....	....	....	....	....	....	....	....	17.4	44.5
280	....	....	....	....	....	....	....	....	....	....	....	....	18.8	51.3
300	....	....	....	....	....	....	....	....	....	....	....	....	20.1	58.5
350	....	....	....	....	....	....	....	....	....	....	....	....	23.5	79.2

U.S. GPM	3 in. Vel.	Frict.	4 in. Vel.	Frict.	5 in Vel.	Frict.	6 in. Vel.	Frict.	8 in. Vel.	Frict.	10 in. Vel.	Frict.	12 in. Vel.	Frict.	14 in. Vel.	Frict.	16 in. Vel.	Frict.	18 in. Vel.	Frict.	20 in. Vel.	Frict.
20	.91	.15																				
40	1.82	.55	1.02	.13																		
50	2.72	1.17	1.53	.28	.96	.08																
80	3.63	2.02	2.04	.48	1.28	.14	.91	.06														
100	4.54	3.10	2.55	.73	1.60	.20	1.13	.10														
120	5.45	4.40	3.06	1.03	1.92	.29	1.36	.13														
140	6.35	5.93	3.57	1.38	2.25	.38	1.59	.18														
160	7.26	7.71	4.08	1.78	2.57	.49	1.82	.23														
180	8.17	9.73	4.60	2.24	2.89	.61	2.04	.28														
200	9.08	11.9	5.11	2.74	3.21	.74	2.27	.35														
220	9.98	14.3	5.62	3.28	3.53	.88	2.50	.42	1.40	.10												
240	10.9	17.0	6.13	3.88	3.85	1.04	2.72	.49	1.53	.12												
260	11.8	19.8	6.64	4.54	4.17	1.20	2.95	.57	1.66	.14												
280	12.7	22.8	7.15	5.25	4.49	1.38	3.18	.66	1.79	.16												
300	13.6	26.1	7.66	6.03	4.81	1.58	3.40	.75	1.91	.18												
350	....	....	8.94	8.22	5.61	2.11	3.97	1.01	2.24	.24												
400	....	....	10.20	10.7	6.41	2.72	4.54	1.30	2.55	.30												
450	....	....	11.45	13.4	7.22	3.41	5.11	1.64	2.87	.38	1.84	.12										
500	....	....	12.8	16.6	8.02	4.16	5.67	2.02	3.19	.46	2.04	.15	1.42	.06								
550	....	....	14.0	19.9	8.82	4.98	6.24	2.42	3.51	.56	2.25	.18	1.56	.07								
600	....	....	....	....	9.62	5.88	6.81	2.84	3.83	.66	2.45	.21	1.70	.08	1.25	.04						
700	....	....	....	....	11.2	7.93	7.94	3.87	4.47	.88	2.86	.29	1.99	.12	1.46	.05						
800	....	....	....	....	12.8	10.22	9.08	5.06	5.11	1.14	3.27	.37	2.27	.15	1.67	.07						
900	....	....	....	....	14.4	12.9	10.2	6.34	5.74	1.44	3.68	.46	2.55	.18	1.88	.09						
1000	....	....	....	....	....	....	11.3	7.73	6.68	1.76	4.09	.57	2.84	.22	2.08	.11						
1100	....	....	....	....	....	....	12.5	9.80	7.02	2.14	4.49	.68	3.12	.27	2.29	.13						
1200	....	....	....	....	....	....	13.6	11.2	7.66	2.53	4.90	.81	3.40	.32	2.50	.15	1.91	.08				
1300	....	....	....	....	....	....	14.7	13.0	8.30	2.94	6.31	.95	3.69	.37	2.71	.17	2.07	.09				
1400	....	....	....	....	....	....	....	....	8.93	3.40	5.72	1.09	3.97	.43	2.92	.20	2.23	.10				
1500	....	....	....	....	....	....	....	....	9.57	3.91	6.13	1.25	4.26	.49	3.13	.23	2.34	.12				
1600	....	....	....	....	....	....	....	....	10.2	4.45	6.54	1.42	4.54	.55	3.33	.25	2.55	.13	2.02	.07		
1700	....	....	....	....	....	....	....	....	10.8	5.00	6.94	1.60	4.87	.62	3.54	.29	2.71	.15	2.15	.08		
1800	....	....	....	....	....	....	....	....	11.5	5.58	7.35	1.78	5.11	.70	3.75	.32	2.87	.16	2.27	.09		
1900	....	....	....	....	....	....	....	....	12.1	6.19	7.76	1.97	5.39	.77	3.96	.35	3.03	.18	2.40	.10		
2000	....	....	....	....	....	....	....	....	12.8	6.84	8.17	2.17	5.67	.86	4.17	.39	3.19	.20	2.52	.11		
2500	....	....	....	....	....	....	....	....	....	....	10.2	3.38	7.10	1.33	5.21	.60	3.99	.31	3.15	.17		
3000	....	....	....	....	....	....	....	....	....	....	12.3	4.79	8.51	1.88	6.25	.86	4.79	.44	3.78	.24	3.06	.14
3500	....	....	....	....	....	....	....	....	....	....	14.3	6.55	9.93	2.56	7.29	1.16	5.58	.58	4.41	.32	3.57	.19
4000	....	....	....	....	....	....	....	....	....	....	....	....	11.3	3.31	8.34	1.50	6.38	.75	5.04	.42	4.08	.24
4500	....	....	....	....	....	....	....	....	....	....	....	....	12.8	4.18	9.38	1.88	7.18	.95	5.67	.53	4.59	.31
5000	....	....	....	....	....	....	....	....	....	....	....	....	14.7	5.13	10.4	2.30	7.98	1.17	6.30	.65	5.11	.38
6000	....	....	....	....	....	....	....	....	....	....	....	....	....	....	12.5	3.31	9.57	1.66	7.56	.92	6.13	.53
7000	....	....	....	....	....	....	....	....	....	....	....	....	....	....	14.6	4.50	11.2	2.26	8.83	1.24	7.15	.72
8000	....	....	....	....	....	....	....	....	....	....	....	....	....	....	....	....	12.8	2.96	10.09	1.61	8.17	.94
9000	....	....	....	....	....	....	....	....	....	....	....	....	....	....	....	....	14.4	3.73	11.3	2.02	9.19	1.18
10000	....	....	....	....	....	....	....	....	....	....	....	....	....	....	....	....	....	....	12.6	2.48	10.2	1.45

No allowance has been made for age, differences in diameter, or any other abnormal condition of interior surface. Any Factor of Safety must be estimated from the local conditions and the requirements of each particular installation. For general purposes, 15% is a responsible Factor of Safety.

Fig. A.2   Friction loss for water in feet per 100 feet of pipe

(Reprinted from the 10th Edition of the Standards of the Hydraulic Institute,
122 East 42nd Street, New York)

A.9   STEPS IN SOLVING PROBLEMS

A.90   Identification of Problem

To solve any problem, you have to identify the problem, determine what kind of answer is needed, and collect the information needed to solve the problem. A good approach to this type of problem is to examine the problem and make a list of *KNOWN* and *UNKNOWN* information.

EXAMPLE: Find the theoretical detention time in a rectangular sedimentation tank 8 feet deep, 30 feet wide, and 60 feet long when the flow is 1.4 MGD.

Known	Unknown
Depth, ft = 8 ft	Detention Time, hours
Width, ft = 30 ft	
Length, ft = 60 ft	
Flow, MGD = 1.4 MGD	

Sometimes a drawing or sketch will help to illustrate a problem and indicate the knowns, unknowns, and possibly additional information needed.

A.91   Selection of Formula

Most problems involving mathematics in water treatment plant operation can be solved by selecting the proper formula, inserting the known information, and calculating the unknown. In our example, we could look in Chapter 5, "Sedimentation," or in Section A.11 of this chapter, "Basic Formulas," to find a formula for calculating detention time.

From Section A.11:

$$\text{Detention Time, hr} = \frac{(\text{Tank Volume, cu ft})(7.48 \text{ gal/cu ft})(24 \text{ hr/day})}{\text{Flow, gal/day}}$$

To convert the known information to fit the terms in a formula sometimes requires extra calculations. The next step is to find the values of any terms in the formula that are not in the list of known values.

$$\text{Flow, gal/day} = 1.4 \text{ MGD}$$
$$= 1,400,000 \text{ gal/day}$$

From Section A.30:

$$\text{Tank Volume, cu ft} = (\text{Length, ft})(\text{Width, ft})(\text{Height, ft})$$
$$= 60 \text{ ft} \times 30 \text{ ft} \times 8 \text{ ft}$$
$$= 14,400 \text{ cu ft}$$

Solution of Problem:

$$\text{Detention Time, hr} = \frac{(\text{Tank Volume, cu ft})(7.48 \text{ gal/cu ft})(24 \text{ hr/day})}{\text{Flow, gal/day}}$$
$$= \frac{(14,400 \text{ cu ft})(7.48 \text{ gal/cu ft})(24 \text{ hr/day})}{1,400,000 \text{ gal/day}}$$
$$= 1.85 \text{ hr}$$

The remainder of this section discusses the details that must be considered in solving this problem.

A.92   Arrangement of Formula

Once the proper formula is selected, you may have to rearrange the terms to solve for the unknown term. From Section A.71, "Flow Rate," we can develop the formula:

$$\text{Velocity, ft/sec} = \frac{\text{Flow Rate, cu ft/sec}}{\text{Cross-Sectional Area, sq ft}}$$

or

$$V = \frac{Q}{A}$$

In this equation if Q and A were given, the equation could be solved for V. If V and A were known, the equation would have to be rearranged to solve for Q. To move terms from one side of an equation to another, use the following rule:

When moving a term or number from one side of an equation to the other, move the numerator (top) of one side to the denominator (bottom) of the other; or from the denominator (bottom) of one side to the numerator (top) of the other.

$$V = \frac{Q}{A} \text{ or } Q = AV \text{ or } A = \frac{Q}{V}$$

If the volume of a sedimentation tank and the desired detention time were given, the detention time formula could be rearranged to calculate the design flow.

$$\text{Detention Time, hr} = \frac{(\text{Tank Vol, cu ft})(7.48 \text{ gal/cu ft})(24 \text{ hr/day})}{\text{Flow, gal/day}}$$

By rearranging the terms

$$\text{Flow, gal/day} = \frac{(\text{Tank Vol, cu ft})(7.48 \text{ gal/cu ft})(24 \text{ hr/day})}{\text{Detention Time, hr}}$$

A.93   Unit Conversions

Each term in a formula or mathematical calculation must be of the correct units. The area of a rectangular clarifier (Area, sq ft = Length, ft x Width, ft) can't be calculated in square feet if the width is given as 246 inches or 20 feet 6 inches. The width must be converted to 20.5 feet. In the example problem, if the tank volume were given in gallons, the 7.48 gal/cu ft would not be needed. *THE UNITS IN A FORMULA MUST ALWAYS BE CHECKED BEFORE ANY CALCULATIONS ARE PERFORMED TO AVOID TIME-CONSUMING MISTAKES.*

$$\text{Detention Time, hr} = \frac{(\text{Tank Volume, cu ft})(7.48 \text{ gal/cu ft})(24 \text{ hr/day})}{\text{Flow, gal/day}}$$
$$= \frac{\cancel{\text{cu ft}}}{1} \times \frac{\cancel{\text{gal}}}{\cancel{\text{cu ft}}} \times \frac{\text{hr}}{\cancel{\text{day}}} \times \frac{\cancel{\text{day}}}{\cancel{\text{gal}}}$$
$$= \text{hr (all other units cancel)}$$

NOTE: We have hours = hr. One should note that the hour unit on both sides of the equation can be cancelled out and nothing would remain. This is one more check that we have the correct units. By rearranging the detention time formula, other unknowns could be determined.

If the design detention time and design flow were known, the required capacity of the tank could be calculated.

$$\text{Tank Volume, cu ft} = \frac{(\text{Detention Time, hr})(\text{Flow, gal/day})}{(7.48 \text{ gal/cu ft})(24 \text{ hr/day})}$$

If the tank volume and design detention time were known, the design flow could be calculated.

$$\text{Flow, gal/day} = \frac{(\text{Tank Volume, cu ft})(7.48 \text{ gal/cu ft})(24 \text{ hr/day})}{\text{Detention Time, hr}}$$

Rearrangement of the detention time formula to find other unknowns illustrates the need to always use the correct units.

A.94 Calculations

Sections A.12, "Multiplication," and A.13 "Division," outline the steps to follow in mathematical calculations. In general, do the calculations inside parentheses ( ) first and brackets [ ] next. Calculations should be done left to right above and below the division line before dividing.

$$\text{Detention Time, hr} = \frac{[(\text{Tank Volume, cu ft})(7.48 \text{ gal/cu ft})(24 \text{ hr/day})]}{\text{Flow, gal/day}}$$

$$= \frac{[(14,400 \text{ cu ft})(7.48 \text{ gal/cu ft})(24 \text{ hr/day})]}{1,400,000 \text{ gal/day}}$$

$$= \frac{2,585,088 \text{ gal-hr/day}}{1,400,000 \text{ gal/day}}$$

$$= 1.85, \text{ or}$$

$$= 1.9 \text{ hr}$$

A.95 Significant Figures

In calculating the detention time in the previous section, the answer is given as 1.9 hr. The answer could have been calculated:

$$\text{Detention Time, hr} = \frac{2,585,088 \text{ gal-hr/day}}{1,400,000 \text{ gal/day}}$$

$$= 1.846491429 \ldots \text{ hours}$$

How does one know when to stop dividing? Common sense and significant figures both help.

First, consider the meaning of detention time and the measurements that were taken to determine the knowns in the formula. Detention time in a tank is a theoretical value and assumes that all particles of water throughout the tank move through the tank at the same velocity. This assumption is not correct; therefore, detention time can only be a representative time for some of the water particles.

Will the flow of 1.4 MGD be constant throughout the 1.9 hours, and is the flow exactly 1.4 MGD, or could it be 1.35 MGD or 1.428 MGD? A carefully calibrated flowmeter may give a reading within 2% of the actual flow rate. Flows into a tank fluctuate and flowmeters do not measure flows extremely accurately; so the detention time again appears to be a representative or typical detention time.

Tank dimensions are probably satisfactory within 0.1 ft. A flowmeter reading of 1.4 MGD is less precise and it could be 1.3 or 1.5 MGD. A 0.1 MGD flowmeter error when the flow is 1.4 MGD is (0.1/1.4) x 100% = 7% error. A detention time of 1.9 hours, based on a flowmeter reading error of plus or minus 7%, also could have the same error or more, even if the flow was constant. Therefore, the detention time error could be 1.9 hours x 0.07 = ±0.13 hour.

In most of the calculations in the operation of water treatment plants, the operator uses measurements determined in

the lab or read from charts, scales, or meters. The accuracy of every measurement depends on the sample being measured, the equipment doing the measuring, and the operator reading or measuring the results. Your estimate is no better than the least precise measurement. Do not retain more than one doubtful number.

To determine how many figures or numbers mean anything in an answer, the approach called "significant figures" is used. In the example the flow was given in two significant figures (1.4 MGD), and the tank dimensions could be considered accurate to the nearest tenth of a foot (depth = 9.0 ft) or two significant figures. Since all measurements and the constants contained two significant figures, the results should be reported as two significant figures or 1.9 hours. The calculations are normally carried out to three significant figures (1.85 hours) and rounded off to two significant figures (1.9 hours).

Decimal points require special attention when determining the number of significant figures in a measurement.

Measurement	Significant Figures
0.00325	3
11.078	5
21,000.	2

EXAMPLE: The distance between two points was divided into three sections, and each section was measured by a different group. What is the distance between the two points if each group reported the distance it measured as follows?

Group	Distance, ft	Significant Figures
A	11,300.	3
B	2,438.9	5
C	87.62	4
Total Distance	13,826.52	

Group A reported the length of the section it measured to three significant figures; therefore, the distance between the two points should be reported as 13,800 feet (3 significant figures).

When adding, subtracting, multiplying, or dividing, the number of significant figures in the answer should not be more than the term in the calculations with the least number of significant figures.

A.96 Check Your Results

After completing your calculations, you should carefully examine your calculations and answer. Does the answer seem reasonable? If possible, have another operator check your calculations before making any operational changes.

A.10 BASIC CONVERSION FACTORS (ENGLISH SYSTEM)

UNITS
1,000,000	= 1 Million	1,000,000/1 Million

LENGTH
12 in	= 1 ft	12 in/ft
3 ft	= 1 yd	3 ft/yd
5,280 ft	= 1 mi	5,280 ft/mi

AREA
144 sq in	= 1 sq ft	144 sq in/sq ft
43,560 sq ft	= 1 acre	43,560 sq ft/ac

VOLUME

7.48 gal	= 1 cu ft	7.48 gal/cu ft
1,000 mL	= 1 liter	1,000 mL/L
3.785 L	= 1 gal	3.785 L/gal
231 cu in	= 1 gal	231 cu in/gal

WEIGHT

1,000 mg	= 1 gm	1,000 mg/gm
1,000 gm	= 1 kg	1,000 gm/kg
454 gm	= 1 lb	454 gm/lb
2.2 lbs	= 1 kg	2.2 lbs/kg

POWER

0.746 kW	= 1 HP	0.746 kW/HP

DENSITY

8.34 lbs	= 1 gal	8.34 lbs/gal
62.4 lbs	= 1 cu ft	62.4 lbs/cu ft

DOSAGE

17.1 mg/L	= 1 grain/gal	17.1 mg/L/gpg
64.7 mg	= 1 grain	64.7 mg/grain

PRESSURE

2.31 ft water	= 1 psi	2.31 ft water/psi
0.433 psi	= 1 ft water	0.433 psi/ft water
1.133 ft water	= 1 in Mercury	1.133 ft water/in Mercury

FLOW

694 GPM	= 1 MGD	694 GPM/MGD
1.55 CFS	= 1 MGD	1.55 CFS/MGD

TIME

60 sec	= 1 min	60 sec/min
60 min	= 1 hr	60 min/hr
24 hr	= 1 day	24 hr/day

NOTE: In our equations the values in the right-hand column may be written either as 24 hr/day or 1 day/24 hours depending on which units we wish to convert to get our desired results.

A.11 BASIC FORMULAS

FLOWS

1. $\text{Flow, MGD} = \dfrac{(\text{Flow, GPM})(60 \text{ min/hr})(24 \text{ hr/day})}{1{,}000{,}000/M}$

 or

 $\text{Flow, GPM} = \dfrac{(\text{Flow, MGD})(1{,}000{,}000/M)}{(60 \text{ min/hr})(24 \text{ hr/day})}$

CHEMICAL DOSES

2. $\text{Chemical Feed, lbs/day} = (\text{Flow, MGD})(\text{Dose, mg/L})(8.34 \text{ lbs/gal})$

3. $\text{Chemical Feeder Setting, mL/min} = \dfrac{(\text{Flow, MGD})(\text{Alum Dose, mg/L})(3.785 \text{ L/gal})(1{,}000{,}000/M)}{(\text{Liquid Alum, mg/mL})(24 \text{ hr/day})(60 \text{ min/hr})}$

4. $\text{Chemical Feeder Setting, gal/day} = \dfrac{(\text{Flow, MGD})(\text{Alum Dose, mg/L})(8.34 \text{ lbs/gal})}{\text{Liquid Alum, lbs/gal}}$

Calibration of a Dry Chemical Feeder

5. $\text{Chemical Feed, lbs/day} = \dfrac{\text{Chemical Applied, lbs}}{\text{Length of Application, day}}$

Calibration of a Solution Chemical Feeder (Chemical Feed Pump or a Hypochlorinator)

6. $\text{Chemical Feed, lbs/day} = \dfrac{(\text{Chem Conc, mg/L})(\text{Vol Pumped, mL})(60 \text{ min/hr})(24 \text{ hr/day})}{(\text{Time Pumped, min})(1{,}000 \text{ mL/L})(1{,}000 \text{ mg/gm})(454 \text{ gm/lb})}$

7. $\text{Chemical Feed, GPM} = \dfrac{\text{Chemical Used, gal}}{(\text{Time, hr})(60 \text{ min/hr})}$

 or $= \dfrac{(\text{Chemical Feed Rate, mL/sec})(60 \text{ sec/min})}{3{,}785 \text{ mL/gal}}$

8a. $\text{Chemical Solution, lbs/gal} = \dfrac{(\text{Chemical Solution, \%})(8.34 \text{ lbs/gal})}{100\%}$

8b. $\text{Feed Pump, GPD} = \dfrac{\text{Chemical Feed, lbs/day}}{\text{Chemical Solution, lbs/gal}}$

8c. $\text{Feeder Setting, \%} = \dfrac{(\text{Desired Feed Pump, GPD})(100\%)}{\text{Maximum Feed Pump, GPD}}$

RESERVOIR MANAGEMENT AND INTAKE STRUCTURES

9. $\text{Reservoir Volume, ac/ft} = \dfrac{\text{Reservoir Volume, cu ft}}{43{,}560 \text{ sq ft/ac}}$

10. $\text{Reservoir Vol, gal} = (\text{Vol, ac-ft})(43{,}560 \text{ sq ft/ac})(7.48 \text{ gal/cu ft})$

11. $\text{Chemical Dose, lbs} = (\text{Surface Area, ac})(\text{Dose, lbs/ac})$

12. $\text{Chemical Dose, lbs} = (\text{Vol, M Gal})(\text{Dose, mg/L})(8.34 \text{ lbs/gal})$

13. $\text{Chemical, lbs} = \dfrac{(\text{Vol, M Gal})(\text{Dose, mg/L})(8.34 \text{ lbs/gal})(100\%)}{\text{Chemical, \%}}$

COAGULATION AND FLOCCULATION

14. $\text{Polymer, lbs} = \dfrac{(\text{Polymer Solution, gal})(8.34 \text{ lbs/gal})(\text{Polymer, \%})(\text{Sp Gr})}{100\%}$

15. $\text{Dose, mg/L} = \dfrac{\text{Chemical Feed, lbs/day}}{(\text{Flow, MGD})(8.34 \text{ lbs/gal})}$

16. $\text{Polymer, \%} = \dfrac{(\text{Dry Polymer, lbs})(100\%)}{(\text{Dry Polymer, lbs} + \text{Water, lbs})}$

17. $\text{Liquid Polymer, gal} = \dfrac{(\text{Polymer Solution, \%})(\text{Volume of Solution, gal})}{(\text{Liquid Polymer, \%})}$

SEDIMENTATION

18. $\text{Detention Time, hr} = \dfrac{(\text{Volume, gal})(24 \text{ hr/day})}{\text{Flow, gal/day}}$

 $= \dfrac{(\text{Vol, cu ft})(7.48 \text{ gal/cu ft})(24 \text{ hr/day})}{\text{Flow, gal/day}}$

19. $\text{Overflow Rate, GPM/sq ft or Surface Loading} = \dfrac{\text{Flow, GPM}}{\text{Surface Area, sq ft}}$

20. $\text{Mean Flow Velocity, ft/min} = \dfrac{\text{Flow, GPM}}{(\text{Cross-Sectional Area, sq ft})(7.48 \text{ gal/cu ft})}$

21. $\dfrac{\text{Weir Loading}}{\text{GPM/ft}} = \dfrac{\text{Flow, GPM}}{\text{Weir Length, ft}}$

FILTRATION

22. $\dfrac{\text{Filtration Rate,}}{\text{GPM/sq ft}} = \dfrac{\text{Flow, GPM}}{\text{Surface Area, sq ft}}$

23. Velocity, ft/min $= \dfrac{\text{Water Drop, ft}}{\text{Time, min}}$

24. Flow, cu ft/min = (Area, sq ft)(Velocity, ft/min)

25. Flow, gal/min = (Flow, cu ft/min)(7.48 gal/cu ft)

or

26. Flow, gal/min $= \dfrac{\text{Total Flow, gal}}{\text{(Filter Run, hr)(60 min/hr)}}$

27. Unit Filter Run Volume (UFRV)

UFRV, gal/sq ft $= \dfrac{\text{Volume Filtered, gal}}{\text{Filter Surface Area, sq ft}}$

28. $\dfrac{\text{UFRV,}}{\text{gal/sq ft}}$ = (Filtration Rate, GPM/sq ft)(Filter Run, hr)(60 min/hr)

29. $\dfrac{\text{Backwash Flow,}}{\text{GPM}}$ = (Filter Area, sq ft)(Backwash Rate, GPM/sq ft)

30. Backwash, in/min $= \dfrac{\text{(Backwash, GPM/sq ft)(12 in/ft)}}{7.48 \text{ gal/cu ft}}$

31. Backwash Water, gal = (Backwash Flow, GPM)(Backwash Time, min)

32. Backwash, % $= \dfrac{\text{(Backwash Water, gal)(100\%)}}{\text{Water Filtered, gal}}$

DISINFECTION

33. $\dfrac{\text{Chlorine}}{\text{Dose, mg/}L} = \dfrac{\text{Chlorine Feed, lbs/day}}{\text{(Flow, MGD)(8.34 lbs/gal)}}$

34. $\dfrac{\text{Chlorine Demand,}}{\text{mg/}L}$ = Chlorine Dose, mg/L – Chlorine Residual, mg/L

35. Chlorine, lbs = (Hypochlorite, gal)(8.34 lbs/gal)$\left(\dfrac{\text{Hypochlorite, \%}}{100\%}\right)$

36. $\dfrac{\text{Hypochlorite}}{\text{Flow, GPD}} = \dfrac{\text{(Container Area, sq ft)(Drop, ft)(7.48 gal/cu ft)(24 hr/day)}}{\text{Time, hr}}$

37. $\dfrac{\text{Hypochlorite}}{\text{Strength, \%}} = \dfrac{\text{(Chlorine Required, lbs/day)(100\%)}}{\text{(Hypochlorinator Flow, gal/day)(8.34 lbs/gal)}}$

38. $\dfrac{\text{Water Added,}}{\text{gal (to Hypo-}}$ chlorite Solution) $= \dfrac{\text{(Hypo, gal)(Hypo, \%)} - \text{(Hypo, gal)(Desired Hypo, \%)}}{\text{Desired Hypo, \%}}$

CORROSION CONTROL

39. $pH_S = A + B - \log(Ca^{2+}) - \log(Alky)$

40. Langelier Index $= pH - pH_S$

PLANT OPERATION

41. $\dfrac{\text{Average Chemical}}{\text{Use, lbs/day}} = \dfrac{\text{Sum of Chemical Used Each Day, lbs}}{\text{Total Time, days}}$

42. $\dfrac{\text{Supply of Chlorine,}}{\text{days}} = \dfrac{\text{(Cylinder Wt, lbs/cyl)(No. of Cylinders)}}{\text{Avg Use, lbs/day}}$

LABORATORY PROCEDURES

43. Temperature, °C $= \dfrac{5}{9}$ (°F – 32°F)

44. Temperature, °F $= \dfrac{9}{5}$ (°C) + 32°F

45. Drop-Dilution Chlorine Residual

$\dfrac{\text{Actual Chlorine}}{\text{Residual, mg/}L} = \dfrac{\text{(Chlorine Residual, mg/}L\text{)(Distilled Water, m}L\text{)}}{\text{(Sample Volume, drops)(0.05 m}L\text{/drop)}}$

46. $\dfrac{\text{Mean or}}{\text{Average}} = \dfrac{\text{Sum of Values or Measurements}}{\text{Number of Values or Measurements}}$

47. Median = Middle Value of a Group of Data

48. Geometric Mean $= [(X_1)(X_2)(X_3) \ldots\ldots (X_n)]^{1/n}$

A.12 HOW TO USE THE BASIC FORMULAS

One clever way of using the basic formulas is to use the Davidson* Pie Method. To apply this method to the basic formula for chemical doses,

1. Chlorine Feed, lbs/day = (Flow, MGD)(Dose, mg/L)(8.34 lbs/gal)

   (a) Draw a circle and draw a horizontal line through the middle of the circle;

   (b) Write the Chemical Feed, lbs/day in the top half;

   (c) Divide the bottom half into three parts; and

   (d) Write Flow, MGD; Dose, mg/L; and 8.34 lbs/gal in the other three parts.

   (e) The line across the middle of the circle represents the line in the equation. The items above the line stay above the line and those below the line stay below the line.

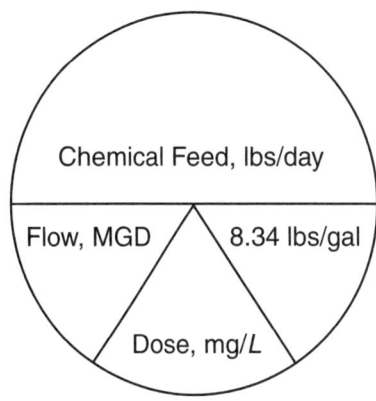

---
* Gerald Davidson, Manager, Clear Lake Oaks Water District, Clear Lake Oaks, California.

If you want to find the Chemical Feed, lbs/day, cover up the Chemical Feed, lbs/day, and what is left uncovered will give you the correct formula.

2. $\dfrac{\text{Chemical Feed,}}{\text{lbs/day}}$ = (Flow, MGD)(Dose, mg/L)(8.34 lbs/gal)

If you know the chlorinator setting in pounds per day and the flow in MGD and would like to know the dose in mg/L, cover up the Dose, mg/L, and what is left uncovered will give you the correct formula.

3. Dose, mg/L $= \dfrac{\text{Chemical Feed, lbs/day}}{(\text{Flow, MGD})(8.34 \text{ lbs/gal})}$

Another approach to using the basic formulas is to memorize the basic formula, for example the detention time formula.

4. Detention Time, hr $= \dfrac{(\text{Tank Volume, gal})(24 \text{ hr/day})}{\text{Flow, gal/day}}$

This formula works fine to solve for the detention time when the Tank Volume, gal, and Flow, gal/day, are given.

If you wish to determine the Flow, gal/day, when the Detention Time, hr, and Tank Volume, gal, are given, you must change the basic formula. You want the Flow, gal/day, on the left of the equal sign and everything else on the right of the equal sign. This is done by moving the terms diagonally (from top to bottom or from bottom to top) past the equal sign.

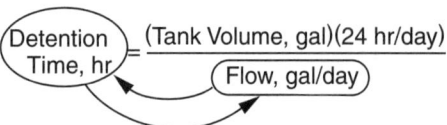

or

Flow, gal/day $= \dfrac{(\text{Tank Volume, gal})(24 \text{ hr/day})}{\text{Detention Time, hr}}$

This same approach can be used if the Tank Volume, gal, was unknown and the Detention Time, hr, and Flow, gal/day, were given. We want Tank Volume, gal, on one side of the equation and everything else on the other side.

Detention Time, hr $= \dfrac{(\text{Tank Volume, gal})(24 \text{ hr/day})}{\text{Flow, gal/day}}$

To Top   To Bottom

or

$\dfrac{(\text{Detention Time, hr})(\text{Flow, gal/day})}{24 \text{ hr/day}}$ = Tank Volume, gal

or

Tank Volume, gal $= \dfrac{(\text{Detention Time, hr})(\text{Flow, gal/day})}{24 \text{ hr/day}}$

One more check is to be sure the units in the rearranged formula cancel out correctly.

For additional information on the use of the basic formulas, refer to Sections:

A.91, "Selection of Formula,"

A.92, "Arrangement of Formula,"

A.93, "Unit Conversions," and

A.94, "Calculations."

A.13   TYPICAL WATER TREATMENT PLANT PROBLEMS (ENGLISH SYSTEM)

A.130   Flows

EXAMPLE 1

Convert a flow of 800 gallons per minute to million gallons per day.

Known	Unknown
Flow, GPM = 800 GPM	Flow, MGD

Convert flow from 800 GPM to MGD.

Flow, MGD $= \dfrac{(\text{Flow, GPM})(60 \text{ min/hr})(24 \text{ hr/day})}{1,000,000/M}$

$\qquad = \dfrac{(800 \text{ GPM})(60 \text{ min/hr})(24 \text{ hr/day})}{1,000,000/M}$

$\qquad = 1.15 \text{ MGD}$

NOTE: When we multiply or divide an equation by 1,000,000/M we do not change anything except the units. This is just like multiplying an equation by 12 inches/foot or 60 min/hr; all we are doing is changing units.

A.131   Chemical Doses

EXAMPLE 2

Determine the chlorinator setting in pounds per 24 hours to treat a flow of 2 MGD with a chlorine dose of 3.0 mg/L.

Known	Unknown
Flow, MGD = 2 MGD	Chlorinator Setting, lbs/24 hours
Chlorine Dose, mg/L = 3.0 mg/L	

Determine the chlorinator setting in pounds per 24 hours or pounds per day.

Chemical Feed, lbs/day $= (\text{Flow, MGD})(\text{Dose, mg/L})(8.34 \text{ lbs/gal})$

$\qquad = (2 \text{ MGD})(3.0 \text{ mg/L})(8.34 \text{ lbs/gal})$

$\qquad = 50 \text{ lbs/day}$

NOTE: For an explanation of the relationship between mg/L and lbs/M lbs, read the *NOTE* in Example 7 on page 352.

EXAMPLE 3

The optimum liquid alum dose from the jar tests is 12 mg/L. Determine the setting on the liquid alum chemical feeder in milliliters per minute when the plant flow is 4.7 MGD. The liquid alum delivered to the plant contains 642.3 milligrams of alum per milliliter of liquid solution.

Known	Unknown
Alum Dose, mg/L = 12 mg/L	Chemical Feeder Setting, mL/min
Flow, MGD = 4.7 MGD	
Liquid Alum, mg/mL = 642.3 mg/mL	

Calculate the liquid alum chemical feeder setting in milliliters per minute.

$$\begin{aligned}\text{Chemical}\atop\text{Feeder}\atop\text{Setting,}\atop\text{m}L\text{/min} &= \frac{(\text{Flow, MGD})(\text{Alum Dose, mg}/L)(3.785\ L/\text{gal})(1{,}000{,}000/\text{M})}{(\text{Liquid Alum, mg/m}L)(24\ \text{hr/day})(60\ \text{min/hr})} \\[2mm] &= \frac{(4.7\ \text{MGD})(12\ \text{mg}/L)(3.785\ L/\text{gal})(1{,}000{,}000/\text{M})}{(642.3\ \text{mg/m}L)(24\ \text{hr/day})(60\ \text{min/hr})}\end{aligned}$$

$$= 231\ \text{m}L\text{/min}$$

EXAMPLE 4

The optimum liquid alum dose from the jar tests is 12 mg/L. Determine the setting on the liquid alum chemical feeder in gallons per day when the flow is 4.7 MGD. The liquid alum delivered to the plant contains 5.36 pounds of alum per gallon of liquid solution.

Known	Unknown
Alum Dose, mg/L = 12 mg/L	Chemical Feeder Setting, GPD
Flow, MGD = 4.7 MGD	
Liquid Alum, lbs/gal = 5.36 lbs/gal	

Calculate the liquid alum chemical feeder setting in gallons per day.

$$\begin{aligned}\text{Chemical Feeder}\atop\text{Setting, GPD} &= \frac{(\text{Flow, MGD})(\text{Alum Dose, mg}/L)(8.34\ \text{lbs/gal})}{\text{Liquid Alum, lbs/gal}} \\[2mm] &= \frac{(4.7\ \text{MGD})(12\ \text{mg}/L)(8.34\ \text{lbs/gal})}{5.36\ \text{lbs/gal}}\end{aligned}$$

$$= 88\ \text{GPD}$$

EXAMPLE 5

Determine the actual chemical feed in pounds per day from a dry chemical feeder. A bucket placed under the chemical feeder weighed 0.3 pound empty and 2.1 pounds after 30 minutes.

Known	Unknown
Empty Bucket, lbs = 0.3 lb	Chemical Feed, lbs/day
Full Bucket, lbs = 2.1 lbs	
Time to Fill, min = 30 min	

Determine the chemical feed in pounds of chemical applied per day.

$$\begin{aligned}\text{Chemical Feed,}\atop\text{lbs/day} &= \frac{(\text{Chemical Applied, lbs})(60\ \text{min/hr})(24\ \text{hr/day})}{\text{Length of Application, min}} \\[2mm] &= \frac{(2.1\ \text{lbs} - 0.3\ \text{lb})(60\ \text{min/hr})(24\ \text{hr/day})}{30\ \text{min}}\end{aligned}$$

$$= 86\ \text{lbs/day}$$

EXAMPLE 6

Determine the chemical feed in pounds of polymer per day from a chemical feed pump. The polymer solution is 1.5 percent or 15,000 mg polymer per liter. During a test run the chemical feed pump delivered 800 mL of polymer solution during five minutes.

Known	Unknown
Polymer Solution, % = 1.5%	Polymer Feed, lbs/day
Polymer Conc, mg/L = 15,000 mg/L	
Polymer Sp Gr = 1.0	
Volume Pumped, mL = 800 mL	
Time Pumped, min = 5 min	

Calculate the polymer fed by the chemical feed pump in pounds of polymer per day.

$$\begin{aligned}\text{Polymer}\atop\text{Feed,}\atop\text{lbs/day} &= \frac{(\text{Poly Conc, mg}/L)(\text{Vol Pumped, m}L)(60\ \text{min/hr})(24\ \text{hr/day})}{(\text{Time Pumped, min})(1{,}000\ \text{m}L/L)(1{,}000\ \text{mg/gm})(454\ \text{gm/lb})} \\[2mm] &= \frac{(15{,}000\ \text{mg}/L)(800\ \text{m}L)(60\ \text{min/hr})(24\ \text{hr/day})}{(5\ \text{min})(1{,}000\ \text{m}L/L)(1{,}000\ \text{mg/gm})(454\ \text{gm/lb})}\end{aligned}$$

$$= 7.6\ \text{lbs polymer/day}$$

EXAMPLE 7

A small chemical feed pump lowered the chemical solution in a three-foot diameter tank one foot and seven inches during an eight-hour period. Estimate the flow delivered by the pump in gallons per minute and gallons per day.

Known	Unknown
Tank Diameter, ft = 3 feet	1. Flow, GPM
Chemical Drop, ft = 1 ft 7 in	2. Flow, GPD
Time, hr = 8 hr	

1. Convert the tank drop from one foot seven inches to feet.

    Tank Drop, ft = 1 ft + 7 inches

    $$= 1\ \text{ft} + \frac{7\ \text{in}}{12\ \text{in/ft}}$$

    $$= 1\ \text{ft} + 0.58\ \text{ft}$$

    $$= 1.58\ \text{ft}$$

2. Determine the gallons of solution pumped.

    $$\begin{aligned}\text{Volume Pumped,}\atop\text{gal} &= (\text{Area, sq ft})(\text{Drop, ft})(7.48\ \text{gal/cu ft}) \\[2mm] &= (0.785)(3\ \text{ft})^2(1.58\ \text{ft})(7.48\ \text{gal/cu ft})\end{aligned}$$

    $$= 83.5\ \text{gal}$$

3. Estimate the flow delivered by the pump in gallons per minute and gallons per day.

    $$\begin{aligned}\text{Flow, GPM} &= \frac{\text{Volume Pumped, gal}}{(\text{Time, hr})(60\ \text{min/hr})} \\[2mm] &= \frac{83.5\ \text{gal}}{(8\ \text{hr})(60\ \text{min/hr})}\end{aligned}$$

    $$= 0.17\ \text{GPM}$$

    or

    $$\begin{aligned}\text{Flow, GPD} &= \frac{(\text{Volume Pumped, gal})(24\ \text{hr/day})}{\text{Time, hr}} \\[2mm] &= \frac{(83.5\ \text{gal})(24\ \text{hr/day})}{8\ \text{hr}}\end{aligned}$$

    $$= 251\ \text{GPD}$$

EXAMPLE 8

Determine the settings in percent stroke on a chemical feed pump for various doses of a chemical in milligrams per liter. (The chemical could be chlorine, polymer, potassium permanganate, or any other chemical solution fed by a pump.) The pump delivering the water to be treated pumps at a flow rate of 400 GPM. The solution strength of the chemical being pumped is 4.8 percent. The chemical feed pump has a maximum capacity of 92 gallons per day at a setting of 100 percent capacity.

Known	Unknown
Pump Flow, GPM = 400 GPM	Settings, % stroke for various doses in mg/L
Solution Strength, % = 4.8%	
Feed Pump, GPD (100% stroke) = 92 GPD	

1. Convert the pump flow from gallons per minute to million gallons per day.

$$\text{Pump Flow, MGD} = (\text{Pump Flow, GPM})(60 \text{ min/hr})(24 \text{ hr/day})$$

$$= (400 \text{ gal/min})(60 \text{ min/hr})(24 \text{ hr/day})$$

$$= 576{,}000 \text{ gal/day}$$

$$= 0.576 \text{ MGD}$$

2. Change the chemical solution strength from a percent to pounds of chemical per gallon of solution. A 4.8 percent solution means we have 4.8 pounds of chemical in a solution of water and chemical weighing 100 pounds.

$$\text{Chemical Solution, lbs/gal} = \frac{(\text{Chemical, lbs})(8.34 \text{ lbs/gal})}{\text{Chemical and Water, lbs}}$$

$$= \frac{(4.8 \text{ lbs})(8.34 \text{ lbs/gal})}{100 \text{ lbs}}$$

$$= 0.4 \text{ lb Chemical/Gallon Solution}$$

3. Calculate the chemical feed in pounds per day for a chemical dose of 0.5 milligrams per liter. We are going to assume various chemical doses of 0.5, 1.0, 1.5, 2.0, 2.5 mg/L, and upward so that if we know the desired chemical dose, we can easily determine the setting (percent stroke) on the chemical feed pump.

$$\text{Chemical Feed, lbs/day} = (\text{Flow, MGD})(\text{Dose, mg/}L)(8.34 \text{ lbs/gal})$$

$$= (0.576 \text{ MGD})(0.5 \text{ mg/}L)(8.34 \text{ lbs/gal})$$

$$= 2.4 \text{ lbs/day}$$

4. Determine the desired flow from the chemical feed pump in gallons per day.

$$\text{Feed Pump, GPD} = \frac{\text{Chemical Feed, lbs/day}}{\text{Chemical Solution, lbs/gal}}$$

$$= \frac{2.4 \text{ lbs/day}}{0.4 \text{ lbs/gal}}$$

$$= 6 \text{ GPD}$$

5. Determine the setting on the chemical feed pump as a percent. In this case we want to know the setting as a percent of the pump stroke.

$$\text{Setting, \%} = \frac{(\text{Desired Feed Pump, GPD})(100\%)}{\text{Maximum Feed Pump, GPD}}$$

$$= \frac{(6 \text{ GPD})(100\%)}{92 \text{ GPD}}$$

$$= 6.5\%$$

6. If we changed the chemical dose in Step 3 from 0.5 mg/L to 1.0 mg/L and other higher doses and repeated the remainder of the steps, we could calculate the data in Table A.1.

7. Plot the data in Table A.1 (Chemical Dose, mg/L vs. Pump Setting, % stroke) to obtain Figure A.3. Only three points were needed since the data plotted a straight line. For any desired chemical dose in milligrams per liter, you can use Figure A.3 to determine the necessary chemical feed pump setting.

TABLE A.1   SETTINGS FOR CHEMICAL FEED PUMP

PUMP FLOW, GPM   = 400 GPM
SOLUTION STRENGTH, % = 4.8%

Chemical Dose, mg/L	Chemical Feed, lbs/day	Feed Pump, GPD	Pump Setting, % stroke
0.5	2.4	6.0	6.5
1.0	4.8	12.0	13.0
1.5	7.2	18.0	19.5
2.0	9.6	24.0	26.1
2.5	12.0	30.0	32.6
3.0	14.4	36.0	39.1
3.5	16.8	42.0	45.6
4.0	19.2	48.0	52.2
4.5	21.6	54.0	58.7
5.0	24.0	60.0	65.2
5.5	26.4	66.0	71.7
6.0	28.8	72.0	78.2
6.5	31.2	78.0	84.8
7.0	33.6	84.0	91.3
7.5	36.0	90.0	97.8

A.132   Reservoir Management and Intake Structures

EXAMPLE 9

The volume of a reservoir is estimated to be 581,000 cubic feet. Estimate the volume in acre-feet.

Known	Unknown
Reservoir Vol, cu ft = 581,000 cu ft	Reservoir Vol, ac-ft

Estimate the reservoir volume in acre-feet.

$$\text{Reservoir Vol, ac-ft} = \frac{\text{Reservoir Vol, cu ft}}{43{,}560 \text{ sq ft/ac}}$$

$$= \frac{581{,}000 \text{ cu ft}}{43{,}560 \text{ sq ft/ac}}$$

$$= 13.3 \text{ ac-ft}$$

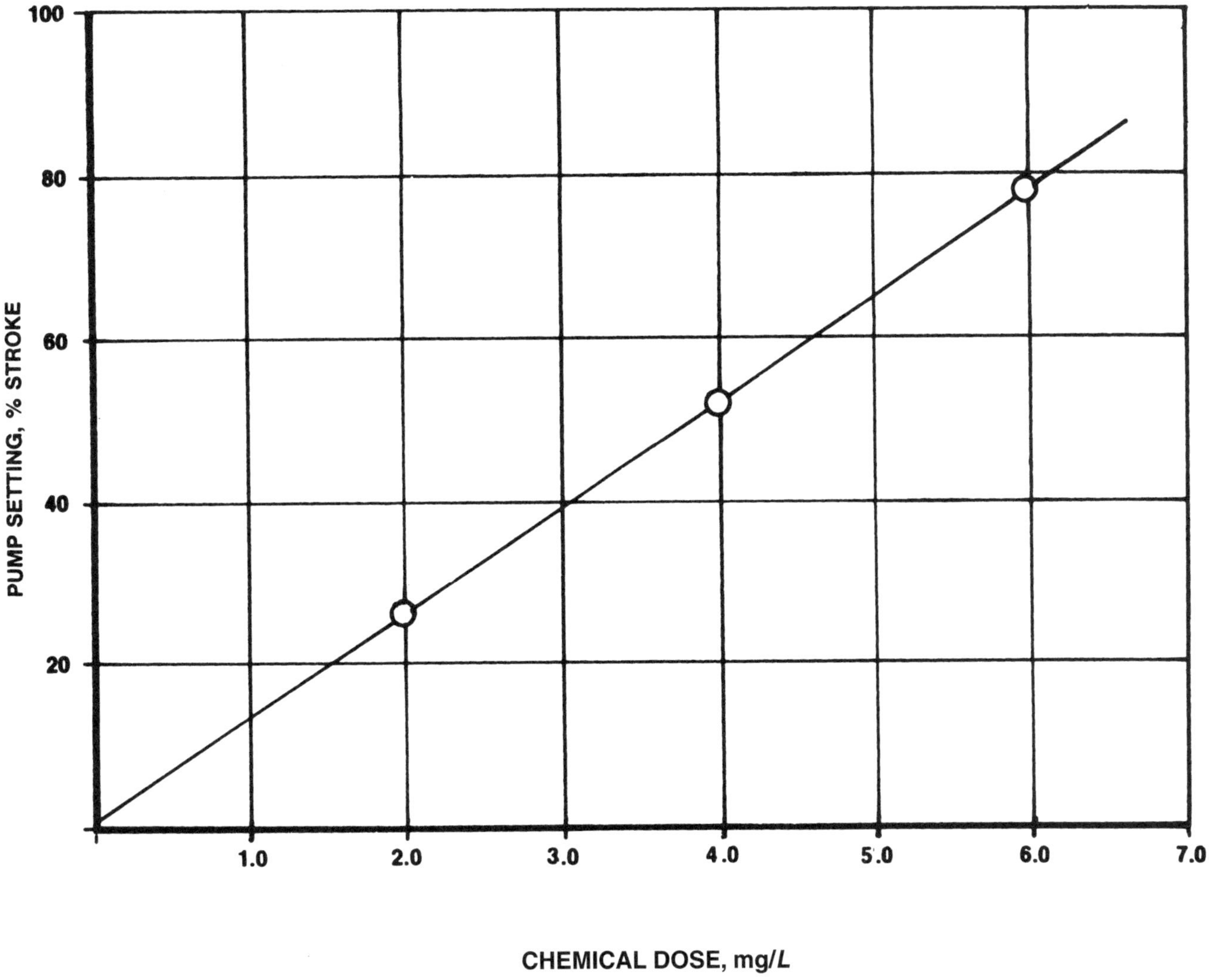

Fig. A.3   Chemical feed pump settings for various chemical doses

EXAMPLE 10

A reservoir has a volume of 6.8 acre-feet. What is the reservoir volume in gallons and million gallons?

Known	Unknown
Reservoir Vol, ac-ft = 6.8 ac-ft	1. Reservoir Vol, gal
	2. Reservoir Vol, MG

Convert reservoir volume from acre-feet to gallons and million gallons.

$$\text{Reservoir Vol, gal} = (\text{Volume, ac-ft})(43{,}560 \text{ sq ft/ac})(7.48 \text{ gal/cu ft})$$

$$= (6.8 \text{ ac-ft})(43{,}560 \text{ sq ft/ac})(7.48 \text{ gal/cu ft})$$

$$= 2{,}215{,}636 \text{ gal}$$

$$= 2.2 \text{ M Gal}$$

EXAMPLE 11

A reservoir has a surface area of 51,200 square feet and the desired dose of copper sulfate is six pounds per acre. How many pounds of copper sulfate will be needed?

Known	Unknown
Surface Area, sq ft = 51,200 sq ft	Copper Sulfate, lbs
Dose, lbs/ac = 6 lbs/ac	

1. Convert the surface area from square feet to acres.

$$\text{Surface Area, ac} = \frac{\text{Surface Area, sq ft}}{43{,}560 \text{ sq ft/ac}}$$

$$= \frac{51{,}200 \text{ sq ft}}{43{,}560 \text{ sq ft/ac}}$$

$$= 1.18 \text{ ac}$$

2. Calculate the pounds of copper sulfate needed.

$$\text{Copper Sulfate, lbs} = (\text{Surface Area, ac})(\text{Dose, lbs/ac})$$

$$= (1.18 \text{ ac})(6 \text{ lbs/ac})$$

$$= 7.1 \text{ lbs Copper Sulfate}$$

EXAMPLE 12

The volume of a reservoir is estimated to be five million gallons. The desired chemical dose is 0.5 mg/L. Estimate the chemical dose in pounds.

Known	Unknown
Reservoir Vol, M Gal = 5 M Gal	Chemical Dose, lbs
Chemical Dose, mg/L = 0.5 mg/L	

Estimate the chemical dose in pounds.

$$\text{Chemical Dose, lbs} = (\text{Vol, M Gal})(\text{Dose, mg/}L)(8.34 \text{ lbs/gal})$$

$$= (5 \text{ M Gal})(0.5 \text{ mg/}L)(8.34 \text{ lbs/gal})$$

$$= 20.9 \text{ lbs}$$

EXAMPLE 13

The volume of a reservoir is estimated to be 581,000 cubic feet. The desired dose of copper is 0.5 mg/L and the copper content of the copper sulfate to be used is 25 percent. How many pounds of copper sulfate will be needed?

Known	Unknown
Reservoir Volume, cu ft = 581,000 cu ft	Copper Sulfate, lbs
Copper Dose, mg/L = 0.5 mg/L	
Copper, % = 25%	

1. Convert the reservoir volume from cubic feet to million gallons.

$$\text{Reservoir Volume, M Gal} = \frac{(\text{Reservoir Vol, cu ft})(7.48 \text{ gal/cu ft})}{1{,}000{,}000/1 \text{ Million}}$$

$$= \frac{(581{,}000 \text{ cu ft})(7.48 \text{ gal/cu ft})}{1{,}000{,}000/1 \text{ Million}}$$

$$= 4.35 \text{ M Gal}$$

2. Calculate the pounds of copper sulfate that will be needed.

$$\text{Copper Sulfate, lbs} = \frac{(\text{Vol, M Gal})(\text{Dose, mg/}L)(8.34 \text{ lbs/gal})(100\%)}{\text{Copper, \%}}$$

$$= \frac{(4.35 \text{ M Gal})(0.5 \text{ mg/}L)(8.34 \text{ lbs/gal})(100\%)}{25\%}$$

$$= 72.6 \text{ lbs Copper Sulfate}$$

A.133   Coagulation and Flocculation

EXAMPLE 14

A polymer feed pump delivers a flow of 200 gallons per day containing a five percent polymer solution with a specific gravity of 1.02. Estimate the polymer delivered in pounds per day.

Known	Unknown
Flow, gal/day = 200 gal/day	Polymer, lbs/day
Flow, MGD = 0.0002 MGD	
Polymer, % = 5%	
Polymer, mg/L = 50,000 mg/L	
Sp Gr = 1.02	

Estimate the polymer delivered in pounds per day.

SOLUTION 1

$$\text{Polymer, lbs/day} = \frac{(\text{Polymer Flow, gal/day})(8.34 \text{ lbs/gal})(\text{Polymer, \%})(\text{Sp Gr})}{100\%}$$

$$= \frac{(200 \text{ gal/day})(8.34 \text{ lbs/gal})(5\%)(1.02)}{100\%}$$

$$= 85 \text{ lbs Polymer/day}$$

SOLUTION 2

$$\begin{array}{rl}\text{Polymer,}\\ \text{lbs/day} \end{array} = \text{(Polymer Flow, MGD)(Polymer, mg/}L\text{)(8.34 lbs/gal)(Sp Gr)}$$

$$= \text{(0.0002 MGD)(50,000 mg/}L\text{)(8.34 lbs/gal)(1.02)}$$

$$= \text{85 lbs Polymer/day} \qquad \textit{SAME ANSWER!}$$

EXAMPLE 15

Estimate the actual alum dose in milligrams per liter if a plant treats a raw water with a flow of 1.8 MGD. The alum feed rate is 135 pounds per day.

Known	**Unknown**
Flow, MGD = 1.8 MGD	Alum Dose, mg/L
Alum Feed, lbs/day = 135 lbs/day	

Calculate the alum dose in milligrams per liter.

BASIC FORMULA

$$\begin{array}{rl}\text{Chemical Feed,}\\ \text{lbs/day} \end{array} = \text{(Flow, MGD)(Dose, mg/}L\text{)(8.34 lbs/gal)}$$

or

$$\begin{aligned}\text{Alum Dose,}\\ \text{mg/}L \end{aligned} &= \frac{\text{Chemical Feed, lbs/day}}{\text{(Flow, MGD)(8.34 lbs/gal)}}\\[2mm] &= \frac{\text{135 lbs/day}}{\text{(1.8 MGD)(8.34 lbs/gal)}}\\[2mm] &= 9.0 \text{ mg/}L\end{aligned}$$

EXAMPLE 16

Determine the strength of a polymer solution as a percent if 80 grams (80 gm/454 gm/lb = 0.176 lb) of dry polymer are mixed with four gallons of water.

Known	**Unknown**
Dry Polymer, lbs = 0.176 lb	Polymer Solution, %
Volume Water, gal = 4 gal	

1. Convert the four gallons of water to pounds.

$$\begin{aligned}\text{Water, lbs} &= \text{(Volume Water, gal)(8.34 lbs/gal)}\\ &= \text{(4 gal)(8.34 lb/gal)}\\ &= 33.36 \text{ lbs}\end{aligned}$$

2. Calculate the polymer solution as a percent.

$$\begin{aligned}\text{Polymer, \%} &= \frac{\text{(Dry Polymer, lbs)(100\%)}}{\text{(Dry Polymer, lbs + Water, lbs)}}\\[2mm] &= \frac{\text{(0.176 lb)(100\%)}}{\text{(0.176 lb + 33.36 lbs)}}\\[2mm] &= 0.52\%\end{aligned}$$

EXAMPLE 17

Liquid polymer is supplied to a water treatment plant as a ten percent solution. How many gallons of liquid polymer should be mixed in a tank with water to produce 200 gallons of 0.6 percent polymer solution?

Known	**Unknown**
Liquid Polymer, % = 10%	Volume of Liquid Polymer, gal
Polymer Solution, % = 0.6%	
Volume of Polymer Solution, gal = 200 gal	

Calculate the volume of liquid polymer in gallons.

$$\begin{aligned}\begin{array}{l}\text{Liquid}\\\text{Polymer,}\\\text{gal}\end{array} &= \frac{\text{(Polymer Solution, \%)(Vol of Solution, gal)}}{\text{Liquid Polymer, \%}}\\[2mm] &= \frac{\text{(0.6\%)(200 gal)}}{10\%}\\[2mm] &= 12 \text{ gallons}\end{aligned}$$

A.134   Sedimentation

EXAMPLE 18

Estimate the detention time in hours for a 30-foot diameter circular clarifier when the flow is 0.5 MGD. The clarifier is 8 feet deep.

Known	**Unknown**
Diameter, ft = 30 ft	Detention Time, hr
Depth, ft = 8 ft	
Flow, MGD = 0.5 MGD	
Flow, gal/day = 500,000 gal/day	

1. Calculate the clarifier volume in cubic feet.

$$\begin{aligned}\text{Volume, cu ft} &= \text{(Area, sq ft)(Depth, ft)}\\ &= (0.785)(30 \text{ ft})^2(8 \text{ ft})\\ &= 5,652 \text{ cu ft}\end{aligned}$$

2. Estimate the detention time of the clarifier in hours.

$$\begin{aligned}\begin{array}{l}\text{Detention}\\\text{Time, hr}\end{array} &= \frac{\text{(Volume, cu ft)(7.48 gal/cu ft)(24 hr/day)}}{\text{Flow, gal/day}}\\[2mm] &= \frac{\text{(5,652 cu ft)(7.48 gal/cu ft)(24 hr/day)}}{\text{500,000 gal/day}}\\[2mm] &= 2.0 \text{ hr}\end{aligned}$$

EXAMPLE 19

Estimate the overflow rate in gallons per minute per square foot for a rectangular sedimentation basin 20 feet wide and 40 feet long when the flow is 0.5 MGD.

Known	Unknown
Width, ft = 20 ft	Overflow Rate, GPM/sq ft
Length, ft = 40 ft	
Flow, MGD = 0.5 MGD	

1. Determine the surface area of the basin in square feet.

   Surface Area, sq ft = (Length, ft)(Width, ft)

   $$= (40 \text{ ft})(20 \text{ ft})$$

   $$= 800 \text{ sq ft}$$

2. Convert the flow from million gallons per day to gallons per minute.

   $$\text{Flow, GPM} = \frac{(\text{Flow, MGD})(1,000,000/M)}{(60 \text{ min/hr})(24 \text{ hr/day})}$$

   $$= \frac{(0.5 \text{ MGD})(1,000,000/M)}{(60 \text{ min/hr})(24 \text{ hr/day})}$$

   $$= 347 \text{ GPM}$$

3. Estimate the overflow rate in gallons per minute per square foot of surface area.

   $$\text{Overflow Rate, GPM/sq ft} = \frac{\text{Flow, GPM}}{\text{Surface Area, sq ft}}$$

   $$= \frac{347 \text{ GPM}}{800 \text{ sq ft}}$$

   $$= 0.43 \text{ GPM/sq ft}$$

EXAMPLE 20

Estimate the flow velocity in feet per minute through a rectangular sedimentation basin 20 feet wide and 8 feet deep when the flow is 350 GPM.

Known	Unknown
Width, ft = 20 ft	Flow Velocity, ft/min
Depth, ft = 8 ft	
Flow, GPM = 350 GPM	

Estimate the flow velocity in feet per minute.

$$\text{Flow Velocity, ft/min} = \frac{\text{Flow, GPM}}{(\text{Cross-Sectional Area, sq ft})(7.48 \text{ gal/cu ft})}$$

$$= \frac{350 \text{ GPM}}{(20 \text{ ft})(8 \text{ ft})(7.48 \text{ gal/cu ft})}$$

$$= 0.29 \text{ ft/min}$$

EXAMPLE 21

Estimate the weir loading in gallons per minute per foot of weir length for a 30-foot diameter circular clarifier treating a flow of 350 GPM. The weir is located on the water edge of the clarifier.

Known	Unknown
Weir Diameter, ft = 30 ft	Weir Loading, GPM/ft
Flow, GPM = 350 GPM	

1. Calculate the weir length in feet.

   Weir Length, ft = π(Diameter, ft)

   $$= (3.14)(30 \text{ ft})$$

   $$= 94.2 \text{ ft}$$

2. Estimate the weir loading in gallons per minute per foot of weir.

   $$\text{Weir Loading, GPM/ft} = \frac{\text{Flow, GPM}}{\text{Weir Length, ft}}$$

   $$= \frac{350 \text{ GPM}}{94.2 \text{ ft}}$$

   $$= 3.7 \text{ GPM/ft}$$

A.135   Filtration

EXAMPLE 22

A 25-foot wide by 30-foot long rapid sand filter treats a flow of 2,000 gallons per minute. Calculate the filtration rate in gallons per minute per square foot of filter area.

Known	Unknown
Width, ft = 25 ft	Filtration Rate, GPM/sq ft
Length, ft = 30 ft	
Flow, GPM = 2,000 GPM	

Calculate the filtration rate in gallons per minute per square foot of filter surface area.

$$\text{Filtration Rate, GPM/sq ft} = \frac{\text{Flow, GPM}}{\text{Surface Area, sq ft}}$$

$$= \frac{2,000 \text{ GPM}}{(25 \text{ ft})(30 \text{ ft})}$$

$$= 2.7 \text{ GPM/sq ft}$$

EXAMPLE 23

With the inflow water to a rapid sand filter shut off, the water is observed to drop 20 inches in nine minutes. What is the velocity of the water dropping in feet per minute?

Known	Unknown
Water Drop, in = 20 in	Velocity of Drop, ft/min
Time of Drop, min = 9 min	

Calculate the velocity of the water drop in feet per minute.

$$\text{Velocity, ft/min} = \frac{\text{Water Drop, in}}{(12 \text{ in/ft})(\text{Time of Drop, min})}$$

$$= \frac{20 \text{ in}}{(12 \text{ in/ft})(9 \text{ min})}$$

$$= 0.185 \text{ ft/min}$$

EXAMPLE 24

Estimate the flow through a rapid sand filter in cubic feet per minute when the velocity of the water dropping is 0.18 feet per minute and the filter is 25 feet wide and 30 feet long.

Known		Unknown
Velocity of Drop, ft/min	= 0.18 ft/min	Flow, cu ft/min
Width, ft	= 25 ft	
Length, ft	= 30 ft	

Estimate the flow through the filter in cubic feet per minute.

$$\text{Flow, cu ft/min} = (\text{Area, sq ft})(\text{Velocity, ft/min})$$

$$= (25 \text{ ft})(30 \text{ ft})(0.18 \text{ ft/min})$$

$$= 135 \text{ cu ft/min}$$

EXAMPLE 25

Calculate the flow through a rapid sand filter in gallons per minute when the flow is 135 cu ft per minute.

Known	Unknown
Flow, cu ft/min = 135 cu ft/min	Flow, GPM

Calculate the flow through the filter in gallons per minute.

$$\text{Flow, GPM} = (\text{Flow, cu ft/min})(7.48 \text{ gal/cu ft})$$

$$= (135 \text{ cu ft/min})(7.48 \text{ gal/cu ft})$$

$$= 1,010 \text{ GPM}$$

EXAMPLE 26

Calculate the flow through a rapid sand filter in gallons per minute when 1.5 million gallons flowed through the filter during a 24-hour filter run.

Known	Unknown
Flow, M Gal = 1.5 M Gal	Flow, GPM
Time, hr = 24 hr	

Calculate the flow through the filter in gallons per minute.

$$\text{Flow, GPM} = \frac{\text{Total Flow, gal}}{(\text{Filter Run, hr})(60 \text{ min/hr})}$$

$$= \frac{1,500,000 \text{ gal}}{(24 \text{ hr})(60 \text{ min/hr})}$$

$$= 1,040 \text{ GPM}$$

EXAMPLE 27

Determine the Unit Filter Run Volume (UFRV) in gallons per square foot for a filter 20 feet long and 16 feet wide if the volume of water filtered between backwash cycles is 2.2 million gallons.

Known		Unknown
Length, ft	= 20 ft	UFRV, gal/sq ft
Width, ft	= 16 ft	
Volume Filtered, gal	= 2,200,000 gal	

Calculate the Unit Filter Run Volume in gallons per square foot of filter surface area.

$$\text{UFRV, gal/sq ft} = \frac{\text{Volume Filtered, gal}}{\text{Filter Surface Area, sq ft}}$$

$$= \frac{2,200,000 \text{ gal}}{(20 \text{ ft})(16 \text{ ft})}$$

$$= 6,875 \text{ gal/sq ft}$$

EXAMPLE 28

Determine the Unit Filter Run Volume (UFRV) in gallons per square foot for a filter if the filtration rate was 2.3 GPM/sq ft during a 46-hour filter run.

Known		Unknown
Filtration Rate, GPM/sq ft	= 2.3 GPM/sq ft	UFRV, gal/sq ft
Filter Run, hr	= 46 hr	

Calculate the Unit Filter Run Volume in gallons per square foot of filter surface area.

$$\text{UFRV, gal/sq ft} = (\text{Filtration Rate, GPM/sq ft})(\text{Filter Run, hr})(60 \text{ min/hr})$$

$$= (2.3 \text{ GPM/sq ft})(46 \text{ hr})(60 \text{ min/hr})$$

$$= 6,348 \text{ gal/sq ft}$$

EXAMPLE 29

Calculate the backwash flow required in gallons per minute to backwash a 25-foot wide by 30-foot long filter if the desired backwash flow rate is 20 gallons per minute per square foot.

Known		Unknown
Width, ft	= 25 ft	Backwash Flow, GPM
Length, ft	= 30 ft	
Backwash Rate, GPM/sq ft	= 20 GPM/sq ft	

Calculate the backwash flow in gallons per minute.

$$\text{Backwash Flow, GPM} = (\text{Filter Area, sq ft})(\text{Backwash Rate, GPM/sq ft})$$

$$= (25 \text{ ft})(30 \text{ ft})(20 \text{ GPM/sq ft})$$

$$= 15,000 \text{ GPM}$$

EXAMPLE 30

Convert a filter backwash rate of 23 gallons per minute per square foot to inches per minute of rise.

Known	Unknown
Backwash, GPM/sq ft = 23 GPM/sq ft	Backwash, in/min

Convert the backwash rate from GPM/sq ft to inches/minute.

$$\text{Backwash, in/min} = \frac{(\text{Backwash, GPM/sq ft})(12 \text{ in/ft})}{7.48 \text{ gal/cu ft}}$$

$$= \frac{(23 \text{ GPM/sq ft})(12 \text{ in/ft})}{7.48 \text{ gal/cu ft}}$$

$$= 37 \text{ in/min}$$

EXAMPLE 31

Determine the volume or amount of water in gallons required to backwash a filter if the backwash flow rate is 9,500 GPM when the backwash time is seven minutes.

Known	Unknown
Backwash Flow Rate, GPM = 9,500 GPM	Backwash Water, gallons
Backwash Time, min = 7 min	

Calculate the volume of backwash water required in gallons.

$$\text{Backwash Water, gallons} = (\text{Backwash Flow, GPM})(\text{Backwash Time, min})$$

$$= (9,500 \text{ gal/min})(7 \text{ min})$$

$$= 66,500 \text{ gallons}$$

EXAMPLE 32

During a filter run the total volume of water filtered was 13.0 million gallons. When the filter was backwashed, 66,500 gallons of water was used. Calculate the percent of the product or finished water used for backwashing.

Known	Unknown
Water Filtered, gal = 13,000,000 gal	Backwash, %
Backwash Water, gal = 66,500 gal	

Calculate the percent of water used for backwashing.

$$\text{Backwash, \%} = \frac{(\text{Backwash Water, gal})(100\%)}{\text{Water Filtered, gal}}$$

$$= \frac{(66,500 \text{ gal})(100\%)}{13,000,000 \text{ gal}}$$

$$= 0.5\%$$

A.136   Disinfection

EXAMPLE 33

Calculate the chlorine dose in mg/L when a chlorinator is set to feed 18 pounds of chlorine in 24 hours. The flow is 570,000 gallons per day.

Known	Unknown
Chlorinator Setting, lbs/24 hr = 18 lbs Cl/24 hr	Chlorine Dose, mg/L
Flow, MGD = 0.57 MGD	

Calculate the chlorine dose in milligrams per liter.

$$\text{Chlorine Dose, mg/L} = \frac{\text{Chlorine Feed, lbs/day}}{(\text{Flow, MGD})(8.34 \text{ lbs/gal})}$$

$$= \frac{18 \text{ lbs/day}}{(0.57 \text{ MGD})(8.34 \text{ lbs/gal})}$$

$$= 3.8 \text{ mg/L}$$

EXAMPLE 34

Estimate the chlorine demand for a water in milligrams per liter if the chlorine dose is 2.9 mg/L and the chlorine residual is 0.6 mg/L.

Known	Unknown
Chlorine Dose, mg/L = 2.9 mg/L	Chlorine Demand, mg/L
Chlorine Residual, mg/L = 0.6 mg/L	

Estimate the chlorine demand of the water in milligrams per liter.

$$\text{Chlorine Demand, mg/L} = \text{Chlorine Dose, mg/L} - \text{Chlorine Residual, mg/L}$$

$$= 2.9 \text{ mg/L} - 0.6 \text{ mg/L}$$

$$= 2.3 \text{ mg/L}$$

EXAMPLE 35

Calculate the pounds of chlorine used to disinfect water if 150 gallons of hypochlorite as a 2.5 percent chlorine solution was used.

Known	Unknown
Hypochlorite, gal = 150 gal	Chlorine, lbs
Hypochlorite, % = 2.5%	

Calculate the pounds of chlorine used.

$$\text{Chlorine, lbs} = \frac{(\text{Hypochlorite, gal})(8.34 \text{ lbs/gal})(\text{Hypochlorite, \%})}{100\%}$$

$$= \frac{(150 \text{ gal})(8.34 \text{ lbs/gal})(2.5\%)}{100\%}$$

$$= 31.3 \text{ lbs Chlorine}$$

EXAMPLE 36

Estimate the flow pumped by a hypochlorinator in gallons per day if the hypochlorite solution is in a container with a diameter of 30 inches (2.5 feet) and the hypochlorite level drops 14 inches during a nine-hour period. The hypochlorinator operated continuously during the nine-hour period.

Known	Unknown
Diameter, ft = 2.5 ft	Hypochlorinator Flow, GPD
Drop, in = 14 in	
Time, hr = 9 hr	

Calculate the hypochlorinator flow in gallons per day.

$$\text{Flow, GPD} = \frac{(\text{Container Area, sq ft})(\text{Drop, in})(7.48 \text{ gal/cu ft})(24 \text{ hr/day})}{(\text{Time, hr})(12 \text{ in/ft})}$$

$$= \frac{(0.785)(2.5 \text{ ft})^2(14 \text{ in})(7.48 \text{ gal/cu ft})(24 \text{ hr/day})}{(9 \text{ hr})(12 \text{ in/ft})}$$

$$= 114 \text{ GPD}$$

EXAMPLE 37

Estimate the desired strength (as a percent chlorine) of a hypochlorite solution which is pumped by a hypochlorinator that delivers 115 gallons per day. The water being treated requires a chlorine dose of twelve pounds of chlorine per day.

Known	Unknown
Hypochlorinator Flow, GPD = 115 GPD	Hypochlorite Strength, %
Chlorine Required, lbs/day = 12 lbs/day	

Estimate the desired hypochlorite strength as a percent chlorine.

$$\text{Hypochlorite Strength, \%} = \frac{(\text{Chlorine Required, lbs/day})(100\%)}{(\text{Hypochlorinator Flow, gal/day})(8.34 \text{ lbs/gal})}$$

$$= \frac{(12 \text{ lbs/day})(100\%)}{(115 \text{ GPD})(8.34 \text{ lbs/gal})}$$

$$= 1.25\%$$

EXAMPLE 38

How many gallons of water must be added to fifteen gallons of five percent hypochlorite solution to produce a 1.25 percent hypochlorite solution?

Known	Unknown
Hypochlorite, gal = 15 gal	Water Added, gal
Desired Hypo, % = 1.25%	
Actual Hypo, % = 5%	

Calculate the gallons of water that must be added to produce a 1.25 percent hypochlorite solution.

$$\text{Water Added, gal (to hypochlorite solution)} = \frac{(\text{Hypo, gal})(\text{Hypo, \%}) - (\text{Hypo, gal})(\text{Desired Hypo, \%})}{\text{Desired Hypo, \%}}$$

$$= \frac{(15 \text{ gal})(5\%) - (15 \text{ gal})(1.25\%)}{1.25\%}$$

$$= \frac{75 - 18.75}{1.25}$$

$$= 45 \text{ gallons}$$

A.137   Corrosion Control

EXAMPLE 39

Find the pH_S of a water at 10°C having a TDS of 100 mg/L, alkalinity of 40 mg/L, and a calcium hardness of 60 mg/L.

Known	Unknown
Water Temp, °C = 10°C	pH_S
TDS, mg/L = 100 mg/L	
Alky, mg/L = 40 mg/L	
Ca Hardness, mg/L = 60 mg/L as $CaCO_3$	

1. Find the formula values from the tables in Chapter 8, "Corrosion Control."

   From Table 8.2 for a water temperature of 10°C,

   A = 2.20

   From Table 8.3 for a TDS of 100 mg/L,

   B = 9.75

   From Table 8.4 for Ca of 60 mg/L and Alky of 40 mg/L,

   $\log(Ca^{2+}) = 1.78$

   $\log(\text{Alky}) = 1.60$

2. Calculate pH_S.

   $$pH_S = A + B - \log(Ca^{2+}) - \log(\text{Alky})$$

   $$= 2.20 + 9.75 - 1.78 - 1.60$$

   $$= 8.57$$

EXAMPLE 40

Calculate the Langelier Index for a water with a calculated pH_S value of 8.69 and an actual pH of 8.5.

Known	Unknown
pH_S = 8.69	Langelier Index
pH = 8.5	

Calculate the Langelier Index.

$$\text{Langelier Index} = pH - pH_S$$

$$= 8.5 - 8.69$$

$$= -0.19$$

Since the Langelier Index is negative, the water is corrosive.

A.138   Plant Operation

EXAMPLE 41

Estimate the average use of chlorine in pounds per day based on actual use of chlorine for one week as shown below.

Day	Sun	Mon	Tue	Wed	Thu	Fri	Sat
Chlorine Used, lbs	23	37	35	31	32	36	24

Known	Unknown
Chlorine Used, lbs/day	Average Chlorine Used, lbs/day

Estimate the average chlorine use in pounds of chlorine per day.

$$\text{Average Chlorine Use, lbs/day} = \frac{\text{Sum of Chlorine Used Each Day, lbs}}{\text{Total Time, days}}$$

$$= \frac{23 \text{ lbs} + 37 \text{ lbs} + 35 \text{ lbs} + 31 \text{ lbs} + 32 \text{ lbs} + 36 \text{ lbs} + 24 \text{ lbs}}{7 \text{ days}}$$

$$= \frac{218 \text{ lbs}}{7 \text{ days}}$$

$$= 31.1 \text{ lbs/day}$$

EXAMPLE 42

A water treatment plant has five 150-pound chlorine cylinders in storage. The plant uses an average of 28 pounds of chlorine per day. How many days' supply of chlorine is in storage?

Known	Unknown
Chlorine Cylinders = 5 cylinders	Supply of Chlorine, days
Cylinder Wt, lbs/cyl = 150 lbs/cyl	
Avg Use, lbs/day = 28 lbs/day	

Calculate the available supply of chlorine in storage in days.

$$\text{Supply of Chlorine, days} = \frac{(\text{Cylinder Wt, lbs/cyl})(\text{No. of Cylinders})}{\text{Avg Use, lbs/day}}$$

$$= \frac{(150 \text{ lbs/cyl})(5 \text{ Cylinders})}{28 \text{ lbs/day}}$$

$$= 27 \text{ days}$$

A.139   Laboratory Procedures

EXAMPLE 43

Convert the temperature of water from 65° Fahrenheit to degrees Celsius.

Known	Unknown
Temp, °F = 65°F	Temp, °C

Change 65°F to degrees Celsius.

$$\text{Temperature, } °C = \frac{5}{9}(°F - 32°F)$$

$$= \frac{5}{9}(65°F - 32°F)$$

$$= 18.3°C$$

EXAMPLE 44

Convert a water temperature of 12° Celsius to degrees Fahrenheit.

Known	Unknown
Temp, °C = 12°C	Temp, °F

Change 12°C to degrees Fahrenheit.

$$\text{Temperature, } °F = \frac{9}{5}(°C) + 32°F$$

$$= \frac{9}{5}(12°C) + 32°F$$

$$= 53.6°F$$

EXAMPLE 45

In the determination of a chlorine residual by the drop-dilution method, three drops of sample produced a chlorine residual of 0.2 mg/L in 10 mL of distilled water. Assume 0.05 mL per drop.

Known	Unknown
Chlorine Residual, mg/L = 0.2 mg/L	Actual Chlorine Residual, mg/L
Sample Volume, drops = 3 drops	
Distilled Water, mL = 10 mL	

Calculate the actual chlorine residual in milligrams per liter.

$$\text{Actual Chlorine Residual, mg/L} = \frac{(\text{Chlorine Residual, mg/L})(\text{Distilled Water, mL})}{(\text{Sample Volume, drops})(0.05 \text{ mL/drop})}$$

$$= \frac{(0.2 \text{ mg/L})(10 \text{ mL})}{(3 \text{ drops})(0.05 \text{ mL/drop})}$$

$$= 13 \text{ mg/L}$$

EXAMPLE 46

Results from the MPN tests during one week were as follows:

Day	S	M	T	W	T	F	S
MPN/100 mL	2	4	6	7	9	5	2

Estimate the (1) mean, (2) median, and (3) geometric mean of the data in MPN/100 mL.

1. Calculate the mean.

$$\text{Mean, MPN/100 mL} = \frac{\text{Sum of All MPNs}}{\text{Number of MPNs}}$$

$$= \frac{2 + 4 + 6 + 7 + 9 + 5 + 2}{7}$$

$$= \frac{35}{7}$$

$$= 5 \text{ MPN/100 mL}$$

2. Determine the median. Rearrange the data in ascending (increasing) order and select the middle value (three will be smaller and three will be larger in this example).

Order	1	2	3	4	5	6	7
MPN/100 mL	2	2	4	5	6	7	9
				↑			

Median, MPN/100 mL = Middle value of a group of data

= 5 MPN/100 mL

3. Calculate the geometric mean for the given data.

$$\begin{array}{l}\text{Geometric Mean,} \\ \text{MPN/100 m}L\end{array} = [(X_1)(X_2)(X_3)(X_4)(X_5)(X_6)(X_7)]^{1/7}$$

$$= [(2)(4)(6)(7)(9)(5)(2)]^{1/7}$$

$$= [30{,}240]^{0.143}$$

$$= 4.4 \text{ MPN/100 m}L$$

A.14   BASIC CONVERSION FACTORS (METRIC SYSTEM)

LENGTH

100 cm	= 1 m	100 cm/m
3.281 ft	= 1 m	3.281 ft/m

AREA

2.4711 ac	= 1 ha*	2.4711 ac/ha
10,000 sq m	= 1 ha	10,000 sq m/ha

VOLUME

1,000 mL	= 1 liter	1,000 mL/L
1,000 L	= 1 cu m	1,000 L/cu m
3.785 L	= 1 gal	3.785 L/gal

WEIGHT

1,000 mg	= 1 gm	1,000 mg/gm
1,000 gm	= 1 kg	1,000 gm/kg

DENSITY

1 kg	= 1 liter	1 kg/L

PRESSURE

10.015 m	= 1 kg/sq cm	10.015 m/kg/sq cm
1 Pascal	= 1 N/sq m	1 Pa/N/sq m
1 psi	= 6,895 Pa	1 psi/6,895 Pa

FLOW

3,785 cu m/day	= 1 MGD	3,785 cu m/day/MGD
3.785 ML/day	= 1 MGD	3.785 ML/day/MGD

---

* hectare

A.15   TYPICAL WATER TREATMENT PLANT PROBLEMS (METRIC SYSTEM)

A.150   Flows

EXAMPLE 1

Convert a flow of 500 gallons per minute to liters per second and cubic meters per day.

Known	Unknown
Flow, GPM = 500 GPM	1. Flow, liters/sec
	2. Flow, cu m/day

1. Convert the flow from 500 GPM to liters per second.

$$\text{Flow, liters/sec} = \frac{(\text{Flow, gal/min})(3.785 \text{ liters/gal})}{60 \text{ sec/min}}$$

$$= \frac{(500 \text{ gal/min})(3.785 \text{ liters/gal})}{60 \text{ sec/min}}$$

$$= 31.5 \text{ liters/sec}$$

2. Convert the flow from 500 GPM to cubic meters per day.

$$\begin{array}{l}\text{Flow,} \\ \text{cu m/day}\end{array} = \frac{(\text{Flow, GPM})(3.785 \text{ }L\text{/gal})(60 \text{ min/hr})(24 \text{ hr/day})}{1{,}000 \text{ }L\text{/cu m}}$$

$$= \frac{(500 \text{ GPM})(3.785 \text{ }L\text{/gal})(60 \text{ min/hr})(24 \text{ hr/day})}{1{,}000 \text{ }L\text{/cu m}}$$

$$= 2{,}725 \text{ cu m/day}$$

A.151   Chemical Doses

EXAMPLE 2

Determine the chlorinator setting in kilograms per 24 hours if 4,000 cubic meters of water per day are to be treated with a desired chlorine dose of 2.5 mg/L.

Known	Unknown
Flow, cu m/day = 4,000 cu m/day	Chlorinator Setting, kg/24 hours
Chlorine Dose, mg/L = 2.5 mg/L	

Determine the chlorinator setting in kilograms per 24 hours.

$$\begin{array}{l}\text{Chlorinator} \\ \text{Setting,} \\ \text{kg/day}\end{array} = \frac{(\text{Flow, cu m/day})(\text{Dose, mg/}L)(1{,}000 \text{ }L\text{/cu m})}{(1{,}000 \text{ mg/gm})(1{,}000 \text{ gm/kg})}$$

$$= \frac{(4{,}000 \text{ cu m/day})(2.5 \text{ mg/}L)(1{,}000 \text{ }L\text{/cu m})}{(1{,}000 \text{ mg/gm})(1{,}000 \text{ gm/kg})}$$

$$= 10 \text{ kg/day}$$

EXAMPLE 3

The optimum liquid alum dose from the jar tests is 12 mg/L. Determine the setting on the liquid alum chemical feeder in milliliters per minute when the plant flow is 15 megaliters per day (or million liters per day). The liquid alum delivered to the plant contains 642.3 milligrams of alum per milliliter of solution.

Known	Unknown
Alum Dose, mg/L = 12 mg/L	Chemical Feeder Setting, mL/min
Flow, ML/day = 15 ML/day	
Liquid Alum, mg/mL = 642.3 mg/mL	

Calculate the liquid alum chemical feeder setting in milliliters per minute.

$$\begin{array}{l}\text{Chemical Feeder} \\ \text{Setting, m}L\text{/min}\end{array} = \frac{(\text{Flow, M}L\text{/day})(\text{Alum Dose, mg/}L)(1{,}000{,}000\text{/M})}{(\text{Liquid Alum, mg/m}L)(24 \text{ hr/day})(60 \text{ min/hr})}$$

$$= \frac{(15 \text{ M}L\text{/day})(12 \text{ mg/}L)(1{,}000{,}000\text{/M})}{(642.3 \text{ mg/m}L)(24 \text{ hr/day})(60 \text{ min/hr})}$$

$$= 195 \text{ m}L\text{/min}$$

EXAMPLE 4

The optimum liquid alum dose from the jar tests is 8 mg/L. Determine the setting on the liquid alum chemical feeder in milliliters per minute when the flow is 12 megaliters per day.

The liquid alum delivered to the plant contains 5.36 pounds of alum per gallon of liquid solution.

Known	Unknown
Alum Dose, mg/L = 8 mg/L	Chemical Feeder Setting, mL/min
Flow, ML/day = 12 ML/day	
Liquid Alum, lbs/gal = 5.36 lbs/gal	

Calculate the liquid alum chemical feeder setting in milliliters per minute.

$$\text{Chemical Feeder Setting, mL/min} = \frac{(\text{Flow, ML/day})(\text{Alum Dose, mg/L})(3.785\ L/\text{gal})(1{,}000\ mL/L)(1{,}000{,}000/M)}{(\text{Liquid Alum, lbs/gal})(454\ gm/lb)(1{,}000\ mg/gm)(24\ hr/day)(60\ min/hr)}$$

$$= \frac{(12\ ML/day)(8\ mg/L)(3.785\ L/\text{gal})(1{,}000\ mL/L)(1{,}000{,}000/M)}{(5.36\ lbs/gal)(454\ gm/lb)(1{,}000\ mg/gm)(24\ hr/day)(60\ min/hr)}$$

$$= 104\ mL/min$$

EXAMPLE 5

Determine the actual chemical dose or chemical feed in kilograms per day from a dry chemical feeder. A bucket placed under the chemical feeder weighed 150 grams empty and 1,800 grams after 12 minutes.

Known	Unknown
Empty Bucket, gm = 150 gm	Chemical Feed, kg/day
Full Bucket, gm = 1,800 gm	
Time to Fill, min = 12 min	

Calculate the actual chemical dose in kg/day.

$$\text{Chemical Feed, kg/day} = \frac{(\text{Chemical Applied, gm})(60\ min/hr)(24\ hr/day)}{(1{,}000\ gm/kg)(\text{Length of Application, min})}$$

$$= \frac{(1{,}800\ gm - 150\ gm)(60\ min/hr)(24\ hr/day)}{(1{,}000\ gm/kg)(12\ min)}$$

$$= 198\ kg/day$$

EXAMPLE 6

Determine the chemical feed in kilograms of polymer per day from a chemical feed pump. The polymer solution is 1.5 percent or 15,000 mg polymer per liter. Assume a specific gravity of the polymer solution of 1.0. During a test run the chemical feed pump delivered 800 mL of polymer solution during five minutes.

Known	Unknown
Polymer Solution, % = 1.5%	Polymer Feed, kg/day
Polymer Conc, mg/L = 15,000 mg/L	
Polymer Sp Gr = 1.0	
Vol Pumped, mL = 800 mL	
Time Pumped, min = 5 min	

Calculate the polymer fed by the chemical feed pump in kilograms of polymer per day.

$$\text{Polymer Feed, kg/day} = \frac{(\text{Poly Conc, mg/L})(\text{Vol Pumped, mL})(60\ min/hr)(24\ hr/day)}{(\text{Time Pumped, min})(1{,}000\ mL/L)(1{,}000\ mg/gm)(1{,}000\ gm/kg)}$$

$$= \frac{(15{,}000\ mg/L)(800\ mL)(60\ min/hr)(24\ hr/day)}{(5\ min)(1{,}000\ mL/L)(1{,}000\ mg/gm)(1{,}000\ gm/kg)}$$

$$= 3.5\ kg/day$$

EXAMPLE 7

A small chemical feed pump lowered the chemical solution in an 80-centimeter diameter tank 35 centimeters during an eight-hour period. Estimate the flow delivered by the pump in liters per minute.

Known	Unknown
Tank Diameter, cm = 80 cm	Flow, liters/min
Chemical Drop, cm = 35 cm	
Time, hr = 8 hr	

Calculate the flow in liters per minute.

$$\text{Flow, liters/min} = \frac{(\text{Area, sq m})(\text{Drop, m})(1{,}000\ \text{liters/cu m})}{(\text{Time, hr})(60\ min/hr)}$$

$$= \frac{(0.785)(0.8\ m)^2(0.35\ m)(1{,}000\ \text{liters/cu m})}{(8\ hr)(60\ min/hr)}$$

$$= 0.37\ \text{liters/min}$$

EXAMPLE 8

Determine the settings in percent stroke on a chemical feed pump for various doses of a chemical in milligrams per liter. (The chemical could be chlorine, polymer, potassium permanganate, or any other chemical solution fed by a pump.) The pump delivering water to be treated pumps at a flow rate of 25 liters per second. The solution strength of the chemical being pumped is five percent. The chemical feed pump has a maximum capacity of 250 milliliters per minute at a setting of 100 percent capacity.

Known	Unknown
Pump Flow, L/sec = 25 L/sec	Settings, % stroke for various doses in mg/L
Solution Strength, % = 5%	
Feed Pump, mL/min = 250 mL/min (100% stroke)	

1. Change the chemical solution strength from a percent to milligrams of chemical per liter of solution. A 5 percent solution means we have 5 milligrams of chemical in a solution of water and chemical weighing 100 milligrams.

$$\text{Chemical Solution, mg/L} = \frac{(\text{Chemical, mg})(1{,}000{,}000\ mg/L)}{\text{Chemical and Water, mg}}$$

$$= \frac{(5\ mg)(1{,}000{,}000\ mg/L)}{100\ mg}$$

$$= 50{,}000\ mg/L$$

2. Calculate the chemical feed in kilograms per day for a chemical dose of 0.5 milligrams per liter. We are going to assume various chemical doses of 0.5, 1.0, 1.5, 2.0, 2.5 mg/L, and upward so that if we know the desired chemical dose, we can easily determine the setting (percent stroke) on the chemical feed pump.

$$\text{Chemical Feed, kg/day} = \frac{(\text{Flow, L/sec})(\text{Dose, mg/L})(60\ sec/min)(60\ min/hr)(24\ hr/day)}{1{,}000{,}000\ mg/kg}$$

$$= \frac{(25\ L/sec)(0.5\ mg/L)(60\ sec/min)(60\ min/hr)(24\ hr/day)}{1{,}000{,}000\ mg/kg}$$

$$= 1.1\ kg/day$$

3. Determine the desired flow from the chemical feed pump in milliliters per minute.

$$\text{Feed Pump,} \atop \text{m}L\text{/min} = \frac{\text{Chemical Feed, kg/day}}{\text{Chemical Solution, mg/}L}$$

$$= \frac{(1.1 \text{ kg/day})(1,000 \text{ m}L/L)(1,000,000 \text{ mg/kg})}{(50,000 \text{ mg/}L)(24 \text{ hr/day})(60 \text{ min/hr})}$$

$$= 15.3 \text{ m}L\text{/min}$$

4. Determine the setting in the chemical feed pump as a percent. In this case we want to know the setting as a percent of the pump stroke.

$$\text{Setting, \%} = \frac{(\text{Desired Feed Pump, m}L\text{/min})(100\%)}{\text{Maximum Feed Pump, m}L\text{/min}}$$

$$= \frac{(15.3 \text{ m}L\text{/min})(100\%)}{250 \text{ m}L\text{/min}}$$

$$= 6.1\%$$

5. If we changed the chemical dose in Step 2 from 0.5 mg/L to 1.0 mg/L and other higher doses and repeated the remainder of the steps, we could calculate the data in Table A.2.

TABLE A.2 SETTINGS FOR CHEMICAL FEED PUMP

PUMP FLOW, L/sec   = 25 L/sec
SOLUTION STRENGTH, % = 5.0%

Chemical Dose, mg/L	Chemical Feed, kg/day	Feed Pump, mL/min	Pump Setting, % stroke
0.5	1.1	15.3	6.1
1.0	2.2	30.6	12.2
2.0	4.3	59.7	23.9
4.0	8.6	119.4	47.8
6.0	13.0	180.6	72.2
8.0	17.3	240.3	96.0

6. The data in Table A.2 could be plotted to produce a chart similar to Figure A.3 (page 619). For any desired chemical dose in milligrams per liter, the chart could be used to determine the necessary chemical feed pump setting.

A.152 Reservoir Management and Intake Structures

EXAMPLE 9

The volume of a reservoir is estimated to be 581,000 cubic feet. Estimate the volume in cubic meters.

Known	Unknown
Reservoir Vol, cu ft = 581,000 cu ft	Reservoir Vol, cu m

Estimate the reservoir volume in cubic meters.

$$\text{Reservoir Vol,} \atop \text{cu m} = \frac{\text{Reservoir Vol, cu ft}}{(3.281 \text{ ft/m})^3}$$

$$= \frac{581,000 \text{ cu ft}}{(3.281 \text{ ft/m})^3}$$

$$= 16,450 \text{ cu m}$$

EXAMPLE 10

A reservoir has a surface area of 0.75 hectare and an average depth of seven meters. Estimate the volume of the reservoir in cubic meters.

Known	Unknown
Surface Area, ha = 0.75 ha	Reservoir Vol, cu m
Depth, m = 7 m	

Estimate the reservoir volume in cubic meters.

$$\text{Reservoir Vol, cu m} = (\text{Surface Area, ha})(\text{Depth, m})(10,000 \text{ sq m/ha})$$

$$= (0.75 \text{ ha})(7 \text{ m})(10,000 \text{ sq m/ha})$$

$$= 52,500 \text{ cu m}$$

EXAMPLE 11

A reservoir has a surface area of 0.48 hectare and the desired dose of copper sulfate is three kilograms per hectare. How many kilograms of copper sulfate will be needed?

Known	Unknown
Surface Area, ha = 0.48 ha	Copper Sulfate, kg
Dose, kg/ha = 3 kg/ha	

Calculate the kilograms of copper sulfate needed.

$$\text{Copper Sulfate, kg} = (\text{Surface Area, ha})(\text{Dose, kg/ha})$$

$$= (0.48 \text{ ha})(3 \text{ kg/ha})$$

$$= 1.44 \text{ kg Copper Sulfate}$$

EXAMPLE 12

The volume of a reservoir is estimated to be 16,000 cubic meters. The desired chemical dose is 0.5 mg/L. Estimate the chemical dose in kilograms.

Known	Unknown
Reservoir Vol, cu m = 16,000 cu m	Chemical Dose, kg
Chemical Dose, mg/L = 0.5 mg/L	

Estimate the chemical dose in kilograms.

$$\text{Chemical Dose, kg} = \frac{(\text{Volume, cu m})(\text{Dose, mg/}L)(1,000 \text{ }L\text{/cu m})}{(1,000 \text{ mg/gm})(1,000 \text{ gm/kg})}$$

$$= \frac{(16,000 \text{ cu m})(0.5 \text{ mg/}L)(1,000 \text{ }L\text{/cu m})}{(1,000 \text{ mg/gm})(1,000 \text{ gm/kg})}$$

$$= 8 \text{ kg}$$

EXAMPLE 13

The volume of a reservoir is estimated to be 16,000 cubic meters. The desired dose of copper is 0.5 mg/L and the copper content of the copper sulfate to be used is 25 percent. How many kilograms of copper sulfate will be needed?

Known	Unknown
Reservoir Volume, cu m = 16,000 cu m	Copper Sulfate, kg
Copper Dose, mg/L = 0.5 mg/L	
Copper, % = 25%	

Calculate the kilograms of copper sulfate that will be needed.

$$\text{Copper Sulfate, kg} = \frac{(\text{Vol, cu m})(\text{Dose, mg/}L)(1{,}000 \text{ }L/\text{cu m})(100\%)}{(1{,}000 \text{ mg/gm})(1{,}000 \text{ gm/kg})(\text{Copper, \%})}$$

$$= \frac{(16{,}000 \text{ cu m})(0.5 \text{ mg/}L)(1{,}000 \text{ }L/\text{cu m})(100\%)}{(1{,}000 \text{ mg/gm})(1{,}000 \text{ gm/kg})(25\%)}$$

$$= 32 \text{ kg}$$

A.153 Coagulation and Flocculation

EXAMPLE 14

A polymer feed pump delivers 600 liters per day containing a five percent polymer solution with a specific gravity of 1.02. Estimate the polymer delivered in kilograms per day.

Known	Unknown
Flow, liters/day = 600 liters/day	Polymer, kg/day
Polymer, % = 5%	
Polymer, mg/L = 50,000 mg/L	
Sp Gr = 1.02	

Estimate the polymer delivered in kilograms per day.

SOLUTION 1

$$\text{Polymer, kg/day} = \frac{(\text{Polymer Flow, }L/\text{day})(1 \text{ kg/}L)(\text{Polymer, \%})(\text{Sp Gr})}{100\%}$$

$$= \frac{(600 \text{ }L/\text{day})(1 \text{ kg/}L)(5\%)(1.02)}{100\%}$$

$$= 30.6 \text{ kg/day}$$

SOLUTION 2

$$\text{Polymer, kg/day} = \frac{(\text{Polymer Flow, }L/\text{day})(\text{Polymer, mg/}L)(\text{Sp Gr})}{(1{,}000 \text{ mg/gm})(1{,}000 \text{ gm/kg})}$$

$$= \frac{(600 \text{ }L/\text{day})(50{,}000 \text{ mg/}L)(1.02)}{(1{,}000 \text{ mg/gm})(1{,}000 \text{ gm/kg})}$$

$$= 30.6 \text{ kg/day} \qquad \textit{SAME ANSWER!}$$

EXAMPLE 15

Estimate the actual alum dose in milligrams per liter if a plant treats a raw water with a flow of 7.0 ML/day. The alum feed rate is 60 kg per day.

Known	Unknown
Flow, ML/day = 7.0 ML/day	Alum Dose, mg/L
Alum Feed, kg/day = 60 kg/day	

Calculate the alum dose in milligrams per liter.

BASIC FORMULA

$$\text{Chemical Feed, kg/day} = (\text{Flow, M}L/\text{day})(\text{Dose, mg/}L)(1 \text{ kg/M mg})$$

or

$$\text{Alum Dose, mg/}L = \frac{\text{Alum Feed, kg/day}}{(\text{Flow, M}L/\text{day})(1 \text{ kg/M mg})}$$

$$= \frac{60 \text{ kg/day}}{(7.0 \text{ M}L/\text{day})(1 \text{ kg/M mg})}$$

$$= 8.6 \text{ mg/}L$$

EXAMPLE 16

Determine the strength of a polymer solution as a percent if 80 grams of dry polymer are mixed with 16 liters of water.

Known	Unknown
Dry Polymer, gm = 80 gm	Polymer Solution, %
Volume Water, L = 16 L	

1. Convert the sixteen liters of water to grams.

$$\text{Water, gm} = (\text{Volume Water, }L)(1{,}000 \text{ gm/}L)$$

$$= (16 \text{ }L)(1{,}000 \text{ gm/}L)$$

$$= 16{,}000 \text{ gm}$$

2. Calculate the polymer solution as a percent.

$$\text{Polymer, \%} = \frac{(\text{Dry Polymer, gm})(100\%)}{(\text{Dry Polymer, gm + Water, gm})}$$

$$= \frac{(80 \text{ gm})(100\%)}{(80 \text{ gm} + 16{,}000 \text{ gm})}$$

$$= 0.50\%$$

EXAMPLE 17

Liquid polymer is supplied to a water treatment plant as a ten percent solution. How many liters of liquid polymer should be mixed in a tank with water to produce 750 liters at 0.8 percent polymer solution?

Known

Liquid Polymer, %   = 10%

Polymer Solution, % = 0.8%

Volume of Polymer
Solution, L   = 750 L

Unknown

Volume of Liquid
Polymer, L

Calculate the volume of liquid polymer in liters.

$$\text{Liquid Polymer, } L = \frac{(\text{Polymer Solution, \%})(\text{Vol of Solution, } L)}{\text{Liquid Polymer, \%}}$$

$$= \frac{(0.8\%)(750\ L)}{10\%}$$

$$= 60 \text{ liters}$$

A.154   Sedimentation

EXAMPLE 18

Estimate the detention time in hours for a 10-meter diameter circular clarifier when the flow is 2,000 cubic meters per day. The clarifier is three meters deep.

Known

Diameter, m   = 10 m

Depth, m   = 3 m

Flow, cu m/day = 2,000 cu m/day

Unknown

Detention Time, hr

1. Calculate the clarifier volume in cubic meters.

Volume, cu m = (Area, sq m)(Depth, m)

$$= (0.785)(10\ m)^2(3\ m)$$

$$= 235.5 \text{ cu m}$$

2. Estimate the detention time of the clarifier in hours.

$$\text{Detention Time, hr} = \frac{(\text{Volume, cu m})(24 \text{ hr/day})}{\text{Flow, cu m/day}}$$

$$= \frac{(235.5 \text{ cu m})(24 \text{ hr/day})}{2,000 \text{ cu m/day}}$$

$$= 2.8 \text{ hr}$$

EXAMPLE 19

Estimate the overflow rate in millimeters per second for a rectangular sedimentation basin 6 meters wide and 12 meters long when the flow is 2,000 cubic meters per day.

Known

Width, m   = 6 m

Length, m   = 12 m

Flow, cu m/day = 2,000 cu m/day

Unknown

Overflow Rate, mm/sec

Determine the overflow rate for the basin in millimeters per second.

$$\text{Overflow Rate, mm/sec} = \frac{(\text{Flow, cu m/day})(1,000 \text{ mm/m})}{(\text{Area, sq m})(24 \text{ hr/day})(60 \text{ min/hr})(60 \text{ sec/min})}$$

$$= \frac{(2,000 \text{ cu m/day})(1,000 \text{ mm/m})}{(6\ m)(12\ m)(24 \text{ hr/day})(60 \text{ min/hr})(60 \text{ sec/min})}$$

$$= 0.32 \text{ mm/sec}$$

EXAMPLE 20

Estimate the flow velocity in meters per minute through a rectangular sedimentation basin 6 meters wide and 2.5 meters deep when the flow is 20 liters per second.

Known

Width, m   = 6 m

Depth, m   = 2.5 m

Flow, L/sec = 20 L/sec

Unknown

Flow Velocity, m/min

Estimate the flow velocity in meters per minute.

$$\text{Flow Velocity, m/min} = \frac{(\text{Flow, } L/\text{sec})(60 \text{ sec/min})}{(\text{Cross-Sectional Area, sq m})}$$

$$= \frac{(20\ L/\text{sec})(60 \text{ sec/min})}{(6\ m)(2.5\ m)(1,000\ L/\text{cu m})}$$

$$= 0.08 \text{ m/min}$$

EXAMPLE 21

Estimate the weir loading in cubic meters per day per meter of weir length for a ten-meter diameter circular clarifier treating a flow of 20 liters per second. The weir is located on the outside edge of the clarifier.

Known

Weir Diameter, m = 10 m

Flow, L/sec   = 20 L/sec

Unknown

Weir Loading, cu m/day/m

1. Calculate the weir length in meters.

Weir Length, m = π(Diameter, m)

$$= (3.14)(10\ m)$$

$$= 31.4 \text{ m}$$

2. Estimate the weir loading in cubic meters per day per meter.

$$\text{Weir Loading, cu m/day/m} = \frac{(\text{Flow, } L/\text{sec})(60 \text{ sec/min})(60 \text{ min/hr})(24 \text{ hr/day})}{(\text{Weir Length, m})(1,000\ L/\text{cu m})}$$

$$= \frac{(20\ L/\text{sec})(60 \text{ sec/min})(60 \text{ min/hr})(24 \text{ hr/day})}{(31.4\ m)(1,000\ L/\text{cu m})}$$

$$= 55 \text{ cu m/day/m}$$

A.155   Filtration

EXAMPLE 22

A 7.5-meter wide by 9-meter long rapid sand filter treats a flow of 125 liters per second. Calculate the filtration rate in liters per second per square meter and also in millimeters per second.

Known	Unknown
Width, m = 7.5 m	1. Filtration Rate, L/sec/sq m
Length, m = 9 m	2. Filtration Rate, mm/sec
Flow, L/sec = 125 L/sec	

1. Calculate the filtration rate in liters per second per square meter of filter surface area.

$$\text{Filtration Rate,} \atop L\text{/sec/sq m} = \frac{\text{Flow, liters/sec}}{\text{Surface Area, sq m}}$$

$$= \frac{125 \ L\text{/sec}}{(7.5 \text{ m})(9 \text{ m})}$$

$$= 1.85 \ L\text{/sec/sq m}$$

2. Calculate the filtration rate in millimeters per second.

$$\text{Filtration Rate,} \atop \text{mm/sec} = \frac{(\text{Flow,} L\text{/sec})(1,000 \text{ mm/m})}{(\text{Surface Area, sq m})(1,000 \ L\text{/cu m})}$$

$$= \frac{(125 \ L\text{/sec})(1,000 \text{ mm/m})}{(7.5 \text{ m})(9 \text{ m})(1,000 \ L\text{/cu m})}$$

$$= 1.85 \text{ mm/sec}$$

EXAMPLE 23

With the inflow water to a rapid sand filter shut off, the water is observed to drop 50 centimeters in nine minutes. What is the velocity of the water dropping in meters per minute?

Known	Unknown
Water Drop, cm = 50 cm	Velocity of Drop, m/min
Time of Drop, min = 9 min	

Calculate the velocity of the water drop in meters per minute.

$$\text{Velocity, m/min} = \frac{\text{Water Drop, cm}}{(100 \text{ cm/m})(\text{Time of Drop, min})}$$

$$= \frac{50 \text{ cm}}{(100 \text{ cm/m})(9 \text{ min})}$$

$$= 0.056 \text{ m/min}$$

EXAMPLE 24

Estimate the flow through a rapid sand filter in cubic meters per minute when the velocity of the water dropping is 0.05 meter per minute and the filter is 7.5 meters wide and 9 meters long.

Known	Unknown
Velocity of Drop, m/min = 0.05 m/min	Flow, cu m/min
Width, m = 7.5 m	
Length, m = 9 m	

Estimate the flow through the filter in cubic meters per minute.

$$\text{Flow, cu m/min} = (\text{Area, sq ft})(\text{Velocity, m/min})$$

$$= (7.5 \text{ m})(9 \text{ m})(0.05 \text{ m/min})$$

$$= 3.38 \text{ cu m/min}$$

EXAMPLE 25

Calculate the flow through a rapid sand filter in liters per second when the flow is 3.5 cubic meters per minute.

Known	Unknown
Flow, cu m/min = 3.5 cu m/min	Flow, L/sec

Calculate the flow through the filter in liters per second.

$$\text{Flow,} L\text{/sec} = \frac{(\text{Flow, cu m/min})(1,000 \ L\text{/cu m})}{60 \text{ sec/min}}$$

$$= \frac{(3.5 \text{ cu m/min})(1,000 \ L\text{/cu m})}{60 \text{ sec/min}}$$

$$= 58.3 \ L\text{/sec}$$

EXAMPLE 26

Calculate the flow through a rapid sand filter in liters per second when 6,000 cubic meters flowed through the filter during a 30-hour filter run.

Known	Unknown
Flow, cu m = 6,000 cu m	Flow, L/sec
Time, hr = 30 hr	

Calculate the flow through the filter in liters per second.

$$\text{Flow,} L\text{/sec} = \frac{(\text{Total Flow, cu m})(1,000 \ L\text{/cu m})}{(\text{Filter Run, hr})(60 \text{ min/hr})(60 \text{ sec/min})}$$

$$= \frac{(6,000 \text{ cu m})(1,000 \ L\text{/cu m})}{(30 \text{ hr})(60 \text{ min/hr})(60 \text{ sec/min})}$$

$$= 55.6 \ L\text{/sec}$$

EXAMPLE 27

Determine the Unit Filter Run Volume (UFRV) in liters per square meter for a filter eight meters long and five meters wide if the volume of water filtered between backwash cycles is eight megaliters.

Known	Unknown
Length, m = 8 m	UFRV, L/sq m
Width, m = 5 m	
Volume Filtered, ML = 8 ML	

Calculate the Unit Filter Run Volume in liters per square meter of filter surface area.

$$\text{UFRV,} L\text{/sq m} = \frac{\text{Volume Filtered,} L}{\text{Filter Surface Area, sq m}}$$

$$= \frac{8,000,000 \ L}{(8 \text{ m})(5 \text{ m})}$$

$$= 200,000 \ L\text{/sq m}$$

or $= 200 \text{ cu m/sq m}$

EXAMPLE 28

Determine the Unit Filter Run Volume (UFRV) in liters per square meter for a filter if the filtration rate was two millimeters per second during a 46-hour filter run.

Known		Unknown
Filtration Rate, mm/sec	= 2 mm/sec	UFRV, L/sq m
Filter Run, hr	= 46 hr	

Calculate the Unit Filter Run Volume in liters per square meter of filter surface area.

$$\text{UFRV,} \atop L/\text{sq m} = \frac{\text{(Filtration Rate, mm/sec)(Filter Run, hr)}}{\frac{(60 \text{ sec/min})(60 \text{ min/hr})(1{,}000 \text{ } L/\text{cu m})}{1{,}000 \text{ mm/m}}}$$

$$= \frac{(2 \text{ mm/sec})(46 \text{ hr})(60 \text{ sec/min})(60 \text{ min/hr})(1{,}000 \text{ } L/\text{cu m})}{1{,}000 \text{ mm/m}}$$

$$= 331{,}200 \text{ } L/\text{sq m}$$

EXAMPLE 29

Calculate the backwash flow required in cubic meters per day to backwash a 7.5-meter wide by 9-meter long filter if the desired backwash flow rate is 0.8 cubic meter per day per square meter.

Known		Unknown
Width, m	= 7.5 m	Backwash Flow, cu m/day
Length, m	= 9 m	
Backwash Rate, cu m/day/sq m	= 0.8 cu m/day/sq m	

Calculate the backwash flow in cubic meters per day.

$$\text{Backwash Flow,} \atop \text{cu m/day} = \text{(Filter Area, sq m)(Backwash Rate, cu m/day/sq m)}$$

$$= (7.5 \text{ m})(9 \text{ m})(0.8 \text{ cu m/day/sq m})$$

$$= 54 \text{ cu m/day}$$

EXAMPLE 30

Convert a filter backwash rate of 25 liters per minute per square meter to millimeters per minute of rise.

Known		Unknown
Backwash, L/min/sq m	= 25 L/min/sq m	Backwash, mm/min

Convert the backwash rate from L/min/sq m to mm/min.

$$\text{Backwash,} \atop \text{mm/min} = \frac{(\text{Backwash, } L/\text{min/sq m})(1{,}000 \text{ mm/m})}{1{,}000 \text{ } L/\text{cu m}}$$

$$= \frac{(25 \text{ } L/\text{min/sq m})(1{,}000 \text{ mm/m})}{1{,}000 \text{ } L/\text{cu m}}$$

$$= 25 \text{ mm/min}$$

EXAMPLE 31

Determine the volume or amount of water, in liters, required to backwash a filter if the backwash flow rate is 600 liters per second when the backwash time is seven minutes.

Known		Unknown
Backwash Flow Rate, L/sec	= 600 L/sec	Backwash Water, L
Backwash Time, min	= 7 min	

Calculate the volume of backwash water required in liters.

$$\text{Backwash} \atop \text{Water, } L = \text{(Backwash Flow, } L/\text{sec)(Backwash Time, min)(60 sec/min)}$$

$$= (600 \text{ } L/\text{sec})(7 \text{ min})(60 \text{ sec/min})$$

$$= 252{,}000 \text{ } L$$

or $\quad = 0.252 \text{ M}L$

EXAMPLE 32

During a filter run the total volume of water filtered was 50 megaliters. When the filter was backwashed, 250,000 liters of water was used. Calculate the percent of the product or finished water used for backwashing.

Known		Unknown
Water Filtered, ML	= 50 ML	Backwash, %
Backwash Water, L	= 250,000 L	

Calculate the percent of water used for backwashing.

$$\text{Backwash, \%} = \frac{(\text{Backwash Water, } L)(100\%)}{(\text{Water Filtered, M}L)(1{,}000{,}000/\text{M})}$$

$$= \frac{(250{,}000 \text{ } L)(100\%)}{(50 \text{ M}L)(1{,}000{,}000/\text{M})}$$

$$= 0.5\%$$

A.156 Disinfection

EXAMPLE 33

Calculate the chlorine dose in mg/L when a chlorinator is set to feed eight kilograms of chlorine in 24 hours. The flow is two megaliters per day.

Known		Unknown
Chlorinator Setting, kg/24 hr	= 8 kg/day	Chlorine Dose, mg/L
Flow, ML/day	= 2 ML/day	

Calculate the chlorine dose in milligrams per liter.

$$\text{Chlorine Dose,} \atop \text{mg/}L = \frac{(\text{Chlorine Feed, kg/day})(1 \text{ M mg/kg})}{\text{Flow, M}L/\text{day}}$$

$$= \frac{(8 \text{ kg/day})(1 \text{ M mg/kg})}{2 \text{ M}L/\text{day}}$$

$$= 4 \text{ mg/}L$$

EXAMPLE 34

Estimate the chlorine demand for a water in milligrams per liter if the chlorine dose is 2.9 mg/L and the chlorine residual is 0.6 mg/L.

Known	Unknown
Chlorine Dose, mg/L = 2.9 mg/L	Chlorine Demand, mg/L
Chlorine Residual, mg/L = 0.6 mg/L	

Estimate the chlorine demand of the water in milligrams per liter.

$$\text{Chlorine Demand, mg/L} = \text{Chlorine Dose, mg/L} - \text{Chlorine Residual, mg/L}$$

$$= 2.9 \text{ mg/L} - 0.6 \text{ mg/L}$$

$$= 2.3 \text{ mg/L}$$

EXAMPLE 35

Calculate the kilograms of chlorine used to disinfect water if 600 liters of hypochlorite as a 2.5 percent chlorine solution was used.

Known	Unknown
Hypochlorite, L = 600 L	Chlorine, kg
Hypochlorite, % = 2.5%	

Calculate the kilograms of chlorine used.

$$\text{Chlorine, kg} = \frac{(\text{Hypochlorite, } L)(1 \text{ kg/}L)(\text{Hypochlorite, \%})}{100\%}$$

$$= \frac{(600 \text{ } L)(1 \text{ kg/}L)(2.5\%)}{100\%}$$

$$= 15 \text{ kg Chlorine}$$

EXAMPLE 36

Estimate the flow pumped by a hypochlorinator in liters per minute if the hypochlorite solution is in a container with a diameter of 0.8 meter and the hypochlorite level drops 35 centimeters during a nine-hour period. The hypochlorinator operated continuously during the nine-hour period.

Known	Unknown
Diameter, m = 0.8 m	Hypochlorinator Flow, L/min
Drop, cm = 35 cm	
Time, hr = 9 hr	

Calculate the hypochlorinator flow in liters per minute.

$$\text{Flow, } L/\text{min} = \frac{(\text{Container Area, sq m})(\text{Drop, cm})(1,000 \text{ } L/\text{cu m})}{(\text{Time, hr})(100 \text{ cm/m})(60 \text{ min/hr})}$$

$$= \frac{(0.785)(0.8 \text{ m})^2(35 \text{ cm})(1,000 \text{ } L/\text{cu m})}{(9 \text{ hr})(100 \text{ cm/m})(60 \text{ min/hr})}$$

$$= 0.33 \text{ } L/\text{min}$$

EXAMPLE 37

Estimate the desired strength (as a percent chlorine) of a hypochlorite solution which is pumped by a hypochlorinator that delivers 0.45 cubic meter per day. The water being treated requires a chlorine dose of six kilograms of chlorine per day.

Known	Unknown
Hypochlorinator Flow, cu m/day = 0.45 cu m/day	Hypochlorite Strength, %
Chlorine Required, kg/day = 6 kg/day	

Estimate the desired hypochlorite strength as a percent chlorine.

$$\text{Hypochlorite Strength, \%} = \frac{(\text{Chlorine Required, kg/day})(100\%)}{(\text{Hypochlorinator Flow, cu m/day})(1,000 \text{ } L/\text{cu m})(1 \text{ kg/}L)}$$

$$= \frac{(6 \text{ kg/day})(100\%)}{(0.45 \text{ cu m/day})(1,000 \text{ } L/\text{cu m})(1 \text{ kg/}L)}$$

$$= 1.33\%$$

EXAMPLE 38

How many liters of water must be added to 40 liters of five percent hypochlorite solution to produce a 1.33 percent hypochlorite solution?

Known	Unknown
Hypochlorite, L = 40 L	Water Added, liters
Desired Hypo, % = 1.33%	
Actual Hypo, % = 5%	

Calculate the liters of water that must be added to produce a 1.33 percent hypochlorite solution.

$$\text{Water Added, } L \text{ (to hypochlorite solution)} = \frac{(\text{Hypo, } L)(\text{Hypo, \%}) - (\text{Hypo, } L)(\text{Desired Hypo, \%})}{\text{Desired Hypo, \%}}$$

$$= \frac{(40 \text{ } L)(5\%) - (40 \text{ } L)(1.33\%)}{1.33\%}$$

$$= \frac{200 - 53.2}{1.33}$$

$$= 110 \text{ liters}$$

A.157   Corrosion Control

EXAMPLE 39

Find the pH_S of a water at 10°C having a TDS of 100 mg/L, alkalinity of 40 mg/L, and a calcium hardness of 60 mg/L.

Known		Unknown
Water Temp, °C	= 10°C	pH_S
TDS, mg/L	= 100 mg/L	
Alky, mg/L	= 40 mg/L	
Ca Hardness, mg/L	= 60 mg/L as $CaCO_3$	

1. Find the formula values from the tables in Chapter 8, "Corrosion Control."

   From Table 8.2 for a water temperature of 10°C,

   $A = 2.20$

   From Table 8.3 for a TDS of 100 mg/L,

   $B = 9.75$

   From Table 8.4 for Ca of 60 mg/L and Alky of 40 mg/L,

   $\log(Ca^{2+}) = 1.78$

   $\log(Alky) = 1.60$

2. Calculate pH_S.

   $pH_S = A + B - \log(Ca^{2+}) - \log(Alky)$

   $\quad\ = 2.20 + 9.75 - 1.78 - 1.60$

   $\quad\ = 8.57$

EXAMPLE 40

Calculate the Langelier Index for a water with a calculated pH_S value of 8.69 and an actual pH of 8.5.

Known	Unknown
pH_S = 8.69	Langelier Index
pH = 8.5	

Calculate the Langelier Index.

Langelier Index $= pH - pH_S$

$\qquad\qquad\qquad = 8.5 - 8.69$

$\qquad\qquad\qquad = -0.19$

Since the Langelier Index is negative, the water is corrosive.

A.158   Plant Operation

EXAMPLE 41

Estimate the average use of chlorine in kilograms per day based on actual use of chlorine for one week as shown below.

Day	Sun	Mon	Tue	Wed	Thu	Fri	Sat
Chlorine Used, kg	23	37	35	31	32	36	24

Known	Unknown
Chlorine Used, kg/day	Average Chlorine Used, kg/day

Estimate the average chlorine use in kilograms of chlorine per day.

$$\begin{aligned}\text{Average}\atop\text{Chlorine}\atop\text{Use, kg/day} &= \frac{\text{Sum of Chlorine Used Each Day, kg}}{\text{Total Time, days}}\\[4pt]
&= \frac{23\ kg + 37\ kg + 35\ kg + 31\ kg + 32\ kg + 36\ kg + 24\ kg}{7\ \text{days}}\\[4pt]
&= \frac{218\ kg}{7\ \text{days}}\\[4pt]
&= 31.1\ \text{kg/day}\end{aligned}$$

EXAMPLE 42

A water treatment plant has five 68-kilogram chlorine cylinders in storage. The plant uses an average of 12 kilograms of chlorine per day. How many days' supply of chlorine is in storage?

Known	Unknown
Chlorine Cylinders = 5 cylinders	Supply of Chlorine, days
Cylinder Wt, kg/cyl = 68 kg/cyl	
Avg Use, kg/day = 12 kg/day	

Calculate the available supply of chlorine in storage in days.

$$\begin{aligned}\text{Supply of Chlorine,}\atop\text{days} &= \frac{(\text{Cylinder Wt, kg/cyl})(\text{No. of Cylinders})}{\text{Avg Use, kg/day}}\\[4pt]
&= \frac{(68\ \text{kg/cyl})(5\ \text{Cylinders})}{12\ \text{kg/day}}\\[4pt]
&= 28\ \text{days}\end{aligned}$$

A.159   Laboratory Procedures

EXAMPLE 43

Convert the temperature of water from 65° Fahrenheit to degrees Celsius.

Known	Unknown
Temp, °F = 65°F	Temp, °C

Change 65°F to degrees Celsius.

$$\begin{aligned}\text{Temperature, °C} &= \frac{5}{9}(\text{°F} - 32\text{°F})\\[4pt]
&= \frac{5}{9}(65\text{°F} - 32\text{°F})\\[4pt]
&= 18.3\text{°C}\end{aligned}$$

EXAMPLE 44

Convert a water temperature of 12° Celsius to degrees Fahrenheit.

Known	Unknown
Temp, °C = 12°C	Temp, °F

Change 12°C to degrees Fahrenheit.

$$\text{Temperature, °F} = \frac{9}{5}\,(\text{°C}) + 32\text{°F}$$

$$= \frac{9}{5}\,(12\text{°C}) + 32\text{°F}$$

$$= 53.6\text{°F}$$

EXAMPLE 45

Results from a five-tube three dilution most probable number (MPN) fermentation-tube test are given below. Calculate the MPN per 100 mL.

Sample, mL	10	1	0.1
Number of Positive	3	1	0

Known	Unknown
Results of Fermentation Tube Test	MPN/100 mL

Calculate the MPN per 100 mL.

$$\text{MPN/100 m}L = \frac{(\text{No. of positive tubes})(100)}{\sqrt{\left(\begin{array}{c}\text{m}L\text{ sample in}\\\text{negative tubes}\end{array}\right) \times \left(\begin{array}{c}\text{m}L\text{ sample in}\\\text{all tubes}\end{array}\right)}}$$

$$= \frac{(4)(100)}{\sqrt{\begin{array}{c}[(2)(10\text{ m}L)+(4)(1\text{ m}L)+(5)(0.1\text{ m}L)]\\ [(5)(10\text{ m}L)+(5)(1\text{ m}L)+(5)(0.1\text{ m}L)]\end{array}}}$$

$$= \frac{400}{\sqrt{(24.5)(55.5)}}$$

$$= 10.8$$

or

$$= 11 \text{ MPN/100 m}L$$

NOTE: See Example 46 of Section A.139, "Laboratory Procedures," for examples of how to calculate the mean, median, and geometric mean.

A.16 CALCULATION OF CT VALUES

The Surface Water Treatment Rule (SWTR) requires all surface water systems and all systems using groundwater under the influence of surface water to achieve 3-log removal of *Giardia* and 4-log removal or inactivation of viruses. This level of treatment can be accomplished by disinfection alone if the source water is relatively free of turbidity and the system meets other specific conditions. Or, the treatment requirements can be met by a combination of filtration and disinfection.

When both filtration and disinfection are used, it is first necessary to know the effectiveness of filtration in removing *Giardia* and viruses before it will be possible to determine the level of disinfection needed to reach the 3-log and 4-log treatment levels. Research studies of conventional, direct, slow sand, and diatomaceous earth filtration have measured the ability of each of these systems to remove *Giardia* cysts. For example, a well-operated conventional filtration system should be able to achieve 2.5-log removal of *Giardia* and 2.0-log removal/inactivation of viruses. Table A.3 lists the treatment removal efficiencies of each of these four types of filtration systems. Using the example above, if filtration achieves 2.5-log removal, disinfection processes will need to achieve at least 0.5-log *Giardia* removal to meet the SWTR requirement of 3-log *Giardia* removal.[1]

TABLE A.3 LOG REMOVAL EFFICIENCY

Type of Filtration	Treatment Removal Credit (Log Removals)		Required Disinfection (Log Inactivations*)	
	Giardia	Viruses	*Giardia*	Viruses
Conventional	2.5	2.0	0.5	2.0
Direct	2.0	1.0	1.0	3.0
Slow Sand	2.0	2.0	1.0	2.0
Diatomaceous Earth	2.0	1.0	1.0	3.0

* Assumed 3-log *Giardia* and 4-log virus removal and/or inactivation (can be increased based on sanitary hazards in watershed).

Not all disinfectants are equally effective at inactivating *Giardia* and viruses. Research has shown that a free chlorination system that achieves 3-log *Giardia* inactivation also achieves 4-log virus inactivation. On the other hand, a system using chloramination must disinfect to the 4-log level for virus inactivation before the water will be considered safe and free of harmful organisms.

The effectiveness of disinfection by free chlorination also depends on pH and temperature. The required CT value for any given free chlorine residual increases with increasing pH and decreasing temperature. Tables A.4 through A.9 (pages 639-644) provide the CT values for inactivation of *Giardia* cysts by free chlorine, and the remainder of this section provides examples of how to calculate disinfection CT (concentration x contact time) values using free chlorination.

EXAMPLE 46

A 10-MGD direct filtration plant applies free chlorine as a disinfectant. The disinfectant has a contact time of 26 minutes under peak flow conditions. The pH of the water is 7.5 and the temperature is 10°C. The free chlorine residual is 2.0 mg/L. Table A.3 indicates that direct filtration is capable of achieving 2-log (99 percent) removal of *Giardia* cysts. Therefore 1-log

---

[1] *The 1996 amendments to the Safe Drinking Water Act require EPA to develop additional regulations to provide greater protection of the public from microorganisms such as Cryptosporidium in drinking water. Log removal levels listed in this section are current as of publication of this manual (2004) but may change in the next few years as new rules are promulgated. Please refer to Volume II, Chapter 22, "Drinking Water Regulations," for additional information about the types of regulations being developed and EPA's schedule for completion. Also, drinking water supply and distribution utilities and operators are strongly urged to stay in close contact with their state and federal regulatory agencies to keep informed of new regulations such as the Interim Enhanced Surface Water Treatment Rule (IESWTR), the Disinfectants/Disinfection By-Products (D/DBP) Rule, and new filter backwash recycling rules.*

(90 percent or CT 90) *Giardia* inactivation must be achieved by disinfection.

SOLUTION

1. Using Table A.6 with a pH of 7.5, a temperature of 10°C, 1.0-log inactivation, and a free chlorine residual of 2.0 mg/L, find the required CT value of 50 mg/L-min.

2. Calculate the CT provided.

CT Provided, mg/L-min = (C, mg/L)(T, min)

$$= (2 \text{ mg/L})(26 \text{ min})$$

$$= 52 \text{ mg/L-min}$$

Since 52 mg/L-min is greater than the required 50 mg/L-min, the free chlorine residual is adequate for the peak flow conditions.

3. Using the above procedure, assume appropriate free chlorine residuals for lower flows and determine if the CT values provided are greater than the required CT 90 (1-log) values.

Flow, MGD	Contact Time, min	Free Chlorine Residual (assumed), mg/L	CT Provided, mg/L-min	CT 90 Required,* mg/L-min
10	26	2	52	50
7.5	39	1.4	55	47
5	52	1	52	45
2.5	104	0.5	52	43

* From Table A.6 (CT 90 = 1-log inactivation).

Changing the disinfectant dose to meet the required CT values as the flow changes is not a practical operational procedure. The suggested operational procedure is to adjust the disinfectant dose to meet the highest flow requirement with various flow ranges.

Flow Range, MGD	Free Chlorine Residual, mg/L
7.5 – 10	2.0
5.0 – 7.5	1.4
2.5 – 5.0	1.0

EXAMPLE 47

The 10-MGD plant in Example 46 is a conventional filtration plant and only needs to achieve 0.5-log *Giardia* inactivation (CT 70). Rework Example 46 for 0.5-log *Giardia* inactivation assuming a free chlorine residual of 1.0 mg/L for peak flow conditions.

Flow, MGD	Contact Time, min	Free Chlorine Residual (assumed), mg/L	CT Provided, mg/L-min	CT 70 Required, mg/L-min
10	26	1.0	26	22
7.5	39	0.7	27	22
5	52	0.5	26	21
2.5	104	0.4	42	21

EXAMPLE 48

A conventional water treatment plant consists of flocculation, sedimentation, filtration, and a clear well. Prechlorination is applied before flocculation and postchlorination after the clear well. A dye tracer study has been done to determine the chlorine contact times (T_{10}, time for 10 percent of the tracer to pass through a tank). Assume a water temperature of 15°C and a pH of 7.5. The plant needs to achieve 0.5-log inactivation of *Giardia* cysts. Determine the fraction of the CT value (CTF) the plant provides the first customer in the distribution system given the following information:

Process	Cl$_2$ Free Residual, mg/L	T_{10}, min
Flocculation	0.8	15
Sedimentation	0.4	75
Filtration	0.2	15
Clear Well	0.6	120
Pipe to Customer	0.4	20

SOLUTION

Calculate the actual CT value for each process, determine the CT required from Table A.7, calculate the CT actual/CT required ratio and sum the ratios. The sum of the ratios is the CTF value and should be greater than 1.0.

Process	Cl Conc, mg/L	T_{10}, min	CT Actual	CT Required*	CT Ratio
Floc.	0.8 mg/L	15 min	12	15	0.80
Sed.	0.4 mg/L	75 min	30	14	2.14
Filt.	0.2 mg/L	15 min	3	14	0.21
CW	0.6 mg/L	120 min	72	14	5.14
Pipe	0.4 mg/L	20 min	8	14	0.57
					8.86

* From Table A.7.

EXAMPLE 49

If the sum of the CT actual/CT required ratios for the flocculation, sedimentation, and filtration processes for Example 48 was 0.72, determine the free chlorine residual from the clear well necessary to achieve a total CTF (CT fraction) value greater than 1.0 for the plant only (neglect pipe CT). Assume the T_{10} value (actual detention time for the clear well) is 30 minutes.

SOLUTION

Determine the minimum CT ratio for the clear well that will produce a total CTF value greater than 1.0 for the plant.

Min. Clear Well CT Ratio Needed = Min. CTF – CTF Provided

$$= 1.0 - 0.72$$

$$= 0.28$$

Determine the CT ratios (CT actual/CT required) for various chlorine residuals in the clear well. Start with the minimum free chlorine residual for the clear well.

Cl$_2$ Free Residual, mg/L	T_{10}, min	CT Actual	CT Required	CT Ratio
0.2 mg/L	30 min	6	14	0.43
0.4 mg/L	30 min	12	14	0.86
0.6 mg/L	30 min	18	14	1.29
0.8 mg/L	30 min	24	15	1.60

Since the free chlorine residual of 0.2 mg/L produces a CT ratio of 0.43, which is greater than 0.28, the free chlorine residual of 0.2 mg/L will be adequate according to these calculations.

EXAMPLE 50

If prechlorination were not practiced in Example 49, what free chlorine residual would be necessary for the clear well alone to provide adequate disinfection?

SOLUTION

The required CT value is 14 mg/L-min.

$$\text{Free Chlorine Residual, mg}/L = \frac{CT, \text{ mg}/L\text{-min}}{T_{10}, \text{ min}}$$

$$= \frac{14 \text{ mg}/L\text{-min}}{30 \text{ min}}$$

$$= 0.5 \text{ mg}/L \text{ or greater}$$

A.17   CALCULATION OF LOG REMOVALS

Regulations may require the calculation of log removals for inactivation of *Giardia* cysts, viruses, or particle counts. How are log removals calculated? The following example illustrates two methods of calculating log removals.

EXAMPLE 51

Calculate the log removal of 5- to 15-micron particles per milliliter if the influent particle count to a water treatment filter reported 2,100 particles in the 5- to 15-microns range per milliliter of water and the filter effluent reported 30 particles in the 5- to 15-microns range per milliliter of filtered water.

Known	Unknown
Influent, particles/mL = 2,100 particles/mL	Log Removal, particles/mL
Effluent, particles/mL = 30 particles/mL	

Calculate the log removal of particles in the 5- to 15-microns range per mL by the filter.

PROCEDURE 1

$$\begin{aligned} \text{Log Removal, particles/m}L &= \text{Log Influent, particles/m}L - \text{Log Effluent, particles/m}L \\ &= \text{Log 2,100 particles/m}L - \text{Log 30 particles/m}L \\ &= 3.3 - 1.5 \\ &= 1.8 \end{aligned}$$

PROCEDURE 2

$$\begin{aligned} \text{Log Removal, particles/m}L &= \text{Log}\left[\frac{\text{Influent, particles/m}L}{\text{Effluent, particles/m}L}\right] \\ &= \text{Log}\left[\frac{2{,}100 \text{ particles/m}L}{30 \text{ particles/m}L}\right] \\ &= \text{Log } 70 \\ &= 1.8 \end{aligned}$$

TABLE A.4 CT VALUES FOR INACTIVATION OF *GIARDIA* CYSTS BY FREE CHLORINE AT 0.5° C OR LOWER (1)

CHLORINE CONCENTRATION (mg/L)	pH<=6 Log Inactivations						pH=6.5 Log Inactivations						pH=7.0 Log Inactivations						pH=7.5 Log Inactivations					
	0.5	1.0	1.5	2.0	2.5	3.0	0.5	1.0	1.5	2.0	2.5	3.0	0.5	1.0	1.5	2.0	2.5	3.0	0.5	1.0	1.5	2.0	2.5	3.0
<=0.4	23	46	69	91	114	137	27	54	82	109	136	163	33	65	98	130	163	195	40	79	119	158	198	237
0.6	24	47	71	94	118	141	28	56	84	112	140	168	33	67	100	133	167	200	40	80	120	159	199	239
0.8	24	48	73	97	121	145	29	57	86	115	143	172	34	68	103	137	171	205	41	82	123	164	205	246
1	25	49	74	99	123	148	29	59	88	117	147	176	35	70	105	140	175	210	42	84	127	169	211	253
1.2	25	51	76	101	127	152	30	60	90	120	150	180	36	72	108	143	179	215	43	86	130	173	216	259
1.4	26	52	78	103	129	155	31	61	92	123	153	184	37	74	111	147	184	221	44	89	133	177	222	266
1.6	26	52	79	105	131	157	31	63	95	126	158	189	38	75	113	151	188	226	46	91	137	182	228	273
1.8	27	54	81	108	135	162	32	64	97	129	161	193	39	77	116	154	193	231	47	93	140	186	233	279
2	28	55	83	110	138	165	33	66	99	131	164	197	39	79	118	157	197	236	48	95	143	191	238	286
2.2	28	56	85	113	141	169	34	67	101	134	168	201	40	81	121	161	202	242	50	99	149	198	248	297
2.4	29	57	86	115	143	172	34	68	103	137	171	205	41	82	124	165	206	247	50	99	149	199	248	298
2.6	29	58	88	117	146	175	35	70	105	139	174	209	42	84	126	168	210	252	51	101	152	203	253	304
2.8	30	59	89	119	148	178	36	71	107	142	178	213	43	86	129	171	214	257	52	103	155	207	258	310
3	30	60	91	121	151	181	36	72	109	145	181	217	44	87	131	174	218	261	53	105	158	211	263	316

CHLORINE CONCENTRATION (mg/L)	pH=8.0 Log Inactivations						pH=8.5 Log Inactivations						pH<=9.0 Log Inactivations					
	0.5	1.0	1.5	2.0	2.5	3.0	0.5	1.0	1.5	2.0	2.5	3.0	0.5	1.0	1.5	2.0	2.5	3.0
<=0.4	46	92	139	185	231	277	55	110	165	219	274	329	65	130	195	260	325	390
0.6	48	95	143	191	238	286	57	114	171	228	285	342	68	136	204	271	339	407
0.8	49	98	148	197	246	295	59	118	177	236	295	354	70	141	211	281	352	422
1	51	101	152	203	253	304	61	122	183	243	304	365	73	146	219	291	364	437
1.2	52	104	157	209	261	313	63	125	188	251	313	376	75	150	226	301	376	451
1.4	54	107	161	214	268	321	65	129	194	258	323	387	77	155	232	309	387	464
1.6	55	110	165	219	274	329	66	132	199	265	331	397	80	159	239	318	398	477
1.8	56	113	169	225	282	338	68	136	204	271	339	407	82	163	245	326	408	489
2	58	115	173	231	288	346	70	139	209	278	348	417	83	167	250	333	417	500
2.2	59	118	177	235	294	353	71	142	213	284	355	426	85	170	256	341	426	511
2.4	60	120	181	241	301	361	73	145	218	290	363	435	87	174	261	348	435	522
2.6	61	123	184	245	307	368	74	148	222	296	370	444	89	178	267	355	444	533
2.8	63	125	188	250	313	375	75	151	226	301	377	452	91	181	272	362	453	543
3	64	127	191	255	318	382	77	153	230	307	383	460	92	184	276	368	460	552

Notes:

(1) CT = CT for 3-log inactivation

99.9

TABLE A.5   CT VALUES FOR INACTIVATION OF *GIARDIA* CYSTS BY FREE CHLORINE AT 5° C (1)

CHLORINE CONCENTRATION (mg/L)	pH ≤ 6 Log Inactivations						pH = 6.5 Log Inactivations						pH = 7.0 Log Inactivations						pH = 7.5 Log Inactivations					
	0.5	1.0	1.5	2.0	2.5	3.0	0.5	1.0	1.5	2.0	2.5	3.0	0.5	1.0	1.5	2.0	2.5	3.0	0.5	1.0	1.5	2.0	2.5	3.0
<=0.4	16	32	49	65	81	97	20	39	59	78	98	117	23	46	70	93	116	139	28	55	83	111	138	166
0.6	17	33	50	67	83	100	20	40	60	80	100	120	24	48	72	95	119	143	29	57	86	114	143	171
0.8	17	34	52	69	86	103	20	41	61	81	102	122	24	49	73	97	122	146	29	58	88	117	146	175
1	18	35	53	70	88	105	21	42	63	83	104	125	25	50	75	99	124	149	30	60	90	119	149	179
1.2	18	36	54	71	89	107	21	42	64	85	106	127	25	51	76	101	127	152	31	61	92	122	153	183
1.4	18	36	55	73	91	109	22	43	65	87	108	130	26	52	78	103	129	155	31	62	94	125	156	187
1.6	19	37	56	74	93	111	22	44	66	88	110	132	26	53	79	105	132	158	32	64	96	128	160	192
1.8	19	38	57	76	95	114	23	45	68	90	113	135	27	54	81	108	135	162	33	65	98	131	163	196
2	19	39	58	77	97	116	23	46	69	92	115	138	28	55	83	110	138	165	33	67	100	133	167	200
2.2	20	39	59	79	98	118	23	47	70	93	117	140	28	56	85	113	141	169	34	68	102	136	170	204
2.4	20	40	60	80	100	120	24	48	72	95	119	143	29	57	86	115	143	172	35	70	105	139	174	209
2.6	20	41	61	81	102	122	24	49	73	97	122	146	29	58	88	117	146	175	36	71	107	142	178	213
2.8	21	41	62	83	103	124	25	49	74	99	123	148	30	59	89	119	148	178	36	72	109	145	181	217
3	21	42	63	84	105	126	25	50	76	101	126	151	30	61	91	121	152	182	37	74	111	147	184	221

CHLORINE CONCENTRATION (mg/L)	pH = 8.0 Log Inactivations						pH = 8.5 Log Inactivations						pH ≤ 9.0 Log Inactivations					
	0.5	1.0	1.5	2.0	2.5	3.0	0.5	1.0	1.5	2.0	2.5	3.0	0.5	1.0	1.5	2.0	2.5	3.0
<=0.4	33	66	99	132	165	198	39	79	118	157	197	236	47	93	140	186	233	279
0.6	34	68	102	136	170	204	41	81	122	163	203	244	49	97	146	194	243	291
0.8	35	70	105	140	175	210	42	84	126	168	210	252	50	100	151	201	251	301
1	36	72	108	144	180	216	43	87	130	173	217	260	52	104	156	208	260	312
1.2	37	74	111	147	184	221	45	89	134	178	223	267	53	107	160	213	267	320
1.4	38	76	114	151	189	227	46	91	137	183	228	274	55	110	165	219	274	329
1.6	39	77	116	155	193	232	47	94	141	187	234	281	56	112	169	225	281	337
1.8	40	79	119	159	198	238	48	96	144	191	239	287	58	115	173	230	288	345
2	41	81	122	162	203	243	49	98	147	196	245	294	59	118	177	235	294	353
2.2	41	83	124	165	207	248	50	100	150	200	250	300	60	120	181	241	301	361
2.4	42	84	127	169	211	253	51	102	153	204	255	306	61	123	184	245	307	368
2.6	43	86	129	172	215	258	52	104	156	208	260	312	63	125	188	250	313	375
2.8	44	88	132	175	219	263	53	106	159	212	265	318	64	127	191	255	318	382
3	45	89	134	179	223	268	54	108	162	216	270	324	65	130	195	259	324	389

Notes:

(1) CT   = CT for 3-log inactivation
99.9

Reprinted from *GUIDANCE MANUAL FOR COMPLIANCE WITH THE FILTRATION AND DISINFECTION REQUIREMENTS FOR PUBLIC WATER SYSTEMS USING SURFACE WATER SOURCES*, a publication of the American Water Works Association. Reproduced with permission. Copyright © 1991 by American Water Works Association.

TABLE A.6   CT VALUES FOR INACTIVATION OF *GIARDIA* CYSTS BY FREE CHLORINE AT 10° C (1)

CHLORINE CONCENTRATION (mg/L)	pH <= 6 Log Inactivations						pH = 6.5 Log Inactivations						pH = 7.0 Log Inactivations						pH = 7.5 Log Inactivations					
	0.5	1.0	1.5	2.0	2.5	3.0	0.5	1.0	1.5	2.0	2.5	3.0	0.5	1.0	1.5	2.0	2.5	3.0	0.5	1.0	1.5	2.0	2.5	3.0
<=0.4	12	24	37	49	61	73	15	29	44	59	73	88	17	35	52	69	87	104	21	42	63	83	104	125
0.6	13	25	38	50	63	75	15	30	45	60	75	90	18	36	54	71	89	107	21	43	64	85	107	128
0.8	13	26	39	52	65	78	15	31	46	61	77	92	18	37	55	73	92	110	22	44	66	87	109	131
1	13	26	40	53	66	79	16	31	47	63	78	94	19	37	56	75	93	112	22	45	67	89	112	134
1.2	13	27	40	53	67	80	16	32	48	63	79	95	19	38	57	76	95	114	23	46	69	91	114	137
1.4	14	27	41	55	68	82	16	33	49	65	82	98	19	39	58	77	97	116	23	47	70	93	117	140
1.6	14	28	42	55	69	83	17	33	50	66	83	99	20	40	60	79	99	119	24	48	72	96	120	144
1.8	14	29	43	57	72	86	17	34	51	67	84	101	20	41	61	81	102	122	25	49	74	98	123	147
2	15	29	44	58	73	87	17	35	52	69	87	104	21	41	62	83	103	124	25	50	75	100	125	150
2.2	15	30	45	59	74	89	18	35	53	70	88	105	21	42	64	85	106	127	26	51	77	102	128	153
2.4	15	30	45	60	75	90	18	36	54	71	89	107	22	43	65	86	108	129	26	52	79	105	131	157
2.6	15	31	46	61	77	92	18	37	55	73	92	110	22	44	66	87	109	131	27	53	80	107	133	160
2.8	16	31	47	62	78	93	19	37	56	74	93	111	22	45	67	89	112	134	27	54	82	109	136	163
3	16	32	48	63	79	95	19	38	57	75	94	113	23	46	69	91	114	137	28	55	83	111	138	166

CHLORINE CONCENTRATION (mg/L)	pH = 8.0 Log Inactivations						pH = 8.5 Log Inactivations						pH <= 9.0 Log Inactivations					
	0.5	1.0	1.5	2.0	2.5	3.0	0.5	1.0	1.5	2.0	2.5	3.0	0.5	1.0	1.5	2.0	2.5	3.0
<=0.4	25	50	75	99	124	149	30	59	89	118	148	177	35	70	105	139	174	209
0.6	26	51	77	102	128	153	31	61	92	122	153	183	36	73	109	145	182	218
0.8	26	53	79	105	132	158	32	63	95	126	158	189	38	75	113	151	188	226
1	27	54	81	108	135	162	33	65	98	130	163	195	39	78	117	156	195	234
1.2	28	55	83	111	138	166	33	67	100	133	167	200	40	80	120	160	200	240
1.4	28	57	85	113	142	170	34	69	103	137	172	206	41	82	124	165	206	247
1.6	29	58	87	116	145	174	35	70	106	141	176	211	42	84	127	169	211	253
1.8	30	60	90	119	149	179	36	72	108	143	179	215	43	86	130	173	216	259
2	30	61	91	121	152	182	37	74	111	147	184	221	44	88	133	177	221	265
2.2	31	62	93	124	155	186	38	75	113	150	188	225	45	90	136	181	226	271
2.4	32	63	95	127	158	190	38	77	115	153	192	230	46	92	138	184	230	276
2.6	32	65	97	129	162	194	39	78	117	156	195	234	47	94	141	187	234	281
2.8	33	66	99	131	164	197	40	80	120	159	199	239	48	96	144	191	239	287
3	34	67	101	134	168	201	41	81	122	162	203	243	49	97	146	195	243	292

Notes:
(1) CT = CT for 3-log inactivation
     99.9

Reprinted from *GUIDANCE MANUAL FOR COMPLIANCE WITH THE FILTRATION AND DISINFECTION REQUIREMENTS FOR PUBLIC WATER SYSTEMS USING SURFACE WATER SOURCES*, a publication of the American Water Works Association. Reproduced with permission. Copyright © 1991 by American Water Works Association.

TABLE A.7 CT VALUES FOR INACTIVATION OF *GIARDIA* CYSTS BY FREE CHLORINE AT 15° C (1)

CHLORINE CONCENTRATION (mg/L)	pH <=6 Log Inactivations						pH=6.5 Log Inactivations						pH=7.0 Log Inactivations						pH=7.5 Log Inactivations					
	0.5	1.0	1.5	2.0	2.5	3.0	0.5	1.0	1.5	2.0	2.5	3.0	0.5	1.0	1.5	2.0	2.5	3.0	0.5	1.0	1.5	2.0	2.5	3.0
<=0.4	8	16	25	33	41	49	10	20	30	39	49	59	12	23	35	47	58	70	14	28	42	55	69	83
0.6	8	17	25	33	42	50	10	20	30	40	50	60	12	24	36	48	60	72	14	29	43	57	72	86
0.8	9	17	26	35	43	52	10	20	31	41	51	61	12	24	37	49	61	73	15	29	44	59	73	88
1	9	18	27	35	44	53	11	21	32	42	53	63	13	25	38	50	63	75	15	30	45	60	75	90
1.2	9	18	27	36	45	54	11	21	32	43	53	64	13	25	38	51	63	76	15	31	46	61	77	92
1.4	9	18	28	37	46	55	11	22	33	43	54	65	13	26	39	52	65	78	16	31	47	63	78	94
1.6	9	19	28	37	47	56	11	22	33	44	55	66	13	26	40	53	66	79	16	32	48	64	80	96
1.8	10	19	29	38	48	57	11	23	34	45	57	68	14	27	41	54	68	81	16	33	49	65	82	98
2	10	19	29	39	48	58	12	23	35	46	58	69	14	28	42	55	69	83	17	33	50	67	83	100
2.2	10	20	30	39	49	59	12	23	35	47	58	70	14	28	43	57	71	85	17	34	51	68	85	102
2.4	10	20	30	40	50	60	12	24	36	48	60	72	14	29	43	57	72	86	18	35	53	70	88	105
2.6	10	20	31	41	51	61	12	24	37	49	61	73	15	29	44	59	73	88	18	36	54	71	89	107
2.8	10	21	31	41	52	62	12	25	37	49	62	74	15	30	45	59	74	89	18	36	55	73	91	109
3	11	21	32	42	53	63	13	25	38	51	63	76	15	30	46	61	76	91	19	37	56	74	93	111

CHLORINE CONCENTRATION (mg/L)	pH=8.0 Log Inactivations						pH=8.5 Log Inactivations						pH <= 9.0 Log Inactivations					
	0.5	1.0	1.5	2.0	2.5	3.0	0.5	1.0	1.5	2.0	2.5	3.0	0.5	1.0	1.5	2.0	2.5	3.0
<=0.4	17	33	50	66	83	99	20	39	59	79	98	118	23	47	70	93	117	140
0.6	17	34	51	68	85	102	20	41	61	81	102	122	24	49	73	97	122	146
0.8	18	35	53	70	88	105	21	42	63	84	105	126	25	50	76	101	126	151
1	18	36	54	72	90	108	22	43	65	87	108	130	26	52	78	104	130	156
1.2	19	37	56	74	93	111	22	45	67	89	112	134	27	53	80	107	133	160
1.4	19	38	57	76	95	114	23	46	69	91	114	137	28	55	83	110	138	165
1.6	19	39	58	77	97	116	24	47	71	94	118	141	28	56	85	113	141	169
1.8	20	40	60	79	99	119	24	48	72	96	120	144	29	58	87	115	144	173
2	20	41	61	81	102	122	25	49	74	98	123	147	30	59	89	118	148	177
2.2	21	41	62	83	103	124	25	50	75	100	125	150	30	60	91	121	151	181
2.4	21	42	64	85	106	127	26	51	77	102	128	153	31	61	92	123	153	184
2.6	22	43	65	86	108	129		52	78	104	130	156	31	63	94	125	157	188
2.8	22	44	66	88	110	132	27	53	80	106	133	159	32	64	96	127	159	191
3	22	45	67	89	112	134	27	54	81	108	135	162	33	65	98	130	163	195

Notes:

(1) $CT = CT_{99.9}$ for 3-log inactivation

TABLE A.8 CT VALUES FOR INACTIVATION OF *GIARDIA* CYSTS BY FREE CHLORINE AT 20° C (1)

CHLORINE CONCENTRATION (mg/L)	pH <= 6 Log Inactivations						pH = 6.5 Log Inactivations						pH = 7.0 Log Inactivations						pH = 7.5 Log Inactivations					
	0.5	1.0	1.5	2.0	2.5	3.0	0.5	1.0	1.5	2.0	2.5	3.0	0.5	1.0	1.5	2.0	2.5	3.0	0.5	1.0	1.5	2.0	2.5	3.0
<=0.4	6	12	18	24	30	36	7	15	22	29	37	44	9	17	26	35	43	52	10	21	31	41	52	62
0.6	6	13	19	25	32	38	8	15	23	30	38	45	9	18	27	36	45	54	11	21	32	43	53	64
0.8	7	13	20	26	33	39	8	15	23	31	38	46	9	18	28	37	46	55	11	22	33	44	55	66
1	7	13	20	26	33	39	8	16	24	31	39	47	9	19	28	37	47	56	11	22	34	45	56	67
1.2	7	13	20	27	33	40	8	16	24	32	40	48	10	19	29	38	48	57	12	23	35	46	58	69
1.4	7	14	21	27	34	41	8	16	25	33	41	49	10	19	29	39	48	58	12	23	35	47	58	70
1.6	7	14	21	28	35	42	8	17	25	33	42	50	10	20	30	39	49	59	12	24	36	48	60	72
1.8	7	14	22	29	36	43	9	17	26	34	43	51	10	20	31	41	51	61	12	25	37	49	62	74
2	7	15	22	29	37	44	9	17	26	35	43	52	10	21	31	41	52	62	13	25	38	50	63	75
2.2	7	15	22	29	37	44	9	18	27	35	44	53	11	21	32	42	53	63	13	26	39	51	64	77
2.4	8	15	23	30	38	45	9	18	27	36	45	54	11	22	33	43	54	65	13	26	39	52	65	78
2.6	8	15	23	31	38	46	9	18	28	37	46	55	11	22	33	44	55	66	13	27	40	53	67	80
2.8	8	16	24	31	39	47	9	19	28	37	47	56	11	22	34	45	56	67	14	27	41	54	68	81
3	8	16	24	31	39	47	10	19	29	38	48	57	11	23	34	45	57	68	14	28	42	55	69	83

CHLORINE CONCENTRATION (mg/L)	pH = 8.0 Log Inactivations						pH = 8.5 Log Inactivations						pH <= 9.0 Log Inactivations					
	0.5	1.0	1.5	2.0	2.5	3.0	0.5	1.0	1.5	2.0	2.5	3.0	0.5	1.0	1.5	2.0	2.5	3.0
<=0.4	12	25	37	49	62	74	15	30	45	59	74	89	18	35	53	70	88	105
0.6	13	26	39	51	64	77	15	31	46	61	77	92	18	36	55	73	91	109
0.8	13	26	40	53	66	79	16	32	48	63	79	95	19	38	57	75	94	113
1	14	27	41	54	68	81	16	33	49	65	82	98	20	39	59	78	98	117
1.2	14	28	42	55	69	83	17	33	50	67	83	100	20	40	60	80	100	120
1.4	14	28	43	57	71	85	17	34	52	69	86	103	21	41	62	82	103	123
1.6	15	29	44	58	73	87	18	35	53	70	88	105	21	42	63	84	105	126
1.8	15	30	45	59	74	89	18	36	54	72	90	108	22	43	65	86	108	129
2	15	30	46	61	76	91	18	37	55	73	92	110	22	44	66	88	110	132
2.2	16	31	47	62	78	93	19	38	57	75	94	113	23	45	68	90	113	135
2.4	16	32	48	63	79	95	19	38	58	77	96	115	23	46	69	92	115	138
2.6	16	32	49	65	81	97	20	39	59	78	98	117	24	47	71	94	118	141
2.8	17	33	50	66	83	99	20	40	60	79	99	119	24	48	72	95	119	143
3	17	34	51	67	84	101	20	41	61	81	102	122	24	49	73	97	122	146

Notes:

(1) CT = CT for 3-log inactivation
99.9

TABLE A.9   CT VALUES FOR INACTIVATION OF *GIARDIA* CYSTS BY FREE CHLORINE AT 25° C (1)

CHLORINE CONCENTRATION (mg/L)	pH <= 6 Log Inactivations						pH = 6.5 Log Inactivations						pH = 7.0 Log Inactivations						pH = 7.5 Log Inactivations					
	0.5	1.0	1.5	2.0	2.5	3.0	0.5	1.0	1.5	2.0	2.5	3.0	0.5	1.0	1.5	2.0	2.5	3.0	0.5	1.0	1.5	2.0	2.5	3.0
<=0.4	4	8	12	16	20	24	5	10	15	19	24	29	6	12	18	23	29	35	7	14	21	28	35	42
0.6	4	8	13	17	21	25	5	10	15	20	25	30	6	12	18	24	30	36	7	14	22	29	36	43
0.8	4	9	13	17	22	26	5	10	16	21	26	31	6	12	19	25	31	37	7	15	22	29	37	44
1	4	9	13	17	22	26	5	10	16	21	26	31	6	12	19	25	31	37	8	15	23	30	38	45
1.2	5	9	14	18	23	27	5	11	16	21	27	32	6	13	19	25	32	38	8	15	23	31	38	46
1.4	5	9	14	18	23	27	6	11	17	22	28	33	7	13	20	26	33	39	8	16	24	31	39	47
1.6	5	9	14	19	23	28	6	11	17	22	28	33	7	13	20	27	33	40	8	16	24	32	40	48
1.8	5	10	15	19	24	29	6	11	17	23	28	34	7	14	21	27	34	41	8	16	25	33	41	49
2	5	10	15	19	24	29	6	12	18	23	29	35	7	14	21	27	34	41	8	17	25	33	42	50
2.2	5	10	15	20	25	30	6	12	18	23	29	35	7	14	21	28	35	42	9	17	26	34	43	51
2.4	5	10	15	20	25	30	6	12	18	24	30	36	7	14	22	29	36	43	9	17	26	35	43	52
2.6	5	10	16	21	26	31	6	12	19	25	31	37	7	15	22	29	37	44	9	18	27	35	44	53
2.8	5	10	16	21	26	31	6	12	19	25	31	37	8	15	23	30	38	45	9	18	27	36	45	54
3	5	11	16	21	27	32	6	13	19	25	32	38	8	15	23	31	38	46	9	18	28	37	46	55

CHLORINE CONCENTRATION (mg/L)	pH = 8.0 Log Inactivations						pH = 8.5 Log Inactivations						pH <= 9.0 Log Inactivations					
	0.5	1.0	1.5	2.0	2.5	3.0	0.5	1.0	1.5	2.0	2.5	3.0	0.5	1.0	1.5	2.0	2.5	3.0
<=0.4	8	17	25	33	42	50	10	20	30	39	49	59	12	23	35	47	58	70
0.6	9	17	26	34	43	51	10	20	31	41	51	61	12	24	37	49	61	73
0.8	9	18	27	35	44	53	11	21	32	42	53	63	13	25	38	50	63	75
1	9	18	27	36	45	54	11	22	33	43	54	65	13	26	39	52	65	78
1.2	9	18	28	37	46	55	11	22	34	45	56	67	13	27	40	53	67	80
1.4	10	19	29	38	48	57	12	23	35	46	58	69	14	27	41	55	68	82
1.6	10	19	29	39	48	58	12	23	35	47	58	70	14	28	42	56	70	84
1.8	10	20	30	40	50	60	12	24	36	48	60	72	14	29	43	57	72	86
2	10	20	31	41	51	61	12	25	37	49	62	74	15	29	44	59	73	88
2.2	10	21	31	41	52	62	13	25	38	50	63	75	15	30	45	60	75	90
2.4	11	21	32	42	53	63	13	26	39	51	64	77	15	31	46	61	77	92
2.6	11	22	33	43	54	65	13	26	39	52	65	78	16	31	47	63	78	94
2.8	11	22	33	44	55	66	13	27	40	53	67	80	16	32	48	64	80	96
3	11	22	34	45	56	67	14	27	41	54	68	81	16	32	49	65	81	97

Notes:

(1) CT$_{99.9}$ = CT for 3-log inactivation

WATER ABBREVIATIONS

ac	acre		km	kilometer
ac-ft	acre-feet		kN	kilonewton
af	acre feet		kW	kilowatt
amp	ampere		kWh	kilowatt-hour
°C	degrees Celsius		*L*	liter
CFM	cubic feet per minute		lb	pound
CFS	cubic feet per second		lbs/sq in	pounds per square inch
Ci	Curie		m	meter
cm	centimeter		M	Mega
cu ft	cubic feet		M	million
cu in	cubic inch		mg	milligram
cu m	cubic meter		MGD	million gallons per day
cu yd	cubic yard		mg/*L*	milligram per liter
°F	degrees Fahrenheit		min	minute
ft	feet or foot		m*L*	milliliter
ft-lb/min	foot-pounds per minute		mm	millimeter
g	gravity		N	Newton
gal	gallon		ohm	ohm
gal/day	gallons per day		Pa	Pascal
GFD	gallons of flux per square foot per day		pCi	picoCurie
gm	gram		ppb	parts per billion
GPD	gallons per day		ppm	parts per million
gpg	grains per gallon		psf	pounds per square foot
GPM	gallons per minute		psi	pounds per square inch
gr	grain		psig	pounds per square inch gage
ha	hectare		RPM	revolutions per minute
HP	horsepower		sec	second
hr	hour		sq ft	square feet
in	inch		sq in	square inches
k	kilo		W	watt
kg	kilogram			

WATER WORDS

A Summary of the Words Defined

in

WATER TREATMENT PLANT OPERATION,

WATER DISTRIBUTION SYSTEM OPERATION AND MAINTENANCE,

and

SMALL WATER SYSTEM OPERATION AND MAINTENANCE

PROJECT PRONUNCIATION KEY

by Warren L. Prentice

The Project Pronunciation Key is designed to aid you in the pronunciation of new words. While this key is based primarily on familiar sounds, it does not attempt to follow any particular pronunciation guide. This key is designed solely to aid operators in this program.

You may find it helpful to refer to other available sources for pronunciation help. Each current standard dictionary contains a guide to its own pronunciation key. Each key will be different from each other and from this key. Examples of the difference between the key used in this program and the *WEBSTER'S NEW WORLD COLLEGE DICTIONARY*[1] "Key" are shown below.

In using this key, you should accent (say louder) the syllable that appears in capital letters. The following chart is presented to give examples of how to pronounce words using the Project Key.

| WORD | SYLLABLE | | | | |
	1st	2nd	3rd	4th	5th
acid	AS	id			
coliform	COAL	i	form		
biological	BUY	o	LODGE	ik	cull

The first word, *ACID*, has its first syllable accented. The second word, *COLIFORM*, has its first syllable accented. The third word, *BIOLOGICAL*, has its first and third syllables accented.

We hope you will find the key useful in unlocking the pronunciation of any new word.

Term	Project Key	Webster Key
acid	AS-id	aś id
coliform	COAL-i-form	kō′ lə fôrm
biological	BUY-o-LODGE-ik-cull	bī ə läj′ i kəl

---

[1] *The WEBSTER'S NEW WORLD COLLEGE DICTIONARY, Fourth Edition, 1999, was chosen rather than an unabridged dictionary because of its availability to the operator. Other editions may be slightly different.*

WATER WORDS

>GREATER THAN

DO >5 mg/*L* would be read as DO GREATER THAN 5 mg/*L*.

<LESS THAN

DO <5 mg/*L* would be read as DO LESS THAN 5 mg/*L*.

A

ABC

See **A**SSOCIATION OF **B**OARDS OF **C**ERTIFICATION.

ACEOPS

See **A**LLIANCE OF **C**ERTIFIED **O**PERATORS, LAB ANALYSTS, INSPECTORS, AND SPECIALISTS (ACEOPS).

atm

The abbreviation for atmosphere. One atmosphere is equal to 14.7 psi or 100 kPa.

AWWA

See **A**MERICAN **W**ATER **W**ORKS **A**SSOCIATION.

ABSORPTION (ab-SORP-shun)

The taking in or soaking up of one substance into the body of another by molecular or chemical action (as tree roots absorb dissolved nutrients in the soil).

ACCOUNTABILITY

When a manager gives power/responsibility to an employee, the employee ensures that the manager is informed of results or events.

ACCURACY

How closely an instrument measures the true or actual value of the process variable being measured or sensed.

ACID RAIN

Precipitation which has been rendered (made) acidic by airborne pollutants.

ACIDIC (uh-SID-ick)

The condition of water or soil which contains a sufficient amount of acid substances to lower the pH below 7.0.

ACIDIFIED (uh-SID-uh-FIE-d)

The addition of an acid (usually nitric or sulfuric) to a sample to lower the pH below 2.0. The purpose of acidification is to "fix" a sample so it won't change until it is analyzed.

ACRE-FOOT

A volume of water that covers one acre to a depth of one foot, or 43,560 cubic feet (1,233.5 cubic meters).

ACTIVATED CARBON

Adsorptive particles or granules of carbon usually obtained by heating carbon (such as wood). These particles or granules have a high capacity to selectively remove certain trace and soluble materials from water.

ACUTE HEALTH EFFECT ACUTE HEALTH EFFECT

An adverse effect on a human or animal body, with symptoms developing rapidly.

ADSORBATE (add-SORE-bait) ADSORBATE

The material being removed by the adsorption process.

ADSORBENT (add-SORE-bent) ADSORBENT

The material (activated carbon) that is responsible for removing the undesirable substance in the adsorption process.

ADSORPTION (add-SORP-shun) ADSORPTION

The gathering of a gas, liquid, or dissolved substance on the surface or interface zone of another material.

AERATION (air-A-shun) AERATION

The process of adding air to water. Air can be added to water by either passing air through water or passing water through air.

AEROBIC (AIR-O-bick) AEROBIC

A condition in which atmospheric or dissolved molecular oxygen is present in the aquatic (water) environment.

AESTHETIC (es-THET-ick) AESTHETIC

Attractive or appealing.

AGE TANK AGE TANK

A tank used to store a known concentration of chemical solution for feed to a chemical feeder. Also called a DAY TANK.

AIR BINDING AIR BINDING

The clogging of a filter, pipe or pump due to the presence of air released from water. Air entering the filter media is harmful to both the filtration and backwash processes. Air can prevent the passage of water during the filtration process and can cause the loss of filter media during the backwash process.

AIR GAP AIR GAP

An open vertical drop, or vertical empty space, that separates a drinking (potable) water supply to be protected from another water system in a water treatment plant or other location. This open gap prevents the contamination of drinking water by backsiphonage or backflow because there is no way raw water or any other water can reach the drinking water supply.

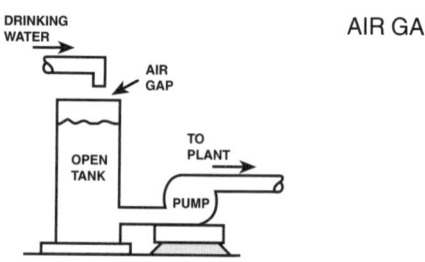

AIR PADDING AIR PADDING

Pumping dry air (dew point −40°F) into a container to assist with the withdrawal of a liquid or to force a liquified gas such as chlorine out of a container.

AIR STRIPPING AIR STRIPPING

A treatment process used to remove dissolved gases and volatile substances from water. Large volumes of air are bubbled through the water being treated to remove (strip out) the dissolved gases and volatile substances.

ALARM CONTACT ALARM CONTACT

A switch that operates when some preset low, high or abnormal condition exists.

ALGAE (AL-gee) ALGAE

Microscopic plants which contain chlorophyll and live floating or suspended in water. They also may be attached to structures, rocks or other submerged surfaces. Excess algal growths can impart tastes and odors to potable water. Algae produce oxygen during sunlight hours and use oxygen during the night hours. Their biological activities appreciably affect the pH, alkalinity, and dissolved oxygen of the water.

ALGAL (AL-gull) BLOOM ALGAL BLOOM

Sudden, massive growths of microscopic and macroscopic plant life, such as green or blue-green algae, which can, under the proper conditions, develop in lakes and reservoirs.

ALGICIDE (AL-juh-SIDE)   ALGICIDE

Any substance or chemical specifically formulated to kill or control algae.

ALIPHATIC (AL-uh-FAT-ick) HYDROXY ACIDS   ALIPHATIC HYDROXY ACIDS

Organic acids with carbon atoms arranged in branched or unbranched open chains rather than in rings.

ALIQUOT (AL-li-kwot)   ALIQUOT

Representative portion of a sample. Often an equally divided portion of a sample.

ALKALI (AL-ka-lie)   ALKALI

Any of certain soluble salts, principally of sodium, potassium, magnesium, and calcium, that have the property of combining with acids to form neutral salts and may be used in chemical water treatment processes.

ALKALINE (AL-ka-LINE)   ALKALINE

The condition of water or soil which contains a sufficient amount of alkali substances to raise the pH above 7.0.

ALKALINITY (AL-ka-LIN-it-tee)   ALKALINITY

The capacity of water to neutralize acids. This capacity is caused by the water's content of carbonate, bicarbonate, hydroxide, and occasionally borate, silicate, and phosphate. Alkalinity is expressed in milligrams per liter of equivalent calcium carbonate. Alkalinity is not the same as pH because water does not have to be strongly basic (high pH) to have a high alkalinity. Alkalinity is a measure of how much acid must be added to a liquid to lower the pH to 4.5.

ALLIANCE OF CERTIFIED OPERATORS,   ALLIANCE OF CERTIFIED OPERATORS,
  LAB ANALYSTS, INSPECTORS,   LAB ANALYSTS, INSPECTORS,
  AND SPECIALISTS (ACEOPS)   AND SPECIALISTS (ACEOPS)

A professional organization for operators, lab analysts, inspectors, and specialists dedicated to improving professionalism; expanding training, certification, and job opportunities; increasing information exchange; and advocating the importance of certified operators, lab analysts, inspectors, and specialists. For information on membership, contact ACEOPS, 1810 Bel Air Drive, Ames, IA 50010-5125, phone (515) 663-4128 or e-mail: Info@aceops.org.

ALLUVIAL (uh-LOU-vee-ul)   ALLUVIAL

Relating to mud and/or sand deposited by flowing water. Alluvial deposits may occur after a heavy rainstorm.

ALTERNATING CURRENT (A.C.)   ALTERNATING CURRENT (A.C.)

An electric current that reverses its direction (positive/negative values) at regular intervals.

ALTITUDE VALVE   ALTITUDE VALVE

A valve that automatically shuts off the flow into an elevated tank when the water level in the tank reaches a predetermined level. The valve automatically opens when the pressure in the distribution system drops below the pressure in the tank.

AMBIENT (AM-bee-ent) TEMPERATURE   AMBIENT TEMPERATURE

Temperature of the surrounding air (or other medium). For example, temperature of the room where a gas chlorinator is installed.

AMERICAN WATER WORKS ASSOCIATION   AMERICAN WATER WORKS ASSOCIATION

A professional organization for all persons working in the water utility field. This organization develops and recommends goals, procedures and standards for water utility agencies to help them improve their performance and effectiveness. For information on AWWA membership and publications, contact AWWA, 6666 W. Quincy Avenue, Denver, CO 80235. Phone (303) 794-7711.

AMPERAGE (AM-purr-age)   AMPERAGE

The strength of an electric current measured in amperes. The amount of electric current flow, similar to the flow of water in gallons per minute.

AMPERE (AM-peer)   AMPERE

The unit used to measure current strength. The current produced by an electromotive force of one volt acting through a resistance of one ohm.

AMPEROMETRIC (am-PURR-o-MET-rick)   AMPEROMETRIC

A method of measurement that records electric current flowing or generated, rather than recording voltage. Amperometric titration is a means of measuring concentrations of certain substances in water.

AMPEROMETRIC (am-PURR-o-MET-rick) TITRATION AMPEROMETRIC TITRATION

A means of measuring concentrations of certain substances in water (such as strong oxidizers) based on the electric current that flows during a chemical reaction. Also see TITRATE.

AMPLITUDE AMPLITUDE

The maximum strength of an alternating current during its cycle, as distinguished from the mean or effective strength.

ANAEROBIC (AN-air-O-bick) ANAEROBIC

A condition in which atmospheric or dissolved molecular oxygen is *NOT* present in the aquatic (water) environment.

ANALOG ANALOG

The readout of an instrument by a pointer (or other indicating means) against a dial or scale.

ANALYZER ANALYZER

A device which conducts periodic or continuous measurement of some factor such as chlorine, fluoride or turbidity. Analyzers operate by any of several methods including photocells, conductivity or complex instrumentation.

ANGSTROM (ANG-strem) ANGSTROM

A unit of length equal to one-tenth of a nanometer or one-tenbillionth of a meter (1 Angstrom = 0.000 000 000 1 meter). One Angstrom is the approximate diameter of an atom.

ANION (AN-EYE-en) ANION

A negatively charged ion in an electrolyte solution, attracted to the anode under the influence of a difference in electrical potential. Chloride ion (Cl⁻) is an anion.

ANIONIC (AN-eye-ON-ick) POLYMER ANIONIC POLYMER

A polymer having negatively charged groups of ions; often used as a filter aid and for dewatering sludges.

ANNULAR (AN-you-ler) SPACE ANNULAR SPACE

A ring-shaped space located between two circular objects, such as two pipes.

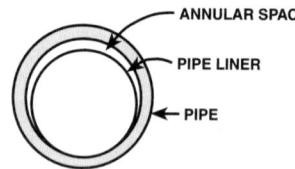

ANODE (an-O-d) ANODE

The positive pole or electrode of an electrolytic system, such as a battery. The anode attracts negatively charged particles or ions (anions).

APPARENT COLOR APPARENT COLOR

Color of the water that includes not only the color due to substances in the water but suspended matter as well.

APPROPRIATIVE APPROPRIATIVE

Water rights to or ownership of a water supply which is acquired for the beneficial use of water by following a specific legal procedure.

APPURTENANCE (uh-PURR-ten-nans) APPURTENANCE

Machinery, appliances, structures and other parts of the main structure necessary to allow it to operate as intended, but not considered part of the main structure.

AQUEOUS (A-kwee-us) AQUEOUS

Something made up of, similar to, or containing water; watery.

AQUIFER (ACK-wi-fer)
AQUIFER

A natural underground layer of porous, water-bearing materials (sand, gravel) usually capable of yielding a large amount or supply of water.

ARCH
ARCH

(1)   The curved top of a sewer pipe or conduit.

(2)   A bridge or arch of hardened or caked chemical which will prevent the flow of the chemical.

ARTESIAN (are-TEE-zhun)
ARTESIAN

Pertaining to groundwater, a well, or underground basin where the water is under a pressure greater than atmospheric and will rise above the level of its upper confining surface if given an opportunity to do so.

ASEPTIC (a-SEP-tick)
ASEPTIC

Free from the living germs of disease, fermentation, or putrefaction. Sterile.

ASSOCIATION OF BOARDS OF CERTIFICATION (ABC)
ASSOCIATION OF BOARDS OF CERTIFICATION (ABC)

An international organization representing over 150 boards which certify the operators of waterworks and wastewater facilities. For information on ABC publications regarding the preparation of and how to study for operator certification examinations, contact ABC, 208 Fifth Street, Ames, IA 50010-6259. Phone (515) 232-3623.

ASYMMETRIC (A-see-MET-rick)
ASYMMETRIC

Not similar in size, shape, form or arrangement of parts on opposite sides of a line, point or plane.

ATOM
ATOM

The smallest unit of a chemical element; composed of protons, neutrons and electrons.

AUDIT, WATER
AUDIT, WATER

A thorough examination of the accuracy of water agency records or accounts (volumes of water) and system control equipment. Water managers can use audits to determine their water distribution system efficiency. The overall goal is to identify and verify water and revenue losses in a water system.

AUTHORITY
AUTHORITY

The power and resources to do a specific job or to get that job done.

AVAILABLE CHLORINE
AVAILABLE CHLORINE

A measure of the amount of chlorine available in chlorinated lime, hypochlorite compounds, and other materials that are used as a source of chlorine when compared with that of elemental (liquid or gaseous) chlorine.

AVAILABLE EXPANSION
AVAILABLE EXPANSION

The vertical distance from the sand surface to the underside of a trough in a sand filter. This distance is also called FREEBOARD.

AVERAGE
AVERAGE

A number obtained by adding quantities or measurements and dividing the sum or total by the number of quantities or measurements. Also called the arithmetic mean.

$$\text{Average} = \frac{\text{Sum of Measurements}}{\text{Number of Measurements}}$$

AVERAGE DEMAND
AVERAGE DEMAND

The total demand for water during a period of time divided by the number of days in that time period. This is also called the average daily demand.

AXIAL TO IMPELLER
AXIAL TO IMPELLER

The direction in which material being pumped flows around the impeller or flows parallel to the impeller shaft.

AXIS OF IMPELLER
AXIS OF IMPELLER

An imaginary line running along the center of a shaft (such as an impeller shaft).

B

BOD (pronounce as separate letters)   BOD

Biochemical **O**xygen **D**emand. The rate at which organisms use the oxygen in water while stabilizing decomposable organic matter under aerobic conditions. In decomposition, organic matter serves as food for the bacteria and energy results from its oxidation. BOD measurements are used as a measure of the organic strength of wastes in water.

BACK PRESSURE   BACK PRESSURE

A pressure that can cause water to backflow into the water supply when a user's water system is at a higher pressure than the public water system.

BACKFLOW   BACKFLOW

A reverse flow condition, created by a difference in water pressures, which causes water to flow back into the distribution pipes of a potable water supply from any source or sources other than an intended source. Also see BACKSIPHONAGE.

BACKSIPHONAGE   BACKSIPHONAGE

A form of backflow caused by a negative or below atmospheric pressure within a water system. Also see BACKFLOW.

BACKWASHING   BACKWASHING

The process of reversing the flow of water back through the filter media to remove the entrapped solids.

BACTERIA (back-TEAR-e-ah)   BACTERIA

Bacteria are living organisms, microscopic in size, which usually consist of a single cell. Most bacteria use organic matter for their food and produce waste products as a result of their life processes.

BAFFLE   BAFFLE

A flat board or plate, deflector, guide or similar device constructed or placed in flowing water or slurry systems to cause more uniform flow velocities, to absorb energy, and to divert, guide, or agitate liquids (water, chemical solutions, slurry).

BAILER (BAY-ler)   BAILER

A 10- to 20-foot-long pipe equipped with a valve at the lower end. A bailer is used to remove slurry from the bottom or the side of a well as it is being drilled.

BASE-EXTRA CAPACITY METHOD   BASE-EXTRA CAPACITY METHOD

A cost allocation method used by water utilities to determine water rates for various water user groups. This method considers base costs (O & M expenses and capital costs), extra capacity costs (additional costs for maximum day and maximum hour demands), customer costs (meter maintenance and reading, billing, collection, accounting) and fire protection costs.

BASE METAL   BASE METAL

A metal (such as iron) which reacts with dilute hydrochloric acid to form hydrogen. Also see NOBLE METAL.

BATCH PROCESS   BATCH PROCESS

A treatment process in which a tank or reactor is filled, the water is treated or a chemical solution is prepared, and the tank is emptied. The tank may then be filled and the process repeated.

BENCH SCALE TESTS   BENCH SCALE TESTS

A method of studying different ways or chemical doses for treating water on a small scale in a laboratory.

BIOCHEMICAL OXYGEN DEMAND (BOD)   BIOCHEMICAL OXYGEN DEMAND (BOD)

The rate at which organisms use the oxygen in water while stabilizing decomposable organic matter under aerobic conditions. In decomposition, organic matter serves as food for the bacteria and energy results from its oxidation. BOD measurements are used as a measure of the organic strength of wastes in water.

BIOLOGICAL GROWTH   BIOLOGICAL GROWTH

The activity and growth of any and all living organisms.

BLANK   BLANK

A bottle containing only dilution water or distilled water; the sample being tested is not added. Tests are frequently run on a *SAMPLE* and a *BLANK* and the differences are compared. The procedure helps to eliminate or reduce test result errors that could be caused when the dilution water or distilled water used is contaminated.

BOND

BOND

(1) A written promise to pay a specified sum of money (called the face value) at a fixed time in the future (called the date of maturity). A bond also carries interest at a fixed rate, payable periodically. The difference between a note and a bond is that a bond usually runs for a longer period of time and requires greater formality. Utility agencies use bonds as a means of obtaining large amounts of money for capital improvements.

(2) A warranty by an underwriting organization, such as an insurance company, guaranteeing honesty, performance, or payment by a contractor.

BONNET (BON-it)

BONNET

The cover on a gate valve.

BOWL, PUMP

BOWL, PUMP

The submerged pumping unit in a well, including the shaft, impellers and housing.

BRAKE HORSEPOWER

BRAKE HORSEPOWER

(1) The horsepower required at the top or end of a pump shaft (input to a pump).

(2) The energy provided by a motor or other power source.

BREAKPOINT CHLORINATION

BREAKPOINT CHLORINATION

Addition of chlorine to water until the chlorine demand has been satisfied. At this point, further additions of chlorine will result in a free chlorine residual that is directly proportional to the amount of chlorine added beyond the breakpoint.

BREAKTHROUGH

BREAKTHROUGH

A crack or break in a filter bed allowing the passage of floc or particulate matter through a filter. This will cause an increase in filter effluent turbidity. A breakthrough can occur (1) when a filter is first placed in service, (2) when the effluent valve suddenly opens or closes, and (3) during periods of excessive head loss through the filter (including when the filter is exposed to negative heads).

BRINELLING (bruh-NEL-ing)

BRINELLING

Tiny indentations (dents) high on the shoulder of the bearing race or bearing. A type of bearing failure.

BUFFER

BUFFER

A solution or liquid whose chemical makeup neutralizes acids or bases without a great change in pH.

BUFFER CAPACITY

BUFFER CAPACITY

A measure of the capacity of a solution or liquid to neutralize acids or bases. This is a measure of the capacity of water for offering a resistance to changes in pH.

C

C FACTOR

C FACTOR

A factor or value used to indicate the smoothness of the interior of a pipe. The higher the C Factor, the smoother the pipe, the greater the carrying capacity, and the smaller the friction or energy losses from water flowing in the pipe. To calculate the C Factor, measure the flow, pipe diameter, distance between two pressure gages, and the friction or energy loss of the water between the gages.

$$\text{C Factor} = \frac{\text{Flow, GPM}}{193.75(\text{Diameter, ft})^{2.63}(\text{Slope})^{0.54}}$$

CT VALUE

CT VALUE

Residual concentration of a given disinfectant in mg/L times the disinfectant's contact time in minutes.

CAISSON (KAY-sawn)

CAISSON

A structure or chamber which is usually sunk or lowered by digging from the inside. Used to gain access to the bottom of a stream or other body of water.

CALCIUM CARBONATE EQUILIBRIUM

CALCIUM CARBONATE EQUILIBRIUM

A water is considered stable when it is just saturated with calcium carbonate. In this condition the water will neither dissolve nor deposit calcium carbonate. Thus, in this water the calcium carbonate is in equilibrium with the hydrogen ion concentration.

CALCIUM CARBONATE (CaCO₃) EQUIVALENT

An expression of the concentration of specified constituents in water in terms of their equivalent value to calcium carbonate. For example, the hardness in water which is caused by calcium, magnesium and other ions is usually described as calcium carbonate equivalent. Alkalinity test results are usually reported as mg/L CaCO₃ equivalents. To convert chloride to CaCO₃ equivalents, multiply the concentration of chloride ions in mg/L by 1.41, and for sulfate, multiply by 1.04.

CALIBRATION

A procedure which checks or adjusts an instrument's accuracy by comparison with a standard or reference.

CALL DATE

First date a bond can be paid off.

CAPILLARY ACTION

The movement of water through very small spaces due to molecular forces.

CAPILLARY FORCES

The molecular forces which cause the movement of water through very small spaces.

CAPILLARY FRINGE

The porous material just above the water table which may hold water by capillarity (a property of surface tension that draws water upward) in the smaller void spaces.

CARCINOGEN (CAR-sin-o-JEN)

Any substance which tends to produce cancer in an organism.

CATALYST (CAT-uh-LIST)

A substance that changes the speed or yield of a chemical reaction without being consumed or chemically changed by the chemical reaction.

CATALYZE (CAT-uh-LIZE)

To act as a catalyst. Or, to speed up a chemical reaction.

CATALYZED (CAT-uh-LIZED)

To be acted upon by a catalyst.

CATHODE (KA-thow-d)

The negative pole or electrode of an electrolytic cell or system. The cathode attracts positively charged particles or ions (cations).

CATHODIC (ca-THOD-ick) PROTECTION

An electrical system for prevention of rust, corrosion, and pitting of metal surfaces which are in contact with water or soil. A low-voltage current is made to flow through a liquid (water) or a soil in contact with the metal in such a manner that the external electromotive force renders the metal structure cathodic. This concentrates corrosion on auxiliary anodic parts which are deliberately allowed to corrode instead of letting the structure corrode.

CATION (CAT-EYE-en)

A positively charged ion in an electrolyte solution, attracted to the cathode under the influence of a difference in electrical potential. Sodium ion (Na⁺) is a cation.

CATIONIC POLYMER

A polymer having positively charged groups of ions; often used as a coagulant aid.

CAUTION

This word warns against potential hazards or cautions against unsafe practices. Also see DANGER, NOTICE, and WARNING.

CAVITATION (CAV-uh-TAY-shun)

The formation and collapse of a gas pocket or bubble on the blade of an impeller or the gate of a valve. The collapse of this gas pocket or bubble drives water into the impeller or gate with a terrific force that can cause pitting on the impeller or gate surface. Cavitation is accompanied by loud noises that sound like someone is pounding on the impeller or gate with a hammer.

CENTRATE

The water leaving a centrifuge after most of the solids have been removed.

CENTRIFUGAL (sen-TRIF-uh-gull) PUMP

A pump consisting of an impeller fixed on a rotating shaft that is enclosed in a casing, and having an inlet and discharge connection. As the rotating impeller whirls the liquid around, centrifugal force builds up enough pressure to force the water through the discharge outlet.

CENTRIFUGE

A mechanical device that uses centrifugal or rotational forces to separate solids from liquids.

CERTIFICATION EXAMINATION

An examination administered by a state agency that operators take to indicate a level of professional competence. In most plants the Chief Operator of the plant must be "certified" (successfully pass a certification examination). In the United States, certification of operators of water treatment plants and wastewater treatment plants is mandatory.

CERTIFIED OPERATOR

A person who has the education and experience required to operate a specific class of treatment facility as indicated by possessing a certificate of professional competence given by a state agency or professional association.

CHARGE CHEMISTRY

A branch of chemistry in which the destabilization and neutralization reactions occur between stable negatively charged and stable positively charged particles.

CHECK SAMPLING

Whenever an initial or routine sample analysis indicates that a Maximum Contaminant Level (MCL) has been exceeded, *CHECK SAMPLING* is required to confirm the routine sampling results. Check sampling is in addition to the routine sampling program.

CHECK VALVE

A special valve with a hinged disc or flap that opens in the direction of normal flow and is forced shut when flows attempt to go in the reverse or opposite direction of normal flows.

CHELATING (key-LAY-ting) AGENT

A chemical used to prevent the precipitation of metals (such as copper).

CHELATION (key-LAY-shun)

A chemical complexing (forming or joining together) of metallic cations (such as copper) with certain organic compounds, such as EDTA (ethylene diamine tetracetic acid). Chelation is used to prevent the precipitation of metals (copper). Also see SEQUESTRATION.

CHLORAMINATION (KLOR-ah-min-NAY-shun)

The application of chlorine and ammonia to water to form chloramines for the purpose of disinfection.

CHLORAMINES (KLOR-uh-means)

Compounds formed by the reaction of hypochlorous acid (or aqueous chlorine) with ammonia.

CHLORINATION (KLOR-uh-NAY-shun)

The application of chlorine to water, generally for the purpose of disinfection, but frequently for accomplishing other biological or chemical results (aiding coagulation and controlling tastes and odors).

CHLORINATOR (KLOR-uh-NAY-ter)

A metering device which is used to add chlorine to water.

CHLORINE DEMAND

Chlorine demand is the difference between the amount of chlorine added to water and the amount of residual chlorine remaining after a given contact time. Chlorine demand may change with dosage, time, temperature, pH, and nature and amount of the impurities in the water.

Chlorine Demand, mg/L = Chlorine Applied, mg/L − Chlorine Residual, mg/L

CHLORINE REQUIREMENT CHLORINE REQUIREMENT

The amount of chlorine which is needed for a particular purpose. Some reasons for adding chlorine are reducing the number of coliform bacteria (Most Probable Number), obtaining a particular chlorine residual, or oxidizing some substance in the water. In each case a definite dosage of chlorine will be necessary. This dosage is the chlorine requirement.

CHLORINE RESIDUAL CHLORINE RESIDUAL

The concentration of chlorine present in water after the chlorine demand has been satisfied. The concentration is expressed in terms of the total chlorine residual, which includes both the free and combined or chemically bound chlorine residuals.

CHLOROPHENOLIC (klor-o-FEE-NO-lick) CHLOROPHENOLIC

Chlorophenolic compounds are phenolic compounds (carbolic acid) combined with chlorine.

CHLOROPHENOXY (KLOR-o-fuh-KNOX-ee) CHLOROPHENOXY

A class of herbicides that may be found in domestic water supplies and cause adverse health effects. Two widely used chlorophenoxy herbicides are 2,4-D (2,4-Dichlorophenoxy acetic acid) and 2,4,5-TP (2,4,5-Trichlorophenoxy propionic acid (silvex)).

CHLORORGANIC (klor-or-GAN-ick) CHLORORGANIC

Organic compounds combined with chlorine. These compounds generally originate from, or are associated with, life processes such as those of algae in water.

CHRONIC HEALTH EFFECT CHRONIC HEALTH EFFECT

An adverse effect on a human or animal body with symptoms that develop slowly over a long period of time or that recur frequently.

CIRCLE OF INFLUENCE CIRCLE OF INFLUENCE

The circular outer edge of a depression produced in the water table by the pumping of water from a well. Also see CONE OF INFLUENCE and CONE OF DEPRESSION.

[SEE DRAWING ON PAGE 659]

CIRCUIT CIRCUIT

The complete path of an electric current, including the generating apparatus or other source; or, a specific segment or section of the complete path.

CIRCUIT BREAKER CIRCUIT BREAKER

A safety device in an electric circuit that automatically shuts off the circuit when it becomes overloaded. The device can be manually reset.

CISTERN (SIS-turn) CISTERN

A small tank (usually covered) or a storage facility used to store water for a home or farm. Often used to store rainwater.

CLARIFIER (KLAIR-uh-fire) CLARIFIER

A large circular or rectangular tank or basin in which water is held for a period of time during which the heavier suspended solids settle to the bottom. Clarifiers are also called settling basins and sedimentation basins.

CLASS, PIPE AND FITTINGS CLASS, PIPE AND FITTINGS

The working pressure rating, including allowances for surges, of a specific pipe for use in water distribution systems. The term is used for cast iron, ductile iron, asbestos cement and some plastic pipe.

CLEAR WELL CLEAR WELL

A reservoir for the storage of filtered water of sufficient capacity to prevent the need to vary the filtration rate with variations in demand. Also used to provide chlorine contact time for disinfection.

COAGULANT AID COAGULANT AID

Any chemical or substance used to assist or modify coagulation.

COAGULANTS (co-AGG-you-lents) COAGULANTS

Chemicals that cause very fine particles to clump (floc) together into larger particles. This makes it easier to separate the solids from the water by settling, skimming, draining or filtering.

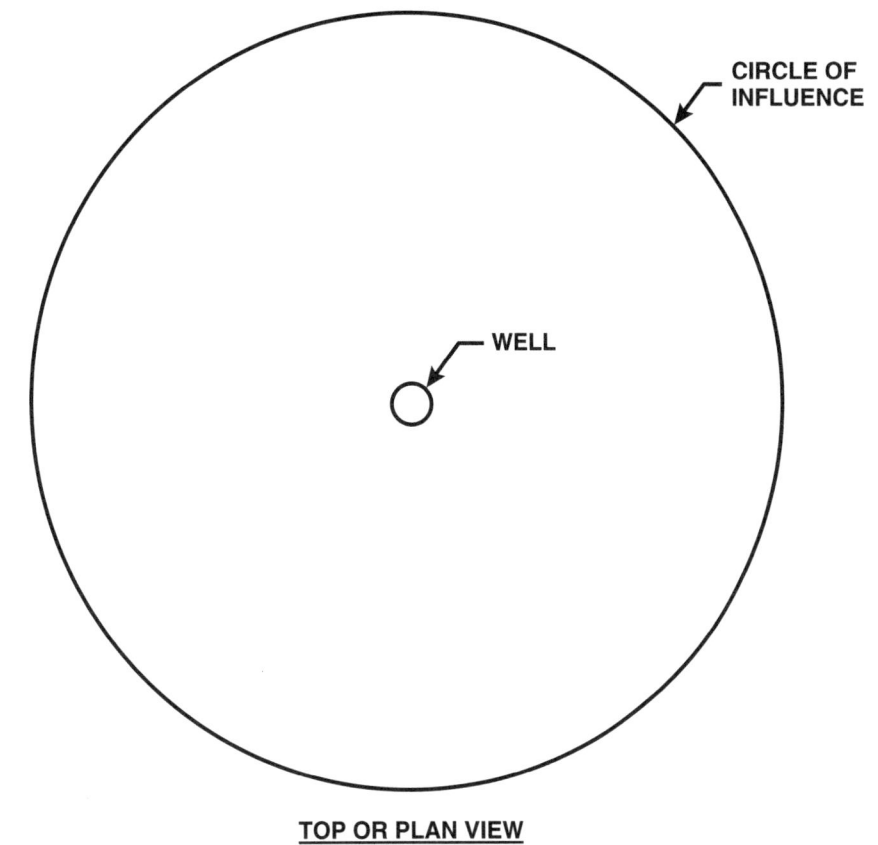

TOP OR PLAN VIEW

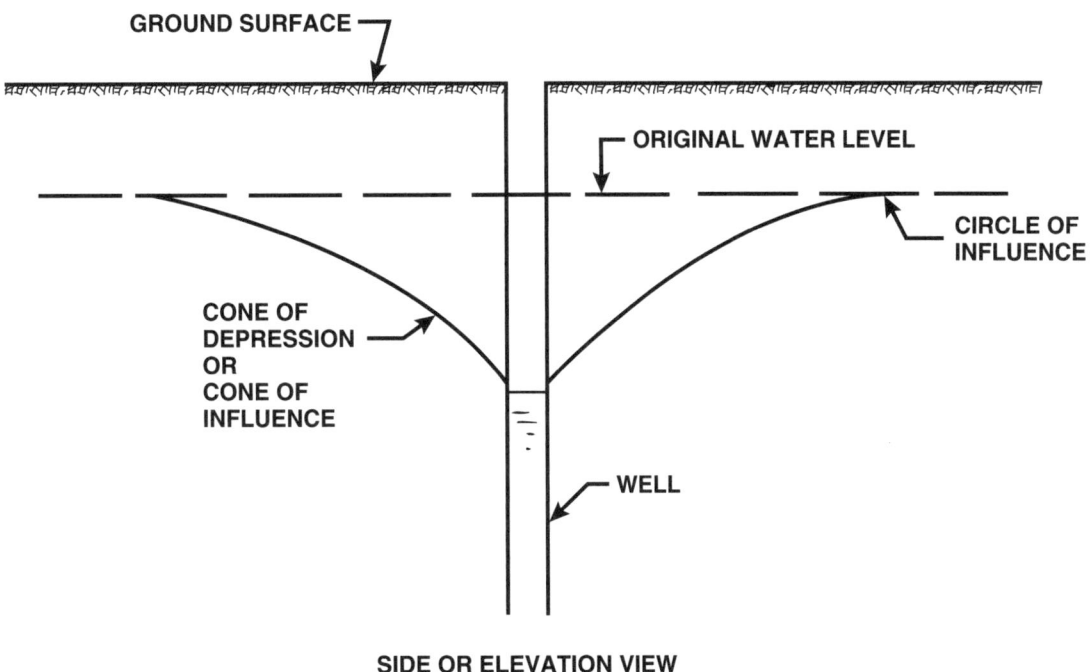

SIDE OR ELEVATION VIEW

CIRCLE OF INFLUENCE and CONE OF DEPRESSION/CONE OF INFLUENCE

COAGULATION (co-AGG-you-LAY-shun)   COAGULATION

The clumping together of very fine particles into larger particles (floc) caused by the use of chemicals (coagulants). The chemicals neutralize the electrical charges of the fine particles, allowing them to come closer and form larger clumps. This clumping together makes it easier to separate the solids from the water by settling, skimming, draining or filtering.

CODE OF FEDERAL REGULATIONS (CFR)   CODE OF FEDERAL REGULATIONS (CFR)

A publication of the United States Government which contains all of the proposed and finalized federal regulations, including environmental regulations.

COLIFORM (COAL-i-form)   COLIFORM

A group of bacteria found in the intestines of warm-blooded animals (including humans) and also in plants, soil, air and water. Fecal coliforms are a specific class of bacteria which only inhabit the intestines of warm-blooded animals. The presence of coliform bacteria is an indication that the water is polluted and may contain pathogenic (disease-causing) organisms.

COLLOIDS (CALL-loids)   COLLOIDS

Very small, finely divided solids (particles that do not dissolve) that remain dispersed in a liquid for a long time due to their small size and electrical charge. When most of the particles in water have a negative electrical charge, they tend to repel each other. This repulsion prevents the particles from clumping together, becoming heavier, and settling out.

COLOR   COLOR

The substances in water that impart a yellowish-brown color to the water. These substances are the result of iron and manganese ions, humus and peat materials, plankton, aquatic weeds, and industrial waste present in the water. Also see TRUE COLOR.

COLORIMETRIC MEASUREMENT   COLORIMETRIC MEASUREMENT

A means of measuring unknown chemical concentrations in water by measuring a sample's color intensity. The specific color of the sample, developed by addition of chemical reagents, is measured with a photoelectric colorimeter or is compared with "color standards" using, or corresponding with, known concentrations of the chemical.

COMBINED AVAILABLE CHLORINE   COMBINED AVAILABLE CHLORINE

The total chlorine, present as chloramine or other derivatives, that is present in a water and is still available for disinfection and for oxidation of organic matter. The combined chlorine compounds are more stable than free chlorine forms, but they are somewhat slower in disinfection action.

COMBINED AVAILABLE CHLORINE RESIDUAL   COMBINED AVAILABLE CHLORINE RESIDUAL

The concentration of residual chlorine that is combined with ammonia, organic nitrogen, or both in water as a chloramine (or other chloro derivative) and yet is still available to oxidize organic matter and help kill bacteria.

COMBINED CHLORINE   COMBINED CHLORINE

The sum of the chlorine species composed of free chlorine and ammonia, including monochloramine, dichloramine, and trichloramine (nitrogen trichloride). Dichloramine is the strongest disinfectant of these chlorine species, but it has less oxidative capacity than free chlorine.

COMBINED RESIDUAL CHLORINATION   COMBINED RESIDUAL CHLORINATION

The application of chlorine to water to produce combined available chlorine residual. This residual can be made up of monochloramines, dichloramines, and nitrogen trichloride.

COMMODITY-DEMAND METHOD   COMMODITY-DEMAND METHOD

A cost allocation method used by water utilities to determine water rates for the various water user groups. This method considers the commodity costs (water, chemicals, power, amount of water use), demand costs (treatment, storage, distribution), customer costs (meter maintenance and reading, billing, collection, accounting) and fire protection costs.

COMPETENT PERSON   COMPETENT PERSON

A competent person is defined by OSHA as a person capable of identifying existing and predictable hazards in the surroundings, or working conditions which are unsanitary, hazardous or dangerous to employees, and who has authorization to take prompt corrective measures to eliminate the hazards.

COMPLETE TREATMENT   COMPLETE TREATMENT

A method of treating water which consists of the addition of coagulant chemicals, flash mixing, coagulation-flocculation, sedimentation and filtration. Also called CONVENTIONAL FILTRATION.

COMPOSITE (come-PAH-zit) (PROPORTIONAL) SAMPLE   COMPOSITE (PROPORTIONAL) SAMPLE

A composite sample is a collection of individual samples obtained at regular intervals, usually every one or two hours during a 24-hour time span. Each individual sample is combined with the others in proportion to the rate of flow when the sample was collected. The resulting mixture (composite sample) forms a representative sample and is analyzed to determine the average conditions during the sampling period.

COMPOUND   COMPOUND

A pure substance composed of two or more elements whose composition is constant. For example, table salt (sodium chloride, $NaCl$) is a compound.

CONCENTRATION POLARIZATION   CONCENTRATION POLARIZATION

(1) A buildup of retained particles on the membrane surface due to dewatering of the feed closest to the membrane. The thickness of the concentration polarization layer is controlled by the flow velocity across the membrane.

(2) Used in corrosion studies to indicate a depletion of ions near an electrode.

(3) The basis for chemical analysis by a polarograph.

CONDITIONING   CONDITIONING

Pretreatment of sludge to facilitate removal of water in subsequent treatment processes.

CONDUCTANCE   CONDUCTANCE

A rapid method of estimating the dissolved solids content of a water supply. The measurement indicates the capacity of a sample of water to carry an electric current, which is related to the concentration of ionized substances in the water. Also called SPECIFIC CONDUCTANCE.

CONDUCTIVITY   CONDUCTIVITY

A measure of the ability of a solution (water) to carry an electric current.

CONDUCTOR   CONDUCTOR

A substance, body, device or wire that readily conducts or carries electric current.

CONDUCTOR CASING   CONDUCTOR CASING

The outer casing of a well. The purpose of this casing is to prevent contaminants from surface waters or shallow groundwaters from entering a well.

CONE OF DEPRESSION   CONE OF DEPRESSION

The depression, roughly conical in shape, produced in the water table by the pumping of water from a well. Also called the CONE OF INFLUENCE. Also see CIRCLE OF INFLUENCE.

[SEE DRAWING ON PAGE 659]

CONE OF INFLUENCE   CONE OF INFLUENCE

The depression, roughly conical in shape, produced in the water table by the pumping of water from a well. Also called the CONE OF DEPRESSION. Also see CIRCLE OF INFLUENCE.

[SEE DRAWING ON PAGE 659]

CONFINED SPACE   CONFINED SPACE

Confined space means a space that:

A. Is large enough and so configured that an employee can bodily enter and perform assigned work; and

B. Has limited or restricted means for entry or exit (for example, manholes, tanks, vessels, silos, storage bins, hoppers, vaults, and pits are spaces that may have limited means of entry); and

C. Is not designed for continuous employee occupancy.

(Definition from the Code of Federal Regulations (CFR) Title 29 Part 1910.146.)

CONFINED SPACE, CLASS "A"   CONFINED SPACE, CLASS "A"

A confined space that presents a situation that is immediately dangerous to life or health (IDLH). These include but are not limited to oxygen deficiency, explosive or flammable atmospheres, and/or concentrations of toxic substances.

(Definition from NIOSH, "Criteria for a Recommended Standard: Working in Confined Spaces.")

CONFINED SPACE, CLASS "B"   CONFINED SPACE, CLASS "B"

A confined space that has the potential for causing injury and illness, if preventive measures are not used, but not immediately dangerous to life and health.

(Definition from NIOSH, "Criteria for a Recommended Standard: Working in Confined Spaces.")

CONFINED SPACE, CLASS "C"   CONFINED SPACE, CLASS "C"

A confined space in which the potential hazard would not require any special modification of the work procedure.

(Definition from NIOSH, "Criteria for a Recommended Standard: Working in Confined Spaces.")

CONFINED SPACE, NON-PERMIT   CONFINED SPACE, NON-PERMIT

A non-permit confined space is a confined space that does not contain or, with respect to atmospheric hazards, have the potential to contain any hazard capable of causing death or serious physical harm.

CONFINED SPACE, PERMIT-REQUIRED   CONFINED SPACE, PERMIT-REQUIRED
  (PERMIT SPACE)   (PERMIT SPACE)

A confined space that has one or more of the following characteristics:

- Contains or has a potential to contain a hazardous atmosphere,
- Contains a material that has the potential for engulfing an entrant,
- Has an internal configuration such that an entrant could be trapped or asphyxiated by inwardly converging walls or by a floor which slopes downward and tapers to a smaller cross section, or
- Contains any other recognized serious safety or health hazard.

(Definition from the Code of Federal Regulations (CFR) Title 29 Part 1910.146.)

CONFINING UNIT   CONFINING UNIT

A layer of rock or soil of very low hydraulic conductivity that hampers the movement of groundwater in and out of an aquifer.

CONSOLIDATED FORMATION   CONSOLIDATED FORMATION

A geologic material whose particles are stratified (layered), cemented or firmly packed together (hard rock); usually occurring at a depth below the ground surface. Also see UNCONSOLIDATED FORMATION.

CONSUMER CONFIDENCE REPORTS   CONSUMER CONFIDENCE REPORTS

An annual report prepared by a water utility to communicate with its consumers. The report provides consumers with information on the source and quality of their drinking water. The report is an opportunity for positive communication with consumers and to convey the importance of paying for good quality drinking water.

CONTACTOR   CONTACTOR

An electric switch, usually magnetically operated.

CONTAMINATION   CONTAMINATION

The introduction into water of microorganisms, chemicals, toxic substances, wastes, or wastewater in a concentration that makes the water unfit for its next intended use.

CONTINUOUS SAMPLE   CONTINUOUS SAMPLE

A flow of water from a particular place in a plant to the location where samples are collected for testing. This continuous stream may be used to obtain grab or composite samples. Frequently, several taps (faucets) will flow continuously in the laboratory to provide test samples from various places in a water treatment plant.

CONTROL LOOP   CONTROL LOOP

The path through the control system between the sensor, which measures a process variable, and the controller, which controls or adjusts the process variable.

CONTROL SYSTEM   CONTROL SYSTEM

An instrumentation system which senses and controls its own operation on a close, continuous basis in what is called proportional (or modulating) control.

CONTROLLER   CONTROLLER

A device which controls the starting, stopping, or operation of a device or piece of equipment.

CONVENTIONAL FILTRATION   CONVENTIONAL FILTRATION

A method of treating water which consists of the addition of coagulant chemicals, flash mixing, coagulation-flocculation, sedimentation and filtration. Also called COMPLETE TREATMENT. Also see DIRECT FILTRATION and IN-LINE FILTRATION.

CONVENTIONAL TREATMENT   CONVENTIONAL TREATMENT

See CONVENTIONAL FILTRATION. Also called COMPLETE TREATMENT.

CORPORATION STOP   CORPORATION STOP

A water service shutoff valve located at a street water main. This valve cannot be operated from the ground surface because it is buried and there is no valve box. Also called a corporation cock.

CORROSION   CORROSION

The gradual decomposition or destruction of a material by chemical action, often due to an electrochemical reaction. Corrosion may be caused by (1) stray current electrolysis, (2) galvanic corrosion caused by dissimilar metals, or (3) differential-concentration cells. Corrosion starts at the surface of a material and moves inward.

CORROSION INHIBITORS   CORROSION INHIBITORS

Substances that slow the rate of corrosion.

CORROSIVE GASES   CORROSIVE GASES

In water, dissolved oxygen reacts readily with metals at the anode of a corrosion cell, accelerating the rate of corrosion until a film of oxidation products such as rust forms. At the cathode where hydrogen gas may form a coating on the cathode and slow the corrosion rate, oxygen reacts rapidly with hydrogen gas forming water, and again increases the rate of corrosion.

CORROSIVITY   CORROSIVITY

An indication of the corrosiveness of a water. The corrosiveness of a water is described by the water's pH, alkalinity, hardness, temperature, total dissolved solids, dissolved oxygen concentration, and the Langelier Index.

COULOMB (COO-lahm)   COULOMB

A measurement of the amount of electrical charge carried by an electric current of one ampere in one second. One coulomb equals about 6.25×10^{18} electrons (6,250,000,000,000,000,000 electrons).

COUPON   COUPON

A steel specimen inserted into water to measure the corrosiveness of water. The rate of corrosion is measured as the loss of weight of the coupon (in milligrams) per surface area (in square decimeters) exposed to the water per day. 10 decimeters = 1 meter = 100 centimeters.

COVERAGE RATIO   COVERAGE RATIO

The coverage ratio is a measure of the ability of the utility to pay the principal and interest on loans and bonds (this is known as "debt service") in addition to any unexpected expenses.

CROSS CONNECTION   CROSS CONNECTION

A connection between a drinking (potable) water system and an unapproved water supply. For example, if you have a pump moving nonpotable water and hook into the drinking water system to supply water for the pump seal, a cross connection or mixing between the two water systems can occur. This mixing may lead to contamination of the drinking water.

CRYPTOSPORIDIUM (CRIP-toe-spo-RID-ee-um)   CRYPTOSPORIDIUM

A waterborne intestinal parasite that causes a disease called cryptosporidiosis (CRIP-toe-spo-rid-ee-O-sis) in infected humans. Symptoms of the disease include diarrhea, cramps, and weight loss. *Cryptosporidium* contamination is found in most surface waters and some groundwaters. Commonly referred to as "crypto."

CURB STOP   CURB STOP

A water service shutoff valve located in a water service pipe near the curb and between the water main and the building. This valve is usually operated by a wrench or valve key and is used to start or stop flows in the water service line to a building. Also called a curb cock.

CURIE

A measure of radioactivity. One Curie of radioactivity is equivalent to 3.7 x 10^{10} or 37,000,000,000 nuclear disintegrations per second.

CURRENT

A movement or flow of electricity. Water flowing in a pipe is measured in gallons per second past a certain point, not by the number of water molecules going past a point. Electric current is measured by the number of coulombs per second flowing past a certain point in a conductor. A coulomb is equal to about 6.25 x 10^{18} electrons (6,250,000,000,000,000,000 electrons). A flow of one coulomb per second is called one ampere, the unit of the rate of flow of current.

CYCLE

A complete alternation of voltage and/or current in an alternating current (A.C.) circuit.

D

DBP

See **DISINFECTION BY-PRODUCT.**

DPD (pronounce as separate letters)

A method of measuring the chlorine residual in water. The residual may be determined by either titrating or comparing a developed color with color standards. DPD stands for N,N-diethyl-p-phenylene-diamine.

DANGER

The word *DANGER* is used where an immediate hazard presents a threat of death or serious injury to employees. Also see CAUTION, NOTICE, and WARNING.

DANGEROUS AIR CONTAMINATION

An atmosphere presenting a threat of causing death, injury, acute illness, or disablement due to the presence of flammable and/or explosive, toxic or otherwise injurious or incapacitating substances.

A. Dangerous air contamination due to the flammability of a gas or vapor is defined as an atmosphere containing the gas or vapor at a concentration greater than 10 percent of its lower explosive (lower flammable) limit.

B. Dangerous air contamination due to a combustible particulate is defined as a concentration greater than 10 percent of the minimum explosive concentration of the particulate.

C. Dangerous air contamination due to the toxicity of a substance is defined as the atmospheric concentration immediately hazardous to life or health.

DATEOMETER (day-TOM-uh-ter)

A small calendar disc attached to motors and equipment to indicate the year in which the last maintenance service was performed.

DATUM LINE

A line from which heights and depths are calculated or measured. Also called a datum plane or a datum level.

DAY TANK

A tank used to store a chemical solution of known concentration for feed to a chemical feeder. A day tank usually stores sufficient chemical solution to properly treat the water being treated for at least one day. Also called an AGE TANK.

DEAD END

The end of a water main which is not connected to other parts of the distribution system by means of a connecting loop of pipe.

DEBT SERVICE

The amount of money required annually to pay the (1) interest on outstanding debts; or (2) funds due on a maturing bonded debt or the redemption of bonds.

DECANT

To draw off the upper layer of liquid (water) after the heavier material (a solid or another liquid) has settled.

DECANT WATER

Water that has separated from sludge and is removed from the layer of water above the sludge.

DECHLORINATION (dee-KLOR-uh-NAY-shun)   DECHLORINATION

The deliberate removal of chlorine from water. The partial or complete reduction of residual chlorine by any chemical or physical process.

DECIBEL (DES-uh-bull)   DECIBEL

A unit for expressing the relative intensity of sounds on a scale from zero for the average least perceptible sound to about 130 for the average level at which sound causes pain to humans. Abbreviated dB.

DECOMPOSITION, DECAY   DECOMPOSITION, DECAY

The conversion of chemically unstable materials to more stable forms by chemical or biological action. If organic matter decays when there is no oxygen present (anaerobic conditions or putrefaction), undesirable tastes and odors are produced. Decay of organic matter when oxygen is present (aerobic conditions) tends to produce much less objectionable tastes and odors.

DEFLUORIDATION (de-FLOOR-uh-DAY-shun)   DEFLUORIDATION

The removal of excess fluoride in drinking water to prevent the mottling (brown stains) of teeth.

DEGASIFICATION (DEE-GAS-if-uh-KAY-shun)   DEGASIFICATION

A water treatment process which removes dissolved gases from the water. The gases may be removed by either mechanical or chemical treatment methods or a combination of both.

DELEGATION   DELEGATION

The act in which power is given to another person in the organization to accomplish a specific job.

DEMINERALIZATION (DEE-MIN-er-al-uh-ZAY-shun)   DEMINERALIZATION

A treatment process which removes dissolved minerals (salts) from water.

DENSITY (DEN-sit-tee)   DENSITY

A measure of how heavy a substance (solid, liquid or gas) is for its size. Density is expressed in terms of weight per unit volume, that is, grams per cubic centimeter or pounds per cubic foot. The density of water (at 4°C or 39°F) is 1.0 gram per cubic centimeter or about 62.4 pounds per cubic foot.

DEPOLARIZATION   DEPOLARIZATION

The removal or depletion of ions in the thin boundary layer adjacent to a membrane or pipe wall.

DEPRECIATION   DEPRECIATION

The gradual loss in service value of a facility or piece of equipment due to all the factors causing the ultimate retirement of the facility or equipment. This loss can be caused by sudden physical damage, wearing out due to age, obsolescence, inadequacy or availability of a newer, more efficient facility or equipment. The value cannot be restored by maintenance.

DESALINIZATION (DEE-SAY-leen-uh-ZAY-shun)   DESALINIZATION

The removal of dissolved salts (such as sodium chloride, NaCl) from water by natural means (leaching) or by specific water treatment processes.

DESICCANT (DESS-uh-kant)   DESICCANT

A drying agent which is capable of removing or absorbing moisture from the atmosphere in a small enclosure.

DESICCATION (DESS-uh-KAY-shun)   DESICCATION

A process used to thoroughly dry air; to remove virtually all moisture from air.

DESICCATOR (DESS-uh-KAY-tor)   DESICCATOR

A closed container into which heated weighing or drying dishes are placed to cool in a dry environment in preparation for weighing. The dishes may be empty or they may contain a sample. Desiccators contain a substance, such as anhydrous calcium chloride, which absorbs moisture and keeps the relative humidity near zero so that the dish or sample will not gain weight from absorbed moisture.

DESTRATIFICATION (de-STRAT-uh-fuh-KAY-shun)   DESTRATIFICATION

The development of vertical mixing within a lake or reservoir to eliminate (either totally or partially) separate layers of temperature, plant, or animal life. This vertical mixing can be caused by mechanical means (pumps) or through the use of forced air diffusers which release air into the lower layers of the reservoir.

DETECTION LAG DETECTION LAG

The time period between the moment a process change is made and the moment when such a change is finally sensed by the associated measuring instrument.

DETENTION TIME DETENTION TIME

(1) The theoretical (calculated) time required for a small amount of water to pass through a tank at a given rate of flow.

(2) The actual time in hours, minutes or seconds that a small amount of water is in a settling basin, flocculating basin or rapid-mix chamber. In storage reservoirs, detention time is the length of time entering water will be held before being drafted for use (several weeks to years, several months being typical).

$$\text{Detention Time, hr} = \frac{(\text{Basin Volume, gal})(24\ \text{hr/day})}{\text{Flow, gal/day}}$$

DEW POINT DEW POINT

The temperature to which air with a given quantity of water vapor must be cooled to cause condensation of the vapor in the air.

DEWATER DEWATER

(1) To remove or separate a portion of the water present in a sludge or slurry. To dry sludge so it can be handled and disposed of.

(2) To remove or drain the water from a tank or a trench.

DIATOMACEOUS (DYE-uh-toe-MAY-shus) EARTH DIATOMACEOUS EARTH

A fine, siliceous (made of silica) "earth" composed mainly of the skeletal remains of diatoms.

DIATOMS (DYE-uh-toms) DIATOMS

Unicellular (single cell), microscopic algae with a rigid (box-like) internal structure consisting mainly of silica.

DIELECTRIC (DIE-ee-LECK-trick) DIELECTRIC

Does not conduct an electric current. An insulator or nonconducting substance.

DIGITAL READOUT DIGITAL READOUT

Use of numbers to indicate the value or measurement of a variable. The readout of an instrument by a direct, numerical reading of the measured value. The signal sent to such readouts is usually an analog signal.

DILUTE SOLUTION DILUTE SOLUTION

A solution that has been made weaker usually by the addition of water.

DIMICTIC (die-MICK-tick) DIMICTIC

Lakes and reservoirs which freeze over and normally go through two stratification and two mixing cycles within a year.

DIRECT CURRENT (D.C.) DIRECT CURRENT (D.C.)

Electric current flowing in one direction only and essentially free from pulsation.

DIRECT FILTRATION DIRECT FILTRATION

A method of treating water which consists of the addition of coagulant chemicals, flash mixing, coagulation, minimal flocculation, and filtration. The flocculation facilities may be omitted, but the physical-chemical reactions will occur to some extent. The sedimentation process is omitted. Also see CONVENTIONAL FILTRATION and IN-LINE FILTRATION.

DIRECT RUNOFF DIRECT RUNOFF

Water that flows over the ground surface or through the ground directly into streams, rivers, or lakes.

DISCHARGE HEAD DISCHARGE HEAD

The pressure (in pounds per square inch or psi) measured at the centerline of a pump discharge and very close to the discharge flange, converted into feet. The pressure is measured from the centerline of the pump to the hydraulic grade line of the water in the discharge pipe.

$$\text{Discharge Head, ft} = (\text{Discharge Pressure, psi})(2.31\ \text{ft/psi})$$

DISINFECTION (dis-in-FECT-shun)   DISINFECTION

The process designed to kill or inactivate most microorganisms in water, including essentially all pathogenic (disease-causing) bacteria. There are several ways to disinfect, with chlorination being the most frequently used in water treatment. Compare with STERILIZATION.

DISINFECTION BY-PRODUCT (DBP)   DISINFECTION BY-PRODUCT (DBP)

A contaminant formed by the reaction of disinfection chemicals (such as chlorine) with other substances in the water being disinfected.

DISTILLATE (DIS-tuh-late)   DISTILLATE

In the distillation of a sample, a portion is collected by evaporation and recondensation; the part that is recondensed is the distillate.

DIVALENT (die-VAY-lent)   DIVALENT

Having a valence of two, such as the ferrous ion, Fe^{2+}. Also called bivalent.

DIVERSION   DIVERSION

Use of part of a stream flow as a water supply.

DRAFT   DRAFT

(1) The act of drawing or removing water from a tank or reservoir.

(2) The water which is drawn or removed from a tank or reservoir.

DRAWDOWN   DRAWDOWN

(1) The drop in the water table or level of water in the ground when water is being pumped from a well.

(2) The amount of water used from a tank or reservoir.

(3) The drop in the water level of a tank or reservoir.

DRIFT   DRIFT

The difference between the actual value and the desired value (or set point); characteristic of proportional controllers that do not incorporate reset action. Also called OFFSET.

DYNAMIC PRESSURE   DYNAMIC PRESSURE

When a pump is operating, the vertical distance (in feet) from a reference point (such as a pump centerline) to the hydraulic grade line is the dynamic head. Also see ENERGY GRADE LINE, STATIC HEAD, STATIC PRESSURE, and TOTAL DYNAMIC HEAD.

$$\text{Dynamic Pressure, psi} = (\text{Dynamic Head, ft})(0.433 \text{ psi/ft})$$

E

EPA   EPA

United States **E**nvironmental **P**rotection **A**gency. A regulatory agency established by the U.S. Congress to administer the nation's environmental laws. Also called the U.S. EPA.

EDUCTOR (e-DUCK-ter)   EDUCTOR

A hydraulic device used to create a negative pressure (suction) by forcing a liquid through a restriction, such as a Venturi. An eductor or aspirator (the hydraulic device) may be used in the laboratory in place of a vacuum pump. As an injector, it is used to produce vacuum for chlorinators. Sometimes used instead of a suction pump.

EFFECTIVE RANGE   EFFECTIVE RANGE

That portion of the design range (usually from 10 to 90+ percent) in which an instrument has acceptable accuracy. Also see RANGE and SPAN.

EFFECTIVE SIZE (E.S.)   EFFECTIVE SIZE (E.S.)

The diameter of the particles in a granular sample (filter media) for which 10 percent of the total grains are smaller and 90 percent larger on a weight basis. Effective size is obtained by passing granular material through sieves with varying dimensions of mesh and weighing the material retained by each sieve. The effective size is also approximately the average size of the grains.

EFFLUENT

EFFLUENT

Water or other liquid—raw (untreated), partially or completely treated—flowing *FROM* a reservoir, basin, treatment process, or treatment plant.

EJECTOR

EJECTOR

A device used to disperse a chemical solution into water being treated.

ELECTROCHEMICAL REACTION

ELECTROCHEMICAL REACTION

Chemical changes produced by electricity (electrolysis) or the production of electricity by chemical changes (galvanic action). In corrosion, a chemical reaction is accompanied by the flow of electrons through a metallic path. The electron flow may come from an external source and cause the reaction, such as electrolysis caused by a D.C. (direct current) electric railway or the electron flow may be caused by a chemical reaction as in the galvanic action of a flashlight dry cell.

ELECTROCHEMICAL SERIES

ELECTROCHEMICAL SERIES

A list of metals with the standard electrode potentials given in volts. The size and sign of the electrode potential indicates how easily these elements will take on or give up electrons, or corrode. Hydrogen is conventionally assigned a value of zero.

ELECTROLYSIS (ee-leck-TRAWL-uh-sis)

ELECTROLYSIS

The decomposition of material by an outside electric current.

ELECTROLYTE (ee-LECK-tro-LITE)

ELECTROLYTE

A substance which dissociates (separates) into two or more ions when it is dissolved in water.

ELECTROLYTIC (ee-LECK-tro-LIT-ick) CELL

ELECTROLYTIC CELL

A device in which the chemical decomposition of material causes an electric current to flow. Also, a device in which a chemical reaction occurs as a result of the flow of electric current. Chlorine and caustic (NaOH) are made from salt (NaCl) in electrolytic cells.

ELECTROMOTIVE FORCE (E.M.F.)

ELECTROMOTIVE FORCE (E.M.F.)

The electrical pressure available to cause a flow of current (amperage) when an electric circuit is closed. Also called VOLTAGE.

ELECTROMOTIVE SERIES

ELECTROMOTIVE SERIES

A list of metals and alloys presented in the order of their tendency to corrode (or go into solution). Also called the GALVANIC SERIES. This is a practical application of the theoretical ELECTROCHEMICAL SERIES.

ELECTRON

ELECTRON

(1) A very small, negatively charged particle which is practically weightless. According to the electron theory, all electrical and electronic effects are caused either by the movement of electrons from place to place or because there is an excess or lack of electrons at a particular place.

(2) The part of an atom that determines its chemical properties.

ELEMENT

ELEMENT

A substance which cannot be separated into its constituent parts and still retain its chemical identity. For example, sodium (Na) is an element.

END BELLS

END BELLS

Devices used to hold the rotor and stator of a motor in position.

END POINT

END POINT

Samples of water or wastewater are titrated to the end point. This means that a chemical is added, drop by drop, to a sample until a certain color change (blue to clear, for example) occurs. This is called the *END POINT* of the titration. In addition to a color change, an end point may be reached by the formation of a precipitate or the reaching of a specified pH. An end point may be detected by the use of an electronic device such as a pH meter. The completion of a desired chemical reaction.

ENDEMIC (en-DEM-ick)

ENDEMIC

Something peculiar to a particular people or locality, such as a disease which is always present in the population.

ENDRIN (EN-drin)

ENDRIN

A pesticide toxic to freshwater and marine aquatic life that produces adverse health effects in domestic water supplies.

ENERGY GRADE LINE (EGL)

ENERGY GRADE LINE (EGL)

A line that represents the elevation of energy head (in feet) of water flowing in a pipe, conduit or channel. The line is drawn above the hydraulic grade line (gradient) a distance equal to the velocity head ($V^2/2g$) of the water flowing at each section or point along the pipe or channel. Also see HYDRAULIC GRADE LINE.

[SEE DRAWING ON PAGE 670]

ENTERIC

ENTERIC

Of intestinal origin, especially applied to wastes or bacteria.

ENTRAIN

ENTRAIN

To trap bubbles in water either mechanically through turbulence or chemically through a reaction.

ENZYMES (EN-zimes)

ENZYMES

Organic substances (produced by living organisms) which cause or speed up chemical reactions. Organic catalysts and/or bio-chemical catalysts.

EPIDEMIC (EP-uh-DEM-ick)

EPIDEMIC

A disease that occurs in a large number of people in a locality at the same time and spreads from person to person.

EPIDEMIOLOGY (EP-uh-DE-me-ALL-o-gee)

EPIDEMIOLOGY

A branch of medicine which studies epidemics (diseases which affect significant numbers of people during the same time period in the same locality). The objective of epidemiology is to determine the factors that cause epidemic diseases and how to prevent them.

EPILIMNION (EP-uh-LIM-knee-on)

EPILIMNION

The upper layer of water in a thermally stratified lake or reservoir. This layer consists of the warmest water and has a fairly uniform (constant) temperature. The layer is readily mixed by wind action.

EQUILIBRIUM, CALCIUM CARBONATE

EQUILIBRIUM, CALCIUM CARBONATE

A water is considered stable when it is just saturated with calcium carbonate. In this condition the water will neither dissolve nor deposit calcium carbonate. Thus, in this water the calcium carbonate is in equilibrium with the hydrogen ion concentration.

EQUITY

EQUITY

The value of an investment in a facility.

EQUIVALENT WEIGHT

EQUIVALENT WEIGHT

That weight which will react with, displace or is equivalent to one gram atom of hydrogen.

ESTER

ESTER

A compound formed by the reaction between an acid and an alcohol with the elimination of a molecule of water.

EUTROPHIC (you-TRO-fick)

EUTROPHIC

Reservoirs and lakes which are rich in nutrients and very productive in terms of aquatic animal and plant life.

EUTROPHICATION (you-TRO-fi-KAY-shun)

EUTROPHICATION

The increase in the nutrient levels of a lake or other body of water; this usually causes an increase in the growth of aquatic animal and plant life.

EVAPORATION

EVAPORATION

The process by which water or other liquid becomes a gas (water vapor or ammonia vapor).

EVAPOTRANSPIRATION (ee-VAP-o-TRANS-purr-A-shun)

EVAPOTRANSPIRATION

(1) The process by which water vapor passes into the atmosphere from living plants. Also called TRANSPIRATION.

(2) The total water removed from an area by transpiration (plants) and by evaporation from soil, snow and water surfaces.

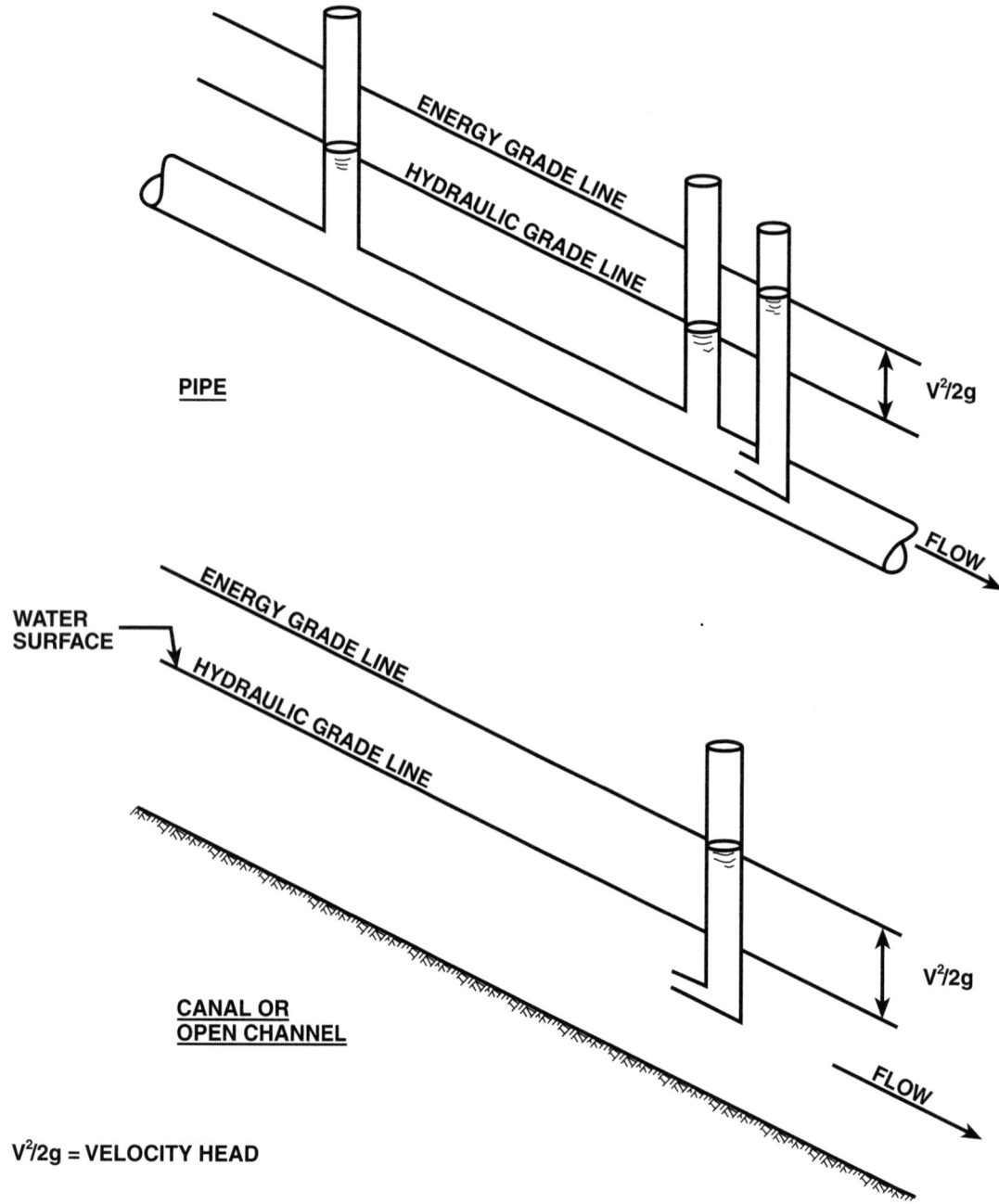

PIPE

$V^2/2g$

FLOW

WATER
SURFACE

ENERGY GRADE LINE

HYDRAULIC GRADE LINE

CANAL OR
OPEN CHANNEL

$V^2/2g$

FLOW

$V^2/2g$ = VELOCITY HEAD

ENERGY GRADE LINE and HYDRAULIC GRADE LINE

F

FACULTATIVE (FACK-ul-TAY-tive)   FACULTATIVE

Facultative bacteria can use either dissolved molecular oxygen or oxygen obtained from food materials such as sulfate or nitrate ions. In other words, facultative bacteria can live under aerobic, anoxic, or anaerobic conditions.

FEEDBACK   FEEDBACK

The circulating action between a sensor measuring a process variable and the controller which controls or adjusts the process variable.

FEEDWATER   FEEDWATER

The water that is fed to a treatment process; the water that is going to be treated.

FINISHED WATER   FINISHED WATER

Water that has passed through a water treatment plant; all the treatment processes are completed or "finished." This water is ready to be delivered to consumers. Also called PRODUCT WATER.

FIX, SAMPLE   FIX, SAMPLE

A sample is "fixed" in the field by adding chemicals that prevent the water quality indicators of interest in the sample from changing before final measurements are performed later in the lab.

FIXED COSTS   FIXED COSTS

Costs that a utility must cover or pay even if there is no demand for water or no water to sell to customers. Also see VARIABLE COSTS.

FLAGELLATES (FLAJ-el-LATES)   FLAGELLATES

Microorganisms that move by the action of tail-like projections.

FLAME POLISHED   FLAME POLISHED

Melted by a flame to smooth out irregularities. Sharp or broken edges of glass (such as the end of a glass tube) are rotated in a flame until the edge melts slightly and becomes smooth.

FLOAT ON SYSTEM   FLOAT ON SYSTEM

A method of operating a water storage facility. Daily flow into the facility is approximately equal to the average daily demand for water. When consumer demands for water are low, the storage facility will be filling. During periods of high demand, the facility will be emptying.

FLOC   FLOC

Clumps of bacteria and particulate impurities that have come together and formed a cluster. Found in flocculation tanks and settling or sedimentation basins.

FLOCCULATION (FLOCK-you-LAY-shun)   FLOCCULATION

The gathering together of fine particles after coagulation to form larger particles by a process of gentle mixing.

FLUIDIZED (FLEW-id-I-zd)   FLUIDIZED

A mass of solid particles that is made to flow like a liquid by injection of water or gas is said to have been fluidized. In water treatment, a bed of filter media is fluidized by backwashing water through the filter.

FLUORIDATION (FLOOR-uh-DAY-shun)   FLUORIDATION

The addition of a chemical to increase the concentration of fluoride ions in drinking water to a predetermined optimum limit to reduce the incidence (number) of dental caries (tooth decay) in children. Defluoridation is the removal of excess fluoride in drinking water to prevent the mottling (brown stains) of teeth.

FLUSHING   FLUSHING

A method used to clean water distribution lines. Hydrants are opened and water with a high velocity flows through the pipes, removes deposits from the pipes, and flows out the hydrants.

FLUX   FLUX

A flowing or flow.

FOOT VALVE FOOT VALVE

A special type of check valve located at the bottom end of the suction pipe on a pump. This valve opens when the pump operates to allow water to enter the suction pipe but closes when the pump shuts off to prevent water from flowing out of the suction pipe.

FREE AVAILABLE RESIDUAL CHLORINE FREE AVAILABLE RESIDUAL CHLORINE

That portion of the total available residual chlorine composed of dissolved chlorine gas (Cl_2), hypochlorous acid (HOCl), and/or hypochlorite ion (OCl⁻) remaining in water after chlorination. This does not include chlorine that has combined with ammonia, nitrogen, or other compounds.

FREE RESIDUAL CHLORINATION FREE RESIDUAL CHLORINATION

The application of chlorine to water to produce a free available chlorine residual equal to at least 80 percent of the total residual chlorine (sum of free and combined available chlorine residual).

FREEBOARD FREEBOARD

(1) The vertical distance from the normal water surface to the top of the confining wall.

(2) The vertical distance from the sand surface to the underside of a trough in a sand filter. This distance is also called AVAILABLE EXPANSION.

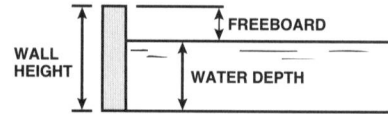

FRICTION LOSSES FRICTION LOSSES

The head, pressure or energy (they are the same) lost by water flowing in a pipe or channel as a result of turbulence caused by the velocity of the flowing water and the roughness of the pipe, channel walls, or restrictions caused by fittings. Water flowing in a pipe loses head, pressure or energy as a result of friction losses. Also see HEAD LOSS.

FUNGI (FUN-ji) FUNGI

Mushrooms, molds, mildews, rusts, and smuts that are small non-chlorophyll-bearing plants lacking roots, stems and leaves. They occur in natural waters and grow best in the absence of light. Their decomposition may cause objectionable tastes and odors in water.

FUSE FUSE

A protective device having a strip or wire of fusible metal which, when placed in a circuit, will melt and break the electric circuit if heated too much. High temperatures will develop in the fuse when a current flows through the fuse in excess of that which the circuit will carry safely.

G

GIS GIS

Geographic **I**nformation **S**ystem. A computer program that combines mapping with detailed information about the physical locations of structures such as pipes, valves, and manholes within geographic areas. The system is used to help operators and maintenance personnel locate utility system features or structures and to assist with the scheduling and performance of maintenance activities.

GAGE PRESSURE GAGE PRESSURE

The pressure within a closed container or pipe as measured with a gage. In contrast, absolute pressure is the sum of atmospheric pressure (14.7 lbs/sq in) *PLUS* pressure within a vessel (as measured with a gage). Most pressure gages read in "gage pressure" or psig (**p**ounds per **s**quare **i**nch **g**age pressure).

GALVANIC CELL GALVANIC CELL

An electrolytic cell capable of producing electric energy by electrochemical action. The decomposition of materials in the cell causes an electric (electron) current to flow from cathode to anode.

GALVANIC SERIES GALVANIC SERIES

A list of metals and alloys presented in the order of their tendency to corrode (or go into solution). Also called the ELECTROMOTIVE SERIES. This is a practical application of the theoretical ELECTROCHEMICAL SERIES.

GALVANIZE GALVANIZE

To coat a metal (especially iron or steel) with zinc. Galvanization is the process of coating a metal with zinc.

GARNET GARNET

A group of hard, reddish, glassy, mineral sands made up of silicates of base metals (calcium, magnesium, iron and manganese). Garnet has a higher density than sand.

GAUGE, PIPE GAUGE, PIPE

A number that defines the thickness of the sheet used to make steel pipe. The larger the number, the thinner the pipe wall.

GEOGRAPHIC INFORMATION SYSTEM (GIS) GEOGRAPHIC INFORMATION SYSTEM (GIS)

A computer program that combines mapping with detailed information about the physical locations of structures such as pipes, valves, and manholes within geographic areas. The system is used to help operators and maintenance personnel locate utility system features or structures and to assist with the scheduling and performance of maintenance activities.

GEOLOGICAL LOG GEOLOGICAL LOG

A detailed description of all underground features discovered during the drilling of a well (depth, thickness and type of formations).

GEOPHYSICAL LOG GEOPHYSICAL LOG

A record of the structure and composition of the earth encountered when drilling a well or similar type of test hole or boring.

GERMICIDE (GERM-uh-SIDE) GERMICIDE

A substance formulated to kill germs or microorganisms. The germicidal properties of chlorine make it an effective disinfectant.

GIARDIA (GEE-ARE-dee-ah) GIARDIA

A waterborne intestinal parasite that causes a disease called giardiasis (GEE-are-DIE-uh-sis) in infected humans. Symptoms of the disease include diarrhea, cramps, and weight loss. *Giardia* contamination is found in most surface waters and some groundwaters.

GIARDIASIS (GEE-are-DIE-uh-sis) GIARDIASIS

Intestinal disease caused by an infestation of *Giardia* flagellates.

GRAB SAMPLE GRAB SAMPLE

A single sample of water collected at a particular time and place which represents the composition of the water only at that time and place.

GRADE GRADE

(1) The elevation of the invert (or bottom) of a pipeline, canal, culvert, or similar conduit.

(2) The inclination or slope of a pipeline, conduit, stream channel, or natural ground surface; usually expressed in terms of the ratio or percentage of number of units of vertical rise or fall per unit of horizontal distance. A 0.5 percent grade would be a drop of one-half foot per hundred feet of pipe.

GRAVIMETRIC GRAVIMETRIC

A means of measuring unknown concentrations of water quality indicators in a sample by *WEIGHING* a precipitate or residue of the sample.

GRAVIMETRIC FEEDER GRAVIMETRIC FEEDER

A dry chemical feeder which delivers a measured weight of chemical during a specific time period.

GREENSAND GREENSAND

A mineral (glauconite) material that looks like ordinary filter sand except that it is green in color. Greensand is a natural ion exchange material which is capable of softening water. Greensand which has been treated with potassium permanganate ($KMnO_4$) is called manganese greensand; this product is used to remove iron, manganese and hydrogen sulfide from groundwaters.

GROUND GROUND

An expression representing an electrical connection to earth or a large conductor which is at the earth's potential or neutral voltage.

H

HTH (pronounce as separate letters) HTH

High **T**est **H**ypochlorite. Calcium hypochlorite or $Ca(OCl)_2$.

HARD WATER HARD WATER

Water having a high concentration of calcium and magnesium ions. A water may be considered hard if it has a hardness greater than the typical hardness of water from the region. Some textbooks define hard water as water with a hardness of more than 100 mg/*L* as calcium carbonate.

HARDNESS, WATER HARDNESS, WATER

A characteristic of water caused mainly by the salts of calcium and magnesium, such as bicarbonate, carbonate, sulfate, chloride and nitrate. Excessive hardness in water is undesirable because it causes the formation of soap curds, increased use of soap, deposition of scale in boilers, damage in some industrial processes, and sometimes causes objectionable tastes in drinking water.

HEAD HEAD

The vertical distance (in feet) equal to the pressure (in psi) at a specific point. The pressure head is equal to the pressure in psi times 2.31 ft/psi.

HEAD LOSS HEAD LOSS

The head, pressure or energy (they are the same) lost by water flowing in a pipe or channel as a result of turbulence caused by the velocity of the flowing water and the roughness of the pipe, channel walls, or restrictions caused by fittings. Water flowing in a pipe loses head, pressure or energy as a result of friction losses. The head loss through a filter is due to friction losses caused by material building up on the surface or in the top part of a filter. Also see FRICTION LOSSES.

HEADER HEADER

A large pipe to which the ends of a series of smaller pipes are connected. Also called a MANIFOLD.

HEAT SENSOR HEAT SENSOR

A device that opens and closes a switch in response to changes in the temperature. This device might be a metal contact, or a thermocouple which generates a minute electric current proportional to the difference in heat, or a variable resistor whose value changes in response to changes in temperature. Also called a TEMPERATURE SENSOR.

HECTARE (HECK-tar) HECTARE

A measure of area in the metric system similar to an acre. One hectare is equal to 10,000 square meters and 2.4711 acres.

HEPATITIS (HEP-uh-TIE-tis) HEPATITIS

Hepatitis is an inflammation of the liver caused by an acute viral infection. Yellow jaundice is one symptom of hepatitis.

HERBICIDE (HERB-uh-SIDE) HERBICIDE

A compound, usually a manmade organic chemical, used to kill or control plant growth.

HERTZ HERTZ

The number of complete electromagnetic cycles or waves in one second of an electric or electronic circuit. Also called the frequency of the current. Abbreviated Hz.

HETEROTROPHIC (HET-er-o-TROF-ick) HETEROTROPHIC

Describes organisms that use organic matter for energy and growth. Animals, fungi and most bacteria are heterotrophs.

HIGH-LINE JUMPERS HIGH-LINE JUMPERS

Pipes or hoses connected to fire hydrants and laid on top of the ground to provide emergency water service for an isolated portion of a distribution system.

HOSE BIB HOSE BIB

Faucet. A location in a water line where a hose is connected.

HYDRATED LIME HYDRATED LIME

Limestone that has been "burned" and treated with water under controlled conditions until the calcium oxide portion has been converted to calcium hydroxide ($Ca(OH)_2$). Hydrated lime is quicklime combined with water. $CaO + H_2O \rightarrow Ca(OH)_2$. Also called slaked lime. Also see QUICKLIME.

HYDRAULIC CONDUCTIVITY (K) HYDRAULIC CONDUCTIVITY (K)

A coefficient describing the relative ease with which groundwater can move through a permeable layer of rock or soil. Typical units of hydraulic conductivity are feet per day, gallons per day per square foot, or meters per day (depending on the unit chosen for the total discharge and the cross-sectional area).

HYDRAULIC GRADE LINE (HGL) HYDRAULIC GRADE LINE (HGL)

The surface or profile of water flowing in an open channel or a pipe flowing partially full. If a pipe is under pressure, the hydraulic grade line is at the level water would rise to in a small vertical tube connected to the pipe. Also see ENERGY GRADE LINE.

[SEE DRAWING ON PAGE 670]

HYDRAULIC GRADIENT   HYDRAULIC GRADIENT

The slope of the hydraulic grade line. This is the slope of the water surface in an open channel, the slope of the water surface of the groundwater table, or the slope of the water pressure for pipes under pressure.

HYDROGEOLOGIST (HI-dro-gee-ALL-uh-gist)   HYDROGEOLOGIST

A person who studies and works with groundwater.

HYDROLOGIC (HI-dro-LOJ-ick) CYCLE   HYDROLOGIC CYCLE

The process of evaporation of water into the air and its return to earth by precipitation (rain or snow). This process also includes transpiration from plants, groundwater movement, and runoff into rivers, streams and the ocean. Also called the WATER CYCLE.

HYDROLYSIS (hi-DROLL-uh-sis)   HYDROLYSIS

(1) A chemical reaction in which a compound is converted into another compound by taking up water.

(2) Usually a chemical degradation of organic matter.

HYDROPHILIC (HI-dro-FILL-ick)   HYDROPHILIC

Having a strong affinity (liking) for water. The opposite of HYDROPHOBIC.

HYDROPHOBIC (HI-dro-FOE-bick)   HYDROPHOBIC

Having a strong aversion (dislike) for water. The opposite of HYDROPHILIC.

HYDROPNEUMATIC (HI-dro-new-MAT-ick)   HYDROPNEUMATIC

A water system, usually small, in which a water pump is automatically controlled (started and stopped) by the air pressure in a compressed-air tank.

HYDROSTATIC (HI-dro-STAT-ick) PRESSURE   HYDROSTATIC PRESSURE

(1) The pressure at a specific elevation exerted by a body of water at rest, or

(2) In the case of groundwater, the pressure at a specific elevation due to the weight of water at higher levels in the same zone of saturation.

HYGROSCOPIC (HI-grow-SKOP-ick)   HYGROSCOPIC

Absorbing or attracting moisture from the air.

HYPOCHLORINATION (HI-poe-KLOR-uh-NAY-shun)   HYPOCHLORINATION

The application of hypochlorite compounds to water for the purpose of disinfection.

HYPOCHLORINATORS (HI-poe-KLOR-uh-NAY-tors)   HYPOCHLORINATORS

Chlorine pumps, chemical feed pumps or devices used to dispense chlorine solutions made from hypochlorites such as bleach (sodium hypochlorite) or calcium hypochlorite into the water being treated.

HYPOCHLORITE (HI-poe-KLOR-ite)   HYPOCHLORITE

Chemical compounds containing available chlorine; used for disinfection. They are available as liquids (bleach) or solids (powder, granules, and pellets) in barrels, drums, and cans. Salts of hypochlorous acid.

HYPOLIMNION (HI-poe-LIM-knee-on)   HYPOLIMNION

The lowest layer in a thermally stratified lake or reservoir. This layer consists of colder, more dense water, has a constant temperature and no mixing occurs.

I

ICR   ICR

The Information Collection Rule (ICR) specifies the requirements for monitoring microbial contaminants and disinfection by-products (DBPs) by large public water systems (PWSs). It also requires large PWSs to conduct either bench- or pilot-scale testing of advanced treatment techniques.

IDLH   IDLH

Immediately **D**angerous to **L**ife or **H**ealth. The atmospheric concentration of any toxic, corrosive or asphyxiant substance that poses an immediate threat to life or would cause irreversible or delayed adverse health effects or would interfere with an individual's ability to escape from a dangerous atmosphere.

IMHOFF CONE   IMHOFF CONE

A clear, cone-shaped container marked with graduations. The cone is used to measure the volume of settleable solids in a specific volume (usually one liter) of water.

IMPELLER   IMPELLER

A rotating set of vanes in a pump or compressor designed to pump or move water or air.

IMPERMEABLE (im-PURR-me-uh-BULL)   IMPERMEABLE

Not easily penetrated. The property of a material or soil that does not allow, or allows only with great difficulty, the movement or passage of water.

INDICATOR (CHEMICAL)   INDICATOR (CHEMICAL)

A substance that gives a visible change, usually of color, at a desired point in a chemical reaction, generally at a specified end point.

INDICATOR (INSTRUMENT)   INDICATOR (INSTRUMENT)

A device which indicates the result of a measurement. Most indicators in the water utility field use either a fixed scale and movable indicator (pointer) such as a pressure gage or a movable scale and movable indicator like those used on a circular flow-recording chart. Also called a RECEIVER.

INFILTRATION (IN-fill-TRAY-shun)   INFILTRATION

The seepage of groundwater into a sewer system, including service connections. Seepage frequently occurs through defective or cracked pipes, pipe joints, connections or manhole walls.

INFLUENT   INFLUENT

Water or other liquid—raw (untreated) or partially treated—flowing *INTO* a reservoir, basin, treatment process, or treatment plant.

INFORMATION COLLECTION RULE (ICR)   INFORMATION COLLECTION RULE (ICR)

The Information Collection Rule (ICR) specifies the requirements for monitoring microbial contaminants and disinfection by-products (DBPs) by large public water systems (PWSs). It also requires large PWSs to conduct either bench- or pilot-scale testing of advanced treatment techniques.

INITIAL SAMPLING   INITIAL SAMPLING

The very first sampling conducted under the Safe Drinking Water Act (SDWA) for each of the applicable contaminant categories.

INJECTOR WATER   INJECTOR WATER

Service water in which chlorine is added (injected) to form a chlorine solution.

IN-LINE FILTRATION   IN-LINE FILTRATION

The addition of chemical coagulants directly to the filter inlet pipe. The chemicals are mixed by the flowing water. Flocculation and sedimentation facilities are eliminated. This pretreatment method is commonly used in pressure filter installations. Also see CONVENTIONAL FILTRATION and DIRECT FILTRATION.

INORGANIC   INORGANIC

Material such as sand, salt, iron, calcium salts and other mineral materials. Inorganic substances are of mineral origin, whereas organic substances are usually of animal or plant origin. Also see ORGANIC.

INORGANIC WASTE   INORGANIC WASTE

Waste material such as sand, salt, iron, calcium, and other mineral materials which are only slightly affected by the action of organisms. Inorganic wastes are chemical substances of mineral origin; whereas organic wastes are chemical substances of an animal or plant origin.

INPUT HORSEPOWER   INPUT HORSEPOWER

The total power used in operating a pump and motor.

$$\text{Input Horsepower, HP} = \frac{(\text{Brake Horsepower, HP})(100\%)}{\text{Motor Efficiency, \%}}$$

INSECTICIDE   INSECTICIDE

Any substance or chemical formulated to kill or control insects.

INSOLUBLE (in-SAWL-you-bull)   INSOLUBLE

Something that cannot be dissolved.

INTEGRATOR   INTEGRATOR

A device or meter that continuously measures and calculates (adds) a process rate variable in cumulative fashion; for example, total flows displayed in gallons, million gallons, cubic feet, or some other unit of volume measurement. Also called a TOTALIZER.

INTERFACE   INTERFACE

The common boundary layer between two substances such as water and a solid (metal); or between two fluids such as water and a gas (air); or between a liquid (water) and another liquid (oil).

INTERLOCK   INTERLOCK

An electric switch, usually magnetically operated. Used to interrupt all (local) power to a panel or device when the door is opened or the circuit is exposed to service.

INTERNAL FRICTION   INTERNAL FRICTION

Friction within a fluid (water) due to cohesive forces.

INTERSTICE (in-TUR-stuhz)   INTERSTICE

A very small open space in a rock or granular material. Also called a PORE, VOID, or void space. Also see VOID.

INVERT (IN-vert)   INVERT

The lowest point of the channel inside a pipe, conduit, or canal.

ION   ION

An electrically charged atom, radical (such as SO_4^{2-}), or molecule formed by the loss or gain of one or more electrons.

ION EXCHANGE   ION EXCHANGE

A water treatment process involving the reversible interchange (switching) of ions between the water being treated and the solid resin. Undesirable ions in the water are switched with acceptable ions on the resin.

ION EXCHANGE RESINS   ION EXCHANGE RESINS

Insoluble polymers, used in water treatment, that are capable of exchanging (switching or giving) acceptable cations or anions to the water being treated for less desirable ions.

IONIC CONCENTRATION   IONIC CONCENTRATION

The concentration of any ion in solution, usually expressed in moles per liter.

IONIZATION (EYE-on-uh-ZAY-shun)   IONIZATION

The splitting or dissociation (separation) of molecules into negatively and positively charged ions.

J

JAR TEST   JAR TEST

A laboratory procedure that simulates a water treatment plant's coagulation/flocculation units with differing chemical doses and also energy of rapid mix, energy of slow mix, and settling time. The purpose of this procedure is to *ESTIMATE* the minimum or ideal coagulant dose required to achieve certain water quality goals. Samples of water to be treated are commonly placed in six jars. Various amounts of chemicals are added to each jar, stirred and the settling of solids is observed. The dose of chemicals that provides satisfactory settling, removal of turbidity and/or color is the dose used to treat the water being taken into the plant at that time. When evaluating the results of a jar test, the operator should also consider the floc quality in the flocculation area and the floc loading on the filter.

JOGGING   JOGGING

The frequent starting and stopping of an electric motor.

JOULE (jewel)   JOULE

A measure of energy, work or quantity of heat. One joule is the work done when the point of application of a force of one newton is displaced a distance of one meter in the direction of the force. Approximately equal to 0.7375 ft-lbs (0.1022 m-kg).

K

KELLY   KELLY

The square section of a rod which causes the rotation of the drill bit. Torque from a drive table is applied to the square rod to cause the rotary motion. The drive table is chain or gear driven by an engine.

KILO   KILO

(1) Kilogram.

(2) Kilometer.

(3) A prefix meaning "thousand" used in the metric system and other scientific systems of measurement.

KINETIC ENERGY   KINETIC ENERGY

Energy possessed by a moving body of matter, such as water, as a result of its motion.

KJELDAHL (KELL-doll) NITROGEN   KJELDAHL NITROGEN

Nitrogen in the form of organic proteins or their decomposition product ammonia, as measured by the Kjeldahl Method.

L

LANGELIER INDEX (L.I.)   LANGELIER INDEX (L.I.)

An index reflecting the equilibrium pH of a water with respect to calcium and alkalinity. This index is used in stabilizing water to control both corrosion and the deposition of scale.

$$\text{Langelier Index} = pH - pH_S$$
$$\text{where } pH = \text{actual pH of the water, and}$$
$$pH_S = pH \text{ at which water having the same alkalinity and calcium content is just saturated with calcium carbonate.}$$

LAUNDERING WEIR (LAWN-der-ing weer)   LAUNDERING WEIR

Sedimentation basin overflow weir. A plate with V-notches along the top to ensure a uniform flow rate and avoid short-circuiting.

LAUNDERS (LAWN-ders)   LAUNDERS

Sedimentation basin and filter discharge channels consisting of overflow weir plates (in sedimentation basins) and conveying troughs.

LEAD (LEE-d)   LEAD

A wire or conductor that can carry electric current.

LEATHERS   LEATHERS

O-rings or gaskets used with piston pumps to provide a seal between the piston and the side wall.

LEVEL CONTROL   LEVEL CONTROL

A float device (or pressure switch) which senses changes in a measured variable and opens or closes a switch in response to that change. In its simplest form, this control might be a floating ball connected mechanically to a switch or valve such as is used to stop water flow into a toilet when the tank is full.

LINDANE (LYNN-dane)   LINDANE

A pesticide that causes adverse health effects in domestic water supplies and also is toxic to freshwater and marine aquatic life.

LINEARITY (LYNN-ee-AIR-it-ee)   LINEARITY

How closely an instrument measures actual values of a variable through its effective range; a measure used to determine the accuracy of an instrument.

LITTORAL (LIT-or-al) ZONE   LITTORAL ZONE

(1) That portion of a body of fresh water extending from the shoreline lakeward to the limit of occupancy of rooted plants.

(2) The strip of land along the shoreline between the high and low water levels.

LOGARITHM (LOG-a-rith-m)   LOGARITHM

The exponent that indicates the power to which a number must be raised to produce a given number. For example: if $B^2 = N$, the 2 is the logarithm of N (to the base B), or $10^2 = 100$ and $\log_{10} 100 = 2$. Also abbreviated to "log."

LOGGING, ELECTRICAL   LOGGING, ELECTRICAL

A procedure used to determine the porosity (spaces or voids) of formations in search of water-bearing formations (aquifers). Electrical probes are lowered into wells, an electric current is induced at various depths, and the resistance measured of various formations indicates the porosity of the material.

LOWER EXPLOSIVE LIMIT (LEL)   LOWER EXPLOSIVE LIMIT (LEL)

The lowest concentration of gas or vapor (percent by volume in air) that explodes if an ignition source is present at ambient temperature. At temperatures above 250°F the LEL decreases because explosibility increases with higher temperature.

M

M or MOLAR   *M* or MOLAR

A molar solution consists of one gram molecular weight of a compound dissolved in enough water to make one liter of solution. A gram molecular weight is the molecular weight of a compound in grams. For example, the molecular weight of sulfuric acid (H_2SO_4) is 98. A one *M* solution of sulfuric acid would consist of 98 grams of H_2SO_4 dissolved in enough distilled water to make one liter of solution.

MBAS   MBAS

Methylene - **B**lue - **A**ctive **S**ubstances. These substances are used in surfactants or detergents.

MCL   MCL

Maximum **C**ontaminant **L**evel. The largest allowable amount. MCLs for various water quality indicators are specified in the National Primary Drinking Water Regulations (NPDWR).

MCLG   MCLG

Maximum **C**ontaminant **L**evel **G**oal. MCLGs are health goals based entirely on health effects. They are a preliminary standard set but not enforced by EPA. MCLs consider health effects, but also take into consideration the feasibility and cost of analysis and treatment of the regulated MCL. Although often less stringent than the corresponding MCLG, the MCL is set to protect health.

mg/*L*   mg/*L*

See MILLIGRAMS PER LITER, mg/*L*.

MPN (pronounce as separate letters)   MPN

MPN is the **M**ost **P**robable **N**umber of coliform-group organisms per unit volume of sample water. Expressed as a density or population of organisms per 100 m*L* of sample water.

MSDS   MSDS

See **M**ATERIAL **S**AFETY **D**ATA **S**HEET.

MACROSCOPIC (MACK-row-SKAWP-ick) ORGANISMS   MACROSCOPIC ORGANISMS

Organisms big enough to be seen by the eye without the aid of a microscope.

MANDREL (MAN-drill)   MANDREL

A special tool used to push bearings in or to pull sleeves out.

MANIFOLD   MANIFOLD

A large pipe to which the ends of a series of smaller pipes are connected. Also called a HEADER.

MANOMETER (man-NAH-mut-ter)   MANOMETER

An instrument for measuring pressure. Usually, a manometer is a glass tube filled with a liquid that is used to measure the difference in pressure across a flow measuring device such as an orifice or a Venturi meter. The instrument used to measure blood pressure is a type of manometer.

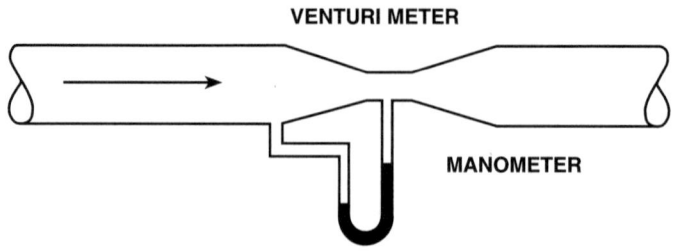

MATERIAL SAFETY DATA SHEET (MSDS)   MATERIAL SAFETY DATA SHEET (MSDS)

A document which provides pertinent information and a profile of a particular hazardous substance or mixture. An MSDS is normally developed by the manufacturer or formulator of the hazardous substance or mixture. The MSDS is required to be made available to employees and operators whenever there is the likelihood of the hazardous substance or mixture being introduced into the workplace. Some manufacturers are preparing MSDSs for products that are not considered to be hazardous to show that the product or substance is *NOT* hazardous.

MAXIMUM CONTAMINANT LEVEL (MCL)   MAXIMUM CONTAMINANT LEVEL (MCL)

See MCL.

MEASURED VARIABLE   MEASURED VARIABLE

A characteristic or component part that is sensed and quantified (reduced to a reading of some kind) by a primary element or sensor.

MECHANICAL JOINT   MECHANICAL JOINT

A flexible device that joins pipes or fittings together by the use of lugs and bolts.

MEG   MEG

A procedure used for checking the insulation resistance on motors, feeders, bus bar systems, grounds, and branch circuit wiring. Also see MEGGER.

MEGGER (from megohm)   MEGGER

An instrument used for checking the insulation resistance on motors, feeders, bus bar systems, grounds, and branch circuit wiring. Also see MEG.

MEGOHM   MEGOHM

Meg means one million, so 5 megohms means 5 million ohms. A megger reads in millions of ohms.

MENISCUS (meh-NIS-cuss)   MENISCUS

The curved surface of a column of liquid (water, oil, mercury) in a small tube. When the liquid wets the sides of the container (as with water), the curve forms a valley. When the confining sides are not wetted (as with mercury), the curve forms a hill or upward bulge. When a meniscus forms in a measuring device, the top of the liquid level of the sample is determined by the bottom of the meniscus.

MESH   MESH

One of the openings or spaces in a screen or woven fabric. The value of the mesh is usually given as the number of openings per inch. This value does not consider the diameter of the wire or fabric; therefore, the mesh number does not always have a definite relationship to the size of the hole.

MESOTROPHIC (MESS-o-TRO-fick)   MESOTROPHIC

Reservoirs and lakes which contain moderate quantities of nutrients and are moderately productive in terms of aquatic animal and plant life.

METABOLISM (meh-TAB-uh-LIZ-um)   METABOLISM

(1)   The biochemical processes in which food is used and wastes are formed by living organisms.

(2)   All biochemical reactions involved in cell formation and growth.

METALIMNION (MET-uh-LIM-knee-on)   METALIMNION

The middle layer in a thermally stratified lake or reservoir. In this layer there is a rapid decrease in temperature with depth. Also called the THERMOCLINE.

METHOXYCHLOR (meth-OXY-klor)   METHOXYCHLOR

A pesticide which causes adverse health effects in domestic water supplies and is also toxic to freshwater and marine aquatic life. The chemical name for methoxychlor is 2,2-bis(p-methoxyphenol)-1,1,1-trichloroethane.

METHYL ORANGE ALKALINITY   METHYL ORANGE ALKALINITY

A measure of the total alkalinity in a water sample. The alkalinity is measured by the amount of standard sulfuric acid required to lower the pH of the water to a pH level of 4.5, as indicated by the change in color of methyl orange from orange to pink. Methyl orange alkalinity is expressed as milligrams per liter equivalent calcium carbonate.

MICROBIAL (my-KROW-bee-ul) GROWTH   MICROBIAL GROWTH

The activity and growth of microorganisms such as bacteria, algae, diatoms, plankton and fungi.

MICRON (MY-kron)   MICRON

μm, Micrometer or Micron. A unit of length. One millionth of a meter or one thousandth of a millimeter. One micron equals 0.00004 of an inch.

MICROORGANISMS (MY-crow-OR-gan-IS-zums)   MICROORGANISMS

Living organisms that can be seen individually only with the aid of a microscope.

MIL   MIL

A unit of length equal to 0.001 of an inch. The diameter of wires and tubing is measured in mils, as is the thickness of plastic sheeting.

MILLIGRAMS PER LITER, mg/L MILLIGRAMS PER LITER, mg/L

A measure of the concentration by weight of a substance per unit volume. For practical purposes, one mg/L of a substance in fresh water is equal to one part per million parts (ppm). Thus a liter of water with a specific gravity of 1.0 weighs one million milligrams. If water contains 10 milligrams of calcium, the concentration is 10 milligrams per million milligrams, or 10 milligrams per liter (10 mg/L), or 10 parts of calcium per million parts of water, or 10 parts per million (10 ppm).

MILLIMICRON (MILL-uh-MY-kron)   MILLIMICRON

A unit of length equal to $10^{-3}\mu$ (one thousandth of a micron), 10^{-6} millimeters, or 10^{-9} meters; correctly called a nanometer, nm.

MOLAR   MOLAR

See *M* for MOLAR.

MOLE   MOLE

The molecular weight of a substance, usually expressed in grams.

MOLECULAR WEIGHT   MOLECULAR WEIGHT

The molecular weight of a compound in grams is the sum of the atomic weights of the elements in the compound. The molecular weight of sulfuric acid (H_2SO_4) in grams is 98.

Element	Atomic Weight	Number of Atoms	Molecular Weight
H	1	2	2
S	32	1	32
O	16	4	64
			98

MOLECULE (MOLL-uh-KULE)   MOLECULE

The smallest division of a compound that still retains or exhibits all the properties of the substance.

MONOMER (MON-o-MER)   MONOMER

A molecule of low molecular weight capable of reacting with identical or different monomers to form polymers.

MONOMICTIC (mo-no-MICK-tick)   MONOMICTIC

Lakes and reservoirs which are relatively deep, do not freeze over during the winter months, and undergo a single stratification and mixing cycle during the year. These lakes and reservoirs usually become destratified during the mixing cycle, usually in the fall of the year.

MONOVALENT   MONOVALENT

Having a valence of one, such as the cuprous (copper) ion, Cu^+.

MOST PROBABLE NUMBER (MPN)   MOST PROBABLE NUMBER (MPN)

See MPN.

MOTILE (MO-till)   MOTILE

Capable of self-propelled movement. A term that is sometimes used to distinguish between certain types of organisms found in water.

MOTOR EFFICIENCY   MOTOR EFFICIENCY

The ratio of energy delivered by a motor to the energy supplied to it during a fixed period or cycle. Motor efficiency ratings will vary depending upon motor manufacturer and usually will be near 90.0 percent.

MUDBALLS   MUDBALLS

Material that is approximately round in shape and varies from pea-sized up to two or more inches in diameter. This material forms in filters and gradually increases in size when not removed by the backwashing process.

MULTI-STAGE PUMP   MULTI-STAGE PUMP

A pump that has more than one impeller. A single-stage pump has one impeller.

N

N or NORMAL   *N* or NORMAL

A normal solution contains one gram equivalent weight of reactant (compound) per liter of solution. The equivalent weight of an acid is that weight which contains one gram atom of ionizable hydrogen or its chemical equivalent. For example, the equivalent weight of sulfuric acid (H_2SO_4) is 49 (98 divided by 2 because there are two replaceable hydrogen ions). A one *N* solution of sulfuric acid would consist of 49 grams of H_2SO_4 dissolved in enough water to make one liter of solution.

NESHTA (formerly NETA)   NESHTA

See **NATIONAL ENVIRONMENTAL, SAFETY & HEALTH TRAINING ASSOCIATION (NESHTA)**.

NETA   NETA

See **NATIONAL ENVIRONMENTAL, SAFETY & HEALTH TRAINING ASSOCIATION (NESHTA)**.

NIOSH (NYE-osh)   NIOSH

The **N**ational **I**nstitute of **O**ccupational **S**afety and **H**ealth is an organization that tests and approves safety equipment for particular applications. NIOSH is the primary federal agency engaged in research in the national effort to eliminate on-the-job hazards to the health and safety of working people. The NIOSH Publications Catalog, Seventh Edition, NIOSH Pub. No. 87-115, lists the NIOSH publications concerning industrial hygiene and occupational health. To obtain a copy of the catalog, write to National Technical Information Service (NTIS), 5285 Port Royal Road, Springfield, VA 22161. NTIS Stock No. PB88-175013, price, $141.00, plus $5.00 shipping and handling per order.

NOM (NATURAL ORGANIC MATTER)   NOM (NATURAL ORGANIC MATTER)

Humic substances composed of humic and fulvic acids that come from decayed vegetation.

NPDES PERMIT

NPDES PERMIT

National **P**ollutant **D**ischarge **E**limination **S**ystem permit is the regulatory agency document issued by either a federal or state agency which is designed to control all discharges of potential pollutants from point sources and storm water runoff into U.S. waterways. NPDES permits regulate discharges into navigable waters from all point sources of pollution, including industries, municipal wastewater treatment plants, sanitary landfills, large agricultural feedlots and return irrigation flows.

NPDWR

NPDWR

National **P**rimary **D**rinking **W**ater **R**egulations.

NSDWR

NSDWR

National **S**econdary **D**rinking **W**ater **R**egulations.

NTU

NTU

Nephelometric **T**urbidity **U**nits. See TURBIDITY UNITS (TU).

NAMEPLATE

NAMEPLATE

A durable metal plate found on equipment which lists critical operating conditions for the equipment.

NATIONAL ENVIRONMENTAL, SAFETY & HEALTH TRAINING ASSOCIATION (NESHTA) (formerly NATIONAL ENVIRONMENTAL TRAINING ASSOCIATION (NETA))

NATIONAL ENVIRONMENTAL, SAFETY & HEALTH TRAINING ASSOCIATION (NESHTA)

A professional organization devoted to serving the environmental trainer and promoting better operation of waterworks and pollution control facilities. For information on NESHTA membership and publications, contact NESHTA, 5320 North 16th Street, Suite 114, Phoenix, AZ 85016-3241. Phone (602) 956-6099.

NATIONAL ENVIRONMENTAL TRAINING ASSOCIATION (NETA)

NATIONAL ENVIRONMENTAL TRAINING ASSOCIATION (NETA)

See **N**ATIONAL **E**NVIRONMENTAL, **S**AFETY & **H**EALTH **T**RAINING **A**SSOCIATION (NESHTA).

NATIONAL INSTITUTE OF OCCUPATIONAL SAFETY AND HEALTH

NATIONAL INSTITUTE OF OCCUPATIONAL SAFETY AND HEALTH

See NIOSH.

NATIONAL PRIMARY DRINKING WATER REGULATIONS

NATIONAL PRIMARY DRINKING WATER REGULATIONS

Commonly referred to as NPDWR.

NATIONAL SECONDARY DRINKING WATER REGULATIONS

NATIONAL SECONDARY DRINKING WATER REGULATIONS

Commonly referred to as NSDWR.

NEPHELOMETRIC (NEFF-el-o-MET-rick)

NEPHELOMETRIC

A means of measuring turbidity in a sample by using an instrument called a nephelometer. A nephelometer passes light through a sample and the amount of light deflected (usually at a 90-degree angle) is then measured.

NEWTON

NEWTON

A force which, when applied to a body having a mass of one kilogram, gives it an acceleration of one meter per second per second.

NITRIFICATION (NYE-truh-fuh-KAY-shun)

NITRIFICATION

An aerobic process in which bacteria reduce the ammonia and organic nitrogen in water into nitrite and then nitrate.

NITROGENOUS (nye-TRAH-jen-us)

NITROGENOUS

A term used to describe chemical compounds (usually organic) containing nitrogen in combined forms. Proteins and nitrate are nitrogenous compounds.

NOBLE METAL

NOBLE METAL

A chemically inactive metal (such as gold). A metal that does not corrode easily and is much scarcer (and more valuable) than the so-called useful or base metals. Also see BASE METAL.

NOMINAL DIAMETER NOMINAL DIAMETER

An approximate measurement of the diameter of a pipe. Although the nominal diameter is used to describe the size or diameter of a pipe, it is usually not the exact inside diameter of the pipe.

NONIONIC (NON-eye-ON-ick) POLYMER NONIONIC POLYMER

A polymer that has no net electrical charge.

NON-PERMIT CONFINED SPACE NON-PERMIT CONFINED SPACE

See CONFINED SPACE, NON-PERMIT.

NONPOINT SOURCE NONPOINT SOURCE

A runoff or discharge from a field or similar source. A point source refers to a discharge that comes out the end of a pipe.

NONPOTABLE (non-POE-tuh-bull) NONPOTABLE

Water that may contain objectionable pollution, contamination, minerals, or infective agents and is considered unsafe and/or unpalatable for drinking.

NONVOLATILE MATTER NONVOLATILE MATTER

Material such as sand, salt, iron, calcium, and other mineral materials which are only slightly affected by the actions of organisms and are not lost on ignition of the dry solids at 550°C. Volatile materials are chemical substances usually of animal or plant origin. Also see INORGANIC WASTE and VOLATILE MATTER or VOLATILE SOLIDS.

NORMAL NORMAL

See *N* for NORMAL.

NOTICE NOTICE

This word calls attention to information that is especially significant in understanding and operating equipment or processes safely. Also see CAUTION, DANGER, and WARNING.

NUTRIENT NUTRIENT

Any substance that is assimilated (taken in) by organisms and promotes growth. Nitrogen and phosphorus are nutrients which promote the growth of algae. There are other essential and trace elements which are also considered nutrients.

O

ORP (pronounce as separate letters) ORP

Oxidation-**R**eduction **P**otential. The electrical potential required to transfer electrons from one compound or element (the oxidant) to another compound or element (the reductant); used as a qualitative measure of the state of oxidation in water treatment systems. ORP is measured in millivolts, with negative values indicating a tendency to reduce compounds or elements and positive values indicating a tendency to oxidize compounds or elements.

OSHA (O-shuh) OSHA

The Williams-Steiger **O**ccupational **S**afety and **H**ealth **A**ct of 1970 (OSHA) is a federal law designed to protect the health and safety of industrial workers and also the operators of water supply systems and treatment plants. The Act regulates the design, construction, operation and maintenance of water supply systems and water treatment plants. OSHA also refers to the federal and state agencies which administer the OSHA regulations.

OCCUPATIONAL SAFETY AND HEALTH ACT OF 1970 OCCUPATIONAL SAFETY AND HEALTH ACT OF 1970

See OSHA.

ODOR THRESHOLD ODOR THRESHOLD

The minimum odor of a water sample that can just be detected after successive dilutions with odorless water. Also called THRESHOLD ODOR.

OFFSET OFFSET

The difference between the actual value and the desired value (or set point); characteristic of proportional controllers that do not incorporate reset action. Also called DRIFT.

OHM OHM

The unit of electrical resistance. The resistance of a conductor in which one volt produces a current of one ampere.

OLFACTORY (ol-FAK-tore-ee) FATIGUE OLFACTORY FATIGUE

A condition in which a person's nose, after exposure to certain odors, is no longer able to detect the odor.

OLIGOTROPHIC (AH-lig-o-TRO-fick) OLIGOTROPHIC

Reservoirs and lakes which are nutrient poor and contain little aquatic plant or animal life.

OPERATING PRESSURE DIFFERENTIAL OPERATING PRESSURE DIFFERENTIAL

The operating pressure range for a hydropneumatic system. For example, when the pressure drops below 40 psi the pump will come on and stay on until the pressure builds up to 60 psi. When the pressure reaches 60 psi the pump will shut off.

OPERATING RATIO OPERATING RATIO

The operating ratio is a measure of the total revenues divided by the total operating expenses.

ORGANIC ORGANIC

Substances that come from animal or plant sources. Organic substances always contain carbon. (Inorganic materials are chemical substances of mineral origin.) Also see INORGANIC.

ORGANICS ORGANICS

(1) A term used to refer to chemical compounds made from carbon molecules. These compounds may be natural materials (such as animal or plant sources) or manmade materials (such as synthetic organics). Also see ORGANIC.

(2) Any form of animal or plant life. Also see BACTERIA.

ORGANISM ORGANISM

Any form of animal or plant life. Also see BACTERIA.

ORGANIZING ORGANIZING

Deciding who does what work and delegating authority to the appropriate persons.

ORIFICE (OR-uh-fiss) ORIFICE

An opening (hole) in a plate, wall, or partition. An orifice flange or plate placed in a pipe consists of a slot or a calibrated circular hole smaller than the pipe diameter. The difference in pressure in the pipe above and at the orifice may be used to determine the flow in the pipe.

ORTHOTOLIDINE (or-tho-TOL-uh-dine) ORTHOTOLIDINE

Orthotolidine is a colorimetric indicator of chlorine residual. If chlorine is present, a yellow-colored compound is produced. This reagent is no longer approved for chemical analysis to determine chlorine residual.

OSMOSIS (oz-MOE-sis) OSMOSIS

The passage of a liquid from a weak solution to a more concentrated solution across a semipermeable membrane. The membrane allows the passage of the water (solvent) but not the dissolved solids (solutes). This process tends to equalize the conditions on either side of the membrane.

OUCH PRINCIPLE OUCH PRINCIPLE

This principle says that as a manager when you delegate job tasks you must be **O**bjective, **U**niform in your treatment of employees, **C**onsistent with utility policies, and **H**ave job relatedness.

OVERALL EFFICIENCY, PUMP OVERALL EFFICIENCY, PUMP

The combined efficiency of a pump and motor together. Also called the WIRE-TO-WATER EFFICIENCY.

OVERDRAFT OVERDRAFT

The pumping of water from a groundwater basin or aquifer in excess of the supply flowing into the basin. This pumping results in a depletion or "mining" of the groundwater in the basin.

OVERFLOW RATE   OVERFLOW RATE

One of the guidelines for the design of settling tanks and clarifiers in treatment plants. Used by operators to determine if tanks and clarifiers are hydraulically (flow) over- or underloaded. Also called SURFACE LOADING.

$$\text{Overflow Rate, GPD/sq ft} = \frac{\text{Flow, gallons/day}}{\text{Surface Area, sq ft}}$$

OVERHEAD   OVERHEAD

Indirect costs necessary for a water utility to function properly. These costs are not related to the actual production and delivery of water to consumers, but include the costs of rent, lights, office supplies, management and administration.

OVERTURN   OVERTURN

The almost spontaneous mixing of all layers of water in a reservoir or lake when the water temperature becomes similar from top to bottom. This may occur in the fall/winter when the surface waters cool to the same temperature as the bottom waters and also in the spring when the surface waters warm after the ice melts. This is also called "turnover."

OXIDATION (ox-uh-DAY-shun)   OXIDATION

Oxidation is the addition of oxygen, removal of hydrogen, or the removal of electrons from an element or compound. In the environment, organic matter is oxidized to more stable substances. The opposite of REDUCTION.

OXIDATION-REDUCTION POTENTIAL (ORP)   OXIDATION-REDUCTION POTENTIAL (ORP)

The electrical potential required to transfer electrons from one compound or element (the oxidant) to another compound or element (the reductant); used as a qualitative measure of the state of oxidation in water treatment systems. ORP is measured in millivolts, with negative values indicating a tendency to reduce compounds or elements and positive values indicating a tendency to oxidize compounds or elements.

OXIDIZING AGENT   OXIDIZING AGENT

Any substance, such as oxygen (O_2) or chlorine (Cl_2), that will readily add (take on) electrons. The opposite is a REDUCING AGENT.

OXYGEN DEFICIENCY   OXYGEN DEFICIENCY

An atmosphere containing oxygen at a concentration of less than 19.5 percent by volume.

OXYGEN ENRICHMENT   OXYGEN ENRICHMENT

An atmosphere containing oxygen at a concentration of more than 23.5 percent by volume.

OZONATION (O-zoe-NAY-shun)   OZONATION

The application of ozone to water for disinfection or for taste and odor control.

P

PCBs   PCBs
See POLYCHLORINATED BIPHENYLS.

pCi/L   pCi/L

picoCurie per liter. A picoCurie is a measure of radioactivity. One picoCurie of radioactivity is equivalent to 0.037 nuclear disintegrations per second.

pcu (PLATINUM COBALT UNITS)   pcu (PLATINUM COBALT UNITS)

Platinum cobalt units are a measure of color using platinum cobalt standards by visual comparison.

PMCL   PMCL

Primary Maximum Contaminant Level. Primary MCLs for various water quality indicators are established to protect public health.

PPM   PPM
See PARTS PER MILLION.

PSIG

PSIG ·

Pounds per **S**quare **I**nch **G**age pressure. The pressure within a closed container or pipe measured with a gage in pounds per square inch. See GAGE PRESSURE.

PACKER ASSEMBLY

PACKER ASSEMBLY

An inflatable device used to seal the tremie pipe inside the well casing to prevent the grout from entering the inside of the conductor casing.

PALATABLE (PAL-uh-tuh-bull)

PALATABLE

Water at a desirable temperature that is free from objectionable tastes, odors, colors, and turbidity. Pleasing to the senses.

PARSHALL FLUME

PARSHALL FLUME

A device used to measure the flow in an open channel. The flume narrows to a throat of fixed dimensions and then expands again. The rate of flow can be calculated by measuring the difference in head (pressure) before and at the throat of the flume.

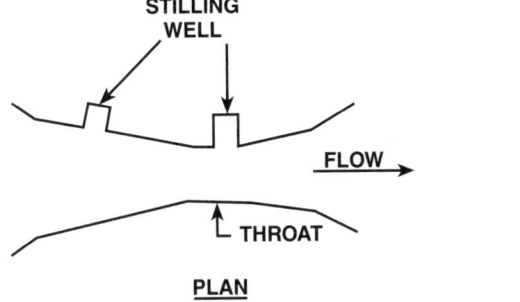

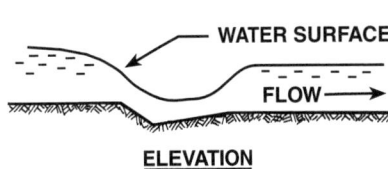

PLAN   ELEVATION

PARTICLE COUNT

PARTICLE COUNT

The results of a microscopic examination of treated water with a special particle counter which classifies suspended particles by number and size.

PARTICLE COUNTER

PARTICLE COUNTER

A device which counts and measures the size of individual particles in water. Particles are divided into size ranges and the number of particles is counted in each of these ranges. The results are reported in terms of the number of particles in different particle diameter size ranges per milliliter of water sampled.

PARTICLE COUNTING

PARTICLE COUNTING

A procedure for counting and measuring the size of individual particles in water. Particles are divided into size ranges and the number of particles is counted in each of these ranges. The results are reported in terms of the number of particles in different particle diameter size ranges per milliliter of water sampled.

PARTICULATE (par-TICK-you-let)

PARTICULATE

A very small solid suspended in water which can vary widely in size, shape, density, and electrical charge. Colloidal and dispersed particulates are artificially gathered together by the processes of coagulation and flocculation.

PARTS PER MILLION (PPM)

PARTS PER MILLION (PPM)

Parts per million parts, a measurement of concentration on a weight or volume basis. This term is equivalent to milligrams per liter (mg/L) which is the preferred term.

PASCAL

PASCAL

The pressure or stress of one newton per square meter. Abbreviated Pa.

$$1 \text{ psi} = 6,895 \text{ Pa} = 6.895 \text{ kN/sq m} = 0.0703 \text{ kg/sq cm}$$

PATHOGENIC (PATH-o-JEN-ick) ORGANISMS

PATHOGENIC ORGANISMS

Organisms, including bacteria, viruses or cysts, capable of causing diseases (giardiasis, cryptosporidiosis, typhoid, cholera, dysentery) in a host (such as a person). There are many types of organisms which do *NOT* cause disease. These organisms are called non-pathogenic.

PATHOGENS (PATH-o-jens)

PATHOGENS

Pathogenic or disease-causing organisms.

PEAK DEMAND

PEAK DEMAND

The maximum momentary load placed on a water treatment plant, pumping station or distribution system. This demand is usually the maximum average load in one hour or less, but may be specified as the instantaneous load or the load during some other short time period.

PERCENT SATURATION

PERCENT SATURATION

The amount of a substance that is dissolved in a solution compared with the amount dissolved in the solution at saturation, expressed as a percent.

$$\text{Percent Saturation, \%} = \frac{\text{Amount of Substance That Is Dissolved x 100\%}}{\text{Amount Dissolved in Solution at Saturation}}$$

PERCOLATING (PURR-co-LAY-ting) WATER

PERCOLATING WATER

Water that passes through soil or rocks under the force of gravity.

PERCOLATION (PURR-co-LAY-shun)

PERCOLATION

The slow passage of water through a filter medium; or, the gradual penetration of soil and rocks by water.

PERIPHYTON (pair-e-FI-tawn)

PERIPHYTON

Microscopic plants and animals that are firmly attached to solid surfaces under water such as rocks, logs, pilings and other structures.

PERMEABILITY (PURR-me-uh-BILL-uh-tee)

PERMEABILITY

The property of a material or soil that permits considerable movement of water through it when it is saturated.

PERMEATE (PURR-me-ate)

PERMEATE

(1) To penetrate and pass through, as water penetrates and passes through soil and other porous materials.

(2) The liquid (demineralized water) produced from the reverse osmosis process that contains a *LOW* concentration of dissolved solids.

PERMIT-REQUIRED CONFINED SPACE (PERMIT SPACE)

PERMIT-REQUIRED CONFINED SPACE (PERMIT SPACE)

See CONFINED SPACE, PERMIT-REQUIRED (PERMIT SPACE).

PESTICIDE

PESTICIDE

Any substance or chemical designed or formulated to kill or control animal pests. Also see INSECTICIDE and RODENTICIDE.

PET COCK

PET COCK

A small valve or faucet used to drain a cylinder or fitting.

pH (pronounce as separate letters)

pH

pH is an expression of the intensity of the basic or acidic condition of a liquid. Mathematically, pH is the logarithm (base 10) of the reciprocal of the hydrogen ion activity.

$$pH = Log \frac{1}{[H^+]}$$

The pH may range from 0 to 14, where 0 is most acidic, 14 most basic, and 7 neutral. Natural waters usually have a pH between 6.5 and 8.5.

PHENOLIC (fee-NO-lick) COMPOUNDS

PHENOLIC COMPOUNDS

Organic compounds that are derivatives of benzene.

PHENOLPHTHALEIN (FEE-nol-THAY-leen) ALKALINITY

PHENOLPHTHALEIN ALKALINITY

The alkalinity in a water sample measured by the amount of standard acid required to lower the pH to a level of 8.3, as indicated by the change in color of phenolphthalein from pink to clear. Phenolphthalein alkalinity is expressed as milligrams per liter of equivalent calcium carbonate.

PHOTOSYNTHESIS (foe-toe-SIN-thuh-sis)

PHOTOSYNTHESIS

A process in which organisms, with the aid of chlorophyll (green plant enzyme), convert carbon dioxide and inorganic substances into oxygen and additional plant material, using sunlight for energy. All green plants grow by this process.

PHYTOPLANKTON (FI-tow-PLANK-ton)   PHYTOPLANKTON

Small, usually microscopic plants (such as algae), found in lakes, reservoirs, and other bodies of water.

PICO   PICO

A prefix used in the metric system and other scientific systems of measurement which means 10^{-12} or 0.000 000 000 001.

PICOCURIE   PICOCURIE

A measure of radioactivity. One picoCurie of radioactivity is equivalent to 0.037 nuclear disintegrations per second.

PITLESS ADAPTER   PITLESS ADAPTER

A fitting which allows the well casing to be extended above ground while having a discharge connection located below the frost line. Advantages of using a pitless adapter include the elimination of the need for a pit or pump house and it is a watertight design, which helps maintain a sanitary water supply.

PLAN VIEW   PLAN VIEW

A diagram or photo showing a facility as it would appear when looking down on top of it.

PLANKTON   PLANKTON

(1)   Small, usually microscopic, plants (phytoplankton) and animals (zooplankton) in aquatic systems.

(2)   All of the smaller floating, suspended or self-propelled organisms in a body of water.

PLANNING   PLANNING

Management of utilities to build the resources and financial capability to provide for future needs.

PLUG FLOW   PLUG FLOW

A type of flow that occurs in tanks, basins or reactors when a slug of water moves through a tank without ever dispersing or mixing with the rest of the water flowing through the tank.

POINT SOURCE   POINT SOURCE

A discharge that comes out the end of a pipe. A nonpoint source refers to runoff or a discharge from a field or similar source.

POLARIZATION   POLARIZATION

The concentration of ions in the thin boundary layer adjacent to a membrane or pipe wall.

POLE SHADER   POLE SHADER

A copper bar circling the laminated iron core inside the coil of a magnetic starter.

POLLUTION   POLLUTION

The impairment (reduction) of water quality by agricultural, domestic, or industrial wastes (including thermal and radioactive wastes) to a degree that has an adverse effect on any beneficial use of water.

POLYANIONIC (poly-AN-eye-ON-ick)   POLYANIONIC

Characterized by many active negative charges especially active on the surface of particles.

POLYCHLORINATED BIPHENYLS   POLYCHLORINATED BIPHENYLS

A class of organic compounds that cause adverse health effects in domestic water supplies.

POLYELECTROLYTE (POLY-ee-LECK-tro-lite)   POLYELECTROLYTE

A high-molecular-weight (relatively heavy) substance having points of positive or negative electrical charges that is formed by either natural or manmade processes. Natural polyelectrolytes may be of biological origin or derived from starch products and cellulose derivatives. Manmade polyelectrolytes consist of simple substances that have been made into complex, high-molecular-weight substances. Used with other chemical coagulants to aid in binding small suspended particles to larger chemical flocs for their removal from water. Often called a POLYMER.

POLYMER (POLY-mer)   POLYMER

A long chain molecule formed by the union of many monomers (molecules of lower molecular weight). Polymers are used with other chemical coagulants to aid in binding small suspended particles to larger chemical flocs for their removal from water.

PORE

PORE

A very small open space in a rock or granular material. Also called an INTERSTICE, VOID, or void space. Also see VOID.

POROSITY

POROSITY

(1) A measure of the spaces or voids in a material or aquifer.

(2) The ratio of the volume of spaces in a rock or soil to the total volume. This ratio is usually expressed as a percentage.

$$\text{Porosity, \%} = \frac{(\text{Volume of Spaces})(100\%)}{\text{Total Volume}}$$

POSITIVE BACTERIOLOGICAL SAMPLE

POSITIVE BACTERIOLOGICAL SAMPLE

A water sample in which gas is produced by coliform organisms during incubation in the multiple tube fermentation test. See Chapter 11, Laboratory Procedures, "Coliform Bacteria."

POSITIVE DISPLACEMENT PUMP

POSITIVE DISPLACEMENT PUMP

A type of piston, diaphragm, gear or screw pump that delivers a constant volume with each stroke. Positive displacement pumps are used as chemical solution feeders.

POSTCHLORINATION

POSTCHLORINATION

The addition of chlorine to the plant effluent, *FOLLOWING* plant treatment, for disinfection purposes.

POTABLE (POE-tuh-bull) WATER

POTABLE WATER

Water that does not contain objectionable pollution, contamination, minerals, or infective agents and is considered satisfactory for drinking.

POWER FACTOR

POWER FACTOR

The ratio of the true power passing through an electric circuit to the product of the voltage and amperage in the circuit. This is a measure of the lag or lead of the current with respect to the voltage. In alternating current the voltage and amperes are not always in phase; therefore, the true power may be slightly less than that determined by the direct product.

PRECHLORINATION

PRECHLORINATION

The addition of chlorine at the headworks of the plant *PRIOR TO* other treatment processes mainly for disinfection and control of tastes, odors and aquatic growths. Also applied to aid in coagulation and settling.

PRECIPITATE (pre-SIP-uh-TATE)

PRECIPITATE

(1) An insoluble, finely divided substance which is a product of a chemical reaction within a liquid.

(2) The separation from solution of an insoluble substance.

PRECIPITATION (pre-SIP-uh-TAY-shun)

PRECIPITATION

(1) The process by which atmospheric moisture falls onto a land or water surface as rain, snow, hail, or other forms of moisture.

(2) The chemical transformation of a substance in solution into an insoluble form (precipitate).

PRECISION

PRECISION

The ability of an instrument to measure a process variable and repeatedly obtain the same result. The ability of an instrument to reproduce the same results.

PRECURSOR, THM (pre-CURSE-or)

PRECURSOR, THM

Natural organic compounds found in all surface and groundwaters. These compounds *MAY* react with halogens (such as chlorine) to form trihalomethanes (tri-HAL-o-METH-hanes) (THMs); they *MUST* be present in order for THMs to form.

PRESCRIPTIVE (pre-SKRIP-tive)

PRESCRIPTIVE

Water rights which are acquired by diverting water and putting it to use in accordance with specified procedures. These procedures include filing a request (with a state agency) to use unused water in a stream, river or lake.

PRESENT WORTH

PRESENT WORTH

The value of a long-term project expressed in today's dollars. Present worth is calculated by converting (discounting) all future benefits and costs over the life of the project to a single economic value at the start of the project. Calculating the present worth of alternative projects makes it possible to compare them and select the one with the largest positive (beneficial) present worth or minimum present cost.

PRESSURE CONTROL   PRESSURE CONTROL

A switch which operates on changes in pressure. Usually this is a diaphragm pressing against a spring. When the force on the diaphragm overcomes the spring pressure, the switch is actuated (activated).

PRESSURE HEAD   PRESSURE HEAD

The vertical distance (in feet) equal to the pressure (in psi) at a specific point. The pressure head is equal to the pressure in psi times 2.31 ft/psi.

PRESTRESSED   PRESTRESSED

A prestressed pipe has been reinforced with wire strands (which are under tension) to give the pipe an active resistance to loads or pressures on it.

PREVENTIVE MAINTENANCE UNITS   PREVENTIVE MAINTENANCE UNITS

Crews assigned the task of cleaning sewers (for example, balling or high-velocity cleaning crews) to prevent stoppages and odor complaints. Preventive maintenance is performing the most effective cleaning procedure, in the area where it is most needed, at the proper time in order to prevent failures and emergency situations.

PRIMARY ELEMENT   PRIMARY ELEMENT

(1)   A device that measures (senses) a physical condition or variable of interest. Floats and thermocouples are examples of primary elements. Also called a SENSOR.

(2)   The hydraulic structure used to measure flows. In open channels, weirs and flumes are primary elements or devices. Venturi meters and orifice plates are the primary elements in pipes or pressure conduits.

PRIME   PRIME

The action of filling a pump casing with water to remove the air. Most pumps must be primed before start-up or they will not pump any water.

PROCESS VARIABLE   PROCESS VARIABLE

A physical or chemical quantity which is usually measured and controlled in the operation of a water treatment plant or an industrial plant.

PRODUCT WATER   PRODUCT WATER

Water that has passed through a water treatment plant. All the treatment processes are completed or finished. This water is the product from the water treatment plant and is ready to be delivered to the consumers. Also called FINISHED WATER.

PROFILE   PROFILE

A drawing showing elevation plotted against distance, such as the vertical section or *SIDE* view of a pipeline.

PRUSSIAN BLUE   PRUSSIAN BLUE

A blue paste or liquid (often on a paper like carbon paper) used to show a contact area. Used to determine if gate valve seats fit properly.

PUMP BOWL   PUMP BOWL

The submerged pumping unit in a well, including the shaft, impellers and housing.

PUMPING WATER LEVEL   PUMPING WATER LEVEL

The vertical distance in feet from the centerline of the pump discharge to the level of the free pool while water is being drawn from the pool.

PURVEYOR (purr-VAY-or), WATER   PURVEYOR, WATER

An agency or person that supplies water (usually potable water).

PUTREFACTION (PEW-truh-FACK-shun)   PUTREFACTION

Biological decomposition of organic matter, with the production of foul-smelling and -tasting products, associated with anaerobic (no oxygen present) conditions.

Q

QUICKLIME   QUICKLIME

A material that is mostly calcium oxide (CaO) or calcium oxide in natural association with a lesser amount of magnesium oxide. Quicklime is capable of combining with water, that is, becoming slaked. Also see HYDRATED LIME.

R

RADIAL TO IMPELLER   RADIAL TO IMPELLER

Perpendicular to the impeller shaft. Material being pumped flows at a right angle to the impeller.

RADICAL   RADICAL

A group of atoms that is capable of remaining unchanged during a series of chemical reactions. Such combinations (radicals) exist in the molecules of many organic compounds; sulfate (SO_4^{2-}) is an inorganic radical.

RANGE   RANGE

The spread from minimum to maximum values that an instrument is designed to measure. Also see EFFECTIVE RANGE and SPAN.

RANNEY COLLECTOR   RANNEY COLLECTOR

This water collector is constructed as a dug well from 12 to 16 feet (3.5 to 5 m) in diameter that has been sunk as a caisson near the bank of a river or lake. Screens are driven radially and approximately horizontally from this well into the sand and the gravel deposits underlying the river.

[SEE DRAWING ON PAGE 693]

RATE OF RETURN   RATE OF RETURN

A value which indicates the return of funds received on the basis of the total equity capital used to finance physical facilities. Similar to the interest rate on savings accounts or loans.

RAW WATER   RAW WATER

(1) Water in its natural state, prior to any treatment.

(2) Usually the water entering the first treatment process of a water treatment plant.

REAERATION (RE-air-A-shun)   REAERATION

The introduction of air through forced air diffusers into the lower layers of the reservoir. As the air bubbles form and rise through the water, oxygen from the air dissolves into the water and replenishes the dissolved oxygen. The rising bubbles also cause the lower waters to rise to the surface where oxygen from the atmosphere is transferred to the water. This is sometimes called surface reaeration.

REAGENT (re-A-gent)   REAGENT

A pure chemical substance that is used to make new products or is used in chemical tests to measure, detect, or examine other substances.

RECARBONATION (re-CAR-bun-NAY-shun)   RECARBONATION

A process in which carbon dioxide is bubbled into the water being treated to lower the pH. The pH may also be lowered by the addition of acid. Recarbonation is the final stage in the lime-soda ash softening process. This process converts carbonate ions to bicarbonate ions and stabilizes the solution against the precipitation of carbonate compounds.

RECEIVER   RECEIVER

A device which indicates the result of a measurement. Most receivers in the water utility field use either a fixed scale and movable indicator (pointer) such as a pressure gage or a movable scale and movable indicator like those used on a circular flow-recording chart. Also called an INDICATOR.

RECORDER   RECORDER

A device that creates a permanent record, on a paper chart or magnetic tape, of the changes in a measured variable.

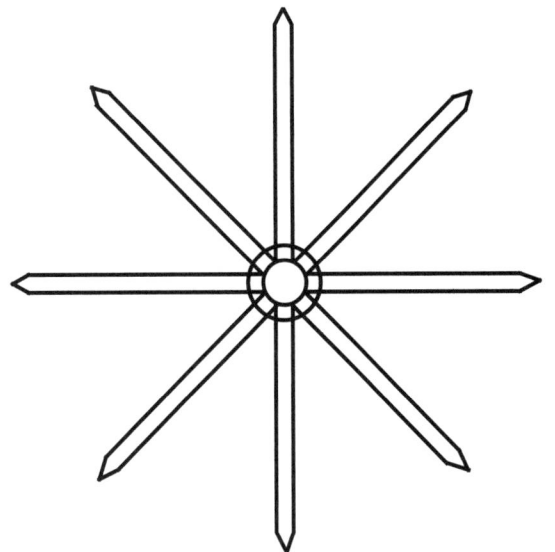

PLAN VIEW OF COLLECTOR PIPES

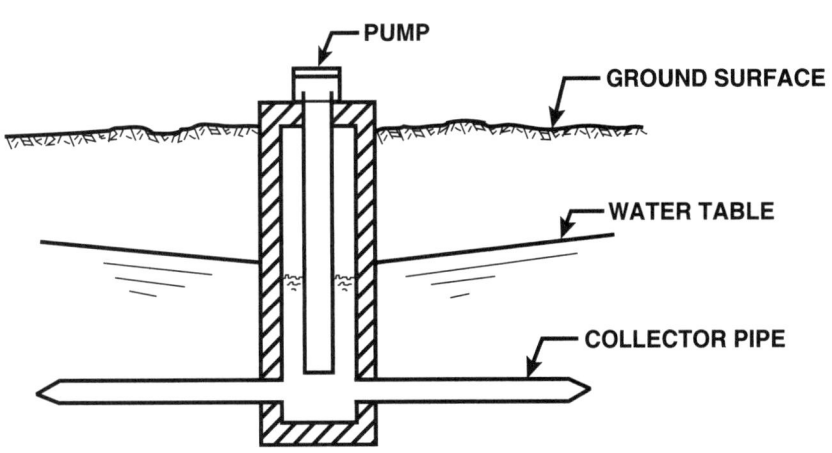

ELEVATION VIEW

RANNEY COLLECTOR

REDUCING AGENT   REDUCING AGENT

Any substance, such as base metal (iron) or the sulfide ion (S^{2-}), that will readily donate (give up) electrons. The opposite is an OXI-DIZING AGENT.

REDUCTION (re-DUCK-shun)   REDUCTION

Reduction is the addition of hydrogen, removal of oxygen, or the addition of electrons to an element or compound. Under anaerobic conditions (no dissolved oxygen present), sulfur compounds are reduced to odor-producing hydrogen sulfide (H_2S) and other compounds. The opposite of OXIDATION.

REFERENCE   REFERENCE

A physical or chemical quantity whose value is known exactly, and thus is used to calibrate instruments or standardize measurements. Also called a STANDARD.

REGULATORY NEGOTIATION   REGULATORY NEGOTIATION

A process whereby the U.S. Environmental Protection Agency acts on an equal basis with outside parties to reach consensus on the content of a proposed rule. If the group reaches consensus, the US EPA commits to propose the rule with the agreed upon content.

RELIQUEFACTION (re-LICK-we-FACK-shun)   RELIQUEFACTION

The return of a gas to the liquid state; for example, a condensation of chlorine gas to return it to its liquid form by cooling.

REPRESENTATIVE SAMPLE   REPRESENTATIVE SAMPLE

A sample portion of material or water that is as nearly identical in content and consistency as possible to that in the larger body of material or water being sampled.

RESIDUAL CHLORINE   RESIDUAL CHLORINE

The concentration of chlorine present in water after the chlorine demand has been satisfied. The concentration is expressed in terms of the total chlorine residual, which includes both the free and combined or chemically bound chlorine residuals.

RESIDUE   RESIDUE

The dry solids remaining after the evaporation of a sample of water or sludge. Also see TOTAL DISSOLVED SOLIDS.

RESINS   RESINS

See ION EXCHANGE RESINS.

RESISTANCE   RESISTANCE

That property of a conductor or wire that opposes the passage of a current, thus causing electric energy to be transformed into heat.

RESPIRATION   RESPIRATION

The process in which an organism uses oxygen for its life processes and gives off carbon dioxide.

RESPONSIBILITY   RESPONSIBILITY

Answering to those above in the chain of command to explain how and why you have used your authority.

REVERSE OSMOSIS (oz-MOE-sis)   REVERSE OSMOSIS

The application of pressure to a concentrated solution which causes the passage of a liquid from the concentrated solution to a weaker solution across a semipermeable membrane. The membrane allows the passage of the water (solvent) but not the dissolved solids (solutes). The liquid produced is a demineralized water. Also see OSMOSIS.

RIPARIAN (ri-PAIR-ee-an)   RIPARIAN

Water rights which are acquired together with title to the land bordering a source of surface water. The right to put to beneficial use surface water adjacent to your land.

RODENTICIDE (row-DENT-uh-SIDE)   RODENTICIDE

Any substance or chemical used to kill or control rodents.

ROTAMETER (RODE-uh-ME-ter)   ROTAMETER

A device used to measure the flow rate of gases and liquids. The gas or liquid being measured flows vertically up a tapered, calibrated tube. Inside the tube is a small ball or bullet-shaped float (it may rotate) that rises or falls depending on the flow rate. The flow rate may be read on a scale behind or on the tube by looking at the middle of the ball or at the widest part or top of the float.

ROTOR

ROTOR

The rotating part of a machine. The rotor is surrounded by the stationary (non-moving) parts (stator) of the machine.

ROUTINE SAMPLING

ROUTINE SAMPLING

Sampling repeated on a regular basis.

S

SCADA (ss-KAY-dah) SYSTEM

SCADA SYSTEM

Supervisory **C**ontrol **A**nd **D**ata **A**cquisition system. A computer-monitored alarm, response, control and data acquisition system used by drinking water facilities to monitor their operations.

SCFM

SCFM

Cubic **F**eet of air per **M**inute at **S**tandard conditions of temperature, pressure, and humidity (0°C, 14.7 psia, and 50% relative humidity).

SDWA

SDWA

See **S**AFE **D**RINKING **W**ATER **A**CT.

SMCL

SMCL

Secondary **M**aximum **C**ontaminant **L**evel. Secondary MCLs for various water quality indicators are established to protect public welfare.

SNARL

SNARL

Suggested **N**o **A**dverse **R**esponse **L**evel. The concentration of a chemical in water that is expected not to cause an adverse health effect.

SACRIFICIAL ANODE

SACRIFICIAL ANODE

An easily corroded material deliberately installed in a pipe or tank. The intent of such an installation is to give up (sacrifice) this anode to corrosion while the water supply facilities remain relatively corrosion free.

SAFE DRINKING WATER ACT

SAFE DRINKING WATER ACT

Commonly referred to as SDWA. An Act passed by the U.S. Congress in 1974. The Act establishes a cooperative program among local, state and federal agencies to ensure safe drinking water for consumers. The Act has been amended several times, including the 1980, 1986, and 1996 Amendments.

SAFE WATER

SAFE WATER

Water that does not contain harmful bacteria, or toxic materials or chemicals. Water may have taste and odor problems, color and certain mineral problems and still be considered safe for drinking.

SAFE YIELD

SAFE YIELD

The annual quantity of water that can be taken from a source of supply over a period of years without depleting the source permanently (beyond its ability to be replenished naturally in "wet years").

SALINITY

SALINITY

(1) The relative concentration of dissolved salts, usually sodium chloride, in a given water.

(2) A measure of the concentration of dissolved mineral substances in water.

SANITARY SURVEY

SANITARY SURVEY

A detailed evaluation and/or inspection of a source of water supply and all conveyances, storage, treatment and distribution facilities to ensure protection of the water supply from all pollution sources.

SAPROPHYTES (SAP-row-FIGHTS)

SAPROPHYTES

Organisms living on dead or decaying organic matter. They help natural decomposition of organic matter in water.

SATURATION

SATURATION

The condition of a liquid (water) when it has taken into solution the maximum possible quantity of a given substance at a given temperature and pressure.

SATURATOR (SAT-you-RAY-tore) SATURATOR

A device which produces a fluoride solution for the fluoridation process. The device is usually a cylindrical container with granular sodium fluoride on the bottom. Water flows either upward or downward through the sodium fluoride to produce the fluoride solution.

SCHEDULE, PIPE SCHEDULE, PIPE

A sizing system of numbers that specifies the I.D. (inside diameter) and O.D. (outside diameter) for each diameter pipe. The schedule number is the ratio of internal pressure in psi divided by the allowable fiber stress multiplied by 1,000. Typical schedules of iron and steel pipe are schedules 40, 80, and 160. Other forms of piping are divided into various classes with their own schedule schemes.

SCHMUTZDECKE (sh-moots-DECK-ee) SCHMUTZDECKE

A layer of trapped matter at the surface of a slow sand filter in which a dense population of microorganisms develops. These microorganisms within the film or mat feed on and break down incoming organic material trapped in the mat. In doing so the microorganisms both remove organic matter and add mass to the mat, further developing the mat and increasing the physical straining action of the mat.

SECCHI (SECK-key) DISC SECCHI DISC

A flat, white disc lowered into the water by a rope until it is just barely visible. At this point, the depth of the disc from the water surface is the recorded Secchi disc transparency.

SEDIMENTATION (SED-uh-men-TAY-shun) SEDIMENTATION

A water treatment process in which solid particles settle out of the water being treated in a large clarifier or sedimentation basin.

SEIZE UP SEIZE UP

Seize up occurs when an engine overheats and a part expands to the point where the engine will not run. Also called "freezing."

SENSITIVITY (PARTICLE COUNTERS) SENSITIVITY (PARTICLE COUNTERS)

The smallest particle a particle counter will measure and count.

SENSOR SENSOR

A device that measures (senses) a physical condition or variable of interest. Floats and thermocouples are examples of sensors. Also called a PRIMARY ELEMENT.

SEPTIC (SEP-tick) SEPTIC

A condition produced by bacteria when all oxygen supplies are depleted. If severe, the bottom deposits produce hydrogen sulfide, the deposits and water turn black, give off foul odors, and the water has a greatly increased chlorine demand.

SEQUESTRATION (SEE-kwes-TRAY-shun) SEQUESTRATION

A chemical complexing (forming or joining together) of metallic cations (such as iron) with certain inorganic compounds, such as phosphate. Sequestration prevents the precipitation of the metals (iron). Also see CHELATION.

SERVICE PIPE SERVICE PIPE

The pipeline extending from the water main to the building served or to the consumer's system.

SET POINT SET POINT

The position at which the control or controller is set. This is the same as the desired value of the process variable. For example, a thermostat is set to maintain a desired temperature.

SEWAGE SEWAGE

The used household water and water-carried solids that flow in sewers to a wastewater treatment plant. The preferred term is WASTEWATER.

SHEAVE (SHE-v) SHEAVE

V-belt drive pulley which is commonly made of cast iron or steel.

SHIM SHIM

Thin metal sheets which are inserted between two surfaces to align or space the surfaces correctly. Shims can be used anywhere a spacer is needed. Usually shims are 0.001 to 0.020 inch thick.

SHOCK LOAD

The arrival at a water treatment plant of raw water containing unusual amounts of algae, colloidal matter, color, suspended solids, turbidity, or other pollutants.

SHORT-CIRCUITING

A condition that occurs in tanks or basins when some of the flowing water entering a tank or basin flows along a nearly direct pathway from the inlet to the outlet. This is usually undesirable since it may result in shorter contact, reaction, or settling times in comparison with the theoretical (calculated) or presumed detention times.

SIMULATE

To reproduce the action of some process, usually on a smaller scale.

SINGLE-STAGE PUMP

A pump that has only one impeller. A multi-stage pump has more than one impeller.

SLAKE

To mix with water so that a true chemical combination (hydration) takes place, such as in the slaking of lime.

SLAKED LIME

See HYDRATED LIME.

SLOPE

The slope or inclination of a trench bottom or a trench side wall is the ratio of the vertical distance to the horizontal distance or "rise over run." Also see GRADE (2).

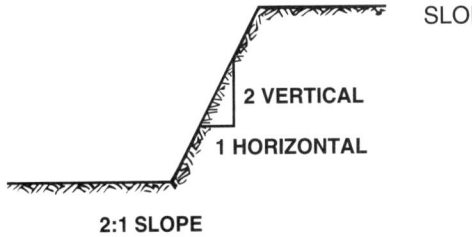

2 VERTICAL

1 HORIZONTAL

2:1 SLOPE

SLUDGE (sluj)

The settleable solids separated from water during processing.

SLURRY (SLUR-e)

A watery mixture or suspension of insoluble (not dissolved) matter; a thin, watery mud or any substance resembling it (such as a grit slurry or a lime slurry).

SOFT WATER

Water having a low concentration of calcium and magnesium ions. According to U.S. Geological Survey guidelines, soft water is water having a hardness of 60 milligrams per liter or less.

SOFTWARE PROGRAMS

Computer programs; the list of instructions that tell a computer how to perform a given task or tasks. Some software programs are designed and written to monitor and control water distribution systems and treatment processes.

SOLENOID (SO-luh-noid)

A magnetically (electric coil) operated mechanical device. Solenoids can operate small valves or electric switches.

SOLUTION

A liquid mixture of dissolved substances. In a solution it is impossible to see all the separate parts.

SOUNDING TUBE

A pipe or tube used for measuring the depths of water.

SPAN

The scale or range of values an instrument is designed to measure. Also see RANGE.

SPECIFIC CAPACITY

A measurement of well yield per unit depth of drawdown after a specific time has passed, usually 24 hours. Typically expressed as gallons per minute per foot (GPM/ft or cu m/day/m).

SHOCK LOAD

SHORT-CIRCUITING

SIMULATE

SINGLE-STAGE PUMP

SLAKE

SLAKED LIME

SLOPE

SLUDGE

SLURRY

SOFT WATER

SOFTWARE PROGRAMS

SOLENOID

SOLUTION

SOUNDING TUBE

SPAN

SPECIFIC CAPACITY

SPECIFIC CAPACITY TEST SPECIFIC CAPACITY TEST

A testing method used to determine the adequacy of an aquifer or well by measuring the specific capacity.

SPECIFIC CONDUCTANCE SPECIFIC CONDUCTANCE

A rapid method of estimating the dissolved solids content of a water supply. The measurement indicates the capacity of a sample of water to carry an electric current, which is related to the concentration of ionized substances in the water. Also called CONDUCTANCE.

SPECIFIC GRAVITY SPECIFIC GRAVITY

(1) Weight of a particle, substance, or chemical solution in relation to the weight of an equal volume of water. Water has a specific gravity of 1.000 at 4°C (39°F). Particulates in raw water may have a specific gravity of 1.005 to 2.5.

(2) Weight of a particular gas in relation to the weight of an equal volume of air at the same temperature and pressure (air has a specific gravity of 1.0). Chlorine has a specific gravity of 2.5 as a gas.

SPECIFIC YIELD SPECIFIC YIELD

The quantity of water that a unit volume of saturated permeable rock or soil will yield when drained by gravity. Specific yield may be expressed as a ratio or as a percentage by volume.

SPOIL SPOIL

Excavated material such as soil from the trench of a water main.

SPORE SPORE

The reproductive body of an organism which is capable of giving rise to a new organism either directly or indirectly. A viable (able to live and grow) body regarded as the resting stage of an organism. A spore is usually more resistant to disinfectants and heat than most organisms. Gangrene and tetanus bacteria are common spore-forming organisms.

SPRING LINE SPRING LINE

Theoretical center of a pipeline. Also, the guideline for laying a course of bricks.

STALE WATER STALE WATER

Water which has not flowed recently and may have picked up tastes and odors from distribution lines or storage facilities.

STANDARD STANDARD

A physical or chemical quantity whose value is known exactly, and thus is used to calibrate instruments or standardize measurements. Also called a REFERENCE.

STANDARD DEVIATION STANDARD DEVIATION

A measure of the spread or dispersion of data.

STANDARD METHODS STANDARD METHODS

STANDARD METHODS FOR THE EXAMINATION OF WATER AND WASTEWATER, 20th Edition. A joint publication of the American Public Health Association (APHA), American Water Works Association (AWWA), and the Water Environment Federation (WEF) which outlines the accepted laboratory procedures used to analyze the impurities in water and wastewater. Available from American Water Works Association, Bookstore, 6666 West Quincy Avenue, Denver, CO 80235. Order No. 10079. Price to members, $167.00; nonmembers, $212.00; price includes cost of shipping and handling.

STANDARD SOLUTION STANDARD SOLUTION

A solution in which the exact concentration of a chemical or compound is known.

STANDARDIZE STANDARDIZE

To compare with a standard.

(1) In wet chemistry, to find out the exact strength of a solution by comparing it with a standard of known strength. This information is used to adjust the strength by adding more water or more of the substance dissolved.

(2) To set up an instrument or device to read a standard. This allows you to adjust the instrument so that it reads accurately, or enables you to apply a correction factor to the readings.

STARTERS (MOTOR) STARTERS (MOTOR)

Devices used to start up large motors gradually to avoid severe mechanical shock to a driven machine and to prevent disturbance to the electrical lines (causing dimming and flickering of lights).

STATIC HEAD STATIC HEAD

When water is not moving, the vertical distance (in feet) from a specific point to the water surface is the static head. (The static pressure in psi is the static head in feet times 0.433 psi/ft.) Also see DYNAMIC PRESSURE and STATIC PRESSURE.

STATIC PRESSURE STATIC PRESSURE

When water is not moving, the vertical distance (in feet) from a specific point to the water surface is the static head. The static pressure in psi is the static head in feet times 0.433 psi/ft. Also see DYNAMIC PRESSURE and STATIC HEAD.

STATIC WATER DEPTH STATIC WATER DEPTH

The vertical distance in feet from the centerline of the pump discharge down to the surface level of the free pool while no water is being drawn from the pool or water table.

STATIC WATER LEVEL STATIC WATER LEVEL

(1) The elevation or level of the water table in a well when the pump is not operating.

(2) The level or elevation to which water would rise in a tube connected to an artesian aquifer, basin, or conduit under pressure.

STATOR STATOR

That portion of a machine which contains the stationary (non-moving) parts that surround the moving parts (rotor).

STERILIZATION (STARE-uh-luh-ZAY-shun) STERILIZATION

The removal or destruction of all microorganisms, including pathogenic and other bacteria, vegetative forms and spores. Compare with DISINFECTION.

STETHOSCOPE STETHOSCOPE

An instrument used to magnify sounds and convey them to the ear.

STORATIVITY (S) STORATIVITY (S)

The volume of groundwater an aquifer releases from or takes into storage per unit surface area of the aquifer per unit change in head. Also called the storage coefficient.

STRATIFICATION (STRAT-uh-fuh-KAY-shun) STRATIFICATION

The formation of separate layers (of temperature, plant, or animal life) in a lake or reservoir. Each layer has similar characteristics such as all water in the layer has the same temperature. Also see THERMAL STRATIFICATION.

STRAY CURRENT CORROSION STRAY CURRENT CORROSION

A corrosion activity resulting from stray electric current originating from some source outside the plumbing system such as D.C. grounding on phone systems.

SUBMERGENCE SUBMERGENCE

The distance between the water surface and the media surface in a filter.

SUBSIDENCE (sub-SIDE-ence) SUBSIDENCE

The dropping or lowering of the ground surface as a result of removing excess water (overdraft or overpumping) from an aquifer. After excess water has been removed, the soil will settle, become compacted and the ground surface will drop and can cause the settling of underground utilities.

SUCTION LIFT SUCTION LIFT

The *NEGATIVE* pressure [in feet (meters) of water or inches (centimeters) of mercury vacuum] on the suction side of the pump. The pressure can be measured from the centerline of the pump *DOWN TO* (lift) the elevation of the hydraulic grade line on the suction side of the pump.

SUPERCHLORINATION (SUE-per-KLOR-uh-NAY-shun) SUPERCHLORINATION

Chlorination with doses that are deliberately selected to produce free or combined residuals so large as to require dechlorination.

SUPERNATANT (sue-per-NAY-tent) SUPERNATANT

Liquid removed from settled sludge. Supernatant commonly refers to the liquid between the sludge on the bottom and the scum on the water surface of a basin or container.

SUPERSATURATED
SUPERSATURATED

An unstable condition of a solution (water) in which the solution contains a substance at a concentration greater than the saturation concentration for the substance.

SURFACE LOADING
SURFACE LOADING

One of the guidelines for the design of settling tanks and clarifiers in treatment plants. Used by operators to determine if tanks and clarifiers are hydraulically (flow) over- or underloaded. Also called OVERFLOW RATE.

$$\text{Surface Loading, GPD/sq ft} = \frac{\text{Flow, gallons/day}}{\text{Surface Area, sq ft}}$$

SURFACTANT (sir-FAC-tent)
SURFACTANT

Abbreviation for surface-active agent. The active agent in detergents that possesses a high cleaning ability.

SURGE CHAMBER
SURGE CHAMBER

A chamber or tank connected to a pipe and located at or near a valve that may quickly open or close or a pump that may suddenly start or stop. When the flow of water in a pipe starts or stops quickly, the surge chamber allows water to flow into or out of the pipe and minimize any sudden positive or negative pressure waves or surges in the pipe.

[SEE DRAWING ON PAGE 701]

SUSPENDED SOLIDS
SUSPENDED SOLIDS

(1) Solids that either float on the surface or are suspended in water, wastewater, or other liquids, and which are largely removable by laboratory filtering.

(2) The quantity of material removed from water in a laboratory test, as prescribed in *STANDARD METHODS FOR THE EXAMINATION OF WATER AND WASTEWATER*, and referred to as Total Suspended Solids Dried at 103–105°C.

T

TCE
TCE

See **TRICHLOROETHANE**.

TDS
TDS

See **TOTAL DISSOLVED SOLIDS**.

THM
THM

See **TRIHALOMETHANES**.

THM PRECURSOR
THM PRECURSOR

See PRECURSOR, THM.

TAILGATE SAFETY MEETING
TAILGATE SAFETY MEETING

Brief (10 to 20 minutes) safety meetings held every 7 to 10 working days. The term *TAILGATE* comes from the safety meetings regularly held by the construction industry around the tailgate of a truck.

TELEMETRY (tel-LEM-uh-tree)
TELEMETRY

The electrical link between the transmitter and the receiver. Telephone lines are commonly used to serve as the electrical line.

TEMPERATURE SENSOR
TEMPERATURE SENSOR

A device that opens and closes a switch in response to changes in the temperature. This device might be a metal contact, or a thermocouple that generates minute electric current proportional to the difference in heat, or a variable resistor whose value changes in response to changes in temperature. Also called a HEAT SENSOR.

THERMAL STRATIFICATION (STRAT-uh-fuh-KAY-shun)
THERMAL STRATIFICATION

The formation of layers of different temperatures in a lake or reservoir. Also see STRATIFICATION.

THERMOCLINE (THUR-moe-KLINE)
THERMOCLINE

The middle layer in a thermally stratified lake or reservoir. In this layer there is a rapid decrease in temperature with depth. Also called the METALIMNION.

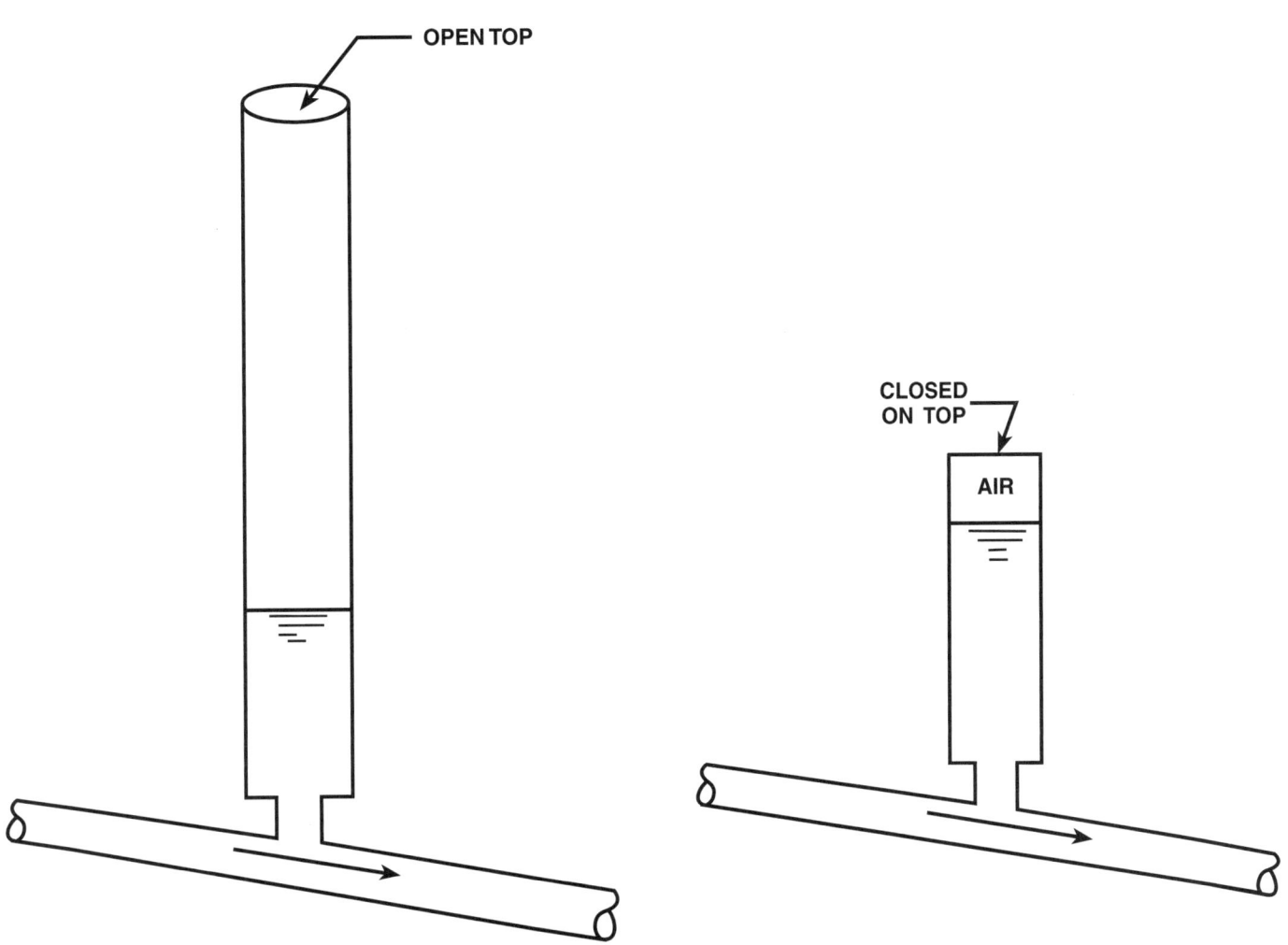

TYPES OF SURGE CHAMBERS

THERMOCOUPLE

A heat-sensing device made of two conductors of different metals joined at their ends. An electric current is produced when there is a difference in temperature between the ends.

THICKENING

Treatment to remove water from the sludge mass to reduce the volume that must be handled.

THRESHOLD ODOR

The minimum odor of a water sample that can just be detected after successive dilutions with odorless water. Also called ODOR THRESHOLD.

THRESHOLD ODOR NUMBER (TON)

The greatest dilution of a sample with odor-free water that still yields a just-detectable odor.

THRUST BLOCK

A mass of concrete or similar material appropriately placed around a pipe to prevent movement when the pipe is carrying water. Usually placed at bends and valve structures.

TIME LAG

The time required for processes and control systems to respond to a signal or to reach a desired level.

TIMER

A device for automatically starting or stopping a machine or other device at a given time.

TITRATE (TIE-trate)

To *TITRATE* a sample, a chemical solution of known strength is added drop by drop until a certain color change, precipitate, or pH change in the sample is observed (end point). Titration is the process of adding the chemical reagent in small increments (0.1 – 1.0 milliliter) until completion of the reaction, as signaled by the end point.

TOPOGRAPHY (toe-PAH-gruh-fee)

The arrangement of hills and valleys in a geographic area.

TOTAL CHLORINE

The total concentration of chlorine in water, including the combined chlorine (such as inorganic and organic chloramines) and the free available chlorine.

TOTAL CHLORINE RESIDUAL

The total amount of chlorine residual (value for residual chlorine, including both free chlorine and chemically bound chlorine) present in a water sample after a given contact time.

TOTAL DISSOLVED SOLIDS (TDS)

All of the dissolved solids in a water. TDS is measured on a sample of water that has passed through a very fine mesh filter to remove suspended solids. The water passing through the filter is evaporated and the residue represents the dissolved solids. Also see SPECIFIC CONDUCTANCE.

TOTAL DYNAMIC HEAD (TDH)

When a pump is lifting or pumping water, the vertical distance (in feet) from the elevation of the energy grade line on the suction side of the pump to the elevation of the energy grade line on the discharge side of the pump.

TOTAL ORGANIC CARBON (TOC)

TOC measures the amount of organic carbon in water.

TOTALIZER

A device or meter that continuously measures and calculates (adds) a process rate variable in cumulative fashion; for example, total flows displayed in gallons, million gallons, cubic feet, or some other unit of volume measurement. Also called an INTEGRATOR.

TOXAPHENE (TOX-uh-FEEN)

A chemical that causes adverse health effects in domestic water supplies and also is toxic to freshwater and marine aquatic life.

TOXIC (TOX-ick)

TOXIC

A substance which is poisonous to a living organism.

TRANSDUCER (trans-DUE-sir)

TRANSDUCER

A device that senses some varying condition measured by a primary sensor and converts it to an electrical or other signal for transmission to some other device (a receiver) for processing or decision making.

TRANSMISSION LINES

TRANSMISSION LINES

Pipelines that transport raw water from its source to a water treatment plant. After treatment, water is usually pumped into pipelines (transmission lines) that are connected to a distribution grid system.

TRANSMISSIVITY (TRANS-miss-SIV-it-tee)

TRANSMISSIVITY

A measure of the ability to transmit (as in the ability of an aquifer to transmit water).

TRANSPIRATION (TRAN-spur-RAY-shun)

TRANSPIRATION

The process by which water vapor is released to the atmosphere by living plants. This process is similar to people sweating. Also see EVAPOTRANSPIRATION.

TREMIE (TREH-me)

TREMIE

A device used to place concrete or grout under water.

TRICHLOROETHANE (TCE) (try-KLOR-o-ETH-hane)

TRICHLOROETHANE (TCE)

An organic chemical used as a cleaning solvent that causes adverse health effects in domestic water supplies.

TRIHALOMETHANES (THMs) (tri-HAL-o-METH-hanes)

TRIHALOMETHANES (THMs)

Derivatives of methane, CH_4, in which three halogen atoms (chlorine or bromine) are substituted for three of the hydrogen atoms. Often formed during chlorination by reactions with natural organic materials in the water. The resulting compounds (THMs) are suspected of causing cancer.

TRUE COLOR

TRUE COLOR

Color of the water from which turbidity has been removed. The turbidity may be removed by double filtering the sample through a Whatman No. 40 filter when using the visual comparison method.

TUBE SETTLER

TUBE SETTLER

A device that uses bundles of small-bore (2 to 3 inches or 50 to 75 mm) tubes installed on an incline as an aid to sedimentation. The tubes may come in a variety of shapes including circular and rectangular. As water rises within the tubes, settling solids fall to the tube surface. As the sludge (from the settled solids) in the tube gains weight, it moves down the tubes and settles to the bottom of the basin for removal by conventional sludge collection means. Tube settlers are sometimes installed in sedimentation basins and clarifiers to improve particle removal.

TUBERCLE (TOO-burr-cull)

TUBERCLE

A protective crust of corrosion products (rust) which builds up over a pit caused by the loss of metal due to corrosion.

TUBERCULATION (too-BURR-cue-LAY-shun)

TUBERCULATION

The development or formation of small mounds of corrosion products (rust) on the inside of iron pipe. These mounds (tubercles) increase the roughness of the inside of the pipe thus increasing resistance to water flow (decreases the C Factor).

TURBID

TURBID

Having a cloudy or muddy appearance.

TURBIDIMETER

TURBIDIMETER

See TURBIDITY METER.

TURBIDITY (ter-BID-it-tee)

TURBIDITY

The cloudy appearance of water caused by the presence of suspended and colloidal matter. In the waterworks field, a turbidity measurement is used to indicate the clarity of water. Technically, turbidity is an optical property of the water based on the amount of light reflected by suspended particles. Turbidity cannot be directly equated to suspended solids because white particles reflect more light than dark-colored particles and many small particles will reflect more light than an equivalent large particle.

TURBIDITY METER TURBIDITY METER

An instrument for measuring and comparing the turbidity of liquids by passing light through them and determining how much light is reflected by the particles in the liquid. The normal measuring range is 0 to 100 and is expressed as Nephelometric Turbidity Units (NTUs).

TURBIDITY UNITS (TU) TURBIDITY UNITS (TU)

Turbidity units are a measure of the cloudiness of water. If measured by a nephelometric (deflected light) instrumental procedure, turbidity units are expressed in nephelometric turbidity units (NTU) or simply TU. Those turbidity units obtained by visual methods are expressed in Jackson Turbidity Units (JTU) which are a measure of the cloudiness of water; they are used to indicate the clarity of water. There is no real connection between NTUs and JTUs. The Jackson turbidimeter is a visual method and the nephelometer is an instrumental method based on deflected light.

TURN-DOWN RATIO TURN-DOWN RATIO

The ratio of the design range to the range of acceptable accuracy or precision of an instrument. Also see EFFECTIVE RANGE.

U

UNCONSOLIDATED FORMATION UNCONSOLIDATED FORMATION

A sediment that is loosely arranged or unstratified (not in layers) or whose particles are not cemented together (soft rock); occurring either at the ground surface or at a depth below the surface. Also see CONSOLIDATED FORMATION.

UNIFORMITY COEFFICIENT (U.C.) UNIFORMITY COEFFICIENT (U.C.)

The ratio of (1) the diameter of a grain (particle) of a size that is barely too large to pass through a sieve that allows 60 percent of the material (by weight) to pass through, to (2) the diameter of a grain (particle) of a size that is barely too large to pass through a sieve that allows 10 percent of the material (by weight) to pass through. The resulting ratio is a measure of the degree of uniformity in a granular material such as filter media.

$$\text{Uniformity Coefficient} = \frac{\text{Particle Diameter}_{60\%}}{\text{Particle Diameter}_{10\%}}$$

UPPER EXPLOSIVE LIMIT (UEL) UPPER EXPLOSIVE LIMIT (UEL)

The point at which the concentration of a gas in air becomes too great to allow an explosion upon ignition due to insufficient oxygen present.

V

VARIABLE COSTS VARIABLE COSTS

Costs that a utility must cover or pay that are associated with the production and delivery of water. The costs vary or fluctuate on the basis of the volume of water treated and delivered to customers (water production). Also see FIXED COSTS.

VARIABLE FREQUENCY DRIVE VARIABLE FREQUENCY DRIVE

A control system that allows the frequency of the current applied to a motor to be varied. The motor is connected to a low-frequency source while standing still; the frequency is then increased gradually until the motor and pump (or other driven machine) are operating at the desired speed.

VARIABLE, MEASURED VARIABLE, MEASURED

A factor (flow, temperature) that is sensed and quantified (reduced to a reading of some kind) by a primary element or sensor.

VARIABLE, PROCESS VARIABLE, PROCESS

A physical or chemical quantity which is usually measured and controlled in the operation of a water treatment plant or an industrial plant.

VELOCITY HEAD VELOCITY HEAD

The energy in flowing water as determined by a vertical height (in feet or meters) equal to the square of the velocity of flowing water divided by twice the acceleration due to gravity ($V^2/2g$).

VENTURI METER   VENTURI METER

A flow measuring device placed in a pipe. The device consists of a tube whose diameter gradually decreases to a throat and then gradually expands to the diameter of the pipe. The flow is determined on the basis of the difference in pressure (caused by different velocity heads) between the entrance and throat of the Venturi meter.

NOTE: Most Venturi meters have pressure sensing taps rather than a manometer to measure the pressure difference. The upstream tap is the high pressure tap or side of the manometer.

VENTURI METER

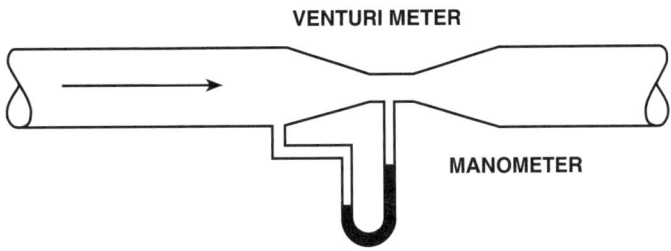

MANOMETER

VISCOSITY (vis-KOSS-uh-tee)   VISCOSITY

A property of water, or any other fluid, which resists efforts to change its shape or flow. Syrup is more viscous (has a higher viscosity) than water. The viscosity of water increases significantly as temperatures decrease. Motor oil is rated by how thick (viscous) it is; 20 weight oil is considered relatively thin while 50 weight oil is relatively thick or viscous.

VOID   VOID

A pore or open space in rock, soil or other granular material, not occupied by solid matter. The pore or open space may be occupied by air, water, or other gaseous or liquid material. Also called an INTERSTICE, PORE, or void space.

VOLATILE (VOL-uh-tull)   VOLATILE

(1)   A volatile substance is one that is capable of being evaporated or changed to a vapor at relatively low temperatures. Volatile substances also can be partially removed by air stripping.

(2)   In terms of solids analysis, volatile refers to materials lost (including most organic matter) upon ignition in a muffle furnace for 60 minutes at 550°C. Natural volatile materials are chemical substances usually of animal or plant origin. Manufactured or synthetic volatile materials such as ether, acetone, and carbon tetrachloride are highly volatile and not of plant or animal origin. Also see NONVOLATILE MATTER.

VOLATILE ACIDS   VOLATILE ACIDS

Fatty acids produced during digestion which are soluble in water and can be steam-distilled at atmospheric pressure. Also called organic acids. Volatile acids are commonly reported as equivalent to acetic acid.

VOLATILE LIQUIDS   VOLATILE LIQUIDS

Liquids which easily vaporize or evaporate at room temperature.

VOLATILE MATTER   VOLATILE MATTER

Matter in water, wastewater, or other liquids that is lost on ignition of the dry solids at 550°C.

VOLATILE SOLIDS   VOLATILE SOLIDS

Those solids in water or other liquids that are lost on ignition of the dry solids at 550°C.

VOLTAGE   VOLTAGE

The electrical pressure available to cause a flow of current (amperage) when an electric circuit is closed. Also called ELECTROMOTIVE FORCE (E.M.F.).

VOLUMETRIC   VOLUMETRIC

A measurement based on the volume of some factor. Volumetric titration is a means of measuring unknown concentrations of water quality indicators in a sample *BY DETERMINING THE VOLUME* of titrant or liquid reagent needed to complete particular reactions.

VOLUMETRIC FEEDER   VOLUMETRIC FEEDER

A dry chemical feeder which delivers a measured volume of chemical during a specific time period.

VORTEX VORTEX

A revolving mass of water which forms a whirlpool. This whirlpool is caused by water flowing out of a small opening in the bottom of a basin or reservoir. A funnel-shaped opening is created downward from the water surface.

W

WARNING WARNING

The word *WARNING* is used to indicate a hazard level between *CAUTION* and *DANGER.* Also see CAUTION, DANGER, and NOTICE.

WASTEWATER WASTEWATER

A community's used water and water-carried solids (including used water from industrial processes) that flow to a treatment plant. Storm water, surface water, and groundwater infiltration also may be included in the wastewater that enters a wastewater treatment plant. The term "sewage" usually refers to household wastes, but this word is being replaced by the term "wastewater."

WATER AUDIT WATER AUDIT

A thorough examination of the accuracy of water agency records or accounts (volumes of water) and system control equipment. Water managers can use audits to determine their water distribution system efficiency. The overall goal is to identify and verify water and revenue losses in a water system.

WATER CYCLE WATER CYCLE

The process of evaporation of water into the air and its return to earth by precipitation (rain or snow). This process also includes transpiration from plants, groundwater movement, and runoff into rivers, streams and the ocean. Also called the HYDROLOGIC CYCLE.

WATER HAMMER WATER HAMMER

The sound like someone hammering on a pipe that occurs when a valve is opened or closed very rapidly. When a valve position is changed quickly, the water pressure in a pipe will increase and decrease back and forth very quickly. This rise and fall in pressures can cause serious damage to the system.

WATER PURVEYOR (purr-VAY-or) WATER PURVEYOR

An agency or person that supplies water (usually potable water).

WATER TABLE WATER TABLE

The upper surface of the zone of saturation of groundwater in an unconfined aquifer.

WATERSHED WATERSHED

The region or land area that contributes to the drainage or catchment area above a specific point on a stream or river.

WATT WATT

A unit of power equal to one joule per second. The power of a current of one ampere flowing across a potential difference of one volt.

WEIR (weer) WEIR

(1) A wall or plate placed in an open channel and used to measure the flow of water. The depth of the flow over the weir can be used to calculate the flow rate, or a chart or conversion table may be used to convert depth to flow.

(2) A wall or obstruction used to control flow (from settling tanks and clarifiers) to ensure a uniform flow rate and avoid short-circuiting.

WEIR (weer) DIAMETER WEIR DIAMETER

Many circular clarifiers have a circular weir within the outside edge of the clarifier. All the water leaving the clarifier flows over this weir. The diameter of the weir is the length of a line from one edge of a weir to the opposite edge and passing through the center of the circle formed by the weir.

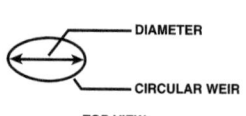

WEIR LOADING   WEIR LOADING

A guideline used to determine the length of weir needed on settling tanks and clarifiers in treatment plants. Used by operators to determine if weirs are hydraulically (flow) overloaded.

$$\text{Weir Loading, GPM/ft} = \frac{\text{Flow, GPM}}{\text{Length of Weir, ft}}$$

WELL ISOLATION ZONE   WELL ISOLATION ZONE

The surface or zone surrounding a water well or well field, supplying a public water system, with restricted land uses to prevent contaminants from a not permitted land use to move toward and reach such water well or well field. Also see WELLHEAD PROTECTION AREA (WHPA).

WELL LOG   WELL LOG

A record of the thickness and characteristics of the soil, rock and water-bearing formations encountered during the drilling (sinking) of a well.

WELLHEAD PROTECTION AREA (WHPA)   WELLHEAD PROTECTION AREA (WHPA)

The surface and subsurface area surrounding a water well or well field, supplying a public water system, through which contaminants are reasonably likely to move toward and reach such water well or well field. Also see WELL ISOLATION ZONE.

WET CHEMISTRY   WET CHEMISTRY

Laboratory procedures used to analyze a sample of water using liquid chemical solutions (wet) instead of, or in addition to, laboratory instruments.

WHOLESOME WATER   WHOLESOME WATER

A water that is safe and palatable for human consumption.

WIRE-TO-WATER EFFICIENCY   WIRE-TO-WATER EFFICIENCY

The combined efficiency of a pump and motor together. Also called the OVERALL EFFICIENCY.

WYE STRAINER   WYE STRAINER

A screen shaped like the letter Y. The water flows in at the top of the Y and the debris in the water is removed in the top part of the Y.

X

(NO LISTINGS)

Y

YIELD   YIELD

The quantity of water (expressed as a rate of flow—GPM, GPH, GPD, or total quantity per year) that can be collected for a given use from surface or groundwater sources. The yield may vary with the use proposed, with the plan of development, and also with economic considerations. Also see SAFE YIELD.

Z

ZEOLITE   ZEOLITE

A type of ion exchange material used to soften water. Natural zeolites are siliceous compounds (made of silica) which remove calcium and magnesium from hard water and replace them with sodium. Synthetic or organic zeolites are ion exchange materials which remove calcium or magnesium and replace them with either sodium or hydrogen. Manganese zeolites are used to remove iron and manganese from water.

ZETA POTENTIAL   ZETA POTENTIAL

In coagulation and flocculation procedures, the difference in the electrical charge between the dense layer of ions surrounding the particle and the charge of the bulk of the suspended fluid surrounding this particle. The zeta potential is usually measured in millivolts.

ZONE OF AERATION ZONE OF AERATION

The comparatively dry soil or rock located between the ground surface and the top of the water table.

ZONE OF SATURATION ZONE OF SATURATION

The soil or rock located below the top of the groundwater table. By definition, the zone of saturation is saturated with water. Also see WATER TABLE.

ZOOPLANKTON (ZOE-PLANK-ton) ZOOPLANKTON

Small, usually microscopic animals (such as protozoans), found in lakes and reservoirs.

SUBJECT INDEX

NOTES

NOTES